Advances in Topological Quantum Field Theory

NATO Science Series

A Series presenting the results of scientific meetings supported under the NATO Science Programme.

The Series is published by IOS Press, Amsterdam, and Kluwer Academic Publishers in conjunction with the NATO Scientific Affairs Division

Sub-Series

I. Life and Behavioural Sciences	IOS Press
II. Mathematics, Physics and Chemistry	Kluwer Academic Publishers
III. Computer and Systems Science	IOS Press
IV. Earth and Environmental Sciences	Kluwer Academic Publishers
V. Science and Technology Policy	IOS Press

The NATO Science Series continues the series of books published formerly as the NATO ASI Series.

The NATO Science Programme offers support for collaboration in civil science between scientists of countries of the Euro-Atlantic Partnership Council. The types of scientific meeting generally supported are "Advanced Study Institutes" and "Advanced Research Workshops", although other types of meeting are supported from time to time. The NATO Science Series collects together the results of these meetings. The meetings are co-organized bij scientists from NATO countries and scientists from NATO's Partner countries – countries of the CIS and Central and Eastern Europe.

Advanced Study Institutes are high-level tutorial courses offering in-depth study of latest advances in a field.
Advanced Research Workshops are expert meetings aimed at critical assessment of a field, and identification of directions for future action.

As a consequence of the restructuring of the NATO Science Programme in 1999, the NATO Science Series has been re-organised and there are currently Five Sub-series as noted above. Please consult the following web sites for information on previous volumes published in the Series, as well as details of earlier Sub-series.

http://www.nato.int/science
http://www.wkap.nl
http://www.iospress.nl
http://www.wtv-books.de/nato-pco.htm

Series II: Mathematics, Physics and Chemistry – Vol. 179

Advances in Topological Quantum Field Theory

edited by

John M. Bryden
Southern Illinois University,
Edwardsville, IL, U.S.A.

Kluwer Academic Publishers

Dordrecht / Boston / London

Published in cooperation with NATO Scientific Affairs Division

Proceedings of the NATO Advanced Research Workshop on
New Techniques in Topological Quantum Field Theory
Kananaskis Village, Canada
22–26 August 2001

A C.I.P. Catalogue record for this book is available from the Library of Congress.

ISBN 1-4020-2771-0 (PB)
ISBN 1-4020-2770-2 (HB)
ISBN 1-4020-2772-9 (e-book)

Published by Kluwer Academic Publishers,
P.O. Box 17, 3300 AA Dordrecht, The Netherlands.

Printed on acid-free paper

Printed in the Netherlands.

Table of Contents

Editorial

During the summer of 1999 Vladimir Turaev, Research Director of IRMA, Université Louis Pasteur, Strasbourg, and Florian Deloup of Université Paul Sabatier in Toulouse were my guests at the University of Calgary. Our prime interest was the relationship between quantum invariants and classical topological invariants. During this time we also had an advanced research workshop in honor of Turaev's visit. This workshop was funded by the Pacific Institute for the Mathematical Sciences (PIMS) and was entitled "Invariants of 3-manifolds". It was held at Nakoda Lodge in Morely, Alberta, Canada from July 18-22, 1999. The theme of this workshop was provided by our interest in the possible relationships between quantum invariants and classical topological invariants.

In 2001 once again both Turaev and Deloup were my guests during the summer months at the University of Calgary. During this visit Turaev was the PIMS Distinguished Visiting Research Chair at the University of Calgary. Furthermore during this time we held the second workshop on invariants of 3-manifolds. This was a NATO Advanced Research Workshop entitled "New Techniques in Topological Quantum Field Theory". We also received funding for this programme from PIMS. Our programme took place at two sites. The first was The University of Calgary, Calgary, Canada on August 23-24, 2001. The second was at the Delta Lodge at Kananaskis, Kananaskis Village, Canada from August 25-27, 2001.

As in 1999 this workshop was very research oriented and its theme was to attempt to find relationships between classical and quantum topology. To this end we invited specialists from different areas of classical and quantum topology. Many leading mathematicians from different fields, from both NATO countries and NATO partner countries, attended our workshop. The programme of the workshop was divided into three sections:

(i) lectures by graduate students and postdoctoral fellows,

(ii) invited lectures on important recent work in the subject areas,

(iii) research discussion sessions on possible new directions in research.

This volume is the collection of invited talks that were submitted and refereed for the proceedings.

I would like to thank PIMS and NATO for funding this project. I would also like to thank my co-director Victor Vassiliev from the Steklov Institute in Moscow and the organizing committee consisting of F. Deloup, D. Rolfsen, V. Turaev and P. Zvengrowski for their help. Finally I would like to thank Marian Miles, the administrative assistant for PIMS at the University of Calgary for making sure the workshop ran smoothly.

This past winter Heiner Zieschang, a valued colleague and collaborator, passed away while visiting Moscow State University. He was a participant at both workshops on invariants of 3-manifolds and contributed to both proceedings. I would also like to note the passing of my mother, Edna F. Bryden, on November 17, 2002. This volume is dedicated to their memory.

John Bryden
Southern Illinois University, Edwardsville, IL, 62026 U.S.A.
email: jbryden@siue.edu

COMBINATORIAL FORMULAS FOR COHOMOLOGY OF SPACES OF KNOTS

V.A. VASSILIEV

ABSTRACT. An algorithmic method of finding combinatorial formulas for knot invariants and other cohomology classes of spaces of knots in $\mathbb{R}^n$, $n \geq 3$, is described. In the case of invariants of knots in $\mathbb{R}^3$, we have a formal algorithm whose input is an arbitrary *weight system*, i.e. a possible principal part of a finite type invariant, and the output is either a proof of the fact that this weight system actually does not correspond to any knot invariant or an effective description of some invariant with this principal part, namely a finite collection of easy subvarieties of full dimension in the space of spatial curves such that the value of this invariant on any generic knot is equal to the sum of multiplicities of these varieties in a neighborhood of the knot. (In examples, the former possibility never occurs.) This algorithm is formally realized over $\mathbb{Z}_2$, but its generalization to the case of arbitrary coefficients is just a technical task.

A similar method of realizing higher dimensional cohomology classes of spaces of knots is not completely formalized yet, however it always works successively in particular examples of comparatively low dimensional cohomology classes; it has proved the existence of several positive-dimensional classes predicted by algebraic computations.

This method is based on a deep analogy between the knot theory and the theory of plane arrangements,

Key words: knot invariant, combinatorial formula, knot space, plane arrangement, simplicial resolution, discriminant

1. INTRODUCTION

The study of knot invariants is only a subproblem of the more natural problem on the cohomology ring of the space of knots. Indeed, the numerical invariants of knots in M^3 are just the 0-dimensional cohomology classes of this space. Recent works by V. Turchin show that the finite type cohomology ring of the space of knots in $\mathbb{R}^3$ has a beautiful algebraic structure; probably the easiest description of the group of finite type invariants can be derived from it by an obvious factorization, see [25], [26], and § 7 below.

It is very convenient to consider simultaneously the spaces of knots in all spaces $\mathbb{R}^n$, $n \geq 3$. If $n > 3$ then all their cohomology classes are of finite type (filtration) in the sense of [29], [32] and can be calculated by the spectral sequence introduced in these works.

We shall mainly consider the spaces of *long knots* in $\mathbb{R}^n$, i.e. of embeddings $\mathbb{R}^1 \to \mathbb{R}^n$ coinciding with a fixed linear embedding outside some segment in $\mathbb{R}^1$, see Fig. 1. The cohomology rings of spaces of long knots and usual "compact" knots (i.e. smooth embeddings $S^1 \to \mathbb{R}^n$) are closely related. The invariants of knots of both types in $\mathbb{R}^3$ are in a natural one to one correspondence, but in higher dimensions the cohomology ring of the space of compact knots is more complicated than that for long knots.

J.M. Bryden (ed.), Advances in Topological Quantum Field Theory, 1–21.

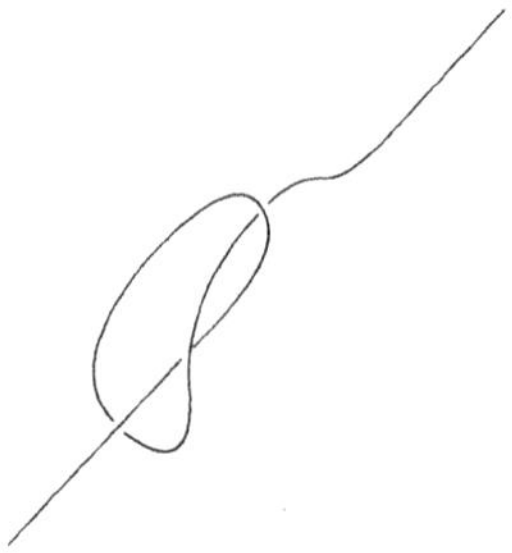

FIGURE 1. A long knot

Any cohomology class of the space of knots (both long or compact) can be realized by the linking number with some cycle (of infinite dimension but finite codimension) in the *discriminant space* of all maps $\mathbb{R}^1 \to \mathbb{R}^n$ or $S^1 \to \mathbb{R}^n$ which are not the knots, i.e. have singular or intersection points.

Definition 1 (see §2 below). Given a cohomology class of the space of knots in $\mathbb{R}^n$ (e.g. a knot invariant if $n = 3$), a *combinatorial formula* for it is an arbitrary *easy subalgebraic relative cycle* in the space of curves modulo the discriminant space Σ, such that our class is equal to the linking number with the boundary of this cycle.

There is a deep analogy between the knot theory and theory of *affine plane arrangements*, i.e. of finite families of affine planes of arbitrary (maybe different) dimensions in $\mathbb{R}^N$. This analogy is determined by the fact that the discriminant Σ in the space $\mathcal{K}$ of curves also is swept out by a family of planes in $\mathcal{K}$. This family is not discrete: it is parametrized by all unordered pairs of points $a, b \in \mathbb{R}^1$. Namely, to any such pair the plane $L(a, b)$ corresponds consisting of maps $\mathbb{R}^1 \to \mathbb{R}^n$ such that $f(a) = f(b)$ if $a \neq b$ or $f'(a) = 0$ if $a = b$.

Short lists of parallel notions of both theories are summarized in [34], [35].

In both theories, it is extremely useful to consider the *simplicial resolutions* of discriminants (respectively, arrangements), and in particular the related spectral sequences. Our method of constructing the combinatorial formulas for cohomology of the knot space $\mathcal{K} \setminus \Sigma$ is nothing else than the direct calculation of such a spectral sequence.[1]

All other methods known to the author of finding combinatorial formulas of all knot invariants deal in an essential way with spatial pictures, drawing the knot diagrams and watching their homotopies. By making a priori geometric choices, our method is purely combinatorial and deals only with easily encodable events such as chord diagrams and their natural generalizations.

[1]By the calculation of a homological spectral sequence one usually means the calculation of its isomorphism class, in particular the existence theorem claiming that any element of the calculated group $E^\infty_{p,q}$ can be extended to a cycle with this principal part; the proof of this theorem usually is implicit and follows from vanishing of all homological obstructions to such an extension. By the *direct* calculation I mean an explicit step by step construction of such cycles.

FIGURE 2. Polyak-Viro formula for the Casson invariant

I shall show here how does this method calculate the simplest non-trivial knot invariant (of order 2). Also we present the results of its work for the next invariant (of order 3), for the Teiblum–Turchin $(3n-8)$-dimensional cocycle (also of order 3) of the space of long knots in $\mathbb{R}^n$, and for all cohomology classes of order ≤ 2 of the spaces of compact knots $S^1 \hookrightarrow \mathbb{R}^n$.

A spectral sequence providing cohomology classes of the space of long knots in $\mathbb{R}^3$ was introduced in [29]. It defines some natural filtration (degree) of these classes. In the simplest case of 0-dimensional classes this filtration has an easy characterization in terms of induced indices of knots with finitely many transverse intersection points, see e.g. §0.2 in [29] and also [2]. The simplest positive-dimensional cohomology class following from this spectral sequence was discovered by D.M. Teiblum and V.E. Turchin in 1995. Again, this was rather an existence theorem: the calculation of the term $E_1^{-3,4}$ of the spectral sequence together with the fact that all the groups into which the differentials of this term can act are trivial. Moreover, it was not known whether this class actually is nontrivial.

Immediate generalizations $E_r^{p,q}(n)$ of this spectral sequence calculate all cohomology groups of spaces of knots in $\mathbb{R}^n$, $n > 3$. These sequences have a natural periodicity property: if m and n are of the same parity or we calculate $\mathbb{Z}_2$-cohomology, then $E_1^{p,q} \sim E_1^{p,q+p(m-n)}$; for m and n of different parities these groups are "super"analogues of one another. Some first calculations, including the extension of the Teiblum–Turchin class to the case of even n and the calculation of all cohomology classes of degree ≤ 2 of the space of compact knots, were given in [33], [31].

A special part of the classes arising from these spectral sequences was then studied in [7]: it are exactly the classes occurring as stabilizations of knot invariants. Also, in [7] a different filtration on the cohomology ring of the space of knots was introduced under the name "Vassiliev-order". Its value on a cohomology class always is no greater than our filtration, but often is strictly smaller: for instance, the "Vassiliev-order" of the Teiblum–Turchin class is equal to zero.

The talk follows my papers [34] and [36].

I thank A.B. Merkov and referee very much for many remarks. This work was supported in part by grants RFBR–01-01-00660, INTAS–00-0259, and NWO–047-008-005.

2. What is a combinatorial formula for a cohomology class of the space of knots

This is a comment to the above Definition 1. First, let us consider an example.

Probably the first non-trivial combinatorial formulas for certain finite type knot invariants were given by J. Lannes in [17]. For some other approaches to combinatorial formulas see e.g. [6], [20], [4], [24], [5]. The best-known and convenient combinatorial expressions for such invariants are the *Polyak–Viro formulas* [21] represented by pictures like Fig. 2 or linear combinations of similar pictures. Fig. 2 should be read as follows. Consider a generic long knot $f : \mathbb{R}^1 \to \mathbb{R}^3$. A *representation* of the picture of Fig. 2 in this knot is any collection of points $a < b < c < d \subset \mathbb{R}^1$ such that $f(a)$ lies below $f(c)$ and $f(d)$ lies below $f(b)$ (with respect to a chosen direction in $\mathbb{R}^3$). The value of this picture on our knot is equal to the number of its representations counted with appropriate signs. An immediate calculation shows that this number is a knot invariant of order 2. Moreover, for any $n > 3$ the same diagram describes a $2(n-3)$-dimensional cohomology class of the space of long knots in $\mathbb{R}^n$, the unique basic class of filtration 2. General Polyak–Viro diagrams consist of several oriented arcs connecting different points of $\mathbb{R}^1$. M. Goussarov has proved that any finite type invariant of long knots can be represented by a linear combination of such diagrams, see [13].

Let us understand in which sense the above diagram defines a subalgebraic chain.

Consider the Cartesian product $\mathcal{K} \times \mathbb{R}^4$ of the space of curves $f : \mathbb{R}^1 \to \mathbb{R}^n$ and the space of quadruples of points $a, b, c, d \in \mathbb{R}^1$. Then all the above conditions obviously define a collection of linear conditions in this enlarged space: 5 conditions of inequality type and $2(n-1)$ conditions of equality type. The relative cycle expressed by the Polyak–Viro formula is just the direct image of the fundamental cycle of the variety distinguished by these conditions under the projection to $\mathcal{K}$.

The proof of the fact that this direct image actually is a relative cycle, all of whose boundary lies in the discriminant, is slightly more complicated, cf. section 5 below.

In the finite dimensional algebraic geometry, the projections of semialgebraic sets or chains remain in the same class of objects by the Tarski-Seidenberg lemma. In our functional space, we do not have a similar fact, therefore we use the word "subalgebraic" for objects obtained as projections of semialgebraic objects from slightly greater spaces. The word "easy" in Definition 1 means that our cycle is a finite linear combination of subalgebraic chains, any of which is the projection of a semialgebraic one defined by no more inequalities than the filtration of the cohomology class; for infinite type cohomology classes easy formulas probably do not exist.

2.1. **Example: a realization of the Teiblum–Turchin cocycle mod 2.** Combinatorial formulas for cohomology classes of positive dimension can be more complicated: I do not see a general expression for them as easy as the Polyak–Viro formulas. Now we give a formula for the simplest such cohomology class, the Teiblum–Turchin class, found by D.M. Teiblum and V.E. Turchin in 1995 and described in [31], [33]. This class is of dimension $3n - 8$ (in particular of dimension 1 if $n = 3$) and of degree 3 in the sense of the natural filtration in the resolved discriminant.

Our combinatorial formula for this class is shown in Fig. 3. The sense of its three pictures is explained in three items of the following theorem.

Denote by $\mathbb{R}^{n-1}$ the quotient space of $\mathbb{R}^n$ by the chosen direction (defining the notions "above" and "below"). Choose a direction "to the east" in this space $\mathbb{R}^{n-1}$.

Theorem 1 (see [34]). *For any $n \geq 3$, the value of the reduced mod 2 Teiblum–Turchin class on any generic $(3n - 8)$-dimensional singular cycle in the space of long knots in*

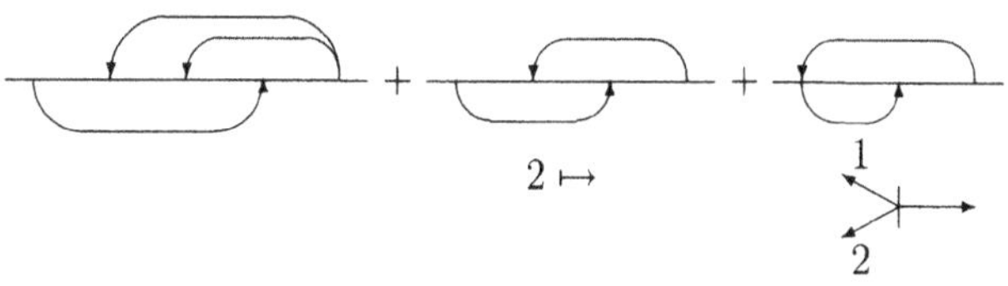

FIGURE 3. Combinatorial formula for the Teiblum–Turchin cocycle

$\mathbb{R}^n$ is equal to the parity of the number of points of this cycle corresponding to knots $f : \mathbb{R}^1 \to \mathbb{R}^n$ such that one of three holds:

1) there are five points $a < b < c < d < e$ in $\mathbb{R}^1$ such that $f(a)$ is above $f(d)$, and $f(e)$ is above both $f(c)$ and $f(b)$;

2) there are four points $a < b < c < d$ in $\mathbb{R}^1$ such that $f(a)$ is above $f(c)$, $f(b)$ is below $f(d)$, and the projection of the derivative $f'(b)$ to $\mathbb{R}^{n-1}$ is directed "to the east";

3) there are three points $a < b < c$ in $\mathbb{R}^1$such that $f(a)$ is above $f(b)$ but below $f(c)$, and the "exterior" angle in $\mathbb{R}^{n-1}$ formed by projections of $f'(a)$ and $f'(b)$ contains the direction "to the east" (i.e. this direction is equal to a linear combination of these projections, and at least one of coefficients in this combination is non-positive).

If for some knot f there exist several point configurations $a < b < \cdots$ satisfying these conditions, then the point f of the cycle should be counted for with the corresponding multiplicity.

Corollary 1. *The Teiblum–Turchin cocycle defines a non-trivial cohomology class in the most interesting case $n = 3$.*

(For $n > 3$ the same follows from dimensional reasons.)

Indeed, let us consider the connected sum of two equal (long) trefoil knots in $\mathbb{R}^3$ and a path in the space of knots connecting this knot with itself as in the proof of the commutativity of the knot semigroup: we shrink the first summand, move it "through" the second, and then blow up again. It was proved in [34] that this closed path in the space of long knots has an odd number of intersection points (counted with multiplicities) with the union of three varieties indicated in items 1, 2 and 3 of Theorem 1. But the Teiblum–Turchin cocycle is a well-defined integral cohomology class, see e.g. [31], [33]. We obtain that its reduction mod 2 takes the non-zero value on the reduction mod 2 of a well-defined integral homology class; hence also the pairing of these integral classes is not equal to zero.

Remark 1. There is an unpublished conjecture by R. Budney, F. Cohen and A. Hatcher that the space of long knots in $\mathbb{R}^3$ is a two-fold loop space which is, up to homotopy equivalence, freely generated over the little squares operad by the spaces of prime knots (whose homotopy types are discussed in [14]). The above proof of Corollary 1 shows that this little square action is non-trivial. I thank the Referee for communicating this remark.

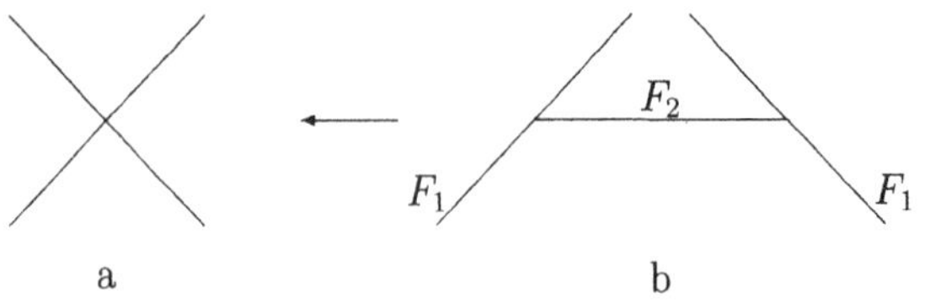

FIGURE 4. Resolution of the cross

3. SIMPLICIAL RESOLUTIONS OF PLANE ARRANGEMENTS AND DISCRIMINANTS

The most convenient way of calculating homology groups of plane arrangements (and also of discriminants in the spaces of curves) is provided by the method of *simplicial resolutions* that is a continuous analog of the combinatorial formula of inclusions and exclusions.

Our main example will be the line arrangement $\mathbf{X} \subset \mathbb{R}^2$ consisting of two crossing lines, see Fig. 4a. Suppose that we need to calculate its *Borel–Moore homology group* $\bar{H}_*(\mathbf{X})$, i.e. the homology group of its one-point compactification $\bar{\mathbf{X}}$ reduced modulo the added point. This group is related via Alexander duality with the usual (reduced modulo a point) cohomology group of the complementary space:

$$\bar{H}_i(\mathbf{X}) \sim \tilde{H}^{N-i-1}(\mathbb{R}^N \setminus \mathbf{X})$$

(in our case $N = 2$). The simplicial resolution of $\mathbf{X}$ is shown in Fig. 4b. Namely, we first take two lines forming $\mathbf{X}$ separately, and then join their common point by a segment. The resulting space $\mathbf{X}!$ admits a natural proper projection to $\mathbf{X}$ defining an isomorphism of Borel–Moore homology groups of these spaces (and moreover extending to a homotopy equivalence of their one-point compactifications). It admits a natural increasing filtration: its term F_1 consists of two divorced lines, and F_2 coincides with entire $\mathbf{X}!$.

In the case of a general arrangement Ψ in $\mathbb{R}^N$ one also takes first all planes forming it separately, and then inserts simplices spanning their common points in such a way that the resulting space $\Psi!$ admits a proper projection to Ψ with contractible (although maybe different) fibers. There are several constructions of the simplicial resolution; we use the one defined in terms of the *order complex* of our arrangement. For its definition, see e.g. [12], [38], or [35]. This resolved space $\Psi!$ always admits a natural increasing filtration of length $\leq n-1$: its ith term is the union of all *proper preimages* under the projection $\Psi! \to \Psi$ of all intersection planes of our arrangement having codimension $\leq i$ in $\mathbb{R}^N$.

In the case of the arrangement $\mathbf{X}$ of Fig. 4, the corresponding spectral sequence calculating the Borel–Moore homology group is as follows: its unique two non-zero terms

$$E^1_{p,q} \equiv \bar{H}_{p+q}(F_p \setminus F_{p-1})$$

are $E^1_{1,0} = \bar{H}_1(F_1) \cong \mathbb{Z}^2$ and $E^1_{2,-1} = \bar{H}_1(F_2 \setminus F_1) \cong \mathbb{Z}$. The reduced cohomology classes of $\mathbb{R}^2 \setminus \mathbf{X}$ of degree (= filtration) one are exactly the linear combinations of linking numbers with either of two lines forming $\mathbf{X}$.

In particular we see that the homological spectral sequence calculating the group $\bar{H}_*(\mathbf{X})$ stabilizes at the first term, and we have

(1) $$\bar{H}_i(\mathbf{X}) \equiv \bar{H}_i(\mathbf{X}!) \cong E^1_{1,i-1} \oplus E^1_{2,i-2}.$$

This is the general fact taking place for arbitrary affine plane arrangements. Moreover, there is the *homotopy splitting* formula [38], [30]: the one-point compactification of any affine plane arrangement is homotopy equivalent to the wedge of one-point compactifications of spaces $F_k \setminus F_{k-1}$ of the natural filtration of the simplicial resolution of this arrangement. The homological version of this splitting is the Goresky-MacPherson formula [12] for the cohomology of the complement of an arrangement.

The simplicial resolution of the discriminant set Σ in the space $\mathcal{K}$ of parametric curves $f : \mathbb{R}^1 \to \mathbb{R}^n$ (with fixed behavior at the infinity) can be constructed in precisely the same way. First we take the *tautological normalization* of Σ, i.e. the total space of the affine bundle, whose base is the configuration space $\overline{B(\mathbb{R}^1, 2)}$ of all unordered pairs of points $a, b \in \mathbb{R}^1$, and the fiber over such a point is the affine subspace $L(a, b) \subset \mathcal{K}$ consisting of all maps $f : \mathbb{R}^1 \to \mathbb{R}^n$ such that $f(a) = f(b)$ if $a \neq b$ or $f'(a) = 0$ if $a = b$. This normalized space is supplied with the obvious projection onto Σ and is the natural analog of the "union of lines taken separately", i.e. the set F_1 in Fig. 4b. However such spaces $L(a, b)$ with different pairs (a, b) intersect in $\mathcal{K}$; therefore we need to span their corresponding points by segments, triangles, etc. in such a way that the resulting space σ admits a natural projection onto Σ, all whose fibers are contractible.

The exact construction of this space σ can be formulated in the terms of the (naturally topologized) order complex of the space of all affine subspaces in $\mathcal{K}$ equal to intersections of several spaces of type $L(a, b)$. It also admits a natural increasing filtration $\sigma_1 \subset \sigma_2 \subset \ldots$, whose first term coincides with the tautological normalization and the common term σ_i is the union of proper preimages of all planes $L(a_1, b_1) \cap L(a_2, b_2) \cap \ldots$ of codimensions $\leq i$.

The resulting space σ is very similar to Σ. If $n > 3$ then their Borel–Moore homology groups of finite codimension are well defined and isomorphic to each other (and are Alexander dual to the cohomology group of the space of knots). If $n = 3$ then the situation is more complicated. A priori only a part of Borel–Moore homology classes of Σ (= cohomology classes of the space of knots in $\mathbb{R}^3$) can be represented by images of cycles from σ. These are exactly the *finite type* cohomology classes; their *degrees* (or *orders*) are defined by our filtration in σ. However this subgroup is quite ample: at this time, no non-trivial homology class of the space of knots in $\mathbb{R}^3$ is known on which all the finite type cohomology classes vanish.

The first term of the filtration, σ_1, is homologically trivial: it is the total space of an affine bundle over the half-plane $\mathbb{R}^2/\{(a, b) = (b, a)\}$, whose fiber is an affine space of codimension n in $\mathcal{K}$. Therefore $\bar{H}_*(\sigma_1) \equiv 0$, and the first column $E^r_{1,q}$ of the related spectral sequence identically vanishes. It is convenient to split the space σ_1 into two cells: one is the affine bundle over the open half-plane $\{(a < b)\}$, and the second is equal to its boundary and is fibered over the line $\{(a, a)\}$. Similar natural decompositions into open cells exist for all terms $\sigma_i \setminus \sigma_{i-1}$ of our spectral sequence.

Further, easy calculations show that the second column $E^1_{2,q} \equiv \bar{H}_{2+q}(\sigma_2 \setminus \sigma_1)$ is generated by the space of a fiber bundle over a 4-dimensional open cell with fiber equal to the product of an open interval and an affine subspace of codimension $2n$ in $\mathcal{K}$. By dimensional reasons, this column survives up to E^∞; moreover, the corresponding cycles survive the projection $\sigma \to \Sigma$ and form a subgroup isomorphic to $\mathbb{Z}$ in the Borel–Moore homology group of codimension $2n-5$ of Σ. By the Alexander duality this means that the unique non-trivial group of cohomology classes of degree 2 of the space of long knots lies in dimension $n-3$ and is isomorphic to $\mathbb{Z}$; for $n=3$ it is generated by the Casson knot invariant v_2.

We shall discuss this basic class v_2 in parallel with the class A generating the group of degree 2 Borel–Moore homology classes of the line arrangement from Fig. 4 reduced modulo the group of classes of degree 1.

The *principal part* of A is the homology class generating the group $\bar{H}_1(F_2(\mathbf{X}!)\setminus F_1(\mathbf{X}!))$, i.e. simply the fundamental class of the horizontal interval in Fig. 4b.

In a similar way, the principal part of the class v_2 is the cycle generating the group $\bar{H}_*(\sigma_2 \setminus \sigma_1)$. This cycle in $\sigma_2 \setminus \sigma_1$ is swept out by the triples of the form

(2) $$\left((a_1 < a_2 < b_1 < b_2) \subset \mathbb{R}^1, f, t\right),$$

where f is a map $\mathbb{R}^1 \to \mathbb{R}^n$ such that $f(a_1) = f(b_1)$, $f(a_2) = f(b_2)$, and $t \in (-1,1)$ is the parameter along an inserted interval arising in the construction of the simplicial resolution and analogous to the horizontal interval in Fig. 4b. The endpoints of any such interval lie in the bigger cell $\breve{\sigma}_1$ of the term σ_1, i.e. in the space of pairs

(3) $$\left((a < b) \subset \mathbb{R}^1, f\right)$$

such that $f(a) = f(b)$. Namely, these endpoints sweep out the sets of points (3) in $\breve{\sigma}_1$ such that additionally $f(a') = f(b')$ for some pair of points $a' < b' \in \mathbb{R}^1$ where either

(4) $$a' < a < b' < b$$

or

(5) $$a < a' < b < b'.$$

Our basic cycle (2) in $\sigma_2 \setminus \sigma_1$ is naturally depicted by the "chord diagram"

(6) 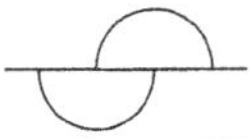 ,

which indicates the mutual disposition of possible pairs of points (a_i, b_i) glued together by the maps f participating in its definition.

Two summands of its boundary in σ_1 corresponding to two possible dispositions (4) and (5) will be denoted respectively by two parts of the expression

(7) 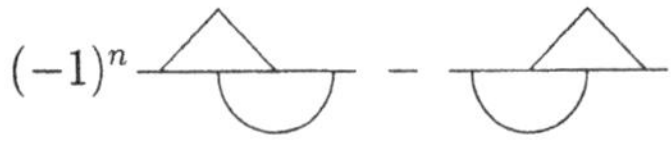

(the signs in all such expressions depend on the orientation conventions; here we use the ones from [34]).

These two summands are exact analogues of the two endpoints of the horizontal segment in Fig. 4b.

4. "COMBINATORIAL FORMULAS" FOR COHOMOLOGY OF PLANE ARRANGEMENTS

Consider again the relative cycle A in the resolved cross $\mathbf{X}!$ generating the group $E^1_{2,-1} \equiv \bar{H}_1(F_2 \setminus F_1)$. Although we have the splitting (1) for the group

$$\bar{H}_1(\mathbf{X}) \cong \tilde{H}^0(\mathbb{R}^2 \setminus \mathbf{X}),$$

this cycle A itself does not define any cohomology class and cannot take values on the zero-homologous 0-cycles in $\mathbb{R}^2 \setminus \mathbf{X}$. The formula (1) says only that this relative cycle *can be* extended to a Borel–Moore cycle in all of $\mathbf{X}$, and hence to define such a homology class. However, in order to define such a class correctly we need an *explicit construction* of such an extension.

Namely, we need first to consider the boundary of this relative cycle in F_1. Formula (1) says us that it can be *spanned* there, i.e. represented as a boundary of a locally finite chain in F_1. We need to choose such a chain A_1, then the difference $A - A_1$ will be a cycle in all of $\mathbf{X}!$. We take the direct image of this cycle in $\mathbf{X}$ and choose a relative cycle in the pair $(\mathbb{R}^2, \mathbf{X})$ spanning this cycle in $\mathbb{R}^2$. This cycle can already take values on particular points of the space $\mathbb{R}^2 \setminus \mathbf{X}$, defining thus a "combinatorial formula". In the case of a general arrangement in $\mathbb{R}^N$ we, by the very definition of the spectral sequence of a filtered space, need to do the same, but with more steps. Given an element of the group $E^1_{p,q}$, we start from a locally finite cycle γ in F_p mod F_{p-1} realizing it, then take its boundary $d^1(\gamma)$ in $F_{p-1} \setminus F_{p-2}$, span it there by a chain γ_1, consider the boundary $d^2(\gamma)$ of the cycle $\gamma - \gamma_1$ in $F_{p-2} \setminus F_{p-3}$, etc.

In the case of plane arrangements in $\mathbb{R}^N$, there is an obvious way to make all of these choices. Indeed, let us choose an arbitrary constant vector field in $\mathbb{R}^N$ generic with respect to our arrangement. It is convenient to imagine it as the gradient of a generic linear function $\mathcal{L}$. All terms $F_i \setminus F_{i-1}$ of the resolution are smooth manifolds, and we always can lift the function $\mathcal{L}$ to these manifolds and span the cycles $d^k(\gamma)$ by trajectories of gradients of these lifted functions issuing from them; the genericity condition implies that these trajectories are always transversal to these cycles.

In particular, for the arrangement $\mathbf{X}$ this procedure is shown in Fig. 5. We assume that the vector field is directed down. In the left-hand top picture we mark by a thick line only the initial relative cycle $A \subset F_2 \setminus F_1$; in the right-hand top picture we add the segments in F_1 spanning its boundary; in the right-hand bottom picture we see the projection of the resulting cycle to $\mathbf{X}$, and in the left-hand bottom picture we shadow the chain spanning this cycle in $\mathbb{R}^2$ and swept out by the trajectories of our vector field issuing from its points.

In fact, in the case of plane arrangements it is not necessary to accomplish all step of this calculation, because its result can be predicted from the very beginning: it is the realization of the Goresky–MacPherson formula given in [38].

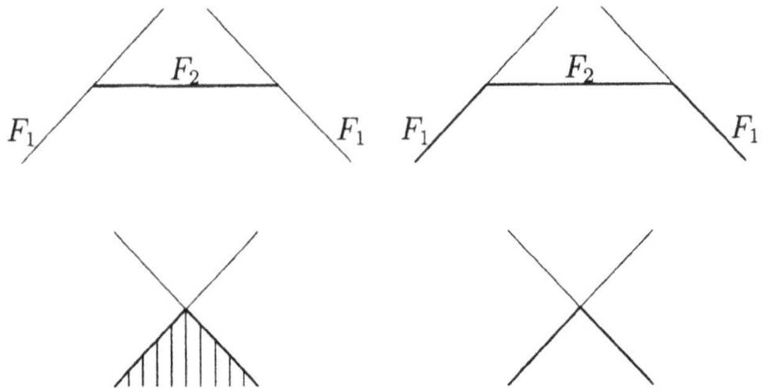

FIGURE 5. "Combinatorial formula" for the degree 2 homology class

Of course, another choice of the direction in $\mathbb{R}^N$ can give a different combinatorial formula, however the difference of these formulas will be a relative cycle of strictly lower filtration.

5. How to calculate the combinatorial formulas for knot spaces

All the same can be done with the homology classes of the discriminant in the space of knots. A formal algorithm performing this calculation of $\mathbb{Z}_2$-valued invariants of knots in $\mathbb{R}^3$ is described and justified in [36]. Its generalization to the case of integral coefficients is straightforward and also will be written soon. Similar formal algorithms for higher-dimensional cohomology classes of spaces of knots in all $\mathbb{R}^n$, $n \geq 3$, also have no chance not to exist, however a precise elaboration of such an algorithm is a more complicated problem: it involves the study of more refined degenerations of knots. Nevertheless, in the first examples (including the Turchin–Teiblum class and all cohomology classes of degree 2 of the space of compact knots) this non-formalized method proved to be efficient: for its results see Theorems 1 and 4 of this paper.

In this section we give the first illustration of our algorithm, calculating a combinatorial formula for the unique degree 2 knot invariant v_2 reduced mod 2.

Theorem 2 (see [36]). *The value of v_2 on a generic long knot $f : \mathbb{R}^1 \to \mathbb{R}^3$ is equal (mod 2) to the sum of three numbers:*

a) the number of configurations $\{a < b < c < b\} \subset \mathbb{R}^1$ such that $f(c)$ is above $f(a)$ and $f(d)$ is above $f(b)$;

b) the number of configurations $\{a < b < c\}$ such that $f(c)$ is above $f(a)$ and the projection of $f(b)$ to $\mathbb{R}^2$ lies to the east of the (common) projection of $f(a)$ and $f(c)$;

c) the number of configurations $\{a < b\}$ such that $f(b)$ is above $f(a)$ and the direction "to the east" in $\mathbb{R}^2$ is a linear combination of projections of derivatives $f'(a)$ and $f'(b)$, such that the first of these projections participates in this linear combination with a positive coefficient, and the second with a negative one.

These three numbers can be depicted by three summands in the next formula:

(8) 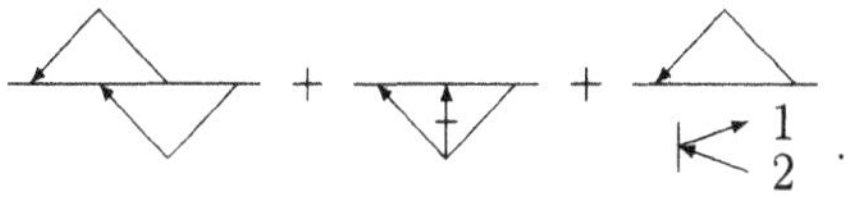

Remark 2. This formula is obviously more complicated than the Polyak–Viro formula of Fig. 2 for the same invariant. It is easy to improve the forthcoming calculation to obtain exactly the Polyak–Viro formula, see Remark 4 below. I cannot yet formalize this improvement in such a way that my algorithm itself will select the most economical computation.

The combinatorial formula for the third degree invariant, provided by our algorithm, consist of fifteen terms.

Theorem 3 (see [36]). *A combinatorial formula for the third degree invariant (mod 2) is given by the sum of fifteen subvarieties indicated in (9).*

(9) 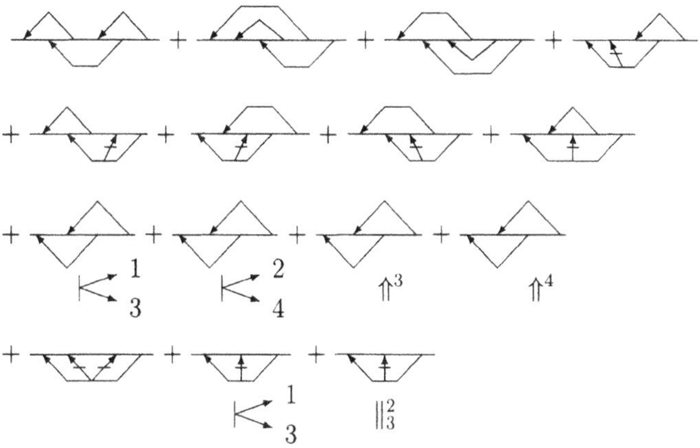

Here the notation $\Uparrow^3$ means that the projection to the blackboard plane $\mathbb{R}^2$ of the derivative of f at the third active point of the picture should be directed into the upper half-plane. The notation $\|_3^2$ means that the similar projections of derivatives at the second and the third active points should be directed into one and the same half-plane of $\mathbb{R}^2$. Other elements of this formula can be decoded exactly as (2). For instance, the first summand in the last row of (9) denotes the number of configurations $(a < b < c < d)$ such that $f(d)$ is above $f(a)$, and the projections of $f(b)$ and $f(c)$ to $\mathbb{R}^2$ are to the east of the projection of $f(a)$ or $f(d)$.

The rest of this section is occupied by a proof of Theorem 2.

5.1. **Principal part, first differential, and its homology to zero.** The principal part of v_2 in $\sigma_2 \setminus \sigma_1$ is expressed by the chord diagram shown in formula (6). The boundary of this principal part in σ_1 is shown in (7) (where we can now forget about the signs because the passage to the $\mathbb{Z}_2$-homology).

Let us span this boundary by a chain in the cell $\breve{\sigma}_1$. Recall that the first summand in (7) is the union of all points $((a,b),f) \in \sigma_1$ such that additionally there exist some two points a', b' with $a' < a < b' < b$ such that $f(a') = f(b')$.

It is natural to try to span this chain in $\breve{\sigma}_1$ by (i.e. to represent it as a piece of the boundary of) the set of points $((a,b),f)$ satisfying all the same conditions, but with the equality $f(a') = f(b')$ replaced by the condition that the projections of $f(a')$ and $f(b')$ to the "blackboard plane" $\mathbb{R}^2$ coincide, and the projection of $f(a')$ to its orthogonal line lies below the projection of $f(b')$.

The latter condition is depicted by the "broken arrow" as in the left-hand part of (10).

In a similar way, we try to span the second summand in (7) by the variety depicted in the left-hand side of (11).

Unfortunately, these two varieties contain additional pieces of boundary, so that their sum does not span the entire chain (7). These pieces correspond to possible degenerations of the configurations of four points a, b, a' and b' participating in the definition of these varieties.

Namely, the full boundaries of these two varieties are described by the right-hand sides of (10) and (11). Let us analyze for instance the first of them.

(10) 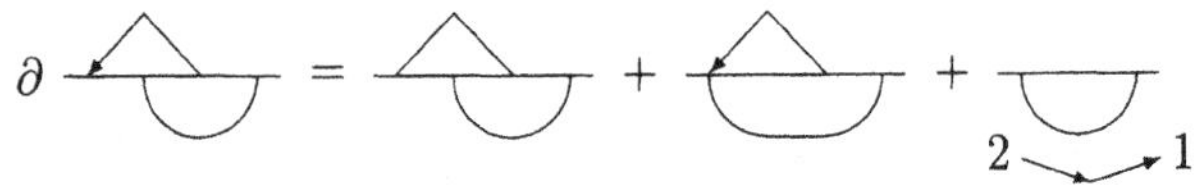

(11) 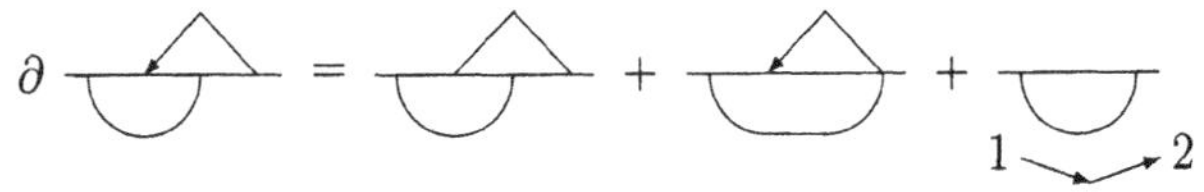

The second summand in its right-hand side appears when the first point a' and the second point a participating in the definition of our variety coincide.

In fact, this formula should have two summands more, arising when the third point b' tends to either a or b. However these two summands coincide since $f(a) = f(b)$ and cancel each other. (This happens also in the similar integral homology calculation: these two summands appear in the integral boundary of our variety with opposite orientations.)

Finally, the last summand arises when a' tends to a and simultaneously b' tends to b. The spatial picture of the corresponding degeneration is shown in Fig. 6a. The labeled arrows in the notation of this summand express the following condition: the projections to the blackboard plane $\mathbb{R}^2$ of derivatives of the knot at the points $a < b$ are co-directed there, but the direction of $f'(a)$ in $\mathbb{R}^3$ goes "above" that of $f'(b)$. The second formula (11) can be analyzed in exactly the same way.

So, we obtain that the cycle (7) is homologous in $\breve{\sigma}_1$ to the sum of second and third summands of right-hand sides of formulas (10) and (11). The sum of their second (respectively, third) summands can be depicted by the first (respectively, the second) summand in the next formula (12).

(12) 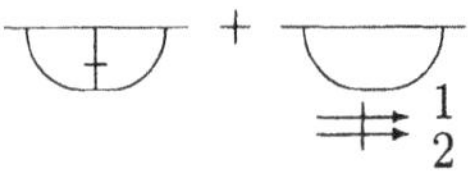

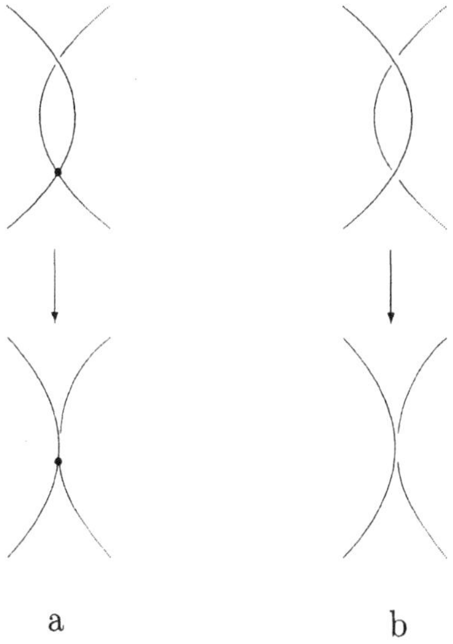

FIGURE 6. Some standard degenerations of singular knot diagrams

By definition, the first summand in (12) consists of points $((a,b),f) \in \breve{\sigma}_1$ which, in addition to the usual condition $f(a) = f(b)$, satisfy the following one: there is a point $c \in (a,b)$ such that the projection of $f(c)$ to the blackboard plane coincides with that of the point $f(a) \equiv f(b)$. Similarly, the additional condition expressed by two crossed arrows under the second summand in (12) says that projections to the blackboard plane of the derivatives $f'(a)$ and $f'(b)$ should be co-directed.

So, the sum of varieties in left-hand sides of formulas (10), (11) provides the homology between the cycles (7) and (12), and we need to find a zero-homology of the latter cycle. This cycle is simpler than the initial one: it consists of two varieties, each of which is defined by certain conditions on the behavior of the function f at 3 or 2 points of $\mathbb{R}^1$, while for both summands of the initial cycle the number of such "active" points was equal to 4. This situation occurs systematically. On any step of the algorithm we need to span a cycle consisting of several varieties, any of which is defined by certain conditions on the behavior of f at several points. At least one of these conditions should be an equality (e.g. the images in $\mathbb{R}^3$ of some two points should coincide, or some two vectors should be co-directed, etc.) Then we try to span such a variety by a similar variety in whose definition this equality is replaced by the inequality relating the same quantities. The latter variety usually has additional pieces of boundary, but these pieces appear at degenerations of the configurations of active points, hence are simpler than the initial cycle. By induction on the number of active points, the algorithm necessarily terminates. The algorithm inventory thus includes a list of varieties (subalgebraic chains) of codimension 1 (i.e. of dimension equal to that of the discriminant) in the space of curves and in the resolution of its discriminant, of which all our spanning chains can consist, plus the list of all irreducible varieties of codimension 2 of which the boundaries of the previous chains consist. These varieties are listed in §2 of [36]. All the same holds for the similar (not written yet) algorithm over the integers, which will differ from the existing mod 2 one by defining coorientations of all these varieties and calculating their incidence coefficients. The similar inventory for the algorithmic calculation of combinatorial formulas for r-dimensional cohomology classes should include similar lists of varieties of codimensions $r+1$ and $r+2$, thus these algorithms will be more complicated.

Let us apply this process to the cycle (12). We replace the equality-type condition distinguishing its first summand, i.e. the condition "there is a point $c \in (a,b)$ such that the projections of $f(c)$ and $f(a) \equiv f(b)$ to $\mathbb{R}^2$ coincide" by the inequality-type condition: "there is a point $c \in (a,b)$ such that the projection of $f(c)$ to $\mathbb{R}^2$ lies *to the east* of the projection of $f(a) = f(b)$". This condition is depicted by the once crossed arrow as in the left-hand side of (13).

Also we replace the equality-type condition distinguishing the second summand of (12), i.e. the condition "projections of vectors $f'(a)$ and $f'(b)$ to $\mathbb{R}^2$ are co-directed" by the inequality type condition: "the chosen direction "to the east" in $\mathbb{R}^2$ lies in the angle between the projections of two vectors $f'(a)$ and $-f'(b)$". The latter condition is expressed by the subscript under the left-hand side of (14).

(13) 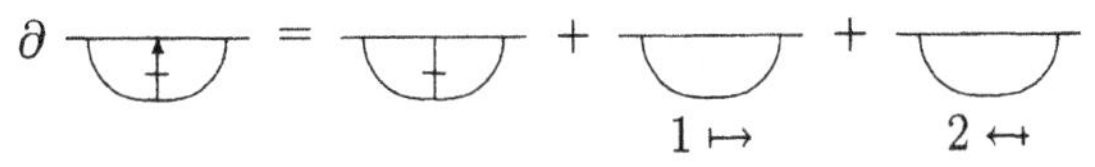

(14) 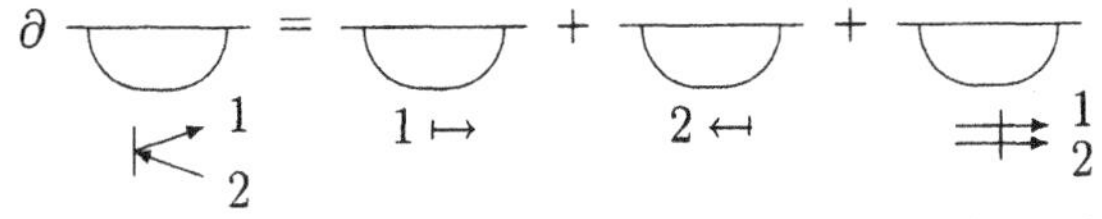

Again, the chains shown in the left-hand sides of equalities (13), (14) contain some additional pieces of boundary described in right-hand sides of these equalities. For instance, the condition $1 \mapsto$ means that the projection to $\mathbb{R}^2$ of the derivative f' at the first active point is directed to the east. The complexities of these additional terms are lower than these of terms which we try to span. For instance, the homology given by the formula (14) does not reduce the number of all active points of its very right-hand picture (i.e. the second picture in (12)), but it replaces a condition involving two such points (expressed by the subscript under this picture) by conditions involving only one point each, thus the complexity of the chain again decreases.

These additional summands in left-hand sides of equalities (13), (14) cancel each other; therefore the sum of all their summands is equal to the cycle (12). The same will necessarily hold also in the similar calculation over the integers (with proper signs before all participating terms) because of the homological conditions: all right-hand parts of our equations and also the cycle which we are going to kill have no boundaries.

Finally, we obtain that the desired chain spanning the cycle (7) in $\check{\sigma}_1$ is equal to the sum of four varieties indicated in left-hand sides of equations (10), (11), (13), and (14)). This sum establishes a homology between the cycle (7) and some cycle in the smaller cell of σ_1. By dimensional reasons, the latter cycle can be equal only to the fundamental cycle of the latter cell taken with some coefficient. Also, it is easy to see that the boundary positions of our four varieties form subvarieties of positive codimension in this cell. Therefore this coefficient is equal to zero, and the sum of our four varieties forms a zero-homology of the cycle (7) in all of σ_1; this sum together with the initial relative cycle (6) forms the cycle in σ_2 generating its group $\bar{H}_*(\sigma_2)$.

Remark 3. This is the unique dangerous instant in the integration of weight systems, i.e. the instant when this integration can fail and prove that there is no knot invariant corresponding to our initial weight system γ. Namely, if the degree of γ is equal to k then the (successful) algorithm consists of k steps. Its starting point is the Borel–Moore cycle of maximal dimension in $\sigma_k \setminus \sigma_{k-1}$ encoded by γ (the maximal cells of $\sigma_k \setminus \sigma_{k-1}$ are in the canonical one-to-one correspondence with the k-chord diagrams, while the vice-maximal cells correspond to one-term and four-term relations). On the rth step, the algorithm considers (a geometric realization of) the rth differential $d^r(\gamma)$, which is a Borel–Moore cycle of codimension 1 in $\sigma_{k-r} \setminus \sigma_{k-r-1}$, and tries to span it there, i.e. to represent as the boundary of a subalgebraic chain of the maximal dimension. If this works then the boundary of this spanning chain in $\sigma_{k-r-1} \setminus \sigma_{k-r-2}$ is the initial data $d^{r+1}(\gamma)$ for the next step.

On any such step, our algorithm no problem establishes a homology between $d^r(\gamma)$ and a cycle in the union of vice-maximal cells, the latter cycle can be only a linear combination of these cells. If the homology class of this cycle is equal to zero, i.e. the cycle is equal to the boundary of a linear combination of maximal cells, then we subtract this linear combination from the above homology and obtain the desired zero-homology of $d^r(\gamma)$. However, if this homology class is not equal to zero then we obtain an obstruction to the integration. (By the Kontsevich theorem, this is impossible over the rational numbers, but the obstruction can be a torsion element and thus prevent the integration over the integers.) In principle, it can happen that this obstruction is not fatal: if exactly the same nontrivial homology class arose previously as an obstruction to the integration of a weight system γ' of a lower degree, then the difference of chains obtained in these calculations for γ and γ' is a relative cycle in $(\sigma_k, \sigma_{k-r-1})$ which can be integrated one step further; however the system γ' will remain an example of a non-integrable system.

Fortunately, in all known examples all these obstructions are trivial, thus it remains to formulate the corresponding general conjecture. In fact, in all these examples the triviality of the obstruction follows from the fact that the boundary positions of our homologies (spanning the cycles $d^r(\gamma)$ in maximal cells of $\sigma_{k-r} \setminus \sigma_{k-r-1}$) form subvarieties of positive codimensions in the vice-maximal cells, thus the homological boundaries of these homologies in the union of the latter cells are not only homologous but even equal to zero. A stronger conjecture says that this also is a general situation.

5.2. **Second differential and its homology to zero.** Now we consider the projection of the obtained cycle from σ to Σ. The projection of its part (6) is of codimension 2 and does not participate in cycles of codimension 1 responsible for knot invariants. On the other hand, the projections to Σ of four chains in σ_1 found in the previous subsection (i.e. the left-hand sides of (10), (11), (13) and (14)) are depicted by four summands of the following formula:

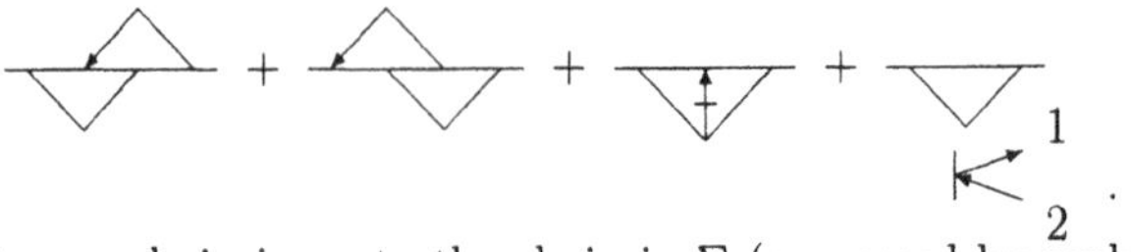

(15)

The passage from a chain in σ_1 to the chain in Σ (expressed by replacing round arcs by broken arcs without arrows) consist in posing the quantifier $\exists$: we replace the set of points $((a, b), f)$ satisfying the condition $f(a) = f(b)$ plus some other conditions by the set of maps f such that there exist points a, b such that $f(a) = f(b)$ plus all the same conditions; if f satisfies these conditions several times then we take it with the corresponding multiplicity.

Again, it is natural to try to span the obtained chain by another one, in whose definition the condition $f(a) = f(b)$ is replaced by "$f(a)$ is below $f(b)$ in $\mathbb{R}^3$". In the language of pictures, this variety is obtained from the initial one by putting an arrow at an endpoint of the broken arc. Then we obtain three pictures indicated in left-hand sides of equalities (16)–(18); note that both the first and the second summands in (15) appear as the pieces of the boundary of the left-hand variety of (16).

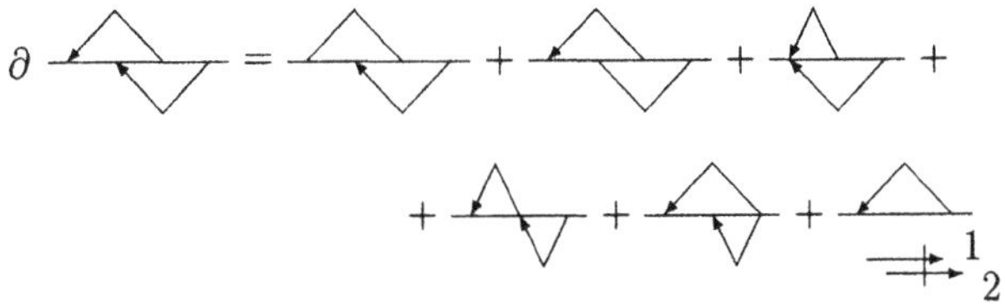

(16)

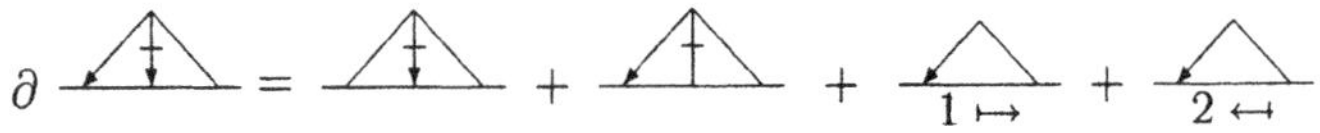

(17)

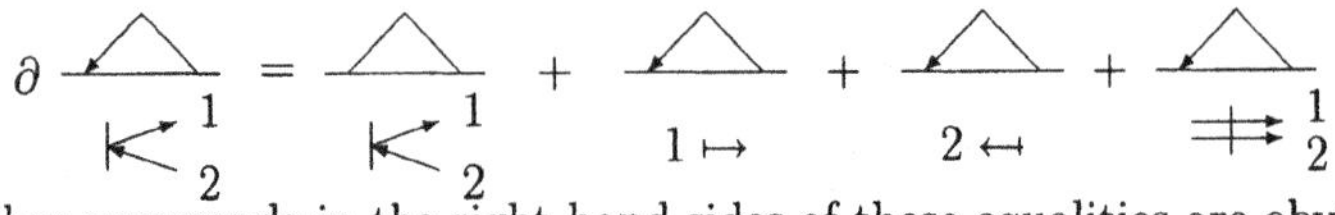

(18)

All the other summands in the right-hand sides of these equalities are obvious, except maybe for the last summand in (16) which is analogous to the last summands in (10), (11), and reflects the second Reidemeister degeneration shown in Fig. 6b.

The sum of the third, fourth, and fifth terms in the right-hand side of (16) equals the second summand in (17). Therefore the sum of right-hand sides of (16)–(18) is equal to the cycle (15), and the sum of chains indicated in left-hand sides of (16)–(18) is the desired combinatorial formula, i.e. the relative cycle of the space of curves modulo Σ, whose boundary coincides with the cycle generating the degree two homology group of Σ. This sum coincides with formula (8), and Theorem 2 is proved.

Remark 4. We could try to kill two summands of (7) not by the left parts of the equalities (10) and (11) but by similar pictures with reversed orientations of broken arcs. If we make such a switch for exactly one of these summands, then the additional summands in the right-hand parts of resulting versions of (10) and (11) cancel each other, so that the sum of their left-hand parts spans the cycle (7). Continuing our algorithm, we obtain in this case exactly the Polyak–Viro formula (2) consisting of a single term, and not of three terms given in Theorem 2. Unfortunately at this stage we have not found a way to make the process both algorithmic and sensitive to such optimal choices.

6. Results for compact knots

In this section we, following [34], describe combinatorial formulas for all cohomology classes of filtration ≤ 2 of the space of compact knots $S^1 \hookrightarrow \mathbb{R}^n$ for any $n \geq 3$. First we, following [33], list all such classes.

We assume that a cyclic coordinate in S^1, i.e. an identification $S^1 \simeq \mathbb{R}^1/2\pi\mathbb{Z}$, is fixed.

Proposition 1 (see [32], [33]). *For any $n \geq 3$ the group of $\mathbb{Z}_2$-cohomology classes of degree 1 of the space of compact knots in $\mathbb{R}^n$ is nontrivial only in dimensions $n-2$ and $n-1$, and is isomorphic to $\mathbb{Z}_2$ in these dimensions. Moreover, for (only) even n similar integral cohomology groups in these dimensions are isomorphic to $\mathbb{Z}$. The generator of the $(n-2)$-dimensional group is Alexander dual to the set of discriminant maps $S^1 \to \mathbb{R}^n$ gluing together some two* opposite *points of S^1, and the $(n-1)$-dimensional one is dual to the set of maps gluing some* chosen *opposite points, say 0 and π.*

Indeed, the first degree cohomology classes ere exactly the ones that can be realized by linking numbers with direct images of some Borel–Moore homology classes of the tautological resolution σ_1 of the discriminant. In the case of compact knots, σ_1 is the space of a vector bundle over the closed Moebius band (i.e. the configuration space of unordered pairs of points $(a, b) \subset S^1$), the fiber over the point (a, b) consisting of all maps $f : S^1 \to \mathbb{R}^n$ such that $f(a) = f(b)$ if $a \neq b$ or $f'(a)$ if $a = b$. This bundle is (co)orientable if and only if n is even. This gives us the column $E_1^{-1,*}$. It is obvious that the dual cohomology classes are well-defined (i.e. all differentials acting from this column are trivial). The fact that these classes are nontrivial follows from their very easy realization: for any $n \geq 3$ they are nontrivial already in the restriction to the space of all naturally parametrized big circles of the unit sphere in $\mathbb{R}^n$ (equal to the Stiefel variety $V(n, 2)$). For $n > 3$ this fact follows also immediately from the shape of the spectral sequence.

Proposition 2 (see [31], [33]). *Additional classes of degree 2 exist in exactly two dimensions: $2n-6$ and $2n-3$. In dimension $2n-6$ for any n they form a group isomorphic to $\mathbb{Z}$ (for $n = 3$ it is generated by the simplest knot invariant). The group in dimension $2n-3$ is isomorphic to $\mathbb{Z}$ for $n > 3$ and is cyclic for $n = 3$; its generator is Alexander dual to the cycle in the discriminant, whose* principal part *(i.e. the analog of the cycle (6) in the double selfintersection of Σ) is swept out by maps $f : S^1 \to \mathbb{R}^n$ such that for some $\alpha \in S^1$ we have $f(\alpha) = f(\alpha + \pi)$, $f(\alpha + \pi/2) = f(\alpha + 3\pi/2)$.*

Again, the proof of this proposition consists in the direct calculation of the column $E_1^{-2,q}$ of the spectral sequence (which has exactly two non-trivial groups, both isomorphic to $\mathbb{Z}$), and dimensional considerations assuring that all these classes (except maybe for some elements of the group $E_1^{-2,5}$ in the case $n = 3$) survive and define non-trivial elements of the term E_∞.

Below we show in particular that for $n = 3$ the latter group also survives and defines a free cyclic subgroup in the cohomology of the space of knots, see Corollary 3. Now we give explicit combinatorial formulas for all classes mentioned in Propositions 1 and 2.

Theorem 4 (see [34]). *For any $n \geq 3$, the values of any of these four cohomology classes on any generic cycle of corresponding dimension in the space $\mathcal{K}_n \setminus \Sigma$ of compact knots in $\mathbb{R}^n$ is equal to the number of points of this cycle, corresponding to knots satisfying the following conditions (and in the case of integer coefficients taken with appropriate signs).*

A. For the $(n-1)$-dimensional class of degree 1: projections of $f(0)$ and $f(\pi)$ into the plane $\mathbb{R}^{n-1}$ coincide, and $f(0)$ is "higher" than $f(\pi)$.

B. For the $(n-2)$-dimensional class of degree 1, one of the following two conditions:

a) there is a point $\alpha \in [0, \pi)$ such that the projections of $f(\alpha)$ and $f(\alpha + \pi)$ to $\mathbb{R}^{n-1}$ coincide, and moreover $f(\alpha)$ is "higher" than $f(\alpha + \pi)$;

b) the projection of the point $f(0)$ to $\mathbb{R}^{n-1}$ lies "to the east" of the projection of $f(\pi)$.

C. For the $(2n-3)$-dimensional class of degree 2, one of the following two conditions:

a) there is a point $\alpha \in [0, \pi/2)$ such that projections of $f(\alpha)$ and $f(\alpha + \pi)$ to $\mathbb{R}^{n-1}$ coincide, projections of $f(\alpha + \pi/2)$ and $f(\alpha + 3\pi/2)$ to $\mathbb{R}^{n-1}$ coincide, and additionally $f(\alpha + \pi)$ is "higher" than $f(\alpha)$ and $f(\alpha + \pi/2)$ is "higher" than $f(\alpha + 3\pi/2)$;

b) projections of $f(0)$ and $f(\pi)$ to $\mathbb{R}^{n-1}$ coincide, $f(\pi)$ is "higher" than $f(0)$, and the projection of $f(\pi/2)$ to $\mathbb{R}^{n-1}$ is "to the east" of the projection of $f(3\pi/2)$.

D. For the $(2n-6)$-dimensional class of degree 2, one of two conditions:

a) there are four distinct points $\alpha, \beta, \gamma, \delta \in S^1$ (whose cyclic coordinates satisfy the inequalities $0 \leq \alpha < \beta < \gamma < \delta < 2\pi$) such that projections of $f(\alpha)$ and $f(\gamma)$ to $\mathbb{R}^{n-1}$ coincide, projections of $f(\beta)$ and $f(\delta)$ to $\mathbb{R}^{n-1}$ coincide, and additionally $f(\gamma)$ is "higher" than $f(\alpha)$ and $f(\beta)$ is "higher" than $f(\delta)$.

b) If $n = 3$ then the second condition is void (and we have only the first one coinciding with the Polyak–Viro formula), but for $n > 3$ we have an additional condition: there are three distinct points β, γ, δ (whose cyclic coordinates satisfy the inequalities $0 < \beta < \gamma < \delta < 2\pi$) such that projections of $f(\gamma)$ and $f(0)$ to $\mathbb{R}^{n-1}$ coincide, $f(\gamma)$ is "higher" than $f(0)$, and the projection of $f(\delta)$ to $\mathbb{R}^{n-1}$ is "to the east" of the projection of $f(\beta)$.

Corollary 2. *For any $n \geq 3$, the basic class of degree 2 and dimension $2n - 3$ takes value ± 1 on the fundamental cycle of the submanifold of the space of knots, consisting of all naturally parametrized great circles of the unit sphere in $\mathbb{R}^n$.*

Indeed, the variety a) of statement C does not intersect this submanifold, and variety b) has with it exactly one intersection point.

In the case of even n the fact that the fundamental class of this submanifold is not homologous to zero in the space of knots was proved in [7] by very different methods.

Corollary 3. *The group of $(2n-3)$-dimensional cohomology classes of degree 2 is free cyclic for $n = 3$ as well.*

Our proofs of Theorems 1 and 4 are very similar to that of Theorem 2. In particular, in the notation of [36] the computation of the combinatorial formula for the Teiblum–Turchin cocycle starts from the generalized weight system

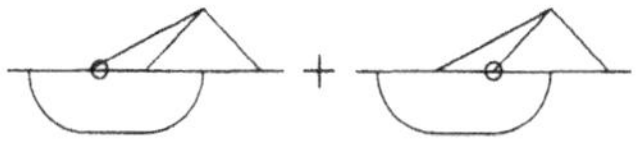

describing some cellular cycle of codimension one in $\sigma_3 \setminus \sigma_2$ and contained in the resolution of the space of curves having one double and one triple point.

Remark 5. In §8 of [7] the map of cohomology rings induced by the inclusion of the space of compact knots into that of immersions $S^1 \to \mathbb{R}^n$ was considered. Our spectral sequence is situated very well for the study of this map (and similar maps for knots in other manifolds), because the "small" discriminant space of maps that are not immersions is a subspace of our discriminant of non-inclusions, and the simplicial resolution of the former discriminant is a subspace of that of the latter. Therefore we arrive at an easy comparison of spectral sequences; in particular these maps in cohomology should respect the filtrations. All multiplicative generators of the rational cohomology ring of the space of immersions are of filtration 1 (if n is even) or 1 and 2 (if n is odd), and the above two propositions say us everything on the cohomology classes of these degrees for the knot space.

Remark 6. In any term $\sigma_k \setminus \sigma_{k-1}$ of the resolution of the discriminant in the space of curves $S^1 \to \mathbb{R}^3$ there is the important open subset $\Box_k$ related with k-fold selfintersections of these curves: it is the space of all triples

$$(C, f, t),$$

where C is a k-chord configuration (i.e. a collection of $2k$ points in S^1 matched into pairs), f is a map $S^1 \to \mathbb{R}^n$ gluing together the points of any pair, and t a point of some $(k-1)$-dimensional simplex arising in the construction of the resolution. For any k the connected components of this set are in the obvious one-to-one correspondence with k-chord diagrams.

The "Vassiliev order" defined in [7] for any cohomology class of the space of knots (or, equivalently, for its Alexander dual Borel–Moore homology class γ of Σ) is the greatest k such that the restriction of γ defines a non-zero class in $\bar{H}_*(\Box_k)$. This index obviously does not exceed the filtration from [29], [33] (defined in exactly the same way but with entire $\sigma_k \setminus \sigma_{k-1}$ instead of $\Box_k$) and sometimes is strictly below it.

7. Appendix: V. Turchin's calculation

The theory of finite type invariants of knots has led to many beautiful algebraic objects, such as the Hopf algebra of chord diagrams and graph-complex of trees, see e.g. [16], [3].

It was shown recently by V. Turchin [25], [26] that these structures are non-separable parts of more general theories, related with entire cohomology rings of spaces of knots and formulated in terms of generalized chord diagrams listed in [29]. The corresponding multiplicative structures resemble the shuffle multiplication in the cohomology of complements of plane arrangements, see [37], [9], [10], but are, of course, much more complicated. It was proved in [25] and [26] that the first term of the main spectral sequence calculating the rational homology of the space of long knots in $\mathbb{R}^n$, $n \geq 3$, can be described in terms of the Hochschild homology of the Poisson algebras operad if n is odd (respectively, of the Gerstenhaber algebras operad if n is even). Namely, the Hochschild homology of these operads is in both cases a polynomial algebra in infinitely many even and odd variables. To obtain the first term of the spectral sequence in the case of even n we need to factor the corresponding polynomial algebra by one odd generator $[x_1, x_2]$. In the case of odd n we need to factor by two generators: one even (equal to $[x_1, x_2]$) and one odd (equal to $[[x_1, x_3], x_2]$).

In particular, the standard bialgebra of chord diagrams factored through the 4-term relations (see [15], [3]) is some subspace in the Hochschild homology of the Poisson algebras operad. To obtain the algebra of finite degree invariants (i.e. cohomology of degree zero in the case $n = 3$) we should factor this bialgebra by one generator $[x_1, x_2]$.

The Poisson operad occurs naturally also in the Goodwillie–Sinha approach to the topology of knot spaces, see [23].

REFERENCES

[1] V.I. Arnold, On some topological invariants of algebraic functions. *Transact. (Trudy) Mosc. Mat. Soc.* **21** (1970), 27–46.

[2] Bar-Natan, D. (1994–) *Bibliography of Vassiliev Invariants.* Web publication **http://www.ma.huji.ac.il/~drorbn/VasBib/VasBib.html**

[3] D. Bar-Natan, On the Vassiliev knot invariants, *Topology,* **34** (1995), 423–472.

[4] R.Bott and C.H.Taubes, On the self-linking of knots, *J. of Mathematical Physics,* 35(10), 1994, 5247–5287.

[5] R. Budney, J. Conant, K. Scannell, and D. Sinha. *New perspectives of self-linking.* **math.GT/0303034**, 2003.

[6] P. Cartier, Construction combinatoire des invariants de Vassiliev, *C.R.Acad. Sci. Paris, Série I,* **316** (1993), 1205–1210.

[7] A.S. Cattaneo, P. Cotta-Ramusino, and R. Longoni, Configuration spaces and Vassiliev classes in any dimension, *Algebr. Geom. Topol.,* **2** (2002), 949–1000.

[8] F. Cohen and S. Gitler, Loop spaces of configuration spaces, *Cohomological methods in homotopy theory* (Bellaterra 1998), 59–78, *Progr. Math.,* Birkhauser, Basel, 2001.

[9] P. Deligne, M. Goresky, and R. MacPherson, L'algebre de cohomologie du complément, dans un espace affine, d'une famille finie de sous-espaces affines, *Michigan J. Math.* **48** (2000), 121–136.

[10] M. de Longueville and C. Schultz, The cohomology rings of complements of subspace arrangements, Math. Annalen. 319 (2000), 625–646.

[11] T. Goodwillie, J. Klein, and M. Weiss, Spaces of smooth embeddings, disjunction and surgery. Surveys on surgery theory, Vol. 2, 221–284, *Ann. of Math. Studies,* **149**, Princeton Univ. Press, Princeton, NJ, 2001.

[12] M. Goresky and R. MacPherson, *Stratified Morse Theory,* Springer, Berlin a.o., 1988

[13] M. Goussarov, M. Polyak, and O. Viro, Finite type invariants of classical and virtual knots, *Topology* **39**:5 (2000), 1045–1068.

[14] A. Hatcher, Spaces of knots, **http://math.cornell.edu/~hatcher**

[15] Kontsevich, M. (1993) Vassiliev's knot invariants, in *Adv. in Sov. Math.,* **16:2,** AMS, Providence RI, 137–150.

[16] M. Kontsevich, Formal (non-)commutative symplectic geometry. In: L. Corvin, I. Gel'fand, J. Lepovsky (eds.), *The I.M. Gel'fand's mathematical seminars 1990-1992*, 1993, Birkhäuser, Basel, 173–187.

[17] J. Lannes, Sur les invariants de Vassiliev de degreé inferieur ou égal à 3. *L'Enseignement Mathématique* **39** (1993), 295–316.

[18] A.B. Merkov, Vassiliev invariants classify plane curves and doodles, Math. Sbornik 194:9 (2003), see also at **http://www.pdmi.ras.ru/˜arnsem/papers**.

[19] A.B. Merkov, Segment-arrow diagrams and invariants of ornaments, *Mat. Sbornik* **191** (2001), 1635–1666.

[20] S.A. Piunikhin, *Combinatorial expression for universal Vassiliev link invariant*, Harvard Univ. preprint **hep-th/9302084**, 1993.

[21] M. Polyak and O. Viro, Gauss diagram formulas for Vassiliev invariants, *Internat. Math. Res. Notes* **11** (1994), 445–453.

[22] M. Polyak and O. Viro, On the Casson knot invariant, *Tel Aviv University and Uppsala University preprint,* 1996, **math.GT/9903158**

[23] D. Sinha, The topology of spaces of knots, **math.AT/0202287**, 2002.

[24] D. Thurston, Integral expressions for the Vassiliev knot invariants, **math.QA/9901110**, 1995.

[25] V. Tourtchine, Sur l'homologie des espaces des nœuds non-compacts, **math.QA/0010017**, 2000.

[26] V. Tourtchine, Sur les questions combinatoires de la théorie spectrale des nœuds, These de Doctorate, Universite Paris-7, 2002.

[27] S.D. Tyurina, On the Lannes and Viro-Polyak type formulas for finite type invariants, *Math. Notes,* **66** (1999), No.3–4, 525–530.

[28] V.A. Vassiliev, *Complements of discriminants of smooth maps: topology and applications, Revised ed.*, Translations of Math. Monographs, AMS, Providence RI, 1994.

[29] V.A. Vassiliev, Cohomology of knot spaces, Theory of Singularities and its Applications (V. I. Arnold, ed.), Advances in Soviet Math. Vol. **1** (1990), p. 23–69 (AMS, Providence, RI).

[30] V.A. Vassiliev, Complexes of connected graphs, in: L. Corvin, I. Gel'fand, J. Lepovsky (eds.), *The I. M. Gel'fand's mathematical seminars 1990–1992,* 1993, Birkhäuser, Basel, 223–235.

[31] V.A. Vassiliev, *Topology of complements of discriminants,* Phasis, Moscow (in Russian), 1997.

[32] V.A. Vassiliev, On invariants and homology of spaces of knots in arbitrary manifolds, in: B. Feigin and V. Vassiliev, eds. *Topics in Quantum Groups and Finite-Type Invariants. Mathematics at the Independent University of Moscow.* AMS Translations. Ser. 2. Vol. **185**. Advances in the Mathematical sciences. AMS, Providence RI, 1998, p. 155–182.

[33] V.A. Vassiliev, Topology of two-connected graphs and homology of spaces of knots, in: S. L. Tabachnikov (ed.), *Differential and Symplectic Topology of Knots and Curves,* AMS Transl., Ser. 2, **190**, AMS, Providence RI, 1999, 253–286.

[34] V.A. Vassiliev, On combinatorial formulas for cohomology of spaces of knots, *Moscow Math. J.*, **1**:1 (2001), 91–123.

[35] V.A. Vassiliev, Topology of plane arrangements and their complements, Russian Math. Surveys, **56**:2 (2001), 167–203.

[36] V.A. Vassiliev, Combinatorial computation of combinatorial formulas for knot invariants, available at **http://www.pdmi.ras.ru/˜arnsem/papers**, to appear in Transact. Moscow Math. Soc.

[37] S. Yuzvinsky, Small rational model of subspace complement, Translations AMS; **http://xxx.lanl.gov/abs/math.CO/9806143**, 1999.

[38] G.M. Ziegler and R.T. Živaljević, Homotopy type of arrangements via diagrams of spaces, *Math. Ann.* **295** (1993), 527–548.

On the Homology of the Spaces of Long Knots

V. Tourtchine*

Abstract. This paper is a more detailed version of [38], where the first term of the Vassiliev spectral sequence (computing the homology of the space of long knots in $\mathbb{R}^d$, $d \geq 3$) was described in terms of the Hochschild homology of the Poisson algebras operad for d odd, and of the Gerstenhaber algebras operad for d even. In particular, the bialgebra of chord diagrams arises as some subspace of this homology. The homology in question is the space of characteristic classes for Hochschild cohomology of Poisson (resp. Gerstenhaber) algebras considered as associative algebras. The paper begins with necessary preliminaries on operads.

Also we give a method to simplify the computations of the first term of the Vassiliev spectral sequence. We do not give proofs of the results.

Keywords: discriminant of the space of knots, bialgebra of chord diagrams, Hochschild complex, operads of Poisson – Gerstenhaber – Batalin-Vilkovisky algebras

Mathematics Subject Classification 2000:-: -Primary: 57Q45 : Secondary: 57Q35, 18D50, 16D03, 16E40, 55P48

0. Introduction

First we recall some known facts on the Vassiliev spectral sequence and then proceed to explaining of the main idea of the work.

0.1.

Let us fix a non-trivial linear map $l : \mathbb{R}^1 \hookrightarrow \mathbb{R}^d$. We will consider the *space of long knots, i. e.*, of injective smooth non-singular maps $\mathbb{R}^1 \hookrightarrow \mathbb{R}^d$, that coincide with the map l outside some compact set (this set is not fixed). The long knots form an open everywhere dense subset in the affine space $\mathcal{K}$ of all smooth maps $\mathbb{R}^1 \to \mathbb{R}^d$ with the same behavior at infinity. The complement $\Sigma \subset \mathcal{K}$ of this dense subset is called the *discriminant space.* It consists of the maps having self-intersections or singularities. Any cohomology class $\gamma \in H^i(\mathcal{K}\backslash\Sigma)$ of the knot space can be realized as the linking coefficient with an appropriate chain in Σ of codimension $i+1$ in $\mathcal{K}$.

Following [46] we will assume that the space $\mathcal{K}$ has a very large but finite dimension ω. A partial justification of this assumption uses finite dimensional approximations of $\mathcal{K}$. Below we indicate by quotes non-rigorous assertions using this assumption and needing a reference to [42, 43] for such a justification.

The main tool of Vassiliev's approach to computation of the (co)homology of the knot space is the *simplicial resolution* σ (constructed in [42]) of the discriminant Σ. This resolution is also called the *resolved discriminant.* The natural projection $\Pi : \bar\sigma \to \bar\Sigma$ is a "homotopy equivalence" between the "one-point compactifications" of the spaces σ and Σ. By the "Alexander duality", the reduced homology groups $\tilde H_*(\bar\sigma, \Bbbk) \equiv \tilde H_*(\bar\Sigma, \Bbbk)$ of these compactifications "coincide" (up to a change of dimension) with the cohomology groups of the space of knots:

* Partially supported by the grants NSh-1972.2003.01, RFBR-00-15-96084, MK-451.2003.01

J.M. Bryden (ed.), Advances in Topological Quantum Field Theory, 23–52.

$$\tilde{H}^i(\mathcal{K}\backslash\Sigma, \mathbb{k}) \simeq \tilde{H}_{\omega-i-1}(\bar{\Sigma}, \mathbb{k}) \equiv \tilde{H}_{\omega-i-1}(\bar{\sigma}, \mathbb{k}), \tag{0.1}$$

where $\mathbb{k}$ is a commutative ring of coefficients.

In the space σ there is a natural filtration

$$\emptyset = \sigma_0 \subset \sigma_1 \subset \sigma_2 \subset \dots . \tag{0.2}$$

CONJECTURE 0.3. *The spectral sequence (called Vassiliev's main spectral sequence) associated with the filtration (0.2) and computing the "Borel-Moore homology groups of the resolution σ" stabilizes over $\mathbb{Q}$ in the first term.* □

CONJECTURE 0.4. (due to Vassiliev) *Filtration (0.2) "homotopically splits", in other words, $\bar{\sigma}$ is "homotopy equivalent" to the wedge $\bigvee_{i=1}^{+\infty}(\bar{\sigma}_i/\bar{\sigma}_{i-1})$.* □

This conjecture would imply the stabilization of our *main spectral sequence* in the first term over any commutative ring $\mathbb{k}$ of coefficients.

Due to the Alexander duality, the filtration (0.2) induces the filtrations

$$H^*_{(0)}(\mathcal{K}\backslash\Sigma) \subset H^*_{(1)}(\mathcal{K}\backslash\Sigma) \subset H^*_{(2)}(\mathcal{K}\backslash\Sigma) \subset \dots, \tag{0.5}$$

$$H_*^{(0)}(\mathcal{K}\backslash\Sigma) \supset H_*^{(1)}(\mathcal{K}\backslash\Sigma) \supset H_*^{(2)}(\mathcal{K}\backslash\Sigma) \supset \dots \tag{0.6}$$

in respectively the cohomology and homology groups of the space of knots. For $d \geq 4$ the filtrations (0.5), (0.6) are finite for any dimension $*$. The Vassiliev spectral sequence in this case computes the graded quotient associated with the above filtrations.

In the most intriguing case $d = 3$ almost nothing is clear. For the dimension $* = 0$ the filtration (0.5) does not exhaust the whole cohomology of degree zero. The knot invariants obtained by this method are called the Vassiliev invariants, or invariants of finite type. One can define them in a more simple and geometrical way, see [42]. The dual space to the graded quotient of the space of finite type knot invariants is the *bialgebra of chord diagrams.* The invariants and the bialgebra in question were intensively studied in the last decade, see [1, 4, 9, 11, 12, 23, 26, 27, 30, 34, 37, 41, 47, 48, 49]. The completeness conjecture for the Vassiliev knot invariants is the question about the convergence of the filtration (0.6) to zero for $d = 3$, $* = 0$. The realization theorem of M. Kontsevich [27] proves that the Vassiliev spectral sequence over $\mathbb{Q}$ for $d = 3$, $* = 0$ also computes the corresponding associated quotient (for positive dimensions $*$ in the case $d = 3$ even this is not for sure) and does stabilize in the first term. The groups of the associated graded quotient to filtrations (0.5), (0.6) in the case $d = 3$, $* > 0$ are some quotient groups of the groups calculated by Vassiliev's main spectral sequence.

To compute the first term of the main spectral sequence, V. A. Vassiliev introduced an *auxiliary filtration* in the spaces $\sigma_i\backslash\sigma_{i-1}$, see [42, 45]. The *auxiliary spectral sequence* associated to this filtration degenerates in the second term, because its first term (for any i) is concentrated at only one line. Therefore *the second term of the auxiliary spectral sequence is isomorphic to the first term of the main spectral sequence.* The term $E_0^{*,*}$ of the auxiliary spectral sequence together with its differential (of degree zero) is a direct sum of tensor products of complexes of connected graphs. The homology groups of the complex of connected graphs with m labelled vertices are concentrated in the dimension

$(m-2)$ only, and the only non-trivial group is isomorphic to $\mathbb{Z}^{(m-1)!}$, see [44, 45]. This homology group has a nice description as the quotient by the 3-term relations of the space spanned by trees (with m labelled vertices), see [39, 40].

0.2.

In fact V. A. Vassiliev considered only cohomological case of the main and auxiliary spectral sequences, *i. e.*, the case corresponding to the homology of the discriminant and (by the Alexander duality) to the cohomology of the knot space $\mathcal{K}\backslash\Sigma$. This was because a convenient description only of the homology of complexes of connected graphs was known. It was noticed in [39, 40] that the cohomology of complexes of connected graphs (with m vertices) admits also a very nice description as the m-th component of the Lie algebras operad. This isomorphism comes from the following observation.

Let us consider the space $F(\mathbb{R}^d, M)$ of injective maps from a finite set M of cardinality m into $\mathbb{R}^d$, $d \geq 2$. This space can be viewed as a finitedimensional analogue of the knot spaces. The corresponding discriminant (consisting of non-injective maps $M \to \mathbb{R}^d$) has also a simplicial resolution, whose filtration analogous to (0.2) does split homotopically, see [43, 45]. The superior non-trivial term $\sigma_{m-1}\backslash\sigma_{m-2}$ of the filtration provides exactly the complex of connected graphs with m vertices labelled by the elements of M. Its only non-trivial homology group is isomorphic by the Alexander duality to the cohomology group in the maximal degree of the space $F(\mathbb{R}^d, M)$. On the other hand the above space is homotopy equivalent to the m-th space of the little cubes operads, see Section 4. The homology operad of this topological operad is well known, see [13] (and Section 4). For different d of the same parity this homology operad is the same up to a change of grading and for odd (resp. even) d it is the Poisson (resp. Gerstenhaber) algebras operad containing in the maximal degree the operad of Lie algebras with even (resp. odd) bracket. (So, we have that the only non-trivial homology group of the complex of connected graphes with m vertices is isomorphic to the m-th component of the Lie algebras operad.)

An analogous periodicity takes place for the spaces of knots. The degree zero term of the main spectral sequence together with its differential (of degree zero) depends up to a change of grading on the parity of d only. Obviously the same is true for the whole auxiliary spectral sequence.

The above description of the cohomology of complexes of connected graphs allows one to describe easily the Vassiliev spectral sequence in the homological case. The main results of these computations are explained in Section 5 (see Theorems 5.4, 5.8, 5.9, 5.10). The proofs in full detail are given in [39], (see also the Russian version [40]).

0.3.

The paper is organized as follows.

In Sections 1, 2, 3 we give some preliminaries on linear graded operads. We give a short definition and examples (that will be useful for us) of linear graded operads (Section 1). We construct a graded Lie algebra structure on the space of any linear graded operad (Section 2). We define a Hochschild complex for any graded linear operad endowed with a morphism from the associative algebras operad to this operad (Section 3). The content of Sections 2 and 3 was borrowed (up to a slightly different definition of signs) from [24].

Section 4 is devoted to the May operad of little cubes and to its homology. As it was already mentioned the m-th component of this operad is a space homotopy equvalent to the space $F(\mathbb{R}^d, m)$ of injective maps $\{1, 2, \ldots, m\} \hookrightarrow \mathbb{R}^d$.

In Section 4 we explain how the stratification in the discriminant set of non-injective maps $\{1, 2, \ldots, m\} \to \mathbb{R}^d$ corresponds to a direct sum decomposition of the homology $H_*(F(\mathbb{R}^d, m))$.

In Section 5 we describe a natural stratification in the discriminant set Σ of long singular "knots". This stratification provides a direct sum decomposition of the first term of the auxiliary spectral sequence. In the case of even d the first term of Vassiliev's auxiliary spectral sequence is completely described by the following theorem:

Theorem 5.4. *The first term of Vassiliev's homological auxiliary spectral sequence together with its first differential is isomorphic to the normalized Hochschild complex of the Batalin-Vilkovisky algebras operad.* □

Unfortunately in the case of odd d it is not possible to describe the corresponding complex in terms of the Hochschild complex for some graded linear operad. A description of this complex is given in Section 8. Nevertheless the homology of the complex in question (*i. e.*, the first term of the main spectral sequence) over $\mathbb{Q}$ can be defined in terms of the Hochschild homology of the Poisson algebras operad in the case of odd d (and of the Gerstenhaber algebras operad in the case of even d). A precise statement is given by Theorem 5.10.

In Section 5 we also introduce complexes homologically equivalent (for any commutative ring $\Bbbk$ of coefficients) to the first term of the auxiliary spectral sequence. These complexes simplify a lot the computations of the second term.

In Section 6 we explain how the bialgebra of chord diagrams arises in our construction. We formulate some problems concerning it.

Section 7 does not contain any new results and serves rather to explain one remark of M. Kontsevich. We study there the homology operads of some topological operads that we call *operads of turning balls*. These homology operads make more clear the difference that we have in the cases of odd and even d.

In Section 8 we describe the first term of the auxiliary spectral sequence together with the degree 1 differential both for d even and odd. The corresponding complex is called *Complex of bracket star-diagrams*.

In Section 9 we construct a differential bialgebra structure on this complex. We conjecture that this differential bialgebra structure is compatible with the homology bialgebra structure of the space of long knots.

In Section 10 we give an intuitive geometric inerpretation of the algebraic structures on the first term of the Vassiliev main spectral sequence (*i.e.* multiplication, comultiplication, Gerstenhaber bracket).

0.4.

Let us mention two alternative approaches studying the ((co)homology of the) spaces of knots.

The first one consists in constructing of real cohomology classes of the knot spaces by means of configuration space integrals. This approach generalizes in a non-trivial way the Vassiliev knot invariants obtained in three dimensions from the Chern-Simons perturbation theory, see [1, 5, 9, 10, 23, 34]. A recent result in this direction is due to

A. S. Cattaneo, P. Cotta-Ramusino, R. Longoni, *cf.* [10]. They constructed a map from some graph-complex endowed with a structure of differential Hopf algebra to the de Rham complex of the spaces of long knots. They have shown that this map of complexes respects multiplication and respects commultiplication at the homology level. As in our case the corresponding graph-complex depends only on the parity of the dimension d. In fact these graph-complexes are quasi-isomorphic to the complexes dual to the complexes of bracket star-diagrams (a description of the complexes dual to $\mathcal{BSD}$ is given in [39, 40]) as differential Hopf algebras. (The proof of the last assertion is elementary, however we have not seen it nowhere.) So, may be this approach leads to a proof of Conjecture 0.3.

The second approach is to study the spaces of knots (or more generally spaces of embeddings of arbitrary varieties) by means of the calculus of analitic functors developped by T. Goodwillie, *cf.* [20]. In this approach the spaces of long knots are replaced by their simplicial or cosimplicial models, see [21, 35]. This approach leads also to a complex quasiisomorphic to $\mathcal{BSD}$, *cf.* [36].

Acknowledgements. I would like to thank my scientific advisor V. Vassiliev for his consultations and for his support during the work. Also I would like to express my gratitude to Ecole Normale Supérieur and Institut des Hautes Etudes Scientifiques for hospitality. I would like to thank M. Kontsevich for his consultations and his attention to this work.

Also I am grateful to P. Cartier, A. V. Chernavsky, M. Deza, D. Panov, M. Finkelberg, S. Loktev, I. Marin, G. Racinet, A. Stoyanovsky, O. Boulanov, Y.-J. Lee, J.-O. Moussafir.

1. Linear operads

The definition of many algebraic structures on a linear space (such as the commutative, associative, Lie algebra structures) consists of setting several polylinear operations (in these three cases, only one binary operation), that should satisfy some composition identities (in our example, associativity or Jacoby identity). Instead of doing this one can consider the spaces of all polylinear n-ary operations, for all $n \geq 0$, and the composition rules, that arise from the corresponding algebraic structure. The natural formalization of this object is given by the notion of operad.

Definition

Let $\Bbbk$ be a commutative ring of coefficients. A *graded* $\Bbbk$*-linear operad* $\mathcal{O}$ is a collection $\{\mathcal{O}(n),\, n \geq 0\}$ of graded $\Bbbk$-vector spaces equipped with the following set of data:

(i) An action of the symmetric group S_n on $\mathcal{O}(n)$ for each $n \geq 2$.

(ii) Linear maps (called *compositions*), preserving the grading,

$$\gamma_{m_1,\dots,m_\ell} : \mathcal{O}(\ell) \otimes (\mathcal{O}(m_1) \otimes \cdots \otimes \mathcal{O}(m_\ell)) \to \mathcal{O}(m_1 + \cdots + m_\ell) \tag{1.1}$$

for all $m_1, \dots, m_\ell \geq 0$. We write $\mu(\nu_1, \dots, \nu_\ell)$ instead of $\gamma_{m_1,\dots,m_\ell}(\mu \otimes \nu_1 \cdots \otimes \nu_\ell)$.

(iii) An element $id \in \mathcal{O}(1)$, called the *unit*, such that $id(\mu) = \mu(id, \dots, id) = \mu$ for any non-negative ℓ and any $\mu \in \mathcal{O}(\ell)$.

It is required that these data satisfy some conditions of associativity and equivariance with respect to the symmetric group actions, see [32, Chapter 1], [25], [31].

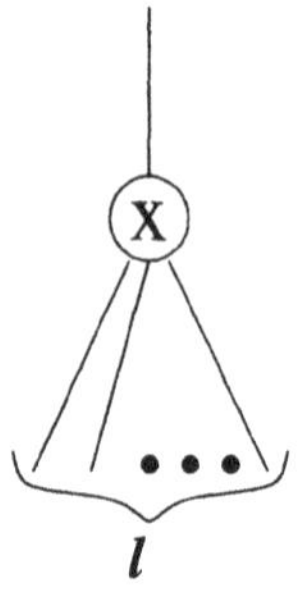

Figure 1. Element $x \in \mathcal{O}(\ell)$.

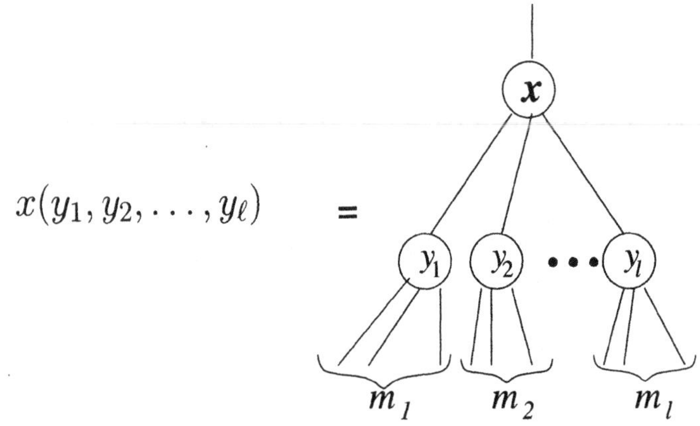

Figure 2. Composiotion operation.

One can consider any element $x \in \mathcal{O}(\ell)$ as something that has ℓ inputs and 1 output, see Figure 1.

The composition operation (1.1) of $x \in \mathcal{O}(\ell)$ with $y_1 \in \mathcal{O}(m_1), \ldots, y_\ell \in \mathcal{O}(m_\ell)$ is the substitution of $y_1, \ldots, y_\ell$ into ℓ inputs of x, see Figure 2.

The resulting element has $m_1 + \cdots + m_\ell$ inputs and 1 output, *i. e.*, it belongs to $\mathcal{O}(m_1 + \cdots + m_\ell)$.

Examples

We will give several well known examples of graded linear operads, see [13, 15, 19].

a) Let V be a graded $\mathbb{k}$-vector space. We define the *endomorphism operad* $\mathcal{END}(V) := \{Hom(V^{\otimes n}, V), n \geq 0\}$. The unit element $id \in Hom(V, V)$ is put to be the identical map $V \to V$. The composition operations and the symmetric group actions are defined in the canonical way.

This operad is very important due to the following definition.

DEFINITION 1.2. Let $\mathcal{O}$ be a graded linear operad. By an *$\mathcal{O}$-algebra* (or *algebra over $\mathcal{O}$*) we call any couple (V, ρ), where V is a graded vector space and ρ is a morphism $\rho : \mathcal{O} \to \mathcal{END}(V)$ of operads. $\square$

In other words, the theory of algebras over an operad $\mathcal{O}$ is the representation theory of $\mathcal{O}$.

b) The *operad $\mathcal{LIE}$ of Lie algebras.* The component $\mathcal{LIE}(0)$ of this operad is trivial. The n-th component $\mathcal{LIE}(n)$ is defined as the subspace of the free Lie algebra with generators $x_1, x_2, \ldots, x_n$, that is spanned by the brackets containing each generator exactly once.

EXAMPLE 1.3. For $n = 5$ one can take the bracket $[[[x_5, x_3], [x_1, x_2]], x_4]$ as an element of $\mathcal{LIE}(5)$. □

Since we work in the category of graded vector spaces, we need to define a grading on each considered space. The grading of the spaces $\mathcal{LIE}(n)$, $n \geq 1$, is put to be zero. It is well known that these spaces are free $\Bbbk$-modules, $\mathcal{LIE}(n) \simeq \Bbbk^{(n-1)!}$. The S_n-action is defined by permutations of $x_1, \ldots, x_n$.

Let $A(x_1, \ldots, x_\ell)$, $B_1(x_1, \ldots, x_{m_1})$, $\ldots$, $B_\ell(x_1, \ldots, x_{m_\ell})$ be brackets respectively from $\mathcal{LIE}(\ell)$, $\mathcal{LIE}(m_1)$, $\ldots$, $\mathcal{LIE}(m_\ell)$. We define the composition operations (1.1) as follows.

$$A(B_1, \ldots, B_\ell)(x_1, \ldots, x_{m_1+\cdots+m_\ell}) := A(B_1(x_1, \ldots, x_{m_1}),$$

$$B_2(x_{m_1+1}, \ldots, x_{m_1+m_2}), \ldots, B_l(x_{m_1+\cdots+m_{\ell-1}+1}, \ldots, x_{m_1+\cdots+m_\ell})). \tag{1.4}$$

The element $x_1 \in \mathcal{LIE}(1)$ is the unit element for this operad.

Note that a $\mathcal{LIE}$-algebra structure in the sense of Definition 1.2 is exactly the same as a (graded) Lie algebra structure in the usual sense. Indeed, the element $\rho([x_1, x_2]) \in Hom(V^{\otimes 2}, V)$ always defines a Lie bracket. The converse is also true. It is easy to see that the unit $id = x_1 \in \mathcal{LIE}(1)$ and the element $[x_1, x_2] \in \mathcal{LIE}(2)$ generate by means of the compositions (1.1) the operad $\mathcal{LIE}$, so if we put $\rho([x_1, x_2])$ equal to our Lie bracket, then we immediately obtain a map $\rho : \mathcal{LIE} \to \mathcal{END}(V)$ of the whole operad $\mathcal{LIE}$.

c) By analogy with the operad $\mathcal{LIE}$ one defines the *operad $\mathcal{COMM}$* (resp. $\mathcal{ASSOC}$) *of commutative* (resp. *associative*) *algebras.* The space $\mathcal{COMM}(n)$ (resp. $\mathcal{ASSOC}(n)$), for $n \geq 1$, is put to be one-dimensional (resp. $n!$-dimensional) free $\Bbbk$-module defined as the subspace of the free commutative (resp. associative) algebra with generators $x_1, \ldots, x_n$, that is spanned by the monomials containing each generator exactly once. The S_n-actions and the composition operations are defined in the same way as for the operad $\mathcal{LIE}$. The element $x_1 \in \mathcal{COMM}(1)$ (resp. $\mathcal{ASSOC}(1)$) is the unit element *id*. There are two different ways to define the space $\mathcal{COMM}(0)$ (resp. $\mathcal{ASSOC}(0)$). We can put this space to be trivial or one-dimensional. In the first case we get the operad of commutative (resp. associative) algebras without unit, in the second case – the operad of commutative (resp. associative) algebras with unit. Below we will consider the second situation.

REMARK 1.5. A commutative (resp. associative) algebra structure is exactly the same as a structure of $\mathcal{COMM}$-algebra (resp. $\mathcal{ASSOC}$-algebra) in the sense of Definition 1.2. □

d) The *operads $\mathcal{POISS}$, $\mathcal{GERST}$, $\mathcal{POISS}_d$ of Poisson, Gerstenhaber* and *d-Poisson algebras.* First of all, let us recall the definition of Poisson, Gerstenhaber and d-Poisson algebras.

DEFINITION 1.6. A graded commutative algebra A is called a *d-Poisson algebra*, if it has a Lie bracket

$$[.,.] : A \otimes A \to A$$

of degree $-d$. The bracket is supposed to be compatible with the multiplication. This means that for any elements $x, y, z \in A$

$$[x, yz] = [x, y]z + (-1)^{\tilde{y}(\tilde{x}-d)}y[x, z]. \square \tag{1.7}$$

0-Poisson (resp. 1-Poisson) algebras are called simply *Poisson* (resp. *Gerstenhaber*) algebras. 1-Poisson algebras are called Gerstenhaber algebras in honor of Murray Gerstenhaber, who discovered this structure on the Hochschild cohomology of associative algebras, see [16] and also Sections 2 and 3.

EXAMPLE 1.8. Let $\mathfrak{g}$ be a graded Lie algebra with the bracket of degree $-d$. Then the symmetric (in the graded sense) algebra $S^*\mathfrak{g}$ has a natural structure of a d-Poisson algebra with the usual multiplication of a symmetric algebra and with the bracket defined by the following formula:

$$[A_1 \cdot A_2 \ldots A_k, B_1 \cdot B_2 \ldots B_l] = \sum_{i,j}(-1)^{\epsilon} A_1 \ldots \widehat{A_i} \ldots A_k \cdot [A_i, B_j] \cdot B_1 \ldots \widehat{B_j} \ldots B_l, \tag{1.9}$$

where $\epsilon = \tilde{A}_i(\sum_{i'=i+1}^{k} \tilde{A}_{i'}) + \tilde{B}_j(\sum_{j'=1}^{j-1} \tilde{B}_{j'})$. $\square$

REMARK 1.10. For a graded Lie algebra $\mathfrak{g}$ with the bracket of degree 0 the d-tuple suspension $\mathfrak{g}[d]$ is also a graded Lie algebra with the bracket of degree $-d$. Thus, the space $S^*(\mathfrak{g}[d])$ is a d-Poisson algebra. $\square$

Note, that for commutative, associative, or Lie algebras the n-th component of the operad is defined as the space of all natural polylinear n-ary operations, that come from the corresponding algebra structure. Now, let us describe the spaces $\mathcal{POISS}_d(n)$ of all natural polylinear n-ary operations of d-Poisson algebras. Consider a free graded Lie algebra $Lie_d(x_1, \ldots, x_n)$ with the bracket of degree $-d$ and with the generators $x_1, \ldots, x_n$ of degree zero, and consider (following Example 1.8) the d-Poisson algebra $Poiss_d(x_1, \ldots, x_n) := S^*(Lie_d(x_1, \ldots, x_n))$. This is a free d-Poisson algebra. We will define the space $\mathcal{POISS}_d(n)$ as the subspace of $Poiss_d(x_1, \ldots, x_n)$ spanned by the products (of brackets) containing each generator x_i exactly once. For instance for $n = 5$ we will take the product $[x_1, x_3] \cdot [[x_2, x_5]x_4]$ as an element of $\mathcal{POISS}_d(5)$. The unit element id is $x_1 \in \mathcal{POISS}_d(1)$. The symmetric group actions and the composition operations are defined analogously to the case of the operad $\mathcal{LIE}$.

The space $\mathcal{POISS}_d(n)$ can be decomposed into a direct sum with the summands numbered by partitions of the set $\{1, \ldots, n\}$:

$$\mathcal{POISS}_d(n) = \bigoplus_A \mathcal{POISS}_d(A, n). \tag{1.11}$$

For a partition $A = \{\bar{A}_1, \ldots, \bar{A}_{\#A}\}$ of the set $\{1, \ldots, n\} = \coprod_{i=1}^{\#A} \bar{A}_i$ we define the space $\mathcal{POISS}_d(A, n) \subset \mathcal{POISS}_d(n)$ to be linearly spanned by products of $\#A$ brackets, such that the i-th bracket contains generators only from the set $\bar{A}_i$ (thus, each generator from $\bar{A}_i$ is presented exactly once in the i-th bracket).

Let $\bar{A}_1, \dots \bar{A}_{\#A}$ be of cardinalities $a_1, \dots, a_{\#A}$ respectively, then

$$\mathcal{POISS}_d(A, n) \simeq \otimes_{i=1}^{\#A} \mathbb{k}^{(a_i - 1)!}. \tag{1.12}$$

This implies that the space $\mathcal{POISS}_d(n)$ is isomorphic to $\mathbb{k}^{n!}$, and its Poincaré polynomial is $(1+t^{-d})(1+2t^{-d})\dots(1+(n-1)t^{-d})$ (use the induction over n).

e) The operad $\mathcal{BV}$ (resp. $\mathcal{BV}_d$, d being odd) of Batalin-Vilkovisky (resp. d-Batalin-Vilkovisky) algebras.

DEFINITION 1.13. A Gerstenhaber algebra (resp. d-Poisson algebra, for odd d) A is called a *Batalin-Vilkovisky algebra* (resp. *d-Batalin-Vilkovisky algebra*), if A is endowed with a linear map δ of degree -1 (resp. $-d$)

$$\delta : A \to A,$$

such that
(i) $\delta^2 = 0$,
(ii) $\delta(ab) = \delta(a)b + (-1)^{\tilde{a}} a\delta(b) + (-1)^{\tilde{a}}[a, b]$. □

Note that (i) and (ii) imply
(iii) $\delta([a, b]) = [\delta(a), b] + (-1)^{\tilde{a}+1}[a, \delta(b)]$.

EXAMPLE 1.14. Let $\mathfrak{g}$ be a graded Lie algebra with the bracket of degree zero. Then the exterior algebra $\Lambda^*\mathfrak{g} := S^*(\mathfrak{g}[1])$ is a Batalin-Vilkovisky algebra, where the structure of a Gerstenhaber algebra is from Remark 1.10; the operator δ is the standard differential on the chain-complex $\Lambda^*\mathfrak{g}$:

$$\delta(A_1 \wedge \dots \wedge A_k) = \sum_{i<j} (-1)^{\epsilon} [A_i, A_j] \wedge A_1 \dots \widehat{A_i} \dots \widehat{A_j} \dots \wedge A_k, \tag{1.15}$$

where $A_1, \dots, A_k \in \mathfrak{g}$, $\epsilon = \tilde{A}_i + (\tilde{A}_i + 1)(\tilde{A}_1 + \dots + \tilde{A}_{i-1} + i - 1) + (\tilde{A}_j + 1)(\tilde{A}_1 + \dots + \widehat{\tilde{A}_i} + \dots + \tilde{A}_{j-1} + j - 2)$. In the same way $\mathfrak{g}$ defines the d-Batalin-Vilkovisky algebra $S^*(\mathfrak{g}[d])$ for any odd d. □

Let us describe the n-th component of the corresponding operad, denoted by $\mathcal{BV}$ and $\mathcal{BV}_d$ (d being always odd). Since the spaces $\mathcal{BV}(n)$ and $\mathcal{BV}_d(n)$ are isomorphic (in the super-sense) up to a change of grading, we will consider now only the case $d = 1$. Obviously, the space $\mathcal{BV}(n)$ of all natural polylinear n-ary operations for such algebras contains $\mathcal{GERST}(n)$. Consider the symmetric algebra of the free graded Lie algebra $Lie_1(x_1, \dots, x_n, \delta(x_1), \dots, \delta(x_n))$ with the bracket of degree -1 and with the generators $x_1, \dots, x_n$ of degree zero and the generators $\delta(x_1), \dots, \delta(x_n)$ of degree -1. This space has a structure of a Batalin-Vilkovisky algebra. (In fact it is a free $\mathcal{BV}$-algebra with generators $x_1, \dots, x_n$). In $S^*(Lie_1(x_1, \dots, x_n, \delta(x_1), \dots, \delta(x_n)))$ we will take the subspace $\mathcal{BV}(n)$ linearly spanned by all the products (of brackets), containing each index $i \in \{1, \dots, n\}$ exactly once. For instance $[\delta(x_1), x_3] \cdot [x_2, [\delta(x_4), \delta(x_5)]]$ belongs to $\mathcal{BV}(5)$. Due to relations (i), (ii), (iii) for δ, this subspace is exactly the space of all natural polylinear n-ary operations on Batalin-Vilkovisly algebras.

Analogously to (1.11) the space $\mathcal{BV}(n)$ can be decomposed into the direct sum:

$$\mathcal{BV}(n) = \bigoplus_{A,S} \mathcal{BV}(A, S, n), \tag{1.16}$$

where A is a partition of the set $\{1,\ldots,n\}$, S is a subset of $\{1,\ldots,n\}$ corresponding to indices i presented by $\delta(x_i)$ in products of brackets of $\mathcal{BV}(n)$.

Note that the space $\mathcal{BV}(n)$ is isomorphic to $\Bbbk^{2^n n!}$. The Poincaré polynomial of this graded space is $(1+t^{-d})^{n+1}(1+2t^{-d})(1+3t^{-d})\ldots(1+(n-1)t^{-d})$.

In the sequel we will use the following definition.

DEFINITION 1.17. For a finite set M any pair (A,S) consisting of a partition A of M and of a subset $S\subset M$ will be called a *star-partition* of the set M. (Each point $i\in S$ will be called a *star*.) □

Note that we have the following natural morphisms of operads:

$$\mathcal{ASSOC}\to\mathcal{COMM}\to\mathcal{POISS}_d \dashrightarrow \mathcal{BV}_d \tag{1.18}$$

The last arrow of (1.18) is defined only if d is odd.

2. Graded Lie algebra structure on graded linear operads

In this section we define a graded Lie algebra structure on an arbitrary graded linear operad. In the next section we use this structure and define the Hochschild complex for a graded linear operad equipped with a morphism from the operad $\mathcal{ASSOC}$ to our operad. Both these constructions, which generalize the Hochschild cochain complex for associative algebras, were introduced in [24]. The only difference between the operations (2.1), (2.5), (3.5) given below and those of [24] is in signs. First of all, in the paper [24] M. Gerstenhaber and A. Voronov considered only linear (non-graded) operads, hence our case is more general. But even for the case of purely even gradings the signs are slightly different. The difference can be easily obtained by conjugation of the operations (2.1), (2.5), (3.5) by means of the linear operator that maps any element $x\in\mathcal{O}(n)$, $n\geq 0$, to $(-1)^{\frac{n(n-1)}{2}}x$.

Let $\mathcal{O}=\{\mathcal{O}(n), n\geq 0\}$ be a graded $\Bbbk$-linear operad. By abuse of the language the space $\bigoplus_{n\geq 0}\mathcal{O}(n)$ will be also denoted by $\mathcal{O}$. A tilde over an element will always designate its grading. For any element $x\in\mathcal{O}(n)$ we put $n_x:=n-1$. The numbers n and 1 here correspond to n inputs and to 1 output respectively.

Define a new grading $|\,.\,|$ on the space $\mathcal{O}$. For an element $x\in\mathcal{O}(n)$ we put $|x|:=\tilde{x}+n_x=\tilde{x}+n-1$. It turns out that $\mathcal{O}$ is a graded Lie algebra with respect to the grading $|\,.\,|$. Note that the composition operations (1.1) respect this grading. Define the following collection of multilinear operations on the space $\mathcal{O}$.

$$x\{x_1,\ldots,x_n\}:=\sum(-1)^{\epsilon}x(id,\ldots,id,x_1,id,\ldots,id,x_n,id,\ldots,id) \tag{2.1}$$

for $x,x_1,\ldots,x_n\in\mathcal{O}$, where the summation runs over all possible substitutions of $x_1,\ldots,x_n$ into x in the prescribed order, $\epsilon:=\sum_{p=1}^n n_{x_p}r_p+n_x\sum_{p=1}^n\tilde{x}_p+\sum_{p<q}n_{x_p}\tilde{x}_q$, r_p being the total number of inputs in x going after x_p. For instance, for $x\in\mathcal{O}(2)$ and arbitrary $x_1,x_2\in\mathcal{O}$

$$x\{x_1,x_2\}=(-1)^{n_{x_1}+(\tilde{x}_1+\tilde{x}_2)+n_{x_1}\tilde{x}_2}x(x_1,x_2).$$

We will also adopt the following convention:

$$x\{\}:=x.$$

One can check immediately the following identities:

$$x\{x_1,\dots,x_m\}\{y_1,\dots,y_n\} =$$

$$\sum_{0\le i_1\le j_1\le\dots\le i_m\le j_m\le n} (-1)^{\epsilon} x\{y_1,\dots,y_{i_1},x_1\{y_{i_1+1},\dots,y_{j_1}\},y_{j_1+1},\dots,y_{i_m},$$

$$x_m\{y_{i_m+1},\dots,y_{j_m}\},y_{j_m+1},\dots,y_n\}, \tag{2.2}$$

where $\epsilon=\sum_{p=1}^{m}(|x_p|\sum_{q=1}^{i_p}|y_q|)$. (These signs are the same as in [24]).

Define a bilinear operation (respecting the grading $|\,.\,|$) $\circ$ on the space $\mathcal{O}$:

$$x\circ y := x\{y\}, \tag{2.3}$$

for $x,y\in\mathcal{O}$.

DEFINITION 2.4. A graded vector space A with a bilinear operation

$$\circ : A\otimes A\to A$$

is called a *Pre-Lie algebra*, if for any $x,y,z\in A$ the following holds:

$$(x\circ y)\circ z - x\circ(y\circ z) = (-1)^{|y||z|}((x\circ z)\circ y - x\circ(z\circ y)).\ \square$$

Any graded Pre-Lie algebra A can be considered as a graded Lie algebra with the bracket

$$[x,y] := x\circ y - (-1)^{|x||y|} y\circ x. \tag{2.5}$$

The description of the operad of Pre-Lie algebras is given in [14].

The following lemma is a corollary of the identity (2.2) applied to the case $m=n=1$.

LEMMA 2.6. *The operation* (2.3) *defines a graded Pre-Lie algebra structure on the space* $\mathcal{O}$. $\square$

In particular this lemma means that any graded linear operad $\mathcal{O}$ can be considered as a graded Lie algebra with the bracket (2.5).

3. Hochschild complexes

Let $\mathcal{O}=\bigoplus_{n\ge 0}\mathcal{O}(n)$ be a graded linear operad equipped with a morphism

$$\Pi : \mathcal{ASSOC}\to\mathcal{O}$$

from the operad $\mathcal{ASSOC}$. This morphism defines the element $m=\Pi(m_2)\in\mathcal{O}(2)$, where the element $m_2=x_1x_2\in\mathcal{ASSOC}(2)$ is the operation of multiplication. Note that the elements m_2, m are odd with respect to the new grading $|\,.\,|$ ($|m|=|m_2|=1$) and $[m,m]=[\Pi(m_2),\Pi(m_2)]=2\Pi(m_2\circ m_2)=0$. (One has $m_2\circ m_2=-(x_1\cdot x_2)\cdot x_3+x_1\cdot(x_2\cdot x_3)=x_1\cdot x_2\cdot x_3-x_1\cdot\cdot x_2\cdot x_3=0$.) Thus $\mathcal{O}$ becomes a differential graded Lie algebra with the differential ∂:

$$\partial x := [m,x] = m\circ x - (-1)^{|x|} x\circ m, \tag{3.1}$$

for $x \in \mathcal{O}$.

We will call the complex $(\mathcal{O}, \partial)$ *Hochschild complex* for the operad $\mathcal{O}$. Actually a better name would be *Hochschild complex* of the *morphism* $\Pi : \mathcal{ASSOC} \to \mathcal{O}$ since this complex is in fact the deformation complex of the morphism Π, see [39]. We prefered here the first name for its shortness.

EXAMPLE 3.2. If $\mathcal{O}$ is the endomorphism operad $\mathcal{END}(A)$ of a vector space A, and we have a morphism

$$\Pi : \mathcal{ASSOC} \to \mathcal{END}(A),$$

that defines an associative algebra structure on A, then the corresponding complex $(\bigoplus_{n=0}^{+\infty} Hom(A^{\otimes n}, A), \partial)$ is the standard Hochschild cochain complex $C^*(A, A)$ of the associative algebra A. □

EXAMPLE 3.3. Due to the morphisms (1.18) we have the Hochschild complexes $(\mathcal{ASSOC}, \partial)$, $(\mathcal{COMM}, \partial)$, $(\mathcal{POISS}_d, \partial)$, $(\mathcal{BV}_d, \partial)$. It can be shown that the complexes $(\mathcal{ASSOC}, \partial)$ and $(\mathcal{COMM}, \partial)$ are acyclic. □

Define another grading

$$deg := |\,.\,| + 1 \tag{3.4}$$

on the space $\mathcal{O}$. With respect to this grading the bracket $[.,.]$ is homogeneous of degree -1.

It is easy to see that the product $*$, defined as follows

$$x * y := (-1)^{|x|} m\{x, y\} = (-1)^{\tilde{y}(n_x+1)} m(x, y), \tag{3.5}$$

for $x, y \in \mathcal{O}$, together with the differential ∂ defines a differential graded associative algebra structure on O with respect to the grading $deg = |\,.\,| + 1$.

THEOREM 3.6. [24] *The multiplication $*$ and the bracket $[.,.]$ induce a Gerstenhaber (or what is the same 1-Poisson) algebra structure on the homology of the Hochschild complex $(\mathcal{O}, \partial)$.* □

Proof of Theorem 3.6: The proof is deduced from the following homotopy formulas.

$$x * y - (-1)^{deg(x)deg(y)} y * x = (-1)^{deg(x)} (\partial(x \circ y) - \partial x \circ y - (-1)^{deg(x)-1} x \circ \partial y). \tag{3.7}$$

The above formula proves the graded commutativity of the multiplication $*$.

$$[x, y * z] - [x, y] * z - (-1)^{(deg(x)-1)deg(y)} y * [x, z] =$$

$$= (-1)^{deg(x)+deg(y)} \big(\partial(x\{y, z\}) - (\partial x)\{y, z\} - (-1)^{|x|} x\{\partial y, z\} - (-1)^{|x|+|y|} x\{y, \partial z\}\big). \tag{3.8}$$

This formula proves the compatibility of the bracket with the multiplication. □

4. The little cubes operad

Analogously to linear operads one can define topological operads, *i. e.*, collections $\{\mathcal{O}(n), n \geq 0\}$ of topological sets with

(i) an S_n-action on each $\mathcal{O}(n)$;

(ii) compositions

$$\gamma : \mathcal{O}(\ell) \times (\mathcal{O}(m_1) \times \cdots \times \mathcal{O}(m_\ell)) \to \mathcal{O}(m_1 + \cdots + m_\ell);$$

(iii) a unit element $id \in \mathcal{O}(1)$.

We assume the same associativity and symmetric group equivariance requirements.

The homology spaces $\{H_*(\mathcal{O}(n), \Bbbk), n \geq 0\}$ over any commutative ring $\Bbbk$ form a graded $\Bbbk$-linear operad.

Usually, when one considers the homology of a topological operad, one inverses the grading, supposing that the i-th homology group $H_i(\mathcal{O}(n), \Bbbk)$ has the degree $-i$.

Historically one of the first examples of topological operads are the little cubes operads $\mathcal{LC}_d = \{\mathcal{LC}_d(n), n \geq 0\}$, $d \geq 1$, see [7], [8, Chapter 2], [32, Chapter 4]. Here $\mathcal{LC}_d(n)$ denotes the configuration space of n disjoint cubes labelled by $\{1, \ldots, n\}$ in a unit cube. It is supposed that the faces of the cubes are parallel to the corresponding faces of the unit cube. The group S_n acts by permutations of the cubes in configurations. The element $id \in \mathcal{LC}_d(1)$ is the configuration of one cube coinciding with the unit cube. The composition operations (ii) are insertions of ℓ configurations respectively of $n_1, \ldots, n_\ell$ cubes into the corresponding ℓ cubes of a configuration of ℓ cubes.

The following result is due to F. Cohen. In [13] one can find its implicit formulation and proof.

THEOREM 4.1. [13] *The homology groups of the operad $\mathcal{LC}_d$, $d \geq 1$, have no torsion and form the following graded linear operad*

1) if $d = 1$, $\{H_{-*}(\mathcal{LC}_1(n), \Bbbk), n \geq 0\}$ is isomorpic to the associative algebras operad $\mathcal{ASSOC}$,

2) if $d \geq 2$, $\{H_{-*}(\mathcal{LC}_d(n), \Bbbk), n \geq 0\}$ is isomorphic to the $(d-1)$-Posson algebras operad $\mathcal{POISS}_{(d-1)}$. □

REMARK 4.2. The operad of associative algebras has a natural filtration compatible with the operad structure. The graded quotient associated with this filtration is the Poisson algebras operad $\mathcal{POISS}$. This assertion follows from the Poincaré-Birkhoff-Witt theorem and from the fact that the universal envelopping algebra of a free Lie algebra is canonically isomorphic to a free associative algebra. □

We will not prove Theorem 4.1, however we will make some explanations.

First of all, note that the space $\mathcal{LC}_d(1)$ is contractible for any $d \geq 1$. The homology classe of one point in this space corresponds to $id = x_1 \in \mathcal{POISS}_{(d-1)}(1)$. The space $\mathcal{LC}_d(2)$ is homotopy equivalent to the $(d-1)$-dimensional sphere S^{d-1}. For $d \geq 2$ this gives us one operation of degree zero and one operation of degree $1-d$. The first operation corresponds to the multiplication $x_1 \cdot x_2 \in \mathcal{POISS}_{(d-1)}(2)$, the second one to the bracket $[x_1, x_2] \in \mathcal{POISS}_{(d-1)}(2)$.

Obviously, the space $\mathcal{LC}_d(n)$ is homotopy equivalent to the configuration space of collections of n distinct points in $\mathbb{R}^d$, *i. e.*, to the space of injective maps $\{1, \ldots, n\} \hookrightarrow \mathbb{R}^d$. The latter space is an open everywhere dense subset of the vector space $\mathbb{R}^{nd}$ of all maps $\{1, \ldots, n\} \to \mathbb{R}^d$. The complement (called the *discriminant*) $\Delta_d(n) \subset \mathbb{R}^{nd}$ is a union of $\frac{n(n-1)}{2}$ vector subspaces of codimension d. Each of these subspaces $L_{i,j}$ corresponds to a pair of distinct points $i, j \in \{1, \ldots, n\}$ and consists of maps $\psi : \{1, \ldots, n\} \to \mathbb{R}^d$, such that $\psi(i) = \psi(j)$.

The non-complete partitions of the set $\{1,\dots,n\}$ (by the *complete* partition we mean the partition into n singletons) are in one-to-one correspondence with the strata of the arrangement $\Delta_d(n) = \bigcup_{i<j} L_{i,j}$. To a partition we assign a vector subspace consisting of maps $\{1,\dots,n\} \to \mathbb{R}^d$ that glue the points of each set in the partition. Following the general theory of arrangements the reduced homology groups $\tilde{H}_*(\mathbb{R}^{nd}\backslash\Delta_d(n),\Bbbk)$ can be decomposed into a direct sum, each summand being assigned to some stratum of the arrangement, see [22, 43, 45, 50]. In the case $d \geq 2$ this decomposition is canonical (in the case $d = 1$ it depends on the choice of a component of $\mathbb{R}^n\backslash\Delta_1(n)$). Let us assign to the complete partition the degree zero homology group $H_0(\mathbb{R}^{nd}\backslash\Delta_d(n),\Bbbk) \simeq \Bbbk$.

PROPOSITION 4.3. *The above decomposition of the homology groups $H_{-*}(\mathbb{R}^{nd}\backslash\Delta_d(n),\Bbbk)$, $d \geq 2$, coincides with the decomposition (1.11).* □

5. The first term of the Vassiliev auxiliary spectral sequence

In the same way as the homology groups of the space of injective maps $\{1,\dots,n\} \hookrightarrow \mathbb{R}^d$ are decomposed into a direct sum by strata of the discriminant $\Delta_d(n)$ (see Section 4 — Proposition 4.3), the first term of the Vassiliev auxiliary spectral sequence is naturally decomposed into a direct sum in which the summands are numbered by equivalence classes of so called *(A,b)-configurations* defined below.

Let A be a non-ordered finite collection of natural numbers $A = (a_1,\dots,a_{\#A})$, any of which is not less than 2, and let b be a non-negative integer. Set $|A| := a_1 + \cdots + a_{\#A}$. An *$(A,b)$-configuration* is a collection of $|A|$ distinct points in $\mathbb{R}^1$ separated into $\#A$ groups of cardinalities $a_1,\dots,a_{\#A}$, plus a collection of b distinct points in $\mathbb{R}^1$ (some of which can coincide with the above $|A|$ points). For short $(A,0)$-configuration are called simply *A-configurations.* A map $\phi : \mathbb{R}^1 \to \mathbb{R}^d$ *respects* an (A,b)-configuration, if it glues together all points inside any of its groups of cardinalities $a_1,\dots,a_{\#A}$, and the derivative ϕ' is equal to 0 at all the b last points of this configuration. For any (A,b)-configuration the set of maps respecting it is an affine subspace in $\mathcal{K}$ of codimension $d(|A| - \#A + b)$; the number $|A| - \#A + b$ is called the *complexity* of the configuration. Two (A,b)-configurations are called *equivalent* if they can be transformed into one another by an orientation-preserving homeomorphism $\mathbb{R}^1 \to \mathbb{R}^1$.

Consider any (A,b)-configuration J of complexity i and with j geometrically distinct points in $\mathbb{R}^1$. The stratum consisting of all mappings $\mathbb{R}^1 \to \mathbb{R}^d$ respecting at least one (A,b)-configuration J' equivalent to J, can be parametrized by the space $S(J)$ of affine fiber bundle, whose base space is the space E^j of (A,b)-configurations J' equivalent to J, and the fiber over J' is the affine space $\mathbb{R}^{\omega-di}$ of maps respecting J'. Note that E^j is contractible being an open cell of dimension j. Therefore this fiber bundle can be trivialized:

$$S(J) \simeq E^j \times \mathbb{R}^{\omega-di}. \tag{5.1}$$

REMARK 5.2. The corresponding stratum is not homeomorphic to $S(J)$: one map $\mathbb{R}^1 \to \mathbb{R}^d$ may respect two different (A,b)-configurations equivalent to J, and therefore the stratum has self-intersections. □

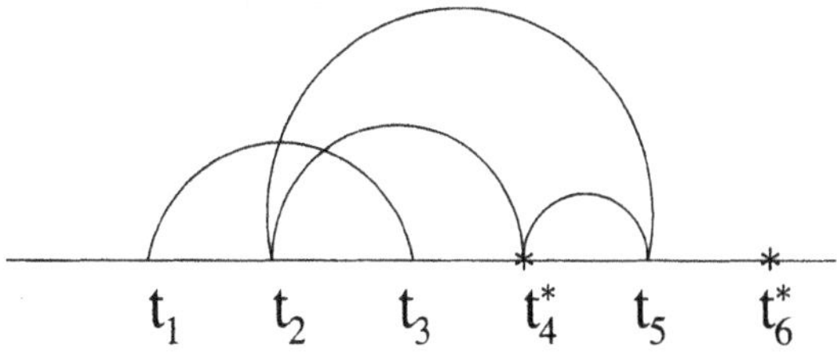

Figure 3. Example of (A, b)-configuration.

The auxiliary spectral sequence computing the homology groups of the term $\sigma_i \backslash \sigma_{i-1}$ in our filtration (0.2) uses those and only those (A, b)-configurations, that are of complexity i. The geometrical meaning of the first differential in the auxiliary spectral sequence is in how the strata (or rather the base spaces of the corresponding affine bundles (5.1)) coresponding to equivalence classes of (A, b)-configurations of complexity i bound to each other.

Let first d be even. For any equivalence class of (A, b)-configurations of complexity i and with j geometrically distinct points we will assign a star-partition $(\bar{A}, S)$ of the set $\{1, \dots, j\}$ (see Definition 1.17) and therefore a subspace $\mathcal{BV}_{(d-1)}(\bar{A}, S, j)$ of the operad $\mathcal{BV}_{(d-1)}$ (see Decomposition 1.16), where the partition $\bar{A}$ of the set $\{1, \dots, j\}$ and the subset $S \subset \{1, \dots, j\}$ are defined below.

DEFINITION 5.3. A *minimal component* of an (A, b)-configuration is either one of its b points, which does not coincide with none of the $|A|$ points, or one of the $\#A$ groups of points with all the stars contained there. □

For instance, the (A, b)-configuration on the Figure 3 has 3 minimal components consisting respectively of the following groups of points: 1) t_1, t_3; 2) t_2, t_4^*, t_5; 3) t_6^*.

To any (A, b)-configuration with j geometrically distinct points we can assign a star-partition of the set $\{1, \dots, j\}$. Let $t_1 < t_2 < \dots < t_j$ be the points in $\mathbb{R}^1$ of our (A, b)-configuration (b of them are marked by stars). The set of minimal components of this (A, b)-configuration defines a partition $\bar{A}$ of the set $\{1, \dots, j\}$. We also have the subset $S \subset \{1, \dots, j\}$ (of cardinality b) of indices, corresponding to the points marked by stars.

Note that different equivalence classes of (A, b)-configurations correspond to different star-partitions. But the correspondence is far from being bijective. The star-partitions not corresponding to (A, b)-configurations are those and only those, which contain singletons not marked by a star.

THEOREM 5.4. see [39]. *The first term of the Vassiliev auxiliary spectral sequence for even d together with its first diferenrial is isomorphic to the subcomplex of the Hochschild complex $(\mathcal{BV}_{(d-1)}, \partial)$, linearly spanned by the summands of the decomposition (1.16), corresponding to star-partitions, which don't contain singletons not marked by a star. The grading corresponding to the homology of the knot space $\mathcal{K} \backslash \Sigma$ is minus the grading "deg" defined by (3.4).* □

In particular, the theorem claims that the subspace in $\mathcal{BV}_{(d-1)}$ spanned by the summands in question is invariant with respect to the differential ∂.

Consider the decomposition of the space $\mathcal{BV}_{(d-1)} = \bigoplus_{i\geq 0} E_i$, where E_i is the sum over all star-partitions having exactly i singletons not marked by a star. The filtration $F_0 \supset F_1 \supset F_2 \ldots$, with $F_i := \bigoplus_{j\geq i} E_j$, is compatible with the differential ∂. Note, that the complex (E_0, ∂) is exactly the complex from Theorem 5.4.

PROPOSITION 5.5. [39] *The Hochschild complex $(\mathcal{BV}_{(d-1)}, \partial)$ (d is even) is a direct sum of the complexes (E_0, ∂) and (F_1, ∂). The first complex (E_0, ∂) is homology (and even homotopy) equivalent to $(\mathcal{BV}_{(d-1)}, \partial)$; the second one (F_1, ∂) is acyclic (and even contractible).* □

An analogous statement holds for the complexes $(\mathcal{POISS}_{(d-1)}, \partial)$, d being any integer number. The subcomplexes spanned by the summands of the decomosition (1.11) corresponding to partitions non-containing singletons will be called the *normalized Hochschild complexes* and denoted by $(\mathcal{POISS}^{Norm}_{(d-1)}, \partial)$. The *normalized Hochschild complexes* (E_0, ∂) for the operads $\mathcal{BV}_{(d-1)}$, d being even, will be denoted also by $(\mathcal{BV}^{Norm}_{(d-1)}, \partial)$.

COROLLARY 5.6. The first term of the main spectral sequence (for even d and for any commutative ring $\Bbbk$ of coefficients) is isomorphic to the Hochschild homology (with inversed grading) of the Batalin-Vilkovisky operad $\mathcal{BV}_{(d-1)}$. □

REMARK 5.7. (due to M. Kontsevich) The operator $\delta \in \mathcal{BV}_{(d-1)}(1)$ has a natural geometrical interpretation as the Euler class $\delta^E_{d-1} \in H_{d-1}(SO(d))$ of the special orthogonal group $SO(d)$, see Section 7. □

If d is odd, then the first term of the auxiliary spectral sequence is very similar to the normalized Hochschild complex $(\mathcal{BV}^{Norm}, \partial)$, but it does not correspond to any operad. A description of the obtained complex, which I called the *complex of bracket star-diagrams*, see in Section 8.

In [39] the following theorem is proved. For even d this is an immediate corollary of Theorem 5.4.

THEOREM 5.8. *For any $d \geq 3$ the subspace of the first term of the auxiliary spectral sequence linearly spanned only by the summands corresponding to A-configurations forms a subcomplex isomorphic to the normalized Hochschild complex $(\mathcal{POISS}^{Norm}_{(d-1)}, \partial)$ with inversed grading.* □

There is a very nice way to simplify the computations of the second term of the auxiliary spectral sequence. This construction works both for d even and odd. Consider the normalized Hochschild complex $(\mathcal{POISS}^{Norm}_{(d-1)}, \partial)$ and consider the quotient of the space of this complex by the "neighboring commutativity relations" — in other words, for any $i = 1, 2, \ldots$ we set x_i and x_{i+1} to commute. For example the element $[x_1, x_3] \cdot [x_2, [x_4, x_5]] \in \mathcal{POISS}^{Norm}_{(d-1)}(5)$ is equal to zero modulo these relations, because it contains $[x_4, x_5]$. The space of relations is invariant with respect to the differential, thus the quotient space has the structure of a quotient complex. Denote this quotient complex by $(\mathcal{POISS}^{zero}_{(d-1)}, \partial)$.

THEOREM 5.9. [39, 40] *For any commutative ring $\Bbbk$ of coefficients the space $\mathcal{POISS}^{zero}_{(d-1)}$ is a free $\Bbbk$-module. The homology space of the complex $(\mathcal{POISS}^{zero}_{(d-1)}, \partial)$ is isomorphic to the first term of the main spectral sequence (=to the second term of the auxiliary spectral sequence).* □

Idea of the proof: Consider the filtration in the first term (of the auxiliary spectral sequence) by the number of minimal components of corresponding (A, b)-configurations. The associated spectral sequence degenerates in the second term, because its first term is concentrated on the only line corresponding to A-configurations. In the proof it was important that we used a good basis in the complex of bracket star-diagrams $\mathcal{BSD}$ of products of so called "monotone brackets". □

Before describing the relation of the Hochschild homology space of $\mathcal{POISS}_{(d-1)}$ defined over $\mathbb{Q}$ with the first term of the main spectral sequence also defined over $\mathbb{Q}$ (see Theorem 5.10) we will give some explanations.

In Section 8 a structure of a differential graded cocommutative bialgebra on the first term of the auxiliary spectral sequence is defined. According to Theorems 5.4 and 5.8 such a structure is defined also on the normalized Hochschild complexes $(\mathcal{POISS}^{Norm}, \partial)$, $(\mathcal{GERST}^{Norm}, \partial)$, $(\mathcal{BV}^{Norm}, \partial)$. A motivation of the existence of such a structure is that the space of long knots is an H-space (has a homotopy associative multiplication), see [38, 39, 40]; therefore over any field $\mathbb{k}$ its homology spaces $H_*(\mathcal{K}\backslash\Sigma, \mathbb{k})$ form a graded cocommutative bialgebra, and its cohomology spaces $H^*(\mathcal{K}\backslash\Sigma, \mathbb{k})$ form a graded commutative bialgebra dual to the previous one.

Applying Theorem 3.6 to the operads of Poisson, Gerstenhaber or Batalin-Vilkovisky algebras over any field $\mathbb{k}$, we obtain that their Hochschild homology bialgebras are graded commutative (and therefore *bicommutative* – commutative and cocommutative). It follows from the Milnor theorem, see [33], that for $\mathbb{k} = \mathbb{Q}$ these bialgebras are polynomial. The spaces of primitive elements are the spaces of their generators. Unfortunately the bracket 2.5 for these operads does not preserve the spaces of primitive elements (note, in previous papers [38, 39, 40] I stated erroneously the opposite). However I guess the bracket does preserve the spaces of primitive elements at the homology level.

Let us consider for even d the element $\alpha_{even} = [x_1, x_2] \in \mathcal{POISS}_{d-1}(2)$ and for odd d the elements $\alpha_{odd} = [x_1, x_2] \in \mathcal{POISS}_{d-1}(2)$ and $\beta_{odd} = [[x_1, x_3], x_2] \in \mathcal{POISS}_{d-1}(3)$. It is an easy check that $\partial\alpha_{even} = 0$, $\partial\alpha_{odd} = 0$, $\partial\beta_{odd} = 0$:

d is even, bracket is odd:

$$\partial\alpha_{even} = \partial([x_1, x_2]) = (x_1 \cdot x_2) \circ [x_1, x_2] - [x_1, x_2] \circ (x_1 \cdot x_2) =$$

$$([x_1, x_2] \cdot x_3 - x_1 \cdot [x_2, x_3]) - (-[x_1 \cdot x_2, x_3] + [x_1, x_2 \cdot x_3]) =$$

$$[x_1, x_2] \cdot x_3 - x_1 \cdot [x_2, x_3] + x_1 \cdot [x_2, x_3] + x_2 \cdot [x_1, x_3] - [x_1, x_2] \cdot x_3 - [x_1, x_3] \cdot x_2 = 0.$$

d is odd, bracket is even:

$$\partial\alpha_{odd} = \partial([x_1, x_2]) = (x_1 \cdot x_2) \circ [x_1, x_2] + [x_1, x_2] \circ (x_1 \cdot x_2) =$$

$$(-[x_1, x_2] \cdot x_3 + x_1 \cdot [x_2, x_3]) + (-[x_1 \cdot x_2, x_3] + [x_1, x_2 \cdot x_3]) =$$

$$-[x_1, x_2] \cdot x_3 + x_1 \cdot [x_2, x_3] - x_1 \cdot [x_2, x_3] - x_2 \cdot [x_1, x_3] + [x_1, x_2] \cdot x_3 + [x_1, x_3] \cdot x_2 = 0.$$

$$\partial\beta_{odd} = \partial([[x_1, x_3], x_2]) = (x_1 \cdot x_2) \circ ([[x_1, x_3], x_2]) - ([[x_1, x_3], x_2]) \circ (x_1 \cdot x_2) =$$

$$[[x_1, x_3], x_2] \cdot x_4 + x_1 \cdot [[x_2, x_4], x_3] - [[x_1 \cdot x_2, x_4], x_3] + [[x_1, x_4], x_2 \cdot x_3] -$$

$$[[x_1, x_3 \cdot x_4], x_2] =$$

$$[[x_1,x_3],x_2]\cdot x_4 + x_1\cdot[[x_2,x_4],x_3] - x_1\cdot[[x_2,x_4],x_3] - x_2\cdot[[x_1,x_4],x_3] -$$
$$[x_1,x_4]\cdot[x_2,x_3] - [x_1,x_3]\cdot[x_2,x_4] + [[x_1,x_4],x_2]\cdot x_3 + [[x_1,x_4],x_3]\cdot x_2 -$$
$$x_4\cdot[[x_1,x_3],x_2] - x_3\cdot[[x_1,x_4],x_2] - [x_1,x_3]\cdot[x_4,x_2] - [x_1,x_4]\cdot[x_3,x_2] = 0.$$

By abuse of the language we denote by α_{even}, α_{odd}, β_{odd} the corresponding classes in the Hochschild homology.

THEOREM 5.10. [39, 40] *As a graded bialgebra the first term of the main spectral sequence over* $\mathbb{Q}$ *is isomorphic (up to an inversing of grading) to the following quotient of the Hochschild homology bialgebra of the* $\mathbb{Q}$*-linear operad* $\mathcal{POISS}_{d-1}$ *:*

1) for even d: we quotient by one odd primitive generator α_{even} *(of degree* $3-d$*);*

2) for odd d: we quotient by one even primitive generator α_{odd} *(of degree* $3-d$*) and one odd primitive generator* β_{odd} *(of degree* $5-2d$*).* □

REMARK 5.11. Note that

$$\begin{aligned}\alpha_{odd}\circ\alpha_{odd} = \; & [x_1,x_2]\circ[x_1,x_2] = -[[x_1,x_2],x_3] + [x_1,[x_2,x_3]] = \\ & -[[x_1,x_3],x_2] = -\beta_{odd}.\end{aligned}$$

We get $[\alpha_{odd},\alpha_{odd}] = -2\beta$, where $[.,.]$ designates the Gerstenhaber bracket (2.5). □

REMARK 5.12. From the above theorem follows that if $\Bbbk = \mathbb{Q}$ then the multiplication on the first term of the Vassiliev spectral sequence is allways commutative whenever d is even or odd. This fact supports Conjecture 0.3. Really, in [39], [40] it is proven that the homology bialgebra $H_*(\mathcal{K}\backslash\Sigma,\mathbb{Q})$ is graded bicommutative (and in the proof it was important that $\Bbbk=\mathbb{Q}$). □

REMARK 5.13. Multiplication and commultiplication respect the "neighboring commutative relations". Therefore $(\mathcal{POISS}^{zero}_{(d-1)},\partial)$ is also a diferential bialgebra. □

6. The bialgebra of chord diagrams

The *bialgebra of chord diagrams*, the dual to the associated quotient bialgebra of Vassiliev knot invariants, was intensively studied in the last decade. In this section we give an interpretation of the bialgebra of chord diagrams as a part of the Hochschild homology algebra of the Poisson algebras operad.

Consider the normalized Hochschild complex $(\mathcal{POISS}^{Norm}_{(d-1)},\partial)$. In this complex one can define a bigrading by the complexity i and by the number $j=|A|$ of geometrically distinct points of the corresponding A-configurations. The differential ∂ is of bidegree (0,1). If the first grading i is fixed, then the number j varies from $i+1$ to $2i$. So any element of the bigrading $(i,2i)$ belongs to the kernel of ∂. The case $j=2i$ corresponds to the minimal possible dimension of non-trivial homology classes of the space of long knots for the complexity i fixed. For example, if $d=3$, then this dimension is equal to $(d-1)i-j=2i-j=0$. The part of the Hochschild homology groups, that lies in the bigradings $(i,2i)$, $i\geq 0$, will be called the *bialgebra of chord diagrams* if d is

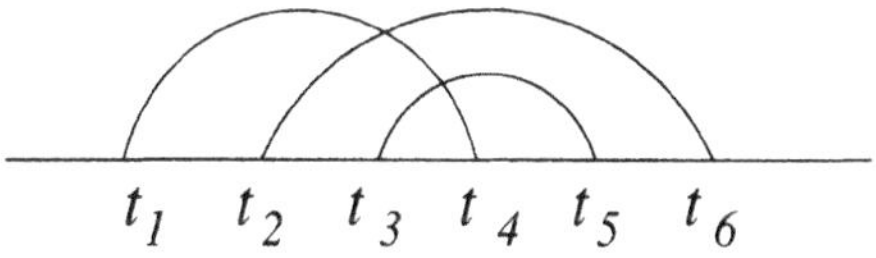

Figure 4. Chord diagram assigned to $[x_3, x_5] \cdot [x_4, x_1] \cdot [x_2, x_6]$

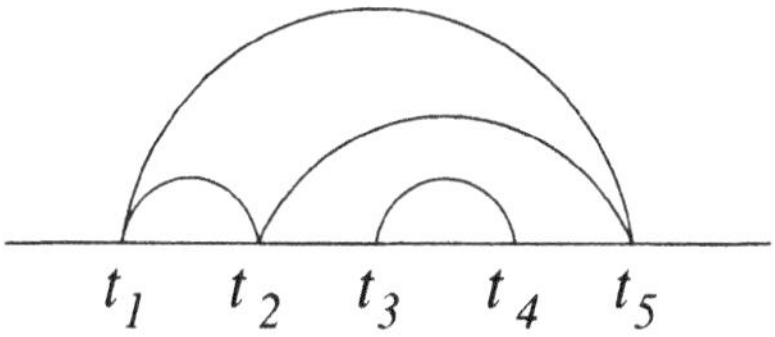

Figure 5. A $(3, 2)$-configuration.

odd, and the *bialgebra of chord superdiagrams* if d is even. Any product of brackets in $\mathcal{POISS}^{Norm}_{(d-1)}(2i) \subset Poiss_{(d-1)}(x_1, \dots, x_{2i}) = S^*Lie_{(d-1)}(x_1, \dots, x_{2i})$ of bidegree $(i, 2i)$ is the product of i brackets, each of wich contains exactly 2 generators. Thus, any such product of brackets can be depicted as $2i$ points on the line $\mathbb{R}^1$, that are decomposed into i pairs and connected by a chord inside each pair. For example, $[x_3, x_5] \cdot [x_4, x_1] \cdot [x_2, x_6]$ is assigned to the diagram of the Figure 4.

The so called *4-term relations* arise as the differential ∂ of products of brackets, in which all brackets except one are "chords", and the only non-chord is a bracket on three elements. In other words, these products of brackets correspond to A-configurations, with $A = (3, 2, 2, \dots, 2)$.

REMARK 6.1. Sometimes one considers the bialgebra of chord diagrams factorized not only by 4-term relations, but also by 1-term relations. We quotient by the diagrams with chords that connect neighboring points. There are two things to say about it. First, these relations arise (when we consider the whole first term of the auxiliary spectral sequence) as the differential of diagrams corresponding to (A, b)-configurations, with $A = (2, \dots, 2)$, $b = 1$, and the only star does not coincide with none of $|A|$ points. Second, it is a manifestation of the "neighboring commutativity relations", see the previous section. □

In the case of odd d we need to take into account orientations of chords (since $[x_{i_1}, x_{i_2}] = -[x_{i_2}, x_{i_1}]$). This definition does not coincide with the standard one, where these orientations are not important. This discordance of definitions can be easily eliminated. Consider the Hochschild complex $(\mathcal{POISS}_{(d-1)}, \partial) = (\bigoplus_{n \geq 0} \mathcal{POISS}_{(d-1)}(n), \partial)$, and replace each space $\mathcal{POISS}_{(d-1)}(n)$ by its tensor product with the one-dimensional sign representation *sign* of the symmetric group S_n. This can be done in the following way. We take a free $(d-1)$-Poisson algebra $Poiss_{(d-1)}(x'_1, \dots, x'_n) = S^*Lie(x'_1, \dots, x'_n)$ with generators $x'_1, \dots, x'_n$ of degree one instead of the analogous algebra $Poiss_{(d-1)}(x_1, \dots, x_n) \supset \mathcal{POISS}_{(d-1)}(n)$ with generators $x_1, \dots, x_n$ of degree zero. Afterwards we consider the subspace $\mathcal{POISS}'_{(d-1)}(n) \subset Poiss_{(d-1)}(x'_1, \dots, x'_n)$ spanned by products of brackets containing each generator x'_i exactly once. Obviously, the S_n-module $\mathcal{POISS}'_{(d-1)}(n)$ is isomorphic to $\mathcal{POISS}_{(d-1)}(n) \otimes sign$. Defining properly the differential, see Section 8, we obtain another version of the Hochschild complex for the operads $\mathcal{POISS}_{(d-1)}$. This new version is interpreted in the geometry of the discriminant Σ as introducing an orientation of the spaces E^j of the strata (5.1) not according to the usual order $t_1 < t_2 < \cdots < t_j$ of the points on the line $\mathbb{R}^1$, but according to the order, that was in our product of brackets. For instance, for the element $[x_3, x_4] \cdot [x_5, [x_2, x_1]]$ the orientation of the corresponding space $E^5 = \{t_1 < t_2 < t_3 < t_4 < t_5\}$ of $(3,2)$-configurations equivalent to configuration on the Figure 5 is according to the order $(t_3, t_4, t_5, t_2, t_1)$.

An advantage of the new Hochschild complexes (for operads $\mathcal{POISS}_{(d-1)}$) is a simpler rule of signs in the definition of the differential, see Section 8.

Let $\mathcal{O}$ be a graded linear operad, equipped with a morphism from the operad $\mathcal{ASSOC}$, then any $\mathcal{O}$-algebra A is an associative algebra because of the following morphisms:

$$\mathcal{ASSOC} \to \mathcal{O} \to \mathcal{END}(A).$$

On the other hand, the map $\mathcal{O} \to \mathcal{END}(A)$ defines a morphism of Hochschild complexes:

$$(\mathcal{O}, \partial) \to (\mathcal{END}(A), \partial).$$

Therefore the classes in the Hochschild homology of $\mathcal{O}$ can be considered as characteristic classes of the Hochschild cohomology of $\mathcal{O}$-algebras (considered as associative algebras). An interesting question is whether all the classes in the Hochschild homology of $\mathcal{BV}$, $\mathcal{POISS}$, $\mathcal{GERST}$, $\mathcal{BV}_{(d-1)}$, $\mathcal{POISS}_{(d-1)}$ have a non-trivial realization as characteristic classes.

Consider Example 1.8, where we take $S^*(\mathfrak{g}[d-1])$ as a $(d-1)$-Poisson algebra. The Hochschild homology space of a polynomial algebra is well known. It is the space of polynomial polyvector fields on the space of generators. In our case the space of generators is $\mathfrak{g}[d-1]$. According to the grading rule, we get that the Hochschild homology space of $\mathcal{POISS}_{(d-1)}$ of bigrading (i,j) is mapped to the space of homogenous degree i j-polyvector fields, *i. e.*, of expressions of the form

$$\sum_{\substack{q_1,\ldots,q_j \\ p_1,\ldots,p_i}} A^{q_1\ldots q_j}_{p_1\ldots p_i} x^{p_1} \ldots x^{p_i} \frac{\partial}{\partial x^{q_1}} \wedge \cdots \wedge \frac{\partial}{\partial x^{q_j}}. \tag{6.2}$$

If d is odd (resp. even), the tensor $A^{q_1\ldots q_j}_{p_1\ldots p_i}$ is symmetric (resp. antisymmetric) with respect to the indices $p_1, \ldots, p_i$ and antisymmetric (resp. symmetric) with respect to the indices $q_1, \ldots, q_j$.

Problem 6.3. Find explicitly $A^{q_1\ldots q_j}_{p_1\ldots p_i}$ via the bracket on $\mathfrak{g}$, for example, in the case $j = 2i$ of chord diagrams. □

The answer to Problem 6.3 can be related with the invariant tensors, defined for Casimir Lie algebras by chord diagrams, see [4, 26, 41].

Also a problem which looks interesting is to find the Gerstenhaber subalgebra in the Hochschild homology of the operads $\mathcal{POISS}$, $\mathcal{GERST}$ generated by the space of chord (super)diagrams. For instance, in the case of the operad $\mathcal{GERST}$ there is an element of bigrading (3,5), that cannot be obtained from chord diagrams by means of the multiplication and the bracket (2.5), see [38, 39]. In the case of the operad $\mathcal{POISS}$ we do not know such an example.

7. Operads of turning balls

In this section we clarify the difference between the cases of odd and even d and give a geometrical interpretation of the operad $\mathcal{BV}_{(d-1)}$ (for d even). In particular we explain Remark 5.7 (due to M. Kontsevich). Theorems 7.1, 7.2 given below are classical, see [15], [2].

For any $d \geq 2$ let us introduce the topological *operad* $\mathcal{TB} = \{\mathcal{TB}_d(n), n \geq 0\}$ *of turning balls.* The space $\mathcal{TB}_d(n)$ is put to be the configuration space of n mappings of the unit ball $B^d \subset \mathbb{R}^d$ into itself. These n mappings are supposed

1) to be injective, preserving the orientation and the ratio of distances;

2) to have disjoint images.

The space $\mathcal{TB}_d(n)$ can be evidently identified with the direct product of the n-th power $(SO(d))^n$ of the special orthogonal group with the configuration space of n disjoint balls in the unit ball B^d. The last space is homotopy equivalent to the n-th component $\mathcal{LC}_d(n)$ of the little cubes operad. The composition operations, the symmetric group actions and the unit element are defined analogously to the case of the little cubes operad.

THEOREM 7.1. [15] *The homology* $\{H_{-*}(\mathcal{TB}_2(n), \mathbb{Z}), n \geq 0\}$ *of the turning discs operad (balls of dimension 2) is the Batalin-Vilkovisky algebras operad* $\mathcal{BV}$. □

Let us describe the operad $\{H_{-*}(\mathcal{TB}_d(n), \mathbb{Q}), n \geq 0\}$ for any $d \geq 2$. Note, that the space of unary operations is the homology algebra $H_{-*}(SO(d), \mathbb{Q})$.

THEOREM 7.2. [2] *The homology bialgebra* $H_*(SO(d), \mathbb{Q})$ *is the exterior algebra on the following primitive generators:*

1) case $d = 2k+1$*: generators* $\delta_3, \delta_7, \ldots, \delta_{4k-1}$ *of degree* $3, 7, \ldots, 4k-1$ *respectively;*

2) case $d = 2k$*: generators* $\delta_3, \delta_7, \ldots, \delta_{4k-5}, \delta^E_{2k-1}$ *of degree* $3, 7, \ldots, 4k-5$ *and* $2k-1$ *respectively.* □

The generators $\delta_{4i-1} \in H_{4i-1}(SO(d), \mathbb{Q})$, $d \geq 2i+1$, are called the *Pontriagin classes*, the generator $\delta^E_{2k-1} \in H_{2k-1}(SO(2k), \mathbb{Q})$ is called the *Euler class*. The Pontriagin classes (and the subalgebra generated by them) lie in the kernel of the map in homology $H_*(SO(d), \mathbb{Q}) \to H_*(S^{d-1}, \mathbb{Q})$ induced by the natural projection

$$SO(d) \overset{SO(d-1)}{\longrightarrow} S^{d-1}.$$

The Euler class is sent to the canonical class of dimension $d-1$ in the homology of the sphere S^{d-1}.

Now we are ready to describe the operads $\{H_{-*}(\mathcal{TB}_d(n), \mathbb{Q}), n \geq 0\}$, $d \geq 2$. We will say which objects are algebras over these operads.

THEOREM 7.3. [39] *Algebras over the operad* $\{H_{-*}(\mathcal{TB}_d(n), \mathbb{Q}), n \geq 0\}$ *are*

1) for even d: $(d-1)$*-Batalin-Vilkovisky algebras, where the operator* δ *is* δ^E_{d-1}*;*

2) for odd d: $(d-1)$*-Poisson algebras.*

Furthermore these algebras are supposed to have $[\frac{d-1}{2}]$ *(where "*$[.]$*" denotes the integral part) mutually super-comuting differentials* $\delta_3, \delta_7, \ldots, \delta_{4[\frac{d-1}{2}]-1}$ *(of degree* $-3, -7, \ldots, 1-4[\frac{d-1}{2}]$*) of the Batalin-Vilkovisky (resp. Poisson) algebra structure. This means, that for any elements* a, b *of the algebra and* $1 \leq i \leq [\frac{d-1}{2}]$*, one has*

(i) $\delta_{4i-1}([a,b]) = [\delta_{4i-1}a, b] + (-1)^{\tilde{a}+d-1}[a, \delta_{4i-1}b]$;
(ii) $\delta_{4i-1}(a \cdot b) = (\delta_{4i-1}a) \cdot b + (-1)^{\tilde{a}} a \cdot (\delta_{4i-1}b)$;
(iii) only for even d: $\delta_{4i-1}\delta^E_{d-1}a = -\delta^E_{d-1}\delta_{4i-1}a$. □

8. Complexes of bracket star-diagrams

In this section we describe the first term of the auxiliary spectral sequence together with its first differential. The corresponding complex will be called *complex of bracket star-diagrams*. In the case of even d this complex is isomorphic to the normolized Hochschild complex of the Batalin-Vilkovisky algebras operad $(\mathcal{BV}^{Norm}, \partial)$.

8.1. Case of odd dimension

In this subsection we consider d to be odd.

Let us fix an (A, b)-configuration J. Let t_α, $\alpha \in \mathfrak{A}$ (resp. t^*_β, $\beta \in \mathfrak{B}$) be the points of J on the line $\mathbb{R}^1$ that do not have stars (resp. that do have stars). Consider the free Lie super-algebra with the even bracket and with odd generators x_{t_α}, $\alpha \in \mathfrak{A}$, $x_{t^*_\beta}$, $\beta \in \mathfrak{B}$. We will take the symmetric algebra (in the super-sens) of the space of this Lie super-algebra. In the obtained space we will consider the subspace $\mathcal{BSD}(J)$ spanned by the products of brackets, where each minimal component of J is presented by one bracket, containing only generators indexed by the points of this minimal component and containing each such generator exactly once.

Such products of brackets will be called *bracket star-diagrams*.

EXAMPLE 8.1. The space $\mathcal{BSD}(J)$ of the bracket star-diagrams corresponding to the (A, b)-configuration J of the Figure 3 is two-dimensional. The diagrams

$$[x_{t_1}, x_{t_3}] \cdot [[x_{t_2}, x_{t^*_4}], x_{t_5}] \cdot x_{t^*_6}, \quad [x_{t_1}, x_{t_3}] \cdot [[x_{t_2}, x_{t_5}]x_{t^*_4}] \cdot x_{t^*_6}$$

form a basis in this space. □

If two bracket star-diagrams can be transformed into one another by an orientation preserving homeomorphism $\mathbb{R}^1 \to \mathbb{R}^1$, then they are set to be equal. For any equivalence class $\mathbf{J}$ of (A, b)-configurations we define the space $\mathcal{BSD}(\mathbf{J})$ as the space $\mathcal{BSD}(J)$, where J is any element of $\mathbf{J}$.

The *space of bracket star-diagrams* is defined as the direct sum of the spaces $\mathcal{BSD}(\mathbf{J})$ over all equivalence classes $\mathbf{J}$ of (A, b)-configurations.

The complexity i and the number j of geometrically distinct points of the corresponding (A, b)-configurations define the bigrading (i, j) on the space of bracket star-diagrams. Remind that the complex of bracket star-diagrams (both for d odd and even) is supposed to compute the first term $E^1_{p,q}$ of the main spectal sequence, whose (p, q) coordinates are expressed as follows:

$$p = -i,$$
$$q = id - j.$$

The corresponding homology degree of the space of knots $\mathcal{K}\backslash\Sigma$ is

$$p + q = i(d-1) - j. \tag{8.2}$$

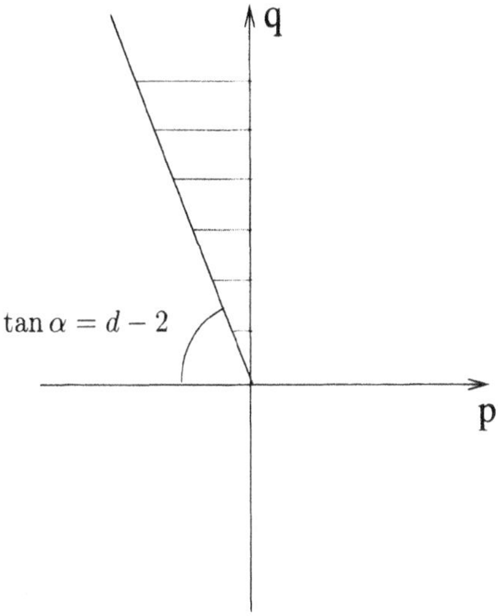

Figure 6. First term of the main spectral sequence.

Note that the first term of the main spectral sequence is non-trivial only in the second quadrant $p \leq 0$. The inequality $j \leq 2i$ for (A, b)-configurations provides the condition $q \geq (2-d)p$, see Figure 6.

The bialgebra of chord (super)diagrams occupies the diagonal $q = (2-d)p$.

To describe the differential ∂ on the space of bracket star-diagrams we will need some complementary definitions and notations.

DEFINITION 8.3. Let us permit to (A, b)-configurations, with $A = (a_1, \ldots, a_{\#A})$, to have $a_i = 1$; we demand also that one-element sets should never coincide with stars. These (A, b)-configurations will be called *generalized (A, b)-configurations.* The generalized (A, b)-configurations that are not (A, b)-configurations in the usual sens (that have $a_i = 1$) will be called *special* generalized (A, b)-configurations. □

Analogously we define the *space of (special) generalized bracket star-diagrams.*

EXAMPLE 8.4. $[x_{t_1}, x_{t_3}] \cdot x_{t_2}$ *is a special generalized star-diagram.* □

Note that the space of generalized bracket star-diagrams is the direct sum of the space of bracket star-diagrams with the space of special generalized bracket star-diagrams.

DEFINITION 8.5. We say that a (generalized) (A, b)-configuration J can be *inserted* in a point $t_0^{(*)}$ of another (generalized) (A', b')-configuration J', if J does not have common points with J' except possibly the point $t_0^{(*)}$. □

DEFINITION 8.6. We say that a (generalized) bracket star-diagram can be *inserted* in the point $t_0^{(*)}$ of another (generalized) bracket star-diagram, if it is the case for their (A, b)-configurations. □

Let A and B be two generalized bracket star-diagrams, such that A can be inserted in the point t_0 (or t_0^*) of B, define the element $B|_{x_{t_0}=A}$ (resp. $B|_{x_{t_0^*}=A}$) of the space of generalized bracket star-diagrams. Up to a sign $B|_{x_{t_0^{(*)}}=A}$ is defined by replacing $x_{t_0^{(*)}}$ (in the diagram B) for A. The sign is defined as $(-1)^{(\tilde{A}-1)\times n}$, where $\tilde{A}$ is the parity of A (the parity of the number of geometrically distinct points), n is the number of generators of the form $x_{t_\alpha}, x_{t_\beta^*}$ before $x_{t_0^{(*)}}$ in B. In other words : we put the bracket containing $x_{t_0^{(*)}}$ on the first place, then by means of antisymmetry relations we put $x_{t_0^{(*)}}$ on the first place in the bracket (and therefore in the diagram); we replace $x_{t_0^{(*)}}$ for A; and we do all these manipulations in the inverse order. It is easy to see that these two definitions give the same sign.

EXAMPLE 8.7.

$$[x_{t_2}x_{t_3^*}]\cdot[x_{t_1^*}x_{t_0}]|_{x_{t_0}=[x_{t_4}x_{t_5}]\cdot x_{t_6^*}} = (-1)^{(3-1)\cdot 3}[x_{t_2}x_{t_3^*}]\cdot[x_{t_1^*},[x_{t_4}x_{t_5}]\cdot x_{t_6^*}]. \square$$

Note that if A has more than 1 minimal components, then the element $B|_{x_{t_0^{(*)}}=A}$ contains multiplications inside brackets. Therefore it is no more a (generalized) bracket star-diagram. To express this element as a sum of (generalized) bracket star-diagrams we will use the formula (1.9).

Now we are ready to define the differential ∂ on the space of bracket star-diagrams.

Let A be a bracket star-diagram, and let t_α be one of its points without a star, then we define

$$\partial_{t_\alpha}A := P(A|_{x_{t_\alpha}=x_{t_{\alpha-}}\cdot x_{t_{\alpha+}}}), \tag{8.8}$$

where P is the projection of the space of generalized bracket star-diagrams on the space of bracket star-diagrams, that sends the space of special generalized bracket star-diagrams to zero; the points $t_{\alpha_-}, t_{\alpha_+} \in \mathbb{R}^1$ are respectively $t_\alpha - \epsilon$ and $t_\alpha + \epsilon$ for a very small $\epsilon > 0$.

REMARK 8.9. The formula (8.8) can be made more precise:

$$\partial_{t_\alpha}A + (x_{t_{\alpha-}} - x_{t_{\alpha+}})\cdot A = A|_{x_{t_\alpha}=x_{t_{\alpha-}}\cdot x_{t_{\alpha+}}}. \square \tag{8.10}$$

Let now t_β^* be one of the points of A having a star. We define

$$\partial_{t_\beta^*}A := P(A|_{x_{t_\beta^*}=x_{t_{\beta-}}\cdot x_{t_{\beta+}^*}+x_{t_{\beta-}^*}\cdot x_{t_{\beta+}}+[x_{t_{\beta-}},x_{t_{\beta+}}]}), \tag{8.11}$$

where P is the same projection, the points $t_{\beta-}^{(*)}, t_{\beta+}^{(*)}$ are respectively $t_\beta^* - \epsilon$ and $t_\beta^* + \epsilon$ for a very small $\epsilon > 0$.

REMARK 8.12. The formula (8.10) can be made more precise:

$$\partial_{t_\beta^*}A + (x_{t_{\beta-}} - x_{t_{\beta+}})\cdot A = A|_{x_{t_\beta^*}=x_{t_{\beta-}}\cdot x_{t_{\beta+}^*}+x_{t_{\beta-}^*}\cdot x_{t_{\beta+}}+[x_{t_{\beta-}},x_{t_{\beta+}}]}. \square \tag{8.13}$$

The differential ∂ on the space of bracket star-diagrams is the sum of the operators ∂_{t_α} and $\partial_{t_\beta^*}$ over all points t_α, $\alpha \in \mathcal{A}$, and t_β^*, $\beta \in \mathcal{B}$, of the corresponding (A,b)-configurations:

$$\partial = \sum_{\alpha\in\mathcal{A}} \partial_{t_\alpha} + \sum_{\beta\in\mathcal{B}} \partial_{t_\beta^*}. \tag{8.14}$$

It is easy to see that $\partial^2 = 0$.

REMARK 8.15.

$$\partial A = \left(\sum_{\alpha \in \mathcal{A}} A|_{x_{t_\alpha} = x_{t_{\alpha-}} \cdot x_{t_{\alpha+}}} \right) +$$

$$+ \left(\sum_{\beta \in \mathcal{B}} A|_{x_{t^*_\beta} = x_{t_{\beta-}} \cdot x_{t^*_{\beta+}} + x_{t^*_{\beta-}} \cdot x_{t_{\beta+}} + [x_{t_{\beta-}}, x_{t_{\beta+}}]} \right) - (x_{t_-} - x_{t_+}) \cdot A, \qquad (8.16)$$

where t_- (resp. t_+) is less (resp. greater) than all the points of the diagram A on the line $\mathbb{R}^1$. □

8.2. Case of even dimension

In this subsection we consider d to be even.

Let us fix an (A, b)-configuration J and consider the free Lie super-algebra with the even bracket and with the even generators x_{t_α}, $\alpha \in \mathcal{A}$, and the odd generators $x_{t^*_\beta}$, $\beta \in \mathcal{B}$, where t_α, $\alpha \in \mathcal{A}$, and t^*_β, $\beta \in \mathcal{B}$, are the points of our (A, b)-configuration J. Let us take the exterior algebra (in the super-sens) of the space of this Lie super-algebra. By convention the parity of an element $A = A_1 \wedge \ldots \wedge A_k$ is $\tilde{A} = \tilde{A}_1 + \ldots + \tilde{A}_k + k - 1$ the sum of the parities of A_i, $1 \le i \le k$, plus the number $k-1$ of the exterior product signs. In the obtained space we will consider the subspace $\mathcal{BSD}(J)$ spanned by the analogous products of brackets (see the previous subsection 8.1). These products of brackets will be also called *bracket star-diagrams.* The *space of bracket star-diagrams* is defined analogously to the case of odd d. We also accept Definitions 8.3, 8.5, 8.6. Note that the parity of a (generalized) bracket star-diagram is the number of stars plus the number of the exterior product signs. This parity is opposite to the parity of the corresponding homology degree (8.2).

Let A and B be two generalized bracket star-diagrams, such that A can be inserted in the point $t_0^{(*)}$ of B. Let us define $B|_{x_{t_0^{(*)}} = A}$. To do this we replace $x_{t_0^{(*)}}$ in B by A, an we multiply the obtained expression by $(-1)^{(\tilde{A} - \epsilon_0) \times (n_1 + n_2)}$, where ϵ_0 is equal to zero (resp. to one) if the point t_0 has no star (resp. if the point t_0^* has a star); n_1 (resp. n_2) is the number of the exterior product signs (resp. of the generators corresponding to stars) before $x_{t_0^{(*)}}$ in B.

EXAMPLE 8.17.

$$[x_{t_2} x_{t^*_3}] \wedge [x_{t^*_1} x_{t_0}]|_{x_{t_0} = [x_{t_4} x_{t_5}] \wedge x_{t^*_6}} = (-1)^{(2-0)\cdot(1+2)} [x_{t_2} x_{t^*_3}] \wedge [x_{t^*_1}, [x_{t_4} x_{t_5}] \wedge x_{t^*_6}]. \ \square$$

Let A be a bracket star-diagram, let t_α be a point of A without a star. Define

$$\partial_{t_\alpha} A := P\left(A|_{x_{t_\alpha} = x_{t_{\alpha-}} \wedge x_{t_{\alpha+}}} \right), \qquad (8.18)$$

where P is the projection from the space of generalized bracket star-diagrams to the space of bracket star-diagrams.

REMARK 8.19. The formula (8.18) can be made more precise:

$$\partial_{t_\alpha} A + (x_{t_{\alpha-}} - x_{t_{\alpha+}}) \wedge A = A|_{x_{t_\alpha} = x_{t_{\alpha-}} \wedge x_{t_{\alpha+}}}. \ \square \qquad (8.20)$$

Figure 7. Multiplication in the space of long knots.

Let t^*_β be a point of A having a star, we define

$$\partial_{t^*_\beta} A := P\left(A|_{x_{t^*_\beta}=x_{t_{\beta-}}\wedge x_{t^*_{\beta+}}-x_{t^*_{\beta-}}\wedge x_{t_{\beta+}}-[x_{t_{\beta-}},x_{t_{\beta+}}]}\right)\cdot \square \tag{8.21}$$

REMARK 8.22. The formula (8.21) can be made more precise:

$$\partial_{t^*_\beta} A + (x_{t_{\beta-}} - x_{t_{\beta+}})\wedge A = A|_{x_{t^*_\beta}=x_{t_{\beta-}}\wedge x_{t^*_{\beta+}}-x_{t^*_{\beta-}}\wedge x_{t_{\beta+}}-[x_{t_{\beta-}},x_{t_{\beta+}}]}.\ \square \tag{8.23}$$

The differential ∂ on the space of bracket star-diagrams is defined by the formula (8.14) analogously to the case of odd d.

REMARK 8.24.

$$\partial A = \left(\sum_{\alpha\in\alpha} A|_{x_{t_\alpha}=x_{t_{\alpha-}}\wedge x_{t_{\alpha+}}}\right)+$$

$$+\left(\sum_{\beta\in\beta} A|_{x_{t^*_\beta}=x_{t_{\beta-}}\wedge x_{t^*_{\beta+}}-x_{t^*_{\beta-}}\wedge x_{t_{\beta+}}-[x_{t_{\beta-}},x_{t_{\beta+}}]}\right)-(x_{t_-}-x_{t_+})\wedge A. \tag{8.25}$$

□

The constructed complex is isomorphic to the normalized Hochschild complex of the Batalin-Vilkovisky algebras operad. This isomorphism is given in [39, Théorème 40.5]. A product of brackets on generators $x_{t_1^{(*)}}, x_{t_2^{(*)}}, \ldots, x_{t_n^{(*)}}$, with $t_1^{(*)} < t_2^{(*)} < \cdots < t_n^{(*)}$, via this isomorphism is mapped to the same (up to a sign) product of brackets, where the generators x_{t_i} without star are replaced by the generators x_i, the generators $x_{t_i^*}$ containing a star are replaced by $\delta(x_i)$.

9. Differential bialgebra of bracket star-diagrams

In this section we define the structure of differential bialgebras on the complexes of bracket star-diagrams. We conjecture that this structure is compatible with the corresponding bialgebra structure on the homology space of the long knots space; the corresponding conjectures are given in [38, 39, 40].

The cases of odd and even d will be considered simultaneously.

Let D be a bracket star-diagram, T be a real number. Define a digram D^T as the diagram obtained from D by the translation of $\mathbb{R}^1$ (D^T is equal to D):

$$t \mapsto t + T. \tag{9.1}$$

Let now A and B be two bracket star-diagrams, we define their product $A * B$ as the diagram $A^{-T} \cdot B^T$ in the case of odd d, and as the diagram $A^{-T} \wedge B^T$ in the case of even d, T being a very large positive number. This product resembles the product in the space of long knots, see Figure 7.

It follows from Theorem 3.6 and Corollary 5.6 that the homology algebra of the differential algebra of bracket star-diagrams is commutative in the case of even d for any commutative ring of coefficients. This is also true over $\mathbb{Q}$ in the case of odd d, see Remark 5.12 and Theorem 5.10.

CONJECTURE 9.2. *Over $\mathbb{Z}$ in the case of odd d, the homology algebra of the differential algebra of the bracket star-diagrams is not commutative.* □

Now we will define a comultiplication on the space of bracket star-diagrams. In the case of odd d the coproduct Δ of any diagram $A = A_1 \cdot A_2 \dots A_k$, where A_i, $1 \le i \le k$, are brackets, is defined as follows:

$$\Delta A = \Delta(A_1 \cdot A_2 \dots A_k) := \sum_{\substack{I \sqcup J = \{1,\dots,k\} \\ I = \{i_1 < \dots < i_l\} \\ J = \{j_1 < \dots < j_{k-l}\}}} (-1)^{\epsilon} A_{i_1} \cdot \dots \cdot A_{i_l} \otimes A_{j_1} \cdot \dots \cdot A_{j_{k-l}}, \tag{9.3}$$

Where $\epsilon = \sum_{i_p > j_q} \tilde{A}_{i_p} \tilde{A}_{j_q}$.

In the case of even d

$$\Delta(A) = \Delta(A_1 \wedge \dots \wedge A_k) := \sum_{\substack{I \sqcup J = \{1,\dots,k\} \\ I = \{i_1 < \dots < i_l\} \\ J = \{j_1 < \dots < j_{k-l}\}}} (-1)^{\epsilon} A_{i_1} \wedge \dots \wedge A_{i_l} \otimes A_{j_1} \wedge \dots \wedge A_{j_{k-l}}, \tag{9.4}$$

where $\epsilon = \sum_{i_p > j_q} (\tilde{A}_{i_p} + 1)(\tilde{A}_{j_q} + 1)$.

Note that the coproduct Δ is graded cocommutative.

It can be easily verified that the operations Δ, $*$, ∂ define a differential bialgebra structure on the space of bracket star-diagrams.

10. Geometric interpretation of the algebraic structures

Reminder. The first term $E^1_{*,*}$ of the Vassiliev main spectral sequence is computed by means of the auxiliary spectral sequence. The auxiliary spectral sequence degenerates on the second term because its first term is concentrated on only one line. In other words the term of degree zero of the main spectral together with its differential is quasiisomorphic to the first term of the auxiliary spectral sequence (also considered with its differential).

Let us remind all the algebraic structures that we have on the first term $E^1_{*,*}$ of the Vassiliev main spectral sequence. The situation depends on the parity of the dimension d.

10.1. Case of even dimension d

By Corollary 5.6 and by Theorem 3.6 the first term of the main spectral sequence is a Gerstenhaber algebra for any commutative ring $\mathbb{k}$ of coefficients. In particular it means that multiplication on it is commutative. If $\mathbb{k}$ is a field then by the previous section $E^1_{*,*}$ is a graded bicommutative bialgebra. By the Milnor-Moore theorem for $\mathbb{k} = \mathbb{Q}$ this bialgebra is polynomial where the space of primitive elements is exactly the space of generators. We conjecture that the Gerstenhaber bracket preserves the space of primitive elements.

10.2. Case of odd dimension d

The case off odd d is different. For any commutative ring $\mathbb{k}$ of coefficients one allways has an algebra structure on $E^1_{*,*}$. But this structure might be non-commutative, see Conjectue 9.2. Due to the previous section, $E^1_{*,*}$ becomes a graded cocommutative bialgebra when k is a field. Moreover for $\mathbb{k} = Q$ this bialgebra is graded bicommutative, see 5.12, and therefore polynomial. On the contrary, $E^1_{*,*}$ has no more Lie algebra structure (even for $\mathbb{k} = \mathbb{Q}$). Really, we have an inclusion of complexes

$$(\mathcal{POISS}, \partial) \hookrightarrow (BSD, \partial),$$

that is surjective on homology (for $\mathbb{k} = \mathbb{Q}$), but we have no warranty that the kernel is a Lie algebra ideal.

10.3. Geometric interpretation of comultiplication

The term $E^1_{*,*}$ is the first term of the Vassiliev main spectral sequence computing the homology groups of the space of long knots. For any field $\mathbb{k}$ of coefficients these groups $H_*(\mathcal{K}\backslash\Sigma, \mathbb{k})$ form a graded cocommutative coalgebra. The corresponding structure on $E^1_{*,*}$ is conjectured to arise from this one, see [38, 39, 40].

10.4. Geometric interpretation of multiplication

As it was already mentioned, the space $\mathcal{K}\backslash\Sigma$ of long knots is an H-space. Therefore its homology groups $H_*(\mathcal{K}\backslash\Sigma, \mathbb{k})$ form an associative algebra for any commutative ring of coefficients $\mathbb{k}$. It is proved in [39, 40] that for $\mathbb{k} = \mathbb{Q}$ the multiplication is commutative (taking $\mathbb{k} = \mathbb{Q}$ is important here). Geometrical (and rather heuristic) reason for this is that the space of long knots is closely related to the space of long framed knots. (Framing for a knot means trivialization of the normal vector bundle. The trivialization should coincide with a fixed trivialization outside some compact subset of $\mathbb{R}$.) On the other hand, it is not difficult to show that the space of long framed knots is a homotopy commutative H-space. The idea of the proof is to shrink the second knot B of the product, see figure 7, to a very small knot; then to pull it on the left through the knot A by means of the connexion unduced by the framing – trivialization of the normal vector bundle of A; and at the end to stretch B to its initial size. Note that this proof does not work for the space $\mathcal{K}\backslash\Sigma$ of long knots.

The precise conjecture how the multiplication on $H_*(\mathcal{K}\backslash\Sigma, \mathbb{k})$ is related to those on $E^1_{*,*}$ is formulated in [38, 39, 40].

10.5. Geometric interpretation of the Gerstenhaber bracket

R. Budney and F. Cohen have announced a proof, that the spaces of long framed knots admit an action of the little squares operad, cf [3]. At the homology level it means that the homology groups of these spaces form a Gerstenhaber algebra. They conjecture that the resulting bracket on homology is compatible with the bracket implied by this paper.

References

1. D.Altschuler, L.Freidel, *Vassiliev knot invariants and Chern-Simons perturbation theory to all orders*, Comm. Math. Phys. 187(1997), pp. 261-287. q-alg/9603010.
2. A.Borel, *Sur la cohomologie des espaces fibrés principaux et des espaces homogènes de groupes de Lie compacts*, Ann. of math. 57 (1953), 115-207.
3. R.Budney, F.Cohen. *Private communication*, 2003.
4. D.Bar-Natan, *On the Vassiliev knot invariants*, Topology, 34 (1995), p.p. 423–472.
5. D.Bar-Natan, *Perturbative Chern-Simons Theory*, Journal of Knot Theory and its Ramifications **4–4** (1995) 503–548.
6. R.Bott and C.Taubes, *On the self-linking of knots*, Jour. Math. Phys. **35** (10) (1994), 5247–5287.
7. J.M.Boardman, R.M.Vogt, *Homotopy-everything H-spaces*. Bull. Amer. Math. Soc. **74** (1968), p.p.1117–1122.
8. J.M.Boardman, R.M.Vogt, *Homotopy invariant algebraic structures on topological spaces*. Heidelberg; New York NY: Springer, 1973 (Lecture notes in mathematics).
9. A.S.Cattaneo, P.Cotta-Ramusino, R.Longoni *Configuration space integrals and Vassiliev classes in any dimension*, Agebr. Geom. Topol. **2**, 949-1000 (2002). GT/9910139.
10. A.S.Cattaneo, P.Cotta-Ramusino, R.Longoni Algebraic structures on graph cohomology, to appear.
11. S.V.Chmutov, S.V.Duzhin, *An upper bound for the number of Vassiliev knot invariants*. J. of Knot Theory and its Ramifications 3 (1994), p.p.141-151.
12. S.V.Chmutov, S.V.Duzhin, S.K.Lando, *Vassiliev knot invariants*. I. Introduction, In: Singularities and Bifurcations, Providence, RI: AMS, 1994, p.p. 117–126 (Adv. in Sov Math. 21).
13. F.R.Cohen, *The homology of C_{n+1}-spaces, $n \geq 0$, The homology of iterated loop spaces*, Lecture Notes in Mathematics 533, Springer–Verlag, Berlin, 1976, p.p.207–351.
14. F.Chapoton, M.Livernet, *Pre-Lie algebras and the rooted trees*, Int. Math. Res. Notices, vol. 2001, Issue 8, p. 395-408. q-alg/0002069.
15. E.Getzler. *Batalin-Vilkovisky algebras and two-dimensional topological field theories*, Comm. Math. Phys. 159 (1994), 265-285.
16. M.Gerstenhaber. *The cohomology structure of an associative ring*. Ann. of Math. **78** (1963), 267–288.
17. V.Ginzburg, *Resolution of diagonals and moduli spaces*, Prog. in Math. 129, pp. 231-266, 1993.
18. E.Getzler, J.D.S.Jones, *Operads, homotopy algebra and iterated integrals for double loop spaces*, preprint hep-th/9403055, Departement of Mathematics, Massachusetts Institute of Technology, March 1994.
19. V.Ginzburg, M.Kapranov, *Koszul duality for operads*, Duke Math. J. 76 (1994), p.p.203–272.
20. T.Goodwillie. *Calculus II. Analytic functors*. K-Theory **5** (1991/92), no. 4, 295-332.
21. T.G.Goodwillie, M.Weiss: *Embeddings from the point of view of immersion theory*: Part II, Geometry & Topology **3**, 103-118 (1999).
22. M.Goresky, R.MacPherson, *Stratified Morse theory*, Ergebnisse der Mathematik und ihrer Grenzgebiete. Vol. 14, Springer-Verlag, Berlin/Heidelberg/New York, 1988.
23. E.Guadagnini, M.Martellini and M.Mintchev, *Chern-Simons field theory and link invariants*, Nucl. Phys **B330** (1990) 575–607.
24. M.Gerstenhaber, A.Voronov, *Homotopy G-algebras and moduli space operad*, Intern. Math. Res. Notices (1995), No.3, p.p. 141–153.
25. V.Hinich, V.Schechtman, *Homotopy Lie algebras*, Adv. in Soviet Math., **16**, Part 2, (1993) 1-28.
26. V.Hinich, A.Vaintrob, *Cyclic operads and algebra of chord diagrams*, Selecta Math., New Ser. Vol. 8, Issue 2 (2002), pp.237-282. q-alg/0005197.
27. M.Kontsevich, *Vassiliev's knot invariants*, Adv. in Sov. Math., vol.16, part 2, AMS, Providence, RI, 1993, p.p.137–150.
28. J.A.Kneissler, *The number of primitive Vassiliev invariants up to degree twelve*, University of Bonn preprint, June 1997. q-alg/97060222.
29. T.Kimura, J.Stasheff, A.A.Voronov, *On operad structures of moduli spaces and string theory*, hep-th/9307114.

30. S.K.Lando. *On primitive elements in the bialgebra of chord diagrams*, AMS Translations (2), vol.180, 1997, p.p.167–174.
31. J.-L. Loday. *La renaissance des opérades*, Séminaire BOURBAKI, 47ème année, 1994-95, n°792.
32. P.May. *The geometry of iterated loop spaces.* Lecture Notes in Math. **271** (1972)
33. J.Milnor, J.Moore, *On the structure of Hopf algebras*, Ann. Math. 81 (1965), pp. 211–264.
34. S.Poirier. *The configuration space integral for links and tangles in* $\mathbb{R}^3$, geom-top/0005085.
35. D.Sinha, *On the topology of spaces of knots*, math. AT/0202287.
36. D.Sinha, *Private communication,* 2002.
37. A.Stoimenow, *Enumeration of chord diagrams and an upper bound for Vassiliev invariants.* J. of Knot Theory and its Ramifications 7 (1998) 93-114.
38. V.Tourtchine, *Sur l'homologie des espaces de nœuds non-compacts*, preprint IHES M/00/66, math. q-alg/0010017
39. V.Tourtchine, *Sur les questions combinatoires de la théorie spectrale des nœuds*, PHD thesis, Université Paris 7, (Mai 2002).
40. V.Tourtchine, *Nekotoryje kombinatornyje voprosy spectral'noj teorii uzlov*, in Russian, thesis (2000), Moscow State University.
41. A.Vaintrob, *Vassiliev knot invariants and Lie S-algebras*, Math. Res. Lett. **1** (1991), 579–595.
42. V.A.Vassiliev, *Cohomology of knot spaces.* In: Adv. in Sov. Math.; Theory of Singularities and its Applications (ed. V.I.Arnol'd). AMS, Providence, R.I., 1990, p.p.23–69.
43. V.A.Vassiliev. *Stable homotopy type of the complement to affine plane arrangements.* Preprint 1991.
44. V.A.Vassiliev. *Complexes of connected graphs.* In: Gelfand's Mathematical Seminars, 1990–1992. L.Corwin, I.Gelfand, J.Lepovsky, eds. Basel: Birkhäuser, 1993, p.p.223–235.
45. V.A.Vassiliev. *Complements of Discriminants of Smooth Maps: Topology and Applications.* Revised ed. Providence, R.I.: AMS, 1994 (Translation of Mathem. Monographs, 98).
46. V.A.Vassiliev, *Topology of two-connected graphs and homology of spaces of knots*, in Differential and Symplectic Topology of Knots and Curves (S.Tabachnikov, ed.), AMS translations, Ser. 2, **190**, AMS, Providence RI, 253–286, 1999.
47. P.Vogel, *Algebraic structures on modules of diagrams*, July 1995 (revised 1997), to appear in Invent. Mathematicae. www.math.jussieu.fr/ vogel/
48. P.Vogel, *The universal Lie algebra*, June 1999, www.math.jussieu.fr/ vogel/
49. D.Zagier, *Vassiliev invariants and a strange identity related to the Dedekind eta-function*, Topology 40, 945-960, 2001 (preprint 1999).
50. G.M.Ziegler, R.Živalević, *Homotopy types of subspace arrangements via diagrams of spaces*, Math. Ann. **295** (1993), 527–548.

Address for Offprints: Independent University of Moscow
Moscow 121002, B.Vlasevsky Pereulok 11, RUSSIA
e-mail: turchin@mccme.ru

Some computations of Ohtsuki series

Nori Jacoby (nori_jacoby@hotmail.com) and Ruth Lawrence (ruthel@ma.huji.ac.il)

Einstein Institute of Mathematics, Hebrew University, Givat Ram 91904 Jerusalem, ISRAEL

Abstract. We present some computational data on Ohtsuki series for a two parameter family of integer homology spheres obtained by surgery around what we call '2–strand knots', closures of the simplest rational tangles. This data allows us to make certain conjectures about the growth rate of the coefficients in Ohtsuki series generally, based on which we introduce an invariant which we call the *slope* $\sigma(M)$ of a manifold M (not to be confused with slopes in hyperbolic geometry). For Seifert fibred manifolds, M, the conjectures are known to hold while $\pi^2\sigma(M) \in \mathbf{Q}$; furthermore if M is also an integer homology sphere, $\pi^2\sigma(M) \in \mathbf{Z}$. Assuming the conjectures, the numerical data enables us to give an example of a $\mathbf{Z}$HS for which $\pi^2\sigma(M) \notin \mathbf{Z}$. This paper is based on the first author's M.Sc. thesis.

Keywords: Ohtsuki series, Seifert fibred manifold, rational tangle, quantum invariants

Mathematics Subject Classification 2000: Primary 57M27, Secondary 05A30 11B65 17B37 57R56

1. Introduction

Suppose that M is a compact oriented 3–manifold without boundary. The sl_2 Witten-Reshetikhin-Turaev invariant (see (Wi), (RT)) is a complex number invariant $Z_K(M, L)$ of embeddings of links L in M, dependent on an integer K. It is known that for links in S^3, $Z_K(S^3, L)$ is a polynomial in $q = \exp\frac{2\pi i}{K}$ and q^{-1}, namely the generalised Jones polynomial of the link L.

Now assume that M is a rational homology sphere with $H = |H^1(M, \mathbf{Z})|$. In this paper we will only consider the case of invariants where the link L is empty. From its algebraic definition, e.g. via quantum groups, $Z_K(M, \emptyset)$ can be written as a rational (or polynomial) function of q at K^{th} roots of unity. In the normalization for which $Z_K(S^3, \emptyset) = 1$, denote the invariant for the pair $(M, \emptyset)$, by $Z_K(M)$. For odd prime K, $Z_K(M) \in \mathbf{Z}[q]$ (see (M1), (M2)), so that for some $a_{m,K}(M) \in \mathbf{Z}$, one

J.M. Bryden (ed.), Advances in Topological Quantum Field Theory, 53–70.

has

$$Z_K(M) = \sum_{m=0}^{\infty} a_{m,K}(M)(q-1)^m . \qquad (1)$$

Although the $a_{m,K}$ are not uniquely determined from this relation since $\frac{q^K-1}{q-1} = 0$, however they are determined modulo K for $m \le K-2$ when K is prime. It is known from (O1) and (O2) that there exist rational numbers $\lambda_m(M)$ such that,

$$a_{m,K}(M) \equiv \left(\frac{H}{K}\right)\lambda_m(M) \qquad (2)$$

as elements of $\mathbf{Z}/K\mathbf{Z}$ for all sufficiently primes $K \ge 2m+3$. The formal power series

$$Z_\infty(M) = \sum_{m=0}^{\infty} \lambda_m(q-1)^m ,$$

is known as the *Ohtsuki series* of M and by (R3) $Z_\infty(M) \in \mathbf{Z}[\frac{1}{2H}][[h]]$ where $q = 1 + h$. For integer homology spheres ($H = 1$) we have $Z_\infty(M) \in \mathbf{Z}[[h]]$, while $\lambda_0(M) = 1$ and $\lambda_1(M) = 6\lambda(M)$ where $\lambda(M)$ denotes the Casson invariant of M. In general Z_∞ is expected to be the asymptotic expansion of the trivial connection contribution to $Z_K(M)$ in Chern-Simons theory.

Relatively few computations of Z_∞ have been carried out. For Lens spaces, Jeffrey's closed formula (Je) for Z_K gives a formula for Z_∞,

$$Z_\infty(L(P,Q)) = q^{\pm 3s(Q,P)} \frac{q^{\frac{1}{2P}} - q^{-\frac{1}{2P}}}{q^{1/2} - q^{-1/2}} ,$$

where $s(Q,P)$ is a modified Dedekind sum. For Seifert fibred manifolds, Z_K can be written as the asymptotic expansion around $q = 1$ of a holomorphic function of q expressed as a complex (contour) integral (LR). Thus for a Seifert fibred manifold which is an **ZHS**,

$$Z_\infty\left(\Sigma\left(\frac{Q_1}{P_1}, \ldots, \frac{Q_N}{P_N}\right)\right) = c.q^{-\frac{\phi}{4}} \int e^{\frac{iKy^2}{8\pi P}} \frac{\prod_{j=1}^N \sinh \frac{y}{2P_j}}{(\sinh \frac{y}{2})^{N-2}} \, dy ,$$

where $P = \prod P_j$, c is a constant (dependent on P) and ϕ is a rational number,

$$\phi = 3\mathrm{sign}P + \sum_{j=1}^{N}\left(12s(Q_j, P_j) - \frac{Q_j}{P_j}\right) .$$

For special cases some other formulae are known, for example for ± 1 surgery around the trefoil, that is for the Poincaré homology sphere

$\Sigma(2,3,5)$ and for $\Sigma(2,3,7)$, (see (LZ), (Le))

$$q(q-1)Z_\infty(\Sigma(2,3,5))(q) = 1 - 1/2 \sum_{n=1}^{\infty} \chi(n) q^{(n^2-1)/120}$$
$$(q-1)Z_\infty(\Sigma(2,3,7))(q) = \sum_{n=0}^{\infty} q^{-\frac{1}{2}n(n+1)}(q^{n+1}-1)\dots(q^{2n+1}-1)$$

where the odd function $\chi : \mathbf{Z}/60\mathbf{Z} \longrightarrow \{-1,0,1\}$ takes value $+1$ precisely at 1, 11, 19 and 29 (and nowhere else).

Such computations have been obtained using three approaches. All use a surgery presentation of the manifold via a link.

- One can write a state sum for Z_K in which a state is a labeling of the regions and components of a diagram of the link (see (KL)) and the local weights are quantum dimensions, θ-nets and quantum $6j$-symbols on the different elements of the diagram (components/regions, edges, crossings, respectively). One can use the method of *recombination* here to rewrite the state sum obtained in a possibly simpler way. To extract a formula for Z_∞ now requires some manipulations of the form of the sums involved in Z_K which only work for a small class of manifolds. This was used in (L) and (LZ).

- Using conformal field theory, one can write Z_K as a sum of products of S and T matrix elements which take a particular simple form for special manifolds. This was used to obtain Z_∞ in (Je), (LR).

- Using the formulation of $J_L(\rho_1,\dots,\rho_c)$, the coloured Jones polynomial of a link L coloured by representations ρ_i on its c components, as the image of the link (or rather of a 1-tangle whose closure is the link) under a representation of the category of tangles, one can write J_L in terms of the universal R-matrix, as the scalar value taken by an element of ${U_q sl_2}^{\otimes c}$ in the representations ρ_i. From this universal invariant of the link an expression for $Z_\infty(M)$ can be obtained as an infinite sum with the property that coefficients of $(q-1)^n$ (for any n) arise only form a finite number of terms in the sum. This was the method used by Le, (Le).

To obtain closed forms for Z_∞ using the first two approaches, relies on the manifold being obtainable from a particularly simple link, since in the end the combinatorial formulae for Z_K depend on quantum $6j$-symbols (or what is the same thing, R-matrix entries) which enter at each crossing in a link diagram presentation. These have complicated formulae even for $U_q sl_2$ and so only for those manifolds for which the quantum $6j$ symbols can be manipulated to cancel can a closed form be expected sufficient to find Z_∞. This occurs for Seifert fibred manifolds.

In this paper we will consider a special (two parameter) class of manifolds (see §2) for which the state sum formula is relatively simple, although still complicated enough that a closed formula for Z_∞ is unknown. In (Ja), computer computations were carried out for Z_∞ on this family; we outline the method and some peculiarities relating to numerical precision, in §3. In §4, some of the results are given graphically along with conjectures that they support. Finally, in §5, some possible generalizations are suggested.

2. A two-parameter family of manifolds

2.1. Definition of the manifolds

Let $K_{((\frac{S}{T}))}$ be the knot obtained by connecting two tangles each of which consists of two strands simply braided, with S and T crossings respectively, the two tangles being connected as shown in Figure 1.

Negative S and/or T result in crossings of opposite orientation. Here, $K_{((\frac{S}{T}))}$ will be a knot (and not a link) when S and T are not both odd. Note that for different parities of S and T, the relative orientations of the different parts of $K_{((\frac{S}{T}))}$ will be different. Replacing (S,T) by $(-T,-S)$ doesn't change the knot, while reversing the signs of both S and T (or interchanging S and T) changes the knot to its mirror image. Some special cases are, $S=0$ (unknot), $S=\pm 1$ ($(2,T\mp 1)$ torus knot) along with the examples in the table in table below, where the knots have been given with their Conway names.

Table .

S	2	3	4	5	6	7	8	4	6	4	5	6
T	2	2	2	2	2	2	2	3	3	4	4	4
$K_{((\frac{S}{T}))}$	4_1	5_2	6_1	7_2	8_1	9_2	10_1	7_3	9_3	8_3	9_4	10_3

In this paper we consider the manifolds $M_{((\frac{S}{T}))}$ obtained from S^3 by surgery on the *framed* knot $K_{((\frac{S}{T}))}$ of Figure 1. The number of twists U in $K_{((\frac{S}{T}))}$ is chosen so that the resulting blackboard framed knot will

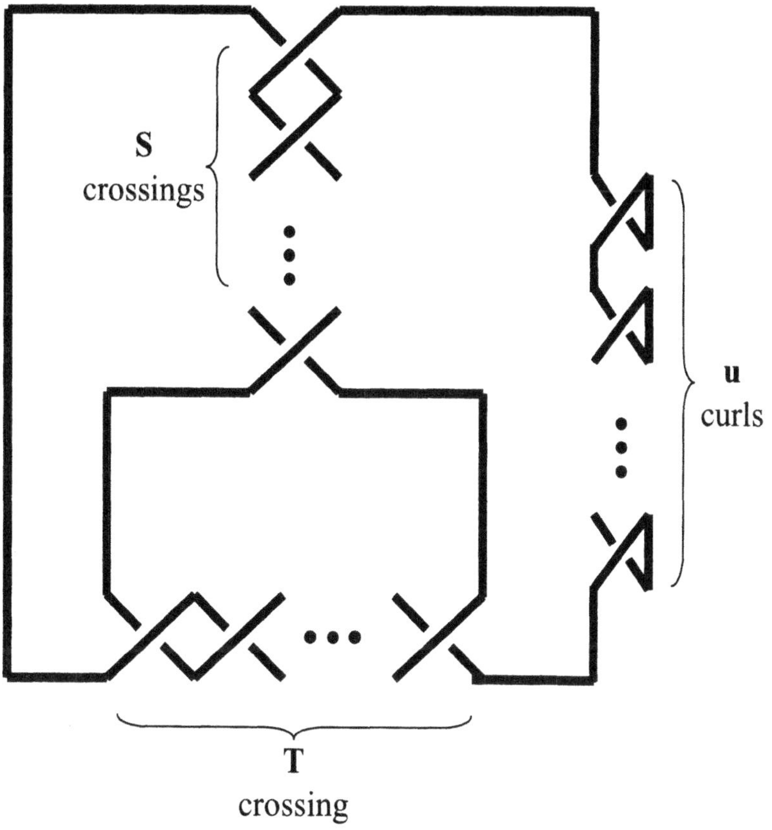

Figure 1. Knot family $K_{(\binom{S}{T})}$

have framing ± 1, that is, so that $M_{(\binom{S}{T})}$ will be an integer homology sphere. This requires $U = (-1)^T S - (-1)^S T \pm 1$, where the signs depend on the parities of S and T (since the relative orientations of different sections of the knot depend upon the parities of S and T), so in fact for each S and T we obtain two manifolds $M^{\pm}_{(\binom{S}{T})}$.

2.2. Calculation of Z_K

The Kauffman-Lins state sum formulation (see (KL)) of the WRT invariant $Z_K(M^{\pm}_{(\binom{S}{T})})$ uses a sum over states which are colorings of the one component of $K_{(\binom{S}{T})}$ and of the regions into which the knot diagram of $K_{(\binom{S}{T})}$ in Figure 1 cuts the plane, by colors from the set $\{0, 1, \ldots, K-2\}$. The only admissible colorings are those which satisfy the constraint that for each edge of the knot diagram, the triple

The $\begin{bmatrix} \begin{smallmatrix} S \\ T \end{smallmatrix} \end{bmatrix}$ Knot Family WRT Formula

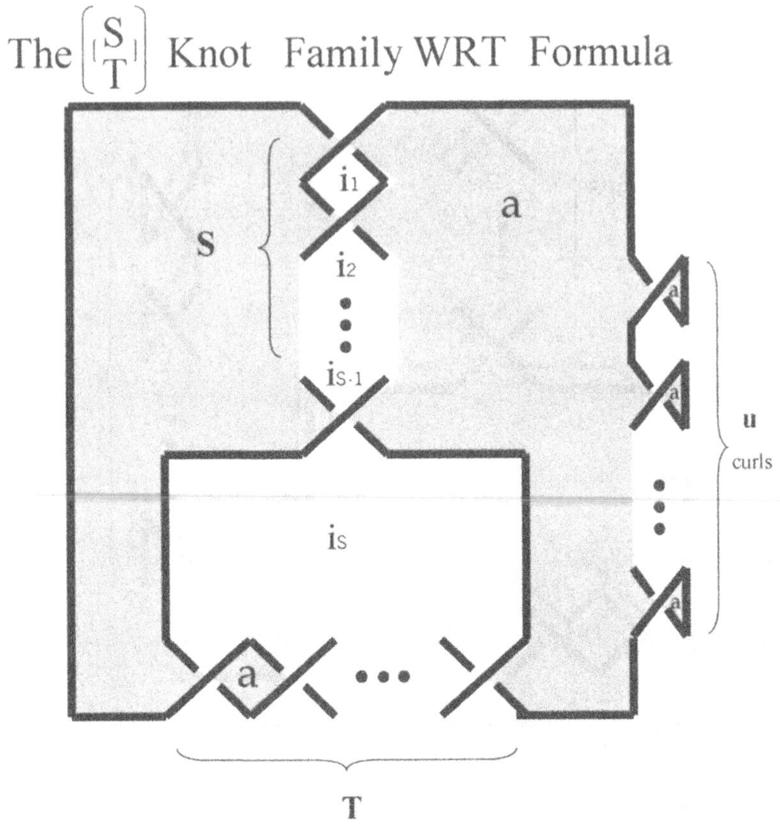

Figure 2. Allowed states for $K_{((\frac{S}{T}))}$

of colors coming from the component and the two regions on either side of the edge, satisfies the Clebsch-Gordan constraint. The possible colorings in our case are shown in Figure 2. This would give a formula for $Z_K(M^\pm_{((\frac{S}{T}))})$ as a sum over $S+1$ indices in which the summand contains $S-1$ nontrivial quantum $6j$ symbols.

Using recombination, each crossing can be written as a sum over the label on the internal edge, of the evaluation of an 'H' type graph with two trivalent vertices (Figure 3(1)). This leads to the configuration in Figure 3(2) which can be simplified using Figure 3(3) to a network whose size is independent of S and T, although with a non-standard weighting. It has just four trivalent vertices (that is, it is a tetrahedral

---

A triple $\{a, b, c\}$ is said to satify the Clebsch-Gordan constraint if they satisfy the triangle inequality $|a - b| \le c \le a + b$ while $a + b + c$ is even and at most $2K - 4$.

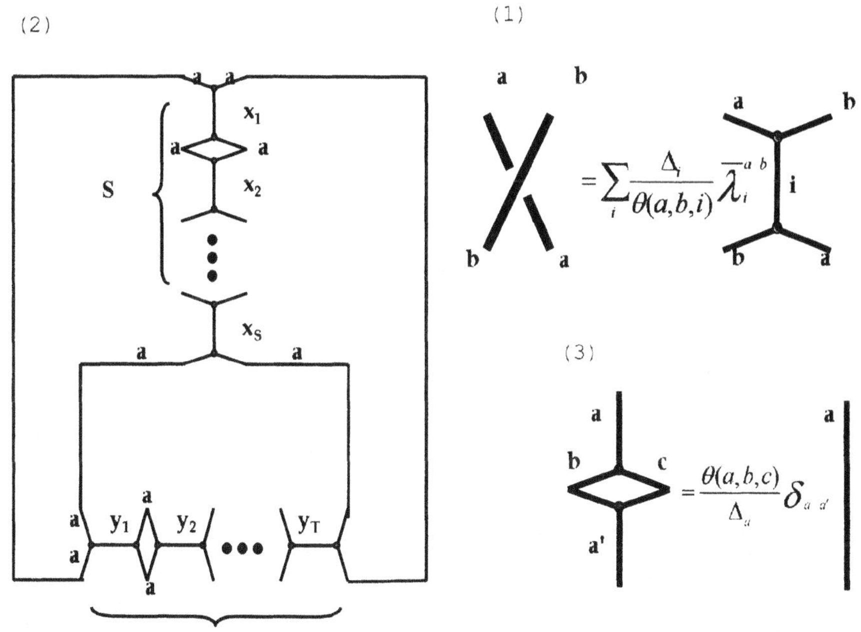

Figure 3. Recombination

network) and the resulting formula for the WRT invariant is

$$Z_K(M^{\pm}_{((\frac{S}{T}))}) = G_{\pm}^{-1} \sum_{a,x,y} \Delta_x \Delta_y (\overline{\lambda}_x^{a\ a})^S (\lambda_y^{a\ a})^T (\lambda_0^{a\ a})^U \begin{Bmatrix} a & x & a \\ a & y & a \end{Bmatrix} \quad (3)$$

where the sum is over colors a, x and y for which $\{a, a, x\}$ and $\{a, a, y\}$ are admissible triples, while $G_{\pm}$ are the Gauss sums $\sum_a \Delta_a^2 (\lambda_0^{a\ a})^{\pm 1}$ obtained by evaluation on the unknot with framing ± 1. We have used notation,

$$\begin{aligned} \Delta_n &= (-1)^n \frac{A^{2(n+1)} - q^{-2(n+1)}}{q^2 - q^{-2}} \\ \lambda_c^{a\ b} &= (-1)^{(a+b+c)/2} A^{(a(a+2)+b(b+2)-c(c+2))/2} \end{aligned}$$

where A is a fourth root of q.

2.3. Properties of Ohtsuki series

The Ohtsuki series $\sum_{m=0}^{\infty} \lambda_m (q-1)^m$ will depend on S and T (and U, or equivalently on the sign of the surgery). To include this dependence clearly in the notation, we will denote the coefficients by $\lambda_m^{\pm}(S,T)$.

Theorem *For fixed parity and signs for S and T, λ_m is a polynomial in S and T of degree $2m$ with rational coefficients.*

Proof. It is known that $M \longrightarrow \lambda_m(M)$ is a finite type invariant of manifolds of order m. On the other hand, the map $K \longrightarrow \lambda_m(S^3_K)$ (doing $+1$ surgery on the knot) gives a Vassiliev invariant of order $2m$. (This follows for example from the Melvin-Morton-Rozansky conjecture (B-NG) and Le (Le).)

However if $|S|+|T| > 2m$, then let us pick any $2m+1$ of the crossings in the knot diagram of $K_{\left(\binom{S}{T}\right)}$ in Figure 1, say s of these crossings will be from the ones that were counted by S and t from the crossings that were counted by T, where $s+t = 2m+1$. One can consider the 2^{2m+1} knots obtained by variously altering the orientations of these crossings in all possible ways. Since λ_m is Vassiliev of order $2m$, the alternating sum of the values of λ_m on these knots vanishes. When the orientation of a crossing (say of type S) in $K_{\left(\binom{S}{T}\right)}$ is flipped, the resulting knot is of the same type, by with S changed to $S-2\mathrm{sign}S$. Without loss of generality, S and T are positive. Then the result of changing the orientations of i crossings of type S and j crossings of type T, is $K_{\left(\binom{S-2i}{T-2j}\right)}$. There are $\binom{s}{i}\binom{t}{j}$ ways in which to pick the subsets of crossings to flip, so that

$$\sum_{i=0}^{s}\sum_{j=0}^{t}(-1)^{i+j}\binom{s}{i}\binom{t}{j}\lambda_m^{\pm}(S-2i,T-2j) = 0\,, \tag{4}$$

for all S and T positive and all s, t with $s+t = 2m+1$. The left hand side of (4) is the order $s+t$ finite difference partial derivative of $\lambda_m^{\pm}$, s times in the S direction and t times in the T direction. Observe also that all the terms in (4) involve S and T of fixed parities. This condition is sufficient to guarantee that (for fixed parities) $\lambda_m^{\pm}(S,T)$ is expressible as a polynomial in S and T of order at most $2m$, as required. Since $\lambda_m \in \mathbf{Z}$, these polynomials must have rational coefficients, in fact $2^{2m}((2m)!!)^2\lambda_m(S,T) \in \mathbf{Z}[S,T]$ where $n!! \equiv \prod_{r=1}^{n} r!$. □

3. Computer calculations of Ohtsuki series

Using (3), $Z_K(M^{\pm}_{((\frac{S}{T}))})$ can be computed for all the $K-1$ different K^{th} roots of unity, $q = q_s = \exp \frac{2\pi i s}{K}$ $(1 \leq s < K)$ for fixed S, T and prime K. From these values, the coefficients $a_{m,K}(M^{\pm}_{((\frac{S}{T}))}) \in \mathbf{Z}$ can be found in (1) where we assume that they vanish past the $(K-2)^{\text{th}}$. Namely, one solves the following linear system of size $K-1$,

$$Z_K|_{q=q_s} = \sum_{m=0}^{K-2} a_{m,K}\cdot(q_s - 1)^m \,, \tag{5}$$

where the dependence on the manifold has been omitted from the notation. The integrality of the solutions provides a check on the errors in the computations of the individual WRT invariants at different roots of unity.

There a number of computational challenges here.

3.1. Growth of number of states

As was mentioned in the previous section, the sum obtained from naively applying, say the Kauffman-Lins prescription, to the knot diagram Figure 1 leads to a sum over a number of states which grows exponentially with S, since there would be a summation over S indices i_j, and the non-trivial quantum $6j$-symbols effectively introduce a further $S-1$ summations, where the summand is now a product (or quotient of products) of quantum numbers and factorials. This problem is obviated with the used of recombination; thus (3) only involves a sum over 4 indices (a, x, y and an additional index from the single quantum $6j$-symbol) so that there are the order of K^3 states to be summed over (or K^4 with a summand which is a quotient of products of quantum numbers and factorials), and this is independent of S and T.

3.2. Numerical precision and summand size

Since q-factorials can be numerically very large, the individual terms in (3) can be many orders of magnitude larger than their sum and this necessitates using very high precision arithmetic.

We give the example of $K = 97$ for the manifold $M^{+}_{((\frac{3}{2}))}$ obtained by surgery around the knot 5_2 (obtained from $S = 3$, $T = 2$, $U = 6$).

For $s = 1$, $[90]! \approx 10^{114}$ and the order of summands in Z_K for $s = 2$ is $\sim 10^{26}$, while their sum (Z_{97}) is $\sim 10^6$. To obtain an accuracy of d significant figures in the result then requires $20 + d + e$ figures in the summands, where the number of states is of order 10^e. See Figure 4 which shows a plot of the numerical sizes of Z_K and the individual terms in the sum for Z_K (more precisely, the sum of absolute values of the real part of terms in the sum for Z_K) on a log scale, against s.

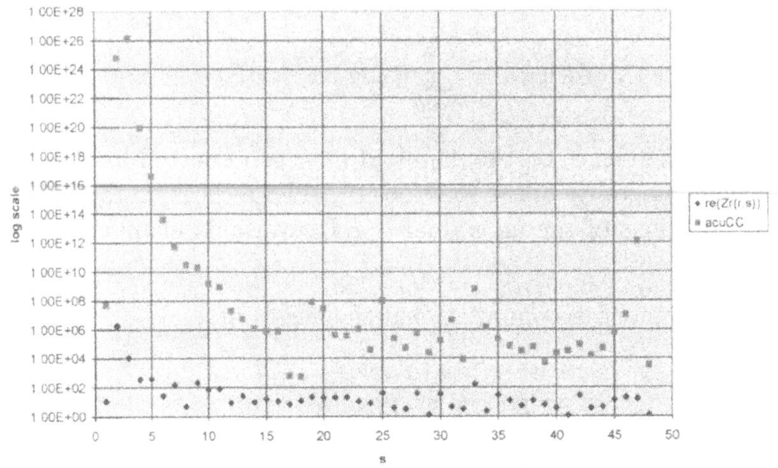

Figure 4. Relative sizes of Z_K and its summands against s

Because of the very large numerical values of the quantum factorials, despite the smaller sizes of terms in the sums (10^{100} versus 10^{26} in the above example), it is necessary to be able to work with numbers with large exponent (say 300) though maybe only 30 significant figures. It is very costly in time to use higher precision than absolutely necessary, so some special routines were written to easily manipulate sum numbers with large exponents, but (comparitively) not so high precision. For example, calculating Z_{97} for all s using 30-digits precision took 596 seconds on a Pentium 600MHz computer but it takes 1740 seconds to compute it for only one s using 80-digits precision!

Because of the large discrepancy in orders of magnitude of the individual terms and the total sum, which as can be see from Figure 4 only occurs for some values of s (near 0 and K), it is necessary for those values of s to perform the computations with specially higher precision. Therefore two different routines were written, and used on different values of s. For details, see (Ja).

The result of these two factors is to practically limit the size of K (though not the complexity of the knot, that is S and T, since the difficulty in calculation is practically independent of S and T). For $M^+_{((\frac{3}{2}))}$ these techniques allow computation for all primes K less then 137 with a total running time less than 1.5 days (using a 586 1GHz computer).

Knowing $a_{m,K}$ for primes $K = K_i \geq m+2$ determines λ_m modulo $\prod K_i$, according to their definition in (2) (recall that here $H = 1$). From Ohtsuki's theorem, it is known that this class, as an integer in $[-\frac{1}{2}\prod K_i, \frac{1}{2}\prod K_i]$, stabilizes as the number of primes used increases. This gives an algorithm for obtaining (the first few terms in) Ohtsuki series. The practical limitation on K above gives a limitation on how many terms can be calculated. The full implementation of this algorithm for our manifolds $M^{\pm}_{((\frac{S}{T}))}$ can be found at `http://www.ma.huji.ac.il/~ruthel/nori/index.html`.

4. Results and conjectures

4.1. Individual Ohtsuki series

For small K (up to 100), it turns out that the number of states is $\approx K^{2.9}$ (rather than K^3) while the running time is $\approx K^{4.2}$ (rather than K^4) due to the increase in computational time of each summand and initialization procedures, for the larger values of K. For $M^+_{((\frac{3}{2}))}$ discussed above ($(S,T,U) = (-2,-3,6)$), $K < 137$ was sufficient to compute the first 21 coefficients of Z_∞,

$$
\begin{aligned}
&Z_\infty(M^+_{((\frac{3}{2}))}) \\
&= 1 + 12h + 258h^2 + 7756h^3 + 300055h^4 + 14192892h^5 \\
&\ +793556722h^6 + 51201783488h^7 + 3744412949224h^8 \\
&\ +306062634843942h^9 + 27651533457983745h^{10} \\
&\ +2736207255879667844h^{11} + 294306807889008940143h^{12} \\
&\ +34188707473104409330168h^{13} \\
&\ +4265845139103716469762268h^{14} \\
&\ +568978507509845435699024672h^{15} \\
&\ +80787229265313530505892175542h^{16} \\
&\ +12165972894589961487357113418955h^{17} \\
&\ +1936811327962748352514940775515283h^{18} \\
&\ +325007156713501796302801741846095206h^{19} \\
&\ +57334985329655520887251821186176103843h^{20} \\
&\ +10607981215487793536113323249915379712259h^{21} \\
&\ +2053956644731187123340443541756436810603354h^{22} + \cdots
\end{aligned}
$$

where $h = q - 1$.

4.2. Polynomiality of coefficients

We already know that for fixed m, the dependence on S and T (of fixed parity) of $\lambda^\pm_m(S,T)$ is polynomial of degree $2m$. Since for $S = 0$ (or $T = 0$), $K_{((\frac{S}{T}))}$ is the unknot for which $Z_\infty = 1$, thus $\lambda^\pm_m(S,T)$ is divisible by S (for S even) and by T (for T even) and by ST (for S, T both even).

Since the WRT invariants of the mirror image manifold $\bar{M}$ of M are given by $Z_K(\bar{M})(q) = Z_K(M)(q^{-1})$ at all roots of unity, thus $Z_\infty(\bar{M})$ can be obtained from $Z_\infty(M)$ by replacing q by q^{-1}, that is by substituting $\frac{1}{1+h} - 1 = -\frac{h}{1+h}$ for h. This gives (complicated) relations between $\lambda^\pm_m(S,T)$ and $\lambda^\mp_m(T,S)$, while $\lambda^\pm_m(-T,-S) = \lambda^\pm_m(S,T)$. Hence it is only necessary to compute the polynomials $\lambda^+_m(S,T)$ for the two parity combinations, (even, even), (even, odd). Knowing the degrees of the polynomials, numerical results for a large enough number of pairs (S,T) are sufficient to determine these polynomials. The results for the first few coefficients are, for S and T both even,

$$
\begin{aligned}
\lambda^+_0 &= 1 \\
\lambda^+_1 &= -\tfrac{3}{2}ST \\
\lambda^+_2 &= \tfrac{3}{4}ST(5ST + T - S + 3)
\end{aligned}
$$

and for S even, T odd,

$$\begin{aligned}\lambda_0^+ &= 1\\ \lambda_1^+ &= \tfrac{3}{4}S(S+2T)\\ \lambda_2^+ &= \tfrac{25}{32}S^4+\tfrac{25}{8}S^3T+\tfrac{15}{4}S^2T^2-\tfrac{1}{2}S^3-\tfrac{3}{2}S^2T-\tfrac{3}{4}ST^2-\tfrac{1}{2}S^2+\tfrac{1}{4}ST-\tfrac{1}{4}S\end{aligned}$$

Observe, for example, that λ_1 is divisible by 6, and indeed the Casson invariant is

$$\lambda(M^+_{((\frac{S}{T}))}) = \frac{1}{6}\lambda_1^+(S,T) = \begin{cases} -\frac{1}{4}ST & \text{for } S \text{ and } T \text{ both even}\\ \frac{1}{8}S(S+2T) & \text{for } S \text{ even and } T \text{ odd}\end{cases}$$

4.3. Slopes

For fixed S and T, the dependence on m of $\lambda_m^{\pm}(S,T)$ is more interesting. See Figure 5 for graphs of the ratios $\frac{\lambda_n^+(S,T)}{\lambda_{n-1}^+(S,T)}$ for $S=-2$ and several odd values of T. From this figure there is an obvious conjecture.

Conjecture *If M is a rational homology sphere, then the ratio $\frac{\lambda_n(M)}{\lambda_{n-1}(M)}$ is asymptotically linear in n, for large n.*

When this conjecture holds, set

$$\sigma(M) = \lim_{n\longrightarrow\infty}\frac{\lambda_n(M)}{n\lambda_{n-1}(M)};$$

this will be called the *slope* of M (not to be confused with slopes in hyperbolic geometry).

It is known from (LR) that the slope exists for Seifert fibred manifolds and is

$$\sigma\left(\Sigma\left(\frac{Q_1}{P_1},\ldots,\frac{Q_N}{P_N}\right)\right) = \pm\frac{\prod P_i}{\pi^2} = \pm\left(\pi^2\sum_{i=1}^{N}\frac{Q_i}{P_i}\right)^{-1}.$$

As a corollary $\pi^2\sigma(M)\in\mathbf{Q}$ for such manifolds, while if in addition the manifold is an integer homology sphere then $\pi^2\sigma(M)\in\mathbf{Z}$. Since the WRT invariant (and therefore also the Ohtsuki series) is multiplicative under connect sum,

$$|\sigma(M\#N)| = \max(|\sigma(M)|,|\sigma(N)|)$$

when $|\sigma(M)|\neq|\sigma(N)|$, while under mirror image, the sign of the slope is reversed ($\sigma(\overline{M}) = -\sigma(M)$).

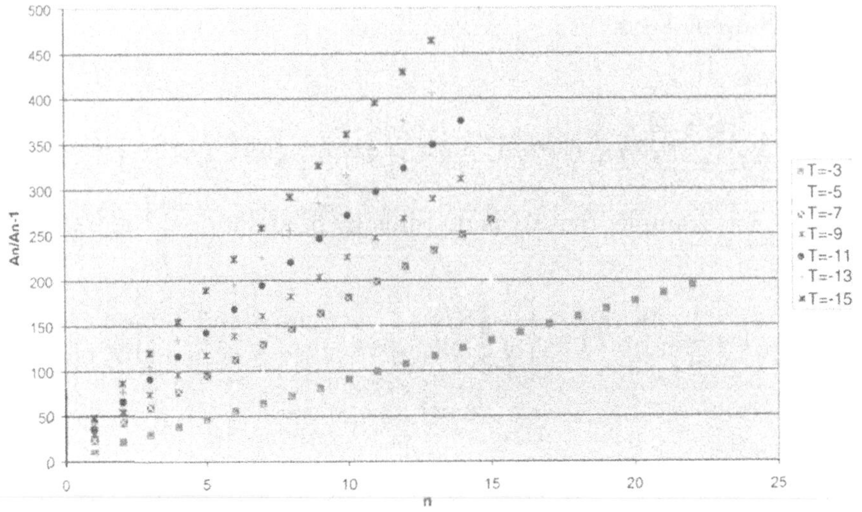

Figure 5. $\lambda_n^+(S,T)/\lambda_{n-1}^+(S,T)$ for $S=-2$ against n

Assuming that slopes exist for our family of manifolds, from the polynomiality of the terms $\lambda_m^\pm(S,T)$, one can make a further conjecture.

Conjecture *The slope for the manifolds $M^\pm_{((\frac{S}{T}))}$ is polynomial in S and T of degree at most 2.*

Observe that this conjecture **does not** follow from the previous conjecture and the theorem giving $\lambda_m^\pm(S,T)$ as a polynomial of degree $2m$ in S and T, since we have no control on the growth rate of the coefficients in such polynomials with m (it is conceivable that the lower order coefficients may grow more rapidly that the leading one). This conjecture is backed by numerical data. See for example Figure 6 which gives the slopes for $S=-3$ plotted against T (even); and Figure 7 which gives slopes for S and T even, plotted against T for various values of S.

Again assuming this conjecture, we can theoretically determine what its form should be. For the case of S odd and T even, we know that the knot for $S=\pm1$ is the $(2,T\mp1)$ torus knot, surgery around which gives for $M^\delta_{((\frac{S}{T}))}$ the Seifert fibred manifold on three fibers $\Sigma(2,T-S,2T-2S-\delta)$, and hence its slope is $\frac{2}{\pi^2}(T-S)((2T-2S)\delta-1)$. Hence for general (odd) S and even T, the quadratic giving the slope must still have coefficient $\frac{4}{\pi^2}$ for T^2.

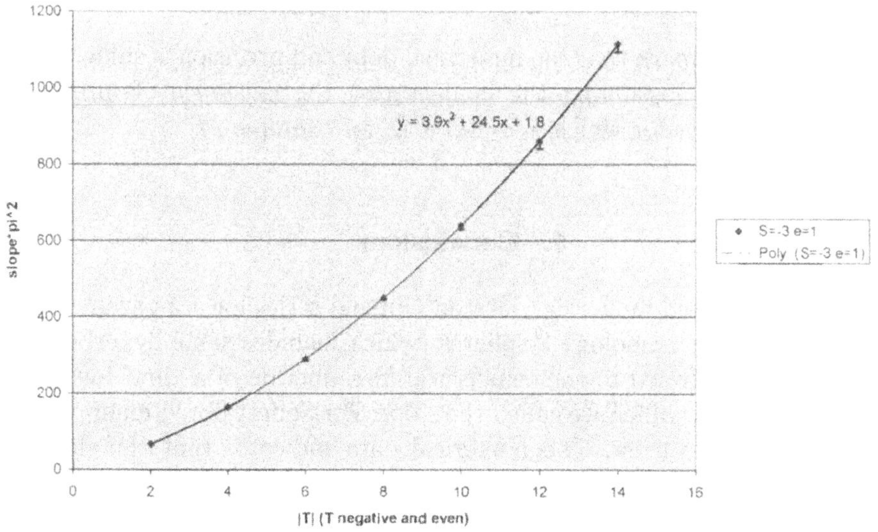

Figure 6. Graph of slopes for $S = -3$ and T even

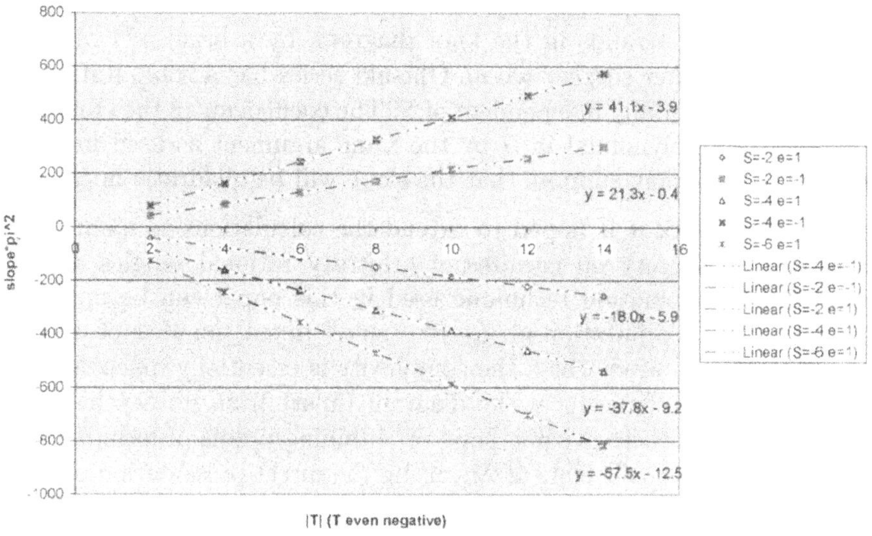

Figure 7. Graph of slopes for S and T even, against T

For the case of S and T both even, we have no similar data from special cases, since although we know that for $S = 0$, the manifold obtained is S^3 and $Z_\infty = 1$, there is no well-defined slope!

Finally we remark that the numerical data and precision is sufficient to demonstrate (assuming the conjectures) the existence of integer homology spheres for which $\pi^2\sigma(M) \notin \mathbf{Z}$, for example $M^+_{(\binom{3}{2})}$.

5. Conclusions

We have presented numerical data on Ohtsuki series for a 2-parameter family of integer homology 3-spheres (which includes some hyperbolic manifolds), which led to the conjectural introduction of a 'new' invariant of manifolds, which we called the *slope*. For Seifert fibred manifolds this is known to exist. The numerical data indicates that the slope (assuming it exists) is quadratic in the two parameters of the manifold.

It may be noted that the computer program may be used to compute the Ohtsuki series for surgery around an arbitrary knot, though the complexity of the calculation is likely to be prohibitively high for more general manifolds. However, for an arbitrary knot diagram, adding a 2-strand (that is, replacing any disc which contains within it just two non-intersecting strands in the knot diagram, by a braid σ_1^S) results in a manifold after surgery whose Ohtsuki series has a computational complexity essentially independent of S. The coefficients in the Ohtsuki series will be polynomial in S by the same argument as used in this paper, and one may suppose that the slope will be quadratic in S.

In future work it is hoped to extend the calculations to manifolds obtained by surgery on closures of arbitrary rational tangles, since then the recombination technique used in this paper can be applied to reduce the computation of Z_K to a small (fixed) number of quantum $6j$ symbols, after which the complexity is essentially independent of the numbers of twists in the diagram (apart from its depth). We also remark the there is some hope of obtaining proofs of formulae at least for leading coefficients of $\lambda^\pm_m$, using the direct presentation of the coloured Jones polynomial via universal R-matrices (see Le's approach to Ohtsuki series (Le)).

It does not seem that Witten-Chern-Simons theory sheds any light on the slope, or its physical meaning. In fact a somewhat 'orthogonal' way of viewing the Ohtsuki series of manifold invariants is via an asymptotic expansion of the ratio of adjacent coefficents, in order

words not only the leading coefficent (that of n, which we have called the slope) but also other coeffients may be of interest.

6. Acknowledgements

The authors would like to thank Dror Bar-Natan, Thang Le, Lev Rozansky and Dylan Thurston for many conversations on the subject of computations of 3-manifold invariants. They would also like to thank John Bryden for the invitation to the very stimulating NATO-PIMS Advanced Research Workshop in Calgary during August 2001, which he organised on the subject of 'New Techniques in Topological Field Theory', where this work was first presented. The second author would like to acknowledge partial support from BSF Grant 1998119.

References

[B-NG] D. Bar-Natan, S. Garoufalidis, *On the Melvin-Morton-Rozansky Conjecture*, Invent. Math. **125**, 103-133 (1996).

[Ja] N. Jacoby, *Computations of Ohtsuki series for surgery on 2-strand knots*, M.Sc. thesis, Hebrew University (Jerusalem, 2002).

[Je] L.C. Jeffrey, *Chern-Simons-Witten Invariants of Lens Spaces and Torus Bundles, and the Semiclassical Approximation*, Commun. Math. Phys. **147**, 563–604 (1992).

[KL] L.H. Kauffman, S. Lins, *Temperley–Lieb recoupling theory and invariants of 3–manifolds*, Princeton University Press (Princeton NJ, 1994).

[L] R. Lawrence, *Asymptotic expansions of Witten-Reshetikhin-Turaev invariants for some simple 3–manifolds*, J. Math. Phys. **36**, 6106–6129 (1995).

[LR] R. Lawrence & L. Rozansky, *Witten-Reshetikhin-Turaev invariants for Seifert manifolds*, Commun. Math. Phys. **205**, 287–314 (1999).

[LZ] R. Lawrence & D. Zagier, *Modular forms and quantum invariants of 3-manifolds*, Asian J. Math. **3**, 93–107 (1999).

[Le] T. Le, *private communication* (1999).

[M1] H. Murakami, *Quantum $SU(2)$–invariants dominate Casson's $SU(2)$–invariant*, Math. Proc. Camb. Phil. Soc. **115**, 253–281 (1993).

[M2] H. Murakami, *Quantum $SO(3)$–invariants dominate the $SU(2)$–invariant of Casson and Walker*, Math. Proc. Camb. Phil. Soc. **117**, 237–249 (1995).

[O1] T. Ohtsuki, *A polynomial invariant of integral homology 3–spheres*, Math. Proc. Camb. Phil. Soc. **117**, 83–112 (1995).

[O2] T. Ohtsuki, *A polynomial invariant of rational homology 3–spheres*, Invent. Math. **123**, 241–257 (1996).

[RT] N.Yu. Reshetikhin & V.G. Turaev, *Invariants of 3-manifolds via link polynomials and quantum groups*, Invent. Math. **103**, 547–597 (1991).

[R1] L. Rozansky, *A contribution of the trivial connection to the Jones polynomial and Witten's invariant of 3d manifolds I, II*, Commun. Math. Phys. **175**, 275–296, 297–318 (1996).

[R2] L. Rozansky, *Witten's invariant of rational homology spheres at prime values of K and trivial connection contribution*, Commun. Math. Phys. **180**, 297–324 (1996).

[R3] L. Rozansky, *On p–adic convergence of perturbative invariants of some rational homology spheres*, Duke Math. J. **91**, 353–379 (1998).

[Wa] K. Walker, *An Extension of Casson's Invariant*, Annals of Mathematical Studies **126**, Princeton University Press (Princeton NJ, 1992).

[We] E. Witten, *Quantum Field Theory and the Jones Polynomial*, Commun. Math. Phys. **121**, 351–399 (1989).

From 3-moves to Lagrangian tangles and cubic skein modules

Józef H. Przytycki

Abstract. We present an expanded version of four talks describing recent developments in Knot Theory to which the author contributed[1]. We discuss several open problems in classical Knot Theory and we develop techniques that allow us to study them: Lagrangian tangles, skein modules and Burnside groups. The method of Burnside groups of links was discovered and developed only half a year after the last talk was delivered in Kananaskis[2].

Keywords: knot, 3-move, rational move, Lagrangian tangle, skein module, Burnside groups of links

Mathematics Subject Classification 2000:- Primary: 57M25: Secondary: 57M27, 20D99

1. Open problems in Knot Theory that everyone can try to solve

When did Knot Theory start?

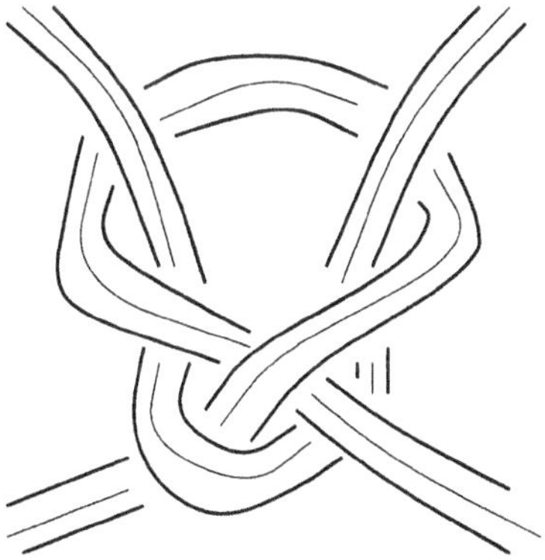

Fig. 1.1; Gauss' meshing knot from 1794

[1] Containing several results that are not yet published elsewhere.

[2] The first three talks were delivered at International Workshop on Graphs – Operads – Logic; Cuautitlán, Mexico, March 12-16, 2001 and the fourth talk "Symplectic Structures on Colored Tangles" at the workshop New Techniques in Topological Quantum Field Theory; Calgary/Kananaskis, August 22-27, 2001.

J.M. Bryden (ed.), Advances in Topological Quantum Field Theory, 71–125.

Was it in 1794 when C. F. Gauss[1] (1777-1855) copied figures of knots from a book written in English (Fig.1.1)?

Fig. 1.2; Framed tangle from Gauss'notebook (Ga)

Or was it before that, in 1771, when A-T. Vandermonde (1735-1796) considered knots and braids as a part of Leibniz's *analysis situs*?

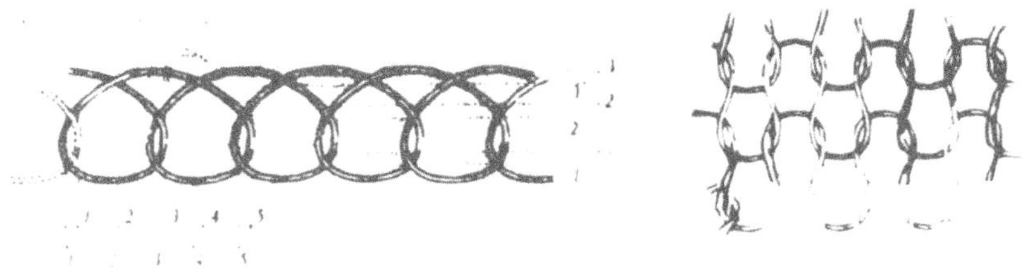

Fig. 1.3; Vandermonde drawings of 1771

[1] Gauss' notebooks contain several drawings related to knots, for example a braid with complex coordinate description (see (Ep-1; P-14)) or the mysterious "framed tangle" which is published here for the first time, see Fig.1.2. (Ga).

Perhaps engravings by Leonardo da Vinci[2] (1452-1519) (Mac) and woodcuts by Albrecht Dürer[3] (1471-1528) (Dur-1; Ha) should also be taken into account, Fig.1.4.

Fig. 1.4; A knot by Dürer (Ku); c. 1505-1507

[2] Giorgio Vasari writes in (Va): "[Leonardo da Vinci] spent much time in making a regular design of a series of knots so that the cord may be traced from one end to the other, the whole filling a round space. There is a fine engraving of this most difficult design, and in the middle are the words: Leonardus Vinci Academia."

[3] "Another great artist with whose works Dürer now became acquainted was Leonardo da Vinci. It does not seem likely that the two artists ever met, but he may have been brought into relation with him through Luca Pacioli, the author of the book De Divina Proportione, which appeared at Venice in 1509, and an intimate friend of the great Leonardo. Dürer would naturally be deeply interested in the proportion theories of Leonardo and Pacioli. He was certainly acquainted with some engravings of Leonardo's school, representing a curious circle of concentric scrollwork on a black ground, one of them entitled Accademia Leonardi Vinci; for he himself executed six woodcuts in imitation, the Six Knots, as he calls them himself. Dürer was amused by and interested in all scientific or mathematical problems..." From: http://www.cwru.edu/edocs/7/258.pdf, compare (Dur-2).

One can go back in time even further to ancient Greece where surgeons considered sling knots, Fig.1.5 (Da; P-14).

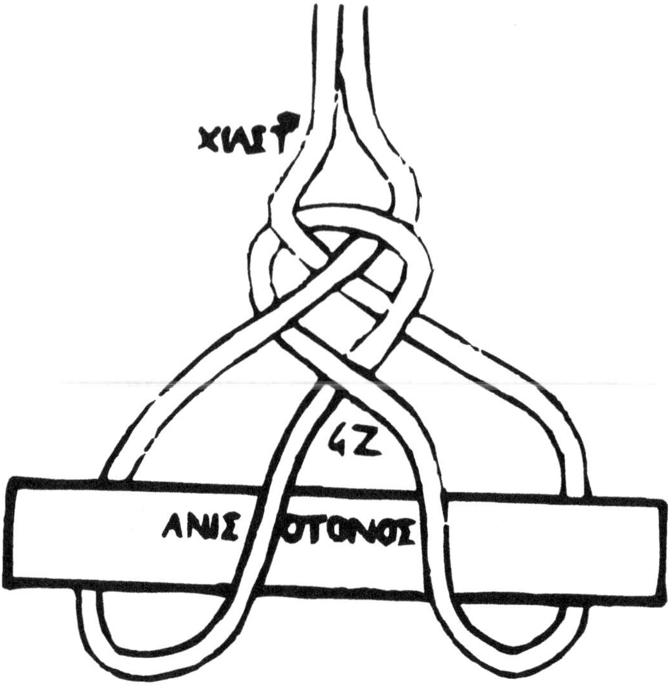

Fig. 1.5; A sling knot of Heraclas

Moreover, we can appreciate ancient stamps and seals with knots and links as their motifs. The oldest examples that I am aware of are from the pre-Hellenic Greece. Excavations at Lerna by the American School of Classical Studies under the direction of Professor J. L. Caskey (1952-1958) discovered two rich deposits of clay seal-impressions. The second deposit dated from about 2200 BC contains several impressions of knots and links[4] (Hig; Hea; Wie) (see Fig.1.6).

[4] The early Bronze Age in Greece is divided, as in Crete and the Cyclades, into three phases. The second phase lasted from 2500 to 2200 BC, and was marked by a considerable increase in prosperity. There were palaces at Lerna, and Tiryns, and probably elsewhere, in contact with the Second City of Troy. The end of this phase (in the Peloponnese) was brought about by invasion and mass burnings. The invaders are thought to be the first speakers of the Greek language to arrive in Greece.

Fig. 1.6; A seal-impression from the House of the Tiles in Lerna (Hig).

As we see Knot Theory has a long history but despite this, or maybe because of this, one still can find inspiring elementary open problems. These problems are not just interesting puzzles but they lead to an interesting theory.

In this section our presentation is absolutely elementary. Links are circles embedded in our space, R^3, up to topological deformation, that is, two links are equivalent if one can be deformed into the other in space without cutting and pasting. We represent links using their plane diagrams.

First we introduce the concept of an n move on a link.

Definition 1.1. *An* n-move *on a link is a local transformation of the link illustrated in Fig.1.7.*

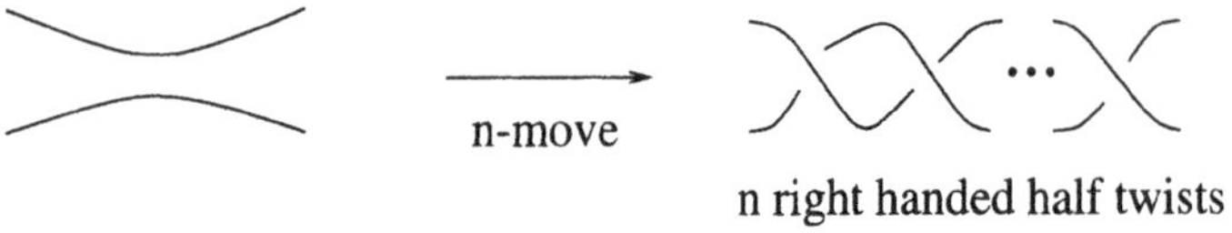

Fig. 1.7; n-move

In our convention, the part of the link outside of the disk in which the move takes place, remains unchanged. One should stress that an n-move can change the topology of the link. For example  —  illustrates a 3-move.

Definition 1.2. *We say that two links, L_1 and L_2, are* n-move equivalent *if one can obtain L_2 from L_1 by a finite number of n-moves and $(-n)$-moves (inverses of n-moves).*

If we work with diagrams of links then the topological deformation of links is captured by Reidemeister moves, that is, two diagrams represent the same link in space if and only if one can obtain one of them from the other by a sequence of Reidemeister moves:

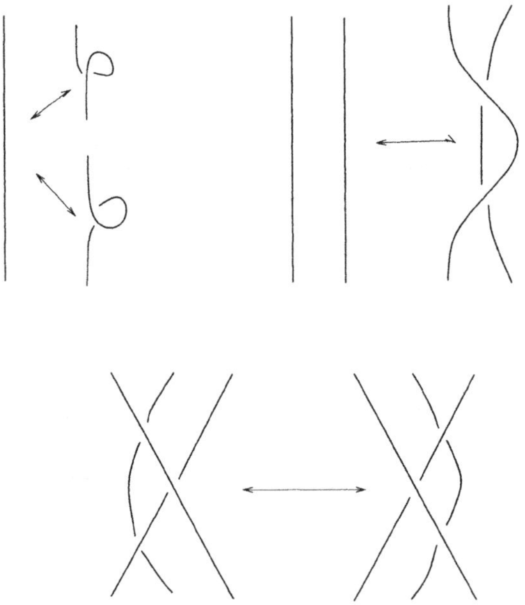

Fig. 1.8; Reidemeister moves

Thus, we say that two diagrams, D_1 and D_2, are n-move equivalent if one can be obtained from the other by a sequence of n-moves, their inverses and Reidemeister moves. To illustrate this, we show that the move 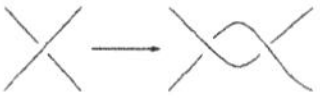 is the result of an application of a 3-move followed by the second Reidemeister move (Fig.1.9).

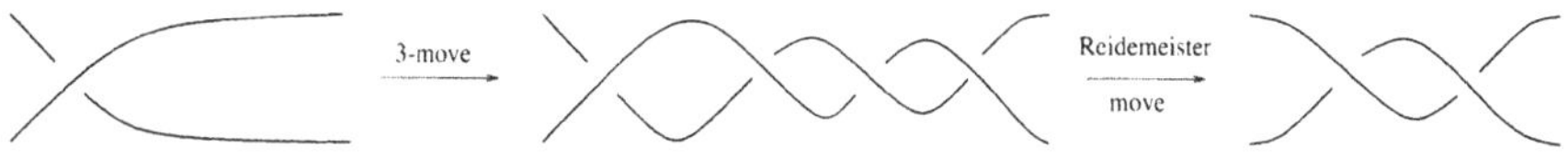

Fig. 1.9

Conjecture 1.3 (Montesinos-Nakanishi).
Every link is 3-move equivalent to a trivial link.

Yasutaka Nakanishi proposed this conjecture in 1981. José Montesinos analyzed 3-moves before, in connection with 3-fold dihedral branch coverings, and asked a related but different question[5] (Mo-2).

Examples 1.4. (i) *Trefoil knots (left- and right-handed) are 3-move equivalent to the trivial link of two components:*

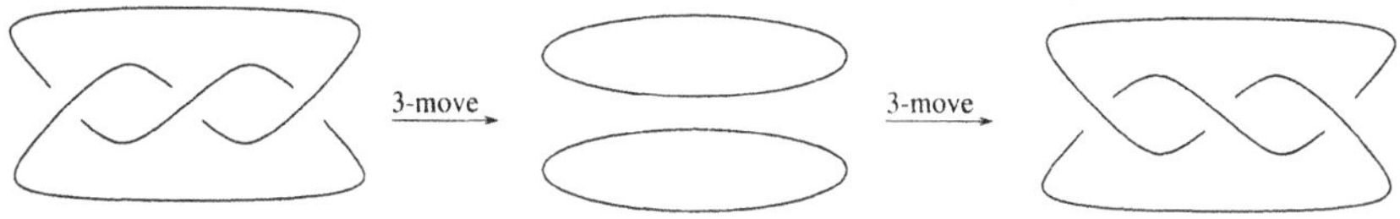

Fig. 1.10

(ii) *The figure eight knot* (4_1) *and the knot* 5_2 *are 3-move equivalent to the trivial knot:*

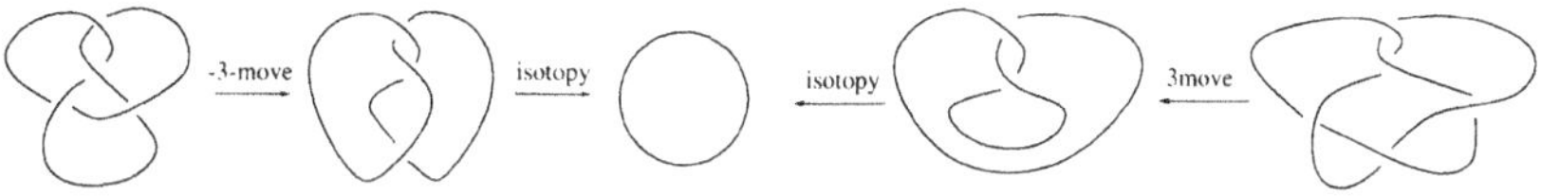

Fig. 1.11

(iii) *The knot* 5_1 *and the Hopf link are 3-move equivalent to the trivial knot*[6]*:*

[5] "Is there a set of moves which do not change the covering manifold and such that if two colored links have the same covering they are related by a finite sequence of those moves?"

[6] One can show that the knot 5_1 cannot be reduced to the trivial knot by one ± 3-move. To see this, one can use the Goeritz matrix approach to the classical signature $(|\sigma(5_1)| = 4)$, see (Go; G-L).

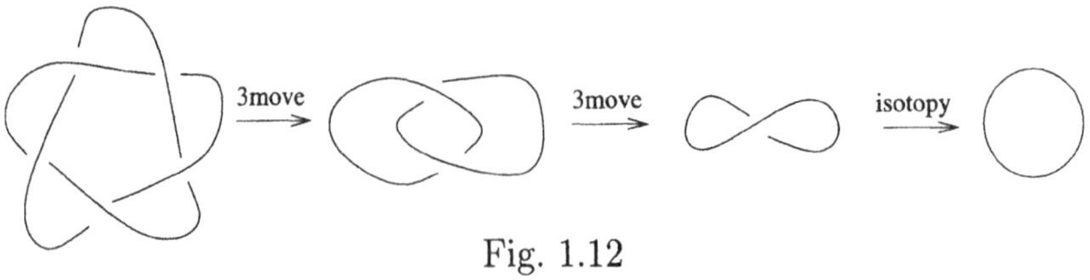

Fig. 1.12

We will show later, in this section, that different trivial links are not 3-move equivalent. However, in order to achieve this conclusion we need an invariant of links preserved by 3-moves and differentiating trivial links. Fox 3-coloring is such an invariant. We will introduce it later (today in its simplest form and, in the second lecture, in a more general context of Fox n-colorings and Alexander-Burau-Fox colorings).

Now let us present some other related conjectures.

Conjecture 1.5.
Any 2-tangle is 3-move equivalent to one of the four 2-tangles shown in Fig.1.13 with possible additional trivial components.

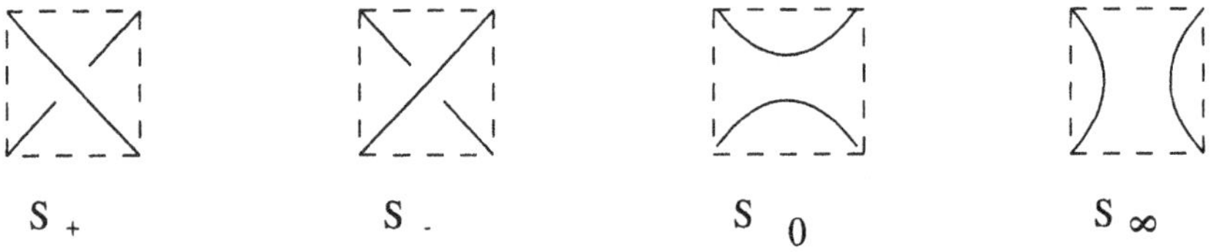

Fig. 1.13

The Montesinos-Nakanishi conjecture follows from Conjecture 1.5. More generally if Conjecture 1.5 holds for some class of 2-tangles, then Conjecture 1.3 holds for any link obtained by closing elements of this class, without introducing any new crossing. The simplest interesting class of tangles for which Conjecture 1.5 holds are algebraic tangles in the sense of Conway (I call them 2-algebraic tangles and in the next section present a generalization). Conjecture 1.5 can be proved by induction for 2-algebraic tangles. I will leave the proof to you as a pleasant exercise (compare Proposition 1.9). The definition you need is as follows

Definition 1.6 ((Co; B-S)). *The family of* 2-algebraic tangles *is the smallest family of 2-tangles satisfying*
(i) *Any 2-tangle with* 0 *or* 1 *crossing is 2-algebraic.*
(ii) *If A and B are 2-algebraic tangles then the 2-tangle $r^i(A) * r^j(B)$*

is also 2-algebraic, where r denotes the counterclockwise rotation of a tangle by 90° along the z-axis, and $*$ denotes the horizontal composition of tangles (see the figure below).

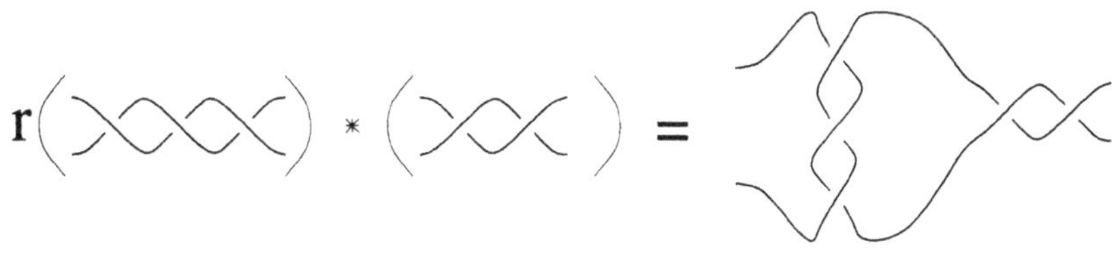

A link is called 2-algebraic *if it can be obtained from a 2-algebraic tangle by closing its ends without introducing crossings*[7].

The Montesinos-Nakanishi 3-move conjecture has been proved by my students Qi Chen and Tatsuya Tsukamoto for many special families of links (Che; Tsu; P-Ts). In particular, Chen proved that the conjecture holds for all 5-braid links except possibly one family, containing the square of the center of the 5-braid group, $\Delta_5^4 = (\sigma_1\sigma_2\sigma_3\sigma_4)^{10}$. He also found a reduction by $\pm$3-moves of Δ_5^4 to the 5-braid link, $(\sigma_1^{-1}\sigma_2\sigma_3\sigma_4^{-1}\sigma_3)^4$, with 20 crossings[8], Fig.1.14. It is now the smallest known possible counterexample to the Montesinos-Nakanishi 3-move conjecture[9].

---

[7] By joining the top ends and then bottom ends of a tangle T one obtains the link $N(T)$, the *numerator* of T, Fig.1.22, 1.23. Joining the left-hand ends and then right-hand ends produces the *denominator* closure $D(T)$.

[8] In the group $B_5/(\sigma_i^3)$ the calculation is as follows: $(\sigma_1\sigma_2\sigma_3\sigma_4)^{10} = (\sigma_1\sigma_2\sigma_3\sigma_4^2\sigma_3\sigma_2\sigma_1)^2(\sigma_2\sigma_3\sigma_4^2\sigma_3\sigma_2)^2(\sigma_3\sigma_4^2\sigma_3)^2 \stackrel{3}{=} (\sigma_1\sigma_2\sigma_3\sigma_4^2\sigma_3\sigma_2\sigma_1)^2(\sigma_2\sigma_3\sigma_4^2\sigma_3\sigma_2)^2 \stackrel{3}{=} (\sigma_1\sigma_2\sigma_3\sigma_4^2\sigma_3\sigma_2\sigma_1)^2(\sigma_2^{-1}\sigma_3\sigma_4^{-1}\sigma_3)^2 \stackrel{3}{=} (\sigma_1\sigma_2\sigma_3\sigma_4^{-1}\sigma_3\sigma_2\sigma_1\sigma_2^{-1}\sigma_3\sigma_4^{-1}\sigma_3)^2 = \sigma_1\sigma_2\sigma_3\sigma_4^{-1}\sigma_3\sigma_1^{-1}\sigma_2\sigma_3\sigma_4^{-1}\sigma_3\sigma_1)^2 \stackrel{3}{=} (\sigma_1^{-1}\sigma_2\sigma_3\sigma_4^{-1}\sigma_3)^4$.

[9] We proved in (D-P-1) that Chen's link is in fact the counterexample to the Montesinos-Nakanishi 3-move conjecture; see Section 4. We think that it is the smallest such counterexample. We also demonstrated that the 2-parallel of the Borromean rings is not 3-move equivalent to a trivial link. It is still possible that Chen's link with an additional trivial component is 3-move equivalent to the 2-parallel of the Borromean rings.

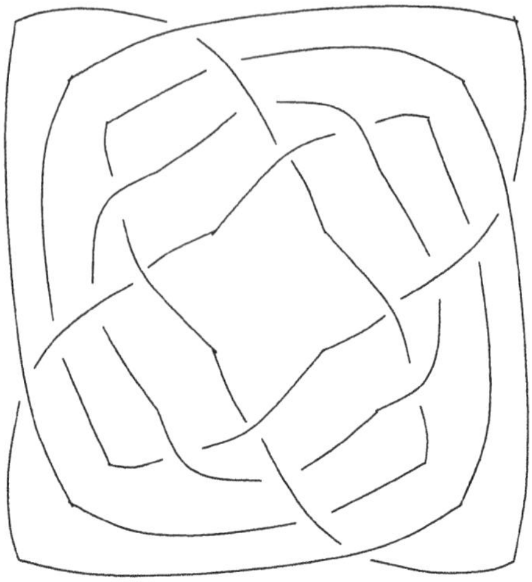

Fig. 1.14

Previously Nakanishi suggested in 1994 (see (Kir)), the 2-parallel of the Borromean rings (a 6-braid with 24 crossings) as a possible counterexample (Fig.1.15).

Fig. 1.15

We will return to the discussion of theories motivated by 3-moves tomorrow. Now we will state some conjectures that employ other elementary moves.

Conjecture 1.7 (Nakanishi, 1979).
Every knot is 4-move equivalent to the trivial knot.

Examples 1.8. *Reduction of the trefoil and the figure eight knots is illustrated in Fig.1.16.*

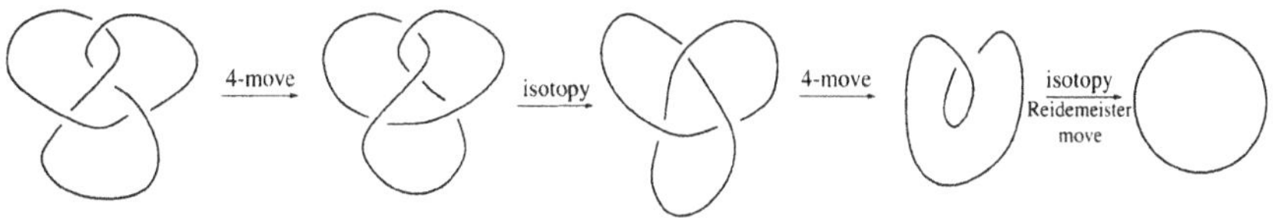

Fig. 1.16

Proposition 1.9 ((P-12)). (i) *Every 2-algebraic tangle without a closed component can be reduced by ±4-moves to one of the six basic 2-tangles shown in Fig.1.17.*

(ii) *Every 2-algebraic knot can be reduced by ±4-moves to the trivial knot.*

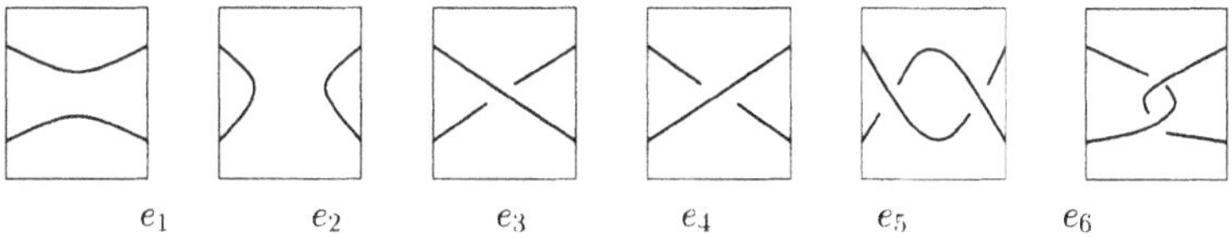

Fig. 1.17

Proof. To prove (i) it suffices to show that every composition (with possible rotation) of tangles presented in Fig.1.17 can be reduced by ±4-moves back to one of the tangles in Fig.1.17 or it has a closed component. These can be easily verified by inspection. Fig.1.18 is the multiplication table for basic tangles. We have chosen our basic tangles to be invariant under the rotation r, so it suffices to be able to reduce every tangle of the table to a basic tangle. One example of such a reduction is shown in Fig.1.19. Part (ii) follows from (i). □

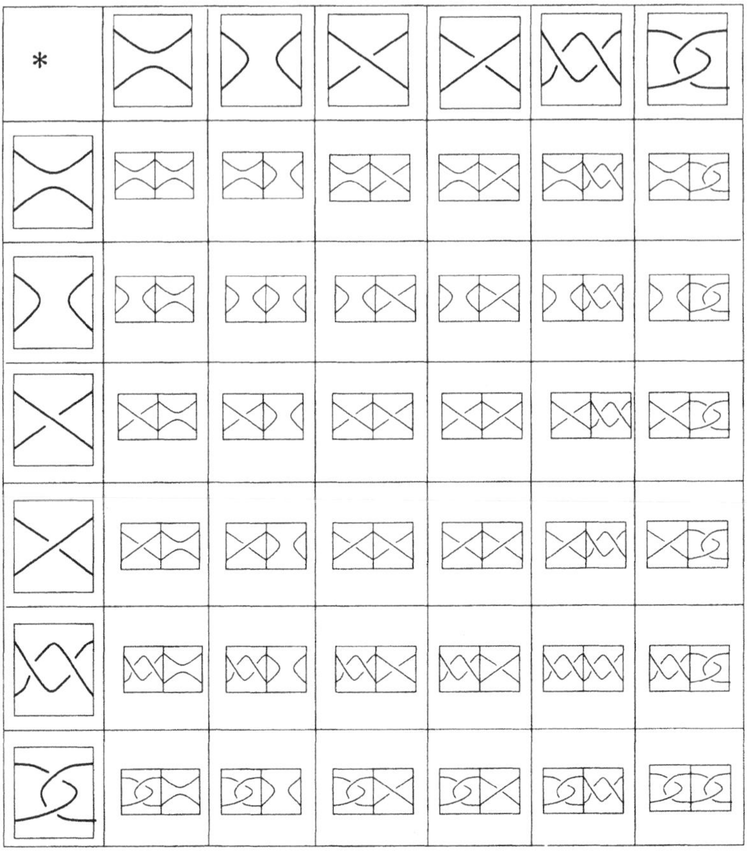

Fig. 1.18

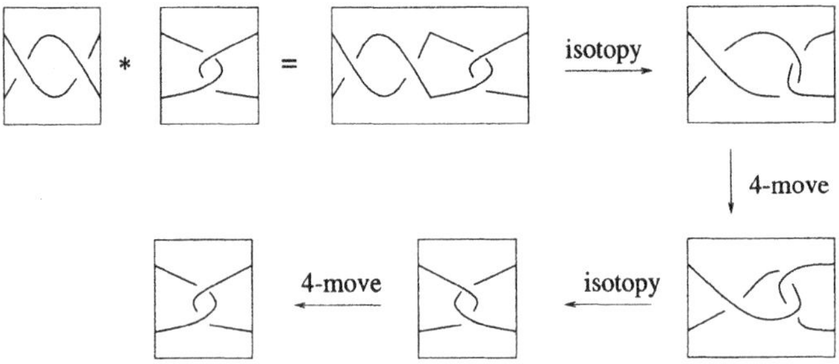

Fig. 1.19

In 1994, Nakanishi began to suspect that a 2-cable of the trefoil knot cannot be simplified by 4-moves (Kir). However, Nikolaos Askitas was able to simplify this knot (Ask). Askitas, in turn, suspects that the (2, 1)-cable of the figure eight knot (with 17 crossings) to be the simplest counterexample to the Nakanishi 4-move conjecture.

Not every link can be reduced to a trivial link by 4-moves. In particular, the linking matrix modulo 2 is preserved by 4-moves. Furthermore, Nakanishi and Suzuki demonstrated that the Borromean rings cannot be reduced to the trivial link of three components (Na-Su).

In 1985, after the seminar talk given by Nakanishi in Osaka, there was discussion about possible generalization of the Nakanishi 4-move conjecture for links. Akio Kawauchi formulated the following question for links

Problem 1.10 ((Kir)).

(i) *Is it true that if two links are link-homotopic*[10] *then they are 4-move equivalent?*

(ii) *In particular, is it true that every two component link is 4-move equivalent to the trivial link of two components or to the Hopf link?*

We can extend the argument used in Proposition 1.9 to show:

Theorem 1.11. *Any two component 2-algebraic link is 4-move equivalent to the trivial link of two components or to the Hopf link.*

Proof. Let L be a 2-algebraic link of two components. Therefore, L is built inductively as in Definition 1.6. Consider the first tangle, T, in the construction, which has a closed component (if it happens). The complement T' of T in the link L is also a 2-algebraic tangle but without a closed component. Therefore it can be reduced to one of the 6 basic tangles shown in Fig.1.17, say e_i. Consider the product $T * e_i$. The only nontrivial tangle T to be considered is $e_6 * e_6$ (the last tangle in Fig.1.18). The compositions $(e_6 * e_6) * e_i$ are illustrated in Fig.1.20. The closure of each of these product tangles (the numerator or the denominator) has two components because it is 4-move equivalent to L. We can easily check that it reduces to the trivial link of two components. □

[10] Two links L_1 and L_2 are *link-homotopic* if one can obtain L_2 from L_1 by a finite number of crossing changes involving only self-crossings of the components.

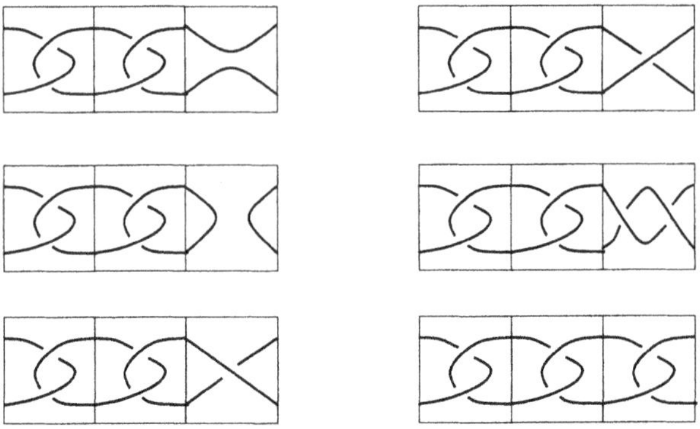

Fig. 1.20

Problem 1.12. (i) *Find a (reasonably small) family of 2-tangles with one closed component so that every 2-tangle with one closed component is 4-move equivalent to one of its elements.*

(ii) *Solve the above problem for 2-algebraic tangles with one closed component.*

Nakanishi (Nak-2) pointed out that the "half" 2-cabling of the Whitehead link, $\mathcal{W}$, Fig.1.21, was the simplest link which he could not reduce to a trivial link by ± 4-moves but which was link-homotopic to a trivial link[11].

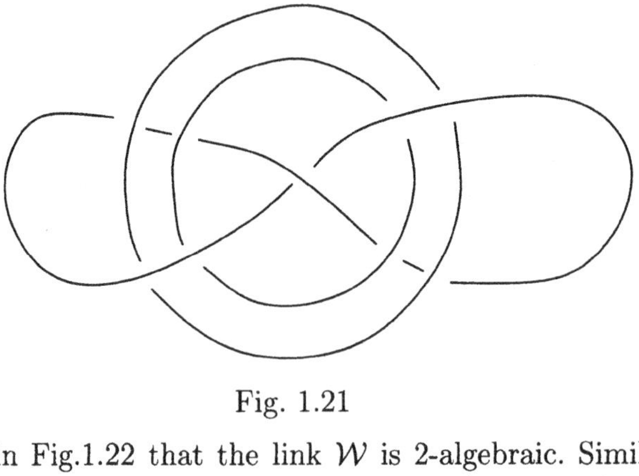

Fig. 1.21

It is shown in Fig.1.22 that the link $\mathcal{W}$ is 2-algebraic. Similarly, the Borromean rings, BR, are 2-algebraic, Fig.1.23 (compare Fig.1.30).

[11] In fact, in June of 2002 we showed that this example cannot be reduced by ± 4-moves (D-P-2).

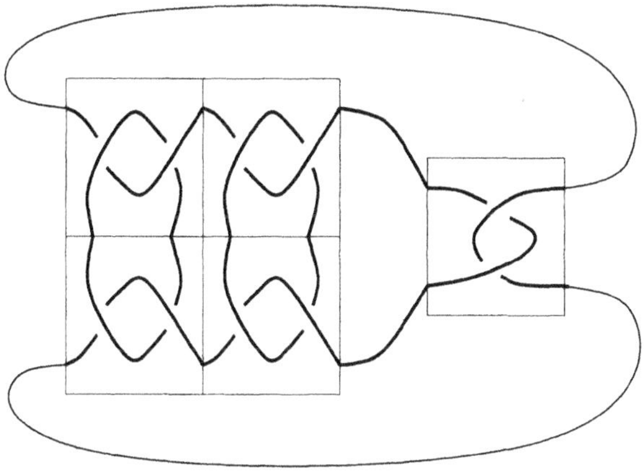

Fig. 1.22; $\mathcal{W} = N(r(r(e_3 * e_3) * r(e_4 * e_4)) * r(r(e_3 * e_3) * r(e_4 * e_4)) * r(e_3 * e_3))$

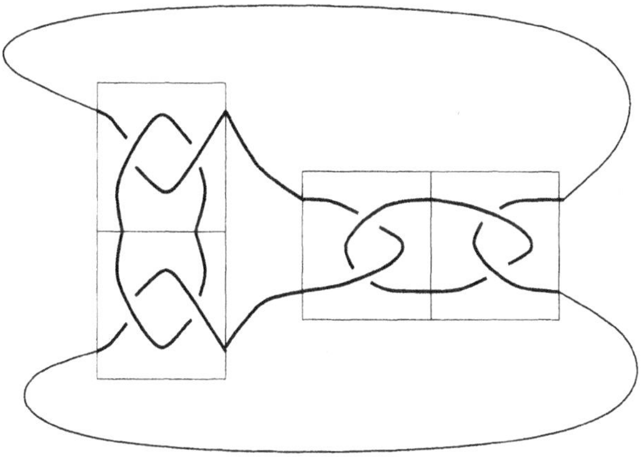

Fig. 1.23; $BR = N(r(r(e_3 * e_3) * r(e_4 * e_4)) * r(e_3 * e_3) * r(e_4 * e_4))$

Problem 1.13. *Is the link $\mathcal{W}$ the only 2-algebraic link of three components (up to 4-move equivalence) which is homotopically trivial but which is not 4-move equivalent to the trivial link of three components?*

We can also prove that the answer to Kawauchi's question is affirmative for closed 3-braids.

Theorem 1.14. (i) *Every knot which is a closed 3-braid is 4-move equivalent to the trivial knot.*

(ii) *Every link of two components which is a closed 3-braid is 4-move equivalent to the trivial link of two components or to the Hopf link.*

(iii) *Every link of three components which is a closed 3-braid is 4-move equivalent either to the trivial link of three components, or to the Hopf link with the additional trivial component, or to the connected sum of two Hopf links, or to the (3,3)-torus link, $\bar{6}^3_1$, represented by $(\sigma_1\sigma_2)^3$ (all linking numbers are equal to 1), or to the Borromean rings (represented by $(\sigma_1\sigma_2^{-1})^3$).*

Proof. Our proof is based on the Coxeter theorem that the quotient group $B_3/(\sigma_i^4)$ is finite with 96 elements, (Cox). Furthermore, $B_3/(\sigma_i^4)$ has 16 conjugacy classes[12]: 9 of them can be easily identified as representing trivial links (up to 4-move equivalence), and 2 of them represent the Hopf link ($\sigma_1^2\sigma_2$ and $\sigma_1^2\sigma_2^{-1}$), and σ_1^2 represents the Hopf link with an additional trivial component. We also have the connected sums of Hopf links ($\sigma_1^2\sigma_2^2$). Finally, we are left with two representatives of the link $\bar{6}^3_1$ ($\sigma_1\sigma_2^2\sigma_1\sigma_2^2$ and $\sigma_1^{-1}\sigma_2^2\sigma_1^{-1}\sigma_2^2$) and the Borromean rings. □

Proposition 1.9 and Theorems 1.11, and 1.14 can be used to analyze 4-move equivalence classes of links with small number of crossings.

Theorem 1.15. (i) *Every knot of no more than* 9 *crossings is* 4 *move equivalent to the trivial knot.*

(ii) *Every two component link of no more than* 9 *crossings is 4-move equivalent to the trivial link of two components or to the Hopf link.*

Proof. Part (ii) follows immediately as the only 2-component links with up to 9 crossings which are not 2-algebraic are $9^2_{40}, 9^2_{41}, 9^2_{42}$ and 9^2_{61} and all these links are closed 3-braids. There are at most 6 knots with up to 9 crossings which are neither 2-algebraic nor 3-braid knots. They are: $9_{34}, 9_{39}, 9_{40}, 9_{41}, 9_{47}$ and 9_{49}. We reduced three of them, $9_{39}, 9_{41}$ and 9_{49} at my Fall 2003 Dean's Seminar. The knot 9_{40} was reduced in December of 2003 by Slavik Jablan and Radmila Sazdanovic. Soon after, my student Maciej Niebrzydowski simplified the remaining pair 9_{34} and 9_{47}, Fig.1.24. □

---

[12] $Id, \sigma_1, \sigma_1^{-1}, \sigma_1^2, \sigma_1\sigma_2, \sigma_1^{-1}\sigma_2, \sigma_1^{-1}\sigma_2^{-1}, \sigma_1^2\sigma_2, \sigma_1^2\sigma_2^{-1}, \sigma_1^2\sigma_2^2, \sigma_1\sigma_2^{-1}\sigma_1\sigma_2^{-1}, \sigma_1\sigma_2^2\sigma_1\sigma_2^{-1}, \sigma_1\sigma_2^{-1}\sigma_1^2\sigma_2^{-1}, \sigma_1\sigma_2^2\sigma_1\sigma_2^2, \sigma_1^{-1}\sigma_2^2\sigma_1^{-1}\sigma_2^2, (\sigma_1\sigma_2^{-1})^3$ (checked by M. Dąbkowski using the *GAP* program).

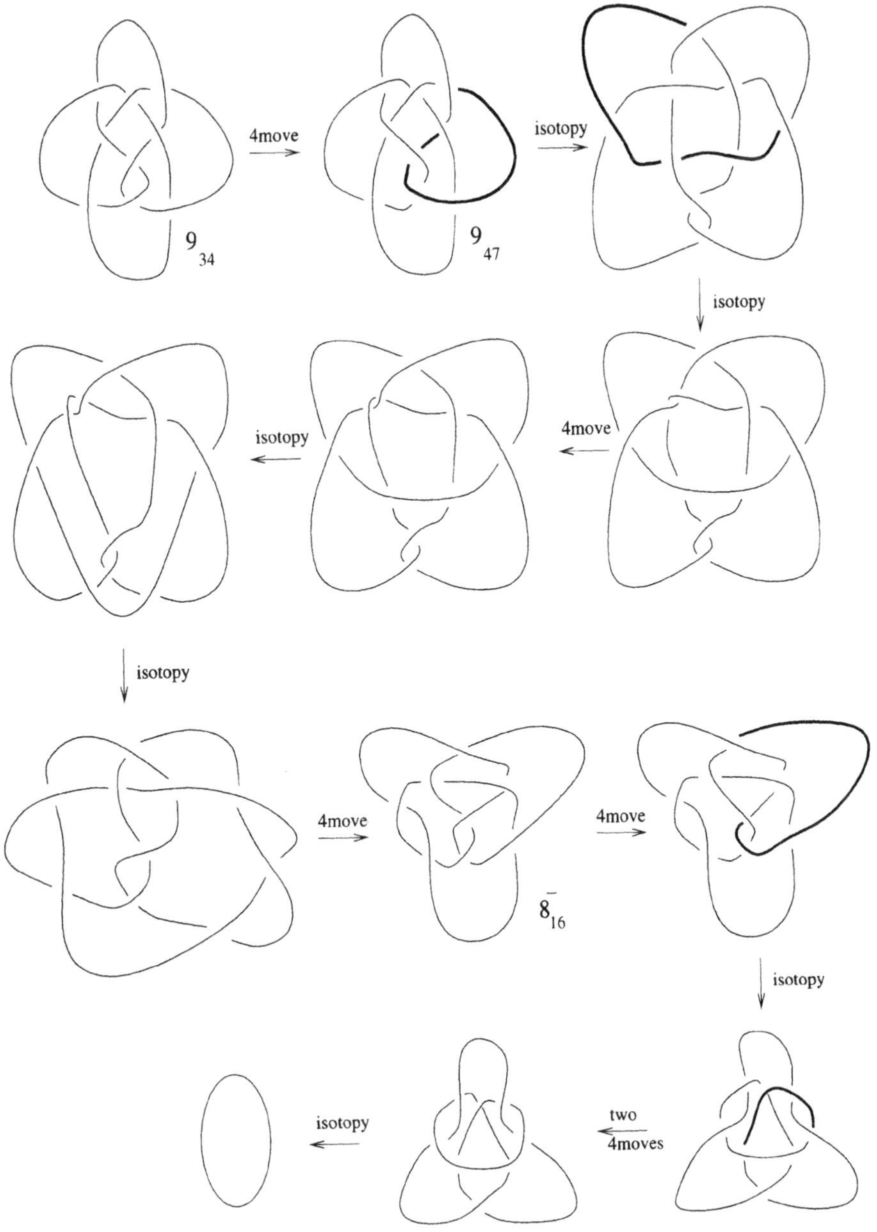
4move
9 34
9 47
isotopy
isotopy
4move
isotopy
isotopy
4move
4move
8̄ 16
isotopy
isotopy
two
4moves

Fig. 1.24

A weaker version of the Kawauchi question has been answered by Nakanishi in 1989, (Nak-1). If $\gamma \in B_n$ then the *γ-move* is the n-tangle move in which the trivial n-braid is replaced by the braid γ.

Theorem 1.16 (Nakanishi). *If two links L_1 and L_2 have the same linking matrix modulo 2, then L_2 can be obtained from L_1 by a finite number of ± 4-moves and Δ_3^4-moves.*

Proof. The square of the center, $\Delta_3^4 = (\sigma_1\sigma_2)^6$, of the 3-braid group B_3 and the Borromean braid, $(\sigma_1\sigma_2^{-1})^3$, are equal[13] in $B_3/(\sigma_i^4)$. From this it also follows that Δ_3^4 and Δ_3^{-4} are equal in $B_3/(\sigma_i^4)$. Furthermore, the $(\sigma_1\sigma_2^{-1})^3$-move is equivalent to Δ-move of Nakanishi in which $\sigma_1\sigma_2^{-1}\sigma_1$ is replaced by $\sigma_2\sigma_1^{-1}\sigma_2$ (we can think of this move as a "false" braid relation or a "false" third Reidemeister move). Nakanishi proved that two oriented links are Δ-move equivalent if and only if their linking matrices are equivalent (Nak-1). Theorem 1.16 follows. □

Selman Akbulut used Nakanishi's theorem to prove John Nash's conjecture for 3-dimensional manifolds (A-K)[14].

It is not true that every link is 5-move equivalent to a trivial link. One can show, using the Jones polynomial, that the figure eight knot is not 5-move equivalent to any trivial link[15]. One can, however, introduce

---

[13] We have in $B_3/(\sigma_i^4)$: $(\sigma_1\sigma_2)^6 = (\sigma_1^2\sigma_2\sigma_1^2\sigma_2)(\sigma_1\sigma_2^2\sigma_1\sigma_2^2) = \sigma_1^2\sigma_2\sigma_1^2(\sigma_1\sigma_2^2\sigma_1\sigma_2^2)\sigma_2 \overset{4}{=} \sigma_1^2\sigma_2\sigma_1^{-1}\sigma_2^2\sigma_1\sigma_2^{-1} = \sigma_1\sigma_2^{-1}\sigma_1\sigma_2\sigma_2^2\sigma_1\sigma_2^{-1} \overset{4}{=} (\sigma_1\sigma_2^{-1})^3$. This calculation can be interpreted as an illustration of Fig.28 in (A-K).

[14] The conjecture that "any two closed smooth connected manifolds of the same dimension can be made diffeomorphic after blowing them up along submanifolds" is an interpretation of the Nash question "Is there an algebraic structure on a any given smooth manifold which is birational to RP^n?" (Nash; A-K). The conjecture is only loosely related to the question mentioned in the book "A beautiful mind" were in Chapter "The 'Blowing Up' Problem", it is written: "Nash seemed, as the Fall [1963] unfolded, to be in far better shape than he had been during his previous interlude at the Institute [IAS]. As he said in his Madrid lecture, he "had had an idea which is referred to as Nash Blowing UP which I discussed with an eminent mathematician named Hironaka." [Letter from J.Nash to V.Nash, 1.9.66] (Hironaka eventually wrote a conjecture up.)" (Nas).

[15] A 5-move preserves the absolute value of the Jones polynomial at $t = e^{\pi i/5}$ (P-1). However, the Jones polynomial $V_{4_1}(e^{\pi i/5}) = 0$ but for any trivial link, T_n, we have $V_{T_n}(e^{\pi i/5}) = (-e^{\pi i/10} - e^{-\pi i/10})^{n-1} \neq 0$.

a more delicate move, called $(2,2)$-move (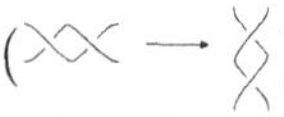), such that the 5-move is a combination of a $(2,2)$-move and its mirror image $(-2,-2)$-move (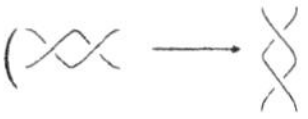), as it is illustrated in Fig.1.25 (H-U; P-3).

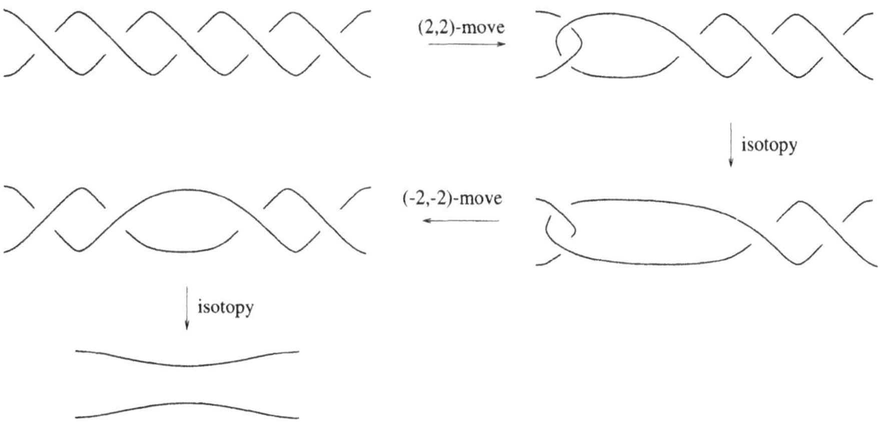

Fig. 1.25

Conjecture 1.17 (Harikae, Nakanishi, Uchida, 1992).
Every link is $(2,2)$*-move equivalent to a trivial link.*

As in the case of 3-moves, an elementary induction shows that the conjecture holds for 2-algebraic links. It is also known that the conjecture holds for all links up to 8 crossings. The key element of the argument in the proof is the observation (going back to Conway (Co)) that any link with up to 8 crossings (different from 8_{18}; see footnote before Fig.1.31) is 2-algebraic. The reduction of the 8_{18} knot to the trivial link of two components by my students, Jarek Buczyński and Mike Veve, is illustrated in Fig.1.26.

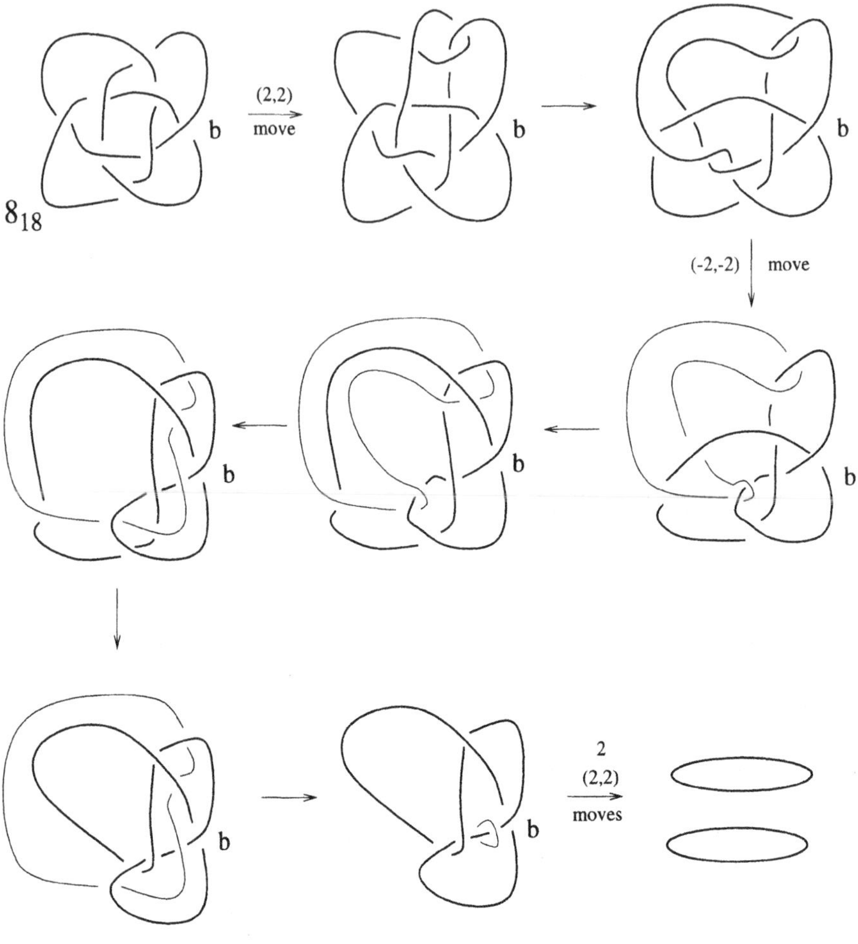

Fig. 1.26; Reduction of the 8_{18} knot

The smallest knots that are not reduced yet are 9_{40} and 9_{49}, Fig.1.27. Possibly you can reduce them![16]

[16] We showed with M. Dąbkowski that the knots 9_{40} and 9_{49} are not $(2,2)$-move equivalent to trivial links (D-P-2). Possibly you can prove that they are in the same $(2,2)$-move equivalence class! If I had to guess, I would say that it is a likely possibility.

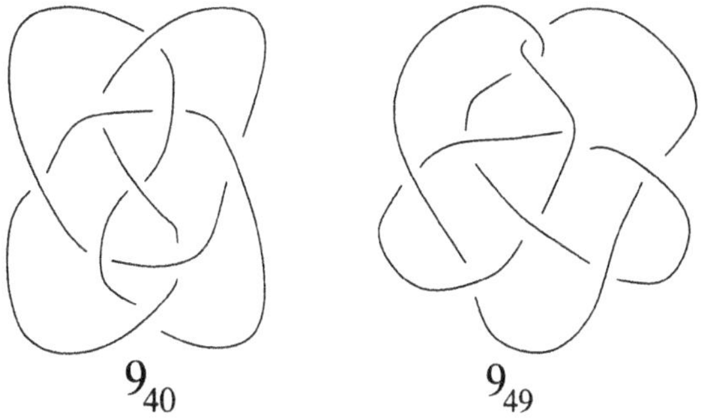

Fig. 1.27

I am much less convinced that the answer to the next open question is positive, so I will not call it a "conjecture". First let us define a (p,q)-move to be a local modification of a link as shown in Fig.1.28. We say that two links, L_1 and L_2, are (p,q)-equivalent if one can obtain one from the other by a finite number of (p,q)-,(q,p)-,$(-p,-q)$- and $(-q,-p)$-moves.

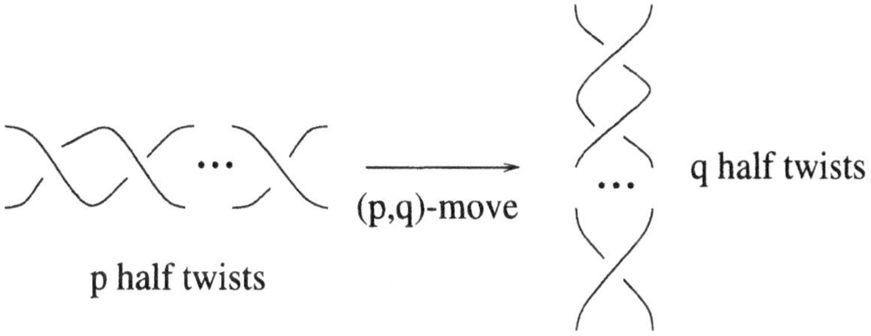

Fig. 1.28

Problem 1.18 ((Kir); Problem 1.59(7), 1995). *Is it true that any link is* $(2,3)$*-move equivalent to a trivial link?*

Example 1.19. *Reduction of the trefoil and the figure eight knots is illustrated in Fig.1.29. Reduction of the Borromean rings is shown in Fig.1.30.*

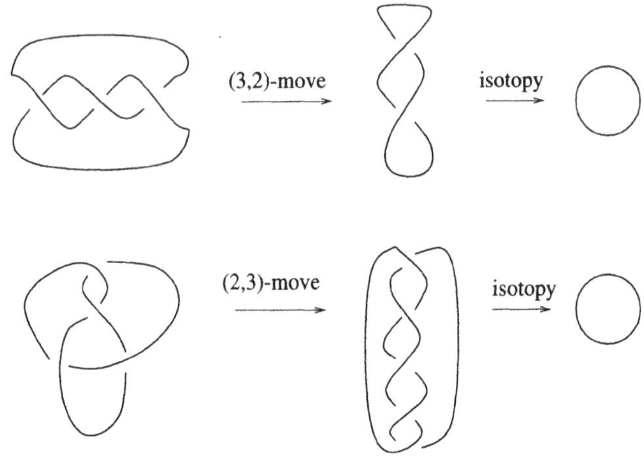

Fig. 1.29

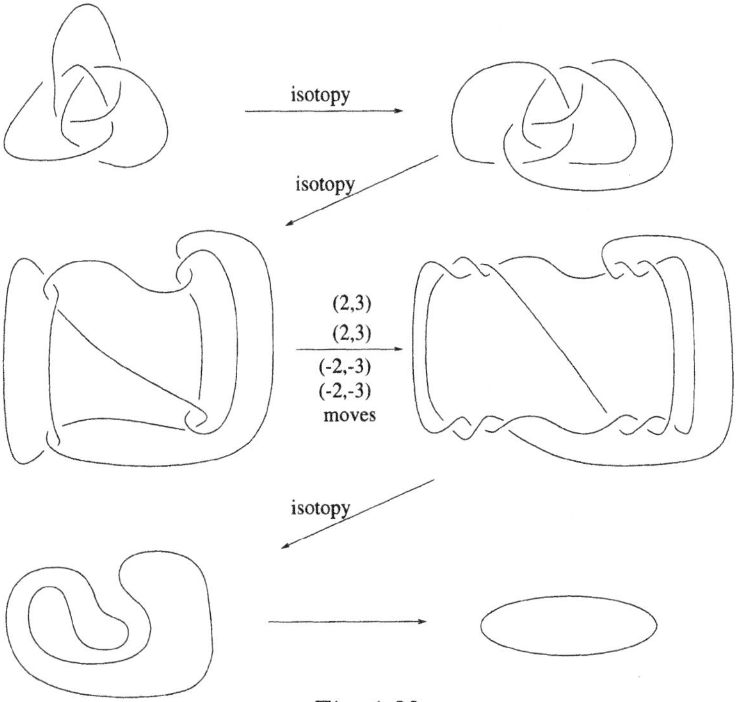

Fig. 1.30

As in the case of Proposition 1.9, simple inductive argument shows that 2-algebraic links are $(2,3)$-move equivalent to trivial links. Fig.1.31 illustrates why the Borromean rings are 2-algebraic. By a proper filling of black dots one can also show that all links with up to 8 crossings, except 8_{18}, are 2-algebraic. Thus, as in the case of $(2,2)$-equivalence, the only link with up to 8 crossings which still should be checked is the 8_{18} knot[17]. Nobody really worked on this problem seriously, so maybe somebody in the audience will try this puzzle.

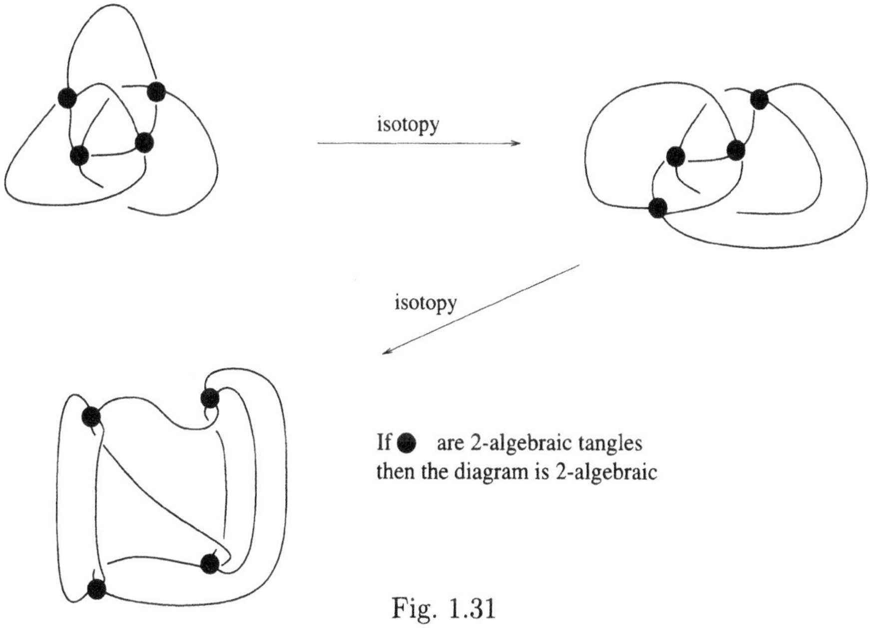

Fig. 1.31

[17] To prove that the knot 8_{18} is not 2-algebraic one considers the 2-fold branched cover of S^3 branched along the knot, $M^{(2)}_{8_{18}}$. Montesinos proved that algebraic knots are covered by Waldhausen graph manifolds (Mo-1). Bonahon and Siebenmann showed ((B-S), Chapter 5) that $M^{(2)}_{8_{18}}$ is a hyperbolic 3-manifold so it cannot be a graph manifold. This manifold is interesting from the point of view of hyperbolic geometry because it is a closed manifold with its volume equal to the volume of the complement of figure eight knot (M-V-1). The knot 9_{49} of Fig.1.27 is not 2-algebraic either because its 2-fold branched cover is a hyperbolic 3-manifold. In fact, it is the manifold I suspected from 1983 to have the smallest volume among oriented hyperbolic 3-manifolds (I-MPT; Kir; M-V-2). In February of 2002 we (my student M.Dąbkowski and myself) found unexpected connection between Knot Theory and the theory of Burnside groups. This has allowed us to present simple combinatorial proof that the knots 9_{40} and 9_{49} are not 2-algebraic. However, our method does not work for the knot 8_{18} (D-P-1; D-P-2; D-P-3).

Fox colorings

The 3-coloring invariant which we are going to use to show that different trivial links are not 3-move equivalent, was introduced by R. H. Fox[18] around 1956 when he was explaining Knot Theory to undergraduate students at Haverford College ("in an attempt to make the subject accessible to everyone" (C-F)). It is a pleasant method of coding representations of the fundamental group of a link complement into the group of symmetries of an equilateral triangle, however this interpretation is not needed for the definition and most of applications of 3-colorings (compare (Cr; C-F; Fo-1; Fo-2)).

Definition 1.20. *(Fox 3-coloring of a link diagram).*
Consider a coloring of a link diagram using colors r (red), y (yellow), and b (blue) in such a way that an arc of the diagram (from an undercrossing to an undercrossing) is colored by one color and at each crossing one uses either only one or all three colors. Such a coloring is called a Fox 3-coloring. *If the whole diagram is colored by just one color we say that we have a* trivial coloring. *The number of different Fox 3-colorings of D is denoted by tri(D).*

Example 1.21. (i) *tri*(◯) = 3 *as the trivial link diagram has only trivial colorings.*

(ii) *tri* (◯ ◯) = 9, *and more generally, for the trivial link diagram of n components,* T_n, *one has* $tri(T_n) = 3^n$.

(iii) *For the standard diagram of the right-handed trefoil knot we have three trivial colorings and six nontrivial colorings. One of them is presented in Fig.1.32 (all the others differ from this one by permutations of colors). Thus, tri*() = 3+6=9.

[18] Ralph Hartzler Fox was born March 24, 1913. A native of Morrisville, Pa., he attended Swarthmore College for two years while studying piano at the Leefson Conservatory of Music in Philadelphia. He was mostly home schooled and later he was a witness in a court case in Virginia, certifying soundness of home schooling. He received his master's degree from the Johns Hopkins University and his Ph.D. from the Princeton University in 1939 under the supervision of Solomon Lefschetz. Fox was married, when he was still a student, to Cynthia Atkinson. They had one son, Robin. After receiving his Princeton doctorate, he spent the following year at Institute for Advanced Study in Princeton. He taught at the University of Illinois and Syracuse University before returning to join the Princeton University faculty in 1945 and staying there until his death. He was giving a series of lectures at the Instituto de Matemáticas de la Universidad Nacional Autónoma de México in the summer of 1951. He was lecturing to American Mathematical Society (1949), to the Summer Seminar of the Canadian Mathematical Society (1953), and at the Universities of Delft and Stockholm, while on a Fulbright grant (1952). He died December 23, 1973 in the University of Pennsylvania Graduate Hospital, where he had undergone open-heart surgery (P-12).

Fig. 1.32; Different colors are marked by lines of different thickness.

Fox 3-colorings were defined for link diagrams. They are, however, invariants of links. One only needs to show that $tri(D)$ is unchanged by Reidemeister moves.

The invariance under R_1 and R_2 is illustrated in Fig.1.33 and the invariance under R_3 is illustrated in Fig.1.34.

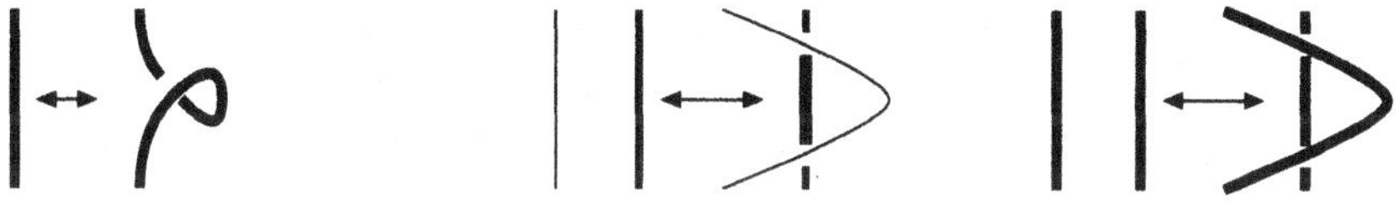

Fig. 1.33

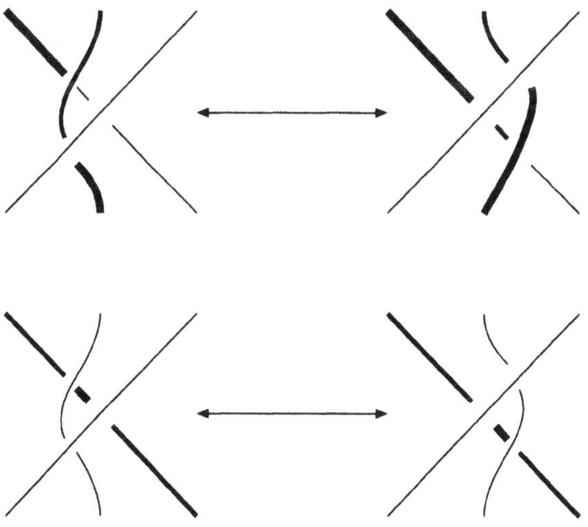

Fig. 1.34

The next property of Fox 3-colorings is the key in proving that different trivial links are not 3-move equivalent.

Lemma 1.22 ((P-1)). *3-moves do not change* $tri(D)$.

The proof of the lemma is illustrated in Figure 1.35.

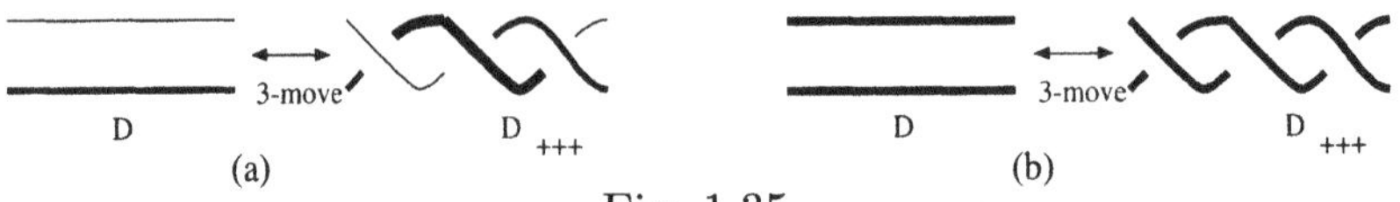

Fig. 1.35

The lemma also explains the fact that the trefoil knot has nontrivial Fox 3-colorings: the trefoil knot is 3-move equivalent to the trivial link of two components (Example 1.4(i)).

Tomorrow, I will place the theory of Fox colorings in a more general (sophisticated) context, and apply it to the analysis of 3-moves (and $(2,2)$- and $(2,3)$-moves) on n-tangles. Interpretation of tangle colorings as Lagrangians in symplectic spaces is our main (and new) tool. In the third section, I will discuss another motivation for studying 3-moves: understanding skein modules based on their deformation.

2. Lagrangian approximation of Fox p-colorings of tangles

We just had the opportunity to listen to a beautiful and elementary talk by Lou Kauffman. I hope to follow this example by making my talk elementary and deep at the same time. I will use several results introduced by Lou, like classification of rational tangles, and also I am going to build on my yesterday's talk. I will culminate today's talk with introduction of the symplectic structure on the boundary of a tangle in such a way that tangles will yield Lagrangians in the associated symplectic space. I could not dream of this connection a year ago; however, now, 10 months after, I see the symplectic structure as a natural development.

Let us start our discussion slowly using my personal perspective and motivation. In the Spring of 1986, I was analyzing behavior of Jones type invariants of links when modified by k-moves (or t_k-, $\bar{t}_{2k}$-moves in the oriented case). My interest had its roots in the fundamental paper by Conway (Co). In July of 1986, I gave a talk at the "Braids" conference in Santa Cruz. After my talk, I was told by Kunio Murasugi and Hitoshi Murakami about the Nakanishi's 3-move conjecture. It was suggested to me by R. Campbell (Rob Kirby's student in 1986) to consider the effect of 3-moves on Fox colorings. Several years later, when writing (P-3) in 1993, I realized that Fox colorings can be successfully used to analyze moves on tangles by considering not only the space of colorings but also the induced colorings of boundary points. More of this later, but let us now define Fox k-colorings first.

Definition 2.1. (i) *We say that a link (or a tangle) diagram is k-colored if every arc is colored by one of the numbers $0, 1, ..., k-1$ (elements of the group $\mathbb{Z}_k$) in such a way that at each crossing the sum of the colors of the undercrossings equals twice the color of the overcrossing modulo k; see Fig.2.1.*

(ii) *The set of k-colorings forms an abelian group, denoted by $Col_k(D)$ (we can also think of $Col_k(D)$ as a module over $\mathbb{Z}_k$). The cardinality of the group will be denoted by $col_k(D)$. For an n-tangle T each Fox k-coloring of T yields a coloring of boundary points of T and we have the homomorphism $\psi : Col_k(T) \to \mathbb{Z}_k^{2n}$*

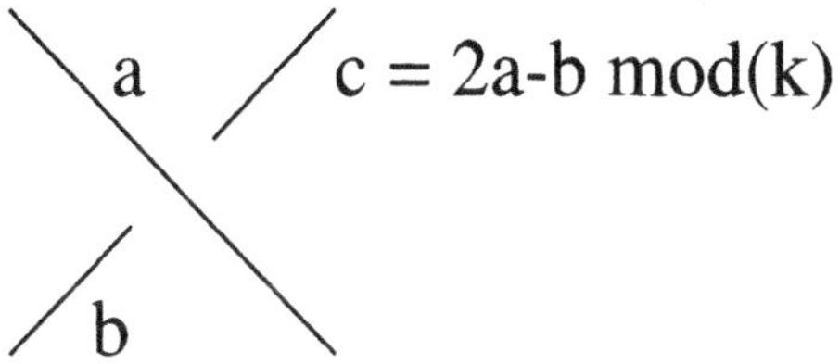

Fig. 2.1

It is a pleasant exercise to show that $Col_k(D)$ is unchanged by Reidemeister moves, so I am going to leave it for you. The invariance under k-moves is explained in Fig.2.2.

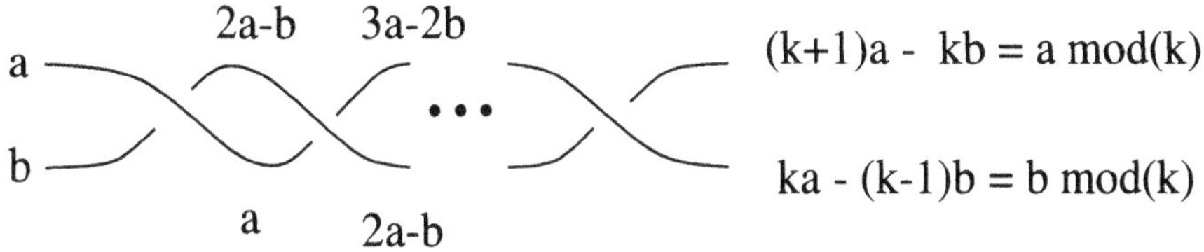

Fig. 2.2

Having observed that k-moves preserve the space of Fox k-colorings, let us take a closer look at the unlinking conjectures described before. We discussed the 3-move conjecture, the 4-move conjecture for knots, and the Kawauchi's question for links. As I mentioned yesterday, not every link can be simplified using 5-moves, but the 5-move is a combination of $\pm(2,2)$-moves and these moves might be sufficient to reduce every link to trivial links. Similarly not every link can be reduced via 7-moves, but again each 7-move is a combination of $(2,3)$-moves[19] which still might be sufficient for reduction. We stopped at this point yesterday, but what could be used instead of general k-moves? Let us consider the case of p-moves, where p is a prime number. I suggest (and state publicly for the first time) that possibly one should consider *rational moves* instead, that is, moves in which a rational $\frac{p}{q}$-tangle of Conway is substituted in place of the identity tangle[20]. The most important observation for us is that $Col_p(D)$ is preserved by $\frac{p}{q}$-moves. Fig.2.3 illustrates, for example, the fact that $Col_{13}(D)$ is unchanged by a $\frac{13}{5}$-move.

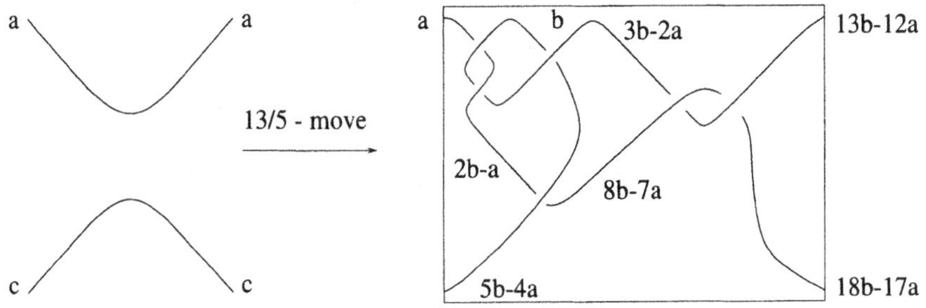

Fig. 2.3

We also should note that (m,q)-moves are equivalent to $\frac{mq+1}{q}$-moves (Fig.2.4) so the space of Fox $(mq+1)$-colorings is preserved by them too.

[19] To be precise, a 7-move is a combination of $(-3,-2)$- and $(2,3)$-moves; compare Fig.1.25.

[20] The move was first considered by J. M. Montesinos (Mo-2); compare also Y. Uchida (Uch).

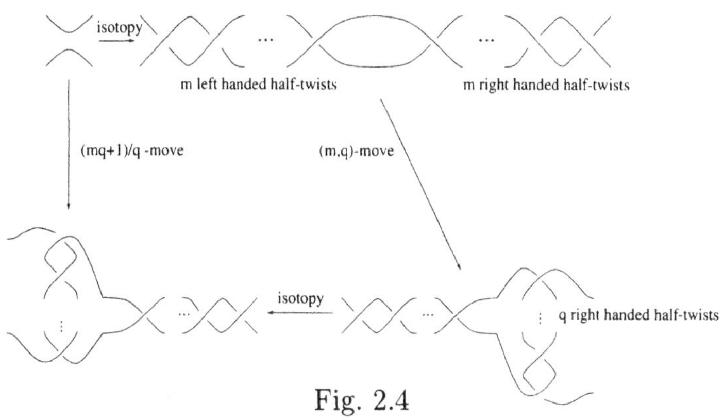

Fig. 2.4

We have just heard about Conway's classification of rational tangles in Lou's talk[21], so I will just briefly sketch necessary definitions and introduce basic notation. The 2-tangles shown in Fig.2.5 are called rational tangles – in Conway's notation, $T(a_1, a_2, ..., a_n)$. A rational tangle is $\frac{p}{q}$-tangle if $\frac{p}{q} = a_n + \frac{1}{a_{n-1}+...+\frac{1}{a_1}}$.[22] Conway proved that two rational tangles are ambient isotopic (with boundary points fixed) if and only if their slopes are equal (compare (Kaw)).

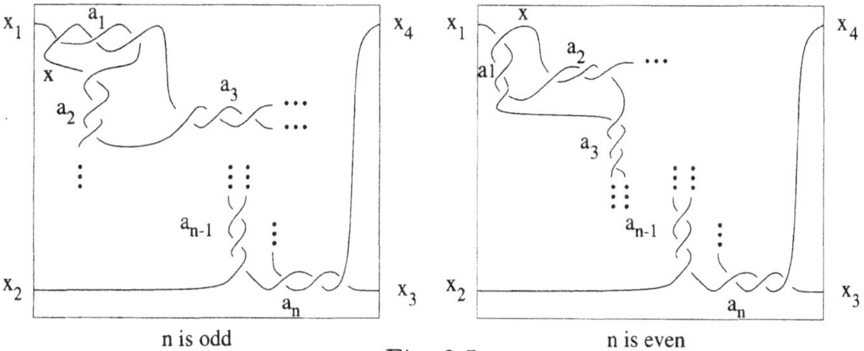

Fig. 2.5

For a given Fox coloring of the rational $\frac{p}{q}$-tangle with boundary colors x_1, x_2, x_3, x_4 (Fig.2.5), one has $x_4 - x_1 = p(x - x_1)$, $x_2 - x_1 = q(x - x_1)$ and $x_3 = x_2 + x_4 - x_1$. If a coloring is nontrivial ($x_1 \neq x$) then $\frac{x_4 - x_1}{x_2 - x_1} = \frac{p}{q}$ as it has been explained by Lou.

[21] L.Kauffman's talk in Cuautitlan , March, 2001; compare (K-L).

[22] $\frac{p}{q}$ is called the slope of the tangle and can be easily identified with the slope of the meridian disk of the solid torus being the branched double cover of the rational tangle.

Conjecture 2.2.
Let p be a fixed prime number, then[23]

(i) *Every link can be reduced to a trivial link by rational $\frac{p}{q}$-moves (q any integer).*

(ii) *There is a function $f(n,p)$ such that any n-tangle can be reduced to one of "basic" $f(n,p)$ n-tangles (allowing additional trivial components) by rational $\frac{p}{q}$-moves.*

First we observe that it suffices to use $\frac{p}{q}$-moves with $|q| \leq \frac{p}{2}$, as they generate all the other $\frac{p}{q}$-moves. Namely, we have $\frac{p}{p-q} = 1 + \frac{1}{-1+\frac{p}{q}}$ and $\frac{p}{-(p+q)} = -1 + \frac{1}{1+\frac{p}{q}}$. Thus $\frac{p}{q}$-moves generate $\frac{p}{-q\pm p}$-moves (e.g., $\frac{p}{p-q}$ tangle is reduced by an inverse of a $\frac{p}{q}$-move to the 0-tangle, $1 + \frac{1}{-1+0} = 0$). Furthermore, we know that for odd p the $\frac{p}{1}$-move is a combination of $\frac{p}{2}$ and $\frac{p}{-2}$-moves (compare Fig.1.25). Thus, in fact, 3-move, $(2,2)$-move and $(2,3)$-move conjectures are special cases of Conjecture 2.2(i). For $p = 11$ we have $\frac{11}{2} = 5 + \frac{1}{2}$, $\frac{11}{3} = 4 - \frac{1}{3}$, $\frac{11}{4} = 3 - \frac{1}{4}$, $\frac{11}{5} = 2 + \frac{1}{5}$. Thus:

Conjecture 2.3.
Every link can be reduced to a trivial link (with the same space of 11-colorings) by $(2,5)$- and $(4,-3)$-moves, their inverses and their mirror images[24].

What about the number $f(n,p)$? We know that because $\frac{p}{q}$-moves preserve p-colorings, therefore $f(n,p)$ is bounded from below by the number of subspaces of p-colorings of the $2n$ boundary points induced by Fox p-colorings of n-tangles (that is by the number of subspaces $\psi(Col_p(T))$ in $\mathbb{Z}_p^{2n}$). I noted in (P-3) that for 2-tangles this number is equal to $p+1$ (even in this special case my argument was complicated). For $p = 3$ and $n = 4$ the number of subspaces followed from the work of my student Tatsuya Tsukamoto and is equal to 40 (P-Ts). The combined effort of Mietek Dąbkowski and Tsukamoto gave the number 1120 for subspaces $\psi(Col_3(T))$ and 4-tangles. That was my knowledge at the early Spring of 2000. On May 2nd and 3rd I attended

[23] I decided to keep the word "Conjecture" as it was used in my talk. However, in Spring of 2002, we disproved it for any p, (D-P-1; D-P-2; D-P-3). The talks in Mexico and Canada were essential for clarifying ideas and finally in constructing counterexamples.

[24] As mentioned in the footnote 23, Conjecture 2.3 does not hold. The closure of the 3-braid Δ_3^4 provides the simplest counterexample (D-P-2; D-P-3). However, it holds for 2-algebraic links; see Proposition 2.7.

talks on Tits buildings (at the Banach Center in Warsaw) by Janek Dymara and Tadek Januszkiewicz. I realized that the topic may have some connection to my work. I asked Januszkiewicz whether he sees relation and I gave him numbers 4, 40, 1120 for $p = 3$. He immediately answered that most likely I was counting the number of Lagrangians in $\mathbb{Z}_3^{2n-2}$ symplectic space, and that the number of Lagrangians in $\mathbb{Z}_p^{2n-2}$ is known to be equal to $\prod_{i=1}^{n-1}(p^i+1)$. Soon I constructed the appropriate symplectic form (as did Dymara). I will spend most of this talk on this construction and end with discussion of classes of tangles for which it has been proved that $f(n,p) = \prod_{i=1}^{n-1}(p^i+1)$.

Consider $2n$ points on a circle (or a square) and a field $\mathbb{Z}_p$ of p-colorings of a point. The colorings of $2n$ points form the linear space $\mathbb{Z}_p^{2n}$. Let $e_1, \ldots, e_{2n}$ be the standard basis of $\mathbb{Z}_p^{2n}$,

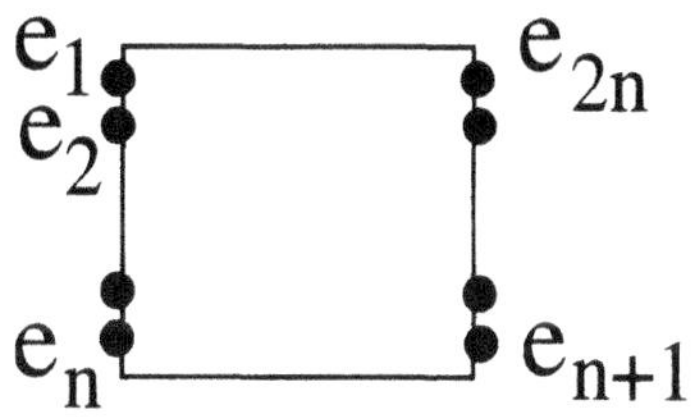

Fig. 2.6

$e_i = (0, \ldots, 1, \ldots, 0)$, where 1 occurs in the i-th position. Let $\mathbb{Z}_p^{2n-1} \subset \mathbb{Z}_p^{2n}$ be the subspace of vectors $\sum a_i e_i$ satisfying $\sum (-1)^i a_i = 0$ (alternating condition). Consider the basis $f_1, \ldots, f_{2n-1}$ of $\mathbf{Z}_p^{2n-1}$ where $f_k = e_k + e_{k+1}$.

Let

$$\phi = \begin{pmatrix} 0 & 1 & \ldots & \ldots \\ -1 & 0 & 1 & \ldots \\ \ldots & \ldots & \ldots & \ldots \\ \ldots & \ldots & -1 & 0 \end{pmatrix}$$

be a skew-symmetric form ϕ on $\mathbb{Z}_p^{2n-1}$ of nullity 1, that is,

$$\phi(f_i, f_j) = \begin{cases} 0 & \text{if } |j-i| \neq 1 \\ 1 & \text{if } j = i+1 \\ -1 & \text{if } j = i-1. \end{cases}$$

Notice that the vector $e_1 + e_2 + \ldots + e_{2n}$ $(= f_1 + f_3 + \ldots + f_{2n-1} = f_2 + f_4 + \ldots + f_{2n})$ is ϕ-orthogonal to any other vector. If we consider $\mathbb{Z}_p^{2n-2} = \mathbb{Z}_p^{2n-1}/\mathbb{Z}_p$, where the subspace $\mathbb{Z}_p$ is generated by $e_1 + \ldots + e_{2n}$, that is, $\mathbb{Z}_p$ consists of monochromatic (i.e., trivial) colorings, then ϕ descends to the symplectic form $\hat{\phi}$ on $\mathbb{Z}_p^{2n-2}$. Now we can analyze isotropic subspaces of $(\mathbb{Z}_p^{2n-2}, \hat{\phi})$, that is subspaces on which $\hat{\phi}$ is 0 ($W \subset \mathbb{Z}_p^{2n-2}$, where $\hat{\phi}(w_1, w_2) = 0$ for all $w_1, w_2 \in W$). The maximal isotropic subspaces of $\mathbb{Z}_p^{2n-2}$ are $(n-1)$-dimensional and they are called Lagrangian subspaces (or maximal totally degenerated subspaces). There are $\prod_{i=1}^{n-1}(p^i + 1)$ of them.

Our local condition on Fox colorings (Fig.2.1) guarantees that for any tangle T, $\psi(Col_pT) \subset \mathbb{Z}_p^{2n-1}$. Furthermore, the space of trivial colorings, $\mathbb{Z}_p$ is always in Col_pT. Thus ψ descends to $\hat{\psi} : Col_pT/\mathbb{Z}_p \to \mathbb{Z}_p^{2n-2} = \mathbb{Z}_p^{2n-1}/\mathbb{Z}_p$. Now we answer the fundamental question: Which subspaces of $\mathbb{Z}_p^{2n-2}$ are yielded by n-tangles?

Theorem 2.4.
$\hat{\psi}(Col_pT/\mathbb{Z}_p)$ is a Lagrangian subspace of $\mathbb{Z}_p^{2n-2}$ with the symplectic form $\hat{\phi}$.

The natural question is whether every Lagrangian subspace can be realized as a space of induced colorings on the boundary for some tangle. The answer is negative for $p = 2$ and positive for $p > 2$.

Theorem 2.5 ((D-J-P)).

(i) *For an odd prime number p, every Lagrangian in $(\mathbb{Z}_p^{2n-2}, \hat{\phi})$ is realized as $\hat{\psi}(Col_pT/\mathbb{Z}_p)$ for some n-tangle T. Furthermore, T can be chosen to be a rational n-tangle*[25]*.*

---

[25] An n-tangle is a rational (or n-bridge) tangle if it is homeomorphic to a tangle without crossing and trivial components (we allow homeomorphism moving the

(ii) *For* $p = 2$, $n > 2$, *not every Lagrangian is realized as* $\hat{\psi}(Col_2T/\mathbb{Z}_2)$. *We have* $f(n,2) = \prod_{i=1}^{n-1}(2i+1)$ *(a 2-coloring is unchanged by a crossing change) but the number of Lagrangians is equal to* $\prod_{i=1}^{n-1}(2^i+1)$.

As a corollary we obtain a fact which was considered to be difficult before, even for 2-tangles (compare (P-3; J-P).

Corollary 2.6.
For any p-coloring $\mathbf{x}$ *of a tangle boundary points satisfying the alternating property (i.e.,* $\mathbf{x} \in \mathbb{Z}_p^{2n-1}$*) there is an n-tangle and p-coloring of it that yields* $\mathbf{x}$*. In other words:* $\mathbb{Z}_p^{2n-1} = \bigcup_T \psi_T(Col_pT)$*. Furthermore, the space* $\psi_T(Col_pT)$ *is n-dimensional for any T.*

We can say that we understand the lower bound for the function $f(n,p)$, but when does Conjecture 2.2 holds with $f(n,p) = \prod_{i=1}^{n-1}(p^i+1)$?

In (D-I-P) we discuss Conjecture 2.2 for 2-algebraic tangles. Here we sketch a proof of a simpler fact.

Proposition 2.7.
Let p be a fixed prime number and let H_p *be the family of 2-tangles:* $\frac{1-p}{2}, \frac{3-p}{2}, \ldots, 0, \ldots, \frac{p-3}{2}, \frac{p-1}{2}$ *and* ∞ *(horizontal family), and let* V_p *be the vertical family of 2-tangles,* $V_p = r(H_p)$*; then*

(i) *Every 2-algebraic tangle can be reduced to a 2-tangle from the family* H_p *(resp.* V_p*) with possible additional trivial components by* $(\frac{sp}{q})$*-moves, where s and q are any integers such that sp and q are relatively prime. Furthermore, for* $p \leq 13$ *one can assume that* $s = 1$.

(ii) *Every 2-algebraic link can be reduced to a trivial link by* $(\frac{sp}{q})$*-moves, where s and q are any integers such that sp and q are relatively prime. Furthermore, for* $p \leq 13$ *one can assume that* $s = 1$.

Outline of the proof. We use the structure of 2-algebraic tangles to perform an inductive proof similar to that of Proposition 1.9. The main problem in the proof is to show that the family V_p can be reduced to the family H_p by our moves. Consider the vertical tangle $r(k)$ where k

---

boundary of the 3-ball). Alternatively, we can use an inductive definition modifying Definition 2.9 in such a way that we start from a tangle without a crossing and a trivial component and we assume in condition (i)(1) that B has exactly one crossing (which is not nugatory, that is, it cannot be eliminated by a first Reidemeister move).

is relatively prime to p. There are integers k' and s such that $kk'+1 = sp$ or equivalently $k' + \frac{1}{k} = \frac{sp}{k}$. Therefore the $\frac{sp}{k}$-move (equivalently (k, k')-move) is changing $r(k)$ to the horizontal tangle k'. In this reasoning we do not have a control over s. Consider now the case of $p = 13$ and $s = 1$. By considering fractions $\frac{13}{2} = 6 + \frac{1}{2}$, $\frac{13}{3} = 4 + \frac{1}{3}$, $\frac{13}{4} = 3 + \frac{1}{4}$, $\frac{13}{6} = 2 + \frac{1}{6}$, we are able to work with all $r(k)$ except $k = 5$. $5 + \frac{1}{5} = \frac{26}{5}$ so $s = 2$ in this case. We can, however, realize $\frac{26}{5}$-move as a combination of $\frac{13}{3}$-move and $\frac{13}{2}$-move as illustrated in Fig.2.7 (we start by presenting $\frac{26}{5}$ as $6 + \frac{1}{-1-\frac{1}{4}}$).

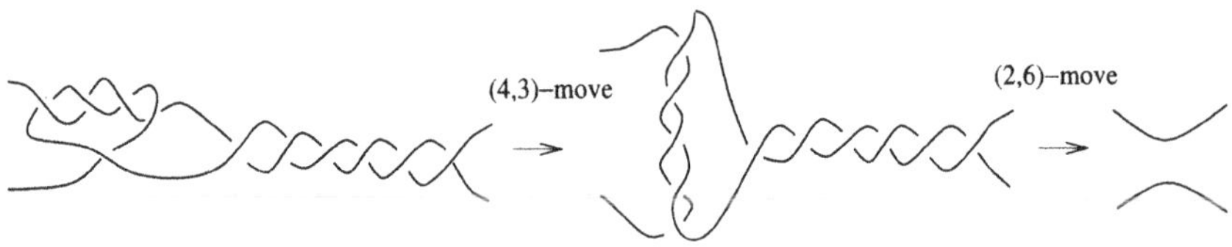

Fig. 2.7

Corollary 2.8.
Every $(\frac{sp}{q})$-rational tangle, p odd prime, can be reduced to the 0 2-tangle by (k, k')-moves where $|k| < \frac{p}{2}$ and $kk' + 1 = sp$ for some s.

In order to be able to use induction for n-tangles with $n > 2$, we generalize the notion of the algebraic tangle.

Definition 2.9.

(i) *The family of n-algebraic tangles is the smallest family of n-tangles which satisfies:*
(0) *Any n-tangle with 0 or 1 crossing is n-algebraic.*
(1) *If A and B are n-algebraic tangles then $r^i(A) * r^j(B)$ is n-algebraic, where r denotes here the rotation of a tangle by $\frac{2\pi}{2n}$ angle, and $*$ denotes horizontal composition of tangles.*

(ii) *If in the condition (1), B is restricted to tangles with no more than k crossings, we obtain the family of (n, k)-algebraic tangles.*

(iii) *If an m-tangle, T, is obtained from an (n, k)-algebraic tangle (resp. n-algebraic tangle) by partially closing its endpoints ($2n - 2m$ of them) without introducing any new crossings, then T is called an (n, k)-algebraic (resp. n-algebraic) m-tangle. For $m = 0$ we obtain an (n, k)-algebraic (resp. n-algebraic) link.*

Conjecture 2.2, for $p = 3$, has been proved for 3-algebraic tangles (P-Ts) ($f(3,3) = 40$) and $(4,5)$-algebraic tangles (Tsu) ($f(4,3) = 1120$). In particular the Montesinos-Nakanishi 3-move conjecture holds for 3-algebraic and $(4,5)$-algebraic links. 40 "basic" 3-tangles are shown in Fig. 2.8.

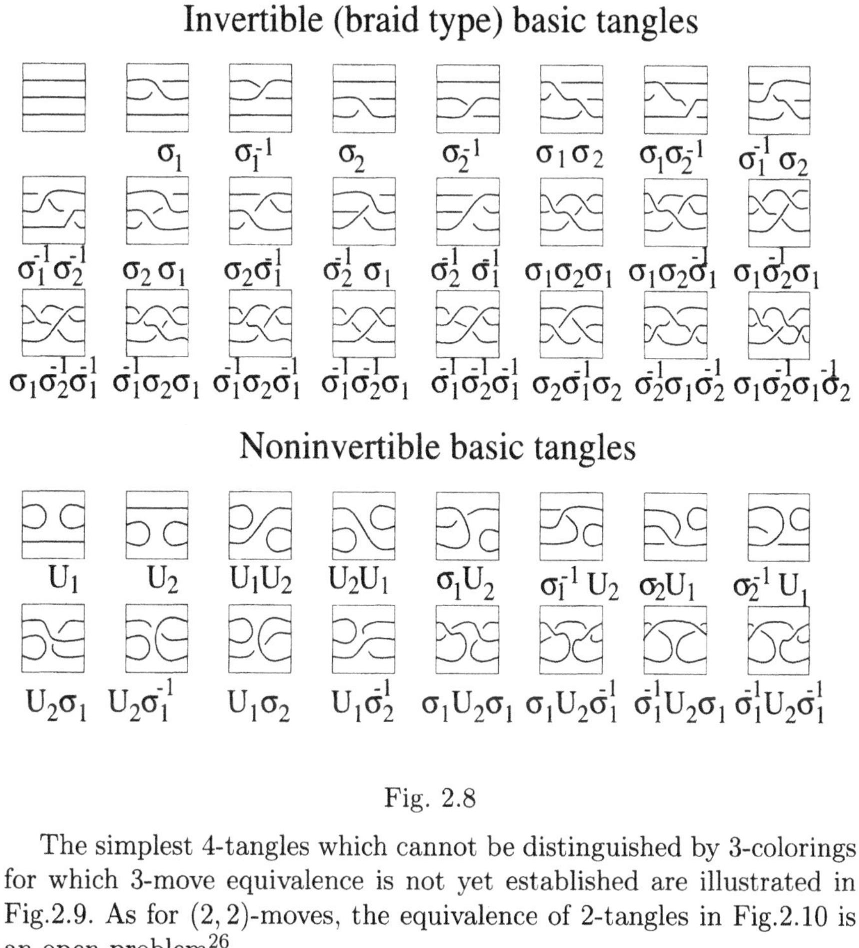

Fig. 2.8

The simplest 4-tangles which cannot be distinguished by 3-colorings for which 3-move equivalence is not yet established are illustrated in Fig.2.9. As for $(2,2)$-moves, the equivalence of 2-tangles in Fig.2.10 is an open problem[26].

[26] The 4-tangles in Fig.2.9 are not 3-move equivalent. This follows from the fact that the Borromean rings and the Chen's link are not 3-move equivalent to trivial links (D-P-1; D-P-3). Similarly, the fact that 2-tangles of Fig.2.10 are not $(2,2)$-move equivalent follows from the result proven in (D-P-2; D-P-3) that the knot 9_{49} and the link 9^2_{40} are not $(2,2)$-move equivalent to the trivial link of three components.

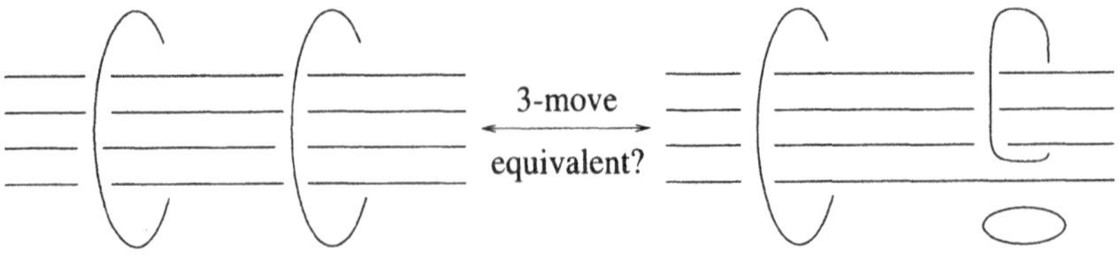

Fig. 2.9

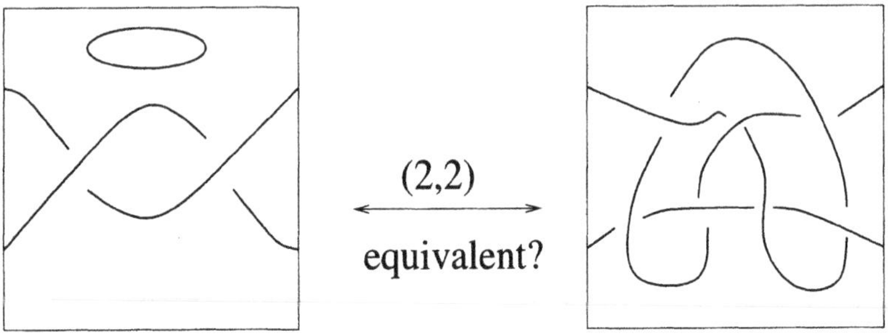

Fig. 2.10

A weaker version of the Montesinos-Nakanishi 3-move conjecture has been proved by Bronek Wajnryb in 1985 (Wa-1; Wa-2) (compare Theorem 1.16).

Theorem 2.10 (Wajnryb). *Every link can be reduced to a trivial link by a finite number of* ± 3*-moves and* Δ_5^4*-moves.*

Let me complete this talk by mentioning two generalizations of the Fox k-colorings.

In the first generalization we consider any commutative ring with the identity, $\mathcal{R}$, instead of $\mathbb{Z}_k$. We construct $Col_{\mathcal{R}}T$ in the same way as before with the relation at each crossing, Fig.2.1, having the form $c = 2a - b$ in $\mathcal{R}$. The skew-symmetric form ϕ on $\mathcal{R}^{2n-1}$, the symplectic form $\hat{\phi}$ on $\mathcal{R}^{2n-2}$ and the homomorphisms ψ and $\hat{\psi}$ are defined in the same manner as before. Theorem 2.4 generalizes as follows ((D-J-P)):

Theorem 2.11. *Let* $\mathcal{R}$ *be a Principal Ideal Domain (PID). Then,* $\hat{\psi}(Col_{\mathcal{R}}T/\mathcal{R})$ *is a virtual Lagrangian submodule of* $\mathcal{R}^{2n-2}$ *with the symplectic form* $\hat{\phi}$. *That is,* $\hat{\psi}(Col_{\mathcal{R}}T/\mathcal{R})$ *is a finite index submodule of a Lagrangian in* $\mathcal{R}^{2n-2}$.

This result can be used to analyze embeddability of tangles in links. It gives an alternative proof of Theorem 2.2 in (P-S-W) in the case of the 2-fold cyclic cover of B^3 branched over a tangle.

Example 2.12. *Consider the pretzel tangle $T = (p, -p)$, Fig.2.11, and the ring $\mathcal{R} = \mathbb{Z}$. Then the virtual Lagrangian $\hat{\psi}(Col_{\mathbb{Z}}T/Z)$ has index p. Namely, coloring of T, as illustrated in Fig.2.11, forces us to have $a = b$ and modulo trivial colorings the image $\hat{\psi}(Col_{\mathbb{Z}}T/\mathbb{Z})$ is generated by the vector $(0, p, p, 0) = p(e_2 + e_3)$. The symplectic space $(\mathbb{Z}^{4-2}, \hat{\phi})$ has a basis $(e_1 + e_2, e_2 + e_3)$. Thus, $\hat{\psi}(Col_{\mathbb{Z}}T/\mathbb{Z})$ is a virtual Lagrangian of index p.*

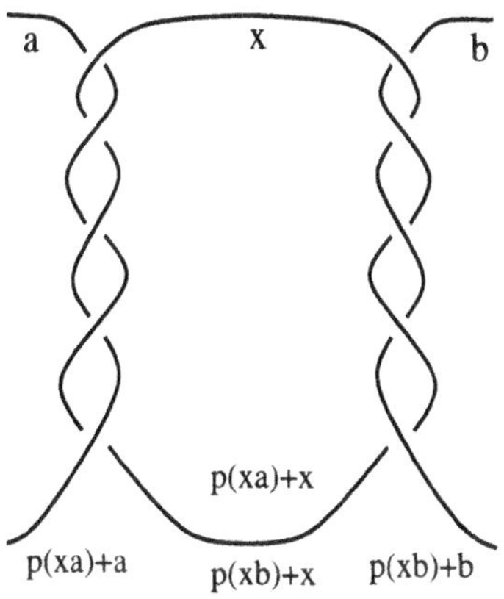

Fig. 2.11; $p = 5$

Corollary 2.13.
If $\hat{\psi}(Col_{\mathbb{Z}}T/\mathbb{Z})$ is a virtual Lagrangian of index $p > 1$, then T does not embed in the trivial knot.

The second generalization leads to racks and quandles (Joy; F-R), but we restrict our remarks to the abelian case – Alexander-Burau-Fox colorings[27]. An ABF-coloring uses colors from a ring $\mathcal{R}$ with an invertible element t (e.g., $\mathcal{R} = \mathbb{Z}[t^{\pm 1}]$). The relation in Fig.2.1 is modified to the relation $c = (1 - t)a + tb$ in $\mathcal{R}$ at each crossing of an oriented link diagram; see Fig. 2.12.

[27] The related approach was first outlined in the letter of J. W. Alexander to O. Veblen, 1919 (A-V). Alexander was probably influenced by Poul Heegaard's dissertation, 1898, which he reviewed for the French translation (Heeg). Burau was considering a braid representation, but locally his relation was the same as that of Fox. According to J. Birman, Burau learned about the representation from Reidemeister or Artin (Ep), p.330.

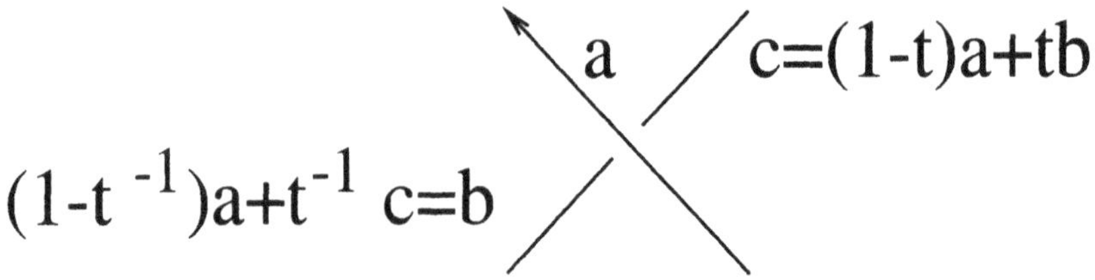

Fig. 2.12

The space $\mathcal{R}^{2n-2}$ has a natural Hermitian structure (Sq), but one can also find a symplectic structure and prove a version of Theorem 2.11 in this setting (D-J-P).

3. Historical Introduction to Skein Modules

In my last talk of the conference, I will discuss *skein modules*, or as I prefer to say more generally, *algebraic topology based on knots*. It was my mind's child, even if the idea was also conceived by other people (most notably Vladimir Turaev), and was envisioned by John H. Conway (as "linear skein") a decade earlier. Skein modules have their origin in the observation made by Alexander (Al), that his polynomials (*Alexander polynomials*) of three links, L_+, L_- and L_0 in R^3 are linearly related (Fig.3.2).

For me it started in Norman, Oklahoma in April of 1987, when I was enlightened to see that the multivariable version of the Jones-Conway (Homflypt) polynomial analyzed by Jim Hoste and Mark Kidwell is really a module of links in a solid torus (or more generally, in the connected sum of solid tori).

I would like to discuss today, in more detail, skein modules related to the (deformations) of 3-moves and the Montesinos-Nakanishi 3-move conjecture, but first I will give the general definition and I will make a short tour of the world of skein modules.

Skein Module is an algebraic object associated with a manifold, usually constructed as a formal linear combination of embedded (or immersed) submanifolds, modulo locally defined relations. In a more restricted setting a skein module is a module associated with a 3-dimensional manifold, by considering linear combinations of links in the manifold, modulo properly chosen (skein) relations. It is the main object of the *algebraic topology based on knots*. When choosing relations one takes into account several factors:

(i) Is the module we obtain accessible (computable)?

(ii) How precise are our modules in distinguishing 3-manifolds and links in them?

(iii) Does the module reflect topology/geometry of a 3-manifold (e.g., surfaces in a manifold, geometric decomposition of a manifold)?

(iv) Does the module admit some additional structure (e.g., filtration, gradation, multiplication, Hopf algebra structure)? Is it leading to a Topological Quantum Field Theory (TQFT) by taking a finite dimensional quotient?

One of the simplest skein modules is a q-deformation of the first homology group of a 3-manifold M, denoted by $\mathcal{S}_2(M;q)$. It is based on the skein relation (between oriented framed links in M): $L_+ = qL_0$; it also satisfies the framing relation $L^{(1)} - qL$, where $L^{(1)}$ denote a link obtained from L by twisting the framing of L once in the positive direction. This easily defined skein module "sees" already nonseparating surfaces in M. These surfaces are responsible for torsion part of our skein module (P-10).

There is a more general pattern: most of the analyzed skein modules reflect various surfaces embedded in a manifold.

The best studied skein modules use skein relations which worked successfully in the classical Knot Theory (when defining polynomial invariants of links in R^3).

(1) The Kauffman bracket skein module, KBSM.
The skein module based on the *Kauffman bracket skein relation*, $L_+ = AL_- + A^{-1}L_\infty$, and denoted by $S_{2,\infty}(M)$, is the best understood among the Jones type skein modules. It can be interpreted as a quantization of the co-ordinate ring of the character variety of $SL(2,\mathbb{C})$ representations of the fundamental group of the manifold M, (Bu-2; B-F-K; P-S). For $M = F \times [0,1]$, KBSM is an algebra (usually noncommutative). It is finitely generated algebra for a compact F (Bu-1), and has no zero divisors (P-S). The center of the algebra is generated by boundary components of F (B-P; P-S). Incompressible tori and 2-spheres in M yield torsion in KBSM; it is the question of fundamental importance whether other surfaces could yield torsion as well. The Kauffman bracket skein modules of the exteriors of 2-bridge knots have been recently (April 2004) computed by Thang Le (Le). For a 2-bridge (rational) knot $K_{\frac{p}{m}}$

the skein module is the free $\mathbb{Z}[A^{\pm 1}]$ module with the basis $\{x^i y^j\}$, $0 \le i$, $0 \le j \le \frac{p-1}{2}$, where $x^i y^j$ denotes the element of the skein module represented by the link composed of i parallel copies of the meridian curve x and j parallel copies of the curve y; see Fig.3.1. Le's theorem generalizes results in (Bu-3) and (B-L).

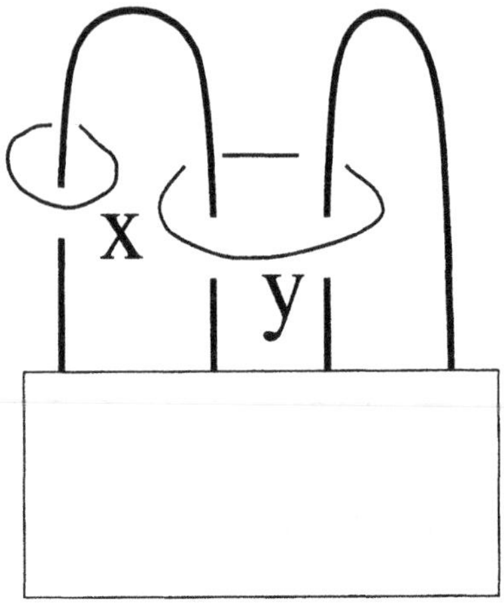

Fig. 3.1

(2) Skein modules based on the Jones-Conway (Homflypt) relation. $v^{-1}L_+ - vL_- = zL_0$, where L_+, L_-, L_0 are oriented links (Fig.3.2). These skein modules are denoted by $S_3(M)$ and generalize skein modules based on Conway relation which were hinted at by Conway. For $M = F \times [0,1]$, $S_3(M)$ is a *Hopf algebra* (usually neither commutative nor co-commutative), (Tu-2; P-6). $S_3(F \times [0,1])$ is a free module and can be interpreted as a quantization (H-K; Tu-1; P-5; Tu-2). $S_3(M)$ is related to the algebraic set of $SL(n, \mathbb{C})$ representations of the fundamental group of the manifold M, (Si).

(3) Skein modules based on the *Kauffman polynomial* relation. $L_{+1} + L_{-1} = x(L_0 + L_\infty)$ (see Fig.3.3) and the framing relation $L^{(1)} - aL$. This module is denoted by $S_{3,\infty}$ and is known to be free for $M = F \times [0,1]$.

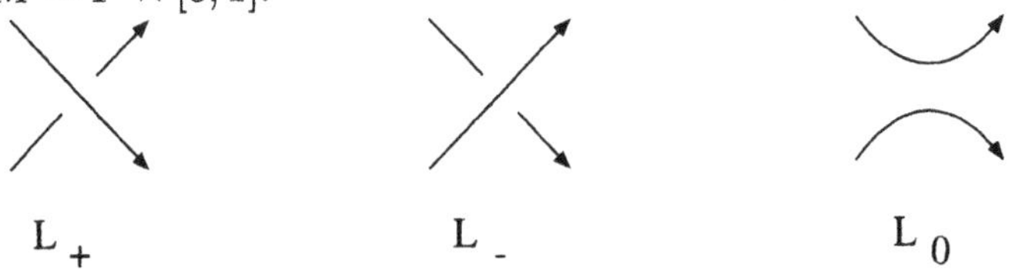

Fig. 3.2

(4) Homotopy skein modules.
In these skein modules, $L_+ = L_-$ for self-crossings. The best studied example is the q-homotopy skein module with the skein relation $q^{-1}L_+ - qL_- = zL_0$ for mixed crossings. For $M = F \times [0,1]$ it is a quantization, (H-P-1; Tu-2; P-11), and as noted by Uwe Kaiser they can be almost completely understood using singular tori technique introduced by Xiao-Song Lin.

(5) Skein modules based on Vassiliev-Gusarov filtration.
We extend the family of knots, $\mathcal{K}$, by allowing singular knots, and resolve a singular crossing by $K_{cr} = K_+ - K_-$. These allow us to define the Vassiliev-Gusarov filtration: $... \subset C_3 \subset C_2 \subset C_1 \subset C_0 = R\mathcal{K}$, where C_k is generated by knots with k singular crossings. The kth Vassiliev-Gusarov skein module is defined to be a quotient: $W_k(M) = R\mathcal{K}/C_{k+1}$. The completion of the space of knots with respect to the Vassiliev-Gusarov filtration, $\hat{R\mathcal{K}}$, is a *Hopf algebra* (for $M = S^3$). Functions dual to Vassiliev-Gusarov skein modules are called *finite type* or *Vassiliev invariants* of knots, (P-7).

(6) Skein modules based on relations deforming n-moves.
$\mathcal{S}_n(M) = R\mathcal{L}/(b_0L_0+b_1L_1+b_2L_2+...+b_{n-1}L_{n-1})$. In the unoriented case, we can add to the relation the term $b_\infty L_\infty$ to get $\mathcal{S}_{n,\infty}(M)$, and also, possibly, a framing relation. The case $n = 4$, on which I am working with my students will be described, in greater detail in a moment.

Examples (1)-(5) gave a short description of skein modules studied extensively until now. I will now spend more time on two other examples which only recently have been considered in more detail. The first example is based on a deformation of the 3-move and the second on the deformation of the $(2,2)$-move. The first one has been studied by my students Tsukamoto and Veve. I denote the skein module described in this example by $\mathcal{S}_{4,\infty}$ since it involves (in the skein relation) 4 horizontal positions and the vertical (∞) smoothing.

Definition 3.1. *Let M be an oriented 3-manifold and let $\mathcal{L}_{fr}$ be the set of unoriented framed links in M (including the empty link, $\emptyset$), and let R be any commutative ring with identity. Then we define the $(4,\infty)$ skein module as: $\mathcal{S}_{4,\infty}(M;R) = R\mathcal{L}_{fr}/I_{(4,\infty)}$, where $I_{(4,\infty)}$ is the submodule of $R\mathcal{L}_{fr}$ generated by the skein relation:*
$b_0L_0 + b_1L_1 + b_2L_2 + b_3L_3 + b_\infty L_\infty = 0$ *and the framing relation:*

$L^{(1)} = aL$ *where* a, b_0, b_3 *are invertible elements in* R *and* b_1, b_2, b_∞ *are any fixed elements of* R *(see Fig.3.3).*

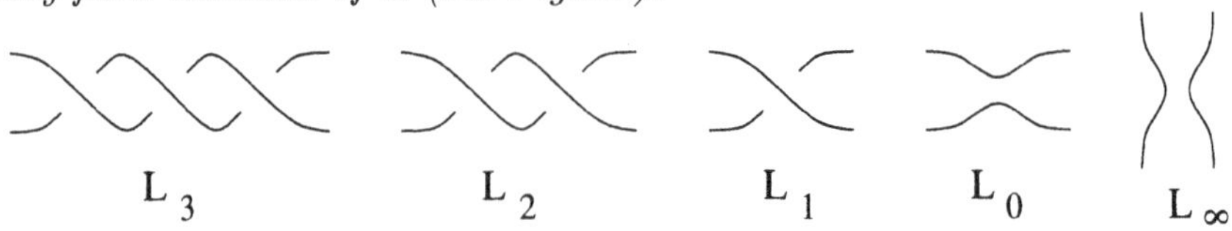

Fig. 3.3

The generalization of the Montesinos-Nakanishi 3-move conjecture says that $S_{4,\infty}(S^3, R)$ is generated by trivial links[28] and that for n-tangles our skein module is generated by $f(n,3)$ basic tangles (with possible trivial components). This would give a generating set for our skein module of S^3 or D^3 with $2n$ boundary points (an n-tangle). In (P-Ts) we analyzed extensively the possibility that trivial links, T_n, are linearly independent. This may happen if $b_\infty = 0$ and $b_0 b_1 = b_2 b_3$. These lead to the following conjecture:

Conjecture 3.2. (1) *There is a polynomial invariant of unoriented links in* S^3, $P_1(L) \in Z[x,t]$, *which satisfies:*

(i) *Initial conditions:* $P_1(T_n) = t^n$, *where* T_n *is a trivial link of* n *components.*

(ii) *Skein relation:* $P_1(L_0) + xP_1(L_1) - xP_1(L_2) - P_1(L_3) = 0$, *where* L_0, L_1, L_2, L_3 *is a standard, unoriented skein quadruple* (L_{i+1} *is obtained from* L_i *by a right-handed half-twist on two arcs involved in* L_i*; compare Fig.3.3).*

(2) *There is a polynomial invariant of unoriented framed links,* $P_2(L) \in Z[A^{\pm 1}, t]$ *which satisfies:*

(i) *Initial conditions:* $P_2(T_n) = t^n$,

(ii) *Framing relation:* $P_2(L^{(1)}) = -A^3 P_2(L)$ *where* $L^{(1)}$ *is obtained from a framed link* L *by a positive half twist on its framing.*

(iii) *Skein relation:* $P_2(L_0) + A(A^2 + A^{-2})P_2(L_1) + (A^2 + A^{-2})P_2(L_2) + AP_2(L_3) = 0$.

---

[28] The counterexamples to the Montesinos-Nakanishi 3-move conjecture, (D-P-1), can be used to show that trivial links "generically" do not generate $S_{4,\infty}(S^3, R)$. This happen, for example, if there is a proper ideal $\mathcal{I} \in R$ such that b_1, b_2 and b_∞ are in $\mathcal{I}$.

The above conjectures assume that $b_\infty = 0$ in our skein relation. Let us consider, for a moment, the possibility that b_∞ is invertible in R. Using the "denominator" of our skein relation (Fig.3.4) we obtain the relation which allows us to compute the effect of adding a trivial component to a link L (we write t^n for the trivial link T_n):

$$(*) \quad (a^{-3}b_3 + a^{-2}b_2 + a^{-1}b_1 + b_0 + b_\infty t)L = 0$$

When considering the "numerator" of the relation and its mirror image (Fig.3.4) we obtain formulas for Hopf link summands, and because the unoriented Hopf link is amphicheiral we can eliminate it from our equations to get the following formula (**):

$$b_3(L\#H) + (ab_2 + b_1 t + a^{-1}b_0 + ab_\infty)L = 0.$$

$$b_0(L\#H) + (a^{-1}b_1 + b_2 t + ab_3 + a^2 b_\infty)L = 0.$$

$$(**) \; ((b_0b_1 - b_2b_3)t + (a^{-1}b_0^2 - ab_3^2) + (ab_0b_2 - a^{-1}b_1b_3) + b_\infty(ab_0 - a^2b_3))L = 0.$$

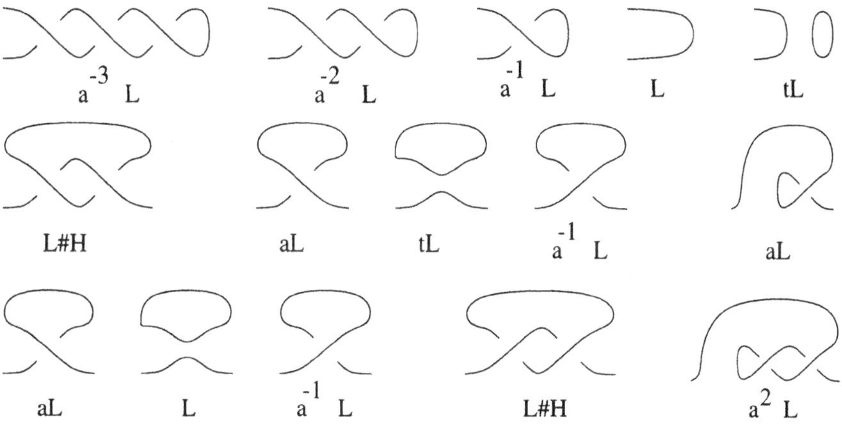

Fig. 3.4

It is possible that $(*)$ and $(**)$ are the only relations in the module. More precisely, we ask whether $\mathcal{S}_{4,\infty}(S^3; R)$ is the quotient ring $R[t]/(\mathcal{I})$ where t^i represents the trivial link of i components and $\mathcal{I}$ is the ideal generated by $(*)$ and $(**)$ for $L = t$. The interesting substitution which satisfies the relations is $b_0 = b_3 = a = 1$, $b_1 = b_2 = x$, $b_\infty = y$. This may lead to a new polynomial invariant (in $\mathbb{Z}[x, y]$) of unoriented links in S^3 satisfying the skein relation $L_3 + xL_2 + xL_1 + L_0 + yL_\infty = 0$.[29]

[29] This speculation should be modified keeping in mind the fact that the Montesinos-Nakanishi 3-move conjecture does not hold (D-P-1).

What about the relations to the Fox colorings? One such a relation, that was already mentioned, is the use of 3-colorings to estimate the number of basic n-tangles (by $\prod_{i=1}^{n-1}(3^i+1)$) for the skein module $\mathcal{S}_{4,\infty}$. I am also convinced that $\mathcal{S}_{4,\infty}(S^3;R)$ contains full information about the space of Fox 7-colorings. It would be a generalization of the fact that the Kauffman bracket polynomial contains information about 3-colorings and the Kauffman polynomial contains information about 5-colorings. In fact, François Jaeger told me that he knew how to form a short skein relation (of the type $(\frac{p+1}{2},\infty)$) involving spaces of p-colorings. Unfortunately, François died prematurely in 1997 and I do not know how to prove his statement[30].

Finally, let me shortly describe the skein module related to the deformation of (2,2)-moves. Because a (2,2)-move is equivalent to the rational $\frac{5}{2}$-move, I will denote the skein module by $\mathcal{S}_{\frac{5}{2}}(M;R)$.

Definition 3.3. *Let M be an oriented 3-manifold. Let $\mathcal{L}_{fr}$ be the set of unoriented framed links in M (including the empty link, $\emptyset$) and let R be any commutative ring with identity. We define the $\frac{5}{2}$-skein module as $\mathcal{S}_{\frac{5}{2}}(M;R) = R\mathcal{L}_{fr}/(I_{\frac{5}{2}})$ where $I_{\frac{5}{2}}$ is the submodule of $R\mathcal{L}_{fr}$ generated by the skein relation:*
(i) $b_2L_2 + b_1L_1 + b_0L_0 + b_\infty L_\infty + b_{-1}L_{-1} + b_{-\frac{1}{2}}L_{-\frac{1}{2}} = 0,$
its mirror image:
($\bar{i}$) $b'_2L_2 + b'_1L_1 + b'_0L_0 + b'_\infty L_\infty + b'_{-1}L_{-1} + b'_{-\frac{1}{2}}L_{-\frac{1}{2}} = 0$
and the framing relation:
$L^{(1)} = aL$, *where $a, b_2, b'_2, b_{-\frac{1}{2}}, b'_{-\frac{1}{2}}$ are invertible elements in R and $b_1, b'_1, b_0, b'_0, b_{-1}, b'_{-1}, b_\infty$, and b'_∞ are any fixed elements of R. The links $L_2, L_1, L_0, L_\infty, L_{-1}$, $L_{\frac{1}{2}}$ and $L_{-\frac{1}{2}}$ are illustrated in Fig.3.5.*[31]

[30] If $col_p(L) = |Col_p(L)|$ denotes the order of the space of Fox p-colorings of the link L, then among $p+1$ links $L_0, L_1, ..., L_{p-1}$, and L_∞, p of them has the same order $col_p(L)$ and one has its order p times larger (P-3). This leads to the relation of type (p,∞). The relation between Jones polynomial (or the Kauffman bracket) and $col_3(L)$ has the form: $col_3(L) = 3|V(e^{\pi i/3})|^2$ and the formula relating the Kauffman polynomial and $col_5(L)$ has the form: $col_5(L) = 5|F(1, e^{2\pi i/5} + e^{-2\pi i/5})|^2$. This seems to suggest that the formula discovered by Jaeger involved Gaussian sums.

[31] Our notation is based on Conway's notation for rational tangles. However, it differs from it by a sign change. The reason is that the Conway convention for a positive crossing is generally not used in the setting of skein relations.

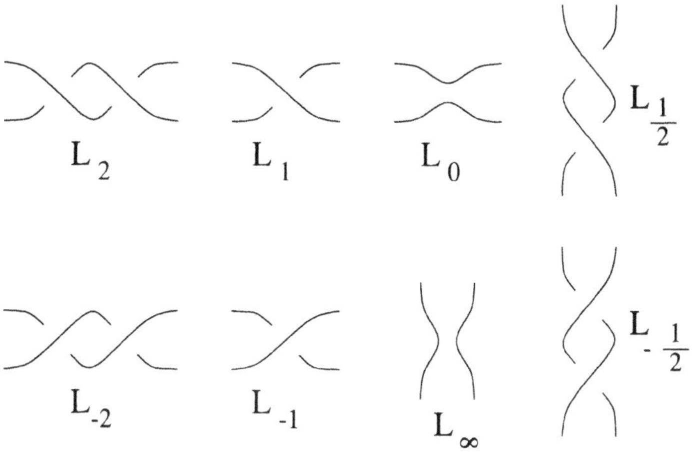

Fig. 3.5

If we rotate the figure from the relation (i) we obtain:
(i') $b_{-\frac{1}{2}}L_2 + b_{-1}L_1 + b_{\infty}L_0 + b_0L_{\infty} + b_1L_{-1} + b_2L_{-\frac{1}{2}} = 0$
One can use (i) and (i') to eliminate $L_{-\frac{1}{2}}$ and to get the relation:
$(b_2^2 - b^2_{-\frac{1}{2}})L_2 + (b_1b_2 - b_{-1}b_{-\frac{1}{2}})L_1 + ((b_0b_2 - b_{\infty}b_{-\frac{1}{2}})L_0 + (b_{-1}b_2 - b_1b_{-\frac{1}{2}})L_{-1} + (b_{\infty}b_2 - b_0b_{-\frac{1}{2}})L_{\infty} = 0.$
Thus, either we deal with the shorter relation (essentially the one in the fourth skein module described before) or all coefficients are equal to 0 and therefore (assuming that there are no zero divisors in R) $b_2 = \varepsilon b_{-\frac{1}{2}}$, $b_1 = \varepsilon b_{-1}$, and $b_0 = \varepsilon b_{\infty}$. Similarly, we would get: $b'_2 = \varepsilon b'_{-\frac{1}{2}}$, $b'_1 = \varepsilon b'_{-1}$, and $b'_0 = \varepsilon b'_{\infty}$, where $\varepsilon = \pm 1$. Assume, for simplicity, that $\varepsilon = 1$. Further relations among coefficients follow from the computation of the Hopf link component using the amphicheirality of the unoriented Hopf link. Namely, by comparing diagrams in Figure 3.6 and their mirror images we get

$$L\#H = -b_2^{-1}(b_1(a + a^{-1}) + a^{-2}b_2 + b_0(1 + T_1))L$$

$$L\#H = -b'^{-1}_2(b'_1(a + a^{-1}) + a^2b'_2 + b'_0(1 + T_1))L.$$

Possibly, the above equalities give the only other relations among coefficients (in the case of S^3). I would present below the simpler question (assuming $a = 1, b_x = b'_x$ and writing t^n for T_n).

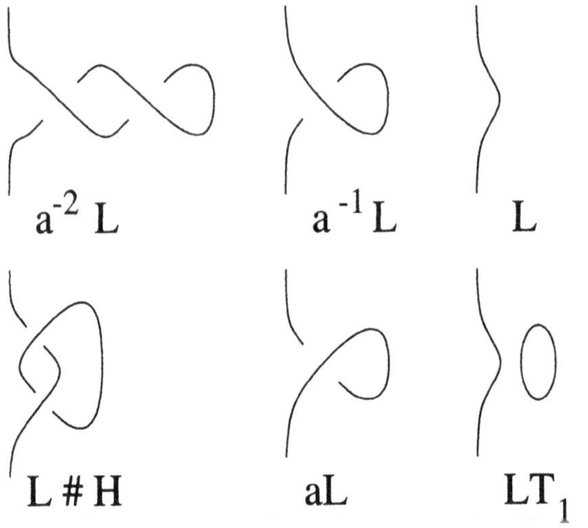

Fig. 3.6

Question 3.4. *Is there a polynomial invariant of unoriented links in* S^3, $P_{\frac{5}{2}}(L) \in \mathbb{Z}[b_0, b_1, t]$, *which satisfies the following conditions?*

(i) *Initial conditions:* $P_{\frac{5}{2}}(T_n) = t^n$, *where* T_n *is a trivial link of* n *components.*

(ii) *Skein relations*

$$P_{\frac{5}{2}}(L_2)+b_1P_{\frac{5}{2}}(L_1)+b_0P_{\frac{5}{2}}(L_0)+b_0P_{\frac{5}{2}}(L_\infty)+b_1P_{\frac{5}{2}}(L_{-1})+P_{\frac{5}{2}}(L_{-\frac{1}{2}}) = 0.$$

$$P_{\frac{5}{2}}(L_{-2})+b_1P_{\frac{5}{2}}(L_{-1})+b_0P_{\frac{5}{2}}(L_0)+b_0P_{\frac{5}{2}}(L_\infty)+b_1P_{\frac{5}{2}}(L_1)+P_{\frac{5}{2}}(L_{\frac{1}{2}}) = 0.$$

Notice that by taking the difference of our skein relations one gets the interesting identity:

$$P_{\frac{5}{2}}(L_2) - P_{\frac{5}{2}}(L_{-2}) = P_{\frac{5}{2}}(L_{\frac{1}{2}}) - P_{\frac{5}{2}}(L_{-\frac{1}{2}}).$$

Nobody has yet studied the skein module $\mathcal{S}_{\frac{5}{2}}(M;R)$ seriously so everything that you can find will be a new research, even a table of the polynomial $P_{\frac{5}{2}}(L)$ for small links, L.

I wish you luck!

4. Added in proof – the Montesinos-Nakanishi 3-move conjecture

A preliminary calculation performed by my student Mietek Dąbkowski (February 21, 2002) shows that the Montesinos-Nakanishi 3-move conjecture does not hold for the Chen link (Fig.1.14). Below is the text of the abstract we have sent for the Knots in Montreal conference organized by Steve Boyer and Adam Sikora in April 2002.

Authors: Mieczysław Dąbkowski, Józef H. Przytycki (GWU)
Title: Obstructions to the Montesinos-Nakanishi 3-move conjecture.
Yasutaka Nakanishi asked in 1981 whether a 3-move is an unknotting operation. This question is called, in the Kirby's problem list, *the Montesinos-Nakanishi Conjecture.* Various partial results have been obtained by Q.Chen, Y.Nakanishi, J.Przytycki and T.Tsukamoto. Nakanishi and Chen presented examples which they couldn't reduce (the Borromean rings and the closure of the square of the center of the fifth braid group, $\bar{\gamma}$, respectively). The only tool, to analyze 3-move equivalence, till 1999, was the Fox 3-coloring (the number of Fox 3-colorings is unchanged by a 3-move). It allowed to distinguish different trivial links but didn't separate Nakanishi and Chen examples from trivial links. The group of 3-colorings of a link L corresponds to the first homology group with Z_3 coefficients of the double branched cover of a link L, $M_L^{(2)}$, i.e.

$$Tri(L) = H_1(M_L^{(2)}, Z_3) \oplus Z_3$$

We find more delicate invariants of 3-moves using homotopy in place homology and we consider the fundamental group of $M_L^{(2)}$.

We define an nth Burnside group of a link as the quotient of the fundamental group of the double branched cover of the link divided by all relations of the form $a^n = 1$. For $n = 2, 3, 4, 6$ the quotient group is finite[32].

---

[32] Burnside groups of links are instances of groups of finite exponents. Our method of analysis of tangle moves rely on the well developed theory of classical Burnside groups and the associated Lie rings. A group G is of a finite exponent if there is a finite integer n such that $g^n = e$ for all $g \in G$. If, in addition, there is no positive integer $m < n$ such that $g^m = e$ for all $g \in G$, then we say that G has an exponent n. Groups of finite exponents were considered for the first time by Burnside in 1902 (Bur). In particular, Burnside himself was interested in the case when G is a finitely generated group of a fixed exponent. He asked the question, known as the Burnside Problem, whether there exist infinite and finitely generated groups G of finite exponents.

The third Burnside group of a link is unchanged by 3-moves[33].

In the proof we use the "core" presentation of the group from the diagram; that is arcs are generators and each crossing gives a relation $c = ab^{-1}a$ where a corresponds to the overcrossing and b and c to undercrossings.

The Montesinos-Nakanishi 3-move conjecture does not hold for Chen's example $\hat{\gamma}$.

To show that $\hat{\gamma}$ has different third Burnside group than any trivial link it suffices to show that the following element, P, of the Burnside free group $B(4,3) = \{x, y, z, t : (a)^3\}$ is nontrivial: $P = uwtu^{-1}w^{-1}t^{-1}$ where
$u = xy^{-1}zt^{-1}$ and $w = x^{-1}yz^{-1}t$.

With the help of GAP it has been achieved!! (Feb. 21, 2002).

We have confirmed our calculation using also computer algebra system Magnus.

References

J. W. Alexander, Letter to Oswald Veblen, 1919, Papers of Oswald Veblen, 1881-1960 (bulk 1920-1960), Archival Manuscript Material (Collection), Library of Congress. (I would like to thank Jim Hoste for providing me with a copy of the letter).
To have a taste of Alexander letter, here is the quotation from the beginning of the interesting part: "When looking over Tait on knots among other things, He really doesn't get very far. He merely writes down all the plane projections of knots with a limited number of crossings, tries out a few transformations that he happen to think of and assumes without

Let $F_r = \langle x_1, x_2, \ldots, x_r \mid - \rangle$ be the free group of rank r and let $B(r,n) = F_r/N$, where N is the normal subgroup of F_r generated by $\{g^n \mid g \in F_r\}$. The group $B(r,n)$ is known as the rth generator Burnside group of exponent n. In this notation, Burnside's question can be rephrased as follows. For what values of r and n is the Burnside group $B(r,n)$ finite? $B(1,n)$ is a cyclic group Z_n. $B(r,2) = Z_2^n$. Burnside proved that $B(r,3)$ is finite for all r and that $B(2,4)$ is finite. In 1940 Sanov proved that $B(r,4)$ is finite for all r, and in 1958 M.Hall proved that $B(r,6)$ finite for all r. However, it was proved by Novikov and Adjan in 1968 that $B(r,n)$ is infinite whenever $r > 1$, n is odd and $n \geq 4381$ (this result was later improved by Adjan, who showed that $B(r,n)$ is infinite if $r > 1$, n is odd and $n \geq 665$). Sergei Ivanov proved that for $k \geq 48$ the group $B(2,2^k)$ is infinite. Lysënok found that $B(2,2^k)$ is infinite for $k \geq 13$. It is still an open problem though whether, for example, $B(2,5)$, $B(2,7)$ or $B(2,8)$ are infinite or finite (VL; D-P-3).

[33] pth Burnside group is preserved by $\frac{ps}{q}$-moves. This fact allows us to disprove Conjecture 2.2 (D-P-3).

proof that if he is unable to reduce one knot to another with a reasonable number of tries, the two are distinct. His invariant, the generalization of the Gaussian invariant ... for links is an invariant merely of the particular projection of the knot that you are dealing with, - the very thing I kept running up against in trying to get an integral that would apply. The same is true of his "Beknottednes".
Here is a genuine and rather jolly invariant: take a plane projection of the knot and color alternate regions light blue (or if you prefer, baby pink). Walk all the way around the knot and ..."

S. Akbulut, H. King, Rational structures on 3-manifolds, *Pacific J. Math.*, 150, 1991, 201-214.

J. W. Alexander, Topological invariants of knots and links, *Trans. Amer. Math. Soc.* 30, 1928, 275-306.

N. Askitas, A note on 4-equivalence, *J. Knot Theory Ramifications*, 8(3), 1999, 261–263.

S. Betley, J. H. Przytycki, T.Żukowski, Hyperbolic structures on Dehn fillings of some punctured-torus bundles over S^1, *Kobe J. Math.*, 3(2), 1986, 117-147.

F. Bonahon, L. Siebenmann, Geometric splittings of classical knots and the algebraic knots of Conway, to appear(?) in L.M.S. Lecture Notes Series, 75.

D. Bullock, A finite set of generators for the Kauffman bracket skein algebra, *Math.Z.*, 231(1), 1999, 91-101.

D. Bullock, Rings of $Sl_2(C)$-characters and the Kauffman bracket skein module, *Comment. Math. Helv.* 72, 1997, 521-542.
http://front.math.ucdavis.edu/q-alg/9604014

D. Bullock, The $(2,\infty)$–skein module of the complement of a $(2, 2p+1)$ torus knot, *Journal of Knot Theory*, 4(4) 1995, 619-632.

D. Bullock, C. Frohman, J. Kania-Bartoszyńska, Understanding the Kauffman bracket skein module, *J. Knot Theory Ramifications* 8(3), 1999, 265–277.
http://front.math.ucdavis.edu/q-alg/9604013

D. Bullock, W. Lofaro, The Kauffman bracket skein module of a twist knot exterior, e-print, Feb. 2004,
http://front.math.ucdavis.edu/math.QA/0402102

D. Bullock, J. H. Przytycki, Multiplicative structure of Kauffman bracket skein module quantizations, *Proc. Amer. Math. Soc.*, 128(3), 2000, 923–931.
http://front.math.ucdavis.edu/math.QA/9902117

W. Burnside, On an Unsettled Question in the Theory of Discontinuous Groups, *Quart. J. Pure Appl. Math.* 33, 1902, 230-238.

Q. Chen, The 3-move conjecture for 5-braids, Knots in Hellas' 98; The Proceedings of the International Conference on Knot Theory and its Ramifications; Volume 1. In the Series on Knots and Everything, Vol. 24, September 2000, pp. 36-47.

J. H. Conway, An enumeration of knots and links, *Computational problems in abstract algebra* (ed. J.Leech), Pergamon Press (1969) 329 - 358.

H. S . M. Coxeter, Factor groups of the braid group, Proc. Fourth Canadian Math. Congress, Banff, 1957, 95-122.

R. H. Crowell, Knots and Wheels, National Council of Teachers of Mathematics (N.C.T.M.) Yearbook, 1961.

R. H. Crowell, R. H. Fox, *An introduction to knot theory*, Ginn and Co., 1963.

C. L. Day, *Quipus and Witches' Knots, With a Translation and Analysis of "Oribasius De Laqueis"*, The University of Kansas Press, Lawrence 1967.

M. K. Dąbkowski, M. Ishiwata, J. H. Przytycki, $(2,2)$-move equivalence for 3-braids and 3-bridge links, in preparation.

M. K. Dąbkowski, J. H. Przytycki, Burnside obstructions to the Montesinos-Nakanishi 3-move conjecture, *Geometry and Topology*, June, 2002, 335-360.
http://front.math.ucdavis.edu/math.GT/0205040

M. K. Dąbkowski, J. H. Przytycki, Unexpected connection between Burnside groups and Knot Theory, preprint 2003.
http://front.math.ucdavis.edu/math.GT/0309140

M. K. Dąbkowski, J. H. Przytycki, Burnside groups in knot theory, preprint 2004.

Albrecht Durer Master Printmaker, Department of Prints and Drawings, Boston Museum of Fine Arts, Hacker Art Books, New York, 1988.

Albrecht Durer, Diary of his Journey to the Netherlands, 1520-1521, translated by P.Trou, edited by J.A.Goris and G.Marlier, London, Lund Humphries, 1971.

J. Dymara, T. Januszkiewicz, J. H. Przytycki, Symplectic structure on Colorings, Lagrangian tangles and Tits buildings, preprint, May 2001.

M. Epple, Geometric aspects in the development of knot theory, History of topology (ed. I. M. James) , 301–357, North-Holland, Amsterdam, 1999.

M. Epple, Die Entstehung der Knotentheorie (Braunschweig, 1999).

R. Fenn, C. Rourke, Racks and links in codimension two, *Journal of Knot Theory and its Ramifications*, 1(4) 1992, 343-406.

R. H. Fox, A quick trip through knot theory, *In: Top. 3-manifolds*, Proc. 1961 Top.Inst.Univ. Georgia (ed. M. K. Fort, jr), 120-167. Englewood Cliffs. N.J.: Princeton-Hall, 1962.

R. H. Fox, Metacyclic invariants of knots and links, Canadian J. Math., XXII(2) 1970, 193-201.

C. F. Gauss, Notebooks, Old library in Göttingen.

C. McA. Gordon, Some aspects of classical knot theory, *In: Knot theory*, L.N.M. 685, 1978, 1-60.

C. McA. Gordon, R. A. Litherland, On the signature of a link, *Invent. Math.*, 47(1978), 53-69.

F. Harary, The knots and links of Albrecht Dürer, *Atti Accad. Pontaniana (N.S.)* 34, 1985, 97-106.

T. Harikae, Y. Uchida, Irregular dihedral branched coverings of knots, in *Topics in knot theory*, N.A.T.O. A.S.I. series C, 399, (ed. M.Bozhüyük) Kluwer Academic Publisher (1993), 269-276.

P. Heegaard, Forstudier til en Topologisk Teori for de algebraiske Fladers Sammenhæng, København, 1898, Filosofiske Doktorgrad; French translation: Sur l'Analysis situs, *Soc. Math. France Bull.*, 44 (1916), 161-242. English translation, by Agata Przybyszewska, of the topological part of Poul Heegaard Dissertation is in: J. H. Przytycki, Knot theory from Vandermonde to Jones, Preprint 43, Odense University 1993.

M. C. Heath, "Early Helladic Clay Sealings from the House of the Tiles at Lerna," *Hesperia* 27(1958) 81-120.

R. Higgins, Minoan and Mycean art, Thames and Hudson, 1997 (Revised edition).

J. Hoste, M. Kidwell, Dichromatic link invariants, *Trans. Amer. Math. Soc.*, 321(1), 1990, 197-229; see also the preliminary version of this paper: "Invariants of colored links", preprint, March 1987.

J. Hoste, J. H. Przytycki, Homotopy skein modules of oriented 3-manifolds, *Math. Proc. Cambridge Phil. Soc.*, 1990, 108, 475-488.

J. Hoste, J. H. Przytycki. A survey of skein modules of 3-manifolds. in Knots 90, Proceedings of the International Conference on Knot Theory and Related Topics, Osaka (Japan), August 15-19, 1990), Editor A. Kawauchi,Walter de Gruyter 1992, 363-379.

An interview with William Thurston, Warsaw, August 1983. A. Mednykh could not come to the International Congress of Mathematicians in Warsaw (August 1983) so he sent me a list of questions to W. Thurston. I met with Thurston and asked him the questions to which he kindly replied in length. Then I wrote his answers (sometimes with my comments). In particular, Thurston conjectured that the smallest volume hyperbolic manifold is a surgery on the figure eight knot. I commented that possibly a manifold we discussed in (B-P-Z) has the smallest volume. I gave expositions about my conjecture at talks at Zaragoza University, May 1984, and at Durham conference in July of 1984 ("The smallest volume hyperbolic 3-manifolds"). Later extensive tabulations by J. Weeks (W) and V. Matveev and A. Fomenko (M-F) gave credibility to the conjecture (which is still open).

F. Jaeger, J. H. Przytycki, A non-commutative version of the Goeritz matrix of a link, in preparation.

D. Joyce, A classifying invariant of knots: the knot quandle, *Jour. Pure Appl. Alg.*, 23, 1982, 37-65.

L. Kauffman, S. Lambropoulou, On the classification of rational tangles, http://front.math.ucdavis.edu/math.GT/0311499

A. Kawauchi, *A survey of Knot Theory*, Birkhäusen Verlag, Basel-Boston-Berlin, 1996.

R. Kirby, Problems in low-dimensional topology; Geometric Topology (Proceedings of the Georgia International Topology Conference, 1993), Studies in Advanced Mathematics, Volume 2 part 2., Ed. W.Kazez, AMS/IP, 1997, 35-473.

S. Kinoshita, *On Wendt's theorem of knots, I*, Osaka Math. J. 9(1), 1957, 61-66.

The complete woodcuts of Albrecht Dürer, Edited by Dr. Willi Kurth, Dover publications, Inc. New York, 1983 (translated from 1927 German edition).

T. Le, The colored Jones polynomial and the A-polynomial of 2-bridge knots, presentation at the *Mini-Conference in Logic and Topology*, GWU, April 9, 2004.

E. MacCurdy, *The Notebooks of Leonardo*, New York: Reynal and Hitchcook, 1938, vol.2, p.588.

V. S. Matveev, A. T. Fomenko, Isoenergetic surfaces of Hamiltonian systems, the enumeration of three-dimensional manifolds in order of growth of their complexity, and the calculation of the volumes of closed hyperbolic manifolds (Russian), *Uspekhi Mat. Nauk* 43 (1988), no. 1(259), 5–22, 247; translation in *Russian Math. Surveys* 43 (1988), no. 1, 3–24.

A. D. Mednykh, A. Vesnin, On the Fibonacci groups, the Turk's head links and hyperbolic 3-manifolds, Groups—Korea '94 (Pusan), 231–239, de Gruyter, Berlin, 1995.

A. D. Mednykh, A. Vesnin, Covering properties of small volume hyperbolic 3-manifolds. J. Knot Theory Ramifications 7(3), 1998, 381–392.

J. M. Montesinos, Variedades de Seifert que son cubiertas ciclicas ramificadas de dos hojas, *Bol. Soc. Mat. Mexicana (2)*, 18, 1973, 1–32.

J. M. Montesinos, Lectures on 3-fold simple coverings and 3-manifolds, Combinatorial methods in topology and algebraic geome-

a more delicate move, called $(2,2)$-move (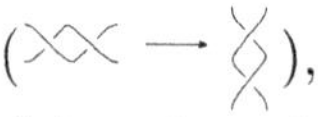), such that the 5-move is a combination of a $(2,2)$-move and its mirror image $(-2,-2)$-move (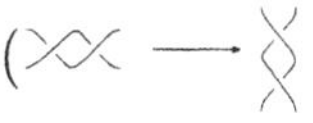), as it is illustrated in Fig.1.25 (H-U; P-3).

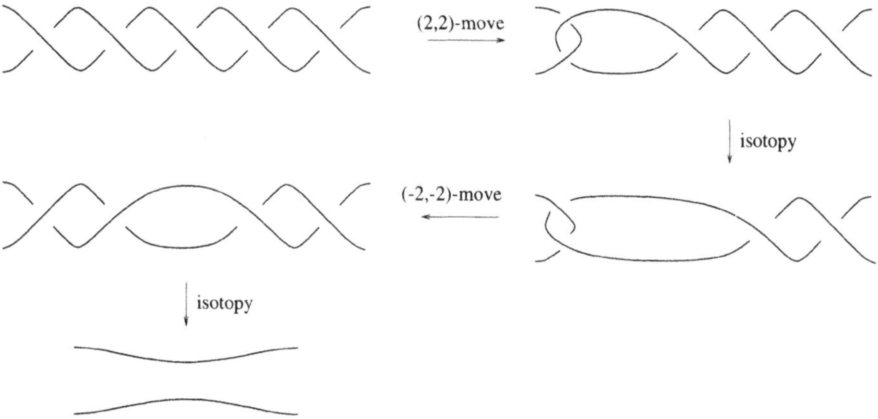

Fig. 1.25

Conjecture 1.17 (Harikae, Nakanishi, Uchida, 1992).
Every link is $(2,2)$*-move equivalent to a trivial link.*

As in the case of 3-moves, an elementary induction shows that the conjecture holds for 2-algebraic links. It is also known that the conjecture holds for all links up to 8 crossings. The key element of the argument in the proof is the observation (going back to Conway (Co)) that any link with up to 8 crossings (different from 8_{18}; see footnote before Fig.1.31) is 2-algebraic. The reduction of the 8_{18} knot to the trivial link of two components by my students, Jarek Buczyński and Mike Veve, is illustrated in Fig.1.26.

tization (Proceedings of the joint AMS-IMS-SIAM conference on Quantization, Mount Holyoke College, 1996); Ed. L.A.Coburn, M.A.Rieffel, AMS 1998, 135-144.

J. H. Przytycki, Homotopy and q-homotopy skein modules of 3-manifolds: an example in Algebra Situs, In *Knots, Braids, and Mapping Class Groups – Papers Dedicated to Joan S.Birman*, Ed.: J.Gilman, W.W.Menasco, X-S.Lin, AMS/IP Studies in Advanced Mathematics, Volume 24, January 2002; (http://front.math.ucdavis.edu/math.GT/0402304).

J. H. Przytycki, The interrelation of the Development of Mathematical Topology in Japan, Poland and USA: Notes to the early history of the Knot Theory in Japan, *Annals of the Institute for Comparative Studies of Culture*, Vol. 63, 2002, 61-86. http://front.math.ucdavis.edu/math.HO/0108072

J. H. Przytycki, Czy coś zostało dla nas? – 10 elementarych zawęźlonych problemów (Is there anything left for us? – 10 elementary knotted problems), *Delta* 5, May, 2002, p.V-VIII.
In this paper, for "Little Delta", the problem of finding unknotting moves was formulated, in a free translation, as follows: *Make a magnificent multi-move minimizing multitude of multi-knots.*

J. H. Przytycki, Classical roots of Knot Theory, Chaos, Solitons and Fractals, Vol. 9 (No. 4-5), 1998, 531-545.

J. H. Przytycki, A. S. Sikora, On Skein Algebras and $Sl_2(C)$-Character Varieties, *Topology*, 39(1), 2000, 115-148. http://front.math.ucdavis.edu/q-alg/9705011

J. H. Przytycki, D. S. Silver, S. G. Williams, 3-manifolds, tangles and persistent invariants, *Math. Proc. Cambridge Phil. Soc.*, to appear. http://front.math.ucdavis.edu/math.GT/0405465

J. H. Przytycki, T. Tsukamoto, The fourth skein module and the Montesinos-Nakanishi conjecture for 3-algebraic links, *J. Knot Theory Ramifications*, 10(7), November 2001, 959-982. http://front.math.ucdavis.edu/math.GT/0010282

A. S. Sikora, PSL_n-character varieties as spaces of graphs, *Trans. Amer. Math. Soc.*, 353, 2001, 2773-2804. http://front.math.ucdavis.edu/math.RT/9806016

C. Squier, The Burau representation is unitary, *Proc. Amer. Math. Soc.*, 90(2), 1984, 199-202.

T. Tsukamoto, Ph. D. Thesis: The fourth Skein module for 4-algebraic links George Washington University, 2000.

V. G. Turaev, The Conway and Kauffman modules of the solid torus, *Zap. Nauchn. Sem. Lomi* 167 (1988), 79-89. English translation: *J. Soviet Math.*, 52, 1990, 2799-2805.

V. G. Turaev, Skein quantization of Poisson algebras of loops on surfaces, *Ann. Scient. Éc. Norm. Sup.*, 4(24), 1991, 635-704.

Y. Uchida, Proc. of Knots 96 ed. S. Suzuki 1997, 109-113 *Knots 96*, Proceedings of the Fifth International Research Institute of MSJ, edited by Shin'ichi Suzuki, 1997 World Scientific Publishing Co., 109-113.

Giorgio Vasari, *Lives of the Most Eminent Italian Architects, Painters, and Sculptors'* (published first in 1550 in Florence). The Lives of the Artists (Oxford World's Classics) by Giorgio Vasari, Julia Conaway Bondanella (Translator), Peter Bondanella (Translator), 1998.

M. Vaughan-Lee, *The restricted Burnside problem*; Second edition. London Mathematical Society Monographs. New Series, 8. The Clarendon Press, Oxford University Press, New York, 1993. xiv+256 pp.

B. Wajnryb, Markov classes in certain finite symplectic representations of braid groups, in: Braids (Santa Cruz, CA, 1986), Contemp. Math., 78, Amer. Math. Soc., Providence, RI, 1988, 687–695.

B. Wajnryb, A braid-like presentation of $Sp(n, p)$, *Israel J. Math.*, 76, 1991, 265-288.

J. Weeks, Hyperbolic structures on 3-manifolds, Princeton Univ. Ph.D.Thesis, 1985.

M. H. Wiencke, "Further Seals and Sealings from Lerna," *Hesperia* 38(1969) 500-521.

Address for Offprints:

Józef H. Przytycki
Department of Mathematics
George Washington University
email: przytyck@gwu.edu

ON SPIN AND COMPLEX SPIN BORROMEAN SURGERIES

FLORIAN DELOUP

ABSTRACT. The Borromean surgery, which was defined by S. Matveev and is the basic building block of a theory of finite-type invariants of 3-manifolds, admits two refinements, spin and complex spin. We define two invariants of spin and complex spin 3-manifolds and we show that two spin 3-manifolds are related by a finite sequence of spin Borromean surgeries if and only if they have the same invariant. We show that the corresponding statement for complex spin fails to hold and suggests an alternative definition.

1. INTRODUCTION

The notion of Borromean surgery was introduced by S. Matveev [16] about fifteen years ago and is the elementary move of the Goussarov-Habiro finite-type invariants for 3-manifolds [6].

Spin structures and complex spin structures (the latter also denoted by spin^c) are classical additional structures of homotopical type which a manifold may be endowed with. Every oriented closed 3-manifold M admit spin (resp. spin^c) structures [12] [7]. The set of spin (resp. spin^c) structures on M is in bijective correspondence with $H^1(M;\mathbb{Z}/2)$ (resp. with $H^2(M;\mathbb{Z})$).

In [2], we introduced an invariant τ of oriented closed 3-manifolds. While this invariant shares many properties with quantum invariants, it is considerably simpler because of its Abelian nature. It is implicit in [3] that this invariant τ induces naturally an invariant τ^{spin} (resp. τ^{spin^c}) of oriented closed spin (resp. spin^c) 3-manifolds.

The Goussarov-Habiro theory was recently refined by G. Massuyeau to spin 3-manifolds [15] and by G. Massuyeau and the author to spin^c 3-manifolds [5]. In particular, there is a well-defined notion of spin (resp. spin^c) Borromean surgery. Consider the equivalence relation generated by spin (resp. spin^c) Borromean surgeries. It was proved that this equivalence relation is characterized by the isomorphism class of a certain quadratic function canonically associated to the spin (resp. spin^c structure). In the terminology of finite-type invariants, the isomorphism class of the quadratic function determines the degree 0 invariants (for the spin and complex spin Goussarov-Habiro theories).

It is the purpose of this article to define explicitly τ^{spin} and τ^{spin^c} and investigate to which extent they determine the degree 0 invariants in the respective theories. First, we show that τ^{spin} determines and is determined by the set of all the degree 0 invariants of the spin Goussarov-Habiro theory (this generalizes the main result of [4]). The case of complex spin refinement is a little more surprising: we show that the analogous statement for complex spin 3-manifolds does not hold. We suggest an alternative definition.

2. AN INVARIANT FOR SPIN 3-MANIFOLDS

In this section, we describe an invariant for spin 3-manifolds (§§2.1, 2.2 and 2.4) and prove (Theorems 1 and 2) that it characterizes the spin Borromean equivalence,

J.M. Bryden (ed.), Advances in Topological Quantum Field Theory, 127–133.

which we recall in §2.3.

2.1. Let M be a closed oriented 3-manifold. Recall that for every 3-manifold M, the torsion subgroup of its first homology group carries the linking pairing $\lambda_M : TH_1(M) \times TH_1(M) \to \mathbb{Q}/\mathbb{Z}$ which is a nondegenerate symmetric bilinear pairing. Denote by $\mathrm{Spin}(M)$ the set of spin structures of M. It is a classical fact (see e.g., [13]) that to $\sigma \in \mathrm{Spin}(M)$, one can associate a quadratic form $\phi_\sigma : TH_1(M) \to \mathbb{Q}/\mathbb{Z}$ over the linking pairing λ_M. In other words, the map ϕ_σ satisfies the quadratic equality

$$\phi_\sigma(x+y) - \phi_\sigma(x) - \phi_\sigma(y) = \lambda_M(x,y) \quad \text{for all } x,\ y \in TH_1(M). \tag{2.1}$$

and the homogeneous condition

$$\phi_\sigma(nx) = n^2\phi_\sigma(x) \quad \text{for all} \quad x, y \in TH_1(M) \text{ and } n \in \mathbb{Z}. \tag{2.2}$$

The set of all quadratic forms – maps satisfying (2.1) and (2.2) – will be denoted $\mathrm{Quad}^0(\lambda_M)$. The map

$$\phi : \mathrm{Spin}(M) \to \mathrm{Quad}^0(\lambda_M),\ \sigma \mapsto \phi_\sigma$$

is affine over the homomorphism $H^1(M;\mathbb{Z}_2) \to \mathrm{Hom}(TH_1(M);\mathbb{Z}_2)$. In particular, ϕ is bijective if M is a rational homology 3-sphere.

2.2. We now extract relevant elements of the construction of the invariant τ of 3-manifolds in [2]. Let $\lambda : T \times T \to \mathbb{Q}/\mathbb{Z}$ be a symmetric bilinear pairing on a finite Abelian group T. Let b be a nonnegative integer. Let $q : G \to \mathbb{Q}/\mathbb{Z}$ be a quadratic form on a finite Abelian group G. Let $\mathfrak{M}$ (resp. $\mathfrak{Q}^0$) be the monoid of isomorphism classes of symmetric bilinear pairings (resp. quadratic forms) on finite Abelian groups. Let also $\mathcal{M}$, respectively $\mathcal{M}^+$, be the monoid of isomorphism classes of nondegenerate symmetric bilinear pairings on finitely generated free Abelian groups, respectively the monoid of isomorphism classes of pairs (a nondegenerate symmetric bilinear pairing λ on a finitely generated free Abelian group H, an element $h \in H$ (considered mod $2H$) such that $f(x,x) - f(h,x) \in 2\mathbb{Z}$ for all $x \in H$). The operation on each of these monoids is induced by the orthogonal sum. From the discriminant construction (see e.g., [3]), there are well-defined homomorphisms $\partial : \mathcal{M} \to \mathfrak{M}$, $\partial : \mathcal{M}^+ \to \mathfrak{Q}^0$ such that the diagram

$$\begin{array}{ccc} \mathcal{M}^+ & \xrightarrow{\text{forget}} & \mathcal{M} \\ \downarrow{\scriptstyle \partial} & & \downarrow{\scriptstyle \partial} \\ \mathfrak{Q}^0 & \longrightarrow & \mathfrak{M} \end{array}$$

(where the lower horizontal arrow is induced by taking the associated bilinear pairing) is commutative. Let $\mathfrak{M}^+ = \mathfrak{M} \times \mathbb{N}$. Note that it is a monoid for the usual direct product operation. Now define a map $\langle \cdot,\cdot \rangle : \mathfrak{M}^+ \times \mathfrak{Q}^0 \to \mathbb{C}$ by

$$\langle(\lambda,n),q\rangle = \gamma(G,q)^{-\mathrm{sign}(\tilde{\lambda})}\gamma(G\otimes H, q\otimes\tilde{\lambda}) \cdot |G|^{-\frac{n}{2}} \tag{2.3}$$

where $\tilde{\lambda} \in \mathcal{M}$ is any pairing satisfying $\partial\tilde{\lambda} = \lambda$; $\mathrm{sign}(\tilde{\lambda})$ denotes the signature of $\tilde{\lambda}\otimes\mathbb{R}$. By [3, §1.4], this map is well-defined and is a bilinear pairing. (With the notations of [2], setting $\lambda = \lambda_M, b = b_1(M)$, where M is a closed oriented 3-manifold, we have $\tau(M;G,q) = \langle(\lambda_M, b_1(M)), q\rangle$.)

The main result of [4] can be reformulated as follows.

Proposition 1. *The bilinear map, $\langle \cdot,\cdot \rangle : \mathfrak{M}^+ \times \mathfrak{Q}^0 \to \mathbb{C}$ is left nondegenerate.*

Since the following particular point will be of importance later, let us rephrase precisely Proposition 1: the equality $\langle(\lambda, n), q\rangle = \langle(\lambda', n'), q'\rangle$ for all $q \in \mathfrak{Q}^0$ implies $([\lambda], n) = ([\lambda'], n')$ in $\mathfrak{M}^+$. In particular, as bilinear pairings, λ and λ' are isomorphic. Caveat: it does not imply that they are equal.

In this note, we shall prove

Theorem 1. *The bilinear map* $\tau = \langle\cdot,\cdot\rangle : \mathfrak{M}^+ \times \mathfrak{Q}^0 \to \mathbb{C}$ *is nondegenerate.*

An equivalent topological version of Theorem 1, Theorem 2 below, will be proved later at the end of this section. Before we state it, we shall briefly review, for the convenience of the reader, Borromean surgery equivalence (in fact, a surgery generating the same equivalence relation, due to H. Murakami and Y. Nakanishi [17]).

2.3. Suppose that M is presented as a surgery link $L = L_1 \cup \cdots \cup L_n$ in the 3-sphere S^3. A Δ-move is a modification of the surgery involving at most three components of L, see Fig. 2.1. Two closed oriented 3-manifolds M and M' presented by surgery on L and L' are diffeomorphic if and only if L and L' can be joined by a finite sequence of Kirby moves ([12]). The Borromean equivalence is the equivalence relation generated by the Kirby moves and the additional Δ-move. Suppose now that M is equipped with a spin structure σ. It is known that (M, σ) can be presented by surgery on a framed link $L = L_1 \cup \cdots \cup L_n$ in S^3 with a residue s_j mod 2 attached to each component L_j. The spin structure is presented as an element $s = (s_1, \ldots, s_n) \in (\mathbb{Z}_2)^n$. Two closed oriented spin 3-manifolds presented by surgery on (L, s) and (L', s') are spin- diffeomorphic if and only if (L, s) and (L', s') are related by a finite sequence of spin Kirby moves (see e.g., [9, §5.7]). The spin Δ-move is a move on pairs (L, s) (a framed link $L = L_1 \cup \cdots \cup L_n$, equipped with an element $s \in (\mathbb{Z}_2)^n$): it acts as the usual Δ-move on L and trivially on s, see Fig. 2.1. The *spin Borromean equivalence* is now defined as the equivalence relation on pairs (L, s) generated by the spin Kirby moves and the spin Δ-move.

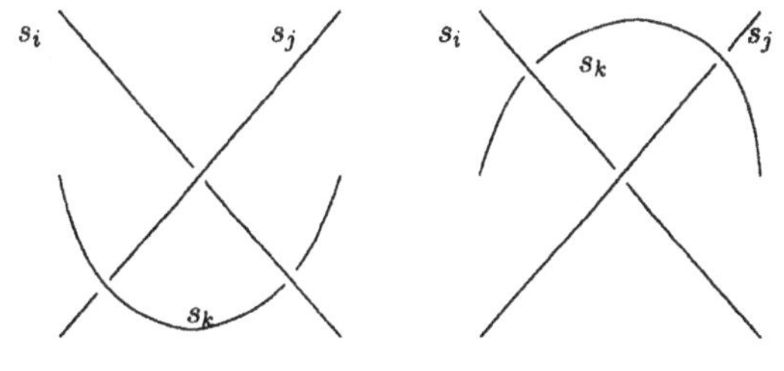

FIGURE 2.1. The Δ-move.

2.4. Paragraph §2.1 suggests that one can define a invariant τ^{spin} of closed oriented spin 3-manifolds as follows. Let $\sigma \in \mathrm{Spin}(M)$ and $(\lambda, n) \in \mathfrak{M}^+$. Set

$$\tau^{\mathrm{spin}}(M, \sigma; \lambda, n) = \langle(\lambda, n); \phi_\sigma\rangle \in \mathbb{C}. \tag{2.4}$$

For simplicity, we denote by $\tau^{\mathrm{spin}}(M, \sigma) \in \mathrm{Hom}(\mathfrak{M}^+, \mathbb{C})$ the map $(\lambda, n) \mapsto \langle(\lambda, n); \phi_\sigma\rangle$.

Theorem 2. *Let* (M, σ) *and* (M', σ') *two closed oriented spin 3-manifolds with same first Betti number. They are spin Borromean equivalent if and only if* $\tau^{\mathrm{spin}}(M, \sigma) = \tau^{\mathrm{spin}}(M', \sigma')$.

Proof. By the main Theorem of [15], the spin Borromean equivalence class of (M, σ) is determined by the first Betti number and the isomorphism class of ϕ_σ. Necessity is therefore obvious. To prove the converse, we show that $\tau^{\text{spin}}(M, \sigma)$ determines the isomorphism class of $\phi_\sigma : \mathrm{T}H_1(M) \to \mathbb{Q}/\mathbb{Z}$. First, it follows from (2.3) that

$$|\langle(\lambda, 0); \phi_\sigma\rangle| = |\mathrm{T}H_1(M) \otimes T|^{\frac{1}{2}}$$

for any symmetric bilinear pairing $\lambda : T \times T \to \mathbb{Q}/\mathbb{Z}$. In particular, by allowing T to vary, we recover all p-components of $\mathrm{T}H_1(M)$ and thus $\mathrm{T}H_1(M)$ itself. Secondly, let k be greater than the exponent of $\mathrm{T}H_1(M)$ (the smallest integer N such that $Nx = 0$ for all $x \in \mathrm{T}H_1(M)$). Choose the pairing $\lambda(x, y) = \frac{xy}{2k}$ mod 1, for $x, y \in \mathbb{Z}/2k$. We may then choose $\tilde{\lambda}(x, y) = 2kxy$, for $x, y \in \mathbb{Z}$. In particular, $\mathrm{sign}(\tilde{\lambda}) = 1$ and $\phi_\sigma \otimes \lambda = 0$. From (2.3), we see that

$$\langle(\lambda, n); \phi_\sigma\rangle = \gamma(\mathrm{T}H_1(M), \phi_\sigma)^{-1}.$$

Hence we recover the value of the Gauss sum $\gamma(\mathrm{T}H_1(M), \phi_\sigma)$. Then it follows again from (2.3) that we also recover the Gauss sum $\gamma(\mathrm{T}H_1(M), 2m\phi_\sigma)$ for *any* $m \in \mathbb{Z}$. Since ϕ_σ is homogeneous, we have $2\phi_\sigma(x) = \lambda_M(x, x)$, for $x \in \mathrm{T}H_1(M)$. Since the values of all Gauss sums $\gamma(\mathrm{T}H_1(M), m\lambda_M)$ are known, it follows from [4, Theorem 2] or [10, Theorem 4.1] that $\tau^{\text{spin}}(M, \sigma)$ determines the isomorphism class of λ_M. Since in addition the Gauss sum $\gamma(\mathrm{T}H_1(M), \phi_\sigma)$ is determined by $\tau^{\text{spin}}(M, \sigma)$, by [15, Theorem 2], $\tau^{\text{spin}}(M, \sigma)$ also determines the isomorphism class of ϕ_σ. ■

Remark. The equivalence between Theorem 1 and Theorem 2 should be clear from [15, Theorem 1] and the following lemma.

Lemma 1. *The quadratic forms associated to spin structures on 3-manifolds exhausts all possible quadratic forms on finite Abelian groups.*

Proof. By [10, Theorem 6.1], linking pairings on oriented rational homology 3-spheres exhaust all possible symmetric bilinear pairings on finite Abelian groups. Let M be a closed oriented 3-manifold. The map $\phi : \mathrm{Spin}(M) \to \mathrm{Quad}^0(\lambda_M), \sigma \mapsto \phi_\sigma$ is affine over the map $H^1(M; \mathbb{Z}_2) \to \mathrm{Hom}(\mathrm{T}H_1(M), \mathbb{Z}_2)$ and therefore is surjective. The claim follows. ■

3. An invariant for complex spin 3-manifolds

In this section, we adapt the previous invariant to the setting of complex spin 3-manifolds (§§3.1 and 3.2 below), recall the basics of complex spin Borromean surgery (§3.2) and briefly note that a statement analogous to Theorem 2 does not hold for complex spin 3-manifolds (§).

3.1. For simplicity, we shall only consider rational homology 3-spheres in this section. To each $\sigma \in \mathrm{Spin}^c(M)$, one can associate in a canonical fashion a map $\phi_\sigma : \mathrm{T}H_1(M) \to \mathbb{Q}/\mathbb{Z}$ which satisfies (2.1), which we call a *quadratic function.* The map

$$\phi : \sigma \mapsto \phi_\sigma$$

is an affine bijection from the set of complex spin structures on an oriented rational homology 3-sphere to the set of quadratic *functions* onto the linking pairing λ_M. The main difference with the previous setup is that ϕ is affine now over the homomorphism $H^2(M) \to \mathrm{Hom}(H_1(M), \mathbb{Q}/\mathbb{Z})$. Denote by $\mathrm{Quad}(\lambda_M)$ the set of all quadratic functions over the linking pairing λ_M. Since M is a rational homology sphere, the natural map $\mathrm{Spin}(M) \to \mathrm{Spin}^c(M)$, affine over the Bockstein homomorphism $H^1(M;\mathbb{Z}_2) \to H^2(M)$, is injective, so that the diagram

$$\begin{array}{ccc} \mathrm{Spin}(M) & \longrightarrow & \mathrm{Spin}^c(\mathrm{M}) \\ \downarrow & & \downarrow \\ \mathrm{Quad}^q(\lambda_M) & \xrightarrow{\text{inclusion}} & \mathrm{Quad}(\lambda_M) \end{array}$$

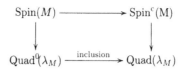

is commutative.

3.2. Denote by $\mathfrak{Q}$ the monoid of all isomorphism classes of quadratic functions on finite Abelian groups. We wish to carry over the argument in §2.2, resp. in §2.4, to define a bilinear map $\langle\cdot,\cdot\rangle : \mathfrak{M}^+ \times \mathfrak{Q} \to \mathbb{C}$, resp. a topological invariant τ^{spin^c} of complex spin oriented rational homology 3-spheres, by the (same) formulas (2.3) and (2.4) respectively. The difficulty consists in defining properly the tensor product of a general quadratic function $q : G \to \mathbb{Q}/\mathbb{Z}$ with a symmetric bilinear pairing $f : H \times H \to \mathbb{Z}$ on a lattice. One possibility is to proceed as follows. (Another possibility will be briefly mentioned at the end of this paper.) Denote by $\lambda_q : G\times G \to \mathbb{Q}/\mathbb{Z}$ the symmetric bilinear pairing associated to q. First suppose that q is homogeneous. Then there is a unique quadratic function $q \otimes f : G \otimes H \to \mathbb{Q}/\mathbb{Z}$ such that the symmetric bilinear pairing is $\lambda_q \otimes f : (G \otimes H) \times (G \otimes H) \to \mathbb{Q}/\mathbb{Z}$ and such that $(q \otimes f)(x \otimes y) = q(x)f(y,y)$ for all $(x,y) \in G \times H$. (Proof: choose a basis (e_j) for H, set $(q \otimes f)(\sum_j x_j \otimes e_j) = \sum_j q(x_j)f(e_j,e_j) + \sum_{j<k} \lambda_q(x_j,x_k)f(e_j,e_k)$ and check that this is well-defined using the definition of the tensor product $G\otimes H$.) If q is not homogeneous, there exists $\alpha \in G$ and a homogeneous quadratic form $q_0 : G \to \mathbb{Q}/\mathbb{Z}$ such that $q(x) = q_0(x)+\lambda_q(\alpha,x)$ for all $x \in G$. (Such a decomposition is not unique in general.) Then we define

$$(q \otimes f)(x) = (q_0 \otimes f)(x) + (\lambda_q \otimes f)(\alpha \otimes v, x) \quad \text{for all } x \in G \otimes H$$

where v is a characteristic element for f. We leave it to the reader to check that this definition makes sense, generalizes the previous definition and is independent of the decomposition of q. It does depend in general of the characteristic element.

Now let us suppose we want to proceed exactly as in the previous argument, that is, use formula (2.4). We would define

$$\tau^{\mathrm{spin}^c}(M,\sigma;\lambda,n) = \langle(\lambda,n);\phi_\sigma\rangle \in \mathbb{C}$$

where σ is a complex spin structure on M, and ϕ_σ is the associated quadratic function. We still have to see how this formula depends on the characteristic element v for the lift $\tilde{\lambda}$. For this we use the

Lemma 2. *Let $q : G \to \mathbb{Q}/\mathbb{Z}$ be a quadratic function and $\alpha \in G$. Denote by q' the quadratic function defined by $q'(x) = q(x) + \lambda_q(x,\alpha)$, $x \in G$. Then*

$$\gamma(G,q') = e^{-2i\pi q(\alpha)}\, \gamma(G,q). \tag{3.1}$$

Set $\mathrm{T} = \mathrm{T}H_1(M)$. Choose $\sigma_0 \in \mathrm{spin}(M)$ and $\alpha \in T$ so that $\phi_\sigma(x) = \phi_{\sigma_0}(x) + \lambda_M(x,\alpha)$ for all $x \in T$. We have $\gamma(T,\phi_\sigma) = e^{-2i\pi\phi_{\sigma_0}(\alpha)}\,\gamma(T,\phi_{\sigma_0})$ and

$$\gamma(T \otimes H, \phi_\sigma \otimes \tilde{\lambda}) = e^{-2i\pi\phi_{\sigma_0}(\alpha)\tilde{\lambda}_M(\alpha,v)}\, \gamma(T \otimes H, \phi_{\sigma_0} \otimes \tilde{\lambda}).$$

It follows that

$$\tau^{\mathrm{spin}^c}(M,\sigma;\lambda,n) = e^{2i\pi\phi_{\sigma_0}(\alpha)(\mathrm{sign}(\tilde{\lambda})-\tilde{\lambda}(v,v))}.$$

The dependency on v is apparent. Replacing the characteristic element v par an another arbitrary one, $v' = v + 2z$, we find that the indeterminacy is

$$\exp\Big(2i\pi\phi_{\sigma_0}(\alpha)(\tilde{\lambda}(v,z) - \tilde{\lambda}(z,z))\Big) \in \exp(4i\pi\phi_{\sigma_0}(\alpha)\mathbb{Z}).$$

We note that this indeterminacy is not trivial in general.

3.3. There is also a notion of complex spin Borromean equivalence. Suppose that (M,σ) is a complex spin 3-manifold. It is known (see e.g. [5, Theorem 2.2]) that (M,σ) can be presented by surgery on a framed link $L = L_1 \cup \cdots \cup L_n$ with an *integer* s_j attached to each component L_j. The complex spin Kirby moves can be defined in this context as well. The spin^c Δ-move is now a move on pairs (L, s) (a framed link $L = L_1 \cup \cdots \cup L_n$, equipped with an element $s \in (\mathbb{Z})^n$): it acts as the usual Δ-move on L and trivially on s, see Fig. 2.1. Finally, the spin^c Borromean equivalence is similarly as before the equivalence relation generated by the spin^c Kirby moves and the spin^c Δ-move. Using this equivalence, it is proved in [5] that there exists a nontrivial spin^c refinement of the Goussarov-Habiro theory. The spin^c Borromean equivalence class of a spin^c rational homology 3-sphere (M,σ) is determined by the isomorphism class of ϕ_σ. One can ask whether a complex spin version of Theorem 2 (i.e., with spin^c instead of spin) holds. We shall show that the answer is negative.

3.4. We give an example of two complex spin Borromean nonequivalent complex spin 3-manifolds (M,σ), (M',σ') such that $\tau^{\text{spin}^c}(M,\sigma) = \tau^{\text{spin}^c}(M',\sigma')$ (modulo the indeterminacy). Set $M = M' = L(32,1)$ (a lens space) and choose complex spin structures σ, σ' on M by letting $\phi_\sigma(x) = \frac{x^2}{64} + \frac{x}{4}$ and $\phi_\sigma(x) = \frac{x^2}{64} + \frac{x}{2}$ for all $x \in H_1(M) = \mathbb{Z}_{32}$ respectively. We have $\langle(\lambda, n), \phi_\sigma\rangle = \langle(\lambda, n), \phi_{\sigma'}\rangle$ for all $(\lambda, n) \in \mathcal{M}^+$. Thus $\tau^{\text{spin}^c}(M,\sigma) = \tau^{\text{spin}^c}(M,\sigma')$. But ϕ_σ and $\phi_{\sigma'}$ are inequivalent quadratic functions since $\phi_{\sigma'}$ is homogeneous whereas ϕ_σ is not. By [5, Theorem 2], (M,σ) and (M',σ') are not complex spin Borromean equivalent.

For completeness, we need to include the following observation.

Lemma 3. *The quadratic functions associated to* spin^c *structures on* 3*-manifolds exhausts all possible quadratic functions on finite Abelian groups.*

Proof. Actually oriented rational homology 3-spheres suffice. By Lemma 1, the result is true for quadratic forms (homogeneous quadratic functions). Now use $H^2(M)$-equivariance of the map $\phi : \text{spin}^c(M) \to \text{Quad}(\lambda_M)$. ■

The very fact that there is an indeterminacy in the definition of $\tau^{\text{spin}^c}(M,\sigma)$ suggests that there ought to exist a better definition. Such a definition in fact exists. It consists in replacing the tensor product $\phi_\sigma \otimes \tilde{\lambda}$ by the quadratic function associated to $\tilde{\lambda}_M \otimes \tilde{\lambda}$ with characteristic element $w \otimes v$ via the discriminant construction. Here w is a (rational) characteristic element for $\tilde{\lambda}_M$ such that $\partial(\tilde{\lambda}_M, w) = \phi_\sigma$ (cf. [3, 1.2.1]) and v is an arbitrary characteristic element for $\tilde{\lambda}$. This defines a (stronger) invariant of complex spin borromean equivalence which contains no indeterminacy. We conjecture that it characterizes the spin^c Borromean equivalence.

References

[1] C. Blanchet, *Invariants on three-manifolds with spin-structure*, Comment. Math. Helvetici **67** (1992), 406–427.

[2] F. Deloup, *Linking forms, reciprocity for Gauss sums and invariants of* 3*-manifolds*, Trans. Amer. Math. Soc. 351 (1999), no. 5, 1895–1918.

[3] F. Deloup, *On Abelian quantum invariants of links in* 3*-manifolds*, Math. Ann. **319** (2001), 759–795.

[4] F. Deloup, C. Gille, *Abelian quantum invariants indeed classify linking pairings*, Knots in Hellas '98, Vol. 2 (Delphi), J. Knot Theory Ramifications 10 (2001), no. 2, 295–302.

[5] F. Deloup, G. Massuyeau, *Quadratic functions and complex spin structures on* 3*-manifolds*, Topology, to appear.

[6] S. Garoufalidis, M. Goussarov, M. Polyak, *Calculus of clovers and FTI of 3-manifolds*, Geometry and Topology **5** (2001), 75–108.

[7] R. Gompf, Spin^c*-structures and homotopy equivalences*, Geom. Topol. 1 (1997), 41–50.

[8] M. Goussarov, *Finite type invariants and n-equivalence of 3-manifolds*, Compt. Rend. Ac. Sc. Paris **329** Série I (1999), 517–522.

[9] R. Gompf, A. Stipsicz, *4-manifolds and Kirby calculus*, Graduate Studies in Mathematics, AMS, Providence, 1999.

[10] A. Kawauchi, S. Kojima, *Algebraic classification of linking pairings on 3-manifolds*, Math. Ann. 253 (1980), no. 1, 29–42.

[11] K. Habiro, *Claspers and finite type invariants of links*, Geometry and Topology **4** (2000), 1–83.

[12] R.C. Kirby, The topology of 4-manifolds, Springer-Verlag LNM **1374**, 1991.

[13] J. Lannes, F. Latour, *Signature modulo 8 des variétés de dimension 4k dont le bord est stablement parallélisé*, Compt. Rend. Ac. Sc. Paris **279** Série A (1974), 705–707.

[14] E. Looijenga, J. Wahl, *Quadratic functions and smoothing surface singularities*, Topology **25** no. 3 (1986), 261–291.

[15] G. Massuyeau, *Spin Borromean surgeries*, Trans. of the A.M.S. **355** (2003), no. 10, 3991–4017.

[16] S.V. Matveev, *Generalized Surgery of three-dimensional manifolds and representations of homology spheres*, Mat. Zametki **42** n°2 (1987), 268–278 (English translation in Math. Notices Acad. Sci. USSR, **42** : **2**).

[17] H. Murakami, Y. Nakanishi, *On a certain move generating link homology*, Math. Ann. **284** (1989), 75–89.

Florian Deloup. Einstein Institute of Mathematics, Edmond J. Safra Campus, Givat Ram, The Hebrew University of Jerusalem, 91904 Jerusalem, Israel. *and*

Laboratoire Emile Picard, UMR 5580 CNRS/Université Paul Sabatier, 118 route de Narbonne, 31062 Toulouse, France.

E-mail address: deloup@math.huji.ac.il

Khovanov homology: torsion and thickness

Marta M. Asaeda[†] and Józef H. Przytycki

Abstract. We partially solve the conjecture by A.Shumakovitch that the Khovanov homology of a nontrivial, prime, non-split link (different from the Hopf link) in S^3 has a non-trivial torsion part. We give a size restriction on the Khovanov homology of almost alternating links. We relate the Khovanov homology of the connected sum of a link diagram with the Hopf link to the Khovanov homology of the diagram via a short exact sequence of homology and prove that this sequence splits. Finally, we show that our results can be adapted to reduced Khovanov homology and that there is a long exact sequence connecting reduced Khovanov homology with unreduced homology.

Introduction

Khovanov homology offers a nontrivial generalization of the Jones polynomials of links in S^3 (and of the Kauffman bracket skein modules of some 3-manifolds). In this paper we use Viro's approach to construction of Khovanov homology, and utilize the fact that one works with unoriented diagrams (unoriented framed links) in which case there is a long exact sequence of Khovanov homology. Khovanov homology, over the field $\mathbb{Q}$, is a categorification of the Jones polynomial (i.e. we represent the Jones polynomial as the generating function of Euler characteristics). However, for integral coefficients Khovanov homology almost always has torsion. The first part of the paper is devoted to the construction of torsion in Khovanov homology. In the second part of the paper we analyze the thickness of Khovanov homology and reduced Khovanov homology.

The paper is organized as follows. In the first section we recall the definition of Khovanov homology and its basic properties.

In the second section we prove that adequate link diagrams with an odd cycle property have $\mathbb{Z}_2$-torsion in Khovanov homology.

In the third section we discuss torsion in the Khovanov homology of an adequate link diagram with an even cycle property.

In the fourth section we prove Shumakovitch's theorem that prime, non-split alternating links different from the trivial knot and the Hopf link have $\mathbb{Z}_2$-torsion in Khovanov homology. We generalize this result to a class of adequate links.

[†] Partially sponsored by the NSF grant #DMS 0202613.

J.M. Bryden (ed.), Advances in Topological Quantum Field Theory, 135–166.

In the fifth section we generalize result of E.S.Lee about the Khovanov homology of alternating links (they are H-thin[1]). We do not assume rational coefficients in this generalization and we allow alternating adequate links on a surface. We use Viro's exact sequence of Khovanov homology to extend Lee's results to almost alternating diagrams and H-k-thick links.

In the sixth section we compute the Khovanov homology for a connected sum of n copies of Hopf links and construct a short exact sequence of Khovanov homology involving a link and its connected sum with the Hopf link. By showing that this sequence splits, we answer the question asked by Shumakovitch.

In the seventh section we notice that the results of sections 5 and 6 can be adapted to reduced Khovanov homology. Finally, we show that there is a long exact sequence connecting reduced Khovanov homology with unreduced homology.

1. Basic properties of Khovanov homology

The first spectacular application of the Jones polynomial (via Kauffman bracket relation) was the solution of Tait conjectures on alternating diagrams and their generalizations to adequate diagrams. Our method of analyzing torsion in Khovanov homology has its root in work related to solutions of Tait conjectures (Ka; Mu; Thi).

Recall that the Kauffman bracket polynomial $< D >$ of a link diagram D is defined by the skein relations $< \times > = A < \asymp > + A^{-1} <)(>$ and $< D \sqcup \bigcirc > = (-A^2 - A^{-2}) < D >$ and the normalization $< \bigcirc > = 1$. The categorification of this invariant (named by Khovanov *reduced homology*) is discussed in Section 7. For the Khovanov homology we use the version of the Kauffman bracket polynomial normalized to be 1 for the empty link (we use the notation $[D]$ in this case).

Definition 1.1 (Kauffman States).
Let D be a diagram[2] of an unoriented, framed link in a 3-ball B^3. A

[1] We also outline a simple proof of Lee's result(Lee-1; Lee-2) that for alternating links Khovanov homology yields the classical signature, see Remark 1.6.

[2] We think of the 3-ball B^3 as $D^2 \times I$ and the diagram is drawn on the disc D^2. In (APS), we have proved that the theory of Khovanov homology can be extended to links in an oriented 3-manifold M that is the bundle over a surface F ($M = F \tilde{\times} I$). If F is orientable then $M = F \times I$. If F is unorientable then M is a twisted I bundle over F (denoted by $F \hat{\times} I$). Several results of the paper are valid for the Khovanov homology of links in $M = F \hat{\times} I$.

Kauffman state s of D is a function from the set of crossings of D to the set $\{+1, -1\}$. Equivalently, we assign to each crossing of D a marker according to the following convention:

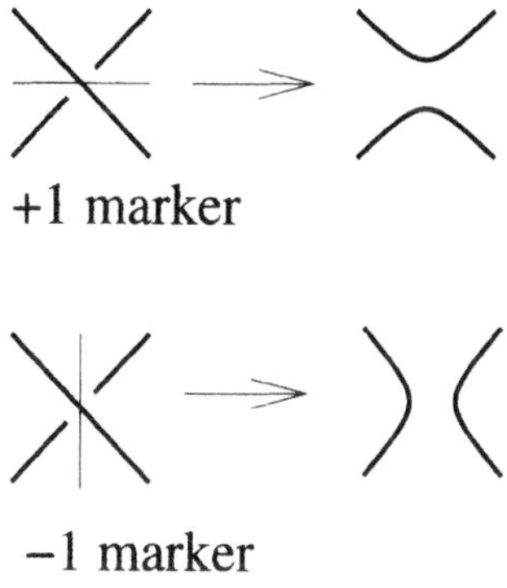

Fig. 1.1; markers and associated smoothings

By D_s we denote the system of circles in the diagram obtained by smoothing all crossings of D according to the markers of the state s, Fig. 1.1.
By $|s|$ we denote the number of components of D_s.

Using this notation we have the Kauffman bracket polynomial given by the state sum formula: $[D] = (-A^2 - A^{-2}) < D >= \sum_s A^{\sigma(s)}(-A^2 - A^{-2})^{|s|}$, where $\sigma(s)$ is the number of positive markers minus the number of negative markers in the state s.

To define Khovanov homology it is convenient (as noticed by Viro) to consider enhanced Kauffman states[3].

Definition 1.2. *An enhanced Kauffman state S of an unoriented framed link diagram D is a Kauffman state s with an additional assignment of + or − sign to each circle of D_s.*

Using enhanced states we express the Kauffman bracket polynomial as a (state) sum of monomials which is important in the definition of Khovanov homology we use. We have $[D] = (-A^2 - A^{-2}) < D >= \sum_S (-1)^{\tau(S)} A^{\sigma(s)+2\tau(S)}$, where $\tau(S)$ is the number of positive circles minus the number of negative circles in the enhanced state S.

---

[3] In Khovanov's original approach every circle of a Kauffman state was decorated by a 2-dimensional module A (with basis $\mathbf{1}$ and X) with the additional structure of Frobenius algebra. As an algebra $A = \mathbb{Z}[X]/(X^2)$ and comultiplication is given by $\Delta(\mathbf{1}) = X \otimes \mathbf{1} + \mathbf{1} \otimes X$ and $\Delta(X) = X \otimes X$. Viro uses − and + in place of $\mathbf{1}$ and X.

Definition 1.3 (Khovanov chain complex).

(i) Let $\mathcal{S}(D)$ denote the set of enhanced Kauffman states of a diagram D, and let $\mathcal{S}_{i,j}(D)$ denote the set of enhanced Kauffman states S such that $\sigma(S) = i$ and $\sigma(S) + 2\tau(S) = j$, The group $\mathcal{C}(D)$ (resp. $\mathcal{C}_{i,j}(D)$) is defined to be the free abelian group spanned by $\mathcal{S}(D)$ (resp. $\mathcal{S}_{i,j}(D)$). $\mathcal{C}(D) = \bigoplus_{i,j\in\mathbb{Z}} \mathcal{C}_{i,j}(D)$ is a free abelian group with (bi)-gradation.

(ii) For a link diagram D with ordered crossings, we define the chain complex $(\mathcal{C}(D), d)$ where $d = \{d_{i,j}\}$ and the differential $d_{i,j} : \mathcal{C}_{i,j}(D) \to \mathcal{C}_{i-2,j}(D)$ satisfies $d(S) = \sum_{S'}(-1)^{t(S:S')}[S : S']S'$ with $S \in \mathcal{S}_{i,j}(D)$, $S' \in \mathcal{S}_{i-2,j}(D)$, and $[S : S']$ equal to 0 or 1. $[S : S'] = 1$ if and only if markers of S and S' differ exactly at one crossing, call it c, and all the circles of D_S and $D_{S'}$ not touching c have the same sign[4]. Furthermore, $t(S : S')$ is the number of negative markers assigned to crossings in S bigger than c in the chosen ordering.

(iii) The Khovanov homology of the diagram D is defined to be the homology of the chain complex $(\mathcal{C}(D), d)$, that is $H_{i,j}(D) = ker(d_{i,j})/d_{i+2,j}(\mathcal{C}_{i+2,j}(D))$. The Khovanov cohomology of the diagram D is defined to be the cohomology of the chain complex $(\mathcal{C}(D), d)$.

Below we list a few elementary properties of Khovanov homology, following from properties of Kauffman states used in the proof of Tait conjectures (Ka; Mu; Thi).

The positive state $s_+ = s_+(D)$ (respectively the negative state $s_- = s_-(D)$) is the state with all positive markers (resp. negative markers). The alternating diagrams without nugatory crossings (i.e. crossings in a diagram of the form 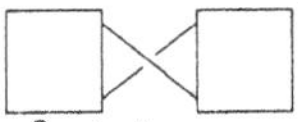) are generalized to adequate diagrams using properties of states s_+ and s_-. Namely, the diagram D is +-adequate (resp. −-adequate) if the state of positive (resp. negative) markers, s_+ (resp. s_-), cuts the diagram to the collection of circles, so that every crossing is connecting different circles. D is an adequate diagram if it is +- and −-adequate (L-T).

Property 1.4.
If D is a diagram of n crossings and its positive state s_+ has $|s_+|$ circles

[4] From our conditions it follows that at the crossing c the marker of S is positive, the marker of S' is negative, and that $\tau(S') = \tau(S) + 1$.

then the highest term (in both grading indexes) of Khovanov chain complex is $C_{n,n+2|s_+|}(D)$*; we have* $C_{n,n+2|s_+|}(D) = \mathbb{Z}$*. Furthermore, if* D *is a* $+$*-adequate diagram, then the whole group* $C_{*,n+2|s_+|}(D) = \mathbb{Z}$*. Similarly the lowest term in the Khovanov chain complex is* $C_{-n,-n-2|s_-|}(D)$*. Assume that* D *is a non-split diagram then* $|s_+| + |s_-| \leq n+2$ *and the equality holds if and only if* D *is an alternating diagram or a connected sum of such diagrams (Wu's dual state lemma (Wu)).*

Property 1.5. *Let* $\sigma(L)$ *be the classical (Trotter-Murasugi) signature*[5] *of an oriented link* L *and* $\hat{\sigma}(L) = \sigma(L) + lk(L)$*, where* $lk(L)$ *is the global linking number of* L*, its Murasugi's version which does not depend on an orientation of* L*. Then*

(i) *[Traczyk's local property] If* D_0^v *is a link diagram obtained from an oriented alternating link diagram* D *by smoothing its crossing* v *and* D_0^v *has the same number of (graph) components as* D*, then* $\sigma(D) = \sigma(D_0^v) - sgn(v)$*. One defines the sign of a crossing* v *as* $sgn(v) = \pm 1$ *according to the convention* $sgn($$) = 1$ *and* $sgn($$) = -1$.

(ii) *[Traczyk Theorem (Tra; Pr)]*
The signature, $\sigma(D)$*, of the non-split alternating oriented link diagram* D *is equal to* $n^- - |s_-| + 1 = -n^+ + |s_+| - 1 = -\frac{1}{2}(n^+ - n^- - (|s_+| - |s_-|)) = n^- - n^+ + d^+ - d^-$*, where* $n^+(D)$ *(resp.* $n^-(D)$*) is the number of positive (resp. negative) crossings of* D *and* d^+ *(resp.* d^-*) is the number of positive (resp. negative) edges in a spanning forest of the Seifert graph*[6] *of* D*.*

(iii) *[Murasugi's Theorem (Mu-1; Mu-2)]*
Let D *be a non-split alternating oriented diagram without nugatory crossings or a connected sum of such diagrams. Let* $V_D(t)$ *be its Jones polynomial*[7]*, then the maximal degree* $\max V_D(t) = n^+(D) - \frac{\sigma(D)}{2}$ *and the minimal degree* $\min V_D(t) = -n^-(D) - \frac{\sigma(D)}{2}$*.*

---

[5] One should not mix the signature $\sigma(L)$ with $\sigma(s)$ which is the signed sum of markers of the state s of a link diagram.

[6] The Seifert graph, $GS(D)$, of an oriented link diagram D is a signed graph whose vertices are in bijection with Seifert circles of D and edges are in a natural bijection with crossings of D. For an alternating diagram the 2-connected components (blocks) of $GS(D)$ have edges of the same sign which makes d^+ and d^- well defined.

[7] Recall that if $\vec{D}$ is an oriented diagram (any orientation put on the unoriented diagram D), and $w(\vec{D})$ is its writhe or Tait number, $w(\vec{D}) = n^+ - n^-$, then $V_{\vec{D}}(t) = A^{-3w(\vec{D})} < D >$ for $t = A^{-4}$. P.G.Tait (1831-1901) was the first to consider the number $w(\vec{D})$ and it is often called the Tait number of the diagram $\vec{D}$ and denoted by $Tait(\vec{D})$.

(iv) [Murasugi's Theorem for unoriented link diagrams]. Let D be a non-split alternating unoriented diagram without nugatory crossings or a connected sum of such diagrams.
Then the maximal degree $\max < D >= \max\, [D] - 2 = n + 2|s_+| - 2 = 2n + sw(D) + 2\hat{\sigma}(D)$
and the minimal degree $\min < D >= \min\, [D] + 2 = -n - 2|s_-| + 2 = -2n + sw(D) + 2\hat{\sigma}(D)$. *The self-twist number of a diagram* $sw(D) = \sum_v sgn(v)$, *where the sum is taken over all self-crossings of D. A self-crossing involves arcs from the same component of a link. sw(D) does not depend on orientation of D.*

Remark 1.6.
In Section 5 we reprove the result of Lee (Lee-1) that the Khovanov homology of non-split alternating links is supported by two adjacent diagonals of slope 2, that is $H_{i,j}(D)$ *can be nontrivial only for two values of* $j - 2i$ *which differ by 4 (Corollary 5.5). One can combine Murasugi-Traczyk result with Viro's long exact sequence of Khovanov homology and Theorem 7.3 to recover Lee's result ((Lee-2)) that for alternating links Khovanov homology has the same information as the Jones polynomial and the classical signature*[8] *(see Chapter 10 of (Pr)). From properties 1.4 and 1.5 it follows that for non-split alternating diagram without nugatory crossings* $H_{n,2n+sw+2\hat{\sigma}+2}(D) = H_{-n,-2n+sw+2\hat{\sigma}-2}(D) = \mathbb{Z}$. $H_{n,2n+sw+2\hat{\sigma}+2}(D) = H_{-n,-2n+sw+2\hat{\sigma}-2}(D) = \mathbb{Z}$. *Thus diagonals which support nontrivial* $H_{i,j}(D)$ *satisfy* $j - 2i = sw(D) + 2\hat{\sigma}(D) \pm 2$. *If we consider Khovanov cohomology* $H^{i',j'}(D)$, *as considered in (Kh-1; BN-1), then* $H^{i',j'}(D) = H_{i,j}(D)$ *for* $i' = \frac{w(D)-i}{2}$, $j' = \frac{3w(D)-j}{2}$ *and thus* $j' - 2i' = \frac{-1}{2}(j - 2i - w(D)) = \sigma(D) \mp 1$ *as in Lee's Theorem.*

Remark 1.7.
The definition of Khovanov homology extends to links in I-bundles over surfaces F ($F \neq RP^2$) *(APS). In the definition we must differentiate between trivial curves, curves bounding a Möbius band, and other non-trivial curves. Namely, we define* $\tau(S)$ *as the sum of signs of circles of* D_S *taken over all trivial circles of* D_S. *Furthermore, to have* $[S : S'] = 1$, *we assume additionally that the sum of signs of circles of* D_S *taken over all nontrivial circles which do not bound a Möbius band is the same for S and S'.*

[8] The beautiful paper by Jacob Rasmussen (Ras) generalizes Lee's results and fulfill our dream (with Paweł Traczyk) of constructing a "supersignature" from Jones type construction (Pr).

2. Diagrams with odd cycle property

In the next few sections we use the concept of a graph, $G_s(D)$, associated to a link diagram D and its state s. The graphs corresponding to states s_+ and s_- are of particular interest. If D is an alternating diagram then $G_{s_+}(D)$ and $G_{s_-}(D)$ are the plane graphs first constructed by Tait.

Definition 2.1.

(i) Let D be a diagram of a link and s its Kauffman state. We form a graph, $G_s(D)$, associated to D and s as follows. Vertices of $G_s(D)$ correspond to circles of D_s. Edges of $G_s(D)$ are in bijection with crossings of D and an edge connects given vertices if the corresponding crossing connects circles of D_s corresponding to the vertices[9].

(ii) In the language of associated graphs we can state the definition of adequate diagrams as follows: the diagram D is $+$-adequate (resp. $-$-adequate) if the graph $G_{s_+}(D)$ (resp. $G_{s_-}(D)$) has no loops.

In this language we can formulate our first result about torsion in Khovanov homology.

Theorem 2.2.
Consider a link diagram D of N crossings. Then

(+) If D is $+$-adequate and $G_{s_+}(D)$ has a cycle of odd length, then the Khovanov homology has $\mathbb{Z}_2$ torsion. More precisely we show that $H_{N-2,N+2|s_+|-4}(D)$ has $\mathbb{Z}_2$ torsion.

(-) If D is $-$-adequate and $G_{s_-}(D)$ has a cycle of odd length, then $H_{-N,-N-2|s_-|+4}(D)$ has $\mathbb{Z}_2$ torsion.

Proof. (+) It suffices to show that the group $\mathcal{C}_{N-2,N+2|s_+|-4}(D)/d(\mathcal{C}_{N,N+2|s_+|-4}(D))$ has 2-torsion.

Consider first the diagram D of the left handed torus knot $T_{-2,n}$ (Fig.2.1 illustrates the case of $n = 5$). The associated graph $G_n =$

[9] If S is an enhanced Kauffman state of D then, in a similar manner, we associate to D and S the graph $G_S(D)$ with signed vertices. Furthermore, we can additionally equip $G_S(D)$ with a cyclic ordering of edges at every vertex following the ordering of crossings at any circle of D_s. The sign of each edge is the label of the corresponding crossing. In short, we can assume that $G_S(D)$ is a ribbon (or framed) graph. We do not use this additional data in this paper but we plan to utilize this in a sequel paper.

$G_{s_+}(T_{-2,n})$ is an n-gon.

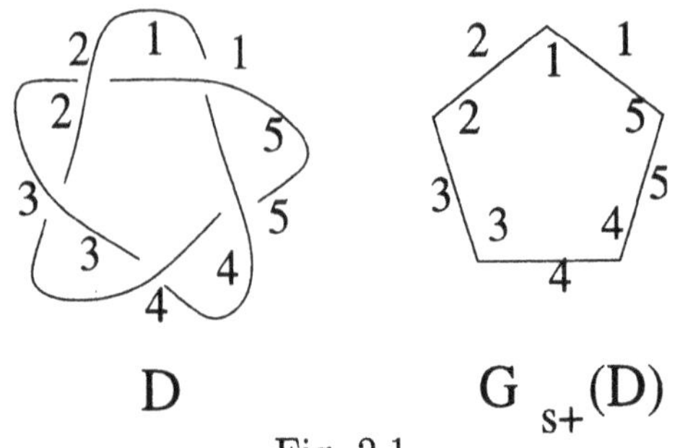

Fig. 2.1

For this diagram we have $\mathcal{C}_{n,n+2|s_+|}(D) = \mathbb{Z}$, $\mathcal{C}_{n,n+2|s_+|-4}(D) = \mathbb{Z}^n$ and $\mathcal{C}_{n-2,n+2|s_+|-4}(D) = \mathbb{Z}^n$, where enhanced states generating $\mathcal{C}_{n,n+2|s_+|-4}(D)$ have all markers positive and exactly one circle (of D_S) negative[10]. Enhanced states generating $\mathcal{C}_{n-2,n+2|s_+|-4}(D)$ have exactly one negative marker and all positive circles of D_S. The differential $d : \mathcal{C}_{n,n+2|s_+|-4}(D) \to \mathcal{C}_{n-2,n+2|s_+|-4}(D)$ can be described by an $n \times n$ circulant matrix (for the ordering of states corresponding to the ordering of crossings and regions as in Fig. 2.1)).

$$\begin{pmatrix} 1 & 1 & 0 & \ldots & 0 & 0 \\ 0 & 1 & 1 & \ldots & 0 & 0 \\ \ldots & \ldots & \ldots & \ldots & \ldots & \ldots \\ 0 & 0 & \ldots & 0 & 1 & 1 \\ 1 & 0 & \ldots & 0 & 0 & 1 \end{pmatrix}$$

Clearly the determinant of the matrix is equal to 2 (because n is odd; for n even the determinant is equal to 0 because the alternating sum of columns gives the zero column). To see this one can consider for example the first row expansion[11]. Therefore the group described by the matrix is equal to $\mathbb{Z}_2$ (for an even n one would get $\mathbb{Z}$). One more observation (which will be used later). The sum of rows of the matrix is equal to the row vector $(2, 2, 2, ..., 2, 2)$ but the row vector $(1, 1, 1, ..., 1, 1)$ is not an integral linear combination of rows of the matrix. In fact the element $(1, 1, 1, ..., 1, 1)$ is the generator of $\mathbb{Z}_2$ group represented by the matrix. This can be easily checked because if $S_1, S_2, ...S_n$ are states

---

[10] In this case $s_+ = n$ but we keep the general notation so the generalization which follows is natural.

[11] Because the matrix is a circulant one we know furthermore that its eigenvalues are equal to $1 + \omega$, where ω is any n'th root of unity ($\omega^n = 1$), and that $\prod_{\omega^n=1}(1 + \omega) = 0$ for n even and 2 for n odd.

freely generating $\mathcal{C}_{n-2,n+2|s_+|-4}(D)$ then relations given by the image of $\mathcal{C}_{n,n+2|s_+|-4}(D)$ are $S_2 = -S_1, S_3 = -S_2 = S_1, ..., S_1 = -S_n = ... = -S_1$ thus $S_1 + S_2 + ... + S_n$ is the generator of the quotient group $\mathcal{C}_{n-2,n+2|s_+|-4}(D)/d(\mathcal{C}_{n,n+2|s_+|-4}(D)) = \mathbb{Z}_2$. In fact we have proved that any sum of the odd number of states S_i represents the generator of $\mathbb{Z}_2$.

Now consider the general case in which $G_{s_+}(D)$ is a graph without a loop and with an odd polygon. Again, we build a matrix presenting the group $\mathcal{C}_{N-2,N+2|s_+|-4}(D)/d(\mathcal{C}_{N,N+2|s_+|-4}(D))$ with the north-west block corresponding to the odd n-gon. This block is exactly the matrix described previously. Furthermore, the submatrix of the full matrix below this block is the zero matrix, as every column has exactly two nonzero entries (both equal to 1). This is the case because each edge of the graph (generator) has two endpoints (belongs to exactly two relations). If we add all rows of the matrix we get the row of all two's. On the other hand the row of one's cannot be created, even in the first block. Thus the row of all one's representing the sum of all enhanced states in $\mathcal{C}_{N-2,N+2|s_+|-4}(D)$ is $\mathbb{Z}_2$-torsion element in the quotient group (presented by the matrix) so also in $H_{N-2,N+2|s_+|-4}(D)$.
(-) This part follows from the fact that the mirror image of D, the diagram $\bar{D}$, satisfies the assumptions of the part (+) of the theorem. Therefore the quotient $\mathcal{C}_{N-2,N+2|s_+|-4}(\bar{D})/d(\mathcal{C}_{N,N+2|s_+|-4}(\bar{D}))$ has $\mathbb{Z}_2$ torsion. Furthermore, the matrix describing the map $d : \mathcal{C}_{-N+2,-N-2|s_-|+4}(D) \to \mathcal{C}_{-N,-N-2|s_-|+4}(D)$ is (up to sign of every row) equal to the transpose of the matrix describing the map $d : \mathcal{C}_{N,N+2|s_+|-4}(\bar{D}) \to \mathcal{C}_{N-2,N+2|s_+|-4}(\bar{D})$. Therefore the torsion of the group $\mathcal{C}_{-N,-N-2|s_-|+4}(D)/d(\mathcal{C}_{-N+2,-N-2|s_-|+4}(D)$ is the same as the torsion of the group $\mathcal{C}_{N-2,N+2|s_+|-4}(\bar{D})/d(\mathcal{C}_{N,N+2|s_+|-4}(\bar{D}))$ and, in conclusion, $H_{-N,-N-2|s_-|+4}(D)$ has $\mathbb{Z}_2$ torsion[12].

□

Remark 2.3.
Notice that the torsion part of the homology, $T_{N-2,N+2|s_+|-4}(D)$, depends only on the graph $G_{s_+}(D)$. Furthermore if $G_{s_+}(D)$ has no 2-gons then $H_{N-2,N+2|s_+|-4}(D) = \mathcal{C}_{N-2,N+2|s_+|-4}(D)/d(\mathcal{C}_{N,N+2|s_+|-4}(D))$ and depends only on the graph $G_{s_+}(D)$. See a generalization in Remark 3.6

[12] Our reasoning reflects a more general fact observed by Khovanov (Kh-1) (see (APS) for the case of $F \times I$) that Khovanov homology satisfies "duality theorem", namely $H^{ij}(D) = H_{-i,-j}(\bar{D})$. This combined with the Universal Coefficients Theorem saying that $H^{ij}(D) = H_{ij}(D)/T_{ij}(D) \oplus T_{i-2,j}(D)$, where $T_{ij}(D)$ denote the torsion part of $H_{ij}(D)$ gives: $T_{-N,-N-2|s_-|+4}(D) = T_{N-2,N+2|s_+|-4}(\bar{D})$ (notice that $|s_-|$ for D equals to $|s_+|$ for $\bar{D}$).

3. Diagrams with an even cycle property

If every cycle of the graph $G_{s_+}(D)$ is even (i.e. the graph is a bipartite graph) we cannot expect that $H_{N-2,N+|s_+|-4}(D)$ always has nontrivial torsion. The simplest link diagram without an odd cycle in $G_{s_+}(D)$ is the left handed torus link diagram $T_{-2,n}$ for n even. As mentioned before, in this case $\mathcal{C}_{n-2,n+2|s_+|-4}(D)/d(\mathcal{C}_{n,n+2|s_+|-4}(D)) = \mathbb{Z}$, and, in fact $H_{n-2,n+2|s_+|-4}(D) = \mathbb{Z}$ except $n = 2$, i.e. the Hopf link, in which case $H_{0,2}(D) = 0$.

To find torsion we have to look "deeper" into the homology. We will find a condition for which $H_{N-4,N+2|s_+|-8}(D)$ has $\mathbb{Z}_2$ torsion, where N is the number of crossings of D.

Analogously to the odd case, we will start from the left handed torus link $T_{-2,n}$ and associated graph $G_{s_+}(D)$ being an n-gon with even $n \geq 4$; Fig.3.1.

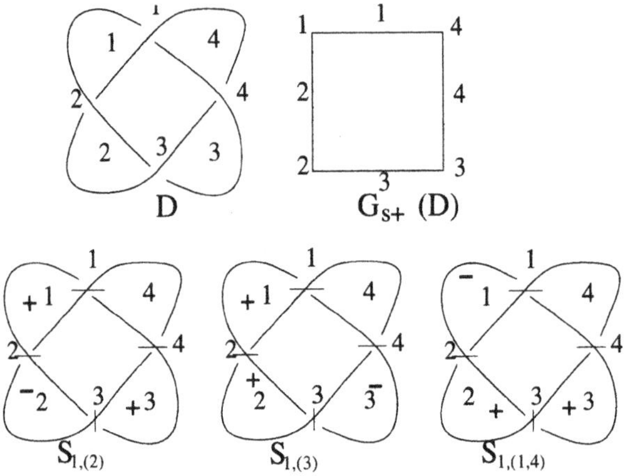

Fig. 3.1

Lemma 3.1.
Let D be the diagram of the left-handed torus link of type $(-2,n)$ with n even, $n \geq 4$. Then $H_{n-4,n+2|s_+|-8}(D) = H_{n-4,3n-8}(D) = \mathcal{C}_{n-4,3n-8}(D)/d(\mathcal{C}_{n-2,3n-8}(D)) = \mathbb{Z}_2$. Furthermore, every enhanced state from the basis of $\mathcal{C}_{n-4,3n-8}(D)$ (or an odd sum of such states) is the generator of $\mathbb{Z}_2$.

Proof. We have $n = |s_+|$. The chain group $C_{n-4,3n-8}(D) = \mathbb{Z}^{\frac{n(n-1)}{2}}$ is freely generated by enhanced states $S_{i,j}$, where exactly ith and jth crossings have negative markers, and all the circles of $D_{S_{i,j}}$ are positive (crossings of D and circles of D_{s_+} are ordered in Fig. 3.1). We have to understand the differential $d : C_{n-2,3n-8}(D) \to C_{n-4,3n-8}(D)$. The chain group $C_{n-2,3n-8}(D) = \mathbb{Z}^{n(n-1)}$ is freely generated by enhanced states with ith negative marker and one negative circle of D_S. In our notation we will write $S_{i,(i-1,i)}$ if the negative circle is obtained by connecting circles $i-1$ and i in D_{s_+} by a negative marker. Notation $S_{i,(j)}$ is used if we have jth negative circle, $j \neq i-1$, $j \neq i$. The states $S_{1,(2)}$, $S_{1,(3)}$ and $S_{1,(4,1)}$ are shown in Fig. 3.1 ($n = 4$ in the figure). The quotient group $C_{n-4,3n-8}(D)/d(C_{n-2,3n-8}(D))$ can be presented by a $n(n-1) \times \frac{n(n-1)}{2}$ matrix, E_n. One should just understand the images of enhanced states of $C_{n-2,3n-8}(D)$. In fact, for a fixed crossing i the corresponding $n-1 \times n-1$ block is (up to sign of columns[13]) the circulant matrix discussed in Section 2. Our goal is to understand the matrix E_n, to show that it represents the group $\mathbb{Z}_2$ and to find natural representatives of the generator of the group. For $n = 4$, $d : \mathbb{Z}^{12} \to \mathbb{Z}^6$ and it is given by: $d(S_{1,(2)}) = S_{1,2} + S_{1,3}$, $d(S_{1,(3)}) = S_{1,3} + S_{1,4}$, $d(S_{1,(1,4)}) = S_{1,2}+S_{1,4}$, $d(S_{2,(1,2)}) = -S_{2,1}+S_{2,3}$, $d(S_{2,(3)}) = S_{2,3}+S_{2,4}$, $d(S_{2,(4)}) = -S_{2,1} + S_{2,4}$, $d(S_{3,(1)}) = -S_{3,1} - S_{3,2}$, $d(S_{3,(2,3)}) = -S_{3,2} + S_{3,4}$, $d(S_{3,(4)}) = -S_{3,1} + S_{3,4}$, $d(S_{4,(1)}) = -S_{4,1} - S_{4,2}$, $d(S_{4,(2)}) = -S_{4,2} - S_{4,3}$, $d(S_{4,(4,3)}) = -S_{4,1} - S_{4,3}$,

Therefore d can be described by the 12×6 matrix. States are ordered lexicographically, e.g. $S_{i,j}$ $(i < j)$ is before $S_{i',j'}$ $(i' < j')$ if $i < i'$ or $i = i'$ and $j < j'$.

$$\begin{pmatrix} 1 & 1 & 0 & 0 & 0 & 0 \\ 0 & 1 & 1 & 0 & 0 & 0 \\ 1 & 0 & 1 & 0 & 0 & 0 \\ -1 & 0 & 0 & 1 & 0 & 0 \\ 0 & 0 & 0 & 1 & 1 & 0 \\ -1 & 0 & 0 & 0 & 1 & 0 \\ 0 & -1 & 0 & -1 & 0 & 0 \\ 0 & 0 & 0 & -1 & 0 & 1 \\ 0 & -1 & 0 & 0 & 0 & 1 \\ 0 & 0 & -1 & 0 & -1 & 0 \\ 0 & 0 & 0 & 0 & -1 & -1 \\ 0 & 0 & -1 & 0 & 0 & -1 \end{pmatrix},$$

In our example the rows correspond to $S_{1,(2)}$, $S_{1,(3)}$, $S_{1,(1,4)}$, $S_{2,(1,2)}$, $S_{2,(3)}$, $S_{2,(4)}$, $S_{3,(1)}$, $S_{3,(2,3)}$, $S_{3,(4)}$, $S_{4,(1)}$, $S_{4,(2)}$, and $S_{4,(4,3)}$, the columns correspond to $S_{1,2}$, $S_{1,3}$, $S_{1,4}$, $S_{2,3}$, $S_{2,4}$, $S_{3,4}$ in this order. Notice that the sum

[13] In the $(n-1) \times (n-1)$ block corresponding to the ith crossing (i.e. we consider only states in which ith crossing has a negative marker), the column under the generator $S_{i,j}$ of $C_{n-4,3n-8}$ has $+1$ entries if $i < j$ and -1 entries if $i > j$.

of columns of the matrix gives the non-zero column of all ± 2 or 0. Therefore over $\mathbb{Z}_2$ our matrix represents a nontrivial group. On the other hand, over $\mathbb{Q}$, the matrix represent the trivial group. Thus over $\mathbb{Z}$ the group represented by the matrix has $\mathbb{Z}_2$ torsion. More precisely, we can see that the group is $\mathbb{Z}_2$ as follows: The row relations can be expressed as: $S_{1,2} = -S_{1,3} = S_{1,4} = -S_{1,2}$, $S_{2,1} = S_{2,3} = -S_{2,4} = -S_{2,1}$, $S_{3,1} = -S_{3,2} = -S_{3,4} = -S_{3,1}$ and $S_{4,1} = -S_{4,2} = S_{4,3} = -S_{4,1}$. $S_{i,j} = S_{j,i}$ in our notation. In particular, it follows from these equalities that the group given by the matrix is equal to $\mathbb{Z}_2$ and is generated by any basic enhanced state $S_{i,j}$ or the sum of odd number of $S_{i,j}$'s.

Similar reasoning works for any even $n \geq 4$ (not only $n = 4$).

Furthermore, the chain group $\mathcal{C}_{n-6,3n-8} = 0$, therefore $H_{n-4,3n-8} = \mathcal{C}_{n-4,3n-8}/d(\mathcal{C}_{n-2,3n-8}) = \mathbb{Z}_2$. □

We are ready now to use Lemma 3.1 in the general case of an even cycle.

Theorem 3.2.
Let D be a connected diagram of a link of N crossings such that the associated graph $G_{s_+}(D)$ has no loops (i.e. D is +-adequate) and the graph has an even n-cycle with a singular edge (i.e. not a part of a 2-gon). Then $H_{N-4,N+2|s_+|-8}(D)$ has $\mathbb{Z}_2$ torsion.

Proof. Consider an ordering of crossings of D such that $e_1, e_2, ..., e_n$ are crossings (edges) of the n-cycle. The chain group $\mathcal{C}_{N-2,N+2|s_+|-8}(D)$ is freely generated by $N(V-1)$ enhanced states,$S_{i,(c)}$, where N is the number of crossings of D (edges of $G_{s_+}(D)$) and $V = |s_+|$ is the number of circles of D_{s_+} (vertices of $G_{s_+}(D)$). $S_{i,(c)}$ is the enhanced state in which the crossing e_i has the negative marker and the circle c of D_{s_i} is negative, where s_i is the state which has all positive markers except at e_i. The chain group $\mathcal{C}_{N-4,N+2|s_+|-8}(D)$ is freely generated by enhanced states which we can partition into two groups.
(i) States $S_{i,j}$, where crossings e_i, e_j have negative markers and corresponding edges of $G_{s_+}(D)$ do not form part of a multi-edge (i.e. e_i and e_j do not have the same endpoints). All circles of the state $S_{i,j}$ are positive.
(ii) States $S'_{i,j}$ and $S''_{i,j}$, where crossings e_i, e_j have negative markers and corresponding edges of $G_{s_+}(D)$ are parts of a multi-edge (i.e. e_i, e_j have the same endpoints). All but one circle of $S'_{i,j}$ and $S''_{i,j}$ are positive and we have two choices for a negative circle leading to $S'_{i,j}$ and $S''_{i,j}$, i.e. the crossings e_i, e_j touch two circles, and we give negative sign to one of them.

In our proof we will make the essential use of the assumption that the edge (crossing) e_1 is a singular edge.

We analyze the matrix presenting the group $\mathcal{C}_{N-4,N+2|s_+|-8}(D)/d(\mathcal{C}_{N-2,N+2|s_+|-8}(D))$.

By Lemma 3.1, we understand already the $n(n-1) \times \frac{1}{2}n(n-1)$ block corresponding to the even n-cycle. In this block every column has 4 non-zero entries (two $+1$ and two -1), therefore columns of the full matrix corresponding to states $S_{i,j}$, where e_i and e_j are in the n-gon, have zeros outside our block. We use this property later.

We now analyze another block represented by rows and columns associated to states having the first crossing e_1 with the negative marker. This $(V-1)\times(N-1)$ block has entries equal to 0 or 1. If we add rows in this block we obtain the vector row of two's $(2, 2, ..., 2)$, following from the fact that every edge of $G_{s_+}(D)$ and of $G_{s_1}(D)$ has 2 endpoints (we use the fact that D is $+$ adequate and e_1 is a singular edge). Consider now the bigger submatrix of the full matrix composed of the same rows as our block but without restriction on columns. All additional columns are 0 columns as our row relations involve only states with negative marker at e_1. Thus the sum of these rows is equal to the row vector $(2, 2, ..., 2, 0, ..., 0)$. We will argue now that the half of this vector, $(1, 1, ..., 1, 0, ..., 0)$, is not an integral linear combination of rows of the full matrix and so represents $\mathbb{Z}_2$-torsion element of the group $\mathcal{C}_{N-4,N+2|s_+|-8}(D)/d(\mathcal{C}_{N-2,N+2|s_+|-8}(D))$. For simplicity assume that $n = 4$ (but the argument holds for any even $n \geq 4$). Consider the columns indexed by $S_{1,2}, S_{1,3}, S_{1,4}, S_{2,3}, S_{2,4}$ and $S_{3,4}$. The integral linear combination of rows restricted to this columns cannot give a row with odd number of one's, as proven in Lemma 3.1. In particular we cannot get the row vector $(1, 1, 1, 0, 0, 0)$. This excludes the row $(1, 1, ..., 1, 0, ..., 0)$, as an integral linear combination of rows of the full matrix. Therefore the sum of enhanced states with the marker of e_1 negative is 2-torsion element in $\mathcal{C}_{N-4,N+2|s_+|-8}(D)/d(\mathcal{C}_{N-2,N+2|s_+|-8}(D))$ and therefore in $H_{N-4,N+2|s_+|-8}(D)$. □

Similarly, using duality, we can deal with $^-$-adequate diagrams.

Corollary 3.3.
Let D be a connected, $-$-adequate diagram of a link and the graph $G_{s_-}(D)$ has an even n-cycle, $n \geq 4$, with a singular edge.
Then $H_{-N+2,-N-2|s_-|+8}(D)$ has $\mathbb{Z}_2$ torsion.

Remark 3.4.
The restriction on D to be a connected diagram is not essential (it just

simplifies the proof) as for a non-connected diagram, $D = D_1 \sqcup D_2$ *we have "Künneth formula"* $H_*(D) = H_*(D_1) \otimes H_*(D_2)$ *so if any of* $H_*(D_i)$ *has torsion then* $H_*(D)$ *has torsion as well.*

We say that a link diagram is doubly +-adequate if its graph $G_{s_+}(D)$ has no loops and 2-gons. In other words, if a state s differs from the state s_+ by two markers then $|s| = |s_+| - 2$. We say that a link diagram is doubly −-adequate if its mirror image is doubly +-adequate.

Corollary 3.5.
Let D *be a connected doubly +-adequate diagram of a link of* N *crossings, then either* D *represents the trivial knot or one of the groups* $H_{N-2,N+2|s_+|-4}(D)$ *and* $H_{N-4,N+2|s_+|-8}(D)$ *has* $\mathbb{Z}_2$ *torsion.*

Proof. The associated graph $G_{s_+}(D)$ has no loops and 2-gons. If $G_{s_+}(D)$ has an odd cycle then by Theorem 2.2 $H_{N-2,N+2|s_+|-4}(D)$ has $\mathbb{Z}_2$ torsion. If $G_{s_+}(D)$ has an even n-cycle, $n \geq 4$ then $H_{N-4,N+2|s_+|-8}(D)$ has $\mathbb{Z}_2$ torsion by Theorem 3.2 (every edge of $G_{s_+}(D)$ is a singular edge as $G_{s_+}(D)$ has no 2-gons). Otherwise $G_{s_+}(D)$ is a tree, each crossing of D is a nugatory crossing and D represents the trivial knot. □

We can generalize and interpret Remark 2.3 as follows.

Remark 3.6.
Assume that the associated graph $G_{s_+}(D)$ *has no* k*-gons, for every* $k \leq m$. *Then the torsion part of Khovanov homology,* $T_{N-2m,N+2|s_+|-4m}(D)$ *depends only on the graph* $G_{s_+}(D)$. *Furthermore,* $H_{N-2m+2,N+2|s_+|-4m+4}(D) = C_{N-2m+2,N+2|s_+|-4m+4}(D)/d$ $(_{N-2m+4,N+2|s_+|-4m+4}(D))$ *and it depends only on the graph* $G_{s_+}(D)$. *On a more philosophical level* [14] *our observation is related to the fact that if the edge* e_c *in* $G_{s_+}(D)$ *corresponding to a crossing* c *in* D *is not a loop then for the crossing* c *the graphs* $G_{s_+}(D_0)$ *and* $G_{s_+}(D_\infty)$

[14] In order to be able to recover the full Khovanov homology from the graph G_{s_+} we would have to equip the graph with additional data: ordering of signed edges adjacent to every vertex. This allows us to construct a closed surface and the link diagram D on it so that $G_{s_+} = G_{s_+}(D)$. The construction imitates the 2-cell embedding of Edmonds (but every vertex corresponds to a circle and signs of edges regulate whether an edge is added inside or outside of the circle). If the surface we obtain is equal to S^2 we get the classical Khovanov homology. If we get a higher genus surface we have to use (APS) theory. This can also be utilized to construct Khovanov homology of virtual links (via Kuperberg minimal genus embedding theory (Ku)). For example, if the graph G_{s_+} is a loop with adjacent edge(s) ordered $e, -e$ then the diagram is composed of a meridian and a longitude on the torus.

are the graphs obtained from $G_{s_+}(D)$ *by deleting* $(G_{s_+}(D) - e_c)$ *and contracting* $(G_{s_+}(D)/e_c)$*, respectively, the edge* e_c *(compare Fig.3.2).*

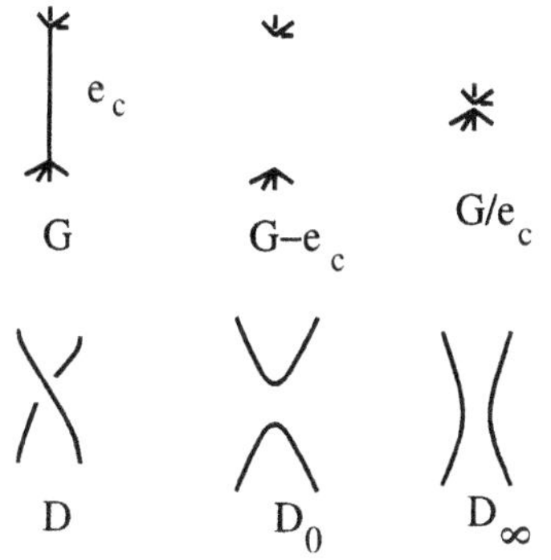

Fig. 3.2

Example 3.7.
Consider the 2-component alternating link 6^2_2 *(*$\frac{10}{3}$ *rational link), with* $G_{s_+}(D) = G_{s_-}(D)$ *being a square with one edge tripled (this is a self-dual graph); see Fig 3.3. Corollary 3.5 does not apply to this case but Theorem 3.2 guarantees* $\mathbb{Z}_2$ *torsion at* $H_{2,6}(D)$ *and* $H_{-4,-6}(D)$.
In fact, the KhoHo (Sh-2) computation gives the following Khovanov homology[15]*:* $H_{6,14} = H_{6,10} = H_{4,10} = \mathbb{Z}$, $H_{2,6} = \mathbb{Z} \oplus \mathbb{Z}_2$, $H_{2,2} = \mathbb{Z}$, $H_{0,2} = \mathbb{Z} \oplus \mathbb{Z}_2$, $H_{0,-2} = \mathbb{Z}$, $H_{-2,-2} = \mathbb{Z} \oplus \mathbb{Z}_2$, $H_{-2,-6} = \mathbb{Z}$, $H_{-4,-6} = \mathbb{Z}_2$, $H_{-4,-10} = \mathbb{Z}$, $H_{-6,-10} = H_{-6,-14} = \mathbb{Z}$.

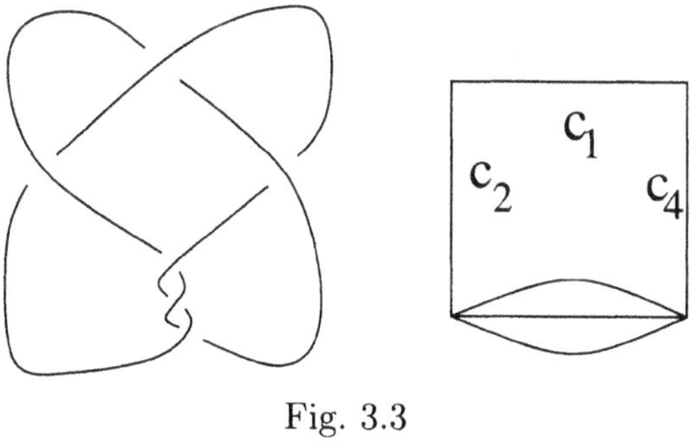

Fig. 3.3

[15] Tables and programs by Bar-Natan and Shumakovitch (BN-3; Sh-2) use the version of Khovanov homology for oriented diagrams, and the variable $q = A^{-2}$, therefore their monomial $q^a t^b$ corresponds to the free part of the group $H_{i,j}(D;\mathbb{Z})$ for $j = -2b + 3w(D)$, $i = -2a + w(D)$ and the monomial $Q^a t^b$ corresponds to the $\mathbb{Z}_2$ part of the group again with $j = -2b + 3w(D)$, $i = -2a + w(D)$. KhoHo gives the torsion part of the polynomial for the oriented link 6^2_2, with $w(D) = -6$, as $Q^{-6}t^{-1} + Q^{-8}t^{-2} + Q^{-10}t^{-3} + Q^{-12}t^{-4}$.

4. Torsion in the Khovanov homology of alternating and adequate links

We show in this section how to use technical results from the previous sections to prove Shumakovitch's result on torsion in the Khovanov homology of alternating links and the analogous result for a class of adequate diagrams.

Theorem 4.1 (Shumakovitch). *The alternating link has torsion free Khovanov homology if and only if it is the trivial knot, the Hopf link or the connected or split sum of copies of them. The nontrivial torsion always contains the $\mathbb{Z}_2$ subgroup.*

The fact that the Khovanov homology of the connected sum of Hopf links is a free group, is discussed in Section 6 (Corollary 6.6).

We start with the "only if" part of the proof by showing the following geometric fact.

Lemma 4.2. *Assume that D is a link diagram which contains a clasp: either $T_{[-2]} =$*  *or $T_{[2]} =$* 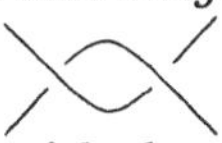 *. Assume additionally that the clasp is not a part of the Hopf link summand of D. Then if the clasp is of $T_{[-2]}$ type then the associated graph $G_{s_+}(D)$ has a singular edge. If the clasp is of $T_{[2]}$ type then the associated graph $G_{s_-}(D)$ has a singular edge. Furthermore the singular edge is not a loop.*

Proof. Consider the case of the clasp $T_{[-2]}$, the case of $T_{[2]}$ being similar. The region bounded by the clasp corresponds to the vertex of degree 2 in $G_{s_+}(D)$. The two edges adjacent to this vertex are not loops and they are not singular edges only if they share the second endpoint as well. In that case our diagram looks like on the Fig. 4.1 so it clearly has a Hopf link summand (possibly it is just a Hopf link) as the north part is separated by a clasp from the south part of the diagram. □

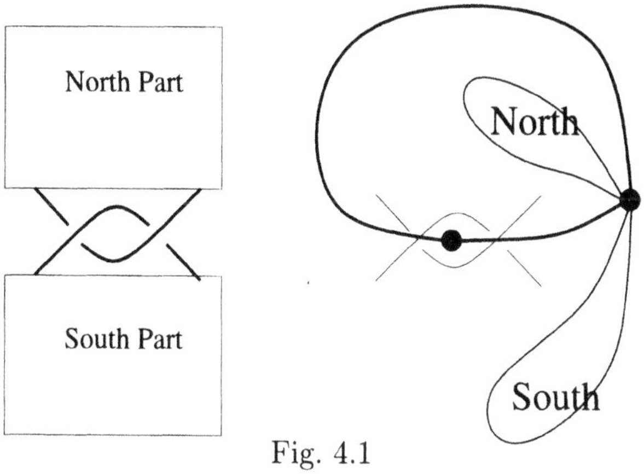

Fig. 4.1

Corollary 4.3. *If D is a $+$-adequate diagram (resp. $-$-adequate diagram) with a clasp of type $T_{[-2]}$ (resp. $T_{[2]}$), then Khovanov homology contains $\mathbb{Z}_2$-torsion or $T_{[-2]}$ (resp. $T_{[2]}$)) is a part of a Hopf link summand of D.*

Proof. Assume that $T_{[-2]}$ is not a part of Hopf link summand of D. By Lemma 4.2 the graph $G_{s_+}(D)$ has a singular edge. Furthermore, the graph $G_{s_+}(D)$ has no loops as D is $+$-adequate. If the graph has an odd cycle then $H_{N-2,N+2|s_+|-4}(D)$ has $\mathbb{Z}_2$ torsion by Theorem 2.2. If $G_{s_+}(D)$ is bipartite (i.e. it has only even cycles), then consider the cycle containing the singular edge. It is an even cycle of length at least 4, so by Theorem 3.2 $H_{N-4,N+2|s_+|-8}(D)$ has $\mathbb{Z}_2$ torsion. A similar proof works in $-$-adequate case. □

With this preliminary result we can complete our proof of Theorem 4.1.

Proof. First we prove the theorem for non-split, prime alternating links. Let D be a diagram of such a link without a nugatory crossing. D is an adequate diagram (i.e. it is $+$ and $-$ adequate diagram), so it is enough to show that if $G_{s_+}(D)$ (or $G_{s_-}(D)$) has a double edge then D can be modified by Tait flypes into a diagram with $T_{[-2]}$ (resp. $T_{[2]}$) clasp. This is a standard fact, justification of which is illustrated in Fig.4.2[16].

[16] For alternating diagrams, $G_{s_+}(D)$ and $G_{s_-}(D)$ are Tait graphs of D. These graphs are plane graphs and the only possibilities when multiple edges are not "parallel" is if our graphs are not 3-connected (as D is not a split link, graphs are connected, and because D is a prime link, the graphs are 2-connected). Tait flype corresponds to the special case of change of the graph in its 2-isomorphic class as illustrated in Fig.4.2.

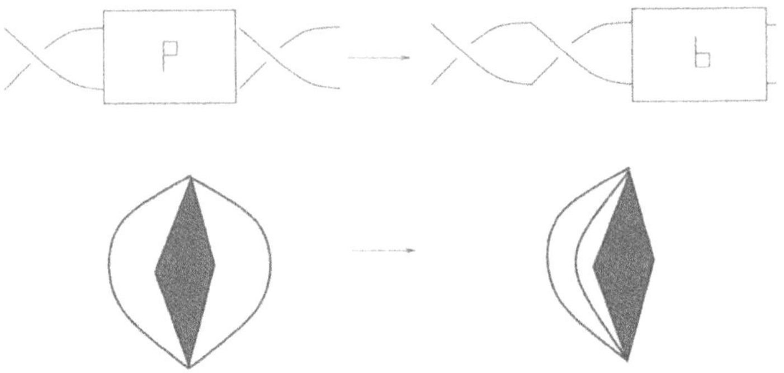

Fig. 4.2

If we do not assume that D is a prime link then we use the theorem by Menasco (Me) that prime decomposition is visible on the level of a diagram. In particular the Tait graphs $G_{s_+}(D)$ and $G_{s_-}(D)$ have block structure, where each block (2-connected component) corresponds to prime factor of a link. Using the previous results we see that the only situation when we didn't find torsion is if every block represents a Hopf link so D represents the sum of Hopf links (including the possibility that the graph is just one vertex representing the trivial knot).

If we relax condition that D is a non-split link then we use the fact, mentioned before, that for $D = D_1 \sqcup D_2$, Khovanov homology satisfies Künneth's formula, $H(D) = H(D_1) \otimes H(D_2)$. □

Example 4.4 (The 8_{19} knot).
The first entry in the knot tables which is not alternating is the $(3,4)$ torus knot, 8_{19}. It is $+$-adequate as it is a positive 3-braid, the closure of $(\sigma_1\sigma_2)^4$. Every positive braid is $+$-adequate but its associated graph $G_{s_+}(D)$ is composed of 2-gons. Furthermore the diagram D of 8_{19} is not $-$-adequate, Fig.4.3. KhoHo shows that the Khovanov homology of 8_{19} has torsion, namely $H_{2,2} = \mathbb{Z}_2$. This torsion is hidden deeply inside the homology spectrum[17], which starts from maximum $H_{8,14}(D) = \mathbb{Z}$ and ends on the minimum $H_{-2,-10}(D) = \mathbb{Z}$.

17 The full graded homology group is: $H_{8,14}(D) = H_{8,10}(D) = H_{4,6}(D) = \mathbb{Z}$, $H_{2,2} = \mathbb{Z}_2$, $H_{0,2}(D) = H_{2,-2}(D) = H_{0,-2}(D) = H_{-2,-4}(D) = H_{-2,-10}(D) = \mathbb{Z}$.

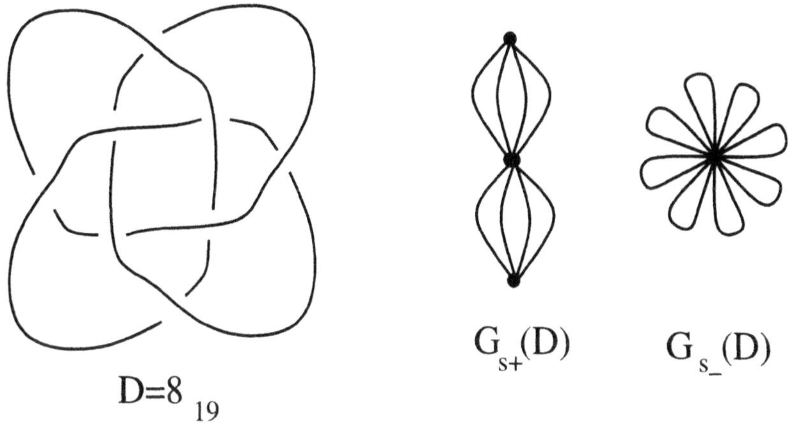

Fig. 4.3

The simplest alternating link which satisfies all conditions of Theorem 3.2 except for the existence of a singular edge, is the four component alternating link of 8 crossings 8^4_1 (Rol); Fig.4.4. We know that $H_{**}(8^4_1)$ has torsion (by using duality) but Theorem 3.2 does not guarantee torsion in $H_{N-4,N+2|s_+|-8}(D) = H_{4,8}(D)$, the graph $G_{s_+}(D)$ is a square with every edge doubled; Fig.4.3. We checked, however using KhoHo the torsion part and in fact $T_{4,8}(D) = \mathbb{Z}_2$. This suggests that Theorem 3.2 can be improved[18].

[18] In (BN-2) the figure describes, by mistake, the mirror image of 8^4_1. The full homology is as follows: $H_{8,16} = \mathbb{Z}$, $H_{8,12} = \mathbb{Z} = H_{6,12}$, $H_{4,8} = \mathbb{Z}_2 \oplus \mathbb{Z}^4$, $H_{4,4} = \mathbb{Z}$, $H_{2,4} = \mathbb{Z}_2^4$, $H_{2,0} = \mathbb{Z}^4$, $H_{0,0} = \mathbb{Z}^7$, $H_{0,-4} = \mathbb{Z}^6$, $H_{-2,-4} = \mathbb{Z}_2^3 \oplus \mathbb{Z}^3$, $H_{-2,-8} = \mathbb{Z}$, $H_{-4,-8} = \mathbb{Z}^3$, $H_{-4,-12} = \mathbb{Z}^3$, $H_{-6,-12} = \mathbb{Z}_2^3$, $H_{-6,-16} = \mathbb{Z}^3$, $H_{-8,-16} = \mathbb{Z}$, $H_{-8,-20} = \mathbb{Z}$. In KhoHo the generating polynomials, assuming $w(8^4_1) = -8$, are: KhPol("8a",21)= $[((q^{18}+q^{16})*t^8+3*q^{16}*t^7+(3*q^{14}+3*q^{12})*t^6+(q^{12}+3*q^{10})*t^5+(6*q^{10}+7*q^8)*t^4+4*q^8*t^3+(q^6+4*q^4)*t^2+q^2*t+(q^2+1))/(q^{20}*t^8), (3*Q^{10}*t^5+3*Q^8*t^4+Q^6*t^3+4*Q^2*t+1)/(Q^{16}*t^6)]$.

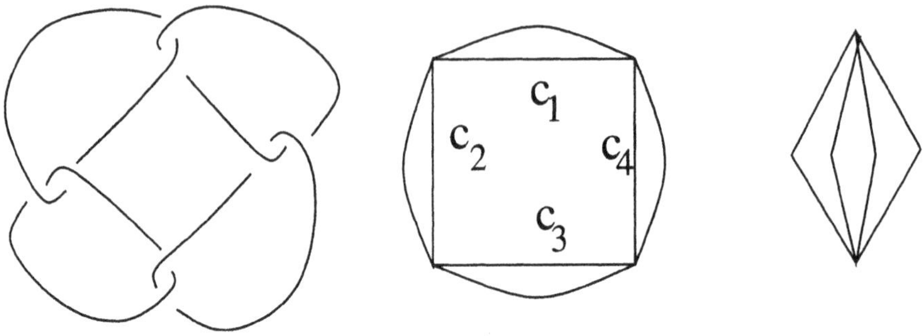

Fig. 4.4

5. Thickness of Khovanov homology and almost alternating links

We define, in this section, the notion of an H-k-thick link diagram and relate it to $(k-1)$-almost alternating diagrams. In particular we give a short proof of Lee's theorem (Lee-1) (conjectured by Khovanov, Bar-Natan, and Garoufalidis) that alternating non-split links are H-1-thick (H-thin in Khovanov terminology).

Definition 5.1. *We say that a link is k-almost alternating if it has a diagram which becomes alternating after changing k of its crossings.*

As noted in Property 1.4 the "extreme" terms of Khovanov chain comples are $C_{N,N+2|s_+|}(D) = C_{-N,-N-2|s_-|}(D) = \mathbb{Z}$. In the following definition of a H-(k_1,k_2)-thick diagram we compare indices of actual Khovanov homology of D with lines of slope 2 going through the points $(N, N+2|s_+|)$ and $(-N, -N-2|s_-|)$.

Definition 5.2. *(i) We say that a link diagram, D of N crossings is H-(k_1,k_2)-thick if $H_{i,j}(D) = 0$ with a possible exception of i and j satisfying:*

$$N - 2|s_-| - 4k_2 \leq j - 2i \leq 2|s_+| - N + 4k_1.$$

(ii) We say that a link diagram of N crossings is H-k-thick[19] *if, it is H-(k_1, k_2)-thick where k_1 and k_2 satisfy:*

$$k \geq k_1 + k_2 + \frac{1}{2}(|s_+| + |s_-| - N).$$

(iii) We define also (k_1, k_2)-thickness (resp. k-thickness) of Khovanov homology separately for the torsion part (we use the notation TH-(k_1, k_2)-thick diagram), and for the free part (we use the notation FH-(k_1, k_2)-thick diagram).

Our FH-1-thick diagram is a H-thin diagram in (Kh-2; Lee-1; BN-1; Sh-1).

With the above notation we are able to formulate our main result of this section.

Theorem 5.3.
If the diagram $D_\infty = D($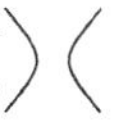*$)$ is H-$(k_1(D_\infty), k_2(D_\infty))$-thick and the diagram $D_0 = D($**$)$ is H-$(k_1(D_0), k_2(D_0))$-thick, then the diagram $D_+ = D($**$)$ is H-$(k_1(D_+), k_2(D_+))$-thick where*
$k_1(D_+) = \max(k_1(D_\infty) + \frac{1}{2}(|s_+(D_\infty)| - |s_+(D_+)| + 1), k_1(D_0))$ *and* .
$k_2(D_+) = \max(k_2(D_\infty), k_2(D_0)) + \frac{1}{2}(|s_-(D_0)| - |s_-(D_+)| + 1))$.
In particular

(i) if $|s_+(D_+)| - |s_+(D_\infty)| = 1$, as is always the case for a $+$-adequate diagram, then $k_1(D_+) = \max(k_1(D_\infty), k_1(D_0))$,

(ii) if $|s_-(D_+)| - |s_-(D_0| = 1$, as is always the case for a $-$-adequate diagram, then $k_2(D_+) = \max(k_2(D_\infty), k_2(D_0))$.

Proof. We formulated our definitions so that our proof follows almost immediately via the Viro's long exact sequence of Khovanov homology:

$$\dots \to H_{i+1,j-1}(D_0) \xrightarrow{\partial} H_{i+1,j+1}(D_\infty) \xrightarrow{\alpha} H_{i,j}(D_+) \xrightarrow{\beta}$$

$$H_{i-1,j-1}(D_0) \xrightarrow{\partial} H_{i-1,j+1}(D_\infty) \to \dots$$

If $0 \neq h \in H_{i,j}(D_+)$ then either $h = \alpha(h')$ for $0 \neq h' \in H_{i+1,j+1}(D_\infty)$ or $0 \neq \beta(h) \in H_{i-1,j-1}(D_0)$. Thus if $H_{i,j}(D_+) \neq 0$ theneither $H_{i+1,j+1}(D_\infty) \neq$

---

[19] Possibly, the more appropriate name would be H-k-thin diagram, as the width of Khovanov homology is bounded from above by k. Khovanov ((Kh-2), page 7) suggests the term homological width; $hw(D) = k$ if homology of D lies on k adjacent diagonals (in our terminology, D is $k-1$ thick).

0 or $H_{i-1,j-1}(D_0) \neq 0$. The first possibility gives the inequalities involving $(j+1) - 2(i+1)$:

$$N(D_\infty)-2|s_-(D_\infty)|-4k_2(D_\infty) \leq j-2i-1 \leq 2|s_+(D_\infty)|-N(D_\infty)+4k_1(D_\infty)$$

which, after observing that $|s_-(D_+)| = |s_-(D_\infty)|$, leads to:

$$N(D_+) - 2|s_-(D_+)| - 4k_2(D_\infty) \leq j - 2i \leq$$
$$2|s_+(D_+)| - N(D_+) + 4k_1(D_\infty) + 2(|s_+(D_\infty)| - |s_+(D_+)| + 1).$$

The second possibility gives the inequalities involving $(j-1)-2(i-1)$:

$$N(D_0)-2|s_-(D_0)|-4k_2(D_0) \leq j-2i+1 \leq 2|s_+(D_0)|-N(D_0)+4k_1(D_0)$$

which, after observing that $|s_+(D_+)| = |s_+(D_0)|$, leads to:

$$N(D_+) - 2|s_-(D_+)| - 4k_2(D_0) - 2(|s_-(D_0)| - |s_-(D_+)| + 1) \leq$$
$$j - 2i \leq 2|s_+(D_+)| - N(D_+) + 4k_1(D_0).$$

Combining these two cases we obtain the conclusion of Theorem 5.3. □

Corollary 5.4. *If D is an adequate diagram such that, for some crossing of D, the diagrams D_0 and D_∞ are H-(k_1, k_2)-thick (resp. H-k-thick) then D is H-(k_1, k_2)-thick (resp. H-k-thick).*

Corollary 5.5. *Every alternating non-split diagram without a nugatory crossing is H-$(0,0)$-thick and H-1-thick.*

Proof. The H-(k_1, k_2)-thickness in Corollary 5.4 follows immediately from Theorem 5.3. To show H-k-thickness we observe additionally that for an adequate diagram D_+ one has $|s_+(D_0)| + |s_-(D_0)| - N(D_0) = |s_+(D_+)| + |s_-(D_+)| - N(D_+) = |s_+(D_\infty)| + |s_-(D_\infty)| - N(D_\infty)$.

We prove Corollary 5.5 using induction on the number of crossings a slightly more general statement allowing nugatory crossings.
(+) If D is an alternating non-split +-adequate diagram then $H_{i,j}(D) \neq 0$ can happen only for $j - 2i \leq 2|s_+(D)| - N(D)$.
(−) If D is an alternating non-split −-adequate diagram then $H_{i,j}(D) \neq 0$ can happen only for $N(D) - 2|s_-(D)| \leq j - 2i$.
If the diagram D from (+) has only nugatory crossings then it represents the trivial knot and its nontrivial Khovanov homology are $H_{N,3N-2}(D) = H_{N,3N+2}(D) = \mathbb{Z}$. Because $|s_+(D)| = N(D) + 1$ in this case, the inequality (+) holds. In the inductive step we use the property

of a non-nugatory crossing of a non-split +-adequate diagram, namely D_0 is also an alternating non-split +-adequate diagram and inductive step follows from Theorem 5.3.
Similarly one proves the condition (–). Because the non-split alternating diagram without nugatory crossings is an adequate diagram, therefore Corollary 5.5 follows from Conditions (+) and (–). □

The conclusion of the theorem is the same if we are interested only in the free part of Khovanov homology (or work over a field). In the case of the torsion part of the homology we should take into account the possibility that torsion "comes" from the free part of the homology, that is $H_{i+1,j+1}(D_\infty)$ may be torsion free but its image under α may have torsion element.

Theorem 5.6. *If* $T_{i,j}(D_+) \neq 0$ *then either*
(1) $T_{i+1,j+1}(D_\infty) \neq 0$ *or* $T_{i-1,j-1}(D_0) \neq 0$,
or
(2) $FH_{i+1,j+1}(D_\infty) \neq 0$ *and* $FH_{i+1,j-1}(D_0) \neq 0$.

Proof. From the long exact sequence of Khovanov homology it follows that the only way the torsion is not related to the torsion of $H_{i+1,j+1}(D_\infty)$ or $H_{i-1,j-1}(D_0)$ is the possibility of torsion created by taking the quotient
$FH_{i+1,j+1}(D_\infty)/\partial(FH_{i+1,j-1}(D_0))$ and in this case both groups $FH_{i+1,j+1}(D_\infty)$ and $FH_{i+1,j-1}(D_0)$ have to be nontrivial. □

Corollary 5.7. *If D is an alternating non-split diagram without a nugatory crossing then D is TH-$(0,-1)$-thick and TH-0-thick. In other words if* $T_{i,j}(D) \neq 0$ *then* $j - 2i = 2|s_+(D)| - N(D) = N(D) - 2|s_-(D)| + 4$.

Proof. We proceed in the same (inductive) manner as in the proof of Corollary 5.5, using Theorem 5.7 and Corollary 5.5. In the first step of the induction we use the fact that the trivial knot has no torsion in Khovanov homology. □

The interest in H-thin diagrams was motivated by the observation (proved by Lee) that diagrams of non-split alternating links are H-thin (see Corollary 5.5). Our approach allows the straightforward generalization to k-almost alternating diagrams.

Corollary 5.8. *Let D be a non-split k-almost alternating diagram without a nugatory crossing. Then D is H-(k,k)-thick and TH-$(k,k-1)$-thick.*

Proof. The corollary holds for $k = 0$ (alternating diagrams) and we use an induction on the number of crossings needed to change the diagram D to an alternating digram, using Theorem 5.3 in each step. □

We were assuming throughout the section that our diagrams are non-split. This assumption was not always necessary. In particular even the split alternating diagram without nugatory crossings is H-$(0,0)$-thick as follows from the following observation.

Lemma 5.9. *If the diagrams D' and D'' are H-(k_1', k_2')-thick and H-(k_1'', k_2'')-thick, respectively, then the diagram $D = D' \sqcup D''$ is H-$(k_1' + k_1'', k_2' + k_2'')$-thick.*

Proof. Lemma 5.9 follows from the obvious fact that in the split sum $D = D' \sqcup D''$ we have $N(D) = N(D') + N(D'')$, $|s_+(D)| = |s_+(D')| + |s_+(D'')|$ and $|s_-(D)| = |s_-(D')| + |s_-(D'')|$. □

Khovanov observed ((Kh-2), Proposition 7) that adequate non-alternating knots are not H-1-thick. We are able to proof the similar result about torsion of adequate non-alternating links.

Theorem 5.10.
Let D be a connected adequate diagram which does not represent an alternating link and such that $G_{s_+}(D)$ and $G_{s_-}(D)$ have either an odd cycle or an even cycle with a singular edge, then D is not TH-0-thick diagram. More generally, D is at best TH-$\frac{1}{2}(N+2-(|s_+(D)|+|s_-(D)|)$-thick.

Proof. The first part of Theorem 5.10 follows from the second part because by Proposition 1.4 (Wu's Lemma), $\frac{1}{2}(N + 2 - (|s_+(D)| + |s_+(D)|) > 0$ for a diagram which is not a connected sum of alternating diagrams. By Theorems 2.2, 3.2 and Corollary 3.3, $TH_{i,j}(D)$ is nontrivial on slope 2 diagonals $j - 2i = 2|s_+| - N$ and $N - 2|s_-| + 4$. The j distance between these diagonals is $N - 2|s_-| + 4 - (2|s_+| - N) = 2(N + 2 - (|s_+(D)| + |s_+(D)|)$, so the theorem follows. □

Example 5.11.
Consider the knot 10_{153} (in the notation of (Rol)). It is an adequate non-alternating knot. Its associated graphs $G_{s_+}(10_{153})$ and $G_{s_-}(10_{153})$ have triangles (Fig.5.1) so Theorem 5.10 applies. Here $|s_+| = 6$, $|s_-| = 4$ and by Theorem 2.2, $H_{8,18}(10_{153})$ and $H_{-10,-14}(10_{153})$ have $\mathbb{Z}_2$ torsion. Thus support of torsion requires at least 2 adjacent diagonals[20]

[20] Checking (Sh-2), gives the full torsion of the Khovanov homology of 10_{153} as: $T_{8,18} = T_{4,10} = T_{2,6} = T_{0,6} = T_{-2,-2} = T_{-4,-2} = T_{-6,-6} = T_{-10,-14} = \mathbb{Z}_2$.

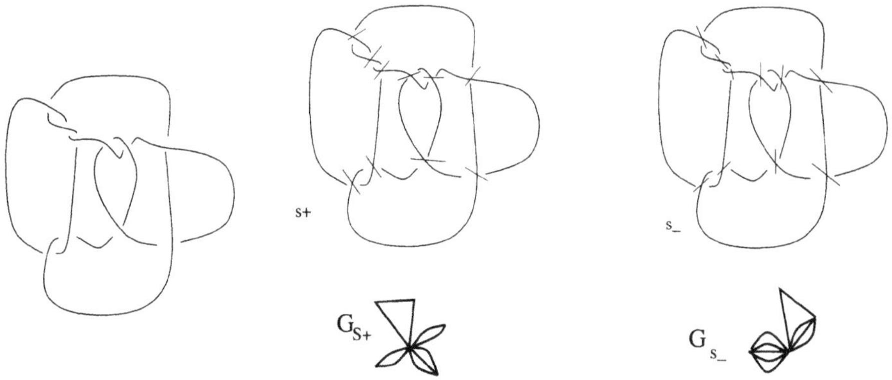

Fig. 5.1

Corollary 5.12. *Any doubly adequate link which is not an alternating link is not TH-0-thick.*

6. Hopf link addition

We find, in this section, the structure of the Khovanov homology of connected sum of n copies of the Hopf link, as promised in Section 5. As a byproduct of our method, we are able to compute Khovanov homology of a connected sum of a diagram D and the Hopf link D_h, Fig 6.1, confirming a conjecture by A.Shumakovitch that the Khovanov homology of the connected sum of D with the Hopf link, is the double of the Khovanov homology of D.

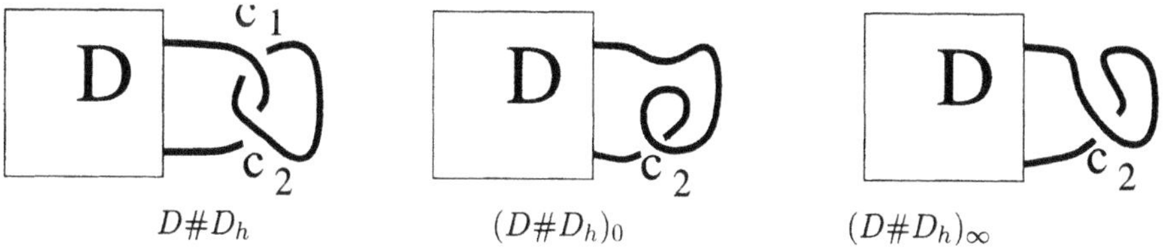

Fig. 6.1

Theorem 6.1. *For every diagram D we have the short exact sequence of Khovanov homology*[21]

$$0 \to H_{i+2,j+4}(D) \xrightarrow{\alpha_h} H_{i,j}(D\#D_h) \xrightarrow{\beta_h} H_{i-2,j-4}(D) \to 0$$

where α_h is given on a state S by Fig.6.2(a) and β_h is a projection given by Fig.6.2(b) (and 0 on other states). The theorem holds for any ring of coefficients, $\mathcal{R}$, not just $\mathcal{R} = \mathbb{Z}$.

(a) α_h (b) β_h

Fig. 6.2

Theorem 6.2.
The short exact sequence of homology from Theorem 6.1 splits, so we have

$$H_{i,j}(D\#D_h) = H_{i+2,j+4}(D) \oplus H_{i-2,j-4}(D).$$

Proof. To prove Theorem 6.1 we consider the long exact sequence of the Khovanov homology of the diagram $D\#D_h$ with respect to the first crossing of the diagram, e_1 (Fig.6.1). To simplify the notation we assume that $\mathcal{R} = \mathbb{Z}$ but our proof works for any ring of coefficients.

$$\ldots \to H_{i+1,j-1}((D\#D_h)_0) \xrightarrow{\partial} H_{i+1,j+1}((D\#D_h)_\infty) \xrightarrow{\alpha} H_{i,j}(D\#D_h) \xrightarrow{\beta}$$
$$H_{i-1,j-1}((D\#D_h)_0) \xrightarrow{\partial} H_{i-1,j+1}((D\#D_h)_\infty) \to \ldots$$

We show that the homomorphism ∂ is the zero map. We use the fact that $(D\#D_h)_0$ differs from D by a positive first Reidemeister move R_{+1} and that $(D\#D_h)_\infty$ differs from D by a negative first Reidemeister move R_{-1}; Fig.6.1. We know, see (APS) for example, that the chain map
$r_{-1} : \mathcal{C}(D) \to \mathcal{C}(R_{-1}(D))$ given by $r_{-1}($  $) =$  yields the isomorphism of homology:

$$r_{-1*} : H_{i,j}(D) \to H_{i-1,j-3}(R_{-1}(D))$$

[21] Theorems 6.1 and 6.2 hold for a diagram D on any surface F and for any ring of coefficients $\mathcal{R}$ with the restriction that for $F = RP^2$ we need $2\mathcal{R} = 0$. In this more general case of a manifold being I-bundle over a surface, we use definitions and setting of (APS).

and the chain map $\bar{r}_{+1}(\mathcal{C}(R_{+1}(D)) = \mathcal{C}((D)$ given by the projection with $\bar{r}_{+1}($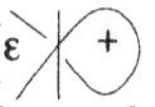 $) = ($  $)$ and 0 otherwise, induces the isomorphism of homology:

$$\bar{r}_{+1*} : H_{i+1,j+3}(R_{+1}(D)) \to H_{i,j}(D).$$

From these we get immediately that the composition homomorphism:

$$r_{-1*}^{-1} \partial \bar{r}_{+1*}^{-1} : H_{i,j-4}(D) \to H_{i+2,j+4}(D)$$

is the zero map by considering the composition of homomorphisms

$$H_{i,j-4}\,(D) \overset{\bar{r}_{+1*}^{-1}}{\to} H_{i+1.j-1}((D\#D_h)_0) \overset{\partial}{\to} H_{i+1.j+1}\,(D\#D_h)_\infty) \overset{r_{-1*}^{-1}}{\to} H_{i+2,j+4}(D).$$

□

Let $h(a,b)(D)$ (resp. $h_{\mathcal{F}}(a,b)(D)$ for a field $\mathcal{F}$) be the generating polynomial of the free part of $H_{**}(D)$ (resp. $H_{**}(D;\mathcal{F})$), where kb^ia^j (resp. $k_{\mathcal{F}}b^ia^j$) represents the fact that the free part of $H_{i,j}(D)$, $FH_{i,j}(D) = \mathbb{Z}^k$ (resp. $H_{i,j}(D;\mathcal{F}) = \mathcal{F}^k$).

Theorem 6.2 will be proved in several steps.

Lemma 6.3.
If the module $H_{i-2,j-4}(D;\mathcal{R})$ is free (e.g. $\mathcal{R}$ is a field) then the sequence from Theorem 6.1 splits and $H_{i,j}((D\#D_h);\mathcal{R}) = H_{i-2,j-4}(D;\mathcal{R}) \oplus H_{i+2,j+4}(D;\mathcal{R})$ or shortly $H_{**}(D\#D_h;\mathcal{R}) = H_{**}(D;\mathcal{R})(b^2a^4+b^{-2}a^{-4})$.

For the free part we have always $FH_{i,j}\,(D\#D_h) = FH_{i+2,j+4}(D) \oplus FH_{i-2,j-4}(D)$ or in the language of generating functions: $h(a,b)(D\#D_h) = (b^2a^4 + b^{-2}a^{-4})h(a,b)(D)$.

Proof. The first part of the lemma follows immediately from Theorem 6.1 which holds for any ring of coefficients, in particular $rank(FH_{i,j}(D\# D_h)) = rank(FH_{i+2,j+4}(D)) + rank(FH_{i-2,j-4}(D))$. □

Lemma 6.4.
There is the exact sequence of $\mathbb{Z}_p$ linear spaces:

$$0 \to H_{i+2,j+4}(D) \otimes \mathbb{Z}_p \to H_{i,j}(D\#D_h) \otimes \mathbb{Z}_p \to H_{i-2,j-4}(D) \otimes \mathbb{Z}_p \to 0.$$

Proof. Our main tool is the universal coefficients theorem (see, for example, (Ha); Theorem 3A.3) combined with Lemma 6.3. By the second part of Lemma 6.3 it suffices to prove that:

$$T_{i,j}(D\#D_h) \otimes \mathbb{Z}_p = T_{i+2,j+4}(D) \otimes \mathbb{Z}_p \oplus T_{i-2,j-4}(D) \otimes \mathbb{Z}_p.$$

From the universal coefficients theorem we have:

$$H_{i,j}((D\#D_h);\mathbb{Z}_p) = H_{i,j}(D\#D_h)\otimes\mathbb{Z}_p \oplus Tor(H_{i-2,j}(D\#D_h),\mathbb{Z}_p)$$

and $Tor(H_{i-2,j}(D\#D_h),\mathbb{Z}_p) = T_{i-2,j}(D\#D_h)\otimes\mathbb{Z}_p$ and the analogous formulas for the Khovanov homology of D. Combining this with both parts of Lemma 6.3, we obtain:

$$T_{i,j}(D\#D_h)\otimes\mathbb{Z}_p \oplus T_{i-2,j}(D\#D_h)\otimes\mathbb{Z}_p =$$

$$(T_{i+2,j+4}(D)\otimes\mathbb{Z}_p\oplus T_{i,j+4}(D)\otimes\mathbb{Z}_p)\oplus(T_{i-2,j-4}(D)\otimes\mathbb{Z}_p\oplus T_{i-4,j-4}(D)\otimes\mathbb{Z}_p).$$

We can express this in the language of generating functions assuming that $t(b,a)(D)$ is the generating function of dimensions of $T_{i,j}(D)\otimes\mathbb{Z}_p$:

$$(1+b^{-2})t(b,a)(D\#D_h) = (1+b^{-2})(b^2a^4+b^{-2}a^{-4}t(b,a)(D)).$$

Therefore $t(b,a)(D\#D_h) = (b^2a^4+b^{-2}a^{-4})t(b,a)(D)$ and $dim(T_{i,j}(D\#D_h)\otimes\mathbb{Z}_p) = dim(T_{i+2,j+4}(D)\otimes\mathbb{Z}_p)+dim(T_{i,j+4}(D)\otimes\mathbb{Z}_p)$. The lemma follows by observing that the short exact sequence with $\mathbb{Z}$ coefficients leads to the sequence

$$0\to ker(\alpha_p)\to H_{i+2,j+4}(D)\otimes\mathbb{Z}_p \overset{\alpha_p}{\to}$$

$$H_{i,j}(D\#D_h)\otimes\mathbb{Z}_p\to H_{i-2,j-4}(D)\otimes\mathbb{Z}_p\to 0.$$

By the previous computation $dim(ker(\alpha_p)) = 0$ and the proof is completed. □

To finish our proof of Theorem 6.2 we only need the following lemma.

Lemma 6.5. *Consider a short exact sequence of finitely generated abelian groups:*

$$0\to A\to B\to C\to 0.$$

If for every prime number p we have also the exact sequence:

$$0\to A\otimes\mathbb{Z}_p\to B\otimes\mathbb{Z}_p\to C\otimes\mathbb{Z}_p\to 0$$

then the exact sequence $0\to A\to B\to C\to 0$ *splits and* $B = A\oplus C$.

Proof. Assume, for contradiction, that the sequence $0\to A\overset{\alpha}{\to} B\to C\to 0$ does not split. Then there is an element $a\in A$ such that $\alpha(a)$ is not p-primitive in B, that is $\alpha(a) = pb$ for $b\in B$ and p a prime number and b does not lies in the subgroup of B span by $\alpha(a)$ (to see that such an a exists one can use the maximal decomposition of A and B into cyclic subgroups (e.g. $A = \mathbb{Z}^k\oplus_{p,i}\mathbb{Z}_{p^i}^{k_{p,i}}$). Now comparing dimensions of linear spaces $A\otimes\mathbb{Z}_p$, $B\otimes\mathbb{Z}_p$, $C\otimes\mathbb{Z}_p$ (e.g. $dim(A\otimes\mathbb{Z}_p) = k+k_{p,1}+k_{p,2}+\ldots$) we see that the sequence $0\to A\otimes\mathbb{Z}_p\to B\otimes\mathbb{Z}_p\to C\otimes\mathbb{Z}_p\to 0$ is not exact, a contradiction. □

Corollary 6.6. *For the connected sum of n copies of the Hopf link we get*[22]
$H_{*,*}(D_h\#...\#D_h) = h(a,b)(D) = (a^2+a^{-2})(a^4b^2+a^{-4}b^{-2})^n$

Remark 6.7. *Notice that* $h(a,b)(D_h)-h(a,b)(OO) = (a^2+a^{-2})(a^4b^2+a^{-4}b^{-2})-(a^2+a^{-2})^2 = b^{-2}a^{-4}(a^2+a^{-2})(1+ba)(1-ba)(1+ba^3)(1-ba^3)$. *This equality may serve as a starting point to formulate a conjecture for links, analogous to Bar-Natan-Garoufalidis-Khovanov conjecture (Kh-2; Ga),(BN-1) (Conjecture 1), formulated for knots and proved for alternating knots by Lee (Lee-1).*

7. Reduced Khovanov homology

Most of the results of Sections 5 and 6 can be adjusted to the case of reduced Khovanov homology[23]. We introduce the concept of H^r-(k_1,k_2)-thick diagram and formulate the result analogous to Theorem 5.3. The highlight of this section is the exact sequence connecting reduced and unreduced Khovanov homology.

Choose a base point, b, on a link diagram D. Enhanced states, $S(D)$ can be decomposed into disjoint union of enhanced states $S_+(D)$ and $S_-(D)$, where the circle containing the base point is positive, respectively negative. The Khovanov abelian group $\mathcal{C}(D) = \mathcal{C}_+(D) \oplus \mathcal{C}_-(D)$ where $\mathcal{C}_+(D)$ is spanned by $S_+(D)$ and $\mathcal{C}_-(D)$ is spanned by $S_-(D)$. $\mathcal{C}_+(D)$ is a chain subcomplex of $\mathcal{C}(D)$. Its homology, $H^r(D)$, is called the reduced Khovanov homology of D, or more precisely of (D,b) (it may depends on the component on which the base point lies). Using the long exact sequence of reduced Khovanov homology we can reformulate most of the results of Sections 5 and 6.

Definition 7.1.
We say that a link diagram, D of N crossings is H^r-(k_1,k_2)-*thick if* $H^r_{i,j}(D) = 0$ *with a possible exception of i and j satisfying:*

$$N - 2|s_-| - 4k_2 + 4 \leq j - 2i \leq 2|s_+| - N + 4k_1.$$

With this definition we have

[22] In the oriented version (with the linking number equal to n, so the writhe number $w = 2n$) and with Bar-Natan notation one gets: $q^{3n}t^n(q+q^{-1})(q^2t+q^{-2}t^{-1})^n$, as computed first by Shumakovitch.

[23] Introduced by Khovanov; we follow here Shumakovitch's approach adjusted to the framed link version.

Theorem 7.2.
If the diagram D_∞ *is* H^r*-*$(k_1(D_\infty), k_2(D_\infty))$*-thick and the diagram* D_0 *is* H^r*-*$(k_1(D_0), k_2(D_0))$*-thick, then the diagram* D_+ *is* H^r*-*$(k_1(D_+), k_2(D_+))$*-thick where*
$k_1(D_+) = \max(k_1(D_\infty) + \frac{1}{2}(|s_+(D_\infty)| - |s_+(D_+)| + 1), k_1(D_0))$ *and*
$k_2(D_+) = k_2(D_+) = \max(k_2(D_\infty), k_2(D_0) + \frac{1}{2}(|s_-(D_0)| - |s_-(D_+)| + 1))$.
Every alternating non-split diagram D *without a nugatory crossing is* H^r*-*$(0,0)$*-thick, and* $H^r_{**}(D)$ *is torsion free (Lee-1; Sh-1).*

The graded abelian group $\mathcal{C}_-(D) = \bigoplus_{i,j} \mathcal{C}_{i,j;-}(D)$ is not a sub-chain complex of $\mathcal{C}(D)$, as $d(S)$ is not necessary in $\mathcal{C}_-(D)$, for $S \in S_-(D)$. However the quotient $\mathcal{C}^-(D) = \mathcal{C}(D)/\mathcal{C}_+(D)$ is a graded chain complex and as a graded abelian group it can be identified with $\mathcal{C}_-(D)$.

Theorem 7.3. *(i) We have the following short exact sequence of chain complexes:*

$$0 \to \mathcal{C}_+(D) \xrightarrow{\phi} \mathcal{C}(D) \xrightarrow{\psi} \mathcal{C}^-(D) \to 0.$$

(ii) We have the following long exact sequence of homology:

$$\ldots \to H^r_{i,j}(D) \xrightarrow{\phi_*} H_{i,j}(D) \xrightarrow{\psi_*} H^{\bar{r}}_{i,j}(D) \xrightarrow{\partial} H^r_{i-2,j}(D) \to \ldots$$

where $H^{\bar{r}}_{i,j}(D)$ *is the homology of* $C^-(D)$*, called co-reduced Khovanov homology. The boundary map can be roughly interpreted for a state* $S \in S_-(D)$ *as* $d(S)$ *restricted to* $\mathcal{C}_+(D)$.

Applications of Theorem 7.3 will be the topic of a sequel paper, here we only mention that if D is an alternating non-split diagram without a nugatory crossing then $H^r_{i,j}(D) = H^{\bar{r}}_{i,j-4}(D)$.

8. Acknowledgments

We would like to thank Alexander Shumakovitch for inspiration and very helpful discussion.

References

M. M. Asaeda, J. H. Przytycki, A. S. Sikora, Khovanov homology of links in I-bundles over surfaces, preprint 2003 (AGT, to appear).
e-print: http://arxiv.org/abs/math/0402402

D. Bar-Natan, On Khovanov's categorification of the Jones polynomial, *Algebraic and Geometric Topology* 2, 2002, 337-370.
http://xxx.lanl.gov/abs/math.QA/0201043

D. Bar-Natan, Introduction to Khovanov homology; http://at.yorku.ca/cgi-bin/amca/calg-65

D. Bar-Natan, http://www.math.toronto.edu/ drorbn/KAtlas/

S. Garoufalidis, A conjecture on Khovanov's invariants, University of Warwick preprint, October 2001 (to appear in Fundamenta Mathematicae).
http://www.math.gatech.edu/ stavros/publications/khovanov.pdf

A. Hatcher, Algebraic Topology, Cambridge University Press, 2002.
http://www.math.cornell.edu/ hatcher/AT/ATch3.pdf

M.Jakobsson, An invariant of link cobordisms,
http://xxx.lanl.gov/abs/math.GT/0206303

L.H. Kauffman, An invariant of regular isotopy, *Trans. Amer. Math. Soc.*, 318(2), 1990, 417–471.

M. Khovanov, A categorification of the Jones polynomial, *Duke Math. J.* 101 (2000), no. 3, 359–426,
http://xxx.lanl.gov/abs/math.QA/9908171

M. Khovanov, Patterns in knot cohomology I, *Experiment. Math.* 12(3), 2003, 365-374,
http://arxiv.org/abs/math/0201306

M. Khovanov, Categorifications of the colored Jones polynomial, preprint,
http://xxx.lanl.gov/abs/math.QA/0302060

M. Khovanov, An invariant of tangle cobordisms, preprint, 2002
http://arxiv.org/abs/math.QA/0207264

G. Kuperberg, What is a virtual link? *Algebraic and Geometric Topology*, 3, 2003, 587-591,
http://www.maths.warwick.ac.uk/agt/AGTVol3/agt-3-20.abs.html

E. S. Lee, The support of the Khovanov's invariants for alternating knots,
http://arxiv.org/abs/math.GT/0201105

E. S. Lee, On Khovanov invariant for alternating links,
http://arxiv.org/abs/math.GT/0210213

W. B. R. Lickorish, M. B. Thistlethwaite, Some links with non-trivial polynomials and their crossing-numbers, *Comment. Math. Helv.*, 63, 1988, 527-539.

J. B. Listing, *Vorstudien zur Topologie*, Göttinger Studien (Abtheilung 1) 1 (1847), 811-875.

W. M. Menasco, Closed incompressible surfaces in alternating knot and link complements, *Topology* 23, 1984, 37-44.

K. Murasugi, Jones polynomial and classical conjectures in knot theory, *Topology*, 26(2), 1987, 187-194.

K. Murasugi, Jones polynomial and classical conjectures in knot theory, II, *Math. Proc. Camb. Phil. Soc.*, 102, 1987, 317-318.

K. Murasugi, On invariants of graphs with application to knot theory, *Trans. Amer. Math. Soc.*, 314 (1989) 1-49.

J. H. Przytycki, **Knots;** – From combinatorics of knot diagrams to the combinatorial topology based on knots, Cambridge University Press, to appear, 2005.

J. Rasmussen, Khovanov homology and the slice genus,
http://xxx.lanl.gov/abs/math.GT/0402131

D. Rolfsen, *Knots and links*. Publish or Perish, 1976 (second edition, 1990).

A. Shumakovitch, Torsion of the Khovanov Homology, Presentations at the conferences: Knots in Poland (July, 2003) and Knots in Washington XVII, December

20, 2003.
http://at.yorku.ca/cgi-bin/amca/camw-15
e-print (May 2004) of the related paper, Torsion of the Khovanov homology, http://arxiv.org/abs/math.GT/0405474

A. Shumakovitch, KhoHo: http://www.geometrie.ch/KhoHo/

M.B.Thistlethwaite, A spanning tree expansion of the Jones polynomial, *Topology*, 26 (1987), 297-309.

P. Traczyk, A combinatorial formula for the signature of alternating links, preprint, 1987 (to appear in Fundamenta Mathematicae).

O. Viro, Remarks on definition of Khovanov homology (to appear in Fundamenta Mathematicae),
http://arxiv.org/abs/math.GT/0202199

Y.-Q. Wu, Jones polynomial and the crossing number of links, Differential geometry and topology (Tjanjin, 1986-87), Lectures Notes in Math., 1369, Springer, Berlin - New York, 1989, 286-288.

Address for Offprints:

Józef H. Przytycki
Department of Mathematics
George Washington University
email: przytyck@gwu.edu

Marta M. Asaeda
Department of Mathematics
University of Iowa
email: asaeda@math.uiowa.edu

KHOVANOV HOMOLOGY FOR KNOTS AND LINKS WITH UP TO 11 CROSSINGS

DROR BAR-NATAN

ABSTRACT. We provide tables of the ranks of the Khovanov homology of all prime knots and links with up to 11 crossings.

CONTENTS

Date: First edition: May 2003. This edition: Aug. 17, 2004.
The research presented here was supported by BSF grant #1998-119 and NSERC grant RGPIN 262178.
This report is available from `http://www.math.toronto.edu/~drorbn/papers/`. The programs with which it was produced are at `http://www.math.toronto.edu/~drorbn/papers/Categorification/` and/or at `http://www.math.toronto.edu/~drorbn/KAtlas/Manual/`.

J.M. Bryden (ed.), Advances in Topological Quantum Field Theory, 167–241.

In his seminal paper [Kh1] Khovanov (see also [BN1]) describes a chain complex $\mathcal{C}(L)$ of graded $\mathbb{Z}$-modules, associated to any knot- or link-diagram L. The Khovanov complex has the following properties, which make it interesting and which suggest that it is but the tip of an iceberg, and that many further results are likely to follow:

- The homology of $\mathcal{C}(L)$ (indeed the homotopy class of $\mathcal{C}(L)$) is an invariant of the knot/link described by L. Below we call it $KH(L)$.
- The graded Euler characteristic of $\mathcal{C}(L)$ is the Jones polynomial $J(L)$ of L.
- $KH(L)$ is a finer knot/link invariant than $J(L)$.
- $\mathcal{C}(L)$ and $KH(L)$ can be extended to tangles [Kh2, BN2].
- As is often the case with homology, $KH(L)$ is functorial in the appropriate sense — it extends to a (projective) functor from the category of knot cobordisms to the category of $\mathbb{Z}$-modules [Ja, Kh3, BN2].
- In particular, there is an associated invariant of 2-knots and 2-links in 4-space.
- Much is known about generalizations and relations of the Jones polynomial, but little is known about generalizations and relations of its "categorification" the Khovanov homology. Thus little is known about categorification in the context of other Lie algebras, of 3-manifolds and/or knots and links inside 3-manifolds, of finite type invariants and of virtual knots, and we don't yet know if Khovanov homology has a rich relation with quantum field theory as does the Jones polynomial. Some promising signs are in [OS, Kh4].

The only purpose of this report is to provide tables of ranks of the Khovanov homology for all prime knots and links with up to 11 crossings, as computed by the program described in [BN1].

We believe our notation is mostly self-explanatory. Our only comments are:

- Knots with up to 10 crossings are enumerated as in Rolfsen's table [Ro], except that the duplication in the "Perko pair" ($10_{161} = 10_{162}$) was removed, and hence Rolfsen's 10_{163} through 10_{166} are ours 10_{162} through 10_{165}. The pictures below (for these knots) were generated using R. Scharein's program KnotPlot [Sc].
- Knots with 11 crossings and all links are enumerated as by J. Hoste and M. Thistlethwaite program Knotscape [HT], which also produced their pictures. As Knotscape doesn't keep track of the orientation of $\mathbb{R}^3$, the picture of each of these knots or links may in fact be a picture of its mirror image.
- To save space underline negative numbers; thus $\underline{1} = -1$.
- A symbol like 2^3_4 means "2 homology classes in KH^3 at degree 4", and thus a line like "3_1: $1^{\underline{3}}_{\underline{9}}1^{\underline{2}}_{\underline{5}}1^{0}_{\underline{3}}1^{0}_{\underline{1}}$" means that the total Khovanov homology of 3_1 is four dimensional, with one class of degree -9 in $KH^{-3}(3_1)$, one class of degree -5 in $KH^{-2}(3_1)$, and two classes in $KH^0(3_1)$, one each of degrees -3 and -1.

2.1. **3–7 Crossing Knots.**

K: KH	1st line	2nd line	3rd line
3_1: $1^{\underline{3}}_{\underline{9}}1^{\underline{2}}_{\underline{5}}1^{0}_{\underline{3}}1^{0}_{\underline{1}}$ 4_1: $1^{\underline{2}}_{\underline{5}}1^{\underline{1}}_{\underline{1}}1^{0}_{\underline{1}}1^{0}_{1}1^{1}_{1}1^{2}_{5}$ 5_1: $1^{\underline{5}}_{\underline{15}}1^{\underline{4}}_{\underline{11}}1^{\underline{3}}_{\underline{11}}1^{\underline{2}}_{\underline{7}}1^{0}_{\underline{5}}1^{0}_{\underline{3}}$	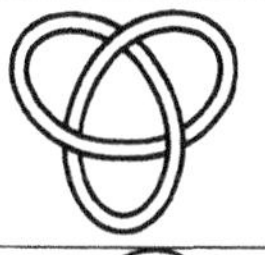		
5_2: $1^{\underline{5}}_{\underline{13}}1^{\underline{4}}_{\underline{9}}1^{\underline{3}}_{\underline{9}}1^{\underline{2}}_{\underline{7}}1^{\underline{2}}_{\underline{5}}1^{\underline{1}}_{\underline{3}}1^{0}_{\underline{3}}1^{0}_{\underline{1}}$ 6_1: $1^{\underline{4}}_{\underline{9}}1^{\underline{3}}_{\underline{5}}1^{\underline{2}}_{\underline{5}}1^{\underline{1}}_{\underline{3}}1^{\underline{1}}_{\underline{1}}1^{0}_{\underline{1}}2^{0}_{1}1^{1}_{1}1^{2}_{5}$ 6_2: $1^{\underline{4}}_{\underline{11}}1^{\underline{3}}_{\underline{9}}1^{\underline{3}}_{\underline{7}}1^{\underline{2}}_{\underline{7}}1^{\underline{2}}_{\underline{5}}1^{\underline{1}}_{\underline{5}}1^{\underline{1}}_{\underline{3}}1^{0}_{\underline{3}}2^{0}_{\underline{1}}1^{1}_{\underline{1}}1^{2}_{3}$			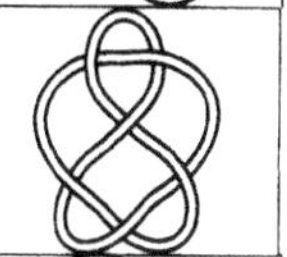
6_3: $1^{\underline{3}}_{\underline{7}}1^{\underline{2}}_{\underline{5}}1^{\underline{2}}_{\underline{3}}1^{\underline{1}}_{\underline{3}}1^{\underline{1}}_{\underline{1}}2^{0}_{\underline{1}}2^{0}_{1}1^{1}_{1}1^{1}_{3}1^{2}_{3}1^{2}_{5}1^{3}_{7}$ 7_1: $1^{\underline{7}}_{\underline{21}}1^{\underline{6}}_{\underline{17}}1^{\underline{5}}_{\underline{17}}1^{\underline{4}}_{\underline{13}}1^{\underline{3}}_{\underline{13}}1^{\underline{2}}_{\underline{9}}1^{0}_{\underline{7}}1^{0}_{\underline{5}}$ 7_2: $1^{\underline{7}}_{\underline{17}}1^{\underline{6}}_{\underline{13}}1^{\underline{5}}_{\underline{13}}1^{\underline{4}}_{\underline{11}}1^{\underline{4}}_{\underline{9}}1^{\underline{3}}_{\underline{9}}1^{\underline{3}}_{\underline{7}}1^{\underline{2}}_{\underline{7}}1^{\underline{2}}_{\underline{5}}1^{\underline{1}}_{\underline{3}}1^{0}_{\underline{3}}1^{0}_{\underline{1}}$		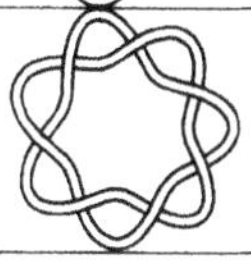	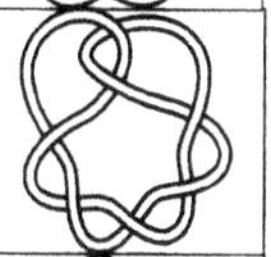
7_3: $1^{0}_{3}1^{0}_{5}1^{1}_{5}1^{2}_{7}1^{2}_{9}1^{3}_{9}1^{3}_{11}2^{4}_{11}1^{4}_{13}2^{5}_{15}1^{6}_{15}1^{7}_{19}$ 7_4: $1^{0}_{1}1^{0}_{3}2^{1}_{3}1^{2}_{5}2^{2}_{7}1^{3}_{7}1^{3}_{9}2^{4}_{9}1^{4}_{11}2^{5}_{13}1^{6}_{13}1^{7}_{17}$ 7_5: $1^{\underline{7}}_{\underline{19}}1^{\underline{6}}_{\underline{17}}1^{\underline{6}}_{\underline{15}}2^{\underline{5}}_{\underline{15}}1^{\underline{5}}_{\underline{13}}1^{\underline{4}}_{\underline{13}}2^{\underline{4}}_{\underline{11}}2^{\underline{3}}_{\underline{11}}1^{\underline{3}}_{\underline{9}}1^{\underline{2}}_{\underline{9}}2^{\underline{2}}_{\underline{7}}1^{\underline{1}}_{\underline{5}}1^{0}_{\underline{5}}1^{0}_{\underline{3}}$	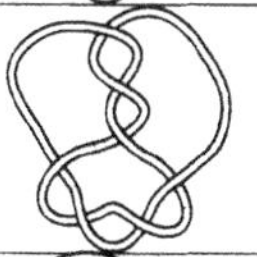	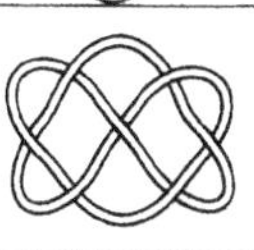	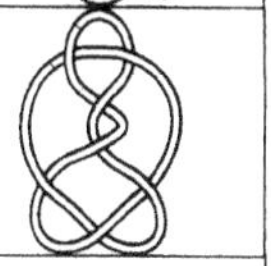
7_6: $1^{\underline{5}}_{\underline{13}}1^{\underline{4}}_{\underline{11}}1^{\underline{4}}_{\underline{9}}2^{\underline{3}}_{\underline{9}}1^{\underline{3}}_{\underline{7}}2^{\underline{2}}_{\underline{7}}2^{\underline{2}}_{\underline{5}}1^{\underline{1}}_{\underline{5}}2^{\underline{1}}_{\underline{3}}2^{0}_{\underline{3}}2^{0}_{\underline{1}}1^{1}_{\underline{1}}1^{1}_{1}1^{2}_{3}$ 7_7: $1^{\underline{3}}_{\underline{7}}2^{\underline{2}}_{\underline{5}}1^{\underline{2}}_{\underline{3}}1^{\underline{1}}_{\underline{3}}2^{\underline{1}}_{\underline{1}}3^{0}_{\underline{1}}2^{0}_{1}2^{1}_{1}2^{1}_{3}1^{2}_{3}2^{2}_{5}1^{3}_{5}1^{3}_{7}1^{4}_{9}$			

(All are alternating).

2.2. **8 Crossing Knots.**

K: KH	1st line	2nd line	3rd line
8_1: $1^{\underline{6}}_{\underline{13}}1^{\underline{5}}_{\underline{9}}1^{\underline{4}}_{\underline{9}}1^{\underline{3}}_{\underline{7}}1^{\underline{3}}_{\underline{5}}1^{\underline{2}}_{\underline{5}}1^{\underline{2}}_{\underline{3}}1^{\underline{1}}_{\underline{3}}1^{\underline{1}}_{\underline{1}}1^{0}_{\underline{1}}2^{0}_{1}1^{1}_{1}1^{2}_{5}$ 8_2: $1^{\underline{6}}_{\underline{17}}1^{\underline{5}}_{\underline{15}}1^{\underline{5}}_{\underline{13}}1^{\underline{4}}_{\underline{13}}1^{\underline{4}}_{\underline{11}}2^{\underline{3}}_{\underline{11}}1^{\underline{3}}_{\underline{9}}1^{\underline{2}}_{\underline{9}}2^{\underline{2}}_{\underline{7}}1^{\underline{1}}_{\underline{7}}1^{\underline{1}}_{\underline{5}}1^{0}_{\underline{5}}2^{0}_{\underline{3}}1^{1}_{\underline{3}}1^{2}_{1}$ 8_3: $1^{\underline{4}}_{\underline{9}}1^{\underline{3}}_{\underline{5}}2^{\underline{2}}_{\underline{5}}1^{\underline{1}}_{\underline{3}}2^{\underline{1}}_{\underline{1}}2^{0}_{\underline{1}}2^{0}_{1}2^{1}_{1}1^{1}_{3}2^{2}_{5}1^{3}_{5}1^{4}_{9}$	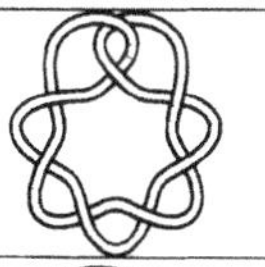		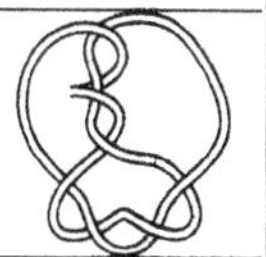
8_4: $1^{\underline{4}}_{\underline{11}}1^{\underline{3}}_{\underline{9}}1^{\underline{3}}_{\underline{7}}2^{\underline{2}}_{\underline{7}}1^{\underline{2}}_{\underline{5}}1^{\underline{1}}_{\underline{5}}2^{\underline{1}}_{\underline{3}}2^{0}_{\underline{3}}2^{0}_{\underline{1}}2^{1}_{\underline{1}}1^{1}_{1}2^{2}_{3}1^{3}_{3}1^{4}_{7}$ 8_5: $1^{\underline{2}}_{\underline{1}}1^{\underline{1}}_{3}3^{0}_{3}1^{0}_{5}1^{1}_{5}2^{1}_{7}2^{2}_{7}1^{2}_{9}2^{3}_{9}2^{3}_{11}1^{4}_{11}2^{4}_{13}1^{5}_{13}1^{5}_{15}1^{6}_{17}$ 8_6: $1^{\underline{6}}_{\underline{15}}1^{\underline{5}}_{\underline{13}}1^{\underline{5}}_{\underline{11}}2^{\underline{4}}_{\underline{11}}1^{\underline{4}}_{\underline{9}}2^{\underline{3}}_{\underline{9}}2^{\underline{3}}_{\underline{7}}2^{\underline{2}}_{\underline{7}}2^{\underline{2}}_{\underline{5}}2^{\underline{1}}_{\underline{5}}2^{\underline{1}}_{\underline{3}}1^{0}_{\underline{3}}3^{0}_{\underline{1}}1^{1}_{\underline{1}}1^{2}_{3}$			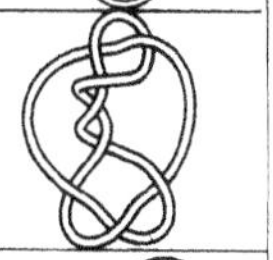
8_7: $1^{\underline{3}}_{\underline{5}}1^{\underline{2}}_{\underline{3}}1^{\underline{2}}_{\underline{1}}1^{\underline{1}}_{\underline{1}}1^{\underline{1}}_{1}3^{0}_{1}2^{0}_{3}2^{1}_{3}2^{1}_{5}2^{2}_{5}2^{2}_{7}1^{3}_{7}2^{3}_{9}1^{4}_{9}1^{4}_{11}1^{5}_{13}$ 8_8: $1^{\underline{3}}_{\underline{7}}1^{\underline{2}}_{\underline{5}}1^{\underline{2}}_{\underline{3}}2^{\underline{1}}_{\underline{3}}1^{\underline{1}}_{\underline{1}}3^{0}_{\underline{1}}3^{0}_{1}2^{1}_{1}2^{1}_{3}2^{2}_{3}2^{2}_{5}1^{3}_{5}2^{3}_{7}1^{4}_{7}1^{4}_{9}1^{5}_{11}$ 8_9: $1^{\underline{4}}_{\underline{9}}1^{\underline{3}}_{\underline{7}}1^{\underline{3}}_{\underline{5}}2^{\underline{2}}_{\underline{5}}1^{\underline{2}}_{\underline{3}}2^{\underline{1}}_{\underline{3}}2^{\underline{1}}_{\underline{1}}3^{0}_{\underline{1}}3^{0}_{1}2^{1}_{1}2^{1}_{3}1^{2}_{3}2^{2}_{5}1^{3}_{5}1^{3}_{7}1^{4}_{9}$			

K: KH	1st line	2nd line	3rd line
8_{10}: $1^{\underline{3}}_{\underline{5}}1^{\underline{2}}_{\underline{3}}1^{\underline{2}}_{\underline{1}}2^{\underline{1}}_{\underline{1}}1^{\underline{1}}_{1}3^{0}_{1}3^{0}_{3}2^{1}_{3}2^{1}_{5}3^{2}_{5}2^{2}_{7}1^{3}_{7}3^{3}_{9}1^{4}_{9}1^{4}_{11}1^{5}_{13}$			
8_{11}: $1^{\underline{6}}_{\underline{15}}1^{\underline{5}}_{\underline{13}}1^{\underline{5}}_{\underline{11}}2^{\underline{4}}_{\underline{11}}1^{\underline{4}}_{\underline{9}}3^{\underline{3}}_{\underline{9}}2^{\underline{3}}_{\underline{7}}2^{\underline{2}}_{\underline{7}}3^{\underline{2}}_{\underline{5}}2^{\underline{1}}_{\underline{5}}2^{\underline{1}}_{\underline{3}}2^{0}_{\underline{3}}3^{0}_{\underline{1}}1^{1}_{\underline{1}}1^{1}_{1}1^{2}_{3}$			
8_{12}: $1^{\underline{4}}_{\underline{9}}1^{\underline{3}}_{\underline{7}}1^{\underline{3}}_{\underline{5}}3^{\underline{2}}_{\underline{5}}1^{\underline{2}}_{\underline{3}}2^{\underline{1}}_{\underline{3}}3^{\underline{1}}_{\underline{1}}3^{0}_{\underline{1}}3^{0}_{1}3^{1}_{1}2^{1}_{3}1^{2}_{3}3^{2}_{5}1^{3}_{5}1^{3}_{7}1^{4}_{9}$			
8_{13}: $1^{\underline{3}}_{\underline{7}}2^{\underline{2}}_{\underline{5}}1^{\underline{2}}_{\underline{3}}2^{\underline{1}}_{\underline{3}}2^{\underline{1}}_{\underline{1}}3^{0}_{\underline{1}}3^{0}_{1}3^{1}_{1}2^{1}_{3}2^{2}_{3}3^{2}_{5}1^{3}_{5}2^{3}_{7}1^{4}_{7}1^{4}_{9}1^{5}_{11}$			
8_{14}: $1^{\underline{6}}_{\underline{15}}2^{\underline{5}}_{\underline{13}}1^{\underline{5}}_{\underline{11}}2^{\underline{4}}_{\underline{11}}2^{\underline{4}}_{\underline{9}}3^{\underline{3}}_{\underline{9}}2^{\underline{3}}_{\underline{7}}3^{\underline{2}}_{\underline{7}}3^{\underline{2}}_{\underline{5}}2^{\underline{1}}_{\underline{5}}3^{\underline{1}}_{\underline{3}}2^{0}_{\underline{3}}3^{0}_{\underline{1}}1^{1}_{\underline{1}}1^{1}_{1}1^{2}_{3}$			
8_{15}: $1^{\underline{8}}_{\underline{21}}2^{\underline{7}}_{\underline{19}}1^{\underline{7}}_{\underline{17}}2^{\underline{6}}_{\underline{17}}2^{\underline{6}}_{\underline{15}}4^{\underline{5}}_{\underline{15}}2^{\underline{5}}_{\underline{13}}2^{\underline{4}}_{\underline{13}}4^{\underline{4}}_{\underline{11}}3^{\underline{3}}_{\underline{11}}2^{\underline{3}}_{\underline{9}}2^{\underline{2}}_{\underline{9}}3^{\underline{2}}_{\underline{7}}2^{\underline{1}}_{\underline{5}}1^{0}_{\underline{5}}1^{0}_{\underline{3}}$			
8_{16}: $1^{\underline{5}}_{\underline{13}}2^{\underline{4}}_{\underline{11}}1^{\underline{4}}_{\underline{9}}3^{\underline{3}}_{\underline{9}}2^{\underline{3}}_{\underline{7}}3^{\underline{2}}_{\underline{7}}3^{\underline{2}}_{\underline{5}}3^{\underline{1}}_{\underline{5}}3^{\underline{1}}_{\underline{3}}3^{0}_{\underline{3}}4^{0}_{\underline{1}}2^{1}_{\underline{1}}2^{1}_{1}1^{2}_{1}2^{2}_{3}1^{3}_{5}$			
8_{17}: $1^{\underline{4}}_{\underline{9}}2^{\underline{3}}_{\underline{7}}1^{\underline{3}}_{\underline{5}}3^{\underline{2}}_{\underline{5}}2^{\underline{2}}_{\underline{3}}3^{\underline{1}}_{\underline{3}}3^{\underline{1}}_{\underline{1}}4^{0}_{\underline{1}}4^{0}_{1}3^{1}_{1}3^{1}_{3}2^{2}_{3}3^{2}_{5}1^{3}_{5}2^{3}_{7}1^{4}_{9}$			
8_{18}: $1^{\underline{4}}_{\underline{9}}3^{\underline{3}}_{\underline{7}}1^{\underline{3}}_{\underline{5}}3^{\underline{2}}_{\underline{5}}3^{\underline{2}}_{\underline{3}}4^{\underline{1}}_{\underline{3}}3^{\underline{1}}_{\underline{1}}5^{0}_{\underline{1}}5^{0}_{1}3^{1}_{1}4^{1}_{3}3^{2}_{3}3^{2}_{5}1^{3}_{5}3^{3}_{7}1^{4}_{9}$			
8_{19}: $1^{0}_{5}1^{0}_{7}1^{2}_{9}1^{3}_{13}1^{4}_{11}1^{4}_{13}1^{5}_{15}1^{5}_{17}$			
8_{20}: $1^{\underline{5}}_{\underline{11}}1^{\underline{4}}_{\underline{7}}1^{\underline{3}}_{\underline{7}}1^{\underline{2}}_{\underline{5}}1^{\underline{2}}_{\underline{3}}1^{\underline{1}}_{\underline{1}}2^{0}_{\underline{1}}1^{0}_{1}1^{1}_{3}$			
8_{21}: $1^{\underline{6}}_{\underline{15}}1^{\underline{5}}_{\underline{13}}1^{\underline{5}}_{\underline{11}}1^{\underline{4}}_{\underline{11}}1^{\underline{4}}_{\underline{9}}2^{\underline{3}}_{\underline{9}}1^{\underline{3}}_{\underline{7}}1^{\underline{2}}_{\underline{7}}2^{\underline{2}}_{\underline{5}}1^{\underline{1}}_{\underline{5}}1^{\underline{1}}_{\underline{3}}1^{0}_{\underline{3}}2^{0}_{\underline{1}}$			

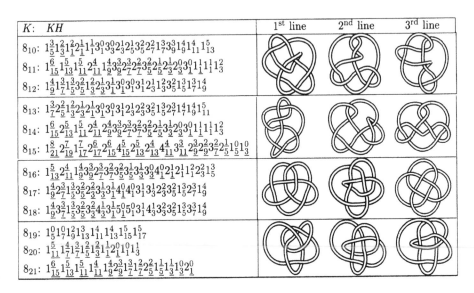

(8_1–8_{18} are alternating, the rest are not).

2.3. **9 Crossing Knots.**

K: KH	1st line	2nd line	3rd line
9_{1}: $1^{\underline{9}}_{\underline{27}}1^{\underline{8}}_{\underline{23}}1^{\underline{7}}_{\underline{23}}1^{\underline{6}}_{\underline{19}}1^{\underline{5}}_{\underline{19}}1^{\underline{4}}_{\underline{15}}1^{\underline{3}}_{\underline{15}}1^{\underline{2}}_{\underline{11}}1^{0}_{\underline{9}}1^{0}_{\underline{7}}$	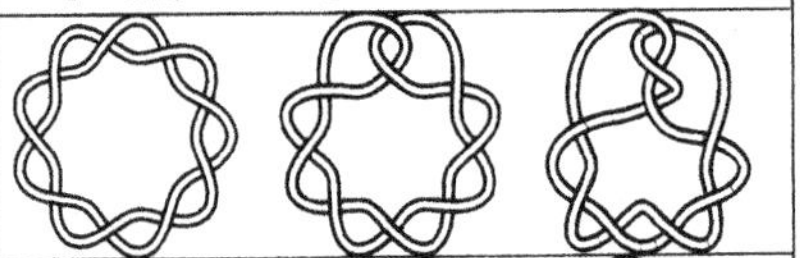		
9_{2}: $1^{\underline{9}}_{\underline{21}}1^{\underline{8}}_{\underline{17}}1^{\underline{7}}_{\underline{17}}1^{\underline{6}}_{\underline{15}}1^{\underline{6}}_{\underline{13}}1^{\underline{5}}_{\underline{13}}1^{\underline{5}}_{\underline{11}}1^{\underline{4}}_{\underline{11}}1^{\underline{4}}_{\underline{9}}1^{\underline{3}}_{\underline{9}}1^{\underline{3}}_{\underline{7}}1^{\underline{2}}_{\underline{7}}1^{\underline{2}}_{\underline{5}}1^{\underline{1}}_{\underline{3}}1^{0}_{\underline{3}}1^{0}_{\underline{1}}$			
9_{3}: $1^{0}_{5}1^{0}_{7}1^{1}_{7}1^{2}_{9}1^{2}_{11}1^{3}_{11}1^{3}_{13}2^{4}_{13}1^{4}_{15}1^{5}_{15}2^{5}_{17}2^{6}_{17}1^{6}_{19}2^{7}_{21}1^{8}_{21}1^{9}_{25}$			
9_{4}: $1^{\underline{9}}_{\underline{23}}1^{\underline{8}}_{\underline{19}}2^{\underline{7}}_{\underline{19}}1^{\underline{6}}_{\underline{17}}2^{\underline{6}}_{\underline{15}}2^{\underline{5}}_{\underline{15}}1^{\underline{5}}_{\underline{13}}2^{\underline{4}}_{\underline{13}}2^{\underline{4}}_{\underline{11}}1^{\underline{3}}_{\underline{11}}2^{\underline{3}}_{\underline{9}}1^{\underline{2}}_{\underline{9}}1^{\underline{2}}_{\underline{7}}1^{\underline{1}}_{\underline{5}}1^{0}_{\underline{5}}1^{0}_{\underline{3}}$			
9_{5}: $1^{0}_{1}1^{0}_{3}2^{1}_{3}1^{2}_{5}2^{2}_{7}2^{3}_{7}1^{3}_{9}2^{4}_{9}2^{4}_{11}1^{5}_{11}2^{5}_{13}2^{6}_{13}1^{6}_{15}2^{7}_{17}1^{8}_{17}1^{9}_{21}$			
9_{6}: $1^{\underline{9}}_{\underline{25}}1^{\underline{8}}_{\underline{23}}1^{\underline{8}}_{\underline{21}}2^{\underline{7}}_{\underline{21}}1^{\underline{7}}_{\underline{19}}2^{\underline{6}}_{\underline{19}}2^{\underline{6}}_{\underline{17}}3^{\underline{5}}_{\underline{17}}2^{\underline{5}}_{\underline{15}}1^{\underline{4}}_{\underline{15}}3^{\underline{4}}_{\underline{13}}2^{\underline{3}}_{\underline{13}}1^{\underline{3}}_{\underline{11}}1^{\underline{2}}_{\underline{11}}2^{\underline{2}}_{\underline{9}}1^{\underline{1}}_{\underline{7}}1^{0}_{\underline{7}}1^{0}_{\underline{5}}$			
9_{7}: $1^{\underline{9}}_{\underline{23}}1^{\underline{8}}_{\underline{21}}1^{\underline{8}}_{\underline{19}}2^{\underline{7}}_{\underline{19}}1^{\underline{7}}_{\underline{17}}2^{\underline{6}}_{\underline{17}}2^{\underline{6}}_{\underline{15}}3^{\underline{5}}_{\underline{15}}2^{\underline{5}}_{\underline{13}}2^{\underline{4}}_{\underline{13}}3^{\underline{4}}_{\underline{11}}2^{\underline{3}}_{\underline{11}}2^{\underline{3}}_{\underline{9}}1^{\underline{2}}_{\underline{9}}2^{\underline{2}}_{\underline{7}}1^{\underline{1}}_{\underline{5}}1^{0}_{\underline{5}}1^{0}_{\underline{3}}$			
9_{8}: $1^{\underline{5}}_{\underline{13}}1^{\underline{4}}_{\underline{11}}1^{\underline{4}}_{\underline{9}}2^{\underline{3}}_{\underline{9}}1^{\underline{3}}_{\underline{7}}3^{\underline{2}}_{\underline{7}}2^{\underline{2}}_{\underline{5}}2^{\underline{1}}_{\underline{5}}3^{\underline{1}}_{\underline{3}}3^{0}_{\underline{3}}3^{0}_{\underline{1}}2^{1}_{\underline{1}}2^{1}_{1}1^{2}_{1}2^{2}_{3}1^{3}_{3}1^{3}_{5}1^{4}_{7}$			
9_{9}: $1^{\underline{9}}_{\underline{25}}1^{\underline{8}}_{\underline{23}}1^{\underline{8}}_{\underline{21}}3^{\underline{7}}_{\underline{21}}1^{\underline{7}}_{\underline{19}}2^{\underline{6}}_{\underline{19}}3^{\underline{6}}_{\underline{17}}3^{\underline{5}}_{\underline{17}}2^{\underline{5}}_{\underline{15}}2^{\underline{4}}_{\underline{15}}3^{\underline{4}}_{\underline{13}}2^{\underline{3}}_{\underline{13}}2^{\underline{3}}_{\underline{11}}1^{\underline{2}}_{\underline{11}}2^{\underline{2}}_{\underline{9}}1^{\underline{1}}_{\underline{7}}1^{0}_{\underline{7}}1^{0}_{\underline{5}}$			
9_{10}: $1^{0}_{3}1^{0}_{5}2^{1}_{5}2^{2}_{7}2^{2}_{9}3^{3}_{9}2^{3}_{11}3^{4}_{11}3^{4}_{13}2^{5}_{13}3^{5}_{15}3^{6}_{15}2^{6}_{17}3^{7}_{19}1^{8}_{19}1^{9}_{23}$			
9_{11}: $1^{\underline{2}}_{\underline{1}}1^{\underline{1}}_{1}1^{\underline{1}}_{3}2^{0}_{3}2^{0}_{5}3^{1}_{5}1^{1}_{7}3^{2}_{7}3^{2}_{9}2^{3}_{9}3^{3}_{11}3^{4}_{11}2^{4}_{13}1^{5}_{13}3^{5}_{15}1^{6}_{15}1^{6}_{17}1^{7}_{19}$			
9_{12}: $1^{\underline{7}}_{\underline{17}}1^{\underline{6}}_{\underline{15}}1^{\underline{6}}_{\underline{13}}2^{\underline{5}}_{\underline{13}}1^{\underline{5}}_{\underline{11}}3^{\underline{4}}_{\underline{11}}2^{\underline{4}}_{\underline{9}}3^{\underline{3}}_{\underline{9}}3^{\underline{3}}_{\underline{7}}3^{\underline{2}}_{\underline{7}}3^{\underline{2}}_{\underline{5}}2^{\underline{1}}_{\underline{5}}3^{\underline{1}}_{\underline{3}}2^{0}_{\underline{3}}3^{0}_{\underline{1}}1^{1}_{\underline{1}}1^{1}_{1}1^{2}_{3}$			
9_{13}: $1^{0}_{3}1^{0}_{5}2^{1}_{5}2^{2}_{7}2^{2}_{9}3^{3}_{9}2^{3}_{11}4^{4}_{11}3^{4}_{13}2^{5}_{13}4^{5}_{15}3^{6}_{15}2^{6}_{17}1^{7}_{17}3^{7}_{19}1^{8}_{19}1^{8}_{21}1^{9}_{23}$	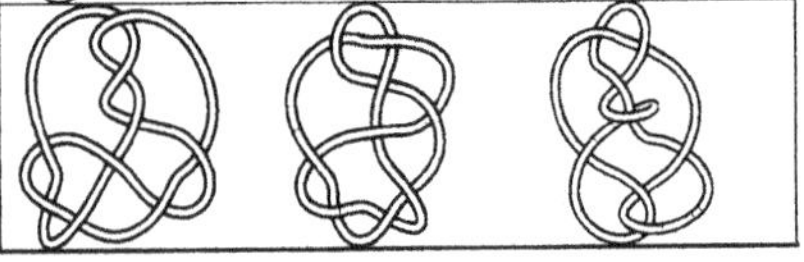		
9_{14}: $1^{\underline{3}}_{\underline{7}}2^{\underline{2}}_{\underline{5}}1^{\underline{2}}_{\underline{3}}2^{\underline{1}}_{\underline{3}}2^{\underline{1}}_{\underline{1}}4^{0}_{\underline{1}}3^{0}_{1}3^{1}_{1}3^{1}_{3}3^{2}_{3}3^{2}_{5}2^{3}_{5}3^{3}_{7}1^{4}_{7}2^{4}_{9}1^{5}_{9}1^{5}_{11}1^{6}_{13}$			
9_{15}: $1^{\underline{2}}_{\underline{3}}1^{\underline{1}}_{\underline{1}}1^{\underline{1}}_{1}3^{0}_{1}2^{0}_{3}4^{1}_{3}2^{1}_{5}3^{2}_{5}4^{2}_{7}3^{3}_{7}3^{3}_{9}3^{4}_{9}3^{4}_{11}1^{5}_{11}3^{5}_{13}1^{6}_{13}1^{6}_{15}1^{7}_{17}$			

K: KH	1st line	2nd line	3rd line
9_{16}: $1^{0}_{5}1^{0}_{7}1^{1}_{7}3^{2}_{9}1^{2}_{11}2^{3}_{11}3^{3}_{13}4^{4}_{13}2^{4}_{15}3^{5}_{15}4^{5}_{17}3^{6}_{17}3^{6}_{19}2^{7}_{19}3^{7}_{21}1^{8}_{21}2^{8}_{23}1^{9}_{25}$ 9_{17}: $1^{\underline{5}}_{\underline{13}}2^{\underline{4}}_{\underline{11}}1^{\underline{4}}_{\underline{9}}2^{\underline{3}}_{\underline{9}}2^{\underline{3}}_{\underline{7}}4^{\underline{2}}_{\underline{7}}2^{\underline{2}}_{\underline{5}}3^{\underline{1}}_{\underline{5}}4^{\underline{1}}_{\underline{3}}3^{0}_{\underline{3}}4^{0}_{\underline{1}}3^{1}_{\underline{1}}2^{1}_{1}1^{2}_{1}3^{2}_{3}1^{3}_{3}1^{3}_{5}1^{4}_{7}$ 9_{18}: $1^{\underline{9}}_{\underline{23}}1^{\underline{8}}_{\underline{21}}1^{\underline{8}}_{\underline{19}}3^{\underline{7}}_{\underline{19}}1^{\underline{7}}_{\underline{17}}3^{\underline{6}}_{\underline{17}}3^{\underline{6}}_{\underline{15}}4^{\underline{5}}_{\underline{15}}3^{\underline{5}}_{\underline{13}}3^{\underline{4}}_{\underline{13}}4^{\underline{4}}_{\underline{11}}3^{\underline{3}}_{\underline{11}}3^{\underline{3}}_{\underline{9}}2^{\underline{2}}_{\underline{9}}3^{\underline{2}}_{\underline{7}}2^{\underline{1}}_{\underline{5}}1^{0}_{\underline{5}}1^{0}_{\underline{3}}$	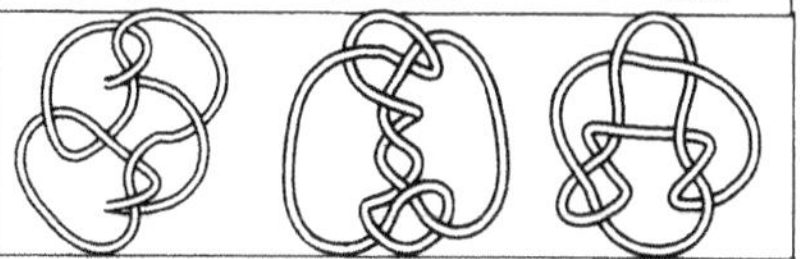		
9_{19}: $1^{\underline{5}}_{\underline{11}}2^{\underline{4}}_{\underline{9}}1^{\underline{4}}_{\underline{7}}2^{\underline{3}}_{\underline{7}}2^{\underline{3}}_{\underline{5}}4^{\underline{2}}_{\underline{5}}2^{\underline{2}}_{\underline{3}}3^{\underline{1}}_{\underline{3}}4^{\underline{1}}_{\underline{1}}4^{0}_{\underline{1}}4^{0}_{1}3^{1}_{1}3^{1}_{3}1^{2}_{3}3^{2}_{5}1^{3}_{5}1^{3}_{7}1^{4}_{9}$ 9_{20}: $1^{\underline{7}}_{\underline{19}}2^{\underline{6}}_{\underline{17}}1^{\underline{6}}_{\underline{15}}3^{\underline{5}}_{\underline{15}}2^{\underline{5}}_{\underline{13}}3^{\underline{4}}_{\underline{13}}3^{\underline{4}}_{\underline{11}}4^{\underline{3}}_{\underline{11}}3^{\underline{3}}_{\underline{9}}3^{\underline{2}}_{\underline{9}}4^{\underline{2}}_{\underline{7}}2^{\underline{1}}_{\underline{7}}3^{\underline{1}}_{\underline{5}}2^{0}_{\underline{5}}3^{0}_{\underline{3}}1^{1}_{\underline{3}}1^{1}_{\underline{1}}1^{2}_{1}$ 9_{21}: $1^{\underline{2}}_{\underline{3}}2^{\underline{1}}_{\underline{1}}1^{\underline{1}}_{1}3^{0}_{1}3^{0}_{3}4^{1}_{3}2^{1}_{5}4^{2}_{5}4^{2}_{7}3^{3}_{7}4^{3}_{9}3^{4}_{9}3^{4}_{11}1^{5}_{11}3^{5}_{13}1^{6}_{13}1^{6}_{15}1^{7}_{17}$			
9_{22}: $1^{\underline{4}}_{\underline{7}}1^{\underline{3}}_{\underline{5}}1^{\underline{3}}_{\underline{3}}3^{\underline{2}}_{\underline{3}}1^{\underline{2}}_{\underline{1}}3^{\underline{1}}_{\underline{1}}3^{\underline{1}}_{1}4^{0}_{1}4^{0}_{3}4^{1}_{3}3^{1}_{5}3^{2}_{5}4^{2}_{7}2^{3}_{7}3^{3}_{9}1^{4}_{9}2^{4}_{11}1^{5}_{13}$ 9_{23}: $1^{\underline{9}}_{\underline{23}}2^{\underline{8}}_{\underline{21}}1^{\underline{8}}_{\underline{19}}3^{\underline{7}}_{\underline{19}}2^{\underline{7}}_{\underline{17}}3^{\underline{6}}_{\underline{17}}3^{\underline{6}}_{\underline{15}}5^{\underline{5}}_{\underline{15}}3^{\underline{5}}_{\underline{13}}3^{\underline{4}}_{\underline{13}}5^{\underline{4}}_{\underline{11}}3^{\underline{3}}_{\underline{11}}3^{\underline{3}}_{\underline{9}}2^{\underline{2}}_{\underline{9}}3^{\underline{2}}_{\underline{7}}2^{\underline{1}}_{\underline{5}}1^{0}_{\underline{5}}1^{0}_{\underline{3}}$ 9_{24}: $1^{\underline{5}}_{\underline{11}}1^{\underline{4}}_{\underline{9}}1^{\underline{4}}_{\underline{7}}3^{\underline{3}}_{\underline{7}}1^{\underline{3}}_{\underline{5}}4^{\underline{2}}_{\underline{5}}3^{\underline{2}}_{\underline{3}}3^{\underline{1}}_{\underline{3}}4^{\underline{1}}_{\underline{1}}5^{0}_{\underline{1}}4^{0}_{1}3^{1}_{1}4^{1}_{3}2^{2}_{3}3^{2}_{5}1^{3}_{5}2^{3}_{7}1^{4}_{9}$			
9_{25}: $1^{\underline{7}}_{\underline{17}}2^{\underline{6}}_{\underline{15}}1^{\underline{6}}_{\underline{13}}3^{\underline{5}}_{\underline{13}}2^{\underline{5}}_{\underline{11}}4^{\underline{4}}_{\underline{11}}3^{\underline{4}}_{\underline{9}}4^{\underline{3}}_{\underline{9}}4^{\underline{3}}_{\underline{7}}4^{\underline{2}}_{\underline{7}}4^{\underline{2}}_{\underline{5}}3^{\underline{1}}_{\underline{5}}4^{\underline{1}}_{\underline{3}}2^{0}_{\underline{3}}4^{0}_{\underline{1}}1^{1}_{\underline{1}}1^{1}_{1}1^{2}_{3}$ 9_{26}: $1^{\underline{3}}_{\underline{5}}2^{\underline{2}}_{\underline{3}}1^{\underline{2}}_{\underline{1}}2^{\underline{1}}_{\underline{1}}2^{\underline{1}}_{1}5^{0}_{1}3^{0}_{3}4^{1}_{3}4^{1}_{5}4^{2}_{5}4^{2}_{7}3^{3}_{7}4^{3}_{9}2^{4}_{9}3^{4}_{11}1^{5}_{11}2^{5}_{13}1^{6}_{15}$ 9_{27}: $1^{\underline{5}}_{\underline{11}}2^{\underline{4}}_{\underline{9}}1^{\underline{4}}_{\underline{7}}3^{\underline{3}}_{\underline{7}}2^{\underline{3}}_{\underline{5}}4^{\underline{2}}_{\underline{5}}3^{\underline{2}}_{\underline{3}}4^{\underline{1}}_{\underline{3}}4^{\underline{1}}_{\underline{1}}5^{0}_{\underline{1}}5^{0}_{1}3^{1}_{1}4^{1}_{3}2^{2}_{3}3^{2}_{5}1^{3}_{5}2^{3}_{7}1^{4}_{9}$			
9_{28}: $1^{\underline{6}}_{\underline{15}}2^{\underline{5}}_{\underline{13}}1^{\underline{5}}_{\underline{11}}3^{\underline{4}}_{\underline{11}}2^{\underline{4}}_{\underline{9}}5^{\underline{3}}_{\underline{9}}3^{\underline{3}}_{\underline{7}}4^{\underline{2}}_{\underline{7}}5^{\underline{2}}_{\underline{5}}4^{\underline{1}}_{\underline{5}}4^{\underline{1}}_{\underline{3}}4^{0}_{\underline{3}}5^{0}_{\underline{1}}2^{1}_{\underline{1}}3^{1}_{1}1^{2}_{1}2^{2}_{3}1^{3}_{5}$ 9_{29}: $1^{\underline{5}}_{\underline{13}}2^{\underline{4}}_{\underline{11}}1^{\underline{4}}_{\underline{9}}4^{\underline{3}}_{\underline{9}}2^{\underline{3}}_{\underline{7}}4^{\underline{2}}_{\underline{7}}4^{\underline{2}}_{\underline{5}}4^{\underline{1}}_{\underline{5}}4^{\underline{1}}_{\underline{3}}5^{0}_{\underline{3}}5^{0}_{\underline{1}}3^{1}_{\underline{1}}4^{1}_{1}2^{2}_{1}3^{2}_{3}1^{3}_{3}2^{3}_{5}1^{4}_{7}$ 9_{30}: $1^{\underline{5}}_{\underline{11}}2^{\underline{4}}_{\underline{9}}1^{\underline{4}}_{\underline{7}}3^{\underline{3}}_{\underline{7}}2^{\underline{3}}_{\underline{5}}5^{\underline{2}}_{\underline{5}}3^{\underline{2}}_{\underline{3}}4^{\underline{1}}_{\underline{3}}5^{\underline{1}}_{\underline{1}}5^{0}_{\underline{1}}5^{0}_{1}4^{1}_{1}4^{1}_{3}2^{2}_{3}4^{2}_{5}1^{3}_{5}2^{3}_{7}1^{4}_{9}$			
9_{31}: $1^{\underline{6}}_{\underline{15}}3^{\underline{5}}_{\underline{13}}1^{\underline{5}}_{\underline{11}}3^{\underline{4}}_{\underline{11}}3^{\underline{4}}_{\underline{9}}5^{\underline{3}}_{\underline{9}}3^{\underline{3}}_{\underline{7}}5^{\underline{2}}_{\underline{7}}5^{\underline{2}}_{\underline{5}}4^{\underline{1}}_{\underline{5}}5^{\underline{1}}_{\underline{3}}4^{0}_{\underline{3}}5^{0}_{\underline{1}}2^{1}_{\underline{1}}3^{1}_{1}1^{2}_{1}2^{2}_{3}1^{3}_{5}$ 9_{32}: $1^{\underline{3}}_{\underline{5}}3^{\underline{2}}_{\underline{3}}1^{\underline{2}}_{\underline{1}}3^{\underline{1}}_{\underline{1}}3^{\underline{1}}_{1}6^{0}_{1}4^{0}_{3}5^{1}_{3}5^{1}_{5}5^{2}_{5}5^{2}_{7}4^{3}_{7}5^{3}_{9}2^{4}_{9}4^{4}_{11}1^{5}_{11}2^{5}_{13}1^{6}_{15}$ 9_{33}: $1^{\underline{5}}_{\underline{11}}2^{\underline{4}}_{\underline{9}}1^{\underline{4}}_{\underline{7}}4^{\underline{3}}_{\underline{7}}2^{\underline{3}}_{\underline{5}}5^{\underline{2}}_{\underline{5}}4^{\underline{2}}_{\underline{3}}5^{\underline{1}}_{\underline{3}}5^{\underline{1}}_{\underline{1}}6^{0}_{\underline{1}}6^{0}_{1}4^{1}_{1}5^{1}_{3}3^{2}_{3}4^{2}_{5}1^{3}_{5}3^{3}_{7}1^{4}_{9}$	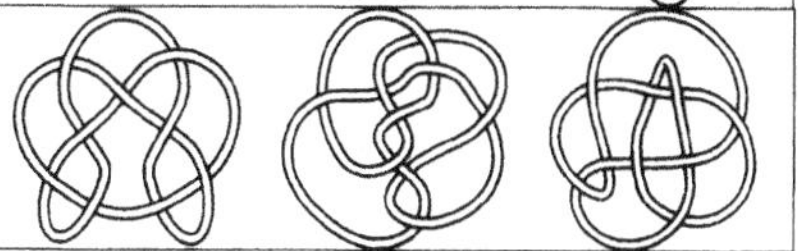		
9_{34}: $1^{\underline{5}}_{\underline{11}}3^{\underline{4}}_{\underline{9}}1^{\underline{4}}_{\underline{7}}4^{\underline{3}}_{\underline{7}}3^{\underline{3}}_{\underline{5}}6^{\underline{2}}_{\underline{5}}4^{\underline{2}}_{\underline{3}}6^{\underline{1}}_{\underline{3}}6^{\underline{1}}_{\underline{1}}6^{0}_{\underline{1}}7^{0}_{1}5^{1}_{1}5^{1}_{3}3^{2}_{3}5^{2}_{5}1^{3}_{5}3^{3}_{7}1^{4}_{9}$ 9_{35}: $1^{\underline{9}}_{\underline{21}}1^{\underline{8}}_{\underline{17}}3^{\underline{7}}_{\underline{17}}1^{\underline{6}}_{\underline{15}}3^{\underline{6}}_{\underline{13}}2^{\underline{5}}_{\underline{13}}1^{\underline{5}}_{\underline{11}}3^{\underline{4}}_{\underline{11}}2^{\underline{4}}_{\underline{9}}1^{\underline{3}}_{\underline{9}}3^{\underline{3}}_{\underline{7}}2^{\underline{2}}_{\underline{7}}1^{\underline{2}}_{\underline{5}}2^{\underline{1}}_{\underline{3}}1^{0}_{\underline{3}}1^{0}_{\underline{1}}$ 9_{36}: $1^{\underline{2}}_{\underline{1}}1^{\underline{1}}_{1}1^{\underline{1}}_{3}3^{0}_{3}2^{0}_{5}3^{1}_{5}2^{1}_{7}3^{2}_{7}3^{2}_{9}3^{3}_{9}3^{3}_{11}3^{4}_{11}3^{4}_{13}1^{5}_{13}3^{5}_{15}1^{6}_{15}1^{6}_{17}1^{7}_{19}$			
9_{37}: $1^{\underline{5}}_{\underline{11}}2^{\underline{4}}_{\underline{9}}1^{\underline{4}}_{\underline{7}}2^{\underline{3}}_{\underline{7}}2^{\underline{3}}_{\underline{5}}5^{\underline{2}}_{\underline{5}}2^{\underline{2}}_{\underline{3}}3^{\underline{1}}_{\underline{3}}5^{\underline{1}}_{\underline{1}}4^{0}_{\underline{1}}4^{0}_{1}4^{1}_{1}3^{1}_{3}1^{2}_{3}4^{2}_{5}1^{3}_{5}1^{3}_{7}1^{4}_{9}$ 9_{38}: $1^{\underline{9}}_{\underline{23}}2^{\underline{8}}_{\underline{21}}1^{\underline{8}}_{\underline{19}}4^{\underline{7}}_{\underline{19}}2^{\underline{7}}_{\underline{17}}4^{\underline{6}}_{\underline{17}}4^{\underline{6}}_{\underline{15}}6^{\underline{5}}_{\underline{15}}4^{\underline{5}}_{\underline{13}}4^{\underline{4}}_{\underline{13}}6^{\underline{4}}_{\underline{11}}4^{\underline{3}}_{\underline{11}}4^{\underline{3}}_{\underline{9}}3^{\underline{2}}_{\underline{9}}4^{\underline{2}}_{\underline{7}}3^{\underline{1}}_{\underline{5}}1^{0}_{\underline{5}}1^{0}_{\underline{3}}$ 9_{39}: $1^{\underline{2}}_{\underline{3}}2^{\underline{1}}_{\underline{1}}1^{\underline{1}}_{1}4^{0}_{1}3^{0}_{3}5^{1}_{3}3^{1}_{5}5^{2}_{5}5^{2}_{7}4^{3}_{7}5^{3}_{9}4^{4}_{9}4^{4}_{11}2^{5}_{11}4^{5}_{13}1^{6}_{13}2^{6}_{15}1^{7}_{17}$			
9_{40}: $1^{\underline{6}}_{\underline{15}}3^{\underline{5}}_{\underline{13}}1^{\underline{5}}_{\underline{11}}5^{\underline{4}}_{\underline{11}}3^{\underline{4}}_{\underline{9}}6^{\underline{3}}_{\underline{9}}5^{\underline{3}}_{\underline{7}}7^{\underline{2}}_{\underline{7}}6^{\underline{2}}_{\underline{5}}6^{\underline{1}}_{\underline{5}}7^{\underline{1}}_{\underline{3}}5^{0}_{\underline{3}}7^{0}_{\underline{1}}4^{1}_{\underline{1}}4^{1}_{1}1^{2}_{1}4^{2}_{3}1^{3}_{5}$ 9_{41}: $1^{\underline{6}}_{\underline{13}}2^{\underline{5}}_{\underline{11}}1^{\underline{5}}_{\underline{9}}3^{\underline{4}}_{\underline{9}}2^{\underline{4}}_{\underline{7}}4^{\underline{3}}_{\underline{7}}3^{\underline{3}}_{\underline{5}}4^{\underline{2}}_{\underline{5}}4^{\underline{2}}_{\underline{3}}4^{\underline{1}}_{\underline{3}}4^{\underline{1}}_{\underline{1}}4^{0}_{\underline{1}}5^{0}_{1}2^{1}_{1}3^{1}_{3}1^{2}_{3}2^{2}_{5}1^{3}_{7}$ 9_{42}: $1^{\underline{4}}_{\underline{7}}1^{\underline{3}}_{\underline{3}}1^{\underline{2}}_{\underline{3}}1^{\underline{1}}_{\underline{1}}1^{\underline{1}}_{1}1^{0}_{\underline{1}}1^{0}_{1}1^{0}_{3}1^{1}_{3}1^{2}_{7}$			
9_{43}: $1^{\underline{2}}_{\underline{1}}1^{\underline{1}}_{3}2^{0}_{3}1^{0}_{5}1^{1}_{5}1^{1}_{7}1^{2}_{7}1^{2}_{9}1^{3}_{9}1^{3}_{11}1^{4}_{11}1^{4}_{13}1^{5}_{15}$ 9_{44}: $1^{\underline{5}}_{\underline{11}}1^{\underline{4}}_{\underline{9}}1^{\underline{4}}_{\underline{7}}1^{\underline{3}}_{\underline{7}}1^{\underline{3}}_{\underline{5}}2^{\underline{2}}_{\underline{5}}1^{\underline{2}}_{\underline{3}}1^{\underline{1}}_{\underline{3}}2^{\underline{1}}_{\underline{1}}2^{0}_{\underline{1}}2^{0}_{1}1^{1}_{1}1^{1}_{3}1^{2}_{5}$ 9_{45}: $1^{\underline{7}}_{\underline{17}}1^{\underline{6}}_{\underline{15}}1^{\underline{6}}_{\underline{13}}2^{\underline{5}}_{\underline{13}}1^{\underline{5}}_{\underline{11}}2^{\underline{4}}_{\underline{11}}2^{\underline{4}}_{\underline{9}}2^{\underline{3}}_{\underline{9}}2^{\underline{3}}_{\underline{7}}2^{\underline{2}}_{\underline{7}}2^{\underline{2}}_{\underline{5}}1^{\underline{1}}_{\underline{5}}2^{\underline{1}}_{\underline{3}}1^{0}_{\underline{3}}2^{0}_{\underline{1}}$			

K: KH	1st line	2nd line	3rd line
9_{46}: $1^{\underline{6}}_{\underline{13}}1^{\underline{5}}_{\underline{9}}1^{\underline{4}}_{\underline{9}}1^{\underline{3}}_{\underline{7}}1^{\underline{3}}_{\underline{5}}1^{\underline{2}}_{\underline{3}}1^{\underline{1}}_{\underline{3}}1^{0}_{\underline{1}}2^{0}_{1}$ 9_{47}: $1^{\underline{3}}_{\underline{5}}2^{\underline{2}}_{\underline{3}}1^{\underline{2}}_{\underline{1}}1^{\underline{1}}_{\underline{1}}2^{\underline{1}}_{1}4^{0}_{1}2^{0}_{3}2^{1}_{3}3^{1}_{5}2^{2}_{5}2^{2}_{7}2^{3}_{7}2^{3}_{9}2^{4}_{11}$ 9_{48}: $1^{\underline{2}}_{\underline{3}}2^{\underline{1}}_{\underline{1}}1^{\underline{1}}_{1}2^{0}_{1}3^{0}_{3}3^{1}_{3}1^{1}_{5}3^{2}_{5}3^{2}_{7}1^{3}_{7}3^{3}_{9}2^{4}_{9}1^{4}_{11}2^{5}_{13}$			
9_{49}: $1^{0}_{3}1^{0}_{5}2^{1}_{5}2^{2}_{7}2^{2}_{9}2^{3}_{9}2^{3}_{11}3^{4}_{11}2^{4}_{13}1^{5}_{13}3^{5}_{15}2^{6}_{15}1^{6}_{17}2^{7}_{19}$			

(9_1–9_{41} are alternating, the rest are not).

2.4. **10 Crossing Knots.**

K: KH	1st line	2nd line	3rd line
10_{1}: $1^{\underline{8}}_{\underline{17}}1^{\underline{7}}_{\underline{13}}1^{\underline{6}}_{\underline{13}}1^{\underline{5}}_{\underline{11}}1^{\underline{5}}_{\underline{9}}1^{\underline{4}}_{\underline{9}}1^{\underline{4}}_{\underline{7}}1^{\underline{3}}_{\underline{7}}1^{\underline{3}}_{\underline{5}}1^{\underline{2}}_{\underline{5}}1^{\underline{2}}_{\underline{3}}1^{\underline{1}}_{\underline{3}}1^{\underline{1}}_{\underline{1}}1^{0}_{\underline{1}}2^{0}_{1}1^{1}_{1}1^{2}_{5}$ 10_{2}: $1^{\underline{8}}_{\underline{23}}1^{\underline{7}}_{\underline{21}}1^{\underline{7}}_{\underline{19}}1^{\underline{6}}_{\underline{19}}1^{\underline{6}}_{\underline{17}}2^{\underline{5}}_{\underline{17}}1^{\underline{5}}_{\underline{15}}1^{\underline{4}}_{\underline{15}}2^{\underline{4}}_{\underline{13}}2^{\underline{3}}_{\underline{13}}1^{\underline{3}}_{\underline{11}}1^{\underline{2}}_{\underline{11}}2^{\underline{2}}_{\underline{9}}1^{\underline{1}}_{\underline{9}}1^{\underline{1}}_{\underline{7}}1^{0}_{\underline{7}}2^{0}_{\underline{5}}1^{1}_{\underline{5}}1^{2}_{\underline{1}}$ 10_{3}: $1^{\underline{6}}_{\underline{13}}1^{\underline{5}}_{\underline{9}}2^{\underline{4}}_{\underline{9}}1^{\underline{3}}_{\underline{7}}2^{\underline{3}}_{\underline{5}}2^{\underline{2}}_{\underline{5}}1^{\underline{2}}_{\underline{3}}2^{\underline{1}}_{\underline{3}}2^{\underline{1}}_{\underline{1}}2^{0}_{\underline{1}}3^{0}_{1}2^{1}_{1}1^{1}_{3}2^{2}_{5}1^{3}_{5}1^{4}_{9}$	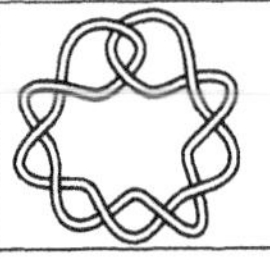	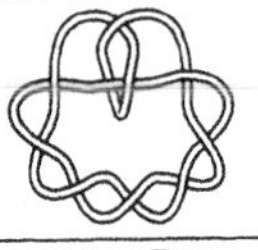	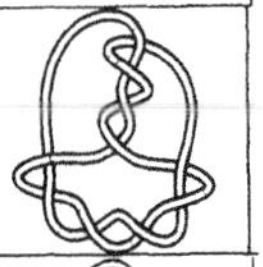
10_{4}: $1^{\underline{4}}_{\underline{11}}1^{\underline{3}}_{\underline{9}}1^{\underline{3}}_{\underline{7}}2^{\underline{2}}_{\underline{7}}1^{\underline{2}}_{\underline{5}}1^{\underline{1}}_{\underline{5}}2^{\underline{1}}_{\underline{3}}3^{0}_{\underline{3}}2^{0}_{\underline{1}}2^{1}_{\underline{1}}2^{1}_{1}1^{2}_{1}2^{2}_{3}2^{3}_{3}1^{3}_{5}2^{4}_{7}1^{5}_{7}1^{6}_{11}$ 10_{5}: $1^{\underline{3}}_{\underline{3}}1^{\underline{2}}_{\underline{1}}1^{\underline{2}}_{1}1^{\underline{1}}_{1}1^{\underline{1}}_{3}3^{0}_{3}2^{0}_{5}2^{1}_{5}2^{1}_{7}3^{2}_{7}2^{2}_{9}2^{3}_{9}3^{3}_{11}2^{4}_{11}2^{4}_{13}1^{5}_{13}2^{5}_{15}1^{6}_{15}1^{6}_{17}1^{7}_{19}$ 10_{6}: $1^{\underline{8}}_{\underline{21}}1^{\underline{7}}_{\underline{19}}1^{\underline{7}}_{\underline{17}}2^{\underline{6}}_{\underline{17}}1^{\underline{6}}_{\underline{15}}3^{\underline{5}}_{\underline{15}}2^{\underline{5}}_{\underline{13}}3^{\underline{4}}_{\underline{13}}3^{\underline{4}}_{\underline{11}}3^{\underline{3}}_{\underline{11}}3^{\underline{3}}_{\underline{9}}2^{\underline{2}}_{\underline{9}}3^{\underline{2}}_{\underline{7}}2^{\underline{1}}_{\underline{7}}2^{\underline{1}}_{\underline{5}}1^{0}_{\underline{5}}3^{0}_{\underline{3}}1^{1}_{\underline{3}}1^{2}_{1}$	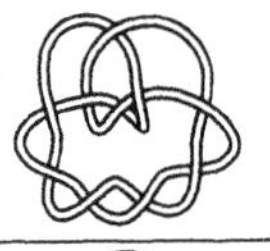		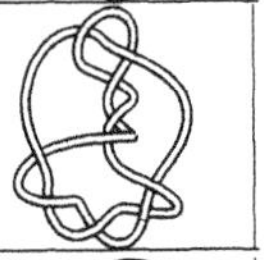
10_{7}: $1^{\underline{8}}_{\underline{19}}1^{\underline{7}}_{\underline{17}}1^{\underline{7}}_{\underline{15}}2^{\underline{6}}_{\underline{15}}1^{\underline{6}}_{\underline{13}}3^{\underline{5}}_{\underline{13}}2^{\underline{5}}_{\underline{11}}3^{\underline{4}}_{\underline{11}}3^{\underline{4}}_{\underline{9}}4^{\underline{3}}_{\underline{9}}3^{\underline{3}}_{\underline{7}}3^{\underline{2}}_{\underline{7}}4^{\underline{2}}_{\underline{5}}2^{\underline{1}}_{\underline{5}}3^{\underline{1}}_{\underline{3}}2^{0}_{\underline{3}}3^{0}_{\underline{1}}1^{1}_{\underline{1}}1^{1}_{1}1^{2}_{3}$ 10_{8}: $1^{\underline{6}}_{\underline{17}}1^{\underline{5}}_{\underline{15}}1^{\underline{5}}_{\underline{13}}2^{\underline{4}}_{\underline{13}}1^{\underline{4}}_{\underline{11}}2^{\underline{3}}_{\underline{11}}2^{\underline{3}}_{\underline{9}}2^{\underline{2}}_{\underline{9}}2^{\underline{2}}_{\underline{7}}2^{\underline{1}}_{\underline{7}}2^{\underline{1}}_{\underline{5}}2^{0}_{\underline{5}}3^{0}_{\underline{3}}2^{1}_{\underline{3}}1^{1}_{\underline{1}}2^{2}_{1}1^{3}_{1}1^{4}_{5}$ 10_{9}: $1^{\underline{4}}_{\underline{7}}1^{\underline{3}}_{\underline{5}}1^{\underline{3}}_{\underline{3}}2^{\underline{2}}_{\underline{3}}1^{\underline{2}}_{\underline{1}}2^{\underline{1}}_{\underline{1}}2^{\underline{1}}_{1}4^{0}_{1}3^{0}_{3}3^{1}_{3}3^{1}_{5}3^{2}_{5}3^{2}_{7}2^{3}_{7}3^{3}_{9}1^{4}_{9}2^{4}_{11}1^{5}_{11}1^{5}_{13}1^{6}_{15}$	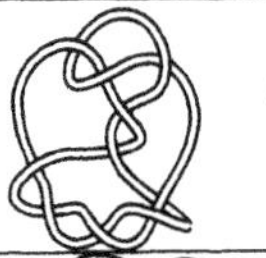	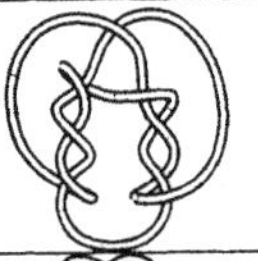	
10_{10}: $1^{\underline{3}}_{\underline{7}}2^{\underline{2}}_{\underline{5}}1^{\underline{2}}_{\underline{3}}2^{\underline{1}}_{\underline{3}}2^{\underline{1}}_{\underline{1}}4^{0}_{\underline{1}}3^{0}_{1}4^{1}_{1}3^{1}_{3}3^{2}_{3}4^{2}_{5}3^{3}_{5}3^{3}_{7}2^{4}_{7}3^{4}_{9}1^{5}_{9}2^{5}_{11}1^{6}_{11}1^{6}_{13}1^{7}_{15}$ 10_{11}: $1^{\underline{6}}_{\underline{15}}1^{\underline{5}}_{\underline{13}}1^{\underline{5}}_{\underline{11}}3^{\underline{4}}_{\underline{11}}1^{\underline{4}}_{\underline{9}}3^{\underline{3}}_{\underline{9}}3^{\underline{3}}_{\underline{7}}4^{\underline{2}}_{\underline{7}}3^{\underline{2}}_{\underline{5}}3^{\underline{1}}_{\underline{5}}4^{\underline{1}}_{\underline{3}}3^{0}_{\underline{3}}4^{0}_{\underline{1}}3^{1}_{\underline{1}}2^{1}_{1}3^{2}_{3}1^{3}_{3}1^{4}_{7}$ 10_{12}: $1^{\underline{3}}_{\underline{5}}1^{\underline{2}}_{\underline{3}}1^{\underline{2}}_{\underline{1}}2^{\underline{1}}_{\underline{1}}1^{\underline{1}}_{1}4^{0}_{1}3^{0}_{3}4^{1}_{3}3^{1}_{5}4^{2}_{5}4^{2}_{7}3^{3}_{7}4^{3}_{9}3^{4}_{9}3^{4}_{11}1^{5}_{11}3^{5}_{13}1^{6}_{13}1^{6}_{15}1^{7}_{17}$		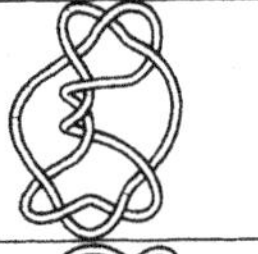	
10_{13}: $1^{\underline{6}}_{\underline{13}}1^{\underline{5}}_{\underline{11}}1^{\underline{5}}_{\underline{9}}3^{\underline{4}}_{\underline{9}}1^{\underline{4}}_{\underline{7}}3^{\underline{3}}_{\underline{7}}3^{\underline{3}}_{\underline{5}}5^{\underline{2}}_{\underline{5}}3^{\underline{2}}_{\underline{3}}4^{\underline{1}}_{\underline{3}}5^{\underline{1}}_{\underline{1}}4^{0}_{\underline{1}}5^{0}_{1}4^{1}_{1}3^{1}_{3}1^{2}_{3}4^{2}_{5}1^{3}_{5}1^{3}_{7}1^{4}_{9}$ 10_{14}: $1^{\underline{8}}_{\underline{21}}2^{\underline{7}}_{\underline{19}}1^{\underline{7}}_{\underline{17}}3^{\underline{6}}_{\underline{17}}2^{\underline{6}}_{\underline{15}}5^{\underline{5}}_{\underline{15}}3^{\underline{5}}_{\underline{13}}4^{\underline{4}}_{\underline{13}}5^{\underline{4}}_{\underline{11}}5^{\underline{3}}_{\underline{11}}4^{\underline{3}}_{\underline{9}}4^{\underline{2}}_{\underline{9}}5^{\underline{2}}_{\underline{7}}2^{\underline{1}}_{\underline{7}}4^{\underline{1}}_{\underline{5}}2^{0}_{\underline{5}}3^{0}_{\underline{3}}1^{1}_{\underline{3}}1^{1}_{\underline{1}}1^{2}_{1}$ 10_{15}: $1^{\underline{5}}_{\underline{9}}1^{\underline{4}}_{\underline{7}}1^{\underline{4}}_{\underline{5}}2^{\underline{3}}_{\underline{5}}1^{\underline{3}}_{\underline{3}}3^{\underline{2}}_{\underline{3}}2^{\underline{2}}_{\underline{1}}3^{\underline{1}}_{\underline{1}}3^{\underline{1}}_{1}4^{0}_{1}4^{0}_{3}3^{1}_{3}3^{1}_{5}3^{2}_{5}3^{2}_{7}1^{3}_{7}3^{3}_{9}1^{4}_{9}1^{4}_{11}1^{5}_{13}$	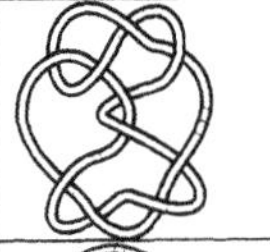		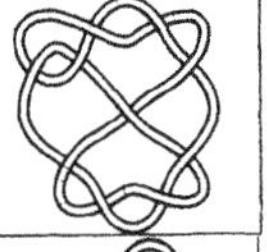
10_{16}: $1^{\underline{4}}_{\underline{7}}1^{\underline{3}}_{\underline{5}}1^{\underline{3}}_{\underline{3}}3^{\underline{2}}_{\underline{3}}1^{\underline{2}}_{\underline{1}}2^{\underline{1}}_{\underline{1}}3^{\underline{1}}_{1}5^{0}_{1}3^{0}_{3}4^{1}_{3}4^{1}_{5}3^{2}_{5}4^{2}_{7}3^{3}_{7}3^{3}_{9}1^{4}_{9}3^{4}_{11}1^{5}_{11}1^{5}_{13}1^{6}_{15}$ 10_{17}: $1^{\underline{5}}_{\underline{11}}1^{\underline{4}}_{\underline{9}}1^{\underline{4}}_{\underline{7}}2^{\underline{3}}_{\underline{7}}1^{\underline{3}}_{\underline{5}}3^{\underline{2}}_{\underline{5}}2^{\underline{2}}_{\underline{3}}3^{\underline{1}}_{\underline{3}}3^{\underline{1}}_{\underline{1}}4^{0}_{\underline{1}}4^{0}_{1}3^{1}_{1}3^{1}_{3}2^{2}_{3}3^{2}_{5}1^{3}_{5}2^{3}_{7}1^{4}_{7}1^{4}_{9}1^{5}_{11}$ 10_{18}: $1^{\underline{6}}_{\underline{15}}2^{\underline{5}}_{\underline{13}}1^{\underline{5}}_{\underline{11}}3^{\underline{4}}_{\underline{11}}2^{\underline{4}}_{\underline{9}}4^{\underline{3}}_{\underline{9}}3^{\underline{3}}_{\underline{7}}5^{\underline{2}}_{\underline{7}}4^{\underline{2}}_{\underline{5}}4^{\underline{1}}_{\underline{5}}5^{\underline{1}}_{\underline{3}}4^{0}_{\underline{3}}5^{0}_{\underline{1}}3^{1}_{\underline{1}}3^{1}_{1}1^{2}_{1}3^{2}_{3}1^{3}_{3}1^{3}_{5}1^{4}_{7}$		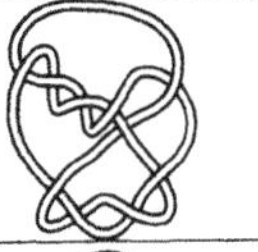	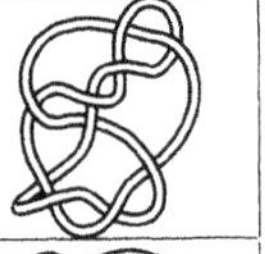
10_{19}: $1^{\underline{5}}_{\underline{13}}2^{\underline{4}}_{\underline{11}}1^{\underline{4}}_{\underline{9}}3^{\underline{3}}_{\underline{9}}2^{\underline{3}}_{\underline{7}}4^{\underline{2}}_{\underline{7}}3^{\underline{2}}_{\underline{5}}4^{\underline{1}}_{\underline{5}}4^{\underline{1}}_{\underline{3}}4^{0}_{\underline{3}}5^{0}_{\underline{1}}4^{1}_{\underline{1}}3^{1}_{1}2^{2}_{1}4^{2}_{3}1^{3}_{3}2^{3}_{5}1^{4}_{5}1^{4}_{7}1^{5}_{9}$ 10_{20}: $1^{\underline{8}}_{\underline{19}}1^{\underline{7}}_{\underline{17}}1^{\underline{7}}_{\underline{15}}2^{\underline{6}}_{\underline{15}}1^{\underline{6}}_{\underline{13}}2^{\underline{5}}_{\underline{13}}2^{\underline{5}}_{\underline{11}}3^{\underline{4}}_{\underline{11}}2^{\underline{4}}_{\underline{9}}3^{\underline{3}}_{\underline{9}}3^{\underline{3}}_{\underline{7}}2^{\underline{2}}_{\underline{7}}3^{\underline{2}}_{\underline{5}}2^{\underline{1}}_{\underline{5}}2^{\underline{1}}_{\underline{3}}1^{0}_{\underline{3}}3^{0}_{\underline{1}}1^{1}_{\underline{1}}1^{2}_{3}$ 10_{21}: $1^{\underline{8}}_{\underline{21}}1^{\underline{7}}_{\underline{19}}1^{\underline{7}}_{\underline{17}}2^{\underline{6}}_{\underline{17}}1^{\underline{6}}_{\underline{15}}4^{\underline{5}}_{\underline{15}}2^{\underline{5}}_{\underline{13}}3^{\underline{4}}_{\underline{13}}4^{\underline{4}}_{\underline{11}}4^{\underline{3}}_{\underline{11}}3^{\underline{3}}_{\underline{9}}3^{\underline{2}}_{\underline{9}}4^{\underline{2}}_{\underline{7}}2^{\underline{1}}_{\underline{7}}3^{\underline{1}}_{\underline{5}}2^{0}_{\underline{5}}3^{0}_{\underline{3}}1^{1}_{\underline{3}}1^{1}_{\underline{1}}1^{2}_{1}$	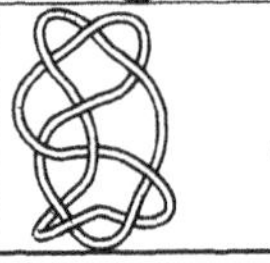		

K: KH	1st line	2nd line	3rd line
10_{22}: $1^{\underline{4}}_{\underline{9}}1^{\underline{3}}_{\underline{7}}1^{\underline{3}}_{\underline{5}}3^{\underline{2}}_{\underline{5}}1^{\underline{2}}_{\underline{3}}3^{\underline{1}}_{\underline{3}}3^{\underline{1}}_{\underline{1}}5^{0}_{\underline{1}}4^{0}_{1}4^{1}_{1}4^{1}_{3}3^{2}_{3}4^{2}_{5}3^{3}_{5}3^{3}_{7}1^{4}_{7}3^{4}_{9}1^{5}_{9}1^{5}_{11}1^{6}_{13}$ 10_{23}: $1^{\underline{3}}_{\underline{5}}2^{\underline{2}}_{\underline{3}}1^{\underline{2}}_{\underline{1}}3^{\underline{1}}_{\underline{1}}2^{\underline{1}}_{1}5^{0}_{1}4^{0}_{3}5^{1}_{3}4^{1}_{5}5^{2}_{5}5^{2}_{7}4^{3}_{7}5^{3}_{9}3^{4}_{9}4^{4}_{11}1^{5}_{11}3^{5}_{13}1^{6}_{13}1^{6}_{15}1^{7}_{17}$ 10_{24}: $1^{\underline{8}}_{\underline{19}}1^{\underline{7}}_{\underline{17}}1^{\underline{7}}_{\underline{15}}3^{\underline{6}}_{\underline{15}}1^{\underline{6}}_{\underline{13}}4^{\underline{5}}_{\underline{13}}3^{\underline{5}}_{\underline{11}}4^{\underline{4}}_{\underline{11}}4^{\underline{4}}_{\underline{9}}5^{\underline{3}}_{\underline{9}}4^{\underline{3}}_{\underline{7}}4^{\underline{2}}_{\underline{7}}5^{\underline{2}}_{\underline{5}}3^{\underline{1}}_{\underline{5}}4^{\underline{1}}_{\underline{3}}2^{0}_{\underline{3}}4^{0}_{\underline{1}}1^{1}_{\underline{1}}1^{1}_{1}1^{2}_{3}$	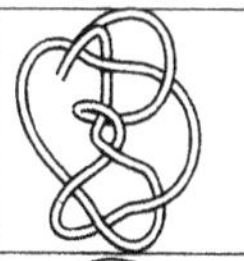		
10_{25}: $1^{\underline{8}}_{\underline{21}}2^{\underline{7}}_{\underline{19}}1^{\underline{7}}_{\underline{17}}4^{\underline{6}}_{\underline{17}}2^{\underline{6}}_{\underline{15}}5^{\underline{5}}_{\underline{15}}4^{\underline{5}}_{\underline{13}}5^{\underline{4}}_{\underline{13}}5^{\underline{4}}_{\underline{11}}6^{\underline{3}}_{\underline{11}}5^{\underline{3}}_{\underline{9}}4^{\underline{2}}_{\underline{9}}6^{\underline{2}}_{\underline{7}}3^{\underline{1}}_{\underline{7}}4^{\underline{1}}_{\underline{5}}2^{0}_{\underline{5}}4^{0}_{\underline{3}}1^{1}_{\underline{3}}1^{1}_{\underline{1}}1^{2}_{1}$ 10_{26}: $1^{\underline{4}}_{\underline{9}}2^{\underline{3}}_{\underline{7}}1^{\underline{3}}_{\underline{5}}4^{\underline{2}}_{\underline{5}}2^{\underline{2}}_{\underline{3}}4^{\underline{1}}_{\underline{3}}4^{\underline{1}}_{\underline{1}}6^{0}_{\underline{1}}5^{0}_{1}5^{1}_{1}5^{1}_{3}4^{2}_{3}5^{2}_{5}3^{3}_{5}4^{3}_{7}1^{4}_{7}3^{4}_{9}1^{5}_{9}1^{5}_{11}1^{6}_{13}$ 10_{27}: $1^{\underline{7}}_{\underline{17}}2^{\underline{6}}_{\underline{15}}1^{\underline{6}}_{\underline{13}}4^{\underline{5}}_{\underline{13}}2^{\underline{5}}_{\underline{11}}5^{\underline{4}}_{\underline{11}}4^{\underline{4}}_{\underline{9}}6^{\underline{3}}_{\underline{9}}5^{\underline{3}}_{\underline{7}}6^{\underline{2}}_{\underline{7}}6^{\underline{2}}_{\underline{5}}5^{\underline{1}}_{\underline{5}}6^{\underline{1}}_{\underline{3}}4^{0}_{\underline{3}}6^{0}_{\underline{1}}2^{1}_{\underline{1}}3^{1}_{1}1^{2}_{1}2^{2}_{3}1^{3}_{5}$			
10_{28}: $1^{\underline{3}}_{\underline{7}}2^{\underline{2}}_{\underline{5}}1^{\underline{2}}_{\underline{3}}3^{\underline{1}}_{\underline{3}}2^{\underline{1}}_{\underline{1}}4^{0}_{\underline{1}}4^{0}_{1}5^{1}_{1}3^{1}_{3}4^{2}_{3}5^{2}_{5}3^{3}_{5}4^{3}_{7}3^{4}_{7}3^{4}_{9}1^{5}_{9}3^{5}_{11}1^{6}_{11}1^{6}_{13}1^{7}_{15}$ 10_{29}: $1^{\underline{6}}_{\underline{15}}2^{\underline{5}}_{\underline{13}}1^{\underline{5}}_{\underline{11}}4^{\underline{4}}_{\underline{11}}2^{\underline{4}}_{\underline{9}}4^{\underline{3}}_{\underline{9}}4^{\underline{3}}_{\underline{7}}6^{\underline{2}}_{\underline{7}}4^{\underline{2}}_{\underline{5}}5^{\underline{1}}_{\underline{5}}6^{\underline{1}}_{\underline{3}}4^{0}_{\underline{3}}6^{0}_{\underline{1}}4^{1}_{\underline{1}}3^{1}_{1}1^{2}_{1}4^{2}_{3}1^{3}_{3}1^{3}_{5}1^{4}_{7}$ 10_{30}: $1^{\underline{8}}_{\underline{19}}2^{\underline{7}}_{\underline{17}}1^{\underline{7}}_{\underline{15}}3^{\underline{6}}_{\underline{15}}2^{\underline{6}}_{\underline{13}}5^{\underline{5}}_{\underline{13}}3^{\underline{5}}_{\underline{11}}5^{\underline{4}}_{\underline{11}}5^{\underline{4}}_{\underline{9}}6^{\underline{3}}_{\underline{9}}5^{\underline{3}}_{\underline{7}}5^{\underline{2}}_{\underline{7}}6^{\underline{2}}_{\underline{5}}3^{\underline{1}}_{\underline{5}}5^{\underline{1}}_{\underline{3}}3^{0}_{\underline{3}}4^{0}_{\underline{1}}1^{1}_{\underline{1}}2^{1}_{1}1^{2}_{3}$			
10_{31}: $1^{\underline{5}}_{\underline{11}}2^{\underline{4}}_{\underline{9}}1^{\underline{4}}_{\underline{7}}3^{\underline{3}}_{\underline{7}}2^{\underline{3}}_{\underline{5}}4^{\underline{2}}_{\underline{5}}3^{\underline{2}}_{\underline{3}}5^{\underline{1}}_{\underline{3}}4^{\underline{1}}_{\underline{1}}5^{0}_{\underline{1}}6^{0}_{1}4^{1}_{1}4^{1}_{3}3^{2}_{3}4^{2}_{5}1^{3}_{5}3^{3}_{7}1^{4}_{7}1^{4}_{9}1^{5}_{11}$ 10_{32}: $1^{\underline{6}}_{\underline{13}}2^{\underline{5}}_{\underline{11}}1^{\underline{5}}_{\underline{9}}3^{\underline{4}}_{\underline{9}}2^{\underline{4}}_{\underline{7}}5^{\underline{3}}_{\underline{7}}3^{\underline{3}}_{\underline{5}}6^{\underline{2}}_{\underline{5}}5^{\underline{2}}_{\underline{3}}5^{\underline{1}}_{\underline{3}}6^{\underline{1}}_{\underline{1}}6^{0}_{\underline{1}}6^{0}_{1}4^{1}_{1}5^{1}_{3}2^{2}_{3}4^{2}_{5}1^{3}_{5}2^{3}_{7}1^{4}_{9}$ 10_{33}: $1^{\underline{5}}_{\underline{11}}2^{\underline{4}}_{\underline{9}}1^{\underline{4}}_{\underline{7}}3^{\underline{3}}_{\underline{7}}2^{\underline{3}}_{\underline{5}}5^{\underline{2}}_{\underline{5}}3^{\underline{2}}_{\underline{3}}5^{\underline{1}}_{\underline{3}}5^{\underline{1}}_{\underline{1}}6^{0}_{\underline{1}}6^{0}_{1}5^{1}_{1}5^{1}_{3}3^{2}_{3}5^{2}_{5}2^{3}_{5}3^{3}_{7}1^{4}_{7}2^{4}_{9}1^{5}_{11}$			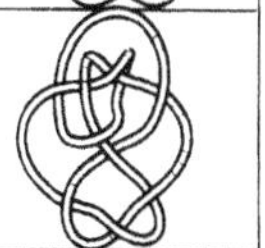
10_{34}: $1^{\underline{3}}_{\underline{7}}1^{\underline{2}}_{\underline{5}}1^{\underline{2}}_{\underline{3}}2^{\underline{1}}_{\underline{3}}1^{\underline{1}}_{\underline{1}}3^{0}_{\underline{1}}3^{0}_{1}3^{1}_{1}2^{1}_{3}3^{2}_{3}3^{2}_{5}2^{3}_{5}3^{3}_{7}2^{4}_{7}2^{4}_{9}1^{5}_{9}2^{5}_{11}1^{6}_{11}1^{6}_{13}1^{7}_{15}$ 10_{35}: $1^{\underline{4}}_{\underline{9}}1^{\underline{3}}_{\underline{7}}1^{\underline{3}}_{\underline{5}}3^{\underline{2}}_{\underline{5}}1^{\underline{2}}_{\underline{3}}3^{\underline{1}}_{\underline{3}}3^{\underline{1}}_{\underline{1}}5^{0}_{\underline{1}}4^{0}_{1}4^{1}_{1}4^{1}_{3}3^{2}_{3}4^{2}_{5}3^{3}_{5}3^{3}_{7}1^{4}_{7}3^{4}_{9}1^{5}_{9}1^{5}_{11}1^{6}_{13}$ 10_{36}: $1^{\underline{8}}_{\underline{19}}2^{\underline{7}}_{\underline{17}}1^{\underline{7}}_{\underline{15}}2^{\underline{6}}_{\underline{15}}2^{\underline{6}}_{\underline{13}}4^{\underline{5}}_{\underline{13}}2^{\underline{5}}_{\underline{11}}4^{\underline{4}}_{\underline{11}}4^{\underline{4}}_{\underline{9}}4^{\underline{3}}_{\underline{9}}4^{\underline{3}}_{\underline{7}}4^{\underline{2}}_{\underline{7}}4^{\underline{2}}_{\underline{5}}2^{\underline{1}}_{\underline{5}}4^{\underline{1}}_{\underline{3}}2^{0}_{\underline{3}}3^{0}_{\underline{1}}1^{1}_{\underline{1}}1^{1}_{1}1^{2}_{3}$			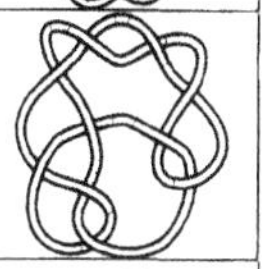
10_{37}: $1^{\underline{5}}_{\underline{11}}1^{\underline{4}}_{\underline{9}}1^{\underline{4}}_{\underline{7}}3^{\underline{3}}_{\underline{7}}1^{\underline{3}}_{\underline{5}}4^{\underline{2}}_{\underline{5}}3^{\underline{2}}_{\underline{3}}4^{\underline{1}}_{\underline{3}}4^{\underline{1}}_{\underline{1}}5^{0}_{\underline{1}}5^{0}_{1}4^{1}_{1}4^{1}_{3}3^{2}_{3}4^{2}_{5}1^{3}_{5}3^{3}_{7}1^{4}_{7}1^{4}_{9}1^{5}_{11}$ 10_{38}: $1^{\underline{8}}_{\underline{19}}2^{\underline{7}}_{\underline{17}}1^{\underline{7}}_{\underline{15}}3^{\underline{6}}_{\underline{15}}2^{\underline{6}}_{\underline{13}}4^{\underline{5}}_{\underline{13}}3^{\underline{5}}_{\underline{11}}5^{\underline{4}}_{\underline{11}}4^{\underline{4}}_{\underline{9}}5^{\underline{3}}_{\underline{9}}5^{\underline{3}}_{\underline{7}}4^{\underline{2}}_{\underline{7}}5^{\underline{2}}_{\underline{5}}3^{\underline{1}}_{\underline{5}}4^{\underline{1}}_{\underline{3}}2^{0}_{\underline{3}}4^{0}_{\underline{1}}1^{1}_{\underline{1}}1^{1}_{1}1^{2}_{3}$ 10_{39}: $1^{\underline{8}}_{\underline{21}}2^{\underline{7}}_{\underline{19}}1^{\underline{7}}_{\underline{17}}3^{\underline{6}}_{\underline{17}}2^{\underline{6}}_{\underline{15}}5^{\underline{5}}_{\underline{15}}3^{\underline{5}}_{\underline{13}}5^{\underline{4}}_{\underline{13}}5^{\underline{4}}_{\underline{11}}5^{\underline{3}}_{\underline{11}}5^{\underline{3}}_{\underline{9}}4^{\underline{2}}_{\underline{9}}5^{\underline{2}}_{\underline{7}}3^{\underline{1}}_{\underline{7}}4^{\underline{1}}_{\underline{5}}2^{0}_{\underline{5}}4^{0}_{\underline{3}}1^{1}_{\underline{3}}1^{1}_{\underline{1}}1^{2}_{1}$	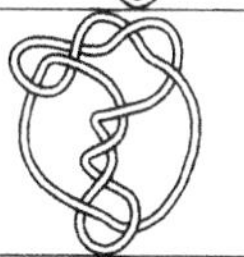		
10_{40}: $1^{\underline{3}}_{\underline{5}}2^{\underline{2}}_{\underline{3}}1^{\underline{2}}_{\underline{1}}4^{\underline{1}}_{\underline{1}}2^{\underline{1}}_{1}6^{0}_{1}5^{0}_{3}6^{1}_{3}5^{1}_{5}7^{2}_{5}6^{2}_{7}5^{3}_{7}7^{3}_{9}4^{4}_{9}5^{4}_{11}2^{5}_{11}4^{5}_{13}1^{6}_{13}2^{6}_{15}1^{7}_{17}$ 10_{41}: $1^{\underline{6}}_{\underline{15}}2^{\underline{5}}_{\underline{13}}1^{\underline{5}}_{\underline{11}}4^{\underline{4}}_{\underline{11}}2^{\underline{4}}_{\underline{9}}5^{\underline{3}}_{\underline{9}}4^{\underline{3}}_{\underline{7}}6^{\underline{2}}_{\underline{7}}5^{\underline{2}}_{\underline{5}}6^{\underline{1}}_{\underline{5}}6^{\underline{1}}_{\underline{3}}5^{0}_{\underline{3}}7^{0}_{\underline{1}}4^{1}_{\underline{1}}4^{1}_{1}2^{2}_{1}4^{2}_{3}1^{3}_{3}2^{3}_{5}1^{4}_{7}$ 10_{42}: $1^{\underline{5}}_{\underline{11}}3^{\underline{4}}_{\underline{9}}1^{\underline{4}}_{\underline{7}}4^{\underline{3}}_{\underline{7}}3^{\underline{3}}_{\underline{5}}6^{\underline{2}}_{\underline{5}}4^{\underline{2}}_{\underline{3}}7^{\underline{1}}_{\underline{3}}6^{\underline{1}}_{\underline{1}}7^{0}_{\underline{1}}8^{0}_{1}6^{1}_{1}6^{1}_{3}4^{2}_{3}6^{2}_{5}2^{3}_{5}4^{3}_{7}1^{4}_{7}2^{4}_{9}1^{5}_{11}$	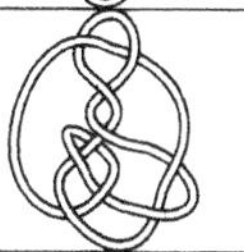		
10_{43}: $1^{\underline{5}}_{\underline{11}}2^{\underline{4}}_{\underline{9}}1^{\underline{4}}_{\underline{7}}4^{\underline{3}}_{\underline{7}}2^{\underline{3}}_{\underline{5}}5^{\underline{2}}_{\underline{5}}4^{\underline{2}}_{\underline{3}}6^{\underline{1}}_{\underline{3}}5^{\underline{1}}_{\underline{1}}7^{0}_{\underline{1}}7^{0}_{1}5^{1}_{1}6^{1}_{3}4^{2}_{3}5^{2}_{5}2^{3}_{5}4^{3}_{7}1^{4}_{7}2^{4}_{9}1^{5}_{11}$ 10_{44}: $1^{\underline{6}}_{\underline{15}}3^{\underline{5}}_{\underline{13}}1^{\underline{5}}_{\underline{11}}4^{\underline{4}}_{\underline{11}}3^{\underline{4}}_{\underline{9}}6^{\underline{3}}_{\underline{9}}4^{\underline{3}}_{\underline{7}}7^{\underline{2}}_{\underline{7}}6^{\underline{2}}_{\underline{5}}6^{\underline{1}}_{\underline{5}}7^{\underline{1}}_{\underline{3}}6^{0}_{\underline{3}}7^{0}_{\underline{1}}4^{1}_{\underline{1}}5^{1}_{1}2^{2}_{1}4^{2}_{3}1^{3}_{3}2^{3}_{5}1^{4}_{7}$ 10_{45}: $1^{\underline{5}}_{\underline{11}}3^{\underline{4}}_{\underline{9}}1^{\underline{4}}_{\underline{7}}4^{\underline{3}}_{\underline{7}}3^{\underline{3}}_{\underline{5}}7^{\underline{2}}_{\underline{5}}4^{\underline{2}}_{\underline{3}}7^{\underline{1}}_{\underline{3}}7^{\underline{1}}_{\underline{1}}8^{0}_{\underline{1}}8^{0}_{1}7^{1}_{1}7^{1}_{3}4^{2}_{3}7^{2}_{5}3^{3}_{5}4^{3}_{7}1^{4}_{7}3^{4}_{9}1^{5}_{11}$			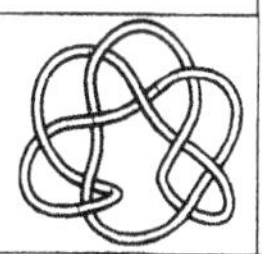
10_{46}: $1^{\underline{2}}_{1}1^{\underline{1}}_{5}3^{0}_{5}1^{0}_{7}1^{1}_{7}2^{1}_{9}3^{2}_{9}1^{2}_{11}2^{3}_{11}3^{3}_{13}2^{4}_{13}2^{4}_{15}2^{5}_{15}2^{5}_{17}1^{6}_{17}2^{6}_{19}1^{7}_{19}1^{7}_{21}1^{8}_{23}$ 10_{47}: $1^{\underline{3}}_{\underline{3}}1^{\underline{2}}_{\underline{1}}1^{\underline{2}}_{1}2^{\underline{1}}_{1}1^{\underline{1}}_{3}3^{0}_{3}3^{0}_{5}3^{1}_{5}2^{1}_{7}4^{2}_{7}3^{2}_{9}2^{3}_{9}4^{3}_{11}3^{4}_{11}2^{4}_{13}1^{5}_{13}3^{5}_{15}1^{6}_{15}1^{6}_{17}1^{7}_{19}$ 10_{48}: $1^{\underline{5}}_{\underline{11}}1^{\underline{4}}_{\underline{9}}1^{\underline{4}}_{\underline{7}}3^{\underline{3}}_{\underline{7}}1^{\underline{3}}_{\underline{5}}3^{\underline{2}}_{\underline{5}}3^{\underline{2}}_{\underline{3}}4^{\underline{1}}_{\underline{3}}3^{\underline{1}}_{\underline{1}}5^{0}_{\underline{1}}5^{0}_{1}3^{1}_{1}4^{1}_{3}3^{2}_{3}3^{2}_{5}1^{3}_{5}3^{3}_{7}1^{4}_{7}1^{4}_{9}1^{5}_{11}$			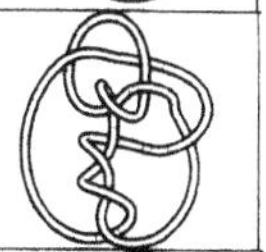
10_{49}: $1^{\underline{10}}_{\underline{27}}2^{\underline{9}}_{\underline{25}}1^{\underline{9}}_{\underline{23}}3^{\underline{8}}_{\underline{23}}2^{\underline{8}}_{\underline{21}}5^{\underline{7}}_{\underline{21}}3^{\underline{7}}_{\underline{19}}4^{\underline{6}}_{\underline{19}}5^{\underline{6}}_{\underline{17}}6^{\underline{5}}_{\underline{17}}4^{\underline{5}}_{\underline{15}}3^{\underline{4}}_{\underline{15}}6^{\underline{4}}_{\underline{13}}3^{\underline{3}}_{\underline{13}}3^{\underline{3}}_{\underline{11}}2^{\underline{2}}_{\underline{11}}3^{\underline{2}}_{\underline{9}}2^{\underline{1}}_{\underline{7}}1^{0}_{\underline{7}}1^{0}_{\underline{5}}$ 10_{50}: $1^{\underline{2}}_{\underline{1}}1^{\underline{1}}_{1}1^{\underline{1}}_{3}4^{0}_{3}2^{0}_{5}3^{1}_{5}3^{1}_{7}5^{2}_{7}3^{2}_{9}4^{3}_{9}5^{3}_{11}4^{4}_{11}4^{4}_{13}3^{5}_{13}4^{5}_{15}1^{6}_{15}3^{6}_{17}1^{7}_{17}1^{7}_{19}1^{8}_{21}$ 10_{51}: $1^{\underline{3}}_{\underline{5}}2^{\underline{2}}_{\underline{3}}1^{\underline{2}}_{\underline{1}}4^{\underline{1}}_{\underline{1}}2^{\underline{1}}_{1}5^{0}_{1}5^{0}_{3}6^{1}_{3}4^{1}_{5}6^{2}_{5}6^{2}_{7}4^{3}_{7}6^{3}_{9}4^{4}_{9}4^{4}_{11}1^{5}_{11}4^{5}_{13}1^{6}_{13}1^{6}_{15}1^{7}_{17}$	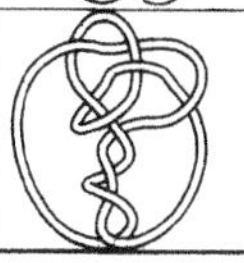		

K: KH	1st line	2nd line	3rd line
10_{52}: $1^{\underline{5}}_{\underline{9}}1^{\underline{4}}_{\underline{7}}1^{\underline{4}}_{\underline{5}}3^{\underline{3}}_{\underline{5}}1^{\underline{3}}_{\underline{3}}4^{\underline{2}}_{\underline{3}}3^{\underline{2}}_{\underline{1}}4^{\underline{1}}_{\underline{1}}4^{\underline{1}}_{1}6^{0}_{1}5^{0}_{3}4^{1}_{3}5^{1}_{5}4^{2}_{5}4^{2}_{7}2^{3}_{7}4^{3}_{9}1^{4}_{9}2^{4}_{11}1^{5}_{13}$ 10_{53}: $1^{\underline{10}}_{\underline{25}}2^{\underline{9}}_{\underline{23}}1^{\underline{9}}_{\underline{21}}3^{\underline{8}}_{\underline{21}}2^{\underline{8}}_{\underline{19}}6^{\underline{7}}_{\underline{19}}3^{\underline{7}}_{\underline{17}}5^{\underline{6}}_{\underline{17}}6^{\underline{6}}_{\underline{15}}7^{\underline{5}}_{\underline{15}}5^{\underline{5}}_{\underline{13}}5^{\underline{4}}_{\underline{13}}7^{\underline{4}}_{\underline{11}}4^{\underline{3}}_{\underline{11}}5^{\underline{3}}_{\underline{9}}3^{\underline{2}}_{\underline{9}}4^{\underline{2}}_{\underline{7}}3^{\underline{1}}_{\underline{5}}1^{0}_{\underline{5}}1^{0}_{\underline{3}}$ 10_{54}: $1^{\underline{5}}_{\underline{9}}1^{\underline{4}}_{\underline{7}}1^{\underline{4}}_{\underline{5}}3^{\underline{3}}_{\underline{5}}1^{\underline{3}}_{\underline{3}}3^{\underline{2}}_{\underline{3}}3^{\underline{2}}_{\underline{1}}3^{\underline{1}}_{\underline{1}}3^{\underline{1}}_{1}5^{0}_{1}4^{0}_{3}3^{1}_{3}4^{1}_{5}3^{2}_{5}3^{2}_{7}1^{3}_{7}3^{3}_{9}1^{4}_{9}1^{4}_{11}1^{5}_{13}$			
10_{55}: $1^{\underline{10}}_{\underline{25}}2^{\underline{9}}_{\underline{23}}1^{\underline{9}}_{\underline{21}}3^{\underline{8}}_{\underline{21}}2^{\underline{8}}_{\underline{19}}5^{\underline{7}}_{\underline{19}}3^{\underline{7}}_{\underline{17}}4^{\underline{6}}_{\underline{17}}5^{\underline{6}}_{\underline{15}}6^{\underline{5}}_{\underline{15}}4^{\underline{5}}_{\underline{13}}4^{\underline{4}}_{\underline{13}}6^{\underline{4}}_{\underline{11}}3^{\underline{3}}_{\underline{11}}4^{\underline{3}}_{\underline{9}}2^{\underline{2}}_{\underline{9}}3^{\underline{2}}_{\underline{7}}2^{\underline{1}}_{\underline{5}}1^{0}_{\underline{5}}1^{0}_{\underline{3}}$ 10_{56}: $1^{\underline{2}}_{\underline{1}}1^{\underline{1}}_{1}1^{\underline{1}}_{3}4^{0}_{3}2^{0}_{5}4^{1}_{5}3^{1}_{7}6^{2}_{7}4^{2}_{9}5^{3}_{9}6^{3}_{11}5^{4}_{11}5^{4}_{13}4^{5}_{13}5^{5}_{15}2^{6}_{15}4^{6}_{17}1^{7}_{17}2^{7}_{19}1^{8}_{21}$ 10_{57}: $1^{\underline{3}}_{\underline{5}}2^{\underline{2}}_{\underline{3}}1^{\underline{2}}_{\underline{1}}4^{\underline{1}}_{\underline{1}}2^{\underline{1}}_{1}6^{0}_{1}5^{0}_{3}7^{1}_{3}5^{1}_{5}7^{2}_{5}7^{2}_{7}5^{3}_{7}7^{3}_{9}5^{4}_{9}5^{4}_{11}2^{5}_{11}5^{5}_{13}1^{6}_{13}2^{6}_{15}1^{7}_{17}$		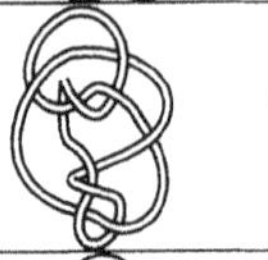	
10_{58}: $1^{\underline{6}}_{\underline{13}}2^{\underline{5}}_{\underline{11}}1^{\underline{5}}_{\underline{9}}4^{\underline{4}}_{\underline{9}}2^{\underline{4}}_{\underline{7}}4^{\underline{3}}_{\underline{7}}4^{\underline{3}}_{\underline{5}}6^{\underline{2}}_{\underline{5}}4^{\underline{2}}_{\underline{3}}5^{\underline{1}}_{\underline{3}}6^{\underline{1}}_{\underline{1}}5^{0}_{\underline{1}}6^{0}_{1}4^{1}_{1}4^{1}_{3}1^{2}_{3}4^{2}_{5}1^{3}_{5}1^{3}_{7}1^{4}_{9}$ 10_{59}: $1^{\underline{4}}_{\underline{7}}2^{\underline{3}}_{\underline{5}}1^{\underline{3}}_{\underline{3}}4^{\underline{2}}_{\underline{3}}2^{\underline{2}}_{\underline{1}}5^{\underline{1}}_{\underline{1}}4^{\underline{1}}_{1}7^{0}_{1}6^{0}_{3}6^{1}_{3}6^{1}_{5}6^{2}_{5}6^{2}_{7}4^{3}_{7}6^{3}_{9}2^{4}_{9}4^{4}_{11}1^{5}_{11}2^{5}_{13}1^{6}_{15}$ 10_{60}: $1^{\underline{6}}_{\underline{13}}2^{\underline{5}}_{\underline{11}}1^{\underline{5}}_{\underline{9}}4^{\underline{4}}_{\underline{9}}2^{\underline{4}}_{\underline{7}}6^{\underline{3}}_{\underline{7}}4^{\underline{3}}_{\underline{5}}7^{\underline{2}}_{\underline{5}}6^{\underline{2}}_{\underline{3}}7^{\underline{1}}_{\underline{3}}7^{\underline{1}}_{\underline{1}}7^{0}_{\underline{1}}8^{0}_{1}5^{1}_{1}6^{1}_{3}3^{2}_{3}5^{2}_{5}1^{3}_{5}3^{3}_{7}1^{4}_{9}$			
10_{61}: $1^{\underline{4}}_{\underline{5}}1^{\underline{3}}_{\underline{1}}3^{\underline{2}}_{\underline{1}}1^{\underline{1}}_{1}3^{\underline{1}}_{3}3^{0}_{3}2^{0}_{5}3^{1}_{5}2^{1}_{7}2^{2}_{7}3^{2}_{9}2^{3}_{9}2^{3}_{11}1^{4}_{11}2^{4}_{13}1^{5}_{13}1^{5}_{15}1^{6}_{17}$ 10_{62}: $1^{\underline{3}}_{\underline{3}}1^{\underline{2}}_{\underline{1}}1^{\underline{2}}_{1}2^{\underline{1}}_{1}1^{\underline{1}}_{3}4^{0}_{3}3^{0}_{5}3^{1}_{5}3^{1}_{7}4^{2}_{7}3^{2}_{9}3^{3}_{9}4^{3}_{11}3^{4}_{11}3^{4}_{13}1^{5}_{13}3^{5}_{15}1^{6}_{15}1^{6}_{17}1^{7}_{19}$ 10_{63}: $1^{\underline{10}}_{\underline{25}}2^{\underline{9}}_{\underline{23}}1^{\underline{9}}_{\underline{21}}2^{\underline{8}}_{\underline{21}}2^{\underline{8}}_{\underline{19}}5^{\underline{7}}_{\underline{19}}2^{\underline{7}}_{\underline{17}}4^{\underline{6}}_{\underline{17}}5^{\underline{6}}_{\underline{15}}5^{\underline{5}}_{\underline{15}}4^{\underline{5}}_{\underline{13}}4^{\underline{4}}_{\underline{13}}5^{\underline{4}}_{\underline{11}}3^{\underline{3}}_{\underline{11}}4^{\underline{3}}_{\underline{9}}2^{\underline{2}}_{\underline{9}}3^{\underline{2}}_{\underline{7}}2^{\underline{1}}_{\underline{5}}1^{0}_{\underline{5}}1^{0}_{\underline{3}}$			
10_{64}: $1^{\underline{4}}_{\underline{7}}1^{\underline{3}}_{\underline{5}}1^{\underline{3}}_{\underline{3}}3^{\underline{2}}_{\underline{3}}1^{\underline{2}}_{\underline{1}}3^{\underline{1}}_{\underline{1}}3^{\underline{1}}_{1}5^{0}_{1}4^{0}_{3}4^{1}_{3}4^{1}_{5}4^{2}_{5}4^{2}_{7}3^{3}_{7}4^{3}_{9}1^{4}_{9}3^{4}_{11}1^{5}_{11}1^{5}_{13}1^{6}_{15}$ 10_{65}: $1^{\underline{3}}_{\underline{5}}2^{\underline{2}}_{\underline{3}}1^{\underline{2}}_{\underline{1}}3^{\underline{1}}_{\underline{1}}2^{\underline{1}}_{1}5^{0}_{1}4^{0}_{3}6^{1}_{3}4^{1}_{5}5^{2}_{5}6^{2}_{7}4^{3}_{7}5^{3}_{9}4^{4}_{9}4^{4}_{11}1^{5}_{11}4^{5}_{13}1^{6}_{13}1^{6}_{15}1^{7}_{17}$ 10_{66}: $1^{\underline{10}}_{\underline{27}}3^{\underline{9}}_{\underline{25}}1^{\underline{9}}_{\underline{23}}4^{\underline{8}}_{\underline{23}}3^{\underline{8}}_{\underline{21}}6^{\underline{7}}_{\underline{21}}4^{\underline{7}}_{\underline{19}}6^{\underline{6}}_{\underline{19}}6^{\underline{6}}_{\underline{17}}7^{\underline{5}}_{\underline{17}}6^{\underline{5}}_{\underline{15}}4^{\underline{4}}_{\underline{15}}7^{\underline{4}}_{\underline{13}}4^{\underline{3}}_{\underline{13}}4^{\underline{3}}_{\underline{11}}2^{\underline{2}}_{\underline{11}}4^{\underline{2}}_{\underline{9}}2^{\underline{1}}_{\underline{7}}1^{0}_{\underline{7}}1^{0}_{\underline{5}}$			
10_{67}: $1^{\underline{8}}_{\underline{19}}2^{\underline{7}}_{\underline{17}}1^{\underline{7}}_{\underline{15}}3^{\underline{6}}_{\underline{15}}2^{\underline{6}}_{\underline{13}}5^{\underline{5}}_{\underline{13}}3^{\underline{5}}_{\underline{11}}5^{\underline{4}}_{\underline{11}}5^{\underline{4}}_{\underline{9}}5^{\underline{3}}_{\underline{9}}5^{\underline{3}}_{\underline{7}}5^{\underline{2}}_{\underline{7}}5^{\underline{2}}_{\underline{5}}3^{\underline{1}}_{\underline{5}}5^{\underline{1}}_{\underline{3}}2^{0}_{\underline{3}}4^{0}_{\underline{1}}1^{1}_{\underline{1}}1^{1}_{1}1^{2}_{3}$ 10_{68}: $1^{\underline{7}}_{\underline{15}}1^{\underline{6}}_{\underline{13}}1^{\underline{6}}_{\underline{11}}3^{\underline{5}}_{\underline{11}}1^{\underline{5}}_{\underline{9}}4^{\underline{4}}_{\underline{9}}3^{\underline{4}}_{\underline{7}}4^{\underline{3}}_{\underline{7}}4^{\underline{3}}_{\underline{5}}5^{\underline{2}}_{\underline{5}}4^{\underline{2}}_{\underline{3}}4^{\underline{1}}_{\underline{3}}5^{\underline{1}}_{\underline{1}}4^{0}_{\underline{1}}5^{0}_{1}2^{1}_{1}3^{1}_{3}1^{2}_{3}2^{2}_{5}1^{3}_{7}$ 10_{69}: $1^{\underline{3}}_{\underline{5}}3^{\underline{2}}_{\underline{3}}1^{\underline{2}}_{\underline{1}}4^{\underline{1}}_{\underline{1}}3^{\underline{1}}_{1}7^{0}_{1}5^{0}_{3}8^{1}_{3}6^{1}_{5}7^{2}_{5}8^{2}_{7}6^{3}_{7}7^{3}_{9}5^{4}_{9}6^{4}_{11}2^{5}_{11}5^{5}_{13}1^{6}_{13}2^{6}_{15}1^{7}_{17}$			
10_{70}: $1^{\underline{4}}_{\underline{7}}1^{\underline{3}}_{\underline{5}}1^{\underline{3}}_{\underline{3}}4^{\underline{2}}_{\underline{3}}1^{\underline{2}}_{\underline{1}}4^{\underline{1}}_{\underline{1}}4^{\underline{1}}_{1}6^{0}_{1}5^{0}_{3}6^{1}_{3}5^{1}_{5}5^{2}_{5}6^{2}_{7}4^{3}_{7}5^{3}_{9}2^{4}_{9}4^{4}_{11}1^{5}_{11}2^{5}_{13}1^{6}_{15}$ 10_{71}: $1^{\underline{5}}_{\underline{11}}2^{\underline{4}}_{\underline{9}}1^{\underline{4}}_{\underline{7}}4^{\underline{3}}_{\underline{7}}2^{\underline{3}}_{\underline{5}}6^{\underline{2}}_{\underline{5}}4^{\underline{2}}_{\underline{3}}6^{\underline{1}}_{\underline{3}}6^{\underline{1}}_{\underline{1}}7^{0}_{\underline{1}}7^{0}_{1}6^{1}_{1}6^{1}_{3}4^{2}_{3}6^{2}_{5}2^{3}_{5}4^{3}_{7}1^{4}_{7}2^{4}_{9}1^{5}_{11}$ 10_{72}: $1^{\underline{2}}_{\underline{1}}1^{\underline{1}}_{1}1^{\underline{1}}_{3}4^{0}_{3}2^{0}_{5}5^{1}_{5}3^{1}_{7}6^{2}_{7}5^{2}_{9}6^{3}_{9}6^{3}_{11}6^{4}_{11}6^{4}_{13}4^{5}_{13}6^{5}_{15}3^{6}_{15}4^{6}_{17}1^{7}_{17}3^{7}_{19}1^{8}_{21}$		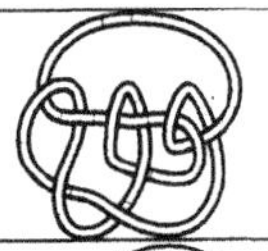	
10_{73}: $1^{\underline{7}}_{\underline{17}}2^{\underline{6}}_{\underline{15}}1^{\underline{6}}_{\underline{13}}4^{\underline{5}}_{\underline{13}}2^{\underline{5}}_{\underline{11}}6^{\underline{4}}_{\underline{11}}4^{\underline{4}}_{\underline{9}}7^{\underline{3}}_{\underline{9}}6^{\underline{3}}_{\underline{7}}7^{\underline{2}}_{\underline{7}}7^{\underline{2}}_{\underline{5}}6^{\underline{1}}_{\underline{5}}7^{\underline{1}}_{\underline{3}}5^{0}_{\underline{3}}7^{0}_{\underline{1}}3^{1}_{\underline{1}}4^{1}_{1}1^{2}_{1}3^{2}_{3}1^{3}_{5}$ 10_{74}: $1^{\underline{8}}_{\underline{19}}1^{\underline{7}}_{\underline{17}}1^{\underline{7}}_{\underline{15}}3^{\underline{6}}_{\underline{15}}1^{\underline{6}}_{\underline{13}}5^{\underline{5}}_{\underline{13}}3^{\underline{5}}_{\underline{11}}4^{\underline{4}}_{\underline{11}}5^{\underline{4}}_{\underline{9}}6^{\underline{3}}_{\underline{9}}4^{\underline{3}}_{\underline{7}}5^{\underline{2}}_{\underline{7}}6^{\underline{2}}_{\underline{5}}3^{\underline{1}}_{\underline{5}}5^{\underline{1}}_{\underline{3}}3^{0}_{\underline{3}}4^{0}_{\underline{1}}1^{1}_{\underline{1}}2^{1}_{1}1^{2}_{3}$ 10_{75}: $1^{\underline{4}}_{\underline{9}}3^{\underline{3}}_{\underline{7}}1^{\underline{3}}_{\underline{5}}4^{\underline{2}}_{\underline{5}}3^{\underline{2}}_{\underline{3}}6^{\underline{1}}_{\underline{3}}4^{\underline{1}}_{\underline{1}}8^{0}_{\underline{1}}7^{0}_{1}6^{1}_{1}7^{1}_{3}6^{2}_{3}6^{2}_{5}4^{3}_{5}6^{3}_{7}2^{4}_{7}4^{4}_{9}1^{5}_{9}2^{5}_{11}1^{6}_{13}$			
10_{76}: $1^{\underline{2}}_{\underline{1}}1^{\underline{1}}_{3}4^{0}_{3}1^{0}_{5}3^{1}_{5}3^{1}_{7}5^{2}_{7}3^{2}_{9}5^{3}_{9}5^{3}_{11}4^{4}_{11}5^{4}_{13}4^{5}_{13}4^{5}_{15}2^{6}_{15}4^{6}_{17}1^{7}_{17}2^{7}_{19}1^{8}_{21}$ 10_{77}: $1^{\underline{3}}_{\underline{5}}1^{\underline{2}}_{\underline{3}}1^{\underline{2}}_{\underline{1}}3^{\underline{1}}_{\underline{1}}1^{\underline{1}}_{1}5^{0}_{1}4^{0}_{3}5^{1}_{3}4^{1}_{5}6^{2}_{5}5^{2}_{7}4^{3}_{7}6^{3}_{9}4^{4}_{9}4^{4}_{11}2^{5}_{11}4^{5}_{13}1^{6}_{13}2^{6}_{15}1^{7}_{17}$ 10_{78}: $1^{\underline{8}}_{\underline{21}}2^{\underline{7}}_{\underline{19}}1^{\underline{7}}_{\underline{17}}3^{\underline{6}}_{\underline{17}}2^{\underline{6}}_{\underline{15}}6^{\underline{5}}_{\underline{15}}3^{\underline{5}}_{\underline{13}}5^{\underline{4}}_{\underline{13}}6^{\underline{4}}_{\underline{11}}6^{\underline{3}}_{\underline{11}}5^{\underline{3}}_{\underline{9}}5^{\underline{2}}_{\underline{9}}6^{\underline{2}}_{\underline{7}}3^{\underline{1}}_{\underline{7}}5^{\underline{1}}_{\underline{5}}3^{0}_{\underline{5}}4^{0}_{\underline{3}}1^{1}_{\underline{3}}2^{1}_{\underline{1}}1^{2}_{1}$		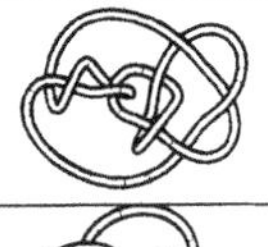	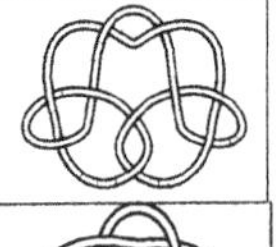
10_{79}: $1^{\underline{5}}_{\underline{11}}1^{\underline{4}}_{\underline{9}}1^{\underline{4}}_{\underline{7}}4^{\underline{3}}_{\underline{7}}1^{\underline{3}}_{\underline{5}}4^{\underline{2}}_{\underline{5}}4^{\underline{2}}_{\underline{3}}5^{\underline{1}}_{\underline{3}}4^{\underline{1}}_{\underline{1}}6^{0}_{\underline{1}}6^{0}_{1}4^{1}_{1}5^{1}_{3}4^{2}_{3}4^{2}_{5}1^{3}_{5}4^{3}_{7}1^{4}_{7}1^{4}_{9}1^{5}_{11}$ 10_{80}: $1^{\underline{10}}_{\underline{27}}2^{\underline{9}}_{\underline{25}}1^{\underline{9}}_{\underline{23}}4^{\underline{8}}_{\underline{23}}2^{\underline{8}}_{\underline{21}}6^{\underline{7}}_{\underline{21}}4^{\underline{7}}_{\underline{19}}5^{\underline{6}}_{\underline{19}}6^{\underline{6}}_{\underline{17}}7^{\underline{5}}_{\underline{17}}5^{\underline{5}}_{\underline{15}}4^{\underline{4}}_{\underline{15}}7^{\underline{4}}_{\underline{13}}4^{\underline{3}}_{\underline{13}}4^{\underline{3}}_{\underline{11}}2^{\underline{2}}_{\underline{11}}4^{\underline{2}}_{\underline{9}}2^{\underline{1}}_{\underline{7}}1^{0}_{\underline{7}}1^{0}_{\underline{5}}$ 10_{81}: $1^{\underline{5}}_{\underline{11}}2^{\underline{4}}_{\underline{9}}1^{\underline{4}}_{\underline{7}}5^{\underline{3}}_{\underline{7}}2^{\underline{3}}_{\underline{5}}6^{\underline{2}}_{\underline{5}}5^{\underline{2}}_{\underline{3}}7^{\underline{1}}_{\underline{3}}6^{\underline{1}}_{\underline{1}}8^{0}_{\underline{1}}8^{0}_{1}6^{1}_{1}7^{1}_{3}5^{2}_{3}6^{2}_{5}2^{3}_{5}5^{3}_{7}1^{4}_{7}2^{4}_{9}1^{5}_{11}$			
10_{82}: $1^{\underline{6}}_{\underline{15}}2^{\underline{5}}_{\underline{13}}1^{\underline{5}}_{\underline{11}}3^{\underline{4}}_{\underline{11}}2^{\underline{4}}_{\underline{9}}5^{\underline{3}}_{\underline{9}}3^{\underline{3}}_{\underline{7}}5^{\underline{2}}_{\underline{7}}5^{\underline{2}}_{\underline{5}}5^{\underline{1}}_{\underline{5}}5^{\underline{1}}_{\underline{3}}5^{0}_{\underline{3}}6^{0}_{\underline{1}}3^{1}_{\underline{1}}4^{1}_{1}2^{2}_{1}3^{2}_{3}1^{3}_{3}2^{3}_{5}1^{4}_{7}$ 10_{83}: $1^{\underline{3}}_{\underline{5}}3^{\underline{2}}_{\underline{3}}1^{\underline{2}}_{\underline{1}}4^{\underline{1}}_{\underline{1}}3^{\underline{1}}_{1}7^{0}_{1}5^{0}_{3}7^{1}_{3}6^{1}_{5}7^{2}_{5}7^{2}_{7}6^{3}_{7}7^{3}_{9}4^{4}_{9}6^{4}_{11}2^{5}_{11}4^{5}_{13}1^{6}_{13}2^{6}_{15}1^{7}_{17}$ 10_{84}: $1^{\underline{3}}_{\underline{5}}2^{\underline{2}}_{\underline{3}}1^{\underline{2}}_{\underline{1}}4^{\underline{1}}_{\underline{1}}2^{\underline{1}}_{1}7^{0}_{1}5^{0}_{3}7^{1}_{3}6^{1}_{5}8^{2}_{5}7^{2}_{7}6^{3}_{7}8^{3}_{9}5^{4}_{9}6^{4}_{11}3^{5}_{11}5^{5}_{13}1^{6}_{13}3^{6}_{15}1^{7}_{17}$			

K: KH	1st line / 2nd line / 3rd line
10_{85}: $1^{\underline{7}}_{\underline{19}} 2^{\underline{6}}_{\underline{17}} 1^{\underline{6}}_{\underline{15}} 3^{\underline{5}}_{\underline{15}} 2^{\underline{5}}_{\underline{13}} 4^{\underline{4}}_{\underline{13}} 3^{\underline{4}}_{\underline{11}} 5^{\underline{3}}_{\underline{11}} 4^{\underline{3}}_{\underline{9}} 4^{\underline{2}}_{\underline{9}} 5^{\underline{2}}_{\underline{7}} 4^{\underline{1}}_{\underline{7}} 4^{\underline{1}}_{\underline{5}} 3^{0}_{\underline{5}} 5^{0}_{\underline{3}} 2^{1}_{\underline{3}} 2^{1}_{\underline{1}} 1^{2}_{\underline{1}} 2^{2}_{1} 1^{3}_{3}$	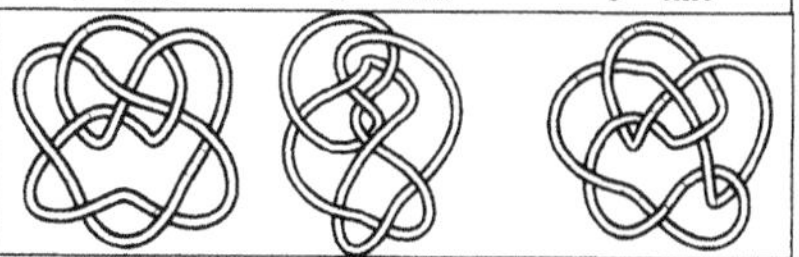
10_{86}: $1^{\underline{4}}_{\underline{9}} 3^{\underline{3}}_{\underline{7}} 1^{\underline{3}}_{\underline{5}} 5^{\underline{2}}_{\underline{5}} 3^{\underline{2}}_{\underline{3}} 6^{\underline{1}}_{\underline{3}} 5^{\underline{1}}_{\underline{1}} 8^{0}_{\underline{1}} 7^{0}_{1} 7^{1}_{1} 7^{1}_{3} 6^{2}_{3} 7^{2}_{5} 4^{3}_{5} 6^{3}_{7} 2^{4}_{7} 4^{4}_{9} 1^{5}_{9} 2^{5}_{11} 1^{6}_{13}$	
10_{87}: $1^{\underline{4}}_{\underline{9}} 2^{\underline{3}}_{\underline{7}} 1^{\underline{3}}_{\underline{5}} 4^{\underline{2}}_{\underline{5}} 2^{\underline{2}}_{\underline{3}} 6^{\underline{1}}_{\underline{3}} 4^{\underline{1}}_{\underline{1}} 7^{0}_{\underline{1}} 7^{0}_{1} 7^{1}_{1} 6^{1}_{3} 6^{2}_{3} 7^{2}_{5} 4^{3}_{5} 6^{3}_{7} 3^{4}_{7} 4^{4}_{9} 1^{5}_{9} 3^{5}_{11} 1^{6}_{13}$	
10_{88}: $1^{\underline{5}}_{\underline{11}} 3^{\underline{4}}_{\underline{9}} 1^{\underline{4}}_{\underline{7}} 5^{\underline{3}}_{\underline{7}} 3^{\underline{3}}_{\underline{5}} 8^{\underline{2}}_{\underline{5}} 5^{\underline{2}}_{\underline{3}} 8^{\underline{1}}_{\underline{3}} 8^{\underline{1}}_{\underline{1}} 9^{0}_{\underline{1}} 9^{0}_{1} 8^{1}_{1} 8^{1}_{3} 5^{2}_{3} 8^{2}_{5} 3^{3}_{5} 5^{3}_{7} 1^{4}_{7} 3^{4}_{9} 1^{5}_{11}$	
10_{89}: $1^{\underline{7}}_{\underline{17}} 2^{\underline{6}}_{\underline{15}} 1^{\underline{6}}_{\underline{13}} 5^{\underline{5}}_{\underline{13}} 2^{\underline{5}}_{\underline{11}} 7^{\underline{4}}_{\underline{11}} 5^{\underline{4}}_{\underline{9}} 8^{\underline{3}}_{\underline{9}} 7^{\underline{3}}_{\underline{7}} 9^{\underline{2}}_{\underline{7}} 8^{\underline{2}}_{\underline{5}} 7^{\underline{1}}_{\underline{5}} 9^{\underline{1}}_{\underline{3}} 6^{0}_{\underline{3}} 8^{0}_{\underline{1}} 4^{1}_{\underline{1}} 5^{1}_{1} 1^{2}_{1} 4^{2}_{3} 1^{3}_{5}$	
10_{90}: $1^{\underline{4}}_{\underline{9}} 2^{\underline{3}}_{\underline{7}} 1^{\underline{3}}_{\underline{5}} 5^{\underline{2}}_{\underline{5}} 2^{\underline{2}}_{\underline{3}} 5^{\underline{1}}_{\underline{3}} 5^{\underline{1}}_{\underline{1}} 7^{0}_{\underline{1}} 6^{0}_{1} 7^{1}_{1} 6^{1}_{3} 5^{2}_{3} 7^{2}_{5} 4^{3}_{5} 5^{3}_{7} 2^{4}_{7} 4^{4}_{9} 1^{5}_{9} 2^{5}_{11} 1^{6}_{13}$	
10_{91}: $1^{\underline{5}}_{\underline{11}} 2^{\underline{4}}_{\underline{9}} 1^{\underline{4}}_{\underline{7}} 4^{\underline{3}}_{\underline{7}} 2^{\underline{3}}_{\underline{5}} 5^{\underline{2}}_{\underline{5}} 4^{\underline{2}}_{\underline{3}} 6^{\underline{1}}_{\underline{3}} 5^{\underline{1}}_{\underline{1}} 7^{0}_{\underline{1}} 7^{0}_{1} 5^{1}_{1} 6^{1}_{3} 4^{2}_{3} 5^{2}_{5} 2^{3}_{5} 4^{3}_{7} 1^{4}_{7} 2^{4}_{9} 1^{5}_{11}$	
10_{92}: $1^{\underline{2}}_{\underline{1}} 2^{\underline{1}}_{1} 1^{\underline{1}}_{3} 5^{0}_{3} 3^{0}_{5} 6^{1}_{5} 4^{1}_{7} 8^{2}_{7} 6^{2}_{9} 7^{3}_{9} 8^{3}_{11} 7^{4}_{11} 7^{4}_{13} 5^{5}_{13} 7^{5}_{15} 3^{6}_{15} 5^{6}_{17} 1^{7}_{17} 3^{7}_{19} 1^{8}_{21}$	
10_{93}: $1^{\underline{5}}_{\underline{13}} 2^{\underline{4}}_{\underline{11}} 1^{\underline{4}}_{\underline{9}} 4^{\underline{3}}_{\underline{9}} 2^{\underline{3}}_{\underline{7}} 5^{\underline{2}}_{\underline{7}} 4^{\underline{2}}_{\underline{5}} 5^{\underline{1}}_{\underline{5}} 5^{\underline{1}}_{\underline{3}} 6^{0}_{\underline{3}} 6^{0}_{\underline{1}} 5^{1}_{\underline{1}} 5^{1}_{1} 3^{2}_{1} 5^{2}_{3} 2^{3}_{3} 3^{3}_{5} 1^{4}_{5} 2^{4}_{7} 1^{5}_{9}$	
10_{94}: $1^{\underline{4}}_{\underline{7}} 2^{\underline{3}}_{\underline{5}} 1^{\underline{3}}_{\underline{3}} 4^{\underline{2}}_{\underline{3}} 2^{\underline{2}}_{\underline{1}} 4^{\underline{1}}_{\underline{1}} 4^{\underline{1}}_{1} 7^{0}_{1} 5^{0}_{3} 6^{1}_{3} 6^{1}_{5} 5^{2}_{5} 6^{2}_{7} 4^{3}_{7} 5^{3}_{9} 2^{4}_{9} 4^{4}_{11} 1^{5}_{11} 2^{5}_{13} 1^{6}_{15}$	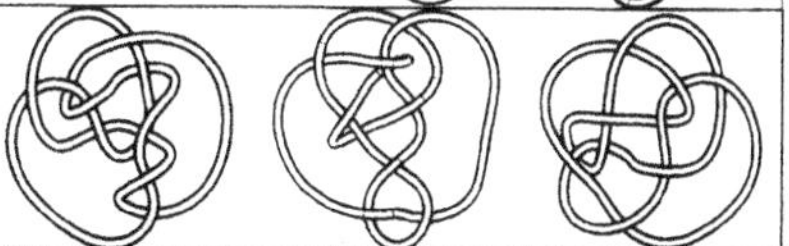
10_{95}: $1^{\underline{3}}_{\underline{5}} 3^{\underline{2}}_{\underline{3}} 1^{\underline{2}}_{\underline{1}} 5^{\underline{1}}_{\underline{1}} 3^{\underline{1}}_{1} 7^{0}_{1} 6^{0}_{3} 8^{1}_{3} 6^{1}_{5} 8^{2}_{5} 8^{2}_{7} 6^{3}_{7} 8^{3}_{9} 5^{4}_{9} 6^{4}_{11} 2^{5}_{11} 5^{5}_{13} 1^{6}_{13} 2^{6}_{15} 1^{7}_{17}$	
10_{96}: $1^{\underline{4}}_{\underline{9}} 3^{\underline{3}}_{\underline{7}} 1^{\underline{3}}_{\underline{5}} 6^{\underline{2}}_{\underline{5}} 3^{\underline{2}}_{\underline{3}} 6^{\underline{1}}_{\underline{3}} 6^{\underline{1}}_{\underline{1}} 9^{0}_{\underline{1}} 7^{0}_{1} 8^{1}_{1} 8^{1}_{3} 6^{2}_{3} 8^{2}_{5} 5^{3}_{5} 6^{3}_{7} 2^{4}_{7} 5^{4}_{9} 1^{5}_{9} 2^{5}_{11} 1^{6}_{13}$	
10_{97}: $1^{\underline{2}}_{\underline{3}} 2^{\underline{1}}_{\underline{1}} 1^{\underline{1}}_{1} 5^{0}_{1} 3^{0}_{3} 7^{1}_{3} 4^{1}_{5} 7^{2}_{5} 7^{2}_{7} 7^{3}_{7} 7^{3}_{9} 7^{4}_{9} 7^{4}_{11} 4^{5}_{11} 7^{5}_{13} 3^{6}_{13} 4^{6}_{15} 1^{7}_{15} 3^{7}_{17} 1^{8}_{19}$	
10_{98}: $1^{\underline{8}}_{\underline{21}} 2^{\underline{7}}_{\underline{19}} 1^{\underline{7}}_{\underline{17}} 5^{\underline{6}}_{\underline{17}} 2^{\underline{6}}_{\underline{15}} 6^{\underline{5}}_{\underline{15}} 5^{\underline{5}}_{\underline{13}} 6^{\underline{4}}_{\underline{13}} 6^{\underline{4}}_{\underline{11}} 8^{\underline{3}}_{\underline{11}} 6^{\underline{3}}_{\underline{9}} 5^{\underline{2}}_{\underline{9}} 8^{\underline{2}}_{\underline{7}} 4^{\underline{1}}_{\underline{7}} 5^{\underline{1}}_{\underline{5}} 3^{0}_{\underline{5}} 5^{0}_{\underline{3}} 1^{1}_{\underline{3}} 2^{1}_{\underline{1}} 1^{2}_{1}$	
10_{99}: $1^{\underline{5}}_{\underline{11}} 2^{\underline{4}}_{\underline{9}} 1^{\underline{4}}_{\underline{7}} 5^{\underline{3}}_{\underline{7}} 2^{\underline{3}}_{\underline{5}} 5^{\underline{2}}_{\underline{5}} 5^{\underline{2}}_{\underline{3}} 7^{\underline{1}}_{\underline{3}} 5^{\underline{1}}_{\underline{1}} 8^{0}_{\underline{1}} 8^{0}_{1} 5^{1}_{1} 7^{1}_{3} 5^{2}_{3} 5^{2}_{5} 2^{3}_{5} 5^{3}_{7} 1^{4}_{7} 2^{4}_{9} 1^{5}_{11}$	
10_{100}: $1^{\underline{7}}_{\underline{19}} 2^{\underline{6}}_{\underline{17}} 1^{\underline{6}}_{\underline{15}} 4^{\underline{5}}_{\underline{15}} 2^{\underline{5}}_{\underline{13}} 4^{\underline{4}}_{\underline{13}} 4^{\underline{4}}_{\underline{11}} 6^{\underline{3}}_{\underline{11}} 4^{\underline{3}}_{\underline{9}} 5^{\underline{2}}_{\underline{9}} 6^{\underline{2}}_{\underline{7}} 4^{\underline{1}}_{\underline{7}} 5^{\underline{1}}_{\underline{5}} 4^{0}_{\underline{5}} 5^{0}_{\underline{3}} 2^{1}_{\underline{3}} 3^{1}_{\underline{1}} 1^{2}_{\underline{1}} 2^{2}_{1} 1^{3}_{3}$	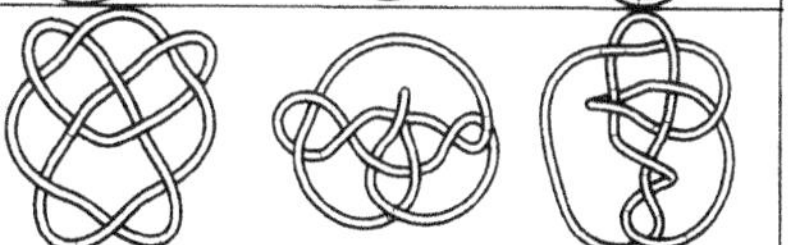
10_{101}: $1^{0}_{3} 1^{0}_{5} 3^{1}_{5} 4^{2}_{7} 3^{2}_{9} 6^{3}_{9} 4^{3}_{11} 8^{4}_{11} 6^{4}_{13} 6^{5}_{13} 8^{5}_{15} 7^{6}_{15} 6^{6}_{17} 4^{7}_{17} 7^{7}_{19} 3^{8}_{19} 4^{8}_{21} 1^{9}_{21} 3^{9}_{23} 1^{10}_{25}$	
10_{102}: $1^{\underline{4}}_{\underline{9}} 2^{\underline{3}}_{\underline{7}} 1^{\underline{3}}_{\underline{5}} 4^{\underline{2}}_{\underline{5}} 2^{\underline{2}}_{\underline{3}} 5^{\underline{1}}_{\underline{3}} 4^{\underline{1}}_{\underline{1}} 7^{0}_{\underline{1}} 6^{0}_{1} 6^{1}_{1} 6^{1}_{3} 5^{2}_{3} 6^{2}_{5} 4^{3}_{5} 5^{3}_{7} 2^{4}_{7} 4^{4}_{9} 1^{5}_{9} 2^{5}_{11} 1^{6}_{13}$	
10_{103}: $1^{\underline{7}}_{\underline{17}} 2^{\underline{6}}_{\underline{15}} 1^{\underline{6}}_{\underline{13}} 4^{\underline{5}}_{\underline{13}} 2^{\underline{5}}_{\underline{11}} 5^{\underline{4}}_{\underline{11}} 4^{\underline{4}}_{\underline{9}} 7^{\underline{3}}_{\underline{9}} 5^{\underline{3}}_{\underline{7}} 6^{\underline{2}}_{\underline{7}} 7^{\underline{2}}_{\underline{5}} 5^{\underline{1}}_{\underline{5}} 6^{\underline{1}}_{\underline{3}} 5^{0}_{\underline{3}} 6^{0}_{\underline{1}} 2^{1}_{\underline{1}} 4^{1}_{1} 1^{2}_{1} 2^{2}_{3} 1^{3}_{5}$	
10_{104}: $1^{\underline{5}}_{\underline{11}} 2^{\underline{4}}_{\underline{9}} 1^{\underline{4}}_{\underline{7}} 4^{\underline{3}}_{\underline{7}} 2^{\underline{3}}_{\underline{5}} 6^{\underline{2}}_{\underline{5}} 4^{\underline{2}}_{\underline{3}} 6^{\underline{1}}_{\underline{3}} 6^{\underline{1}}_{\underline{1}} 7^{0}_{\underline{1}} 7^{0}_{1} 6^{1}_{1} 6^{1}_{3} 4^{2}_{3} 6^{2}_{5} 2^{3}_{5} 4^{3}_{7} 1^{4}_{7} 2^{4}_{9} 1^{5}_{11}$	
10_{105}: $1^{\underline{4}}_{\underline{7}} 2^{\underline{3}}_{\underline{5}} 1^{\underline{3}}_{\underline{3}} 5^{\underline{2}}_{\underline{3}} 2^{\underline{2}}_{\underline{1}} 6^{\underline{1}}_{\underline{1}} 5^{\underline{1}}_{1} 8^{0}_{1} 7^{0}_{3} 8^{1}_{3} 7^{1}_{5} 7^{2}_{5} 8^{2}_{7} 5^{3}_{7} 7^{3}_{9} 3^{4}_{9} 5^{4}_{11} 1^{5}_{11} 3^{5}_{13} 1^{6}_{15}$	
10_{106}: $1^{\underline{4}}_{\underline{7}} 2^{\underline{3}}_{\underline{5}} 1^{\underline{3}}_{\underline{3}} 4^{\underline{2}}_{\underline{3}} 2^{\underline{2}}_{\underline{1}} 5^{\underline{1}}_{\underline{1}} 4^{\underline{1}}_{1} 7^{0}_{1} 6^{0}_{3} 6^{1}_{3} 6^{1}_{5} 6^{2}_{5} 6^{2}_{7} 4^{3}_{7} 6^{3}_{9} 2^{4}_{9} 4^{4}_{11} 1^{5}_{11} 2^{5}_{13} 1^{6}_{15}$	
10_{107}: $1^{\underline{5}}_{\underline{11}} 3^{\underline{4}}_{\underline{9}} 1^{\underline{4}}_{\underline{7}} 5^{\underline{3}}_{\underline{7}} 3^{\underline{3}}_{\underline{5}} 7^{\underline{2}}_{\underline{5}} 5^{\underline{2}}_{\underline{3}} 8^{\underline{1}}_{\underline{3}} 7^{\underline{1}}_{\underline{1}} 8^{0}_{\underline{1}} 9^{0}_{1} 7^{1}_{1} 7^{1}_{3} 5^{2}_{3} 7^{2}_{5} 2^{3}_{5} 5^{3}_{7} 1^{4}_{7} 2^{4}_{9} 1^{5}_{11}$	
10_{108}: $1^{\underline{5}}_{\underline{9}} 2^{\underline{4}}_{\underline{7}} 1^{\underline{4}}_{\underline{5}} 3^{\underline{3}}_{\underline{5}} 2^{\underline{3}}_{\underline{3}} 5^{\underline{2}}_{\underline{3}} 3^{\underline{2}}_{\underline{1}} 4^{\underline{1}}_{\underline{1}} 5^{\underline{1}}_{1} 6^{0}_{1} 5^{0}_{3} 5^{1}_{3} 5^{1}_{5} 3^{2}_{5} 5^{2}_{7} 2^{3}_{7} 3^{3}_{9} 1^{4}_{9} 2^{4}_{11} 1^{5}_{13}$	
10_{109}: $1^{\underline{5}}_{\underline{11}} 2^{\underline{4}}_{\underline{9}} 1^{\underline{4}}_{\underline{7}} 5^{\underline{3}}_{\underline{7}} 2^{\underline{3}}_{\underline{5}} 6^{\underline{2}}_{\underline{5}} 5^{\underline{2}}_{\underline{3}} 7^{\underline{1}}_{\underline{3}} 6^{\underline{1}}_{\underline{1}} 8^{0}_{\underline{1}} 8^{0}_{1} 6^{1}_{1} 7^{1}_{3} 5^{2}_{3} 6^{2}_{5} 2^{3}_{5} 5^{3}_{7} 1^{4}_{7} 2^{4}_{9} 1^{5}_{11}$	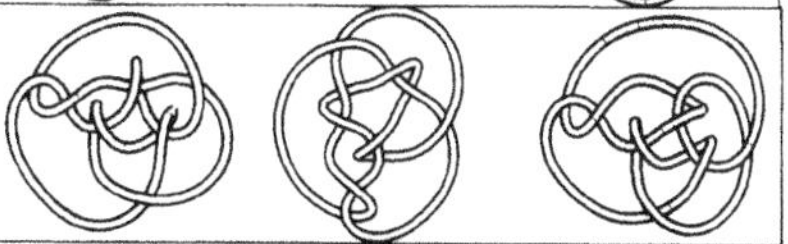
10_{110}: $1^{\underline{6}}_{\underline{15}} 2^{\underline{5}}_{\underline{13}} 1^{\underline{5}}_{\underline{11}} 5^{\underline{4}}_{\underline{11}} 2^{\underline{4}}_{\underline{9}} 6^{\underline{3}}_{\underline{9}} 5^{\underline{3}}_{\underline{7}} 7^{\underline{2}}_{\underline{7}} 6^{\underline{2}}_{\underline{5}} 7^{\underline{1}}_{\underline{5}} 7^{\underline{1}}_{\underline{3}} 6^{0}_{\underline{3}} 8^{0}_{\underline{1}} 5^{1}_{\underline{1}} 5^{1}_{1} 2^{2}_{1} 5^{2}_{3} 1^{3}_{3} 2^{3}_{5} 1^{4}_{7}$	
10_{111}: $1^{\underline{2}}_{\underline{1}} 2^{\underline{1}}_{1} 1^{\underline{1}}_{3} 5^{0}_{3} 3^{0}_{5} 5^{1}_{5} 4^{1}_{7} 7^{2}_{7} 5^{2}_{9} 6^{3}_{9} 7^{3}_{11} 6^{4}_{11} 6^{4}_{13} 4^{5}_{13} 6^{5}_{15} 2^{6}_{15} 4^{6}_{17} 1^{7}_{17} 2^{7}_{19} 1^{8}_{21}$	
10_{112}: $1^{\underline{6}}_{\underline{15}} 3^{\underline{5}}_{\underline{13}} 1^{\underline{5}}_{\underline{11}} 4^{\underline{4}}_{\underline{11}} 3^{\underline{4}}_{\underline{9}} 7^{\underline{3}}_{\underline{9}} 4^{\underline{3}}_{\underline{7}} 7^{\underline{2}}_{\underline{7}} 7^{\underline{2}}_{\underline{5}} 7^{\underline{1}}_{\underline{5}} 7^{\underline{1}}_{\underline{3}} 7^{0}_{\underline{3}} 8^{0}_{\underline{1}} 4^{1}_{\underline{1}} 6^{1}_{1} 3^{2}_{1} 4^{2}_{3} 1^{3}_{3} 3^{3}_{5} 1^{4}_{7}$	
10_{113}: $1^{\underline{3}}_{\underline{5}} 3^{\underline{2}}_{\underline{3}} 1^{\underline{2}}_{\underline{1}} 5^{\underline{1}}_{\underline{1}} 3^{\underline{1}}_{1} 9^{0}_{1} 6^{0}_{3} 9^{1}_{3} 8^{1}_{5} 10^{2}_{5} 9^{2}_{7} 8^{3}_{7} 10^{3}_{9} 6^{4}_{9} 8^{4}_{11} 4^{5}_{11} 6^{5}_{13} 1^{6}_{13} 4^{6}_{15} 1^{7}_{17}$	
10_{114}: $1^{\underline{6}}_{\underline{13}} 3^{\underline{5}}_{\underline{11}} 1^{\underline{5}}_{\underline{9}} 4^{\underline{4}}_{\underline{9}} 3^{\underline{4}}_{\underline{7}} 7^{\underline{3}}_{\underline{7}} 4^{\underline{3}}_{\underline{5}} 8^{\underline{2}}_{\underline{5}} 7^{\underline{2}}_{\underline{3}} 7^{\underline{1}}_{\underline{3}} 8^{\underline{1}}_{\underline{1}} 8^{0}_{\underline{1}} 8^{0}_{1} 5^{1}_{1} 7^{1}_{3} 3^{2}_{3} 5^{2}_{5} 1^{3}_{5} 3^{3}_{7} 1^{4}_{9}$	
10_{115}: $1^{\underline{5}}_{\underline{11}} 3^{\underline{4}}_{\underline{9}} 1^{\underline{4}}_{\underline{7}} 6^{\underline{3}}_{\underline{7}} 3^{\underline{3}}_{\underline{5}} 8^{\underline{2}}_{\underline{5}} 6^{\underline{2}}_{\underline{3}} 9^{\underline{1}}_{\underline{3}} 8^{\underline{1}}_{\underline{1}} 10^{0}_{\underline{1}} 10^{0}_{1} 8^{1}_{1} 9^{1}_{3} 6^{2}_{3} 8^{2}_{5} 3^{3}_{5} 6^{3}_{7} 1^{4}_{7} 3^{4}_{9} 1^{5}_{11}$	
10_{116}: $1^{\underline{6}}_{\underline{15}} 3^{\underline{5}}_{\underline{13}} 1^{\underline{5}}_{\underline{11}} 5^{\underline{4}}_{\underline{11}} 3^{\underline{4}}_{\underline{9}} 7^{\underline{3}}_{\underline{9}} 5^{\underline{3}}_{\underline{7}} 8^{\underline{2}}_{\underline{7}} 7^{\underline{2}}_{\underline{5}} 8^{\underline{1}}_{\underline{5}} 8^{\underline{1}}_{\underline{3}} 7^{0}_{\underline{3}} 9^{0}_{\underline{1}} 5^{1}_{\underline{1}} 6^{1}_{1} 3^{2}_{1} 5^{2}_{3} 1^{3}_{3} 3^{3}_{5} 1^{4}_{7}$	
10_{117}: $1^{\underline{3}}_{\underline{5}} 3^{\underline{2}}_{\underline{3}} 1^{\underline{2}}_{\underline{1}} 5^{\underline{1}}_{\underline{1}} 3^{\underline{1}}_{1} 8^{0}_{1} 6^{0}_{3} 9^{1}_{3} 7^{1}_{5} 9^{2}_{5} 9^{2}_{7} 7^{3}_{7} 9^{3}_{9} 6^{4}_{9} 7^{4}_{11} 3^{5}_{11} 6^{5}_{13} 1^{6}_{13} 3^{6}_{15} 1^{7}_{17}$	

K: KH	1st line	2nd line	3rd line
10_{118}: $1^{\underline{5}}_{\underline{11}}3^{\underline{4}}_{\underline{9}}1^{\underline{4}}_{\underline{7}}5^{\underline{3}}_{\underline{7}}3^{\underline{3}}_{\underline{5}}7^{\underline{2}}_{\underline{5}}5^{\underline{2}}_{\underline{3}}8^{\underline{1}}_{\underline{3}}7^{\underline{1}}_{\underline{1}}9^{0}_{\underline{1}}9^{0}_{1}7^{1}_{1}8^{1}_{3}5^{2}_{3}7^{2}_{5}3^{3}_{5}5^{3}_{7}1^{4}_{7}3^{4}_{9}1^{5}_{11}$	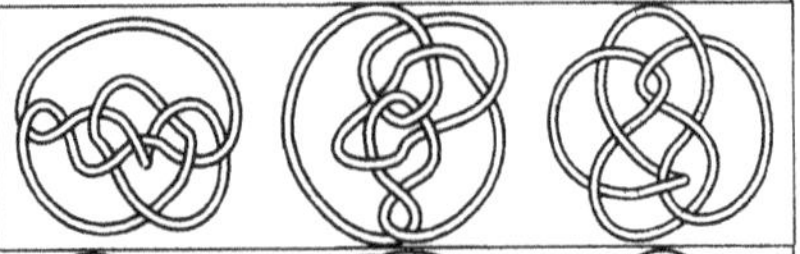		
10_{119}: $1^{\underline{4}}_{\underline{9}}3^{\underline{3}}_{\underline{7}}1^{\underline{3}}_{\underline{5}}6^{\underline{2}}_{\underline{5}}3^{\underline{2}}_{\underline{3}}7^{\underline{1}}_{\underline{3}}6^{\underline{1}}_{\underline{1}}9^{0}_{\underline{1}}8^{0}_{1}9^{1}_{1}8^{1}_{3}7^{2}_{3}9^{2}_{5}5^{3}_{5}7^{3}_{7}3^{4}_{7}5^{4}_{9}1^{5}_{9}3^{5}_{11}1^{6}_{13}$			
10_{120}: $1^{\underline{10}}_{\underline{25}}3^{\underline{9}}_{\underline{23}}1^{\underline{9}}_{\underline{21}}5^{\underline{8}}_{\underline{21}}3^{\underline{8}}_{\underline{19}}8^{\underline{7}}_{\underline{19}}5^{\underline{7}}_{\underline{17}}8^{\underline{6}}_{\underline{17}}8^{\underline{6}}_{\underline{15}}10^{\underline{5}}_{\underline{15}}8^{\underline{5}}_{\underline{13}}7^{\underline{4}}_{\underline{13}}10^{\underline{4}}_{\underline{11}}6^{\underline{3}}_{\underline{11}}7^{\underline{3}}_{\underline{9}}4^{\underline{2}}_{\underline{9}}6^{\underline{2}}_{\underline{7}}4^{\underline{1}}_{\underline{5}}1^{0}_{\underline{5}}1^{0}_{\underline{3}}$			
10_{121}: $1^{\underline{7}}_{\underline{17}}3^{\underline{6}}_{\underline{15}}1^{\underline{6}}_{\underline{13}}6^{\underline{5}}_{\underline{13}}3^{\underline{5}}_{\underline{11}}8^{\underline{4}}_{\underline{11}}6^{\underline{4}}_{\underline{9}}10^{\underline{3}}_{\underline{9}}8^{\underline{3}}_{\underline{7}}10^{\underline{2}}_{\underline{7}}10^{\underline{2}}_{\underline{5}}8^{\underline{1}}_{\underline{5}}10^{\underline{1}}_{\underline{3}}7^{0}_{\underline{3}}9^{0}_{\underline{1}}4^{1}_{\underline{1}}6^{1}_{1}1^{2}_{1}4^{2}_{3}1^{3}_{5}$			
10_{122}: $1^{\underline{4}}_{\underline{9}}3^{\underline{3}}_{\underline{7}}1^{\underline{3}}_{\underline{5}}5^{\underline{2}}_{\underline{5}}3^{\underline{2}}_{\underline{3}}8^{\underline{1}}_{\underline{3}}5^{\underline{1}}_{\underline{1}}9^{0}_{\underline{1}}9^{0}_{1}9^{1}_{1}8^{1}_{3}8^{2}_{3}9^{2}_{5}5^{3}_{5}8^{3}_{7}4^{4}_{7}5^{4}_{9}1^{5}_{9}4^{5}_{11}1^{6}_{13}$			
10_{123}: $1^{\underline{5}}_{\underline{11}}4^{\underline{4}}_{\underline{9}}1^{\underline{4}}_{\underline{7}}6^{\underline{3}}_{\underline{7}}4^{\underline{3}}_{\underline{5}}9^{\underline{2}}_{\underline{5}}6^{\underline{2}}_{\underline{3}}10^{\underline{1}}_{\underline{3}}9^{\underline{1}}_{\underline{1}}11^{0}_{\underline{1}}11^{0}_{1}9^{1}_{1}10^{1}_{3}6^{2}_{3}9^{2}_{5}4^{3}_{5}6^{3}_{7}1^{4}_{7}4^{4}_{9}1^{5}_{11}$			
10_{124}: $1^{0}_{7}1^{0}_{9}1^{2}_{11}1^{3}_{15}1^{4}_{13}1^{4}_{15}1^{5}_{17}1^{5}_{19}1^{6}_{17}1^{7}_{21}$			
10_{125}: $1^{\underline{5}}_{\underline{9}}1^{\underline{4}}_{\underline{5}}1^{\underline{3}}_{\underline{5}}1^{\underline{2}}_{\underline{3}}1^{\underline{2}}_{\underline{1}}1^{\underline{1}}_{1}2^{0}_{1}1^{0}_{3}1^{1}_{5}1^{2}_{5}1^{3}_{9}$			
10_{126}: $1^{\underline{7}}_{\underline{17}}1^{\underline{6}}_{\underline{13}}2^{\underline{5}}_{\underline{13}}1^{\underline{4}}_{\underline{11}}2^{\underline{4}}_{\underline{9}}2^{\underline{3}}_{\underline{9}}1^{\underline{3}}_{\underline{7}}2^{\underline{2}}_{\underline{7}}2^{\underline{2}}_{\underline{5}}2^{\underline{1}}_{\underline{3}}2^{0}_{\underline{3}}1^{0}_{\underline{1}}1^{1}_{1}$			
10_{127}: $1^{\underline{8}}_{\underline{21}}1^{\underline{7}}_{\underline{19}}1^{\underline{7}}_{\underline{17}}2^{\underline{6}}_{\underline{17}}1^{\underline{6}}_{\underline{15}}3^{\underline{5}}_{\underline{15}}2^{\underline{5}}_{\underline{13}}2^{\underline{4}}_{\underline{13}}3^{\underline{4}}_{\underline{11}}3^{\underline{3}}_{\underline{11}}2^{\underline{3}}_{\underline{9}}1^{\underline{2}}_{\underline{9}}3^{\underline{2}}_{\underline{7}}1^{\underline{1}}_{\underline{7}}1^{\underline{1}}_{\underline{5}}1^{0}_{\underline{5}}2^{0}_{\underline{3}}$			
10_{128}: $1^{0}_{5}1^{0}_{7}1^{1}_{7}1^{2}_{9}1^{2}_{11}1^{3}_{11}1^{3}_{13}1^{4}_{11}2^{4}_{13}1^{4}_{15}1^{5}_{15}2^{5}_{17}1^{6}_{17}1^{7}_{21}$			
10_{129}: $1^{\underline{5}}_{\underline{11}}1^{\underline{4}}_{\underline{9}}1^{\underline{4}}_{\underline{7}}2^{\underline{3}}_{\underline{7}}1^{\underline{3}}_{\underline{5}}2^{\underline{2}}_{\underline{5}}2^{\underline{2}}_{\underline{3}}2^{\underline{1}}_{\underline{3}}2^{\underline{1}}_{\underline{1}}3^{0}_{\underline{1}}3^{0}_{1}1^{1}_{1}2^{1}_{3}1^{2}_{3}1^{2}_{5}1^{3}_{7}$			
10_{130}: $1^{\underline{7}}_{\underline{15}}1^{\underline{6}}_{\underline{11}}2^{\underline{5}}_{\underline{11}}1^{\underline{4}}_{\underline{9}}2^{\underline{4}}_{\underline{7}}1^{\underline{3}}_{\underline{7}}1^{\underline{3}}_{\underline{5}}2^{\underline{2}}_{\underline{5}}1^{\underline{2}}_{\underline{3}}2^{\underline{1}}_{\underline{1}}2^{0}_{\underline{1}}1^{0}_{1}1^{1}_{3}$	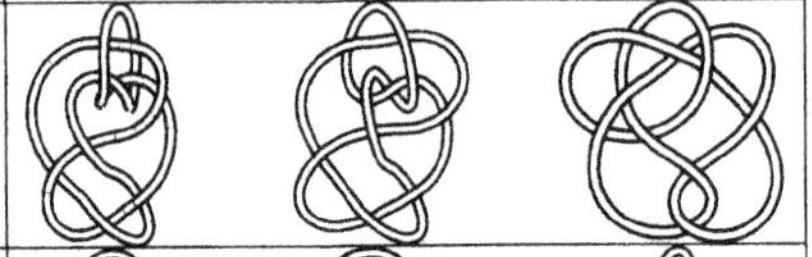		
10_{131}: $1^{\underline{8}}_{\underline{19}}1^{\underline{7}}_{\underline{17}}1^{\underline{7}}_{\underline{15}}2^{\underline{6}}_{\underline{15}}1^{\underline{6}}_{\underline{13}}3^{\underline{5}}_{\underline{13}}2^{\underline{5}}_{\underline{11}}2^{\underline{4}}_{\underline{11}}3^{\underline{4}}_{\underline{9}}3^{\underline{3}}_{\underline{9}}2^{\underline{3}}_{\underline{7}}2^{\underline{2}}_{\underline{7}}3^{\underline{2}}_{\underline{5}}1^{\underline{1}}_{\underline{5}}2^{\underline{1}}_{\underline{3}}1^{0}_{\underline{3}}2^{0}_{\underline{1}}$			
10_{132}: $1^{\underline{7}}_{\underline{15}}1^{\underline{6}}_{\underline{11}}1^{\underline{5}}_{\underline{11}}1^{\underline{4}}_{\underline{9}}1^{\underline{4}}_{\underline{7}}1^{\underline{3}}_{\underline{9}}1^{\underline{3}}_{\underline{5}}2^{\underline{2}}_{\underline{5}}1^{\underline{1}}_{\underline{1}}1^{0}_{\underline{3}}1^{0}_{\underline{1}}$			
10_{133}: $1^{\underline{8}}_{\underline{19}}1^{\underline{7}}_{\underline{17}}1^{\underline{7}}_{\underline{15}}1^{\underline{6}}_{\underline{15}}1^{\underline{6}}_{\underline{13}}2^{\underline{5}}_{\underline{13}}1^{\underline{5}}_{\underline{11}}1^{\underline{4}}_{\underline{11}}2^{\underline{4}}_{\underline{9}}2^{\underline{3}}_{\underline{9}}1^{\underline{3}}_{\underline{7}}1^{\underline{2}}_{\underline{7}}2^{\underline{2}}_{\underline{5}}1^{\underline{1}}_{\underline{3}}1^{0}_{\underline{3}}1^{0}_{\underline{1}}$	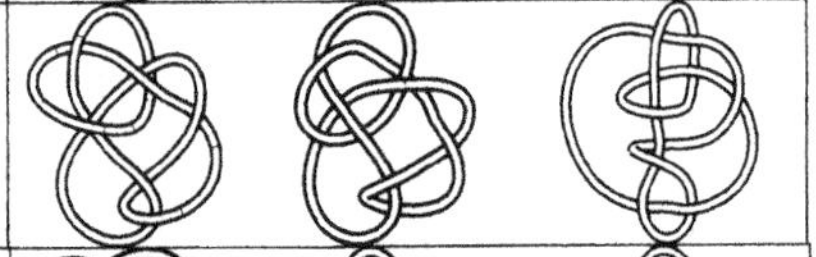		
10_{134}: $1^{0}_{5}1^{0}_{7}1^{1}_{7}2^{2}_{9}1^{2}_{11}1^{3}_{11}2^{3}_{13}3^{4}_{13}1^{4}_{15}1^{5}_{15}3^{5}_{17}2^{6}_{17}1^{6}_{19}1^{7}_{19}2^{7}_{21}1^{8}_{23}$			
10_{135}: $1^{\underline{5}}_{\underline{11}}1^{\underline{4}}_{\underline{9}}1^{\underline{4}}_{\underline{7}}3^{\underline{3}}_{\underline{7}}1^{\underline{3}}_{\underline{5}}3^{\underline{2}}_{\underline{5}}3^{\underline{2}}_{\underline{3}}3^{\underline{1}}_{\underline{3}}3^{\underline{1}}_{\underline{1}}4^{0}_{\underline{1}}4^{0}_{1}2^{1}_{1}3^{1}_{3}2^{2}_{3}2^{2}_{5}2^{3}_{7}$			
10_{136}: $1^{\underline{4}}_{\underline{7}}1^{\underline{3}}_{\underline{5}}1^{\underline{3}}_{\underline{3}}1^{\underline{2}}_{\underline{3}}1^{\underline{2}}_{\underline{1}}2^{\underline{1}}_{\underline{1}}1^{\underline{1}}_{1}1^{0}_{\underline{1}}2^{0}_{1}2^{0}_{3}1^{1}_{3}1^{1}_{5}1^{2}_{5}1^{2}_{7}1^{3}_{9}$			
10_{137}: $1^{\underline{6}}_{\underline{13}}1^{\underline{5}}_{\underline{11}}1^{\underline{5}}_{\underline{9}}2^{\underline{4}}_{\underline{9}}1^{\underline{4}}_{\underline{7}}2^{\underline{3}}_{\underline{7}}2^{\underline{3}}_{\underline{5}}2^{\underline{2}}_{\underline{5}}2^{\underline{2}}_{\underline{3}}2^{\underline{1}}_{\underline{3}}2^{\underline{1}}_{\underline{1}}2^{0}_{\underline{1}}3^{0}_{1}1^{1}_{1}1^{1}_{3}1^{2}_{5}$			
10_{138}: $1^{\underline{4}}_{\underline{7}}1^{\underline{3}}_{\underline{5}}1^{\underline{3}}_{\underline{3}}3^{\underline{2}}_{\underline{3}}1^{\underline{2}}_{\underline{1}}2^{\underline{1}}_{\underline{1}}3^{\underline{1}}_{1}4^{0}_{1}3^{0}_{3}3^{1}_{3}3^{1}_{5}2^{2}_{5}3^{2}_{7}2^{3}_{7}2^{3}_{9}2^{4}_{11}$			
10_{139}: $1^{0}_{7}1^{0}_{9}1^{2}_{11}1^{3}_{15}1^{4}_{13}1^{4}_{15}1^{5}_{15}1^{5}_{17}1^{5}_{19}1^{6}_{17}1^{6}_{19}1^{7}_{21}1^{8}_{21}1^{9}_{25}$			
10_{140}: $1^{\underline{7}}_{\underline{15}}1^{\underline{6}}_{\underline{11}}1^{\underline{5}}_{\underline{11}}1^{\underline{4}}_{\underline{9}}1^{\underline{4}}_{\underline{7}}1^{\underline{3}}_{\underline{5}}1^{\underline{2}}_{\underline{5}}1^{\underline{1}}_{\underline{1}}1^{0}_{\underline{1}}1^{0}_{1}$			
10_{141}: $1^{\underline{6}}_{\underline{13}}1^{\underline{5}}_{\underline{11}}1^{\underline{5}}_{\underline{9}}1^{\underline{4}}_{\underline{9}}1^{\underline{4}}_{\underline{7}}2^{\underline{3}}_{\underline{7}}1^{\underline{3}}_{\underline{5}}2^{\underline{2}}_{\underline{5}}2^{\underline{2}}_{\underline{3}}1^{\underline{1}}_{\underline{3}}2^{\underline{1}}_{\underline{1}}2^{0}_{\underline{1}}2^{0}_{1}1^{1}_{1}1^{1}_{3}1^{2}_{5}$			
10_{142}: $1^{0}_{5}1^{0}_{7}1^{1}_{7}1^{2}_{9}1^{2}_{11}1^{3}_{11}1^{3}_{13}2^{4}_{13}1^{4}_{15}2^{5}_{17}2^{6}_{17}2^{7}_{21}$			
10_{143}: $1^{\underline{7}}_{\underline{17}}1^{\underline{6}}_{\underline{15}}1^{\underline{6}}_{\underline{13}}2^{\underline{5}}_{\underline{13}}1^{\underline{5}}_{\underline{11}}2^{\underline{4}}_{\underline{11}}2^{\underline{4}}_{\underline{9}}3^{\underline{3}}_{\underline{9}}2^{\underline{3}}_{\underline{7}}2^{\underline{2}}_{\underline{7}}3^{\underline{2}}_{\underline{5}}1^{\underline{1}}_{\underline{5}}2^{\underline{1}}_{\underline{3}}2^{0}_{\underline{3}}2^{0}_{\underline{1}}1^{1}_{1}$			
10_{144}: $1^{\underline{6}}_{\underline{15}}2^{\underline{5}}_{\underline{13}}1^{\underline{5}}_{\underline{11}}3^{\underline{4}}_{\underline{11}}2^{\underline{4}}_{\underline{9}}3^{\underline{3}}_{\underline{9}}3^{\underline{3}}_{\underline{7}}4^{\underline{2}}_{\underline{7}}3^{\underline{2}}_{\underline{5}}3^{\underline{1}}_{\underline{5}}4^{\underline{1}}_{\underline{3}}2^{0}_{\underline{3}}4^{0}_{\underline{1}}2^{1}_{\underline{1}}1^{1}_{1}2^{2}_{3}$			
10_{145}: $1^{\underline{9}}_{\underline{21}}1^{\underline{8}}_{\underline{17}}1^{\underline{7}}_{\underline{17}}1^{\underline{6}}_{\underline{15}}1^{\underline{6}}_{\underline{13}}1^{\underline{5}}_{\underline{15}}1^{\underline{5}}_{\underline{11}}2^{\underline{4}}_{\underline{11}}1^{\underline{3}}_{\underline{11}}1^{\underline{3}}_{\underline{7}}1^{\underline{2}}_{\underline{7}}1^{0}_{\underline{5}}1^{0}_{\underline{3}}$			
10_{146}: $1^{\underline{5}}_{\underline{11}}2^{\underline{4}}_{\underline{9}}1^{\underline{4}}_{\underline{7}}2^{\underline{3}}_{\underline{7}}2^{\underline{3}}_{\underline{5}}3^{\underline{2}}_{\underline{5}}2^{\underline{2}}_{\underline{3}}3^{\underline{1}}_{\underline{3}}3^{\underline{1}}_{\underline{1}}3^{0}_{\underline{1}}4^{0}_{1}2^{1}_{1}2^{1}_{3}1^{2}_{3}2^{2}_{5}1^{3}_{7}$			
10_{147}: $1^{\underline{4}}_{\underline{7}}1^{\underline{3}}_{\underline{5}}1^{\underline{3}}_{\underline{3}}2^{\underline{2}}_{\underline{3}}1^{\underline{2}}_{\underline{1}}2^{\underline{1}}_{\underline{1}}2^{\underline{1}}_{1}3^{0}_{1}3^{0}_{3}2^{1}_{3}2^{1}_{5}2^{2}_{5}2^{2}_{7}1^{3}_{7}2^{3}_{9}1^{4}_{11}$			

K: KH	1st line 2nd line 3rd line
10_{148}: $1^{\underline{7}}_{\underline{17}} 1^{\underline{6}}_{\underline{15}} 1^{\underline{6}}_{\underline{13}} 3^{\underline{5}}_{\underline{13}} 1^{\underline{5}}_{\underline{11}} 2^{\underline{4}}_{\underline{11}} 3^{\underline{4}}_{\underline{9}} 3^{\underline{3}}_{\underline{9}} 2^{\underline{3}}_{\underline{7}} 3^{\underline{2}}_{\underline{7}} 3^{\underline{2}}_{\underline{5}} 1^{\underline{1}}_{\underline{5}} 3^{\underline{1}}_{\underline{3}} 2^{0}_{\underline{3}} 2^{0}_{\underline{1}} 1^{1}_{1}$ 10_{149}: $1^{\underline{8}}_{\underline{21}} 2^{\underline{7}}_{\underline{19}} 1^{\underline{7}}_{\underline{17}} 3^{\underline{6}}_{\underline{17}} 2^{\underline{6}}_{\underline{15}} 4^{\underline{5}}_{\underline{15}} 3^{\underline{5}}_{\underline{13}} 3^{\underline{4}}_{\underline{13}} 4^{\underline{4}}_{\underline{11}} 4^{\underline{3}}_{\underline{11}} 3^{\underline{3}}_{\underline{9}} 2^{\underline{2}}_{\underline{9}} 4^{\underline{2}}_{\underline{7}} 1^{\underline{1}}_{\underline{7}} 2^{\underline{1}}_{\underline{5}} 1^{0}_{\underline{5}} 2^{0}_{\underline{3}}$ 10_{150}: $1^{\underline{2}}_{\underline{1}} 1^{\underline{1}}_{1} 1^{\underline{1}}_{3} 3^{0}_{3} 2^{0}_{5} 2^{1}_{5} 2^{1}_{7} 3^{2}_{7} 2^{2}_{9} 2^{3}_{9} 3^{3}_{11} 2^{4}_{11} 2^{4}_{13} 1^{5}_{13} 2^{5}_{15} 1^{6}_{17}$	
10_{151}: $1^{\underline{3}}_{\underline{5}} 2^{\underline{2}}_{\underline{3}} 1^{\underline{2}}_{\underline{1}} 3^{\underline{1}}_{\underline{1}} 2^{\underline{1}}_{1} 4^{0}_{1} 4^{0}_{3} 4^{1}_{3} 3^{1}_{5} 4^{2}_{5} 4^{2}_{7} 2^{3}_{7} 4^{3}_{9} 2^{4}_{9} 2^{4}_{11} 2^{5}_{13}$ 10_{152}: $1^{\underline{10}}_{\underline{27}} 1^{\underline{9}}_{\underline{25}} 1^{\underline{9}}_{\underline{23}} 1^{\underline{8}}_{\underline{23}} 1^{\underline{8}}_{\underline{21}} 2^{\underline{7}}_{\underline{21}} 1^{\underline{7}}_{\underline{19}} 1^{\underline{6}}_{\underline{19}} 2^{\underline{6}}_{\underline{17}} 1^{\underline{5}}_{\underline{19}} 2^{\underline{5}}_{\underline{17}} 1^{\underline{5}}_{\underline{15}} 1^{\underline{4}}_{\underline{15}} 2^{\underline{4}}_{\underline{13}} 1^{\underline{3}}_{\underline{15}} 1^{\underline{2}}_{\underline{11}} 1^{0}_{\underline{9}} 1^{0}_{\underline{7}}$ 10_{153}: $1^{\underline{5}}_{\underline{11}} 1^{\underline{4}}_{\underline{7}} 1^{\underline{3}}_{\underline{7}} 1^{\underline{2}}_{\underline{5}} 1^{\underline{2}}_{\underline{3}} 1^{\underline{1}}_{\underline{5}} 1^{\underline{1}}_{\underline{1}} 3^{0}_{\underline{1}} 1^{0}_{1} 1^{1}_{\underline{1}} 1^{1}_{3} 1^{2}_{1} 1^{2}_{3} 1^{3}_{5} 1^{4}_{5} 1^{5}_{9}$	
10_{154}: $1^{0}_{5} 1^{0}_{7} 1^{2}_{9} 1^{3}_{9} 1^{3}_{13} 2^{4}_{11} 2^{4}_{13} 1^{5}_{13} 2^{5}_{15} 1^{5}_{17} 2^{6}_{15} 1^{6}_{17} 1^{7}_{17} 2^{7}_{19} 1^{8}_{19} 1^{8}_{21} 1^{9}_{21} 1^{9}_{23} 1^{10}_{25}$ 10_{155}: $1^{\underline{2}}_{\underline{5}} 1^{\underline{1}}_{\underline{3}} 1^{\underline{1}}_{\underline{1}} 3^{0}_{\underline{1}} 2^{0}_{1} 2^{1}_{1} 2^{1}_{3} 2^{2}_{3} 2^{2}_{5} 2^{3}_{5} 2^{3}_{7} 1^{4}_{7} 2^{4}_{9} 1^{5}_{9} 1^{5}_{11} 1^{6}_{13}$ 10_{156}: $1^{\underline{5}}_{\underline{13}} 2^{\underline{4}}_{\underline{11}} 1^{\underline{4}}_{\underline{9}} 3^{\underline{3}}_{\underline{9}} 2^{\underline{3}}_{\underline{7}} 3^{\underline{2}}_{\underline{7}} 3^{\underline{2}}_{\underline{5}} 3^{\underline{1}}_{\underline{5}} 3^{\underline{1}}_{\underline{3}} 3^{0}_{\underline{3}} 4^{0}_{\underline{1}} 2^{1}_{\underline{1}} 2^{1}_{1} 1^{2}_{1} 2^{2}_{3} 1^{3}_{5}$	
10_{157}: $2^{0}_{3} 1^{0}_{5} 3^{1}_{5} 1^{1}_{7} 4^{2}_{7} 3^{2}_{9} 4^{3}_{9} 4^{3}_{11} 5^{4}_{11} 4^{4}_{13} 3^{5}_{13} 5^{5}_{15} 3^{6}_{15} 3^{6}_{17} 1^{7}_{17} 3^{7}_{19} 1^{8}_{21}$ 10_{158}: $1^{\underline{4}}_{\underline{9}} 2^{\underline{3}}_{\underline{7}} 1^{\underline{3}}_{\underline{5}} 4^{\underline{2}}_{\underline{5}} 2^{\underline{2}}_{\underline{3}} 3^{\underline{1}}_{\underline{3}} 4^{\underline{1}}_{\underline{1}} 5^{0}_{\underline{1}} 4^{0}_{1} 4^{1}_{1} 4^{1}_{3} 2^{2}_{3} 4^{2}_{5} 2^{3}_{5} 2^{3}_{7} 2^{4}_{9}$ 10_{159}: $1^{\underline{7}}_{\underline{17}} 2^{\underline{6}}_{\underline{15}} 1^{\underline{6}}_{\underline{13}} 3^{\underline{5}}_{\underline{13}} 2^{\underline{5}}_{\underline{11}} 3^{\underline{4}}_{\underline{11}} 3^{\underline{4}}_{\underline{9}} 4^{\underline{3}}_{\underline{9}} 3^{\underline{3}}_{\underline{7}} 3^{\underline{2}}_{\underline{7}} 4^{\underline{2}}_{\underline{5}} 2^{\underline{1}}_{\underline{5}} 3^{\underline{1}}_{\underline{3}} 2^{0}_{\underline{3}} 3^{0}_{\underline{1}} 1^{1}_{1}$	
10_{160}: $1^{\underline{2}}_{\underline{1}} 1^{\underline{1}}_{1} 1^{\underline{1}}_{3} 2^{0}_{3} 2^{0}_{5} 2^{1}_{5} 1^{1}_{7} 2^{2}_{7} 2^{2}_{9} 1^{3}_{9} 2^{3}_{11} 2^{4}_{11} 1^{4}_{13} 2^{5}_{15}$ 10_{161}: $1^{\underline{9}}_{\underline{23}} 1^{\underline{8}}_{\underline{19}} 1^{\underline{7}}_{\underline{19}} 1^{\underline{6}}_{\underline{17}} 1^{\underline{6}}_{\underline{15}} 1^{\underline{5}}_{\underline{17}} 1^{\underline{5}}_{\underline{15}} 1^{\underline{5}}_{\underline{13}} 2^{\underline{4}}_{\underline{13}} 1^{\underline{4}}_{\underline{11}} 1^{\underline{3}}_{\underline{13}} 1^{\underline{3}}_{\underline{9}} 1^{\underline{2}}_{\underline{9}} 1^{0}_{\underline{7}} 1^{0}_{\underline{5}}$ 10_{162}: $1^{\underline{6}}_{\underline{15}} 1^{\underline{5}}_{\underline{13}} 1^{\underline{5}}_{\underline{11}} 3^{\underline{4}}_{\underline{11}} 1^{\underline{4}}_{\underline{9}} 3^{\underline{3}}_{\underline{9}} 3^{\underline{3}}_{\underline{7}} 3^{\underline{2}}_{\underline{7}} 3^{\underline{2}}_{\underline{5}} 3^{\underline{1}}_{\underline{5}} 3^{\underline{1}}_{\underline{3}} 2^{0}_{\underline{3}} 4^{0}_{\underline{1}} 2^{1}_{\underline{1}} 1^{1}_{1} 2^{2}_{3}$	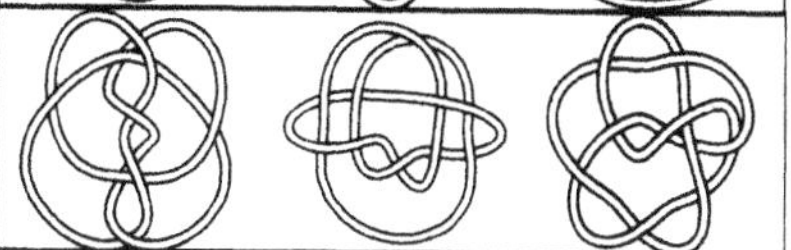
10_{163}: $1^{\underline{3}}_{\underline{5}} 3^{\underline{2}}_{\underline{3}} 1^{\underline{2}}_{\underline{1}} 3^{\underline{1}}_{\underline{1}} 3^{\underline{1}}_{1} 5^{0}_{1} 4^{0}_{3} 5^{1}_{3} 4^{1}_{5} 4^{2}_{5} 5^{2}_{7} 3^{3}_{7} 4^{3}_{9} 2^{4}_{9} 3^{4}_{11} 2^{5}_{13}$ 10_{164}: $1^{\underline{5}}_{\underline{11}} 2^{\underline{4}}_{\underline{9}} 1^{\underline{4}}_{\underline{7}} 3^{\underline{3}}_{\underline{7}} 2^{\underline{3}}_{\underline{5}} 4^{\underline{2}}_{\underline{5}} 3^{\underline{2}}_{\underline{3}} 1^{\underline{1}}_{\underline{3}} 4^{\underline{1}}_{\underline{1}} 4^{0}_{\underline{1}} 5^{0}_{1} 3^{1}_{1} 3^{1}_{3} 2^{2}_{3} 3^{2}_{5} 2^{3}_{7}$ 10_{165}: $2^{0}_{1} 1^{0}_{3} 3^{1}_{3} 1^{1}_{5} 3^{2}_{5} 3^{2}_{7} 3^{3}_{7} 3^{3}_{9} 4^{4}_{9} 3^{4}_{11} 2^{5}_{11} 4^{5}_{13} 2^{6}_{13} 2^{6}_{15} 1^{7}_{15} 2^{7}_{17} 1^{8}_{19}$	

(10_1–10_{123} are alternating, the rest are not).

2.5. **11 Crossing Alternating Knots.**

K: KH	1st line 2nd line 3rd line
11^a_1: $1^{\underline{4}}_{\underline{7}} 2^{\underline{3}}_{\underline{5}} 1^{\underline{3}}_{\underline{3}} 5^{\underline{2}}_{\underline{3}} 2^{\underline{2}}_{\underline{1}} 7^{\underline{1}}_{\underline{1}} 5^{\underline{1}}_{1} 10^{0}_{1} 8^{0}_{3} 11^{1}_{3} 9^{1}_{5} 10^{2}_{5} 11^{2}_{7} 8^{3}_{7} 10^{3}_{9} 6^{4}_{9} 8^{4}_{11} 3^{5}_{11} 6^{5}_{13} 1^{6}_{13} 3^{6}_{15} 1^{7}_{17}$ 11^a_2: $1^{\underline{2}}_{\underline{1}} 2^{\underline{1}}_{1} 1^{\underline{1}}_{3} 6^{0}_{3} 3^{0}_{5} 8^{1}_{5} 5^{1}_{7} 11^{2}_{7} 8^{2}_{9} 11^{3}_{9} 11^{3}_{11} 11^{4}_{11} 11^{4}_{13} 9^{5}_{13} 11^{5}_{15} 6^{6}_{15} 9^{6}_{17} 3^{7}_{17} 6^{7}_{19} 1^{8}_{19} 3^{8}_{21} 1^{9}_{23}$ 11^a_3: $1^{\underline{7}}_{\underline{17}} 2^{\underline{6}}_{\underline{15}} 1^{\underline{6}}_{\underline{13}} 5^{\underline{5}}_{\underline{13}} 2^{\underline{5}}_{\underline{11}} 7^{\underline{4}}_{\underline{11}} 5^{\underline{4}}_{\underline{9}} 9^{\underline{3}}_{\underline{9}} 7^{\underline{3}}_{\underline{7}} 10^{\underline{2}}_{\underline{7}} 9^{\underline{2}}_{\underline{5}} 8^{\underline{1}}_{\underline{5}} 10^{\underline{1}}_{\underline{3}} 8^{0}_{\underline{3}} 9^{0}_{\underline{1}} 5^{1}_{\underline{1}} 7^{1}_{1} 2^{2}_{1} 5^{2}_{3} 1^{3}_{3} 2^{3}_{5} 1^{4}_{7}$	
11^a_4: $1^{\underline{7}}_{\underline{15}} 2^{\underline{6}}_{\underline{13}} 1^{\underline{6}}_{\underline{11}} 4^{\underline{5}}_{\underline{11}} 2^{\underline{5}}_{\underline{9}} 6^{\underline{4}}_{\underline{9}} 4^{\underline{4}}_{\underline{7}} 7^{\underline{3}}_{\underline{7}} 6^{\underline{3}}_{\underline{5}} 8^{\underline{2}}_{\underline{5}} 7^{\underline{2}}_{\underline{3}} 7^{\underline{1}}_{\underline{3}} 8^{\underline{1}}_{\underline{1}} 7^{0}_{\underline{1}} 8^{0}_{1} 4^{1}_{1} 6^{1}_{3} 2^{2}_{3} 4^{2}_{5} 1^{3}_{5} 2^{3}_{7} 1^{4}_{9}$ 11^a_5: $1^{\underline{6}}_{\underline{13}} 2^{\underline{5}}_{\underline{11}} 1^{\underline{5}}_{\underline{9}} 5^{\underline{4}}_{\underline{9}} 2^{\underline{4}}_{\underline{7}} 7^{\underline{3}}_{\underline{7}} 5^{\underline{3}}_{\underline{5}} 10^{\underline{2}}_{\underline{5}} 7^{\underline{2}}_{\underline{3}} 10^{\underline{1}}_{\underline{3}} 10^{\underline{1}}_{\underline{1}} 10^{0}_{\underline{1}} 11^{0}_{1} 9^{1}_{1} 9^{1}_{3} 5^{2}_{3} 9^{2}_{5} 3^{3}_{5} 5^{3}_{7} 1^{4}_{7} 3^{4}_{9} 1^{5}_{11}$ 11^a_6: $1^{\underline{4}}_{\underline{7}} 2^{\underline{3}}_{\underline{5}} 1^{\underline{3}}_{\underline{3}} 5^{\underline{2}}_{\underline{3}} 2^{\underline{2}}_{\underline{1}} 7^{\underline{1}}_{\underline{1}} 5^{\underline{1}}_{1} 11^{0}_{1} 8^{0}_{3} 11^{1}_{3} 10^{1}_{5} 11^{2}_{5} 11^{2}_{7} 9^{3}_{7} 11^{3}_{9} 6^{4}_{9} 9^{4}_{11} 4^{5}_{11} 6^{5}_{13} 1^{6}_{13} 4^{6}_{15} 1^{7}_{17}$	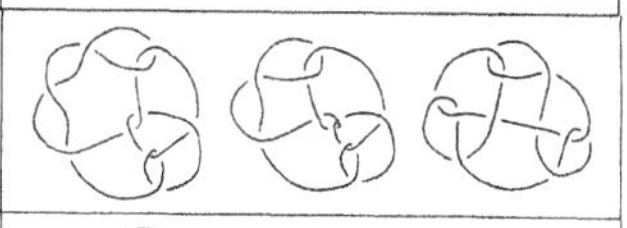
11^a_7: $1^{\underline{7}}_{\underline{17}} 2^{\underline{6}}_{\underline{15}} 1^{\underline{6}}_{\underline{13}} 4^{\underline{5}}_{\underline{13}} 2^{\underline{5}}_{\underline{11}} 6^{\underline{4}}_{\underline{11}} 4^{\underline{4}}_{\underline{9}} 7^{\underline{3}}_{\underline{9}} 6^{\underline{3}}_{\underline{7}} 8^{\underline{2}}_{\underline{7}} 7^{\underline{2}}_{\underline{5}} 7^{\underline{1}}_{\underline{5}} 8^{\underline{1}}_{\underline{3}} 6^{0}_{\underline{3}} 8^{0}_{\underline{1}} 4^{1}_{\underline{1}} 5^{1}_{1} 2^{2}_{1} 4^{2}_{3} 1^{3}_{3} 2^{3}_{5} 1^{4}_{7}$ 11^a_8: $1^{\underline{7}}_{\underline{15}} 2^{\underline{6}}_{\underline{13}} 1^{\underline{6}}_{\underline{11}} 4^{\underline{5}}_{\underline{11}} 2^{\underline{5}}_{\underline{9}} 7^{\underline{4}}_{\underline{9}} 4^{\underline{4}}_{\underline{7}} 8^{\underline{3}}_{\underline{7}} 7^{\underline{3}}_{\underline{5}} 10^{\underline{2}}_{\underline{5}} 8^{\underline{2}}_{\underline{3}} 9^{\underline{1}}_{\underline{3}} 10^{\underline{1}}_{\underline{1}} 8^{0}_{\underline{1}} 10^{0}_{1} 6^{1}_{1} 7^{1}_{3} 3^{2}_{3} 6^{2}_{5} 1^{3}_{5} 3^{3}_{7} 1^{4}_{9}$ 11^a_9: $1^{\underline{4}}_{\underline{5}} 1^{\underline{3}}_{\underline{3}} 1^{\underline{3}}_{\underline{1}} 3^{\underline{2}}_{\underline{1}} 1^{\underline{2}}_{1} 3^{\underline{1}}_{1} 3^{\underline{1}}_{3} 5^{0}_{3} 4^{0}_{5} 5^{1}_{5} 4^{1}_{7} 5^{2}_{7} 5^{2}_{9} 4^{3}_{9} 5^{3}_{11} 3^{4}_{11} 4^{4}_{13} 2^{5}_{13} 3^{5}_{15} 1^{6}_{15} 2^{6}_{17} 1^{7}_{19}$	
11^a_{10}: $1^{\underline{4}}_{\underline{7}} 2^{\underline{3}}_{\underline{5}} 1^{\underline{3}}_{\underline{3}} 5^{\underline{2}}_{\underline{3}} 2^{\underline{2}}_{\underline{1}} 6^{\underline{1}}_{\underline{1}} 5^{\underline{1}}_{1} 9^{0}_{1} 7^{0}_{3} 9^{1}_{3} 8^{1}_{5} 8^{2}_{5} 9^{2}_{7} 7^{3}_{7} 8^{3}_{9} 4^{4}_{9} 7^{4}_{11} 2^{5}_{11} 4^{5}_{13} 1^{6}_{13} 2^{6}_{15} 1^{7}_{17}$ 11^a_{11}: $1^{\underline{5}}_{\underline{11}} 2^{\underline{4}}_{\underline{9}} 1^{\underline{4}}_{\underline{7}} 4^{\underline{3}}_{\underline{7}} 2^{\underline{3}}_{\underline{5}} 7^{\underline{2}}_{\underline{5}} 4^{\underline{2}}_{\underline{3}} 8^{\underline{1}}_{\underline{3}} 7^{\underline{1}}_{\underline{1}} 10^{0}_{\underline{1}} 9^{0}_{1} 9^{1}_{1} 9^{1}_{3} 7^{2}_{3} 9^{2}_{5} 5^{3}_{5} 7^{3}_{7} 3^{4}_{7} 5^{4}_{9} 1^{5}_{9} 3^{5}_{11} 1^{6}_{13}$ 11^a_{12}: $1^{\underline{4}}_{\underline{7}} 1^{\underline{3}}_{\underline{5}} 1^{\underline{3}}_{\underline{3}} 4^{\underline{2}}_{\underline{3}} 1^{\underline{2}}_{\underline{1}} 5^{\underline{1}}_{\underline{1}} 4^{\underline{1}}_{1} 8^{0}_{1} 6^{0}_{3} 9^{1}_{3} 7^{1}_{5} 8^{2}_{5} 9^{2}_{7} 7^{3}_{7} 8^{3}_{9} 5^{4}_{9} 7^{4}_{11} 3^{5}_{11} 5^{5}_{13} 1^{6}_{13} 3^{6}_{15} 1^{7}_{17}$	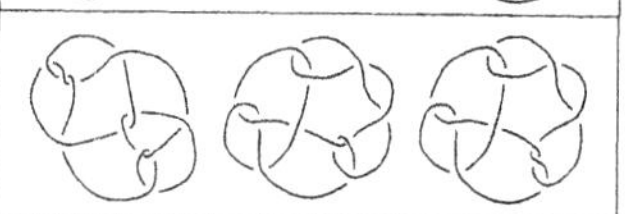

K: KH	1st line	2nd line	3rd line
11^a_{13}: $1^{\underline{4}}_{\underline{9}}1^{\underline{3}}_{\underline{7}}1^{\underline{3}}_{\underline{5}}3^{\underline{2}}_{\underline{5}}1^{\underline{2}}_{\underline{3}}3^{\underline{1}}_{\underline{3}}3^{\underline{1}}_{\underline{1}}5^{0}_{\underline{1}}4^{0}_{1}5^{1}_{1}4^{1}_{3}4^{2}_{3}5^{2}_{5}4^{3}_{5}4^{3}_{7}2^{4}_{7}4^{4}_{9}2^{5}_{9}2^{5}_{11}1^{6}_{11}2^{6}_{13}1^{7}_{15}$			
11^a_{14}: $1^{\underline{5}}_{\underline{11}}2^{\underline{4}}_{\underline{9}}1^{\underline{4}}_{\underline{7}}6^{\underline{3}}_{\underline{7}}2^{\underline{3}}_{\underline{5}}8^{\underline{2}}_{\underline{5}}6^{\underline{2}}_{\underline{3}}10^{\underline{1}}_{\underline{3}}8^{\underline{1}}_{\underline{1}}12^{0}_{\underline{1}}11^{0}_{1}10^{1}_{1}11^{1}_{3}9^{2}_{3}10^{2}_{5}5^{3}_{5}9^{3}_{7}3^{4}_{7}5^{4}_{9}1^{5}_{9}3^{5}_{11}1^{6}_{13}$			
11^a_{15}: $1^{\underline{6}}_{\underline{15}}2^{\underline{5}}_{\underline{13}}1^{\underline{5}}_{\underline{11}}5^{\underline{4}}_{\underline{11}}2^{\underline{4}}_{\underline{9}}7^{\underline{3}}_{\underline{9}}5^{\underline{3}}_{\underline{7}}8^{\underline{2}}_{\underline{7}}7^{\underline{2}}_{\underline{5}}9^{\underline{1}}_{\underline{5}}8^{\underline{1}}_{\underline{3}}8^{0}_{\underline{3}}10^{0}_{\underline{1}}7^{1}_{\underline{1}}7^{1}_{1}4^{2}_{1}7^{2}_{3}2^{3}_{3}4^{3}_{5}1^{4}_{5}2^{4}_{7}1^{5}_{9}$			
11^a_{16}: $1^{\underline{6}}_{\underline{13}}2^{\underline{5}}_{\underline{11}}1^{\underline{5}}_{\underline{9}}5^{\underline{4}}_{\underline{9}}2^{\underline{4}}_{\underline{7}}6^{\underline{3}}_{\underline{7}}5^{\underline{3}}_{\underline{5}}8^{\underline{2}}_{\underline{5}}6^{\underline{2}}_{\underline{3}}9^{\underline{1}}_{\underline{3}}8^{\underline{1}}_{\underline{1}}8^{0}_{\underline{1}}10^{0}_{1}7^{1}_{1}7^{1}_{3}4^{2}_{3}7^{2}_{5}2^{3}_{5}4^{3}_{7}1^{4}_{7}2^{4}_{9}1^{5}_{11}$			
11^a_{17}: $1^{\underline{4}}_{\underline{7}}2^{\underline{3}}_{\underline{5}}1^{\underline{3}}_{\underline{3}}6^{\underline{2}}_{\underline{3}}2^{\underline{2}}_{\underline{1}}7^{\underline{1}}_{\underline{1}}6^{\underline{1}}_{1}10^{0}_{1}8^{0}_{3}11^{1}_{3}9^{1}_{5}9^{2}_{5}11^{2}_{7}8^{3}_{7}9^{3}_{9}5^{4}_{9}8^{4}_{11}2^{5}_{11}5^{5}_{13}1^{6}_{13}2^{6}_{15}1^{7}_{17}$			
11^a_{18}: $1^{\underline{3}}_{\underline{5}}2^{\underline{2}}_{\underline{3}}1^{\underline{2}}_{\underline{1}}5^{\underline{1}}_{\underline{1}}2^{\underline{1}}_{1}8^{0}_{1}6^{0}_{3}10^{1}_{3}7^{1}_{5}11^{2}_{5}10^{2}_{7}9^{3}_{7}11^{3}_{9}9^{4}_{9}9^{4}_{11}5^{5}_{11}9^{5}_{13}3^{6}_{13}5^{6}_{15}1^{7}_{15}3^{7}_{17}1^{8}_{19}$			
11^a_{19}: $1^{\underline{4}}_{\underline{7}}3^{\underline{3}}_{\underline{5}}1^{\underline{3}}_{\underline{3}}7^{\underline{2}}_{\underline{3}}3^{\underline{2}}_{\underline{1}}9^{\underline{1}}_{\underline{1}}7^{\underline{1}}_{1}13^{0}_{1}10^{0}_{3}13^{1}_{3}12^{1}_{5}12^{2}_{5}13^{2}_{7}10^{3}_{7}12^{3}_{9}6^{4}_{9}10^{4}_{11}3^{5}_{11}6^{5}_{13}1^{6}_{13}3^{6}_{15}1^{7}_{17}$			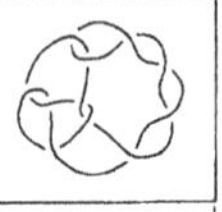
11^a_{20}: $1^{\underline{2}}_{\underline{1}}1^{\underline{1}}_{1}1^{\underline{1}}_{3}5^{0}_{3}2^{0}_{5}6^{1}_{5}4^{1}_{7}9^{2}_{7}6^{2}_{9}9^{3}_{9}9^{3}_{11}9^{4}_{11}9^{4}_{13}8^{5}_{13}9^{5}_{15}5^{6}_{15}8^{6}_{17}3^{7}_{17}5^{7}_{19}1^{8}_{19}3^{8}_{21}1^{9}_{23}$			
11^a_{21}: $1^{\underline{2}}_{3}1^{\underline{1}}_{\underline{1}}1^{\underline{1}}_{1}4^{0}_{1}2^{0}_{3}4^{1}_{3}3^{1}_{5}6^{2}_{5}4^{2}_{7}6^{3}_{7}6^{3}_{9}5^{4}_{9}6^{4}_{11}5^{5}_{11}5^{5}_{13}3^{6}_{13}5^{6}_{15}2^{7}_{15}3^{7}_{17}1^{8}_{17}2^{8}_{19}1^{9}_{21}$			
11^a_{22}: $1^{\underline{3}}_{\underline{3}}2^{\underline{2}}_{\underline{1}}1^{\underline{2}}_{1}4^{\underline{1}}_{1}2^{\underline{1}}_{3}7^{0}_{3}5^{0}_{5}7^{1}_{5}6^{1}_{7}9^{2}_{7}7^{2}_{9}7^{3}_{9}9^{3}_{11}7^{4}_{11}7^{4}_{13}4^{5}_{13}7^{5}_{15}2^{6}_{15}4^{6}_{17}1^{7}_{17}2^{7}_{19}1^{8}_{21}$			
11^a_{23}: $1^{\underline{3}}_{\underline{5}}2^{\underline{2}}_{\underline{3}}1^{\underline{2}}_{\underline{1}}4^{\underline{1}}_{\underline{1}}2^{\underline{1}}_{1}7^{0}_{1}5^{0}_{3}8^{1}_{3}6^{1}_{5}9^{2}_{5}8^{2}_{7}7^{3}_{7}9^{3}_{9}7^{4}_{9}7^{4}_{11}4^{5}_{11}7^{5}_{13}2^{6}_{13}4^{6}_{15}1^{7}_{15}2^{7}_{17}1^{8}_{19}$			
11^a_{24}: $1^{\underline{5}}_{\underline{11}}3^{\underline{4}}_{\underline{9}}1^{\underline{4}}_{\underline{7}}7^{\underline{3}}_{\underline{7}}3^{\underline{3}}_{\underline{5}}10^{\underline{2}}_{\underline{5}}7^{\underline{2}}_{\underline{3}}12^{\underline{1}}_{\underline{3}}10^{\underline{1}}_{\underline{1}}14^{0}_{\underline{1}}13^{0}_{1}12^{1}_{1}13^{1}_{3}10^{2}_{3}12^{2}_{5}6^{3}_{5}10^{3}_{7}3^{4}_{7}6^{4}_{9}1^{5}_{9}3^{5}_{11}1^{6}_{13}$			
11^a_{25}: $1^{\underline{4}}_{\underline{7}}3^{\underline{3}}_{\underline{5}}1^{\underline{3}}_{\underline{3}}7^{\underline{2}}_{\underline{3}}3^{\underline{2}}_{\underline{1}}9^{\underline{1}}_{\underline{1}}7^{\underline{1}}_{1}13^{0}_{1}10^{0}_{3}13^{1}_{3}12^{1}_{5}12^{2}_{5}13^{2}_{7}10^{3}_{7}12^{3}_{9}6^{4}_{9}10^{4}_{11}3^{5}_{11}6^{5}_{13}1^{6}_{13}3^{6}_{15}1^{7}_{17}$			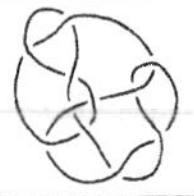
11^a_{26}: $1^{\underline{5}}_{\underline{11}}3^{\underline{4}}_{\underline{9}}1^{\underline{4}}_{\underline{7}}7^{\underline{3}}_{\underline{7}}3^{\underline{3}}_{\underline{5}}10^{\underline{2}}_{\underline{5}}7^{\underline{2}}_{\underline{3}}12^{\underline{1}}_{\underline{3}}10^{\underline{1}}_{\underline{1}}14^{0}_{\underline{1}}13^{0}_{1}12^{1}_{1}13^{1}_{3}10^{2}_{3}12^{2}_{5}6^{3}_{5}10^{3}_{7}3^{4}_{7}6^{4}_{9}1^{5}_{9}3^{5}_{11}1^{6}_{13}$			
11^a_{27}: $1^{\underline{8}}_{\underline{19}}3^{\underline{7}}_{\underline{17}}1^{\underline{7}}_{\underline{15}}6^{\underline{6}}_{\underline{15}}3^{\underline{6}}_{\underline{13}}9^{\underline{5}}_{\underline{13}}6^{\underline{5}}_{\underline{11}}11^{\underline{4}}_{\underline{11}}9^{\underline{4}}_{\underline{9}}12^{\underline{3}}_{\underline{9}}11^{\underline{3}}_{\underline{7}}11^{\underline{2}}_{\underline{7}}12^{\underline{2}}_{\underline{5}}9^{\underline{1}}_{\underline{5}}11^{\underline{1}}_{\underline{3}}6^{0}_{\underline{3}}10^{0}_{\underline{1}}3^{1}_{\underline{1}}5^{1}_{1}1^{2}_{1}3^{2}_{3}1^{3}_{5}$			
11^a_{28}: $1^{\underline{6}}_{\underline{13}}2^{\underline{5}}_{\underline{11}}1^{\underline{5}}_{\underline{9}}5^{\underline{4}}_{\underline{9}}2^{\underline{4}}_{\underline{7}}7^{\underline{3}}_{\underline{7}}5^{\underline{3}}_{\underline{5}}9^{\underline{2}}_{\underline{5}}7^{\underline{2}}_{\underline{3}}10^{\underline{1}}_{\underline{3}}9^{\underline{1}}_{\underline{1}}10^{0}_{\underline{1}}11^{0}_{1}8^{1}_{1}9^{1}_{3}5^{2}_{3}8^{2}_{5}3^{3}_{5}5^{3}_{7}1^{4}_{7}3^{4}_{9}1^{5}_{11}$			
11^a_{29}: $1^{\underline{6}}_{\underline{15}}2^{\underline{5}}_{\underline{13}}1^{\underline{5}}_{\underline{11}}5^{\underline{4}}_{\underline{11}}2^{\underline{4}}_{\underline{9}}7^{\underline{3}}_{\underline{9}}5^{\underline{3}}_{\underline{7}}9^{\underline{2}}_{\underline{7}}7^{\underline{2}}_{\underline{5}}9^{\underline{1}}_{\underline{5}}9^{\underline{1}}_{\underline{3}}9^{0}_{\underline{3}}10^{0}_{\underline{1}}8^{1}_{\underline{1}}8^{1}_{1}4^{2}_{1}8^{2}_{3}3^{3}_{3}4^{3}_{5}1^{4}_{5}3^{4}_{7}1^{5}_{9}$			
11^a_{30}: $1^{\underline{5}}_{\underline{11}}3^{\underline{4}}_{\underline{9}}1^{\underline{4}}_{\underline{7}}6^{\underline{3}}_{\underline{7}}3^{\underline{3}}_{\underline{5}}10^{\underline{2}}_{\underline{5}}6^{\underline{2}}_{\underline{3}}11^{\underline{1}}_{\underline{3}}10^{\underline{1}}_{\underline{1}}13^{0}_{\underline{1}}12^{0}_{1}12^{1}_{1}12^{1}_{3}9^{2}_{3}12^{2}_{5}6^{3}_{5}9^{3}_{7}3^{4}_{7}6^{4}_{9}1^{5}_{9}3^{5}_{11}1^{6}_{13}$			
11^a_{31}: $1^{\underline{2}}_{\underline{1}}2^{\underline{1}}_{1}1^{\underline{1}}_{3}6^{0}_{3}3^{0}_{5}7^{1}_{5}5^{1}_{7}10^{2}_{7}7^{2}_{9}10^{3}_{9}10^{3}_{11}10^{4}_{11}10^{4}_{13}8^{5}_{13}10^{5}_{15}5^{6}_{15}8^{6}_{17}3^{7}_{17}5^{7}_{19}1^{8}_{19}3^{8}_{21}1^{9}_{23}$			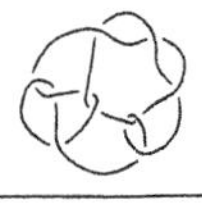
11^a_{32}: $1^{\underline{3}}_{\underline{5}}3^{\underline{2}}_{\underline{3}}1^{\underline{2}}_{\underline{1}}6^{\underline{1}}_{\underline{1}}3^{\underline{1}}_{1}9^{0}_{1}7^{0}_{3}11^{1}_{3}8^{1}_{5}12^{2}_{5}11^{2}_{7}10^{3}_{7}12^{3}_{9}9^{4}_{9}10^{4}_{11}5^{5}_{11}9^{5}_{13}3^{6}_{13}5^{6}_{15}1^{7}_{15}3^{7}_{17}1^{8}_{19}$			
11^a_{33}: $1^{\underline{6}}_{\underline{15}}2^{\underline{5}}_{\underline{13}}1^{\underline{5}}_{\underline{11}}4^{\underline{4}}_{\underline{11}}2^{\underline{4}}_{\underline{9}}6^{\underline{3}}_{\underline{9}}4^{\underline{3}}_{\underline{7}}7^{\underline{2}}_{\underline{7}}6^{\underline{2}}_{\underline{5}}8^{\underline{1}}_{\underline{5}}7^{\underline{1}}_{\underline{3}}7^{0}_{\underline{3}}9^{0}_{\underline{1}}6^{1}_{\underline{1}}6^{1}_{1}4^{2}_{1}6^{2}_{3}2^{3}_{3}4^{3}_{5}1^{4}_{5}2^{4}_{7}1^{5}_{9}$			
11^a_{34}: $1^{\underline{4}}_{\underline{7}}2^{\underline{3}}_{\underline{5}}1^{\underline{3}}_{\underline{3}}5^{\underline{2}}_{\underline{3}}2^{\underline{2}}_{\underline{1}}6^{\underline{1}}_{\underline{1}}5^{\underline{1}}_{1}10^{0}_{1}7^{0}_{3}10^{1}_{3}9^{1}_{5}9^{2}_{5}10^{2}_{7}8^{3}_{7}9^{3}_{9}5^{4}_{9}8^{4}_{11}3^{5}_{11}5^{5}_{13}1^{6}_{13}3^{6}_{15}1^{7}_{17}$			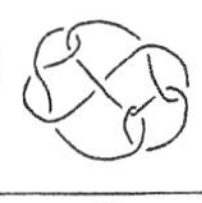
11^a_{35}: $1^{\underline{5}}_{\underline{11}}2^{\underline{4}}_{\underline{9}}1^{\underline{4}}_{\underline{7}}5^{\underline{3}}_{\underline{7}}2^{\underline{3}}_{\underline{5}}7^{\underline{2}}_{\underline{5}}5^{\underline{2}}_{\underline{3}}9^{\underline{1}}_{\underline{3}}7^{\underline{1}}_{\underline{1}}11^{0}_{\underline{1}}10^{0}_{1}9^{1}_{1}10^{1}_{3}8^{2}_{3}9^{2}_{5}5^{3}_{5}8^{3}_{7}3^{4}_{7}5^{4}_{9}1^{5}_{9}3^{5}_{11}1^{6}_{13}$			
11^a_{36}: $1^{\underline{6}}_{\underline{13}}3^{\underline{5}}_{\underline{11}}1^{\underline{5}}_{\underline{9}}5^{\underline{4}}_{\underline{9}}3^{\underline{4}}_{\underline{7}}8^{\underline{3}}_{\underline{7}}5^{\underline{3}}_{\underline{5}}9^{\underline{2}}_{\underline{5}}8^{\underline{2}}_{\underline{3}}10^{\underline{1}}_{\underline{3}}9^{\underline{1}}_{\underline{1}}10^{0}_{\underline{1}}11^{0}_{1}7^{1}_{1}9^{1}_{3}5^{2}_{3}7^{2}_{5}2^{3}_{5}5^{3}_{7}1^{4}_{7}2^{4}_{9}1^{5}_{11}$			
11^a_{37}: $1^{\underline{4}}_{\underline{9}}2^{\underline{3}}_{\underline{7}}1^{\underline{3}}_{\underline{5}}5^{\underline{2}}_{\underline{5}}2^{\underline{2}}_{\underline{3}}5^{\underline{1}}_{\underline{3}}5^{\underline{1}}_{\underline{1}}8^{0}_{\underline{1}}6^{0}_{1}8^{1}_{1}7^{1}_{3}6^{2}_{3}8^{2}_{5}6^{3}_{5}6^{3}_{7}3^{4}_{7}6^{4}_{9}2^{5}_{9}3^{5}_{11}1^{6}_{11}2^{6}_{13}1^{7}_{15}$		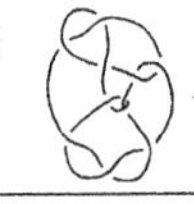	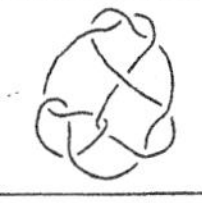
11^a_{38}: $1^{\underline{6}}_{\underline{13}}2^{\underline{5}}_{\underline{11}}1^{\underline{5}}_{\underline{9}}5^{\underline{4}}_{\underline{9}}2^{\underline{4}}_{\underline{7}}7^{\underline{3}}_{\underline{7}}5^{\underline{3}}_{\underline{5}}9^{\underline{2}}_{\underline{5}}7^{\underline{2}}_{\underline{3}}10^{\underline{1}}_{\underline{3}}9^{\underline{1}}_{\underline{1}}9^{0}_{\underline{1}}11^{0}_{1}8^{1}_{1}8^{1}_{3}5^{2}_{3}8^{2}_{5}2^{3}_{5}5^{3}_{7}1^{4}_{7}2^{4}_{9}1^{5}_{11}$			
11^a_{39}: $1^{\underline{4}}_{\underline{9}}2^{\underline{3}}_{\underline{7}}1^{\underline{3}}_{\underline{5}}5^{\underline{2}}_{\underline{5}}2^{\underline{2}}_{\underline{3}}6^{\underline{1}}_{\underline{3}}5^{\underline{1}}_{\underline{1}}8^{0}_{\underline{1}}7^{0}_{1}9^{1}_{1}7^{1}_{3}7^{2}_{3}9^{2}_{5}6^{3}_{5}7^{3}_{7}4^{4}_{7}6^{4}_{9}2^{5}_{9}4^{5}_{11}1^{6}_{11}2^{6}_{13}1^{7}_{15}$			
11^a_{40}: $1^{\underline{3}}_{\underline{3}}2^{\underline{2}}_{\underline{1}}1^{\underline{2}}_{1}3^{\underline{1}}_{1}2^{\underline{1}}_{3}6^{0}_{3}4^{0}_{5}6^{1}_{5}5^{1}_{7}8^{2}_{7}6^{2}_{9}6^{3}_{9}8^{3}_{11}6^{4}_{11}6^{4}_{13}4^{5}_{13}6^{5}_{15}2^{6}_{15}4^{6}_{17}1^{7}_{17}2^{7}_{19}1^{8}_{21}$			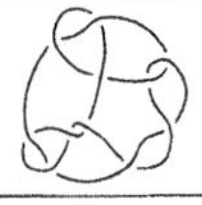
11^a_{41}: $1^{\underline{3}}_{\underline{5}}3^{\underline{2}}_{\underline{3}}1^{\underline{2}}_{\underline{1}}5^{\underline{1}}_{\underline{1}}3^{\underline{1}}_{1}8^{0}_{1}6^{0}_{3}9^{1}_{3}7^{1}_{5}10^{2}_{5}9^{2}_{7}8^{3}_{7}10^{3}_{9}7^{4}_{9}8^{4}_{11}4^{5}_{11}7^{5}_{13}2^{6}_{13}4^{6}_{15}1^{7}_{15}2^{7}_{17}1^{8}_{19}$			
11^a_{42}: $1^{\underline{4}}_{\underline{7}}2^{\underline{3}}_{\underline{5}}1^{\underline{3}}_{\underline{3}}5^{\underline{2}}_{\underline{3}}2^{\underline{2}}_{\underline{1}}6^{\underline{1}}_{\underline{1}}5^{\underline{1}}_{1}9^{0}_{1}7^{0}_{3}9^{1}_{3}8^{1}_{5}8^{2}_{5}9^{2}_{7}7^{3}_{7}8^{3}_{9}4^{4}_{9}7^{4}_{11}2^{5}_{11}4^{5}_{13}1^{6}_{13}2^{6}_{15}1^{7}_{17}$			
11^a_{43}: $1^{0}_{5}1^{0}_{7}3^{1}_{7}6^{2}_{9}3^{2}_{11}7^{3}_{11}6^{3}_{13}12^{4}_{13}7^{4}_{15}10^{5}_{15}12^{5}_{17}11^{6}_{17}10^{6}_{19}9^{7}_{19}11^{7}_{21}5^{8}_{21}9^{8}_{23}3^{9}_{23}5^{9}_{25}1^{10}_{25}3^{10}_{27}1^{11}_{29}$		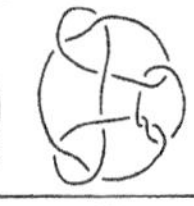	
11^a_{44}: $1^{\underline{5}}_{\underline{11}}2^{\underline{4}}_{\underline{9}}1^{\underline{4}}_{\underline{7}}6^{\underline{3}}_{\underline{7}}2^{\underline{3}}_{\underline{5}}7^{\underline{2}}_{\underline{5}}6^{\underline{2}}_{\underline{3}}9^{\underline{1}}_{\underline{3}}7^{\underline{1}}_{\underline{1}}11^{0}_{\underline{1}}10^{0}_{1}8^{1}_{1}10^{1}_{3}8^{2}_{3}8^{2}_{5}4^{3}_{5}8^{3}_{7}2^{4}_{7}4^{4}_{9}1^{5}_{9}2^{5}_{11}1^{6}_{13}$			
11^a_{45}: $1^{\underline{4}}_{\underline{7}}1^{\underline{3}}_{\underline{5}}1^{\underline{3}}_{\underline{3}}4^{\underline{2}}_{\underline{3}}1^{\underline{2}}_{\underline{1}}4^{\underline{1}}_{\underline{1}}4^{\underline{1}}_{1}7^{0}_{1}5^{0}_{3}8^{1}_{3}6^{1}_{5}6^{2}_{5}8^{2}_{7}6^{3}_{7}6^{3}_{9}4^{4}_{9}6^{4}_{11}2^{5}_{11}4^{5}_{13}1^{6}_{13}2^{6}_{15}1^{7}_{17}$			
11^a_{46}: $1^{\underline{5}}_{\underline{9}}2^{\underline{4}}_{\underline{7}}1^{\underline{4}}_{\underline{5}}4^{\underline{3}}_{\underline{5}}2^{\underline{3}}_{\underline{3}}5^{\underline{2}}_{\underline{3}}4^{\underline{2}}_{\underline{1}}6^{\underline{1}}_{\underline{1}}5^{\underline{1}}_{1}8^{0}_{1}7^{0}_{3}6^{1}_{3}7^{1}_{5}6^{2}_{5}6^{2}_{7}3^{3}_{7}6^{3}_{9}2^{4}_{9}3^{4}_{11}1^{5}_{11}2^{5}_{13}1^{6}_{15}$		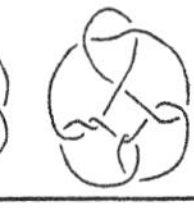	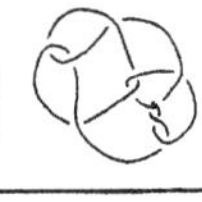
11^a_{47}: $1^{\underline{5}}_{\underline{11}}2^{\underline{4}}_{\underline{9}}1^{\underline{4}}_{\underline{7}}6^{\underline{3}}_{\underline{7}}2^{\underline{3}}_{\underline{5}}7^{\underline{2}}_{\underline{5}}6^{\underline{2}}_{\underline{3}}9^{\underline{1}}_{\underline{3}}7^{\underline{1}}_{\underline{1}}11^{0}_{\underline{1}}10^{0}_{1}8^{1}_{1}10^{1}_{3}8^{2}_{3}8^{2}_{5}4^{3}_{5}8^{3}_{7}2^{4}_{7}4^{4}_{9}1^{5}_{9}2^{5}_{11}1^{6}_{13}$			
11^a_{48}: $1^{\underline{9}}_{\underline{23}}2^{\underline{8}}_{\underline{21}}1^{\underline{8}}_{\underline{19}}5^{\underline{7}}_{\underline{19}}2^{\underline{7}}_{\underline{17}}7^{\underline{6}}_{\underline{17}}5^{\underline{6}}_{\underline{15}}9^{\underline{5}}_{\underline{15}}7^{\underline{5}}_{\underline{13}}9^{\underline{4}}_{\underline{13}}9^{\underline{4}}_{\underline{11}}9^{\underline{3}}_{\underline{11}}9^{\underline{3}}_{\underline{9}}7^{\underline{2}}_{\underline{9}}9^{\underline{2}}_{\underline{7}}4^{\underline{1}}_{\underline{7}}7^{\underline{1}}_{\underline{5}}3^{0}_{\underline{5}}5^{0}_{\underline{3}}1^{1}_{\underline{3}}2^{1}_{\underline{1}}1^{2}_{1}$			

K: KH	1st line 2nd line 3rd line
11^a_{49}: $1^{\underline{2}}_{\underline{1}}1^{\underline{1}}_{1}1^{\underline{1}}_{3}5^{0}_{3}2^{0}_{5}6^{1}_{5}4^{1}_{7}8^{2}_{7}6^{2}_{9}9^{3}_{9}8^{3}_{11}8^{4}_{11}9^{4}_{13}7^{5}_{13}8^{5}_{15}5^{6}_{15}7^{6}_{17}2^{7}_{17}5^{7}_{19}1^{8}_{19}2^{8}_{21}1^{9}_{23}$	
11^a_{50}: $1^{\underline{2}}_{\underline{3}}1^{\underline{1}}_{\underline{1}}1^{\underline{1}}_{1}4^{0}_{1}2^{0}_{3}5^{1}_{3}3^{1}_{5}6^{2}_{5}5^{2}_{7}7^{3}_{7}6^{3}_{9}6^{4}_{9}7^{4}_{11}5^{5}_{11}6^{5}_{13}4^{6}_{13}5^{6}_{15}2^{7}_{15}4^{7}_{17}1^{8}_{17}2^{8}_{19}1^{9}_{21}$	
11^a_{51}: $1^{\underline{7}}_{\underline{17}}2^{\underline{6}}_{\underline{15}}1^{\underline{6}}_{\underline{13}}5^{\underline{5}}_{\underline{13}}2^{\underline{5}}_{\underline{11}}7^{\underline{4}}_{\underline{11}}5^{\underline{4}}_{\underline{9}}9^{\underline{3}}_{\underline{9}}7^{\underline{3}}_{\underline{7}}10^{\underline{2}}_{\underline{7}}9^{\underline{2}}_{\underline{5}}8^{\underline{1}}_{\underline{5}}10^{\underline{1}}_{\underline{3}}8^{0}_{\underline{3}}9^{0}_{\underline{1}}5^{1}_{\underline{1}}7^{1}_{1}2^{2}_{1}5^{2}_{3}1^{3}_{3}2^{3}_{5}1^{4}_{7}$	
11^a_{52}: $1^{\underline{5}}_{\underline{11}}2^{\underline{4}}_{\underline{9}}1^{\underline{4}}_{\underline{7}}5^{\underline{3}}_{\underline{7}}2^{\underline{3}}_{\underline{5}}8^{\underline{2}}_{\underline{5}}5^{\underline{2}}_{\underline{3}}10^{\underline{1}}_{\underline{3}}8^{\underline{1}}_{\underline{1}}12^{0}_{\underline{1}}11^{0}_{1}11^{1}_{1}11^{1}_{3}9^{2}_{3}11^{2}_{5}6^{3}_{5}9^{3}_{7}4^{4}_{7}6^{4}_{9}1^{5}_{9}4^{5}_{11}1^{6}_{13}$	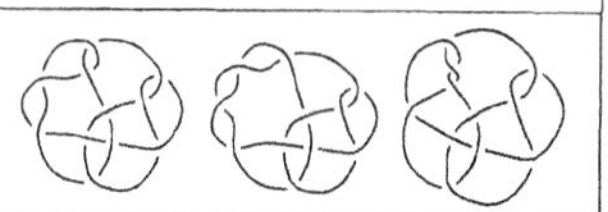
11^a_{53}: $1^{\underline{8}}_{\underline{21}}2^{\underline{7}}_{\underline{19}}1^{\underline{7}}_{\underline{17}}4^{\underline{6}}_{\underline{17}}2^{\underline{6}}_{\underline{15}}6^{\underline{5}}_{\underline{15}}4^{\underline{5}}_{\underline{13}}7^{\underline{4}}_{\underline{13}}6^{\underline{4}}_{\underline{11}}8^{\underline{3}}_{\underline{11}}7^{\underline{3}}_{\underline{9}}7^{\underline{2}}_{\underline{9}}8^{\underline{2}}_{\underline{7}}6^{\underline{1}}_{\underline{7}}7^{\underline{1}}_{\underline{5}}4^{0}_{\underline{5}}7^{0}_{\underline{3}}3^{1}_{\underline{3}}3^{1}_{\underline{1}}1^{2}_{\underline{1}}3^{2}_{1}1^{3}_{3}$	
11^a_{54}: $1^{\underline{8}}_{\underline{19}}2^{\underline{7}}_{\underline{17}}1^{\underline{7}}_{\underline{15}}5^{\underline{6}}_{\underline{15}}2^{\underline{6}}_{\underline{13}}8^{\underline{5}}_{\underline{13}}5^{\underline{5}}_{\underline{11}}10^{\underline{4}}_{\underline{11}}8^{\underline{4}}_{\underline{9}}12^{\underline{3}}_{\underline{9}}10^{\underline{3}}_{\underline{7}}11^{\underline{2}}_{\underline{7}}12^{\underline{2}}_{\underline{5}}9^{\underline{1}}_{\underline{5}}11^{\underline{1}}_{\underline{3}}7^{0}_{\underline{3}}10^{0}_{\underline{1}}4^{1}_{\underline{1}}6^{1}_{1}1^{2}_{1}4^{2}_{3}1^{3}_{5}$	
11^a_{55}: $1^{\underline{6}}_{\underline{15}}2^{\underline{5}}_{\underline{13}}1^{\underline{5}}_{\underline{11}}3^{\underline{4}}_{\underline{11}}2^{\underline{4}}_{\underline{9}}5^{\underline{3}}_{\underline{9}}3^{\underline{3}}_{\underline{7}}5^{\underline{2}}_{\underline{7}}5^{\underline{2}}_{\underline{5}}6^{\underline{1}}_{\underline{5}}5^{\underline{1}}_{\underline{3}}5^{0}_{\underline{3}}7^{0}_{\underline{1}}4^{1}_{\underline{1}}4^{1}_{1}3^{2}_{1}4^{2}_{3}1^{3}_{3}3^{3}_{5}1^{4}_{5}1^{4}_{7}1^{5}_{9}$	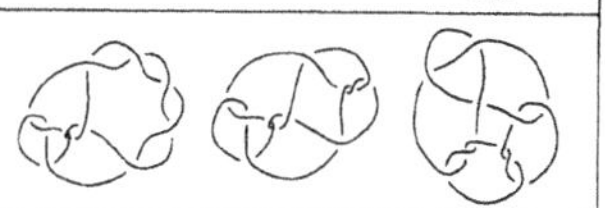
11^a_{56}: $1^{\underline{6}}_{\underline{13}}2^{\underline{5}}_{\underline{11}}1^{\underline{5}}_{\underline{9}}4^{\underline{4}}_{\underline{9}}2^{\underline{4}}_{\underline{7}}7^{\underline{3}}_{\underline{7}}4^{\underline{3}}_{\underline{5}}8^{\underline{2}}_{\underline{5}}7^{\underline{2}}_{\underline{3}}9^{\underline{1}}_{\underline{3}}8^{\underline{1}}_{\underline{1}}9^{0}_{\underline{1}}10^{0}_{1}7^{1}_{1}8^{1}_{3}5^{2}_{3}7^{2}_{5}2^{3}_{5}5^{3}_{7}1^{4}_{7}2^{4}_{9}1^{5}_{11}$	
11^a_{57}: $1^{\underline{6}}_{\underline{15}}2^{\underline{5}}_{\underline{13}}1^{\underline{5}}_{\underline{11}}5^{\underline{4}}_{\underline{11}}2^{\underline{4}}_{\underline{9}}7^{\underline{3}}_{\underline{9}}5^{\underline{3}}_{\underline{7}}7^{\underline{2}}_{\underline{7}}7^{\underline{2}}_{\underline{5}}9^{\underline{1}}_{\underline{5}}7^{\underline{1}}_{\underline{3}}7^{0}_{\underline{3}}10^{0}_{\underline{1}}6^{1}_{\underline{1}}6^{1}_{1}4^{2}_{1}6^{2}_{3}1^{3}_{3}4^{3}_{5}1^{4}_{5}1^{4}_{7}1^{5}_{9}$	
11^a_{58}: $1^{\underline{6}}_{\underline{13}}2^{\underline{5}}_{\underline{11}}1^{\underline{5}}_{\underline{9}}4^{\underline{4}}_{\underline{9}}2^{\underline{4}}_{\underline{7}}5^{\underline{3}}_{\underline{7}}4^{\underline{3}}_{\underline{5}}6^{\underline{2}}_{\underline{5}}5^{\underline{2}}_{\underline{3}}7^{\underline{1}}_{\underline{3}}6^{\underline{1}}_{\underline{1}}6^{0}_{\underline{1}}8^{0}_{1}5^{1}_{1}5^{1}_{3}3^{2}_{3}5^{2}_{5}1^{3}_{5}3^{3}_{7}1^{4}_{7}1^{4}_{9}1^{5}_{11}$	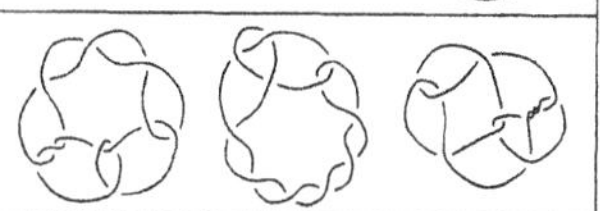
11^a_{59}: $1^{\underline{6}}_{\underline{11}}1^{\underline{5}}_{\underline{9}}1^{\underline{5}}_{\underline{7}}2^{\underline{4}}_{\underline{7}}1^{\underline{4}}_{\underline{5}}2^{\underline{3}}_{\underline{5}}2^{\underline{3}}_{\underline{3}}3^{\underline{2}}_{\underline{3}}2^{\underline{2}}_{\underline{1}}3^{\underline{1}}_{\underline{1}}3^{\underline{1}}_{1}3^{0}_{1}4^{0}_{3}3^{1}_{3}2^{1}_{5}2^{2}_{5}3^{2}_{7}1^{3}_{7}2^{3}_{9}1^{4}_{9}1^{4}_{11}1^{5}_{13}$	
11^a_{60}: $1^{\underline{2}}_{\underline{1}}1^{\underline{1}}_{1}1^{\underline{1}}_{3}4^{0}_{3}2^{0}_{5}5^{1}_{5}3^{1}_{7}6^{2}_{7}5^{2}_{9}7^{3}_{9}6^{3}_{11}7^{4}_{11}7^{4}_{13}5^{5}_{13}7^{5}_{15}4^{6}_{15}5^{6}_{17}2^{7}_{17}4^{7}_{19}1^{8}_{19}2^{8}_{21}1^{9}_{23}$	
11^a_{61}: $1^{\underline{2}}_{\underline{3}}2^{\underline{1}}_{\underline{1}}1^{\underline{1}}_{1}5^{0}_{1}3^{0}_{3}7^{1}_{3}4^{1}_{5}8^{2}_{5}7^{2}_{7}8^{3}_{7}8^{3}_{9}8^{4}_{9}8^{4}_{11}6^{5}_{11}8^{5}_{13}4^{6}_{13}6^{6}_{15}2^{7}_{15}4^{7}_{17}1^{8}_{17}2^{8}_{19}1^{9}_{21}$	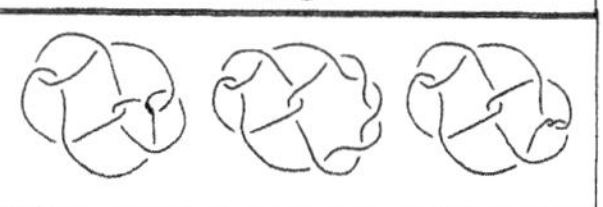
11^a_{62}: $1^{\underline{9}}_{\underline{25}}1^{\underline{8}}_{\underline{23}}1^{\underline{8}}_{\underline{21}}3^{\underline{7}}_{\underline{21}}1^{\underline{7}}_{\underline{19}}3^{\underline{6}}_{\underline{19}}3^{\underline{6}}_{\underline{17}}4^{\underline{5}}_{\underline{17}}3^{\underline{5}}_{\underline{15}}4^{\underline{4}}_{\underline{15}}4^{\underline{4}}_{\underline{13}}4^{\underline{3}}_{\underline{13}}4^{\underline{3}}_{\underline{11}}3^{\underline{2}}_{\underline{11}}4^{\underline{2}}_{\underline{9}}2^{\underline{1}}_{\underline{9}}3^{\underline{1}}_{\underline{7}}2^{0}_{\underline{7}}3^{0}_{\underline{5}}1^{1}_{\underline{5}}1^{1}_{\underline{3}}1^{2}_{\underline{1}}$	
11^a_{63}: $1^{\underline{9}}_{\underline{23}}1^{\underline{8}}_{\underline{21}}1^{\underline{8}}_{\underline{19}}4^{\underline{7}}_{\underline{19}}1^{\underline{7}}_{\underline{17}}5^{\underline{6}}_{\underline{17}}4^{\underline{6}}_{\underline{15}}7^{\underline{5}}_{\underline{15}}5^{\underline{5}}_{\underline{13}}8^{\underline{4}}_{\underline{13}}7^{\underline{4}}_{\underline{11}}7^{\underline{3}}_{\underline{11}}8^{\underline{3}}_{\underline{9}}6^{\underline{2}}_{\underline{9}}7^{\underline{2}}_{\underline{7}}4^{\underline{1}}_{\underline{7}}6^{\underline{1}}_{\underline{5}}3^{0}_{\underline{5}}5^{0}_{\underline{3}}1^{1}_{\underline{3}}2^{1}_{\underline{1}}1^{2}_{1}$	
11^a_{64}: $1^{\underline{9}}_{\underline{23}}2^{\underline{8}}_{\underline{21}}1^{\underline{8}}_{\underline{19}}5^{\underline{7}}_{\underline{19}}2^{\underline{7}}_{\underline{17}}6^{\underline{6}}_{\underline{17}}5^{\underline{6}}_{\underline{15}}8^{\underline{5}}_{\underline{15}}6^{\underline{5}}_{\underline{13}}8^{\underline{4}}_{\underline{13}}8^{\underline{4}}_{\underline{11}}7^{\underline{3}}_{\underline{11}}8^{\underline{3}}_{\underline{9}}6^{\underline{2}}_{\underline{9}}7^{\underline{2}}_{\underline{7}}3^{\underline{1}}_{\underline{7}}6^{\underline{1}}_{\underline{5}}2^{0}_{\underline{5}}4^{0}_{\underline{3}}1^{1}_{\underline{3}}1^{1}_{\underline{1}}1^{2}_{1}$	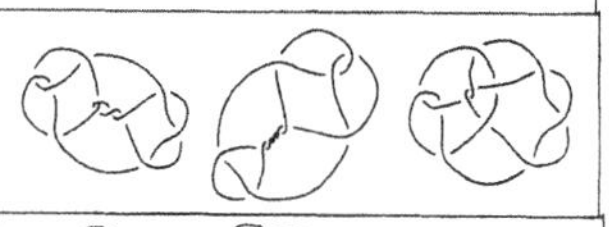
11^a_{65}: $1^{\underline{9}}_{\underline{21}}1^{\underline{8}}_{\underline{19}}1^{\underline{8}}_{\underline{17}}3^{\underline{7}}_{\underline{17}}1^{\underline{7}}_{\underline{15}}3^{\underline{6}}_{\underline{15}}3^{\underline{6}}_{\underline{13}}4^{\underline{5}}_{\underline{13}}3^{\underline{5}}_{\underline{11}}5^{\underline{4}}_{\underline{11}}4^{\underline{4}}_{\underline{9}}4^{\underline{3}}_{\underline{9}}5^{\underline{3}}_{\underline{7}}4^{\underline{2}}_{\underline{7}}4^{\underline{2}}_{\underline{5}}2^{\underline{1}}_{\underline{5}}4^{\underline{1}}_{\underline{3}}2^{0}_{\underline{3}}3^{0}_{\underline{1}}1^{1}_{\underline{1}}1^{1}_{1}1^{2}_{3}$	
11^a_{66}: $1^{\underline{7}}_{\underline{17}}2^{\underline{6}}_{\underline{15}}1^{\underline{6}}_{\underline{13}}5^{\underline{5}}_{\underline{13}}2^{\underline{5}}_{\underline{11}}7^{\underline{4}}_{\underline{11}}5^{\underline{4}}_{\underline{9}}9^{\underline{3}}_{\underline{9}}7^{\underline{3}}_{\underline{7}}10^{\underline{2}}_{\underline{7}}9^{\underline{2}}_{\underline{5}}9^{\underline{1}}_{\underline{5}}10^{\underline{1}}_{\underline{3}}8^{0}_{\underline{3}}10^{0}_{\underline{1}}5^{1}_{\underline{1}}7^{1}_{1}3^{2}_{1}5^{2}_{3}1^{3}_{3}3^{3}_{5}1^{4}_{7}$	
11^a_{67}: $1^{\underline{7}}_{\underline{15}}2^{\underline{6}}_{\underline{13}}1^{\underline{6}}_{\underline{11}}5^{\underline{5}}_{\underline{11}}2^{\underline{5}}_{\underline{9}}7^{\underline{4}}_{\underline{9}}5^{\underline{4}}_{\underline{7}}9^{\underline{3}}_{\underline{7}}7^{\underline{3}}_{\underline{5}}11^{\underline{2}}_{\underline{5}}9^{\underline{2}}_{\underline{3}}9^{\underline{1}}_{\underline{3}}11^{\underline{1}}_{\underline{1}}9^{0}_{\underline{1}}10^{0}_{1}6^{1}_{1}8^{1}_{3}3^{2}_{3}6^{2}_{5}1^{3}_{5}3^{3}_{7}1^{4}_{9}$	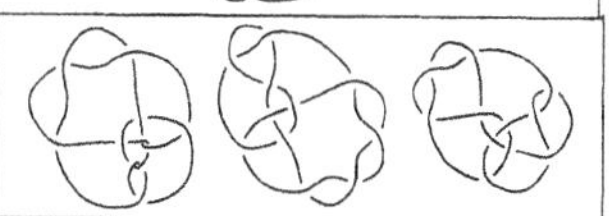
11^a_{68}: $1^{\underline{5}}_{\underline{9}}2^{\underline{4}}_{\underline{7}}1^{\underline{4}}_{\underline{5}}4^{\underline{3}}_{\underline{5}}2^{\underline{3}}_{\underline{3}}6^{\underline{2}}_{\underline{3}}4^{\underline{2}}_{\underline{1}}7^{\underline{1}}_{\underline{1}}6^{\underline{1}}_{1}9^{0}_{1}8^{0}_{3}8^{1}_{3}8^{1}_{5}7^{2}_{5}8^{2}_{7}4^{3}_{7}7^{3}_{9}3^{4}_{9}4^{4}_{11}1^{5}_{11}3^{5}_{13}1^{6}_{15}$	
11^a_{69}: $1^{\underline{5}}_{\underline{11}}3^{\underline{4}}_{\underline{9}}1^{\underline{4}}_{\underline{7}}6^{\underline{3}}_{\underline{7}}3^{\underline{3}}_{\underline{5}}9^{\underline{2}}_{\underline{5}}6^{\underline{2}}_{\underline{3}}11^{\underline{1}}_{\underline{3}}9^{\underline{1}}_{\underline{1}}12^{0}_{\underline{1}}12^{0}_{1}11^{1}_{1}11^{1}_{3}9^{2}_{3}11^{2}_{5}5^{3}_{5}9^{3}_{7}3^{4}_{7}5^{4}_{9}1^{5}_{9}3^{5}_{11}1^{6}_{13}$	
11^a_{70}: $1^{\underline{4}}_{\underline{7}}2^{\underline{3}}_{\underline{5}}1^{\underline{3}}_{\underline{3}}6^{\underline{2}}_{\underline{3}}2^{\underline{2}}_{\underline{1}}8^{\underline{1}}_{\underline{1}}6^{\underline{1}}_{1}12^{0}_{1}9^{0}_{3}13^{1}_{3}11^{1}_{5}12^{2}_{5}13^{2}_{7}10^{3}_{7}12^{3}_{9}7^{4}_{9}10^{4}_{11}4^{5}_{11}7^{5}_{13}1^{6}_{13}4^{6}_{15}1^{7}_{17}$	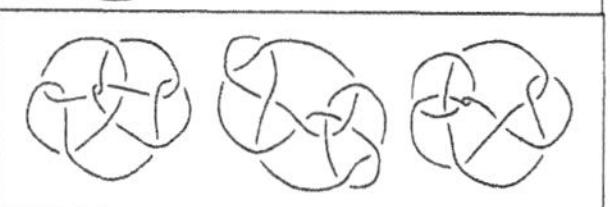
11^a_{71}: $1^{\underline{4}}_{\underline{7}}3^{\underline{3}}_{\underline{5}}1^{\underline{3}}_{\underline{3}}6^{\underline{2}}_{\underline{3}}3^{\underline{2}}_{\underline{1}}9^{\underline{1}}_{\underline{1}}6^{\underline{1}}_{1}13^{0}_{1}10^{0}_{3}13^{1}_{3}12^{1}_{5}13^{2}_{5}13^{2}_{7}10^{3}_{7}13^{3}_{9}7^{4}_{9}10^{4}_{11}4^{5}_{11}7^{5}_{13}1^{6}_{13}4^{6}_{15}1^{7}_{17}$	
11^a_{72}: $1^{\underline{5}}_{\underline{11}}3^{\underline{4}}_{\underline{9}}1^{\underline{4}}_{\underline{7}}6^{\underline{3}}_{\underline{7}}3^{\underline{3}}_{\underline{5}}9^{\underline{2}}_{\underline{5}}6^{\underline{2}}_{\underline{3}}12^{\underline{1}}_{\underline{3}}9^{\underline{1}}_{\underline{1}}13^{0}_{\underline{1}}13^{0}_{1}12^{1}_{1}12^{1}_{3}10^{2}_{3}12^{2}_{5}6^{3}_{5}10^{3}_{7}4^{4}_{7}6^{4}_{9}1^{5}_{9}4^{5}_{11}1^{6}_{13}$	
11^a_{73}: $1^{\underline{6}}_{\underline{13}}4^{\underline{5}}_{\underline{11}}1^{\underline{5}}_{\underline{9}}7^{\underline{4}}_{\underline{9}}4^{\underline{4}}_{\underline{7}}11^{\underline{3}}_{\underline{7}}7^{\underline{3}}_{\underline{5}}14^{\underline{2}}_{\underline{5}}11^{\underline{2}}_{\underline{3}}14^{\underline{1}}_{\underline{3}}14^{\underline{1}}_{\underline{1}}15^{0}_{\underline{1}}15^{0}_{1}11^{1}_{1}14^{1}_{3}7^{2}_{3}11^{2}_{5}4^{3}_{5}7^{3}_{7}1^{4}_{7}4^{4}_{9}1^{5}_{11}$	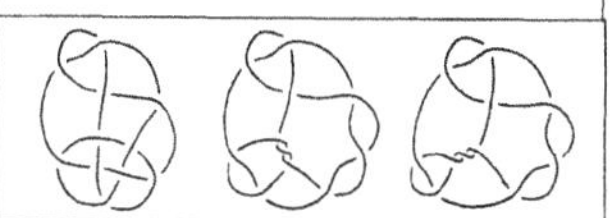
11^a_{74}: $1^{\underline{4}}_{\underline{5}}1^{\underline{3}}_{\underline{3}}1^{\underline{3}}_{\underline{1}}4^{\underline{2}}_{\underline{1}}1^{\underline{2}}_{1}3^{\underline{1}}_{1}4^{\underline{1}}_{3}6^{0}_{3}4^{0}_{5}6^{1}_{5}5^{1}_{7}5^{2}_{7}6^{2}_{9}5^{3}_{9}5^{3}_{11}3^{4}_{11}5^{4}_{13}2^{5}_{13}3^{5}_{15}1^{6}_{15}2^{6}_{17}1^{7}_{19}$	
11^a_{75}: $1^{\underline{4}}_{\underline{7}}1^{\underline{3}}_{\underline{5}}1^{\underline{3}}_{\underline{3}}4^{\underline{2}}_{\underline{3}}1^{\underline{2}}_{\underline{1}}4^{\underline{1}}_{\underline{1}}4^{\underline{1}}_{1}7^{0}_{1}5^{0}_{3}7^{1}_{3}6^{1}_{5}6^{2}_{5}7^{2}_{7}6^{3}_{7}6^{3}_{9}3^{4}_{9}6^{4}_{11}2^{5}_{11}3^{5}_{13}1^{6}_{13}2^{6}_{15}1^{7}_{17}$	
11^a_{76}: $1^{\underline{6}}_{\underline{13}}3^{\underline{5}}_{\underline{11}}1^{\underline{5}}_{\underline{9}}6^{\underline{4}}_{\underline{9}}3^{\underline{4}}_{\underline{7}}9^{\underline{3}}_{\underline{7}}6^{\underline{3}}_{\underline{5}}11^{\underline{2}}_{\underline{5}}9^{\underline{2}}_{\underline{3}}12^{\underline{1}}_{\underline{3}}11^{\underline{1}}_{\underline{1}}12^{0}_{\underline{1}}13^{0}_{1}9^{1}_{1}11^{1}_{3}6^{2}_{3}9^{2}_{5}3^{3}_{5}6^{3}_{7}1^{4}_{7}3^{4}_{9}1^{5}_{11}$	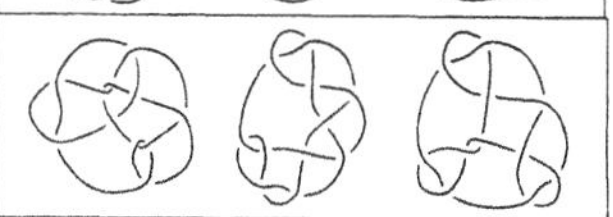
11^a_{77}: $1^{\underline{8}}_{\underline{19}}3^{\underline{7}}_{\underline{17}}1^{\underline{7}}_{\underline{15}}5^{\underline{6}}_{\underline{15}}3^{\underline{6}}_{\underline{13}}8^{\underline{5}}_{\underline{13}}5^{\underline{5}}_{\underline{11}}10^{\underline{4}}_{\underline{11}}8^{\underline{4}}_{\underline{9}}11^{\underline{3}}_{\underline{9}}10^{\underline{3}}_{\underline{7}}10^{\underline{2}}_{\underline{7}}11^{\underline{2}}_{\underline{5}}8^{\underline{1}}_{\underline{5}}10^{\underline{1}}_{\underline{3}}6^{0}_{\underline{3}}9^{0}_{\underline{1}}3^{1}_{\underline{1}}5^{1}_{1}1^{2}_{1}3^{2}_{3}1^{3}_{5}$	
11^a_{78}: $1^{\underline{4}}_{\underline{7}}2^{\underline{3}}_{\underline{5}}1^{\underline{3}}_{\underline{3}}5^{\underline{2}}_{\underline{3}}2^{\underline{2}}_{\underline{1}}7^{\underline{1}}_{\underline{1}}5^{\underline{1}}_{1}10^{0}_{1}8^{0}_{3}10^{1}_{3}9^{1}_{5}10^{2}_{5}10^{2}_{7}8^{3}_{7}10^{3}_{9}5^{4}_{9}8^{4}_{11}3^{5}_{11}5^{5}_{13}1^{6}_{13}3^{6}_{15}1^{7}_{17}$	
11^a_{79}: $1^{\underline{7}}_{\underline{17}}3^{\underline{6}}_{\underline{15}}1^{\underline{6}}_{\underline{13}}6^{\underline{5}}_{\underline{13}}3^{\underline{5}}_{\underline{11}}9^{\underline{4}}_{\underline{11}}6^{\underline{4}}_{\underline{9}}11^{\underline{3}}_{\underline{9}}9^{\underline{3}}_{\underline{7}}12^{\underline{2}}_{\underline{7}}11^{\underline{2}}_{\underline{5}}11^{\underline{1}}_{\underline{5}}12^{\underline{1}}_{\underline{3}}9^{0}_{\underline{3}}12^{0}_{\underline{1}}6^{1}_{\underline{1}}8^{1}_{1}3^{2}_{1}6^{2}_{3}1^{3}_{3}3^{3}_{5}1^{4}_{7}$	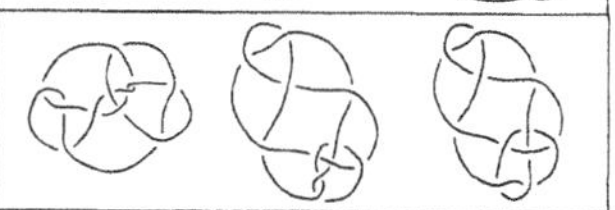
11^a_{80}: $1^{\underline{6}}_{\underline{13}}3^{\underline{5}}_{\underline{11}}1^{\underline{5}}_{\underline{9}}6^{\underline{4}}_{\underline{9}}3^{\underline{4}}_{\underline{7}}8^{\underline{3}}_{\underline{7}}6^{\underline{3}}_{\underline{5}}11^{\underline{2}}_{\underline{5}}8^{\underline{2}}_{\underline{3}}11^{\underline{1}}_{\underline{3}}11^{\underline{1}}_{\underline{1}}11^{0}_{\underline{1}}12^{0}_{1}9^{1}_{1}10^{1}_{3}5^{2}_{3}9^{2}_{5}3^{3}_{5}5^{3}_{7}1^{4}_{7}3^{4}_{9}1^{5}_{11}$	
11^a_{81}: $1^{\underline{5}}_{\underline{9}}3^{\underline{4}}_{\underline{7}}1^{\underline{4}}_{\underline{5}}5^{\underline{3}}_{\underline{5}}3^{\underline{3}}_{\underline{3}}8^{\underline{2}}_{\underline{3}}5^{\underline{2}}_{\underline{1}}9^{\underline{1}}_{\underline{1}}8^{\underline{1}}_{1}11^{0}_{1}10^{0}_{3}10^{1}_{3}10^{1}_{5}8^{2}_{5}10^{2}_{7}5^{3}_{7}8^{3}_{9}3^{4}_{9}5^{4}_{11}1^{5}_{11}3^{5}_{13}1^{6}_{15}$	
11^a_{82}: $1^{\underline{6}}_{\underline{15}}2^{\underline{5}}_{\underline{13}}1^{\underline{5}}_{\underline{11}}4^{\underline{4}}_{\underline{11}}2^{\underline{4}}_{\underline{9}}6^{\underline{3}}_{\underline{9}}4^{\underline{3}}_{\underline{7}}7^{\underline{2}}_{\underline{7}}6^{\underline{2}}_{\underline{5}}8^{\underline{1}}_{\underline{5}}7^{\underline{1}}_{\underline{3}}7^{0}_{\underline{3}}9^{0}_{\underline{1}}6^{1}_{\underline{1}}6^{1}_{1}4^{2}_{1}6^{2}_{3}2^{3}_{3}4^{3}_{5}1^{4}_{5}2^{4}_{7}1^{5}_{9}$	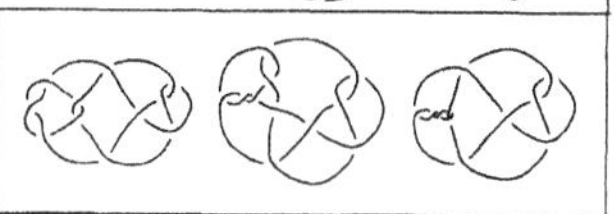
11^a_{83}: $1^{\underline{3}}_{\underline{3}}2^{\underline{2}}_{\underline{1}}1^{\underline{2}}_{1}4^{\underline{1}}_{1}2^{\underline{1}}_{3}7^{0}_{3}5^{0}_{5}8^{1}_{5}6^{1}_{7}10^{2}_{7}8^{2}_{9}8^{3}_{9}10^{3}_{11}8^{4}_{11}8^{4}_{13}5^{5}_{13}8^{5}_{15}3^{6}_{15}5^{6}_{17}1^{7}_{17}3^{7}_{19}1^{8}_{21}$	
11^a_{84}: $1^{\underline{6}}_{\underline{13}}2^{\underline{5}}_{\underline{11}}1^{\underline{5}}_{\underline{9}}4^{\underline{4}}_{\underline{9}}2^{\underline{4}}_{\underline{7}}6^{\underline{3}}_{\underline{7}}4^{\underline{3}}_{\underline{5}}8^{\underline{2}}_{\underline{5}}6^{\underline{2}}_{\underline{3}}8^{\underline{1}}_{\underline{3}}8^{\underline{1}}_{\underline{1}}8^{0}_{\underline{1}}9^{0}_{1}7^{1}_{1}7^{1}_{3}4^{2}_{3}7^{2}_{5}2^{3}_{5}4^{3}_{7}1^{4}_{7}2^{4}_{9}1^{5}_{11}$	

K: KH	1st line	2nd line	3rd line
11^a_{85}: $1^{\underline{3}}_{\underline{5}}2^{\underline{2}}_{\underline{3}}1^{\underline{2}}_{\underline{1}}4^{\underline{1}}_{\underline{1}}2^{\underline{1}}_{1}7^{0}_{1}5^{0}_{3}8^{1}_{3}6^{1}_{5}9^{2}_{5}8^{2}_{7}8^{3}_{7}9^{3}_{9}7^{4}_{9}8^{4}_{11}4^{5}_{11}7^{5}_{13}3^{6}_{13}4^{6}_{15}1^{7}_{15}3^{7}_{17}1^{8}_{19}$	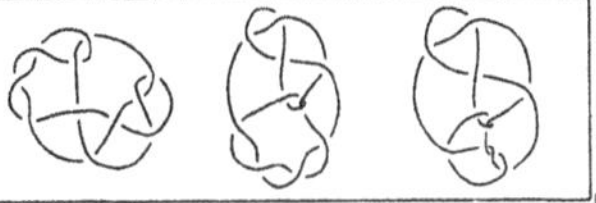		
11^a_{86}: $1^{\underline{5}}_{\underline{9}}2^{\underline{4}}_{\underline{7}}1^{\underline{4}}_{\underline{5}}3^{\underline{3}}_{\underline{5}}2^{\underline{3}}_{\underline{3}}6^{\underline{2}}_{\underline{3}}3^{\underline{2}}_{\underline{1}}6^{\underline{1}}_{\underline{1}}6^{\underline{1}}_{1}8^{0}_{1}7^{0}_{3}7^{1}_{3}7^{1}_{5}6^{2}_{5}7^{2}_{7}4^{3}_{7}6^{3}_{9}2^{4}_{9}4^{4}_{11}1^{5}_{11}2^{5}_{13}1^{6}_{15}$			
11^a_{87}: $1^{\underline{5}}_{\underline{11}}3^{\underline{4}}_{\underline{9}}1^{\underline{4}}_{\underline{7}}5^{\underline{3}}_{\underline{7}}3^{\underline{3}}_{\underline{5}}8^{\underline{2}}_{\underline{5}}5^{\underline{2}}_{\underline{3}}9^{\underline{1}}_{\underline{3}}8^{\underline{1}}_{\underline{1}}11^{0}_{\underline{1}}10^{0}_{1}9^{1}_{1}10^{1}_{3}7^{2}_{3}9^{2}_{5}5^{3}_{5}7^{3}_{7}2^{4}_{7}5^{4}_{9}1^{5}_{9}2^{5}_{11}1^{6}_{13}$			
11^a_{88}: $1^{\underline{6}}_{\underline{13}}2^{\underline{5}}_{\underline{11}}1^{\underline{5}}_{\underline{9}}4^{\underline{4}}_{\underline{9}}2^{\underline{4}}_{\underline{7}}6^{\underline{3}}_{\underline{7}}4^{\underline{3}}_{\underline{5}}8^{\underline{2}}_{\underline{5}}6^{\underline{2}}_{\underline{3}}8^{\underline{1}}_{\underline{3}}8^{\underline{1}}_{\underline{1}}8^{0}_{\underline{1}}9^{0}_{1}7^{1}_{1}7^{1}_{3}4^{2}_{3}7^{2}_{5}2^{3}_{5}4^{3}_{7}1^{4}_{7}2^{4}_{9}1^{5}_{11}$	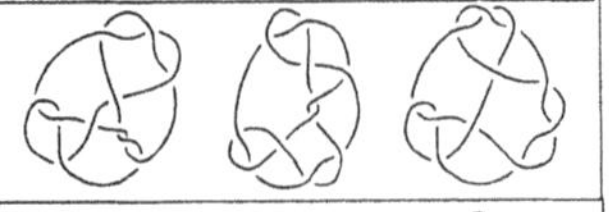		
11^a_{89}: $1^{\underline{4}}_{\underline{7}}2^{\underline{3}}_{\underline{5}}1^{\underline{3}}_{\underline{3}}5^{\underline{2}}_{\underline{3}}2^{\underline{2}}_{\underline{1}}6^{\underline{1}}_{\underline{1}}5^{\underline{1}}_{1}10^{0}_{1}7^{0}_{3}10^{1}_{3}9^{1}_{5}9^{2}_{5}10^{2}_{7}8^{3}_{7}9^{3}_{9}5^{4}_{9}8^{4}_{11}3^{5}_{11}5^{5}_{13}1^{6}_{13}3^{6}_{15}1^{7}_{17}$			
11^a_{90}: $1^{\underline{6}}_{\underline{15}}2^{\underline{5}}_{\underline{13}}1^{\underline{5}}_{\underline{11}}4^{\underline{4}}_{\underline{11}}2^{\underline{4}}_{\underline{9}}5^{\underline{3}}_{\underline{9}}4^{\underline{3}}_{\underline{7}}7^{\underline{2}}_{\underline{7}}5^{\underline{2}}_{\underline{5}}7^{\underline{1}}_{\underline{5}}7^{\underline{1}}_{\underline{3}}6^{0}_{\underline{3}}8^{0}_{\underline{1}}6^{1}_{\underline{1}}5^{1}_{1}3^{2}_{1}6^{2}_{3}2^{3}_{3}3^{3}_{5}1^{4}_{5}2^{4}_{7}1^{5}_{9}$			
11^a_{91}: $1^{\underline{5}}_{\underline{11}}3^{\underline{4}}_{\underline{9}}1^{\underline{4}}_{\underline{7}}5^{\underline{3}}_{\underline{7}}3^{\underline{3}}_{\underline{5}}8^{\underline{2}}_{\underline{5}}5^{\underline{2}}_{\underline{3}}10^{\underline{1}}_{\underline{3}}8^{\underline{1}}_{\underline{1}}11^{0}_{\underline{1}}11^{0}_{1}10^{1}_{1}10^{1}_{3}8^{2}_{3}10^{2}_{5}5^{3}_{5}8^{3}_{7}3^{4}_{7}5^{4}_{9}1^{5}_{9}3^{5}_{11}1^{6}_{13}$	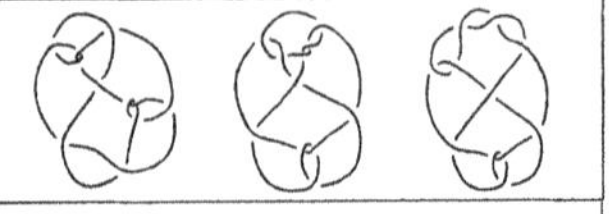		
11^a_{92}: $1^{\underline{7}}_{\underline{17}}2^{\underline{6}}_{\underline{15}}1^{\underline{6}}_{\underline{13}}5^{\underline{5}}_{\underline{13}}2^{\underline{5}}_{\underline{11}}6^{\underline{4}}_{\underline{11}}5^{\underline{4}}_{\underline{9}}8^{\underline{3}}_{\underline{9}}6^{\underline{3}}_{\underline{7}}9^{\underline{2}}_{\underline{7}}8^{\underline{2}}_{\underline{5}}7^{\underline{1}}_{\underline{5}}9^{\underline{1}}_{\underline{3}}7^{0}_{\underline{3}}8^{0}_{\underline{1}}4^{1}_{\underline{1}}6^{1}_{1}2^{2}_{1}4^{2}_{3}1^{3}_{3}2^{3}_{5}1^{4}_{7}$			
11^a_{93}: $1^{\underline{7}}_{\underline{15}}2^{\underline{6}}_{\underline{13}}1^{\underline{6}}_{\underline{11}}4^{\underline{5}}_{\underline{11}}2^{\underline{5}}_{\underline{9}}5^{\underline{4}}_{\underline{9}}4^{\underline{4}}_{\underline{7}}7^{\underline{3}}_{\underline{7}}5^{\underline{3}}_{\underline{5}}8^{\underline{2}}_{\underline{5}}7^{\underline{2}}_{\underline{3}}6^{\underline{1}}_{\underline{3}}8^{\underline{1}}_{\underline{1}}7^{0}_{\underline{1}}7^{0}_{1}4^{1}_{1}6^{1}_{3}2^{2}_{3}4^{2}_{5}1^{3}_{5}2^{3}_{7}1^{4}_{9}$			
11^a_{94}: $1^{0}_{5}1^{0}_{7}2^{1}_{7}4^{2}_{9}2^{2}_{11}5^{3}_{11}4^{3}_{13}9^{4}_{13}5^{4}_{15}8^{5}_{15}9^{5}_{17}9^{6}_{17}8^{6}_{19}7^{7}_{19}9^{7}_{21}5^{8}_{21}7^{8}_{23}3^{9}_{23}5^{9}_{25}1^{10}_{25}3^{10}_{27}1^{11}_{29}$			
11^a_{95}: $1^{0}_{3}1^{0}_{5}2^{1}_{5}3^{2}_{7}2^{2}_{9}4^{3}_{9}3^{3}_{11}6^{4}_{11}4^{4}_{13}5^{5}_{13}6^{5}_{15}6^{6}_{15}5^{6}_{17}4^{7}_{17}6^{7}_{19}3^{8}_{19}4^{8}_{21}2^{9}_{21}3^{9}_{23}1^{10}_{23}2^{10}_{25}1^{11}_{27}$			
11^a_{96}: $1^{\underline{6}}_{\underline{13}}2^{\underline{5}}_{\underline{11}}1^{\underline{5}}_{\underline{9}}5^{\underline{4}}_{\underline{9}}2^{\underline{4}}_{\underline{7}}7^{\underline{3}}_{\underline{7}}5^{\underline{3}}_{\underline{5}}9^{\underline{2}}_{\underline{5}}7^{\underline{2}}_{\underline{3}}10^{\underline{1}}_{\underline{3}}9^{\underline{1}}_{\underline{1}}10^{0}_{\underline{1}}11^{0}_{1}8^{1}_{1}9^{1}_{3}5^{2}_{3}8^{2}_{5}3^{3}_{5}5^{3}_{7}1^{4}_{7}3^{4}_{9}1^{5}_{11}$			
11^a_{97}: $1^{\underline{6}}_{\underline{11}}1^{\underline{5}}_{\underline{9}}1^{\underline{5}}_{\underline{7}}3^{\underline{4}}_{\underline{7}}1^{\underline{4}}_{\underline{5}}4^{\underline{3}}_{\underline{5}}3^{\underline{3}}_{\underline{3}}5^{\underline{2}}_{\underline{3}}4^{\underline{2}}_{\underline{1}}5^{\underline{1}}_{\underline{1}}5^{\underline{1}}_{1}6^{0}_{1}6^{0}_{3}5^{1}_{3}5^{1}_{5}3^{2}_{5}5^{2}_{7}2^{3}_{7}3^{3}_{9}1^{4}_{9}2^{4}_{11}1^{5}_{13}$			
11^a_{98}: $1^{\underline{6}}_{\underline{13}}1^{\underline{5}}_{\underline{11}}1^{\underline{5}}_{\underline{9}}3^{\underline{4}}_{\underline{9}}1^{\underline{4}}_{\underline{7}}4^{\underline{3}}_{\underline{7}}3^{\underline{3}}_{\underline{5}}6^{\underline{2}}_{\underline{5}}4^{\underline{2}}_{\underline{3}}6^{\underline{1}}_{\underline{3}}6^{\underline{1}}_{\underline{1}}6^{0}_{\underline{1}}7^{0}_{1}6^{1}_{1}5^{1}_{3}3^{2}_{3}6^{2}_{5}2^{3}_{5}3^{3}_{7}1^{4}_{7}2^{4}_{9}1^{5}_{11}$			
11^a_{99}: $1^{\underline{5}}_{\underline{9}}3^{\underline{4}}_{\underline{7}}1^{\underline{4}}_{\underline{5}}5^{\underline{3}}_{\underline{5}}3^{\underline{3}}_{\underline{3}}9^{\underline{2}}_{\underline{3}}5^{\underline{2}}_{\underline{1}}9^{\underline{1}}_{\underline{1}}9^{\underline{1}}_{1}12^{0}_{1}10^{0}_{3}11^{1}_{3}11^{1}_{5}8^{2}_{5}11^{2}_{7}6^{3}_{7}8^{3}_{9}3^{4}_{9}6^{4}_{11}1^{5}_{11}3^{5}_{13}1^{6}_{15}$			
11^a_{100}: $1^{\underline{2}}_{\underline{1}}2^{\underline{1}}_{1}1^{\underline{1}}_{3}6^{0}_{3}3^{0}_{5}8^{1}_{5}5^{1}_{7}11^{2}_{7}8^{2}_{9}11^{3}_{9}11^{3}_{11}12^{4}_{11}11^{4}_{13}9^{5}_{13}12^{5}_{15}6^{6}_{15}9^{6}_{17}4^{7}_{17}6^{7}_{19}1^{8}_{19}4^{8}_{21}1^{9}_{23}$			
11^a_{101}: $1^{\underline{3}}_{\underline{5}}3^{\underline{2}}_{\underline{3}}1^{\underline{2}}_{\underline{1}}6^{\underline{1}}_{\underline{1}}3^{\underline{1}}_{1}11^{0}_{1}7^{0}_{3}13^{1}_{3}10^{1}_{5}14^{2}_{5}13^{2}_{7}13^{3}_{7}14^{3}_{9}11^{4}_{9}13^{4}_{11}7^{5}_{11}11^{5}_{13}4^{6}_{13}7^{6}_{15}1^{7}_{15}4^{7}_{17}1^{8}_{19}$			
11^a_{102}: $1^{\underline{4}}_{\underline{7}}1^{\underline{3}}_{\underline{5}}1^{\underline{3}}_{\underline{3}}5^{\underline{2}}_{\underline{3}}1^{\underline{2}}_{\underline{1}}5^{\underline{1}}_{\underline{1}}5^{\underline{1}}_{1}8^{0}_{1}6^{0}_{3}9^{1}_{3}7^{1}_{5}7^{2}_{5}9^{2}_{7}7^{3}_{7}7^{3}_{9}4^{4}_{9}7^{4}_{11}2^{5}_{11}4^{5}_{13}1^{6}_{13}2^{6}_{15}1^{7}_{17}$			
11^a_{103}: $1^{\underline{4}}_{\underline{9}}1^{\underline{3}}_{\underline{7}}1^{\underline{3}}_{\underline{5}}4^{\underline{2}}_{\underline{5}}1^{\underline{2}}_{\underline{3}}4^{\underline{1}}_{\underline{3}}4^{\underline{1}}_{\underline{1}}7^{0}_{\underline{1}}5^{0}_{1}7^{1}_{1}6^{1}_{3}5^{2}_{3}7^{2}_{5}6^{3}_{5}5^{3}_{7}3^{4}_{7}6^{4}_{9}2^{5}_{9}3^{5}_{11}1^{6}_{11}2^{6}_{13}1^{7}_{15}$			
11^a_{104}: $1^{\underline{5}}_{\underline{11}}2^{\underline{4}}_{\underline{9}}1^{\underline{4}}_{\underline{7}}5^{\underline{3}}_{\underline{7}}2^{\underline{3}}_{\underline{5}}8^{\underline{2}}_{\underline{5}}5^{\underline{2}}_{\underline{3}}9^{\underline{1}}_{\underline{3}}8^{\underline{1}}_{\underline{1}}11^{0}_{\underline{1}}10^{0}_{1}10^{1}_{1}10^{1}_{3}8^{2}_{3}10^{2}_{5}5^{3}_{5}8^{3}_{7}3^{4}_{7}5^{4}_{9}1^{5}_{9}3^{5}_{11}1^{6}_{13}$			
11^a_{105}: $1^{\underline{2}}_{\underline{1}}1^{\underline{1}}_{1}1^{\underline{1}}_{3}5^{0}_{3}2^{0}_{5}6^{1}_{5}4^{1}_{7}8^{2}_{7}6^{2}_{9}9^{3}_{9}8^{3}_{11}9^{4}_{11}9^{4}_{13}7^{5}_{13}9^{5}_{15}5^{6}_{15}7^{6}_{17}3^{7}_{17}5^{7}_{19}1^{8}_{19}3^{8}_{21}1^{9}_{23}$			
11^a_{106}: $1^{\underline{5}}_{\underline{11}}2^{\underline{4}}_{\underline{9}}1^{\underline{4}}_{\underline{7}}4^{\underline{3}}_{\underline{7}}2^{\underline{3}}_{\underline{5}}6^{\underline{2}}_{\underline{5}}4^{\underline{2}}_{\underline{3}}7^{\underline{1}}_{\underline{3}}6^{\underline{1}}_{\underline{1}}8^{0}_{\underline{1}}8^{0}_{1}7^{1}_{1}7^{1}_{3}6^{2}_{3}7^{2}_{5}3^{3}_{5}6^{3}_{7}2^{4}_{7}3^{4}_{9}1^{5}_{9}2^{5}_{11}1^{6}_{13}$	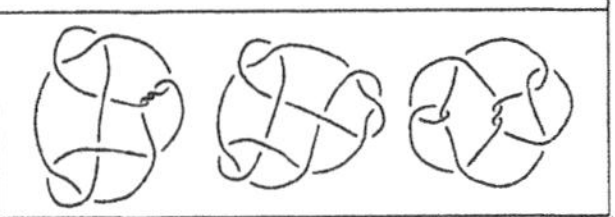		
11^a_{107}: $1^{\underline{5}}_{\underline{9}}2^{\underline{4}}_{\underline{7}}1^{\underline{4}}_{\underline{5}}4^{\underline{3}}_{\underline{5}}2^{\underline{3}}_{\underline{3}}7^{\underline{2}}_{\underline{3}}4^{\underline{2}}_{\underline{1}}7^{\underline{1}}_{\underline{1}}7^{\underline{1}}_{1}10^{0}_{1}8^{0}_{3}9^{1}_{3}9^{1}_{5}7^{2}_{5}9^{2}_{7}5^{3}_{7}7^{3}_{9}3^{4}_{9}5^{4}_{11}1^{5}_{11}3^{5}_{13}1^{6}_{15}$			
11^a_{108}: $1^{\underline{6}}_{\underline{15}}2^{\underline{5}}_{\underline{13}}1^{\underline{5}}_{\underline{11}}4^{\underline{4}}_{\underline{11}}2^{\underline{4}}_{\underline{9}}7^{\underline{3}}_{\underline{9}}4^{\underline{3}}_{\underline{7}}7^{\underline{2}}_{\underline{7}}7^{\underline{2}}_{\underline{5}}8^{\underline{1}}_{\underline{5}}7^{\underline{1}}_{\underline{3}}8^{0}_{\underline{3}}9^{0}_{\underline{1}}6^{1}_{\underline{1}}7^{1}_{1}4^{2}_{1}6^{2}_{3}2^{3}_{3}4^{3}_{5}1^{4}_{5}2^{4}_{7}1^{5}_{9}$			
11^a_{109}: $1^{\underline{5}}_{\underline{11}}2^{\underline{4}}_{\underline{9}}1^{\underline{4}}_{\underline{7}}5^{\underline{3}}_{\underline{7}}2^{\underline{3}}_{\underline{5}}7^{\underline{2}}_{\underline{5}}5^{\underline{2}}_{\underline{3}}9^{\underline{1}}_{\underline{3}}7^{\underline{1}}_{\underline{1}}10^{0}_{\underline{1}}10^{0}_{1}9^{1}_{1}9^{1}_{3}8^{2}_{3}9^{2}_{5}4^{3}_{5}8^{3}_{7}3^{4}_{7}4^{4}_{9}1^{5}_{9}3^{5}_{11}1^{6}_{13}$	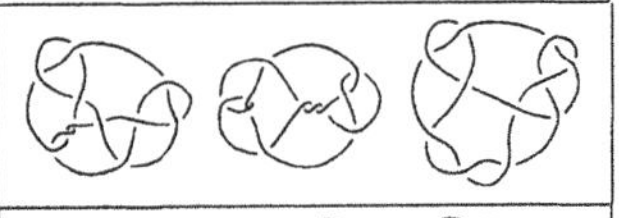		
11^a_{110}: $1^{\underline{6}}_{\underline{13}}2^{\underline{5}}_{\underline{11}}1^{\underline{5}}_{\underline{9}}4^{\underline{4}}_{\underline{9}}2^{\underline{4}}_{\underline{7}}6^{\underline{3}}_{\underline{7}}4^{\underline{3}}_{\underline{5}}7^{\underline{2}}_{\underline{5}}6^{\underline{2}}_{\underline{3}}8^{\underline{1}}_{\underline{3}}7^{\underline{1}}_{\underline{1}}8^{0}_{\underline{1}}9^{0}_{1}6^{1}_{1}7^{1}_{3}4^{2}_{3}6^{2}_{5}2^{3}_{5}4^{3}_{7}1^{4}_{7}2^{4}_{9}1^{5}_{11}$			
11^a_{111}: $1^{\underline{5}}_{\underline{9}}2^{\underline{4}}_{\underline{7}}1^{\underline{4}}_{\underline{5}}4^{\underline{3}}_{\underline{5}}2^{\underline{3}}_{\underline{3}}6^{\underline{2}}_{\underline{3}}4^{\underline{2}}_{\underline{1}}7^{\underline{1}}_{\underline{1}}6^{\underline{1}}_{1}9^{0}_{1}8^{0}_{3}8^{1}_{3}8^{1}_{5}7^{2}_{5}8^{2}_{7}4^{3}_{7}7^{3}_{9}3^{4}_{9}4^{4}_{11}1^{5}_{11}3^{5}_{13}1^{6}_{15}$			
11^a_{112}: $1^{\underline{6}}_{\underline{13}}2^{\underline{5}}_{\underline{11}}1^{\underline{5}}_{\underline{9}}5^{\underline{4}}_{\underline{9}}2^{\underline{4}}_{\underline{7}}7^{\underline{3}}_{\underline{7}}5^{\underline{3}}_{\underline{5}}10^{\underline{2}}_{\underline{5}}7^{\underline{2}}_{\underline{3}}10^{\underline{1}}_{\underline{3}}10^{\underline{1}}_{\underline{1}}10^{0}_{\underline{1}}11^{0}_{1}9^{1}_{1}9^{1}_{3}5^{2}_{3}9^{2}_{5}3^{3}_{5}5^{3}_{7}1^{4}_{7}3^{4}_{9}1^{5}_{11}$	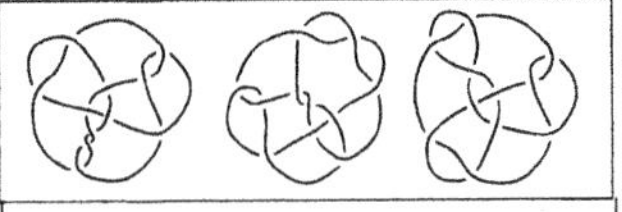		
11^a_{113}: $1^{\underline{8}}_{\underline{21}}2^{\underline{7}}_{\underline{19}}1^{\underline{7}}_{\underline{17}}4^{\underline{6}}_{\underline{17}}2^{\underline{6}}_{\underline{15}}7^{\underline{5}}_{\underline{15}}4^{\underline{5}}_{\underline{13}}8^{\underline{4}}_{\underline{13}}7^{\underline{4}}_{\underline{11}}9^{\underline{3}}_{\underline{11}}8^{\underline{3}}_{\underline{9}}8^{\underline{2}}_{\underline{9}}9^{\underline{2}}_{\underline{7}}7^{\underline{1}}_{\underline{7}}8^{\underline{1}}_{\underline{5}}5^{0}_{\underline{5}}8^{0}_{\underline{3}}3^{1}_{\underline{3}}4^{1}_{\underline{1}}1^{2}_{\underline{1}}3^{2}_{1}1^{3}_{3}$			
11^a_{114}: $1^{\underline{3}}_{\underline{5}}3^{\underline{2}}_{\underline{3}}1^{\underline{2}}_{\underline{1}}6^{\underline{1}}_{\underline{1}}3^{\underline{1}}_{1}10^{0}_{1}7^{0}_{3}12^{1}_{3}9^{1}_{5}13^{2}_{5}12^{2}_{7}11^{3}_{7}13^{3}_{9}10^{4}_{9}11^{4}_{11}6^{5}_{11}10^{5}_{13}3^{6}_{13}6^{6}_{15}1^{7}_{15}3^{7}_{17}1^{8}_{19}$			
11^a_{115}: $1^{\underline{4}}_{\underline{9}}2^{\underline{3}}_{\underline{7}}1^{\underline{3}}_{\underline{5}}5^{\underline{2}}_{\underline{5}}2^{\underline{2}}_{\underline{3}}7^{\underline{1}}_{\underline{3}}5^{\underline{1}}_{\underline{1}}10^{0}_{\underline{1}}8^{0}_{1}10^{1}_{1}9^{1}_{3}9^{2}_{3}10^{2}_{5}8^{3}_{5}9^{3}_{7}5^{4}_{7}8^{4}_{9}3^{5}_{9}5^{5}_{11}1^{6}_{11}3^{6}_{13}1^{7}_{15}$	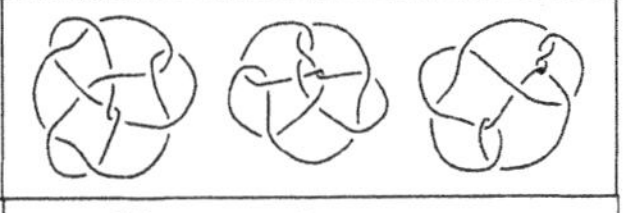		
11^a_{116}: $1^{\underline{2}}_{\underline{1}}2^{\underline{1}}_{1}1^{\underline{1}}_{3}6^{0}_{3}3^{0}_{5}8^{1}_{5}5^{1}_{7}11^{2}_{7}8^{2}_{9}11^{3}_{9}11^{3}_{11}11^{4}_{11}11^{4}_{13}9^{5}_{13}11^{5}_{15}6^{6}_{15}9^{6}_{17}3^{7}_{17}6^{7}_{19}1^{8}_{19}3^{8}_{21}1^{9}_{23}$			
11^a_{117}: $1^{\underline{2}}_{\underline{1}}2^{\underline{1}}_{1}1^{\underline{1}}_{3}5^{0}_{3}3^{0}_{5}7^{1}_{5}4^{1}_{7}9^{2}_{7}7^{2}_{9}9^{3}_{9}9^{3}_{11}10^{4}_{11}9^{4}_{13}7^{5}_{13}10^{5}_{15}5^{6}_{15}7^{6}_{17}3^{7}_{17}5^{7}_{19}1^{8}_{19}3^{8}_{21}1^{9}_{23}$			
11^a_{118}: $1^{\underline{4}}_{\underline{7}}1^{\underline{3}}_{\underline{5}}1^{\underline{3}}_{\underline{3}}4^{\underline{2}}_{\underline{3}}1^{\underline{2}}_{\underline{1}}4^{\underline{1}}_{\underline{1}}4^{\underline{1}}_{1}7^{0}_{1}5^{0}_{3}8^{1}_{3}6^{1}_{5}6^{2}_{5}8^{2}_{7}6^{3}_{7}6^{3}_{9}4^{4}_{9}6^{4}_{11}2^{5}_{11}4^{5}_{13}1^{6}_{13}2^{6}_{15}1^{7}_{17}$	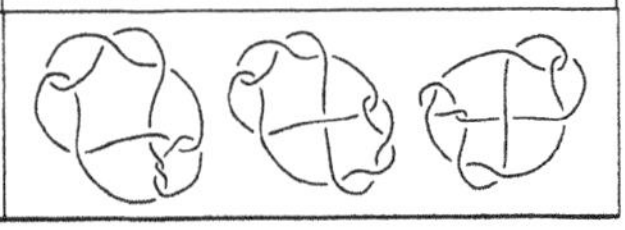		
11^a_{119}: $1^{\underline{4}}_{\underline{9}}1^{\underline{3}}_{\underline{7}}1^{\underline{3}}_{\underline{5}}4^{\underline{2}}_{\underline{5}}1^{\underline{2}}_{\underline{3}}4^{\underline{1}}_{\underline{3}}4^{\underline{1}}_{\underline{1}}6^{0}_{\underline{1}}5^{0}_{1}7^{1}_{1}5^{1}_{3}5^{2}_{3}7^{2}_{5}5^{3}_{5}5^{3}_{7}3^{4}_{7}5^{4}_{9}2^{5}_{9}3^{5}_{11}1^{6}_{11}2^{6}_{13}1^{7}_{15}$			
11^a_{120}: $1^{\underline{9}}_{\underline{23}}2^{\underline{8}}_{\underline{21}}1^{\underline{8}}_{\underline{19}}5^{\underline{7}}_{\underline{19}}2^{\underline{7}}_{\underline{17}}6^{\underline{6}}_{\underline{17}}5^{\underline{6}}_{\underline{15}}9^{\underline{5}}_{\underline{15}}6^{\underline{5}}_{\underline{13}}9^{\underline{4}}_{\underline{13}}9^{\underline{4}}_{\underline{11}}8^{\underline{3}}_{\underline{11}}9^{\underline{3}}_{\underline{9}}7^{\underline{2}}_{\underline{9}}8^{\underline{2}}_{\underline{7}}4^{\underline{1}}_{\underline{7}}7^{\underline{1}}_{\underline{5}}3^{0}_{\underline{5}}5^{0}_{\underline{3}}1^{1}_{\underline{3}}2^{1}_{\underline{1}}1^{2}_{\underline{1}}$			

K: KH	1st line 2nd line 3rd line
11^a_{121}: $1^{\underline{7}}_{\underline{17}} 2^{\underline{6}}_{\underline{15}} 1^{\underline{6}}_{\underline{13}} 5^{\underline{5}}_{\underline{13}} 2^{\underline{5}}_{\underline{11}} 7^{\underline{4}}_{\underline{11}} 5^{\underline{4}}_{\underline{9}} 9^{\underline{3}}_{\underline{9}} 7^{\underline{3}}_{\underline{7}} 10^{\underline{2}}_{\underline{7}} 9^{\underline{2}}_{\underline{5}} 9^{\underline{1}}_{\underline{5}} 10^{\underline{1}}_{\underline{3}} 8^{0}_{\underline{3}} 10^{0}_{\underline{1}} 5^{1}_{\underline{1}} 7^{1}_{1} 3^{2}_{1} 5^{2}_{3} 1^{3}_{3} 3^{3}_{5} 1^{4}_{7}$	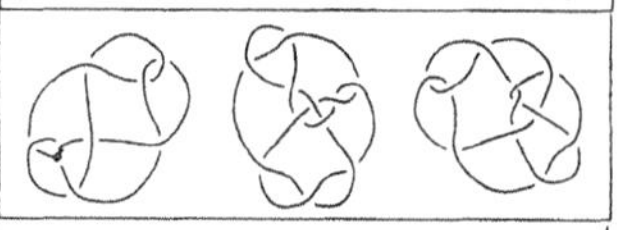
11^a_{122}: $1^{\underline{8}}_{\underline{19}} 2^{\underline{7}}_{\underline{17}} 1^{\underline{7}}_{\underline{15}} 5^{\underline{6}}_{\underline{15}} 2^{\underline{6}}_{\underline{13}} 8^{\underline{5}}_{\underline{13}} 5^{\underline{5}}_{\underline{11}} 9^{\underline{4}}_{\underline{11}} 8^{\underline{4}}_{\underline{9}} 11^{\underline{3}}_{\underline{9}} 9^{\underline{3}}_{\underline{7}} 10^{\underline{2}}_{\underline{7}} 11^{\underline{2}}_{\underline{5}} 8^{\underline{1}}_{\underline{5}} 10^{\underline{1}}_{\underline{3}} 6^{0}_{\underline{3}} 9^{0}_{\underline{1}} 3^{1}_{\underline{1}} 5^{1}_{1} 1^{2}_{1} 3^{2}_{3} 1^{3}_{5}$	
11^a_{123}: $1^{0}_{3} 1^{0}_{5} 3^{1}_{5} 4^{2}_{7} 3^{2}_{9} 7^{3}_{9} 4^{3}_{11} 10^{4}_{11} 7^{4}_{13} 8^{5}_{13} 10^{5}_{15} 10^{6}_{15} 8^{6}_{17} 7^{7}_{17} 10^{7}_{19} 5^{8}_{19} 7^{8}_{21} 3^{9}_{21} 5^{9}_{23} 1^{10}_{23} 3^{10}_{25} 1^{11}_{27}$	
11^a_{124}: $1^{0}_{5} 1^{0}_{7} 3^{1}_{7} 6^{2}_{9} 3^{2}_{11} 8^{3}_{11} 6^{3}_{13} 13^{4}_{13} 8^{4}_{15} 12^{5}_{15} 13^{5}_{17} 13^{6}_{17} 12^{6}_{19} 10^{7}_{19} 13^{7}_{21} 7^{8}_{21} 10^{8}_{23} 4^{9}_{23} 7^{9}_{25} 1^{10}_{25} 4^{10}_{27} 1^{11}_{29}$	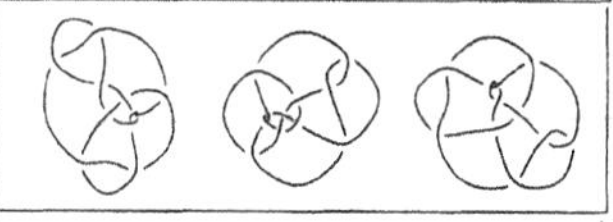
11^a_{125}: $1^{\underline{4}}_{\underline{7}} 3^{\underline{3}}_{\underline{5}} 1^{\underline{3}}_{\underline{3}} 7^{\underline{2}}_{\underline{3}} 3^{\underline{2}}_{\underline{1}} 10^{\underline{1}}_{\underline{1}} 7^{\underline{1}}_{1} 14^{0}_{1} 11^{0}_{3} 15^{1}_{3} 13^{1}_{5} 14^{2}_{5} 15^{2}_{7} 11^{3}_{7} 14^{3}_{9} 8^{4}_{9} 11^{4}_{11} 4^{5}_{11} 8^{5}_{13} 1^{6}_{13} 4^{6}_{15} 1^{7}_{17}$	
11^a_{126}: $1^{\underline{5}}_{\underline{11}} 2^{\underline{4}}_{\underline{9}} 1^{\underline{4}}_{\underline{7}} 6^{\underline{3}}_{\underline{7}} 2^{\underline{3}}_{\underline{5}} 8^{\underline{2}}_{\underline{5}} 6^{\underline{2}}_{\underline{3}} 11^{\underline{1}}_{\underline{3}} 8^{\underline{1}}_{\underline{1}} 13^{0}_{\underline{1}} 12^{0}_{1} 11^{1}_{1} 12^{1}_{3} 10^{2}_{3} 11^{2}_{5} 6^{3}_{5} 10^{3}_{7} 4^{4}_{7} 6^{4}_{9} 1^{5}_{9} 4^{5}_{11} 1^{6}_{13}$	
11^a_{127}: $1^{\underline{3}}_{\underline{3}} 3^{\underline{2}}_{\underline{1}} 1^{\underline{2}}_{1} 5^{\underline{1}}_{1} 3^{\underline{1}}_{3} 9^{0}_{3} 6^{0}_{5} 10^{1}_{5} 8^{1}_{7} 12^{2}_{7} 10^{2}_{9} 10^{3}_{9} 12^{3}_{11} 9^{4}_{11} 10^{4}_{13} 6^{5}_{13} 9^{5}_{15} 3^{6}_{15} 6^{6}_{17} 1^{7}_{17} 3^{7}_{19} 1^{8}_{21}$	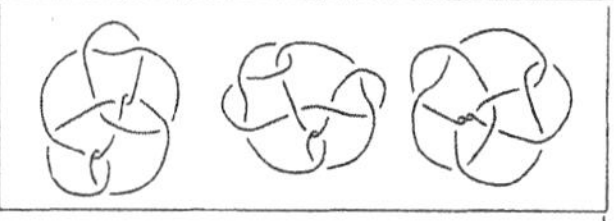
11^a_{128}: $1^{\underline{6}}_{\underline{13}} 2^{\underline{5}}_{\underline{11}} 1^{\underline{5}}_{\underline{9}} 6^{\underline{4}}_{\underline{9}} 2^{\underline{4}}_{\underline{7}} 7^{\underline{3}}_{\underline{7}} 6^{\underline{3}}_{\underline{5}} 10^{\underline{2}}_{\underline{5}} 7^{\underline{2}}_{\underline{3}} 11^{\underline{1}}_{\underline{3}} 10^{\underline{1}}_{\underline{1}} 10^{0}_{\underline{1}} 12^{0}_{1} 9^{1}_{1} 9^{1}_{3} 5^{2}_{3} 9^{2}_{5} 3^{3}_{5} 5^{3}_{7} 1^{4}_{7} 3^{4}_{9} 1^{5}_{11}$	
11^a_{129}: $1^{\underline{8}}_{\underline{21}} 2^{\underline{7}}_{\underline{19}} 1^{\underline{7}}_{\underline{17}} 5^{\underline{6}}_{\underline{17}} 2^{\underline{6}}_{\underline{15}} 7^{\underline{5}}_{\underline{15}} 5^{\underline{5}}_{\underline{13}} 8^{\underline{4}}_{\underline{13}} 7^{\underline{4}}_{\underline{11}} 10^{\underline{3}}_{\underline{11}} 8^{\underline{3}}_{\underline{9}} 8^{\underline{2}}_{\underline{9}} 10^{\underline{2}}_{\underline{7}} 7^{\underline{1}}_{\underline{7}} 8^{\underline{1}}_{\underline{5}} 5^{0}_{\underline{5}} 8^{0}_{\underline{3}} 3^{1}_{\underline{3}} 4^{1}_{\underline{1}} 1^{2}_{\underline{1}} 3^{2}_{1} 1^{3}_{3}$	
11^a_{130}: $1^{\underline{8}}_{\underline{19}} 2^{\underline{7}}_{\underline{17}} 1^{\underline{7}}_{\underline{15}} 5^{\underline{6}}_{\underline{15}} 2^{\underline{6}}_{\underline{13}} 7^{\underline{5}}_{\underline{13}} 5^{\underline{5}}_{\underline{11}} 9^{\underline{4}}_{\underline{11}} 7^{\underline{4}}_{\underline{9}} 11^{\underline{3}}_{\underline{9}} 9^{\underline{3}}_{\underline{7}} 9^{\underline{2}}_{\underline{7}} 11^{\underline{2}}_{\underline{5}} 8^{\underline{1}}_{\underline{5}} 9^{\underline{1}}_{\underline{3}} 6^{0}_{\underline{3}} 9^{0}_{1} 3^{1}_{\underline{1}} 5^{1}_{1} 1^{2}_{1} 3^{2}_{3} 1^{3}_{5}$	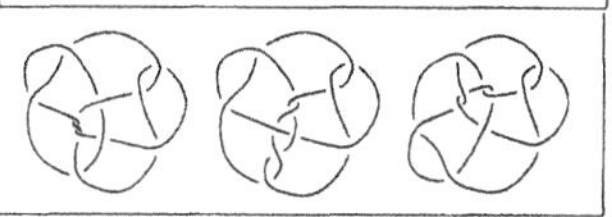
11^a_{131}: $1^{\underline{7}}_{\underline{17}} 2^{\underline{6}}_{\underline{15}} 1^{\underline{6}}_{\underline{13}} 5^{\underline{5}}_{\underline{13}} 2^{\underline{5}}_{\underline{11}} 8^{\underline{4}}_{\underline{11}} 5^{\underline{4}}_{\underline{9}} 10^{\underline{3}}_{\underline{9}} 8^{\underline{3}}_{\underline{7}} 11^{\underline{2}}_{\underline{7}} 10^{\underline{2}}_{\underline{5}} 10^{\underline{1}}_{\underline{5}} 11^{\underline{1}}_{\underline{3}} 9^{0}_{\underline{3}} 11^{0}_{\underline{1}} 6^{1}_{\underline{1}} 8^{1}_{1} 3^{2}_{1} 6^{2}_{3} 1^{3}_{3} 3^{3}_{5} 1^{4}_{7}$	
11^a_{132}: $1^{\underline{4}}_{\underline{7}} 2^{\underline{3}}_{\underline{5}} 1^{\underline{3}}_{\underline{3}} 6^{\underline{2}}_{\underline{3}} 2^{\underline{2}}_{\underline{1}} 7^{\underline{1}}_{\underline{1}} 6^{\underline{1}}_{1} 11^{0}_{1} 8^{0}_{3} 12^{1}_{3} 10^{1}_{5} 10^{2}_{5} 12^{2}_{7} 9^{3}_{7} 10^{3}_{9} 6^{4}_{9} 9^{4}_{11} 3^{5}_{11} 6^{5}_{13} 1^{6}_{13} 3^{6}_{15} 1^{7}_{17}$	
11^a_{133}: $1^{\underline{2}}_{\underline{3}} 1^{\underline{1}}_{\underline{1}} 1^{\underline{1}}_{1} 4^{0}_{1} 2^{0}_{3} 5^{1}_{3} 3^{1}_{5} 6^{2}_{5} 5^{2}_{7} 6^{3}_{7} 6^{3}_{9} 6^{4}_{9} 6^{4}_{11} 5^{5}_{11} 6^{5}_{13} 3^{6}_{13} 5^{6}_{15} 2^{7}_{15} 3^{7}_{17} 1^{8}_{17} 2^{8}_{19} 1^{9}_{21}$	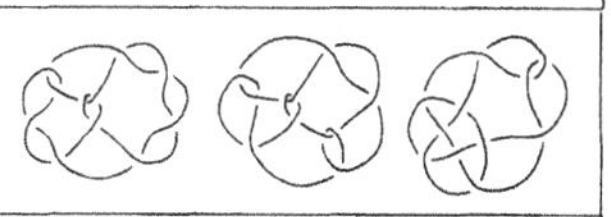
11^a_{134}: $1^{\underline{3}}_{\underline{5}} 2^{\underline{2}}_{\underline{3}} 1^{\underline{2}}_{\underline{1}} 5^{\underline{1}}_{\underline{1}} 2^{\underline{1}}_{1} 8^{0}_{1} 6^{0}_{3} 9^{1}_{3} 7^{1}_{5} 11^{2}_{5} 9^{2}_{7} 9^{3}_{7} 11^{3}_{9} 8^{4}_{9} 9^{4}_{11} 5^{5}_{11} 8^{5}_{13} 3^{6}_{13} 5^{6}_{15} 1^{7}_{15} 3^{7}_{17} 1^{8}_{19}$	
11^a_{135}: $1^{\underline{5}}_{\underline{11}} 3^{\underline{4}}_{\underline{9}} 1^{\underline{4}}_{\underline{7}} 5^{\underline{3}}_{\underline{7}} 3^{\underline{3}}_{\underline{5}} 10^{\underline{2}}_{\underline{5}} 5^{\underline{2}}_{\underline{3}} 11^{\underline{1}}_{\underline{3}} 10^{\underline{1}}_{\underline{1}} 13^{0}_{\underline{1}} 12^{0}_{1} 13^{1}_{1} 12^{1}_{3} 9^{2}_{3} 13^{2}_{5} 7^{3}_{5} 9^{3}_{7} 4^{4}_{7} 7^{4}_{9} 1^{5}_{9} 4^{5}_{11} 1^{6}_{13}$	
11^a_{136}: $1^{\underline{8}}_{\underline{19}} 3^{\underline{7}}_{\underline{17}} 1^{\underline{7}}_{\underline{15}} 6^{\underline{6}}_{\underline{15}} 3^{\underline{6}}_{\underline{13}} 10^{\underline{5}}_{\underline{13}} 6^{\underline{5}}_{\underline{11}} 12^{\underline{4}}_{\underline{11}} 10^{\underline{4}}_{\underline{9}} 14^{\underline{3}}_{\underline{9}} 12^{\underline{3}}_{\underline{7}} 13^{\underline{2}}_{\underline{7}} 14^{\underline{2}}_{\underline{5}} 10^{\underline{1}}_{\underline{5}} 13^{\underline{1}}_{\underline{3}} 8^{0}_{\underline{3}} 11^{0}_{\underline{1}} 4^{1}_{\underline{1}} 7^{1}_{1} 1^{2}_{1} 4^{2}_{3} 1^{3}_{5}$	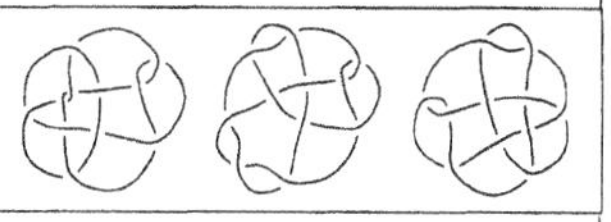
11^a_{137}: $1^{\underline{6}}_{\underline{15}} 2^{\underline{5}}_{\underline{13}} 1^{\underline{5}}_{\underline{11}} 5^{\underline{4}}_{\underline{11}} 2^{\underline{4}}_{\underline{9}} 6^{\underline{3}}_{\underline{9}} 5^{\underline{3}}_{\underline{7}} 9^{\underline{2}}_{\underline{7}} 6^{\underline{2}}_{\underline{5}} 9^{\underline{1}}_{\underline{5}} 9^{\underline{1}}_{\underline{3}} 8^{0}_{\underline{3}} 10^{0}_{\underline{1}} 8^{1}_{\underline{1}} 7^{1}_{1} 4^{2}_{1} 8^{2}_{3} 3^{3}_{3} 4^{3}_{5} 1^{4}_{5} 3^{4}_{7} 1^{5}_{9}$	
11^a_{138}: $1^{\underline{5}}_{\underline{11}} 3^{\underline{4}}_{\underline{9}} 1^{\underline{4}}_{\underline{7}} 6^{\underline{3}}_{\underline{7}} 3^{\underline{3}}_{\underline{5}} 10^{\underline{2}}_{\underline{5}} 6^{\underline{2}}_{\underline{3}} 12^{\underline{1}}_{\underline{3}} 10^{\underline{1}}_{\underline{1}} 14^{0}_{\underline{1}} 13^{0}_{1} 13^{1}_{1} 13^{1}_{3} 10^{2}_{3} 13^{2}_{5} 7^{3}_{5} 10^{3}_{7} 4^{4}_{7} 7^{4}_{9} 1^{5}_{9} 4^{5}_{11} 1^{6}_{13}$	
11^a_{139}: $1^{\underline{4}}_{\underline{7}} 2^{\underline{3}}_{\underline{5}} 1^{\underline{3}}_{\underline{3}} 4^{\underline{2}}_{\underline{3}} 2^{\underline{2}}_{\underline{1}} 6^{\underline{1}}_{\underline{1}} 4^{\underline{1}}_{1} 8^{0}_{1} 7^{0}_{3} 8^{1}_{3} 7^{1}_{5} 8^{2}_{5} 8^{2}_{7} 6^{3}_{7} 8^{3}_{9} 4^{4}_{9} 6^{4}_{11} 2^{5}_{11} 4^{5}_{13} 1^{6}_{13} 2^{6}_{15} 1^{7}_{17}$	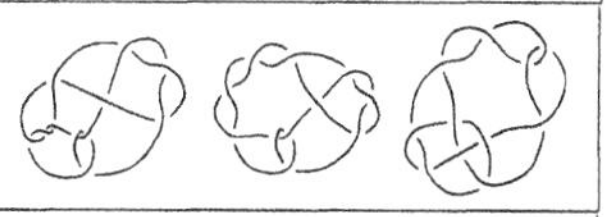
11^a_{140}: $1^{\underline{4}}_{\underline{5}} 1^{\underline{3}}_{\underline{3}} 1^{\underline{3}}_{\underline{1}} 3^{\underline{2}}_{\underline{1}} 1^{\underline{2}}_{1} 3^{\underline{1}}_{1} 3^{\underline{1}}_{3} 5^{0}_{3} 4^{0}_{5} 5^{1}_{5} 4^{1}_{7} 5^{2}_{7} 5^{2}_{9} 4^{3}_{9} 5^{3}_{11} 3^{4}_{11} 4^{4}_{13} 2^{5}_{13} 3^{5}_{15} 1^{6}_{15} 2^{6}_{17} 1^{7}_{19}$	
11^a_{141}: $1^{\underline{6}}_{\underline{15}} 2^{\underline{5}}_{\underline{13}} 1^{\underline{5}}_{\underline{11}} 4^{\underline{4}}_{\underline{11}} 2^{\underline{4}}_{\underline{9}} 6^{\underline{3}}_{\underline{9}} 4^{\underline{3}}_{\underline{7}} 8^{\underline{2}}_{\underline{7}} 6^{\underline{2}}_{\underline{5}} 8^{\underline{1}}_{\underline{5}} 8^{\underline{1}}_{\underline{3}} 8^{0}_{\underline{3}} 9^{0}_{\underline{1}} 7^{1}_{\underline{1}} 7^{1}_{1} 4^{2}_{1} 7^{2}_{3} 3^{3}_{3} 4^{3}_{5} 1^{4}_{5} 3^{4}_{7} 1^{5}_{9}$	
11^a_{142}: $1^{\underline{9}}_{\underline{25}} 1^{\underline{8}}_{\underline{23}} 1^{\underline{8}}_{\underline{21}} 3^{\underline{7}}_{\underline{21}} 1^{\underline{7}}_{\underline{19}} 3^{\underline{6}}_{\underline{19}} 3^{\underline{6}}_{\underline{17}} 5^{\underline{5}}_{\underline{17}} 3^{\underline{5}}_{\underline{15}} 4^{\underline{4}}_{\underline{15}} 5^{\underline{4}}_{\underline{13}} 4^{\underline{3}}_{\underline{13}} 4^{\underline{3}}_{\underline{11}} 4^{\underline{2}}_{\underline{11}} 4^{\underline{2}}_{\underline{9}} 2^{\underline{1}}_{\underline{9}} 4^{\underline{1}}_{\underline{7}} 2^{0}_{\underline{7}} 3^{0}_{\underline{5}} 1^{1}_{\underline{5}} 1^{1}_{\underline{3}} 1^{2}_{\underline{1}}$	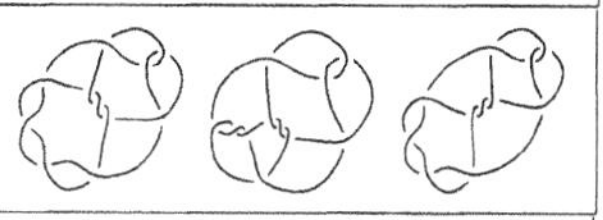
11^a_{143}: $1^{\underline{9}}_{\underline{23}} 1^{\underline{8}}_{\underline{21}} 1^{\underline{8}}_{\underline{19}} 4^{\underline{7}}_{\underline{19}} 1^{\underline{7}}_{\underline{17}} 5^{\underline{6}}_{\underline{17}} 4^{\underline{6}}_{\underline{15}} 7^{\underline{5}}_{\underline{15}} 5^{\underline{5}}_{\underline{13}} 7^{\underline{4}}_{\underline{13}} 7^{\underline{4}}_{\underline{11}} 7^{\underline{3}}_{\underline{11}} 7^{\underline{3}}_{\underline{9}} 6^{\underline{2}}_{\underline{9}} 7^{\underline{2}}_{\underline{7}} 3^{\underline{1}}_{\underline{7}} 6^{\underline{1}}_{\underline{5}} 3^{0}_{\underline{5}} 4^{0}_{\underline{3}} 1^{1}_{\underline{3}} 2^{1}_{\underline{1}} 1^{2}_{1}$	
11^a_{144}: $1^{\underline{9}}_{\underline{23}} 1^{\underline{8}}_{\underline{21}} 1^{\underline{8}}_{\underline{19}} 4^{\underline{7}}_{\underline{19}} 1^{\underline{7}}_{\underline{17}} 4^{\underline{6}}_{\underline{17}} 4^{\underline{6}}_{\underline{15}} 6^{\underline{5}}_{\underline{15}} 4^{\underline{5}}_{\underline{13}} 6^{\underline{4}}_{\underline{13}} 6^{\underline{4}}_{\underline{11}} 5^{\underline{3}}_{\underline{11}} 6^{\underline{3}}_{\underline{9}} 5^{\underline{2}}_{\underline{9}} 5^{\underline{2}}_{\underline{7}} 2^{\underline{1}}_{\underline{7}} 5^{\underline{1}}_{\underline{5}} 2^{0}_{\underline{5}} 3^{0}_{\underline{3}} 1^{1}_{\underline{3}} 1^{1}_{\underline{1}} 1^{2}_{1}$	
11^a_{145}: $1^{\underline{9}}_{\underline{21}} 1^{\underline{8}}_{\underline{19}} 1^{\underline{8}}_{\underline{17}} 4^{\underline{7}}_{\underline{17}} 1^{\underline{7}}_{\underline{15}} 4^{\underline{6}}_{\underline{15}} 4^{\underline{6}}_{\underline{13}} 6^{\underline{5}}_{\underline{13}} 4^{\underline{5}}_{\underline{11}} 7^{\underline{4}}_{\underline{11}} 6^{\underline{4}}_{\underline{9}} 6^{\underline{3}}_{\underline{9}} 7^{\underline{3}}_{\underline{7}} 6^{\underline{2}}_{\underline{7}} 6^{\underline{2}}_{\underline{5}} 3^{\underline{1}}_{\underline{5}} 6^{\underline{1}}_{\underline{3}} 3^{0}_{\underline{3}} 4^{0}_{\underline{1}} 1^{1}_{\underline{1}} 2^{1}_{1} 1^{2}_{3}$	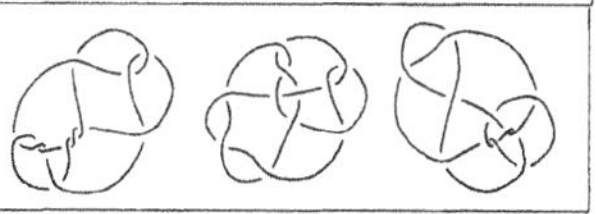
11^a_{146}: $1^{\underline{7}}_{\underline{17}} 2^{\underline{6}}_{\underline{15}} 1^{\underline{6}}_{\underline{13}} 5^{\underline{5}}_{\underline{13}} 2^{\underline{5}}_{\underline{11}} 7^{\underline{4}}_{\underline{11}} 5^{\underline{4}}_{\underline{9}} 10^{\underline{3}}_{\underline{9}} 7^{\underline{3}}_{\underline{7}} 10^{\underline{2}}_{\underline{7}} 10^{\underline{2}}_{\underline{5}} 9^{\underline{1}}_{\underline{5}} 10^{\underline{1}}_{\underline{3}} 9^{0}_{\underline{3}} 10^{0}_{\underline{1}} 5^{1}_{\underline{1}} 8^{1}_{1} 3^{2}_{1} 5^{2}_{3} 1^{3}_{3} 3^{3}_{5} 1^{4}_{7}$	
11^a_{147}: $1^{\underline{4}}_{\underline{7}} 3^{\underline{3}}_{\underline{5}} 1^{\underline{3}}_{\underline{3}} 6^{\underline{2}}_{\underline{3}} 3^{\underline{2}}_{\underline{1}} 9^{\underline{1}}_{\underline{1}} 6^{\underline{1}}_{1} 12^{0}_{1} 10^{0}_{3} 13^{1}_{3} 11^{1}_{5} 12^{2}_{5} 13^{2}_{7} 9^{3}_{7} 12^{3}_{9} 7^{4}_{9} 9^{4}_{11} 3^{5}_{11} 7^{5}_{13} 1^{6}_{13} 3^{6}_{15} 1^{7}_{17}$	
11^a_{148}: $1^{\underline{2}}_{\underline{3}} 2^{\underline{1}}_{\underline{1}} 1^{\underline{1}}_{1} 5^{0}_{1} 3^{0}_{3} 7^{1}_{3} 4^{1}_{5} 9^{2}_{5} 7^{2}_{7} 9^{3}_{7} 9^{3}_{9} 9^{4}_{9} 9^{4}_{11} 7^{5}_{11} 9^{5}_{13} 5^{6}_{13} 7^{6}_{15} 3^{7}_{15} 5^{7}_{17} 1^{8}_{17} 3^{8}_{19} 1^{9}_{21}$	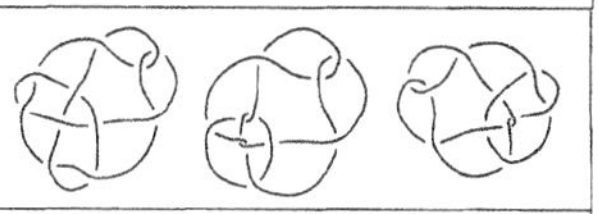
11^a_{149}: $1^{\underline{4}}_{\underline{7}} 2^{\underline{3}}_{\underline{5}} 1^{\underline{3}}_{\underline{3}} 5^{\underline{2}}_{\underline{3}} 2^{\underline{2}}_{\underline{1}} 7^{\underline{1}}_{\underline{1}} 5^{\underline{1}}_{1} 10^{0}_{1} 8^{0}_{3} 11^{1}_{3} 9^{1}_{5} 10^{2}_{5} 11^{2}_{7} 8^{3}_{7} 10^{3}_{9} 6^{4}_{9} 8^{4}_{11} 3^{5}_{11} 6^{5}_{13} 1^{6}_{13} 3^{6}_{15} 1^{7}_{17}$	
11^a_{150}: $1^{\underline{9}}_{\underline{23}} 2^{\underline{8}}_{\underline{21}} 1^{\underline{8}}_{\underline{19}} 5^{\underline{7}}_{\underline{19}} 2^{\underline{7}}_{\underline{17}} 7^{\underline{6}}_{\underline{17}} 5^{\underline{6}}_{\underline{15}} 10^{\underline{5}}_{\underline{15}} 7^{\underline{5}}_{\underline{13}} 10^{\underline{4}}_{\underline{13}} 10^{\underline{4}}_{\underline{11}} 10^{\underline{3}}_{\underline{11}} 10^{\underline{3}}_{\underline{9}} 8^{\underline{2}}_{\underline{9}} 10^{\underline{2}}_{\underline{7}} 5^{\underline{1}}_{\underline{7}} 8^{\underline{1}}_{\underline{5}} 4^{0}_{\underline{5}} 6^{0}_{\underline{3}} 1^{1}_{\underline{3}} 3^{1}_{\underline{1}} 1^{2}_{1}$	
11^a_{151}: $1^{\underline{7}}_{\underline{17}} 2^{\underline{6}}_{\underline{15}} 1^{\underline{6}}_{\underline{13}} 6^{\underline{5}}_{\underline{13}} 2^{\underline{5}}_{\underline{11}} 7^{\underline{4}}_{\underline{11}} 6^{\underline{4}}_{\underline{9}} 10^{\underline{3}}_{\underline{9}} 7^{\underline{3}}_{\underline{7}} 11^{\underline{2}}_{\underline{7}} 10^{\underline{2}}_{\underline{5}} 9^{\underline{1}}_{\underline{5}} 11^{\underline{1}}_{\underline{3}} 9^{0}_{\underline{3}} 10^{0}_{\underline{1}} 5^{1}_{\underline{1}} 8^{1}_{1} 3^{2}_{1} 5^{2}_{3} 1^{3}_{3} 3^{3}_{5} 1^{4}_{7}$	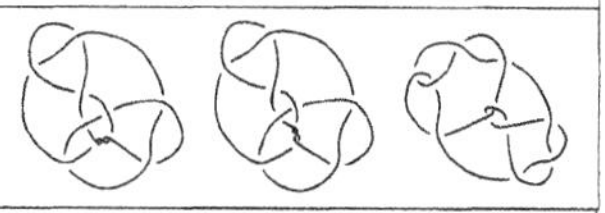
11^a_{152}: $1^{\underline{7}}_{\underline{15}} 2^{\underline{6}}_{\underline{13}} 1^{\underline{6}}_{\underline{11}} 5^{\underline{5}}_{\underline{11}} 2^{\underline{5}}_{\underline{9}} 6^{\underline{4}}_{\underline{9}} 5^{\underline{4}}_{\underline{7}} 9^{\underline{3}}_{\underline{7}} 6^{\underline{3}}_{\underline{5}} 10^{\underline{2}}_{\underline{5}} 9^{\underline{2}}_{\underline{3}} 8^{\underline{1}}_{\underline{3}} 10^{\underline{1}}_{\underline{1}} 9^{0}_{\underline{1}} 9^{0}_{1} 5^{1}_{1} 8^{1}_{3} 3^{2}_{3} 5^{2}_{5} 1^{3}_{5} 3^{3}_{7} 1^{4}_{9}$	
11^a_{153}: $1^{\underline{4}}_{\underline{9}} 2^{\underline{3}}_{\underline{7}} 1^{\underline{3}}_{\underline{5}} 4^{\underline{2}}_{\underline{5}} 2^{\underline{2}}_{\underline{3}} 6^{\underline{1}}_{\underline{3}} 4^{\underline{1}}_{\underline{1}} 7^{0}_{\underline{1}} 7^{0}_{1} 8^{1}_{1} 6^{1}_{3} 6^{2}_{3} 8^{2}_{5} 5^{3}_{5} 6^{3}_{7} 4^{4}_{7} 5^{4}_{9} 1^{5}_{9} 4^{5}_{11} 1^{6}_{11} 1^{6}_{13} 1^{7}_{15}$	
11^a_{154}: $1^{\underline{4}}_{\underline{7}} 1^{\underline{3}}_{\underline{5}} 1^{\underline{3}}_{\underline{3}} 3^{\underline{2}}_{\underline{3}} 1^{\underline{2}}_{\underline{1}} 4^{\underline{1}}_{\underline{1}} 3^{\underline{1}}_{1} 5^{0}_{1} 5^{0}_{3} 6^{1}_{3} 4^{1}_{5} 5^{2}_{5} 6^{2}_{7} 4^{3}_{7} 5^{3}_{9} 3^{4}_{9} 4^{4}_{11} 1^{5}_{11} 3^{5}_{13} 1^{6}_{13} 1^{6}_{15} 1^{7}_{17}$	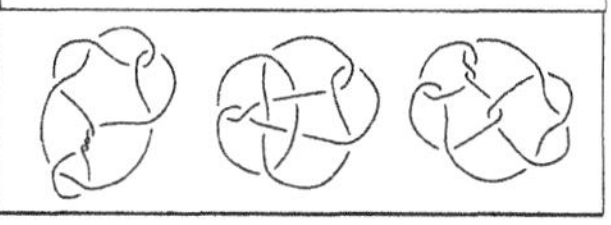
11^a_{155}: $1^{\underline{8}}_{\underline{19}} 3^{\underline{7}}_{\underline{17}} 1^{\underline{7}}_{\underline{15}} 6^{\underline{6}}_{\underline{15}} 3^{\underline{6}}_{\underline{13}} 11^{\underline{5}}_{\underline{13}} 6^{\underline{5}}_{\underline{11}} 12^{\underline{4}}_{\underline{11}} 11^{\underline{4}}_{\underline{9}} 15^{\underline{3}}_{\underline{9}} 12^{\underline{3}}_{\underline{7}} 14^{\underline{2}}_{\underline{7}} 15^{\underline{2}}_{\underline{5}} 10^{\underline{1}}_{\underline{5}} 14^{\underline{1}}_{\underline{3}} 9^{0}_{\underline{3}} 11^{0}_{\underline{1}} 4^{1}_{\underline{1}} 8^{1}_{1} 1^{2}_{1} 4^{2}_{3} 1^{3}_{5}$	
11^a_{156}: $1^{\underline{7}}_{\underline{17}} 1^{\underline{6}}_{\underline{15}} 1^{\underline{6}}_{\underline{13}} 4^{\underline{5}}_{\underline{13}} 1^{\underline{5}}_{\underline{11}} 5^{\underline{4}}_{\underline{11}} 4^{\underline{4}}_{\underline{9}} 7^{\underline{3}}_{\underline{9}} 5^{\underline{3}}_{\underline{7}} 8^{\underline{2}}_{\underline{7}} 7^{\underline{2}}_{\underline{5}} 6^{\underline{1}}_{\underline{5}} 8^{\underline{1}}_{\underline{3}} 7^{0}_{\underline{3}} 7^{0}_{\underline{1}} 4^{1}_{\underline{1}} 6^{1}_{1} 2^{2}_{1} 4^{2}_{3} 1^{3}_{3} 2^{3}_{5} 1^{4}_{7}$	

K: KH	1st line 2nd line 3rd line
11^a_{157}: $1^{\underline{7}}_{\underline{17}}2^{\underline{6}}_{\underline{15}}1^{\underline{6}}_{\underline{13}}6^{\underline{5}}_{\underline{13}}2^{\underline{5}}_{\underline{11}}8^{\underline{4}}_{\underline{11}}6^{\underline{4}}_{\underline{9}}10^{\underline{3}}_{\underline{9}}8^{\underline{3}}_{\underline{7}}12^{\underline{2}}_{\underline{7}}10^{\underline{2}}_{\underline{5}}10^{\underline{1}}_{\underline{5}}12^{\underline{1}}_{\underline{3}}9^{0}_{\underline{3}}11^{0}_{\underline{1}}6^{1}_{\underline{1}}8^{1}_{1}3^{2}_{1}6^{2}_{3}1^{3}_{3}3^{3}_{5}1^{4}_{7}$	
11^a_{158}: $1^{\underline{6}}_{\underline{15}}2^{\underline{5}}_{\underline{13}}1^{\underline{5}}_{\underline{11}}5^{\underline{4}}_{\underline{11}}2^{\underline{4}}_{\underline{9}}7^{\underline{3}}_{\underline{9}}5^{\underline{3}}_{\underline{7}}9^{\underline{2}}_{\underline{7}}7^{\underline{2}}_{\underline{5}}10^{\underline{1}}_{\underline{5}}9^{\underline{1}}_{\underline{3}}9^{0}_{\underline{3}}11^{0}_{\underline{1}}8^{1}_{\underline{1}}8^{1}_{1}5^{2}_{1}8^{2}_{3}3^{3}_{3}5^{3}_{5}1^{4}_{5}3^{4}_{7}1^{5}_{9}$	
11^a_{159}: $1^{\underline{4}}_{\underline{7}}2^{\underline{3}}_{\underline{5}}1^{\underline{3}}_{\underline{3}}5^{\underline{2}}_{\underline{3}}2^{\underline{2}}_{\underline{1}}6^{\underline{1}}_{\underline{1}}5^{\underline{1}}_{1}9^{0}_{1}7^{0}_{3}10^{1}_{3}8^{1}_{5}8^{2}_{5}10^{2}_{7}7^{3}_{7}8^{3}_{9}5^{4}_{9}7^{4}_{11}2^{5}_{11}5^{5}_{13}1^{6}_{13}2^{6}_{15}1^{7}_{17}$	
11^a_{160}: $1^{\underline{5}}_{\underline{11}}3^{\underline{4}}_{\underline{9}}1^{\underline{4}}_{\underline{7}}6^{\underline{3}}_{\underline{7}}3^{\underline{3}}_{\underline{5}}9^{\underline{2}}_{\underline{5}}6^{\underline{2}}_{\underline{3}}11^{\underline{1}}_{\underline{3}}9^{\underline{1}}_{\underline{1}}13^{0}_{\underline{1}}12^{0}_{1}11^{1}_{1}12^{1}_{3}9^{2}_{3}11^{2}_{5}6^{3}_{5}9^{3}_{7}3^{4}_{7}6^{4}_{9}1^{5}_{9}3^{5}_{11}1^{6}_{13}$	
11^a_{161}: $1^{\underline{7}}_{\underline{19}}1^{\underline{6}}_{\underline{17}}1^{\underline{6}}_{\underline{15}}3^{\underline{5}}_{\underline{15}}1^{\underline{5}}_{\underline{13}}3^{\underline{4}}_{\underline{13}}3^{\underline{4}}_{\underline{11}}4^{\underline{3}}_{\underline{11}}3^{\underline{3}}_{\underline{9}}5^{\underline{2}}_{\underline{9}}4^{\underline{2}}_{\underline{7}}3^{\underline{1}}_{\underline{7}}5^{\underline{1}}_{\underline{5}}4^{0}_{\underline{5}}4^{0}_{\underline{3}}3^{1}_{\underline{3}}3^{1}_{\underline{1}}1^{2}_{\underline{1}}3^{2}_{1}1^{3}_{1}1^{3}_{3}1^{4}_{5}$	
11^a_{162}: $1^{\underline{7}}_{\underline{17}}3^{\underline{6}}_{\underline{15}}1^{\underline{6}}_{\underline{13}}7^{\underline{5}}_{\underline{13}}3^{\underline{5}}_{\underline{11}}10^{\underline{4}}_{\underline{11}}7^{\underline{4}}_{\underline{9}}13^{\underline{3}}_{\underline{9}}10^{\underline{3}}_{\underline{7}}14^{\underline{2}}_{\underline{7}}13^{\underline{2}}_{\underline{5}}13^{\underline{1}}_{\underline{5}}14^{\underline{1}}_{\underline{3}}11^{0}_{\underline{3}}14^{0}_{\underline{1}}7^{1}_{\underline{1}}10^{1}_{1}4^{2}_{1}7^{2}_{3}1^{3}_{3}4^{3}_{5}1^{4}_{7}$	
11^a_{163}: $1^{\underline{5}}_{\underline{9}}2^{\underline{4}}_{\underline{7}}1^{\underline{4}}_{\underline{5}}5^{\underline{3}}_{\underline{5}}2^{\underline{3}}_{\underline{3}}7^{\underline{2}}_{\underline{3}}5^{\underline{2}}_{\underline{1}}8^{\underline{1}}_{\underline{1}}7^{\underline{1}}_{1}11^{0}_{1}9^{0}_{3}9^{1}_{3}10^{1}_{5}8^{2}_{5}9^{2}_{7}5^{3}_{7}8^{3}_{9}3^{4}_{9}5^{4}_{11}1^{5}_{11}3^{5}_{13}1^{6}_{15}$	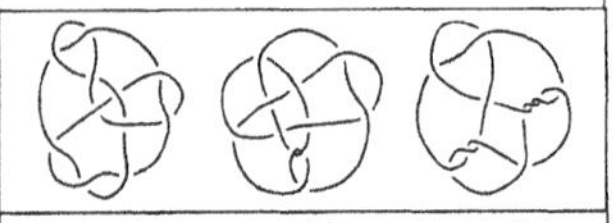
11^a_{164}: $1^{\underline{6}}_{\underline{13}}3^{\underline{5}}_{\underline{11}}1^{\underline{5}}_{\underline{9}}7^{\underline{4}}_{\underline{9}}3^{\underline{4}}_{\underline{7}}10^{\underline{3}}_{\underline{7}}7^{\underline{3}}_{\underline{5}}13^{\underline{2}}_{\underline{5}}10^{\underline{2}}_{\underline{3}}14^{\underline{1}}_{\underline{3}}13^{\underline{1}}_{\underline{1}}14^{0}_{\underline{1}}15^{0}_{1}11^{1}_{1}13^{1}_{3}7^{2}_{3}11^{2}_{5}4^{3}_{5}7^{3}_{7}1^{4}_{7}4^{4}_{9}1^{5}_{11}$	
11^a_{165}: $1^{\underline{4}}_{\underline{9}}2^{\underline{3}}_{\underline{7}}1^{\underline{3}}_{\underline{5}}4^{\underline{2}}_{\underline{5}}2^{\underline{2}}_{\underline{3}}5^{\underline{1}}_{\underline{3}}4^{\underline{1}}_{\underline{1}}7^{0}_{\underline{1}}6^{0}_{1}7^{1}_{1}6^{1}_{3}5^{2}_{3}7^{2}_{5}5^{3}_{5}5^{3}_{7}3^{4}_{7}5^{4}_{9}1^{5}_{9}3^{5}_{11}1^{6}_{11}1^{6}_{13}1^{7}_{15}$	
11^a_{166}: $1^{\underline{4}}_{\underline{7}}1^{\underline{3}}_{\underline{5}}1^{\underline{3}}_{\underline{3}}3^{\underline{2}}_{\underline{3}}1^{\underline{2}}_{\underline{1}}3^{\underline{1}}_{\underline{1}}3^{\underline{1}}_{1}5^{0}_{1}4^{0}_{3}5^{1}_{3}4^{1}_{5}4^{2}_{5}5^{2}_{7}4^{3}_{7}4^{3}_{9}2^{4}_{9}4^{4}_{11}1^{5}_{11}2^{5}_{13}1^{6}_{13}1^{6}_{15}1^{7}_{17}$	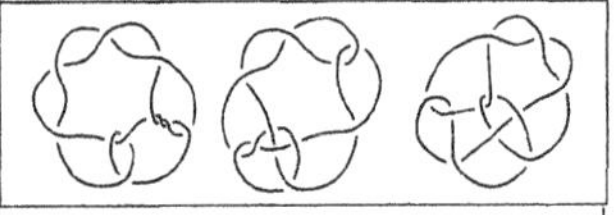
11^a_{167}: $1^{\underline{5}}_{\underline{11}}2^{\underline{4}}_{\underline{9}}1^{\underline{4}}_{\underline{7}}4^{\underline{3}}_{\underline{7}}2^{\underline{3}}_{\underline{5}}7^{\underline{2}}_{\underline{5}}4^{\underline{2}}_{\underline{3}}8^{\underline{1}}_{\underline{3}}7^{\underline{1}}_{\underline{1}}10^{0}_{\underline{1}}9^{0}_{1}9^{1}_{1}9^{1}_{3}7^{2}_{3}9^{2}_{5}5^{3}_{5}7^{3}_{7}3^{4}_{7}5^{4}_{9}1^{5}_{9}3^{5}_{11}1^{6}_{13}$	
11^a_{168}: $1^{\underline{5}}_{\underline{11}}2^{\underline{4}}_{\underline{9}}1^{\underline{4}}_{\underline{7}}5^{\underline{3}}_{\underline{7}}2^{\underline{3}}_{\underline{5}}8^{\underline{2}}_{\underline{5}}5^{\underline{2}}_{\underline{3}}9^{\underline{1}}_{\underline{3}}8^{\underline{1}}_{\underline{1}}11^{0}_{\underline{1}}10^{0}_{1}10^{1}_{1}10^{1}_{3}8^{2}_{3}10^{2}_{5}5^{3}_{5}8^{3}_{7}3^{4}_{7}5^{4}_{9}1^{5}_{9}3^{5}_{11}1^{6}_{13}$	
11^a_{169}: $1^{\underline{7}}_{\underline{15}}2^{\underline{6}}_{\underline{13}}1^{\underline{6}}_{\underline{11}}5^{\underline{5}}_{\underline{11}}2^{\underline{5}}_{\underline{9}}7^{\underline{4}}_{\underline{9}}5^{\underline{4}}_{\underline{7}}9^{\underline{3}}_{\underline{7}}7^{\underline{3}}_{\underline{5}}10^{\underline{2}}_{\underline{5}}9^{\underline{2}}_{\underline{3}}9^{\underline{1}}_{\underline{3}}10^{\underline{1}}_{\underline{1}}9^{0}_{\underline{1}}10^{0}_{1}5^{1}_{1}8^{1}_{3}3^{2}_{3}5^{2}_{5}1^{3}_{5}3^{3}_{7}1^{4}_{9}$	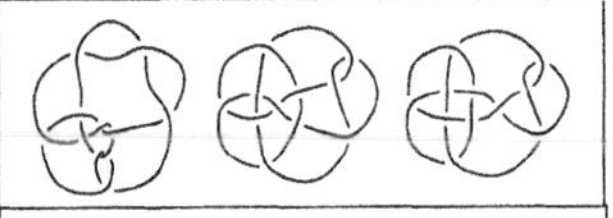
11^a_{170}: $1^{\underline{5}}_{\underline{11}}3^{\underline{4}}_{\underline{9}}1^{\underline{4}}_{\underline{7}}7^{\underline{3}}_{\underline{7}}3^{\underline{3}}_{\underline{5}}11^{\underline{2}}_{\underline{5}}7^{\underline{2}}_{\underline{3}}14^{\underline{1}}_{\underline{3}}11^{\underline{1}}_{\underline{1}}16^{0}_{\underline{1}}15^{0}_{1}15^{1}_{1}15^{1}_{3}12^{2}_{3}15^{2}_{5}8^{3}_{5}12^{3}_{7}5^{4}_{7}8^{4}_{9}1^{5}_{9}5^{5}_{11}1^{6}_{13}$	
11^a_{171}: $1^{\underline{4}}_{\underline{7}}4^{\underline{3}}_{\underline{5}}1^{\underline{3}}_{\underline{3}}7^{\underline{2}}_{\underline{3}}4^{\underline{2}}_{\underline{1}}11^{\underline{1}}_{\underline{1}}7^{\underline{1}}_{1}15^{0}_{1}12^{0}_{3}15^{1}_{3}14^{1}_{5}15^{2}_{5}15^{2}_{7}11^{3}_{7}15^{3}_{9}8^{4}_{9}11^{4}_{11}4^{5}_{11}8^{5}_{13}1^{6}_{13}4^{6}_{15}1^{7}_{17}$	
11^a_{172}: $1^{\underline{4}}_{\underline{7}}2^{\underline{3}}_{\underline{5}}1^{\underline{3}}_{\underline{3}}6^{\underline{2}}_{\underline{3}}2^{\underline{2}}_{\underline{1}}8^{\underline{1}}_{\underline{1}}6^{\underline{1}}_{1}11^{0}_{1}9^{0}_{3}12^{1}_{3}10^{1}_{5}11^{2}_{5}12^{2}_{7}9^{3}_{7}11^{3}_{9}6^{4}_{9}9^{4}_{11}3^{5}_{11}6^{5}_{13}1^{6}_{13}3^{6}_{15}1^{7}_{17}$	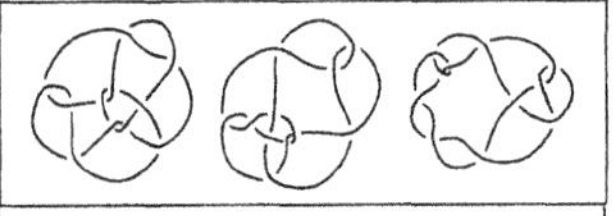
11^a_{173}: $1^{\underline{4}}_{\underline{7}}2^{\underline{3}}_{\underline{5}}1^{\underline{3}}_{\underline{3}}5^{\underline{2}}_{\underline{3}}2^{\underline{2}}_{\underline{1}}8^{\underline{1}}_{\underline{1}}5^{\underline{1}}_{1}10^{0}_{1}9^{0}_{3}12^{1}_{3}9^{1}_{5}11^{2}_{5}12^{2}_{7}8^{3}_{7}11^{3}_{9}7^{4}_{9}8^{4}_{11}3^{5}_{11}7^{5}_{13}1^{6}_{13}3^{6}_{15}1^{7}_{17}$	
11^a_{174}: $1^{\underline{6}}_{\underline{15}}2^{\underline{5}}_{\underline{13}}1^{\underline{5}}_{\underline{11}}3^{\underline{4}}_{\underline{11}}2^{\underline{4}}_{\underline{9}}5^{\underline{3}}_{\underline{9}}3^{\underline{3}}_{\underline{7}}6^{\underline{2}}_{\underline{7}}5^{\underline{2}}_{\underline{5}}6^{\underline{1}}_{\underline{5}}6^{\underline{1}}_{\underline{3}}6^{0}_{\underline{3}}7^{0}_{\underline{1}}5^{1}_{\underline{1}}5^{1}_{1}3^{2}_{1}5^{2}_{3}2^{3}_{3}3^{3}_{5}1^{4}_{5}2^{4}_{7}1^{5}_{9}$	
11^a_{175}: $1^{\underline{5}}_{\underline{11}}2^{\underline{4}}_{\underline{9}}1^{\underline{4}}_{\underline{7}}4^{\underline{3}}_{\underline{7}}2^{\underline{3}}_{\underline{5}}6^{\underline{2}}_{\underline{5}}4^{\underline{2}}_{\underline{3}}8^{\underline{1}}_{\underline{3}}6^{\underline{1}}_{\underline{1}}9^{0}_{\underline{1}}9^{0}_{1}8^{1}_{1}8^{1}_{3}7^{2}_{3}8^{2}_{5}4^{3}_{5}7^{3}_{7}3^{4}_{7}4^{4}_{9}1^{5}_{9}3^{5}_{11}1^{6}_{13}$	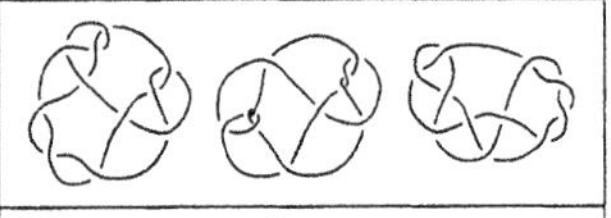
11^a_{176}: $1^{\underline{4}}_{\underline{7}}2^{\underline{3}}_{\underline{5}}1^{\underline{3}}_{\underline{3}}4^{\underline{2}}_{\underline{3}}2^{\underline{2}}_{\underline{1}}6^{\underline{1}}_{\underline{1}}4^{\underline{1}}_{1}9^{0}_{1}7^{0}_{3}9^{1}_{3}8^{1}_{5}9^{2}_{5}9^{2}_{7}7^{3}_{7}9^{3}_{9}5^{4}_{9}7^{4}_{11}3^{5}_{11}5^{5}_{13}1^{6}_{13}3^{6}_{15}1^{7}_{17}$	
11^a_{177}: $1^{\underline{3}}_{\underline{3}}2^{\underline{2}}_{\underline{1}}1^{\underline{2}}_{1}3^{\underline{1}}_{1}2^{\underline{1}}_{3}6^{0}_{3}4^{0}_{5}7^{1}_{5}5^{1}_{7}8^{2}_{7}7^{2}_{9}7^{3}_{9}8^{3}_{11}7^{4}_{11}7^{4}_{13}4^{5}_{13}7^{5}_{15}3^{6}_{15}4^{6}_{17}1^{7}_{17}3^{7}_{19}1^{8}_{21}$	
11^a_{178}: $1^{\underline{3}}_{\underline{5}}3^{\underline{2}}_{\underline{3}}1^{\underline{2}}_{\underline{1}}5^{\underline{1}}_{\underline{1}}3^{\underline{1}}_{1}8^{0}_{1}6^{0}_{3}10^{1}_{3}7^{1}_{5}10^{2}_{5}10^{2}_{7}9^{3}_{7}10^{3}_{9}8^{4}_{9}9^{4}_{11}4^{5}_{11}8^{5}_{13}3^{6}_{13}4^{6}_{15}1^{7}_{15}3^{7}_{17}1^{8}_{19}$	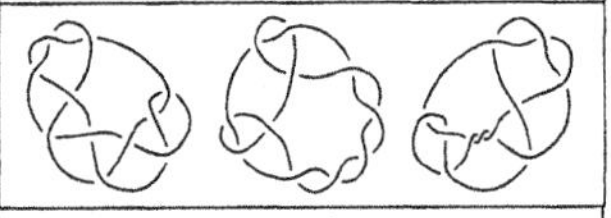
11^a_{179}: $1^{\underline{4}}_{\underline{5}}1^{\underline{3}}_{\underline{3}}1^{\underline{3}}_{\underline{1}}3^{\underline{2}}_{\underline{1}}1^{\underline{2}}_{1}2^{\underline{1}}_{1}3^{\underline{1}}_{3}5^{0}_{3}3^{0}_{5}4^{1}_{5}4^{1}_{7}4^{2}_{7}4^{2}_{9}4^{3}_{9}4^{3}_{11}2^{4}_{11}4^{4}_{13}2^{5}_{13}2^{5}_{15}1^{6}_{15}2^{6}_{17}1^{7}_{19}$	
11^a_{180}: $1^{\underline{6}}_{\underline{13}}2^{\underline{5}}_{\underline{11}}1^{\underline{5}}_{\underline{9}}4^{\underline{4}}_{\underline{9}}2^{\underline{4}}_{\underline{7}}5^{\underline{3}}_{\underline{7}}4^{\underline{3}}_{\underline{5}}7^{\underline{2}}_{\underline{5}}5^{\underline{2}}_{\underline{3}}7^{\underline{1}}_{\underline{3}}7^{\underline{1}}_{\underline{1}}7^{0}_{\underline{1}}8^{0}_{1}6^{1}_{1}6^{1}_{3}3^{2}_{3}6^{2}_{5}2^{3}_{5}3^{3}_{7}1^{4}_{7}2^{4}_{9}1^{5}_{11}$	
11^a_{181}: $1^{\underline{4}}_{\underline{7}}2^{\underline{3}}_{\underline{5}}1^{\underline{3}}_{\underline{3}}5^{\underline{2}}_{\underline{3}}2^{\underline{2}}_{\underline{1}}5^{\underline{1}}_{\underline{1}}5^{\underline{1}}_{1}9^{0}_{1}6^{0}_{3}8^{1}_{3}8^{1}_{5}7^{2}_{5}8^{2}_{7}7^{3}_{7}7^{3}_{9}3^{4}_{9}7^{4}_{11}2^{5}_{11}3^{5}_{13}1^{6}_{13}2^{6}_{15}1^{7}_{17}$	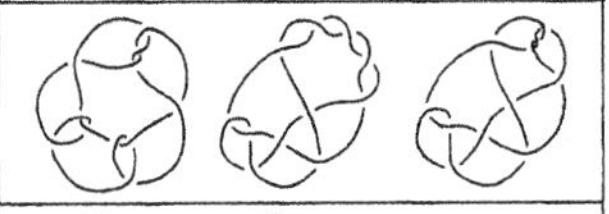
11^a_{182}: $1^{\underline{8}}_{\underline{21}}2^{\underline{7}}_{\underline{19}}1^{\underline{7}}_{\underline{17}}3^{\underline{6}}_{\underline{17}}2^{\underline{6}}_{\underline{15}}5^{\underline{5}}_{\underline{15}}3^{\underline{5}}_{\underline{13}}5^{\underline{4}}_{\underline{13}}5^{\underline{4}}_{\underline{11}}6^{\underline{3}}_{\underline{11}}5^{\underline{3}}_{\underline{9}}5^{\underline{2}}_{\underline{9}}6^{\underline{2}}_{\underline{7}}4^{\underline{1}}_{\underline{7}}5^{\underline{1}}_{\underline{5}}3^{0}_{\underline{5}}5^{0}_{\underline{3}}2^{1}_{\underline{3}}2^{1}_{\underline{1}}1^{2}_{\underline{1}}2^{2}_{1}1^{3}_{3}$	
11^a_{183}: $1^{\underline{8}}_{\underline{19}}2^{\underline{7}}_{\underline{17}}1^{\underline{7}}_{\underline{15}}4^{\underline{6}}_{\underline{15}}2^{\underline{6}}_{\underline{13}}7^{\underline{5}}_{\underline{13}}4^{\underline{5}}_{\underline{11}}8^{\underline{4}}_{\underline{11}}7^{\underline{4}}_{\underline{9}}10^{\underline{3}}_{\underline{9}}8^{\underline{3}}_{\underline{7}}9^{\underline{2}}_{\underline{7}}10^{\underline{2}}_{\underline{5}}7^{\underline{1}}_{\underline{5}}9^{\underline{1}}_{\underline{3}}6^{0}_{\underline{3}}8^{0}_{\underline{1}}3^{1}_{\underline{1}}5^{1}_{1}1^{2}_{1}3^{2}_{3}1^{3}_{5}$	
11^a_{184}: $1^{\underline{7}}_{\underline{17}}2^{\underline{6}}_{\underline{15}}1^{\underline{6}}_{\underline{13}}4^{\underline{5}}_{\underline{13}}2^{\underline{5}}_{\underline{11}}5^{\underline{4}}_{\underline{11}}4^{\underline{4}}_{\underline{9}}7^{\underline{3}}_{\underline{9}}5^{\underline{3}}_{\underline{7}}7^{\underline{2}}_{\underline{7}}7^{\underline{2}}_{\underline{5}}6^{\underline{1}}_{\underline{5}}7^{\underline{1}}_{\underline{3}}6^{0}_{\underline{3}}7^{0}_{\underline{1}}3^{1}_{\underline{1}}5^{1}_{1}2^{2}_{1}3^{2}_{3}1^{3}_{3}2^{3}_{5}1^{4}_{7}$	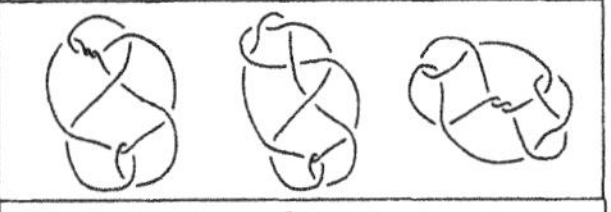
11^a_{185}: $1^{\underline{7}}_{\underline{15}}2^{\underline{6}}_{\underline{13}}1^{\underline{6}}_{\underline{11}}4^{\underline{5}}_{\underline{11}}2^{\underline{5}}_{\underline{9}}6^{\underline{4}}_{\underline{9}}4^{\underline{4}}_{\underline{7}}8^{\underline{3}}_{\underline{7}}6^{\underline{3}}_{\underline{5}}9^{\underline{2}}_{\underline{5}}8^{\underline{2}}_{\underline{3}}8^{\underline{1}}_{\underline{3}}9^{\underline{1}}_{\underline{1}}8^{0}_{\underline{1}}9^{0}_{1}5^{1}_{1}7^{1}_{3}3^{2}_{3}5^{2}_{5}1^{3}_{5}3^{3}_{7}1^{4}_{9}$	
11^a_{186}: $1^{0}_{5}1^{0}_{7}2^{1}_{7}4^{2}_{9}2^{2}_{11}5^{3}_{11}4^{3}_{13}8^{4}_{13}5^{4}_{15}7^{5}_{15}8^{5}_{17}8^{6}_{17}7^{6}_{19}6^{7}_{19}8^{7}_{21}4^{8}_{21}6^{8}_{23}2^{9}_{23}4^{9}_{25}1^{10}_{25}2^{10}_{27}1^{11}_{29}$	
11^a_{187}: $1^{\underline{6}}_{\underline{13}}2^{\underline{5}}_{\underline{11}}1^{\underline{5}}_{\underline{9}}4^{\underline{4}}_{\underline{9}}2^{\underline{4}}_{\underline{7}}7^{\underline{3}}_{\underline{7}}4^{\underline{3}}_{\underline{5}}9^{\underline{2}}_{\underline{5}}7^{\underline{2}}_{\underline{3}}9^{\underline{1}}_{\underline{3}}9^{\underline{1}}_{\underline{1}}10^{0}_{\underline{1}}10^{0}_{1}8^{1}_{1}9^{1}_{3}5^{2}_{3}8^{2}_{5}3^{3}_{5}5^{3}_{7}1^{4}_{7}3^{4}_{9}1^{5}_{11}$	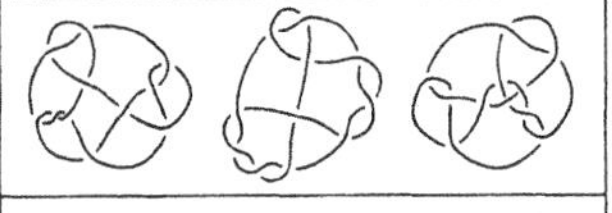
11^a_{188}: $1^{\underline{6}}_{\underline{11}}1^{\underline{5}}_{\underline{9}}1^{\underline{5}}_{\underline{7}}3^{\underline{4}}_{\underline{7}}1^{\underline{4}}_{\underline{5}}3^{\underline{3}}_{\underline{5}}3^{\underline{3}}_{\underline{3}}5^{\underline{2}}_{\underline{3}}3^{\underline{2}}_{\underline{1}}5^{\underline{1}}_{\underline{1}}5^{\underline{1}}_{1}5^{0}_{1}6^{0}_{3}5^{1}_{3}4^{1}_{5}3^{2}_{5}5^{2}_{7}2^{3}_{7}3^{3}_{9}1^{4}_{9}2^{4}_{11}1^{5}_{13}$	
11^a_{189}: $1^{\underline{6}}_{\underline{13}}3^{\underline{5}}_{\underline{11}}1^{\underline{5}}_{\underline{9}}6^{\underline{4}}_{\underline{9}}3^{\underline{4}}_{\underline{7}}9^{\underline{3}}_{\underline{7}}6^{\underline{3}}_{\underline{5}}12^{\underline{2}}_{\underline{5}}9^{\underline{2}}_{\underline{3}}12^{\underline{1}}_{\underline{3}}12^{\underline{1}}_{\underline{1}}12^{0}_{\underline{1}}13^{0}_{1}10^{1}_{1}11^{1}_{3}6^{2}_{3}10^{2}_{5}3^{3}_{5}6^{3}_{7}1^{4}_{7}3^{4}_{9}1^{5}_{11}$	
11^a_{190}: $1^{\underline{4}}_{\underline{9}}2^{\underline{3}}_{\underline{7}}1^{\underline{3}}_{\underline{5}}4^{\underline{2}}_{\underline{5}}2^{\underline{2}}_{\underline{3}}5^{\underline{1}}_{\underline{3}}4^{\underline{1}}_{\underline{1}}7^{0}_{\underline{1}}6^{0}_{1}7^{1}_{1}6^{1}_{3}6^{2}_{3}7^{2}_{5}5^{3}_{5}6^{3}_{7}3^{4}_{7}5^{4}_{9}2^{5}_{9}3^{5}_{11}1^{6}_{11}2^{6}_{13}1^{7}_{15}$	
11^a_{191}: $1^{0}_{5}1^{0}_{7}2^{1}_{7}3^{2}_{9}2^{2}_{11}4^{3}_{11}3^{3}_{13}7^{4}_{13}4^{4}_{15}6^{5}_{15}7^{5}_{17}7^{6}_{17}6^{6}_{19}5^{7}_{19}7^{7}_{21}4^{8}_{21}5^{8}_{23}2^{9}_{23}4^{9}_{25}1^{10}_{25}2^{10}_{27}1^{11}_{29}$	
11^a_{192}: $1^{0}_{3}1^{0}_{5}3^{1}_{5}4^{2}_{7}3^{2}_{9}6^{3}_{9}4^{3}_{11}8^{4}_{11}6^{4}_{13}7^{5}_{13}8^{5}_{15}8^{6}_{15}7^{6}_{17}5^{7}_{17}8^{7}_{19}4^{8}_{19}5^{8}_{21}2^{9}_{21}4^{9}_{23}1^{10}_{23}2^{10}_{25}1^{11}_{27}$	

K: KH	1st line 2nd line 3rd line
11^a_{193}: $1^{\underline{8}}_{\underline{19}}2^{\underline{7}}_{\underline{17}}1^{\underline{7}}_{\underline{15}}4^{\underline{6}}_{\underline{15}}2^{\underline{6}}_{\underline{13}}6^{\underline{5}}_{\underline{13}}4^{\underline{5}}_{\underline{11}}7^{\underline{4}}_{\underline{11}}6^{\underline{4}}_{\underline{9}}8^{\underline{3}}_{\underline{9}}7^{\underline{3}}_{\underline{7}}7^{\underline{2}}_{\underline{7}}8^{\underline{2}}_{\underline{5}}6^{\underline{1}}_{\underline{5}}7^{\underline{1}}_{\underline{3}}4^{0}_{\underline{3}}7^{0}_{\underline{1}}2^{1}_{\underline{1}}3^{1}_{1}1^{2}_{1}2^{2}_{3}1^{3}_{5}$	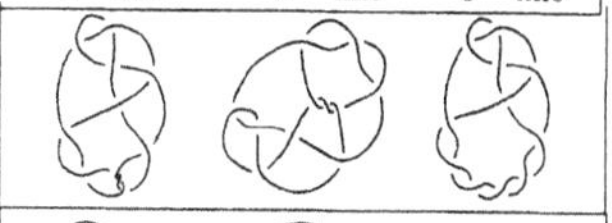
11^a_{194}: $1^{\underline{8}}_{\underline{21}}2^{\underline{7}}_{\underline{19}}1^{\underline{7}}_{\underline{17}}4^{\underline{6}}_{\underline{17}}2^{\underline{6}}_{\underline{15}}6^{\underline{5}}_{\underline{15}}4^{\underline{5}}_{\underline{13}}7^{\underline{4}}_{\underline{13}}6^{\underline{4}}_{\underline{11}}8^{\underline{3}}_{\underline{11}}7^{\underline{3}}_{\underline{9}}6^{\underline{2}}_{\underline{9}}8^{\underline{2}}_{\underline{7}}6^{\underline{1}}_{\underline{7}}6^{\underline{1}}_{\underline{5}}4^{0}_{\underline{5}}7^{0}_{\underline{3}}2^{1}_{\underline{3}}3^{1}_{\underline{1}}1^{2}_{\underline{1}}2^{2}_{1}1^{3}_{3}$	
11^a_{195}: $1^{\underline{8}}_{\underline{17}}1^{\underline{7}}_{\underline{15}}1^{\underline{7}}_{\underline{13}}2^{\underline{6}}_{\underline{13}}1^{\underline{6}}_{\underline{11}}3^{\underline{5}}_{\underline{11}}2^{\underline{5}}_{\underline{9}}3^{\underline{4}}_{\underline{9}}3^{\underline{4}}_{\underline{7}}4^{\underline{3}}_{\underline{7}}3^{\underline{3}}_{\underline{5}}4^{\underline{2}}_{\underline{5}}4^{\underline{2}}_{\underline{3}}3^{\underline{1}}_{\underline{3}}4^{\underline{1}}_{\underline{1}}3^{0}_{\underline{1}}4^{0}_{1}2^{1}_{1}2^{1}_{3}1^{2}_{3}2^{2}_{5}1^{3}_{7}$	
11^a_{196}: $1^{\underline{7}}_{\underline{17}}3^{\underline{6}}_{\underline{15}}1^{\underline{6}}_{\underline{13}}6^{\underline{5}}_{\underline{13}}3^{\underline{5}}_{\underline{11}}9^{\underline{4}}_{\underline{11}}6^{\underline{4}}_{\underline{9}}12^{\underline{3}}_{\underline{9}}9^{\underline{3}}_{\underline{7}}12^{\underline{2}}_{\underline{7}}12^{\underline{2}}_{\underline{5}}11^{\underline{1}}_{\underline{5}}12^{\underline{1}}_{\underline{3}}10^{0}_{\underline{3}}12^{0}_{\underline{1}}6^{1}_{\underline{1}}9^{1}_{1}3^{2}_{1}6^{2}_{3}1^{3}_{3}3^{3}_{5}1^{4}_{7}$	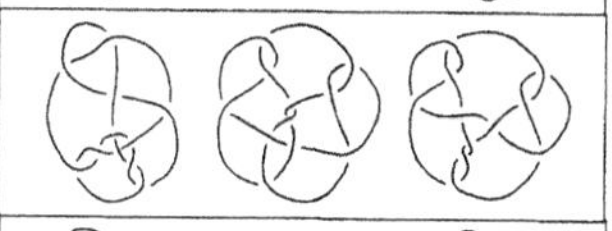
11^a_{197}: $1^{\underline{3}}_{\underline{5}}2^{\underline{2}}_{\underline{3}}1^{\underline{2}}_{\underline{1}}5^{\underline{1}}_{\underline{1}}2^{\underline{1}}_{1}9^{0}_{1}6^{0}_{3}11^{1}_{3}8^{1}_{5}12^{2}_{5}11^{2}_{7}11^{3}_{7}12^{3}_{9}10^{4}_{9}11^{4}_{11}6^{5}_{11}10^{5}_{13}4^{6}_{13}6^{6}_{15}1^{7}_{15}4^{7}_{17}1^{8}_{19}$	
11^a_{198}: $1^{\underline{5}}_{9}2^{\underline{4}}_{7}1^{\underline{4}}_{5}4^{\underline{3}}_{5}2^{\underline{3}}_{\underline{3}}7^{\underline{2}}_{\underline{3}}4^{\underline{2}}_{\underline{1}}8^{\underline{1}}_{\underline{1}}7^{\underline{1}}_{1}10^{0}_{1}9^{0}_{3}9^{1}_{3}9^{1}_{5}8^{2}_{5}9^{2}_{7}5^{3}_{7}8^{3}_{9}3^{4}_{9}5^{4}_{11}1^{5}_{11}3^{5}_{13}1^{6}_{15}$	
11^a_{199}: $1^{\underline{6}}_{\underline{15}}2^{\underline{5}}_{\underline{13}}1^{\underline{5}}_{\underline{11}}5^{\underline{4}}_{\underline{11}}2^{\underline{4}}_{\underline{9}}6^{\underline{3}}_{\underline{9}}5^{\underline{3}}_{\underline{7}}8^{\underline{2}}_{\underline{7}}6^{\underline{2}}_{\underline{5}}8^{\underline{1}}_{\underline{5}}8^{\underline{1}}_{\underline{3}}7^{0}_{\underline{3}}9^{0}_{\underline{1}}7^{1}_{\underline{1}}6^{1}_{1}3^{2}_{1}7^{2}_{3}2^{3}_{3}3^{3}_{5}1^{4}_{5}2^{4}_{7}1^{5}_{9}$	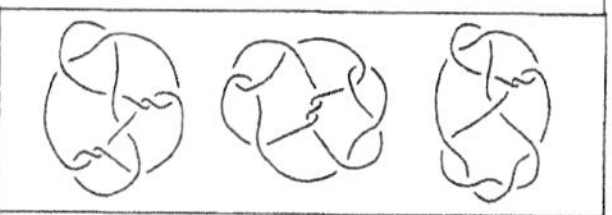
11^a_{200}: $1^{0}_{3}1^{0}_{5}2^{1}_{5}3^{2}_{7}2^{2}_{9}5^{3}_{9}3^{3}_{11}7^{4}_{11}5^{4}_{13}6^{5}_{13}7^{5}_{15}7^{6}_{15}6^{6}_{17}5^{7}_{17}7^{7}_{19}4^{8}_{19}5^{8}_{21}2^{9}_{21}4^{9}_{23}1^{10}_{23}2^{10}_{25}1^{11}_{27}$	
11^a_{201}: $1^{\underline{6}}_{\underline{13}}1^{\underline{5}}_{\underline{11}}1^{\underline{5}}_{\underline{9}}4^{\underline{4}}_{\underline{9}}1^{\underline{4}}_{\underline{7}}4^{\underline{3}}_{\underline{7}}4^{\underline{3}}_{\underline{5}}6^{\underline{2}}_{\underline{5}}4^{\underline{2}}_{\underline{3}}7^{\underline{1}}_{\underline{3}}6^{\underline{1}}_{\underline{1}}6^{0}_{\underline{1}}8^{0}_{1}6^{1}_{1}5^{1}_{3}3^{2}_{3}6^{2}_{5}2^{3}_{5}3^{3}_{7}1^{4}_{7}2^{4}_{9}1^{5}_{11}$	
11^a_{202}: $1^{\underline{4}}_{\underline{7}}1^{\underline{3}}_{\underline{5}}1^{\underline{3}}_{\underline{3}}5^{\underline{2}}_{\underline{3}}1^{\underline{2}}_{\underline{1}}5^{\underline{1}}_{\underline{1}}5^{\underline{1}}_{1}9^{0}_{1}6^{0}_{3}10^{1}_{3}8^{1}_{5}8^{2}_{5}10^{2}_{7}8^{3}_{7}8^{3}_{9}5^{4}_{9}8^{4}_{11}3^{5}_{11}5^{5}_{13}1^{6}_{13}3^{6}_{15}1^{7}_{17}$	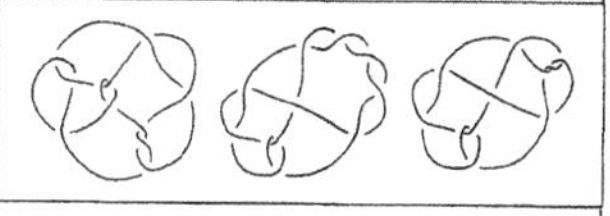
11^a_{203}: $1^{\underline{2}}_{1}1^{\underline{1}}_{3}1^{\underline{1}}_{5}3^{0}_{5}2^{0}_{7}3^{1}_{7}2^{1}_{9}5^{2}_{9}3^{2}_{11}4^{3}_{11}5^{3}_{13}5^{4}_{13}4^{4}_{15}4^{5}_{15}5^{5}_{17}3^{6}_{17}4^{6}_{19}2^{7}_{19}3^{7}_{21}1^{8}_{21}2^{8}_{23}1^{9}_{25}$	
11^a_{204}: $1^{\underline{2}}_{\underline{1}}2^{\underline{1}}_{1}1^{\underline{1}}_{3}5^{0}_{3}3^{0}_{5}6^{1}_{5}4^{1}_{7}8^{2}_{7}6^{2}_{9}8^{3}_{9}8^{3}_{11}8^{4}_{11}8^{4}_{13}6^{5}_{13}8^{5}_{15}4^{6}_{15}6^{6}_{17}2^{7}_{17}4^{7}_{19}1^{8}_{19}2^{8}_{21}1^{9}_{23}$	
11^a_{205}: $1^{\underline{6}}_{\underline{15}}2^{\underline{5}}_{\underline{13}}1^{\underline{5}}_{\underline{11}}4^{\underline{4}}_{\underline{11}}2^{\underline{4}}_{\underline{9}}6^{\underline{3}}_{\underline{9}}4^{\underline{3}}_{\underline{7}}7^{\underline{2}}_{\underline{7}}6^{\underline{2}}_{\underline{5}}7^{\underline{1}}_{\underline{5}}7^{\underline{1}}_{\underline{3}}7^{0}_{\underline{3}}8^{0}_{\underline{1}}6^{1}_{\underline{1}}6^{1}_{1}3^{2}_{1}6^{2}_{3}2^{3}_{3}3^{3}_{5}1^{4}_{5}2^{4}_{7}1^{5}_{9}$	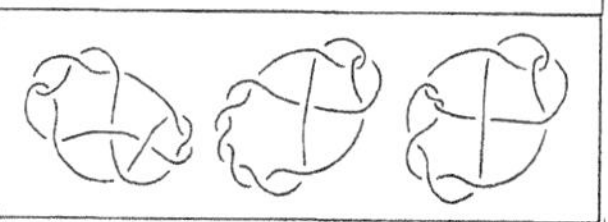
11^a_{206}: $1^{\underline{9}}_{\underline{25}}1^{\underline{8}}_{\underline{23}}1^{\underline{8}}_{\underline{21}}3^{\underline{7}}_{\underline{21}}1^{\underline{7}}_{\underline{19}}2^{\underline{6}}_{\underline{19}}3^{\underline{6}}_{\underline{17}}4^{\underline{5}}_{\underline{17}}2^{\underline{5}}_{\underline{15}}3^{\underline{4}}_{\underline{15}}4^{\underline{4}}_{\underline{13}}3^{\underline{3}}_{\underline{13}}3^{\underline{3}}_{\underline{11}}3^{\underline{2}}_{\underline{11}}3^{\underline{2}}_{\underline{9}}1^{\underline{1}}_{\underline{9}}3^{\underline{1}}_{\underline{7}}2^{0}_{\underline{7}}2^{0}_{\underline{5}}1^{1}_{\underline{5}}1^{1}_{\underline{3}}1^{2}_{\underline{1}}$	
11^a_{207}: $1^{\underline{9}}_{\underline{23}}1^{\underline{8}}_{\underline{21}}1^{\underline{8}}_{\underline{19}}4^{\underline{7}}_{\underline{19}}1^{\underline{7}}_{\underline{17}}4^{\underline{6}}_{\underline{17}}4^{\underline{6}}_{\underline{15}}7^{\underline{5}}_{\underline{15}}4^{\underline{5}}_{\underline{13}}7^{\underline{4}}_{\underline{13}}7^{\underline{4}}_{\underline{11}}6^{\underline{3}}_{\underline{11}}7^{\underline{3}}_{\underline{9}}6^{\underline{2}}_{\underline{9}}6^{\underline{2}}_{\underline{7}}3^{\underline{1}}_{\underline{7}}6^{\underline{1}}_{\underline{5}}3^{0}_{\underline{5}}4^{0}_{\underline{3}}1^{1}_{\underline{3}}2^{1}_{\underline{1}}1^{2}_{1}$	
11^a_{208}: $1^{\underline{9}}_{\underline{23}}2^{\underline{8}}_{\underline{21}}1^{\underline{8}}_{\underline{19}}5^{\underline{7}}_{\underline{19}}2^{\underline{7}}_{\underline{17}}6^{\underline{6}}_{\underline{17}}5^{\underline{6}}_{\underline{15}}9^{\underline{5}}_{\underline{15}}6^{\underline{5}}_{\underline{13}}8^{\underline{4}}_{\underline{13}}9^{\underline{4}}_{\underline{11}}8^{\underline{3}}_{\underline{11}}8^{\underline{3}}_{\underline{9}}7^{\underline{2}}_{\underline{9}}8^{\underline{2}}_{\underline{7}}3^{\underline{1}}_{\underline{7}}7^{\underline{1}}_{\underline{5}}3^{0}_{\underline{5}}4^{0}_{\underline{3}}1^{1}_{\underline{3}}2^{1}_{\underline{1}}1^{2}_{1}$	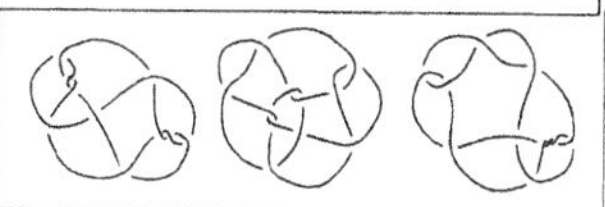
11^a_{209}: $1^{\underline{6}}_{\underline{13}}2^{\underline{5}}_{\underline{11}}1^{\underline{5}}_{\underline{9}}6^{\underline{4}}_{\underline{9}}2^{\underline{4}}_{\underline{7}}8^{\underline{3}}_{\underline{7}}6^{\underline{3}}_{\underline{5}}11^{\underline{2}}_{\underline{5}}8^{\underline{2}}_{\underline{3}}12^{\underline{1}}_{\underline{3}}11^{\underline{1}}_{\underline{1}}11^{0}_{\underline{1}}13^{0}_{1}10^{1}_{1}10^{1}_{3}6^{2}_{3}10^{2}_{5}3^{3}_{5}6^{3}_{7}1^{4}_{7}3^{4}_{9}1^{5}_{11}$	
11^a_{210}: $1^{\underline{6}}_{\underline{13}}1^{\underline{5}}_{\underline{11}}1^{\underline{5}}_{\underline{9}}3^{\underline{4}}_{\underline{9}}1^{\underline{4}}_{\underline{7}}4^{\underline{3}}_{\underline{7}}3^{\underline{3}}_{\underline{5}}5^{\underline{2}}_{\underline{5}}4^{\underline{2}}_{\underline{3}}6^{\underline{1}}_{\underline{3}}5^{\underline{1}}_{\underline{1}}6^{0}_{\underline{1}}7^{0}_{1}5^{1}_{1}5^{1}_{3}3^{2}_{3}5^{2}_{5}2^{3}_{5}3^{3}_{7}1^{4}_{7}2^{4}_{9}1^{5}_{11}$	
11^a_{211}: $1^{\underline{9}}_{\underline{21}}1^{\underline{8}}_{\underline{19}}1^{\underline{8}}_{\underline{17}}3^{\underline{7}}_{\underline{17}}1^{\underline{7}}_{\underline{15}}3^{\underline{6}}_{\underline{15}}3^{\underline{6}}_{\underline{13}}5^{\underline{5}}_{\underline{13}}3^{\underline{5}}_{\underline{11}}5^{\underline{4}}_{\underline{11}}5^{\underline{4}}_{\underline{9}}5^{\underline{3}}_{\underline{9}}5^{\underline{3}}_{\underline{7}}5^{\underline{2}}_{\underline{7}}5^{\underline{2}}_{\underline{5}}2^{\underline{1}}_{\underline{5}}5^{\underline{1}}_{\underline{3}}3^{0}_{\underline{3}}3^{0}_{\underline{1}}1^{1}_{\underline{1}}2^{1}_{1}1^{2}_{3}$	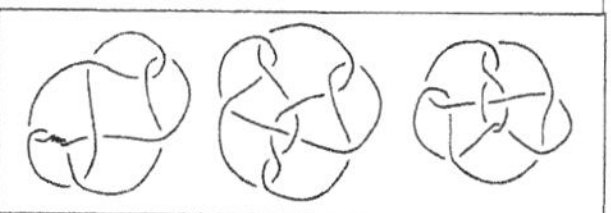
11^a_{212}: $1^{\underline{2}}_{\underline{1}}2^{\underline{1}}_{1}1^{\underline{1}}_{3}6^{0}_{3}3^{0}_{5}9^{1}_{5}5^{1}_{7}12^{2}_{7}9^{2}_{9}12^{3}_{9}12^{3}_{11}13^{4}_{11}12^{4}_{13}10^{5}_{13}13^{5}_{15}7^{6}_{15}10^{6}_{17}4^{7}_{17}7^{7}_{19}1^{8}_{19}4^{8}_{21}1^{9}_{23}$	
11^a_{213}: $1^{\underline{2}}_{\underline{1}}2^{\underline{1}}_{1}1^{\underline{1}}_{3}6^{0}_{3}3^{0}_{5}9^{1}_{5}5^{1}_{7}11^{2}_{7}9^{2}_{9}12^{3}_{9}11^{3}_{11}12^{4}_{11}12^{4}_{13}9^{5}_{13}12^{5}_{15}7^{6}_{15}9^{6}_{17}3^{7}_{17}7^{7}_{19}1^{8}_{19}3^{8}_{21}1^{9}_{23}$	
11^a_{214}: $1^{\underline{4}}_{\underline{9}}1^{\underline{3}}_{\underline{7}}1^{\underline{3}}_{\underline{5}}4^{\underline{2}}_{\underline{5}}1^{\underline{2}}_{\underline{3}}3^{\underline{1}}_{\underline{3}}4^{\underline{1}}_{\underline{1}}6^{0}_{\underline{1}}4^{0}_{1}6^{1}_{1}5^{1}_{3}4^{2}_{3}6^{2}_{5}5^{3}_{5}4^{3}_{7}2^{4}_{7}5^{4}_{9}2^{5}_{9}2^{5}_{11}1^{6}_{11}2^{6}_{13}1^{7}_{15}$	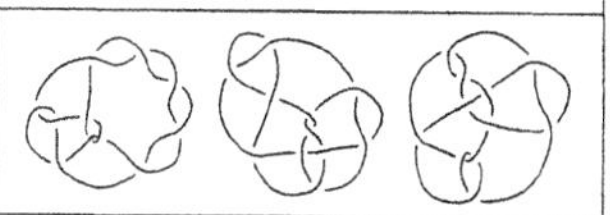
11^a_{215}: $1^{\underline{3}}_{\underline{3}}3^{\underline{2}}_{\underline{1}}1^{\underline{2}}_{1}5^{\underline{1}}_{1}3^{\underline{1}}_{3}9^{0}_{3}6^{0}_{5}10^{1}_{5}8^{1}_{7}11^{2}_{7}10^{2}_{9}10^{3}_{9}11^{3}_{11}9^{4}_{11}10^{4}_{13}5^{5}_{13}9^{5}_{15}3^{6}_{15}5^{6}_{17}1^{7}_{17}3^{7}_{19}1^{8}_{21}$	
11^a_{216}: $1^{\underline{4}}_{\underline{7}}3^{\underline{3}}_{\underline{5}}1^{\underline{3}}_{\underline{3}}6^{\underline{2}}_{\underline{3}}3^{\underline{2}}_{\underline{1}}9^{\underline{1}}_{\underline{1}}6^{\underline{1}}_{1}12^{0}_{1}10^{0}_{3}12^{1}_{3}11^{1}_{5}12^{2}_{5}12^{2}_{7}9^{3}_{7}12^{3}_{9}6^{4}_{9}9^{4}_{11}3^{5}_{11}6^{5}_{13}1^{6}_{13}3^{6}_{15}1^{7}_{17}$	
11^a_{217}: $1^{\underline{7}}_{\underline{17}}2^{\underline{6}}_{\underline{15}}1^{\underline{6}}_{\underline{13}}6^{\underline{5}}_{\underline{13}}2^{\underline{5}}_{\underline{11}}8^{\underline{4}}_{\underline{11}}6^{\underline{4}}_{\underline{9}}11^{\underline{3}}_{\underline{9}}8^{\underline{3}}_{\underline{7}}12^{\underline{2}}_{\underline{7}}11^{\underline{2}}_{\underline{5}}10^{\underline{1}}_{\underline{5}}12^{\underline{1}}_{\underline{3}}10^{0}_{\underline{3}}11^{0}_{\underline{1}}6^{1}_{\underline{1}}9^{1}_{1}3^{2}_{1}6^{2}_{3}1^{3}_{3}3^{3}_{5}1^{4}_{7}$	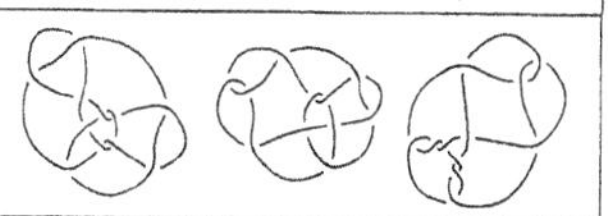
11^a_{218}: $1^{\underline{7}}_{\underline{17}}2^{\underline{6}}_{\underline{15}}1^{\underline{6}}_{\underline{13}}5^{\underline{5}}_{\underline{13}}2^{\underline{5}}_{\underline{11}}8^{\underline{4}}_{\underline{11}}5^{\underline{4}}_{\underline{9}}10^{\underline{3}}_{\underline{9}}8^{\underline{3}}_{\underline{7}}11^{\underline{2}}_{\underline{7}}10^{\underline{2}}_{\underline{5}}10^{\underline{1}}_{\underline{5}}11^{\underline{1}}_{\underline{3}}9^{0}_{\underline{3}}11^{0}_{\underline{1}}6^{1}_{\underline{1}}8^{1}_{1}3^{2}_{1}6^{2}_{3}1^{3}_{3}3^{3}_{5}1^{4}_{7}$	
11^a_{219}: $1^{\underline{9}}_{\underline{21}}1^{\underline{8}}_{\underline{19}}1^{\underline{8}}_{\underline{17}}4^{\underline{7}}_{\underline{17}}1^{\underline{7}}_{\underline{15}}4^{\underline{6}}_{\underline{15}}4^{\underline{6}}_{\underline{13}}7^{\underline{5}}_{\underline{13}}4^{\underline{5}}_{\underline{11}}7^{\underline{4}}_{\underline{11}}7^{\underline{4}}_{\underline{9}}6^{\underline{3}}_{\underline{9}}7^{\underline{3}}_{\underline{7}}7^{\underline{2}}_{\underline{7}}6^{\underline{2}}_{\underline{5}}3^{\underline{1}}_{\underline{5}}7^{\underline{1}}_{\underline{3}}3^{0}_{\underline{3}}4^{0}_{\underline{1}}1^{1}_{\underline{1}}2^{1}_{1}1^{2}_{3}$	
11^a_{220}: $1^{\underline{2}}_{\underline{1}}1^{\underline{1}}_{1}1^{\underline{1}}_{3}4^{0}_{3}2^{0}_{5}5^{1}_{5}3^{1}_{7}6^{2}_{7}5^{2}_{9}7^{3}_{9}6^{3}_{11}7^{4}_{11}7^{4}_{13}5^{5}_{13}7^{5}_{15}4^{6}_{15}5^{6}_{17}2^{7}_{17}4^{7}_{19}1^{8}_{19}2^{8}_{21}1^{9}_{23}$	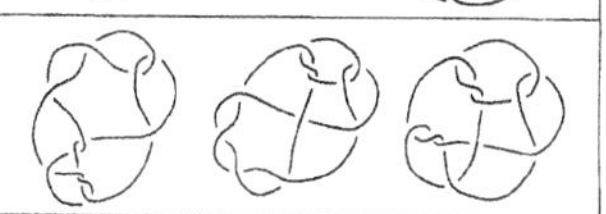
11^a_{221}: $1^{\underline{7}}_{\underline{17}}1^{\underline{6}}_{\underline{15}}1^{\underline{6}}_{\underline{13}}4^{\underline{5}}_{\underline{13}}1^{\underline{5}}_{\underline{11}}4^{\underline{4}}_{\underline{11}}4^{\underline{4}}_{\underline{9}}6^{\underline{3}}_{\underline{9}}4^{\underline{3}}_{\underline{7}}7^{\underline{2}}_{\underline{7}}6^{\underline{2}}_{\underline{5}}5^{\underline{1}}_{\underline{5}}7^{\underline{1}}_{\underline{3}}6^{0}_{\underline{3}}6^{0}_{\underline{1}}3^{1}_{\underline{1}}5^{1}_{1}2^{2}_{1}3^{2}_{3}1^{3}_{3}2^{3}_{5}1^{4}_{7}$	
11^a_{222}: $1^{\underline{7}}_{\underline{15}}1^{\underline{6}}_{\underline{13}}1^{\underline{6}}_{\underline{11}}4^{\underline{5}}_{\underline{11}}1^{\underline{5}}_{\underline{9}}5^{\underline{4}}_{\underline{9}}4^{\underline{4}}_{\underline{7}}7^{\underline{3}}_{\underline{7}}5^{\underline{3}}_{\underline{5}}9^{\underline{2}}_{\underline{5}}7^{\underline{2}}_{\underline{3}}7^{\underline{1}}_{\underline{3}}9^{\underline{1}}_{\underline{1}}8^{0}_{\underline{1}}8^{0}_{1}5^{1}_{1}7^{1}_{3}3^{2}_{3}5^{2}_{5}1^{3}_{5}3^{3}_{7}1^{4}_{9}$	
11^a_{223}: $1^{\underline{2}}_{1}1^{\underline{1}}_{3}1^{\underline{1}}_{5}4^{0}_{5}2^{0}_{7}4^{1}_{7}3^{1}_{9}6^{2}_{9}4^{2}_{11}5^{3}_{11}6^{3}_{13}6^{4}_{13}5^{4}_{15}5^{5}_{15}6^{5}_{17}3^{6}_{17}5^{6}_{19}2^{7}_{19}3^{7}_{21}1^{8}_{21}2^{8}_{23}1^{9}_{25}$	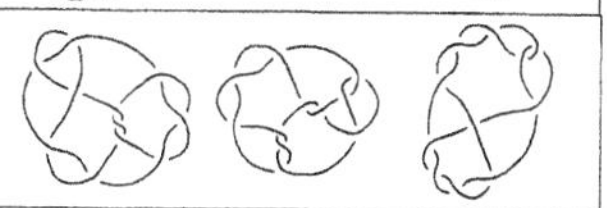
11^a_{224}: $1^{\underline{2}}_{\underline{1}}1^{\underline{1}}_{1}1^{\underline{1}}_{3}4^{0}_{3}2^{0}_{5}5^{1}_{5}3^{1}_{7}7^{2}_{7}5^{2}_{9}7^{3}_{9}7^{3}_{11}7^{4}_{11}7^{4}_{13}6^{5}_{13}7^{5}_{15}4^{6}_{15}6^{6}_{17}2^{7}_{17}4^{7}_{19}1^{8}_{19}2^{8}_{21}1^{9}_{23}$	
11^a_{225}: $1^{\underline{7}}_{\underline{19}}1^{\underline{6}}_{\underline{17}}1^{\underline{6}}_{\underline{15}}3^{\underline{5}}_{\underline{15}}1^{\underline{5}}_{\underline{13}}3^{\underline{4}}_{\underline{13}}3^{\underline{4}}_{\underline{11}}4^{\underline{3}}_{\underline{11}}3^{\underline{3}}_{\underline{9}}4^{\underline{2}}_{\underline{9}}4^{\underline{2}}_{\underline{7}}3^{\underline{1}}_{\underline{7}}4^{\underline{1}}_{\underline{5}}4^{0}_{\underline{5}}4^{0}_{\underline{3}}2^{1}_{\underline{3}}3^{1}_{\underline{1}}1^{2}_{\underline{1}}2^{2}_{1}1^{3}_{1}1^{3}_{3}1^{4}_{5}$	
11^a_{226}: $1^{\underline{7}}_{\underline{17}}1^{\underline{6}}_{\underline{15}}1^{\underline{6}}_{\underline{13}}3^{\underline{5}}_{\underline{13}}1^{\underline{5}}_{\underline{11}}4^{\underline{4}}_{\underline{11}}3^{\underline{4}}_{\underline{9}}5^{\underline{3}}_{\underline{9}}4^{\underline{3}}_{\underline{7}}6^{\underline{2}}_{\underline{7}}5^{\underline{2}}_{\underline{5}}5^{\underline{1}}_{\underline{5}}6^{\underline{1}}_{\underline{3}}5^{0}_{\underline{3}}6^{0}_{\underline{1}}3^{1}_{\underline{1}}4^{1}_{1}2^{2}_{1}3^{2}_{3}1^{3}_{3}2^{3}_{5}1^{4}_{7}$	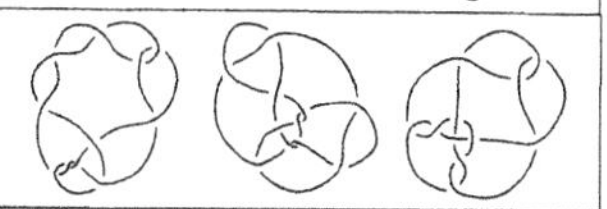
11^a_{227}: $1^{0}_{5}1^{0}_{7}3^{1}_{7}6^{2}_{9}3^{2}_{11}8^{3}_{11}6^{3}_{13}12^{4}_{13}8^{4}_{15}11^{5}_{15}12^{5}_{17}12^{6}_{17}11^{6}_{19}9^{7}_{19}12^{7}_{21}6^{8}_{21}9^{8}_{23}3^{9}_{23}6^{9}_{25}1^{10}_{25}3^{10}_{27}1^{11}_{29}$	
11^a_{228}: $1^{\underline{6}}_{\underline{13}}2^{\underline{5}}_{\underline{11}}1^{\underline{5}}_{\underline{9}}5^{\underline{4}}_{\underline{9}}2^{\underline{4}}_{\underline{7}}8^{\underline{3}}_{\underline{7}}5^{\underline{3}}_{\underline{5}}10^{\underline{2}}_{\underline{5}}8^{\underline{2}}_{\underline{3}}11^{\underline{1}}_{\underline{3}}10^{\underline{1}}_{\underline{1}}11^{0}_{\underline{1}}12^{0}_{1}9^{1}_{1}10^{1}_{3}6^{2}_{3}9^{2}_{5}3^{3}_{5}6^{3}_{7}1^{4}_{7}3^{4}_{9}1^{5}_{11}$	

K: KH	1st line 2nd line 3rd line
11^a_{229}: $1^{\underline{2}}_{\underline{3}}1^{\underline{1}}_{\underline{1}}1^{\underline{1}}_{1}4^{0}_{1}2^{0}_{3}5^{1}_{3}3^{1}_{5}5^{2}_{5}5^{2}_{7}6^{3}_{7}5^{3}_{9}5^{4}_{9}6^{4}_{11}4^{5}_{11}5^{5}_{13}3^{6}_{13}4^{6}_{15}1^{7}_{15}3^{7}_{17}1^{8}_{17}1^{8}_{19}1^{9}_{21}$	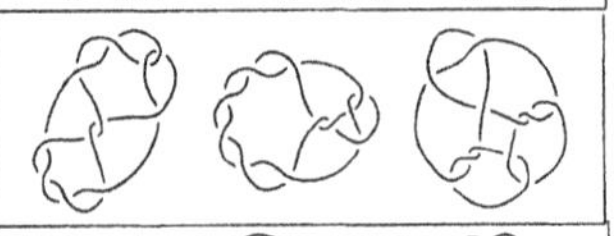
11^a_{230}: $1^{\underline{2}}_{\underline{3}}1^{\underline{1}}_{\underline{1}}1^{\underline{1}}_{1}3^{0}_{1}2^{0}_{3}3^{1}_{3}2^{1}_{5}4^{2}_{5}3^{2}_{7}4^{3}_{7}4^{3}_{9}3^{4}_{9}4^{4}_{11}3^{5}_{11}3^{5}_{13}2^{6}_{13}3^{6}_{15}1^{7}_{15}2^{7}_{17}1^{8}_{17}1^{8}_{19}1^{9}_{21}$	
11^a_{231}: $1^{\underline{6}}_{\underline{15}}2^{\underline{5}}_{\underline{13}}1^{\underline{5}}_{\underline{11}}5^{\underline{4}}_{\underline{11}}2^{\underline{4}}_{\underline{9}}7^{\underline{3}}_{\underline{9}}5^{\underline{3}}_{\underline{7}}7^{\underline{2}}_{\underline{7}}7^{\underline{2}}_{\underline{5}}9^{\underline{1}}_{\underline{5}}7^{\underline{1}}_{\underline{3}}7^{0}_{\underline{3}}10^{0}_{\underline{1}}6^{1}_{\underline{1}}6^{1}_{1}4^{2}_{1}6^{2}_{3}1^{3}_{3}4^{3}_{5}1^{4}_{5}1^{4}_{7}1^{5}_{9}$	
11^a_{232}: $1^{\underline{6}}_{\underline{15}}2^{\underline{5}}_{\underline{13}}1^{\underline{5}}_{\underline{11}}5^{\underline{4}}_{\underline{11}}2^{\underline{4}}_{\underline{9}}8^{\underline{3}}_{\underline{9}}5^{\underline{3}}_{\underline{7}}9^{\underline{2}}_{\underline{7}}8^{\underline{2}}_{\underline{5}}10^{\underline{1}}_{\underline{5}}9^{\underline{1}}_{\underline{3}}10^{0}_{\underline{3}}11^{0}_{\underline{1}}8^{1}_{\underline{1}}9^{1}_{1}5^{2}_{1}8^{2}_{3}3^{3}_{3}5^{3}_{5}1^{4}_{5}3^{4}_{7}1^{5}_{9}$	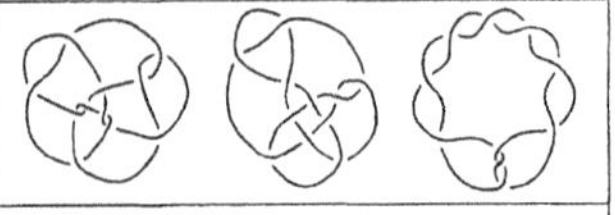
11^a_{233}: $1^{\underline{5}}_{\underline{11}}3^{\underline{4}}_{\underline{9}}1^{\underline{4}}_{\underline{7}}7^{\underline{3}}_{\underline{7}}3^{\underline{3}}_{\underline{5}}11^{\underline{2}}_{\underline{5}}7^{\underline{2}}_{\underline{3}}13^{\underline{1}}_{\underline{3}}11^{\underline{1}}_{\underline{1}}15^{0}_{\underline{1}}14^{0}_{1}14^{1}_{1}14^{1}_{3}11^{2}_{3}14^{2}_{5}7^{3}_{5}11^{3}_{7}4^{4}_{7}7^{4}_{9}1^{5}_{9}4^{5}_{11}1^{6}_{13}$	
11^a_{234}: $1^{0}_{7}1^{0}_{9}1^{1}_{9}2^{2}_{11}1^{2}_{13}1^{3}_{13}2^{3}_{15}3^{4}_{15}1^{4}_{17}2^{5}_{17}3^{5}_{19}3^{6}_{19}2^{6}_{21}2^{7}_{21}3^{7}_{23}2^{8}_{23}2^{8}_{25}1^{9}_{25}2^{9}_{27}1^{10}_{27}1^{10}_{29}1^{11}_{31}$	
11^a_{235}: $1^{0}_{5}1^{0}_{7}2^{1}_{7}3^{2}_{9}2^{2}_{11}4^{3}_{11}3^{3}_{13}6^{4}_{13}4^{4}_{15}5^{5}_{15}6^{5}_{17}6^{6}_{17}5^{6}_{19}4^{7}_{19}6^{7}_{21}3^{8}_{21}4^{8}_{23}1^{9}_{23}3^{9}_{25}1^{10}_{25}1^{10}_{27}1^{11}_{29}$	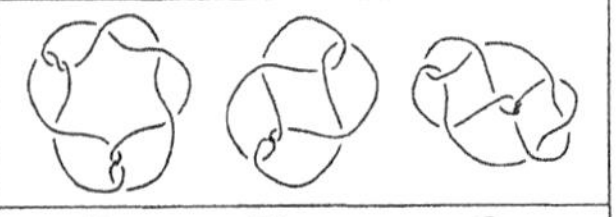
11^a_{236}: $1^{0}_{5}1^{0}_{7}2^{1}_{7}4^{2}_{9}2^{2}_{11}5^{3}_{11}4^{3}_{13}8^{4}_{13}5^{4}_{15}8^{5}_{15}8^{5}_{17}8^{6}_{17}8^{6}_{19}6^{7}_{19}8^{7}_{21}5^{8}_{21}6^{8}_{23}2^{9}_{23}5^{9}_{25}1^{10}_{25}2^{10}_{27}1^{11}_{29}$	
11^a_{237}: $1^{0}_{3}1^{0}_{5}3^{1}_{5}4^{2}_{7}3^{2}_{9}6^{3}_{9}4^{3}_{11}8^{4}_{11}6^{4}_{13}7^{5}_{13}8^{5}_{15}7^{6}_{15}7^{6}_{17}5^{7}_{17}7^{7}_{19}4^{8}_{19}5^{8}_{21}1^{9}_{21}4^{9}_{23}1^{10}_{23}1^{10}_{25}1^{11}_{27}$	
11^a_{238}: $1^{0}_{3}1^{0}_{5}2^{1}_{5}3^{2}_{7}2^{2}_{9}4^{3}_{9}3^{3}_{11}5^{4}_{11}4^{4}_{13}5^{5}_{13}5^{5}_{15}5^{6}_{15}5^{6}_{17}3^{7}_{17}5^{7}_{19}3^{8}_{19}3^{8}_{21}1^{9}_{21}3^{9}_{23}1^{10}_{23}1^{10}_{25}1^{11}_{27}$	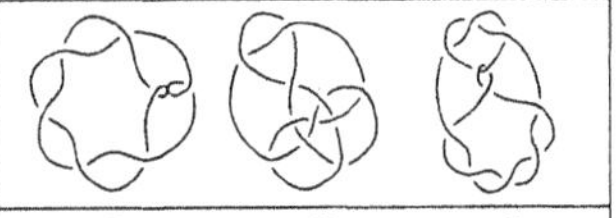
11^a_{239}: $1^{\underline{4}}_{\underline{7}}4^{\underline{3}}_{\underline{5}}1^{\underline{3}}_{\underline{3}}8^{\underline{2}}_{\underline{3}}4^{\underline{2}}_{\underline{1}}12^{\underline{1}}_{\underline{1}}8^{\underline{1}}_{1}16^{0}_{1}13^{0}_{3}16^{1}_{3}15^{1}_{5}16^{2}_{5}16^{2}_{7}12^{3}_{7}16^{3}_{9}8^{4}_{9}12^{4}_{11}4^{5}_{11}8^{5}_{13}1^{6}_{13}4^{6}_{15}1^{7}_{17}$	
11^a_{240}: $1^{0}_{7}1^{0}_{9}1^{1}_{9}3^{2}_{11}1^{2}_{13}2^{3}_{13}3^{3}_{15}5^{4}_{15}2^{4}_{17}4^{5}_{17}5^{5}_{19}5^{6}_{19}4^{6}_{21}4^{7}_{21}5^{7}_{23}3^{8}_{23}4^{8}_{25}2^{9}_{25}3^{9}_{27}1^{10}_{27}2^{10}_{29}1^{11}_{31}$	
11^a_{241}: $1^{0}_{5}1^{0}_{7}2^{1}_{7}4^{2}_{9}2^{2}_{11}5^{3}_{11}4^{3}_{13}8^{4}_{13}5^{4}_{15}7^{5}_{15}8^{5}_{17}8^{6}_{17}7^{6}_{19}6^{7}_{19}8^{7}_{21}4^{8}_{21}6^{8}_{23}2^{9}_{23}4^{9}_{25}1^{10}_{25}2^{10}_{27}1^{11}_{29}$	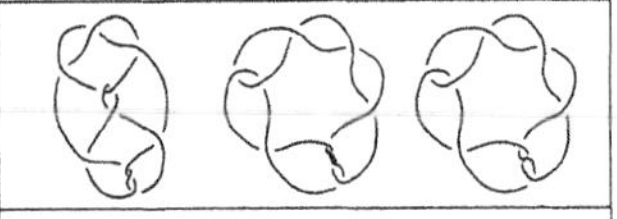
11^a_{242}: $1^{0}_{5}1^{0}_{7}1^{1}_{7}2^{2}_{9}1^{2}_{11}2^{3}_{11}2^{3}_{13}4^{4}_{13}2^{4}_{15}3^{5}_{15}4^{5}_{17}4^{6}_{17}3^{6}_{19}3^{7}_{19}4^{7}_{21}2^{8}_{21}3^{8}_{23}1^{9}_{23}2^{9}_{25}1^{10}_{25}1^{10}_{27}1^{11}_{29}$	
11^a_{243}: $1^{0}_{3}1^{0}_{5}2^{1}_{5}3^{2}_{7}2^{2}_{9}4^{3}_{9}3^{3}_{11}6^{4}_{11}4^{4}_{13}5^{5}_{13}6^{5}_{15}5^{6}_{15}5^{6}_{17}4^{7}_{17}5^{7}_{19}3^{8}_{19}4^{8}_{21}1^{9}_{21}3^{9}_{23}1^{10}_{23}1^{10}_{25}1^{11}_{27}$	
11^a_{244}: $1^{0}_{5}1^{0}_{7}3^{1}_{7}6^{2}_{9}3^{2}_{11}8^{3}_{11}6^{3}_{13}12^{4}_{13}8^{4}_{15}12^{5}_{15}12^{5}_{17}12^{6}_{17}12^{6}_{19}9^{7}_{19}12^{7}_{21}7^{8}_{21}9^{8}_{23}3^{9}_{23}7^{9}_{25}1^{10}_{25}3^{10}_{27}1^{11}_{29}$	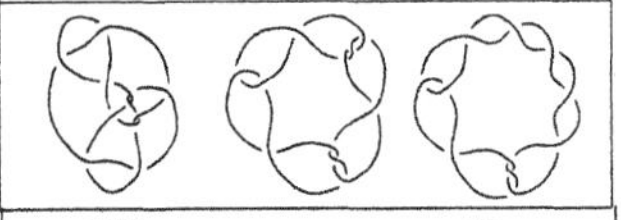
11^a_{245}: $1^{0}_{5}1^{0}_{7}1^{1}_{7}3^{2}_{9}1^{2}_{11}3^{3}_{11}3^{3}_{13}6^{4}_{13}3^{4}_{15}6^{5}_{15}6^{5}_{17}6^{6}_{17}6^{6}_{19}5^{7}_{19}6^{7}_{21}4^{8}_{21}5^{8}_{23}2^{9}_{23}4^{9}_{25}1^{10}_{25}2^{10}_{27}1^{11}_{29}$	
11^a_{246}: $1^{0}_{3}1^{0}_{5}1^{1}_{5}2^{2}_{7}1^{2}_{9}2^{3}_{9}2^{3}_{11}3^{4}_{11}2^{4}_{13}3^{5}_{13}3^{5}_{15}3^{6}_{15}3^{6}_{17}2^{7}_{17}3^{7}_{19}2^{8}_{19}2^{8}_{21}1^{9}_{21}2^{9}_{23}1^{10}_{23}1^{10}_{25}1^{11}_{27}$	
11^a_{247}: $1^{0}_{1}1^{0}_{3}1^{1}_{3}1^{2}_{5}1^{2}_{7}1^{3}_{7}1^{3}_{9}1^{4}_{9}1^{4}_{11}1^{5}_{11}1^{5}_{13}1^{6}_{13}1^{6}_{15}1^{7}_{15}1^{7}_{17}1^{8}_{17}1^{8}_{19}1^{9}_{21}1^{10}_{21}1^{11}_{25}$	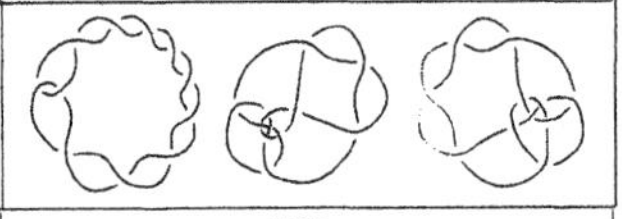
11^a_{248}: $1^{\underline{7}}_{\underline{17}}4^{\underline{6}}_{\underline{15}}1^{\underline{6}}_{\underline{13}}7^{\underline{5}}_{\underline{13}}4^{\underline{5}}_{\underline{11}}10^{\underline{4}}_{\underline{11}}7^{\underline{4}}_{\underline{9}}13^{\underline{3}}_{\underline{9}}10^{\underline{3}}_{\underline{7}}13^{\underline{2}}_{\underline{7}}13^{\underline{2}}_{\underline{5}}12^{\underline{1}}_{\underline{5}}13^{\underline{1}}_{\underline{3}}10^{0}_{\underline{3}}13^{0}_{\underline{1}}6^{1}_{\underline{1}}9^{1}_{1}3^{2}_{1}6^{2}_{3}1^{3}_{3}3^{3}_{5}1^{4}_{7}$	
11^a_{249}: $1^{\underline{7}}_{\underline{15}}3^{\underline{6}}_{\underline{13}}1^{\underline{6}}_{\underline{11}}4^{\underline{5}}_{\underline{11}}3^{\underline{5}}_{\underline{9}}7^{\underline{4}}_{\underline{9}}4^{\underline{4}}_{\underline{7}}9^{\underline{3}}_{\underline{7}}7^{\underline{3}}_{\underline{5}}9^{\underline{2}}_{\underline{5}}9^{\underline{2}}_{\underline{3}}9^{\underline{1}}_{\underline{3}}9^{\underline{1}}_{\underline{1}}8^{0}_{\underline{1}}10^{0}_{1}5^{1}_{1}7^{1}_{3}3^{2}_{3}5^{2}_{5}1^{3}_{5}3^{3}_{7}1^{4}_{9}$	
11^a_{250}: $1^{\underline{4}}_{\underline{5}}1^{\underline{3}}_{\underline{3}}1^{\underline{3}}_{\underline{1}}4^{\underline{2}}_{\underline{1}}1^{\underline{2}}_{1}4^{\underline{1}}_{1}4^{\underline{1}}_{3}7^{0}_{3}5^{0}_{5}7^{1}_{5}6^{1}_{7}6^{2}_{7}7^{2}_{9}6^{3}_{9}6^{3}_{11}4^{4}_{11}6^{4}_{13}2^{5}_{13}4^{5}_{15}1^{6}_{15}2^{6}_{17}1^{7}_{19}$	
11^a_{251}: $1^{\underline{6}}_{\underline{13}}2^{\underline{5}}_{\underline{11}}1^{\underline{5}}_{\underline{9}}5^{\underline{4}}_{\underline{9}}2^{\underline{4}}_{\underline{7}}8^{\underline{3}}_{\underline{7}}5^{\underline{3}}_{\underline{5}}10^{\underline{2}}_{\underline{5}}8^{\underline{2}}_{\underline{3}}11^{\underline{1}}_{\underline{3}}10^{\underline{1}}_{\underline{1}}11^{0}_{\underline{1}}12^{0}_{1}9^{1}_{1}10^{1}_{3}6^{2}_{3}9^{2}_{5}3^{3}_{5}6^{3}_{7}1^{4}_{7}3^{4}_{9}1^{5}_{11}$	
11^a_{252}: $1^{\underline{5}}_{\underline{9}}2^{\underline{4}}_{\underline{7}}1^{\underline{4}}_{\underline{5}}5^{\underline{3}}_{\underline{5}}2^{\underline{3}}_{\underline{3}}8^{\underline{2}}_{\underline{3}}5^{\underline{2}}_{\underline{1}}9^{\underline{1}}_{\underline{1}}8^{\underline{1}}_{1}12^{0}_{1}10^{0}_{3}10^{1}_{3}11^{1}_{5}9^{2}_{5}10^{2}_{7}6^{3}_{7}9^{3}_{9}3^{4}_{9}6^{4}_{11}1^{5}_{11}3^{5}_{13}1^{6}_{15}$	
11^a_{253}: $1^{\underline{6}}_{\underline{13}}2^{\underline{5}}_{\underline{11}}1^{\underline{5}}_{\underline{9}}5^{\underline{4}}_{\underline{9}}2^{\underline{4}}_{\underline{7}}8^{\underline{3}}_{\underline{7}}5^{\underline{3}}_{\underline{5}}10^{\underline{2}}_{\underline{5}}8^{\underline{2}}_{\underline{3}}11^{\underline{1}}_{\underline{3}}10^{\underline{1}}_{\underline{1}}11^{0}_{\underline{1}}12^{0}_{1}9^{1}_{1}10^{1}_{3}6^{2}_{3}9^{2}_{5}3^{3}_{5}6^{3}_{7}1^{4}_{7}3^{4}_{9}1^{5}_{11}$	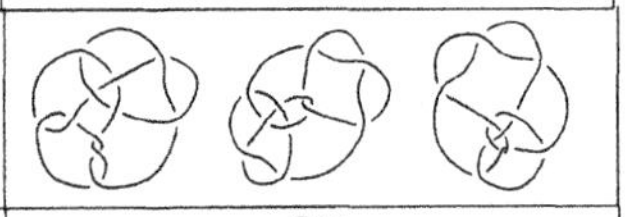
11^a_{254}: $1^{\underline{5}}_{\underline{9}}2^{\underline{4}}_{\underline{7}}1^{\underline{4}}_{\underline{5}}5^{\underline{3}}_{\underline{5}}2^{\underline{3}}_{\underline{3}}8^{\underline{2}}_{\underline{3}}5^{\underline{2}}_{\underline{1}}9^{\underline{1}}_{\underline{1}}8^{\underline{1}}_{1}12^{0}_{1}10^{0}_{3}10^{1}_{3}11^{1}_{5}9^{2}_{5}10^{2}_{7}6^{3}_{7}9^{3}_{9}3^{4}_{9}6^{4}_{11}1^{5}_{11}3^{5}_{13}1^{6}_{15}$	
11^a_{255}: $1^{\underline{7}}_{\underline{17}}3^{\underline{6}}_{\underline{15}}1^{\underline{6}}_{\underline{13}}6^{\underline{5}}_{\underline{13}}3^{\underline{5}}_{\underline{11}}9^{\underline{4}}_{\underline{11}}6^{\underline{4}}_{\underline{9}}11^{\underline{3}}_{\underline{9}}9^{\underline{3}}_{\underline{7}}12^{\underline{2}}_{\underline{7}}11^{\underline{2}}_{\underline{5}}11^{\underline{1}}_{\underline{5}}12^{\underline{1}}_{\underline{3}}9^{0}_{\underline{3}}12^{0}_{\underline{1}}6^{1}_{\underline{1}}8^{1}_{1}3^{2}_{1}6^{2}_{3}1^{3}_{3}3^{3}_{5}1^{4}_{7}$	
11^a_{256}: $1^{\underline{7}}_{\underline{15}}3^{\underline{6}}_{\underline{13}}1^{\underline{6}}_{\underline{11}}5^{\underline{5}}_{\underline{11}}3^{\underline{5}}_{\underline{9}}8^{\underline{4}}_{\underline{9}}5^{\underline{4}}_{\underline{7}}10^{\underline{3}}_{\underline{7}}8^{\underline{3}}_{\underline{5}}11^{\underline{2}}_{\underline{5}}10^{\underline{2}}_{\underline{3}}10^{\underline{1}}_{\underline{3}}11^{\underline{1}}_{\underline{1}}9^{0}_{\underline{1}}11^{0}_{1}6^{1}_{1}8^{1}_{3}3^{2}_{3}6^{2}_{5}1^{3}_{5}3^{3}_{7}1^{4}_{9}$	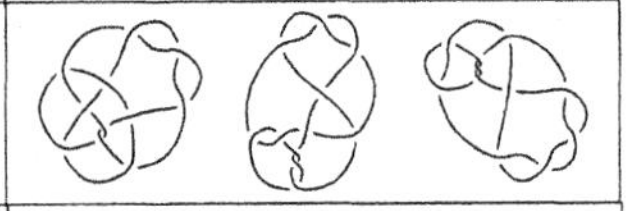
11^a_{257}: $1^{\underline{6}}_{\underline{13}}2^{\underline{5}}_{\underline{11}}1^{\underline{5}}_{\underline{9}}4^{\underline{4}}_{\underline{9}}2^{\underline{4}}_{\underline{7}}6^{\underline{3}}_{\underline{7}}4^{\underline{3}}_{\underline{5}}7^{\underline{2}}_{\underline{5}}6^{\underline{2}}_{\underline{3}}8^{\underline{1}}_{\underline{3}}7^{\underline{1}}_{\underline{1}}8^{0}_{\underline{1}}9^{0}_{1}6^{1}_{1}7^{1}_{3}4^{2}_{3}6^{2}_{5}2^{3}_{5}4^{3}_{7}1^{4}_{7}2^{4}_{9}1^{5}_{11}$	
11^a_{258}: $1^{\underline{6}}_{\underline{11}}1^{\underline{5}}_{\underline{9}}1^{\underline{5}}_{\underline{7}}3^{\underline{4}}_{\underline{7}}1^{\underline{4}}_{\underline{5}}4^{\underline{3}}_{\underline{5}}3^{\underline{3}}_{\underline{3}}5^{\underline{2}}_{\underline{3}}4^{\underline{2}}_{\underline{1}}6^{\underline{1}}_{\underline{1}}5^{\underline{1}}_{1}6^{0}_{1}7^{0}_{3}5^{1}_{3}5^{1}_{5}4^{2}_{5}5^{2}_{7}2^{3}_{7}4^{3}_{9}1^{4}_{9}2^{4}_{11}1^{5}_{13}$	
11^a_{259}: $1^{\underline{2}}_{1}1^{\underline{1}}_{3}1^{\underline{1}}_{5}4^{0}_{5}2^{0}_{7}4^{1}_{7}3^{1}_{9}6^{2}_{9}4^{2}_{11}6^{3}_{11}6^{3}_{13}6^{4}_{13}6^{4}_{15}5^{5}_{15}6^{5}_{17}4^{6}_{17}5^{6}_{19}2^{7}_{19}4^{7}_{21}1^{8}_{21}2^{8}_{23}1^{9}_{25}$	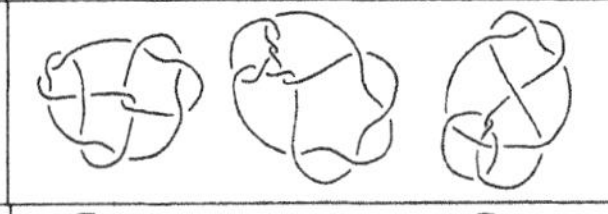
11^a_{260}: $1^{\underline{2}}_{\underline{1}}1^{\underline{1}}_{1}1^{\underline{1}}_{3}4^{0}_{3}2^{0}_{5}4^{1}_{5}3^{1}_{7}5^{2}_{7}4^{2}_{9}6^{3}_{9}5^{3}_{11}5^{4}_{11}6^{4}_{13}4^{5}_{13}5^{5}_{15}3^{6}_{15}4^{6}_{17}1^{7}_{17}3^{7}_{19}1^{8}_{19}1^{8}_{21}1^{9}_{23}$	
11^a_{261}: $1^{\underline{8}}_{\underline{21}}3^{\underline{7}}_{\underline{19}}1^{\underline{7}}_{\underline{17}}6^{\underline{6}}_{\underline{17}}3^{\underline{6}}_{\underline{15}}8^{\underline{5}}_{\underline{15}}6^{\underline{5}}_{\underline{13}}10^{\underline{4}}_{\underline{13}}8^{\underline{4}}_{\underline{11}}11^{\underline{3}}_{\underline{11}}10^{\underline{3}}_{\underline{9}}9^{\underline{2}}_{\underline{9}}11^{\underline{2}}_{\underline{7}}8^{\underline{1}}_{\underline{7}}9^{\underline{1}}_{\underline{5}}5^{0}_{\underline{5}}9^{0}_{\underline{3}}3^{1}_{\underline{3}}4^{1}_{\underline{1}}1^{2}_{\underline{1}}3^{2}_{1}1^{3}_{3}$	
11^a_{262}: $1^{\underline{8}}_{\underline{19}}2^{\underline{7}}_{\underline{17}}1^{\underline{7}}_{\underline{15}}4^{\underline{6}}_{\underline{15}}2^{\underline{6}}_{\underline{13}}6^{\underline{5}}_{\underline{13}}4^{\underline{5}}_{\underline{11}}8^{\underline{4}}_{\underline{11}}6^{\underline{4}}_{\underline{9}}9^{\underline{3}}_{\underline{9}}8^{\underline{3}}_{\underline{7}}8^{\underline{2}}_{\underline{7}}9^{\underline{2}}_{\underline{5}}7^{\underline{1}}_{\underline{5}}8^{\underline{1}}_{\underline{3}}5^{0}_{\underline{3}}8^{0}_{\underline{1}}3^{1}_{\underline{1}}4^{1}_{1}1^{2}_{1}3^{2}_{3}1^{3}_{5}$	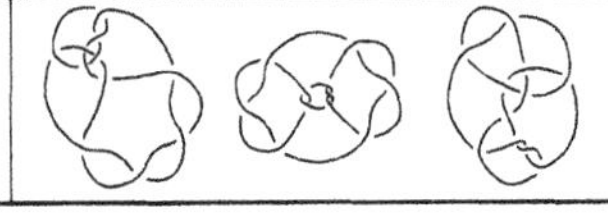
11^a_{263}: $1^{0}_{7}1^{0}_{9}1^{1}_{9}4^{2}_{11}1^{2}_{13}3^{3}_{13}4^{3}_{15}7^{4}_{15}3^{4}_{17}6^{5}_{17}7^{5}_{19}6^{6}_{19}6^{6}_{21}6^{7}_{21}6^{7}_{23}4^{8}_{23}6^{8}_{25}2^{9}_{25}4^{9}_{27}1^{10}_{27}2^{10}_{29}1^{11}_{31}$	
11^a_{264}: $1^{\underline{4}}_{\underline{7}}3^{\underline{3}}_{\underline{5}}1^{\underline{3}}_{\underline{3}}5^{\underline{2}}_{\underline{3}}3^{\underline{2}}_{\underline{1}}8^{\underline{1}}_{\underline{1}}5^{\underline{1}}_{1}11^{0}_{1}9^{0}_{3}11^{1}_{3}10^{1}_{5}11^{2}_{5}11^{2}_{7}8^{3}_{7}11^{3}_{9}6^{4}_{9}8^{4}_{11}3^{5}_{11}6^{5}_{13}1^{6}_{13}3^{6}_{15}1^{7}_{17}$	

K: KH	1st line 2nd line 3rd line
11^a_{265}: $1^{\underline{4}}_{\underline{9}}3^{\underline{3}}_{\underline{7}}1^{\underline{3}}_{\underline{5}}5^{\underline{2}}_{\underline{5}}3^{\underline{2}}_{\underline{3}}7^{\underline{1}}_{\underline{3}}5^{\underline{1}}_{\underline{1}}9^{0}_{\underline{1}}8^{0}_{1}9^{1}_{1}8^{1}_{3}8^{2}_{3}9^{2}_{5}6^{3}_{5}8^{3}_{7}4^{4}_{7}6^{4}_{9}2^{5}_{9}4^{5}_{11}1^{6}_{11}2^{6}_{13}1^{7}_{15}$	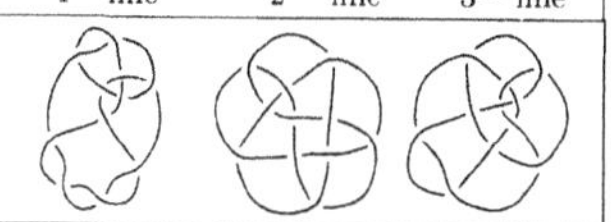
11^a_{266}: $1^{\underline{6}}_{\underline{13}}5^{\underline{5}}_{\underline{11}}1^{\underline{5}}_{\underline{9}}9^{\underline{4}}_{\underline{9}}5^{\underline{4}}_{\underline{7}}13^{\underline{3}}_{\underline{7}}9^{\underline{3}}_{\underline{5}}17^{\underline{2}}_{\underline{5}}13^{\underline{2}}_{\underline{3}}17^{\underline{1}}_{\underline{3}}17^{\underline{1}}_{\underline{1}}17^{0}_{\underline{1}}18^{0}_{1}13^{1}_{1}16^{1}_{3}8^{2}_{3}13^{2}_{5}4^{3}_{5}8^{3}_{7}1^{4}_{7}4^{4}_{9}1^{5}_{11}$	
11^a_{267}: $1^{\underline{7}}_{\underline{17}}4^{\underline{6}}_{\underline{15}}1^{\underline{6}}_{\underline{13}}8^{\underline{5}}_{\underline{13}}4^{\underline{5}}_{\underline{11}}12^{\underline{4}}_{\underline{11}}8^{\underline{4}}_{\underline{9}}15^{\underline{3}}_{\underline{9}}12^{\underline{3}}_{\underline{7}}16^{\underline{2}}_{\underline{7}}15^{\underline{2}}_{\underline{5}}15^{\underline{1}}_{\underline{5}}16^{\underline{1}}_{\underline{3}}12^{0}_{\underline{3}}16^{0}_{\underline{1}}8^{1}_{\underline{1}}11^{1}_{1}4^{2}_{1}8^{2}_{3}1^{3}_{3}4^{3}_{5}1^{4}_{7}$	
11^a_{268}: $1^{\underline{5}}_{\underline{9}}3^{\underline{4}}_{\underline{7}}1^{\underline{4}}_{\underline{5}}5^{\underline{3}}_{\underline{5}}3^{\underline{3}}_{\underline{3}}9^{\underline{2}}_{\underline{3}}5^{\underline{2}}_{\underline{1}}10^{\underline{1}}_{\underline{1}}9^{\underline{1}}_{1}12^{0}_{1}11^{0}_{3}11^{1}_{3}11^{1}_{5}9^{2}_{5}11^{2}_{7}6^{3}_{7}9^{3}_{9}3^{4}_{9}6^{4}_{11}1^{5}_{11}3^{5}_{13}1^{6}_{15}$	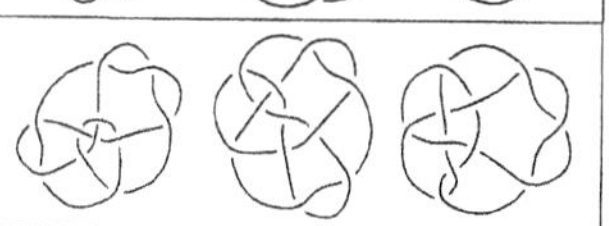
11^a_{269}: $1^{\underline{5}}_{\underline{9}}4^{\underline{4}}_{\underline{7}}1^{\underline{4}}_{\underline{5}}6^{\underline{3}}_{\underline{5}}4^{\underline{3}}_{\underline{3}}10^{\underline{2}}_{\underline{3}}6^{\underline{2}}_{\underline{1}}11^{\underline{1}}_{\underline{1}}10^{\underline{1}}_{1}13^{0}_{1}12^{0}_{3}12^{1}_{3}12^{1}_{5}9^{2}_{5}12^{2}_{7}6^{3}_{7}9^{3}_{9}3^{4}_{9}6^{4}_{11}1^{5}_{11}3^{5}_{13}1^{6}_{15}$	
11^a_{270}: $1^{\underline{5}}_{\underline{11}}3^{\underline{4}}_{\underline{9}}1^{\underline{4}}_{\underline{7}}5^{\underline{3}}_{\underline{7}}3^{\underline{3}}_{\underline{5}}9^{\underline{2}}_{\underline{5}}5^{\underline{2}}_{\underline{3}}10^{\underline{1}}_{\underline{3}}9^{\underline{1}}_{\underline{1}}12^{0}_{\underline{1}}11^{0}_{1}11^{1}_{1}11^{1}_{3}8^{2}_{3}11^{2}_{5}6^{3}_{5}8^{3}_{7}3^{4}_{7}6^{4}_{9}1^{5}_{9}3^{5}_{11}1^{6}_{13}$	
11^a_{271}: $1^{\underline{3}}_{\underline{5}}4^{\underline{2}}_{\underline{3}}1^{\underline{2}}_{\underline{1}}7^{\underline{1}}_{\underline{1}}4^{\underline{1}}_{1}12^{0}_{1}8^{0}_{3}13^{1}_{3}11^{1}_{5}15^{2}_{5}13^{2}_{7}13^{3}_{7}15^{3}_{9}10^{4}_{9}13^{4}_{11}7^{5}_{11}10^{5}_{13}3^{6}_{13}7^{6}_{15}1^{7}_{15}3^{7}_{17}1^{8}_{19}$	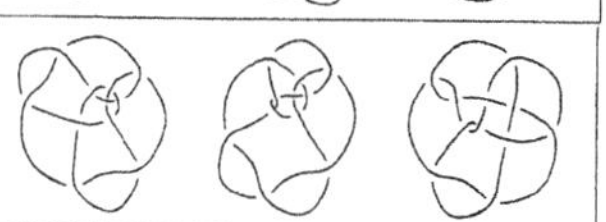
11^a_{272}: $1^{\underline{5}}_{\underline{11}}3^{\underline{4}}_{\underline{9}}1^{\underline{4}}_{\underline{7}}6^{\underline{3}}_{\underline{7}}3^{\underline{3}}_{\underline{5}}10^{\underline{2}}_{\underline{5}}6^{\underline{2}}_{\underline{3}}11^{\underline{1}}_{\underline{3}}10^{\underline{1}}_{\underline{1}}13^{0}_{\underline{1}}12^{0}_{1}12^{1}_{1}12^{1}_{3}9^{2}_{3}12^{2}_{5}6^{3}_{5}9^{3}_{7}3^{4}_{7}6^{4}_{9}1^{5}_{9}3^{5}_{11}1^{6}_{13}$	
11^a_{273}: $1^{\underline{8}}_{\underline{19}}3^{\underline{7}}_{\underline{17}}1^{\underline{7}}_{\underline{15}}7^{\underline{6}}_{\underline{15}}3^{\underline{6}}_{\underline{13}}10^{\underline{5}}_{\underline{13}}7^{\underline{5}}_{\underline{11}}12^{\underline{4}}_{\underline{11}}10^{\underline{4}}_{\underline{9}}14^{\underline{3}}_{\underline{9}}12^{\underline{3}}_{\underline{7}}12^{\underline{2}}_{\underline{7}}14^{\underline{2}}_{\underline{5}}10^{\underline{1}}_{\underline{5}}12^{\underline{1}}_{\underline{3}}7^{0}_{\underline{3}}11^{0}_{\underline{1}}3^{1}_{\underline{1}}6^{1}_{1}1^{2}_{1}3^{2}_{3}1^{3}_{5}$	
11^a_{274}: $1^{\underline{6}}_{\underline{13}}3^{\underline{5}}_{\underline{11}}1^{\underline{5}}_{\underline{9}}7^{\underline{4}}_{\underline{9}}3^{\underline{4}}_{\underline{7}}10^{\underline{3}}_{\underline{7}}7^{\underline{3}}_{\underline{5}}13^{\underline{2}}_{\underline{5}}10^{\underline{2}}_{\underline{3}}14^{\underline{1}}_{\underline{3}}13^{\underline{1}}_{\underline{1}}13^{0}_{\underline{1}}15^{0}_{1}11^{1}_{1}12^{1}_{3}7^{2}_{3}11^{2}_{5}3^{3}_{5}7^{3}_{7}1^{4}_{7}3^{4}_{9}1^{5}_{11}$	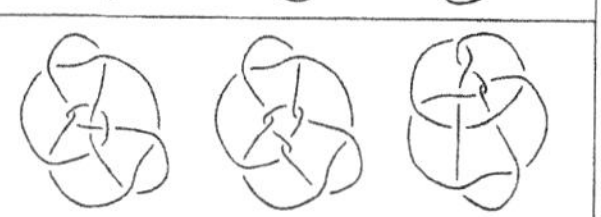
11^a_{275}: $1^{\underline{2}}_{\underline{1}}2^{\underline{1}}_{1}1^{\underline{1}}_{3}6^{0}_{3}3^{0}_{5}8^{1}_{5}5^{1}_{7}10^{2}_{7}8^{2}_{9}11^{3}_{9}10^{3}_{11}10^{4}_{11}11^{4}_{13}8^{5}_{13}10^{5}_{15}6^{6}_{15}8^{6}_{17}2^{7}_{17}6^{7}_{19}1^{8}_{19}2^{8}_{21}1^{9}_{23}$	
11^a_{276}: $1^{\underline{2}}_{\underline{1}}3^{\underline{1}}_{1}1^{\underline{1}}_{3}7^{0}_{3}4^{0}_{5}10^{1}_{5}6^{1}_{7}13^{2}_{7}10^{2}_{9}13^{3}_{9}13^{3}_{11}13^{4}_{11}13^{4}_{13}10^{5}_{13}13^{5}_{15}7^{6}_{15}10^{6}_{17}3^{7}_{17}7^{7}_{19}1^{8}_{19}3^{8}_{21}1^{9}_{23}$	
11^a_{277}: $1^{\underline{4}}_{\underline{7}}3^{\underline{3}}_{\underline{5}}1^{\underline{3}}_{\underline{3}}6^{\underline{2}}_{\underline{3}}3^{\underline{2}}_{\underline{1}}7^{\underline{1}}_{\underline{1}}6^{\underline{1}}_{1}12^{0}_{1}8^{0}_{3}11^{1}_{3}11^{1}_{5}10^{2}_{5}11^{2}_{7}9^{3}_{7}10^{3}_{9}5^{4}_{9}9^{4}_{11}3^{5}_{11}5^{5}_{13}1^{6}_{13}3^{6}_{15}1^{7}_{17}$	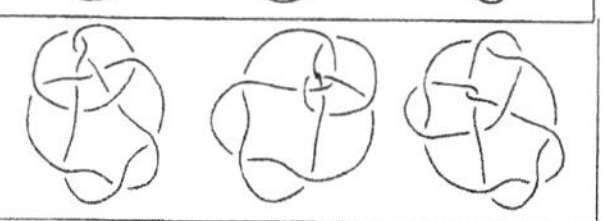
11^a_{278}: $1^{\underline{4}}_{\underline{9}}3^{\underline{3}}_{\underline{7}}1^{\underline{3}}_{\underline{5}}7^{\underline{2}}_{\underline{5}}3^{\underline{2}}_{\underline{3}}8^{\underline{1}}_{\underline{3}}7^{\underline{1}}_{\underline{1}}12^{0}_{\underline{1}}9^{0}_{1}12^{1}_{1}11^{1}_{3}10^{2}_{3}12^{2}_{5}9^{3}_{5}10^{3}_{7}5^{4}_{7}9^{4}_{9}3^{5}_{9}5^{5}_{11}1^{6}_{11}3^{6}_{13}1^{7}_{15}$	
11^a_{279}: $1^{\underline{6}}_{\underline{11}}2^{\underline{5}}_{\underline{9}}1^{\underline{5}}_{\underline{7}}4^{\underline{4}}_{\underline{7}}2^{\underline{4}}_{\underline{5}}5^{\underline{3}}_{\underline{5}}4^{\underline{3}}_{\underline{3}}7^{\underline{2}}_{\underline{3}}5^{\underline{2}}_{\underline{1}}7^{\underline{1}}_{\underline{1}}7^{\underline{1}}_{1}7^{0}_{1}8^{0}_{3}6^{1}_{3}6^{1}_{5}4^{2}_{5}6^{2}_{7}2^{3}_{7}4^{3}_{9}1^{4}_{9}2^{4}_{11}1^{5}_{13}$	
11^a_{280}: $1^{\underline{6}}_{\underline{13}}2^{\underline{5}}_{\underline{11}}1^{\underline{5}}_{\underline{9}}5^{\underline{4}}_{\underline{9}}2^{\underline{4}}_{\underline{7}}6^{\underline{3}}_{\underline{7}}5^{\underline{3}}_{\underline{5}}8^{\underline{2}}_{\underline{5}}6^{\underline{2}}_{\underline{3}}9^{\underline{1}}_{\underline{3}}8^{\underline{1}}_{\underline{1}}8^{0}_{\underline{1}}10^{0}_{1}7^{1}_{1}7^{1}_{3}4^{2}_{3}7^{2}_{5}2^{3}_{5}4^{3}_{7}1^{4}_{7}2^{4}_{9}1^{5}_{11}$	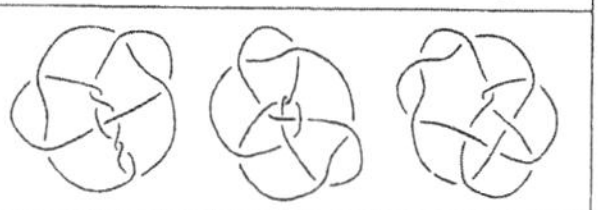
11^a_{281}: $1^{\underline{5}}_{\underline{9}}3^{\underline{4}}_{\underline{7}}1^{\underline{4}}_{\underline{5}}6^{\underline{3}}_{\underline{5}}3^{\underline{3}}_{\underline{3}}10^{\underline{2}}_{\underline{3}}6^{\underline{2}}_{\underline{1}}11^{\underline{1}}_{\underline{1}}10^{\underline{1}}_{1}14^{0}_{1}12^{0}_{3}12^{1}_{3}13^{1}_{5}10^{2}_{5}12^{2}_{7}7^{3}_{7}10^{3}_{9}3^{4}_{9}7^{4}_{11}1^{5}_{11}3^{5}_{13}1^{6}_{15}$	
11^a_{282}: $1^{\underline{6}}_{\underline{15}}3^{\underline{5}}_{\underline{13}}1^{\underline{5}}_{\underline{11}}5^{\underline{4}}_{\underline{11}}3^{\underline{4}}_{\underline{9}}8^{\underline{3}}_{\underline{9}}5^{\underline{3}}_{\underline{7}}10^{\underline{2}}_{\underline{7}}8^{\underline{2}}_{\underline{5}}10^{\underline{1}}_{\underline{5}}10^{\underline{1}}_{\underline{3}}10^{0}_{\underline{3}}11^{0}_{\underline{1}}8^{1}_{\underline{1}}9^{1}_{1}5^{2}_{1}8^{2}_{3}3^{3}_{3}5^{3}_{5}1^{4}_{5}3^{4}_{7}1^{5}_{9}$	
11^a_{283}: $1^{\underline{8}}_{\underline{19}}3^{\underline{7}}_{\underline{17}}1^{\underline{7}}_{\underline{15}}6^{\underline{6}}_{\underline{15}}3^{\underline{6}}_{\underline{13}}9^{\underline{5}}_{\underline{13}}6^{\underline{5}}_{\underline{11}}11^{\underline{4}}_{\underline{11}}9^{\underline{4}}_{\underline{9}}13^{\underline{3}}_{\underline{9}}11^{\underline{3}}_{\underline{7}}11^{\underline{2}}_{\underline{7}}13^{\underline{2}}_{\underline{5}}9^{\underline{1}}_{\underline{5}}11^{\underline{1}}_{\underline{3}}7^{0}_{\underline{3}}10^{0}_{\underline{1}}3^{1}_{\underline{1}}6^{1}_{1}1^{2}_{1}3^{2}_{3}1^{3}_{5}$	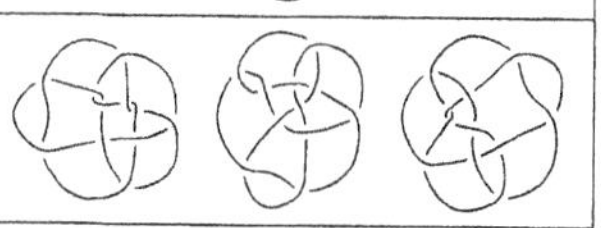
11^a_{284}: $1^{\underline{4}}_{\underline{7}}4^{\underline{3}}_{\underline{5}}1^{\underline{3}}_{\underline{3}}8^{\underline{2}}_{\underline{3}}4^{\underline{2}}_{\underline{1}}11^{\underline{1}}_{\underline{1}}8^{\underline{1}}_{1}15^{0}_{1}12^{0}_{3}15^{1}_{3}14^{1}_{5}14^{2}_{5}15^{2}_{7}11^{3}_{7}14^{3}_{9}7^{4}_{9}11^{4}_{11}3^{5}_{11}7^{5}_{13}1^{6}_{13}3^{6}_{15}1^{7}_{17}$	
11^a_{285}: $1^{\underline{6}}_{\underline{13}}4^{\underline{5}}_{\underline{11}}1^{\underline{5}}_{\underline{9}}7^{\underline{4}}_{\underline{9}}4^{\underline{4}}_{\underline{7}}10^{\underline{3}}_{\underline{7}}7^{\underline{3}}_{\underline{5}}13^{\underline{2}}_{\underline{5}}10^{\underline{2}}_{\underline{3}}13^{\underline{1}}_{\underline{3}}13^{\underline{1}}_{\underline{1}}13^{0}_{\underline{1}}14^{0}_{1}10^{1}_{1}12^{1}_{3}6^{2}_{3}10^{2}_{5}3^{3}_{5}6^{3}_{7}1^{4}_{7}3^{4}_{9}1^{5}_{11}$	
11^a_{286}: $1^{\underline{5}}_{\underline{9}}3^{\underline{4}}_{\underline{7}}1^{\underline{4}}_{\underline{5}}6^{\underline{3}}_{\underline{5}}3^{\underline{3}}_{\underline{3}}9^{\underline{2}}_{\underline{3}}6^{\underline{2}}_{\underline{1}}11^{\underline{1}}_{\underline{1}}9^{\underline{1}}_{1}13^{0}_{1}12^{0}_{3}11^{1}_{3}12^{1}_{5}10^{2}_{5}11^{2}_{7}6^{3}_{7}10^{3}_{9}3^{4}_{9}6^{4}_{11}1^{5}_{11}3^{5}_{13}1^{6}_{15}$	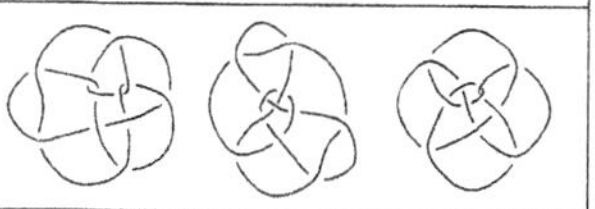
11^a_{287}: $1^{\underline{6}}_{\underline{13}}3^{\underline{5}}_{\underline{11}}1^{\underline{5}}_{\underline{9}}7^{\underline{4}}_{\underline{9}}3^{\underline{4}}_{\underline{7}}11^{\underline{3}}_{\underline{7}}7^{\underline{3}}_{\underline{5}}14^{\underline{2}}_{\underline{5}}11^{\underline{2}}_{\underline{3}}15^{\underline{1}}_{\underline{3}}14^{\underline{1}}_{\underline{1}}15^{0}_{\underline{1}}16^{0}_{1}12^{1}_{1}14^{1}_{3}8^{2}_{3}12^{2}_{5}4^{3}_{5}8^{3}_{7}1^{4}_{7}4^{4}_{9}1^{5}_{11}$	
11^a_{288}: $1^{\underline{5}}_{\underline{11}}4^{\underline{4}}_{\underline{9}}1^{\underline{4}}_{\underline{7}}9^{\underline{3}}_{\underline{7}}4^{\underline{3}}_{\underline{5}}13^{\underline{2}}_{\underline{5}}9^{\underline{2}}_{\underline{3}}16^{\underline{1}}_{\underline{3}}13^{\underline{1}}_{\underline{1}}18^{0}_{\underline{1}}17^{0}_{1}16^{1}_{1}17^{1}_{3}13^{2}_{3}16^{2}_{5}8^{3}_{5}13^{3}_{7}4^{4}_{7}8^{4}_{9}1^{5}_{9}4^{5}_{11}1^{6}_{13}$	
11^a_{289}: $1^{\underline{5}}_{\underline{11}}3^{\underline{4}}_{\underline{9}}1^{\underline{4}}_{\underline{7}}6^{\underline{3}}_{\underline{7}}3^{\underline{3}}_{\underline{5}}9^{\underline{2}}_{\underline{5}}6^{\underline{2}}_{\underline{3}}11^{\underline{1}}_{\underline{3}}9^{\underline{1}}_{\underline{1}}13^{0}_{\underline{1}}12^{0}_{1}11^{1}_{1}12^{1}_{3}9^{2}_{3}11^{2}_{5}6^{3}_{5}9^{3}_{7}3^{4}_{7}6^{4}_{9}1^{5}_{9}3^{5}_{11}1^{6}_{13}$	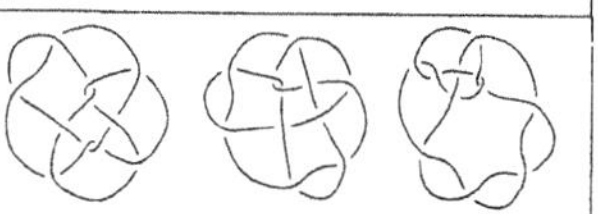
11^a_{290}: $1^{\underline{7}}_{\underline{15}}3^{\underline{6}}_{\underline{13}}1^{\underline{6}}_{\underline{11}}6^{\underline{5}}_{\underline{11}}3^{\underline{5}}_{\underline{9}}8^{\underline{4}}_{\underline{9}}6^{\underline{4}}_{\underline{7}}11^{\underline{3}}_{\underline{7}}8^{\underline{3}}_{\underline{5}}12^{\underline{2}}_{\underline{5}}11^{\underline{2}}_{\underline{3}}10^{\underline{1}}_{\underline{3}}12^{\underline{1}}_{\underline{1}}10^{0}_{\underline{1}}11^{0}_{1}6^{1}_{1}9^{1}_{3}3^{2}_{3}6^{2}_{5}1^{3}_{5}3^{3}_{7}1^{4}_{9}$	
11^a_{291}: $1^{0}_{5}1^{0}_{7}3^{1}_{7}4^{2}_{9}3^{2}_{11}5^{3}_{11}4^{3}_{13}9^{4}_{13}5^{4}_{15}7^{5}_{15}9^{5}_{17}8^{6}_{17}7^{6}_{19}6^{7}_{19}8^{7}_{21}4^{8}_{21}6^{8}_{23}2^{9}_{23}4^{9}_{25}1^{10}_{25}2^{10}_{27}1^{11}_{29}$	
11^a_{292}: $1^{0}_{3}1^{0}_{5}4^{1}_{5}6^{2}_{7}4^{2}_{9}8^{3}_{9}6^{3}_{11}11^{4}_{11}8^{4}_{13}10^{5}_{13}11^{5}_{15}10^{6}_{15}10^{6}_{17}7^{7}_{17}10^{7}_{19}5^{8}_{19}7^{8}_{21}2^{9}_{21}5^{9}_{23}1^{10}_{23}2^{10}_{25}1^{11}_{27}$	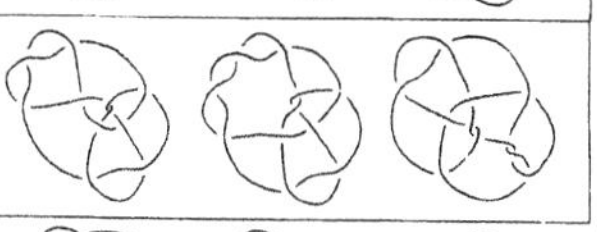
11^a_{293}: $1^{\underline{7}}_{\underline{19}}2^{\underline{6}}_{\underline{17}}1^{\underline{6}}_{\underline{15}}4^{\underline{5}}_{\underline{15}}2^{\underline{5}}_{\underline{13}}4^{\underline{4}}_{\underline{13}}4^{\underline{4}}_{\underline{11}}7^{\underline{3}}_{\underline{11}}4^{\underline{3}}_{\underline{9}}6^{\underline{2}}_{\underline{9}}7^{\underline{2}}_{\underline{7}}5^{\underline{1}}_{\underline{7}}6^{\underline{1}}_{\underline{5}}6^{0}_{\underline{5}}6^{0}_{\underline{3}}3^{1}_{\underline{3}}5^{1}_{\underline{1}}2^{2}_{\underline{1}}3^{2}_{1}1^{3}_{1}2^{3}_{3}1^{4}_{5}$	
11^a_{294}: $1^{\underline{7}}_{\underline{17}}2^{\underline{6}}_{\underline{15}}1^{\underline{6}}_{\underline{13}}5^{\underline{5}}_{\underline{13}}2^{\underline{5}}_{\underline{11}}7^{\underline{4}}_{\underline{11}}5^{\underline{4}}_{\underline{9}}10^{\underline{3}}_{\underline{9}}7^{\underline{3}}_{\underline{7}}10^{\underline{2}}_{\underline{7}}10^{\underline{2}}_{\underline{5}}9^{\underline{1}}_{\underline{5}}10^{\underline{1}}_{\underline{3}}9^{0}_{\underline{3}}10^{0}_{\underline{1}}5^{1}_{\underline{1}}8^{1}_{1}3^{2}_{1}5^{2}_{3}1^{3}_{3}3^{3}_{5}1^{4}_{7}$	
11^a_{295}: $1^{\underline{9}}_{\underline{23}}2^{\underline{8}}_{\underline{21}}1^{\underline{8}}_{\underline{19}}5^{\underline{7}}_{\underline{19}}2^{\underline{7}}_{\underline{17}}6^{\underline{6}}_{\underline{17}}5^{\underline{6}}_{\underline{15}}9^{\underline{5}}_{\underline{15}}6^{\underline{5}}_{\underline{13}}9^{\underline{4}}_{\underline{13}}9^{\underline{4}}_{\underline{11}}8^{\underline{3}}_{\underline{11}}9^{\underline{3}}_{\underline{9}}7^{\underline{2}}_{\underline{9}}8^{\underline{2}}_{\underline{7}}4^{\underline{1}}_{\underline{7}}7^{\underline{1}}_{\underline{5}}3^{0}_{\underline{5}}5^{0}_{\underline{3}}1^{1}_{\underline{3}}2^{1}_{\underline{1}}1^{2}_{1}$	
11^a_{296}: $1^{\underline{9}}_{\underline{21}}2^{\underline{8}}_{\underline{19}}1^{\underline{8}}_{\underline{17}}5^{\underline{7}}_{\underline{17}}2^{\underline{7}}_{\underline{15}}6^{\underline{6}}_{\underline{15}}5^{\underline{6}}_{\underline{13}}9^{\underline{5}}_{\underline{13}}6^{\underline{5}}_{\underline{11}}9^{\underline{4}}_{\underline{11}}9^{\underline{4}}_{\underline{9}}8^{\underline{3}}_{\underline{9}}9^{\underline{3}}_{\underline{7}}8^{\underline{2}}_{\underline{7}}8^{\underline{2}}_{\underline{5}}4^{\underline{1}}_{\underline{5}}8^{\underline{1}}_{\underline{3}}3^{0}_{\underline{3}}5^{0}_{\underline{1}}1^{1}_{\underline{1}}2^{1}_{1}1^{2}_{3}$	
11^a_{297}: $1^{\underline{7}}_{\underline{17}}4^{\underline{6}}_{\underline{15}}1^{\underline{6}}_{\underline{13}}8^{\underline{5}}_{\underline{13}}4^{\underline{5}}_{\underline{11}}11^{\underline{4}}_{\underline{11}}8^{\underline{4}}_{\underline{9}}14^{\underline{3}}_{\underline{9}}11^{\underline{3}}_{\underline{7}}15^{\underline{2}}_{\underline{7}}14^{\underline{2}}_{\underline{5}}13^{\underline{1}}_{\underline{5}}15^{\underline{1}}_{\underline{3}}11^{0}_{\underline{3}}14^{0}_{\underline{1}}7^{1}_{\underline{1}}10^{1}_{1}3^{2}_{1}7^{2}_{3}1^{3}_{3}3^{3}_{5}1^{4}_{7}$	
11^a_{298}: $1^{0}_{5}1^{0}_{7}3^{1}_{7}5^{2}_{9}3^{2}_{11}7^{3}_{11}5^{3}_{13}11^{4}_{13}7^{4}_{15}10^{5}_{15}11^{5}_{17}11^{6}_{17}10^{6}_{19}8^{7}_{19}11^{7}_{21}6^{8}_{21}8^{8}_{23}3^{9}_{23}6^{9}_{25}1^{10}_{25}3^{10}_{27}1^{11}_{29}$	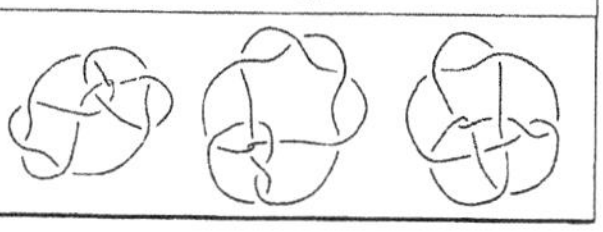
11^a_{299}: $1^{0}_{3}1^{0}_{5}3^{1}_{5}4^{2}_{7}3^{2}_{9}6^{3}_{9}4^{3}_{11}8^{4}_{11}6^{4}_{13}7^{5}_{13}8^{5}_{15}8^{6}_{15}7^{6}_{17}5^{7}_{17}8^{7}_{19}4^{8}_{19}5^{8}_{21}2^{9}_{21}4^{9}_{23}1^{10}_{23}2^{10}_{25}1^{11}_{27}$	
11^a_{300}: $1^{\underline{6}}_{\underline{13}}3^{\underline{5}}_{\underline{11}}1^{\underline{5}}_{\underline{9}}7^{\underline{4}}_{\underline{9}}3^{\underline{4}}_{\underline{7}}9^{\underline{3}}_{\underline{7}}7^{\underline{3}}_{\underline{5}}12^{\underline{2}}_{\underline{5}}9^{\underline{2}}_{\underline{3}}13^{\underline{1}}_{\underline{3}}12^{\underline{1}}_{\underline{1}}12^{0}_{\underline{1}}14^{0}_{1}10^{1}_{1}11^{1}_{3}6^{2}_{3}10^{2}_{5}3^{3}_{5}6^{3}_{7}1^{4}_{7}3^{4}_{9}1^{5}_{11}$	

K: KH	1st line	2nd line	3rd line
11^a_{301}: $1^{\underline{7}}_{\underline{17}}4^{\underline{6}}_{\underline{15}}1^{\underline{6}}_{\underline{13}}9^{\underline{5}}_{\underline{13}}4^{\underline{5}}_{\underline{11}}12^{\underline{4}}_{\underline{11}}9^{\underline{4}}_{\underline{9}}16^{\underline{3}}_{\underline{9}}12^{\underline{3}}_{\underline{7}}17^{\underline{2}}_{\underline{7}}16^{\underline{2}}_{\underline{5}}15^{\underline{1}}_{\underline{5}}17^{\underline{1}}_{\underline{3}}13^{0}_{\underline{3}}16^{0}_{\underline{1}}8^{1}_{\underline{1}}12^{1}_{1}4^{2}_{1}8^{2}_{3}1^{3}_{3}4^{3}_{5}1^{4}_{7}$			
11^a_{302}: $1^{\underline{8}}_{\underline{21}}3^{\underline{7}}_{\underline{19}}1^{\underline{7}}_{\underline{17}}7^{\underline{6}}_{\underline{17}}3^{\underline{6}}_{\underline{15}}10^{\underline{5}}_{\underline{15}}7^{\underline{5}}_{\underline{13}}12^{\underline{4}}_{\underline{13}}10^{\underline{4}}_{\underline{11}}14^{\underline{3}}_{\underline{11}}12^{\underline{3}}_{\underline{9}}12^{\underline{2}}_{\underline{9}}14^{\underline{2}}_{\underline{7}}10^{\underline{1}}_{\underline{7}}12^{\underline{1}}_{\underline{5}}7^{0}_{\underline{5}}11^{0}_{\underline{3}}4^{1}_{\underline{3}}6^{1}_{\underline{1}}1^{2}_{\underline{1}}4^{2}_{1}1^{3}_{3}$			
11^a_{303}: $1^{\underline{4}}_{\underline{9}}3^{\underline{3}}_{\underline{7}}1^{\underline{3}}_{\underline{5}}7^{\underline{2}}_{\underline{5}}3^{\underline{2}}_{\underline{3}}9^{\underline{1}}_{\underline{3}}7^{\underline{1}}_{\underline{1}}12^{0}_{\underline{1}}10^{0}_{1}13^{1}_{1}11^{1}_{3}11^{2}_{3}13^{2}_{5}9^{3}_{5}11^{3}_{7}6^{4}_{7}9^{4}_{9}3^{5}_{9}6^{5}_{11}1^{6}_{11}3^{6}_{13}1^{7}_{15}$			
11^a_{304}: $1^{\underline{2}}_{\underline{1}}2^{\underline{1}}_{1}1^{\underline{1}}_{3}6^{0}_{3}3^{0}_{5}7^{1}_{5}5^{1}_{7}9^{2}_{7}7^{2}_{9}10^{3}_{9}9^{3}_{11}9^{4}_{11}10^{4}_{13}7^{5}_{13}9^{5}_{15}5^{6}_{15}7^{6}_{17}2^{7}_{17}5^{7}_{19}1^{8}_{19}2^{8}_{21}1^{9}_{23}$			
11^a_{305}: $1^{\underline{6}}_{\underline{15}}3^{\underline{5}}_{\underline{13}}1^{\underline{5}}_{\underline{11}}5^{\underline{4}}_{\underline{11}}3^{\underline{4}}_{\underline{9}}9^{\underline{3}}_{\underline{9}}5^{\underline{3}}_{\underline{7}}10^{\underline{2}}_{\underline{7}}9^{\underline{2}}_{\underline{5}}11^{\underline{1}}_{\underline{5}}10^{\underline{1}}_{\underline{3}}11^{0}_{\underline{3}}12^{0}_{\underline{1}}8^{1}_{\underline{1}}10^{1}_{1}6^{2}_{1}8^{2}_{3}3^{3}_{3}6^{3}_{5}1^{4}_{5}3^{4}_{7}1^{5}_{9}$			
11^a_{306}: $1^{\underline{8}}_{\underline{21}}3^{\underline{7}}_{\underline{19}}1^{\underline{7}}_{\underline{17}}5^{\underline{6}}_{\underline{17}}3^{\underline{6}}_{\underline{15}}7^{\underline{5}}_{\underline{15}}5^{\underline{5}}_{\underline{13}}8^{\underline{4}}_{\underline{13}}7^{\underline{4}}_{\underline{11}}9^{\underline{3}}_{\underline{11}}8^{\underline{3}}_{\underline{9}}7^{\underline{2}}_{\underline{9}}9^{\underline{2}}_{\underline{7}}6^{\underline{1}}_{\underline{7}}7^{\underline{1}}_{\underline{5}}4^{0}_{\underline{5}}7^{0}_{\underline{3}}2^{1}_{\underline{3}}3^{1}_{\underline{1}}1^{2}_{\underline{1}}2^{2}_{1}1^{3}_{3}$			
11^a_{307}: $1^{\underline{8}}_{\underline{19}}2^{\underline{7}}_{\underline{17}}1^{\underline{7}}_{\underline{15}}3^{\underline{6}}_{\underline{15}}2^{\underline{6}}_{\underline{13}}5^{\underline{5}}_{\underline{13}}3^{\underline{5}}_{\underline{11}}6^{\underline{4}}_{\underline{11}}5^{\underline{4}}_{\underline{9}}7^{\underline{3}}_{\underline{9}}6^{\underline{3}}_{\underline{7}}6^{\underline{2}}_{\underline{7}}7^{\underline{2}}_{\underline{5}}5^{\underline{1}}_{\underline{5}}6^{\underline{1}}_{\underline{3}}4^{0}_{\underline{3}}6^{0}_{\underline{1}}2^{1}_{\underline{1}}3^{1}_{1}1^{2}_{1}2^{2}_{3}1^{3}_{5}$			
11^a_{308}: $1^{\underline{2}}_{1}1^{\underline{1}}_{3}1^{\underline{1}}_{5}3^{0}_{5}2^{0}_{7}4^{1}_{7}2^{1}_{9}5^{2}_{9}4^{2}_{11}5^{3}_{11}5^{3}_{13}6^{4}_{13}5^{4}_{15}4^{5}_{15}6^{5}_{17}4^{6}_{17}4^{6}_{19}2^{7}_{19}4^{7}_{21}1^{8}_{21}2^{8}_{23}1^{9}_{25}$			
11^a_{309}: $1^{\underline{2}}_{\underline{1}}2^{\underline{1}}_{1}1^{\underline{1}}_{3}4^{0}_{3}3^{0}_{5}6^{1}_{5}3^{1}_{7}7^{2}_{7}6^{2}_{9}7^{3}_{9}7^{3}_{11}8^{4}_{11}7^{4}_{13}5^{5}_{13}8^{5}_{15}4^{6}_{15}5^{6}_{17}2^{7}_{17}4^{7}_{19}1^{8}_{19}2^{8}_{21}1^{9}_{23}$			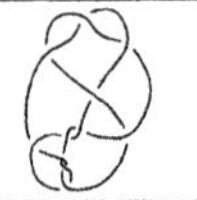
11^a_{310}: $1^{\underline{2}}_{\underline{1}}1^{\underline{1}}_{1}1^{\underline{1}}_{3}3^{0}_{3}2^{0}_{5}4^{1}_{5}2^{1}_{7}4^{2}_{7}4^{2}_{9}5^{3}_{9}4^{3}_{11}5^{4}_{11}5^{4}_{13}3^{5}_{13}5^{5}_{15}3^{6}_{15}3^{6}_{17}1^{7}_{17}3^{7}_{19}1^{8}_{19}1^{8}_{21}1^{9}_{23}$			
11^a_{311}: $1^{\underline{2}}_{\underline{3}}2^{\underline{1}}_{\underline{1}}1^{\underline{1}}_{1}4^{0}_{1}3^{0}_{3}6^{1}_{3}3^{1}_{5}6^{2}_{5}6^{2}_{7}6^{3}_{7}6^{3}_{9}6^{4}_{9}6^{4}_{11}4^{5}_{11}6^{5}_{13}3^{6}_{13}4^{6}_{15}1^{7}_{15}3^{7}_{17}1^{8}_{17}1^{8}_{19}1^{9}_{21}$			
11^a_{312}: $1^{\underline{8}}_{\underline{19}}3^{\underline{7}}_{\underline{17}}1^{\underline{7}}_{\underline{15}}5^{\underline{6}}_{\underline{15}}3^{\underline{6}}_{\underline{13}}8^{\underline{5}}_{\underline{13}}5^{\underline{5}}_{\underline{11}}9^{\underline{4}}_{\underline{11}}8^{\underline{4}}_{\underline{9}}10^{\underline{3}}_{\underline{9}}9^{\underline{3}}_{\underline{7}}9^{\underline{2}}_{\underline{7}}10^{\underline{2}}_{\underline{5}}7^{\underline{1}}_{\underline{5}}9^{\underline{1}}_{\underline{3}}5^{0}_{\underline{3}}8^{0}_{\underline{1}}2^{1}_{\underline{1}}4^{1}_{1}1^{2}_{1}2^{2}_{3}1^{3}_{5}$			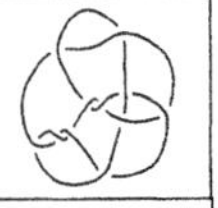
11^a_{313}: $1^{\underline{8}}_{\underline{17}}2^{\underline{7}}_{\underline{15}}1^{\underline{7}}_{\underline{13}}3^{\underline{6}}_{\underline{13}}2^{\underline{6}}_{\underline{11}}5^{\underline{5}}_{\underline{11}}3^{\underline{5}}_{\underline{9}}5^{\underline{4}}_{\underline{9}}5^{\underline{4}}_{\underline{7}}6^{\underline{3}}_{\underline{7}}5^{\underline{3}}_{\underline{5}}6^{\underline{2}}_{\underline{5}}6^{\underline{2}}_{\underline{3}}4^{\underline{1}}_{\underline{3}}6^{\underline{1}}_{\underline{1}}4^{0}_{\underline{1}}5^{0}_{1}2^{1}_{1}3^{1}_{3}1^{2}_{3}2^{2}_{5}1^{3}_{7}$			
11^a_{314}: $1^{\underline{4}}_{\underline{7}}4^{\underline{3}}_{\underline{5}}1^{\underline{3}}_{\underline{3}}8^{\underline{2}}_{\underline{3}}4^{\underline{2}}_{\underline{1}}10^{\underline{1}}_{\underline{1}}8^{\underline{1}}_{1}15^{0}_{1}11^{0}_{3}14^{1}_{3}14^{1}_{5}13^{2}_{5}14^{2}_{7}11^{3}_{7}13^{3}_{9}6^{4}_{9}11^{4}_{11}3^{5}_{11}6^{5}_{13}1^{6}_{13}3^{6}_{15}1^{7}_{17}$		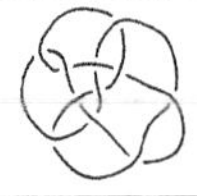	
11^a_{315}: $1^{\underline{5}}_{\underline{11}}3^{\underline{4}}_{\underline{9}}1^{\underline{4}}_{\underline{7}}7^{\underline{3}}_{\underline{7}}3^{\underline{3}}_{\underline{5}}10^{\underline{2}}_{\underline{5}}7^{\underline{2}}_{\underline{3}}12^{\underline{1}}_{\underline{3}}10^{\underline{1}}_{\underline{1}}14^{0}_{\underline{1}}13^{0}_{1}12^{1}_{1}13^{1}_{3}10^{2}_{3}12^{2}_{5}6^{3}_{5}10^{3}_{7}3^{4}_{7}6^{4}_{9}1^{5}_{9}3^{5}_{11}1^{6}_{13}$			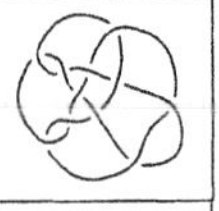
11^a_{316}: $1^{\underline{5}}_{\underline{11}}2^{\underline{4}}_{\underline{9}}1^{\underline{4}}_{\underline{7}}5^{\underline{3}}_{\underline{7}}2^{\underline{3}}_{\underline{5}}7^{\underline{2}}_{\underline{5}}5^{\underline{2}}_{\underline{3}}9^{\underline{1}}_{\underline{3}}7^{\underline{1}}_{\underline{1}}11^{0}_{\underline{1}}10^{0}_{1}9^{1}_{1}10^{1}_{3}8^{2}_{3}9^{2}_{5}5^{3}_{5}8^{3}_{7}3^{4}_{7}5^{4}_{9}1^{5}_{9}3^{5}_{11}1^{6}_{13}$			
11^a_{317}: $1^{\underline{7}}_{\underline{15}}2^{\underline{6}}_{\underline{13}}1^{\underline{6}}_{\underline{11}}5^{\underline{5}}_{\underline{11}}2^{\underline{5}}_{\underline{9}}7^{\underline{4}}_{\underline{9}}5^{\underline{4}}_{\underline{7}}9^{\underline{3}}_{\underline{7}}7^{\underline{3}}_{\underline{5}}11^{\underline{2}}_{\underline{5}}9^{\underline{2}}_{\underline{3}}9^{\underline{1}}_{\underline{3}}11^{\underline{1}}_{\underline{1}}9^{0}_{\underline{1}}10^{0}_{1}6^{1}_{1}8^{1}_{3}3^{2}_{3}6^{2}_{5}1^{3}_{5}3^{3}_{7}1^{4}_{9}$			
11^a_{318}: $1^{0}_{5}1^{0}_{7}3^{1}_{7}6^{2}_{9}3^{2}_{11}8^{3}_{11}6^{3}_{13}11^{4}_{13}8^{4}_{15}11^{5}_{15}11^{5}_{17}11^{6}_{17}11^{6}_{19}8^{7}_{19}11^{7}_{21}6^{8}_{21}8^{8}_{23}2^{9}_{23}6^{9}_{25}1^{10}_{25}2^{10}_{27}1^{11}_{29}$			
11^a_{319}: $1^{0}_{5}1^{0}_{7}3^{1}_{7}5^{2}_{9}3^{2}_{11}7^{3}_{11}5^{3}_{13}10^{4}_{13}7^{4}_{15}9^{5}_{15}10^{5}_{17}10^{6}_{17}9^{6}_{19}7^{7}_{19}10^{7}_{21}5^{8}_{21}7^{8}_{23}2^{9}_{23}5^{9}_{25}1^{10}_{25}2^{10}_{27}1^{11}_{29}$			
11^a_{320}: $1^{0}_{3}1^{0}_{5}3^{1}_{5}4^{2}_{7}3^{2}_{9}7^{3}_{9}4^{3}_{11}9^{4}_{11}7^{4}_{13}8^{5}_{13}9^{5}_{15}9^{6}_{15}8^{6}_{17}6^{7}_{17}9^{7}_{19}5^{8}_{19}6^{8}_{21}2^{9}_{21}5^{9}_{23}1^{10}_{23}2^{10}_{25}1^{11}_{27}$			
11^a_{321}: $1^{\underline{9}}_{\underline{23}}2^{\underline{8}}_{\underline{21}}1^{\underline{8}}_{\underline{19}}6^{\underline{7}}_{\underline{19}}2^{\underline{7}}_{\underline{17}}7^{\underline{6}}_{\underline{17}}6^{\underline{6}}_{\underline{15}}10^{\underline{5}}_{\underline{15}}7^{\underline{5}}_{\underline{13}}10^{\underline{4}}_{\underline{13}}10^{\underline{4}}_{\underline{11}}9^{\underline{3}}_{\underline{11}}10^{\underline{3}}_{\underline{9}}8^{\underline{2}}_{\underline{9}}9^{\underline{2}}_{\underline{7}}4^{\underline{1}}_{\underline{7}}8^{\underline{1}}_{\underline{5}}3^{0}_{\underline{5}}5^{0}_{\underline{3}}1^{1}_{\underline{3}}2^{1}_{\underline{1}}1^{2}_{1}$			
11^a_{322}: $1^{\underline{7}}_{\underline{17}}3^{\underline{6}}_{\underline{15}}1^{\underline{6}}_{\underline{13}}7^{\underline{5}}_{\underline{13}}3^{\underline{5}}_{\underline{11}}9^{\underline{4}}_{\underline{11}}7^{\underline{4}}_{\underline{9}}12^{\underline{3}}_{\underline{9}}9^{\underline{3}}_{\underline{7}}13^{\underline{2}}_{\underline{7}}12^{\underline{2}}_{\underline{5}}11^{\underline{1}}_{\underline{5}}13^{\underline{1}}_{\underline{3}}10^{0}_{\underline{3}}12^{0}_{\underline{1}}6^{1}_{\underline{1}}9^{1}_{1}3^{2}_{1}6^{2}_{3}1^{3}_{3}3^{3}_{5}1^{4}_{7}$			
11^a_{323}: $1^{\underline{5}}_{\underline{13}}2^{\underline{4}}_{\underline{11}}1^{\underline{4}}_{\underline{9}}4^{\underline{3}}_{\underline{9}}2^{\underline{3}}_{\underline{7}}5^{\underline{2}}_{\underline{7}}4^{\underline{2}}_{\underline{5}}6^{\underline{1}}_{\underline{5}}5^{\underline{1}}_{\underline{3}}7^{0}_{\underline{3}}7^{0}_{\underline{1}}6^{1}_{\underline{1}}6^{1}_{1}5^{2}_{1}6^{2}_{3}3^{3}_{3}5^{3}_{5}2^{4}_{5}3^{4}_{7}1^{5}_{7}2^{5}_{9}1^{6}_{11}$			
11^a_{324}: $1^{\underline{9}}_{\underline{21}}2^{\underline{8}}_{\underline{19}}1^{\underline{8}}_{\underline{17}}5^{\underline{7}}_{\underline{17}}2^{\underline{7}}_{\underline{15}}5^{\underline{6}}_{\underline{15}}5^{\underline{6}}_{\underline{13}}8^{\underline{5}}_{\underline{13}}5^{\underline{5}}_{\underline{11}}8^{\underline{4}}_{\underline{11}}8^{\underline{4}}_{\underline{9}}7^{\underline{3}}_{\underline{9}}8^{\underline{3}}_{\underline{7}}7^{\underline{2}}_{\underline{7}}7^{\underline{2}}_{\underline{5}}3^{\underline{1}}_{\underline{5}}7^{\underline{1}}_{\underline{3}}3^{0}_{\underline{3}}4^{0}_{\underline{1}}1^{1}_{\underline{1}}2^{1}_{1}1^{2}_{3}$			
11^a_{325}: $1^{\underline{7}}_{\underline{17}}2^{\underline{6}}_{\underline{15}}1^{\underline{6}}_{\underline{13}}4^{\underline{5}}_{\underline{13}}2^{\underline{5}}_{\underline{11}}6^{\underline{4}}_{\underline{11}}4^{\underline{4}}_{\underline{9}}7^{\underline{3}}_{\underline{9}}6^{\underline{3}}_{\underline{7}}8^{\underline{2}}_{\underline{7}}7^{\underline{2}}_{\underline{5}}7^{\underline{1}}_{\underline{5}}8^{\underline{1}}_{\underline{3}}6^{0}_{\underline{3}}8^{0}_{\underline{1}}4^{1}_{\underline{1}}5^{1}_{1}2^{2}_{1}4^{2}_{3}1^{3}_{3}2^{3}_{5}1^{4}_{7}$			
11^a_{326}: $1^{\underline{5}}_{\underline{11}}3^{\underline{4}}_{\underline{9}}1^{\underline{4}}_{\underline{7}}7^{\underline{3}}_{\underline{7}}3^{\underline{3}}_{\underline{5}}10^{\underline{2}}_{\underline{5}}7^{\underline{2}}_{\underline{3}}13^{\underline{1}}_{\underline{3}}10^{\underline{1}}_{\underline{1}}15^{0}_{\underline{1}}14^{0}_{1}13^{1}_{1}14^{1}_{3}11^{2}_{3}13^{2}_{5}7^{3}_{5}11^{3}_{7}4^{4}_{7}7^{4}_{9}1^{5}_{9}4^{5}_{11}1^{6}_{13}$		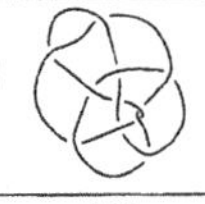	
11^a_{327}: $1^{\underline{3}}_{\underline{5}}4^{\underline{2}}_{\underline{3}}1^{\underline{2}}_{\underline{1}}8^{\underline{1}}_{\underline{1}}4^{\underline{1}}_{1}12^{0}_{1}9^{0}_{3}15^{1}_{3}11^{1}_{5}16^{2}_{5}15^{2}_{7}14^{3}_{7}16^{3}_{9}12^{4}_{9}14^{4}_{11}7^{5}_{11}12^{5}_{13}4^{6}_{13}7^{6}_{15}1^{7}_{15}4^{7}_{17}1^{8}_{19}$			
11^a_{328}: $1^{\underline{2}}_{\underline{1}}3^{\underline{1}}_{1}1^{\underline{1}}_{3}7^{0}_{3}4^{0}_{5}9^{1}_{5}6^{1}_{7}12^{2}_{7}9^{2}_{9}12^{3}_{9}12^{3}_{11}12^{4}_{11}12^{4}_{13}9^{5}_{13}12^{5}_{15}6^{6}_{15}9^{6}_{17}3^{7}_{17}6^{7}_{19}1^{8}_{19}3^{8}_{21}1^{9}_{23}$	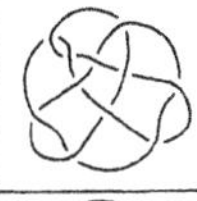		
11^a_{329}: $1^{0}_{3}1^{0}_{5}4^{1}_{5}6^{2}_{7}4^{2}_{9}9^{3}_{9}6^{3}_{11}12^{4}_{11}9^{4}_{13}11^{5}_{13}12^{5}_{15}12^{6}_{15}11^{6}_{17}8^{7}_{17}12^{7}_{19}6^{8}_{19}8^{8}_{21}3^{9}_{21}6^{9}_{23}1^{10}_{23}3^{10}_{25}1^{11}_{27}$		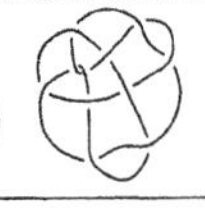	
11^a_{330}: $1^{\underline{7}}_{\underline{19}}2^{\underline{6}}_{\underline{17}}1^{\underline{6}}_{\underline{15}}4^{\underline{5}}_{\underline{15}}2^{\underline{5}}_{\underline{13}}5^{\underline{4}}_{\underline{13}}4^{\underline{4}}_{\underline{11}}7^{\underline{3}}_{\underline{11}}5^{\underline{3}}_{\underline{9}}7^{\underline{2}}_{\underline{9}}7^{\underline{2}}_{\underline{7}}6^{\underline{1}}_{\underline{7}}7^{\underline{1}}_{\underline{5}}6^{0}_{\underline{5}}7^{0}_{\underline{3}}4^{1}_{\underline{3}}5^{1}_{\underline{1}}2^{2}_{\underline{1}}4^{2}_{1}1^{3}_{1}2^{3}_{3}1^{4}_{5}$			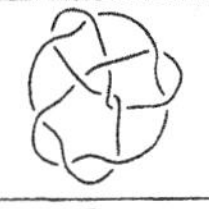
11^a_{331}: $1^{\underline{7}}_{\underline{17}}2^{\underline{6}}_{\underline{15}}1^{\underline{6}}_{\underline{13}}5^{\underline{5}}_{\underline{13}}2^{\underline{5}}_{\underline{11}}7^{\underline{4}}_{\underline{11}}5^{\underline{4}}_{\underline{9}}9^{\underline{3}}_{\underline{9}}7^{\underline{3}}_{\underline{7}}10^{\underline{2}}_{\underline{7}}9^{\underline{2}}_{\underline{5}}8^{\underline{1}}_{\underline{5}}10^{\underline{1}}_{\underline{3}}8^{0}_{\underline{3}}9^{0}_{\underline{1}}5^{1}_{\underline{1}}7^{1}_{1}2^{2}_{1}5^{2}_{3}1^{3}_{3}2^{3}_{5}1^{4}_{7}$	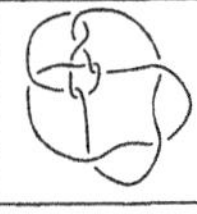		
11^a_{332}: $1^{\underline{5}}_{\underline{11}}4^{\underline{4}}_{\underline{9}}1^{\underline{4}}_{\underline{7}}9^{\underline{3}}_{\underline{7}}4^{\underline{3}}_{\underline{5}}12^{\underline{2}}_{\underline{5}}9^{\underline{2}}_{\underline{3}}15^{\underline{1}}_{\underline{3}}12^{\underline{1}}_{\underline{1}}17^{0}_{\underline{1}}16^{0}_{1}14^{1}_{1}16^{1}_{3}12^{2}_{3}14^{2}_{5}7^{3}_{5}12^{3}_{7}3^{4}_{7}7^{4}_{9}1^{5}_{9}3^{5}_{11}1^{6}_{13}$		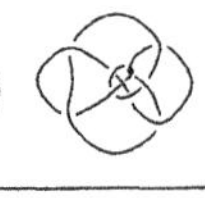	
11^a_{333}: $1^{\underline{8}}_{\underline{17}}1^{\underline{7}}_{\underline{15}}1^{\underline{7}}_{\underline{13}}2^{\underline{6}}_{\underline{13}}1^{\underline{6}}_{\underline{11}}4^{\underline{5}}_{\underline{11}}2^{\underline{5}}_{\underline{9}}4^{\underline{4}}_{\underline{9}}4^{\underline{4}}_{\underline{7}}5^{\underline{3}}_{\underline{7}}4^{\underline{3}}_{\underline{5}}5^{\underline{2}}_{\underline{5}}5^{\underline{2}}_{\underline{3}}4^{\underline{1}}_{\underline{3}}5^{\underline{1}}_{\underline{1}}4^{0}_{\underline{1}}5^{0}_{1}2^{1}_{1}3^{1}_{3}1^{2}_{3}2^{2}_{5}1^{3}_{7}$			
11^a_{334}: $1^{0}_{7}1^{0}_{9}1^{1}_{9}2^{2}_{11}1^{2}_{13}2^{3}_{13}2^{3}_{15}4^{4}_{15}2^{4}_{17}3^{5}_{17}4^{5}_{19}4^{6}_{19}3^{6}_{21}3^{7}_{21}4^{7}_{23}3^{8}_{23}3^{8}_{25}1^{9}_{25}3^{9}_{27}1^{10}_{27}1^{10}_{29}1^{11}_{31}$	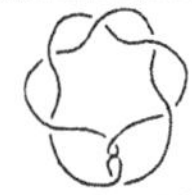		
11^a_{335}: $1^{0}_{5}1^{0}_{7}2^{1}_{7}3^{2}_{9}2^{2}_{11}4^{3}_{11}3^{3}_{13}6^{4}_{13}4^{4}_{15}6^{5}_{15}6^{5}_{17}6^{6}_{17}6^{6}_{19}4^{7}_{19}6^{7}_{21}4^{8}_{21}4^{8}_{23}1^{9}_{23}4^{9}_{25}1^{10}_{25}1^{10}_{27}1^{11}_{29}$		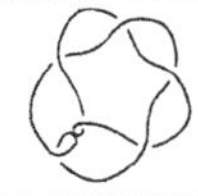	
11^a_{336}: $1^{0}_{5}1^{0}_{7}2^{1}_{7}2^{2}_{9}2^{2}_{11}3^{3}_{11}2^{3}_{13}5^{4}_{13}3^{4}_{15}4^{5}_{15}5^{5}_{17}5^{6}_{17}4^{6}_{19}3^{7}_{19}5^{7}_{21}3^{8}_{21}3^{8}_{23}1^{9}_{23}3^{9}_{25}1^{10}_{25}1^{10}_{27}1^{11}_{29}$			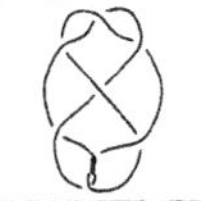

K: KH	1st line	2nd line	3rd line
11^a_{337}: $1^0_3 1^0_5 3^1_5 4^2_7 3^2_9 6^3_9 4^3_{11} 7^4_{11} 6^4_{13} 7^5_{13} 7^5_{15} 7^6_{15} 7^6_{17} 4^7_{17} 7^7_{19} 4^8_{19} 4^8_{21} 1^9_{21} 4^9_{23} 1^{10}_{23} 1^{10}_{25} 1^{11}_{27}$			
11^a_{338}: $1^0_7 1^0_9 1^1_9 3^2_{11} 1^2_{13} 3^3_{13} 3^3_{15} 5^4_{15} 3^4_{17} 5^5_{17} 5^5_{19} 6^6_{19} 5^6_{21} 4^7_{21} 6^7_{23} 4^8_{23} 4^8_{25} 2^9_{25} 4^9_{27} 1^{10}_{27} 2^{10}_{29} 1^{11}_{31}$			
11^a_{339}: $1^0_5 1^0_7 1^1_7 2^2_9 1^2_{11} 3^3_{11} 2^3_{13} 4^4_{13} 3^4_{15} 4^5_{15} 4^5_{17} 5^6_{17} 4^6_{19} 3^7_{19} 5^7_{21} 3^8_{21} 3^8_{23} 1^9_{23} 3^9_{25} 1^{10}_{25} 1^{10}_{27} 1^{11}_{29}$			
11^a_{340}: $1^0_5 1^0_7 2^1_7 3^2_9 2^2_{11} 5^3_{11} 3^3_{13} 7^4_{13} 5^4_{15} 6^5_{15} 7^5_{17} 8^6_{17} 6^6_{19} 5^7_{19} 8^7_{21} 4^8_{21} 5^8_{23} 2^9_{23} 4^9_{25} 1^{10}_{25} 2^{10}_{27} 1^{11}_{29}$	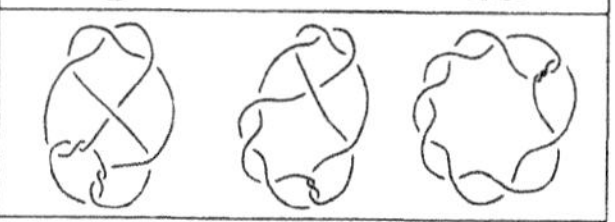		
11^a_{341}: $1^0_3 1^0_5 2^1_5 2^2_7 2^2_9 4^3_9 2^3_{11} 5^4_{11} 4^4_{13} 4^5_{13} 5^5_{15} 5^6_{15} 4^6_{17} 3^7_{17} 5^7_{19} 3^8_{19} 3^8_{21} 1^9_{21} 3^9_{23} 1^{10}_{23} 1^{10}_{25} 1^{11}_{27}$			
11^a_{342}: $1^0_3 1^0_5 1^1_5 1^2_7 1^2_9 2^3_9 1^3_{11} 2^4_{11} 2^4_{13} 2^5_{13} 2^5_{15} 2^6_{15} 2^6_{17} 1^7_{17} 2^7_{19} 2^8_{19} 1^8_{21} 2^9_{23} 1^{10}_{23} 1^{11}_{27}$			
11^a_{343}: $1^0_1 1^0_3 2^1_3 1^2_5 2^2_7 2^3_7 1^3_9 2^4_9 2^4_{11} 2^5_{11} 2^5_{13} 2^6_{13} 2^6_{15} 1^7_{15} 2^7_{17} 2^8_{17} 1^8_{19} 2^9_{21} 1^{10}_{21} 1^{11}_{25}$			
11^a_{344}: $1^{\underline{2}}_{\underline{1}} 2^{\underline{1}}_{1} 1^{\underline{1}}_{3} 5^0_3 3^0_5 8^1_5 4^1_7 10^2_7 8^2_9 10^3_9 10^3_{11} 11^4_{11} 10^4_{13} 8^5_{13} 11^5_{15} 6^6_{15} 8^6_{17} 3^7_{17} 6^7_{19} 1^8_{19} 3^8_{21} 1^9_{23}$			
11^a_{345}: $1^{\underline{2}}_{\underline{3}} 2^{\underline{1}}_{\underline{1}} 1^{\underline{1}}_{1} 4^0_1 3^0_3 6^1_3 3^1_5 7^2_5 6^2_7 7^3_7 7^3_9 7^4_9 7^4_{11} 5^5_{11} 7^5_{13} 4^6_{13} 5^6_{15} 2^7_{15} 4^7_{17} 1^8_{17} 2^8_{19} 1^9_{21}$			
11^a_{346}: $1^{\underline{4}}_{\underline{5}} 2^{\underline{3}}_{\underline{3}} 1^{\underline{3}}_{\underline{1}} 4^{\underline{2}}_{\underline{1}} 2^{\underline{2}}_{1} 5^{\underline{1}}_{1} 4^{\underline{1}}_{3} 8^0_3 6^0_5 7^1_5 7^1_7 7^2_7 7^2_9 6^3_9 7^3_{11} 4^4_{11} 6^4_{13} 2^5_{13} 4^5_{15} 1^6_{15} 2^6_{17} 1^7_{19}$			
11^a_{347}: $1^{\underline{4}}_{\underline{7}} 2^{\underline{3}}_{\underline{5}} 1^{\underline{3}}_{\underline{3}} 5^{\underline{2}}_{\underline{3}} 2^{\underline{2}}_{\underline{1}} 6^{\underline{1}}_{\underline{1}} 5^{\underline{1}}_{1} 9^0_1 7^0_3 10^1_3 8^1_5 8^2_5 10^2_7 7^3_7 8^3_9 5^4_9 7^4_{11} 2^5_{11} 5^5_{13} 1^6_{13} 2^6_{15} 1^7_{17}$			
11^a_{348}: $1^{\underline{3}}_{\underline{3}} 4^{\underline{2}}_{\underline{1}} 1^{\underline{2}}_{1} 5^{\underline{1}}_{1} 4^{\underline{1}}_{3} 10^0_3 6^0_5 11^1_5 9^1_7 12^2_7 11^2_9 11^3_9 12^3_{11} 9^4_{11} 11^4_{13} 6^5_{13} 9^5_{15} 3^6_{15} 6^6_{17} 1^7_{17} 3^7_{19} 1^8_{21}$			
11^a_{349}: $1^{\underline{3}}_{\underline{5}} 4^{\underline{2}}_{\underline{3}} 1^{\underline{2}}_{\underline{1}} 6^{\underline{1}}_{\underline{1}} 4^{\underline{1}}_{1} 11^0_1 7^0_3 12^1_3 10^1_5 13^2_5 12^2_7 12^3_7 13^3_9 9^4_9 12^4_{11} 6^5_{11} 9^5_{13} 3^6_{13} 6^6_{15} 1^7_{15} 3^7_{17} 1^8_{19}$	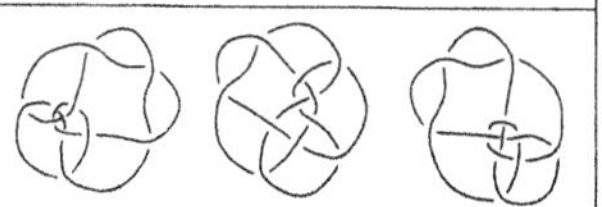		
11^a_{350}: $1^{\underline{5}}_{\underline{11}} 4^{\underline{4}}_{\underline{9}} 1^{\underline{4}}_{\underline{7}} 7^{\underline{3}}_{\underline{7}} 4^{\underline{3}}_{\underline{5}} 12^{\underline{2}}_{\underline{5}} 7^{\underline{2}}_{\underline{3}} 14^{\underline{1}}_{\underline{3}} 12^{\underline{1}}_{\underline{1}} 16^0_{\underline{1}} 15^0_1 15^1_1 15^1_3 11^2_3 15^2_5 8^3_5 11^3_7 4^4_7 8^4_9 1^5_9 4^5_{11} 1^6_{13}$			
11^a_{351}: $1^{\underline{5}}_{\underline{11}} 4^{\underline{4}}_{\underline{9}} 1^{\underline{4}}_{\underline{7}} 7^{\underline{3}}_{\underline{7}} 4^{\underline{3}}_{\underline{5}} 11^{\underline{2}}_{\underline{5}} 7^{\underline{2}}_{\underline{3}} 13^{\underline{1}}_{\underline{3}} 11^{\underline{1}}_{\underline{1}} 14^0_{\underline{1}} 14^0_1 13^1_1 13^1_3 10^2_3 13^2_5 6^3_5 10^3_7 3^4_7 6^4_9 1^5_9 3^5_{11} 1^6_{13}$			
11^a_{352}: $1^{\underline{5}}_{\underline{9}} 4^{\underline{4}}_{\underline{7}} 1^{\underline{4}}_{\underline{5}} 5^{\underline{3}}_{\underline{5}} 4^{\underline{3}}_{\underline{3}} 9^{\underline{2}}_{\underline{3}} 5^{\underline{2}}_{\underline{1}} 10^{\underline{1}}_{\underline{1}} 9^{\underline{1}}_{1} 11^0_1 11^0_3 11^1_3 10^1_5 8^2_5 11^2_7 5^3_7 8^3_9 3^4_9 5^4_{11} 1^5_{11} 3^5_{13} 1^6_{15}$	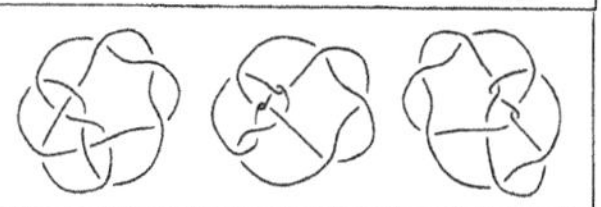		
11^a_{353}: $1^0_5 1^0_7 3^1_7 5^2_9 3^2_{11} 7^3_{11} 5^3_{13} 10^4_{13} 7^4_{15} 10^5_{15} 10^5_{17} 10^6_{17} 10^6_{19} 7^7_{19} 10^7_{21} 6^8_{21} 7^8_{23} 2^9_{23} 6^9_{25} 1^{10}_{25} 2^{10}_{27} 1^{11}_{29}$			
11^a_{354}: $1^0_3 1^0_5 3^1_5 4^2_7 3^2_9 7^3_9 4^3_{11} 8^4_{11} 7^4_{13} 8^5_{13} 8^5_{15} 9^6_{15} 8^6_{17} 5^7_{17} 9^7_{19} 5^8_{19} 5^8_{21} 2^9_{21} 5^9_{23} 1^{10}_{23} 2^{10}_{25} 1^{11}_{27}$			
11^a_{355}: $1^0_7 1^0_9 1^1_9 2^2_{11} 1^2_{13} 2^3_{13} 2^3_{15} 3^4_{15} 2^4_{17} 3^5_{17} 3^5_{19} 4^6_{19} 3^6_{21} 2^7_{21} 4^7_{23} 3^8_{23} 2^8_{25} 1^9_{25} 3^9_{27} 1^{10}_{27} 1^{10}_{29} 1^{11}_{31}$	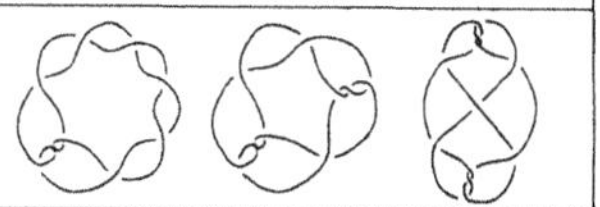		
11^a_{356}: $1^0_5 1^0_7 2^1_7 3^2_9 2^2_{11} 5^3_{11} 3^3_{13} 6^4_{13} 5^4_{15} 6^5_{15} 6^5_{17} 7^6_{17} 6^6_{19} 4^7_{19} 7^7_{21} 4^8_{21} 4^8_{23} 1^9_{23} 4^9_{25} 1^{10}_{25} 1^{10}_{27} 1^{11}_{29}$			
11^a_{357}: $1^0_5 1^0_7 2^1_7 3^2_9 2^2_{11} 5^3_{11} 3^3_{13} 7^4_{13} 5^4_{15} 7^5_{15} 7^5_{17} 8^6_{17} 7^6_{19} 5^7_{19} 8^7_{21} 5^8_{21} 5^8_{23} 2^9_{23} 5^9_{25} 1^{10}_{25} 2^{10}_{27} 1^{11}_{29}$			
11^a_{358}: $1^0_5 1^0_7 1^1_7 1^2_9 1^2_{11} 2^3_{11} 1^3_{13} 2^4_{13} 2^4_{15} 2^5_{15} 2^5_{17} 3^6_{17} 2^6_{19} 1^7_{19} 3^7_{21} 2^8_{21} 1^8_{23} 2^9_{25} 1^{10}_{25} 1^{11}_{29}$	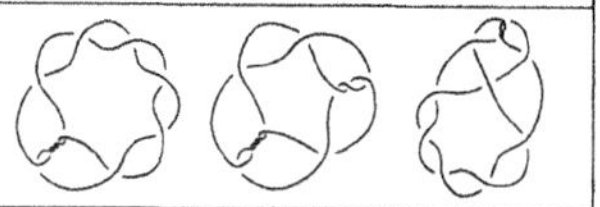		
11^a_{359}: $1^0_3 1^0_5 2^1_5 2^2_7 2^2_9 4^3_9 2^3_{11} 4^4_{11} 4^4_{13} 4^5_{13} 4^5_{15} 4^6_{15} 4^6_{17} 2^7_{17} 4^7_{19} 3^8_{19} 2^8_{21} 3^9_{23} 1^{10}_{23} 1^{11}_{27}$			
11^a_{360}: $1^0_3 1^0_5 2^1_5 2^2_7 2^2_9 4^3_9 2^3_{11} 4^4_{11} 4^4_{13} 4^5_{13} 4^5_{15} 5^6_{15} 4^6_{17} 2^7_{17} 5^7_{19} 3^8_{19} 2^8_{21} 1^9_{21} 3^9_{23} 1^{10}_{23} 1^{10}_{25} 1^{11}_{27}$			
11^a_{361}: $1^0_3 1^0_5 2^1_5 2^2_7 2^2_9 5^3_9 2^3_{11} 5^4_{11} 5^4_{13} 5^5_{13} 5^5_{15} 6^6_{15} 5^6_{17} 3^7_{17} 6^7_{19} 4^8_{19} 3^8_{21} 1^9_{21} 4^9_{23} 1^{10}_{23} 1^{10}_{25} 1^{11}_{27}$	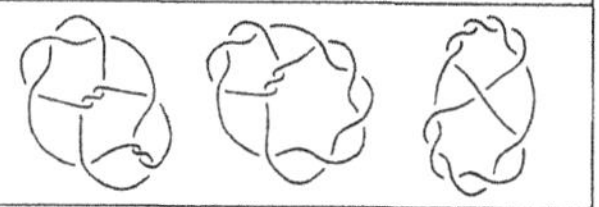		
11^a_{362}: $1^0_1 1^0_3 2^1_3 1^2_5 2^2_7 3^3_7 1^3_9 2^4_9 3^4_{11} 3^5_{11} 2^5_{13} 3^6_{13} 3^6_{15} 1^7_{15} 3^7_{17} 3^8_{17} 1^8_{19} 3^9_{21} 1^{10}_{21} 1^{11}_{25}$			
11^a_{363}: $1^0_1 1^0_3 2^1_3 1^2_5 2^2_7 3^3_7 1^3_9 2^4_9 3^4_{11} 2^5_{11} 2^5_{13} 3^6_{13} 2^6_{15} 1^7_{15} 3^7_{17} 2^8_{17} 1^8_{19} 2^9_{21} 1^{10}_{21} 1^{11}_{25}$			
11^a_{364}: $1^0_7 1^0_9 1^1_9 1^2_{11} 1^2_{13} 1^3_{13} 1^3_{15} 2^4_{15} 1^4_{17} 1^5_{17} 2^5_{19} 2^6_{19} 1^6_{21} 1^7_{21} 2^7_{23} 2^8_{23} 1^8_{25} 2^9_{27} 1^{10}_{27} 1^{11}_{31}$	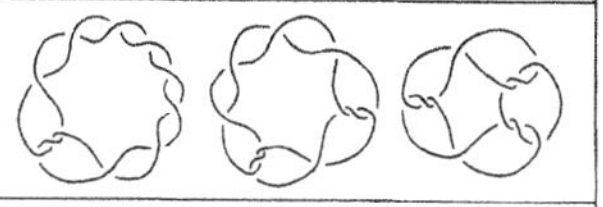		
11^a_{365}: $1^0_5 1^0_7 2^1_7 2^2_9 2^2_{11} 3^3_{11} 2^3_{13} 4^4_{13} 3^4_{15} 4^5_{15} 4^5_{17} 4^6_{17} 4^6_{19} 2^7_{19} 4^7_{21} 3^8_{21} 2^8_{23} 3^9_{25} 1^{10}_{25} 1^{11}_{29}$			
11^a_{366}: $1^0_3 1^0_5 3^1_5 4^2_7 3^2_9 6^3_9 4^3_{11} 6^4_{11} 6^4_{13} 7^5_{13} 6^5_{15} 6^6_{15} 7^6_{17} 3^7_{17} 6^7_{19} 4^8_{19} 3^8_{21} 4^9_{23} 1^{10}_{23} 1^{11}_{27}$			
11^a_{367}: $1^0_9 1^0_{11} 1^2_{13} 1^3_{17} 1^4_{17} 1^5_{21} 1^6_{21} 1^7_{25} 1^8_{25} 1^9_{29} 1^{10}_{29} 1^{11}_{33}$			

2.6. 11 Crossing Non-Alternating Knots.

K: KH	1st line 2nd line 3rd line
11^n_1: $1^{\underline{9}}_{\underline{21}}1^{\underline{8}}_{\underline{19}}1^{\underline{8}}_{\underline{17}}2^{\underline{7}}_{\underline{17}}1^{\underline{7}}_{\underline{15}}2^{\underline{6}}_{\underline{15}}2^{\underline{6}}_{\underline{13}}2^{\underline{5}}_{\underline{13}}2^{\underline{5}}_{\underline{11}}2^{\underline{4}}_{\underline{11}}2^{\underline{4}}_{\underline{9}}2^{\underline{3}}_{\underline{9}}2^{\underline{3}}_{\underline{7}}1^{\underline{2}}_{\underline{7}}2^{\underline{2}}_{\underline{5}}1^{\underline{1}}_{\underline{3}}1^{0}_{\underline{3}}1^{0}_{\underline{1}}$	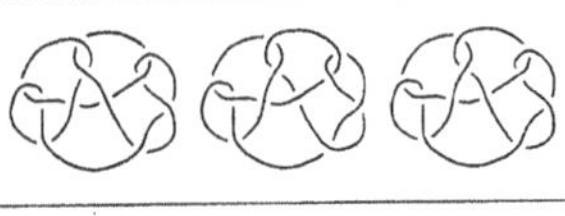
11^n_2: $1^{\underline{2}}_{\underline{1}}1^{\underline{1}}_{1}1^{\underline{1}}_{3}4^{0}_{3}2^{0}_{5}4^{1}_{5}3^{1}_{7}5^{2}_{7}4^{2}_{9}5^{3}_{9}5^{3}_{11}4^{4}_{11}5^{4}_{13}3^{5}_{13}4^{5}_{15}2^{6}_{15}3^{6}_{17}2^{7}_{19}$	
11^n_3: $1^{\underline{5}}_{\underline{13}}2^{\underline{4}}_{\underline{11}}1^{\underline{4}}_{\underline{9}}3^{\underline{3}}_{\underline{9}}2^{\underline{3}}_{\underline{7}}4^{\underline{2}}_{\underline{7}}3^{\underline{2}}_{\underline{5}}3^{\underline{1}}_{\underline{5}}4^{\underline{1}}_{\underline{3}}4^{0}_{\underline{3}}4^{0}_{\underline{1}}3^{1}_{\underline{1}}3^{1}_{1}1^{2}_{1}3^{2}_{3}1^{3}_{3}1^{3}_{5}1^{4}_{7}$	
11^n_4: $1^{\underline{3}}_{\underline{7}}2^{\underline{2}}_{\underline{5}}1^{\underline{2}}_{\underline{3}}3^{\underline{1}}_{\underline{3}}2^{\underline{1}}_{\underline{1}}5^{0}_{\underline{1}}4^{0}_{1}4^{1}_{1}4^{1}_{3}4^{2}_{3}4^{2}_{5}3^{3}_{5}4^{3}_{7}2^{4}_{7}3^{4}_{9}1^{5}_{9}2^{5}_{11}1^{6}_{13}$	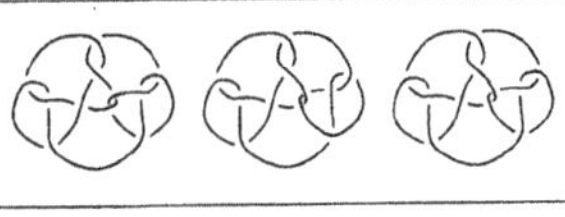
11^n_5: $2^{\underline{2}}_{\underline{3}}2^{\underline{1}}_{\underline{1}}2^{\underline{1}}_{1}6^{0}_{1}3^{0}_{3}6^{1}_{3}5^{1}_{5}6^{2}_{5}6^{2}_{7}6^{3}_{7}6^{3}_{9}4^{4}_{9}6^{4}_{11}3^{5}_{11}4^{5}_{13}1^{6}_{13}3^{6}_{15}1^{7}_{17}$	
11^n_6: $1^{\underline{7}}_{\underline{15}}1^{\underline{6}}_{\underline{13}}1^{\underline{6}}_{\underline{11}}2^{\underline{5}}_{\underline{11}}1^{\underline{5}}_{\underline{9}}2^{\underline{4}}_{\underline{9}}2^{\underline{4}}_{\underline{7}}2^{\underline{3}}_{\underline{7}}2^{\underline{3}}_{\underline{5}}1^{\underline{2}}_{\underline{7}}2^{\underline{2}}_{\underline{5}}2^{\underline{2}}_{\underline{3}}2^{\underline{1}}_{\underline{3}}2^{\underline{1}}_{\underline{1}}1^{0}_{\underline{3}}2^{0}_{\underline{1}}2^{0}_{1}1^{1}_{\underline{1}}1^{1}_{1}1^{1}_{3}1^{2}_{3}1^{3}_{3}1^{4}_{7}$	
11^n_7: $1^{\underline{4}}_{\underline{7}}2^{\underline{3}}_{\underline{5}}1^{\underline{3}}_{\underline{3}}4^{\underline{2}}_{\underline{3}}2^{\underline{2}}_{\underline{1}}5^{\underline{1}}_{\underline{1}}4^{\underline{1}}_{1}6^{0}_{1}6^{0}_{3}6^{1}_{3}5^{1}_{5}5^{2}_{5}6^{2}_{7}3^{3}_{7}5^{3}_{9}2^{4}_{9}3^{4}_{11}2^{5}_{13}$	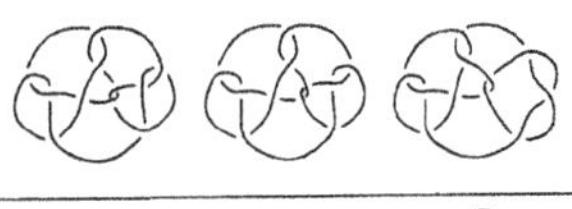
11^n_8: $1^{\underline{7}}_{\underline{19}}2^{\underline{6}}_{\underline{17}}1^{\underline{6}}_{\underline{15}}4^{\underline{5}}_{\underline{15}}2^{\underline{5}}_{\underline{13}}4^{\underline{4}}_{\underline{13}}4^{\underline{4}}_{\underline{11}}5^{\underline{3}}_{\underline{11}}4^{\underline{3}}_{\underline{9}}4^{\underline{2}}_{\underline{9}}5^{\underline{2}}_{\underline{7}}3^{\underline{1}}_{\underline{7}}4^{\underline{1}}_{\underline{5}}3^{0}_{\underline{5}}4^{0}_{\underline{3}}1^{1}_{\underline{3}}2^{1}_{\underline{1}}1^{2}_{1}$	
11^n_9: $1^{\underline{2}}_{1}1^{\underline{1}}_{5}2^{0}_{5}1^{0}_{7}1^{1}_{7}1^{1}_{9}1^{2}_{7}1^{2}_{9}1^{2}_{11}1^{3}_{9}2^{3}_{11}1^{3}_{13}2^{4}_{11}2^{4}_{13}1^{4}_{15}2^{5}_{13}2^{5}_{15}1^{5}_{17}1^{6}_{15}2^{6}_{17}1^{7}_{17}1^{7}_{19}1^{8}_{19}1^{8}_{21}1^{9}_{23}$	
11^n_{10}: $1^{\underline{9}}_{\underline{23}}2^{\underline{8}}_{\underline{21}}1^{\underline{8}}_{\underline{19}}4^{\underline{7}}_{\underline{19}}2^{\underline{7}}_{\underline{17}}5^{\underline{6}}_{\underline{17}}4^{\underline{6}}_{\underline{15}}6^{\underline{5}}_{\underline{15}}5^{\underline{5}}_{\underline{13}}5^{\underline{4}}_{\underline{13}}6^{\underline{4}}_{\underline{11}}5^{\underline{3}}_{\underline{11}}5^{\underline{3}}_{\underline{9}}3^{\underline{2}}_{\underline{9}}5^{\underline{2}}_{\underline{7}}1^{\underline{1}}_{\underline{7}}3^{\underline{1}}_{\underline{5}}1^{0}_{\underline{5}}2^{0}_{\underline{3}}$	
11^n_{11}: $1^{\underline{2}}_{\underline{3}}2^{\underline{1}}_{\underline{1}}1^{\underline{1}}_{1}4^{0}_{1}3^{0}_{3}5^{1}_{3}3^{1}_{5}5^{2}_{5}5^{2}_{7}4^{3}_{7}5^{3}_{9}4^{4}_{9}4^{4}_{11}2^{5}_{11}4^{5}_{13}1^{6}_{13}2^{6}_{15}1^{7}_{17}$	
11^n_{12}: $1^{0}_{\underline{1}}1^{0}_{1}1^{0}_{3}1^{1}_{1}1^{1}_{3}1^{2}_{3}2^{2}_{5}2^{3}_{5}1^{3}_{7}1^{3}_{9}1^{4}_{7}2^{4}_{9}1^{5}_{9}1^{5}_{11}1^{6}_{11}1^{6}_{13}1^{7}_{15}$	
11^n_{13}: $1^{\underline{2}}_{1}1^{\underline{1}}_{5}2^{0}_{5}1^{0}_{7}1^{1}_{7}1^{1}_{9}1^{2}_{9}1^{2}_{11}1^{3}_{11}1^{3}_{13}1^{4}_{13}1^{4}_{15}1^{5}_{17}1^{6}_{17}1^{7}_{21}$	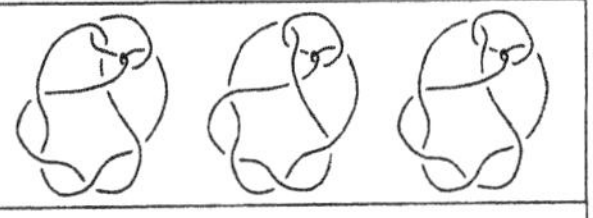
11^n_{14}: $1^{\underline{9}}_{\underline{23}}1^{\underline{8}}_{\underline{21}}1^{\underline{8}}_{\underline{19}}3^{\underline{7}}_{\underline{19}}1^{\underline{7}}_{\underline{17}}3^{\underline{6}}_{\underline{17}}3^{\underline{6}}_{\underline{15}}4^{\underline{5}}_{\underline{15}}3^{\underline{5}}_{\underline{13}}4^{\underline{4}}_{\underline{13}}4^{\underline{4}}_{\underline{11}}3^{\underline{3}}_{\underline{11}}4^{\underline{3}}_{\underline{9}}2^{\underline{2}}_{\underline{9}}3^{\underline{2}}_{\underline{7}}1^{\underline{1}}_{\underline{7}}2^{\underline{1}}_{\underline{5}}1^{0}_{\underline{5}}2^{0}_{\underline{3}}$	
11^n_{15}: $1^{\underline{2}}_{\underline{3}}1^{\underline{1}}_{\underline{1}}1^{\underline{1}}_{1}3^{0}_{1}2^{0}_{3}3^{1}_{3}2^{1}_{5}3^{2}_{5}3^{2}_{7}3^{3}_{7}3^{3}_{9}2^{4}_{9}3^{4}_{11}1^{5}_{11}2^{5}_{13}1^{6}_{13}1^{6}_{15}1^{7}_{17}$	
11^n_{16}: $1^{\underline{2}}_{\underline{1}}1^{\underline{1}}_{1}1^{\underline{1}}_{3}3^{0}_{3}2^{0}_{5}3^{1}_{5}2^{1}_{7}3^{2}_{7}3^{2}_{9}3^{3}_{9}3^{3}_{11}3^{4}_{11}3^{4}_{13}1^{5}_{13}3^{5}_{15}1^{6}_{15}1^{6}_{17}1^{7}_{19}$	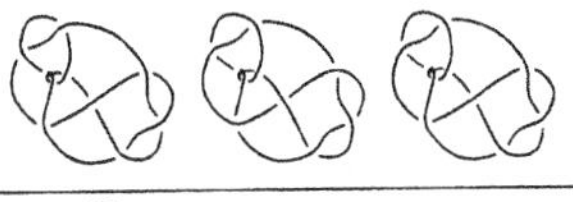
11^n_{17}: $1^{\underline{9}}_{\underline{21}}1^{\underline{8}}_{\underline{19}}1^{\underline{8}}_{\underline{17}}3^{\underline{7}}_{\underline{17}}1^{\underline{7}}_{\underline{15}}3^{\underline{6}}_{\underline{15}}3^{\underline{6}}_{\underline{13}}4^{\underline{5}}_{\underline{13}}3^{\underline{5}}_{\underline{11}}4^{\underline{4}}_{\underline{11}}4^{\underline{4}}_{\underline{9}}3^{\underline{3}}_{\underline{9}}4^{\underline{3}}_{\underline{7}}3^{\underline{2}}_{\underline{7}}3^{\underline{2}}_{\underline{5}}1^{\underline{1}}_{\underline{5}}3^{\underline{1}}_{\underline{3}}1^{0}_{\underline{3}}2^{0}_{\underline{1}}$	
11^n_{18}: $1^{\underline{2}}_{\underline{5}}1^{\underline{1}}_{\underline{3}}1^{\underline{1}}_{\underline{1}}3^{0}_{\underline{1}}2^{0}_{1}3^{1}_{1}2^{1}_{3}2^{2}_{3}3^{2}_{5}3^{3}_{5}2^{3}_{7}2^{4}_{7}3^{4}_{9}1^{5}_{9}2^{5}_{11}1^{6}_{11}1^{6}_{13}1^{7}_{15}$	
11^n_{19}: $1^{\underline{3}}_{\underline{9}}1^{\underline{2}}_{\underline{9}}1^{\underline{2}}_{\underline{5}}1^{\underline{1}}_{\underline{5}}1^{0}_{\underline{5}}1^{0}_{\underline{3}}1^{0}_{\underline{1}}1^{1}_{\underline{3}}1^{1}_{\underline{1}}1^{2}_{1}1^{3}_{1}1^{4}_{5}$	
11^n_{20}: $1^{\underline{4}}_{\underline{11}}1^{\underline{3}}_{\underline{9}}1^{\underline{3}}_{\underline{7}}2^{\underline{2}}_{\underline{7}}1^{\underline{2}}_{\underline{5}}2^{\underline{1}}_{\underline{5}}2^{\underline{1}}_{\underline{3}}2^{0}_{\underline{3}}3^{0}_{\underline{1}}1^{0}_{1}2^{1}_{\underline{1}}2^{1}_{1}1^{2}_{1}2^{2}_{3}1^{3}_{3}1^{3}_{5}1^{4}_{7}$	
11^n_{21}: $1^{\underline{3}}_{\underline{7}}2^{\underline{2}}_{\underline{5}}1^{\underline{2}}_{\underline{3}}3^{\underline{1}}_{\underline{3}}2^{\underline{1}}_{\underline{1}}5^{0}_{\underline{1}}4^{0}_{1}4^{1}_{1}4^{1}_{3}4^{2}_{3}4^{2}_{5}3^{3}_{5}4^{3}_{7}2^{4}_{7}3^{4}_{9}1^{5}_{9}2^{5}_{11}1^{6}_{13}$	
11^n_{22}: $1^{\underline{1}}_{\underline{1}}3^{0}_{1}2^{0}_{3}4^{1}_{3}2^{1}_{5}5^{2}_{5}4^{2}_{7}4^{3}_{7}5^{3}_{9}5^{4}_{9}4^{4}_{11}3^{5}_{11}5^{5}_{13}2^{6}_{13}3^{6}_{15}1^{7}_{15}2^{7}_{17}1^{8}_{19}$	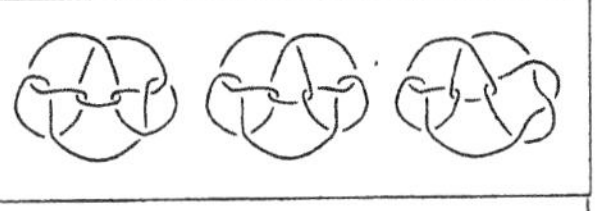
11^n_{23}: $1^{\underline{3}}_{\underline{3}}1^{\underline{2}}_{\underline{1}}1^{\underline{2}}_{1}2^{\underline{1}}_{1}1^{\underline{1}}_{3}3^{0}_{3}3^{0}_{5}2^{1}_{5}2^{1}_{7}3^{2}_{7}2^{2}_{9}1^{3}_{9}3^{3}_{11}2^{4}_{11}1^{4}_{13}2^{5}_{15}$	
11^n_{24}: $1^{\underline{5}}_{\underline{9}}1^{\underline{4}}_{\underline{7}}1^{\underline{4}}_{\underline{5}}2^{\underline{3}}_{\underline{5}}1^{\underline{3}}_{\underline{3}}2^{\underline{2}}_{\underline{3}}2^{\underline{2}}_{\underline{1}}2^{\underline{1}}_{\underline{1}}2^{\underline{1}}_{1}1^{0}_{\underline{1}}3^{0}_{1}2^{0}_{3}1^{1}_{3}2^{1}_{5}1^{2}_{5}1^{2}_{7}1^{3}_{9}$	
11^n_{25}: $1^{\underline{2}}_{\underline{3}}1^{\underline{1}}_{\underline{1}}1^{\underline{1}}_{1}4^{0}_{1}2^{0}_{3}4^{1}_{3}3^{1}_{5}4^{2}_{5}4^{2}_{7}4^{3}_{7}4^{3}_{9}3^{4}_{9}4^{4}_{11}2^{5}_{11}3^{5}_{13}1^{6}_{13}2^{6}_{15}1^{7}_{17}$	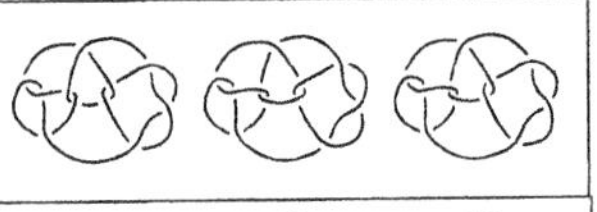
11^n_{26}: $1^{\underline{2}}_{\underline{5}}2^{\underline{1}}_{\underline{3}}1^{\underline{1}}_{\underline{1}}3^{0}_{\underline{1}}3^{0}_{1}4^{1}_{1}2^{1}_{3}3^{2}_{3}4^{2}_{5}3^{3}_{5}3^{3}_{7}3^{4}_{7}3^{4}_{9}1^{5}_{9}3^{5}_{11}1^{6}_{11}1^{6}_{13}1^{7}_{15}$	
11^n_{27}: $1^{\underline{2}}_{1}1^{\underline{1}}_{5}3^{0}_{5}1^{0}_{7}1^{1}_{7}2^{1}_{9}2^{2}_{9}1^{2}_{11}2^{3}_{11}2^{3}_{13}1^{4}_{11}1^{4}_{13}2^{4}_{15}2^{5}_{15}1^{5}_{17}1^{6}_{19}$	
11^n_{28}: $1^{\underline{2}}_{\underline{5}}1^{\underline{1}}_{\underline{1}}2^{0}_{\underline{1}}1^{0}_{1}2^{1}_{1}1^{1}_{3}1^{2}_{3}2^{2}_{5}2^{3}_{5}1^{3}_{7}1^{4}_{7}2^{4}_{9}1^{5}_{9}1^{5}_{11}1^{6}_{11}1^{6}_{13}1^{7}_{15}$	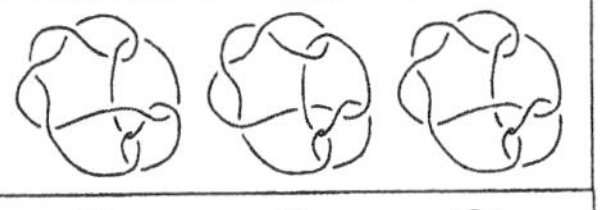
11^n_{29}: $2^{\underline{2}}_{\underline{3}}2^{\underline{1}}_{\underline{1}}2^{\underline{1}}_{1}4^{0}_{1}3^{0}_{3}5^{1}_{3}3^{1}_{5}4^{2}_{5}5^{2}_{7}4^{3}_{7}4^{3}_{9}3^{4}_{9}4^{4}_{11}1^{5}_{11}3^{5}_{13}1^{6}_{13}1^{6}_{15}1^{7}_{17}$	
11^n_{30}: $1^{\underline{2}}_{\underline{1}}1^{\underline{1}}_{3}3^{0}_{3}1^{0}_{5}2^{1}_{5}2^{1}_{7}3^{2}_{7}2^{2}_{9}3^{3}_{9}3^{3}_{11}2^{4}_{11}3^{4}_{13}2^{5}_{13}2^{5}_{15}1^{6}_{15}2^{6}_{17}1^{7}_{19}$	
11^n_{31}: $1^{\underline{2}}_{\underline{1}}1^{\underline{1}}_{3}2^{0}_{3}1^{0}_{5}1^{1}_{5}1^{1}_{7}1^{2}_{5}1^{2}_{7}1^{2}_{9}1^{3}_{7}2^{3}_{9}1^{3}_{11}1^{4}_{9}2^{4}_{11}1^{4}_{13}2^{5}_{11}1^{5}_{13}1^{5}_{15}1^{6}_{13}2^{6}_{15}1^{7}_{15}1^{7}_{17}1^{8}_{17}1^{8}_{19}1^{9}_{21}$	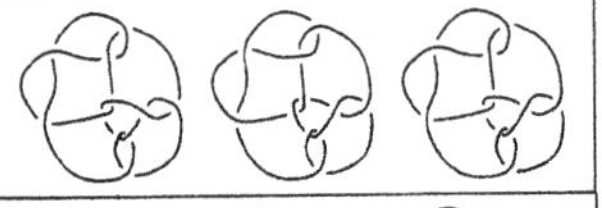
11^n_{32}: $2^{\underline{4}}_{\underline{9}}3^{\underline{3}}_{\underline{7}}2^{\underline{3}}_{\underline{5}}5^{\underline{2}}_{\underline{5}}3^{\underline{2}}_{\underline{3}}6^{\underline{1}}_{\underline{3}}5^{\underline{1}}_{\underline{1}}6^{0}_{\underline{1}}7^{0}_{1}6^{1}_{1}5^{1}_{3}4^{2}_{3}6^{2}_{5}2^{3}_{5}4^{3}_{7}1^{4}_{7}2^{4}_{9}1^{5}_{11}$	
11^n_{33}: $1^{\underline{4}}_{\underline{7}}1^{\underline{3}}_{\underline{5}}1^{\underline{3}}_{\underline{3}}4^{\underline{2}}_{\underline{3}}1^{\underline{2}}_{\underline{1}}3^{\underline{1}}_{\underline{1}}4^{\underline{1}}_{1}5^{0}_{1}4^{0}_{3}5^{1}_{3}4^{1}_{5}3^{2}_{5}5^{2}_{7}3^{3}_{7}3^{3}_{9}1^{4}_{9}3^{4}_{11}1^{5}_{13}$	
11^n_{34}: $1^{\underline{6}}_{\underline{13}}1^{\underline{5}}_{\underline{11}}1^{\underline{5}}_{\underline{9}}1^{\underline{4}}_{\underline{9}}1^{\underline{4}}_{\underline{7}}2^{\underline{3}}_{\underline{7}}1^{\underline{3}}_{\underline{5}}1^{\underline{2}}_{\underline{7}}1^{\underline{2}}_{\underline{5}}2^{\underline{2}}_{\underline{3}}1^{\underline{1}}_{\underline{5}}2^{\underline{1}}_{\underline{3}}1^{\underline{1}}_{\underline{1}}1^{0}_{\underline{3}}3^{0}_{\underline{1}}2^{0}_{1}2^{1}_{\underline{1}}1^{1}_{1}1^{1}_{3}1^{2}_{1}2^{2}_{3}1^{3}_{3}1^{3}_{5}1^{4}_{5}1^{4}_{7}1^{5}_{9}$	
11^n_{35}: $3^{0}_{3}1^{0}_{5}4^{1}_{5}2^{1}_{7}7^{2}_{7}4^{2}_{9}7^{3}_{9}7^{3}_{11}8^{4}_{11}7^{4}_{13}7^{5}_{13}8^{5}_{15}5^{6}_{15}7^{6}_{17}3^{7}_{17}5^{7}_{19}1^{8}_{19}3^{8}_{21}1^{9}_{23}$	
11^n_{36}: $1^{\underline{4}}_{\underline{7}}2^{\underline{3}}_{\underline{5}}1^{\underline{3}}_{\underline{3}}4^{\underline{2}}_{\underline{3}}2^{\underline{2}}_{\underline{1}}5^{\underline{1}}_{\underline{1}}4^{\underline{1}}_{1}6^{0}_{1}6^{0}_{3}6^{1}_{3}5^{1}_{5}5^{2}_{5}6^{2}_{7}3^{3}_{7}5^{3}_{9}2^{4}_{9}3^{4}_{11}2^{5}_{13}$	

K: KH	1st line 2nd line 3rd line
11^n_{37}: $1^{\underline{6}}_{\underline{13}}1^{\underline{5}}_{\underline{11}}1^{\underline{5}}_{\underline{9}}2^{\underline{4}}_{\underline{9}}1^{\underline{4}}_{\underline{7}}2^{\underline{3}}_{\underline{7}}2^{\underline{3}}_{\underline{5}}2^{\underline{2}}_{\underline{5}}2^{\underline{2}}_{\underline{3}}2^{\underline{1}}_{\underline{3}}2^{\underline{1}}_{\underline{1}}2^{0}_{\underline{1}}3^{0}_{1}1^{1}_{1}1^{1}_{3}1^{2}_{5}$ 11^n_{38}: $1^{\underline{6}}_{\underline{11}}1^{\underline{5}}_{\underline{7}}1^{\underline{4}}_{\underline{7}}1^{\underline{3}}_{\underline{5}}1^{\underline{3}}_{\underline{3}}1^{\underline{2}}_{\underline{5}}1^{\underline{2}}_{\underline{1}}2^{\underline{1}}_{\underline{1}}1^{0}_{\underline{1}}1^{0}_{1}1^{0}_{3}1^{1}_{1}1^{2}_{5}$ 11^n_{39}: $1^{\underline{5}}_{9}1^{\underline{4}}_{5}1^{\underline{3}}_{5}1^{\underline{2}}_{5}1^{\underline{2}}_{3}1^{\underline{2}}_{1}2^{\underline{1}}_{3}1^{\underline{1}}_{\underline{1}}1^{\underline{1}}_{1}3^{0}_{\underline{1}}4^{0}_{1}3^{1}_{1}2^{1}_{3}1^{1}_{5}3^{2}_{3}3^{2}_{5}2^{3}_{5}3^{3}_{7}2^{4}_{7}2^{4}_{9}1^{5}_{9}2^{5}_{11}1^{6}_{13}$	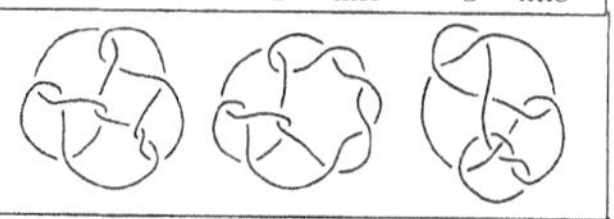
11^n_{40}: $2^{\underline{1}}_{\underline{1}}4^{0}_{1}3^{0}_{3}6^{1}_{3}3^{1}_{5}7^{2}_{5}6^{2}_{7}6^{3}_{7}7^{3}_{9}7^{4}_{9}6^{4}_{11}4^{5}_{11}7^{5}_{13}3^{6}_{13}4^{6}_{15}1^{7}_{15}3^{7}_{17}1^{8}_{19}$ 11^n_{41}: $1^{\underline{3}}_{\underline{3}}2^{\underline{2}}_{\underline{1}}1^{\underline{2}}_{1}3^{\underline{1}}_{1}2^{\underline{1}}_{3}5^{0}_{3}4^{0}_{5}4^{1}_{5}4^{1}_{7}5^{2}_{7}4^{2}_{9}3^{3}_{9}5^{3}_{11}3^{4}_{11}3^{4}_{13}1^{5}_{13}3^{5}_{15}1^{6}_{17}$ 11^n_{42}: $1^{\underline{6}}_{13}1^{\underline{5}}_{\underline{11}}1^{\underline{5}}_{\underline{9}}1^{\underline{4}}_{\underline{9}}1^{\underline{4}}_{\underline{7}}2^{\underline{3}}_{\underline{7}}1^{\underline{3}}_{\underline{5}}1^{\underline{2}}_{\underline{7}}1^{\underline{2}}_{\underline{5}}2^{\underline{2}}_{\underline{3}}1^{\underline{1}}_{\underline{5}}2^{\underline{1}}_{\underline{3}}1^{\underline{1}}_{\underline{1}}1^{0}_{\underline{3}}3^{0}_{\underline{1}}2^{0}_{1}2^{1}_{\underline{1}}1^{1}_{1}1^{1}_{3}1^{2}_{1}2^{2}_{3}1^{3}_{3}1^{3}_{5}1^{4}_{5}1^{4}_{7}1^{5}_{9}$	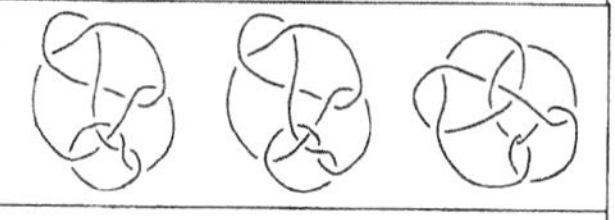
11^n_{43}: $3^{0}_{3}1^{0}_{5}4^{1}_{5}2^{1}_{7}7^{2}_{7}4^{2}_{9}7^{3}_{9}7^{3}_{11}8^{4}_{11}7^{4}_{13}7^{5}_{13}8^{5}_{15}5^{6}_{15}7^{6}_{17}3^{7}_{17}5^{7}_{19}1^{8}_{19}3^{8}_{21}1^{9}_{23}$ 11^n_{44}: $1^{\underline{4}}_{\underline{7}}2^{\underline{3}}_{\underline{5}}1^{\underline{3}}_{\underline{3}}4^{\underline{2}}_{\underline{3}}2^{\underline{2}}_{\underline{1}}5^{\underline{1}}_{\underline{1}}4^{\underline{1}}_{1}6^{0}_{1}6^{0}_{3}6^{1}_{3}5^{1}_{5}5^{2}_{5}6^{2}_{7}3^{3}_{7}5^{3}_{9}2^{4}_{9}3^{4}_{11}2^{5}_{13}$ 11^n_{45}: $1^{\underline{5}}_{\underline{9}}1^{\underline{4}}_{\underline{5}}1^{\underline{3}}_{\underline{5}}1^{\underline{2}}_{\underline{5}}1^{\underline{2}}_{\underline{3}}1^{\underline{2}}_{\underline{1}}2^{\underline{1}}_{\underline{3}}1^{\underline{1}}_{\underline{1}}1^{\underline{1}}_{1}3^{0}_{\underline{1}}4^{0}_{1}3^{1}_{1}2^{1}_{3}1^{1}_{5}3^{2}_{3}3^{2}_{5}2^{3}_{5}3^{3}_{7}2^{4}_{7}2^{4}_{9}1^{5}_{9}2^{5}_{11}1^{6}_{13}$	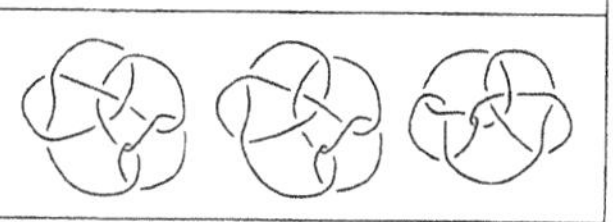
11^n_{46}: $2^{\underline{1}}_{\underline{1}}4^{0}_{1}3^{0}_{3}6^{1}_{3}3^{1}_{5}7^{2}_{5}6^{2}_{7}6^{3}_{7}7^{3}_{9}7^{4}_{9}6^{4}_{11}4^{5}_{11}7^{5}_{13}3^{6}_{13}4^{6}_{15}1^{7}_{15}3^{7}_{17}1^{8}_{19}$ 11^n_{47}: $1^{\underline{3}}_{\underline{3}}2^{\underline{2}}_{\underline{1}}1^{\underline{2}}_{1}3^{\underline{1}}_{1}2^{\underline{1}}_{3}5^{0}_{3}4^{0}_{5}4^{1}_{5}4^{1}_{7}5^{2}_{7}4^{2}_{9}3^{3}_{9}5^{3}_{11}3^{4}_{11}3^{4}_{13}1^{5}_{13}3^{5}_{15}1^{6}_{17}$ 11^n_{48}: $1^{\underline{6}}_{\underline{13}}1^{\underline{5}}_{\underline{11}}1^{\underline{5}}_{\underline{9}}2^{\underline{4}}_{\underline{9}}1^{\underline{4}}_{\underline{7}}2^{\underline{3}}_{\underline{7}}2^{\underline{3}}_{\underline{5}}3^{\underline{2}}_{\underline{5}}2^{\underline{2}}_{\underline{3}}2^{\underline{1}}_{\underline{3}}3^{\underline{1}}_{\underline{1}}2^{0}_{\underline{1}}3^{0}_{1}2^{1}_{1}1^{1}_{3}2^{2}_{5}$	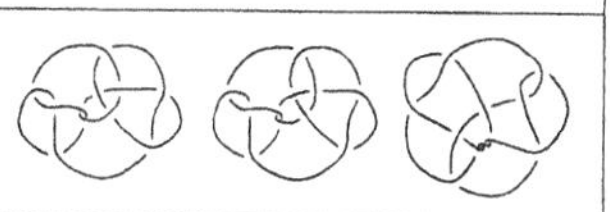
11^n_{49}: $1^{\underline{6}}_{\underline{11}}1^{\underline{5}}_{\underline{7}}1^{\underline{4}}_{\underline{7}}1^{\underline{3}}_{\underline{5}}1^{\underline{3}}_{\underline{3}}1^{\underline{2}}_{\underline{5}}1^{\underline{2}}_{\underline{1}}2^{\underline{1}}_{\underline{1}}2^{0}_{\underline{1}}1^{0}_{1}1^{0}_{3}1^{1}_{1}1^{1}_{3}1^{2}_{5}1^{3}_{5}1^{4}_{9}$ 11^n_{50}: $1^{\underline{7}}_{\underline{15}}1^{\underline{6}}_{\underline{13}}1^{\underline{6}}_{\underline{11}}2^{\underline{5}}_{\underline{11}}1^{\underline{5}}_{\underline{9}}2^{\underline{4}}_{\underline{9}}2^{\underline{4}}_{\underline{7}}2^{\underline{3}}_{\underline{7}}2^{\underline{3}}_{\underline{5}}2^{\underline{2}}_{\underline{5}}2^{\underline{2}}_{\underline{3}}1^{\underline{1}}_{\underline{3}}2^{\underline{1}}_{\underline{1}}2^{0}_{\underline{1}}2^{0}_{1}1^{1}_{3}$ 11^n_{51}: $1^{\underline{2}}_{\underline{5}}1^{\underline{1}}_{\underline{3}}1^{\underline{1}}_{\underline{1}}2^{0}_{\underline{1}}2^{0}_{1}3^{1}_{1}1^{1}_{3}2^{2}_{3}3^{2}_{5}2^{3}_{5}2^{3}_{7}2^{4}_{7}2^{4}_{9}1^{5}_{9}2^{5}_{11}1^{6}_{11}1^{6}_{13}1^{7}_{15}$	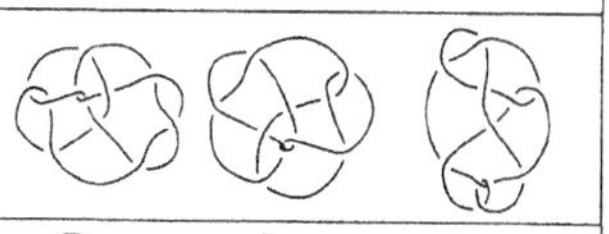
11^n_{52}: $2^{\underline{2}}_{\underline{3}}2^{\underline{1}}_{\underline{1}}2^{\underline{1}}_{1}5^{0}_{1}3^{0}_{3}5^{1}_{3}4^{1}_{5}5^{2}_{5}5^{2}_{7}5^{3}_{7}5^{3}_{9}3^{4}_{9}5^{4}_{11}2^{5}_{11}3^{5}_{13}1^{6}_{13}2^{6}_{15}1^{7}_{17}$ 11^n_{53}: $1^{\underline{6}}_{\underline{13}}1^{\underline{5}}_{\underline{11}}1^{\underline{5}}_{\underline{9}}2^{\underline{4}}_{\underline{9}}1^{\underline{4}}_{\underline{7}}3^{\underline{3}}_{\underline{7}}2^{\underline{3}}_{\underline{5}}3^{\underline{2}}_{\underline{5}}3^{\underline{2}}_{\underline{3}}3^{\underline{1}}_{\underline{3}}3^{\underline{1}}_{\underline{1}}3^{0}_{\underline{1}}4^{0}_{1}2^{1}_{1}2^{1}_{3}1^{2}_{3}2^{2}_{5}1^{3}_{7}$ 11^n_{54}: $1^{\underline{1}}_{\underline{1}}2^{0}_{1}2^{0}_{3}3^{1}_{3}1^{1}_{5}4^{2}_{5}3^{2}_{7}3^{3}_{7}4^{3}_{9}4^{4}_{9}3^{4}_{11}2^{5}_{11}4^{5}_{13}2^{6}_{13}2^{6}_{15}1^{7}_{15}2^{7}_{17}1^{8}_{19}$	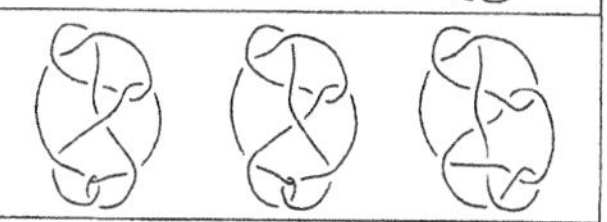
11^n_{55}: $2^{\underline{3}}_{\underline{7}}3^{\underline{2}}_{\underline{5}}2^{\underline{2}}_{\underline{3}}4^{\underline{1}}_{\underline{3}}3^{\underline{1}}_{\underline{1}}6^{0}_{\underline{1}}5^{0}_{1}5^{1}_{1}5^{1}_{3}5^{2}_{3}5^{2}_{5}3^{3}_{5}5^{3}_{7}2^{4}_{7}3^{4}_{9}1^{5}_{9}2^{5}_{11}1^{6}_{13}$ 11^n_{56}: $1^{\underline{5}}_{\underline{9}}1^{\underline{4}}_{\underline{7}}1^{\underline{4}}_{\underline{5}}2^{\underline{3}}_{\underline{5}}1^{\underline{3}}_{\underline{3}}3^{\underline{2}}_{\underline{3}}2^{\underline{2}}_{\underline{1}}2^{\underline{1}}_{\underline{1}}3^{\underline{1}}_{1}4^{0}_{1}3^{0}_{3}2^{1}_{3}3^{1}_{5}2^{2}_{5}2^{2}_{7}1^{3}_{7}2^{3}_{9}1^{4}_{11}$ 11^n_{57}: $1^{\underline{2}}_{1}1^{\underline{1}}_{5}2^{0}_{5}1^{0}_{7}1^{1}_{7}1^{1}_{9}1^{2}_{7}1^{2}_{9}1^{2}_{11}2^{3}_{11}1^{3}_{13}1^{4}_{11}1^{4}_{13}1^{4}_{15}1^{5}_{13}1^{5}_{15}1^{5}_{17}1^{6}_{17}$	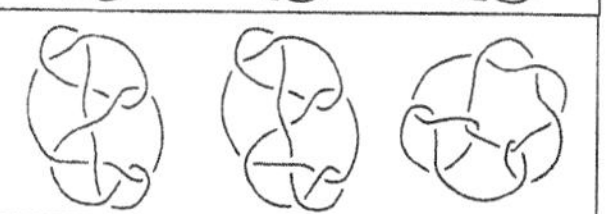
11^n_{58}: $1^{\underline{4}}_{\underline{11}}2^{\underline{3}}_{\underline{9}}1^{\underline{3}}_{\underline{7}}2^{\underline{2}}_{\underline{7}}2^{\underline{2}}_{\underline{5}}3^{\underline{1}}_{\underline{5}}2^{\underline{1}}_{\underline{3}}3^{0}_{\underline{3}}4^{0}_{\underline{1}}3^{1}_{\underline{1}}2^{1}_{1}2^{2}_{1}3^{2}_{3}1^{3}_{3}2^{3}_{5}1^{4}_{5}1^{4}_{7}1^{5}_{9}$ 11^n_{59}: $2^{0}_{3}1^{0}_{5}2^{1}_{5}1^{1}_{7}4^{2}_{7}2^{2}_{9}4^{3}_{9}4^{3}_{11}5^{4}_{11}4^{4}_{13}4^{5}_{13}5^{5}_{15}3^{6}_{15}4^{6}_{17}2^{7}_{17}3^{7}_{19}1^{8}_{19}2^{8}_{21}1^{9}_{23}$ 11^n_{60}: $1^{\underline{4}}_{\underline{7}}1^{\underline{3}}_{\underline{5}}1^{\underline{3}}_{\underline{3}}2^{\underline{2}}_{\underline{3}}1^{\underline{2}}_{\underline{1}}2^{\underline{1}}_{\underline{1}}2^{\underline{1}}_{1}3^{0}_{1}3^{0}_{3}3^{1}_{3}2^{1}_{5}2^{2}_{5}3^{2}_{7}1^{3}_{7}2^{3}_{9}1^{4}_{9}1^{4}_{11}1^{5}_{13}$	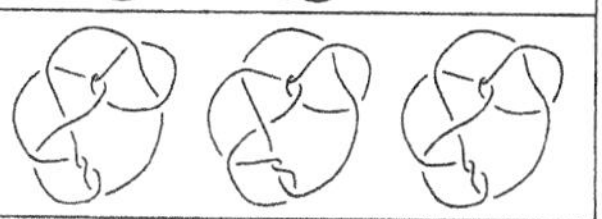
11^n_{61}: $1^{\underline{3}}_{\underline{3}}1^{\underline{2}}_{\underline{1}}1^{\underline{2}}_{1}1^{\underline{1}}_{1}1^{\underline{1}}_{3}3^{0}_{3}2^{0}_{5}1^{1}_{3}1^{1}_{5}2^{1}_{7}3^{2}_{7}1^{2}_{9}1^{3}_{9}2^{3}_{11}1^{4}_{9}1^{4}_{11}1^{4}_{13}1^{5}_{13}1^{5}_{15}$ 11^n_{62}: $1^{\underline{4}}_{\underline{9}}1^{\underline{3}}_{\underline{7}}1^{\underline{3}}_{\underline{5}}2^{\underline{2}}_{\underline{5}}1^{\underline{2}}_{\underline{3}}3^{\underline{1}}_{\underline{3}}2^{\underline{1}}_{\underline{1}}3^{0}_{\underline{1}}4^{0}_{1}3^{1}_{1}2^{1}_{3}2^{2}_{3}3^{2}_{5}1^{3}_{5}2^{3}_{7}1^{4}_{7}1^{4}_{9}1^{5}_{11}$ 11^n_{63}: $2^{0}_{1}1^{0}_{3}2^{1}_{3}1^{1}_{5}3^{2}_{5}2^{2}_{7}3^{3}_{7}3^{3}_{9}3^{4}_{9}3^{4}_{11}3^{5}_{11}3^{5}_{13}2^{6}_{13}3^{6}_{15}1^{7}_{15}2^{7}_{17}1^{8}_{17}1^{8}_{19}1^{9}_{21}$	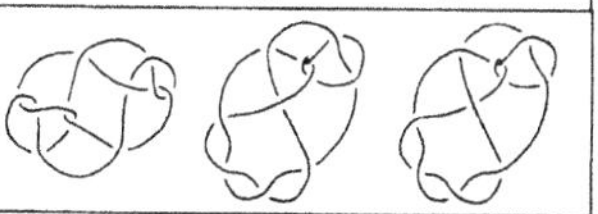
11^n_{64}: $1^{\underline{4}}_{\underline{5}}1^{\underline{3}}_{\underline{1}}2^{\underline{2}}_{\underline{1}}1^{\underline{1}}_{1}2^{\underline{1}}_{3}2^{0}_{3}2^{0}_{5}2^{1}_{5}1^{1}_{7}1^{2}_{7}2^{2}_{9}1^{3}_{9}1^{3}_{11}1^{4}_{11}1^{4}_{13}1^{5}_{15}$ 11^n_{65}: $1^{\underline{7}}_{\underline{15}}1^{\underline{6}}_{\underline{13}}1^{\underline{6}}_{\underline{11}}3^{\underline{5}}_{\underline{11}}1^{\underline{5}}_{\underline{9}}2^{\underline{4}}_{\underline{9}}3^{\underline{4}}_{\underline{7}}3^{\underline{3}}_{\underline{7}}2^{\underline{3}}_{\underline{5}}3^{\underline{2}}_{\underline{5}}3^{\underline{2}}_{\underline{3}}1^{\underline{1}}_{\underline{3}}3^{\underline{1}}_{\underline{1}}3^{0}_{\underline{1}}2^{0}_{1}2^{1}_{3}$ 11^n_{66}: $1^{\underline{3}}_{\underline{5}}3^{\underline{2}}_{\underline{3}}1^{\underline{2}}_{\underline{1}}4^{\underline{1}}_{\underline{1}}3^{\underline{1}}_{1}7^{0}_{1}5^{0}_{3}6^{1}_{3}6^{1}_{5}7^{2}_{5}6^{2}_{7}5^{3}_{7}7^{3}_{9}3^{4}_{9}5^{4}_{11}2^{5}_{11}3^{5}_{13}2^{6}_{15}$	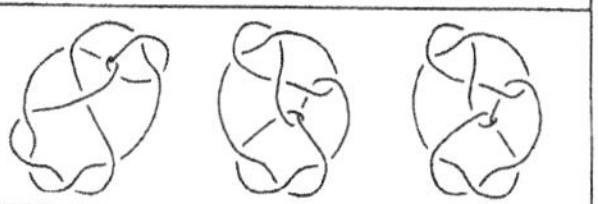
11^n_{67}: $1^{\underline{4}}_{\underline{7}}1^{\underline{3}}_{\underline{3}}1^{\underline{2}}_{\underline{3}}1^{\underline{1}}_{\underline{1}}1^{\underline{1}}_{1}2^{0}_{\underline{1}}1^{0}_{1}1^{0}_{3}1^{1}_{1}2^{1}_{3}1^{2}_{3}1^{2}_{5}1^{2}_{7}2^{3}_{5}1^{3}_{7}1^{4}_{7}2^{4}_{9}1^{5}_{9}1^{5}_{11}1^{6}_{11}1^{6}_{13}1^{7}_{15}$ 11^n_{68}: $2^{\underline{2}}_{\underline{3}}2^{\underline{1}}_{\underline{1}}2^{\underline{1}}_{1}5^{0}_{1}3^{0}_{3}6^{1}_{3}4^{1}_{5}5^{2}_{5}6^{2}_{7}5^{3}_{7}5^{3}_{9}4^{4}_{9}5^{4}_{11}2^{5}_{11}4^{5}_{13}1^{6}_{13}2^{6}_{15}1^{7}_{17}$ 11^n_{69}: $1^{\underline{2}}_{\underline{1}}1^{\underline{1}}_{1}1^{\underline{1}}_{3}4^{0}_{3}2^{0}_{5}3^{1}_{5}3^{1}_{7}4^{2}_{7}3^{2}_{9}4^{3}_{9}4^{3}_{11}3^{4}_{11}4^{4}_{13}2^{5}_{13}3^{5}_{15}1^{6}_{15}2^{6}_{17}1^{7}_{19}$	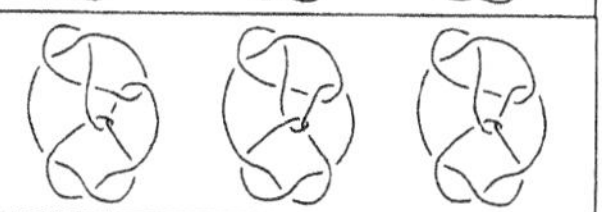
11^n_{70}: $1^{\underline{4}}_{\underline{5}}1^{\underline{3}}_{\underline{1}}2^{\underline{2}}_{\underline{1}}1^{\underline{1}}_{1}2^{\underline{1}}_{3}1^{0}_{1}2^{0}_{3}1^{0}_{5}2^{1}_{5}1^{1}_{7}1^{2}_{5}2^{2}_{9}2^{3}_{9}1^{4}_{13}$ 11^n_{71}: $2^{\underline{1}}_{\underline{1}}3^{0}_{1}3^{0}_{3}5^{1}_{3}2^{1}_{5}6^{2}_{5}5^{2}_{7}4^{3}_{7}6^{3}_{9}6^{4}_{9}4^{4}_{11}3^{5}_{11}6^{5}_{13}2^{6}_{13}3^{6}_{15}1^{7}_{15}2^{7}_{17}1^{8}_{19}$ 11^n_{72}: $3^{0}_{3}1^{0}_{5}3^{1}_{5}2^{1}_{7}7^{2}_{7}3^{2}_{9}6^{3}_{9}7^{3}_{11}7^{4}_{11}6^{4}_{13}7^{5}_{13}7^{5}_{15}4^{6}_{15}7^{6}_{17}3^{7}_{17}4^{7}_{19}1^{8}_{19}3^{8}_{21}1^{9}_{23}$	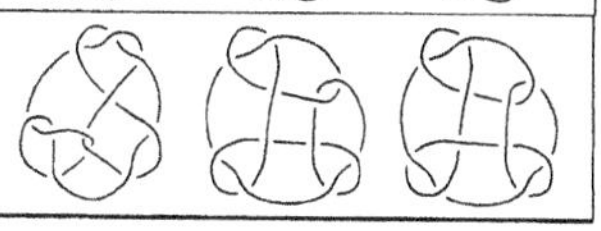

K: KH	1st line 2nd line 3rd line
11^n_{73}: $1^{\underline{5}}_{\underline{9}}1^{\underline{4}}_{\underline{5}}1^{\underline{3}}_{\underline{5}}1^{\underline{2}}_{\underline{3}}1^{\underline{2}}_{\underline{1}}1^{\underline{1}}_{\underline{3}}1^{\underline{1}}_{1}2^{0}_{\underline{1}}3^{0}_{1}1^{1}_{1}1^{1}_{3}1^{1}_{5}2^{2}_{3}1^{2}_{5}1^{3}_{5}2^{3}_{7}1^{4}_{7}1^{4}_{9}1^{5}_{9}1^{5}_{11}1^{6}_{13}$	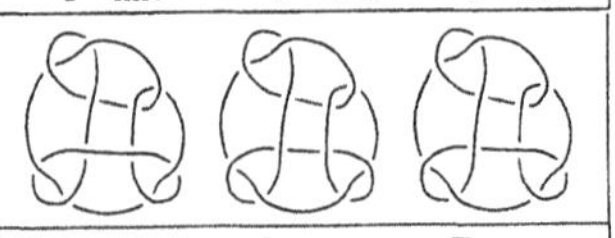
11^n_{74}: $1^{\underline{5}}_{\underline{9}}1^{\underline{4}}_{\underline{5}}1^{\underline{3}}_{\underline{5}}1^{\underline{2}}_{\underline{3}}1^{\underline{2}}_{\underline{1}}1^{\underline{1}}_{\underline{3}}1^{\underline{1}}_{1}2^{0}_{\underline{1}}3^{0}_{1}1^{1}_{1}1^{1}_{3}1^{1}_{5}2^{2}_{3}1^{2}_{5}1^{3}_{5}2^{3}_{7}1^{4}_{7}1^{4}_{9}1^{5}_{9}1^{5}_{11}1^{6}_{13}$	
11^n_{75}: $1^{\underline{8}}_{\underline{19}}2^{\underline{7}}_{\underline{17}}1^{\underline{7}}_{\underline{15}}3^{\underline{6}}_{\underline{15}}2^{\underline{6}}_{\underline{13}}6^{\underline{5}}_{\underline{13}}3^{\underline{5}}_{\underline{11}}4^{\underline{4}}_{\underline{11}}6^{\underline{4}}_{\underline{9}}6^{\underline{3}}_{\underline{9}}4^{\underline{3}}_{\underline{7}}5^{\underline{2}}_{\underline{7}}6^{\underline{2}}_{\underline{5}}2^{\underline{1}}_{\underline{5}}5^{\underline{1}}_{\underline{3}}3^{0}_{\underline{3}}3^{0}_{\underline{1}}2^{1}_{1}$	
11^n_{76}: $1^{\underline{6}}_{\underline{17}}3^{\underline{5}}_{\underline{15}}1^{\underline{5}}_{\underline{13}}2^{\underline{4}}_{\underline{13}}3^{\underline{4}}_{\underline{11}}5^{\underline{3}}_{\underline{11}}2^{\underline{3}}_{\underline{9}}3^{\underline{2}}_{\underline{9}}5^{\underline{2}}_{\underline{7}}3^{\underline{1}}_{\underline{7}}3^{\underline{1}}_{\underline{5}}4^{0}_{\underline{5}}4^{0}_{\underline{3}}1^{1}_{\underline{3}}3^{1}_{\underline{1}}1^{2}_{\underline{1}}1^{2}_{1}1^{3}_{3}$	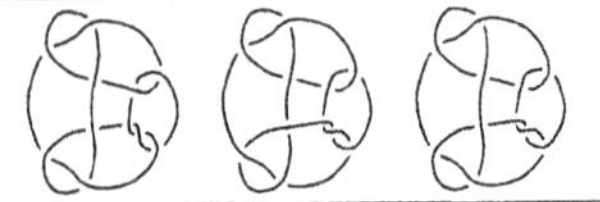
11^n_{77}: $1^{0}_{7}1^{0}_{9}1^{2}_{11}1^{3}_{15}3^{4}_{13}1^{4}_{15}2^{5}_{15}3^{5}_{17}1^{5}_{19}3^{6}_{17}2^{6}_{19}3^{7}_{19}3^{7}_{21}2^{8}_{21}3^{8}_{23}2^{9}_{23}2^{9}_{25}1^{10}_{25}2^{10}_{27}1^{11}_{29}$	
11^n_{78}: $1^{\underline{3}}_{\underline{3}}1^{\underline{2}}_{\underline{1}}1^{\underline{2}}_{1}3^{\underline{1}}_{1}1^{\underline{1}}_{3}4^{0}_{3}4^{0}_{5}3^{1}_{5}3^{1}_{7}5^{2}_{7}3^{2}_{9}2^{3}_{9}5^{3}_{11}3^{4}_{11}2^{4}_{13}1^{5}_{13}3^{5}_{15}1^{6}_{17}$	
11^n_{79}: $1^{\underline{4}}_{\underline{7}}1^{\underline{3}}_{\underline{3}}2^{\underline{2}}_{\underline{3}}1^{\underline{1}}_{\underline{1}}2^{\underline{1}}_{1}1^{0}_{\underline{1}}2^{0}_{1}1^{0}_{3}2^{1}_{3}1^{1}_{5}2^{2}_{7}1^{3}_{7}1^{4}_{11}$	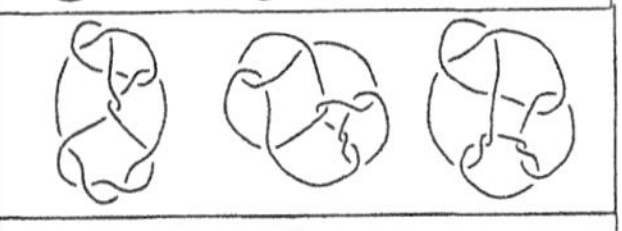
11^n_{80}: $1^{\underline{7}}_{\underline{17}}1^{\underline{6}}_{\underline{15}}1^{\underline{6}}_{\underline{13}}2^{\underline{5}}_{\underline{13}}1^{\underline{5}}_{\underline{11}}2^{\underline{4}}_{\underline{11}}2^{\underline{4}}_{\underline{9}}2^{\underline{3}}_{\underline{9}}2^{\underline{3}}_{\underline{7}}1^{\underline{2}}_{\underline{9}}2^{\underline{2}}_{\underline{7}}2^{\underline{2}}_{\underline{5}}2^{\underline{1}}_{\underline{5}}2^{\underline{1}}_{\underline{3}}1^{0}_{\underline{5}}1^{0}_{\underline{3}}2^{0}_{\underline{1}}1^{1}_{\underline{3}}1^{1}_{\underline{1}}1^{2}_{1}1^{3}_{1}1^{4}_{5}$	
11^n_{81}: $1^{\underline{2}}_{1}1^{\underline{1}}_{5}4^{0}_{5}1^{0}_{7}1^{1}_{7}3^{1}_{9}3^{2}_{9}1^{2}_{11}3^{3}_{11}3^{3}_{13}1^{4}_{11}1^{4}_{13}3^{4}_{15}3^{5}_{15}1^{5}_{17}2^{6}_{19}$	
11^n_{82}: $1^{\underline{5}}_{\underline{9}}1^{\underline{4}}_{\underline{7}}1^{\underline{4}}_{\underline{5}}1^{\underline{3}}_{\underline{5}}1^{\underline{3}}_{\underline{3}}2^{\underline{2}}_{\underline{3}}1^{\underline{2}}_{\underline{1}}1^{\underline{1}}_{\underline{1}}2^{\underline{1}}_{1}2^{0}_{1}2^{0}_{3}1^{1}_{3}1^{1}_{5}1^{2}_{5}1^{2}_{7}1^{3}_{9}$	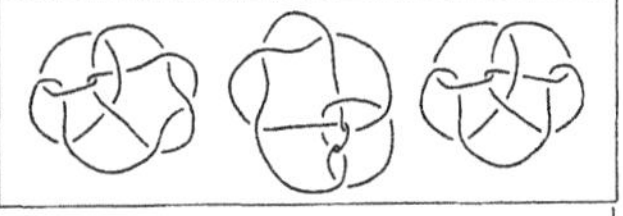
11^n_{83}: $1^{\underline{5}}_{\underline{11}}2^{\underline{4}}_{\underline{9}}1^{\underline{4}}_{\underline{7}}3^{\underline{3}}_{\underline{7}}2^{\underline{3}}_{\underline{5}}4^{\underline{2}}_{\underline{5}}3^{\underline{2}}_{\underline{3}}4^{\underline{1}}_{\underline{3}}4^{\underline{1}}_{\underline{1}}5^{0}_{\underline{1}}5^{0}_{1}3^{1}_{1}4^{1}_{3}2^{2}_{3}3^{2}_{5}1^{3}_{5}2^{3}_{7}1^{4}_{9}$	
11^n_{84}: $1^{\underline{8}}_{\underline{19}}2^{\underline{7}}_{\underline{17}}1^{\underline{7}}_{\underline{15}}2^{\underline{6}}_{\underline{15}}2^{\underline{6}}_{\underline{13}}3^{\underline{5}}_{\underline{13}}2^{\underline{5}}_{\underline{11}}3^{\underline{4}}_{\underline{11}}3^{\underline{4}}_{\underline{9}}3^{\underline{3}}_{\underline{9}}3^{\underline{3}}_{\underline{7}}2^{\underline{2}}_{\underline{7}}3^{\underline{2}}_{\underline{5}}1^{\underline{1}}_{\underline{5}}2^{\underline{1}}_{\underline{3}}1^{0}_{\underline{3}}2^{0}_{\underline{1}}$	
11^n_{85}: $1^{\underline{3}}_{\underline{7}}2^{\underline{2}}_{\underline{5}}1^{\underline{2}}_{\underline{3}}3^{\underline{1}}_{\underline{3}}2^{\underline{1}}_{\underline{1}}4^{0}_{\underline{1}}4^{0}_{1}4^{1}_{1}3^{1}_{3}4^{2}_{3}4^{2}_{5}2^{3}_{5}4^{3}_{7}2^{4}_{7}2^{4}_{9}1^{5}_{9}2^{5}_{11}1^{6}_{13}$	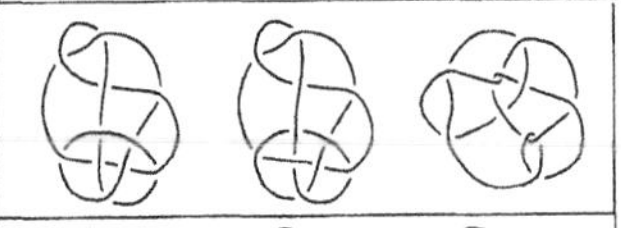
11^n_{86}: $1^{\underline{2}}_{\underline{5}}2^{\underline{1}}_{\underline{3}}1^{\underline{1}}_{\underline{1}}3^{0}_{\underline{1}}3^{0}_{1}3^{1}_{1}2^{1}_{3}3^{2}_{3}3^{2}_{5}2^{3}_{5}3^{3}_{7}2^{4}_{7}2^{4}_{9}1^{5}_{9}2^{5}_{11}1^{6}_{13}$	
11^n_{87}: $1^{\underline{8}}_{\underline{19}}2^{\underline{7}}_{\underline{17}}1^{\underline{7}}_{\underline{15}}3^{\underline{6}}_{\underline{15}}2^{\underline{6}}_{\underline{13}}4^{\underline{5}}_{\underline{13}}3^{\underline{5}}_{\underline{11}}4^{\underline{4}}_{\underline{11}}4^{\underline{4}}_{\underline{9}}5^{\underline{3}}_{\underline{9}}4^{\underline{3}}_{\underline{7}}3^{\underline{2}}_{\underline{7}}5^{\underline{2}}_{\underline{5}}2^{\underline{1}}_{\underline{5}}3^{\underline{1}}_{\underline{3}}2^{0}_{\underline{3}}3^{0}_{\underline{1}}1^{1}_{1}$	
11^n_{88}: $1^{\underline{2}}_{1}1^{\underline{1}}_{5}2^{0}_{5}1^{0}_{7}1^{1}_{7}1^{1}_{9}1^{2}_{9}1^{2}_{11}1^{3}_{11}1^{3}_{13}1^{4}_{11}1^{4}_{13}1^{4}_{15}1^{5}_{15}1^{5}_{17}$	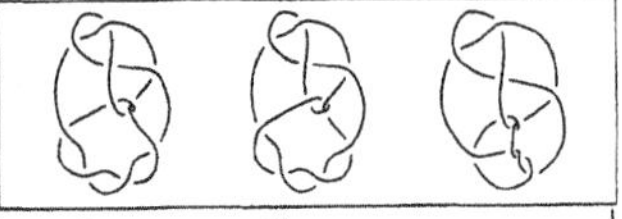
11^n_{89}: $1^{\underline{9}}_{\underline{23}}2^{\underline{8}}_{\underline{21}}1^{\underline{8}}_{\underline{19}}4^{\underline{7}}_{\underline{19}}2^{\underline{7}}_{\underline{17}}4^{\underline{6}}_{\underline{17}}4^{\underline{6}}_{\underline{15}}6^{\underline{5}}_{\underline{15}}4^{\underline{5}}_{\underline{13}}5^{\underline{4}}_{\underline{13}}6^{\underline{4}}_{\underline{11}}4^{\underline{3}}_{\underline{11}}5^{\underline{3}}_{\underline{9}}3^{\underline{2}}_{\underline{9}}4^{\underline{2}}_{\underline{7}}1^{\underline{1}}_{\underline{7}}3^{\underline{1}}_{\underline{5}}1^{0}_{\underline{5}}2^{0}_{\underline{3}}$	
11^n_{90}: $1^{\underline{2}}_{\underline{1}}1^{\underline{1}}_{1}1^{\underline{1}}_{3}3^{0}_{3}2^{0}_{5}3^{1}_{5}2^{1}_{7}4^{2}_{7}3^{2}_{9}3^{3}_{9}4^{3}_{11}3^{4}_{11}3^{4}_{13}2^{5}_{13}3^{5}_{15}1^{6}_{15}2^{6}_{17}1^{7}_{19}$	
11^n_{91}: $1^{\underline{9}}_{\underline{21}}1^{\underline{8}}_{\underline{19}}1^{\underline{8}}_{\underline{17}}2^{\underline{7}}_{\underline{17}}1^{\underline{7}}_{\underline{15}}2^{\underline{6}}_{\underline{15}}2^{\underline{6}}_{\underline{13}}3^{\underline{5}}_{\underline{13}}2^{\underline{5}}_{\underline{11}}2^{\underline{4}}_{\underline{11}}3^{\underline{4}}_{\underline{9}}2^{\underline{3}}_{\underline{9}}2^{\underline{3}}_{\underline{7}}2^{\underline{2}}_{\underline{7}}2^{\underline{2}}_{\underline{5}}2^{\underline{1}}_{\underline{3}}1^{0}_{\underline{3}}1^{0}_{\underline{1}}$	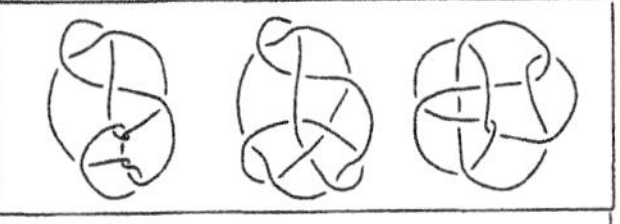
11^n_{92}: $1^{\underline{2}}_{\underline{7}}1^{\underline{1}}_{\underline{5}}1^{\underline{1}}_{\underline{3}}1^{0}_{\underline{3}}2^{0}_{\underline{1}}1^{0}_{1}2^{1}_{\underline{1}}1^{1}_{1}1^{2}_{1}2^{2}_{3}1^{3}_{3}1^{3}_{5}1^{4}_{5}1^{4}_{7}1^{5}_{9}$	
11^n_{93}: $1^{0}_{5}1^{0}_{7}2^{1}_{7}3^{2}_{9}2^{2}_{11}3^{3}_{11}3^{3}_{13}5^{4}_{13}3^{4}_{15}3^{5}_{15}5^{5}_{17}4^{6}_{17}3^{6}_{19}2^{7}_{19}4^{7}_{21}1^{8}_{21}2^{8}_{23}1^{9}_{25}$	
11^n_{94}: $1^{\underline{3}}_{\underline{7}}2^{\underline{2}}_{\underline{5}}1^{\underline{2}}_{\underline{3}}4^{\underline{1}}_{\underline{3}}2^{\underline{1}}_{\underline{1}}5^{0}_{\underline{1}}5^{0}_{1}5^{1}_{1}4^{1}_{3}5^{2}_{3}5^{2}_{5}3^{3}_{5}5^{3}_{7}3^{4}_{7}3^{4}_{9}1^{5}_{9}3^{5}_{11}1^{6}_{13}$	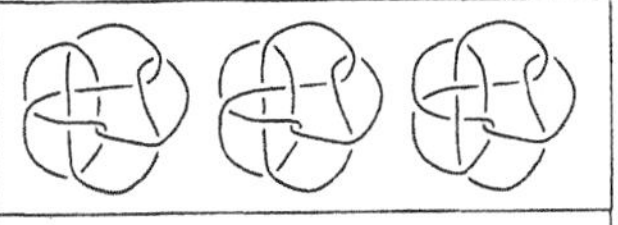
11^n_{95}: $2^{0}_{3}1^{0}_{5}2^{1}_{5}1^{1}_{7}3^{2}_{7}2^{2}_{9}3^{3}_{9}3^{3}_{11}3^{4}_{11}3^{4}_{13}2^{5}_{13}3^{5}_{15}2^{6}_{15}2^{6}_{17}2^{7}_{19}$	
11^n_{96}: $1^{\underline{5}}_{\underline{9}}1^{\underline{4}}_{\underline{7}}1^{\underline{4}}_{\underline{5}}1^{\underline{3}}_{\underline{5}}1^{\underline{3}}_{\underline{3}}2^{\underline{2}}_{\underline{3}}1^{\underline{2}}_{\underline{1}}1^{\underline{1}}_{\underline{3}}1^{\underline{1}}_{\underline{1}}2^{\underline{1}}_{1}1^{0}_{\underline{1}}3^{0}_{1}1^{0}_{3}1^{1}_{1}1^{1}_{3}1^{1}_{5}1^{2}_{3}1^{2}_{5}1^{2}_{7}1^{3}_{7}1^{4}_{7}1^{5}_{11}$	
11^n_{97}: $1^{\underline{5}}_{\underline{11}}1^{\underline{4}}_{\underline{9}}1^{\underline{4}}_{\underline{7}}1^{\underline{3}}_{\underline{7}}1^{\underline{3}}_{\underline{5}}2^{\underline{2}}_{\underline{5}}1^{\underline{2}}_{\underline{3}}1^{\underline{1}}_{\underline{3}}2^{\underline{1}}_{\underline{1}}1^{0}_{\underline{3}}2^{0}_{\underline{1}}2^{0}_{1}2^{1}_{1}1^{1}_{3}1^{2}_{1}1^{2}_{5}1^{3}_{3}1^{3}_{5}1^{4}_{7}1^{5}_{7}1^{6}_{11}$	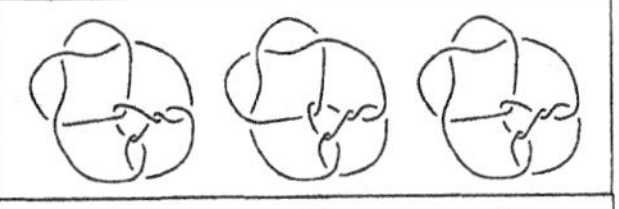
11^n_{98}: $2^{\underline{5}}_{\underline{11}}2^{\underline{4}}_{\underline{9}}2^{\underline{4}}_{\underline{7}}5^{\underline{3}}_{\underline{7}}2^{\underline{3}}_{\underline{5}}6^{\underline{2}}_{\underline{5}}5^{\underline{2}}_{\underline{3}}5^{\underline{1}}_{\underline{3}}6^{\underline{1}}_{\underline{1}}7^{0}_{\underline{1}}6^{0}_{1}4^{1}_{1}6^{1}_{3}3^{2}_{3}4^{2}_{5}1^{3}_{5}3^{3}_{7}1^{4}_{9}$	
11^n_{99}: $1^{\underline{8}}_{\underline{19}}1^{\underline{7}}_{\underline{17}}1^{\underline{7}}_{\underline{15}}3^{\underline{6}}_{\underline{15}}1^{\underline{6}}_{\underline{13}}3^{\underline{5}}_{\underline{13}}3^{\underline{5}}_{\underline{11}}3^{\underline{4}}_{\underline{11}}3^{\underline{4}}_{\underline{9}}4^{\underline{3}}_{\underline{9}}3^{\underline{3}}_{\underline{7}}2^{\underline{2}}_{\underline{7}}4^{\underline{2}}_{\underline{5}}2^{\underline{1}}_{\underline{5}}2^{\underline{1}}_{\underline{3}}1^{0}_{\underline{3}}3^{0}_{\underline{1}}$	
11^n_{100}: $1^{\underline{3}}_{\underline{7}}3^{\underline{2}}_{\underline{5}}1^{\underline{2}}_{\underline{3}}2^{\underline{1}}_{\underline{3}}3^{\underline{1}}_{\underline{1}}5^{0}_{\underline{1}}3^{0}_{1}4^{1}_{1}4^{1}_{3}3^{2}_{3}4^{2}_{5}3^{3}_{5}3^{3}_{7}1^{4}_{7}3^{4}_{9}1^{5}_{9}1^{5}_{11}1^{6}_{13}$	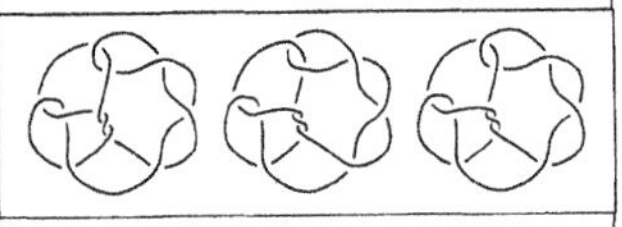
11^n_{101}: $1^{\underline{5}}_{\underline{13}}1^{\underline{4}}_{\underline{11}}1^{\underline{4}}_{\underline{9}}3^{\underline{3}}_{\underline{9}}1^{\underline{3}}_{\underline{7}}3^{\underline{2}}_{\underline{7}}3^{\underline{2}}_{\underline{5}}3^{\underline{1}}_{\underline{5}}3^{\underline{1}}_{\underline{3}}4^{0}_{\underline{3}}4^{0}_{\underline{1}}2^{1}_{\underline{1}}3^{1}_{1}2^{2}_{1}2^{2}_{3}1^{3}_{3}2^{3}_{5}1^{4}_{7}$	
11^n_{102}: $1^{\underline{8}}_{\underline{17}}1^{\underline{7}}_{\underline{13}}1^{\underline{6}}_{\underline{13}}1^{\underline{5}}_{\underline{11}}1^{\underline{5}}_{\underline{9}}1^{\underline{4}}_{\underline{11}}1^{\underline{4}}_{\underline{7}}1^{\underline{3}}_{\underline{9}}2^{\underline{3}}_{\underline{7}}1^{\underline{2}}_{\underline{7}}1^{\underline{2}}_{\underline{5}}1^{\underline{2}}_{\underline{3}}1^{\underline{1}}_{\underline{5}}1^{\underline{1}}_{\underline{3}}1^{0}_{\underline{3}}2^{0}_{\underline{1}}1^{1}_{\underline{1}}1^{2}_{3}$	
11^n_{103}: $2^{\underline{7}}_{\underline{19}}3^{\underline{6}}_{\underline{17}}2^{\underline{6}}_{\underline{15}}5^{\underline{5}}_{\underline{15}}3^{\underline{5}}_{\underline{13}}5^{\underline{4}}_{\underline{13}}5^{\underline{4}}_{\underline{11}}6^{\underline{3}}_{\underline{11}}5^{\underline{3}}_{\underline{9}}5^{\underline{2}}_{\underline{9}}6^{\underline{2}}_{\underline{7}}3^{\underline{1}}_{\underline{7}}5^{\underline{1}}_{\underline{5}}3^{0}_{\underline{5}}4^{0}_{\underline{3}}1^{1}_{\underline{3}}2^{1}_{\underline{1}}1^{2}_{1}$	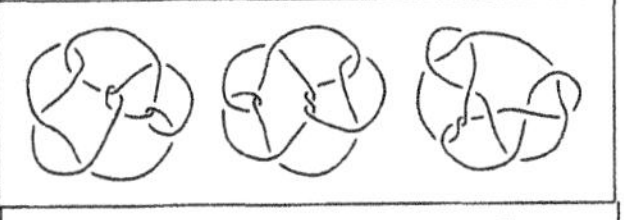
11^n_{104}: $1^{\underline{2}}_{1}1^{\underline{1}}_{5}2^{0}_{5}1^{0}_{7}1^{1}_{7}1^{1}_{9}1^{2}_{7}1^{2}_{9}1^{2}_{11}2^{3}_{11}1^{3}_{13}2^{4}_{11}1^{4}_{13}1^{4}_{15}1^{5}_{13}2^{5}_{15}1^{5}_{17}1^{6}_{17}1^{7}_{17}1^{8}_{21}$	
11^n_{105}: $2^{0}_{3}1^{0}_{5}3^{1}_{5}1^{1}_{7}5^{2}_{7}3^{2}_{9}5^{3}_{9}5^{3}_{11}7^{4}_{11}5^{4}_{13}5^{5}_{13}7^{5}_{15}4^{6}_{15}5^{6}_{17}3^{7}_{17}4^{7}_{19}1^{8}_{19}3^{8}_{21}1^{9}_{23}$	
11^n_{106}: $2^{\underline{3}}_{\underline{9}}1^{\underline{2}}_{\underline{7}}2^{\underline{2}}_{\underline{5}}2^{\underline{1}}_{\underline{5}}1^{\underline{1}}_{\underline{3}}3^{0}_{\underline{3}}3^{0}_{\underline{1}}2^{1}_{\underline{1}}2^{1}_{1}2^{2}_{1}2^{2}_{3}1^{3}_{3}2^{3}_{5}1^{4}_{5}1^{4}_{7}1^{5}_{9}$	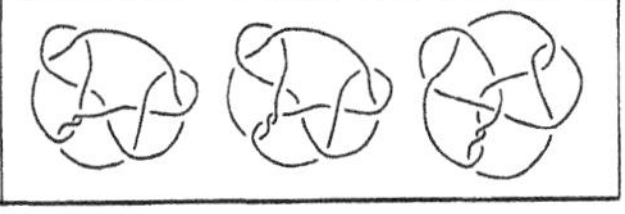
11^n_{107}: $1^{\underline{3}}_{\underline{3}}1^{\underline{2}}_{\underline{1}}1^{\underline{2}}_{1}1^{\underline{1}}_{1}1^{\underline{1}}_{3}3^{0}_{3}2^{0}_{5}1^{1}_{5}2^{1}_{7}2^{2}_{7}1^{2}_{9}1^{3}_{9}2^{3}_{11}1^{4}_{11}1^{4}_{13}1^{5}_{15}$	
11^n_{108}: $1^{\underline{9}}_{\underline{23}}2^{\underline{8}}_{\underline{21}}1^{\underline{8}}_{\underline{19}}5^{\underline{7}}_{\underline{19}}2^{\underline{7}}_{\underline{17}}5^{\underline{6}}_{\underline{17}}5^{\underline{6}}_{\underline{15}}7^{\underline{5}}_{\underline{15}}5^{\underline{5}}_{\underline{13}}6^{\underline{4}}_{\underline{13}}7^{\underline{4}}_{\underline{11}}5^{\underline{3}}_{\underline{11}}6^{\underline{3}}_{\underline{9}}4^{\underline{2}}_{\underline{9}}5^{\underline{2}}_{\underline{7}}1^{\underline{1}}_{\underline{7}}4^{\underline{1}}_{\underline{5}}1^{0}_{\underline{5}}2^{0}_{\underline{3}}$	

K: KH	1st line	2nd line	3rd line
11^n_{109}: $1^{\underline{9}}_{\underline{23}}1^{\underline{8}}_{\underline{21}}1^{\underline{8}}_{\underline{19}}4^{\underline{7}}_{\underline{19}}1^{\underline{7}}_{\underline{17}}4^{\underline{6}}_{\underline{17}}4^{\underline{6}}_{\underline{15}}5^{\underline{5}}_{\underline{15}}4^{\underline{5}}_{\underline{13}}5^{\underline{4}}_{\underline{13}}5^{\underline{4}}_{\underline{11}}4^{\underline{3}}_{\underline{11}}5^{\underline{3}}_{\underline{9}}3^{\underline{2}}_{\underline{9}}4^{\underline{2}}_{\underline{7}}1^{\underline{1}}_{\underline{7}}3^{\underline{1}}_{\underline{5}}1^{0}_{\underline{5}}2^{0}_{\underline{3}}$ 11^n_{110}: $2^{\underline{2}}_{\underline{5}}2^{\underline{1}}_{\underline{3}}2^{\underline{1}}_{\underline{1}}4^{0}_{\underline{1}}3^{0}_{1}4^{1}_{1}3^{1}_{3}3^{2}_{3}4^{2}_{5}3^{3}_{5}3^{3}_{7}2^{4}_{7}3^{4}_{9}1^{5}_{9}2^{5}_{11}1^{6}_{13}$ 11^n_{111}: $1^{\underline{4}}_{\underline{5}}1^{\underline{3}}_{\underline{1}}1^{\underline{2}}_{\underline{1}}1^{\underline{1}}_{1}1^{\underline{1}}_{3}2^{0}_{1}1^{0}_{3}1^{0}_{5}1^{1}_{3}2^{1}_{5}2^{2}_{5}1^{2}_{7}1^{2}_{9}1^{3}_{7}2^{3}_{9}1^{4}_{9}1^{4}_{11}1^{5}_{11}1^{5}_{13}1^{6}_{15}$			
11^n_{112}: $1^{\underline{2}}_{\underline{3}}2^{\underline{1}}_{\underline{1}}1^{\underline{1}}_{1}4^{0}_{1}3^{0}_{3}5^{1}_{3}3^{1}_{5}5^{2}_{5}5^{2}_{7}4^{3}_{7}5^{3}_{9}4^{4}_{9}4^{4}_{11}2^{5}_{11}4^{5}_{13}1^{6}_{13}2^{6}_{15}1^{7}_{17}$ 11^n_{113}: $1^{\underline{9}}_{\underline{21}}1^{\underline{8}}_{\underline{19}}1^{\underline{8}}_{\underline{17}}3^{\underline{7}}_{\underline{17}}1^{\underline{7}}_{\underline{15}}2^{\underline{6}}_{\underline{15}}3^{\underline{6}}_{\underline{13}}3^{\underline{5}}_{\underline{13}}2^{\underline{5}}_{\underline{11}}3^{\underline{4}}_{\underline{11}}3^{\underline{4}}_{\underline{9}}2^{\underline{3}}_{\underline{9}}3^{\underline{3}}_{\underline{7}}2^{\underline{2}}_{\underline{7}}2^{\underline{2}}_{\underline{5}}2^{\underline{1}}_{\underline{3}}1^{0}_{\underline{3}}1^{0}_{\underline{1}}$ 11^n_{114}: $1^{\underline{4}}_{\underline{9}}2^{\underline{3}}_{\underline{7}}1^{\underline{3}}_{\underline{5}}4^{\underline{2}}_{\underline{5}}2^{\underline{2}}_{\underline{3}}4^{\underline{1}}_{\underline{3}}4^{\underline{1}}_{\underline{1}}5^{0}_{\underline{1}}5^{0}_{1}5^{1}_{1}4^{1}_{3}3^{2}_{3}5^{2}_{5}2^{3}_{5}3^{3}_{7}1^{4}_{7}2^{4}_{9}1^{5}_{11}$			
11^n_{115}: $1^{\underline{5}}_{\underline{11}}3^{\underline{4}}_{\underline{9}}1^{\underline{4}}_{\underline{7}}4^{\underline{3}}_{\underline{7}}3^{\underline{3}}_{\underline{5}}7^{\underline{2}}_{\underline{5}}4^{\underline{2}}_{\underline{3}}6^{\underline{1}}_{\underline{3}}7^{\underline{1}}_{\underline{1}}7^{0}_{\underline{1}}7^{0}_{1}6^{1}_{1}6^{1}_{3}3^{2}_{3}6^{2}_{5}2^{3}_{5}3^{3}_{7}2^{4}_{9}$ 11^n_{116}: $1^{\underline{6}}_{\underline{13}}1^{\underline{5}}_{\underline{9}}1^{\underline{4}}_{\underline{9}}1^{\underline{3}}_{\underline{7}}1^{\underline{3}}_{\underline{5}}1^{\underline{2}}_{\underline{7}}1^{\underline{2}}_{\underline{3}}2^{\underline{1}}_{\underline{3}}1^{0}_{\underline{3}}1^{0}_{\underline{1}}2^{0}_{1}1^{1}_{\underline{1}}1^{1}_{1}1^{2}_{3}1^{3}_{3}1^{4}_{7}$ 11^n_{117}: $1^{\underline{4}}_{\underline{7}}1^{\underline{3}}_{\underline{5}}1^{\underline{3}}_{\underline{3}}3^{\underline{2}}_{\underline{3}}1^{\underline{2}}_{\underline{1}}2^{\underline{1}}_{\underline{1}}3^{\underline{1}}_{1}4^{0}_{1}3^{0}_{3}3^{1}_{3}3^{1}_{5}2^{2}_{5}3^{2}_{7}2^{3}_{7}2^{3}_{9}2^{4}_{11}$			
11^n_{118}: $2^{0}_{3}1^{0}_{5}1^{1}_{5}1^{1}_{7}2^{2}_{7}1^{2}_{9}2^{3}_{9}2^{3}_{11}2^{4}_{11}2^{4}_{13}1^{5}_{13}2^{5}_{15}1^{6}_{15}1^{6}_{17}1^{7}_{19}$ 11^n_{119}: $1^{\underline{6}}_{\underline{13}}2^{\underline{5}}_{\underline{11}}1^{\underline{5}}_{\underline{9}}4^{\underline{4}}_{\underline{9}}2^{\underline{4}}_{\underline{7}}5^{\underline{3}}_{\underline{7}}4^{\underline{3}}_{\underline{5}}6^{\underline{2}}_{\underline{5}}5^{\underline{2}}_{\underline{3}}6^{\underline{1}}_{\underline{3}}6^{\underline{1}}_{\underline{1}}5^{0}_{\underline{1}}7^{0}_{1}4^{1}_{1}4^{1}_{3}2^{2}_{3}4^{2}_{5}2^{3}_{7}$ 11^n_{120}: $1^{\underline{4}}_{\underline{7}}2^{\underline{3}}_{\underline{5}}1^{\underline{3}}_{\underline{3}}3^{\underline{2}}_{\underline{3}}2^{\underline{2}}_{\underline{1}}3^{\underline{1}}_{\underline{1}}3^{\underline{1}}_{1}5^{0}_{1}4^{0}_{3}4^{1}_{3}4^{1}_{5}3^{2}_{5}4^{2}_{7}2^{3}_{7}3^{3}_{9}1^{4}_{9}2^{4}_{11}1^{5}_{13}$			
11^n_{121}: $1^{\underline{7}}_{\underline{19}}1^{\underline{6}}_{\underline{17}}1^{\underline{6}}_{\underline{15}}4^{\underline{5}}_{\underline{15}}1^{\underline{5}}_{\underline{13}}3^{\underline{4}}_{\underline{13}}4^{\underline{4}}_{\underline{11}}4^{\underline{3}}_{\underline{11}}3^{\underline{3}}_{\underline{9}}4^{\underline{2}}_{\underline{9}}4^{\underline{2}}_{\underline{7}}2^{\underline{1}}_{\underline{7}}4^{\underline{1}}_{\underline{5}}3^{0}_{\underline{5}}3^{0}_{\underline{3}}1^{1}_{\underline{3}}2^{1}_{\underline{1}}1^{2}_{1}$ 11^n_{122}: $1^{\underline{8}}_{\underline{19}}1^{\underline{7}}_{\underline{17}}1^{\underline{7}}_{\underline{15}}2^{\underline{6}}_{\underline{15}}1^{\underline{6}}_{\underline{13}}2^{\underline{5}}_{\underline{13}}2^{\underline{5}}_{\underline{11}}2^{\underline{4}}_{\underline{11}}2^{\underline{4}}_{\underline{9}}3^{\underline{3}}_{\underline{9}}2^{\underline{3}}_{\underline{7}}1^{\underline{2}}_{\underline{7}}3^{\underline{2}}_{\underline{5}}1^{\underline{1}}_{\underline{5}}1^{\underline{1}}_{\underline{3}}1^{0}_{\underline{3}}2^{0}_{\underline{1}}$ 11^n_{123}: $1^{\underline{6}}_{\underline{13}}2^{\underline{5}}_{\underline{11}}1^{\underline{5}}_{\underline{9}}4^{\underline{4}}_{\underline{9}}2^{\underline{4}}_{\underline{7}}4^{\underline{3}}_{\underline{7}}4^{\underline{3}}_{\underline{5}}5^{\underline{2}}_{\underline{5}}4^{\underline{2}}_{\underline{3}}5^{\underline{1}}_{\underline{3}}5^{\underline{1}}_{\underline{1}}4^{0}_{\underline{1}}6^{0}_{1}3^{1}_{1}3^{1}_{3}1^{2}_{3}3^{2}_{5}1^{3}_{7}$			
11^n_{124}: $1^{\underline{4}}_{\underline{7}}2^{\underline{3}}_{\underline{5}}1^{\underline{3}}_{\underline{3}}4^{\underline{2}}_{\underline{3}}2^{\underline{2}}_{\underline{1}}4^{\underline{1}}_{\underline{1}}4^{\underline{1}}_{1}6^{0}_{1}5^{0}_{3}5^{1}_{3}5^{1}_{5}4^{2}_{5}5^{2}_{7}3^{3}_{7}4^{3}_{9}1^{4}_{9}3^{4}_{11}1^{5}_{13}$ 11^n_{125}: $1^{\underline{2}}_{\underline{3}}2^{\underline{1}}_{\underline{1}}1^{\underline{1}}_{1}5^{0}_{1}3^{0}_{3}5^{1}_{3}4^{1}_{5}6^{2}_{5}5^{2}_{7}5^{3}_{7}6^{3}_{9}4^{4}_{9}5^{4}_{11}3^{5}_{11}4^{5}_{13}1^{6}_{13}3^{6}_{15}1^{7}_{17}$ 11^n_{126}: $1^{0}_{5}1^{0}_{7}2^{1}_{7}2^{2}_{9}2^{2}_{11}2^{3}_{11}2^{3}_{13}1^{4}_{11}4^{4}_{13}2^{4}_{15}2^{5}_{15}4^{5}_{17}2^{6}_{17}1^{6}_{19}1^{7}_{19}2^{7}_{21}1^{8}_{23}$			
11^n_{127}: $1^{\underline{8}}_{\underline{19}}2^{\underline{7}}_{\underline{17}}1^{\underline{7}}_{\underline{15}}3^{\underline{6}}_{\underline{15}}2^{\underline{6}}_{\underline{13}}5^{\underline{5}}_{\underline{13}}3^{\underline{5}}_{\underline{11}}4^{\underline{4}}_{\underline{11}}5^{\underline{4}}_{\underline{9}}5^{\underline{3}}_{\underline{9}}4^{\underline{3}}_{\underline{7}}4^{\underline{2}}_{\underline{7}}5^{\underline{2}}_{\underline{5}}2^{\underline{1}}_{\underline{5}}4^{\underline{1}}_{\underline{3}}2^{0}_{\underline{3}}3^{0}_{\underline{1}}1^{1}_{1}$ 11^n_{128}: $1^{\underline{5}}_{\underline{9}}2^{\underline{4}}_{\underline{7}}1^{\underline{4}}_{\underline{5}}2^{\underline{3}}_{\underline{5}}2^{\underline{3}}_{\underline{3}}4^{\underline{2}}_{\underline{3}}2^{\underline{2}}_{\underline{1}}3^{\underline{1}}_{\underline{1}}4^{\underline{1}}_{1}4^{0}_{1}4^{0}_{3}3^{1}_{3}3^{1}_{5}2^{2}_{5}3^{2}_{7}1^{3}_{7}2^{3}_{9}1^{4}_{11}$ 11^n_{129}: $1^{\underline{7}}_{\underline{17}}1^{\underline{6}}_{\underline{15}}1^{\underline{6}}_{\underline{13}}3^{\underline{5}}_{\underline{13}}1^{\underline{5}}_{\underline{11}}3^{\underline{4}}_{\underline{11}}3^{\underline{4}}_{\underline{9}}4^{\underline{3}}_{\underline{9}}3^{\underline{3}}_{\underline{7}}4^{\underline{2}}_{\underline{7}}4^{\underline{2}}_{\underline{5}}2^{\underline{1}}_{\underline{5}}4^{\underline{1}}_{\underline{3}}3^{0}_{\underline{3}}3^{0}_{\underline{1}}1^{1}_{\underline{1}}2^{1}_{1}1^{2}_{3}$			
11^n_{130}: $1^{\underline{6}}_{\underline{13}}2^{\underline{5}}_{\underline{11}}1^{\underline{5}}_{\underline{9}}3^{\underline{4}}_{\underline{9}}2^{\underline{4}}_{\underline{7}}4^{\underline{3}}_{\underline{7}}3^{\underline{3}}_{\underline{5}}5^{\underline{2}}_{\underline{5}}4^{\underline{2}}_{\underline{3}}4^{\underline{1}}_{\underline{3}}5^{\underline{1}}_{\underline{1}}4^{0}_{\underline{1}}5^{0}_{1}3^{1}_{1}3^{1}_{3}1^{2}_{3}3^{2}_{5}1^{3}_{7}$ 11^n_{131}: $1^{\underline{6}}_{\underline{15}}3^{\underline{5}}_{\underline{13}}1^{\underline{5}}_{\underline{11}}4^{\underline{4}}_{\underline{11}}3^{\underline{4}}_{\underline{9}}6^{\underline{3}}_{\underline{9}}4^{\underline{3}}_{\underline{7}}6^{\underline{2}}_{\underline{7}}6^{\underline{2}}_{\underline{5}}5^{\underline{1}}_{\underline{5}}6^{\underline{1}}_{\underline{3}}5^{0}_{\underline{3}}6^{0}_{\underline{1}}3^{1}_{\underline{1}}4^{1}_{1}1^{2}_{1}3^{2}_{3}1^{3}_{5}$ 11^n_{132}: $1^{\underline{7}}_{\underline{15}}1^{\underline{6}}_{\underline{13}}1^{\underline{6}}_{\underline{11}}2^{\underline{5}}_{\underline{11}}1^{\underline{5}}_{\underline{9}}2^{\underline{4}}_{\underline{9}}2^{\underline{4}}_{\underline{7}}2^{\underline{3}}_{\underline{7}}2^{\underline{3}}_{\underline{5}}2^{\underline{2}}_{\underline{5}}2^{\underline{2}}_{\underline{3}}1^{\underline{1}}_{\underline{3}}2^{\underline{1}}_{\underline{1}}2^{0}_{\underline{1}}2^{0}_{1}1^{1}_{3}$			
11^n_{133}: $1^{\underline{3}}_{\underline{3}}2^{\underline{2}}_{\underline{1}}1^{\underline{2}}_{1}1^{\underline{1}}_{1}2^{\underline{1}}_{3}4^{0}_{3}2^{0}_{5}1^{1}_{3}2^{1}_{5}3^{1}_{7}3^{2}_{7}2^{2}_{9}2^{3}_{9}2^{3}_{11}1^{4}_{9}1^{4}_{11}2^{4}_{13}1^{5}_{13}1^{5}_{15}$ 11^n_{134}: $1^{\underline{8}}_{\underline{19}}2^{\underline{7}}_{\underline{17}}1^{\underline{7}}_{\underline{15}}3^{\underline{6}}_{\underline{15}}2^{\underline{6}}_{\underline{13}}4^{\underline{5}}_{\underline{13}}3^{\underline{5}}_{\underline{11}}4^{\underline{4}}_{\underline{11}}4^{\underline{4}}_{\underline{9}}4^{\underline{3}}_{\underline{9}}4^{\underline{3}}_{\underline{7}}3^{\underline{2}}_{\underline{7}}4^{\underline{2}}_{\underline{5}}2^{\underline{1}}_{\underline{5}}3^{\underline{1}}_{\underline{3}}1^{0}_{\underline{3}}3^{0}_{\underline{1}}$ 11^n_{135}: $1^{\underline{2}}_{\underline{1}}1^{\underline{1}}_{3}2^{0}_{3}1^{0}_{5}1^{1}_{5}1^{1}_{7}1^{2}_{5}1^{2}_{7}1^{2}_{9}2^{3}_{9}1^{3}_{11}1^{4}_{9}1^{4}_{11}1^{4}_{13}1^{5}_{11}1^{5}_{13}1^{5}_{15}1^{6}_{15}1^{7}_{15}1^{8}_{19}$			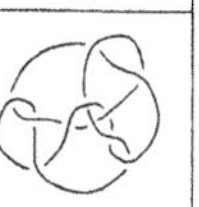
11^n_{136}: $1^{0}_{5}1^{0}_{7}2^{1}_{7}4^{2}_{9}2^{2}_{11}4^{3}_{11}4^{3}_{13}6^{4}_{13}4^{4}_{15}5^{5}_{15}6^{5}_{17}5^{6}_{17}5^{6}_{19}3^{7}_{19}5^{7}_{21}2^{8}_{21}3^{8}_{23}2^{9}_{25}$ 11^n_{137}: $2^{\underline{7}}_{\underline{19}}2^{\underline{6}}_{\underline{17}}2^{\underline{6}}_{\underline{15}}5^{\underline{5}}_{\underline{15}}2^{\underline{5}}_{\underline{13}}4^{\underline{4}}_{\underline{13}}5^{\underline{4}}_{\underline{11}}5^{\underline{3}}_{\underline{11}}4^{\underline{3}}_{\underline{9}}5^{\underline{2}}_{\underline{9}}5^{\underline{2}}_{\underline{7}}2^{\underline{1}}_{\underline{7}}5^{\underline{1}}_{\underline{5}}3^{0}_{\underline{5}}3^{0}_{\underline{3}}1^{1}_{\underline{3}}2^{1}_{\underline{1}}1^{2}_{1}$ 11^n_{138}: $1^{\underline{6}}_{\underline{11}}1^{\underline{5}}_{\underline{7}}2^{\underline{4}}_{\underline{7}}1^{\underline{3}}_{\underline{5}}2^{\underline{3}}_{\underline{3}}1^{\underline{2}}_{\underline{3}}1^{\underline{2}}_{\underline{1}}2^{\underline{1}}_{\underline{1}}1^{\underline{1}}_{1}1^{0}_{\underline{1}}1^{0}_{1}2^{0}_{3}1^{1}_{3}1^{2}_{7}$			
11^n_{139}: $1^{0}_{\underline{1}}1^{0}_{1}1^{2}_{3}1^{3}_{7}1^{4}_{7}1^{5}_{9}1^{5}_{11}1^{6}_{13}1^{7}_{13}1^{8}_{17}$ 11^n_{140}: $2^{\underline{7}}_{\underline{17}}1^{\underline{6}}_{\underline{15}}2^{\underline{6}}_{\underline{13}}4^{\underline{5}}_{\underline{13}}1^{\underline{5}}_{\underline{11}}4^{\underline{4}}_{\underline{11}}4^{\underline{4}}_{\underline{9}}4^{\underline{3}}_{\underline{9}}4^{\underline{3}}_{\underline{7}}5^{\underline{2}}_{\underline{7}}4^{\underline{2}}_{\underline{5}}2^{\underline{1}}_{\underline{5}}5^{\underline{1}}_{\underline{3}}3^{0}_{\underline{3}}3^{0}_{\underline{1}}1^{1}_{\underline{1}}2^{1}_{1}1^{2}_{3}$ 11^n_{141}: $1^{\underline{6}}_{\underline{13}}1^{\underline{5}}_{\underline{9}}2^{\underline{4}}_{\underline{9}}1^{\underline{3}}_{\underline{7}}2^{\underline{3}}_{\underline{5}}2^{\underline{2}}_{\underline{5}}1^{\underline{2}}_{\underline{3}}2^{\underline{1}}_{\underline{3}}2^{\underline{1}}_{\underline{1}}1^{0}_{\underline{1}}3^{0}_{1}2^{1}_{1}2^{2}_{5}$			
11^n_{142}: $1^{\underline{6}}_{\underline{13}}1^{\underline{5}}_{\underline{11}}1^{\underline{5}}_{\underline{9}}3^{\underline{4}}_{\underline{9}}1^{\underline{4}}_{\underline{7}}2^{\underline{3}}_{\underline{7}}3^{\underline{3}}_{\underline{5}}3^{\underline{2}}_{\underline{5}}2^{\underline{2}}_{\underline{3}}3^{\underline{1}}_{\underline{3}}3^{\underline{1}}_{\underline{1}}2^{0}_{\underline{1}}4^{0}_{1}2^{1}_{1}1^{1}_{3}2^{2}_{5}$ 11^n_{143}: $1^{\underline{4}}_{\underline{7}}1^{\underline{3}}_{\underline{3}}1^{\underline{2}}_{\underline{3}}1^{\underline{1}}_{\underline{3}}1^{\underline{1}}_{\underline{1}}1^{\underline{1}}_{1}2^{0}_{\underline{1}}2^{0}_{1}1^{0}_{3}1^{1}_{1}2^{1}_{3}2^{2}_{3}1^{2}_{5}1^{2}_{7}1^{3}_{5}2^{3}_{7}1^{4}_{7}1^{4}_{9}1^{5}_{9}1^{5}_{11}1^{6}_{13}$ 11^n_{144}: $2^{0}_{3}1^{0}_{5}3^{1}_{5}1^{1}_{7}5^{2}_{7}3^{2}_{9}5^{3}_{9}5^{3}_{11}6^{4}_{11}5^{4}_{13}5^{5}_{13}6^{5}_{15}4^{6}_{15}5^{6}_{17}2^{7}_{17}4^{7}_{19}1^{8}_{19}2^{8}_{21}1^{9}_{23}$			

K: KH	1st line 2nd line 3rd line
11^n_{145}: $1^{\underline{5}}_{\underline{9}}1^{\underline{4}}_{\underline{5}}1^{\underline{3}}_{\underline{5}}1^{\underline{2}}_{\underline{5}}1^{\underline{2}}_{\underline{3}}1^{\underline{2}}_{\underline{1}}1^{\underline{1}}_{\underline{3}}1^{\underline{1}}_{\underline{1}}1^{\underline{1}}_{1}2^{0}_{\underline{1}}3^{0}_{1}2^{1}_{1}1^{1}_{3}1^{1}_{5}1^{2}_{3}2^{2}_{5}1^{3}_{5}1^{3}_{7}1^{4}_{7}1^{4}_{9}1^{5}_{11}$	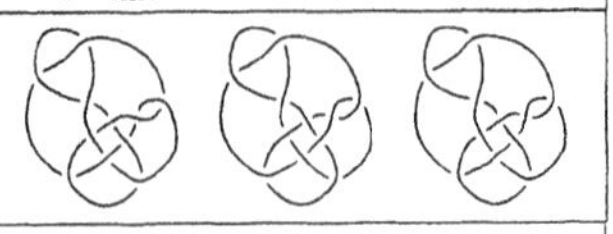
11^n_{146}: $1^{\underline{1}}_{\underline{1}}3^{0}_{1}2^{0}_{3}5^{1}_{3}2^{1}_{5}5^{2}_{5}5^{2}_{7}5^{3}_{7}5^{3}_{9}6^{4}_{9}5^{4}_{11}3^{5}_{11}6^{5}_{13}3^{6}_{13}3^{6}_{15}1^{7}_{15}3^{7}_{17}1^{8}_{19}$	
11^n_{147}: $1^{\underline{3}}_{\underline{3}}2^{\underline{2}}_{\underline{1}}1^{\underline{2}}_{1}2^{\underline{1}}_{1}2^{\underline{1}}_{3}4^{0}_{3}3^{0}_{5}3^{1}_{5}3^{1}_{7}3^{2}_{7}3^{2}_{9}2^{3}_{9}3^{3}_{11}2^{4}_{11}2^{4}_{13}2^{5}_{15}$	
11^n_{148}: $1^{\underline{4}}_{\underline{7}}3^{\underline{3}}_{\underline{5}}1^{\underline{3}}_{\underline{3}}4^{\underline{2}}_{\underline{3}}3^{\underline{2}}_{\underline{1}}6^{\underline{1}}_{\underline{1}}4^{\underline{1}}_{1}7^{0}_{1}7^{0}_{3}6^{1}_{3}6^{1}_{5}6^{2}_{5}6^{2}_{7}3^{3}_{7}6^{3}_{9}2^{4}_{9}3^{4}_{11}2^{5}_{13}$	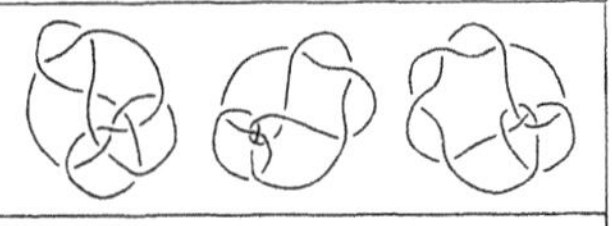
11^n_{149}: $1^{\underline{3}}_{\underline{3}}2^{\underline{2}}_{\underline{1}}1^{\underline{2}}_{1}1^{\underline{1}}_{1}2^{\underline{1}}_{3}4^{0}_{3}2^{0}_{5}2^{1}_{5}3^{1}_{7}3^{2}_{7}2^{2}_{9}2^{3}_{9}3^{3}_{11}1^{4}_{11}2^{4}_{13}1^{5}_{13}1^{5}_{15}1^{6}_{17}$	
11^n_{150}: $1^{\underline{3}}_{\underline{5}}3^{\underline{2}}_{\underline{3}}1^{\underline{2}}_{\underline{1}}4^{\underline{1}}_{\underline{1}}3^{\underline{1}}_{1}7^{0}_{1}5^{0}_{3}6^{1}_{3}6^{1}_{5}7^{2}_{5}6^{2}_{7}5^{3}_{7}7^{3}_{9}3^{4}_{9}5^{4}_{11}2^{5}_{11}3^{5}_{13}2^{6}_{15}$	
11^n_{151}: $1^{\underline{4}}_{\underline{5}}1^{\underline{3}}_{\underline{1}}1^{\underline{2}}_{\underline{1}}1^{\underline{1}}_{1}1^{\underline{1}}_{3}3^{0}_{1}1^{0}_{3}1^{0}_{5}2^{1}_{3}3^{1}_{5}3^{2}_{5}2^{2}_{7}1^{2}_{9}3^{3}_{7}3^{3}_{9}2^{4}_{9}3^{4}_{11}2^{5}_{11}2^{5}_{13}1^{6}_{13}2^{6}_{15}1^{7}_{17}$	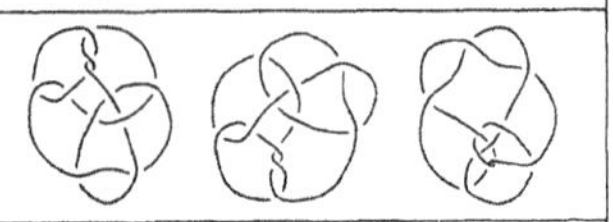
11^n_{152}: $1^{\underline{4}}_{\underline{5}}1^{\underline{3}}_{\underline{1}}1^{\underline{2}}_{\underline{1}}1^{\underline{1}}_{1}1^{\underline{1}}_{3}3^{0}_{1}1^{0}_{3}1^{0}_{5}2^{1}_{3}3^{1}_{5}3^{2}_{5}2^{2}_{7}1^{2}_{9}3^{3}_{7}3^{3}_{9}2^{4}_{9}3^{4}_{11}2^{5}_{11}2^{5}_{13}1^{6}_{13}2^{6}_{15}1^{7}_{17}$	
11^n_{153}: $1^{\underline{5}}_{\underline{11}}2^{\underline{4}}_{\underline{9}}1^{\underline{4}}_{\underline{7}}3^{\underline{3}}_{\underline{7}}2^{\underline{3}}_{\underline{5}}5^{\underline{2}}_{\underline{5}}3^{\underline{2}}_{\underline{3}}4^{\underline{1}}_{\underline{3}}5^{\underline{1}}_{\underline{1}}6^{0}_{\underline{1}}5^{0}_{1}4^{1}_{1}5^{1}_{3}2^{2}_{3}4^{2}_{5}2^{3}_{5}2^{3}_{7}2^{4}_{9}$	
11^n_{154}: $2^{\underline{2}}_{\underline{3}}3^{\underline{1}}_{\underline{1}}2^{\underline{1}}_{1}6^{0}_{1}4^{0}_{3}7^{1}_{3}5^{1}_{5}7^{2}_{5}7^{2}_{7}6^{3}_{7}7^{3}_{9}5^{4}_{9}6^{4}_{11}3^{5}_{11}5^{5}_{13}1^{6}_{13}3^{6}_{15}1^{7}_{17}$	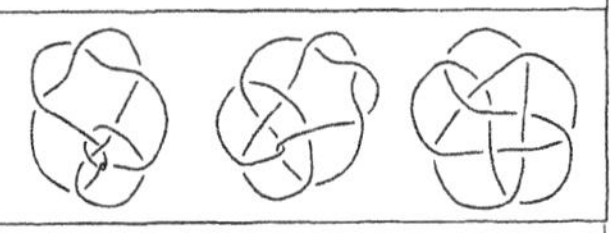
11^n_{155}: $1^{\underline{5}}_{\underline{9}}2^{\underline{4}}_{\underline{7}}1^{\underline{4}}_{\underline{5}}2^{\underline{3}}_{\underline{5}}2^{\underline{3}}_{\underline{3}}5^{\underline{2}}_{\underline{3}}2^{\underline{2}}_{\underline{1}}3^{\underline{1}}_{\underline{1}}5^{\underline{1}}_{1}5^{0}_{1}4^{0}_{3}4^{1}_{3}4^{1}_{5}2^{2}_{5}4^{2}_{7}2^{3}_{7}2^{3}_{9}2^{4}_{11}$	
11^n_{156}: $2^{\underline{3}}_{\underline{7}}4^{\underline{2}}_{\underline{5}}2^{\underline{2}}_{\underline{3}}5^{\underline{1}}_{\underline{3}}4^{\underline{1}}_{\underline{1}}7^{0}_{\underline{1}}6^{0}_{1}7^{1}_{1}6^{1}_{3}6^{2}_{3}7^{2}_{5}4^{3}_{5}6^{3}_{7}3^{4}_{7}4^{4}_{9}1^{5}_{9}3^{5}_{11}1^{6}_{13}$	
11^n_{157}: $1^{\underline{6}}_{\underline{13}}3^{\underline{5}}_{\underline{11}}1^{\underline{5}}_{\underline{9}}4^{\underline{4}}_{\underline{9}}3^{\underline{4}}_{\underline{7}}5^{\underline{3}}_{\underline{7}}4^{\underline{3}}_{\underline{5}}6^{\underline{2}}_{\underline{5}}5^{\underline{2}}_{\underline{3}}5^{\underline{1}}_{\underline{3}}6^{\underline{1}}_{\underline{1}}5^{0}_{\underline{1}}6^{0}_{1}3^{1}_{1}4^{1}_{3}1^{2}_{3}3^{2}_{5}1^{3}_{7}$	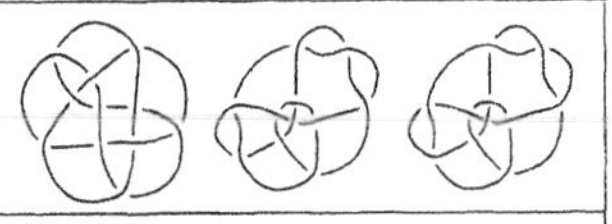
11^n_{158}: $1^{\underline{3}}_{\underline{3}}2^{\underline{2}}_{\underline{1}}1^{\underline{2}}_{1}2^{\underline{1}}_{1}2^{\underline{1}}_{3}5^{0}_{3}3^{0}_{5}3^{1}_{5}4^{1}_{7}4^{2}_{7}3^{2}_{9}3^{3}_{9}4^{3}_{11}2^{4}_{11}3^{4}_{13}1^{5}_{13}2^{5}_{15}1^{6}_{17}$	
11^n_{159}: $1^{\underline{8}}_{\underline{19}}3^{\underline{7}}_{\underline{17}}1^{\underline{7}}_{\underline{15}}4^{\underline{6}}_{\underline{15}}3^{\underline{6}}_{\underline{13}}6^{\underline{5}}_{\underline{13}}4^{\underline{5}}_{\underline{11}}6^{\underline{4}}_{\underline{11}}6^{\underline{4}}_{\underline{9}}6^{\underline{3}}_{\underline{9}}6^{\underline{3}}_{\underline{7}}5^{\underline{2}}_{\underline{7}}6^{\underline{2}}_{\underline{5}}3^{\underline{1}}_{\underline{5}}5^{\underline{1}}_{\underline{3}}2^{0}_{\underline{3}}4^{0}_{\underline{1}}1^{1}_{1}$	
11^n_{160}: $2^{\underline{2}}_{\underline{3}}3^{\underline{1}}_{\underline{1}}2^{\underline{1}}_{1}5^{0}_{1}4^{0}_{3}6^{1}_{3}4^{1}_{5}6^{2}_{5}6^{2}_{7}5^{3}_{7}6^{3}_{9}4^{4}_{9}5^{4}_{11}2^{5}_{11}4^{5}_{13}1^{6}_{13}2^{6}_{15}1^{7}_{17}$	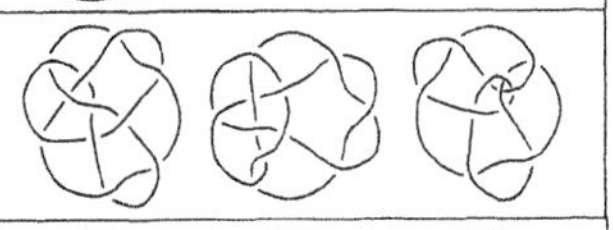
11^n_{161}: $1^{\underline{3}}_{\underline{5}}2^{\underline{2}}_{\underline{3}}1^{\underline{2}}_{\underline{1}}3^{\underline{1}}_{\underline{1}}2^{\underline{1}}_{1}6^{0}_{1}4^{0}_{3}5^{1}_{3}5^{1}_{5}6^{2}_{5}5^{2}_{7}4^{3}_{7}6^{3}_{9}3^{4}_{9}4^{4}_{11}2^{5}_{11}3^{5}_{13}2^{6}_{15}$	
11^n_{162}: $2^{0}_{1}1^{0}_{3}3^{1}_{3}1^{1}_{5}4^{2}_{5}3^{2}_{7}4^{3}_{7}4^{3}_{9}5^{4}_{9}4^{4}_{11}4^{5}_{11}5^{5}_{13}3^{6}_{13}4^{6}_{15}2^{7}_{15}3^{7}_{17}1^{8}_{17}2^{8}_{19}1^{9}_{21}$	
11^n_{163}: $1^{\underline{3}}_{\underline{5}}4^{\underline{2}}_{\underline{3}}1^{\underline{2}}_{\underline{1}}5^{\underline{1}}_{\underline{1}}4^{\underline{1}}_{1}8^{0}_{1}6^{0}_{3}8^{1}_{3}7^{1}_{5}8^{2}_{5}8^{2}_{7}6^{3}_{7}8^{3}_{9}4^{4}_{9}6^{4}_{11}2^{5}_{11}4^{5}_{13}2^{6}_{15}$	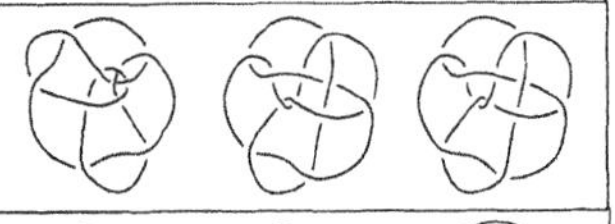
11^n_{164}: $1^{\underline{2}}_{\underline{1}}2^{\underline{1}}_{1}1^{\underline{1}}_{3}4^{0}_{3}3^{0}_{5}3^{1}_{5}3^{1}_{7}5^{2}_{7}3^{2}_{9}3^{3}_{9}5^{3}_{11}3^{4}_{11}3^{4}_{13}2^{5}_{13}3^{5}_{15}2^{6}_{17}$	
11^n_{165}: $2^{\underline{4}}_{\underline{9}}4^{\underline{3}}_{\underline{7}}2^{\underline{3}}_{\underline{5}}6^{\underline{2}}_{\underline{5}}4^{\underline{2}}_{\underline{3}}7^{\underline{1}}_{\underline{3}}6^{\underline{1}}_{\underline{1}}8^{0}_{\underline{1}}8^{0}_{1}7^{1}_{1}7^{1}_{3}5^{2}_{3}7^{2}_{5}3^{3}_{5}5^{3}_{7}1^{4}_{7}3^{4}_{9}1^{5}_{11}$	
11^n_{166}: $1^{\underline{4}}_{\underline{7}}2^{\underline{3}}_{\underline{5}}1^{\underline{3}}_{\underline{3}}4^{\underline{2}}_{\underline{3}}2^{\underline{2}}_{\underline{1}}4^{\underline{1}}_{\underline{1}}4^{\underline{1}}_{1}6^{0}_{1}5^{0}_{3}5^{1}_{3}5^{1}_{5}4^{2}_{5}5^{2}_{7}3^{3}_{7}4^{3}_{9}1^{4}_{9}3^{4}_{11}1^{5}_{13}$	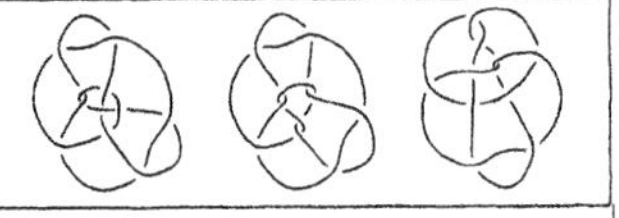
11^n_{167}: $1^{\underline{2}}_{\underline{3}}3^{\underline{1}}_{\underline{1}}1^{\underline{1}}_{1}4^{0}_{1}4^{0}_{3}6^{1}_{3}3^{1}_{5}6^{2}_{5}6^{2}_{7}4^{3}_{7}6^{3}_{9}5^{4}_{9}4^{4}_{11}2^{5}_{11}5^{5}_{13}1^{6}_{13}2^{6}_{15}1^{7}_{17}$	
11^n_{168}: $1^{\underline{4}}_{\underline{7}}3^{\underline{3}}_{\underline{5}}1^{\underline{3}}_{\underline{3}}4^{\underline{2}}_{\underline{3}}3^{\underline{2}}_{\underline{1}}6^{\underline{1}}_{\underline{1}}4^{\underline{1}}_{1}7^{0}_{1}7^{0}_{3}6^{1}_{3}6^{1}_{5}6^{2}_{5}6^{2}_{7}3^{3}_{7}6^{3}_{9}2^{4}_{9}3^{4}_{11}2^{5}_{13}$	
11^n_{169}: $1^{0}_{5}1^{0}_{7}2^{1}_{7}2^{2}_{9}2^{2}_{11}2^{3}_{11}2^{3}_{13}4^{4}_{13}2^{4}_{15}2^{5}_{15}4^{5}_{17}3^{6}_{17}2^{6}_{19}1^{7}_{19}3^{7}_{21}1^{8}_{21}1^{8}_{23}1^{9}_{25}$	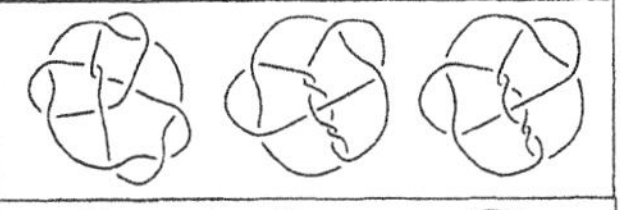
11^n_{170}: $2^{\underline{7}}_{\underline{17}}2^{\underline{6}}_{\underline{15}}2^{\underline{6}}_{\underline{13}}5^{\underline{5}}_{\underline{13}}2^{\underline{5}}_{\underline{11}}5^{\underline{4}}_{\underline{11}}5^{\underline{4}}_{\underline{9}}5^{\underline{3}}_{\underline{9}}5^{\underline{3}}_{\underline{7}}6^{\underline{2}}_{\underline{7}}5^{\underline{2}}_{\underline{5}}3^{\underline{1}}_{\underline{5}}6^{\underline{1}}_{\underline{3}}3^{0}_{\underline{3}}4^{0}_{\underline{1}}1^{1}_{\underline{1}}2^{1}_{1}1^{2}_{3}$	
11^n_{171}: $1^{0}_{3}1^{0}_{5}3^{1}_{5}4^{2}_{7}3^{2}_{9}5^{3}_{9}4^{3}_{11}6^{4}_{11}5^{4}_{13}5^{5}_{13}6^{5}_{15}5^{6}_{15}5^{6}_{17}2^{7}_{17}5^{7}_{19}2^{8}_{19}2^{8}_{21}2^{9}_{23}$	
11^n_{172}: $1^{\underline{7}}_{\underline{15}}2^{\underline{6}}_{\underline{13}}1^{\underline{6}}_{\underline{11}}3^{\underline{5}}_{\underline{11}}2^{\underline{5}}_{\underline{9}}4^{\underline{4}}_{\underline{9}}3^{\underline{4}}_{\underline{7}}4^{\underline{3}}_{\underline{7}}4^{\underline{3}}_{\underline{5}}4^{\underline{2}}_{\underline{5}}4^{\underline{2}}_{\underline{3}}3^{\underline{1}}_{\underline{3}}4^{\underline{1}}_{\underline{1}}3^{0}_{\underline{1}}4^{0}_{1}1^{1}_{1}2^{1}_{3}1^{2}_{5}$	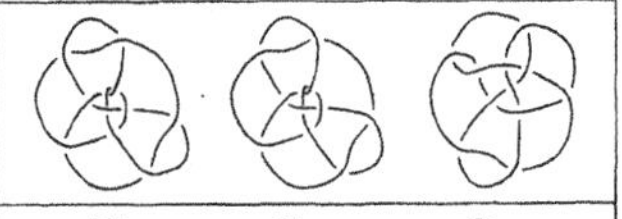
11^n_{173}: $1^{\underline{3}}_{\underline{3}}2^{\underline{2}}_{\underline{1}}1^{\underline{2}}_{1}3^{\underline{1}}_{1}2^{\underline{1}}_{3}4^{0}_{3}4^{0}_{5}4^{1}_{5}3^{1}_{7}4^{2}_{7}4^{2}_{9}2^{3}_{9}4^{3}_{11}3^{4}_{11}2^{4}_{13}3^{5}_{15}$	
11^n_{174}: $3^{0}_{3}1^{0}_{5}5^{1}_{5}2^{1}_{7}7^{2}_{7}5^{2}_{9}8^{3}_{9}7^{3}_{11}9^{4}_{11}8^{4}_{13}7^{5}_{13}9^{5}_{15}6^{6}_{15}7^{6}_{17}3^{7}_{17}6^{7}_{19}1^{8}_{19}3^{8}_{21}1^{9}_{23}$	
11^n_{175}: $1^{\underline{2}}_{\underline{1}}2^{\underline{1}}_{1}1^{\underline{1}}_{3}4^{0}_{3}3^{0}_{5}5^{1}_{5}3^{1}_{7}6^{2}_{7}5^{2}_{9}5^{3}_{9}6^{3}_{11}5^{4}_{11}5^{4}_{13}3^{5}_{13}5^{5}_{15}2^{6}_{15}3^{6}_{17}2^{7}_{19}$	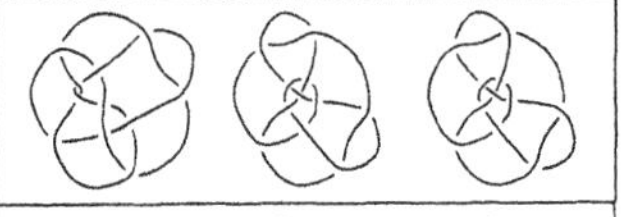
11^n_{176}: $1^{\underline{8}}_{\underline{19}}2^{\underline{7}}_{\underline{17}}1^{\underline{7}}_{\underline{15}}4^{\underline{6}}_{\underline{15}}2^{\underline{6}}_{\underline{13}}5^{\underline{5}}_{\underline{13}}4^{\underline{5}}_{\underline{11}}5^{\underline{4}}_{\underline{11}}5^{\underline{4}}_{\underline{9}}6^{\underline{3}}_{\underline{9}}5^{\underline{3}}_{\underline{7}}4^{\underline{2}}_{\underline{7}}6^{\underline{2}}_{\underline{5}}3^{\underline{1}}_{\underline{5}}4^{\underline{1}}_{\underline{3}}2^{0}_{\underline{3}}4^{0}_{\underline{1}}1^{1}_{1}$	
11^n_{177}: $1^{\underline{4}}_{\underline{7}}3^{\underline{3}}_{\underline{5}}1^{\underline{3}}_{\underline{3}}5^{\underline{2}}_{\underline{3}}3^{\underline{2}}_{\underline{1}}6^{\underline{1}}_{\underline{1}}5^{\underline{1}}_{1}8^{0}_{1}7^{0}_{3}7^{1}_{3}7^{1}_{5}6^{2}_{5}7^{2}_{7}4^{3}_{7}6^{3}_{9}2^{4}_{9}4^{4}_{11}2^{5}_{13}$	
11^n_{178}: $2^{\underline{1}}_{\underline{1}}5^{0}_{1}3^{0}_{3}7^{1}_{3}4^{1}_{5}8^{2}_{5}7^{2}_{7}8^{3}_{7}8^{3}_{9}8^{4}_{9}8^{4}_{11}5^{5}_{11}8^{5}_{13}4^{6}_{13}5^{6}_{15}1^{7}_{15}4^{7}_{17}1^{8}_{19}$	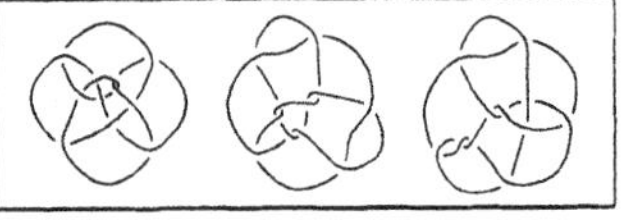
11^n_{179}: $1^{\underline{4}}_{\underline{9}}3^{\underline{3}}_{\underline{7}}1^{\underline{3}}_{\underline{5}}5^{\underline{2}}_{\underline{5}}3^{\underline{2}}_{\underline{3}}6^{\underline{1}}_{\underline{3}}5^{\underline{1}}_{\underline{1}}7^{0}_{\underline{1}}7^{0}_{1}7^{1}_{1}6^{1}_{3}5^{2}_{3}7^{2}_{5}3^{3}_{5}5^{3}_{7}2^{4}_{7}3^{4}_{9}2^{5}_{11}$	
11^n_{180}: $1^{0}_{5}1^{0}_{7}2^{1}_{7}3^{2}_{9}2^{2}_{11}4^{3}_{11}3^{3}_{13}5^{4}_{13}4^{4}_{15}4^{5}_{15}5^{5}_{17}5^{6}_{17}4^{6}_{19}2^{7}_{19}5^{7}_{21}2^{8}_{21}2^{8}_{23}2^{9}_{25}$	

K: KH	1st line 2nd line 3rd line
11^n_{181}: $1^0_3 1^0_5 2^1_5 2^2_7 2^2_9 4^3_9 2^3_{11} 4^4_{11} 4^4_{13} 3^5_{13} 4^5_{15} 4^6_{15} 3^6_{17} 1^7_{17} 4^7_{19} 2^8_{19} 1^8_{21} 2^9_{23}$ 11^n_{182}: $3^{\underline{4}}_{\underline{9}} 4^{\underline{3}}_{\underline{7}} 3^{\underline{3}}_{\underline{5}} 7^{\underline{2}}_{\underline{5}} 4^{\underline{2}}_{\underline{3}} 8^{\underline{1}}_{\underline{3}} 7^{\underline{1}}_{\underline{1}} 8^0_{\underline{1}} 9^0_1 8^1_1 7^1_3 5^2_3 8^2_5 3^3_5 5^3_7 1^4_7 3^4_9 1^5_{11}$ 11^n_{183}: $1^0_5 1^0_7 1^2_9 2^3_9 1^3_{13} 2^4_{11} 3^4_{13} 2^5_{13} 2^5_{15} 1^5_{17} 3^6_{15} 2^6_{17} 1^7_{17} 3^7_{19} 2^8_{19} 1^8_{21} 1^9_{21} 2^9_{23} 1^{10}_{25}$	  
11^n_{184}: $2^{\underline{1}}_{\underline{1}} 5^0_1 3^0_3 6^1_3 4^1_5 8^2_5 6^2_7 7^3_7 8^3_9 7^4_9 7^4_{11} 5^5_{11} 7^5_{13} 3^6_{13} 5^6_{15} 1^7_{15} 3^7_{17} 1^8_{19}$ 11^n_{185}: $3^0_3 1^0_5 5^1_5 2^1_7 8^2_7 5^2_9 8^3_9 8^3_{11} 10^4_{11} 8^4_{13} 8^5_{13} 10^5_{15} 6^6_{15} 8^6_{17} 4^7_{17} 6^7_{19} 1^8_{19} 4^8_{21} 1^9_{23}$	 

3. Links

3.1. 2–5 Crossing Links.

L: KH	1st line	2nd line	3rd line
2^a_1: $1^{\underline{2}}_{\underline{6}}1^{\underline{2}}_{\underline{4}}1^{0}_{\underline{2}}1^{0}_{0}$			
4^a_1: $1^{\underline{4}}_{\underline{10}}1^{\underline{4}}_{\underline{8}}1^{\underline{2}}_{\underline{6}}1^{\underline{1}}_{\underline{2}}1^{0}_{\underline{2}}1^{0}_{0}$		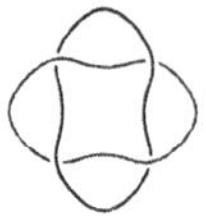	
5^a_1: $1^{\underline{3}}_{\underline{8}}1^{\underline{2}}_{\underline{6}}1^{\underline{2}}_{\underline{4}}1^{\underline{1}}_{\underline{2}}2^{0}_{\underline{2}}2^{0}_{0}1^{1}_{0}1^{2}_{4}$			

(All are alternating).

3.2. 6 Crossing Alternating Links.

L: KH	1st line	2nd line	3rd line
6^a_1: $1^{\underline{4}}_{\underline{10}}1^{\underline{4}}_{\underline{8}}1^{\underline{3}}_{\underline{8}}2^{\underline{2}}_{\underline{6}}1^{\underline{2}}_{\underline{4}}2^{\underline{1}}_{\underline{2}}2^{0}_{\underline{2}}1^{0}_{0}1^{1}_{0}1^{1}_{2}1^{2}_{4}$	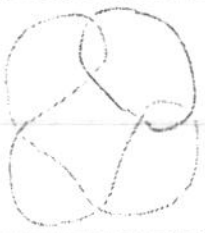		
6^a_2: $1^{\underline{6}}_{\underline{16}}1^{\underline{6}}_{\underline{14}}1^{\underline{5}}_{\underline{14}}1^{\underline{4}}_{\underline{12}}1^{\underline{4}}_{\underline{10}}1^{\underline{3}}_{\underline{10}}1^{\underline{3}}_{\underline{8}}1^{\underline{2}}_{\underline{8}}1^{\underline{2}}_{\underline{6}}1^{\underline{1}}_{\underline{4}}1^{0}_{\underline{4}}1^{0}_{\underline{2}}$			
6^a_3: $1^{\underline{6}}_{\underline{18}}1^{\underline{6}}_{\underline{16}}1^{\underline{5}}_{\underline{16}}1^{\underline{4}}_{\underline{12}}1^{\underline{3}}_{\underline{12}}1^{\underline{2}}_{\underline{8}}1^{0}_{\underline{6}}1^{0}_{\underline{4}}$			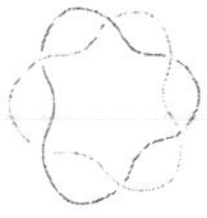
6^a_4: $1^{\underline{3}}_{\underline{7}}2^{\underline{2}}_{\underline{5}}1^{\underline{2}}_{\underline{3}}2^{\underline{1}}_{\underline{1}}4^{0}_{\underline{1}}4^{0}_{1}2^{1}_{1}1^{2}_{3}2^{2}_{5}1^{3}_{7}$	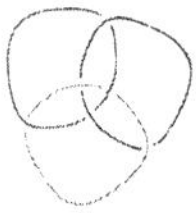		
6^a_5: $1^{\underline{6}}_{\underline{15}}1^{\underline{5}}_{\underline{11}}3^{\underline{4}}_{\underline{11}}3^{\underline{4}}_{\underline{9}}1^{\underline{3}}_{\underline{9}}2^{\underline{2}}_{\underline{7}}1^{\underline{2}}_{\underline{5}}2^{\underline{1}}_{\underline{3}}1^{0}_{\underline{3}}1^{0}_{\underline{1}}$			

3.3. 6 Crossing Non-Alternating Links.

L: KH	1st line	2nd line	3rd line
6^n_1: $2^{0}_{\underline{1}}3^{0}_{1}1^{0}_{3}1^{1}_{1}1^{2}_{5}1^{4}_{7}1^{4}_{9}$			

3.4. 7 Crossing Alternating Links.

L: KH	1st line	2nd line	3rd line
7^a_1: $1^{\underline{3}}_{\underline{6}}2^{\underline{2}}_{\underline{4}}1^{\underline{2}}_{\underline{2}}1^{\underline{1}}_{\underline{2}}2^{\underline{1}}_{0}4^{0}_{0}3^{0}_{2}2^{1}_{2}2^{1}_{4}2^{2}_{4}2^{2}_{6}1^{3}_{6}2^{3}_{8}1^{4}_{10}$	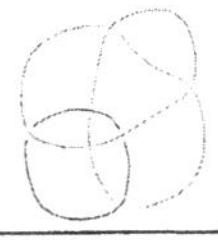		
7^a_2: $1^{\underline{7}}_{\underline{18}}1^{\underline{6}}_{\underline{16}}1^{\underline{6}}_{\underline{14}}2^{\underline{5}}_{\underline{14}}1^{\underline{5}}_{\underline{12}}2^{\underline{4}}_{\underline{12}}3^{\underline{4}}_{\underline{10}}2^{\underline{3}}_{\underline{10}}1^{\underline{3}}_{\underline{8}}2^{\underline{2}}_{\underline{8}}2^{\underline{2}}_{\underline{6}}2^{\underline{1}}_{\underline{4}}1^{0}_{\underline{4}}1^{0}_{\underline{2}}$			
7^a_3: $1^{\underline{2}}_{\underline{2}}1^{\underline{1}}_{2}3^{0}_{2}2^{0}_{4}1^{1}_{4}1^{1}_{6}2^{2}_{6}1^{2}_{8}1^{3}_{8}2^{3}_{10}1^{4}_{10}1^{4}_{12}1^{5}_{14}$			
7^a_4: $1^{\underline{2}}_{\underline{4}}1^{\underline{1}}_{0}3^{0}_{0}2^{0}_{2}2^{1}_{2}1^{1}_{4}1^{2}_{4}2^{2}_{6}1^{3}_{6}1^{3}_{8}1^{4}_{8}1^{4}_{10}1^{5}_{12}$	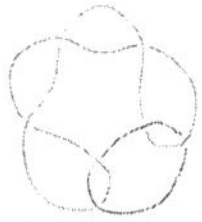		
7^a_5: $1^{\underline{5}}_{\underline{12}}1^{\underline{4}}_{\underline{10}}1^{\underline{4}}_{\underline{8}}1^{\underline{3}}_{\underline{8}}1^{\underline{3}}_{\underline{6}}3^{\underline{2}}_{\underline{6}}2^{\underline{2}}_{\underline{4}}1^{\underline{1}}_{\underline{4}}2^{\underline{1}}_{\underline{2}}2^{0}_{\underline{2}}2^{0}_{0}1^{1}_{0}1^{1}_{2}1^{2}_{4}$			
7^a_6: $1^{\underline{2}}_{\underline{2}}1^{\underline{1}}_{2}2^{0}_{2}1^{0}_{4}1^{1}_{4}1^{1}_{6}2^{2}_{6}2^{2}_{8}1^{3}_{8}1^{3}_{10}1^{4}_{10}1^{4}_{12}1^{5}_{14}$			

L: KH	1st line	2nd line	3rd line
7^a_7: $1^{\underline{4}}_{\underline{9}}1^{\underline{4}}_{\underline{7}}1^{\underline{3}}_{\underline{7}}3^{\underline{2}}_{\underline{5}}1^{\underline{2}}_{\underline{3}}3^{\underline{1}}_{\underline{1}}4^{0}_{\underline{1}}3^{0}_{1}2^{1}_{1}1^{1}_{3}1^{2}_{3}2^{2}_{5}1^{3}_{7}$	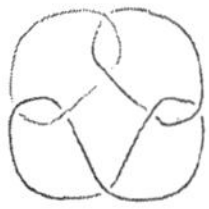		

3.5. **7 Crossing Non-Alternating Links.**

L: KH	1st line	2nd line	3rd line
7^n_1: $1^{\underline{5}}_{\underline{16}}2^{\underline{4}}_{\underline{12}}1^{\underline{4}}_{\underline{10}}1^{\underline{3}}_{\underline{12}}1^{\underline{2}}_{\underline{8}}1^{0}_{\underline{6}}1^{0}_{\underline{4}}$ 7^n_2: $1^{\underline{5}}_{\underline{12}}1^{\underline{4}}_{\underline{8}}1^{\underline{3}}_{\underline{8}}1^{\underline{2}}_{\underline{6}}1^{\underline{2}}_{\underline{4}}1^{\underline{1}}_{\underline{2}}2^{0}_{\underline{2}}2^{0}_{0}$	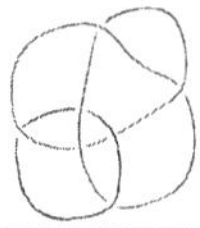	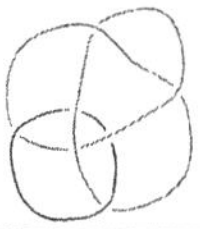	

3.6. **8 Crossing Alternating Links.**

L: KH	1st line	2nd line	3rd line
8^a_1: $1^{\underline{5}}_{\underline{12}}2^{\underline{4}}_{\underline{10}}1^{\underline{4}}_{\underline{8}}3^{\underline{3}}_{\underline{8}}2^{\underline{3}}_{\underline{6}}4^{\underline{2}}_{\underline{6}}3^{\underline{2}}_{\underline{4}}3^{\underline{1}}_{\underline{4}}4^{\underline{1}}_{\underline{2}}4^{0}_{\underline{2}}5^{0}_{0}3^{1}_{0}2^{1}_{2}1^{2}_{2}3^{2}_{4}1^{3}_{6}$ 8^a_2: $1^{\underline{4}}_{\underline{8}}1^{\underline{3}}_{\underline{6}}1^{\underline{3}}_{\underline{4}}3^{\underline{2}}_{\underline{4}}1^{\underline{2}}_{\underline{2}}2^{\underline{1}}_{\underline{2}}3^{\underline{1}}_{0}4^{0}_{0}4^{0}_{2}3^{1}_{2}2^{1}_{4}2^{2}_{4}3^{2}_{6}1^{3}_{6}2^{3}_{8}1^{4}_{10}$ 8^a_3: $1^{\underline{6}}_{\underline{14}}1^{\underline{5}}_{\underline{12}}1^{\underline{5}}_{\underline{10}}3^{\underline{4}}_{\underline{10}}2^{\underline{4}}_{\underline{8}}2^{\underline{3}}_{\underline{8}}2^{\underline{3}}_{\underline{6}}3^{\underline{2}}_{\underline{6}}2^{\underline{2}}_{\underline{4}}2^{\underline{1}}_{\underline{4}}3^{\underline{1}}_{\underline{2}}2^{0}_{\underline{2}}3^{0}_{0}1^{1}_{0}1^{1}_{2}1^{2}_{4}$	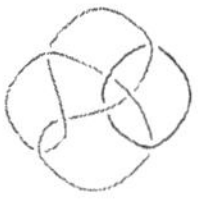	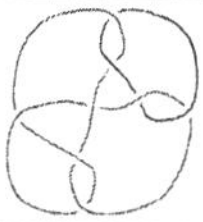	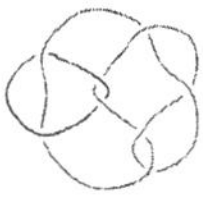
8^a_4: $1^{\underline{5}}_{\underline{12}}1^{\underline{4}}_{\underline{10}}1^{\underline{4}}_{\underline{8}}3^{\underline{3}}_{\underline{8}}1^{\underline{3}}_{\underline{6}}3^{\underline{2}}_{\underline{6}}3^{\underline{2}}_{\underline{4}}2^{\underline{1}}_{\underline{4}}3^{\underline{1}}_{\underline{2}}4^{0}_{\underline{2}}4^{0}_{0}2^{1}_{0}2^{1}_{2}1^{2}_{2}2^{2}_{4}1^{3}_{6}$ 8^a_5: $1^{\underline{4}}_{\underline{8}}1^{\underline{4}}_{\underline{6}}1^{\underline{3}}_{\underline{6}}3^{\underline{2}}_{\underline{4}}1^{\underline{2}}_{\underline{2}}1^{\underline{1}}_{\underline{2}}3^{\underline{1}}_{0}4^{0}_{0}2^{0}_{2}2^{1}_{2}3^{1}_{4}2^{2}_{4}2^{2}_{6}1^{3}_{6}2^{3}_{8}1^{4}_{10}$ 8^a_6: $1^{\underline{4}}_{\underline{10}}1^{\underline{4}}_{\underline{8}}1^{\underline{3}}_{\underline{8}}2^{\underline{2}}_{\underline{6}}1^{\underline{2}}_{\underline{4}}1^{\underline{1}}_{\underline{4}}2^{\underline{1}}_{\underline{2}}3^{0}_{\underline{2}}2^{0}_{0}1^{1}_{0}2^{1}_{2}1^{2}_{2}1^{2}_{4}1^{3}_{4}1^{3}_{6}1^{4}_{8}$		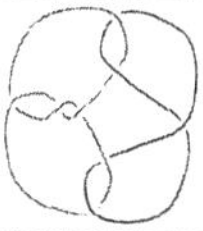	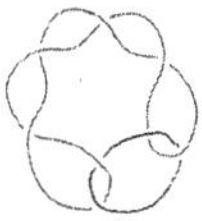
8^a_7: $1^{\underline{8}}_{\underline{20}}2^{\underline{7}}_{\underline{18}}1^{\underline{7}}_{\underline{16}}2^{\underline{6}}_{\underline{16}}2^{\underline{6}}_{\underline{14}}4^{\underline{5}}_{\underline{14}}2^{\underline{5}}_{\underline{12}}3^{\underline{4}}_{\underline{12}}5^{\underline{4}}_{\underline{10}}3^{\underline{3}}_{\underline{10}}2^{\underline{3}}_{\underline{8}}3^{\underline{2}}_{\underline{8}}3^{\underline{2}}_{\underline{6}}3^{\underline{1}}_{\underline{4}}1^{0}_{\underline{4}}1^{0}_{\underline{2}}$ 8^a_8: $1^{\underline{4}}_{\underline{8}}1^{\underline{3}}_{\underline{6}}1^{\underline{3}}_{\underline{4}}3^{\underline{2}}_{\underline{4}}2^{\underline{2}}_{\underline{2}}2^{\underline{1}}_{\underline{2}}2^{\underline{1}}_{0}4^{0}_{0}3^{0}_{2}2^{1}_{2}3^{1}_{4}2^{2}_{4}2^{2}_{6}1^{3}_{6}2^{3}_{8}1^{4}_{10}$ 8^a_9: $1^{\underline{5}}_{\underline{12}}2^{\underline{4}}_{\underline{10}}1^{\underline{4}}_{\underline{8}}2^{\underline{3}}_{\underline{8}}2^{\underline{3}}_{\underline{6}}4^{\underline{2}}_{\underline{6}}3^{\underline{2}}_{\underline{4}}3^{\underline{1}}_{\underline{4}}3^{\underline{1}}_{\underline{2}}3^{0}_{\underline{2}}4^{0}_{0}2^{1}_{0}2^{1}_{2}1^{2}_{2}2^{2}_{4}1^{3}_{6}$	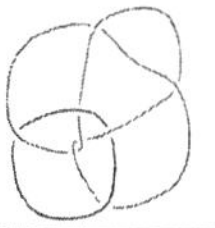	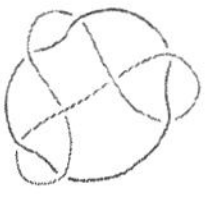	
8^a_{10}: $1^{\underline{8}}_{\underline{20}}1^{\underline{7}}_{\underline{18}}1^{\underline{7}}_{\underline{16}}2^{\underline{6}}_{\underline{16}}2^{\underline{6}}_{\underline{14}}3^{\underline{5}}_{\underline{14}}1^{\underline{5}}_{\underline{12}}2^{\underline{4}}_{\underline{12}}3^{\underline{4}}_{\underline{10}}2^{\underline{3}}_{\underline{10}}2^{\underline{3}}_{\underline{8}}2^{\underline{2}}_{\underline{8}}2^{\underline{2}}_{\underline{6}}2^{\underline{1}}_{\underline{4}}1^{0}_{\underline{4}}1^{0}_{\underline{2}}$ 8^a_{11}: $1^{\underline{8}}_{\underline{22}}1^{\underline{7}}_{\underline{20}}1^{\underline{7}}_{\underline{18}}2^{\underline{6}}_{\underline{18}}2^{\underline{6}}_{\underline{16}}3^{\underline{5}}_{\underline{16}}1^{\underline{5}}_{\underline{14}}1^{\underline{4}}_{\underline{14}}3^{\underline{4}}_{\underline{12}}2^{\underline{3}}_{\underline{12}}1^{\underline{3}}_{\underline{10}}1^{\underline{2}}_{\underline{10}}2^{\underline{2}}_{\underline{8}}1^{\underline{1}}_{\underline{6}}1^{0}_{\underline{6}}1^{0}_{\underline{4}}$ 8^a_{12}: $1^{\underline{8}}_{\underline{22}}1^{\underline{8}}_{\underline{20}}1^{\underline{7}}_{\underline{20}}1^{\underline{6}}_{\underline{18}}1^{\underline{6}}_{\underline{16}}2^{\underline{5}}_{\underline{16}}1^{\underline{5}}_{\underline{14}}1^{\underline{4}}_{\underline{14}}2^{\underline{4}}_{\underline{12}}1^{\underline{3}}_{\underline{12}}1^{\underline{3}}_{\underline{10}}1^{\underline{2}}_{\underline{10}}1^{\underline{2}}_{\underline{8}}1^{\underline{1}}_{\underline{6}}1^{0}_{\underline{6}}1^{0}_{\underline{4}}$	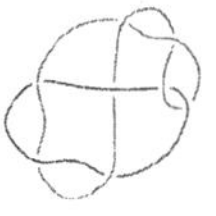	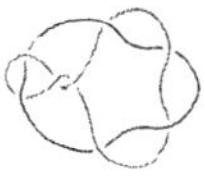	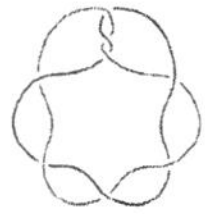
8^a_{13}: $1^{\underline{8}}_{\underline{20}}1^{\underline{8}}_{\underline{18}}1^{\underline{7}}_{\underline{18}}2^{\underline{6}}_{\underline{16}}1^{\underline{6}}_{\underline{14}}2^{\underline{5}}_{\underline{14}}2^{\underline{5}}_{\underline{12}}2^{\underline{4}}_{\underline{12}}2^{\underline{4}}_{\underline{10}}2^{\underline{3}}_{\underline{10}}2^{\underline{3}}_{\underline{8}}2^{\underline{2}}_{\underline{8}}2^{\underline{2}}_{\underline{6}}2^{\underline{1}}_{\underline{4}}1^{0}_{\underline{4}}1^{0}_{\underline{2}}$ 8^a_{14}: $1^{\underline{8}}_{\underline{24}}1^{\underline{8}}_{\underline{22}}1^{\underline{7}}_{\underline{22}}1^{\underline{6}}_{\underline{18}}1^{\underline{5}}_{\underline{18}}1^{\underline{4}}_{\underline{14}}1^{\underline{3}}_{\underline{14}}1^{\underline{2}}_{\underline{10}}1^{0}_{\underline{8}}1^{0}_{\underline{6}}$ 8^a_{15}: $1^{\underline{6}}_{\underline{15}}1^{\underline{5}}_{\underline{11}}4^{\underline{4}}_{\underline{11}}3^{\underline{4}}_{\underline{9}}3^{\underline{3}}_{\underline{9}}1^{\underline{3}}_{\underline{7}}3^{\underline{2}}_{\underline{7}}3^{\underline{2}}_{\underline{5}}1^{\underline{1}}_{\underline{5}}3^{\underline{1}}_{\underline{3}}3^{0}_{\underline{3}}2^{0}_{\underline{1}}1^{1}_{\underline{1}}2^{1}_{1}1^{2}_{3}$	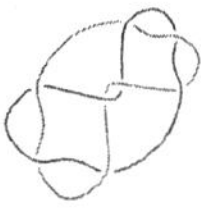	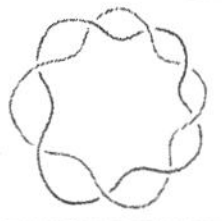	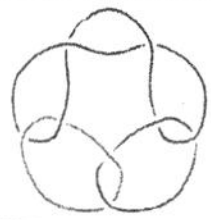
8^a_{16}: $1^{\underline{3}}_{\underline{5}}2^{\underline{2}}_{\underline{3}}1^{\underline{2}}_{\underline{1}}1^{\underline{1}}_{\underline{1}}2^{\underline{1}}_{1}5^{0}_{1}3^{0}_{3}2^{1}_{3}3^{1}_{5}4^{2}_{5}4^{2}_{7}2^{3}_{7}2^{3}_{9}1^{4}_{9}2^{4}_{11}1^{5}_{13}$ 8^a_{17}: $1^{\underline{8}}_{\underline{21}}1^{\underline{7}}_{\underline{19}}1^{\underline{7}}_{\underline{17}}3^{\underline{6}}_{\underline{17}}3^{\underline{6}}_{\underline{15}}3^{\underline{5}}_{\underline{15}}1^{\underline{5}}_{\underline{13}}3^{\underline{4}}_{\underline{13}}4^{\underline{4}}_{\underline{11}}2^{\underline{3}}_{\underline{11}}2^{\underline{3}}_{\underline{9}}2^{\underline{2}}_{\underline{9}}2^{\underline{2}}_{\underline{7}}2^{\underline{1}}_{\underline{5}}1^{0}_{\underline{5}}1^{0}_{\underline{3}}$ 8^a_{18}: $1^{\underline{2}}_{\underline{1}}1^{\underline{1}}_{3}3^{0}_{3}2^{0}_{5}1^{1}_{5}1^{1}_{7}3^{2}_{7}2^{2}_{9}1^{3}_{9}2^{3}_{11}2^{4}_{11}1^{4}_{13}2^{5}_{15}1^{6}_{15}1^{6}_{17}$	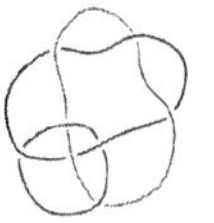	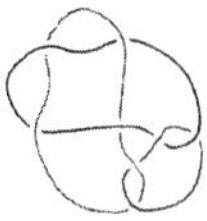	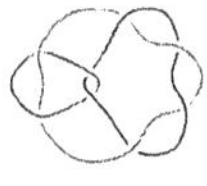
8^a_{19}: $1^{\underline{4}}_{\underline{9}}2^{\underline{3}}_{\underline{7}}1^{\underline{3}}_{\underline{5}}3^{\underline{2}}_{\underline{5}}3^{\underline{2}}_{\underline{3}}3^{\underline{1}}_{\underline{3}}2^{\underline{1}}_{\underline{1}}5^{0}_{\underline{1}}5^{0}_{1}2^{1}_{1}3^{1}_{3}3^{2}_{3}3^{2}_{5}1^{3}_{5}2^{3}_{7}1^{4}_{9}$ 8^a_{20}: $1^{\underline{4}}_{\underline{9}}1^{\underline{4}}_{\underline{7}}2^{\underline{3}}_{\underline{7}}3^{\underline{2}}_{\underline{5}}2^{\underline{2}}_{\underline{3}}2^{\underline{1}}_{\underline{3}}3^{\underline{1}}_{\underline{1}}4^{0}_{\underline{1}}4^{0}_{1}3^{1}_{1}2^{1}_{3}2^{2}_{3}3^{2}_{5}2^{3}_{7}1^{4}_{7}1^{4}_{9}$ 8^a_{21}: $1^{\underline{8}}_{\underline{20}}1^{\underline{8}}_{\underline{18}}1^{\underline{7}}_{\underline{18}}4^{\underline{6}}_{\underline{16}}1^{\underline{6}}_{\underline{14}}4^{\underline{5}}_{\underline{12}}7^{\underline{4}}_{\underline{12}}6^{\underline{4}}_{\underline{10}}3^{\underline{3}}_{\underline{10}}1^{\underline{3}}_{\underline{8}}3^{\underline{2}}_{\underline{8}}3^{\underline{2}}_{\underline{6}}3^{\underline{1}}_{\underline{4}}1^{0}_{\underline{4}}1^{0}_{\underline{2}}$	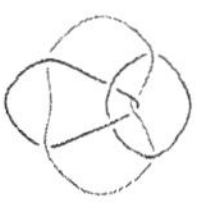	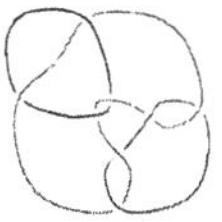	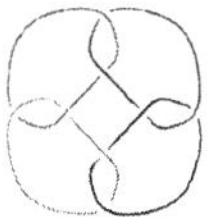

3.7. 8 Crossing Non-Alternating Links.

L: KH	1st line	2nd line	3rd line
8^n_1: $2^{\underline{4}}_{\underline{12}}1^{\underline{4}}_{\underline{10}}1^{\underline{3}}_{\underline{10}}1^{\underline{3}}_{\underline{8}}1^{\underline{2}}_{\underline{8}}1^{\underline{2}}_{\underline{6}}1^{\underline{1}}_{\underline{6}}1^{\underline{1}}_{\underline{4}}1^{0}_{\underline{4}}2^{0}_{\underline{2}}1^{1}_{\underline{2}}1^{2}_{2}$ 8^n_2: $1^{\underline{4}}_{\underline{8}}1^{\underline{3}}_{\underline{4}}1^{\underline{2}}_{\underline{4}}1^{\underline{1}}_{\underline{2}}1^{\underline{1}}_{0}1^{0}_{\underline{2}}2^{0}_{0}2^{0}_{2}1^{1}_{2}1^{2}_{6}$ 8^n_3: $1^{\underline{6}}_{\underline{19}}2^{\underline{6}}_{\underline{17}}1^{\underline{6}}_{\underline{15}}1^{\underline{5}}_{\underline{17}}2^{\underline{4}}_{\underline{13}}1^{\underline{4}}_{\underline{11}}1^{\underline{3}}_{\underline{13}}1^{\underline{2}}_{\underline{9}}1^{0}_{\underline{7}}1^{0}_{\underline{5}}$	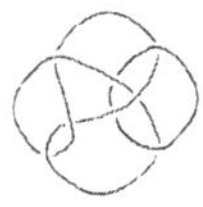	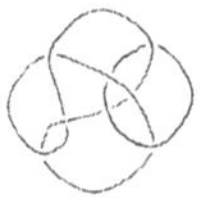	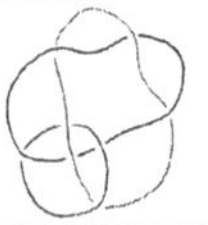
8^n_4: $1^{\underline{6}}_{\underline{15}}1^{\underline{6}}_{\underline{13}}1^{\underline{5}}_{\underline{13}}1^{\underline{4}}_{\underline{11}}1^{\underline{4}}_{\underline{9}}1^{\underline{3}}_{\underline{9}}1^{\underline{3}}_{\underline{7}}2^{\underline{2}}_{\underline{7}}2^{\underline{2}}_{\underline{5}}1^{\underline{1}}_{\underline{3}}2^{0}_{\underline{3}}2^{0}_{\underline{1}}$ 8^n_5: $1^{\underline{6}}_{\underline{15}}1^{\underline{5}}_{\underline{13}}1^{\underline{5}}_{\underline{11}}2^{\underline{4}}_{\underline{11}}3^{\underline{4}}_{\underline{9}}2^{\underline{3}}_{\underline{9}}2^{\underline{2}}_{\underline{7}}2^{\underline{2}}_{\underline{5}}2^{\underline{1}}_{\underline{3}}2^{0}_{\underline{3}}2^{0}_{\underline{1}}$ 8^n_6: $1^{\underline{8}}_{\underline{19}}1^{\underline{8}}_{\underline{17}}1^{\underline{6}}_{\underline{15}}1^{\underline{5}}_{\underline{11}}2^{\underline{4}}_{\underline{13}}3^{\underline{4}}_{\underline{11}}1^{\underline{3}}_{\underline{11}}1^{\underline{3}}_{\underline{7}}1^{\underline{2}}_{\underline{7}}1^{0}_{\underline{5}}1^{0}_{\underline{3}}$	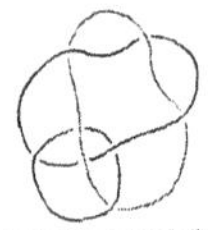	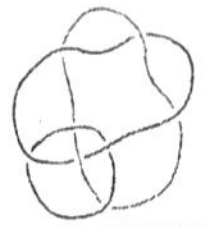	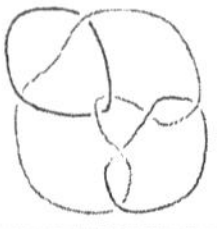
8^n_7: $3^{0}_{0}4^{0}_{2}1^{0}_{4}3^{1}_{2}1^{2}_{4}3^{2}_{6}1^{3}_{8}4^{4}_{8}4^{4}_{10}1^{5}_{10}1^{6}_{14}$ 8^n_8: $1^{\underline{4}}_{\underline{8}}1^{\underline{4}}_{\underline{6}}1^{\underline{2}}_{\underline{4}}1^{\underline{1}}_{0}3^{0}_{\underline{2}}6^{0}_{0}3^{0}_{2}1^{1}_{0}1^{2}_{4}1^{4}_{6}1^{4}_{8}$	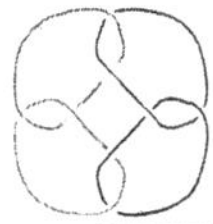	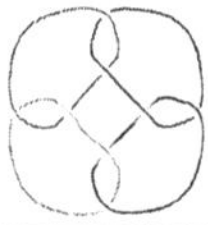	

3.8. 9 Crossing Alternating Links.

L: KH	1st line	2nd line	3rd line
9^a_1: $1^{\underline{3}}_{\underline{6}}3^{\underline{2}}_{\underline{4}}1^{\underline{2}}_{\underline{2}}3^{\underline{1}}_{\underline{2}}3^{\underline{1}}_{0}6^{0}_{0}5^{0}_{2}5^{1}_{2}4^{1}_{4}5^{2}_{4}5^{2}_{6}3^{3}_{6}5^{3}_{8}2^{4}_{8}3^{4}_{10}1^{5}_{10}2^{5}_{12}1^{6}_{14}$ 9^a_2: $1^{\underline{3}}_{\underline{4}}2^{\underline{2}}_{\underline{2}}1^{\underline{2}}_{0}1^{\underline{1}}_{0}2^{\underline{1}}_{2}5^{0}_{2}3^{0}_{4}3^{1}_{4}3^{1}_{6}4^{2}_{6}3^{2}_{8}2^{3}_{8}4^{3}_{10}2^{4}_{10}2^{4}_{12}1^{5}_{12}2^{5}_{14}1^{6}_{16}$ 9^a_3: $1^{\underline{3}}_{\underline{6}}2^{\underline{2}}_{\underline{4}}1^{\underline{2}}_{\underline{2}}3^{\underline{1}}_{\underline{2}}2^{\underline{1}}_{0}6^{0}_{0}5^{0}_{2}5^{1}_{2}4^{1}_{4}5^{2}_{4}5^{2}_{6}3^{3}_{6}5^{3}_{8}3^{4}_{8}3^{4}_{10}1^{5}_{10}3^{5}_{12}1^{6}_{14}$	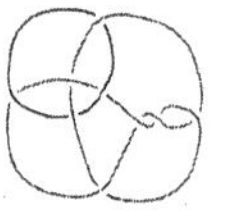	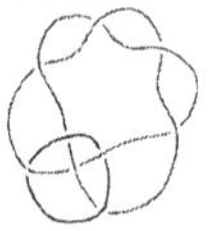	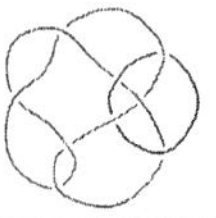
9^a_4: $1^{\underline{2}}_{\underline{2}}1^{\underline{1}}_{0}1^{\underline{1}}_{2}4^{0}_{2}3^{0}_{4}4^{1}_{4}2^{1}_{6}5^{2}_{6}4^{2}_{8}3^{3}_{8}5^{3}_{10}4^{4}_{10}3^{4}_{12}2^{5}_{12}4^{5}_{14}1^{6}_{14}2^{6}_{16}1^{7}_{18}$ 9^a_5: $1^{\underline{4}}_{\underline{8}}1^{\underline{4}}_{\underline{6}}2^{\underline{3}}_{\underline{6}}4^{\underline{2}}_{\underline{4}}2^{\underline{2}}_{\underline{2}}3^{\underline{1}}_{\underline{2}}4^{\underline{1}}_{0}6^{0}_{0}4^{0}_{2}4^{1}_{2}5^{1}_{4}4^{2}_{4}4^{2}_{6}2^{3}_{6}4^{3}_{8}1^{4}_{8}2^{4}_{10}1^{5}_{12}$ 9^a_6: $1^{\underline{9}}_{\underline{24}}1^{\underline{8}}_{\underline{22}}1^{\underline{8}}_{\underline{20}}4^{\underline{7}}_{\underline{20}}1^{\underline{7}}_{\underline{18}}3^{\underline{6}}_{\underline{18}}4^{\underline{6}}_{\underline{16}}4^{\underline{5}}_{\underline{16}}3^{\underline{5}}_{\underline{14}}4^{\underline{4}}_{\underline{14}}5^{\underline{4}}_{\underline{12}}3^{\underline{3}}_{\underline{12}}3^{\underline{3}}_{\underline{10}}2^{\underline{2}}_{\underline{10}}3^{\underline{2}}_{\underline{8}}2^{\underline{1}}_{\underline{6}}1^{0}_{\underline{6}}1^{0}_{\underline{4}}$	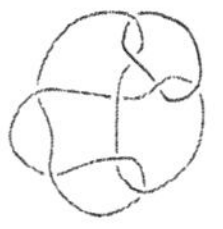	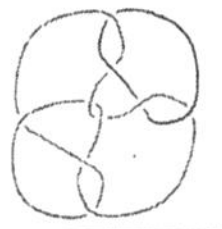	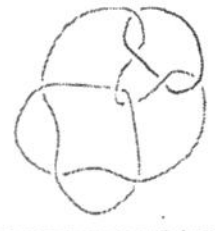
9^a_7: $1^{\underline{9}}_{\underline{22}}1^{\underline{8}}_{\underline{20}}1^{\underline{8}}_{\underline{18}}3^{\underline{7}}_{\underline{18}}1^{\underline{7}}_{\underline{16}}2^{\underline{6}}_{\underline{16}}3^{\underline{6}}_{\underline{14}}3^{\underline{5}}_{\underline{14}}2^{\underline{5}}_{\underline{12}}4^{\underline{4}}_{\underline{12}}4^{\underline{4}}_{\underline{10}}2^{\underline{3}}_{\underline{10}}3^{\underline{3}}_{\underline{8}}2^{\underline{2}}_{\underline{8}}2^{\underline{2}}_{\underline{6}}2^{\underline{1}}_{\underline{4}}1^{0}_{\underline{4}}1^{0}_{\underline{2}}$ 9^a_8: $1^{\underline{4}}_{\underline{8}}1^{\underline{3}}_{\underline{6}}1^{\underline{3}}_{\underline{4}}4^{\underline{2}}_{\underline{4}}1^{\underline{2}}_{\underline{2}}3^{\underline{1}}_{\underline{2}}4^{\underline{1}}_{0}5^{0}_{0}5^{0}_{2}5^{1}_{2}3^{1}_{4}3^{2}_{4}5^{2}_{6}2^{3}_{6}3^{3}_{8}1^{4}_{8}2^{4}_{10}1^{5}_{12}$ 9^a_9: $1^{\underline{5}}_{\underline{12}}2^{\underline{4}}_{\underline{10}}1^{\underline{4}}_{\underline{8}}3^{\underline{3}}_{\underline{8}}2^{\underline{3}}_{\underline{6}}4^{\underline{2}}_{\underline{6}}3^{\underline{2}}_{\underline{4}}4^{\underline{1}}_{\underline{4}}4^{\underline{1}}_{\underline{2}}5^{0}_{\underline{2}}6^{0}_{0}3^{1}_{0}3^{1}_{2}2^{2}_{2}3^{2}_{4}1^{3}_{4}2^{3}_{6}1^{4}_{8}$	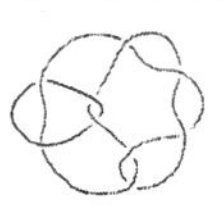	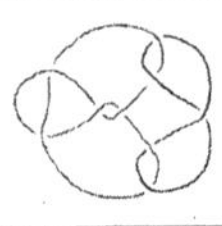	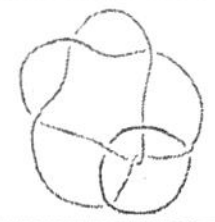
9^a_{10}: $1^{\underline{5}}_{\underline{10}}2^{\underline{4}}_{\underline{8}}1^{\underline{4}}_{\underline{6}}2^{\underline{3}}_{\underline{6}}2^{\underline{3}}_{\underline{4}}5^{\underline{2}}_{\underline{4}}2^{\underline{2}}_{\underline{2}}3^{\underline{1}}_{\underline{2}}5^{\underline{1}}_{0}5^{0}_{0}5^{0}_{2}4^{1}_{2}3^{1}_{4}2^{2}_{4}4^{2}_{6}1^{3}_{6}2^{3}_{8}1^{4}_{10}$ 9^a_{11}: $1^{\underline{6}}_{\underline{14}}1^{\underline{5}}_{\underline{12}}1^{\underline{5}}_{\underline{10}}4^{\underline{4}}_{\underline{10}}2^{\underline{4}}_{\underline{8}}4^{\underline{3}}_{\underline{8}}3^{\underline{3}}_{\underline{6}}5^{\underline{2}}_{\underline{6}}4^{\underline{2}}_{\underline{4}}4^{\underline{1}}_{\underline{4}}5^{\underline{1}}_{\underline{2}}4^{0}_{\underline{2}}5^{0}_{0}3^{1}_{0}3^{1}_{2}1^{2}_{2}3^{2}_{4}1^{3}_{6}$ 9^a_{12}: $1^{\underline{9}}_{\underline{24}}1^{\underline{8}}_{\underline{22}}1^{\underline{8}}_{\underline{20}}3^{\underline{7}}_{\underline{20}}1^{\underline{7}}_{\underline{18}}2^{\underline{6}}_{\underline{18}}3^{\underline{6}}_{\underline{16}}4^{\underline{5}}_{\underline{16}}2^{\underline{5}}_{\underline{14}}3^{\underline{4}}_{\underline{14}}5^{\underline{4}}_{\underline{12}}2^{\underline{3}}_{\underline{12}}2^{\underline{3}}_{\underline{10}}2^{\underline{2}}_{\underline{10}}2^{\underline{2}}_{\underline{8}}2^{\underline{1}}_{\underline{6}}1^{0}_{\underline{6}}1^{0}_{\underline{4}}$	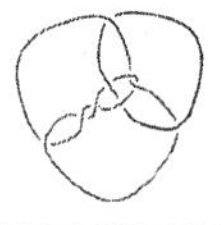	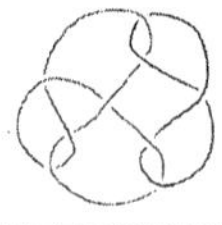	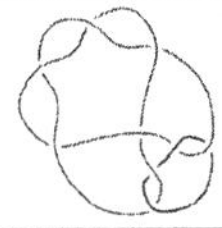
9^a_{13}: $1^{\underline{9}}_{\underline{22}}1^{\underline{8}}_{\underline{20}}1^{\underline{8}}_{\underline{18}}3^{\underline{7}}_{\underline{18}}1^{\underline{7}}_{\underline{16}}3^{\underline{6}}_{\underline{16}}3^{\underline{6}}_{\underline{14}}4^{\underline{5}}_{\underline{14}}3^{\underline{5}}_{\underline{12}}4^{\underline{4}}_{\underline{12}}5^{\underline{4}}_{\underline{10}}3^{\underline{3}}_{\underline{10}}3^{\underline{3}}_{\underline{8}}3^{\underline{2}}_{\underline{8}}3^{\underline{2}}_{\underline{6}}3^{\underline{1}}_{\underline{4}}1^{0}_{\underline{4}}1^{0}_{\underline{2}}$ 9^a_{14}: $1^{\underline{2}}_{0}1^{\underline{1}}_{4}3^{0}_{4}2^{0}_{6}1^{1}_{6}1^{1}_{8}3^{2}_{8}1^{2}_{10}1^{3}_{10}3^{3}_{12}2^{4}_{12}1^{4}_{14}1^{5}_{14}2^{5}_{16}1^{6}_{16}1^{6}_{18}1^{7}_{20}$ 9^a_{15}: $1^{\underline{2}}_{\underline{2}}1^{\underline{1}}_{0}1^{\underline{1}}_{2}4^{0}_{2}3^{0}_{4}3^{1}_{4}2^{1}_{6}4^{2}_{6}3^{2}_{8}3^{3}_{8}4^{3}_{10}3^{4}_{10}3^{4}_{12}1^{5}_{12}3^{5}_{14}1^{6}_{14}1^{6}_{16}1^{7}_{18}$	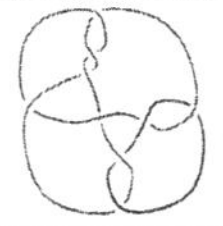	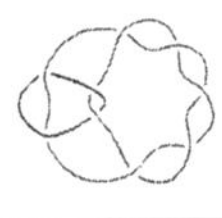	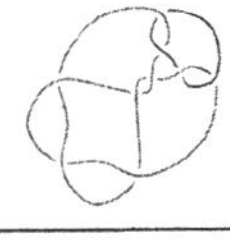
9^a_{16}: $1^{\underline{7}}_{\underline{18}}1^{\underline{6}}_{\underline{16}}1^{\underline{6}}_{\underline{14}}3^{\underline{5}}_{\underline{14}}1^{\underline{5}}_{\underline{12}}4^{\underline{4}}_{\underline{12}}4^{\underline{4}}_{\underline{10}}4^{\underline{3}}_{\underline{10}}3^{\underline{3}}_{\underline{8}}4^{\underline{2}}_{\underline{8}}4^{\underline{2}}_{\underline{6}}2^{\underline{1}}_{\underline{6}}4^{\underline{1}}_{\underline{4}}3^{0}_{\underline{4}}3^{0}_{\underline{2}}1^{1}_{\underline{2}}2^{1}_{0}1^{2}_{2}$ 9^a_{17}: $1^{\underline{2}}_{\underline{2}}1^{\underline{1}}_{2}4^{0}_{2}2^{0}_{4}3^{1}_{4}2^{1}_{6}4^{2}_{6}3^{2}_{8}3^{3}_{8}4^{3}_{10}3^{4}_{10}3^{4}_{12}2^{5}_{12}3^{5}_{14}1^{6}_{14}2^{6}_{16}1^{7}_{18}$ 9^a_{18}: $1^{\underline{2}}_{\underline{4}}1^{\underline{1}}_{0}3^{0}_{0}2^{0}_{2}2^{1}_{2}1^{1}_{4}2^{2}_{4}2^{2}_{6}2^{3}_{6}2^{3}_{8}1^{4}_{8}2^{4}_{10}1^{5}_{10}1^{5}_{12}1^{6}_{12}1^{6}_{14}1^{7}_{16}$	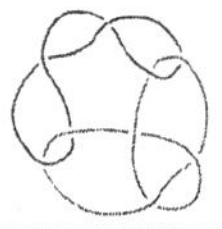	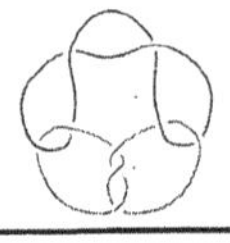	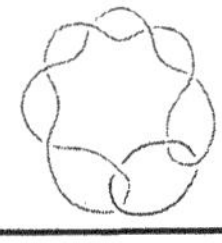

L: KH	1st line	2nd line	3rd line
9^a_{19}: $1^{\underline{6}}_{\underline{14}}2^{\underline{5}}_{\underline{12}}1^{\underline{5}}_{\underline{10}}4^{\underline{4}}_{\underline{10}}3^{\underline{4}}_{\underline{8}}5^{\underline{3}}_{\underline{8}}3^{\underline{3}}_{\underline{6}}6^{\underline{2}}_{\underline{6}}5^{\underline{2}}_{\underline{4}}4^{\underline{1}}_{\underline{4}}6^{\underline{1}}_{\underline{2}}5^{0}_{\underline{2}}5^{0}_{0}3^{1}_{0}4^{1}_{2}1^{2}_{2}3^{2}_{4}1^{3}_{6}$	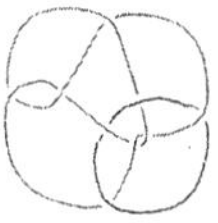		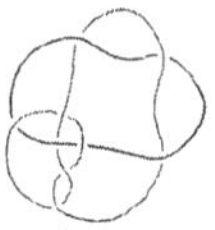
9^a_{20}: $1^{\underline{5}}_{\underline{12}}3^{\underline{4}}_{\underline{10}}1^{\underline{4}}_{\underline{8}}4^{\underline{3}}_{\underline{8}}3^{\underline{3}}_{\underline{6}}6^{\underline{2}}_{\underline{6}}5^{\underline{2}}_{\underline{4}}6^{\underline{1}}_{\underline{4}}5^{\underline{1}}_{\underline{2}}6^{0}_{\underline{2}}7^{0}_{0}4^{1}_{0}5^{1}_{2}3^{2}_{2}4^{2}_{4}1^{3}_{4}3^{3}_{6}1^{4}_{8}$			
9^a_{21}: $1^{\underline{5}}_{\underline{12}}2^{\underline{4}}_{\underline{10}}1^{\underline{4}}_{\underline{8}}3^{\underline{3}}_{\underline{8}}2^{\underline{3}}_{\underline{6}}5^{\underline{2}}_{\underline{6}}4^{\underline{2}}_{\underline{4}}4^{\underline{1}}_{\underline{4}}4^{\underline{1}}_{\underline{2}}5^{0}_{\underline{2}}5^{0}_{0}3^{1}_{0}4^{1}_{2}2^{2}_{2}3^{2}_{4}1^{3}_{4}2^{3}_{6}1^{4}_{8}$			
9^a_{22}: $1^{\underline{6}}_{\underline{16}}2^{\underline{5}}_{\underline{14}}1^{\underline{5}}_{\underline{12}}3^{\underline{4}}_{\underline{12}}2^{\underline{4}}_{\underline{10}}4^{\underline{3}}_{\underline{10}}3^{\underline{3}}_{\underline{8}}4^{\underline{2}}_{\underline{8}}5^{\underline{2}}_{\underline{6}}4^{\underline{1}}_{\underline{6}}3^{\underline{1}}_{\underline{4}}3^{0}_{\underline{4}}5^{0}_{\underline{2}}2^{1}_{\underline{2}}2^{1}_{0}1^{2}_{0}2^{2}_{2}1^{3}_{4}$	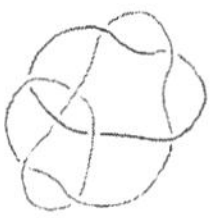		
9^a_{23}: $1^{\underline{6}}_{\underline{14}}1^{\underline{6}}_{\underline{12}}2^{\underline{5}}_{\underline{12}}3^{\underline{4}}_{\underline{10}}2^{\underline{4}}_{\underline{8}}3^{\underline{3}}_{\underline{8}}3^{\underline{3}}_{\underline{6}}5^{\underline{2}}_{\underline{6}}3^{\underline{2}}_{\underline{4}}3^{\underline{1}}_{\underline{4}}5^{\underline{1}}_{\underline{2}}4^{0}_{\underline{2}}4^{0}_{0}2^{1}_{0}3^{1}_{2}1^{2}_{2}2^{2}_{4}1^{3}_{6}$			
9^a_{24}: $1^{\underline{7}}_{\underline{18}}1^{\underline{6}}_{\underline{16}}1^{\underline{6}}_{\underline{14}}3^{\underline{5}}_{\underline{14}}1^{\underline{5}}_{\underline{12}}3^{\underline{4}}_{\underline{12}}3^{\underline{4}}_{\underline{10}}3^{\underline{3}}_{\underline{10}}3^{\underline{3}}_{\underline{8}}4^{\underline{2}}_{\underline{8}}4^{\underline{2}}_{\underline{6}}2^{\underline{1}}_{\underline{6}}3^{\underline{1}}_{\underline{4}}2^{0}_{\underline{4}}3^{0}_{\underline{2}}1^{1}_{\underline{2}}1^{1}_{0}1^{2}_{2}$			
9^a_{25}: $1^{\underline{7}}_{\underline{16}}1^{\underline{6}}_{\underline{14}}1^{\underline{6}}_{\underline{12}}2^{\underline{5}}_{\underline{12}}1^{\underline{5}}_{\underline{10}}3^{\underline{4}}_{\underline{10}}2^{\underline{4}}_{\underline{8}}2^{\underline{3}}_{\underline{8}}3^{\underline{3}}_{\underline{6}}4^{\underline{2}}_{\underline{6}}3^{\underline{2}}_{\underline{4}}2^{\underline{1}}_{\underline{4}}3^{\underline{1}}_{\underline{2}}2^{0}_{\underline{2}}3^{0}_{0}1^{1}_{0}1^{1}_{2}1^{2}_{4}$	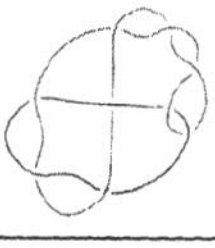	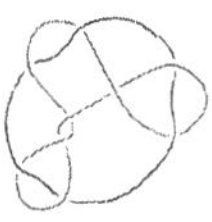	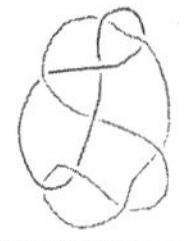
9^a_{26}: $1^{\underline{2}}_{\underline{2}}1^{\underline{1}}_{0}1^{\underline{1}}_{2}4^{0}_{2}2^{0}_{4}3^{1}_{4}3^{1}_{6}5^{2}_{6}4^{2}_{8}4^{3}_{8}4^{3}_{10}3^{4}_{10}4^{4}_{12}2^{5}_{12}3^{5}_{14}1^{6}_{14}2^{6}_{16}1^{7}_{18}$			
9^a_{27}: $1^{\underline{4}}_{\underline{8}}2^{\underline{3}}_{\underline{6}}1^{\underline{3}}_{\underline{4}}3^{\underline{2}}_{\underline{4}}2^{\underline{2}}_{\underline{2}}4^{\underline{1}}_{\underline{2}}3^{\underline{1}}_{0}5^{0}_{0}5^{0}_{2}4^{1}_{2}4^{1}_{4}4^{2}_{4}5^{2}_{6}2^{3}_{6}3^{3}_{8}1^{4}_{8}2^{4}_{10}1^{5}_{12}$			
9^a_{28}: $1^{\underline{6}}_{\underline{16}}1^{\underline{6}}_{\underline{14}}2^{\underline{5}}_{\underline{14}}2^{\underline{4}}_{\underline{12}}2^{\underline{4}}_{\underline{10}}3^{\underline{3}}_{\underline{10}}2^{\underline{3}}_{\underline{8}}3^{\underline{2}}_{\underline{8}}3^{\underline{2}}_{\underline{6}}2^{\underline{1}}_{\underline{6}}3^{\underline{1}}_{\underline{4}}3^{0}_{\underline{4}}3^{0}_{\underline{2}}1^{1}_{\underline{2}}2^{1}_{0}1^{2}_{0}1^{2}_{2}1^{3}_{4}$	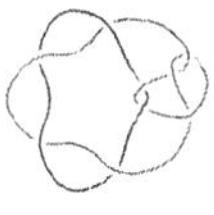	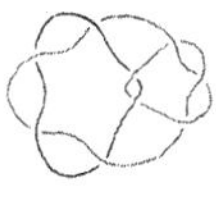	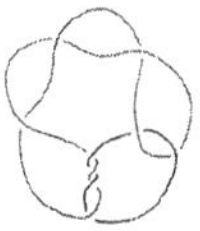
9^a_{29}: $1^{\underline{7}}_{\underline{20}}1^{\underline{6}}_{\underline{18}}1^{\underline{6}}_{\underline{16}}2^{\underline{5}}_{\underline{16}}1^{\underline{5}}_{\underline{14}}2^{\underline{4}}_{\underline{14}}2^{\underline{4}}_{\underline{12}}2^{\underline{3}}_{\underline{12}}2^{\underline{3}}_{\underline{10}}2^{\underline{2}}_{\underline{10}}3^{\underline{2}}_{\underline{8}}2^{\underline{1}}_{\underline{8}}1^{\underline{1}}_{\underline{6}}1^{0}_{\underline{6}}3^{0}_{\underline{4}}1^{1}_{\underline{4}}1^{2}_{0}$			
9^a_{30}: $1^{\underline{7}}_{\underline{18}}1^{\underline{6}}_{\underline{16}}1^{\underline{6}}_{\underline{14}}2^{\underline{5}}_{\underline{14}}1^{\underline{5}}_{\underline{12}}3^{\underline{4}}_{\underline{12}}2^{\underline{4}}_{\underline{10}}2^{\underline{3}}_{\underline{10}}3^{\underline{3}}_{\underline{8}}3^{\underline{2}}_{\underline{8}}3^{\underline{2}}_{\underline{6}}2^{\underline{1}}_{\underline{6}}2^{\underline{1}}_{\underline{4}}1^{0}_{\underline{4}}3^{0}_{\underline{2}}1^{1}_{\underline{2}}1^{2}_{2}$			
9^a_{31}: $1^{\underline{5}}_{\underline{12}}2^{\underline{4}}_{\underline{10}}1^{\underline{4}}_{\underline{8}}4^{\underline{3}}_{\underline{8}}2^{\underline{3}}_{\underline{6}}4^{\underline{2}}_{\underline{6}}4^{\underline{2}}_{\underline{4}}5^{\underline{1}}_{\underline{4}}4^{\underline{1}}_{\underline{2}}5^{0}_{\underline{2}}6^{0}_{0}3^{1}_{0}4^{1}_{2}3^{2}_{2}4^{2}_{4}1^{3}_{4}2^{3}_{6}1^{4}_{8}$	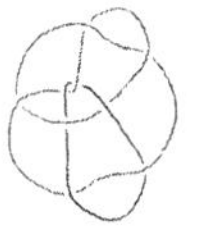	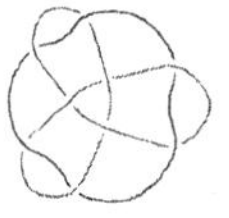	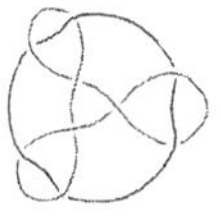
9^a_{32}: $1^{\underline{9}}_{\underline{22}}2^{\underline{8}}_{\underline{20}}1^{\underline{8}}_{\underline{18}}3^{\underline{7}}_{\underline{18}}2^{\underline{7}}_{\underline{16}}4^{\underline{6}}_{\underline{16}}4^{\underline{6}}_{\underline{14}}5^{\underline{5}}_{\underline{14}}3^{\underline{5}}_{\underline{12}}4^{\underline{4}}_{\underline{12}}5^{\underline{4}}_{\underline{10}}3^{\underline{3}}_{\underline{10}}4^{\underline{3}}_{\underline{8}}3^{\underline{2}}_{\underline{8}}3^{\underline{2}}_{\underline{6}}3^{\underline{1}}_{\underline{4}}1^{0}_{\underline{4}}1^{0}_{\underline{2}}$			
9^a_{33}: $1^{\underline{6}}_{\underline{14}}1^{\underline{6}}_{\underline{12}}2^{\underline{5}}_{\underline{12}}4^{\underline{4}}_{\underline{10}}2^{\underline{4}}_{\underline{8}}3^{\underline{3}}_{\underline{8}}4^{\underline{3}}_{\underline{6}}6^{\underline{2}}_{\underline{6}}3^{\underline{2}}_{\underline{4}}4^{\underline{1}}_{\underline{4}}6^{\underline{1}}_{\underline{2}}4^{0}_{\underline{2}}5^{0}_{0}3^{1}_{0}3^{1}_{2}1^{2}_{2}3^{2}_{4}1^{3}_{6}$			
9^a_{34}: $1^{\underline{7}}_{\underline{18}}1^{\underline{6}}_{\underline{16}}1^{\underline{6}}_{\underline{14}}2^{\underline{5}}_{\underline{14}}1^{\underline{5}}_{\underline{12}}4^{\underline{4}}_{\underline{12}}3^{\underline{4}}_{\underline{10}}3^{\underline{3}}_{\underline{10}}3^{\underline{3}}_{\underline{8}}3^{\underline{2}}_{\underline{8}}3^{\underline{2}}_{\underline{6}}2^{\underline{1}}_{\underline{6}}3^{\underline{1}}_{\underline{4}}2^{0}_{\underline{4}}3^{0}_{\underline{2}}1^{1}_{\underline{2}}1^{1}_{0}1^{2}_{2}$	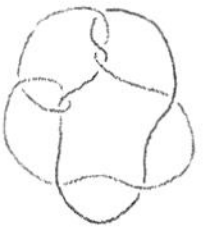		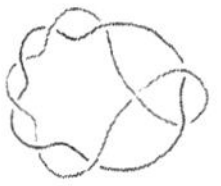
9^a_{35}: $1^{\underline{5}}_{\underline{10}}1^{\underline{4}}_{\underline{8}}1^{\underline{4}}_{\underline{6}}2^{\underline{3}}_{\underline{6}}1^{\underline{3}}_{\underline{4}}4^{\underline{2}}_{\underline{4}}2^{\underline{2}}_{\underline{2}}2^{\underline{1}}_{\underline{2}}4^{\underline{1}}_{0}5^{0}_{0}4^{0}_{2}3^{1}_{2}3^{1}_{4}2^{2}_{4}3^{2}_{6}1^{3}_{6}2^{3}_{8}1^{4}_{10}$			
9^a_{36}: $1^{\underline{2}}_{0}1^{\underline{1}}_{4}2^{0}_{4}1^{0}_{6}1^{1}_{6}1^{1}_{8}2^{2}_{8}1^{2}_{10}1^{3}_{10}2^{3}_{12}2^{4}_{12}2^{4}_{14}1^{5}_{14}1^{5}_{16}1^{6}_{16}1^{6}_{18}1^{7}_{20}$			
9^a_{37}: $1^{\underline{2}}_{\underline{2}}1^{\underline{1}}_{0}1^{\underline{1}}_{2}3^{0}_{2}2^{0}_{4}4^{1}_{4}2^{1}_{6}4^{2}_{6}4^{2}_{8}3^{3}_{8}4^{3}_{10}4^{4}_{10}4^{4}_{12}2^{5}_{12}3^{5}_{14}1^{6}_{14}2^{6}_{16}1^{7}_{18}$	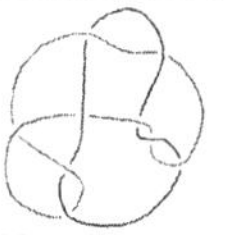	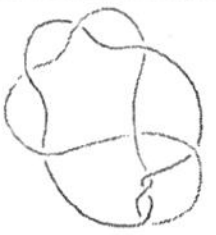	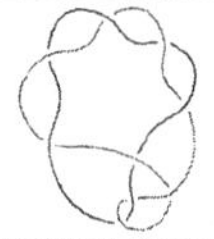
9^a_{38}: $1^{\underline{4}}_{\underline{8}}1^{\underline{3}}_{\underline{6}}1^{\underline{3}}_{\underline{4}}2^{\underline{2}}_{\underline{4}}1^{\underline{2}}_{\underline{2}}2^{\underline{1}}_{\underline{2}}2^{\underline{1}}_{0}4^{0}_{0}4^{0}_{2}3^{1}_{2}2^{1}_{4}2^{2}_{4}3^{2}_{6}1^{3}_{6}2^{3}_{8}1^{4}_{8}1^{4}_{10}1^{5}_{12}$			
9^a_{39}: $1^{\underline{6}}_{\underline{16}}1^{\underline{5}}_{\underline{14}}1^{\underline{5}}_{\underline{12}}2^{\underline{4}}_{\underline{12}}2^{\underline{4}}_{\underline{10}}3^{\underline{3}}_{\underline{10}}1^{\underline{3}}_{\underline{8}}2^{\underline{2}}_{\underline{8}}3^{\underline{2}}_{\underline{6}}2^{\underline{1}}_{\underline{6}}2^{\underline{1}}_{\underline{4}}2^{0}_{\underline{4}}3^{0}_{\underline{2}}1^{1}_{\underline{2}}1^{1}_{0}1^{2}_{0}1^{2}_{2}1^{3}_{4}$			
9^a_{40}: $1^{\underline{5}}_{\underline{14}}1^{\underline{4}}_{\underline{12}}1^{\underline{4}}_{\underline{10}}2^{\underline{3}}_{\underline{10}}1^{\underline{3}}_{\underline{8}}2^{\underline{2}}_{\underline{8}}2^{\underline{2}}_{\underline{6}}1^{\underline{1}}_{\underline{6}}2^{\underline{1}}_{\underline{4}}3^{0}_{\underline{4}}3^{0}_{\underline{2}}2^{1}_{\underline{2}}1^{1}_{0}2^{2}_{2}1^{3}_{2}1^{4}_{6}$	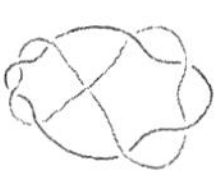	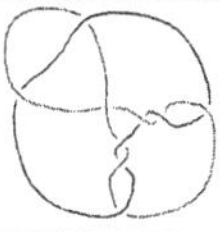	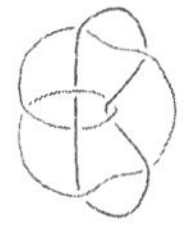
9^a_{41}: $1^{\underline{6}}_{\underline{16}}1^{\underline{5}}_{\underline{14}}1^{\underline{5}}_{\underline{12}}3^{\underline{4}}_{\underline{12}}2^{\underline{4}}_{\underline{10}}4^{\underline{3}}_{\underline{10}}2^{\underline{3}}_{\underline{8}}2^{\underline{2}}_{\underline{8}}4^{\underline{2}}_{\underline{6}}3^{\underline{1}}_{\underline{6}}2^{\underline{1}}_{\underline{4}}3^{0}_{\underline{4}}4^{0}_{\underline{2}}1^{1}_{\underline{2}}2^{1}_{0}1^{2}_{0}1^{2}_{2}1^{3}_{4}$			
9^a_{42}: $1^{\underline{4}}_{\underline{8}}2^{\underline{3}}_{\underline{6}}1^{\underline{3}}_{\underline{4}}4^{\underline{2}}_{\underline{4}}2^{\underline{2}}_{\underline{2}}4^{\underline{1}}_{\underline{2}}4^{\underline{1}}_{0}6^{0}_{0}6^{0}_{2}5^{1}_{2}4^{1}_{4}4^{2}_{4}5^{2}_{6}2^{3}_{6}4^{3}_{8}1^{4}_{8}2^{4}_{10}1^{5}_{12}$			
9^a_{43}: $1^{\underline{9}}_{\underline{23}}1^{\underline{8}}_{\underline{21}}1^{\underline{8}}_{\underline{19}}4^{\underline{7}}_{\underline{19}}1^{\underline{7}}_{\underline{17}}4^{\underline{6}}_{\underline{17}}4^{\underline{6}}_{\underline{15}}4^{\underline{5}}_{\underline{15}}4^{\underline{5}}_{\underline{13}}6^{\underline{4}}_{\underline{13}}7^{\underline{4}}_{\underline{11}}4^{\underline{3}}_{\underline{11}}3^{\underline{3}}_{\underline{9}}3^{\underline{2}}_{\underline{9}}4^{\underline{2}}_{\underline{7}}3^{\underline{1}}_{\underline{5}}1^{0}_{\underline{5}}1^{0}_{\underline{3}}$	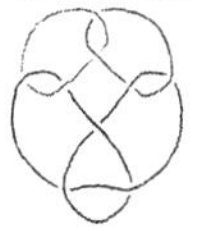	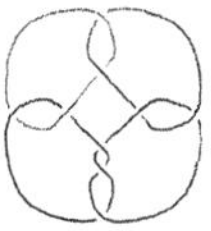	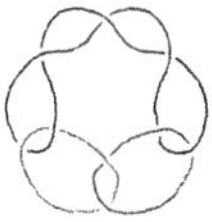
9^a_{44}: $1^{\underline{4}}_{\underline{7}}1^{\underline{4}}_{\underline{5}}1^{\underline{3}}_{\underline{5}}4^{\underline{2}}_{\underline{3}}1^{\underline{2}}_{\underline{1}}1^{\underline{1}}_{\underline{1}}4^{\underline{1}}_{1}7^{0}_{1}4^{0}_{3}3^{1}_{3}4^{1}_{5}4^{2}_{5}3^{2}_{7}2^{3}_{7}4^{3}_{9}1^{4}_{9}2^{4}_{11}1^{5}_{13}$			
9^a_{45}: $1^{\underline{4}}_{\underline{9}}1^{\underline{4}}_{\underline{7}}1^{\underline{3}}_{\underline{7}}3^{\underline{2}}_{\underline{5}}1^{\underline{2}}_{\underline{3}}1^{\underline{1}}_{\underline{3}}3^{\underline{1}}_{\underline{1}}6^{0}_{\underline{1}}4^{0}_{1}3^{1}_{1}3^{1}_{3}2^{2}_{3}3^{2}_{5}2^{3}_{5}2^{3}_{7}1^{4}_{7}2^{4}_{9}1^{5}_{11}$			
9^a_{46}: $1^{\underline{5}}_{\underline{11}}3^{\underline{4}}_{\underline{9}}1^{\underline{4}}_{\underline{7}}3^{\underline{3}}_{\underline{7}}3^{\underline{3}}_{\underline{5}}7^{\underline{2}}_{\underline{5}}5^{\underline{2}}_{\underline{3}}5^{\underline{1}}_{\underline{3}}5^{\underline{1}}_{\underline{1}}7^{0}_{\underline{1}}7^{0}_{1}4^{1}_{1}5^{1}_{3}3^{2}_{3}4^{2}_{5}1^{3}_{5}3^{3}_{7}1^{4}_{9}$	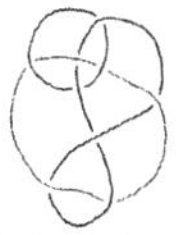		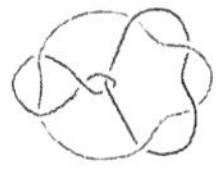
9^a_{47}: $1^{\underline{7}}_{\underline{17}}2^{\underline{6}}_{\underline{15}}1^{\underline{6}}_{\underline{13}}3^{\underline{5}}_{\underline{13}}2^{\underline{5}}_{\underline{11}}5^{\underline{4}}_{\underline{11}}4^{\underline{4}}_{\underline{9}}4^{\underline{3}}_{\underline{9}}4^{\underline{3}}_{\underline{7}}6^{\underline{2}}_{\underline{7}}6^{\underline{2}}_{\underline{5}}3^{\underline{1}}_{\underline{5}}4^{\underline{1}}_{\underline{3}}3^{0}_{\underline{3}}4^{0}_{\underline{1}}1^{1}_{\underline{1}}2^{1}_{1}1^{2}_{3}$			
9^a_{48}: $1^{\underline{4}}_{\underline{7}}1^{\underline{4}}_{\underline{5}}1^{\underline{3}}_{\underline{5}}3^{\underline{2}}_{\underline{3}}1^{\underline{2}}_{\underline{1}}1^{\underline{1}}_{\underline{1}}3^{\underline{1}}_{1}5^{0}_{1}2^{0}_{3}2^{1}_{3}4^{1}_{5}4^{2}_{5}4^{2}_{7}2^{3}_{7}2^{3}_{9}1^{4}_{9}2^{4}_{11}1^{5}_{13}$			

L: KH	1st line	2nd line	3rd line
9^a_{49}: $1^{\underline{4}}_{\underline{9}}1^{\underline{4}}_{\underline{7}}2^{\underline{3}}_{\underline{7}}3^{\underline{2}}_{\underline{5}}2^{\underline{2}}_{\underline{3}}3^{\underline{1}}_{\underline{3}}3^{\underline{1}}_{\underline{1}}5^{0}_{\underline{1}}4^0_1 3^1_1 4^1_3 4^2_3 5^2_5 2^3_5 2^3_7 1^4_7 2^4_9 1^5_{11}$ 9^a_{50}: $1^{\underline{4}}_{\underline{7}}1^{\underline{3}}_{\underline{5}}1^{\underline{3}}_{\underline{3}}4^{\underline{2}}_{\underline{3}}2^{\underline{2}}_{\underline{1}}2^{\underline{1}}_{\underline{1}}3^{\underline{1}}_1 6^0_1 4^0_3 3^1_3 4^1_5 4^2_5 4^2_7 2^3_7 3^3_9 1^4_9 2^4_{11} 1^5_{13}$ 9^a_{51}: $1^{\underline{3}}_{\underline{5}}3^{\underline{2}}_{\underline{3}}1^{\underline{2}}_{\underline{1}}3^{\underline{1}}_{\underline{1}}3^{\underline{1}}_1 6^0_1 5^0_3 6^1_3 4^1_5 5^2_5 7^2_7 4^3_7 4^3_9 3^4_9 4^4_{11} 3^5_{13} 1^6_{13} 1^6_{15}$	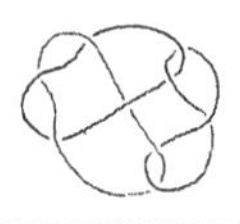		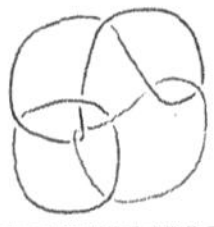
9^a_{52}: $1^{\underline{7}}_{\underline{17}}2^{\underline{6}}_{\underline{15}}1^{\underline{6}}_{\underline{13}}3^{\underline{5}}_{\underline{13}}2^{\underline{5}}_{\underline{11}}5^{\underline{4}}_{\underline{11}}5^{\underline{4}}_{\underline{9}}4^{\underline{3}}_{\underline{9}}3^{\underline{3}}_{\underline{7}}5^{\underline{2}}_{\underline{7}}4^{\underline{2}}_{\underline{5}}2^{\underline{1}}_{\underline{5}}5^{\underline{1}}_{\underline{3}}3^0_{\underline{3}}4^0_{\underline{1}}1^1_{\underline{1}}1^1_1 1^2_3$ 9^a_{53}: $1^{\underline{5}}_{\underline{11}}2^{\underline{4}}_{\underline{9}}1^{\underline{4}}_{\underline{7}}4^{\underline{3}}_{\underline{7}}2^{\underline{3}}_{\underline{5}}6^{\underline{2}}_{\underline{5}}4^{\underline{2}}_{\underline{3}}4^{\underline{1}}_{\underline{3}}6^{\underline{1}}_{\underline{1}}8^0_{\underline{1}}8^0_1 5^1_1 4^1_3 3^2_3 5^2_5 1^3_5 3^3_7 1^4_9$ 9^a_{54}: $1^{\underline{4}}_{\underline{7}}1^{\underline{3}}_{\underline{5}}1^{\underline{3}}_{\underline{3}}4^{\underline{2}}_{\underline{3}}1^{\underline{2}}_{\underline{1}}2^{\underline{1}}_{\underline{1}}4^{\underline{1}}_1 7^0_1 6^0_3 4^1_3 3^1_5 4^2_5 4^2_7 2^3_7 4^3_9 1^4_9 2^4_{11} 1^5_{13}$	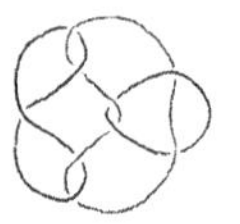	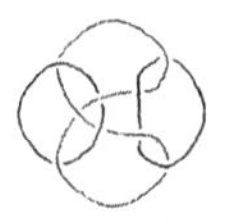	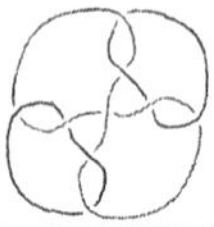
9^a_{55}: $1^{\underline{6}}_{\underline{14}}1^{\underline{5}}_{\underline{10}}5^{\underline{4}}_{\underline{10}}4^{\underline{4}}_{\underline{8}}4^{\underline{3}}_{\underline{8}}1^{\underline{3}}_{\underline{6}}6^{\underline{2}}_{\underline{6}}4^{\underline{2}}_{\underline{4}}1^{\underline{1}}_{\underline{4}}6^{\underline{1}}_{\underline{2}}7^0_{\underline{2}}5^0_0 3^1_0 3^1_2 1^2_2 3^2_4 1^3_6$	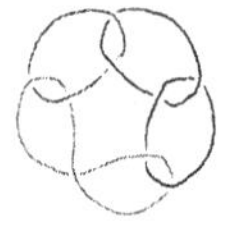		

3.9. 9 Crossing Non-Alternating Links.

L: KH	1st line	2nd line	3rd line
9^n_1: $1^{\underline{7}}_{\underline{20}}1^{\underline{6}}_{\underline{16}}2^{\underline{5}}_{\underline{16}}1^{\underline{4}}_{\underline{14}}3^{\underline{4}}_{\underline{12}}1^{\underline{4}}_{\underline{10}}1^{\underline{3}}_{\underline{12}}1^{\underline{3}}_{\underline{10}}1^{\underline{2}}_{\underline{10}}1^{\underline{2}}_{\underline{8}}1^{\underline{1}}_{\underline{6}}1^0_{\underline{6}}1^0_{\underline{4}}$ 9^n_2: $1^{\underline{7}}_{\underline{16}}1^{\underline{6}}_{\underline{12}}2^{\underline{5}}_{\underline{12}}1^{\underline{4}}_{\underline{10}}2^{\underline{4}}_{\underline{8}}1^{\underline{3}}_{\underline{8}}1^{\underline{3}}_{\underline{6}}2^{\underline{2}}_{\underline{6}}1^{\underline{2}}_{\underline{4}}2^{\underline{1}}_{\underline{2}}2^0_{\underline{2}}2^0_0$ 9^n_3: $1^{\underline{7}}_{\underline{16}}1^{\underline{6}}_{\underline{12}}1^{\underline{5}}_{\underline{12}}1^{\underline{4}}_{\underline{10}}1^{\underline{4}}_{\underline{8}}1^{\underline{3}}_{\underline{10}}1^{\underline{3}}_{\underline{8}}1^{\underline{3}}_{\underline{6}}2^{\underline{2}}_{\underline{6}}1^{\underline{2}}_{\underline{4}}1^{\underline{1}}_{\underline{2}}1^0_{\underline{4}}2^0_{\underline{2}}1^0_0$	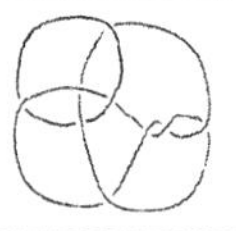	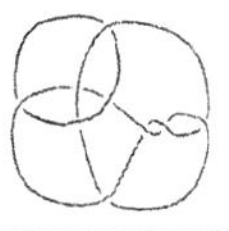	
9^n_4: $1^{\underline{7}}_{\underline{22}}1^{\underline{7}}_{\underline{20}}1^{\underline{6}}_{\underline{18}}1^{\underline{6}}_{\underline{16}}1^{\underline{5}}_{\underline{18}}2^{\underline{4}}_{\underline{14}}1^{\underline{4}}_{\underline{12}}1^{\underline{3}}_{\underline{14}}1^{\underline{2}}_{\underline{10}}1^0_{\underline{8}}1^0_{\underline{6}}$ 9^n_5: $1^{\underline{7}}_{\underline{18}}1^{\underline{6}}_{\underline{14}}2^{\underline{5}}_{\underline{14}}1^{\underline{4}}_{\underline{12}}2^{\underline{4}}_{\underline{10}}2^{\underline{3}}_{\underline{10}}1^{\underline{3}}_{\underline{8}}1^{\underline{2}}_{\underline{8}}2^{\underline{2}}_{\underline{6}}1^{\underline{1}}_{\underline{4}}2^0_{\underline{4}}2^0_{\underline{2}}$ 9^n_6: $1^{\underline{7}}_{\underline{18}}1^{\underline{6}}_{\underline{16}}1^{\underline{6}}_{\underline{14}}3^{\underline{5}}_{\underline{14}}1^{\underline{5}}_{\underline{12}}1^{\underline{4}}_{\underline{12}}3^{\underline{4}}_{\underline{10}}3^{\underline{3}}_{\underline{10}}1^{\underline{3}}_{\underline{8}}2^{\underline{2}}_{\underline{8}}3^{\underline{2}}_{\underline{6}}2^{\underline{1}}_{\underline{4}}2^0_{\underline{4}}2^0_{\underline{2}}$	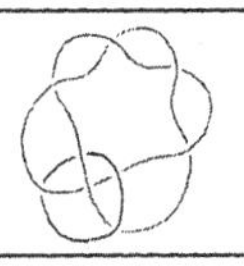	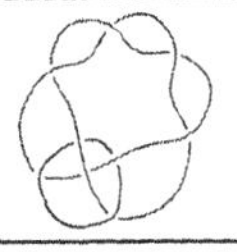	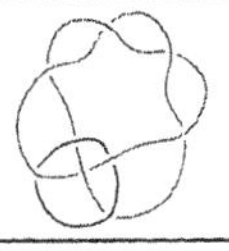
9^n_7: $2^{\underline{7}}_{\underline{20}}1^{\underline{6}}_{\underline{18}}2^{\underline{6}}_{\underline{16}}2^{\underline{5}}_{\underline{16}}1^{\underline{5}}_{\underline{14}}2^{\underline{4}}_{\underline{14}}3^{\underline{4}}_{\underline{12}}2^{\underline{3}}_{\underline{12}}1^{\underline{3}}_{\underline{10}}1^{\underline{2}}_{\underline{10}}2^{\underline{2}}_{\underline{8}}1^{\underline{1}}_{\underline{6}}1^0_{\underline{6}}1^0_{\underline{4}}$ 9^n_8: $2^{\underline{5}}_{\underline{14}}1^{\underline{4}}_{\underline{12}}2^{\underline{4}}_{\underline{10}}3^{\underline{3}}_{\underline{10}}1^{\underline{3}}_{\underline{8}}2^{\underline{2}}_{\underline{8}}3^{\underline{2}}_{\underline{6}}1^{\underline{1}}_{\underline{6}}2^{\underline{1}}_{\underline{4}}3^0_{\underline{4}}3^0_{\underline{2}}1^1_{\underline{2}}1^1_0 1^2_2$ 9^n_9: $1^{\underline{9}}_{\underline{22}}1^{\underline{8}}_{\underline{18}}1^{\underline{7}}_{\underline{18}}1^{\underline{6}}_{\underline{16}}1^{\underline{6}}_{\underline{14}}1^{\underline{5}}_{\underline{16}}1^{\underline{5}}_{\underline{12}}3^{\underline{4}}_{\underline{12}}1^{\underline{4}}_{\underline{10}}1^{\underline{3}}_{\underline{12}}1^{\underline{3}}_{\underline{8}}1^{\underline{2}}_{\underline{8}}1^0_{\underline{6}}1^0_{\underline{4}}$	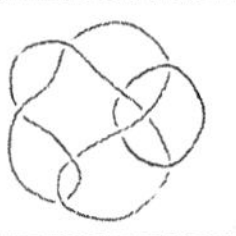	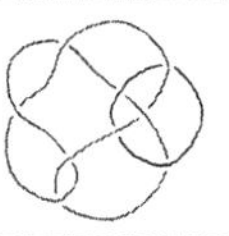	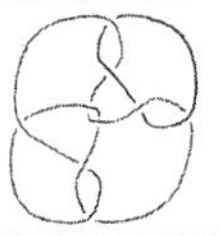
9^n_{10}: $2^{\underline{4}}_{\underline{10}}1^{\underline{4}}_{\underline{8}}2^{\underline{3}}_{\underline{8}}1^{\underline{3}}_{\underline{6}}3^{\underline{2}}_{\underline{6}}2^{\underline{2}}_{\underline{4}}2^{\underline{1}}_{\underline{4}}3^{\underline{1}}_{\underline{2}}3^0_{\underline{2}}3^0_0 2^1_0 2^1_2 1^2_2 2^2_4 1^3_6$ 9^n_{11}: $1^{\underline{5}}_{\underline{14}}2^{\underline{4}}_{\underline{12}}2^{\underline{4}}_{\underline{10}}2^{\underline{3}}_{\underline{10}}1^{\underline{3}}_{\underline{8}}2^{\underline{2}}_{\underline{8}}2^{\underline{2}}_{\underline{6}}1^{\underline{1}}_{\underline{6}}2^{\underline{1}}_{\underline{4}}2^0_{\underline{4}}2^0_{\underline{2}}1^1_{\underline{2}}1^1_0 1^2_2$ 9^n_{12}: $1^{\underline{4}}_{\underline{6}}1^{\underline{4}}_{\underline{4}}1^{\underline{2}}_{\underline{2}}1^{\underline{1}}_2 2^0_0 2^0_2 1^1_2 1^1_4 1^1_6 1^2_4 1^2_6 1^3_8 1^4_8 1^5_{12}$	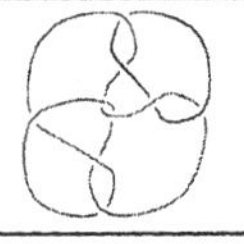	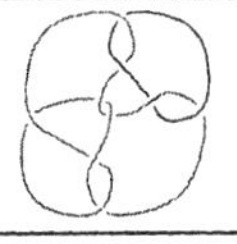	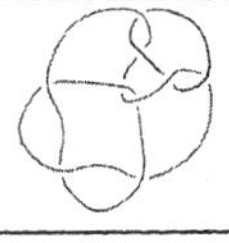
9^n_{13}: $1^{\underline{6}}_{\underline{14}}1^{\underline{5}}_{\underline{12}}1^{\underline{5}}_{\underline{10}}1^{\underline{4}}_{\underline{10}}1^{\underline{4}}_{\underline{8}}2^{\underline{3}}_{\underline{8}}1^{\underline{3}}_{\underline{6}}2^{\underline{2}}_{\underline{6}}3^{\underline{2}}_{\underline{4}}1^{\underline{1}}_{\underline{4}}1^{\underline{1}}_{\underline{2}}2^0_{\underline{2}}2^0_0 1^1_2$ 9^n_{14}: $1^{\underline{2}}_{\underline{6}}1^{\underline{2}}_{\underline{4}}1^{\underline{1}}_{\underline{4}}1^0_{\underline{2}}2^0_0 1^1_0 1^2_2 1^2_4 1^3_6 1^4_6 1^5_{10}$ 9^n_{15}: $1^{\underline{6}}_{\underline{18}}1^{\underline{6}}_{\underline{16}}1^{\underline{5}}_{\underline{18}}1^{\underline{5}}_{\underline{16}}1^{\underline{4}}_{\underline{14}}1^{\underline{4}}_{\underline{12}}1^{\underline{3}}_{\underline{14}}1^{\underline{2}}_{\underline{10}}1^0_{\underline{8}}1^0_{\underline{6}}$	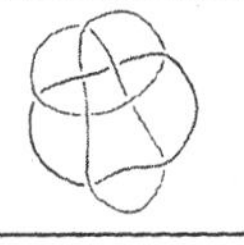	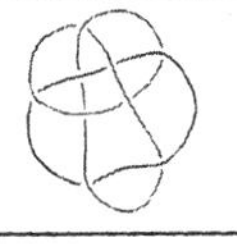	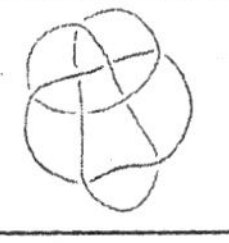
9^n_{16}: $1^{\underline{6}}_{\underline{14}}1^{\underline{6}}_{\underline{12}}1^{\underline{5}}_{\underline{12}}1^{\underline{4}}_{\underline{10}}1^{\underline{4}}_{\underline{8}}1^{\underline{3}}_{\underline{8}}1^{\underline{3}}_{\underline{6}}2^{\underline{2}}_{\underline{6}}1^{\underline{2}}_{\underline{4}}2^{\underline{1}}_{\underline{2}}2^0_{\underline{2}}1^0_0 1^1_2$ 9^n_{17}: $1^{\underline{7}}_{\underline{18}}1^{\underline{6}}_{\underline{16}}1^{\underline{6}}_{\underline{14}}2^{\underline{5}}_{\underline{14}}1^{\underline{5}}_{\underline{12}}2^{\underline{4}}_{\underline{12}}2^{\underline{4}}_{\underline{10}}2^{\underline{3}}_{\underline{10}}2^{\underline{3}}_{\underline{8}}2^{\underline{2}}_{\underline{8}}3^{\underline{2}}_{\underline{6}}1^{\underline{1}}_{\underline{6}}1^{\underline{1}}_{\underline{4}}1^0_{\underline{4}}2^0_{\underline{2}}$ 9^n_{18}: $1^{\underline{8}}_{\underline{22}}1^{\underline{8}}_{\underline{20}}1^{\underline{6}}_{\underline{18}}1^{\underline{5}}_{\underline{18}}1^{\underline{5}}_{\underline{16}}1^{\underline{5}}_{\underline{14}}1^{\underline{4}}_{\underline{14}}1^{\underline{4}}_{\underline{12}}1^{\underline{3}}_{\underline{14}}1^{\underline{2}}_{\underline{10}}1^0_{\underline{8}}1^0_{\underline{6}}$	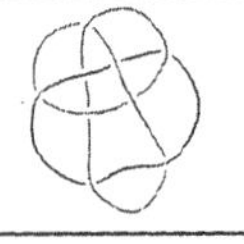	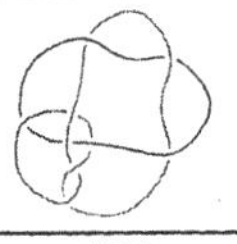	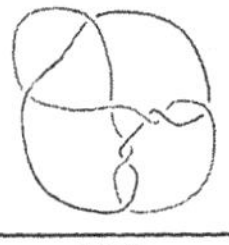
9^n_{19}: $1^{\underline{8}}_{\underline{20}}1^{\underline{8}}_{\underline{18}}1^{\underline{6}}_{\underline{16}}1^{\underline{5}}_{\underline{16}}1^{\underline{5}}_{\underline{12}}2^{\underline{4}}_{\underline{12}}1^{\underline{3}}_{\underline{12}}1^{\underline{3}}_{\underline{8}}1^{\underline{2}}_{\underline{8}}1^0_{\underline{6}}1^0_{\underline{4}}$ 9^n_{20}: $3^0_1 3^0_3 3^1_3 3^2_5 3^2_7 1^3_7 3^3_9 4^4_9 2^4_{11} 1^5_{11} 3^5_{13} 1^6_{13} 1^6_{15} 1^7_{17}$ 9^n_{21}: $1^{\underline{5}}_{\underline{11}}1^{\underline{4}}_{\underline{7}}1^{\underline{3}}_{\underline{7}}1^{\underline{2}}_{\underline{5}}1^{\underline{2}}_{\underline{3}}1^{\underline{1}}_{\underline{5}}1^{\underline{1}}_{\underline{1}}4^0_{\underline{1}}3^0_1 1^1_{\underline{1}}1^2_3 1^4_5 1^4_7$	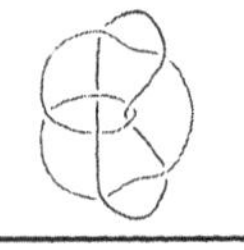	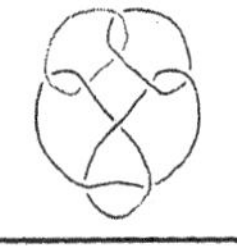	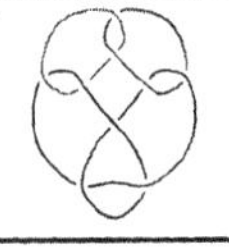

L: KH	1st line	2nd line	3rd line
9^n_{22}: $1^{\underline{6}}_{\underline{13}}1^{\underline{5}}_{\underline{9}}4^{\underline{4}}_{\underline{9}}3^{\underline{4}}_{\underline{7}}1^{\underline{3}}_{\underline{7}}1^{\underline{3}}_{\underline{5}}3^{\underline{2}}_{\underline{5}}1^{\underline{2}}_{\underline{3}}1^{\underline{1}}_{\underline{3}}3^{\underline{1}}_{\underline{1}}1^{0}_{\underline{3}}3^{0}_{\underline{1}}1^{0}_{1}2^{1}_{3}$	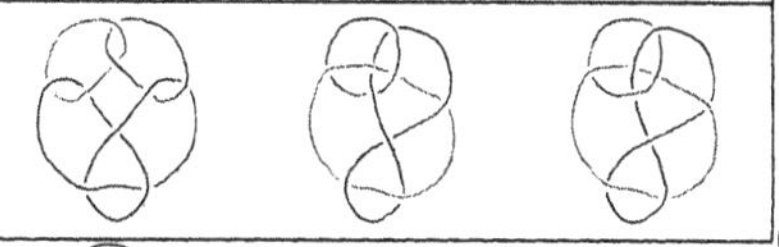		
9^n_{23}: $1^{\underline{5}}_{\underline{15}}1^{\underline{4}}_{\underline{13}}2^{\underline{4}}_{\underline{11}}1^{\underline{4}}_{\underline{9}}1^{\underline{3}}_{\underline{11}}1^{\underline{3}}_{\underline{9}}2^{\underline{2}}_{\underline{9}}3^{\underline{2}}_{\underline{7}}1^{\underline{2}}_{\underline{5}}1^{\underline{1}}_{\underline{7}}1^{\underline{1}}_{\underline{5}}1^{0}_{\underline{5}}2^{0}_{\underline{3}}1^{1}_{\underline{3}}1^{2}_{1}$			
9^n_{24}: $1^{\underline{5}}_{\underline{11}}1^{\underline{4}}_{\underline{9}}1^{\underline{4}}_{\underline{7}}1^{\underline{3}}_{\underline{7}}1^{\underline{3}}_{\underline{5}}3^{\underline{2}}_{\underline{5}}2^{\underline{2}}_{\underline{3}}1^{\underline{1}}_{\underline{3}}2^{\underline{1}}_{\underline{1}}3^{0}_{\underline{1}}3^{0}_{1}1^{1}_{1}1^{1}_{3}1^{2}_{3}2^{2}_{5}$			
9^n_{25}: $1^{\underline{5}}_{\underline{11}}1^{\underline{4}}_{\underline{9}}1^{\underline{4}}_{\underline{7}}1^{\underline{3}}_{\underline{7}}1^{\underline{3}}_{\underline{5}}2^{\underline{2}}_{\underline{5}}1^{\underline{2}}_{\underline{3}}1^{\underline{1}}_{\underline{3}}2^{\underline{1}}_{\underline{1}}1^{0}_{\underline{3}}4^{0}_{\underline{1}}4^{0}_{1}1^{1}_{1}1^{2}_{5}$	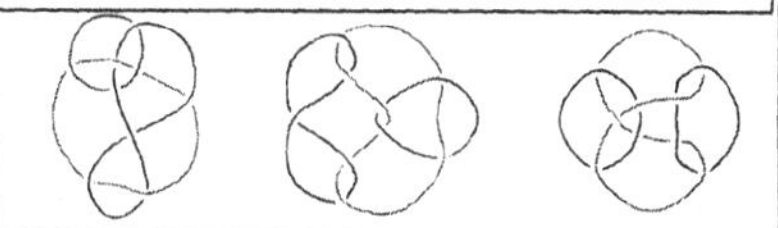		
9^n_{26}: $1^{\underline{6}}_{\underline{13}}1^{\underline{5}}_{\underline{9}}3^{\underline{4}}_{\underline{9}}2^{\underline{4}}_{\underline{7}}1^{\underline{3}}_{\underline{7}}1^{\underline{3}}_{\underline{5}}2^{\underline{2}}_{\underline{5}}1^{\underline{2}}_{\underline{3}}1^{\underline{1}}_{\underline{3}}2^{\underline{1}}_{\underline{1}}1^{0}_{\underline{3}}3^{0}_{\underline{1}}2^{0}_{1}1^{1}_{3}$			
9^n_{27}: $1^{\underline{5}}_{\underline{11}}1^{\underline{4}}_{\underline{7}}1^{\underline{3}}_{\underline{7}}1^{\underline{2}}_{\underline{7}}1^{\underline{2}}_{\underline{5}}1^{\underline{2}}_{\underline{3}}1^{\underline{1}}_{\underline{3}}1^{\underline{1}}_{\underline{1}}2^{0}_{\underline{3}}4^{0}_{\underline{1}}3^{0}_{1}1^{1}_{\underline{1}}1^{1}_{1}1^{2}_{3}1^{3}_{3}1^{4}_{7}$			
9^n_{28}: $1^{\underline{3}}_{\underline{5}}2^{\underline{2}}_{\underline{3}}1^{\underline{2}}_{\underline{1}}2^{\underline{1}}_{\underline{1}}2^{\underline{1}}_{1}4^{0}_{1}4^{0}_{3}3^{1}_{3}2^{1}_{5}3^{2}_{5}3^{2}_{7}1^{3}_{7}3^{3}_{9}2^{4}_{9}3^{4}_{11}$	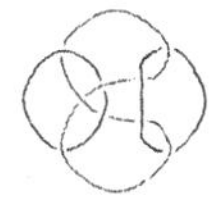		

3.10. 10 Crossing Alternating Links.

L: KH	1st line	2nd line	3rd line
10^a_{1}: $1^{\underline{6}}_{\underline{14}}4^{\underline{5}}_{\underline{12}}1^{\underline{5}}_{\underline{10}}5^{\underline{4}}_{\underline{10}}4^{\underline{4}}_{\underline{8}}8^{\underline{3}}_{\underline{8}}5^{\underline{3}}_{\underline{6}}9^{\underline{2}}_{\underline{6}}8^{\underline{2}}_{\underline{4}}8^{\underline{1}}_{\underline{4}}9^{\underline{1}}_{\underline{2}}9^{0}_{\underline{2}}10^{0}_{0}5^{1}_{0}7^{1}_{2}3^{2}_{2}5^{2}_{4}1^{3}_{4}3^{3}_{6}1^{4}_{8}$	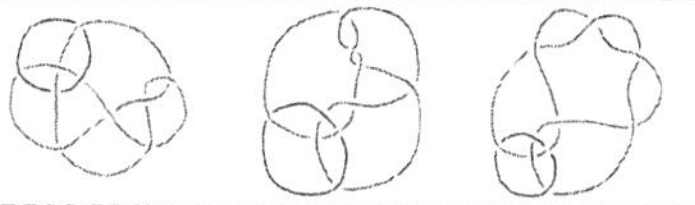		
10^a_{2}: $1^{\underline{7}}_{\underline{18}}3^{\underline{6}}_{\underline{16}}1^{\underline{6}}_{\underline{14}}5^{\underline{5}}_{\underline{14}}3^{\underline{5}}_{\underline{12}}6^{\underline{4}}_{\underline{12}}5^{\underline{4}}_{\underline{10}}8^{\underline{3}}_{\underline{10}}6^{\underline{3}}_{\underline{8}}7^{\underline{2}}_{\underline{8}}8^{\underline{2}}_{\underline{6}}6^{\underline{1}}_{\underline{6}}7^{\underline{1}}_{\underline{4}}5^{0}_{\underline{4}}8^{0}_{\underline{2}}3^{1}_{\underline{2}}3^{1}_{0}1^{2}_{0}3^{2}_{2}1^{3}_{4}$			
10^a_{3}: $1^{\underline{7}}_{\underline{16}}2^{\underline{6}}_{\underline{14}}1^{\underline{6}}_{\underline{12}}3^{\underline{5}}_{\underline{12}}2^{\underline{5}}_{\underline{10}}5^{\underline{4}}_{\underline{10}}3^{\underline{4}}_{\underline{8}}6^{\underline{3}}_{\underline{8}}5^{\underline{3}}_{\underline{6}}6^{\underline{2}}_{\underline{6}}6^{\underline{2}}_{\underline{4}}5^{\underline{1}}_{\underline{4}}6^{\underline{1}}_{\underline{2}}5^{0}_{\underline{2}}7^{0}_{0}3^{1}_{0}3^{1}_{2}1^{2}_{2}3^{2}_{4}1^{3}_{6}$			
10^a_{4}: $1^{\underline{5}}_{\underline{10}}3^{\underline{4}}_{\underline{8}}1^{\underline{4}}_{\underline{6}}4^{\underline{3}}_{\underline{6}}3^{\underline{3}}_{\underline{4}}7^{\underline{2}}_{\underline{4}}4^{\underline{2}}_{\underline{2}}6^{\underline{1}}_{\underline{2}}7^{\underline{1}}_{0}9^{0}_{0}8^{0}_{2}7^{1}_{2}7^{1}_{4}4^{2}_{4}7^{2}_{6}3^{3}_{6}4^{3}_{8}1^{4}_{8}3^{4}_{10}1^{5}_{12}$	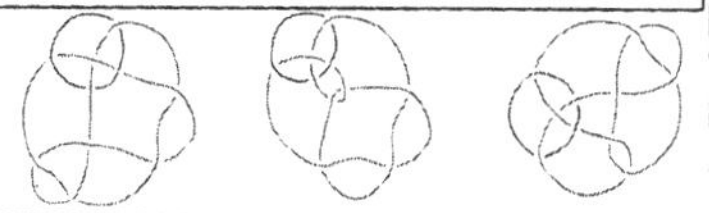		
10^a_{5}: $1^{\underline{5}}_{\underline{10}}2^{\underline{4}}_{\underline{8}}1^{\underline{4}}_{\underline{6}}4^{\underline{3}}_{\underline{6}}2^{\underline{3}}_{\underline{4}}7^{\underline{2}}_{\underline{4}}4^{\underline{2}}_{\underline{2}}6^{\underline{1}}_{\underline{2}}7^{\underline{1}}_{0}9^{0}_{0}8^{0}_{2}7^{1}_{2}7^{1}_{4}5^{2}_{4}7^{2}_{6}3^{3}_{6}5^{3}_{8}1^{4}_{8}3^{4}_{10}1^{5}_{12}$			
10^a_{6}: $1^{\underline{6}}_{\underline{14}}3^{\underline{5}}_{\underline{12}}1^{\underline{5}}_{\underline{10}}5^{\underline{4}}_{\underline{10}}3^{\underline{4}}_{\underline{8}}8^{\underline{3}}_{\underline{8}}5^{\underline{3}}_{\underline{6}}9^{\underline{2}}_{\underline{6}}8^{\underline{2}}_{\underline{4}}8^{\underline{1}}_{\underline{4}}9^{\underline{1}}_{\underline{2}}9^{0}_{\underline{2}}10^{0}_{0}6^{1}_{0}7^{1}_{2}3^{2}_{2}6^{2}_{4}1^{3}_{4}3^{3}_{6}1^{4}_{8}$			
10^a_{7}: $1^{\underline{5}}_{\underline{12}}2^{\underline{4}}_{\underline{10}}1^{\underline{4}}_{\underline{8}}3^{\underline{3}}_{\underline{8}}2^{\underline{3}}_{\underline{6}}6^{\underline{2}}_{\underline{6}}3^{\underline{2}}_{\underline{4}}5^{\underline{1}}_{\underline{4}}6^{\underline{1}}_{\underline{2}}7^{0}_{\underline{2}}7^{0}_{0}6^{1}_{0}5^{1}_{2}3^{2}_{2}6^{2}_{4}3^{3}_{4}3^{3}_{6}1^{4}_{6}3^{4}_{8}1^{5}_{10}$			
10^a_{8}: $1^{\underline{5}}_{\underline{10}}2^{\underline{4}}_{\underline{8}}1^{\underline{4}}_{\underline{6}}4^{\underline{3}}_{\underline{6}}2^{\underline{3}}_{\underline{4}}6^{\underline{2}}_{\underline{4}}4^{\underline{2}}_{\underline{2}}6^{\underline{1}}_{\underline{2}}6^{\underline{1}}_{0}8^{0}_{0}8^{0}_{2}6^{1}_{2}6^{1}_{4}5^{2}_{4}6^{2}_{6}2^{3}_{6}5^{3}_{8}1^{4}_{8}2^{4}_{10}1^{5}_{12}$			
10^a_{9}: $1^{\underline{4}}_{\underline{6}}1^{\underline{3}}_{\underline{4}}1^{\underline{3}}_{\underline{2}}4^{\underline{2}}_{\underline{2}}1^{\underline{2}}_{0}3^{\underline{1}}_{0}4^{\underline{1}}_{2}7^{0}_{2}5^{0}_{4}5^{1}_{4}5^{1}_{6}5^{2}_{6}5^{2}_{8}4^{3}_{8}5^{3}_{10}2^{4}_{10}4^{4}_{12}1^{5}_{12}2^{5}_{14}1^{6}_{16}$			
10^a_{10}: $1^{\underline{4}}_{\underline{8}}1^{\underline{3}}_{\underline{6}}1^{\underline{3}}_{\underline{4}}4^{\underline{2}}_{\underline{4}}1^{\underline{2}}_{\underline{2}}3^{\underline{1}}_{\underline{2}}4^{\underline{1}}_{0}7^{0}_{0}5^{0}_{2}6^{1}_{2}5^{1}_{4}4^{2}_{4}6^{2}_{6}4^{3}_{6}4^{3}_{8}2^{4}_{8}4^{4}_{10}1^{5}_{10}2^{5}_{12}1^{6}_{14}$	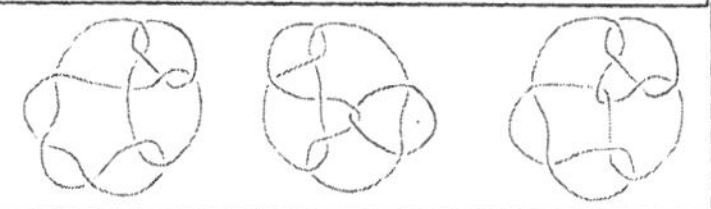		
10^a_{11}: $1^{\underline{8}}_{\underline{20}}2^{\underline{7}}_{\underline{18}}1^{\underline{7}}_{\underline{16}}5^{\underline{6}}_{\underline{16}}2^{\underline{6}}_{\underline{14}}6^{\underline{5}}_{\underline{14}}5^{\underline{5}}_{\underline{12}}7^{\underline{4}}_{\underline{12}}7^{\underline{4}}_{\underline{10}}8^{\underline{3}}_{\underline{10}}6^{\underline{3}}_{\underline{8}}6^{\underline{2}}_{\underline{8}}8^{\underline{2}}_{\underline{6}}4^{\underline{1}}_{\underline{6}}6^{\underline{1}}_{\underline{4}}3^{0}_{\underline{4}}5^{0}_{\underline{2}}1^{1}_{\underline{2}}2^{1}_{0}1^{2}_{2}$			
10^a_{12}: $1^{\underline{6}}_{\underline{14}}1^{\underline{5}}_{\underline{12}}1^{\underline{5}}_{\underline{10}}5^{\underline{4}}_{\underline{10}}2^{\underline{4}}_{\underline{8}}5^{\underline{3}}_{\underline{8}}4^{\underline{3}}_{\underline{6}}7^{\underline{2}}_{\underline{6}}5^{\underline{2}}_{\underline{4}}6^{\underline{1}}_{\underline{4}}7^{\underline{1}}_{\underline{2}}6^{0}_{\underline{2}}7^{0}_{0}5^{1}_{0}5^{1}_{2}2^{2}_{2}5^{2}_{4}1^{3}_{4}2^{3}_{6}1^{4}_{8}$			
10^a_{13}: $1^{\underline{7}}_{\underline{16}}2^{\underline{6}}_{\underline{14}}1^{\underline{6}}_{\underline{12}}4^{\underline{5}}_{\underline{12}}2^{\underline{5}}_{\underline{10}}6^{\underline{4}}_{\underline{10}}5^{\underline{4}}_{\underline{8}}6^{\underline{3}}_{\underline{8}}5^{\underline{3}}_{\underline{6}}7^{\underline{2}}_{\underline{6}}6^{\underline{2}}_{\underline{4}}5^{\underline{1}}_{\underline{4}}7^{\underline{1}}_{\underline{2}}5^{0}_{\underline{2}}6^{0}_{0}2^{1}_{0}4^{1}_{2}1^{2}_{2}2^{2}_{4}1^{3}_{6}$	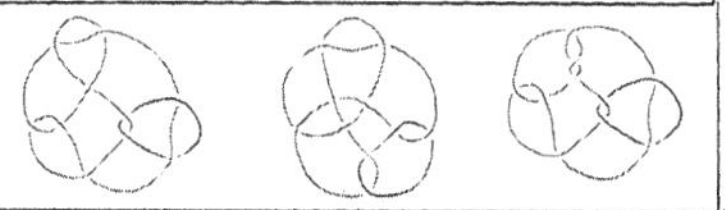		
10^a_{14}: $1^{\underline{6}}_{\underline{14}}2^{\underline{5}}_{\underline{12}}1^{\underline{5}}_{\underline{10}}5^{\underline{4}}_{\underline{10}}2^{\underline{4}}_{\underline{8}}7^{\underline{3}}_{\underline{8}}5^{\underline{3}}_{\underline{6}}8^{\underline{2}}_{\underline{6}}7^{\underline{2}}_{\underline{4}}8^{\underline{1}}_{\underline{4}}8^{\underline{1}}_{\underline{2}}8^{0}_{\underline{2}}10^{0}_{0}6^{1}_{0}6^{1}_{2}3^{2}_{2}6^{2}_{4}1^{3}_{4}3^{3}_{6}1^{4}_{8}$			
10^a_{15}: $1^{\underline{8}}_{\underline{20}}2^{\underline{7}}_{\underline{18}}1^{\underline{7}}_{\underline{16}}4^{\underline{6}}_{\underline{16}}2^{\underline{6}}_{\underline{14}}5^{\underline{5}}_{\underline{14}}4^{\underline{5}}_{\underline{12}}6^{\underline{4}}_{\underline{12}}6^{\underline{4}}_{\underline{10}}6^{\underline{3}}_{\underline{10}}5^{\underline{3}}_{\underline{8}}5^{\underline{2}}_{\underline{8}}6^{\underline{2}}_{\underline{6}}3^{\underline{1}}_{\underline{6}}5^{\underline{1}}_{\underline{4}}2^{0}_{\underline{4}}4^{0}_{\underline{2}}1^{1}_{\underline{2}}1^{1}_{0}1^{2}_{2}$			
10^a_{16}: $1^{\underline{8}}_{\underline{18}}1^{\underline{7}}_{\underline{16}}1^{\underline{7}}_{\underline{14}}3^{\underline{6}}_{\underline{14}}1^{\underline{6}}_{\underline{12}}2^{\underline{5}}_{\underline{12}}3^{\underline{5}}_{\underline{10}}4^{\underline{4}}_{\underline{10}}3^{\underline{4}}_{\underline{8}}4^{\underline{3}}_{\underline{8}}3^{\underline{3}}_{\underline{6}}3^{\underline{2}}_{\underline{6}}4^{\underline{2}}_{\underline{4}}2^{\underline{1}}_{\underline{4}}3^{\underline{1}}_{\underline{2}}2^{0}_{\underline{2}}3^{0}_{0}1^{1}_{0}1^{1}_{2}1^{2}_{4}$	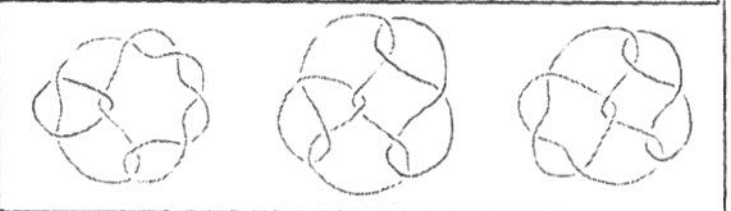		
10^a_{17}: $1^{\underline{2}}_{\underline{2}}1^{\underline{1}}_{0}1^{\underline{1}}_{2}5^{0}_{2}3^{0}_{4}6^{1}_{4}3^{1}_{6}7^{2}_{6}6^{2}_{8}6^{3}_{8}7^{3}_{10}7^{4}_{10}6^{4}_{12}4^{5}_{12}7^{5}_{14}3^{6}_{14}4^{6}_{16}1^{7}_{16}3^{7}_{18}1^{8}_{20}$			
10^a_{18}: $1^{\underline{7}}_{\underline{18}}1^{\underline{6}}_{\underline{16}}1^{\underline{6}}_{\underline{14}}4^{\underline{5}}_{\underline{14}}1^{\underline{5}}_{\underline{12}}4^{\underline{4}}_{\underline{12}}4^{\underline{4}}_{\underline{10}}6^{\underline{3}}_{\underline{10}}4^{\underline{3}}_{\underline{8}}5^{\underline{2}}_{\underline{8}}6^{\underline{2}}_{\underline{6}}4^{\underline{1}}_{\underline{6}}5^{\underline{1}}_{\underline{4}}5^{0}_{\underline{4}}6^{0}_{\underline{2}}2^{1}_{\underline{2}}3^{1}_{0}1^{2}_{0}2^{2}_{2}1^{3}_{4}$			
10^a_{19}: $1^{\underline{7}}_{\underline{16}}1^{\underline{6}}_{\underline{14}}1^{\underline{6}}_{\underline{12}}4^{\underline{5}}_{\underline{12}}1^{\underline{5}}_{\underline{10}}4^{\underline{4}}_{\underline{10}}4^{\underline{4}}_{\underline{8}}5^{\underline{3}}_{\underline{8}}4^{\underline{3}}_{\underline{6}}6^{\underline{2}}_{\underline{6}}5^{\underline{2}}_{\underline{4}}4^{\underline{1}}_{\underline{4}}6^{\underline{1}}_{\underline{2}}5^{0}_{\underline{2}}6^{0}_{0}2^{1}_{0}3^{1}_{2}1^{2}_{2}2^{2}_{4}1^{3}_{6}$	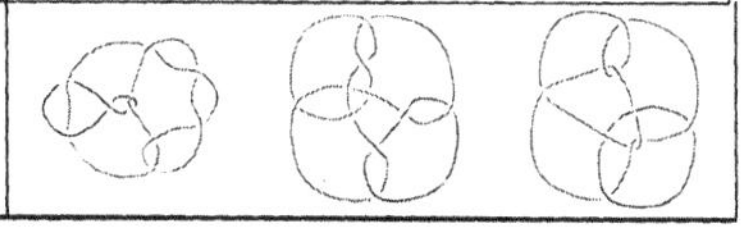		
10^a_{20}: $1^{\underline{5}}_{\underline{10}}2^{\underline{4}}_{\underline{8}}1^{\underline{4}}_{\underline{6}}5^{\underline{3}}_{\underline{6}}2^{\underline{3}}_{\underline{4}}7^{\underline{2}}_{\underline{4}}5^{\underline{2}}_{\underline{2}}7^{\underline{1}}_{\underline{2}}7^{\underline{1}}_{0}10^{0}_{0}9^{0}_{2}7^{1}_{2}8^{1}_{4}6^{2}_{4}7^{2}_{6}3^{3}_{6}6^{3}_{8}1^{4}_{8}3^{4}_{10}1^{5}_{12}$			
10^a_{21}: $1^{\underline{4}}_{\underline{8}}3^{\underline{3}}_{\underline{6}}1^{\underline{3}}_{\underline{4}}5^{\underline{2}}_{\underline{4}}3^{\underline{2}}_{\underline{2}}6^{\underline{1}}_{\underline{2}}5^{\underline{1}}_{0}10^{0}_{0}8^{0}_{2}8^{1}_{2}8^{1}_{4}7^{2}_{4}8^{2}_{6}5^{3}_{6}7^{3}_{8}3^{4}_{8}5^{4}_{10}1^{5}_{10}3^{5}_{12}1^{6}_{14}$			

L: KH	1st line	2nd line	3rd line
10^a_{22}: $1^{\underline{3}}_{\underline{4}}3^{\underline{2}}_{\underline{2}}1^{\underline{2}}_{0}4^{\underline{1}}_{0}3^{\underline{1}}_{2}8^{0}_{2}6^{0}_{4}8^{1}_{4}6^{1}_{6}9^{2}_{6}8^{2}_{8}6^{3}_{8}9^{3}_{10}6^{4}_{10}6^{4}_{12}3^{5}_{12}6^{5}_{14}1^{6}_{14}3^{6}_{16}1^{7}_{18}$	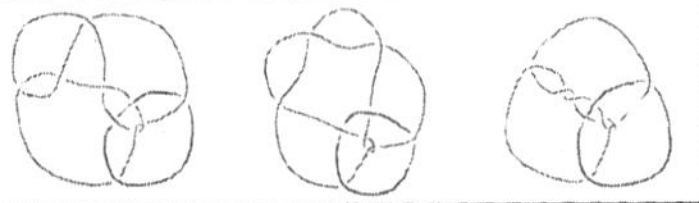		
10^a_{23}: $1^{\underline{7}}_{\underline{18}}2^{\underline{6}}_{\underline{16}}1^{\underline{6}}_{\underline{14}}5^{\underline{5}}_{\underline{14}}2^{\underline{5}}_{\underline{12}}5^{\underline{4}}_{\underline{12}}5^{\underline{4}}_{\underline{10}}7^{\underline{3}}_{\underline{10}}5^{\underline{3}}_{\underline{8}}7^{\underline{2}}_{\underline{8}}7^{\underline{2}}_{\underline{6}}5^{\underline{1}}_{\underline{6}}7^{\underline{1}}_{\underline{4}}5^{0}_{\underline{4}}7^{0}_{\underline{2}}3^{1}_{\underline{2}}3^{1}_{0}1^{2}_{0}3^{2}_{2}1^{3}_{4}$			
10^a_{24}: $1^{\underline{7}}_{\underline{16}}2^{\underline{6}}_{\underline{14}}1^{\underline{6}}_{\underline{12}}4^{\underline{5}}_{\underline{12}}2^{\underline{5}}_{\underline{10}}5^{\underline{4}}_{\underline{10}}4^{\underline{4}}_{\underline{8}}7^{\underline{3}}_{\underline{8}}5^{\underline{3}}_{\underline{6}}7^{\underline{2}}_{\underline{6}}7^{\underline{2}}_{\underline{4}}5^{\underline{1}}_{\underline{4}}7^{\underline{1}}_{\underline{2}}6^{0}_{\underline{2}}7^{0}_{0}3^{1}_{0}4^{1}_{2}1^{2}_{2}3^{2}_{4}1^{3}_{6}$			
10^a_{25}: $1^{\underline{8}}_{\underline{20}}2^{\underline{7}}_{\underline{18}}1^{\underline{7}}_{\underline{16}}4^{\underline{6}}_{\underline{16}}2^{\underline{6}}_{\underline{14}}5^{\underline{5}}_{\underline{14}}4^{\underline{5}}_{\underline{12}}7^{\underline{4}}_{\underline{12}}6^{\underline{4}}_{\underline{10}}7^{\underline{3}}_{\underline{10}}6^{\underline{3}}_{\underline{8}}5^{\underline{2}}_{\underline{8}}7^{\underline{2}}_{\underline{6}}4^{\underline{1}}_{\underline{6}}5^{\underline{1}}_{\underline{4}}3^{0}_{\underline{4}}5^{0}_{\underline{2}}1^{1}_{\underline{2}}2^{1}_{0}1^{2}_{2}$	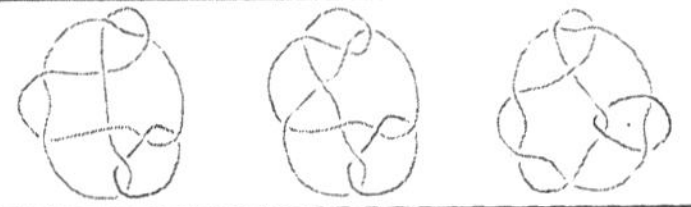		
10^a_{26}: $1^{\underline{7}}_{\underline{16}}2^{\underline{6}}_{\underline{14}}1^{\underline{6}}_{\underline{12}}4^{\underline{5}}_{\underline{12}}2^{\underline{5}}_{\underline{10}}6^{\underline{4}}_{\underline{10}}5^{\underline{4}}_{\underline{8}}7^{\underline{3}}_{\underline{8}}5^{\underline{3}}_{\underline{6}}8^{\underline{2}}_{\underline{6}}7^{\underline{2}}_{\underline{4}}5^{\underline{1}}_{\underline{4}}8^{\underline{1}}_{\underline{2}}6^{0}_{\underline{2}}6^{0}_{0}3^{1}_{0}5^{1}_{2}1^{2}_{2}3^{2}_{4}1^{3}_{6}$			
10^a_{27}: $1^{\underline{4}}_{\underline{6}}1^{\underline{3}}_{\underline{4}}1^{\underline{3}}_{\underline{2}}4^{\underline{2}}_{\underline{2}}1^{\underline{2}}_{0}2^{\underline{1}}_{0}4^{\underline{1}}_{2}6^{0}_{2}4^{0}_{4}5^{1}_{4}4^{1}_{6}4^{2}_{6}5^{2}_{8}3^{3}_{8}4^{3}_{10}2^{4}_{10}3^{4}_{12}1^{5}_{12}2^{5}_{14}1^{6}_{16}$			
10^a_{28}: $1^{\underline{6}}_{\underline{14}}2^{\underline{5}}_{\underline{12}}1^{\underline{5}}_{\underline{10}}4^{\underline{4}}_{\underline{10}}2^{\underline{4}}_{\underline{8}}5^{\underline{3}}_{\underline{8}}4^{\underline{3}}_{\underline{6}}6^{\underline{2}}_{\underline{6}}5^{\underline{2}}_{\underline{4}}6^{\underline{1}}_{\underline{4}}6^{\underline{1}}_{\underline{2}}6^{0}_{\underline{2}}8^{0}_{0}4^{1}_{0}4^{1}_{2}2^{2}_{2}4^{2}_{4}1^{3}_{4}2^{3}_{6}1^{4}_{8}$	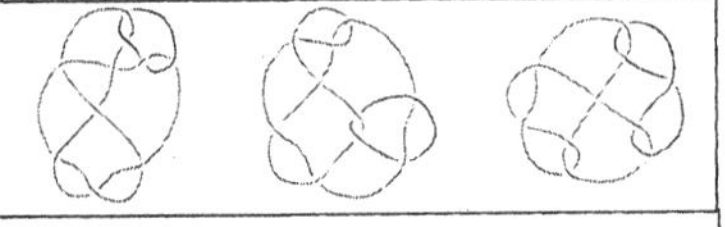		
10^a_{29}: $1^{\underline{5}}_{\underline{10}}2^{\underline{4}}_{\underline{8}}1^{\underline{4}}_{\underline{6}}4^{\underline{3}}_{\underline{6}}2^{\underline{3}}_{\underline{4}}5^{\underline{2}}_{\underline{4}}4^{\underline{2}}_{\underline{2}}5^{\underline{1}}_{\underline{2}}5^{\underline{1}}_{0}8^{0}_{0}7^{0}_{2}5^{1}_{2}6^{1}_{4}4^{2}_{4}5^{2}_{6}2^{3}_{6}4^{3}_{8}1^{4}_{8}2^{4}_{10}1^{5}_{12}$			
10^a_{30}: $1^{\underline{4}}_{\underline{8}}1^{\underline{4}}_{\underline{6}}2^{\underline{3}}_{\underline{6}}4^{\underline{2}}_{\underline{4}}2^{\underline{2}}_{\underline{2}}4^{\underline{1}}_{\underline{2}}4^{\underline{1}}_{0}8^{0}_{0}5^{0}_{2}6^{1}_{2}7^{1}_{4}6^{2}_{4}6^{2}_{6}4^{3}_{6}6^{3}_{8}3^{4}_{8}4^{4}_{10}1^{5}_{10}3^{5}_{12}1^{6}_{14}$			
10^a_{31}: $1^{\underline{6}}_{\underline{12}}1^{\underline{5}}_{\underline{10}}1^{\underline{5}}_{\underline{8}}3^{\underline{4}}_{\underline{8}}1^{\underline{4}}_{\underline{6}}3^{\underline{3}}_{\underline{6}}3^{\underline{3}}_{\underline{4}}5^{\underline{2}}_{\underline{4}}3^{\underline{2}}_{\underline{2}}4^{\underline{1}}_{\underline{2}}5^{\underline{1}}_{0}5^{0}_{0}6^{0}_{2}4^{1}_{2}3^{1}_{4}2^{2}_{4}4^{2}_{6}1^{3}_{6}2^{3}_{8}1^{4}_{10}$			
10^a_{32}: $1^{\underline{4}}_{\underline{8}}1^{\underline{3}}_{\underline{6}}1^{\underline{3}}_{\underline{4}}5^{\underline{2}}_{\underline{4}}1^{\underline{2}}_{\underline{2}}4^{\underline{1}}_{\underline{2}}5^{\underline{1}}_{0}7^{0}_{0}6^{0}_{2}7^{1}_{2}5^{1}_{4}5^{2}_{4}7^{2}_{6}4^{3}_{6}5^{3}_{8}2^{4}_{8}4^{4}_{10}1^{5}_{10}2^{5}_{12}1^{6}_{14}$			
10^a_{33}: $1^{\underline{6}}_{\underline{12}}1^{\underline{5}}_{\underline{10}}1^{\underline{5}}_{\underline{8}}4^{\underline{4}}_{\underline{8}}2^{\underline{4}}_{\underline{6}}3^{\underline{3}}_{\underline{6}}3^{\underline{3}}_{\underline{4}}6^{\underline{2}}_{\underline{4}}3^{\underline{2}}_{\underline{2}}4^{\underline{1}}_{\underline{2}}6^{\underline{1}}_{0}5^{0}_{0}5^{0}_{2}4^{1}_{2}4^{1}_{4}2^{2}_{4}4^{2}_{6}1^{3}_{6}2^{3}_{8}1^{4}_{10}$			
10^a_{34}: $1^{\underline{4}}_{\underline{8}}1^{\underline{3}}_{\underline{6}}1^{\underline{3}}_{\underline{4}}4^{\underline{2}}_{\underline{4}}1^{\underline{2}}_{\underline{2}}4^{\underline{1}}_{\underline{2}}4^{\underline{1}}_{0}7^{0}_{0}6^{0}_{2}7^{1}_{2}5^{1}_{4}5^{2}_{4}7^{2}_{6}4^{3}_{6}5^{3}_{8}3^{4}_{8}4^{4}_{10}1^{5}_{10}3^{5}_{12}1^{6}_{14}$	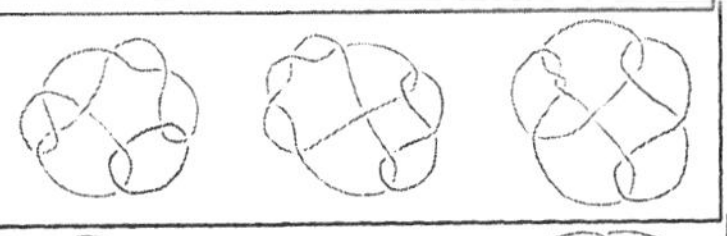		
10^a_{35}: $1^{\underline{7}}_{\underline{18}}1^{\underline{6}}_{\underline{16}}1^{\underline{6}}_{\underline{14}}4^{\underline{5}}_{\underline{14}}1^{\underline{5}}_{\underline{12}}3^{\underline{4}}_{\underline{12}}4^{\underline{4}}_{\underline{10}}5^{\underline{3}}_{\underline{10}}3^{\underline{3}}_{\underline{8}}5^{\underline{2}}_{\underline{8}}5^{\underline{2}}_{\underline{6}}3^{\underline{1}}_{\underline{6}}5^{\underline{1}}_{\underline{4}}4^{0}_{\underline{4}}5^{0}_{\underline{2}}2^{1}_{\underline{2}}2^{1}_{0}1^{2}_{0}2^{2}_{2}1^{3}_{4}$			
10^a_{36}: $1^{\underline{7}}_{\underline{16}}1^{\underline{6}}_{\underline{14}}1^{\underline{6}}_{\underline{12}}4^{\underline{5}}_{\underline{12}}1^{\underline{5}}_{\underline{10}}4^{\underline{4}}_{\underline{10}}4^{\underline{4}}_{\underline{8}}6^{\underline{3}}_{\underline{8}}4^{\underline{3}}_{\underline{6}}7^{\underline{2}}_{\underline{6}}6^{\underline{2}}_{\underline{4}}4^{\underline{1}}_{\underline{4}}7^{\underline{1}}_{\underline{2}}6^{0}_{\underline{2}}6^{0}_{0}3^{1}_{0}4^{1}_{2}1^{2}_{2}3^{2}_{4}1^{3}_{6}$			
10^a_{37}: $1^{\underline{6}}_{\underline{14}}1^{\underline{5}}_{\underline{12}}1^{\underline{5}}_{\underline{10}}4^{\underline{4}}_{\underline{10}}2^{\underline{4}}_{\underline{8}}5^{\underline{3}}_{\underline{8}}3^{\underline{3}}_{\underline{6}}6^{\underline{2}}_{\underline{6}}5^{\underline{2}}_{\underline{4}}5^{\underline{1}}_{\underline{4}}6^{\underline{1}}_{\underline{2}}6^{0}_{\underline{2}}6^{0}_{0}4^{1}_{0}5^{1}_{2}2^{2}_{2}4^{2}_{4}1^{3}_{4}2^{3}_{6}1^{4}_{8}$	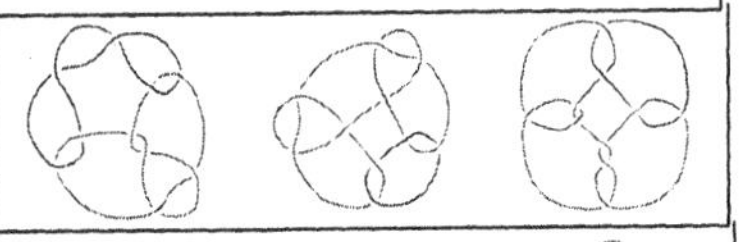		
10^a_{38}: $1^{\underline{10}}_{\underline{26}}2^{\underline{9}}_{\underline{24}}1^{\underline{9}}_{\underline{22}}4^{\underline{8}}_{\underline{22}}2^{\underline{8}}_{\underline{20}}7^{\underline{7}}_{\underline{20}}4^{\underline{7}}_{\underline{18}}6^{\underline{6}}_{\underline{18}}7^{\underline{6}}_{\underline{16}}8^{\underline{5}}_{\underline{16}}6^{\underline{5}}_{\underline{14}}6^{\underline{4}}_{\underline{14}}9^{\underline{4}}_{\underline{12}}5^{\underline{3}}_{\underline{12}}5^{\underline{3}}_{\underline{10}}3^{\underline{2}}_{\underline{10}}5^{\underline{2}}_{\underline{8}}3^{\underline{1}}_{\underline{6}}1^{0}_{\underline{6}}1^{0}_{\underline{4}}$			
10^a_{39}: $1^{\underline{5}}_{\underline{10}}1^{\underline{4}}_{\underline{8}}1^{\underline{4}}_{\underline{6}}4^{\underline{3}}_{\underline{6}}1^{\underline{3}}_{\underline{4}}5^{\underline{2}}_{\underline{4}}4^{\underline{2}}_{\underline{2}}5^{\underline{1}}_{\underline{2}}5^{\underline{1}}_{0}8^{0}_{0}7^{0}_{2}5^{1}_{2}6^{1}_{4}5^{2}_{4}5^{2}_{6}2^{3}_{6}5^{3}_{8}1^{4}_{8}2^{4}_{10}1^{5}_{12}$			
10^a_{40}: $1^{\underline{6}}_{\underline{14}}1^{\underline{5}}_{\underline{12}}1^{\underline{5}}_{\underline{10}}4^{\underline{4}}_{\underline{10}}2^{\underline{4}}_{\underline{8}}3^{\underline{3}}_{\underline{8}}3^{\underline{3}}_{\underline{6}}5^{\underline{2}}_{\underline{6}}3^{\underline{2}}_{\underline{4}}4^{\underline{1}}_{\underline{4}}5^{\underline{1}}_{\underline{2}}4^{0}_{\underline{2}}5^{0}_{0}3^{1}_{0}3^{1}_{2}1^{2}_{2}3^{2}_{4}1^{3}_{4}1^{3}_{6}1^{4}_{8}$	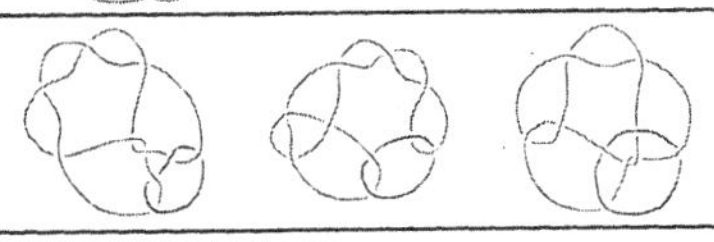		
10^a_{41}: $1^{\underline{5}}_{\underline{12}}1^{\underline{4}}_{\underline{10}}1^{\underline{4}}_{\underline{8}}3^{\underline{3}}_{\underline{8}}1^{\underline{3}}_{\underline{6}}4^{\underline{2}}_{\underline{6}}3^{\underline{2}}_{\underline{4}}4^{\underline{1}}_{\underline{4}}4^{\underline{1}}_{\underline{2}}6^{0}_{\underline{2}}6^{0}_{0}4^{1}_{0}4^{1}_{2}3^{2}_{2}4^{2}_{4}2^{3}_{4}3^{3}_{6}1^{4}_{6}2^{4}_{8}1^{5}_{10}$			
10^a_{42}: $1^{\underline{8}}_{\underline{20}}2^{\underline{7}}_{\underline{18}}1^{\underline{7}}_{\underline{16}}4^{\underline{6}}_{\underline{16}}2^{\underline{6}}_{\underline{14}}6^{\underline{5}}_{\underline{14}}4^{\underline{5}}_{\underline{12}}7^{\underline{4}}_{\underline{12}}7^{\underline{4}}_{\underline{10}}8^{\underline{3}}_{\underline{10}}6^{\underline{3}}_{\underline{8}}6^{\underline{2}}_{\underline{8}}8^{\underline{2}}_{\underline{6}}4^{\underline{1}}_{\underline{6}}6^{\underline{1}}_{\underline{4}}4^{0}_{\underline{4}}5^{0}_{\underline{2}}1^{1}_{\underline{2}}3^{1}_{0}1^{2}_{2}$			
10^a_{43}: $1^{\underline{4}}_{\underline{8}}1^{\underline{4}}_{\underline{6}}3^{\underline{3}}_{\underline{6}}5^{\underline{2}}_{\underline{4}}3^{\underline{2}}_{\underline{2}}6^{\underline{1}}_{\underline{2}}5^{\underline{1}}_{0}10^{0}_{0}7^{0}_{2}8^{1}_{2}9^{1}_{4}8^{2}_{4}8^{2}_{6}5^{3}_{6}8^{3}_{8}4^{4}_{8}5^{4}_{10}1^{5}_{10}4^{5}_{12}1^{6}_{14}$	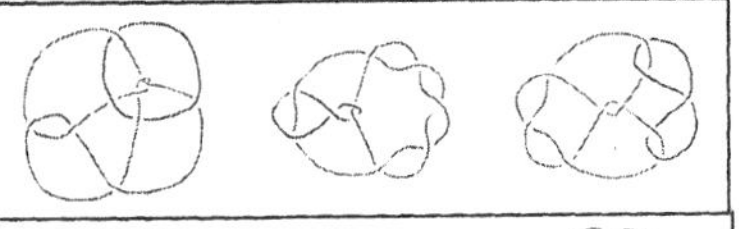		
10^a_{44}: $1^{\underline{4}}_{\underline{6}}1^{\underline{4}}_{\underline{4}}1^{\underline{3}}_{\underline{4}}3^{\underline{2}}_{\underline{2}}1^{\underline{2}}_{0}1^{\underline{1}}_{0}3^{\underline{1}}_{2}5^{0}_{2}2^{0}_{4}3^{1}_{4}4^{1}_{6}4^{2}_{6}3^{2}_{8}2^{3}_{8}4^{3}_{10}2^{4}_{10}2^{4}_{12}1^{5}_{12}2^{5}_{14}1^{6}_{16}$			
10^a_{45}: $1^{\underline{4}}_{\underline{8}}1^{\underline{4}}_{\underline{6}}2^{\underline{3}}_{\underline{6}}4^{\underline{2}}_{\underline{4}}2^{\underline{2}}_{\underline{2}}4^{\underline{1}}_{\underline{2}}4^{\underline{1}}_{0}7^{0}_{0}5^{0}_{2}5^{1}_{2}6^{1}_{4}6^{2}_{4}5^{2}_{6}3^{3}_{6}6^{3}_{8}2^{4}_{8}3^{4}_{10}1^{5}_{10}2^{5}_{12}1^{6}_{14}$			
10^a_{46}: $1^{\underline{4}}_{\underline{6}}1^{\underline{4}}_{\underline{4}}1^{\underline{3}}_{\underline{4}}4^{\underline{2}}_{\underline{2}}1^{\underline{2}}_{0}2^{\underline{1}}_{0}4^{\underline{1}}_{2}7^{0}_{2}3^{0}_{4}4^{1}_{4}6^{1}_{6}5^{2}_{6}4^{2}_{8}4^{3}_{8}5^{3}_{10}2^{4}_{10}4^{4}_{12}1^{5}_{12}2^{5}_{14}1^{6}_{16}$	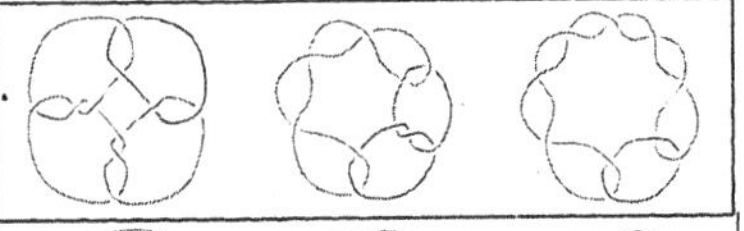		
10^a_{47}: $1^{\underline{4}}_{\underline{8}}1^{\underline{4}}_{\underline{6}}1^{\underline{3}}_{\underline{6}}3^{\underline{2}}_{\underline{4}}1^{\underline{2}}_{\underline{2}}2^{\underline{1}}_{\underline{2}}3^{\underline{1}}_{0}6^{0}_{0}3^{0}_{2}4^{1}_{2}5^{1}_{4}4^{2}_{4}4^{2}_{6}3^{3}_{6}4^{3}_{8}2^{4}_{8}3^{4}_{10}1^{5}_{10}2^{5}_{12}1^{6}_{14}$			
10^a_{48}: $1^{\underline{4}}_{\underline{10}}1^{\underline{4}}_{\underline{8}}1^{\underline{3}}_{\underline{8}}2^{\underline{2}}_{\underline{6}}1^{\underline{2}}_{\underline{4}}1^{\underline{1}}_{\underline{4}}2^{\underline{1}}_{\underline{2}}3^{0}_{\underline{2}}2^{0}_{0}2^{1}_{0}2^{1}_{2}2^{2}_{2}2^{2}_{4}1^{3}_{4}2^{3}_{6}1^{4}_{6}1^{4}_{8}1^{5}_{8}1^{5}_{10}1^{6}_{12}$			
10^a_{49}: $1^{\underline{10}}_{\underline{26}}3^{\underline{9}}_{\underline{24}}1^{\underline{9}}_{\underline{22}}4^{\underline{8}}_{\underline{22}}3^{\underline{8}}_{\underline{20}}7^{\underline{7}}_{\underline{20}}4^{\underline{7}}_{\underline{18}}6^{\underline{6}}_{\underline{18}}7^{\underline{6}}_{\underline{16}}8^{\underline{5}}_{\underline{16}}6^{\underline{5}}_{\underline{14}}6^{\underline{4}}_{\underline{14}}9^{\underline{4}}_{\underline{12}}4^{\underline{3}}_{\underline{12}}5^{\underline{3}}_{\underline{10}}3^{\underline{2}}_{\underline{10}}4^{\underline{2}}_{\underline{8}}3^{\underline{1}}_{\underline{6}}1^{0}_{\underline{6}}1^{0}_{\underline{4}}$	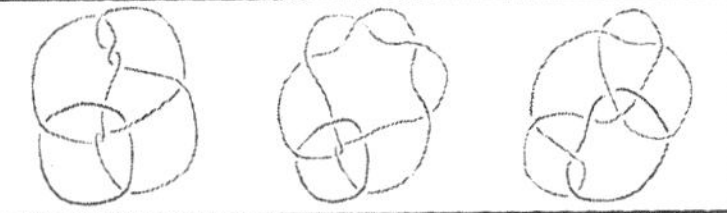		
10^a_{50}: $1^{\underline{10}}_{\underline{24}}2^{\underline{9}}_{\underline{22}}1^{\underline{9}}_{\underline{20}}2^{\underline{8}}_{\underline{20}}2^{\underline{8}}_{\underline{18}}5^{\underline{7}}_{\underline{18}}2^{\underline{7}}_{\underline{16}}4^{\underline{6}}_{\underline{16}}5^{\underline{6}}_{\underline{14}}5^{\underline{5}}_{\underline{14}}4^{\underline{5}}_{\underline{12}}5^{\underline{4}}_{\underline{12}}6^{\underline{4}}_{\underline{10}}3^{\underline{3}}_{\underline{10}}4^{\underline{3}}_{\underline{8}}3^{\underline{2}}_{\underline{8}}3^{\underline{2}}_{\underline{6}}3^{\underline{1}}_{\underline{4}}1^{0}_{\underline{4}}1^{0}_{\underline{2}}$			
10^a_{51}: $1^{\underline{6}}_{\underline{14}}3^{\underline{5}}_{\underline{12}}1^{\underline{5}}_{\underline{10}}5^{\underline{4}}_{\underline{10}}3^{\underline{4}}_{\underline{8}}7^{\underline{3}}_{\underline{8}}5^{\underline{3}}_{\underline{6}}9^{\underline{2}}_{\underline{6}}8^{\underline{2}}_{\underline{4}}8^{\underline{1}}_{\underline{4}}8^{\underline{1}}_{\underline{2}}8^{0}_{\underline{2}}9^{0}_{0}5^{1}_{0}7^{1}_{2}3^{2}_{2}5^{2}_{4}1^{3}_{4}3^{3}_{6}1^{4}_{8}$			
10^a_{52}: $1^{\underline{5}}_{\underline{10}}3^{\underline{4}}_{\underline{8}}1^{\underline{4}}_{\underline{6}}4^{\underline{3}}_{\underline{6}}3^{\underline{3}}_{\underline{4}}8^{\underline{2}}_{\underline{4}}5^{\underline{2}}_{\underline{2}}7^{\underline{1}}_{\underline{2}}7^{\underline{1}}_{0}9^{0}_{0}8^{0}_{2}7^{1}_{2}8^{1}_{4}5^{2}_{4}7^{2}_{6}3^{3}_{6}5^{3}_{8}1^{4}_{8}3^{4}_{10}1^{5}_{12}$	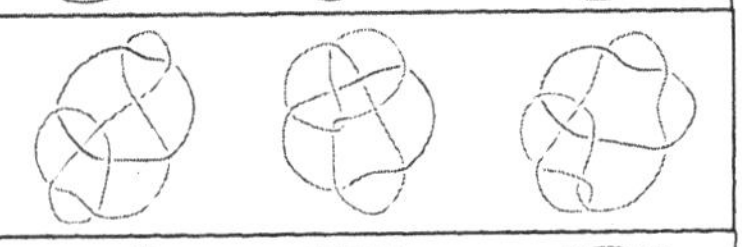		
10^a_{53}: $1^{\underline{5}}_{\underline{12}}3^{\underline{4}}_{\underline{10}}1^{\underline{4}}_{\underline{8}}5^{\underline{3}}_{\underline{8}}3^{\underline{3}}_{\underline{6}}8^{\underline{2}}_{\underline{6}}6^{\underline{2}}_{\underline{4}}8^{\underline{1}}_{\underline{4}}7^{\underline{1}}_{\underline{2}}9^{0}_{\underline{2}}9^{0}_{0}7^{1}_{0}8^{1}_{2}5^{2}_{2}7^{2}_{4}3^{3}_{4}5^{3}_{6}1^{4}_{6}3^{4}_{8}1^{5}_{10}$			
10^a_{54}: $1^{\underline{7}}_{\underline{18}}2^{\underline{6}}_{\underline{16}}1^{\underline{6}}_{\underline{14}}4^{\underline{5}}_{\underline{14}}2^{\underline{5}}_{\underline{12}}6^{\underline{4}}_{\underline{12}}4^{\underline{4}}_{\underline{10}}6^{\underline{3}}_{\underline{10}}6^{\underline{3}}_{\underline{8}}7^{\underline{2}}_{\underline{8}}7^{\underline{2}}_{\underline{6}}6^{\underline{1}}_{\underline{6}}6^{\underline{1}}_{\underline{4}}4^{0}_{\underline{4}}7^{0}_{\underline{2}}3^{1}_{\underline{2}}3^{1}_{0}1^{2}_{0}3^{2}_{2}1^{3}_{4}$			
10^a_{55}: $1^{\underline{5}}_{\underline{12}}3^{\underline{4}}_{\underline{10}}1^{\underline{4}}_{\underline{8}}4^{\underline{3}}_{\underline{8}}3^{\underline{3}}_{\underline{6}}7^{\underline{2}}_{\underline{6}}5^{\underline{2}}_{\underline{4}}7^{\underline{1}}_{\underline{4}}6^{\underline{1}}_{\underline{2}}7^{0}_{\underline{2}}8^{0}_{0}6^{1}_{0}6^{1}_{2}4^{2}_{2}6^{2}_{4}2^{3}_{4}4^{3}_{6}1^{4}_{6}2^{4}_{8}1^{5}_{10}$	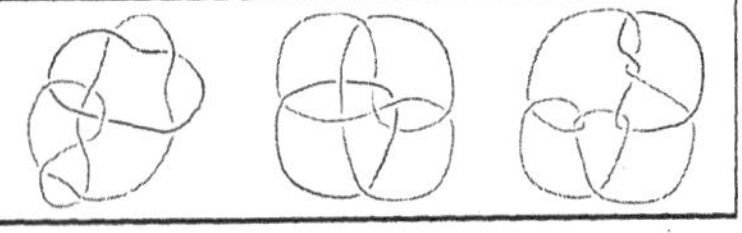		
10^a_{56}: $1^{\underline{6}}_{\underline{14}}4^{\underline{5}}_{\underline{12}}1^{\underline{5}}_{\underline{10}}7^{\underline{4}}_{\underline{10}}4^{\underline{4}}_{\underline{8}}9^{\underline{3}}_{\underline{8}}7^{\underline{3}}_{\underline{6}}12^{\underline{2}}_{\underline{6}}10^{\underline{2}}_{\underline{4}}11^{\underline{1}}_{\underline{4}}11^{\underline{1}}_{\underline{2}}10^{0}_{\underline{2}}12^{0}_{0}7^{1}_{0}9^{1}_{2}4^{2}_{2}7^{2}_{4}1^{3}_{4}4^{3}_{6}1^{4}_{8}$			
10^a_{57}: $1^{\underline{6}}_{\underline{14}}2^{\underline{5}}_{\underline{12}}1^{\underline{5}}_{\underline{10}}4^{\underline{4}}_{\underline{10}}2^{\underline{4}}_{\underline{8}}5^{\underline{3}}_{\underline{8}}4^{\underline{3}}_{\underline{6}}7^{\underline{2}}_{\underline{6}}6^{\underline{2}}_{\underline{4}}6^{\underline{1}}_{\underline{4}}6^{\underline{1}}_{\underline{2}}6^{0}_{\underline{2}}7^{0}_{0}4^{1}_{0}5^{1}_{2}2^{2}_{2}4^{2}_{4}1^{3}_{4}2^{3}_{6}1^{4}_{8}$			

L: KH	1st line 2nd line 3rd line
10^a_{58}: $1^{\underline{8}}_{\underline{20}}1^{\underline{7}}_{\underline{18}}1^{\underline{7}}_{\underline{16}}4^{\underline{6}}_{\underline{16}}2^{\underline{6}}_{\underline{14}}5^{\underline{5}}_{\underline{14}}3^{\underline{5}}_{\underline{12}}5^{\underline{4}}_{\underline{12}}5^{\underline{4}}_{\underline{10}}6^{\underline{3}}_{\underline{10}}5^{\underline{3}}_{\underline{8}}5^{\underline{2}}_{\underline{8}}6^{\underline{2}}_{\underline{6}}3^{\underline{1}}_{\underline{6}}5^{\underline{1}}_{\underline{4}}3^{0}_{\underline{4}}4^{0}_{\underline{2}}1^{1}_{\underline{2}}2^{1}_{0}1^{2}_{2}$	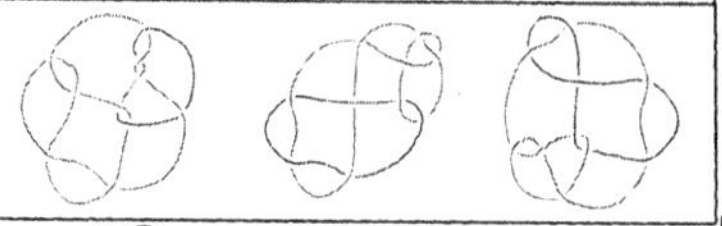
10^a_{59}: $1^{\underline{5}}_{\underline{10}}2^{\underline{4}}_{\underline{8}}1^{\underline{4}}_{\underline{6}}3^{\underline{3}}_{\underline{6}}2^{\underline{3}}_{\underline{4}}6^{\underline{2}}_{\underline{4}}4^{\underline{2}}_{\underline{2}}5^{\underline{1}}_{\underline{2}}5^{\underline{1}}_{0}7^{0}_{0}6^{0}_{2}5^{1}_{2}6^{1}_{4}4^{2}_{4}5^{2}_{6}2^{3}_{6}4^{3}_{8}1^{4}_{8}2^{4}_{10}1^{5}_{12}$	
10^a_{60}: $1^{\underline{5}}_{\underline{10}}2^{\underline{4}}_{\underline{8}}1^{\underline{4}}_{\underline{6}}4^{\underline{3}}_{\underline{6}}2^{\underline{3}}_{\underline{4}}5^{\underline{2}}_{\underline{4}}4^{\underline{2}}_{\underline{2}}6^{\underline{1}}_{\underline{2}}5^{\underline{1}}_{0}7^{0}_{0}7^{0}_{2}5^{1}_{2}6^{1}_{4}5^{2}_{4}6^{2}_{6}2^{3}_{6}4^{3}_{8}1^{4}_{8}2^{4}_{10}1^{5}_{12}$	
10^a_{61}: $1^{\underline{2}}_{\underline{2}}2^{\underline{1}}_{0}1^{\underline{1}}_{2}5^{0}_{2}3^{0}_{4}6^{1}_{4}4^{1}_{6}8^{2}_{6}7^{2}_{8}7^{3}_{8}7^{3}_{10}7^{4}_{10}7^{4}_{12}4^{5}_{12}7^{5}_{14}3^{6}_{14}4^{6}_{16}1^{7}_{16}3^{7}_{18}1^{8}_{20}$	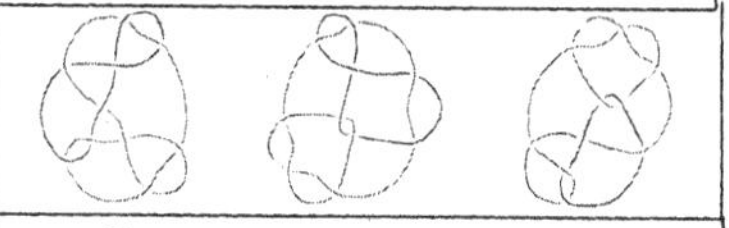
10^a_{62}: $1^{\underline{4}}_{\underline{6}}1^{\underline{3}}_{\underline{4}}1^{\underline{3}}_{\underline{2}}4^{\underline{2}}_{\underline{2}}2^{\underline{2}}_{0}3^{\underline{1}}_{0}3^{\underline{1}}_{2}6^{0}_{2}4^{0}_{4}4^{1}_{4}5^{1}_{6}5^{2}_{6}4^{2}_{8}3^{3}_{8}5^{3}_{10}2^{4}_{10}3^{4}_{12}1^{5}_{12}2^{5}_{14}1^{6}_{16}$	
10^a_{63}: $1^{\underline{7}}_{\underline{18}}2^{\underline{6}}_{\underline{16}}1^{\underline{6}}_{\underline{14}}4^{\underline{5}}_{\underline{14}}2^{\underline{5}}_{\underline{12}}5^{\underline{4}}_{\underline{12}}4^{\underline{4}}_{\underline{10}}6^{\underline{3}}_{\underline{10}}5^{\underline{3}}_{\underline{8}}6^{\underline{2}}_{\underline{8}}7^{\underline{2}}_{\underline{6}}5^{\underline{1}}_{\underline{6}}5^{\underline{1}}_{\underline{4}}4^{0}_{\underline{4}}6^{0}_{\underline{2}}2^{1}_{\underline{2}}3^{1}_{0}1^{2}_{0}2^{2}_{2}1^{3}_{4}$	
10^a_{64}: $1^{\underline{4}}_{\underline{8}}1^{\underline{3}}_{\underline{6}}1^{\underline{3}}_{\underline{4}}4^{\underline{2}}_{\underline{4}}2^{\underline{2}}_{\underline{2}}4^{\underline{1}}_{\underline{2}}3^{\underline{1}}_{0}6^{0}_{0}5^{0}_{2}5^{1}_{2}5^{1}_{4}5^{2}_{4}5^{2}_{6}3^{3}_{6}5^{3}_{8}2^{4}_{8}3^{4}_{10}1^{5}_{10}2^{5}_{12}1^{6}_{14}$	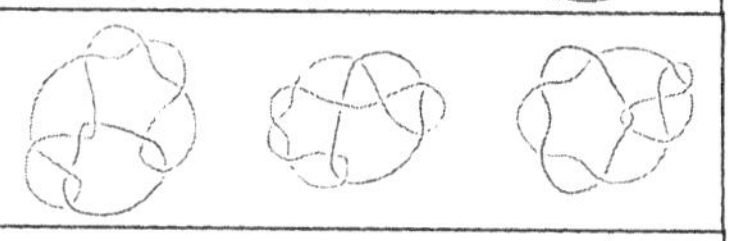
10^a_{65}: $1^{\underline{7}}_{\underline{16}}2^{\underline{6}}_{\underline{14}}1^{\underline{6}}_{\underline{12}}3^{\underline{5}}_{\underline{12}}2^{\underline{5}}_{\underline{10}}5^{\underline{4}}_{\underline{10}}3^{\underline{4}}_{\underline{8}}5^{\underline{3}}_{\underline{8}}5^{\underline{3}}_{\underline{6}}6^{\underline{2}}_{\underline{6}}6^{\underline{2}}_{\underline{4}}5^{\underline{1}}_{\underline{4}}5^{\underline{1}}_{\underline{2}}4^{0}_{\underline{2}}6^{0}_{0}2^{1}_{0}3^{1}_{2}1^{2}_{2}2^{2}_{4}1^{3}_{6}$	
10^a_{66}: $1^{\underline{6}}_{\underline{16}}2^{\underline{5}}_{\underline{14}}1^{\underline{5}}_{\underline{12}}3^{\underline{4}}_{\underline{12}}2^{\underline{4}}_{\underline{10}}4^{\underline{3}}_{\underline{10}}3^{\underline{3}}_{\underline{8}}5^{\underline{2}}_{\underline{8}}5^{\underline{2}}_{\underline{6}}4^{\underline{1}}_{\underline{6}}4^{\underline{1}}_{\underline{4}}4^{0}_{\underline{4}}5^{0}_{\underline{2}}3^{1}_{\underline{2}}3^{1}_{0}1^{2}_{0}3^{2}_{2}1^{3}_{2}1^{3}_{4}1^{4}_{6}$	
10^a_{67}: $1^{\underline{8}}_{\underline{22}}1^{\underline{7}}_{\underline{20}}1^{\underline{7}}_{\underline{18}}3^{\underline{6}}_{\underline{18}}2^{\underline{6}}_{\underline{16}}4^{\underline{5}}_{\underline{16}}2^{\underline{5}}_{\underline{14}}3^{\underline{4}}_{\underline{14}}4^{\underline{4}}_{\underline{12}}4^{\underline{3}}_{\underline{12}}3^{\underline{3}}_{\underline{10}}3^{\underline{2}}_{\underline{10}}4^{\underline{2}}_{\underline{8}}2^{\underline{1}}_{\underline{8}}3^{\underline{1}}_{\underline{6}}2^{0}_{\underline{6}}3^{0}_{\underline{4}}1^{1}_{\underline{4}}1^{1}_{\underline{2}}1^{2}_{0}$	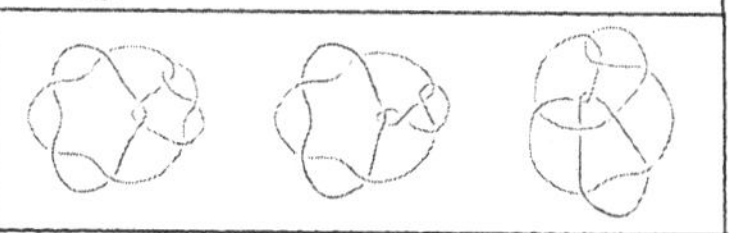
10^a_{68}: $1^{\underline{5}}_{\underline{12}}2^{\underline{4}}_{\underline{10}}1^{\underline{4}}_{\underline{8}}3^{\underline{3}}_{\underline{8}}2^{\underline{3}}_{\underline{6}}5^{\underline{2}}_{\underline{6}}4^{\underline{2}}_{\underline{4}}5^{\underline{1}}_{\underline{4}}4^{\underline{1}}_{\underline{2}}5^{0}_{\underline{2}}6^{0}_{0}4^{1}_{0}4^{1}_{2}3^{2}_{2}4^{2}_{4}1^{3}_{4}3^{3}_{6}1^{4}_{6}1^{4}_{8}1^{5}_{10}$	
10^a_{69}: $1^{\underline{4}}_{\underline{8}}3^{\underline{3}}_{\underline{6}}1^{\underline{3}}_{\underline{4}}5^{\underline{2}}_{\underline{4}}3^{\underline{2}}_{\underline{2}}6^{\underline{1}}_{\underline{2}}5^{\underline{1}}_{0}9^{0}_{0}7^{0}_{2}7^{1}_{2}8^{1}_{4}7^{2}_{4}8^{2}_{6}5^{3}_{6}6^{3}_{8}2^{4}_{8}5^{4}_{10}1^{5}_{10}2^{5}_{12}1^{6}_{14}$	
10^a_{70}: $1^{\underline{5}}_{\underline{12}}3^{\underline{4}}_{\underline{10}}1^{\underline{4}}_{\underline{8}}6^{\underline{3}}_{\underline{8}}3^{\underline{3}}_{\underline{6}}8^{\underline{2}}_{\underline{6}}6^{\underline{2}}_{\underline{4}}9^{\underline{1}}_{\underline{4}}8^{\underline{1}}_{\underline{2}}10^{0}_{\underline{2}}10^{0}_{0}8^{1}_{0}9^{1}_{2}6^{2}_{2}9^{2}_{4}4^{3}_{4}5^{3}_{6}1^{4}_{6}4^{4}_{8}1^{5}_{10}$	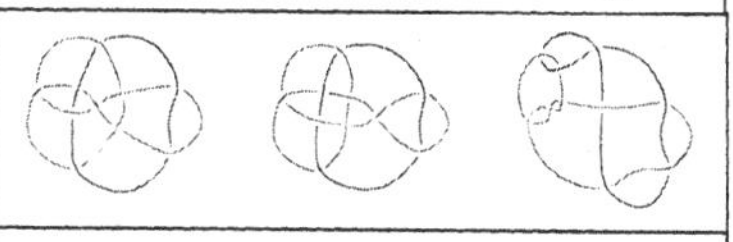
10^a_{71}: $1^{\underline{4}}_{\underline{8}}4^{\underline{3}}_{\underline{6}}1^{\underline{3}}_{\underline{4}}6^{\underline{2}}_{\underline{4}}4^{\underline{2}}_{\underline{2}}8^{\underline{1}}_{\underline{2}}6^{\underline{1}}_{0}11^{0}_{0}9^{0}_{2}9^{1}_{2}10^{1}_{4}9^{2}_{4}10^{2}_{6}6^{3}_{6}8^{3}_{8}3^{4}_{8}6^{4}_{10}1^{5}_{10}3^{5}_{12}1^{6}_{14}$	
10^a_{72}: $1^{\underline{10}}_{\underline{26}}2^{\underline{9}}_{\underline{24}}1^{\underline{9}}_{\underline{22}}3^{\underline{8}}_{\underline{22}}2^{\underline{8}}_{\underline{20}}5^{\underline{7}}_{\underline{20}}3^{\underline{7}}_{\underline{18}}5^{\underline{6}}_{\underline{18}}6^{\underline{6}}_{\underline{16}}6^{\underline{5}}_{\underline{16}}4^{\underline{5}}_{\underline{14}}4^{\underline{4}}_{\underline{14}}6^{\underline{4}}_{\underline{12}}3^{\underline{3}}_{\underline{12}}4^{\underline{3}}_{\underline{10}}2^{\underline{2}}_{\underline{10}}3^{\underline{2}}_{\underline{8}}2^{\underline{1}}_{\underline{6}}1^{0}_{\underline{6}}1^{0}_{\underline{4}}$	
10^a_{73}: $1^{\underline{10}}_{\underline{24}}1^{\underline{9}}_{\underline{22}}1^{\underline{9}}_{\underline{20}}2^{\underline{8}}_{\underline{20}}1^{\underline{8}}_{\underline{18}}3^{\underline{7}}_{\underline{18}}2^{\underline{7}}_{\underline{16}}3^{\underline{6}}_{\underline{16}}4^{\underline{6}}_{\underline{14}}4^{\underline{5}}_{\underline{14}}2^{\underline{5}}_{\underline{12}}3^{\underline{4}}_{\underline{12}}4^{\underline{4}}_{\underline{10}}2^{\underline{3}}_{\underline{10}}3^{\underline{3}}_{\underline{8}}2^{\underline{2}}_{\underline{8}}2^{\underline{2}}_{\underline{6}}2^{\underline{1}}_{\underline{4}}1^{0}_{\underline{4}}1^{0}_{\underline{2}}$	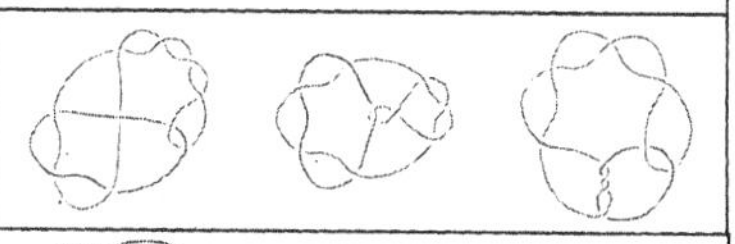
10^a_{74}: $1^{\underline{10}}_{\underline{28}}2^{\underline{9}}_{\underline{26}}1^{\underline{9}}_{\underline{24}}3^{\underline{8}}_{\underline{24}}2^{\underline{8}}_{\underline{22}}4^{\underline{7}}_{\underline{22}}3^{\underline{7}}_{\underline{20}}4^{\underline{6}}_{\underline{20}}5^{\underline{6}}_{\underline{18}}5^{\underline{5}}_{\underline{18}}3^{\underline{5}}_{\underline{16}}2^{\underline{4}}_{\underline{16}}5^{\underline{4}}_{\underline{14}}3^{\underline{3}}_{\underline{14}}2^{\underline{3}}_{\underline{12}}1^{\underline{2}}_{\underline{12}}3^{\underline{2}}_{\underline{10}}1^{\underline{1}}_{\underline{8}}1^{0}_{\underline{8}}1^{0}_{\underline{6}}$	
10^a_{75}: $1^{\underline{10}}_{\underline{26}}1^{\underline{9}}_{\underline{24}}1^{\underline{9}}_{\underline{22}}2^{\underline{8}}_{\underline{22}}1^{\underline{8}}_{\underline{20}}3^{\underline{7}}_{\underline{20}}2^{\underline{7}}_{\underline{18}}3^{\underline{6}}_{\underline{18}}4^{\underline{6}}_{\underline{16}}4^{\underline{5}}_{\underline{16}}2^{\underline{5}}_{\underline{14}}2^{\underline{4}}_{\underline{14}}4^{\underline{4}}_{\underline{12}}2^{\underline{3}}_{\underline{12}}2^{\underline{3}}_{\underline{10}}1^{\underline{2}}_{\underline{10}}2^{\underline{2}}_{\underline{8}}1^{\underline{1}}_{\underline{6}}1^{0}_{\underline{6}}1^{0}_{\underline{4}}$	
10^a_{76}: $1^{\underline{6}}_{\underline{14}}3^{\underline{5}}_{\underline{12}}1^{\underline{5}}_{\underline{10}}5^{\underline{4}}_{\underline{10}}3^{\underline{4}}_{\underline{8}}8^{\underline{3}}_{\underline{8}}5^{\underline{3}}_{\underline{6}}8^{\underline{2}}_{\underline{6}}8^{\underline{2}}_{\underline{4}}9^{\underline{1}}_{\underline{4}}8^{\underline{1}}_{\underline{2}}8^{0}_{\underline{2}}10^{0}_{0}5^{1}_{0}7^{1}_{2}4^{2}_{2}6^{2}_{4}1^{3}_{4}3^{3}_{6}1^{4}_{8}$	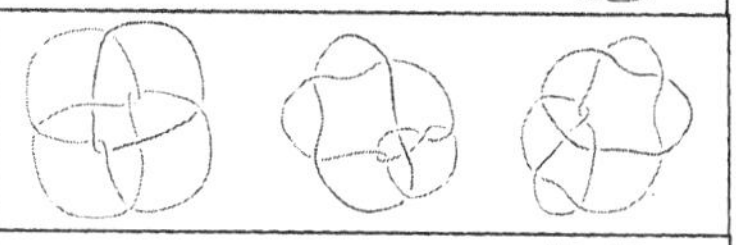
10^a_{77}: $1^{\underline{10}}_{\underline{26}}2^{\underline{9}}_{\underline{24}}1^{\underline{9}}_{\underline{22}}4^{\underline{8}}_{\underline{22}}2^{\underline{8}}_{\underline{20}}6^{\underline{7}}_{\underline{20}}4^{\underline{7}}_{\underline{18}}6^{\underline{6}}_{\underline{18}}7^{\underline{6}}_{\underline{16}}8^{\underline{5}}_{\underline{16}}5^{\underline{5}}_{\underline{14}}5^{\underline{4}}_{\underline{14}}8^{\underline{4}}_{\underline{12}}4^{\underline{3}}_{\underline{12}}5^{\underline{3}}_{\underline{10}}3^{\underline{2}}_{\underline{10}}4^{\underline{2}}_{\underline{8}}3^{\underline{1}}_{\underline{6}}1^{0}_{\underline{6}}1^{0}_{\underline{4}}$	
10^a_{78}: $1^{\underline{6}}_{\underline{16}}1^{\underline{6}}_{\underline{14}}3^{\underline{5}}_{\underline{14}}3^{\underline{4}}_{\underline{12}}3^{\underline{4}}_{\underline{10}}5^{\underline{3}}_{\underline{10}}3^{\underline{3}}_{\underline{8}}6^{\underline{2}}_{\underline{8}}5^{\underline{2}}_{\underline{6}}4^{\underline{1}}_{\underline{6}}6^{\underline{1}}_{\underline{4}}6^{0}_{\underline{4}}5^{0}_{\underline{2}}3^{1}_{\underline{2}}5^{1}_{0}2^{2}_{0}3^{2}_{2}1^{3}_{2}2^{3}_{4}1^{4}_{6}$	
10^a_{79}: $1^{\underline{7}}_{\underline{16}}2^{\underline{6}}_{\underline{14}}1^{\underline{6}}_{\underline{12}}5^{\underline{5}}_{\underline{12}}2^{\underline{5}}_{\underline{10}}6^{\underline{4}}_{\underline{10}}5^{\underline{4}}_{\underline{8}}7^{\underline{3}}_{\underline{8}}6^{\underline{3}}_{\underline{6}}9^{\underline{2}}_{\underline{6}}8^{\underline{2}}_{\underline{4}}6^{\underline{1}}_{\underline{4}}8^{\underline{1}}_{\underline{2}}6^{0}_{\underline{2}}7^{0}_{0}3^{1}_{0}5^{1}_{2}1^{2}_{2}3^{2}_{4}1^{3}_{6}$	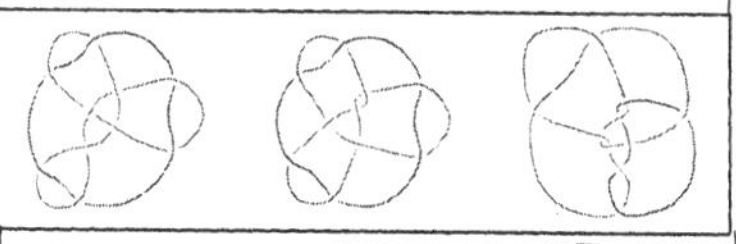
10^a_{80}: $1^{\underline{6}}_{\underline{12}}2^{\underline{5}}_{\underline{10}}1^{\underline{5}}_{\underline{8}}4^{\underline{4}}_{\underline{8}}2^{\underline{4}}_{\underline{6}}4^{\underline{3}}_{\underline{6}}4^{\underline{3}}_{\underline{4}}7^{\underline{2}}_{\underline{4}}5^{\underline{2}}_{\underline{2}}5^{\underline{1}}_{\underline{2}}6^{\underline{1}}_{0}6^{0}_{0}6^{0}_{2}4^{1}_{2}5^{1}_{4}2^{2}_{4}4^{2}_{6}1^{3}_{6}2^{3}_{8}1^{4}_{10}$	
10^a_{81}: $1^{\underline{8}}_{\underline{20}}2^{\underline{7}}_{\underline{18}}1^{\underline{7}}_{\underline{16}}5^{\underline{6}}_{\underline{16}}3^{\underline{6}}_{\underline{14}}6^{\underline{5}}_{\underline{14}}4^{\underline{5}}_{\underline{12}}7^{\underline{4}}_{\underline{12}}6^{\underline{4}}_{\underline{10}}7^{\underline{3}}_{\underline{10}}7^{\underline{3}}_{\underline{8}}6^{\underline{2}}_{\underline{8}}7^{\underline{2}}_{\underline{6}}4^{\underline{1}}_{\underline{6}}6^{\underline{1}}_{\underline{4}}3^{0}_{\underline{4}}5^{0}_{\underline{2}}1^{1}_{\underline{2}}2^{1}_{0}1^{2}_{2}$	
10^a_{82}: $1^{\underline{4}}_{\underline{8}}1^{\underline{3}}_{\underline{6}}1^{\underline{3}}_{\underline{4}}5^{\underline{2}}_{\underline{4}}2^{\underline{2}}_{\underline{2}}5^{\underline{1}}_{\underline{2}}4^{\underline{1}}_{0}7^{0}_{0}6^{0}_{2}7^{1}_{2}6^{1}_{4}6^{2}_{4}7^{2}_{6}4^{3}_{6}6^{3}_{8}3^{4}_{8}4^{4}_{10}1^{5}_{10}3^{5}_{12}1^{6}_{14}$	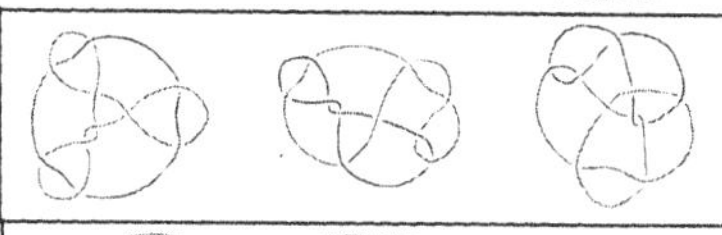
10^a_{83}: $1^{\underline{8}}_{\underline{20}}2^{\underline{7}}_{\underline{18}}1^{\underline{7}}_{\underline{16}}4^{\underline{6}}_{\underline{16}}3^{\underline{6}}_{\underline{14}}6^{\underline{5}}_{\underline{14}}3^{\underline{5}}_{\underline{12}}6^{\underline{4}}_{\underline{12}}6^{\underline{4}}_{\underline{10}}6^{\underline{3}}_{\underline{10}}6^{\underline{3}}_{\underline{8}}6^{\underline{2}}_{\underline{8}}6^{\underline{2}}_{\underline{6}}3^{\underline{1}}_{\underline{6}}6^{\underline{1}}_{\underline{4}}3^{0}_{\underline{4}}4^{0}_{\underline{2}}1^{1}_{\underline{2}}2^{1}_{0}1^{2}_{2}$	
10^a_{84}: $1^{\underline{5}}_{\underline{12}}2^{\underline{4}}_{\underline{10}}1^{\underline{4}}_{\underline{8}}5^{\underline{3}}_{\underline{8}}2^{\underline{3}}_{\underline{6}}6^{\underline{2}}_{\underline{6}}5^{\underline{2}}_{\underline{4}}7^{\underline{1}}_{\underline{4}}6^{\underline{1}}_{\underline{2}}8^{0}_{\underline{2}}8^{0}_{0}6^{1}_{0}7^{1}_{2}5^{2}_{2}7^{2}_{4}3^{3}_{4}4^{3}_{6}1^{4}_{6}3^{4}_{8}1^{5}_{10}$	
10^a_{85}: $1^{\underline{6}}_{\underline{14}}1^{\underline{6}}_{\underline{12}}3^{\underline{5}}_{\underline{12}}5^{\underline{4}}_{\underline{10}}3^{\underline{4}}_{\underline{8}}6^{\underline{3}}_{\underline{8}}5^{\underline{3}}_{\underline{6}}9^{\underline{2}}_{\underline{6}}6^{\underline{2}}_{\underline{4}}7^{\underline{1}}_{\underline{4}}9^{\underline{1}}_{\underline{2}}8^{0}_{\underline{2}}8^{0}_{0}5^{1}_{0}7^{1}_{2}3^{2}_{2}5^{2}_{4}1^{3}_{4}3^{3}_{6}1^{4}_{8}$	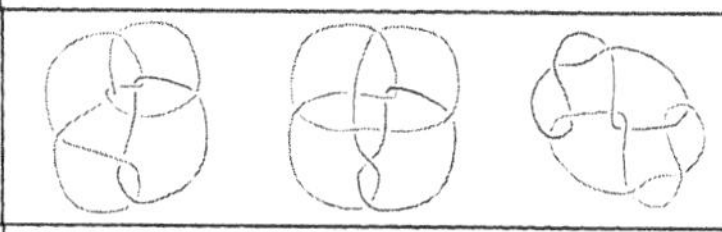
10^a_{86}: $1^{\underline{4}}_{\underline{8}}3^{\underline{3}}_{\underline{6}}1^{\underline{3}}_{\underline{4}}6^{\underline{2}}_{\underline{4}}3^{\underline{2}}_{\underline{2}}7^{\underline{1}}_{\underline{2}}6^{\underline{1}}_{0}10^{0}_{0}8^{0}_{2}9^{1}_{2}9^{1}_{4}8^{2}_{4}10^{2}_{6}6^{3}_{6}7^{3}_{8}3^{4}_{8}6^{4}_{10}1^{5}_{10}3^{5}_{12}1^{6}_{14}$	
10^a_{87}: $1^{\underline{10}}_{\underline{26}}2^{\underline{9}}_{\underline{24}}1^{\underline{9}}_{\underline{22}}4^{\underline{8}}_{\underline{22}}2^{\underline{8}}_{\underline{20}}5^{\underline{7}}_{\underline{20}}4^{\underline{7}}_{\underline{18}}6^{\underline{6}}_{\underline{18}}6^{\underline{6}}_{\underline{16}}7^{\underline{5}}_{\underline{16}}5^{\underline{5}}_{\underline{14}}4^{\underline{4}}_{\underline{14}}7^{\underline{4}}_{\underline{12}}4^{\underline{3}}_{\underline{12}}4^{\underline{3}}_{\underline{10}}2^{\underline{2}}_{\underline{10}}4^{\underline{2}}_{\underline{8}}2^{\underline{1}}_{\underline{6}}1^{0}_{\underline{6}}1^{0}_{\underline{4}}$	
10^a_{88}: $1^{\underline{7}}_{\underline{16}}2^{\underline{6}}_{\underline{14}}1^{\underline{6}}_{\underline{12}}3^{\underline{5}}_{\underline{12}}2^{\underline{5}}_{\underline{10}}6^{\underline{4}}_{\underline{10}}4^{\underline{4}}_{\underline{8}}6^{\underline{3}}_{\underline{8}}5^{\underline{3}}_{\underline{6}}7^{\underline{2}}_{\underline{6}}6^{\underline{2}}_{\underline{4}}5^{\underline{1}}_{\underline{4}}7^{\underline{1}}_{\underline{2}}5^{0}_{\underline{2}}6^{0}_{0}3^{1}_{0}4^{1}_{2}1^{2}_{2}3^{2}_{4}1^{3}_{6}$	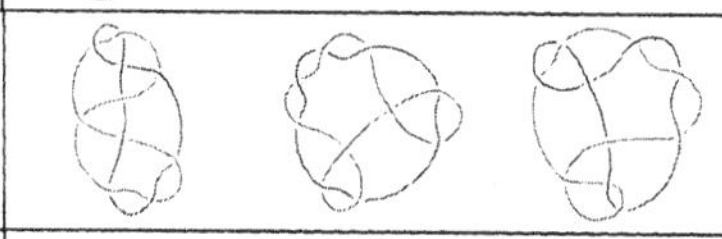
10^a_{89}: $1^{\underline{4}}_{\underline{6}}1^{\underline{3}}_{\underline{4}}1^{\underline{3}}_{\underline{2}}3^{\underline{2}}_{\underline{2}}1^{\underline{2}}_{0}2^{\underline{1}}_{0}3^{\underline{1}}_{2}5^{0}_{2}4^{0}_{4}4^{1}_{4}3^{1}_{6}4^{2}_{6}4^{2}_{8}2^{3}_{8}4^{3}_{10}2^{4}_{10}2^{4}_{12}1^{5}_{12}2^{5}_{14}1^{6}_{16}$	
10^a_{90}: $1^{\underline{5}}_{\underline{10}}2^{\underline{4}}_{\underline{8}}1^{\underline{4}}_{\underline{6}}3^{\underline{3}}_{\underline{6}}2^{\underline{3}}_{\underline{4}}5^{\underline{2}}_{\underline{4}}3^{\underline{2}}_{\underline{2}}4^{\underline{1}}_{\underline{2}}5^{\underline{1}}_{0}7^{0}_{0}6^{0}_{2}5^{1}_{2}5^{1}_{4}3^{2}_{4}5^{2}_{6}2^{3}_{6}3^{3}_{8}1^{4}_{8}2^{4}_{10}1^{5}_{12}$	
10^a_{91}: $1^{\underline{4}}_{\underline{8}}2^{\underline{3}}_{\underline{6}}1^{\underline{3}}_{\underline{4}}4^{\underline{2}}_{\underline{4}}2^{\underline{2}}_{\underline{2}}5^{\underline{1}}_{\underline{2}}4^{\underline{1}}_{0}8^{0}_{0}7^{0}_{2}7^{1}_{2}6^{1}_{4}6^{2}_{4}7^{2}_{6}4^{3}_{6}6^{3}_{8}3^{4}_{8}4^{4}_{10}1^{5}_{10}3^{5}_{12}1^{6}_{14}$	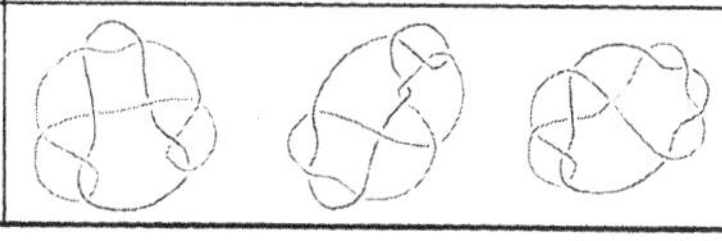
10^a_{92}: $1^{\underline{6}}_{\underline{14}}2^{\underline{5}}_{\underline{12}}1^{\underline{5}}_{\underline{10}}4^{\underline{4}}_{\underline{10}}3^{\underline{4}}_{\underline{8}}5^{\underline{3}}_{\underline{8}}3^{\underline{3}}_{\underline{6}}6^{\underline{2}}_{\underline{6}}5^{\underline{2}}_{\underline{4}}5^{\underline{1}}_{\underline{4}}6^{\underline{1}}_{\underline{2}}6^{0}_{\underline{2}}6^{0}_{0}3^{1}_{0}5^{1}_{2}2^{2}_{2}3^{2}_{4}1^{3}_{4}2^{3}_{6}1^{4}_{8}$	
10^a_{93}: $1^{\underline{7}}_{\underline{18}}2^{\underline{6}}_{\underline{16}}1^{\underline{6}}_{\underline{14}}3^{\underline{5}}_{\underline{14}}2^{\underline{5}}_{\underline{12}}5^{\underline{4}}_{\underline{12}}4^{\underline{4}}_{\underline{10}}5^{\underline{3}}_{\underline{10}}4^{\underline{3}}_{\underline{8}}5^{\underline{2}}_{\underline{8}}5^{\underline{2}}_{\underline{6}}4^{\underline{1}}_{\underline{6}}5^{\underline{1}}_{\underline{4}}3^{0}_{\underline{4}}5^{0}_{\underline{2}}2^{1}_{\underline{2}}2^{1}_{0}1^{2}_{0}2^{2}_{2}1^{3}_{4}$	

L: KH	1st line	2nd line	3rd line
10^a_{94}: $1^{\underline{10}}_{\underline{26}}1^{\underline{9}}_{\underline{24}}1^{\underline{9}}_{\underline{22}}3^{\underline{8}}_{\underline{22}}2^{\underline{8}}_{\underline{20}}5^{\underline{7}}_{\underline{20}}2^{\underline{7}}_{\underline{18}}4^{\underline{6}}_{\underline{18}}5^{\underline{6}}_{\underline{16}}5^{\underline{5}}_{\underline{16}}4^{\underline{5}}_{\underline{14}}4^{\underline{4}}_{\underline{14}}5^{\underline{4}}_{\underline{12}}3^{\underline{3}}_{\underline{12}}4^{\underline{3}}_{\underline{10}}2^{\underline{2}}_{\underline{10}}3^{\underline{2}}_{\underline{8}}2^{\underline{1}}_{\underline{6}}1^{0}_{\underline{6}}1^{0}_{\underline{4}}$	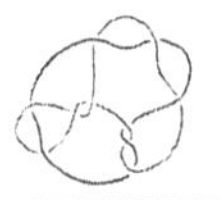	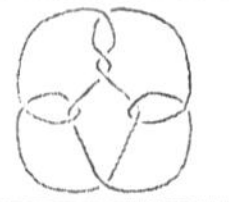	
10^a_{95}: $1^{\underline{6}}_{\underline{14}}2^{\underline{5}}_{\underline{12}}1^{\underline{5}}_{\underline{10}}3^{\underline{4}}_{\underline{10}}2^{\underline{4}}_{\underline{8}}6^{\underline{3}}_{\underline{8}}3^{\underline{3}}_{\underline{6}}6^{\underline{2}}_{\underline{6}}6^{\underline{2}}_{\underline{4}}5^{\underline{1}}_{\underline{4}}6^{\underline{1}}_{\underline{2}}7^{0}_{\underline{2}}7^{0}_{0}4^{1}_{0}5^{1}_{2}2^{2}_{2}4^{2}_{4}1^{3}_{4}2^{3}_{6}1^{4}_{8}$			
10^a_{96}: $1^{\underline{10}}_{\underline{26}}2^{\underline{9}}_{\underline{24}}1^{\underline{9}}_{\underline{22}}4^{\underline{8}}_{\underline{22}}3^{\underline{8}}_{\underline{20}}5^{\underline{7}}_{\underline{20}}3^{\underline{7}}_{\underline{18}}5^{\underline{6}}_{\underline{18}}5^{\underline{6}}_{\underline{16}}6^{\underline{5}}_{\underline{16}}5^{\underline{5}}_{\underline{14}}4^{\underline{4}}_{\underline{14}}6^{\underline{4}}_{\underline{12}}3^{\underline{3}}_{\underline{12}}4^{\underline{3}}_{\underline{10}}2^{\underline{2}}_{\underline{10}}3^{\underline{2}}_{\underline{8}}2^{\underline{1}}_{\underline{6}}1^{0}_{\underline{6}}1^{0}_{\underline{4}}$			
10^a_{97}: $1^{\underline{8}}_{\underline{22}}1^{\underline{8}}_{\underline{20}}2^{\underline{7}}_{\underline{20}}2^{\underline{6}}_{\underline{18}}2^{\underline{6}}_{\underline{16}}3^{\underline{5}}_{\underline{16}}2^{\underline{5}}_{\underline{14}}3^{\underline{4}}_{\underline{14}}3^{\underline{4}}_{\underline{12}}3^{\underline{3}}_{\underline{12}}3^{\underline{3}}_{\underline{10}}3^{\underline{2}}_{\underline{10}}3^{\underline{2}}_{\underline{8}}1^{\underline{1}}_{\underline{8}}3^{\underline{1}}_{\underline{6}}2^{0}_{\underline{6}}2^{0}_{\underline{4}}1^{1}_{\underline{4}}1^{1}_{\underline{2}}1^{2}_{0}$	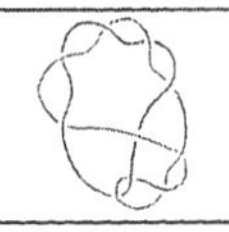	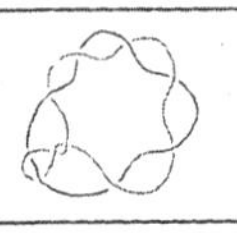	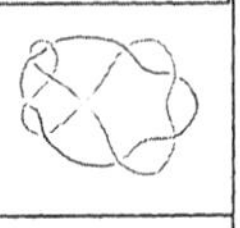
10^a_{98}: $1^{\underline{10}}_{\underline{28}}1^{\underline{9}}_{\underline{26}}1^{\underline{9}}_{\underline{24}}2^{\underline{8}}_{\underline{24}}2^{\underline{8}}_{\underline{22}}3^{\underline{7}}_{\underline{22}}1^{\underline{7}}_{\underline{20}}2^{\underline{6}}_{\underline{20}}3^{\underline{6}}_{\underline{18}}3^{\underline{5}}_{\underline{18}}2^{\underline{5}}_{\underline{16}}1^{\underline{4}}_{\underline{16}}3^{\underline{4}}_{\underline{14}}2^{\underline{3}}_{\underline{14}}1^{\underline{3}}_{\underline{12}}1^{\underline{2}}_{\underline{12}}2^{\underline{2}}_{\underline{10}}1^{\underline{1}}_{\underline{8}}1^{0}_{\underline{8}}1^{0}_{\underline{6}}$			
10^a_{99}: $1^{\underline{6}}_{\underline{12}}1^{\underline{5}}_{\underline{10}}1^{\underline{5}}_{\underline{8}}3^{\underline{4}}_{\underline{8}}2^{\underline{4}}_{\underline{6}}3^{\underline{3}}_{\underline{6}}2^{\underline{3}}_{\underline{4}}5^{\underline{2}}_{\underline{4}}3^{\underline{2}}_{\underline{2}}3^{\underline{1}}_{\underline{2}}5^{\underline{1}}_{0}5^{0}_{0}4^{0}_{2}3^{1}_{2}4^{1}_{4}2^{2}_{4}3^{2}_{6}1^{3}_{6}2^{3}_{8}1^{4}_{10}$			
10^a_{100}: $1^{\underline{10}}_{\underline{26}}1^{\underline{9}}_{\underline{24}}1^{\underline{9}}_{\underline{22}}3^{\underline{8}}_{\underline{22}}2^{\underline{8}}_{\underline{20}}4^{\underline{7}}_{\underline{20}}2^{\underline{7}}_{\underline{18}}3^{\underline{6}}_{\underline{18}}4^{\underline{6}}_{\underline{16}}5^{\underline{5}}_{\underline{16}}3^{\underline{5}}_{\underline{14}}3^{\underline{4}}_{\underline{14}}5^{\underline{4}}_{\underline{12}}2^{\underline{3}}_{\underline{12}}3^{\underline{3}}_{\underline{10}}2^{\underline{2}}_{\underline{10}}2^{\underline{2}}_{\underline{8}}2^{\underline{1}}_{\underline{6}}1^{0}_{\underline{6}}1^{0}_{\underline{4}}$	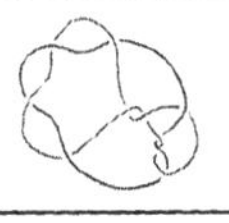	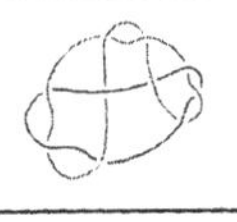	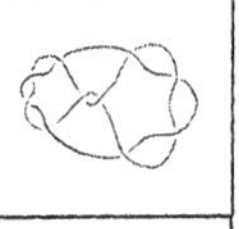
10^a_{101}: $1^{\underline{10}}_{\underline{24}}1^{\underline{9}}_{\underline{22}}1^{\underline{9}}_{\underline{20}}3^{\underline{8}}_{\underline{20}}2^{\underline{8}}_{\underline{18}}4^{\underline{7}}_{\underline{18}}2^{\underline{7}}_{\underline{16}}4^{\underline{6}}_{\underline{16}}4^{\underline{6}}_{\underline{14}}5^{\underline{5}}_{\underline{14}}4^{\underline{5}}_{\underline{12}}4^{\underline{4}}_{\underline{12}}5^{\underline{4}}_{\underline{10}}3^{\underline{3}}_{\underline{10}}4^{\underline{3}}_{\underline{8}}3^{\underline{2}}_{\underline{8}}3^{\underline{2}}_{\underline{6}}3^{\underline{1}}_{\underline{4}}1^{0}_{\underline{4}}1^{0}_{\underline{2}}$			
10^a_{102}: $1^{\underline{10}}_{\underline{28}}1^{\underline{9}}_{\underline{26}}1^{\underline{9}}_{\underline{24}}3^{\underline{8}}_{\underline{24}}2^{\underline{8}}_{\underline{22}}4^{\underline{7}}_{\underline{22}}2^{\underline{7}}_{\underline{20}}2^{\underline{6}}_{\underline{20}}4^{\underline{6}}_{\underline{18}}4^{\underline{5}}_{\underline{18}}2^{\underline{5}}_{\underline{16}}2^{\underline{4}}_{\underline{16}}4^{\underline{4}}_{\underline{14}}2^{\underline{3}}_{\underline{14}}2^{\underline{3}}_{\underline{12}}1^{\underline{2}}_{\underline{12}}2^{\underline{2}}_{\underline{10}}1^{\underline{1}}_{\underline{8}}1^{0}_{\underline{8}}1^{0}_{\underline{6}}$			
10^a_{103}: $1^{\underline{5}}_{\underline{10}}2^{\underline{4}}_{\underline{8}}1^{\underline{4}}_{\underline{6}}3^{\underline{3}}_{\underline{6}}2^{\underline{3}}_{\underline{4}}6^{\underline{2}}_{\underline{4}}3^{\underline{2}}_{\underline{2}}5^{\underline{1}}_{\underline{2}}6^{\underline{1}}_{0}7^{0}_{0}7^{0}_{2}6^{1}_{2}5^{1}_{4}4^{2}_{4}6^{2}_{6}2^{3}_{6}4^{3}_{8}1^{4}_{8}2^{4}_{10}1^{5}_{12}$		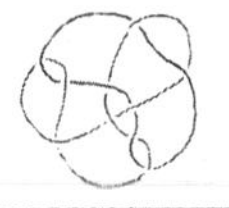	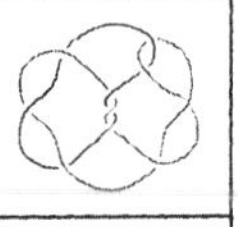
10^a_{104}: $1^{\underline{4}}_{\underline{8}}3^{\underline{3}}_{\underline{6}}1^{\underline{3}}_{\underline{4}}5^{\underline{2}}_{\underline{4}}3^{\underline{2}}_{\underline{2}}7^{\underline{1}}_{\underline{2}}5^{\underline{1}}_{0}8^{0}_{0}8^{0}_{2}8^{1}_{2}7^{1}_{4}7^{2}_{4}8^{2}_{6}4^{3}_{6}7^{3}_{8}3^{4}_{8}5^{4}_{10}1^{5}_{10}2^{5}_{12}1^{6}_{14}$			
10^a_{105}: $1^{\underline{8}}_{\underline{22}}1^{\underline{8}}_{\underline{20}}2^{\underline{7}}_{\underline{20}}2^{\underline{6}}_{\underline{18}}2^{\underline{6}}_{\underline{16}}4^{\underline{5}}_{\underline{16}}2^{\underline{5}}_{\underline{14}}4^{\underline{4}}_{\underline{14}}4^{\underline{4}}_{\underline{12}}3^{\underline{3}}_{\underline{12}}4^{\underline{3}}_{\underline{10}}4^{\underline{2}}_{\underline{10}}3^{\underline{2}}_{\underline{8}}2^{\underline{1}}_{\underline{8}}4^{\underline{1}}_{\underline{6}}2^{0}_{\underline{6}}3^{0}_{\underline{4}}1^{1}_{\underline{4}}1^{1}_{\underline{2}}1^{2}_{0}$			
10^a_{106}: $1^{\underline{7}}_{\underline{18}}2^{\underline{6}}_{\underline{16}}1^{\underline{6}}_{\underline{14}}4^{\underline{5}}_{\underline{14}}2^{\underline{5}}_{\underline{12}}7^{\underline{4}}_{\underline{12}}5^{\underline{4}}_{\underline{10}}7^{\underline{3}}_{\underline{10}}6^{\underline{3}}_{\underline{8}}7^{\underline{2}}_{\underline{8}}7^{\underline{2}}_{\underline{6}}6^{\underline{1}}_{\underline{6}}7^{\underline{1}}_{\underline{4}}5^{0}_{\underline{4}}7^{0}_{\underline{2}}3^{1}_{\underline{2}}4^{1}_{0}1^{2}_{0}3^{2}_{2}1^{3}_{4}$	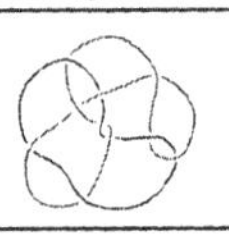	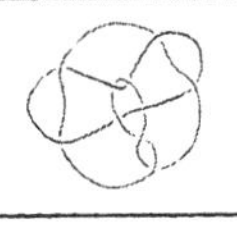	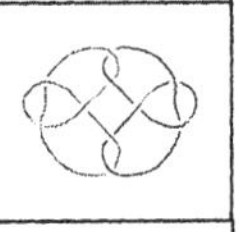
10^a_{107}: $1^{\underline{8}}_{\underline{20}}1^{\underline{8}}_{\underline{18}}3^{\underline{7}}_{\underline{18}}5^{\underline{6}}_{\underline{16}}3^{\underline{6}}_{\underline{14}}6^{\underline{5}}_{\underline{14}}5^{\underline{5}}_{\underline{12}}8^{\underline{4}}_{\underline{12}}6^{\underline{4}}_{\underline{10}}7^{\underline{3}}_{\underline{10}}8^{\underline{3}}_{\underline{8}}7^{\underline{2}}_{\underline{8}}7^{\underline{2}}_{\underline{6}}4^{\underline{1}}_{\underline{6}}7^{\underline{1}}_{\underline{4}}3^{0}_{\underline{4}}5^{0}_{\underline{2}}1^{1}_{\underline{2}}2^{1}_{0}1^{2}_{2}$			
10^a_{108}: $1^{\underline{10}}_{\underline{26}}2^{\underline{9}}_{\underline{24}}1^{\underline{9}}_{\underline{22}}4^{\underline{8}}_{\underline{22}}2^{\underline{8}}_{\underline{20}}7^{\underline{7}}_{\underline{20}}4^{\underline{7}}_{\underline{18}}6^{\underline{6}}_{\underline{18}}7^{\underline{6}}_{\underline{16}}8^{\underline{5}}_{\underline{16}}6^{\underline{5}}_{\underline{14}}6^{\underline{4}}_{\underline{14}}9^{\underline{4}}_{\underline{12}}5^{\underline{3}}_{\underline{12}}5^{\underline{3}}_{\underline{10}}3^{\underline{2}}_{\underline{10}}5^{\underline{2}}_{\underline{8}}3^{\underline{1}}_{\underline{6}}1^{0}_{\underline{6}}1^{0}_{\underline{4}}$			
10^a_{109}: $1^{\underline{5}}_{\underline{10}}1^{\underline{4}}_{\underline{8}}1^{\underline{4}}_{\underline{6}}4^{\underline{3}}_{\underline{6}}1^{\underline{3}}_{\underline{4}}5^{\underline{2}}_{\underline{4}}4^{\underline{2}}_{\underline{2}}5^{\underline{1}}_{\underline{2}}5^{\underline{1}}_{0}8^{0}_{0}7^{0}_{2}5^{1}_{2}6^{1}_{4}5^{2}_{4}5^{2}_{6}2^{3}_{6}5^{3}_{8}1^{4}_{8}2^{4}_{10}1^{5}_{12}$	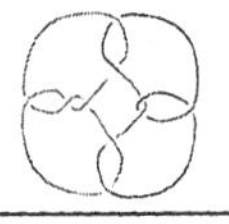	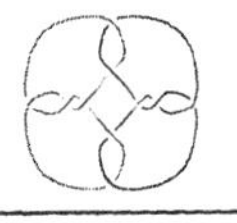	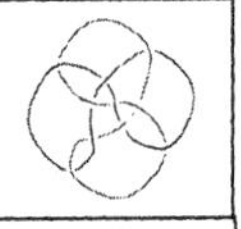
10^a_{110}: $1^{\underline{4}}_{\underline{6}}1^{\underline{4}}_{\underline{4}}1^{\underline{3}}_{\underline{4}}4^{\underline{2}}_{\underline{2}}1^{\underline{2}}_{0}2^{\underline{1}}_{0}4^{\underline{1}}_{2}7^{0}_{2}3^{0}_{4}4^{1}_{4}6^{1}_{6}5^{2}_{6}4^{2}_{8}4^{3}_{8}5^{3}_{10}2^{4}_{10}4^{4}_{12}1^{5}_{12}2^{5}_{14}1^{6}_{16}$			
10^a_{111}: $1^{\underline{6}}_{\underline{14}}3^{\underline{5}}_{\underline{12}}1^{\underline{5}}_{\underline{10}}6^{\underline{4}}_{\underline{10}}3^{\underline{4}}_{\underline{8}}9^{\underline{3}}_{\underline{8}}6^{\underline{3}}_{\underline{6}}10^{\underline{2}}_{\underline{6}}9^{\underline{2}}_{\underline{4}}10^{\underline{1}}_{\underline{4}}10^{\underline{1}}_{\underline{2}}10^{0}_{\underline{2}}12^{0}_{0}7^{1}_{0}8^{1}_{2}4^{2}_{2}7^{2}_{4}1^{3}_{4}4^{3}_{6}1^{4}_{8}$			
10^a_{112}: $1^{\underline{5}}_{\underline{12}}3^{\underline{4}}_{\underline{10}}1^{\underline{4}}_{\underline{8}}6^{\underline{3}}_{\underline{8}}3^{\underline{3}}_{\underline{6}}8^{\underline{2}}_{\underline{6}}6^{\underline{2}}_{\underline{4}}8^{\underline{1}}_{\underline{4}}8^{\underline{1}}_{\underline{2}}10^{0}_{\underline{2}}10^{0}_{0}8^{1}_{0}8^{1}_{2}5^{2}_{2}8^{2}_{4}3^{3}_{4}5^{3}_{6}1^{4}_{6}3^{4}_{8}1^{5}_{10}$	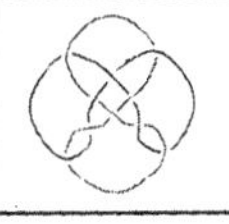	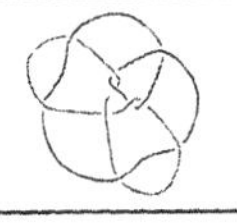	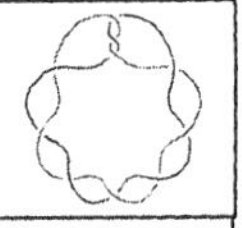
10^a_{113}: $1^{\underline{4}}_{\underline{6}}2^{\underline{3}}_{\underline{4}}1^{\underline{3}}_{\underline{2}}4^{\underline{2}}_{\underline{2}}2^{\underline{2}}_{0}4^{\underline{1}}_{0}4^{\underline{1}}_{2}8^{0}_{2}6^{0}_{4}5^{1}_{4}6^{1}_{6}6^{2}_{6}5^{2}_{8}4^{3}_{8}6^{3}_{10}2^{4}_{10}4^{4}_{12}1^{5}_{12}2^{5}_{14}1^{6}_{16}$			
10^a_{114}: $1^{\underline{10}}_{\underline{28}}1^{\underline{10}}_{\underline{26}}1^{\underline{9}}_{\underline{26}}1^{\underline{8}}_{\underline{24}}1^{\underline{8}}_{\underline{22}}2^{\underline{7}}_{\underline{22}}1^{\underline{7}}_{\underline{20}}1^{\underline{6}}_{\underline{20}}2^{\underline{6}}_{\underline{18}}2^{\underline{5}}_{\underline{18}}1^{\underline{5}}_{\underline{16}}1^{\underline{4}}_{\underline{16}}2^{\underline{4}}_{\underline{14}}1^{\underline{3}}_{\underline{14}}1^{\underline{3}}_{\underline{12}}1^{\underline{2}}_{\underline{12}}1^{\underline{2}}_{\underline{10}}1^{\underline{1}}_{\underline{8}}1^{0}_{\underline{8}}1^{0}_{\underline{6}}$			
10^a_{115}: $1^{\underline{10}}_{\underline{26}}1^{\underline{10}}_{\underline{24}}1^{\underline{9}}_{\underline{24}}2^{\underline{8}}_{\underline{22}}1^{\underline{8}}_{\underline{20}}3^{\underline{7}}_{\underline{20}}2^{\underline{7}}_{\underline{18}}3^{\underline{6}}_{\underline{18}}3^{\underline{6}}_{\underline{16}}4^{\underline{5}}_{\underline{16}}3^{\underline{5}}_{\underline{14}}3^{\underline{4}}_{\underline{14}}4^{\underline{4}}_{\underline{12}}2^{\underline{3}}_{\underline{12}}3^{\underline{3}}_{\underline{10}}2^{\underline{2}}_{\underline{10}}2^{\underline{2}}_{\underline{8}}2^{\underline{1}}_{\underline{6}}1^{0}_{\underline{6}}1^{0}_{\underline{4}}$	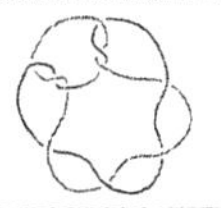	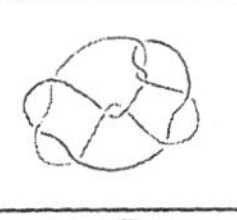	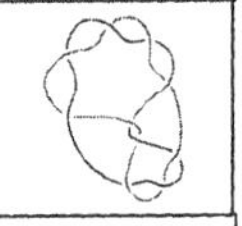
10^a_{116}: $1^{\underline{10}}_{\underline{26}}1^{\underline{10}}_{\underline{24}}2^{\underline{9}}_{\underline{24}}3^{\underline{8}}_{\underline{22}}2^{\underline{8}}_{\underline{20}}4^{\underline{7}}_{\underline{20}}3^{\underline{7}}_{\underline{18}}5^{\underline{6}}_{\underline{18}}4^{\underline{6}}_{\underline{16}}5^{\underline{5}}_{\underline{16}}5^{\underline{5}}_{\underline{14}}4^{\underline{4}}_{\underline{14}}5^{\underline{4}}_{\underline{12}}3^{\underline{3}}_{\underline{12}}4^{\underline{3}}_{\underline{10}}2^{\underline{2}}_{\underline{10}}3^{\underline{2}}_{\underline{8}}2^{\underline{1}}_{\underline{6}}1^{0}_{\underline{6}}1^{0}_{\underline{4}}$			
10^a_{117}: $1^{\underline{10}}_{\underline{28}}1^{\underline{10}}_{\underline{26}}2^{\underline{9}}_{\underline{26}}2^{\underline{8}}_{\underline{24}}2^{\underline{8}}_{\underline{22}}3^{\underline{7}}_{\underline{22}}2^{\underline{7}}_{\underline{20}}3^{\underline{6}}_{\underline{20}}3^{\underline{6}}_{\underline{18}}3^{\underline{5}}_{\underline{18}}3^{\underline{5}}_{\underline{16}}2^{\underline{4}}_{\underline{16}}3^{\underline{4}}_{\underline{14}}2^{\underline{3}}_{\underline{14}}2^{\underline{3}}_{\underline{12}}1^{\underline{2}}_{\underline{12}}2^{\underline{2}}_{\underline{10}}1^{\underline{1}}_{\underline{8}}1^{0}_{\underline{8}}1^{0}_{\underline{6}}$			
10^a_{118}: $1^{\underline{10}}_{\underline{30}}1^{\underline{10}}_{\underline{28}}1^{\underline{9}}_{\underline{28}}1^{\underline{8}}_{\underline{24}}1^{\underline{7}}_{\underline{24}}1^{\underline{6}}_{\underline{20}}1^{\underline{5}}_{\underline{20}}1^{\underline{4}}_{\underline{16}}1^{\underline{3}}_{\underline{16}}1^{\underline{2}}_{\underline{12}}1^{0}_{\underline{10}}1^{0}_{\underline{8}}$	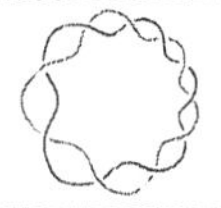	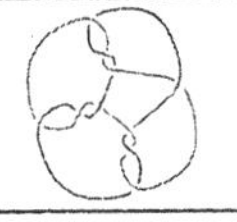	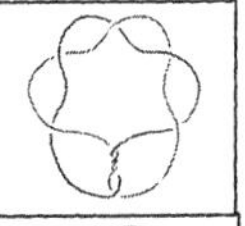
10^a_{119}: $1^{\underline{10}}_{\underline{24}}1^{\underline{10}}_{\underline{22}}1^{\underline{9}}_{\underline{22}}3^{\underline{8}}_{\underline{20}}1^{\underline{8}}_{\underline{18}}3^{\underline{7}}_{\underline{18}}3^{\underline{7}}_{\underline{16}}4^{\underline{6}}_{\underline{16}}3^{\underline{6}}_{\underline{14}}5^{\underline{5}}_{\underline{14}}4^{\underline{5}}_{\underline{12}}4^{\underline{4}}_{\underline{12}}5^{\underline{4}}_{\underline{10}}3^{\underline{3}}_{\underline{10}}4^{\underline{3}}_{\underline{8}}3^{\underline{2}}_{\underline{8}}3^{\underline{2}}_{\underline{6}}3^{\underline{1}}_{\underline{4}}1^{0}_{\underline{4}}1^{0}_{\underline{2}}$			
10^a_{120}: $1^{\underline{10}}_{\underline{26}}1^{\underline{10}}_{\underline{24}}1^{\underline{9}}_{\underline{24}}1^{\underline{8}}_{\underline{22}}1^{\underline{8}}_{\underline{20}}2^{\underline{7}}_{\underline{20}}1^{\underline{7}}_{\underline{18}}2^{\underline{6}}_{\underline{18}}2^{\underline{6}}_{\underline{16}}2^{\underline{5}}_{\underline{16}}2^{\underline{5}}_{\underline{14}}2^{\underline{4}}_{\underline{14}}2^{\underline{4}}_{\underline{12}}1^{\underline{3}}_{\underline{12}}2^{\underline{3}}_{\underline{10}}1^{\underline{2}}_{\underline{10}}1^{\underline{2}}_{\underline{8}}1^{\underline{1}}_{\underline{6}}1^{0}_{\underline{6}}1^{0}_{\underline{4}}$			
10^a_{121}: $1^{\underline{10}}_{\underline{26}}1^{\underline{10}}_{\underline{24}}3^{\underline{9}}_{\underline{24}}5^{\underline{8}}_{\underline{22}}3^{\underline{8}}_{\underline{20}}6^{\underline{7}}_{\underline{20}}5^{\underline{7}}_{\underline{18}}8^{\underline{6}}_{\underline{18}}6^{\underline{6}}_{\underline{16}}8^{\underline{5}}_{\underline{16}}8^{\underline{5}}_{\underline{14}}6^{\underline{4}}_{\underline{14}}8^{\underline{4}}_{\underline{12}}5^{\underline{3}}_{\underline{12}}6^{\underline{3}}_{\underline{10}}3^{\underline{2}}_{\underline{10}}5^{\underline{2}}_{\underline{8}}3^{\underline{1}}_{\underline{6}}1^{0}_{\underline{6}}1^{0}_{\underline{4}}$	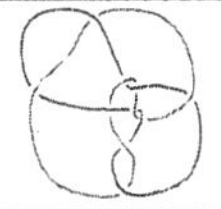	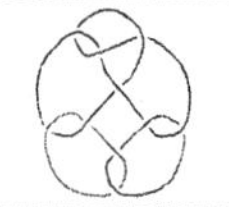	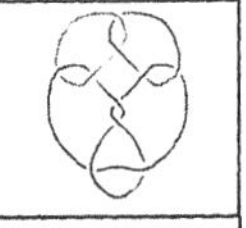
10^a_{122}: $1^{\underline{6}}_{\underline{13}}1^{\underline{5}}_{\underline{11}}1^{\underline{5}}_{\underline{9}}5^{\underline{4}}_{\underline{9}}2^{\underline{4}}_{\underline{7}}5^{\underline{3}}_{\underline{7}}4^{\underline{3}}_{\underline{5}}8^{\underline{2}}_{\underline{5}}5^{\underline{2}}_{\underline{3}}6^{\underline{1}}_{\underline{3}}8^{\underline{1}}_{\underline{1}}8^{0}_{\underline{1}}9^{0}_{1}6^{1}_{1}5^{1}_{3}3^{2}_{3}6^{2}_{5}1^{3}_{5}3^{3}_{7}1^{4}_{9}$			
10^a_{123}: $1^{\underline{8}}_{\underline{19}}1^{\underline{7}}_{\underline{17}}1^{\underline{7}}_{\underline{15}}5^{\underline{6}}_{\underline{15}}1^{\underline{6}}_{\underline{13}}4^{\underline{5}}_{\underline{13}}5^{\underline{5}}_{\underline{11}}8^{\underline{4}}_{\underline{11}}7^{\underline{4}}_{\underline{9}}7^{\underline{3}}_{\underline{9}}5^{\underline{3}}_{\underline{7}}6^{\underline{2}}_{\underline{7}}7^{\underline{2}}_{\underline{5}}4^{\underline{1}}_{\underline{5}}6^{\underline{1}}_{\underline{3}}3^{0}_{\underline{3}}5^{0}_{\underline{1}}1^{1}_{\underline{1}}2^{1}_{1}1^{2}_{3}$			
10^a_{124}: $1^{\underline{7}}_{\underline{17}}1^{\underline{6}}_{\underline{15}}1^{\underline{6}}_{\underline{13}}4^{\underline{5}}_{\underline{13}}1^{\underline{5}}_{\underline{11}}5^{\underline{4}}_{\underline{11}}5^{\underline{4}}_{\underline{9}}7^{\underline{3}}_{\underline{9}}4^{\underline{3}}_{\underline{7}}7^{\underline{2}}_{\underline{7}}7^{\underline{2}}_{\underline{5}}4^{\underline{1}}_{\underline{5}}7^{\underline{1}}_{\underline{3}}7^{0}_{\underline{3}}7^{0}_{\underline{1}}3^{1}_{\underline{1}}4^{1}_{1}1^{2}_{1}3^{2}_{3}1^{3}_{5}$	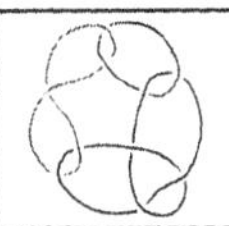	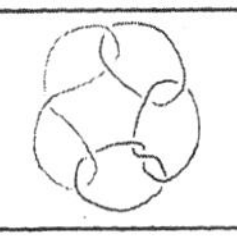	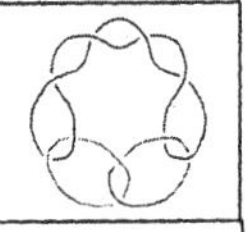
10^a_{125}: $1^{\underline{6}}_{\underline{13}}1^{\underline{5}}_{\underline{9}}5^{\underline{4}}_{\underline{9}}3^{\underline{4}}_{\underline{7}}4^{\underline{3}}_{\underline{7}}2^{\underline{3}}_{\underline{5}}7^{\underline{2}}_{\underline{5}}4^{\underline{2}}_{\underline{3}}4^{\underline{1}}_{\underline{3}}7^{\underline{1}}_{\underline{1}}7^{0}_{\underline{1}}5^{0}_{1}4^{1}_{1}6^{1}_{3}3^{2}_{3}4^{2}_{5}1^{3}_{5}3^{3}_{7}1^{4}_{9}$			
10^a_{126}: $1^{\underline{6}}_{\underline{15}}1^{\underline{5}}_{\underline{11}}4^{\underline{4}}_{\underline{11}}3^{\underline{4}}_{\underline{9}}3^{\underline{3}}_{\underline{9}}1^{\underline{3}}_{\underline{7}}4^{\underline{2}}_{\underline{7}}3^{\underline{2}}_{\underline{5}}3^{\underline{1}}_{\underline{5}}4^{\underline{1}}_{\underline{3}}4^{0}_{\underline{3}}4^{0}_{\underline{1}}2^{1}_{\underline{1}}3^{1}_{1}2^{2}_{1}2^{2}_{3}1^{3}_{3}2^{3}_{5}1^{4}_{7}$			
10^a_{127}: $1^{\underline{3}}_{\underline{5}}3^{\underline{2}}_{\underline{3}}1^{\underline{2}}_{\underline{1}}4^{\underline{1}}_{\underline{1}}3^{\underline{1}}_{1}9^{0}_{1}6^{0}_{3}7^{1}_{3}7^{1}_{5}10^{2}_{5}9^{2}_{7}7^{3}_{7}8^{3}_{9}5^{4}_{9}7^{4}_{11}3^{5}_{11}5^{5}_{13}1^{6}_{13}3^{6}_{15}1^{7}_{17}$	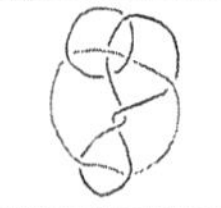	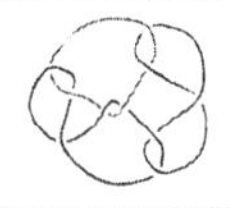	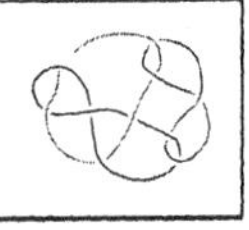
10^a_{128}: $1^{\underline{6}}_{\underline{13}}1^{\underline{5}}_{\underline{11}}1^{\underline{5}}_{\underline{9}}5^{\underline{4}}_{\underline{9}}2^{\underline{4}}_{\underline{7}}4^{\underline{3}}_{\underline{7}}4^{\underline{3}}_{\underline{5}}8^{\underline{2}}_{\underline{5}}6^{\underline{2}}_{\underline{3}}6^{\underline{1}}_{\underline{3}}6^{\underline{1}}_{\underline{1}}7^{0}_{\underline{1}}7^{0}_{1}4^{1}_{1}6^{1}_{3}3^{2}_{3}4^{2}_{5}1^{3}_{5}3^{3}_{7}1^{4}_{9}$			
10^a_{129}: $1^{\underline{7}}_{\underline{17}}2^{\underline{6}}_{\underline{15}}1^{\underline{6}}_{\underline{13}}4^{\underline{5}}_{\underline{13}}2^{\underline{5}}_{\underline{11}}7^{\underline{4}}_{\underline{11}}5^{\underline{4}}_{\underline{9}}6^{\underline{3}}_{\underline{9}}6^{\underline{3}}_{\underline{7}}9^{\underline{2}}_{\underline{7}}8^{\underline{2}}_{\underline{5}}6^{\underline{1}}_{\underline{5}}7^{\underline{1}}_{\underline{3}}5^{0}_{\underline{3}}7^{0}_{\underline{1}}3^{1}_{\underline{1}}4^{1}_{1}1^{2}_{1}3^{2}_{3}1^{3}_{5}$			

L: KH	1st line	2nd line	3rd line
10^a_{130}: $1^{\underline{10}}_{\underline{25}}2^{\underline{9}}_{\underline{23}}1^{\underline{9}}_{\underline{21}}4^{\underline{8}}_{\underline{21}}2^{\underline{8}}_{\underline{19}}5^{\underline{7}}_{\underline{19}}4^{\underline{7}}_{\underline{17}}7^{\underline{6}}_{\underline{17}}7^{\underline{6}}_{\underline{15}}7^{\underline{5}}_{\underline{15}}5^{\underline{5}}_{\underline{13}}6^{\underline{4}}_{\underline{13}}8^{\underline{4}}_{\underline{11}}4^{\underline{3}}_{\underline{11}}5^{\underline{3}}_{\underline{9}}3^{\underline{2}}_{\underline{9}}4^{\underline{2}}_{\underline{7}}3^{\underline{1}}_{\underline{5}}1^{0}_{\underline{5}}1^{0}_{\underline{3}}$	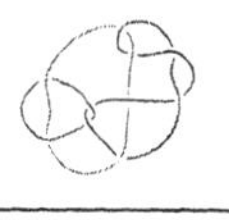	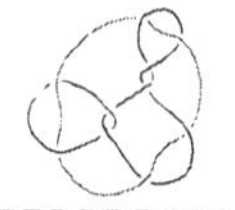	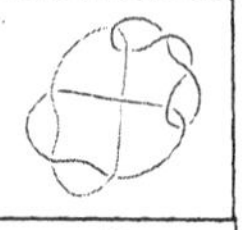
10^a_{131}: $1^{\underline{2}}_{\underline{1}}1^{\underline{1}}_{1}1^{\underline{1}}_{3}5^{0}_{3}3^{0}_{5}4^{1}_{5}3^{1}_{7}7^{2}_{7}5^{2}_{9}5^{3}_{9}6^{3}_{11}6^{4}_{11}5^{4}_{13}3^{5}_{13}6^{5}_{15}3^{6}_{15}4^{6}_{17}1^{7}_{17}2^{7}_{19}1^{8}_{21}$			
10^a_{132}: $1^{\underline{8}}_{\underline{21}}1^{\underline{7}}_{\underline{19}}1^{\underline{7}}_{\underline{17}}4^{\underline{6}}_{\underline{17}}3^{\underline{6}}_{\underline{15}}5^{\underline{5}}_{\underline{15}}2^{\underline{5}}_{\underline{13}}5^{\underline{4}}_{\underline{13}}6^{\underline{4}}_{\underline{11}}5^{\underline{3}}_{\underline{11}}4^{\underline{3}}_{\underline{9}}5^{\underline{2}}_{\underline{9}}5^{\underline{2}}_{\underline{7}}2^{\underline{1}}_{\underline{7}}5^{\underline{1}}_{\underline{5}}3^{0}_{\underline{5}}3^{0}_{\underline{3}}1^{1}_{\underline{3}}2^{1}_{\underline{1}}1^{2}_{1}$			
10^a_{133}: $1^{\underline{2}}_{\underline{1}}1^{\underline{1}}_{3}4^{0}_{3}2^{0}_{5}3^{1}_{5}2^{1}_{7}5^{2}_{7}4^{2}_{9}4^{3}_{9}4^{3}_{11}5^{4}_{11}4^{4}_{13}2^{5}_{13}5^{5}_{15}3^{6}_{15}3^{6}_{17}1^{7}_{17}2^{7}_{19}1^{8}_{21}$	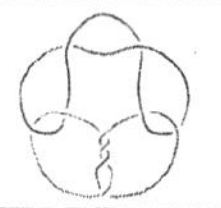	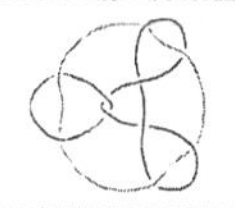	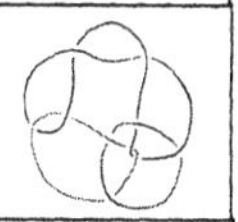
10^a_{134}: $1^{\underline{6}}_{\underline{13}}1^{\underline{6}}_{\underline{11}}2^{\underline{5}}_{\underline{11}}5^{\underline{4}}_{\underline{9}}2^{\underline{4}}_{\underline{7}}4^{\underline{3}}_{\underline{7}}5^{\underline{3}}_{\underline{5}}9^{\underline{2}}_{\underline{5}}5^{\underline{2}}_{\underline{3}}6^{\underline{1}}_{\underline{3}}8^{\underline{1}}_{\underline{1}}8^{0}_{\underline{1}}8^{0}_{1}5^{1}_{1}6^{1}_{3}3^{2}_{3}5^{2}_{5}1^{3}_{5}3^{3}_{7}1^{4}_{9}$			
10^a_{135}: $1^{\underline{4}}_{\underline{9}}2^{\underline{3}}_{\underline{7}}1^{\underline{3}}_{\underline{5}}5^{\underline{2}}_{\underline{5}}3^{\underline{2}}_{\underline{3}}6^{\underline{1}}_{\underline{3}}4^{\underline{1}}_{\underline{1}}8^{0}_{\underline{1}}8^{0}_{1}7^{1}_{1}6^{1}_{3}7^{2}_{3}8^{2}_{5}4^{3}_{5}6^{3}_{7}3^{4}_{7}4^{4}_{9}1^{5}_{9}3^{5}_{11}1^{6}_{13}$			
10^a_{136}: $1^{\underline{5}}_{\underline{11}}3^{\underline{4}}_{\underline{9}}1^{\underline{4}}_{\underline{7}}5^{\underline{3}}_{\underline{7}}3^{\underline{3}}_{\underline{5}}8^{\underline{2}}_{\underline{5}}6^{\underline{2}}_{\underline{3}}8^{\underline{1}}_{\underline{3}}7^{\underline{1}}_{\underline{1}}10^{0}_{\underline{1}}10^{0}_{1}7^{1}_{1}8^{1}_{3}6^{2}_{3}8^{2}_{5}3^{3}_{5}5^{3}_{7}1^{4}_{7}3^{4}_{9}1^{5}_{11}$		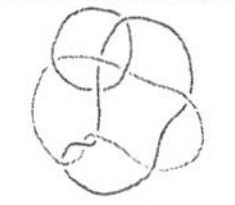	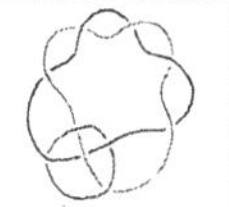
10^a_{137}: $1^{\underline{3}}_{\underline{5}}3^{\underline{2}}_{\underline{3}}1^{\underline{2}}_{\underline{1}}4^{\underline{1}}_{\underline{1}}3^{\underline{1}}_{1}7^{0}_{1}6^{0}_{3}7^{1}_{3}5^{1}_{5}7^{2}_{5}7^{2}_{7}5^{3}_{7}7^{3}_{9}5^{4}_{9}7^{4}_{11}2^{5}_{11}3^{5}_{13}1^{6}_{13}2^{6}_{15}1^{7}_{17}$			
10^a_{138}: $1^{\underline{3}}_{\underline{3}}2^{\underline{2}}_{\underline{1}}1^{\underline{2}}_{1}1^{\underline{1}}_{1}2^{\underline{1}}_{3}5^{0}_{3}3^{0}_{5}3^{1}_{5}3^{1}_{7}5^{2}_{7}3^{2}_{9}2^{3}_{9}5^{3}_{11}4^{4}_{11}4^{4}_{13}2^{5}_{13}2^{5}_{15}1^{6}_{15}2^{6}_{17}1^{7}_{19}$			
10^a_{139}: $1^{\underline{6}}_{\underline{13}}2^{\underline{5}}_{\underline{11}}1^{\underline{5}}_{\underline{9}}4^{\underline{4}}_{\underline{9}}3^{\underline{4}}_{\underline{7}}5^{\underline{3}}_{\underline{7}}3^{\underline{3}}_{\underline{5}}5^{\underline{2}}_{\underline{5}}5^{\underline{2}}_{\underline{3}}5^{\underline{1}}_{\underline{3}}5^{\underline{1}}_{\underline{1}}6^{0}_{\underline{1}}7^{0}_{1}3^{1}_{1}4^{1}_{3}2^{2}_{3}3^{2}_{5}2^{3}_{7}1^{4}_{7}1^{4}_{9}$	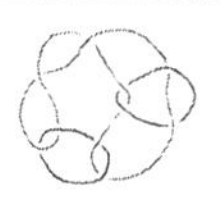	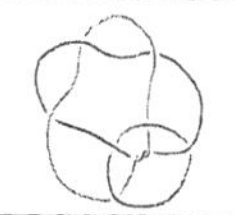	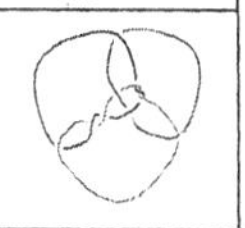
10^a_{140}: $1^{\underline{5}}_{\underline{11}}2^{\underline{4}}_{\underline{9}}1^{\underline{4}}_{\underline{7}}3^{\underline{3}}_{\underline{7}}2^{\underline{3}}_{\underline{5}}5^{\underline{2}}_{\underline{5}}3^{\underline{2}}_{\underline{3}}4^{\underline{1}}_{\underline{3}}5^{\underline{1}}_{\underline{1}}8^{0}_{\underline{1}}8^{0}_{1}5^{1}_{1}4^{1}_{3}3^{2}_{3}5^{2}_{5}2^{3}_{5}3^{3}_{7}1^{4}_{7}2^{4}_{9}1^{5}_{11}$			
10^a_{141}: $1^{\underline{5}}_{\underline{9}}2^{\underline{4}}_{\underline{7}}1^{\underline{4}}_{\underline{5}}2^{\underline{3}}_{\underline{5}}2^{\underline{3}}_{\underline{3}}6^{\underline{2}}_{\underline{3}}2^{\underline{2}}_{\underline{1}}3^{\underline{1}}_{\underline{1}}6^{\underline{1}}_{1}8^{0}_{1}7^{0}_{3}5^{1}_{3}4^{1}_{5}4^{2}_{5}5^{2}_{7}2^{3}_{7}4^{3}_{9}1^{4}_{9}2^{4}_{11}1^{5}_{13}$			
10^a_{142}: $1^{\underline{10}}_{\underline{27}}1^{\underline{9}}_{\underline{25}}1^{\underline{9}}_{\underline{23}}3^{\underline{8}}_{\underline{23}}3^{\underline{8}}_{\underline{21}}4^{\underline{7}}_{\underline{21}}1^{\underline{7}}_{\underline{19}}3^{\underline{6}}_{\underline{19}}4^{\underline{6}}_{\underline{17}}4^{\underline{5}}_{\underline{17}}3^{\underline{5}}_{\underline{15}}3^{\underline{4}}_{\underline{15}}5^{\underline{4}}_{\underline{13}}2^{\underline{3}}_{\underline{13}}2^{\underline{3}}_{\underline{11}}2^{\underline{2}}_{\underline{11}}2^{\underline{2}}_{\underline{9}}2^{\underline{1}}_{\underline{7}}1^{0}_{\underline{7}}1^{0}_{\underline{5}}$	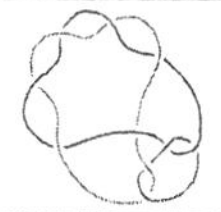	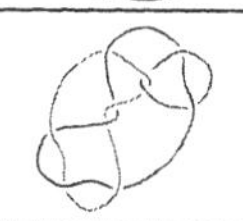	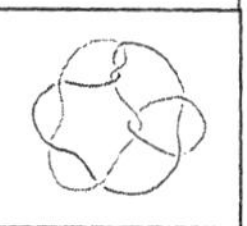
10^a_{143}: $1^{\underline{10}}_{\underline{25}}1^{\underline{9}}_{\underline{23}}1^{\underline{9}}_{\underline{21}}4^{\underline{8}}_{\underline{21}}3^{\underline{8}}_{\underline{19}}5^{\underline{7}}_{\underline{19}}2^{\underline{7}}_{\underline{17}}6^{\underline{6}}_{\underline{17}}5^{\underline{6}}_{\underline{15}}5^{\underline{5}}_{\underline{15}}6^{\underline{5}}_{\underline{13}}6^{\underline{4}}_{\underline{13}}6^{\underline{4}}_{\underline{11}}4^{\underline{3}}_{\underline{11}}5^{\underline{3}}_{\underline{9}}3^{\underline{2}}_{\underline{9}}4^{\underline{2}}_{\underline{7}}3^{\underline{1}}_{\underline{5}}1^{0}_{\underline{5}}1^{0}_{\underline{3}}$			
10^a_{144}: $1^{\underline{2}}_{\underline{1}}1^{\underline{1}}_{1}1^{\underline{1}}_{3}4^{0}_{3}3^{0}_{5}4^{1}_{5}2^{1}_{7}4^{2}_{7}4^{2}_{9}4^{3}_{9}4^{3}_{11}5^{4}_{11}5^{4}_{13}2^{5}_{13}4^{5}_{15}2^{6}_{15}2^{6}_{17}2^{7}_{19}1^{8}_{19}1^{8}_{21}$			
10^a_{145}: $1^{\underline{2}}_{1}1^{\underline{1}}_{5}3^{0}_{5}2^{0}_{7}1^{1}_{7}1^{1}_{9}3^{2}_{9}1^{2}_{11}1^{3}_{11}3^{3}_{13}3^{4}_{13}2^{4}_{15}1^{5}_{15}2^{5}_{17}2^{6}_{17}1^{6}_{19}2^{7}_{21}1^{8}_{21}1^{8}_{23}$	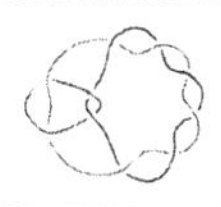	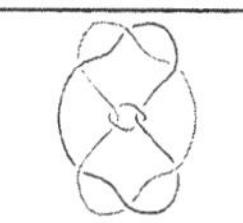	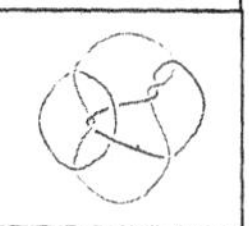
10^a_{146}: $1^{\underline{2}}_{\underline{3}}1^{\underline{1}}_{\underline{1}}1^{\underline{1}}_{1}4^{0}_{1}3^{0}_{3}6^{1}_{3}2^{1}_{5}4^{2}_{5}6^{2}_{7}5^{3}_{7}4^{3}_{9}5^{4}_{9}6^{4}_{11}3^{5}_{11}4^{5}_{13}2^{6}_{13}3^{6}_{15}2^{7}_{17}1^{8}_{17}1^{8}_{19}$			
10^a_{147}: $1^{\underline{4}}_{\underline{9}}3^{\underline{3}}_{\underline{7}}1^{\underline{3}}_{\underline{5}}5^{\underline{2}}_{\underline{5}}3^{\underline{2}}_{\underline{3}}6^{\underline{1}}_{\underline{3}}5^{\underline{1}}_{\underline{1}}8^{0}_{\underline{1}}9^{0}_{1}8^{1}_{1}5^{1}_{3}6^{2}_{3}8^{2}_{5}3^{3}_{5}6^{3}_{7}3^{4}_{7}4^{4}_{9}1^{5}_{9}2^{5}_{11}1^{6}_{13}$			
10^a_{148}: $1^{\underline{4}}_{\underline{7}}2^{\underline{3}}_{\underline{5}}1^{\underline{3}}_{\underline{3}}3^{\underline{2}}_{\underline{3}}2^{\underline{2}}_{\underline{1}}3^{\underline{1}}_{\underline{1}}3^{\underline{1}}_{1}7^{0}_{1}6^{0}_{3}5^{1}_{3}4^{1}_{5}5^{2}_{5}5^{2}_{7}2^{3}_{7}5^{3}_{9}3^{4}_{9}3^{4}_{11}1^{5}_{11}2^{5}_{13}1^{6}_{15}$	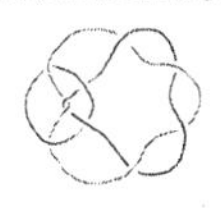	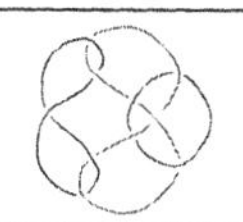	
10^a_{149}: $1^{\underline{3}}_{\underline{5}}2^{\underline{2}}_{\underline{3}}1^{\underline{2}}_{\underline{1}}4^{\underline{1}}_{\underline{1}}2^{\underline{1}}_{1}8^{0}_{1}6^{0}_{3}8^{1}_{3}6^{1}_{5}9^{2}_{5}8^{2}_{7}6^{3}_{7}9^{3}_{9}7^{4}_{9}8^{4}_{11}4^{5}_{11}5^{5}_{13}1^{6}_{13}4^{6}_{15}1^{7}_{17}$			
10^a_{150}: $1^{\underline{2}}_{\underline{1}}1^{\underline{1}}_{1}1^{\underline{1}}_{3}5^{0}_{3}3^{0}_{5}5^{1}_{5}3^{1}_{7}8^{2}_{7}5^{2}_{9}5^{3}_{9}8^{3}_{11}8^{4}_{11}7^{4}_{13}5^{5}_{13}6^{5}_{15}3^{6}_{15}5^{6}_{17}1^{7}_{17}3^{7}_{19}1^{8}_{21}$			
10^a_{151}: $1^{\underline{5}}_{\underline{11}}2^{\underline{4}}_{\underline{9}}1^{\underline{4}}_{\underline{7}}4^{\underline{3}}_{\underline{7}}2^{\underline{3}}_{\underline{5}}7^{\underline{2}}_{\underline{5}}4^{\underline{2}}_{\underline{3}}5^{\underline{1}}_{\underline{3}}7^{\underline{1}}_{\underline{1}}9^{0}_{\underline{1}}9^{0}_{1}7^{1}_{1}5^{1}_{3}4^{2}_{3}7^{2}_{5}2^{3}_{5}4^{3}_{7}1^{4}_{7}2^{4}_{9}1^{5}_{11}$	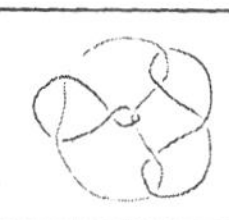	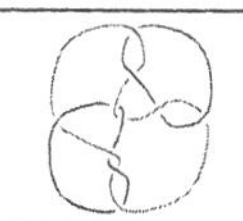	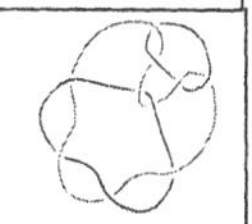
10^a_{152}: $1^{\underline{4}}_{\underline{7}}1^{\underline{4}}_{\underline{5}}2^{\underline{3}}_{\underline{5}}4^{\underline{2}}_{\underline{3}}2^{\underline{2}}_{\underline{1}}4^{\underline{1}}_{\underline{1}}4^{\underline{1}}_{1}7^{0}_{1}5^{0}_{3}6^{1}_{3}6^{1}_{5}6^{2}_{5}7^{2}_{7}4^{3}_{7}5^{3}_{9}3^{4}_{9}4^{4}_{11}3^{5}_{13}1^{6}_{13}1^{6}_{15}$			
10^a_{153}: $1^{\underline{10}}_{\underline{27}}1^{\underline{10}}_{\underline{25}}2^{\underline{9}}_{\underline{25}}3^{\underline{8}}_{\underline{23}}2^{\underline{8}}_{\underline{21}}4^{\underline{7}}_{\underline{21}}3^{\underline{7}}_{\underline{19}}5^{\underline{6}}_{\underline{19}}5^{\underline{6}}_{\underline{17}}5^{\underline{5}}_{\underline{17}}4^{\underline{5}}_{\underline{15}}4^{\underline{4}}_{\underline{15}}6^{\underline{4}}_{\underline{13}}3^{\underline{3}}_{\underline{13}}3^{\underline{3}}_{\underline{11}}2^{\underline{2}}_{\underline{11}}3^{\underline{2}}_{\underline{9}}2^{\underline{1}}_{\underline{7}}1^{0}_{\underline{7}}1^{0}_{\underline{5}}$			
10^a_{154}: $1^{\underline{8}}_{\underline{19}}1^{\underline{8}}_{\underline{17}}2^{\underline{7}}_{\underline{17}}5^{\underline{6}}_{\underline{15}}2^{\underline{6}}_{\underline{13}}4^{\underline{5}}_{\underline{13}}5^{\underline{5}}_{\underline{11}}9^{\underline{4}}_{\underline{11}}6^{\underline{4}}_{\underline{9}}6^{\underline{3}}_{\underline{9}}7^{\underline{3}}_{\underline{7}}7^{\underline{2}}_{\underline{7}}6^{\underline{2}}_{\underline{5}}4^{\underline{1}}_{\underline{5}}7^{\underline{1}}_{\underline{3}}3^{0}_{\underline{3}}5^{0}_{\underline{1}}1^{1}_{\underline{1}}2^{1}_{1}1^{2}_{3}$	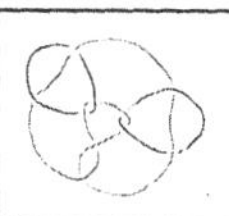	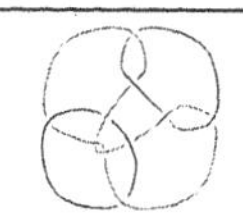	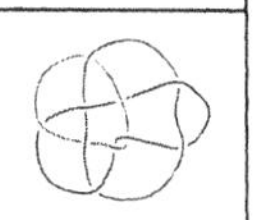
10^a_{155}: $1^{\underline{6}}_{\underline{13}}2^{\underline{5}}_{\underline{11}}1^{\underline{5}}_{\underline{9}}5^{\underline{4}}_{\underline{9}}3^{\underline{4}}_{\underline{7}}7^{\underline{3}}_{\underline{7}}4^{\underline{3}}_{\underline{5}}8^{\underline{2}}_{\underline{5}}7^{\underline{2}}_{\underline{3}}8^{\underline{1}}_{\underline{3}}8^{\underline{1}}_{\underline{1}}8^{0}_{\underline{1}}10^{0}_{1}6^{1}_{1}6^{1}_{3}4^{2}_{3}6^{2}_{5}4^{3}_{7}1^{4}_{7}1^{4}_{9}$			
10^a_{156}: $1^{\underline{5}}_{\underline{11}}3^{\underline{4}}_{\underline{9}}1^{\underline{4}}_{\underline{7}}5^{\underline{3}}_{\underline{7}}3^{\underline{3}}_{\underline{5}}8^{\underline{2}}_{\underline{5}}6^{\underline{2}}_{\underline{3}}8^{\underline{1}}_{\underline{3}}7^{\underline{1}}_{\underline{1}}10^{0}_{\underline{1}}10^{0}_{1}7^{1}_{1}8^{1}_{3}6^{2}_{3}8^{2}_{5}3^{3}_{5}5^{3}_{7}1^{4}_{7}3^{4}_{9}1^{5}_{11}$			
10^a_{157}: $1^{\underline{5}}_{\underline{13}}2^{\underline{4}}_{\underline{11}}1^{\underline{4}}_{\underline{9}}3^{\underline{3}}_{\underline{9}}2^{\underline{3}}_{\underline{7}}5^{\underline{2}}_{\underline{7}}4^{\underline{2}}_{\underline{5}}5^{\underline{1}}_{\underline{5}}4^{\underline{1}}_{\underline{3}}5^{0}_{\underline{3}}7^{0}_{\underline{1}}5^{1}_{\underline{1}}3^{1}_{1}3^{2}_{1}6^{2}_{3}2^{3}_{3}2^{3}_{5}1^{4}_{5}2^{4}_{7}1^{5}_{9}$	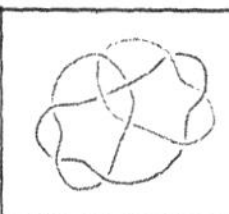	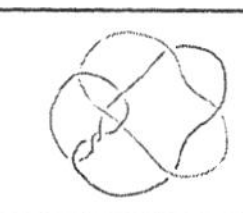	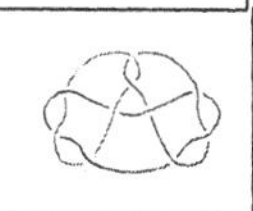
10^a_{158}: $1^{\underline{5}}_{\underline{11}}2^{\underline{4}}_{\underline{9}}1^{\underline{4}}_{\underline{7}}3^{\underline{3}}_{\underline{7}}2^{\underline{3}}_{\underline{5}}6^{\underline{2}}_{\underline{5}}4^{\underline{2}}_{\underline{3}}5^{\underline{1}}_{\underline{3}}5^{\underline{1}}_{\underline{1}}7^{0}_{\underline{1}}7^{0}_{1}5^{1}_{1}5^{1}_{3}4^{2}_{3}6^{2}_{5}2^{3}_{5}3^{3}_{7}1^{4}_{7}2^{4}_{9}1^{5}_{11}$			
10^a_{159}: $1^{\underline{10}}_{\underline{27}}2^{\underline{9}}_{\underline{25}}1^{\underline{9}}_{\underline{23}}4^{\underline{8}}_{\underline{23}}3^{\underline{8}}_{\underline{21}}5^{\underline{7}}_{\underline{21}}3^{\underline{7}}_{\underline{19}}6^{\underline{6}}_{\underline{19}}7^{\underline{6}}_{\underline{17}}6^{\underline{5}}_{\underline{17}}4^{\underline{5}}_{\underline{15}}4^{\underline{4}}_{\underline{15}}6^{\underline{4}}_{\underline{13}}3^{\underline{3}}_{\underline{13}}4^{\underline{3}}_{\underline{11}}2^{\underline{2}}_{\underline{11}}3^{\underline{2}}_{\underline{9}}2^{\underline{1}}_{\underline{7}}1^{0}_{\underline{7}}1^{0}_{\underline{5}}$			
10^a_{160}: $1^{\underline{2}}_{\underline{1}}1^{\underline{1}}_{1}1^{\underline{1}}_{3}3^{0}_{3}2^{0}_{5}3^{1}_{5}2^{1}_{7}5^{2}_{7}5^{2}_{9}4^{3}_{9}3^{3}_{11}4^{4}_{11}4^{4}_{13}2^{5}_{13}4^{5}_{15}2^{6}_{15}2^{6}_{17}2^{7}_{19}1^{8}_{19}1^{8}_{21}$	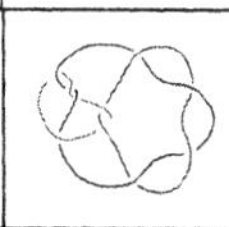	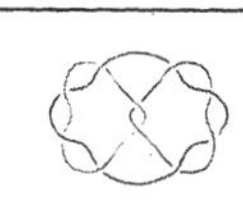	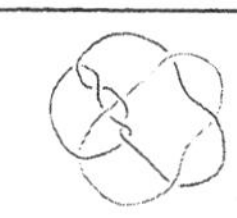
10^a_{161}: $1^{\underline{2}}_{1}1^{\underline{1}}_{5}3^{0}_{5}1^{0}_{7}1^{1}_{7}2^{1}_{9}4^{2}_{9}3^{2}_{11}2^{3}_{11}2^{3}_{13}3^{4}_{13}2^{4}_{15}1^{5}_{15}3^{5}_{17}2^{6}_{17}1^{6}_{19}2^{7}_{21}1^{8}_{21}1^{8}_{23}$			
10^a_{162}: $1^{\underline{4}}_{\underline{7}}2^{\underline{3}}_{\underline{5}}1^{\underline{3}}_{\underline{3}}4^{\underline{2}}_{\underline{3}}3^{\underline{2}}_{\underline{1}}5^{\underline{1}}_{\underline{1}}3^{\underline{1}}_{1}7^{0}_{1}6^{0}_{3}5^{1}_{3}6^{1}_{5}7^{2}_{5}6^{2}_{7}3^{3}_{7}6^{3}_{9}3^{4}_{9}4^{4}_{11}1^{5}_{11}2^{5}_{13}1^{6}_{15}$			
10^a_{163}: $1^{\underline{6}}_{\underline{15}}3^{\underline{5}}_{\underline{13}}1^{\underline{5}}_{\underline{11}}5^{\underline{4}}_{\underline{11}}3^{\underline{4}}_{\underline{9}}7^{\underline{3}}_{\underline{9}}5^{\underline{3}}_{\underline{7}}9^{\underline{2}}_{\underline{7}}9^{\underline{2}}_{\underline{5}}8^{\underline{1}}_{\underline{5}}7^{\underline{1}}_{\underline{3}}8^{0}_{\underline{3}}10^{0}_{\underline{1}}5^{1}_{\underline{1}}6^{1}_{1}3^{2}_{1}5^{2}_{3}1^{3}_{3}3^{3}_{5}1^{4}_{7}$	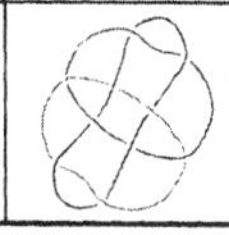	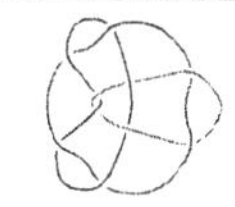	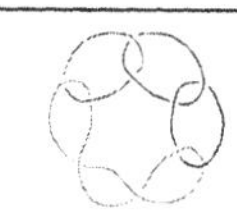
10^a_{164}: $1^{\underline{10}}_{\underline{25}}2^{\underline{9}}_{\underline{23}}1^{\underline{9}}_{\underline{21}}5^{\underline{8}}_{\underline{21}}3^{\underline{8}}_{\underline{19}}5^{\underline{7}}_{\underline{19}}4^{\underline{7}}_{\underline{17}}8^{\underline{6}}_{\underline{17}}7^{\underline{6}}_{\underline{15}}7^{\underline{5}}_{\underline{15}}6^{\underline{5}}_{\underline{13}}6^{\underline{4}}_{\underline{13}}7^{\underline{4}}_{\underline{11}}4^{\underline{3}}_{\underline{11}}6^{\underline{3}}_{\underline{9}}3^{\underline{2}}_{\underline{9}}4^{\underline{2}}_{\underline{7}}3^{\underline{1}}_{\underline{5}}1^{0}_{\underline{5}}1^{0}_{\underline{3}}$			
10^a_{165}: $1^{\underline{8}}_{\underline{20}}1^{\underline{8}}_{\underline{18}}1^{\underline{7}}_{\underline{18}}4^{\underline{6}}_{\underline{16}}1^{\underline{6}}_{\underline{14}}1^{\underline{5}}_{\underline{14}}4^{\underline{5}}_{\underline{12}}10^{\underline{4}}_{\underline{12}}7^{\underline{4}}_{\underline{10}}6^{\underline{3}}_{\underline{10}}4^{\underline{3}}_{\underline{8}}5^{\underline{2}}_{\underline{8}}6^{\underline{2}}_{\underline{6}}3^{\underline{1}}_{\underline{6}}5^{\underline{1}}_{\underline{4}}4^{0}_{\underline{4}}4^{0}_{\underline{2}}1^{1}_{\underline{2}}3^{1}_{0}1^{2}_{2}$			

L: KH	1st line	2nd line	3rd line
10^a_{166}: $1^{\underline{10}}_{\underline{26}}1^{\underline{10}}_{\underline{24}}2^{\underline{9}}_{\underline{24}}4^{\underline{8}}_{\underline{22}}2^{\underline{8}}_{\underline{20}}4^{\underline{7}}_{\underline{20}}4^{\underline{7}}_{\underline{18}}8^{\underline{6}}_{\underline{18}}7^{\underline{6}}_{\underline{16}}6^{\underline{5}}_{\underline{16}}5^{\underline{5}}_{\underline{14}}7^{\underline{4}}_{\underline{14}}9^{\underline{4}}_{\underline{12}}4^{\underline{3}}_{\underline{12}}4^{\underline{3}}_{\underline{10}}3^{\underline{2}}_{\underline{10}}4^{\underline{2}}_{\underline{8}}3^{\underline{1}}_{\underline{6}}1^{0}_{\underline{6}}1^{0}_{\underline{4}}$	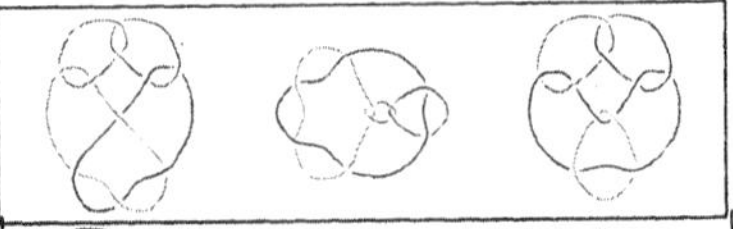		
10^a_{167}: $1^{\underline{4}}_{\underline{6}}1^{\underline{4}}_{\underline{4}}1^{\underline{3}}_{\underline{4}}4^{\underline{2}}_{\underline{2}}1^{\underline{2}}_{0}1^{\underline{1}}_{0}4^{\underline{1}}_{2}8^{0}_{2}4^{0}_{4}3^{1}_{4}5^{1}_{6}7^{2}_{6}6^{2}_{8}3^{3}_{8}4^{3}_{10}3^{4}_{10}3^{4}_{12}3^{5}_{14}1^{6}_{14}1^{6}_{16}$			
10^a_{168}: $1^{\underline{6}}_{\underline{14}}1^{\underline{5}}_{\underline{12}}1^{\underline{5}}_{\underline{10}}5^{\underline{4}}_{\underline{10}}4^{\underline{4}}_{\underline{8}}7^{\underline{3}}_{\underline{8}}2^{\underline{3}}_{\underline{6}}7^{\underline{2}}_{\underline{6}}7^{\underline{2}}_{\underline{4}}5^{\underline{1}}_{\underline{4}}7^{\underline{1}}_{\underline{2}}9^{0}_{\underline{2}}9^{0}_{0}5^{1}_{0}5^{1}_{2}3^{2}_{2}5^{2}_{4}3^{3}_{6}1^{4}_{6}1^{4}_{8}$			
10^a_{169}: $1^{\underline{6}}_{\underline{14}}4^{\underline{5}}_{\underline{12}}1^{\underline{5}}_{\underline{10}}7^{\underline{4}}_{\underline{10}}4^{\underline{4}}_{\underline{8}}8^{\underline{3}}_{\underline{8}}7^{\underline{3}}_{\underline{6}}14^{\underline{2}}_{\underline{6}}12^{\underline{2}}_{\underline{4}}10^{\underline{1}}_{\underline{4}}10^{\underline{1}}_{\underline{2}}12^{0}_{\underline{2}}14^{0}_{0}7^{1}_{0}8^{1}_{2}4^{2}_{2}7^{2}_{4}1^{3}_{4}4^{3}_{6}1^{4}_{8}$	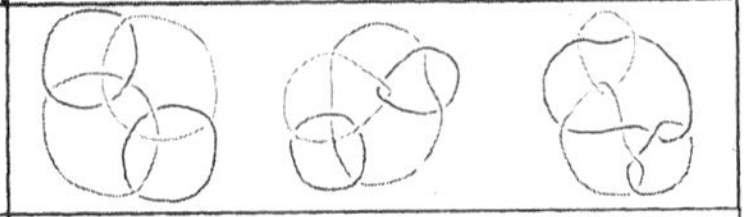		
10^a_{170}: $1^{\underline{5}}_{\underline{10}}3^{\underline{4}}_{\underline{8}}1^{\underline{4}}_{\underline{6}}4^{\underline{3}}_{\underline{6}}3^{\underline{3}}_{\underline{4}}9^{\underline{2}}_{\underline{4}}6^{\underline{2}}_{\underline{2}}6^{\underline{1}}_{\underline{2}}7^{\underline{1}}_{0}12^{0}_{0}10^{0}_{2}6^{1}_{2}8^{1}_{4}7^{2}_{4}8^{2}_{6}3^{3}_{6}5^{3}_{8}1^{4}_{8}3^{4}_{10}1^{5}_{12}$			
10^a_{171}: $1^{\underline{8}}_{\underline{20}}2^{\underline{7}}_{\underline{18}}1^{\underline{7}}_{\underline{16}}5^{\underline{6}}_{\underline{16}}4^{\underline{6}}_{\underline{14}}6^{\underline{5}}_{\underline{14}}3^{\underline{5}}_{\underline{12}}8^{\underline{4}}_{\underline{12}}8^{\underline{4}}_{\underline{10}}6^{\underline{3}}_{\underline{10}}6^{\underline{3}}_{\underline{8}}8^{\underline{2}}_{\underline{8}}8^{\underline{2}}_{\underline{6}}3^{\underline{1}}_{\underline{6}}6^{\underline{1}}_{\underline{4}}4^{0}_{\underline{4}}5^{0}_{\underline{2}}1^{1}_{\underline{2}}2^{1}_{0}1^{2}_{2}$			
10^a_{172}: $1^{\underline{4}}_{\underline{6}}1^{\underline{3}}_{\underline{4}}1^{\underline{3}}_{\underline{2}}5^{\underline{2}}_{\underline{2}}2^{\underline{2}}_{0}2^{\underline{1}}_{0}4^{\underline{1}}_{2}9^{0}_{2}6^{0}_{4}4^{1}_{4}5^{1}_{6}7^{2}_{6}6^{2}_{8}3^{3}_{8}5^{3}_{10}3^{4}_{10}3^{4}_{12}3^{5}_{14}1^{6}_{14}1^{6}_{16}$	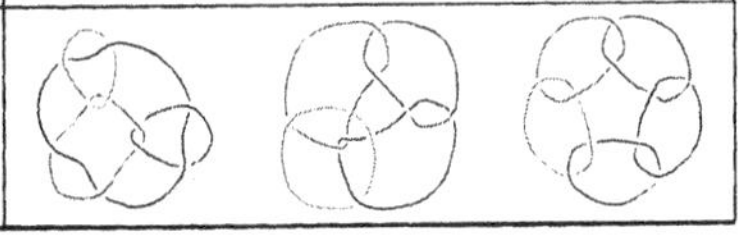		
10^a_{173}: $1^{\underline{8}}_{\underline{20}}1^{\underline{8}}_{\underline{18}}3^{\underline{7}}_{\underline{18}}5^{\underline{6}}_{\underline{16}}3^{\underline{6}}_{\underline{14}}6^{\underline{5}}_{\underline{14}}5^{\underline{5}}_{\underline{12}}10^{\underline{4}}_{\underline{12}}8^{\underline{4}}_{\underline{10}}7^{\underline{3}}_{\underline{10}}8^{\underline{3}}_{\underline{8}}10^{\underline{2}}_{\underline{8}}11^{\underline{2}}_{\underline{6}}5^{\underline{1}}_{\underline{6}}6^{\underline{1}}_{\underline{4}}4^{0}_{\underline{4}}6^{0}_{\underline{2}}1^{1}_{\underline{2}}3^{1}_{0}1^{2}_{2}$			
10^a_{174}: $1^{\underline{10}}_{\underline{25}}1^{\underline{9}}_{\underline{21}}6^{\underline{8}}_{\underline{21}}5^{\underline{8}}_{\underline{19}}5^{\underline{7}}_{\underline{19}}1^{\underline{7}}_{\underline{17}}10^{\underline{6}}_{\underline{17}}5^{\underline{6}}_{\underline{15}}1^{\underline{5}}_{\underline{15}}10^{\underline{5}}_{\underline{13}}14^{\underline{4}}_{\underline{13}}11^{\underline{4}}_{\underline{11}}6^{\underline{3}}_{\underline{11}}4^{\underline{3}}_{\underline{9}}4^{\underline{2}}_{\underline{9}}6^{\underline{2}}_{\underline{7}}4^{\underline{1}}_{\underline{5}}1^{0}_{\underline{5}}1^{0}_{\underline{3}}$			

3.11. 10 Crossing Non-Alternating Links.

L: KH	1st line	2nd line	3rd line
10^n_{1}: $1^{\underline{5}}_{\underline{14}}1^{\underline{4}}_{\underline{12}}2^{\underline{4}}_{\underline{10}}1^{\underline{4}}_{\underline{8}}2^{\underline{3}}_{\underline{10}}1^{\underline{3}}_{\underline{8}}2^{\underline{2}}_{\underline{8}}3^{\underline{2}}_{\underline{6}}2^{\underline{1}}_{\underline{6}}2^{\underline{1}}_{\underline{4}}1^{\underline{1}}_{\underline{2}}2^{0}_{\underline{4}}3^{0}_{\underline{2}}1^{1}_{\underline{2}}1^{1}_{0}1^{2}_{0}1^{2}_{2}1^{3}_{4}$	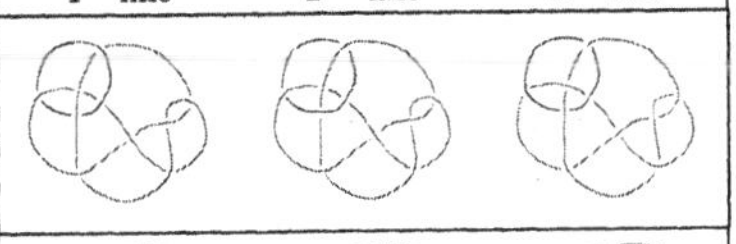		
10^n_{2}: $1^{\underline{5}}_{\underline{10}}1^{\underline{4}}_{\underline{8}}1^{\underline{4}}_{\underline{6}}2^{\underline{3}}_{\underline{6}}1^{\underline{3}}_{\underline{4}}3^{\underline{2}}_{\underline{4}}2^{\underline{2}}_{\underline{2}}2^{\underline{1}}_{\underline{2}}3^{\underline{1}}_{0}4^{0}_{0}4^{0}_{2}2^{1}_{2}2^{1}_{4}2^{2}_{4}2^{2}_{6}2^{3}_{8}$			
10^n_{3}: $1^{\underline{5}}_{\underline{10}}1^{\underline{4}}_{\underline{8}}1^{\underline{4}}_{\underline{6}}2^{\underline{3}}_{\underline{6}}1^{\underline{3}}_{\underline{4}}2^{\underline{2}}_{\underline{4}}2^{\underline{2}}_{\underline{2}}2^{\underline{1}}_{\underline{2}}2^{\underline{1}}_{0}1^{0}_{\underline{2}}4^{0}_{0}3^{0}_{2}1^{1}_{2}2^{1}_{4}1^{2}_{4}1^{2}_{6}1^{3}_{8}$			
10^n_{4}: $2^{\underline{4}}_{\underline{10}}1^{\underline{4}}_{\underline{8}}1^{\underline{3}}_{\underline{8}}1^{\underline{3}}_{\underline{6}}3^{\underline{2}}_{\underline{6}}1^{\underline{2}}_{\underline{4}}2^{\underline{1}}_{\underline{4}}3^{\underline{1}}_{\underline{2}}3^{0}_{\underline{2}}3^{0}_{0}2^{1}_{0}2^{1}_{2}1^{2}_{2}2^{2}_{4}1^{3}_{4}1^{3}_{6}1^{4}_{8}$	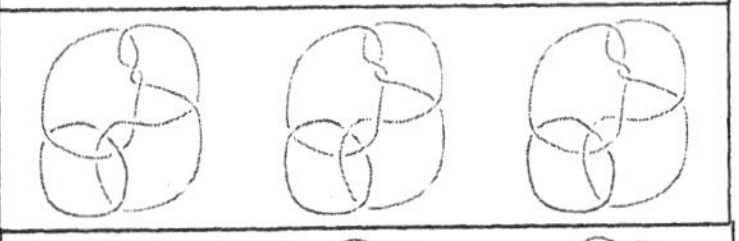		
10^n_{5}: $1^{\underline{4}}_{\underline{6}}1^{\underline{3}}_{\underline{2}}2^{\underline{2}}_{\underline{2}}1^{\underline{1}}_{0}2^{\underline{1}}_{2}1^{0}_{0}3^{0}_{2}2^{0}_{4}2^{1}_{4}1^{1}_{6}1^{2}_{4}1^{2}_{6}2^{2}_{8}2^{3}_{8}1^{3}_{10}1^{4}_{12}$			
10^n_{6}: $1^{\underline{2}}_{\underline{4}}1^{\underline{1}}_{0}4^{0}_{0}2^{0}_{2}2^{1}_{2}2^{1}_{4}2^{2}_{4}2^{2}_{6}2^{3}_{6}2^{3}_{8}1^{4}_{8}2^{4}_{10}1^{5}_{10}1^{5}_{12}1^{6}_{14}$			
10^n_{7}: $2^{\underline{4}}_{\underline{12}}1^{\underline{4}}_{\underline{10}}1^{\underline{3}}_{\underline{10}}1^{\underline{3}}_{\underline{8}}2^{\underline{2}}_{\underline{8}}1^{\underline{2}}_{\underline{6}}1^{\underline{1}}_{\underline{6}}2^{\underline{1}}_{\underline{4}}2^{0}_{\underline{4}}2^{0}_{\underline{2}}2^{1}_{\underline{2}}1^{1}_{0}2^{2}_{2}1^{3}_{2}1^{4}_{6}$	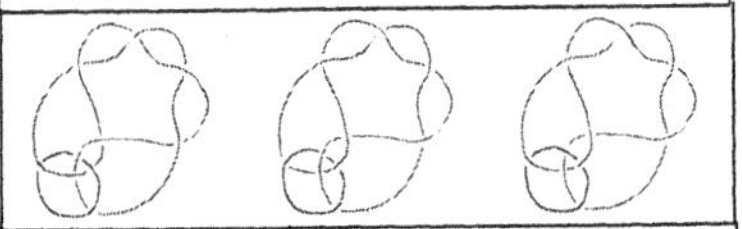		
10^n_{8}: $1^{\underline{4}}_{\underline{8}}1^{\underline{3}}_{\underline{4}}2^{\underline{2}}_{\underline{4}}1^{\underline{1}}_{\underline{2}}2^{\underline{1}}_{0}1^{0}_{\underline{2}}3^{0}_{0}2^{0}_{2}2^{1}_{2}1^{1}_{4}2^{2}_{6}1^{3}_{6}1^{4}_{10}$			
10^n_{9}: $1^{\underline{2}}_{\underline{6}}1^{\underline{2}}_{\underline{4}}1^{\underline{1}}_{\underline{2}}1^{\underline{1}}_{0}2^{0}_{\underline{2}}2^{0}_{0}1^{0}_{2}1^{1}_{0}2^{1}_{2}1^{2}_{2}1^{2}_{4}1^{2}_{6}1^{3}_{4}1^{3}_{6}1^{4}_{8}1^{5}_{8}1^{6}_{12}$			
10^n_{10}: $1^{\underline{6}}_{\underline{18}}1^{\underline{6}}_{\underline{16}}1^{\underline{5}}_{\underline{16}}1^{\underline{5}}_{\underline{14}}1^{\underline{5}}_{\underline{12}}1^{\underline{4}}_{\underline{14}}2^{\underline{4}}_{\underline{12}}1^{\underline{4}}_{\underline{10}}2^{\underline{3}}_{\underline{12}}2^{\underline{3}}_{\underline{10}}1^{\underline{2}}_{\underline{10}}2^{\underline{2}}_{\underline{8}}1^{\underline{2}}_{\underline{6}}1^{\underline{1}}_{\underline{8}}1^{\underline{1}}_{\underline{6}}1^{0}_{\underline{6}}2^{0}_{\underline{4}}1^{1}_{\underline{4}}1^{2}_{0}$	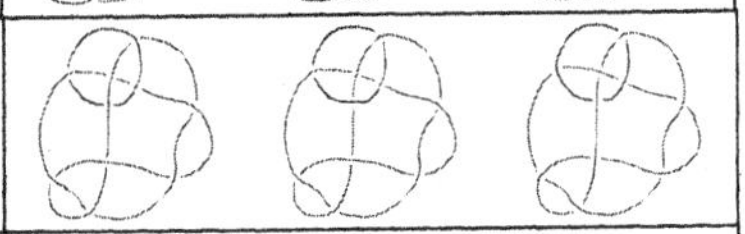		
10^n_{11}: $1^{\underline{6}}_{\underline{14}}1^{\underline{5}}_{\underline{12}}1^{\underline{5}}_{\underline{10}}2^{\underline{4}}_{\underline{10}}1^{\underline{4}}_{\underline{8}}3^{\underline{3}}_{\underline{8}}2^{\underline{3}}_{\underline{6}}3^{\underline{2}}_{\underline{6}}3^{\underline{2}}_{\underline{4}}2^{\underline{1}}_{\underline{4}}3^{\underline{1}}_{\underline{2}}3^{0}_{\underline{2}}4^{0}_{0}2^{1}_{0}1^{1}_{2}2^{2}_{4}$			
10^n_{12}: $1^{\underline{6}}_{\underline{14}}2^{\underline{5}}_{\underline{12}}1^{\underline{5}}_{\underline{10}}3^{\underline{4}}_{\underline{10}}2^{\underline{4}}_{\underline{8}}3^{\underline{3}}_{\underline{8}}3^{\underline{3}}_{\underline{6}}4^{\underline{2}}_{\underline{6}}3^{\underline{2}}_{\underline{4}}3^{\underline{1}}_{\underline{4}}4^{\underline{1}}_{\underline{2}}3^{0}_{\underline{2}}5^{0}_{0}2^{1}_{0}1^{1}_{2}2^{2}_{4}$			
10^n_{13}: $1^{\underline{6}}_{\underline{18}}1^{\underline{5}}_{\underline{16}}1^{\underline{5}}_{\underline{14}}2^{\underline{4}}_{\underline{14}}2^{\underline{4}}_{\underline{12}}1^{\underline{4}}_{\underline{10}}2^{\underline{3}}_{\underline{12}}2^{\underline{3}}_{\underline{10}}1^{\underline{2}}_{\underline{10}}2^{\underline{2}}_{\underline{8}}2^{\underline{1}}_{\underline{8}}1^{\underline{1}}_{\underline{6}}1^{0}_{\underline{6}}3^{0}_{\underline{4}}1^{1}_{\underline{4}}1^{2}_{0}$	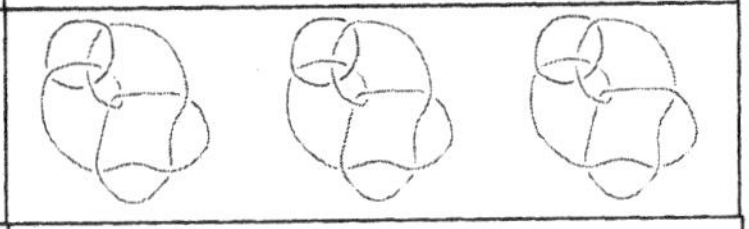		
10^n_{14}: $1^{\underline{6}}_{\underline{14}}1^{\underline{5}}_{\underline{12}}1^{\underline{5}}_{\underline{10}}1^{\underline{4}}_{\underline{10}}1^{\underline{4}}_{\underline{8}}2^{\underline{3}}_{\underline{8}}1^{\underline{3}}_{\underline{6}}1^{\underline{2}}_{\underline{8}}1^{\underline{2}}_{\underline{6}}2^{\underline{2}}_{\underline{4}}2^{\underline{1}}_{\underline{4}}1^{\underline{1}}_{\underline{2}}1^{0}_{\underline{4}}2^{0}_{\underline{2}}3^{0}_{0}1^{1}_{\underline{2}}1^{1}_{0}1^{2}_{2}1^{3}_{2}1^{4}_{6}$			
10^n_{15}: $1^{\underline{3}}_{\underline{4}}2^{\underline{2}}_{\underline{2}}1^{\underline{2}}_{0}2^{\underline{1}}_{0}2^{\underline{1}}_{2}5^{0}_{2}4^{0}_{4}3^{1}_{4}3^{1}_{6}4^{2}_{6}3^{2}_{8}2^{3}_{8}4^{3}_{10}2^{4}_{10}2^{4}_{12}2^{5}_{14}$			
10^n_{16}: $1^{\underline{8}}_{\underline{20}}3^{\underline{7}}_{\underline{18}}1^{\underline{7}}_{\underline{16}}3^{\underline{6}}_{\underline{16}}3^{\underline{6}}_{\underline{14}}6^{\underline{5}}_{\underline{14}}3^{\underline{5}}_{\underline{12}}4^{\underline{4}}_{\underline{12}}6^{\underline{4}}_{\underline{10}}5^{\underline{3}}_{\underline{10}}4^{\underline{3}}_{\underline{8}}4^{\underline{2}}_{\underline{8}}5^{\underline{2}}_{\underline{6}}1^{\underline{1}}_{\underline{6}}4^{\underline{1}}_{\underline{4}}2^{0}_{\underline{4}}3^{0}_{\underline{2}}$	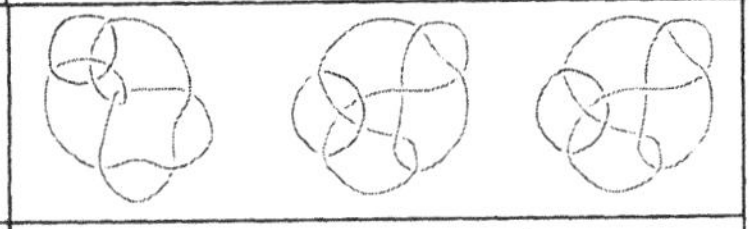		
10^n_{17}: $1^{\underline{5}}_{\underline{14}}2^{\underline{4}}_{\underline{12}}2^{\underline{4}}_{\underline{10}}3^{\underline{3}}_{\underline{10}}1^{\underline{3}}_{\underline{8}}2^{\underline{2}}_{\underline{8}}3^{\underline{2}}_{\underline{6}}2^{\underline{1}}_{\underline{6}}2^{\underline{1}}_{\underline{4}}3^{0}_{\underline{4}}3^{0}_{\underline{2}}1^{1}_{\underline{2}}2^{1}_{0}1^{2}_{0}1^{2}_{2}1^{3}_{4}$			
10^n_{18}: $1^{\underline{5}}_{\underline{10}}1^{\underline{4}}_{\underline{6}}1^{\underline{3}}_{\underline{6}}1^{\underline{2}}_{\underline{6}}1^{\underline{2}}_{\underline{4}}1^{\underline{2}}_{\underline{2}}1^{\underline{1}}_{\underline{4}}1^{\underline{1}}_{\underline{2}}1^{\underline{1}}_{0}2^{0}_{\underline{2}}4^{0}_{0}1^{0}_{2}2^{1}_{0}1^{1}_{2}1^{1}_{4}1^{2}_{2}2^{2}_{4}1^{3}_{4}1^{3}_{6}1^{4}_{6}1^{4}_{8}1^{5}_{10}$			
10^n_{19}: $2^{\underline{6}}_{\underline{16}}1^{\underline{5}}_{\underline{14}}2^{\underline{5}}_{\underline{12}}3^{\underline{4}}_{\underline{12}}2^{\underline{4}}_{\underline{10}}3^{\underline{3}}_{\underline{10}}2^{\underline{3}}_{\underline{8}}2^{\underline{2}}_{\underline{8}}3^{\underline{2}}_{\underline{6}}2^{\underline{1}}_{\underline{6}}2^{\underline{1}}_{\underline{4}}1^{0}_{\underline{4}}3^{0}_{\underline{2}}1^{1}_{\underline{2}}1^{2}_{2}$	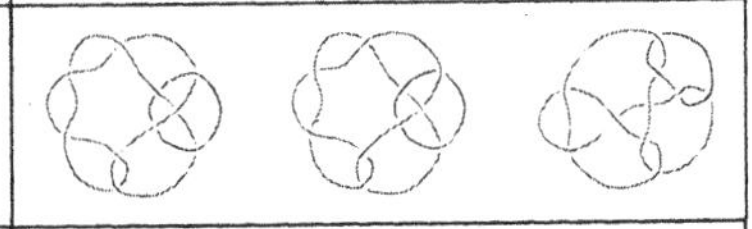		
10^n_{20}: $2^{\underline{4}}_{\underline{10}}2^{\underline{3}}_{\underline{8}}2^{\underline{3}}_{\underline{6}}4^{\underline{2}}_{\underline{6}}2^{\underline{2}}_{\underline{4}}3^{\underline{1}}_{\underline{4}}4^{\underline{1}}_{\underline{2}}4^{0}_{\underline{2}}5^{0}_{0}4^{1}_{0}2^{1}_{2}1^{2}_{2}4^{2}_{4}1^{3}_{4}1^{3}_{6}1^{4}_{8}$			
10^n_{21}: $1^{\underline{2}}_{\underline{4}}1^{\underline{1}}_{\underline{2}}1^{\underline{1}}_{0}4^{0}_{0}3^{0}_{2}3^{1}_{2}2^{1}_{4}3^{2}_{4}3^{2}_{6}2^{3}_{6}3^{3}_{8}2^{4}_{8}2^{4}_{10}1^{5}_{10}2^{5}_{12}1^{6}_{14}$			
10^n_{22}: $1^{\underline{7}}_{\underline{16}}1^{\underline{6}}_{\underline{14}}1^{\underline{6}}_{\underline{12}}3^{\underline{5}}_{\underline{12}}1^{\underline{5}}_{\underline{10}}2^{\underline{4}}_{\underline{10}}3^{\underline{4}}_{\underline{8}}3^{\underline{3}}_{\underline{8}}2^{\underline{3}}_{\underline{6}}3^{\underline{2}}_{\underline{6}}3^{\underline{2}}_{\underline{4}}1^{\underline{1}}_{\underline{4}}3^{\underline{1}}_{\underline{2}}3^{0}_{\underline{2}}3^{0}_{0}1^{1}_{2}$	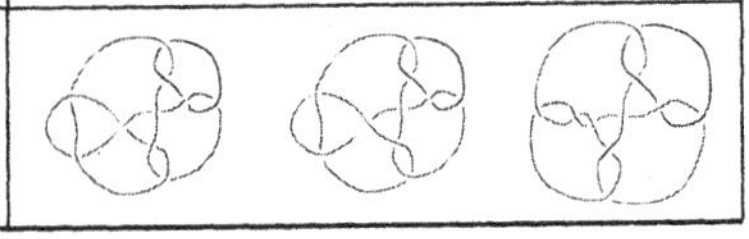		
10^n_{23}: $1^{\underline{2}}_{0}1^{\underline{1}}_{4}3^{0}_{4}2^{0}_{6}1^{1}_{6}1^{1}_{8}2^{2}_{8}1^{2}_{10}1^{3}_{10}2^{3}_{12}1^{4}_{10}2^{4}_{12}1^{4}_{14}1^{5}_{14}2^{5}_{16}$			
10^n_{24}: $1^{\underline{2}}_{\underline{6}}1^{\underline{1}}_{\underline{4}}1^{\underline{1}}_{\underline{2}}2^{0}_{\underline{2}}3^{0}_{0}1^{0}_{2}2^{1}_{0}1^{1}_{2}1^{2}_{2}2^{2}_{4}1^{3}_{4}1^{3}_{6}1^{4}_{6}1^{4}_{8}1^{5}_{10}$			

L: KH	1st line 2nd line 3rd line
10^n_{25}: $1^{\underline{6}}_{\underline{12}}1^{\underline{5}}_{\underline{8}}2^{\underline{4}}_{\underline{8}}1^{\underline{4}}_{\underline{6}}1^{\underline{3}}_{\underline{6}}1^{\underline{3}}_{\underline{4}}1^{\underline{2}}_{\underline{6}}1^{\underline{2}}_{\underline{4}}1^{\underline{2}}_{\underline{2}}2^{\underline{1}}_{\underline{2}}1^{\underline{1}}_{0}2^{0}_{\underline{2}}2^{0}_{0}1^{0}_{2}1^{1}_{0}1^{1}_{2}1^{1}_{4}1^{2}_{4}1^{3}_{4}1^{4}_{8}$	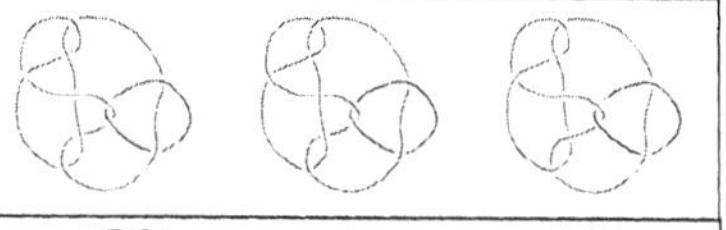
10^n_{26}: $2^{\underline{4}}_{\underline{10}}1^{\underline{4}}_{\underline{8}}3^{\underline{3}}_{\underline{8}}1^{\underline{3}}_{\underline{6}}4^{\underline{2}}_{\underline{6}}3^{\underline{2}}_{\underline{4}}3^{\underline{1}}_{\underline{4}}4^{\underline{1}}_{\underline{2}}5^{0}_{\underline{2}}4^{0}_{0}3^{1}_{0}4^{1}_{2}2^{2}_{2}3^{2}_{4}1^{3}_{4}2^{3}_{6}1^{4}_{8}$	
10^n_{27}: $1^{\underline{6}}_{\underline{16}}2^{\underline{5}}_{\underline{14}}1^{\underline{5}}_{\underline{12}}3^{\underline{4}}_{\underline{12}}3^{\underline{4}}_{\underline{10}}4^{\underline{3}}_{\underline{10}}2^{\underline{3}}_{\underline{8}}3^{\underline{2}}_{\underline{8}}4^{\underline{2}}_{\underline{6}}2^{\underline{1}}_{\underline{6}}3^{\underline{1}}_{\underline{4}}3^{0}_{\underline{4}}3^{0}_{\underline{2}}1^{1}_{\underline{2}}2^{1}_{0}1^{2}_{2}$	
10^n_{28}: $1^{\underline{8}}_{\underline{18}}1^{\underline{7}}_{\underline{14}}1^{\underline{6}}_{\underline{14}}1^{\underline{5}}_{\underline{12}}1^{\underline{5}}_{\underline{10}}2^{\underline{4}}_{\underline{12}}1^{\underline{4}}_{\underline{10}}1^{\underline{4}}_{\underline{8}}1^{\underline{3}}_{\underline{10}}2^{\underline{3}}_{\underline{8}}1^{\underline{2}}_{\underline{8}}1^{\underline{2}}_{\underline{6}}1^{\underline{2}}_{\underline{4}}1^{\underline{1}}_{\underline{6}}1^{\underline{1}}_{\underline{4}}1^{0}_{\underline{4}}2^{0}_{\underline{2}}1^{1}_{\underline{2}}1^{2}_{2}$	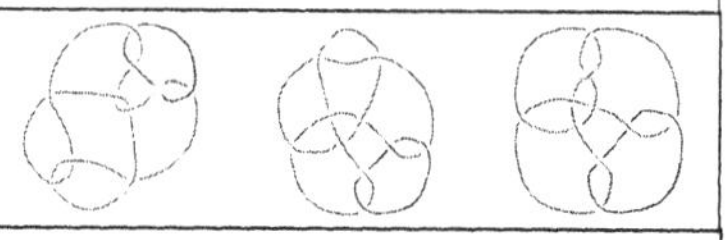
10^n_{29}: $3^{\underline{4}}_{\underline{10}}1^{\underline{4}}_{\underline{8}}3^{\underline{3}}_{\underline{8}}2^{\underline{3}}_{\underline{6}}5^{\underline{2}}_{\underline{6}}3^{\underline{2}}_{\underline{4}}4^{\underline{1}}_{\underline{4}}5^{\underline{1}}_{\underline{2}}5^{0}_{\underline{2}}5^{0}_{0}4^{1}_{0}4^{1}_{2}2^{2}_{2}4^{2}_{4}1^{3}_{4}2^{3}_{6}1^{4}_{8}$	
10^n_{30}: $2^{\underline{5}}_{\underline{14}}3^{\underline{4}}_{\underline{12}}3^{\underline{4}}_{\underline{10}}4^{\underline{3}}_{\underline{10}}2^{\underline{3}}_{\underline{8}}4^{\underline{2}}_{\underline{8}}4^{\underline{2}}_{\underline{6}}3^{\underline{1}}_{\underline{6}}4^{\underline{1}}_{\underline{4}}4^{0}_{\underline{4}}4^{0}_{\underline{2}}2^{1}_{\underline{2}}3^{1}_{0}1^{2}_{0}2^{2}_{2}1^{3}_{4}$	
10^n_{31}: $2^{\underline{6}}_{\underline{16}}1^{\underline{5}}_{\underline{14}}2^{\underline{5}}_{\underline{12}}4^{\underline{4}}_{\underline{12}}2^{\underline{4}}_{\underline{10}}4^{\underline{3}}_{\underline{10}}3^{\underline{3}}_{\underline{8}}2^{\underline{2}}_{\underline{8}}4^{\underline{2}}_{\underline{6}}3^{\underline{1}}_{\underline{6}}2^{\underline{1}}_{\underline{4}}2^{0}_{\underline{4}}4^{0}_{\underline{2}}1^{1}_{\underline{2}}1^{1}_{0}1^{2}_{2}$	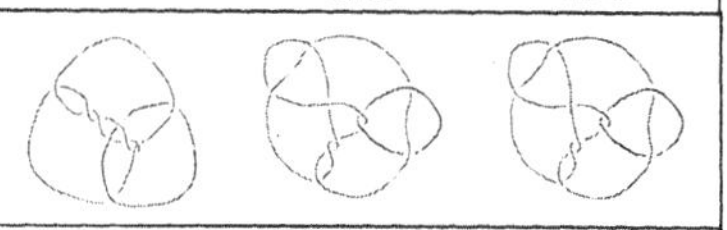
10^n_{32}: $1^{\underline{6}}_{\underline{12}}1^{\underline{5}}_{\underline{8}}1^{\underline{4}}_{\underline{8}}1^{\underline{3}}_{\underline{6}}1^{\underline{3}}_{\underline{4}}1^{\underline{2}}_{\underline{6}}1^{\underline{2}}_{\underline{2}}2^{\underline{1}}_{\underline{2}}2^{0}_{\underline{2}}2^{0}_{0}2^{0}_{2}1^{1}_{0}1^{1}_{2}1^{2}_{4}1^{3}_{4}1^{4}_{8}$	
10^n_{33}: $2^{\underline{5}}_{\underline{12}}2^{\underline{4}}_{\underline{10}}2^{\underline{4}}_{\underline{8}}4^{\underline{3}}_{\underline{8}}2^{\underline{3}}_{\underline{6}}5^{\underline{2}}_{\underline{6}}4^{\underline{2}}_{\underline{4}}3^{\underline{1}}_{\underline{4}}5^{\underline{1}}_{\underline{2}}5^{0}_{\underline{2}}5^{0}_{0}3^{1}_{0}3^{1}_{2}1^{2}_{2}3^{2}_{4}1^{3}_{6}$	
10^n_{34}: $3^{0}_{0}2^{0}_{2}2^{1}_{2}1^{1}_{4}2^{2}_{4}2^{2}_{6}2^{3}_{6}2^{3}_{8}2^{4}_{8}2^{4}_{10}1^{5}_{10}2^{5}_{12}1^{6}_{12}1^{6}_{14}1^{7}_{16}$	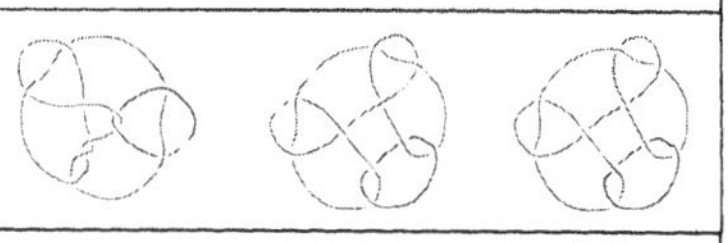
10^n_{35}: $3^{0}_{2}2^{0}_{4}3^{1}_{4}1^{1}_{6}5^{2}_{6}3^{2}_{8}3^{3}_{8}5^{3}_{10}5^{4}_{10}3^{4}_{12}3^{5}_{12}5^{5}_{14}2^{6}_{14}3^{6}_{16}1^{7}_{16}2^{7}_{18}1^{8}_{20}$	
10^n_{36}: $1^{\underline{5}}_{\underline{10}}1^{\underline{4}}_{\underline{6}}1^{\underline{3}}_{\underline{6}}1^{\underline{2}}_{\underline{4}}1^{\underline{2}}_{\underline{2}}1^{\underline{1}}_{\underline{4}}1^{\underline{1}}_{0}1^{0}_{\underline{2}}4^{0}_{0}1^{0}_{2}1^{1}_{0}1^{1}_{4}1^{2}_{2}1^{2}_{4}1^{3}_{6}1^{4}_{6}1^{5}_{10}$	
10^n_{37}: $1^{\underline{4}}_{\underline{6}}1^{\underline{4}}_{\underline{4}}1^{\underline{2}}_{\underline{2}}1^{\underline{1}}_{2}3^{0}_{0}2^{0}_{2}1^{1}_{2}2^{1}_{4}1^{1}_{6}2^{2}_{4}1^{2}_{6}1^{3}_{6}2^{3}_{8}1^{4}_{8}1^{4}_{10}1^{5}_{10}1^{5}_{12}1^{6}_{14}$	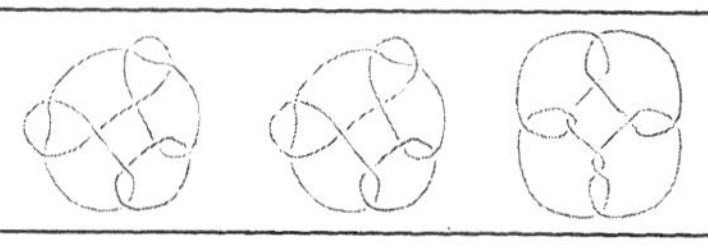
10^n_{38}: $1^{\underline{7}}_{\underline{16}}1^{\underline{6}}_{\underline{14}}1^{\underline{6}}_{\underline{12}}3^{\underline{5}}_{\underline{12}}1^{\underline{5}}_{\underline{10}}3^{\underline{4}}_{\underline{10}}4^{\underline{4}}_{\underline{8}}3^{\underline{3}}_{\underline{8}}2^{\underline{3}}_{\underline{6}}4^{\underline{2}}_{\underline{6}}3^{\underline{2}}_{\underline{4}}1^{\underline{1}}_{\underline{4}}4^{\underline{1}}_{\underline{2}}3^{0}_{\underline{2}}2^{0}_{0}2^{1}_{2}$	
10^n_{39}: $1^{\underline{2}}_{0}1^{\underline{1}}_{4}4^{0}_{4}2^{0}_{6}1^{1}_{6}2^{1}_{8}3^{2}_{8}1^{2}_{10}2^{3}_{10}3^{3}_{12}1^{4}_{10}2^{4}_{12}2^{4}_{14}2^{5}_{14}2^{5}_{16}1^{6}_{18}$	
10^n_{40}: $1^{\underline{3}}_{\underline{8}}1^{\underline{2}}_{\underline{6}}1^{\underline{2}}_{\underline{4}}2^{\underline{1}}_{\underline{4}}1^{\underline{1}}_{\underline{2}}2^{0}_{\underline{2}}3^{0}_{0}2^{1}_{0}1^{1}_{2}2^{2}_{2}3^{2}_{4}1^{3}_{4}1^{3}_{6}1^{4}_{6}1^{4}_{8}1^{5}_{10}$	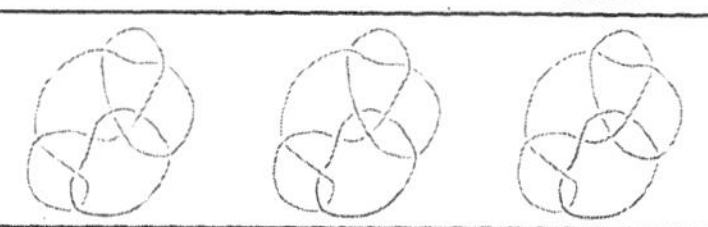
10^n_{41}: $1^{\underline{7}}_{\underline{16}}1^{\underline{6}}_{\underline{14}}1^{\underline{6}}_{\underline{12}}2^{\underline{5}}_{\underline{12}}1^{\underline{5}}_{\underline{10}}2^{\underline{4}}_{\underline{10}}2^{\underline{4}}_{\underline{8}}2^{\underline{3}}_{\underline{8}}2^{\underline{3}}_{\underline{6}}3^{\underline{2}}_{\underline{6}}3^{\underline{2}}_{\underline{4}}1^{\underline{1}}_{\underline{4}}2^{\underline{1}}_{\underline{2}}2^{0}_{\underline{2}}2^{0}_{0}1^{1}_{2}$	
10^n_{42}: $1^{\underline{2}}_{0}1^{\underline{1}}_{4}2^{0}_{4}1^{0}_{6}1^{1}_{6}1^{1}_{8}1^{2}_{6}2^{2}_{8}1^{2}_{10}1^{3}_{10}1^{3}_{12}1^{4}_{10}1^{4}_{12}1^{4}_{14}1^{5}_{14}1^{5}_{16}$	
10^n_{43}: $2^{\underline{2}}_{\underline{4}}1^{\underline{2}}_{\underline{2}}2^{\underline{1}}_{\underline{2}}1^{\underline{1}}_{0}4^{0}_{0}3^{0}_{2}3^{1}_{2}3^{1}_{4}4^{2}_{4}3^{2}_{6}2^{3}_{6}4^{3}_{8}2^{4}_{8}2^{4}_{10}1^{5}_{10}2^{5}_{12}1^{6}_{14}$	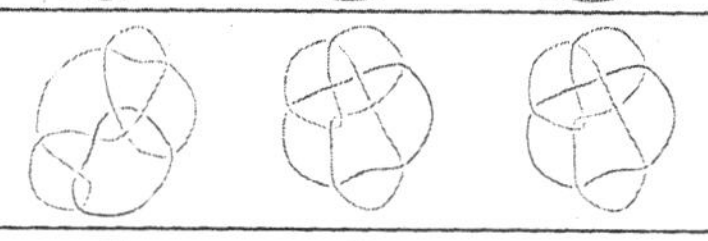
10^n_{44}: $1^{\underline{7}}_{\underline{16}}1^{\underline{6}}_{\underline{14}}1^{\underline{6}}_{\underline{12}}1^{\underline{5}}_{\underline{12}}1^{\underline{5}}_{\underline{10}}2^{\underline{4}}_{\underline{10}}1^{\underline{4}}_{\underline{8}}1^{\underline{3}}_{\underline{8}}2^{\underline{3}}_{\underline{6}}2^{\underline{2}}_{\underline{6}}2^{\underline{2}}_{\underline{4}}1^{\underline{1}}_{\underline{4}}1^{\underline{1}}_{\underline{2}}1^{0}_{\underline{2}}2^{0}_{0}$	
10^n_{45}: $1^{\underline{2}}_{\underline{8}}1^{\underline{2}}_{\underline{6}}1^{\underline{2}}_{\underline{4}}1^{\underline{1}}_{\underline{4}}1^{0}_{\underline{4}}1^{0}_{\underline{2}}1^{0}_{0}1^{1}_{\underline{2}}1^{1}_{0}1^{2}_{2}1^{3}_{2}1^{4}_{6}$	
10^n_{46}: $2^{\underline{6}}_{\underline{16}}1^{\underline{6}}_{\underline{14}}1^{\underline{5}}_{\underline{14}}1^{\underline{5}}_{\underline{12}}2^{\underline{4}}_{\underline{12}}1^{\underline{4}}_{\underline{10}}2^{\underline{3}}_{\underline{10}}2^{\underline{3}}_{\underline{8}}1^{\underline{2}}_{\underline{8}}2^{\underline{2}}_{\underline{6}}1^{\underline{1}}_{\underline{6}}1^{\underline{1}}_{\underline{4}}1^{0}_{\underline{4}}2^{0}_{\underline{2}}$	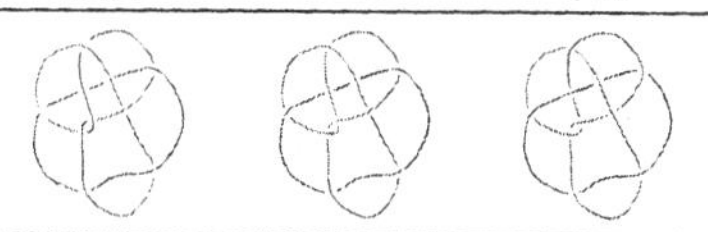
10^n_{47}: $1^{\underline{6}}_{\underline{14}}1^{\underline{6}}_{\underline{12}}2^{\underline{5}}_{\underline{12}}2^{\underline{4}}_{\underline{10}}2^{\underline{4}}_{\underline{8}}2^{\underline{3}}_{\underline{8}}2^{\underline{3}}_{\underline{6}}4^{\underline{2}}_{\underline{6}}2^{\underline{2}}_{\underline{4}}1^{\underline{1}}_{\underline{4}}4^{\underline{1}}_{\underline{2}}3^{0}_{\underline{2}}2^{0}_{0}1^{1}_{0}2^{1}_{2}1^{2}_{4}$	
10^n_{48}: $1^{\underline{6}}_{\underline{14}}1^{\underline{5}}_{\underline{12}}1^{\underline{5}}_{\underline{10}}2^{\underline{4}}_{\underline{10}}1^{\underline{4}}_{\underline{8}}2^{\underline{3}}_{\underline{8}}2^{\underline{3}}_{\underline{6}}3^{\underline{2}}_{\underline{6}}3^{\underline{2}}_{\underline{4}}2^{\underline{1}}_{\underline{4}}2^{\underline{1}}_{\underline{2}}2^{0}_{\underline{2}}3^{0}_{0}1^{1}_{0}1^{1}_{2}1^{2}_{4}$	
10^n_{49}: $1^{\underline{6}}_{\underline{18}}1^{\underline{6}}_{\underline{16}}1^{\underline{5}}_{\underline{16}}1^{\underline{4}}_{\underline{14}}1^{\underline{4}}_{\underline{12}}1^{\underline{3}}_{\underline{12}}1^{\underline{3}}_{\underline{10}}1^{\underline{2}}_{\underline{10}}1^{\underline{2}}_{\underline{8}}1^{\underline{1}}_{\underline{8}}1^{\underline{1}}_{\underline{6}}1^{0}_{\underline{6}}2^{0}_{\underline{4}}1^{1}_{\underline{4}}1^{2}_{0}$	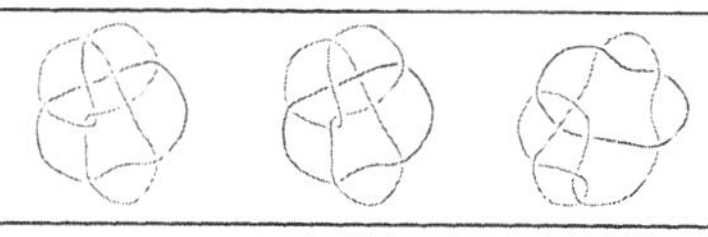
10^n_{50}: $1^{\underline{5}}_{\underline{14}}2^{\underline{4}}_{\underline{12}}1^{\underline{4}}_{\underline{10}}2^{\underline{3}}_{\underline{10}}2^{\underline{3}}_{\underline{8}}3^{\underline{2}}_{\underline{8}}3^{\underline{2}}_{\underline{6}}3^{\underline{1}}_{\underline{6}}2^{\underline{1}}_{\underline{4}}2^{0}_{\underline{4}}4^{0}_{\underline{2}}2^{1}_{\underline{2}}1^{1}_{0}1^{2}_{0}2^{2}_{2}1^{3}_{4}$	
10^n_{51}: $1^{\underline{8}}_{\underline{20}}1^{\underline{7}}_{\underline{18}}1^{\underline{7}}_{\underline{16}}3^{\underline{6}}_{\underline{16}}2^{\underline{6}}_{\underline{14}}3^{\underline{5}}_{\underline{14}}2^{\underline{5}}_{\underline{12}}3^{\underline{4}}_{\underline{12}}3^{\underline{4}}_{\underline{10}}3^{\underline{3}}_{\underline{10}}3^{\underline{3}}_{\underline{8}}2^{\underline{2}}_{\underline{8}}3^{\underline{2}}_{\underline{6}}1^{\underline{1}}_{\underline{6}}2^{\underline{1}}_{\underline{4}}1^{0}_{\underline{4}}2^{0}_{\underline{2}}$	
10^n_{52}: $1^{\underline{6}}_{\underline{14}}2^{\underline{5}}_{\underline{12}}1^{\underline{5}}_{\underline{10}}3^{\underline{4}}_{\underline{10}}2^{\underline{4}}_{\underline{8}}3^{\underline{3}}_{\underline{8}}3^{\underline{3}}_{\underline{6}}5^{\underline{2}}_{\underline{6}}4^{\underline{2}}_{\underline{4}}3^{\underline{1}}_{\underline{4}}4^{\underline{1}}_{\underline{2}}3^{0}_{\underline{2}}4^{0}_{0}2^{1}_{0}2^{1}_{2}2^{2}_{4}$	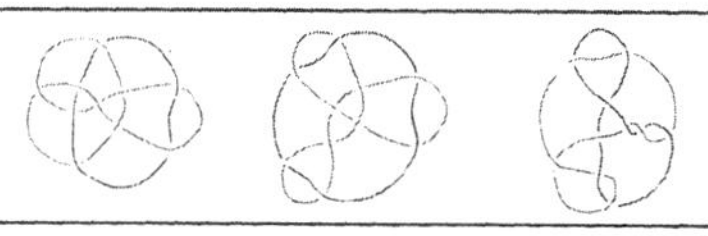
10^n_{53}: $2^{\underline{6}}_{\underline{16}}1^{\underline{6}}_{\underline{14}}3^{\underline{5}}_{\underline{14}}1^{\underline{5}}_{\underline{12}}3^{\underline{4}}_{\underline{12}}3^{\underline{4}}_{\underline{10}}4^{\underline{3}}_{\underline{10}}3^{\underline{3}}_{\underline{8}}4^{\underline{2}}_{\underline{8}}4^{\underline{2}}_{\underline{6}}2^{\underline{1}}_{\underline{6}}4^{\underline{1}}_{\underline{4}}3^{0}_{\underline{4}}3^{0}_{\underline{2}}1^{1}_{\underline{2}}2^{1}_{0}1^{2}_{2}$	
10^n_{54}: $1^{\underline{2}}_{0}1^{\underline{1}}_{4}2^{0}_{4}1^{0}_{6}1^{1}_{6}1^{1}_{8}1^{2}_{8}1^{2}_{10}1^{3}_{10}1^{3}_{12}1^{4}_{10}2^{4}_{12}1^{4}_{14}1^{5}_{16}$	
10^n_{55}: $1^{\underline{8}}_{\underline{20}}1^{\underline{8}}_{\underline{18}}2^{\underline{7}}_{\underline{18}}3^{\underline{6}}_{\underline{16}}2^{\underline{6}}_{\underline{14}}3^{\underline{5}}_{\underline{14}}3^{\underline{5}}_{\underline{12}}4^{\underline{4}}_{\underline{12}}3^{\underline{4}}_{\underline{10}}3^{\underline{3}}_{\underline{10}}4^{\underline{3}}_{\underline{8}}3^{\underline{2}}_{\underline{8}}3^{\underline{2}}_{\underline{6}}1^{\underline{1}}_{\underline{6}}3^{\underline{1}}_{\underline{4}}1^{0}_{\underline{4}}2^{0}_{\underline{2}}$	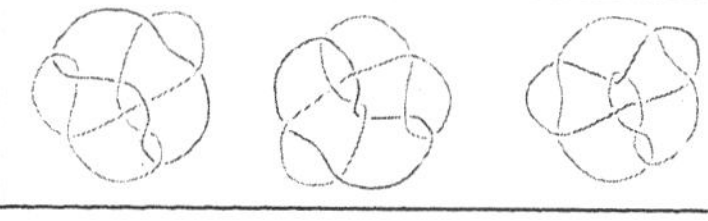
10^n_{56}: $1^{\underline{5}}_{\underline{12}}1^{\underline{4}}_{\underline{8}}1^{\underline{3}}_{\underline{8}}1^{\underline{2}}_{\underline{8}}1^{\underline{2}}_{\underline{6}}1^{\underline{2}}_{\underline{4}}1^{\underline{1}}_{\underline{4}}1^{\underline{1}}_{\underline{2}}1^{0}_{\underline{4}}2^{0}_{\underline{2}}2^{0}_{0}1^{1}_{\underline{2}}1^{1}_{0}1^{2}_{2}1^{3}_{2}1^{4}_{6}$	
10^n_{57}: $1^{\underline{5}}_{\underline{10}}1^{\underline{4}}_{\underline{6}}1^{\underline{3}}_{\underline{6}}1^{\underline{2}}_{\underline{6}}1^{\underline{2}}_{\underline{4}}1^{\underline{2}}_{\underline{2}}1^{\underline{1}}_{\underline{2}}1^{\underline{1}}_{0}2^{0}_{\underline{2}}3^{0}_{0}1^{0}_{2}1^{1}_{0}1^{1}_{2}1^{1}_{4}1^{2}_{4}1^{3}_{4}1^{4}_{8}$	
10^n_{58}: $3^{0}_{2}2^{0}_{4}3^{1}_{4}1^{1}_{6}5^{2}_{6}3^{2}_{8}3^{3}_{8}5^{3}_{10}5^{4}_{10}3^{4}_{12}3^{5}_{12}5^{5}_{14}2^{6}_{14}3^{6}_{16}1^{7}_{16}2^{7}_{18}1^{8}_{20}$	
10^n_{59}: $1^{\underline{5}}_{\underline{10}}1^{\underline{4}}_{\underline{6}}1^{\underline{3}}_{\underline{6}}1^{\underline{2}}_{\underline{4}}1^{\underline{2}}_{\underline{2}}1^{\underline{1}}_{\underline{4}}1^{\underline{1}}_{0}1^{0}_{\underline{2}}4^{0}_{0}1^{0}_{2}1^{1}_{0}1^{1}_{4}1^{2}_{2}1^{2}_{4}1^{3}_{6}1^{4}_{6}1^{5}_{10}$	
10^n_{60}: $1^{\underline{4}}_{\underline{6}}1^{\underline{4}}_{\underline{4}}1^{\underline{2}}_{\underline{2}}1^{\underline{1}}_{2}3^{0}_{0}2^{0}_{2}1^{1}_{2}2^{1}_{4}1^{1}_{6}2^{2}_{4}1^{2}_{6}1^{3}_{6}2^{3}_{8}1^{4}_{8}1^{4}_{10}1^{5}_{10}1^{5}_{12}1^{6}_{14}$	

L: KH	1st line	2nd line	3rd line
10^n_{61}: $1^{\underline{7}}_{\underline{16}}1^{\underline{6}}_{\underline{14}}1^{\underline{6}}_{\underline{12}}3^{\underline{5}}_{\underline{12}}1^{\underline{5}}_{\underline{10}}3^{\underline{4}}_{\underline{10}}4^{\underline{4}}_{\underline{8}}3^{\underline{3}}_{\underline{8}}2^{\underline{3}}_{\underline{6}}4^{\underline{2}}_{\underline{6}}3^{\underline{2}}_{\underline{4}}1^{\underline{1}}_{\underline{4}}4^{\underline{1}}_{\underline{2}}3^{0}_{\underline{2}}2^{0}_{0}2^{1}_{2}$ 10^n_{62}: $1^{\underline{2}}_{0}1^{\underline{1}}_{4}4^{0}_{4}2^{0}_{6}1^{1}_{6}2^{1}_{8}3^{2}_{8}1^{2}_{10}2^{3}_{10}3^{3}_{12}1^{4}_{10}2^{4}_{12}2^{4}_{14}2^{5}_{14}2^{5}_{16}1^{6}_{18}$ 10^n_{63}: $3^{\underline{4}}_{\underline{10}}1^{\underline{4}}_{\underline{8}}4^{\underline{3}}_{\underline{8}}2^{\underline{3}}_{\underline{6}}5^{\underline{2}}_{\underline{6}}4^{\underline{2}}_{\underline{4}}5^{\underline{1}}_{\underline{4}}5^{\underline{1}}_{\underline{2}}6^{0}_{\underline{2}}6^{0}_{0}4^{1}_{0}5^{1}_{2}3^{2}_{2}4^{2}_{4}1^{3}_{4}3^{3}_{6}1^{4}_{8}$	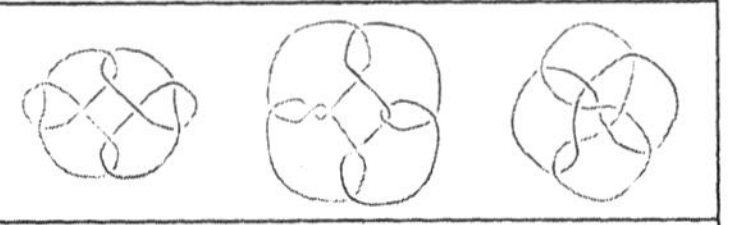		
10^n_{64}: $1^{\underline{7}}_{\underline{16}}2^{\underline{6}}_{\underline{14}}1^{\underline{6}}_{\underline{12}}2^{\underline{5}}_{\underline{12}}2^{\underline{5}}_{\underline{10}}4^{\underline{4}}_{\underline{10}}3^{\underline{4}}_{\underline{8}}3^{\underline{3}}_{\underline{8}}3^{\underline{3}}_{\underline{6}}3^{\underline{2}}_{\underline{6}}3^{\underline{2}}_{\underline{4}}2^{\underline{1}}_{\underline{4}}3^{\underline{1}}_{\underline{2}}2^{0}_{\underline{2}}3^{0}_{0}1^{1}_{2}$ 10^n_{65}: $3^{\underline{4}}_{\underline{11}}1^{\underline{4}}_{\underline{9}}3^{\underline{3}}_{\underline{9}}2^{\underline{3}}_{\underline{7}}4^{\underline{2}}_{\underline{7}}3^{\underline{2}}_{\underline{5}}3^{\underline{1}}_{\underline{5}}4^{\underline{1}}_{\underline{3}}5^{0}_{\underline{3}}6^{0}_{\underline{1}}4^{1}_{\underline{1}}2^{1}_{1}1^{2}_{1}4^{2}_{3}1^{3}_{3}1^{3}_{5}1^{4}_{7}$ 10^n_{66}: $1^{\underline{6}}_{\underline{13}}1^{\underline{5}}_{\underline{9}}2^{\underline{4}}_{\underline{9}}1^{\underline{4}}_{\underline{7}}1^{\underline{3}}_{\underline{7}}1^{\underline{3}}_{\underline{5}}1^{\underline{2}}_{\underline{7}}1^{\underline{2}}_{\underline{5}}1^{\underline{2}}_{\underline{3}}2^{\underline{1}}_{\underline{3}}1^{\underline{1}}_{\underline{1}}2^{0}_{\underline{3}}3^{0}_{\underline{1}}3^{0}_{1}1^{1}_{\underline{1}}1^{1}_{1}1^{2}_{3}1^{3}_{3}1^{4}_{7}$	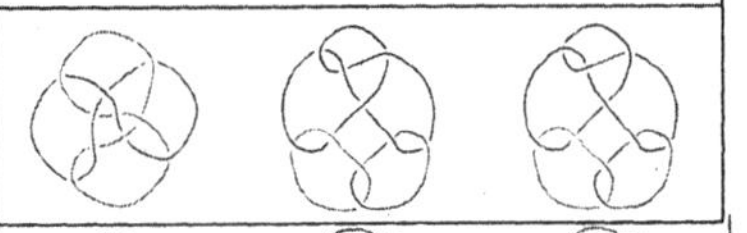		
10^n_{67}: $1^{\underline{6}}_{\underline{13}}1^{\underline{5}}_{\underline{11}}1^{\underline{5}}_{\underline{9}}4^{\underline{4}}_{\underline{9}}2^{\underline{4}}_{\underline{7}}2^{\underline{3}}_{\underline{7}}3^{\underline{3}}_{\underline{5}}4^{\underline{2}}_{\underline{5}}2^{\underline{2}}_{\underline{3}}3^{\underline{1}}_{\underline{3}}4^{\underline{1}}_{\underline{1}}1^{0}_{\underline{3}}4^{0}_{\underline{1}}5^{0}_{1}2^{1}_{1}1^{1}_{3}2^{2}_{5}$ 10^n_{68}: $1^{\underline{8}}_{\underline{23}}1^{\underline{7}}_{\underline{21}}1^{\underline{7}}_{\underline{19}}2^{\underline{6}}_{\underline{19}}3^{\underline{6}}_{\underline{17}}1^{\underline{6}}_{\underline{15}}3^{\underline{5}}_{\underline{17}}1^{\underline{5}}_{\underline{15}}1^{\underline{4}}_{\underline{15}}4^{\underline{4}}_{\underline{13}}1^{\underline{4}}_{\underline{11}}2^{\underline{3}}_{\underline{13}}1^{\underline{3}}_{\underline{11}}1^{\underline{2}}_{\underline{11}}2^{\underline{2}}_{\underline{9}}1^{\underline{1}}_{\underline{7}}1^{0}_{\underline{7}}1^{0}_{\underline{5}}$ 10^n_{69}: $1^{\underline{8}}_{\underline{19}}1^{\underline{7}}_{\underline{17}}1^{\underline{7}}_{\underline{15}}2^{\underline{6}}_{\underline{15}}2^{\underline{6}}_{\underline{13}}3^{\underline{5}}_{\underline{13}}1^{\underline{5}}_{\underline{11}}2^{\underline{4}}_{\underline{11}}3^{\underline{4}}_{\underline{9}}2^{\underline{3}}_{\underline{9}}2^{\underline{3}}_{\underline{7}}3^{\underline{2}}_{\underline{7}}3^{\underline{2}}_{\underline{5}}2^{\underline{1}}_{\underline{3}}2^{0}_{\underline{3}}2^{0}_{\underline{1}}$	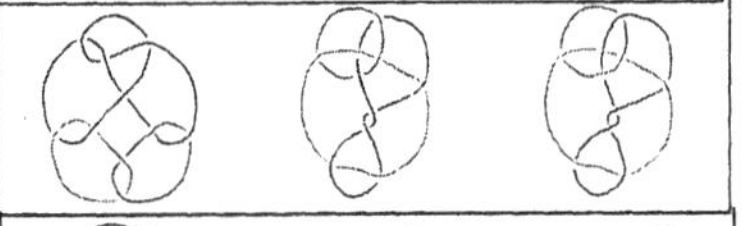		
10^n_{70}: $1^{\underline{8}}_{\underline{19}}1^{\underline{7}}_{\underline{17}}1^{\underline{7}}_{\underline{15}}1^{\underline{6}}_{\underline{15}}1^{\underline{6}}_{\underline{13}}2^{\underline{5}}_{\underline{13}}1^{\underline{5}}_{\underline{11}}1^{\underline{4}}_{\underline{13}}3^{\underline{4}}_{\underline{11}}3^{\underline{4}}_{\underline{9}}1^{\underline{3}}_{\underline{11}}2^{\underline{3}}_{\underline{9}}1^{\underline{3}}_{\underline{7}}2^{\underline{2}}_{\underline{7}}2^{\underline{2}}_{\underline{5}}1^{\underline{1}}_{\underline{3}}1^{0}_{\underline{5}}2^{0}_{\underline{3}}1^{0}_{\underline{1}}$ 10^n_{71}: $2^{\underline{4}}_{\underline{11}}1^{\underline{4}}_{\underline{9}}3^{\underline{3}}_{\underline{9}}1^{\underline{3}}_{\underline{7}}2^{\underline{2}}_{\underline{7}}3^{\underline{2}}_{\underline{5}}4^{\underline{1}}_{\underline{5}}2^{\underline{1}}_{\underline{3}}3^{0}_{\underline{3}}5^{0}_{\underline{1}}2^{1}_{\underline{1}}2^{1}_{1}3^{2}_{1}4^{2}_{3}1^{3}_{3}1^{3}_{5}1^{4}_{7}$ 10^n_{72}: $1^{\underline{4}}_{\underline{7}}1^{\underline{3}}_{\underline{5}}1^{\underline{3}}_{\underline{3}}1^{\underline{2}}_{\underline{3}}1^{\underline{2}}_{\underline{1}}2^{\underline{1}}_{\underline{1}}1^{\underline{1}}_{1}1^{0}_{\underline{1}}3^{0}_{1}3^{0}_{3}1^{1}_{1}1^{1}_{3}1^{1}_{5}3^{2}_{5}2^{2}_{7}1^{3}_{5}1^{3}_{9}1^{4}_{9}1^{6}_{11}1^{6}_{13}$	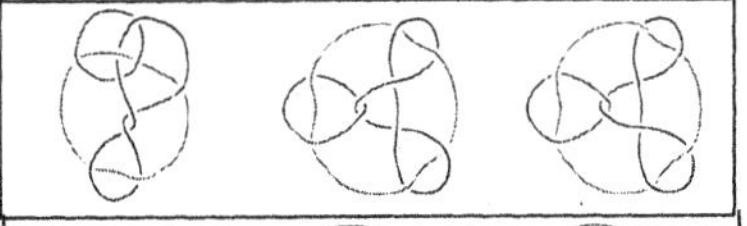		
10^n_{73}: $1^{\underline{3}}_{\underline{7}}3^{\underline{2}}_{\underline{5}}2^{\underline{2}}_{\underline{3}}2^{\underline{1}}_{\underline{3}}2^{\underline{1}}_{\underline{1}}5^{0}_{\underline{1}}4^{0}_{1}3^{1}_{1}3^{1}_{3}3^{2}_{3}4^{2}_{5}2^{3}_{5}2^{3}_{7}1^{4}_{7}2^{4}_{9}1^{5}_{11}$ 10^n_{74}: $1^{\underline{8}}_{\underline{23}}2^{\underline{8}}_{\underline{21}}1^{\underline{8}}_{\underline{19}}1^{\underline{7}}_{\underline{21}}1^{\underline{6}}_{\underline{19}}2^{\underline{6}}_{\underline{17}}2^{\underline{5}}_{\underline{17}}1^{\underline{5}}_{\underline{15}}1^{\underline{5}}_{\underline{13}}1^{\underline{4}}_{\underline{15}}3^{\underline{4}}_{\underline{13}}1^{\underline{4}}_{\underline{11}}1^{\underline{3}}_{\underline{13}}1^{\underline{3}}_{\underline{11}}1^{\underline{2}}_{\underline{11}}1^{\underline{2}}_{\underline{9}}1^{\underline{1}}_{\underline{7}}1^{0}_{\underline{7}}1^{0}_{\underline{5}}$ 10^n_{75}: $1^{\underline{8}}_{\underline{19}}1^{\underline{8}}_{\underline{17}}1^{\underline{7}}_{\underline{17}}2^{\underline{6}}_{\underline{15}}1^{\underline{6}}_{\underline{13}}2^{\underline{5}}_{\underline{13}}2^{\underline{5}}_{\underline{11}}3^{\underline{4}}_{\underline{11}}3^{\underline{4}}_{\underline{9}}2^{\underline{3}}_{\underline{9}}2^{\underline{3}}_{\underline{7}}3^{\underline{2}}_{\underline{7}}2^{\underline{2}}_{\underline{5}}3^{\underline{1}}_{\underline{3}}2^{0}_{\underline{3}}2^{0}_{\underline{1}}$	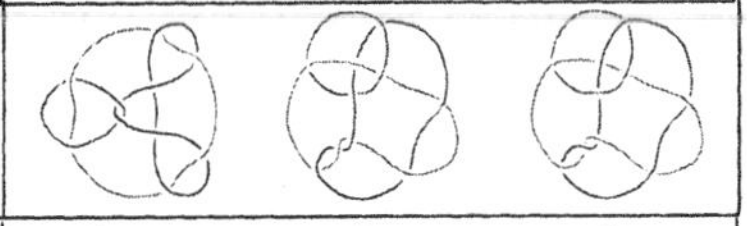		
10^n_{76}: $1^{\underline{8}}_{\underline{19}}1^{\underline{7}}_{\underline{17}}1^{\underline{7}}_{\underline{15}}3^{\underline{6}}_{\underline{15}}3^{\underline{6}}_{\underline{13}}3^{\underline{5}}_{\underline{13}}1^{\underline{5}}_{\underline{11}}3^{\underline{4}}_{\underline{11}}3^{\underline{4}}_{\underline{9}}2^{\underline{3}}_{\underline{9}}3^{\underline{3}}_{\underline{7}}3^{\underline{2}}_{\underline{7}}2^{\underline{2}}_{\underline{5}}3^{\underline{1}}_{\underline{3}}2^{0}_{\underline{3}}2^{0}_{\underline{1}}$ 10^n_{77}: $1^{\underline{8}}_{\underline{25}}2^{\underline{8}}_{\underline{23}}1^{\underline{8}}_{\underline{21}}1^{\underline{7}}_{\underline{23}}1^{\underline{7}}_{\underline{21}}1^{\underline{6}}_{\underline{19}}1^{\underline{6}}_{\underline{17}}1^{\underline{5}}_{\underline{19}}2^{\underline{4}}_{\underline{15}}1^{\underline{4}}_{\underline{13}}1^{\underline{3}}_{\underline{15}}1^{\underline{2}}_{\underline{11}}1^{0}_{\underline{9}}1^{0}_{\underline{7}}$ 10^n_{78}: $1^{\underline{8}}_{\underline{21}}1^{\underline{8}}_{\underline{19}}1^{\underline{7}}_{\underline{19}}1^{\underline{6}}_{\underline{17}}1^{\underline{6}}_{\underline{15}}2^{\underline{5}}_{\underline{15}}1^{\underline{5}}_{\underline{13}}2^{\underline{4}}_{\underline{13}}3^{\underline{4}}_{\underline{11}}2^{\underline{3}}_{\underline{11}}1^{\underline{3}}_{\underline{9}}1^{\underline{2}}_{\underline{9}}2^{\underline{2}}_{\underline{7}}1^{\underline{1}}_{\underline{5}}2^{0}_{\underline{5}}2^{0}_{\underline{3}}$	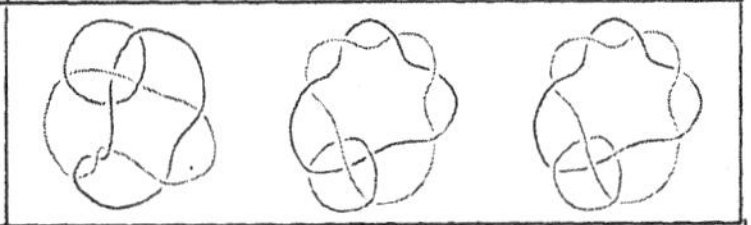		
10^n_{79}: $1^{\underline{8}}_{\underline{21}}1^{\underline{7}}_{\underline{19}}1^{\underline{7}}_{\underline{17}}3^{\underline{6}}_{\underline{17}}3^{\underline{6}}_{\underline{15}}4^{\underline{5}}_{\underline{15}}1^{\underline{5}}_{\underline{13}}2^{\underline{4}}_{\underline{13}}4^{\underline{4}}_{\underline{11}}3^{\underline{3}}_{\underline{11}}2^{\underline{3}}_{\underline{9}}2^{\underline{2}}_{\underline{9}}3^{\underline{2}}_{\underline{7}}2^{\underline{1}}_{\underline{5}}2^{0}_{\underline{5}}2^{0}_{\underline{3}}$ 10^n_{80}: $1^{\underline{8}}_{\underline{19}}1^{\underline{8}}_{\underline{17}}1^{\underline{7}}_{\underline{17}}3^{\underline{6}}_{\underline{15}}1^{\underline{6}}_{\underline{13}}1^{\underline{5}}_{\underline{13}}3^{\underline{5}}_{\underline{11}}5^{\underline{4}}_{\underline{11}}3^{\underline{4}}_{\underline{9}}2^{\underline{3}}_{\underline{9}}3^{\underline{3}}_{\underline{7}}3^{\underline{2}}_{\underline{7}}2^{\underline{2}}_{\underline{5}}1^{\underline{1}}_{\underline{5}}3^{\underline{1}}_{\underline{3}}1^{0}_{\underline{3}}2^{0}_{\underline{1}}$ 10^n_{81}: $1^{\underline{8}}_{\underline{21}}1^{\underline{8}}_{\underline{19}}2^{\underline{7}}_{\underline{19}}3^{\underline{6}}_{\underline{17}}2^{\underline{6}}_{\underline{15}}3^{\underline{5}}_{\underline{15}}3^{\underline{5}}_{\underline{13}}3^{\underline{4}}_{\underline{13}}4^{\underline{4}}_{\underline{11}}4^{\underline{3}}_{\underline{11}}2^{\underline{3}}_{\underline{9}}2^{\underline{2}}_{\underline{9}}4^{\underline{2}}_{\underline{7}}2^{\underline{1}}_{\underline{5}}2^{0}_{\underline{5}}2^{0}_{\underline{3}}$	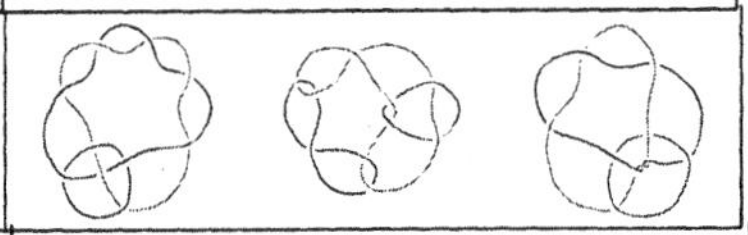		
10^n_{82}: $1^{\underline{8}}_{\underline{17}}1^{\underline{8}}_{\underline{15}}1^{\underline{6}}_{\underline{13}}1^{\underline{5}}_{\underline{9}}2^{\underline{4}}_{\underline{11}}2^{\underline{4}}_{\underline{9}}2^{\underline{3}}_{\underline{9}}1^{\underline{3}}_{\underline{7}}1^{\underline{3}}_{\underline{5}}2^{\underline{2}}_{\underline{7}}2^{\underline{2}}_{\underline{5}}2^{\underline{1}}_{\underline{3}}3^{0}_{\underline{3}}2^{0}_{\underline{1}}1^{1}_{\underline{1}}1^{1}_{1}1^{2}_{3}$ 10^n_{83}: $1^{\underline{6}}_{\underline{17}}2^{\underline{5}}_{\underline{15}}1^{\underline{5}}_{\underline{13}}3^{\underline{4}}_{\underline{13}}4^{\underline{4}}_{\underline{11}}1^{\underline{4}}_{\underline{9}}4^{\underline{3}}_{\underline{11}}2^{\underline{3}}_{\underline{9}}2^{\underline{2}}_{\underline{9}}4^{\underline{2}}_{\underline{7}}2^{\underline{1}}_{\underline{7}}2^{\underline{1}}_{\underline{5}}3^{0}_{\underline{5}}4^{0}_{\underline{3}}1^{1}_{\underline{3}}1^{1}_{\underline{1}}1^{2}_{1}$ 10^n_{84}: $1^{\underline{10}}_{\underline{25}}1^{\underline{10}}_{\underline{23}}1^{\underline{9}}_{\underline{23}}1^{\underline{8}}_{\underline{21}}1^{\underline{8}}_{\underline{19}}1^{\underline{7}}_{\underline{19}}1^{\underline{7}}_{\underline{17}}1^{\underline{6}}_{\underline{19}}3^{\underline{6}}_{\underline{17}}1^{\underline{6}}_{\underline{15}}1^{\underline{5}}_{\underline{17}}2^{\underline{5}}_{\underline{13}}3^{\underline{4}}_{\underline{13}}1^{\underline{4}}_{\underline{11}}1^{\underline{3}}_{\underline{13}}1^{\underline{3}}_{\underline{9}}1^{\underline{2}}_{\underline{9}}1^{0}_{\underline{7}}1^{0}_{\underline{5}}$	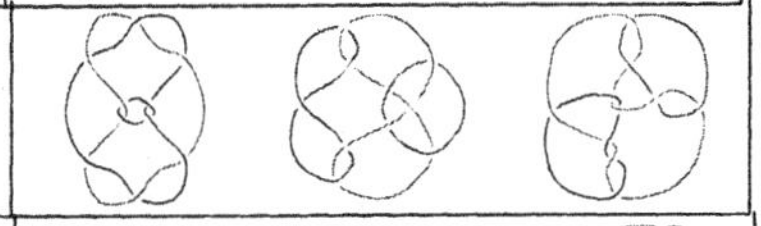		
10^n_{85}: $2^{\underline{4}}_{\underline{9}}1^{\underline{4}}_{\underline{7}}2^{\underline{3}}_{\underline{7}}1^{\underline{3}}_{\underline{5}}4^{\underline{2}}_{\underline{5}}3^{\underline{2}}_{\underline{3}}3^{\underline{1}}_{\underline{3}}3^{\underline{1}}_{\underline{1}}5^{0}_{\underline{1}}4^{0}_{1}2^{1}_{1}4^{1}_{3}3^{2}_{3}3^{2}_{5}1^{3}_{5}2^{3}_{7}1^{4}_{9}$ 10^n_{86}: $1^{\underline{6}}_{\underline{17}}1^{\underline{6}}_{\underline{15}}2^{\underline{5}}_{\underline{15}}2^{\underline{4}}_{\underline{13}}3^{\underline{4}}_{\underline{11}}2^{\underline{3}}_{\underline{11}}1^{\underline{3}}_{\underline{9}}3^{\underline{2}}_{\underline{9}}3^{\underline{2}}_{\underline{7}}1^{\underline{1}}_{\underline{7}}2^{\underline{1}}_{\underline{5}}2^{0}_{\underline{5}}2^{0}_{\underline{3}}1^{1}_{\underline{3}}1^{1}_{\underline{1}}1^{2}_{1}$ 10^n_{87}: $1^{\underline{4}}_{\underline{5}}1^{\underline{4}}_{\underline{3}}1^{\underline{2}}_{\underline{1}}1^{\underline{1}}_{3}2^{0}_{1}2^{0}_{3}1^{1}_{3}1^{1}_{5}1^{1}_{7}2^{2}_{5}2^{2}_{7}1^{3}_{7}1^{3}_{9}1^{4}_{9}1^{4}_{11}1^{5}_{13}1^{6}_{13}1^{6}_{15}$	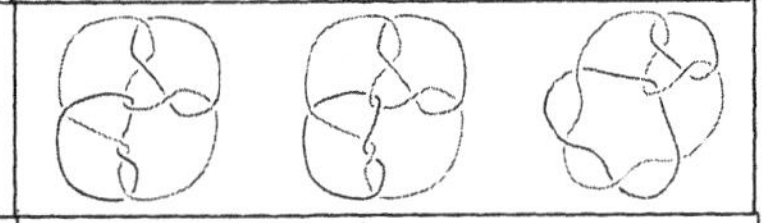		
10^n_{88}: $1^{\underline{6}}_{\underline{13}}1^{\underline{5}}_{\underline{11}}1^{\underline{5}}_{\underline{9}}2^{\underline{4}}_{\underline{9}}2^{\underline{4}}_{\underline{7}}2^{\underline{3}}_{\underline{7}}1^{\underline{3}}_{\underline{5}}2^{\underline{2}}_{\underline{5}}2^{\underline{2}}_{\underline{3}}1^{\underline{1}}_{\underline{5}}1^{\underline{1}}_{\underline{3}}2^{\underline{1}}_{\underline{1}}4^{0}_{\underline{1}}3^{0}_{1}1^{1}_{\underline{1}}1^{1}_{3}1^{2}_{3}1^{4}_{5}1^{4}_{7}$ 10^n_{89}: $2^{\underline{8}}_{\underline{21}}1^{\underline{8}}_{\underline{19}}3^{\underline{7}}_{\underline{19}}1^{\underline{7}}_{\underline{17}}4^{\underline{6}}_{\underline{17}}3^{\underline{6}}_{\underline{15}}4^{\underline{5}}_{\underline{15}}4^{\underline{5}}_{\underline{13}}5^{\underline{4}}_{\underline{13}}6^{\underline{4}}_{\underline{11}}4^{\underline{3}}_{\underline{11}}3^{\underline{3}}_{\underline{9}}3^{\underline{2}}_{\underline{9}}4^{\underline{2}}_{\underline{7}}3^{\underline{1}}_{\underline{5}}1^{0}_{\underline{5}}1^{0}_{\underline{3}}$ 10^n_{90}: $3^{\underline{4}}_{\underline{11}}2^{\underline{4}}_{\underline{9}}4^{\underline{3}}_{\underline{9}}1^{\underline{3}}_{\underline{7}}4^{\underline{2}}_{\underline{7}}4^{\underline{2}}_{\underline{5}}3^{\underline{1}}_{\underline{5}}4^{\underline{1}}_{\underline{3}}6^{0}_{\underline{3}}5^{0}_{\underline{1}}3^{1}_{\underline{1}}4^{1}_{1}2^{2}_{1}3^{2}_{3}1^{3}_{3}2^{3}_{5}1^{4}_{7}$	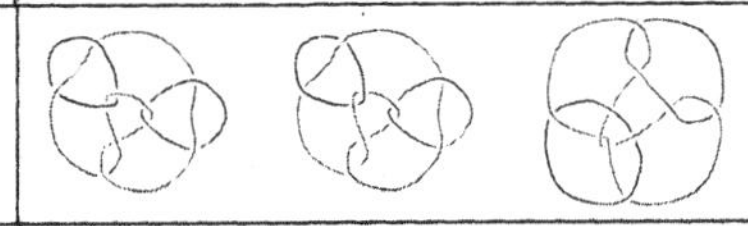		
10^n_{91}: $1^{\underline{8}}_{\underline{19}}1^{\underline{8}}_{\underline{17}}1^{\underline{7}}_{\underline{17}}1^{\underline{6}}_{\underline{15}}1^{\underline{6}}_{\underline{13}}1^{\underline{5}}_{\underline{13}}1^{\underline{5}}_{\underline{11}}3^{\underline{4}}_{\underline{13}}4^{\underline{4}}_{\underline{11}}1^{\underline{4}}_{\underline{9}}1^{\underline{3}}_{\underline{11}}1^{\underline{3}}_{\underline{9}}2^{\underline{3}}_{\underline{7}}1^{\underline{2}}_{\underline{9}}2^{\underline{2}}_{\underline{7}}1^{\underline{1}}_{\underline{7}}1^{\underline{1}}_{\underline{5}}1^{\underline{1}}_{\underline{3}}1^{0}_{\underline{5}}2^{0}_{\underline{3}}1^{1}_{\underline{3}}1^{2}_{1}$ 10^n_{92}: $1^{\underline{6}}_{\underline{13}}1^{\underline{6}}_{\underline{11}}1^{\underline{5}}_{\underline{11}}1^{\underline{4}}_{\underline{9}}1^{\underline{4}}_{\underline{7}}1^{\underline{3}}_{\underline{7}}1^{\underline{3}}_{\underline{5}}2^{\underline{2}}_{\underline{5}}1^{\underline{2}}_{\underline{3}}2^{\underline{1}}_{\underline{1}}3^{0}_{\underline{1}}2^{0}_{1}1^{1}_{3}1^{2}_{3}1^{2}_{5}$ 10^n_{93}: $1^{\underline{8}}_{\underline{23}}1^{\underline{8}}_{\underline{21}}2^{\underline{6}}_{\underline{19}}2^{\underline{6}}_{\underline{17}}1^{\underline{5}}_{\underline{19}}1^{\underline{5}}_{\underline{17}}1^{\underline{4}}_{\underline{15}}1^{\underline{4}}_{\underline{13}}1^{\underline{3}}_{\underline{15}}1^{\underline{2}}_{\underline{11}}1^{0}_{\underline{9}}1^{0}_{\underline{7}}$	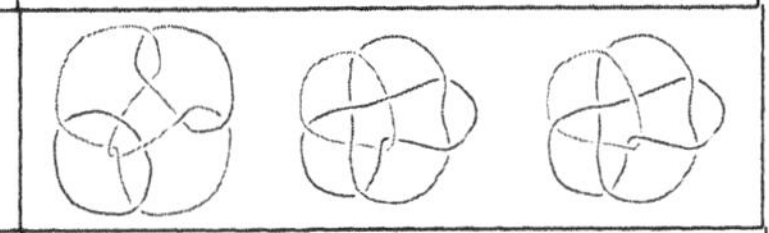		
10^n_{94}: $1^{0}_{3}1^{0}_{5}1^{2}_{5}3^{2}_{7}1^{2}_{9}1^{3}_{7}1^{3}_{11}2^{4}_{11}1^{5}_{11}1^{5}_{15}1^{6}_{15}1^{8}_{17}1^{8}_{19}$ 10^n_{95}: $1^{\underline{6}}_{\underline{15}}1^{\underline{6}}_{\underline{13}}2^{\underline{5}}_{\underline{13}}2^{\underline{4}}_{\underline{11}}2^{\underline{4}}_{\underline{9}}3^{\underline{3}}_{\underline{9}}2^{\underline{3}}_{\underline{7}}3^{\underline{2}}_{\underline{7}}3^{\underline{2}}_{\underline{5}}2^{\underline{1}}_{\underline{5}}3^{\underline{1}}_{\underline{3}}3^{0}_{\underline{3}}4^{0}_{\underline{1}}1^{1}_{\underline{1}}1^{1}_{1}1^{2}_{1}2^{2}_{3}$ 10^n_{96}: $3^{0}_{2}3^{0}_{4}3^{1}_{4}4^{2}_{6}4^{2}_{8}3^{3}_{8}3^{3}_{10}5^{4}_{10}4^{4}_{12}1^{5}_{12}4^{5}_{14}4^{6}_{14}4^{6}_{16}1^{7}_{16}1^{7}_{18}1^{8}_{20}$	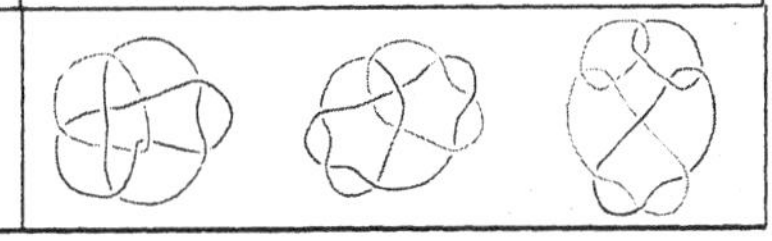		

L: KH	1st line	2nd line	3rd line
10^n_{97}: $1^{\underline{6}}_{\underline{14}} 1^{\underline{6}}_{\underline{12}} 1^{\underline{5}}_{\underline{12}} 1^{\underline{4}}_{\underline{10}} 1^{\underline{4}}_{\underline{8}} 1^{\underline{3}}_{\underline{8}} 1^{\underline{3}}_{\underline{6}} 1^{\underline{2}}_{\underline{8}} 4^{\underline{2}}_{\underline{6}} 3^{\underline{2}}_{\underline{4}} 1^{\underline{1}}_{\underline{6}} 1^{\underline{1}}_{\underline{2}} 4^{0}_{\underline{2}} 3^{0}_{0} 1^{1}_{\underline{2}} 1^{2}_{2} 1^{4}_{4} 1^{4}_{6}$	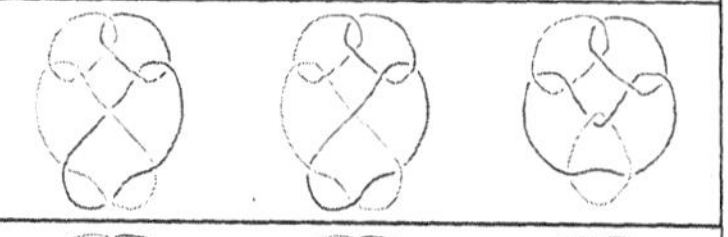		
10^n_{98}: $1^{\underline{6}}_{\underline{12}} 1^{\underline{5}}_{\underline{8}} 4^{\underline{4}}_{\underline{8}} 3^{\underline{4}}_{\underline{6}} 1^{\underline{3}}_{\underline{6}} 1^{\underline{3}}_{\underline{4}} 4^{\underline{2}}_{\underline{4}} 2^{\underline{2}}_{\underline{2}} 1^{\underline{1}}_{\underline{2}} 3^{\underline{1}}_{0} 1^{0}_{\underline{2}} 4^{0}_{0} 1^{0}_{2} 3^{1}_{4} 1^{2}_{2} 3^{2}_{4} 2^{2}_{6}$			
10^n_{99}: $3^{\underline{4}}_{\underline{10}} 3^{\underline{4}}_{\underline{8}} 4^{\underline{3}}_{\underline{8}} 5^{\underline{2}}_{\underline{6}} 4^{\underline{2}}_{\underline{4}} 2^{\underline{1}}_{\underline{4}} 5^{\underline{1}}_{\underline{2}} 7^{0}_{\underline{2}} 6^{0}_{0} 4^{1}_{0} 3^{1}_{2} 2^{2}_{2} 4^{2}_{4} 2^{3}_{6} 1^{4}_{6} 1^{4}_{8}$			
10^n_{100}: $1^{\underline{3}}_{\underline{6}} 3^{\underline{2}}_{\underline{4}} 1^{\underline{2}}_{\underline{2}} 1^{\underline{1}}_{\underline{2}} 3^{\underline{1}}_{0} 6^{0}_{0} 7^{0}_{2} 4^{1}_{2} 3^{2}_{4} 4^{2}_{6} 3^{3}_{8} 2^{4}_{8} 2^{4}_{10}$	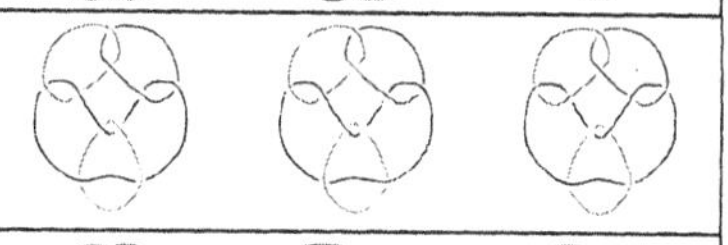		
10^n_{101}: $1^{\underline{2}}_{\underline{2}} 1^{\underline{1}}_{0} 1^{\underline{1}}_{2} 3^{0}_{2} 4^{0}_{4} 2^{1}_{4} 3^{2}_{6} 2^{2}_{8} 1^{3}_{6} 3^{3}_{10} 5^{4}_{10} 4^{4}_{12} 1^{5}_{10} 1^{6}_{14} 1^{8}_{16} 1^{8}_{18}$			
10^n_{102}: $1^{\underline{10}}_{\underline{24}} 1^{\underline{9}}_{\underline{20}} 4^{\underline{8}}_{\underline{20}} 3^{\underline{8}}_{\underline{18}} 1^{\underline{7}}_{\underline{18}} 1^{\underline{7}}_{\underline{16}} 3^{\underline{6}}_{\underline{16}} 1^{\underline{6}}_{\underline{14}} 1^{\underline{5}}_{\underline{14}} 3^{\underline{5}}_{\underline{12}} 3^{\underline{4}}_{\underline{14}} 6^{\underline{4}}_{\underline{12}} 2^{\underline{4}}_{\underline{10}} 1^{\underline{3}}_{\underline{12}} 2^{\underline{3}}_{\underline{8}} 1^{\underline{2}}_{\underline{8}} 1^{0}_{\underline{6}} 1^{0}_{\underline{4}}$			
10^n_{103}: $1^{\underline{6}}_{\underline{16}} 1^{\underline{5}}_{\underline{12}} 6^{\underline{4}}_{\underline{12}} 6^{\underline{4}}_{\underline{10}} 1^{\underline{4}}_{\underline{8}} 4^{\underline{3}}_{\underline{10}} 1^{\underline{3}}_{\underline{8}} 3^{\underline{2}}_{\underline{8}} 4^{\underline{2}}_{\underline{6}} 1^{\underline{1}}_{\underline{6}} 3^{\underline{1}}_{\underline{4}} 4^{0}_{\underline{4}} 3^{0}_{\underline{2}} 1^{1}_{\underline{2}} 2^{1}_{0} 1^{2}_{2}$	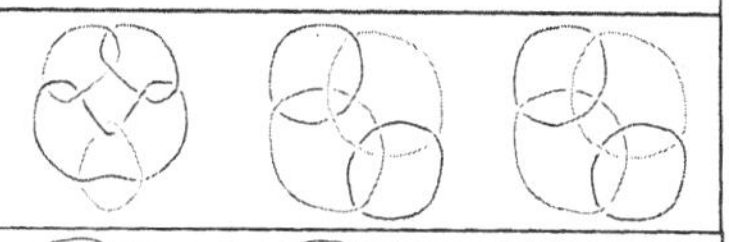		
10^n_{104}: $1^{\underline{8}}_{\underline{18}} 1^{\underline{8}}_{\underline{16}} 1^{\underline{6}}_{\underline{14}} 1^{\underline{5}}_{\underline{10}} 2^{\underline{4}}_{\underline{12}} 3^{\underline{4}}_{\underline{10}} 1^{\underline{3}}_{\underline{6}} 3^{\underline{2}}_{\underline{8}} 4^{\underline{2}}_{\underline{6}} 1^{\underline{2}}_{\underline{4}} 1^{0}_{\underline{4}} 1^{0}_{\underline{2}}$			
10^n_{105}: $2^{\underline{4}}_{\underline{10}} 2^{\underline{4}}_{\underline{8}} 1^{\underline{3}}_{\underline{8}} 3^{\underline{2}}_{\underline{6}} 3^{\underline{2}}_{\underline{4}} 1^{\underline{1}}_{\underline{4}} 1^{\underline{1}}_{\underline{2}} 3^{0}_{\underline{2}} 3^{0}_{0} 1^{1}_{2} 2^{2}_{2} 2^{2}_{4}$			
10^n_{106}: $1^{\underline{5}}_{\underline{14}} 3^{\underline{4}}_{\underline{12}} 3^{\underline{4}}_{\underline{10}} 1^{\underline{4}}_{\underline{8}} 2^{\underline{3}}_{\underline{10}} 2^{\underline{3}}_{\underline{8}} 5^{\underline{2}}_{\underline{8}} 6^{\underline{2}}_{\underline{6}} 1^{\underline{2}}_{\underline{4}} 3^{\underline{1}}_{\underline{6}} 2^{\underline{1}}_{\underline{4}} 3^{0}_{\underline{4}} 5^{0}_{\underline{2}} 2^{1}_{\underline{2}} 1^{1}_{0} 1^{2}_{0} 2^{2}_{2} 1^{3}_{4}$	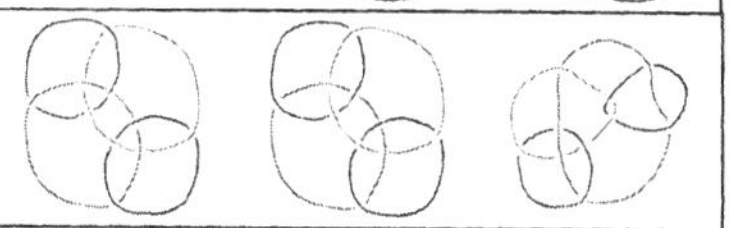		
10^n_{107}: $1^{\underline{5}}_{\underline{10}} 1^{\underline{4}}_{\underline{8}} 1^{\underline{4}}_{\underline{6}} 1^{\underline{3}}_{\underline{6}} 1^{\underline{3}}_{\underline{4}} 1^{\underline{2}}_{\underline{6}} 2^{\underline{2}}_{\underline{4}} 1^{\underline{2}}_{\underline{2}} 2^{\underline{1}}_{\underline{2}} 2^{\underline{1}}_{0} 5^{0}_{\underline{2}} 8^{0}_{0} 5^{0}_{2} 2^{1}_{0} 2^{1}_{2} 1^{2}_{2} 2^{2}_{4} 1^{2}_{6} 1^{3}_{4} 1^{3}_{6} 1^{4}_{6} 1^{4}_{8} 1^{5}_{10}$			
10^n_{108}: $1^{\underline{6}}_{\underline{18}} 2^{\underline{6}}_{\underline{16}} 1^{\underline{6}}_{\underline{14}} 2^{\underline{5}}_{\underline{16}} 1^{\underline{4}}_{\underline{14}} 4^{\underline{4}}_{\underline{12}} 2^{\underline{4}}_{\underline{10}} 2^{\underline{3}}_{\underline{12}} 1^{\underline{3}}_{\underline{10}} 2^{\underline{2}}_{\underline{10}} 4^{\underline{2}}_{\underline{8}} 1^{\underline{2}}_{\underline{6}} 1^{\underline{1}}_{\underline{8}} 1^{\underline{1}}_{\underline{6}} 2^{0}_{\underline{6}} 3^{0}_{\underline{4}} 1^{1}_{\underline{4}} 1^{2}_{0}$			
10^n_{109}: $1^{\underline{6}}_{\underline{14}} 1^{\underline{6}}_{\underline{12}} 2^{\underline{5}}_{\underline{12}} 2^{\underline{4}}_{\underline{10}} 2^{\underline{4}}_{\underline{8}} 2^{\underline{3}}_{\underline{8}} 2^{\underline{3}}_{\underline{6}} 5^{\underline{2}}_{\underline{6}} 4^{\underline{2}}_{\underline{4}} 1^{\underline{1}}_{\underline{4}} 3^{\underline{1}}_{\underline{2}} 5^{0}_{\underline{2}} 5^{0}_{0} 1^{1}_{0} 1^{1}_{2} 1^{2}_{2} 2^{2}_{4}$	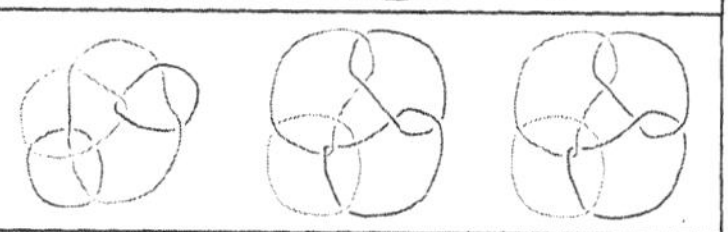		
10^n_{110}: $3^{\underline{2}}_{\underline{4}} 2^{\underline{2}}_{\underline{2}} 3^{\underline{1}}_{\underline{2}} 1^{\underline{1}}_{0} 5^{0}_{0} 5^{0}_{2} 4^{1}_{2} 3^{1}_{4} 6^{2}_{4} 6^{2}_{6} 2^{3}_{6} 4^{3}_{8} 4^{4}_{8} 4^{4}_{10} 1^{5}_{10} 2^{5}_{12} 1^{6}_{14}$			
10^n_{111}: $1^{\underline{6}}_{\underline{12}} 1^{\underline{5}}_{\underline{8}} 2^{\underline{4}}_{\underline{8}} 1^{\underline{4}}_{\underline{6}} 1^{\underline{3}}_{\underline{6}} 1^{\underline{3}}_{\underline{4}} 3^{\underline{2}}_{\underline{6}} 5^{\underline{2}}_{\underline{4}} 2^{\underline{2}}_{\underline{2}} 1^{\underline{1}}_{\underline{4}} 1^{\underline{1}}_{\underline{2}} 1^{\underline{1}}_{0} 3^{0}_{\underline{2}} 4^{0}_{0} 1^{0}_{2} 1^{1}_{0} 1^{1}_{2} 1^{1}_{4} 1^{2}_{2} 1^{2}_{4} 1^{3}_{6} 1^{4}_{6} 1^{4}_{8}$			
10^n_{112}: $4^{0}_{1} 5^{0}_{3} 1^{0}_{5} 6^{1}_{3} 4^{2}_{5} 6^{2}_{7} 1^{3}_{7} 4^{3}_{9} 10^{4}_{9} 11^{4}_{11} 5^{5}_{11} 1^{6}_{13} 5^{6}_{15} 1^{7}_{17} 1^{8}_{17} 1^{8}_{19}$	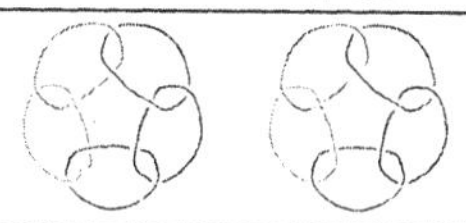		
10^n_{113}: $1^{\underline{4}}_{\underline{7}} 1^{\underline{4}}_{\underline{5}} 1^{\underline{2}}_{\underline{3}} 1^{\underline{1}}_{1} 6^{0}_{\underline{1}} 10^{0}_{1} 4^{0}_{3} 4^{1}_{1} 1^{2}_{3} 4^{2}_{5} 1^{3}_{7} 5^{4}_{7} 5^{4}_{9} 1^{5}_{9} 1^{6}_{13}$			

3.12. 11 Crossing Alternating Links.

L: KH	1st line	2nd line
11^a_{1}: $1^{\underline{5}}_{\underline{10}} 4^{\underline{4}}_{\underline{8}} 1^{\underline{4}}_{\underline{6}} 6^{\underline{3}}_{\underline{6}} 4^{\underline{3}}_{\underline{4}} 10^{\underline{2}}_{\underline{4}} 6^{\underline{2}}_{\underline{2}} 11^{\underline{1}}_{\underline{2}} 10^{\underline{1}}_{0} 14^{0}_{0} 13^{0}_{2} 12^{1}_{2} 12^{1}_{4} 9^{2}_{4} 12^{2}_{6} 6^{3}_{6} 9^{3}_{8} 3^{4}_{8} 6^{4}_{10} 1^{5}_{10} 3^{5}_{12} 1^{6}_{14}$	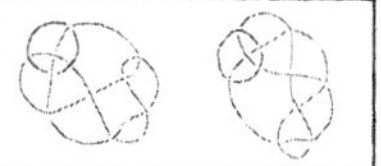	
11^a_{2}: $1^{\underline{6}}_{\underline{14}} 4^{\underline{5}}_{\underline{12}} 1^{\underline{5}}_{\underline{10}} 7^{\underline{4}}_{\underline{10}} 4^{\underline{4}}_{\underline{8}} 10^{\underline{3}}_{\underline{8}} 7^{\underline{3}}_{\underline{6}} 14^{\underline{2}}_{\underline{6}} 10^{\underline{2}}_{\underline{4}} 13^{\underline{1}}_{\underline{4}} 14^{\underline{1}}_{\underline{2}} 14^{0}_{\underline{2}} 15^{0}_{0} 11^{1}_{0} 12^{1}_{2} 6^{2}_{2} 11^{2}_{4} 4^{3}_{4} 6^{3}_{6} 1^{4}_{6} 4^{4}_{8} 1^{5}_{10}$		
11^a_{3}: $1^{\underline{5}}_{\underline{10}} 3^{\underline{4}}_{\underline{8}} 1^{\underline{4}}_{\underline{6}} 6^{\underline{3}}_{\underline{6}} 3^{\underline{3}}_{\underline{4}} 9^{\underline{2}}_{\underline{4}} 6^{\underline{2}}_{\underline{2}} 11^{\underline{1}}_{\underline{2}} 9^{\underline{1}}_{0} 14^{0}_{0} 13^{0}_{2} 12^{1}_{2} 12^{1}_{4} 10^{2}_{4} 12^{2}_{6} 6^{3}_{6} 10^{3}_{8} 4^{4}_{8} 6^{4}_{10} 1^{5}_{10} 4^{5}_{12} 1^{6}_{14}$	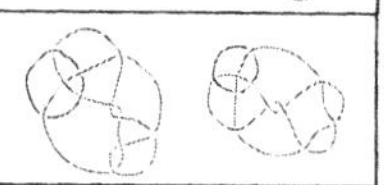	
11^a_{4}: $1^{\underline{3}}_{\underline{4}} 3^{\underline{2}}_{\underline{2}} 1^{\underline{2}}_{0} 4^{\underline{1}}_{0} 3^{\underline{1}}_{2} 10^{0}_{2} 6^{0}_{4} 10^{1}_{4} 8^{1}_{6} 12^{2}_{6} 10^{2}_{8} 10^{3}_{8} 12^{3}_{10} 9^{4}_{10} 10^{4}_{12} 6^{5}_{12} 9^{5}_{14} 3^{6}_{14} 6^{6}_{16} 1^{7}_{16} 3^{7}_{18} 1^{8}_{20}$		
11^a_{5}: $1^{\underline{3}}_{\underline{6}} 3^{\underline{2}}_{\underline{4}} 1^{\underline{2}}_{\underline{2}} 3^{\underline{1}}_{\underline{2}} 3^{\underline{1}}_{0} 7^{0}_{0} 5^{0}_{2} 7^{1}_{2} 5^{1}_{4} 7^{2}_{4} 7^{2}_{6} 6^{3}_{6} 7^{3}_{8} 5^{4}_{8} 6^{4}_{10} 3^{5}_{10} 5^{5}_{12} 2^{6}_{12} 3^{6}_{14} 1^{7}_{14} 2^{7}_{16} 1^{8}_{18}$	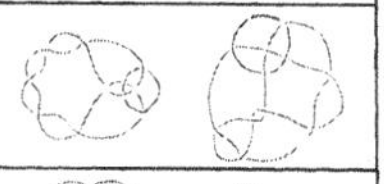	
11^a_{6}: $1^{\underline{3}}_{\underline{4}} 3^{\underline{2}}_{\underline{2}} 1^{\underline{2}}_{0} 5^{\underline{1}}_{0} 3^{\underline{1}}_{2} 10^{0}_{2} 7^{0}_{4} 10^{1}_{4} 8^{1}_{6} 12^{2}_{6} 10^{2}_{8} 10^{3}_{8} 12^{3}_{10} 9^{4}_{10} 10^{4}_{12} 5^{5}_{12} 9^{5}_{14} 3^{6}_{14} 5^{6}_{16} 1^{7}_{16} 3^{7}_{18} 1^{8}_{20}$		
11^a_{7}: $1^{\underline{3}}_{\underline{4}} 3^{\underline{2}}_{\underline{2}} 1^{\underline{2}}_{0} 4^{\underline{1}}_{0} 3^{\underline{1}}_{2} 8^{0}_{2} 6^{0}_{4} 8^{1}_{4} 6^{1}_{6} 9^{2}_{6} 8^{2}_{8} 7^{3}_{8} 9^{3}_{10} 7^{4}_{10} 7^{4}_{12} 3^{5}_{12} 7^{5}_{14} 2^{6}_{14} 3^{6}_{16} 1^{7}_{16} 2^{7}_{18} 1^{8}_{20}$	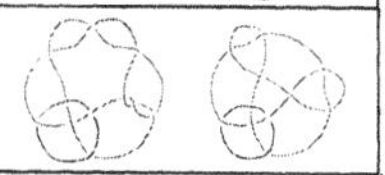	
11^a_{8}: $1^{\underline{7}}_{\underline{16}} 3^{\underline{6}}_{\underline{14}} 1^{\underline{6}}_{\underline{12}} 7^{\underline{5}}_{\underline{12}} 3^{\underline{5}}_{\underline{10}} 10^{\underline{4}}_{\underline{10}} 7^{\underline{4}}_{\underline{8}} 13^{\underline{3}}_{\underline{8}} 10^{\underline{3}}_{\underline{6}} 14^{\underline{2}}_{\underline{6}} 13^{\underline{2}}_{\underline{4}} 13^{\underline{1}}_{\underline{4}} 14^{\underline{1}}_{\underline{2}} 12^{0}_{\underline{2}} 15^{0}_{0} 7^{1}_{0} 10^{1}_{2} 4^{2}_{2} 7^{2}_{4} 1^{3}_{4} 4^{3}_{6} 1^{4}_{8}$		
11^a_{9}: $1^{\underline{3}}_{\underline{6}} 3^{\underline{2}}_{\underline{4}} 1^{\underline{2}}_{\underline{2}} 5^{\underline{1}}_{\underline{2}} 3^{\underline{1}}_{0} 9^{0}_{0} 7^{0}_{2} 10^{1}_{2} 7^{1}_{4} 10^{2}_{4} 10^{2}_{6} 8^{3}_{6} 10^{3}_{8} 8^{4}_{8} 8^{4}_{10} 4^{5}_{10} 8^{5}_{12} 2^{6}_{12} 4^{6}_{14} 1^{7}_{14} 2^{7}_{16} 1^{8}_{18}$	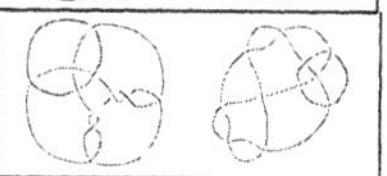	
11^a_{10}: $1^{\underline{3}}_{\underline{6}} 3^{\underline{2}}_{\underline{4}} 1^{\underline{2}}_{\underline{2}} 5^{\underline{1}}_{\underline{2}} 3^{\underline{1}}_{0} 8^{0}_{0} 7^{0}_{2} 10^{1}_{2} 6^{1}_{4} 10^{2}_{4} 10^{2}_{6} 8^{3}_{6} 10^{3}_{8} 8^{4}_{8} 8^{4}_{10} 4^{5}_{10} 8^{5}_{12} 3^{6}_{12} 4^{6}_{14} 1^{7}_{14} 3^{7}_{16} 1^{8}_{18}$		
11^a_{11}: $1^{\underline{7}}_{\underline{16}} 3^{\underline{6}}_{\underline{14}} 1^{\underline{6}}_{\underline{12}} 6^{\underline{5}}_{\underline{12}} 3^{\underline{5}}_{\underline{10}} 10^{\underline{4}}_{\underline{10}} 6^{\underline{4}}_{\underline{8}} 13^{\underline{3}}_{\underline{8}} 10^{\underline{3}}_{\underline{6}} 14^{\underline{2}}_{\underline{6}} 13^{\underline{2}}_{\underline{4}} 13^{\underline{1}}_{\underline{4}} 14^{\underline{1}}_{\underline{2}} 12^{0}_{\underline{2}} 15^{0}_{0} 8^{1}_{0} 10^{1}_{2} 4^{2}_{2} 8^{2}_{4} 1^{3}_{4} 4^{3}_{6} 1^{4}_{8}$	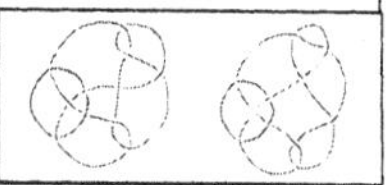	
11^a_{12}: $1^{\underline{8}}_{\underline{20}} 3^{\underline{7}}_{\underline{18}} 1^{\underline{7}}_{\underline{16}} 5^{\underline{6}}_{\underline{16}} 3^{\underline{6}}_{\underline{14}} 10^{\underline{5}}_{\underline{14}} 5^{\underline{5}}_{\underline{12}} 11^{\underline{4}}_{\underline{12}} 10^{\underline{4}}_{\underline{10}} 13^{\underline{3}}_{\underline{10}} 11^{\underline{3}}_{\underline{8}} 12^{\underline{2}}_{\underline{8}} 13^{\underline{2}}_{\underline{6}} 9^{\underline{1}}_{\underline{6}} 12^{\underline{1}}_{\underline{4}} 8^{0}_{\underline{4}} 11^{0}_{\underline{2}} 4^{1}_{\underline{2}} 6^{1}_{0} 1^{2}_{0} 4^{2}_{2} 1^{3}_{4}$		

L: KH	1st line	2nd line
11^a_{13}: $1^{\underline{5}}_{\underline{10}}3^{\underline{4}}_{\underline{8}}1^{\underline{4}}_{\underline{6}}6^{\underline{3}}_{\underline{6}}3^{\underline{3}}_{\underline{4}}10^{\underline{2}}_{\underline{4}}6^{\underline{2}}_{\underline{2}}11^{\underline{1}}_{\underline{2}}10^{\underline{1}}_{0}14^{0}_{0}13^{0}_{2}12^{1}_{2}12^{1}_{4}10^{2}_{4}12^{2}_{6}6^{3}_{6}10^{3}_{8}3^{4}_{8}6^{4}_{10}1^{5}_{10}3^{5}_{12}1^{6}_{14}$ 11^a_{14}: $1^{\underline{3}}_{\underline{4}}2^{\underline{2}}_{\underline{2}}1^{\underline{2}}_{0}4^{\underline{1}}_{0}2^{\underline{1}}_{2}9^{0}_{2}6^{0}_{4}10^{1}_{4}7^{1}_{6}12^{2}_{6}10^{2}_{8}10^{3}_{8}12^{3}_{10}10^{4}_{10}10^{4}_{12}6^{5}_{12}10^{5}_{14}4^{6}_{14}6^{6}_{16}1^{7}_{16}4^{7}_{18}1^{8}_{20}$	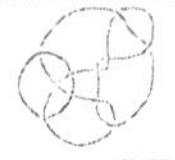	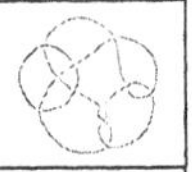
11^a_{15}: $1^{\underline{3}}_{\underline{6}}2^{\underline{2}}_{\underline{4}}1^{\underline{2}}_{\underline{2}}3^{\underline{1}}_{\underline{2}}2^{\underline{1}}_{0}6^{0}_{0}5^{0}_{2}7^{1}_{2}4^{1}_{4}7^{2}_{4}7^{2}_{6}6^{3}_{6}7^{3}_{8}6^{4}_{8}6^{4}_{10}3^{5}_{10}6^{5}_{12}3^{6}_{12}3^{6}_{14}1^{7}_{14}3^{7}_{16}1^{8}_{18}$ 11^a_{16}: $1^{\underline{3}}_{\underline{2}}2^{\underline{2}}_{0}1^{\underline{2}}_{2}1^{\underline{1}}_{2}2^{\underline{1}}_{4}5^{0}_{4}3^{0}_{6}3^{1}_{6}3^{1}_{8}5^{2}_{8}3^{2}_{10}3^{3}_{10}5^{3}_{12}4^{4}_{12}3^{4}_{14}2^{5}_{14}4^{5}_{16}2^{6}_{16}2^{6}_{18}1^{7}_{18}2^{7}_{20}1^{8}_{22}$		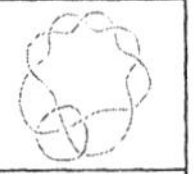
11^a_{17}: $1^{\underline{3}}_{\underline{4}}2^{\underline{2}}_{\underline{2}}1^{\underline{2}}_{0}3^{\underline{1}}_{0}2^{\underline{1}}_{2}7^{0}_{2}5^{0}_{4}8^{1}_{4}5^{1}_{6}9^{2}_{6}8^{2}_{8}7^{3}_{8}9^{3}_{10}8^{4}_{10}7^{4}_{12}4^{5}_{12}8^{5}_{14}3^{6}_{14}4^{6}_{16}1^{7}_{16}3^{7}_{18}1^{8}_{20}$ 11^a_{18}: $1^{\underline{6}}_{\underline{12}}2^{\underline{5}}_{\underline{10}}1^{\underline{5}}_{\underline{8}}5^{\underline{4}}_{\underline{8}}2^{\underline{4}}_{\underline{6}}7^{\underline{3}}_{\underline{6}}5^{\underline{3}}_{\underline{4}}10^{\underline{2}}_{\underline{4}}7^{\underline{2}}_{\underline{2}}10^{\underline{1}}_{\underline{2}}10^{\underline{1}}_{0}11^{0}_{0}12^{0}_{2}9^{1}_{2}9^{1}_{4}6^{2}_{4}9^{2}_{6}3^{3}_{6}6^{3}_{8}1^{4}_{8}3^{4}_{10}1^{5}_{12}$		
11^a_{19}: $1^{\underline{4}}_{\underline{6}}2^{\underline{3}}_{\underline{4}}1^{\underline{3}}_{\underline{2}}5^{\underline{2}}_{\underline{2}}2^{\underline{2}}_{0}6^{\underline{1}}_{0}5^{\underline{1}}_{2}10^{0}_{2}8^{0}_{4}9^{1}_{4}8^{1}_{6}9^{2}_{6}9^{2}_{8}7^{3}_{8}9^{3}_{10}5^{4}_{10}7^{4}_{12}2^{5}_{12}5^{5}_{14}1^{6}_{14}2^{6}_{16}1^{7}_{18}$ 11^a_{20}: $1^{\underline{7}}_{\underline{16}}2^{\underline{6}}_{\underline{14}}1^{\underline{6}}_{\underline{12}}5^{\underline{5}}_{\underline{12}}2^{\underline{5}}_{\underline{10}}7^{\underline{4}}_{\underline{10}}5^{\underline{4}}_{\underline{8}}10^{\underline{3}}_{\underline{8}}7^{\underline{3}}_{\underline{6}}11^{\underline{2}}_{\underline{6}}10^{\underline{2}}_{\underline{4}}9^{\underline{1}}_{\underline{4}}11^{\underline{1}}_{\underline{2}}10^{0}_{\underline{2}}11^{0}_{0}6^{1}_{0}8^{1}_{2}3^{2}_{2}6^{2}_{4}1^{3}_{4}3^{3}_{6}1^{4}_{8}$	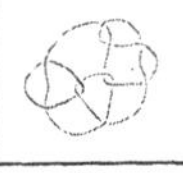	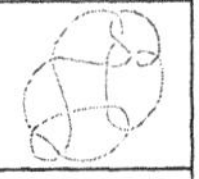
11^a_{21}: $1^{\underline{2}}_{0}1^{\underline{1}}_{2}1^{\underline{1}}_{4}5^{0}_{4}3^{0}_{6}6^{1}_{6}3^{1}_{8}9^{2}_{8}6^{2}_{10}8^{3}_{10}9^{3}_{12}10^{4}_{12}8^{4}_{14}7^{5}_{14}10^{5}_{16}6^{6}_{16}7^{6}_{18}3^{7}_{18}6^{7}_{20}1^{8}_{20}3^{8}_{22}1^{9}_{24}$ 11^a_{22}: $1^{\underline{2}}_{\underline{2}}1^{\underline{1}}_{0}1^{\underline{1}}_{2}4^{0}_{2}3^{0}_{4}5^{1}_{4}2^{1}_{6}6^{2}_{6}5^{2}_{8}6^{3}_{8}6^{3}_{10}7^{4}_{10}6^{4}_{12}4^{5}_{12}7^{5}_{14}4^{6}_{14}4^{6}_{16}2^{7}_{16}4^{7}_{18}1^{8}_{18}2^{8}_{20}1^{9}_{22}$	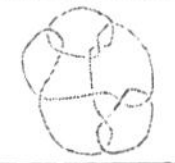	
11^a_{23}: $1^{\underline{11}}_{\underline{28}}2^{\underline{10}}_{\underline{26}}1^{\underline{10}}_{\underline{24}}6^{\underline{9}}_{\underline{24}}2^{\underline{9}}_{\underline{22}}7^{\underline{8}}_{\underline{22}}6^{\underline{8}}_{\underline{20}}10^{\underline{7}}_{\underline{20}}7^{\underline{7}}_{\underline{18}}10^{\underline{6}}_{\underline{18}}10^{\underline{6}}_{\underline{16}}10^{\underline{5}}_{\underline{16}}10^{\underline{5}}_{\underline{14}}8^{\underline{4}}_{\underline{14}}11^{\underline{4}}_{\underline{12}}5^{\underline{3}}_{\underline{12}}7^{\underline{3}}_{\underline{10}}3^{\underline{2}}_{\underline{10}}5^{\underline{2}}_{\underline{8}}3^{\underline{1}}_{\underline{6}}1^{0}_{\underline{6}}1^{0}_{\underline{4}}$ 11^a_{24}: $1^{\underline{9}}_{\underline{22}}2^{\underline{8}}_{\underline{20}}1^{\underline{8}}_{\underline{18}}6^{\underline{7}}_{\underline{18}}2^{\underline{7}}_{\underline{16}}8^{\underline{6}}_{\underline{16}}6^{\underline{6}}_{\underline{14}}10^{\underline{5}}_{\underline{14}}8^{\underline{5}}_{\underline{12}}12^{\underline{4}}_{\underline{12}}11^{\underline{4}}_{\underline{10}}10^{\underline{3}}_{\underline{10}}11^{\underline{3}}_{\underline{8}}9^{\underline{2}}_{\underline{8}}10^{\underline{2}}_{\underline{6}}5^{\underline{1}}_{\underline{6}}9^{\underline{1}}_{\underline{4}}3^{0}_{\underline{4}}6^{0}_{\underline{2}}1^{1}_{\underline{2}}2^{1}_{0}1^{2}_{2}$	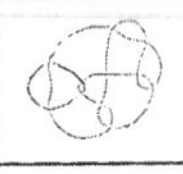	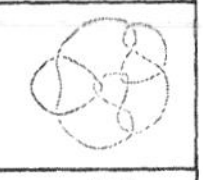
11^a_{25}: $1^{\underline{4}}_{\underline{6}}1^{\underline{4}}_{\underline{4}}2^{\underline{3}}_{\underline{4}}5^{\underline{2}}_{\underline{2}}2^{\underline{2}}_{0}5^{\underline{1}}_{0}5^{\underline{1}}_{2}10^{0}_{2}6^{0}_{4}8^{1}_{4}9^{1}_{6}9^{2}_{6}8^{2}_{8}7^{3}_{8}9^{3}_{10}5^{4}_{10}7^{4}_{12}2^{5}_{12}5^{5}_{14}1^{6}_{14}2^{6}_{16}1^{7}_{18}$ 11^a_{26}: $1^{\underline{4}}_{\underline{8}}1^{\underline{4}}_{\underline{6}}2^{\underline{3}}_{\underline{6}}4^{\underline{2}}_{\underline{4}}2^{\underline{2}}_{\underline{2}}5^{\underline{1}}_{\underline{2}}4^{\underline{1}}_{0}9^{0}_{0}6^{0}_{2}8^{1}_{2}8^{1}_{4}8^{2}_{4}8^{2}_{6}6^{3}_{6}8^{3}_{8}5^{4}_{8}6^{4}_{10}2^{5}_{10}5^{5}_{12}1^{6}_{12}2^{6}_{14}1^{7}_{16}$		
11^a_{27}: $1^{\underline{9}}_{\underline{22}}2^{\underline{8}}_{\underline{20}}1^{\underline{8}}_{\underline{18}}5^{\underline{7}}_{\underline{18}}2^{\underline{7}}_{\underline{16}}7^{\underline{6}}_{\underline{16}}5^{\underline{6}}_{\underline{14}}9^{\underline{5}}_{\underline{14}}7^{\underline{5}}_{\underline{12}}10^{\underline{4}}_{\underline{12}}10^{\underline{4}}_{\underline{10}}9^{\underline{3}}_{\underline{10}}9^{\underline{3}}_{\underline{8}}8^{\underline{2}}_{\underline{8}}9^{\underline{2}}_{\underline{6}}4^{\underline{1}}_{\underline{6}}8^{\underline{1}}_{\underline{4}}3^{0}_{\underline{4}}5^{0}_{\underline{2}}1^{1}_{\underline{2}}2^{1}_{0}1^{2}_{2}$ 11^a_{28}: $1^{\underline{5}}_{\underline{10}}3^{\underline{4}}_{\underline{8}}1^{\underline{4}}_{\underline{6}}6^{\underline{3}}_{\underline{6}}3^{\underline{3}}_{\underline{4}}11^{\underline{2}}_{\underline{4}}6^{\underline{2}}_{\underline{2}}11^{\underline{1}}_{\underline{2}}11^{\underline{1}}_{0}15^{0}_{0}13^{0}_{2}13^{1}_{2}13^{1}_{4}10^{2}_{4}13^{2}_{6}7^{3}_{6}10^{3}_{8}3^{4}_{8}7^{4}_{10}1^{5}_{10}3^{5}_{12}1^{6}_{14}$	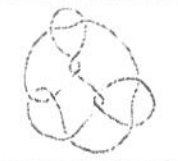	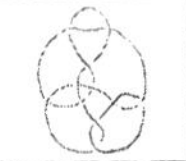
11^a_{29}: $1^{\underline{2}}_{\underline{2}}2^{\underline{1}}_{0}1^{\underline{1}}_{2}6^{0}_{2}4^{0}_{4}10^{1}_{4}4^{1}_{6}11^{2}_{6}10^{2}_{8}11^{3}_{8}11^{3}_{10}13^{4}_{10}11^{4}_{12}8^{5}_{12}13^{5}_{14}7^{6}_{14}8^{6}_{16}3^{7}_{16}7^{7}_{18}1^{8}_{18}3^{8}_{20}1^{9}_{22}$ 11^a_{30}: $1^{\underline{7}}_{\underline{16}}3^{\underline{6}}_{\underline{14}}1^{\underline{6}}_{\underline{12}}5^{\underline{5}}_{\underline{12}}3^{\underline{5}}_{\underline{10}}9^{\underline{4}}_{\underline{10}}6^{\underline{4}}_{\underline{8}}9^{\underline{3}}_{\underline{8}}8^{\underline{3}}_{\underline{6}}11^{\underline{2}}_{\underline{6}}9^{\underline{2}}_{\underline{4}}9^{\underline{1}}_{\underline{4}}11^{\underline{1}}_{\underline{2}}8^{0}_{\underline{2}}10^{0}_{0}5^{1}_{0}7^{1}_{2}2^{2}_{2}5^{2}_{4}1^{3}_{4}2^{3}_{6}1^{4}_{8}$		
11^a_{31}: $1^{\underline{4}}_{\underline{6}}1^{\underline{4}}_{\underline{4}}2^{\underline{3}}_{\underline{4}}4^{\underline{2}}_{\underline{2}}2^{\underline{2}}_{0}4^{\underline{1}}_{0}4^{\underline{1}}_{2}8^{0}_{2}5^{0}_{4}7^{1}_{4}7^{1}_{6}8^{2}_{6}7^{2}_{8}5^{3}_{8}8^{3}_{10}5^{4}_{10}5^{4}_{12}2^{5}_{12}5^{5}_{14}1^{6}_{14}2^{6}_{16}1^{7}_{18}$ 11^a_{32}: $1^{\underline{4}}_{\underline{8}}1^{\underline{4}}_{\underline{6}}3^{\underline{3}}_{\underline{6}}5^{\underline{2}}_{\underline{4}}3^{\underline{2}}_{\underline{2}}7^{\underline{1}}_{\underline{2}}5^{\underline{1}}_{0}10^{0}_{0}8^{0}_{2}9^{1}_{2}9^{1}_{4}10^{2}_{4}9^{2}_{6}6^{3}_{6}10^{3}_{8}5^{4}_{8}6^{4}_{10}2^{5}_{10}5^{5}_{12}1^{6}_{12}2^{6}_{14}1^{7}_{16}$	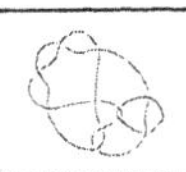	
11^a_{33}: $1^{\underline{11}}_{\underline{30}}1^{\underline{10}}_{\underline{28}}1^{\underline{10}}_{\underline{26}}4^{\underline{9}}_{\underline{26}}1^{\underline{9}}_{\underline{24}}3^{\underline{8}}_{\underline{24}}4^{\underline{8}}_{\underline{22}}6^{\underline{7}}_{\underline{22}}3^{\underline{7}}_{\underline{20}}5^{\underline{6}}_{\underline{20}}6^{\underline{6}}_{\underline{18}}5^{\underline{5}}_{\underline{18}}5^{\underline{5}}_{\underline{16}}4^{\underline{4}}_{\underline{16}}6^{\underline{4}}_{\underline{14}}3^{\underline{3}}_{\underline{14}}3^{\underline{3}}_{\underline{12}}2^{\underline{2}}_{\underline{12}}3^{\underline{2}}_{\underline{10}}2^{\underline{1}}_{\underline{8}}1^{0}_{\underline{8}}1^{0}_{\underline{6}}$ 11^a_{34}: $1^{\underline{11}}_{\underline{28}}1^{\underline{10}}_{\underline{26}}1^{\underline{10}}_{\underline{24}}5^{\underline{9}}_{\underline{24}}1^{\underline{9}}_{\underline{22}}5^{\underline{8}}_{\underline{22}}5^{\underline{8}}_{\underline{20}}9^{\underline{7}}_{\underline{20}}5^{\underline{7}}_{\underline{18}}9^{\underline{6}}_{\underline{18}}9^{\underline{6}}_{\underline{16}}8^{\underline{5}}_{\underline{16}}9^{\underline{5}}_{\underline{14}}8^{\underline{4}}_{\underline{14}}9^{\underline{4}}_{\underline{12}}5^{\underline{3}}_{\underline{12}}7^{\underline{3}}_{\underline{10}}3^{\underline{2}}_{\underline{10}}5^{\underline{2}}_{\underline{8}}3^{\underline{1}}_{\underline{6}}1^{0}_{\underline{6}}1^{0}_{\underline{4}}$	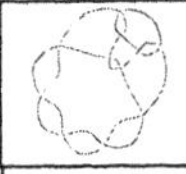	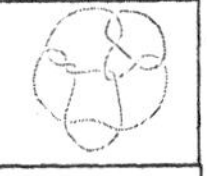
11^a_{35}: $1^{\underline{6}}_{\underline{12}}2^{\underline{5}}_{\underline{10}}1^{\underline{5}}_{\underline{8}}5^{\underline{4}}_{\underline{8}}3^{\underline{4}}_{\underline{6}}6^{\underline{3}}_{\underline{6}}4^{\underline{3}}_{\underline{4}}8^{\underline{2}}_{\underline{4}}6^{\underline{2}}_{\underline{2}}7^{\underline{1}}_{\underline{2}}8^{\underline{1}}_{0}9^{0}_{0}8^{0}_{2}6^{1}_{2}8^{1}_{4}4^{2}_{4}6^{2}_{6}2^{3}_{6}4^{3}_{8}1^{4}_{8}2^{4}_{10}1^{5}_{12}$ 11^a_{36}: $1^{\underline{11}}_{\underline{28}}2^{\underline{10}}_{\underline{26}}1^{\underline{10}}_{\underline{24}}5^{\underline{9}}_{\underline{24}}2^{\underline{9}}_{\underline{22}}5^{\underline{8}}_{\underline{22}}5^{\underline{8}}_{\underline{20}}8^{\underline{7}}_{\underline{20}}5^{\underline{7}}_{\underline{18}}7^{\underline{6}}_{\underline{18}}8^{\underline{6}}_{\underline{16}}7^{\underline{5}}_{\underline{16}}7^{\underline{5}}_{\underline{14}}6^{\underline{4}}_{\underline{14}}8^{\underline{4}}_{\underline{12}}3^{\underline{3}}_{\underline{12}}5^{\underline{3}}_{\underline{10}}2^{\underline{2}}_{\underline{10}}3^{\underline{2}}_{\underline{8}}2^{\underline{1}}_{\underline{6}}1^{0}_{\underline{6}}1^{0}_{\underline{4}}$	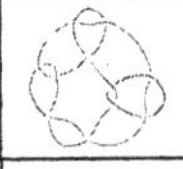	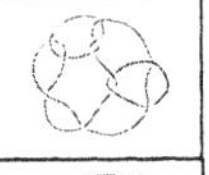
11^a_{37}: $1^{\underline{11}}_{\underline{26}}1^{\underline{10}}_{\underline{24}}1^{\underline{10}}_{\underline{22}}3^{\underline{9}}_{\underline{22}}1^{\underline{9}}_{\underline{20}}2^{\underline{8}}_{\underline{20}}3^{\underline{8}}_{\underline{18}}4^{\underline{7}}_{\underline{18}}2^{\underline{7}}_{\underline{16}}4^{\underline{6}}_{\underline{16}}4^{\underline{6}}_{\underline{14}}3^{\underline{5}}_{\underline{14}}4^{\underline{5}}_{\underline{12}}4^{\underline{4}}_{\underline{12}}4^{\underline{4}}_{\underline{10}}2^{\underline{3}}_{\underline{10}}3^{\underline{3}}_{\underline{8}}2^{\underline{2}}_{\underline{8}}2^{\underline{2}}_{\underline{6}}2^{\underline{1}}_{\underline{4}}1^{0}_{\underline{4}}1^{0}_{\underline{2}}$ 11^a_{38}: $1^{\underline{6}}_{\underline{14}}2^{\underline{5}}_{\underline{12}}1^{\underline{5}}_{\underline{10}}5^{\underline{4}}_{\underline{10}}2^{\underline{4}}_{\underline{8}}8^{\underline{3}}_{\underline{8}}5^{\underline{3}}_{\underline{6}}10^{\underline{2}}_{\underline{6}}8^{\underline{2}}_{\underline{4}}10^{\underline{1}}_{\underline{4}}10^{\underline{1}}_{\underline{2}}11^{0}_{\underline{2}}12^{0}_{0}9^{1}_{0}9^{1}_{2}5^{2}_{2}9^{2}_{4}3^{3}_{4}5^{3}_{6}1^{4}_{6}3^{4}_{8}1^{5}_{10}$	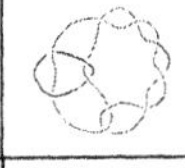	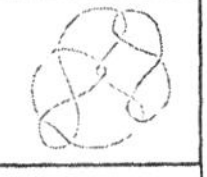
11^a_{39}: $1^{\underline{4}}_{\underline{8}}1^{\underline{3}}_{\underline{6}}1^{\underline{3}}_{\underline{4}}5^{\underline{2}}_{\underline{4}}1^{\underline{2}}_{\underline{2}}5^{\underline{1}}_{\underline{2}}5^{\underline{1}}_{0}10^{0}_{0}7^{0}_{2}10^{1}_{2}8^{1}_{4}8^{2}_{4}10^{2}_{6}8^{3}_{6}8^{3}_{8}5^{4}_{8}8^{4}_{10}3^{5}_{10}5^{5}_{12}1^{6}_{12}3^{6}_{14}1^{7}_{16}$ 11^a_{40}: $1^{\underline{5}}_{\underline{10}}2^{\underline{4}}_{\underline{8}}1^{\underline{4}}_{\underline{6}}5^{\underline{3}}_{\underline{6}}2^{\underline{3}}_{\underline{4}}8^{\underline{2}}_{\underline{4}}5^{\underline{2}}_{\underline{2}}9^{\underline{1}}_{\underline{2}}8^{\underline{1}}_{0}12^{0}_{0}11^{0}_{2}10^{1}_{2}10^{1}_{4}9^{2}_{4}10^{2}_{6}5^{3}_{6}9^{3}_{8}3^{4}_{8}5^{4}_{10}1^{5}_{10}3^{5}_{12}1^{6}_{14}$	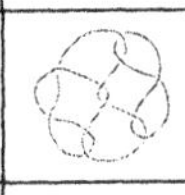	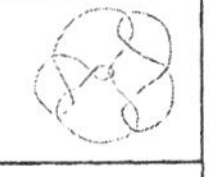
11^a_{41}: $1^{\underline{7}}_{\underline{16}}2^{\underline{6}}_{\underline{14}}1^{\underline{6}}_{\underline{12}}6^{\underline{5}}_{\underline{12}}2^{\underline{5}}_{\underline{10}}9^{\underline{4}}_{\underline{10}}6^{\underline{4}}_{\underline{8}}12^{\underline{3}}_{\underline{8}}9^{\underline{3}}_{\underline{6}}14^{\underline{2}}_{\underline{6}}12^{\underline{2}}_{\underline{4}}12^{\underline{1}}_{\underline{4}}14^{\underline{1}}_{\underline{2}}12^{0}_{\underline{2}}14^{0}_{0}8^{1}_{0}10^{1}_{2}4^{2}_{2}8^{2}_{4}1^{3}_{4}4^{3}_{6}1^{4}_{8}$ 11^a_{42}: $1^{\underline{6}}_{\underline{14}}2^{\underline{5}}_{\underline{12}}1^{\underline{5}}_{\underline{10}}5^{\underline{4}}_{\underline{10}}2^{\underline{4}}_{\underline{8}}7^{\underline{3}}_{\underline{8}}5^{\underline{3}}_{\underline{6}}9^{\underline{2}}_{\underline{6}}7^{\underline{2}}_{\underline{4}}9^{\underline{1}}_{\underline{4}}9^{\underline{1}}_{\underline{2}}9^{0}_{\underline{2}}11^{0}_{0}8^{1}_{0}7^{1}_{2}4^{2}_{2}8^{2}_{4}2^{3}_{4}4^{3}_{6}1^{4}_{6}2^{4}_{8}1^{5}_{10}$	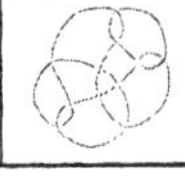	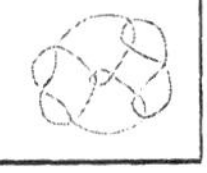

L: KH	1st line	2nd line
11^a_{43}: $1^{\underline{6}}_{\underline{12}}1^{\underline{5}}_{\underline{10}}1^{\underline{5}}_{\underline{8}}4^{\underline{4}}_{\underline{8}}1^{\underline{4}}_{\underline{6}}4^{\underline{3}}_{\underline{6}}4^{\underline{3}}_{\underline{4}}6^{\underline{2}}_{\underline{4}}4^{\underline{2}}_{\underline{2}}6^{\underline{1}}_{\underline{2}}6^{\underline{1}}_{0}7^{0}_{0}8^{0}_{2}6^{1}_{2}5^{1}_{4}3^{2}_{4}6^{2}_{6}2^{3}_{6}3^{3}_{8}1^{4}_{8}2^{4}_{10}1^{5}_{12}$ 11^a_{44}: $1^{\underline{6}}_{\underline{12}}2^{\underline{5}}_{\underline{10}}1^{\underline{5}}_{\underline{8}}6^{\underline{4}}_{\underline{8}}2^{\underline{4}}_{\underline{6}}8^{\underline{3}}_{\underline{6}}6^{\underline{3}}_{\underline{4}}11^{\underline{2}}_{\underline{4}}8^{\underline{2}}_{\underline{2}}12^{\underline{1}}_{\underline{2}}11^{\underline{1}}_{0}12^{0}_{0}14^{0}_{2}10^{1}_{2}10^{1}_{4}7^{2}_{4}10^{2}_{6}3^{3}_{6}7^{3}_{8}1^{4}_{8}3^{4}_{10}1^{5}_{12}$		
11^a_{45}: $1^{\underline{7}}_{\underline{16}}3^{\underline{6}}_{\underline{14}}1^{\underline{6}}_{\underline{12}}6^{\underline{5}}_{\underline{12}}3^{\underline{5}}_{\underline{10}}8^{\underline{4}}_{\underline{10}}6^{\underline{4}}_{\underline{8}}12^{\underline{3}}_{\underline{8}}8^{\underline{3}}_{\underline{6}}12^{\underline{2}}_{\underline{6}}12^{\underline{2}}_{\underline{4}}10^{\underline{1}}_{\underline{4}}12^{\underline{1}}_{\underline{2}}11^{0}_{\underline{2}}12^{0}_{0}6^{1}_{0}9^{1}_{2}3^{2}_{2}6^{2}_{4}1^{3}_{4}3^{3}_{6}1^{4}_{8}$ 11^a_{46}: $1^{\underline{6}}_{\underline{14}}3^{\underline{5}}_{\underline{12}}1^{\underline{5}}_{\underline{10}}7^{\underline{4}}_{\underline{10}}3^{\underline{4}}_{\underline{8}}9^{\underline{3}}_{\underline{8}}7^{\underline{3}}_{\underline{6}}13^{\underline{2}}_{\underline{6}}9^{\underline{2}}_{\underline{4}}13^{\underline{1}}_{\underline{4}}13^{\underline{1}}_{\underline{2}}13^{0}_{\underline{2}}15^{0}_{0}11^{1}_{0}11^{1}_{2}6^{2}_{2}11^{2}_{4}4^{3}_{4}6^{3}_{6}1^{4}_{6}4^{4}_{8}1^{5}_{10}$		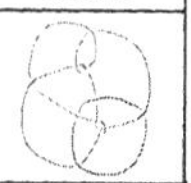
11^a_{47}: $1^{\underline{5}}_{\underline{12}}3^{\underline{4}}_{\underline{10}}1^{\underline{4}}_{\underline{8}}5^{\underline{3}}_{\underline{8}}3^{\underline{3}}_{\underline{6}}8^{\underline{2}}_{\underline{6}}5^{\underline{2}}_{\underline{4}}8^{\underline{1}}_{\underline{4}}8^{\underline{1}}_{\underline{2}}10^{0}_{\underline{2}}10^{0}_{0}9^{1}_{0}8^{1}_{2}6^{2}_{2}9^{2}_{4}4^{3}_{4}6^{3}_{6}2^{4}_{6}4^{4}_{8}1^{5}_{8}2^{5}_{10}1^{6}_{12}$ 11^a_{48}: $1^{\underline{5}}_{\underline{10}}3^{\underline{4}}_{\underline{8}}1^{\underline{4}}_{\underline{6}}5^{\underline{3}}_{\underline{6}}3^{\underline{3}}_{\underline{4}}10^{\underline{2}}_{\underline{4}}5^{\underline{2}}_{\underline{2}}9^{\underline{1}}_{\underline{2}}10^{\underline{1}}_{0}14^{0}_{0}11^{0}_{2}12^{1}_{2}12^{1}_{4}8^{2}_{4}12^{2}_{6}7^{3}_{6}8^{3}_{8}3^{4}_{8}7^{4}_{10}1^{5}_{10}3^{5}_{12}1^{6}_{14}$	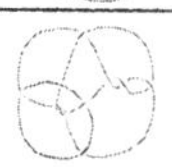	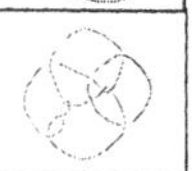
11^a_{49}: $1^{\underline{5}}_{\underline{10}}3^{\underline{4}}_{\underline{8}}1^{\underline{4}}_{\underline{6}}6^{\underline{3}}_{\underline{6}}3^{\underline{3}}_{\underline{4}}10^{\underline{2}}_{\underline{4}}6^{\underline{2}}_{\underline{2}}11^{\underline{1}}_{\underline{2}}10^{\underline{1}}_{0}15^{0}_{0}13^{0}_{2}13^{1}_{2}13^{1}_{4}10^{2}_{4}13^{2}_{6}7^{3}_{6}10^{3}_{8}4^{4}_{8}7^{4}_{10}1^{5}_{10}4^{5}_{12}1^{6}_{14}$ 11^a_{50}: $1^{\underline{5}}_{\underline{10}}3^{\underline{4}}_{\underline{8}}1^{\underline{4}}_{\underline{6}}4^{\underline{3}}_{\underline{6}}3^{\underline{3}}_{\underline{4}}8^{\underline{2}}_{\underline{4}}4^{\underline{2}}_{\underline{2}}7^{\underline{1}}_{\underline{2}}8^{\underline{1}}_{0}11^{0}_{0}9^{0}_{2}9^{1}_{2}9^{1}_{4}6^{2}_{4}9^{2}_{6}5^{3}_{6}6^{3}_{8}2^{4}_{8}5^{4}_{10}1^{5}_{10}2^{5}_{12}1^{6}_{14}$	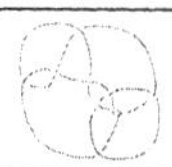	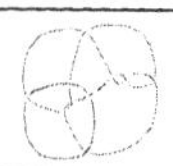
11^a_{51}: $1^{\underline{5}}_{\underline{10}}2^{\underline{4}}_{\underline{8}}1^{\underline{4}}_{\underline{6}}3^{\underline{3}}_{\underline{6}}2^{\underline{3}}_{\underline{4}}5^{\underline{2}}_{\underline{4}}3^{\underline{2}}_{\underline{2}}5^{\underline{1}}_{\underline{2}}5^{\underline{1}}_{0}8^{0}_{0}7^{0}_{2}6^{1}_{2}6^{1}_{4}5^{2}_{4}6^{2}_{6}3^{3}_{6}5^{3}_{8}2^{4}_{8}3^{4}_{10}1^{5}_{10}2^{5}_{12}1^{6}_{14}$ 11^a_{52}: $1^{\underline{5}}_{\underline{8}}2^{\underline{4}}_{\underline{6}}1^{\underline{4}}_{\underline{4}}2^{\underline{3}}_{\underline{4}}2^{\underline{3}}_{\underline{2}}6^{\underline{2}}_{\underline{2}}2^{\underline{2}}_{0}4^{\underline{1}}_{0}6^{\underline{1}}_{2}8^{0}_{2}6^{0}_{4}6^{1}_{4}6^{1}_{6}5^{2}_{6}6^{2}_{8}4^{3}_{8}5^{3}_{10}2^{4}_{10}4^{4}_{12}1^{5}_{12}2^{5}_{14}1^{6}_{16}$		
11^a_{53}: $1^{\underline{5}}_{\underline{12}}2^{\underline{4}}_{\underline{10}}1^{\underline{4}}_{\underline{8}}5^{\underline{3}}_{\underline{8}}2^{\underline{3}}_{\underline{6}}7^{\underline{2}}_{\underline{6}}5^{\underline{2}}_{\underline{4}}8^{\underline{1}}_{\underline{4}}7^{\underline{1}}_{\underline{2}}10^{0}_{\underline{2}}10^{0}_{0}9^{1}_{0}8^{1}_{2}7^{2}_{2}9^{2}_{4}4^{3}_{4}7^{3}_{6}3^{4}_{6}4^{4}_{8}1^{5}_{8}3^{5}_{10}1^{6}_{12}$ 11^a_{54}: $1^{\underline{5}}_{\underline{10}}2^{\underline{4}}_{\underline{8}}1^{\underline{4}}_{\underline{6}}4^{\underline{3}}_{\underline{6}}2^{\underline{3}}_{\underline{4}}7^{\underline{2}}_{\underline{4}}4^{\underline{2}}_{\underline{2}}7^{\underline{1}}_{\underline{2}}7^{\underline{1}}_{0}11^{0}_{0}9^{0}_{2}9^{1}_{2}9^{1}_{4}7^{2}_{4}9^{2}_{6}5^{3}_{6}7^{3}_{8}3^{4}_{8}5^{4}_{10}1^{5}_{10}3^{5}_{12}1^{6}_{14}$	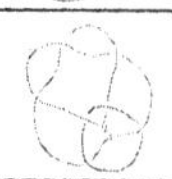	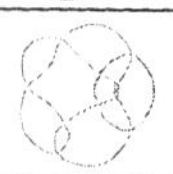
11^a_{55}: $1^{\underline{5}}_{\underline{10}}3^{\underline{4}}_{\underline{8}}1^{\underline{4}}_{\underline{6}}5^{\underline{3}}_{\underline{6}}3^{\underline{3}}_{\underline{4}}9^{\underline{2}}_{\underline{4}}5^{\underline{2}}_{\underline{2}}9^{\underline{1}}_{\underline{2}}9^{\underline{1}}_{0}13^{0}_{0}11^{0}_{2}11^{1}_{2}11^{1}_{4}8^{2}_{4}11^{2}_{6}6^{3}_{6}8^{3}_{8}3^{4}_{8}6^{4}_{10}1^{5}_{10}3^{5}_{12}1^{6}_{14}$ 11^a_{56}: $1^{\underline{2}}_{0}1^{\underline{1}}_{2}1^{\underline{1}}_{4}5^{0}_{4}3^{0}_{6}5^{1}_{6}3^{1}_{8}7^{2}_{8}5^{2}_{10}6^{3}_{10}7^{3}_{12}8^{4}_{12}6^{4}_{14}5^{5}_{14}8^{5}_{16}4^{6}_{16}5^{6}_{18}2^{7}_{18}4^{7}_{20}1^{8}_{20}2^{8}_{22}1^{9}_{24}$	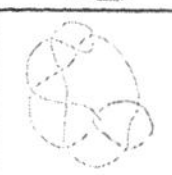	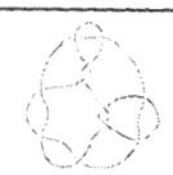
11^a_{57}: $1^{\underline{6}}_{\underline{12}}2^{\underline{5}}_{\underline{10}}1^{\underline{5}}_{\underline{8}}5^{\underline{4}}_{\underline{8}}2^{\underline{4}}_{\underline{6}}7^{\underline{3}}_{\underline{6}}5^{\underline{3}}_{\underline{4}}9^{\underline{2}}_{\underline{4}}7^{\underline{2}}_{\underline{2}}9^{\underline{1}}_{\underline{2}}9^{\underline{1}}_{0}11^{0}_{0}11^{0}_{2}8^{1}_{2}9^{1}_{4}5^{2}_{4}8^{2}_{6}3^{3}_{6}5^{3}_{8}1^{4}_{8}3^{4}_{10}1^{5}_{12}$ 11^a_{58}: $1^{\underline{7}}_{\underline{16}}2^{\underline{6}}_{\underline{14}}1^{\underline{6}}_{\underline{12}}5^{\underline{5}}_{\underline{12}}2^{\underline{5}}_{\underline{10}}8^{\underline{4}}_{\underline{10}}6^{\underline{4}}_{\underline{8}}10^{\underline{3}}_{\underline{8}}7^{\underline{3}}_{\underline{6}}12^{\underline{2}}_{\underline{6}}10^{\underline{2}}_{\underline{4}}9^{\underline{1}}_{\underline{4}}12^{\underline{1}}_{\underline{2}}10^{0}_{\underline{2}}10^{0}_{0}6^{1}_{0}9^{1}_{2}3^{2}_{2}6^{2}_{4}1^{3}_{4}3^{3}_{6}1^{4}_{8}$		
11^a_{59}: $1^{\underline{6}}_{\underline{12}}1^{\underline{5}}_{\underline{10}}1^{\underline{5}}_{\underline{8}}5^{\underline{4}}_{\underline{8}}2^{\underline{4}}_{\underline{6}}5^{\underline{3}}_{\underline{6}}4^{\underline{3}}_{\underline{4}}9^{\underline{2}}_{\underline{4}}5^{\underline{2}}_{\underline{2}}8^{\underline{1}}_{\underline{2}}9^{\underline{1}}_{0}9^{0}_{0}9^{0}_{2}8^{1}_{2}8^{1}_{4}5^{2}_{4}8^{2}_{6}3^{3}_{6}5^{3}_{8}1^{4}_{8}3^{4}_{10}1^{5}_{12}$ 11^a_{60}: $1^{\underline{6}}_{\underline{14}}1^{\underline{5}}_{\underline{12}}1^{\underline{5}}_{\underline{10}}5^{\underline{4}}_{\underline{10}}2^{\underline{4}}_{\underline{8}}5^{\underline{3}}_{\underline{8}}4^{\underline{3}}_{\underline{6}}8^{\underline{2}}_{\underline{6}}5^{\underline{2}}_{\underline{4}}8^{\underline{1}}_{\underline{4}}8^{\underline{1}}_{\underline{2}}8^{0}_{\underline{2}}9^{0}_{0}7^{1}_{0}7^{1}_{2}4^{2}_{2}7^{2}_{4}3^{3}_{4}4^{3}_{6}1^{4}_{6}3^{4}_{8}1^{5}_{10}$	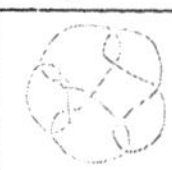	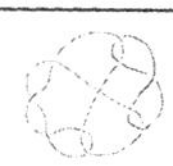
11^a_{61}: $1^{\underline{11}}_{\underline{28}}2^{\underline{10}}_{\underline{26}}1^{\underline{10}}_{\underline{24}}5^{\underline{9}}_{\underline{24}}2^{\underline{9}}_{\underline{22}}6^{\underline{8}}_{\underline{22}}5^{\underline{8}}_{\underline{20}}10^{\underline{7}}_{\underline{20}}6^{\underline{7}}_{\underline{18}}9^{\underline{6}}_{\underline{18}}10^{\underline{6}}_{\underline{16}}9^{\underline{5}}_{\underline{16}}9^{\underline{5}}_{\underline{14}}8^{\underline{4}}_{\underline{14}}10^{\underline{4}}_{\underline{12}}5^{\underline{3}}_{\underline{12}}7^{\underline{3}}_{\underline{10}}3^{\underline{2}}_{\underline{10}}5^{\underline{2}}_{\underline{8}}3^{\underline{1}}_{\underline{6}}1^{0}_{\underline{6}}1^{0}_{\underline{4}}$ 11^a_{62}: $1^{\underline{9}}_{\underline{22}}2^{\underline{8}}_{\underline{20}}1^{\underline{8}}_{\underline{18}}5^{\underline{7}}_{\underline{18}}2^{\underline{7}}_{\underline{16}}7^{\underline{6}}_{\underline{16}}5^{\underline{6}}_{\underline{14}}9^{\underline{5}}_{\underline{14}}7^{\underline{5}}_{\underline{12}}11^{\underline{4}}_{\underline{12}}10^{\underline{4}}_{\underline{10}}10^{\underline{3}}_{\underline{10}}10^{\underline{3}}_{\underline{8}}8^{\underline{2}}_{\underline{8}}10^{\underline{2}}_{\underline{6}}5^{\underline{1}}_{\underline{6}}8^{\underline{1}}_{\underline{4}}4^{0}_{\underline{4}}6^{0}_{\underline{2}}1^{1}_{\underline{2}}3^{1}_{0}1^{2}_{2}$	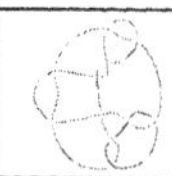	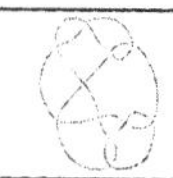
11^a_{63}: $1^{\underline{4}}_{\underline{6}}2^{\underline{3}}_{\underline{4}}1^{\underline{3}}_{\underline{2}}5^{\underline{2}}_{\underline{2}}2^{\underline{2}}_{0}6^{\underline{1}}_{0}5^{\underline{1}}_{2}10^{0}_{2}8^{0}_{4}10^{1}_{4}8^{1}_{6}10^{2}_{6}10^{2}_{8}7^{3}_{8}10^{3}_{10}6^{4}_{10}7^{4}_{12}3^{5}_{12}6^{5}_{14}1^{6}_{14}3^{6}_{16}1^{7}_{18}$ 11^a_{64}: $1^{\underline{2}}_{0}1^{\underline{1}}_{2}1^{\underline{1}}_{4}4^{0}_{4}3^{0}_{6}5^{1}_{6}2^{1}_{8}7^{2}_{8}5^{2}_{10}6^{3}_{10}7^{3}_{12}8^{4}_{12}6^{4}_{14}5^{5}_{14}8^{5}_{16}5^{6}_{16}5^{6}_{18}2^{7}_{18}5^{7}_{20}1^{8}_{20}2^{8}_{22}1^{9}_{24}$	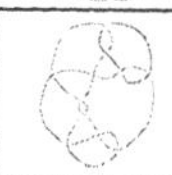	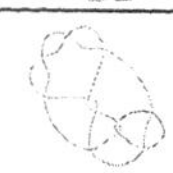
11^a_{65}: $1^{\underline{2}}_{\underline{2}}2^{\underline{1}}_{0}1^{\underline{1}}_{2}5^{0}_{2}4^{0}_{4}7^{1}_{4}3^{1}_{6}8^{2}_{6}7^{2}_{8}8^{3}_{8}8^{3}_{10}9^{4}_{10}8^{4}_{12}5^{5}_{12}9^{5}_{14}5^{6}_{14}5^{6}_{16}2^{7}_{16}5^{7}_{18}1^{8}_{18}2^{8}_{20}1^{9}_{22}$ 11^a_{66}: $1^{\underline{4}}_{\underline{6}}2^{\underline{3}}_{\underline{4}}1^{\underline{3}}_{\underline{2}}5^{\underline{2}}_{\underline{2}}2^{\underline{2}}_{0}5^{\underline{1}}_{0}5^{\underline{1}}_{2}9^{0}_{2}7^{0}_{4}9^{1}_{4}7^{1}_{6}8^{2}_{6}9^{2}_{8}6^{3}_{8}8^{3}_{10}5^{4}_{10}6^{4}_{12}2^{5}_{12}5^{5}_{14}1^{6}_{14}2^{6}_{16}1^{7}_{18}$		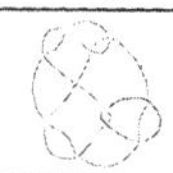
11^a_{67}: $1^{\underline{7}}_{\underline{16}}3^{\underline{6}}_{\underline{14}}1^{\underline{6}}_{\underline{12}}6^{\underline{5}}_{\underline{12}}3^{\underline{5}}_{\underline{10}}12^{\underline{4}}_{\underline{10}}7^{\underline{4}}_{\underline{8}}13^{\underline{3}}_{\underline{8}}11^{\underline{3}}_{\underline{6}}16^{\underline{2}}_{\underline{6}}13^{\underline{2}}_{\underline{4}}14^{\underline{1}}_{\underline{4}}16^{\underline{1}}_{\underline{2}}12^{0}_{\underline{2}}15^{0}_{0}9^{1}_{0}11^{1}_{2}4^{2}_{2}9^{2}_{4}1^{3}_{4}4^{3}_{6}1^{4}_{8}$ 11^a_{68}: $1^{\underline{2}}_{\underline{2}}1^{\underline{1}}_{0}1^{\underline{1}}_{2}5^{0}_{2}3^{0}_{4}7^{1}_{4}3^{1}_{6}8^{2}_{6}7^{2}_{8}8^{3}_{8}8^{3}_{10}9^{4}_{10}8^{4}_{12}6^{5}_{12}9^{5}_{14}5^{6}_{14}6^{6}_{16}2^{7}_{16}5^{7}_{18}1^{8}_{18}2^{8}_{20}1^{9}_{22}$		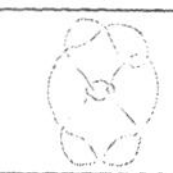
11^a_{69}: $1^{\underline{7}}_{\underline{16}}2^{\underline{6}}_{\underline{14}}1^{\underline{6}}_{\underline{12}}6^{\underline{5}}_{\underline{12}}2^{\underline{5}}_{\underline{10}}8^{\underline{4}}_{\underline{10}}6^{\underline{4}}_{\underline{8}}10^{\underline{3}}_{\underline{8}}8^{\underline{3}}_{\underline{6}}12^{\underline{2}}_{\underline{6}}10^{\underline{2}}_{\underline{4}}10^{\underline{1}}_{\underline{4}}12^{\underline{1}}_{\underline{2}}10^{0}_{\underline{2}}12^{0}_{0}6^{1}_{0}8^{1}_{2}3^{2}_{2}6^{2}_{4}1^{3}_{4}3^{3}_{6}1^{4}_{8}$ 11^a_{70}: $1^{\underline{8}}_{\underline{20}}2^{\underline{7}}_{\underline{18}}1^{\underline{7}}_{\underline{16}}5^{\underline{6}}_{\underline{16}}2^{\underline{6}}_{\underline{14}}8^{\underline{5}}_{\underline{14}}5^{\underline{5}}_{\underline{12}}11^{\underline{4}}_{\underline{12}}9^{\underline{4}}_{\underline{10}}12^{\underline{3}}_{\underline{10}}10^{\underline{3}}_{\underline{8}}11^{\underline{2}}_{\underline{8}}12^{\underline{2}}_{\underline{6}}9^{\underline{1}}_{\underline{6}}11^{\underline{1}}_{\underline{4}}7^{0}_{\underline{4}}10^{0}_{\underline{2}}4^{1}_{\underline{2}}6^{1}_{0}1^{2}_{0}4^{2}_{2}1^{3}_{4}$		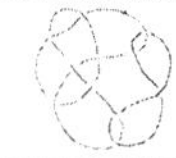
11^a_{71}: $1^{\underline{4}}_{\underline{6}}1^{\underline{4}}_{\underline{4}}2^{\underline{3}}_{\underline{4}}5^{\underline{2}}_{\underline{2}}2^{\underline{2}}_{0}5^{\underline{1}}_{0}5^{\underline{1}}_{2}10^{0}_{2}6^{0}_{4}9^{1}_{4}9^{1}_{6}10^{2}_{6}9^{2}_{8}7^{3}_{8}10^{3}_{10}6^{4}_{10}7^{4}_{12}3^{5}_{12}6^{5}_{14}1^{6}_{14}3^{6}_{16}1^{7}_{18}$ 11^a_{72}: $1^{\underline{11}}_{\underline{30}}1^{\underline{10}}_{\underline{28}}1^{\underline{10}}_{\underline{26}}4^{\underline{9}}_{\underline{26}}1^{\underline{9}}_{\underline{24}}4^{\underline{8}}_{\underline{24}}4^{\underline{8}}_{\underline{22}}7^{\underline{7}}_{\underline{22}}4^{\underline{7}}_{\underline{20}}5^{\underline{6}}_{\underline{20}}7^{\underline{6}}_{\underline{18}}6^{\underline{5}}_{\underline{18}}5^{\underline{5}}_{\underline{16}}5^{\underline{4}}_{\underline{16}}7^{\underline{4}}_{\underline{14}}3^{\underline{3}}_{\underline{14}}4^{\underline{3}}_{\underline{12}}2^{\underline{2}}_{\underline{12}}3^{\underline{2}}_{\underline{10}}2^{\underline{1}}_{\underline{8}}1^{0}_{\underline{8}}1^{0}_{\underline{6}}$	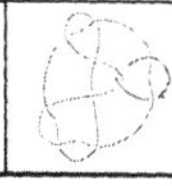	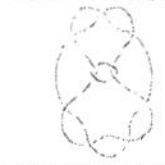

L: KH	1st line	2nd line
11^a_{73}: $1^{\underline{11}}_{\underline{28}}1^{\underline{10}}_{\underline{26}}1^{\underline{10}}_{\underline{24}}5^{\underline{9}}_{\underline{24}}1^{\underline{9}}_{\underline{22}}5^{\underline{8}}_{\underline{22}}5^{\underline{8}}_{\underline{20}}8^{\underline{7}}_{\underline{20}}5^{\underline{7}}_{\underline{18}}8^{\underline{6}}_{\underline{18}}8^{\underline{6}}_{\underline{16}}8^{\underline{5}}_{\underline{16}}8^{\underline{5}}_{\underline{14}}7^{\underline{4}}_{\underline{14}}9^{\underline{4}}_{\underline{12}}4^{\underline{3}}_{\underline{12}}6^{\underline{3}}_{\underline{10}}3^{\underline{2}}_{\underline{10}}4^{\underline{2}}_{\underline{8}}3^{\underline{1}}_{\underline{6}}1^{0}_{\underline{6}}1^{0}_{\underline{4}}$ 11^a_{74}: $1^{\underline{8}}_{\underline{18}}2^{\underline{7}}_{\underline{16}}1^{\underline{7}}_{\underline{14}}5^{\underline{6}}_{\underline{14}}2^{\underline{6}}_{\underline{12}}6^{\underline{5}}_{\underline{12}}5^{\underline{5}}_{\underline{10}}9^{\underline{4}}_{\underline{10}}7^{\underline{4}}_{\underline{8}}10^{\underline{3}}_{\underline{8}}8^{\underline{3}}_{\underline{6}}9^{\underline{2}}_{\underline{6}}10^{\underline{2}}_{\underline{4}}7^{\underline{1}}_{\underline{4}}9^{\underline{1}}_{\underline{2}}6^{0}_{\underline{2}}8^{0}_{0}3^{1}_{0}5^{1}_{2}1^{2}_{2}3^{2}_{4}1^{3}_{6}$		
11^a_{75}: $1^{\underline{11}}_{\underline{28}}1^{\underline{10}}_{\underline{26}}1^{\underline{10}}_{\underline{24}}4^{\underline{9}}_{\underline{24}}1^{\underline{9}}_{\underline{22}}3^{\underline{8}}_{\underline{22}}4^{\underline{8}}_{\underline{20}}6^{\underline{7}}_{\underline{20}}3^{\underline{7}}_{\underline{18}}5^{\underline{6}}_{\underline{18}}6^{\underline{6}}_{\underline{16}}5^{\underline{5}}_{\underline{16}}5^{\underline{5}}_{\underline{14}}5^{\underline{4}}_{\underline{14}}6^{\underline{4}}_{\underline{12}}2^{\underline{3}}_{\underline{12}}4^{\underline{3}}_{\underline{10}}2^{\underline{2}}_{\underline{10}}2^{\underline{2}}_{\underline{8}}2^{\underline{1}}_{\underline{6}}1^{0}_{\underline{6}}1^{0}_{\underline{4}}$ 11^a_{76}: $1^{\underline{11}}_{\underline{26}}1^{\underline{10}}_{\underline{24}}1^{\underline{10}}_{\underline{22}}4^{\underline{9}}_{\underline{22}}1^{\underline{9}}_{\underline{20}}3^{\underline{8}}_{\underline{20}}4^{\underline{8}}_{\underline{18}}6^{\underline{7}}_{\underline{18}}3^{\underline{7}}_{\underline{16}}6^{\underline{6}}_{\underline{16}}6^{\underline{6}}_{\underline{14}}5^{\underline{5}}_{\underline{14}}6^{\underline{5}}_{\underline{12}}6^{\underline{4}}_{\underline{12}}6^{\underline{4}}_{\underline{10}}3^{\underline{3}}_{\underline{10}}5^{\underline{3}}_{\underline{8}}3^{\underline{2}}_{\underline{8}}3^{\underline{2}}_{\underline{6}}3^{\underline{1}}_{\underline{4}}1^{0}_{\underline{4}}1^{0}_{\underline{2}}$		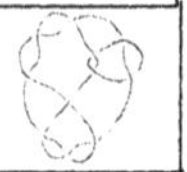
11^a_{77}: $1^{\underline{6}}_{\underline{14}}2^{\underline{5}}_{\underline{12}}1^{\underline{5}}_{\underline{10}}5^{\underline{4}}_{\underline{10}}2^{\underline{4}}_{\underline{8}}7^{\underline{3}}_{\underline{8}}5^{\underline{3}}_{\underline{6}}9^{\underline{2}}_{\underline{6}}7^{\underline{2}}_{\underline{4}}10^{\underline{1}}_{\underline{4}}9^{\underline{1}}_{\underline{2}}10^{0}_{\underline{2}}12^{0}_{0}8^{1}_{0}8^{1}_{2}5^{2}_{2}8^{2}_{4}3^{3}_{4}5^{3}_{6}1^{4}_{6}3^{4}_{8}1^{5}_{10}$ 11^a_{78}: $1^{\underline{5}}_{\underline{10}}2^{\underline{4}}_{\underline{8}}1^{\underline{4}}_{\underline{6}}5^{\underline{3}}_{\underline{6}}2^{\underline{3}}_{\underline{4}}8^{\underline{2}}_{\underline{4}}5^{\underline{2}}_{\underline{2}}9^{\underline{1}}_{\underline{2}}8^{\underline{1}}_{0}13^{0}_{0}11^{0}_{2}11^{1}_{2}11^{1}_{4}9^{2}_{4}11^{2}_{6}6^{3}_{6}9^{3}_{8}4^{4}_{8}6^{4}_{10}1^{5}_{10}4^{5}_{12}1^{6}_{14}$	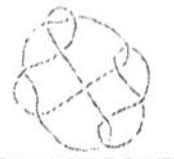	
11^a_{79}: $1^{\underline{4}}_{\underline{8}}1^{\underline{4}}_{\underline{6}}3^{\underline{3}}_{\underline{6}}5^{\underline{2}}_{\underline{4}}3^{\underline{2}}_{\underline{2}}8^{\underline{1}}_{\underline{2}}5^{\underline{1}}_{0}12^{0}_{0}9^{0}_{2}11^{1}_{2}11^{1}_{4}12^{2}_{4}11^{2}_{6}8^{3}_{6}12^{3}_{8}7^{4}_{8}8^{4}_{10}3^{5}_{10}7^{5}_{12}1^{6}_{12}3^{6}_{14}1^{7}_{16}$ 11^a_{80}: $1^{\underline{6}}_{\underline{12}}2^{\underline{5}}_{\underline{10}}1^{\underline{5}}_{\underline{8}}6^{\underline{4}}_{\underline{8}}3^{\underline{4}}_{\underline{6}}8^{\underline{3}}_{\underline{6}}5^{\underline{3}}_{\underline{4}}11^{\underline{2}}_{\underline{4}}8^{\underline{2}}_{\underline{2}}11^{\underline{1}}_{\underline{2}}11^{\underline{1}}_{0}12^{0}_{0}12^{0}_{2}9^{1}_{2}11^{1}_{4}7^{2}_{4}9^{2}_{6}3^{3}_{6}7^{3}_{8}1^{4}_{8}3^{4}_{10}1^{5}_{12}$		
11^a_{81}: $1^{\underline{11}}_{\underline{28}}2^{\underline{10}}_{\underline{26}}1^{\underline{10}}_{\underline{24}}6^{\underline{9}}_{\underline{24}}2^{\underline{9}}_{\underline{22}}8^{\underline{8}}_{\underline{22}}6^{\underline{8}}_{\underline{20}}12^{\underline{7}}_{\underline{20}}8^{\underline{7}}_{\underline{18}}12^{\underline{6}}_{\underline{18}}12^{\underline{6}}_{\underline{16}}12^{\underline{5}}_{\underline{16}}12^{\underline{5}}_{\underline{14}}10^{\underline{4}}_{\underline{14}}13^{\underline{4}}_{\underline{12}}7^{\underline{3}}_{\underline{12}}9^{\underline{3}}_{\underline{10}}4^{\underline{2}}_{\underline{10}}7^{\underline{2}}_{\underline{8}}4^{\underline{1}}_{\underline{6}}1^{0}_{\underline{6}}1^{0}_{\underline{4}}$ 11^a_{82}: $1^{\underline{7}}_{\underline{16}}3^{\underline{6}}_{\underline{14}}1^{\underline{6}}_{\underline{12}}5^{\underline{5}}_{\underline{12}}3^{\underline{5}}_{\underline{10}}9^{\underline{4}}_{\underline{10}}6^{\underline{4}}_{\underline{8}}10^{\underline{3}}_{\underline{8}}8^{\underline{3}}_{\underline{6}}12^{\underline{2}}_{\underline{6}}10^{\underline{2}}_{\underline{4}}9^{\underline{1}}_{\underline{4}}12^{\underline{1}}_{\underline{2}}9^{0}_{\underline{2}}10^{0}_{0}6^{1}_{0}8^{1}_{2}2^{2}_{2}6^{2}_{4}1^{3}_{4}2^{3}_{6}1^{4}_{8}$	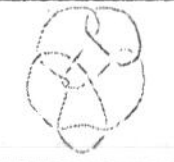	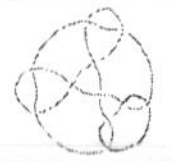
11^a_{83}: $1^{\underline{2}}_{\underline{2}}1^{\underline{1}}_{0}1^{\underline{1}}_{2}5^{0}_{2}3^{0}_{4}6^{1}_{4}3^{1}_{6}8^{2}_{6}6^{2}_{8}8^{3}_{8}8^{3}_{10}9^{4}_{10}8^{4}_{12}6^{5}_{12}9^{5}_{14}5^{6}_{14}6^{6}_{16}3^{7}_{16}5^{7}_{18}1^{8}_{18}3^{8}_{20}1^{9}_{22}$ 11^a_{84}: $1^{\underline{4}}_{\underline{8}}1^{\underline{4}}_{\underline{6}}2^{\underline{3}}_{\underline{6}}4^{\underline{2}}_{\underline{4}}2^{\underline{2}}_{\underline{2}}5^{\underline{1}}_{\underline{2}}4^{\underline{1}}_{0}10^{0}_{0}6^{0}_{2}8^{1}_{2}9^{1}_{4}9^{2}_{4}8^{2}_{6}7^{3}_{6}9^{3}_{8}5^{4}_{8}7^{4}_{10}3^{5}_{10}5^{5}_{12}1^{6}_{12}3^{6}_{14}1^{7}_{16}$	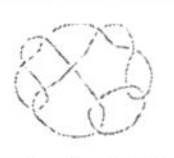	
11^a_{85}: $1^{\underline{9}}_{\underline{24}}1^{\underline{8}}_{\underline{22}}1^{\underline{8}}_{\underline{20}}4^{\underline{7}}_{\underline{20}}1^{\underline{7}}_{\underline{18}}4^{\underline{6}}_{\underline{18}}4^{\underline{6}}_{\underline{16}}7^{\underline{5}}_{\underline{16}}4^{\underline{5}}_{\underline{14}}7^{\underline{4}}_{\underline{14}}8^{\underline{4}}_{\underline{12}}6^{\underline{3}}_{\underline{12}}6^{\underline{3}}_{\underline{10}}6^{\underline{2}}_{\underline{10}}6^{\underline{2}}_{\underline{8}}3^{\underline{1}}_{\underline{8}}6^{\underline{1}}_{\underline{6}}3^{0}_{\underline{6}}4^{0}_{\underline{4}}1^{1}_{\underline{4}}2^{1}_{\underline{2}}1^{2}_{0}$ 11^a_{86}: $1^{\underline{9}}_{\underline{22}}1^{\underline{8}}_{\underline{20}}1^{\underline{8}}_{\underline{18}}4^{\underline{7}}_{\underline{18}}1^{\underline{7}}_{\underline{16}}5^{\underline{6}}_{\underline{16}}4^{\underline{6}}_{\underline{14}}7^{\underline{5}}_{\underline{14}}5^{\underline{5}}_{\underline{12}}8^{\underline{4}}_{\underline{12}}8^{\underline{4}}_{\underline{10}}7^{\underline{3}}_{\underline{10}}7^{\underline{3}}_{\underline{8}}7^{\underline{2}}_{\underline{8}}7^{\underline{2}}_{\underline{6}}3^{\underline{1}}_{\underline{6}}7^{\underline{1}}_{\underline{4}}3^{0}_{\underline{4}}4^{0}_{\underline{2}}1^{1}_{\underline{2}}2^{1}_{0}1^{2}_{2}$		
11^a_{87}: $1^{\underline{7}}_{\underline{16}}2^{\underline{6}}_{\underline{14}}1^{\underline{6}}_{\underline{12}}5^{\underline{5}}_{\underline{12}}2^{\underline{5}}_{\underline{10}}9^{\underline{4}}_{\underline{10}}6^{\underline{4}}_{\underline{8}}11^{\underline{3}}_{\underline{8}}8^{\underline{3}}_{\underline{6}}13^{\underline{2}}_{\underline{6}}11^{\underline{2}}_{\underline{4}}11^{\underline{1}}_{\underline{4}}13^{\underline{1}}_{\underline{2}}11^{0}_{\underline{2}}12^{0}_{0}7^{1}_{0}10^{1}_{2}4^{2}_{2}7^{2}_{4}1^{3}_{4}4^{3}_{6}1^{4}_{8}$ 11^a_{88}: $1^{\underline{4}}_{\underline{8}}1^{\underline{4}}_{\underline{6}}2^{\underline{3}}_{\underline{6}}4^{\underline{2}}_{\underline{4}}2^{\underline{2}}_{\underline{2}}5^{\underline{1}}_{\underline{2}}4^{\underline{1}}_{0}8^{0}_{0}6^{0}_{2}7^{1}_{2}7^{1}_{4}8^{2}_{4}7^{2}_{6}5^{3}_{6}8^{3}_{8}4^{4}_{8}5^{4}_{10}2^{5}_{10}4^{5}_{12}1^{6}_{12}2^{6}_{14}1^{7}_{16}$		
11^a_{89}: $1^{\underline{9}}_{\underline{22}}2^{\underline{8}}_{\underline{20}}1^{\underline{8}}_{\underline{18}}5^{\underline{7}}_{\underline{18}}2^{\underline{7}}_{\underline{16}}6^{\underline{6}}_{\underline{16}}5^{\underline{6}}_{\underline{14}}8^{\underline{5}}_{\underline{14}}6^{\underline{5}}_{\underline{12}}10^{\underline{4}}_{\underline{12}}9^{\underline{4}}_{\underline{10}}8^{\underline{3}}_{\underline{10}}9^{\underline{3}}_{\underline{8}}7^{\underline{2}}_{\underline{8}}8^{\underline{2}}_{\underline{6}}4^{\underline{1}}_{\underline{6}}7^{\underline{1}}_{\underline{4}}3^{0}_{\underline{4}}5^{0}_{\underline{2}}1^{1}_{\underline{2}}2^{1}_{0}1^{2}_{2}$ 11^a_{90}: $1^{\underline{4}}_{\underline{6}}1^{\underline{3}}_{\underline{4}}1^{\underline{3}}_{\underline{2}}5^{\underline{2}}_{\underline{2}}1^{\underline{2}}_{0}4^{\underline{1}}_{0}5^{\underline{1}}_{2}9^{0}_{2}6^{0}_{4}9^{1}_{4}7^{1}_{6}8^{2}_{6}9^{2}_{8}7^{3}_{8}8^{3}_{10}5^{4}_{10}7^{4}_{12}3^{5}_{12}5^{5}_{14}1^{6}_{14}3^{6}_{16}1^{7}_{18}$		
11^a_{91}: $1^{\underline{7}}_{\underline{16}}2^{\underline{6}}_{\underline{14}}1^{\underline{6}}_{\underline{12}}5^{\underline{5}}_{\underline{12}}2^{\underline{5}}_{\underline{10}}8^{\underline{4}}_{\underline{10}}5^{\underline{4}}_{\underline{8}}10^{\underline{3}}_{\underline{8}}8^{\underline{3}}_{\underline{6}}12^{\underline{2}}_{\underline{6}}10^{\underline{2}}_{\underline{4}}10^{\underline{1}}_{\underline{4}}12^{\underline{1}}_{\underline{2}}10^{0}_{\underline{2}}12^{0}_{0}7^{1}_{0}8^{1}_{2}3^{2}_{2}7^{2}_{4}1^{3}_{4}3^{3}_{6}1^{4}_{8}$ 11^a_{92}: $1^{\underline{9}}_{\underline{22}}2^{\underline{8}}_{\underline{20}}1^{\underline{8}}_{\underline{18}}5^{\underline{7}}_{\underline{18}}2^{\underline{7}}_{\underline{16}}8^{\underline{6}}_{\underline{16}}5^{\underline{6}}_{\underline{14}}9^{\underline{5}}_{\underline{14}}8^{\underline{5}}_{\underline{12}}11^{\underline{4}}_{\underline{12}}10^{\underline{4}}_{\underline{10}}10^{\underline{3}}_{\underline{10}}10^{\underline{3}}_{\underline{8}}8^{\underline{2}}_{\underline{8}}10^{\underline{2}}_{\underline{6}}5^{\underline{1}}_{\underline{6}}8^{\underline{1}}_{\underline{4}}3^{0}_{\underline{4}}6^{0}_{\underline{2}}1^{1}_{\underline{2}}2^{1}_{0}1^{2}_{2}$		
11^a_{93}: $1^{\underline{4}}_{\underline{6}}1^{\underline{3}}_{\underline{4}}1^{\underline{3}}_{\underline{2}}5^{\underline{2}}_{\underline{2}}1^{\underline{2}}_{0}5^{\underline{1}}_{0}5^{\underline{1}}_{2}9^{0}_{2}7^{0}_{4}9^{1}_{4}7^{1}_{6}8^{2}_{6}9^{2}_{8}7^{3}_{8}8^{3}_{10}5^{4}_{10}7^{4}_{12}2^{5}_{12}5^{5}_{14}1^{6}_{14}2^{6}_{16}1^{7}_{18}$ 11^a_{94}: $1^{\underline{4}}_{\underline{8}}1^{\underline{3}}_{\underline{6}}1^{\underline{3}}_{\underline{4}}4^{\underline{2}}_{\underline{4}}1^{\underline{2}}_{\underline{2}}4^{\underline{1}}_{\underline{2}}4^{\underline{1}}_{0}8^{0}_{0}6^{0}_{2}8^{1}_{2}6^{1}_{4}6^{2}_{4}8^{2}_{6}6^{3}_{6}6^{3}_{8}4^{4}_{8}6^{4}_{10}2^{5}_{10}4^{5}_{12}1^{6}_{12}2^{6}_{14}1^{7}_{16}$		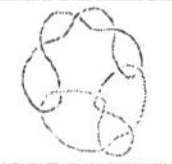
11^a_{95}: $1^{\underline{8}}_{\underline{20}}2^{\underline{7}}_{\underline{18}}1^{\underline{7}}_{\underline{16}}4^{\underline{6}}_{\underline{16}}2^{\underline{6}}_{\underline{14}}8^{\underline{5}}_{\underline{14}}4^{\underline{5}}_{\underline{12}}8^{\underline{4}}_{\underline{12}}8^{\underline{4}}_{\underline{10}}11^{\underline{3}}_{\underline{10}}8^{\underline{3}}_{\underline{8}}9^{\underline{2}}_{\underline{8}}11^{\underline{2}}_{\underline{6}}7^{\underline{1}}_{\underline{6}}9^{\underline{1}}_{\underline{4}}7^{0}_{\underline{4}}9^{0}_{\underline{2}}3^{1}_{\underline{2}}5^{1}_{0}1^{2}_{0}3^{2}_{2}1^{3}_{4}$ 11^a_{96}: $1^{\underline{4}}_{\underline{8}}1^{\underline{3}}_{\underline{6}}1^{\underline{3}}_{\underline{4}}5^{\underline{2}}_{\underline{4}}1^{\underline{2}}_{\underline{2}}4^{\underline{1}}_{\underline{2}}5^{\underline{1}}_{0}8^{0}_{0}6^{0}_{2}8^{1}_{2}6^{1}_{4}6^{2}_{4}8^{2}_{6}6^{3}_{6}6^{3}_{8}3^{4}_{8}6^{4}_{10}2^{5}_{10}3^{5}_{12}1^{6}_{12}2^{6}_{14}1^{7}_{16}$		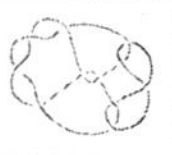
11^a_{97}: $1^{\underline{11}}_{\underline{28}}3^{\underline{10}}_{\underline{26}}1^{\underline{10}}_{\underline{24}}6^{\underline{9}}_{\underline{24}}3^{\underline{9}}_{\underline{22}}9^{\underline{8}}_{\underline{22}}6^{\underline{8}}_{\underline{20}}13^{\underline{7}}_{\underline{20}}9^{\underline{7}}_{\underline{18}}12^{\underline{6}}_{\underline{18}}13^{\underline{6}}_{\underline{16}}13^{\underline{5}}_{\underline{16}}12^{\underline{5}}_{\underline{14}}10^{\underline{4}}_{\underline{14}}14^{\underline{4}}_{\underline{12}}7^{\underline{3}}_{\underline{12}}9^{\underline{3}}_{\underline{10}}4^{\underline{2}}_{\underline{10}}7^{\underline{2}}_{\underline{8}}4^{\underline{1}}_{\underline{6}}1^{0}_{\underline{6}}1^{0}_{\underline{4}}$ 11^a_{98}: $1^{\underline{6}}_{\underline{12}}2^{\underline{5}}_{\underline{10}}1^{\underline{5}}_{\underline{8}}5^{\underline{4}}_{\underline{8}}3^{\underline{4}}_{\underline{6}}8^{\underline{3}}_{\underline{6}}4^{\underline{3}}_{\underline{4}}10^{\underline{2}}_{\underline{4}}8^{\underline{2}}_{\underline{2}}10^{\underline{1}}_{\underline{2}}10^{\underline{1}}_{0}12^{0}_{0}11^{0}_{2}8^{1}_{2}11^{1}_{4}7^{2}_{4}8^{2}_{6}3^{3}_{6}7^{3}_{8}1^{4}_{8}3^{4}_{10}1^{5}_{12}$	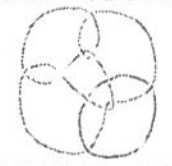	
11^a_{99}: $1^{\underline{6}}_{\underline{14}}2^{\underline{5}}_{\underline{12}}1^{\underline{5}}_{\underline{10}}4^{\underline{4}}_{\underline{10}}3^{\underline{4}}_{\underline{8}}7^{\underline{3}}_{\underline{8}}3^{\underline{3}}_{\underline{6}}8^{\underline{2}}_{\underline{6}}7^{\underline{2}}_{\underline{4}}8^{\underline{1}}_{\underline{4}}8^{\underline{1}}_{\underline{2}}10^{0}_{\underline{2}}9^{0}_{0}6^{1}_{0}9^{1}_{2}5^{2}_{2}6^{2}_{4}3^{3}_{4}5^{3}_{6}1^{4}_{6}3^{4}_{8}1^{5}_{10}$ 11^a_{100}: $1^{\underline{7}}_{\underline{18}}2^{\underline{6}}_{\underline{16}}1^{\underline{6}}_{\underline{14}}3^{\underline{5}}_{\underline{14}}2^{\underline{5}}_{\underline{12}}4^{\underline{4}}_{\underline{12}}3^{\underline{4}}_{\underline{10}}6^{\underline{3}}_{\underline{10}}4^{\underline{3}}_{\underline{8}}5^{\underline{2}}_{\underline{8}}6^{\underline{2}}_{\underline{6}}5^{\underline{1}}_{\underline{6}}5^{\underline{1}}_{\underline{4}}5^{0}_{\underline{4}}7^{0}_{\underline{2}}3^{1}_{\underline{2}}3^{1}_{0}2^{2}_{0}3^{2}_{2}1^{3}_{2}2^{3}_{4}1^{4}_{6}$	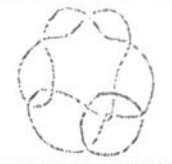	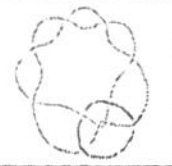
11^a_{101}: $1^{\underline{7}}_{\underline{16}}2^{\underline{6}}_{\underline{14}}1^{\underline{6}}_{\underline{12}}4^{\underline{5}}_{\underline{12}}2^{\underline{5}}_{\underline{10}}7^{\underline{4}}_{\underline{10}}4^{\underline{4}}_{\underline{8}}9^{\underline{3}}_{\underline{8}}7^{\underline{3}}_{\underline{6}}10^{\underline{2}}_{\underline{6}}9^{\underline{2}}_{\underline{4}}9^{\underline{1}}_{\underline{4}}10^{\underline{1}}_{\underline{2}}9^{0}_{\underline{2}}11^{0}_{0}6^{1}_{0}7^{1}_{2}3^{2}_{2}6^{2}_{4}1^{3}_{4}3^{3}_{6}1^{4}_{8}$ 11^a_{102}: $1^{\underline{7}}_{\underline{16}}3^{\underline{6}}_{\underline{14}}1^{\underline{6}}_{\underline{12}}5^{\underline{5}}_{\underline{12}}3^{\underline{5}}_{\underline{10}}8^{\underline{4}}_{\underline{10}}5^{\underline{4}}_{\underline{8}}9^{\underline{3}}_{\underline{8}}8^{\underline{3}}_{\underline{6}}10^{\underline{2}}_{\underline{6}}9^{\underline{2}}_{\underline{4}}9^{\underline{1}}_{\underline{4}}10^{\underline{1}}_{\underline{2}}8^{0}_{\underline{2}}11^{0}_{0}5^{1}_{0}6^{1}_{2}2^{2}_{2}5^{2}_{4}1^{3}_{4}2^{3}_{6}1^{4}_{8}$	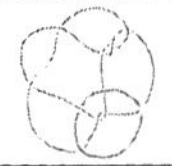	

L: KH	1st line	2nd line
11^a_{103}: $1^{\underline{7}}_{\underline{14}}2^{\underline{6}}_{\underline{12}}1^{\underline{6}}_{\underline{10}}2^{\underline{5}}_{\underline{10}}2^{\underline{5}}_{\underline{8}}5^{\underline{4}}_{\underline{8}}2^{\underline{4}}_{\underline{6}}4^{\underline{3}}_{\underline{6}}5^{\underline{3}}_{\underline{4}}6^{\underline{2}}_{\underline{4}}4^{\underline{2}}_{\underline{2}}5^{\underline{1}}_{\underline{2}}6^{\underline{1}}_{0}5^{0}_{0}7^{0}_{2}4^{1}_{2}3^{1}_{4}2^{2}_{4}4^{2}_{6}1^{3}_{6}2^{3}_{8}1^{4}_{10}$ 11^a_{104}: $1^{\underline{6}}_{\underline{14}}2^{\underline{5}}_{\underline{12}}1^{\underline{5}}_{\underline{10}}6^{\underline{4}}_{\underline{10}}3^{\underline{4}}_{\underline{8}}8^{\underline{3}}_{\underline{8}}5^{\underline{3}}_{\underline{6}}11^{\underline{2}}_{\underline{6}}8^{\underline{2}}_{\underline{4}}11^{\underline{1}}_{\underline{4}}11^{\underline{1}}_{\underline{2}}12^{0}_{\underline{2}}12^{0}_{0}9^{1}_{0}11^{1}_{2}6^{2}_{2}9^{2}_{4}4^{3}_{4}6^{3}_{6}1^{4}_{6}4^{4}_{8}1^{5}_{10}$		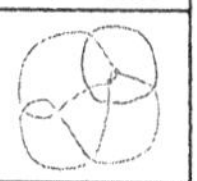
11^a_{105}: $1^{\underline{6}}_{\underline{12}}1^{\underline{5}}_{\underline{10}}1^{\underline{5}}_{\underline{8}}5^{\underline{4}}_{\underline{8}}2^{\underline{4}}_{\underline{6}}5^{\underline{3}}_{\underline{6}}4^{\underline{3}}_{\underline{4}}8^{\underline{2}}_{\underline{4}}5^{\underline{2}}_{\underline{2}}7^{\underline{1}}_{\underline{2}}8^{\underline{1}}_{0}9^{0}_{0}8^{0}_{2}7^{1}_{2}8^{1}_{4}4^{2}_{4}7^{2}_{6}3^{3}_{6}4^{3}_{8}1^{4}_{8}3^{4}_{10}1^{5}_{12}$ 11^a_{106}: $1^{\underline{8}}_{\underline{20}}2^{\underline{7}}_{\underline{18}}1^{\underline{7}}_{\underline{16}}5^{\underline{6}}_{\underline{16}}2^{\underline{6}}_{\underline{14}}7^{\underline{5}}_{\underline{14}}5^{\underline{5}}_{\underline{12}}9^{\underline{4}}_{\underline{12}}8^{\underline{4}}_{\underline{10}}10^{\underline{3}}_{\underline{10}}8^{\underline{3}}_{\underline{8}}9^{\underline{2}}_{\underline{8}}10^{\underline{2}}_{\underline{6}}7^{\underline{1}}_{\underline{6}}9^{\underline{1}}_{\underline{4}}5^{0}_{\underline{4}}8^{0}_{\underline{2}}3^{1}_{\underline{2}}4^{1}_{0}1^{2}_{0}3^{2}_{2}1^{3}_{4}$	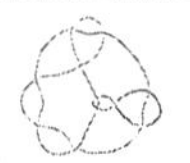	
11^a_{107}: $1^{\underline{11}}_{\underline{30}}1^{\underline{10}}_{\underline{28}}1^{\underline{10}}_{\underline{26}}3^{\underline{9}}_{\underline{26}}1^{\underline{9}}_{\underline{24}}2^{\underline{8}}_{\underline{24}}3^{\underline{8}}_{\underline{22}}5^{\underline{7}}_{\underline{22}}2^{\underline{7}}_{\underline{20}}3^{\underline{6}}_{\underline{20}}5^{\underline{6}}_{\underline{18}}4^{\underline{5}}_{\underline{18}}3^{\underline{5}}_{\underline{16}}3^{\underline{4}}_{\underline{16}}5^{\underline{4}}_{\underline{14}}2^{\underline{3}}_{\underline{14}}2^{\underline{3}}_{\underline{12}}2^{\underline{2}}_{\underline{12}}2^{\underline{2}}_{\underline{10}}2^{\underline{1}}_{\underline{8}}1^{0}_{\underline{8}}1^{0}_{\underline{6}}$ 11^a_{108}: $1^{\underline{11}}_{\underline{28}}1^{\underline{10}}_{\underline{26}}1^{\underline{10}}_{\underline{24}}4^{\underline{9}}_{\underline{24}}1^{\underline{9}}_{\underline{22}}4^{\underline{8}}_{\underline{22}}4^{\underline{8}}_{\underline{20}}8^{\underline{7}}_{\underline{20}}4^{\underline{7}}_{\underline{18}}7^{\underline{6}}_{\underline{18}}8^{\underline{6}}_{\underline{16}}7^{\underline{5}}_{\underline{16}}7^{\underline{5}}_{\underline{14}}7^{\underline{4}}_{\underline{14}}8^{\underline{4}}_{\underline{12}}4^{\underline{3}}_{\underline{12}}6^{\underline{3}}_{\underline{10}}3^{\underline{2}}_{\underline{10}}4^{\underline{2}}_{\underline{8}}3^{\underline{1}}_{\underline{6}}1^{0}_{\underline{6}}1^{0}_{\underline{4}}$	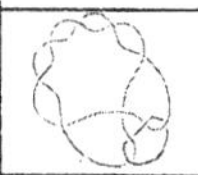	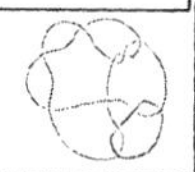
11^a_{109}: $1^{\underline{11}}_{\underline{28}}2^{\underline{10}}_{\underline{26}}1^{\underline{10}}_{\underline{24}}5^{\underline{9}}_{\underline{24}}2^{\underline{9}}_{\underline{22}}6^{\underline{8}}_{\underline{22}}5^{\underline{8}}_{\underline{20}}9^{\underline{7}}_{\underline{20}}6^{\underline{7}}_{\underline{18}}8^{\underline{6}}_{\underline{18}}9^{\underline{6}}_{\underline{16}}9^{\underline{5}}_{\underline{16}}8^{\underline{5}}_{\underline{14}}7^{\underline{4}}_{\underline{14}}10^{\underline{4}}_{\underline{12}}4^{\underline{3}}_{\underline{12}}6^{\underline{3}}_{\underline{10}}3^{\underline{2}}_{\underline{10}}4^{\underline{2}}_{\underline{8}}3^{\underline{1}}_{\underline{6}}1^{0}_{\underline{6}}1^{0}_{\underline{4}}$ 11^a_{110}: $1^{\underline{2}}_{2}1^{\underline{1}}_{6}3^{0}_{6}2^{0}_{8}1^{1}_{8}1^{1}_{10}3^{2}_{10}1^{2}_{12}1^{3}_{12}3^{3}_{14}3^{4}_{14}1^{4}_{16}1^{5}_{16}3^{5}_{18}2^{6}_{18}1^{6}_{20}1^{7}_{20}2^{7}_{22}1^{8}_{22}1^{8}_{24}1^{9}_{26}$	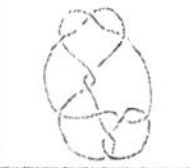	
11^a_{111}: $1^{\underline{2}}_{0}1^{\underline{1}}_{2}1^{\underline{1}}_{4}4^{0}_{4}3^{0}_{6}4^{1}_{6}2^{1}_{8}5^{2}_{8}4^{2}_{10}4^{3}_{10}5^{3}_{12}6^{4}_{12}4^{4}_{14}3^{5}_{14}6^{5}_{16}3^{6}_{16}3^{6}_{18}1^{7}_{18}3^{7}_{20}1^{8}_{20}1^{8}_{22}1^{9}_{24}$ 11^a_{112}: $1^{\underline{2}}_{0}1^{\underline{1}}_{2}1^{\underline{1}}_{4}5^{0}_{4}3^{0}_{6}5^{1}_{6}3^{1}_{8}8^{2}_{8}5^{2}_{10}7^{3}_{10}8^{3}_{12}8^{4}_{12}7^{4}_{14}6^{5}_{14}8^{5}_{16}5^{6}_{16}6^{6}_{18}2^{7}_{18}5^{7}_{20}1^{8}_{20}2^{8}_{22}1^{9}_{24}$	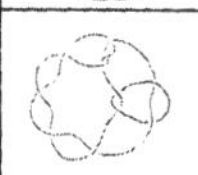	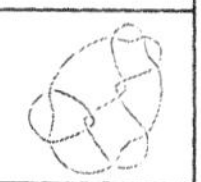
11^a_{113}: $1^{\underline{7}}_{\underline{16}}2^{\underline{6}}_{\underline{14}}1^{\underline{6}}_{\underline{12}}5^{\underline{5}}_{\underline{12}}2^{\underline{5}}_{\underline{10}}8^{\underline{4}}_{\underline{10}}6^{\underline{4}}_{\underline{8}}9^{\underline{3}}_{\underline{8}}7^{\underline{3}}_{\underline{6}}11^{\underline{2}}_{\underline{6}}9^{\underline{2}}_{\underline{4}}9^{\underline{1}}_{\underline{4}}11^{\underline{1}}_{\underline{2}}9^{0}_{\underline{2}}10^{0}_{0}5^{1}_{0}8^{1}_{2}3^{2}_{2}5^{2}_{4}1^{3}_{4}3^{3}_{6}1^{4}_{8}$ 11^a_{114}: $1^{\underline{2}}_{\underline{2}}2^{\underline{1}}_{0}1^{\underline{1}}_{2}5^{0}_{2}4^{0}_{4}7^{1}_{4}3^{1}_{6}8^{2}_{6}7^{2}_{8}7^{3}_{8}8^{3}_{10}8^{4}_{10}7^{4}_{12}5^{5}_{12}8^{5}_{14}4^{6}_{14}5^{6}_{16}1^{7}_{16}4^{7}_{18}1^{8}_{18}1^{8}_{20}1^{9}_{22}$	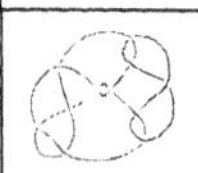	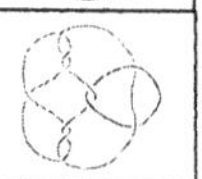
11^a_{115}: $1^{\underline{8}}_{\underline{18}}1^{\underline{7}}_{\underline{16}}1^{\underline{7}}_{\underline{14}}4^{\underline{6}}_{\underline{14}}1^{\underline{6}}_{\underline{12}}4^{\underline{5}}_{\underline{12}}4^{\underline{5}}_{\underline{10}}7^{\underline{4}}_{\underline{10}}5^{\underline{4}}_{\underline{8}}8^{\underline{3}}_{\underline{8}}6^{\underline{3}}_{\underline{6}}7^{\underline{2}}_{\underline{6}}8^{\underline{2}}_{\underline{4}}6^{\underline{1}}_{\underline{4}}7^{\underline{1}}_{\underline{2}}5^{0}_{\underline{2}}7^{0}_{0}3^{1}_{0}4^{1}_{2}1^{2}_{2}3^{2}_{4}1^{3}_{6}$ 11^a_{116}: $1^{\underline{11}}_{\underline{26}}1^{\underline{10}}_{\underline{24}}1^{\underline{10}}_{\underline{22}}3^{\underline{9}}_{\underline{22}}1^{\underline{9}}_{\underline{20}}3^{\underline{8}}_{\underline{20}}3^{\underline{8}}_{\underline{18}}5^{\underline{7}}_{\underline{18}}3^{\underline{7}}_{\underline{16}}5^{\underline{6}}_{\underline{16}}5^{\underline{6}}_{\underline{14}}5^{\underline{5}}_{\underline{14}}5^{\underline{5}}_{\underline{12}}5^{\underline{4}}_{\underline{12}}6^{\underline{4}}_{\underline{10}}3^{\underline{3}}_{\underline{10}}4^{\underline{3}}_{\underline{8}}3^{\underline{2}}_{\underline{8}}3^{\underline{2}}_{\underline{6}}3^{\underline{1}}_{\underline{4}}1^{0}_{\underline{4}}1^{0}_{\underline{2}}$	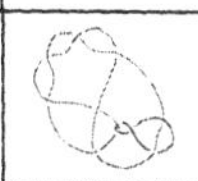	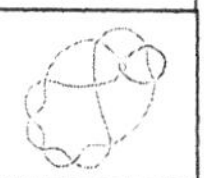
11^a_{117}: $1^{\underline{2}}_{\underline{2}}1^{\underline{1}}_{0}1^{\underline{1}}_{2}4^{0}_{2}3^{0}_{4}4^{1}_{4}2^{1}_{6}5^{2}_{6}4^{2}_{8}5^{3}_{8}5^{3}_{10}5^{4}_{10}5^{4}_{12}3^{5}_{12}5^{5}_{14}3^{6}_{14}3^{6}_{16}1^{7}_{16}3^{7}_{18}1^{8}_{18}1^{8}_{20}1^{9}_{22}$ 11^a_{118}: $1^{\underline{9}}_{\underline{22}}2^{\underline{8}}_{\underline{20}}1^{\underline{8}}_{\underline{18}}6^{\underline{7}}_{\underline{18}}2^{\underline{7}}_{\underline{16}}9^{\underline{6}}_{\underline{16}}6^{\underline{6}}_{\underline{14}}11^{\underline{5}}_{\underline{14}}9^{\underline{5}}_{\underline{12}}13^{\underline{4}}_{\underline{12}}12^{\underline{4}}_{\underline{10}}12^{\underline{3}}_{\underline{10}}12^{\underline{3}}_{\underline{8}}10^{\underline{2}}_{\underline{8}}12^{\underline{2}}_{\underline{6}}6^{\underline{1}}_{\underline{6}}10^{\underline{1}}_{\underline{4}}4^{0}_{\underline{4}}7^{0}_{\underline{2}}1^{1}_{\underline{2}}3^{1}_{0}1^{2}_{2}$	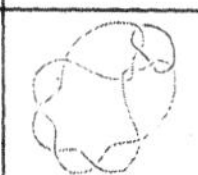	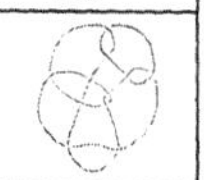
11^a_{119}: $1^{\underline{6}}_{\underline{14}}1^{\underline{5}}_{\underline{12}}1^{\underline{5}}_{\underline{10}}5^{\underline{4}}_{\underline{10}}2^{\underline{4}}_{\underline{8}}6^{\underline{3}}_{\underline{8}}4^{\underline{3}}_{\underline{6}}9^{\underline{2}}_{\underline{6}}6^{\underline{2}}_{\underline{4}}8^{\underline{1}}_{\underline{4}}9^{\underline{1}}_{\underline{2}}9^{0}_{\underline{2}}9^{0}_{0}8^{1}_{0}8^{1}_{2}4^{2}_{2}8^{2}_{4}3^{3}_{4}4^{3}_{6}1^{4}_{6}3^{4}_{8}1^{5}_{10}$ 11^a_{120}: $1^{\underline{11}}_{\underline{26}}1^{\underline{10}}_{\underline{24}}1^{\underline{10}}_{\underline{22}}4^{\underline{9}}_{\underline{22}}1^{\underline{9}}_{\underline{20}}4^{\underline{8}}_{\underline{20}}4^{\underline{8}}_{\underline{18}}7^{\underline{7}}_{\underline{18}}4^{\underline{7}}_{\underline{16}}6^{\underline{6}}_{\underline{16}}7^{\underline{6}}_{\underline{14}}6^{\underline{5}}_{\underline{14}}6^{\underline{5}}_{\underline{12}}7^{\underline{4}}_{\underline{12}}7^{\underline{4}}_{\underline{10}}3^{\underline{3}}_{\underline{10}}6^{\underline{3}}_{\underline{8}}3^{\underline{2}}_{\underline{8}}3^{\underline{2}}_{\underline{6}}3^{\underline{1}}_{\underline{4}}1^{0}_{\underline{4}}1^{0}_{\underline{2}}$	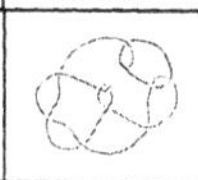	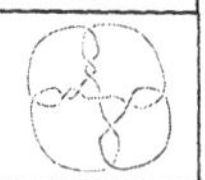
11^a_{121}: $1^{\underline{6}}_{\underline{12}}1^{\underline{5}}_{\underline{10}}1^{\underline{5}}_{\underline{8}}5^{\underline{4}}_{\underline{8}}2^{\underline{4}}_{\underline{6}}6^{\underline{3}}_{\underline{6}}4^{\underline{3}}_{\underline{4}}9^{\underline{2}}_{\underline{4}}6^{\underline{2}}_{\underline{2}}9^{\underline{1}}_{\underline{2}}9^{\underline{1}}_{0}10^{0}_{0}10^{0}_{2}8^{1}_{2}9^{1}_{4}6^{2}_{4}8^{2}_{6}3^{3}_{6}6^{3}_{8}1^{4}_{8}3^{4}_{10}1^{5}_{12}$ 11^a_{122}: $1^{\underline{6}}_{\underline{14}}1^{\underline{5}}_{\underline{12}}1^{\underline{5}}_{\underline{10}}4^{\underline{4}}_{\underline{10}}2^{\underline{4}}_{\underline{8}}5^{\underline{3}}_{\underline{8}}3^{\underline{3}}_{\underline{6}}7^{\underline{2}}_{\underline{6}}5^{\underline{2}}_{\underline{4}}7^{\underline{1}}_{\underline{4}}7^{\underline{1}}_{\underline{2}}8^{0}_{\underline{2}}8^{0}_{0}6^{1}_{0}7^{1}_{2}4^{2}_{2}6^{2}_{4}3^{3}_{4}4^{3}_{6}1^{4}_{6}3^{4}_{8}1^{5}_{10}$	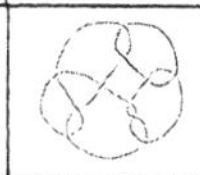	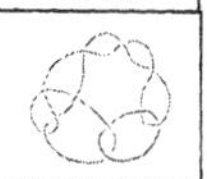
11^a_{123}: $1^{\underline{9}}_{\underline{24}}1^{\underline{8}}_{\underline{22}}1^{\underline{8}}_{\underline{20}}4^{\underline{7}}_{\underline{20}}1^{\underline{7}}_{\underline{18}}4^{\underline{6}}_{\underline{18}}4^{\underline{6}}_{\underline{16}}6^{\underline{5}}_{\underline{16}}4^{\underline{5}}_{\underline{14}}6^{\underline{4}}_{\underline{14}}7^{\underline{4}}_{\underline{12}}6^{\underline{3}}_{\underline{12}}5^{\underline{3}}_{\underline{10}}5^{\underline{2}}_{\underline{10}}6^{\underline{2}}_{\underline{8}}2^{\underline{1}}_{\underline{8}}5^{\underline{1}}_{\underline{6}}3^{0}_{\underline{6}}3^{0}_{\underline{4}}1^{1}_{\underline{4}}2^{1}_{\underline{2}}1^{2}_{0}$ 11^a_{124}: $1^{\underline{9}}_{\underline{22}}1^{\underline{8}}_{\underline{20}}1^{\underline{8}}_{\underline{18}}4^{\underline{7}}_{\underline{18}}1^{\underline{7}}_{\underline{16}}5^{\underline{6}}_{\underline{16}}4^{\underline{6}}_{\underline{14}}7^{\underline{5}}_{\underline{14}}5^{\underline{5}}_{\underline{12}}9^{\underline{4}}_{\underline{12}}8^{\underline{4}}_{\underline{10}}8^{\underline{3}}_{\underline{10}}8^{\underline{3}}_{\underline{8}}7^{\underline{2}}_{\underline{8}}8^{\underline{2}}_{\underline{6}}4^{\underline{1}}_{\underline{6}}7^{\underline{1}}_{\underline{4}}4^{0}_{\underline{4}}5^{0}_{\underline{2}}1^{1}_{\underline{2}}3^{1}_{0}1^{2}_{2}$	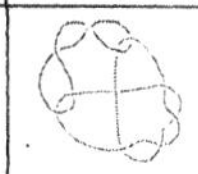	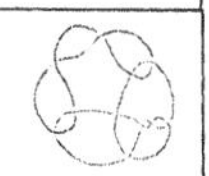
11^a_{125}: $1^{\underline{2}}_{0}1^{\underline{1}}_{4}4^{0}_{4}2^{0}_{6}3^{1}_{6}2^{1}_{8}5^{2}_{8}3^{2}_{10}4^{3}_{10}5^{3}_{12}6^{4}_{12}4^{4}_{14}4^{5}_{14}6^{5}_{16}3^{6}_{16}4^{6}_{18}2^{7}_{18}3^{7}_{20}1^{8}_{20}2^{8}_{22}1^{9}_{24}$ 11^a_{126}: $1^{\underline{2}}_{\underline{2}}1^{\underline{1}}_{0}1^{\underline{1}}_{2}5^{0}_{2}3^{0}_{4}6^{1}_{4}3^{1}_{6}8^{2}_{6}6^{2}_{8}7^{3}_{8}8^{3}_{10}8^{4}_{10}7^{4}_{12}6^{5}_{12}8^{5}_{14}4^{6}_{14}6^{6}_{16}2^{7}_{16}4^{7}_{18}1^{8}_{18}2^{8}_{20}1^{9}_{22}$	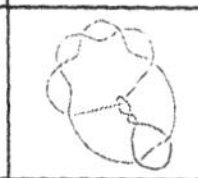	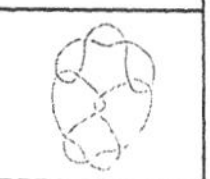
11^a_{127}: $1^{\underline{8}}_{\underline{18}}2^{\underline{7}}_{\underline{16}}1^{\underline{7}}_{\underline{14}}6^{\underline{6}}_{\underline{14}}2^{\underline{6}}_{\underline{12}}8^{\underline{5}}_{\underline{12}}6^{\underline{5}}_{\underline{10}}11^{\underline{4}}_{\underline{10}}9^{\underline{4}}_{\underline{8}}12^{\underline{3}}_{\underline{8}}10^{\underline{3}}_{\underline{6}}11^{\underline{2}}_{\underline{6}}12^{\underline{2}}_{\underline{4}}9^{\underline{1}}_{\underline{4}}11^{\underline{1}}_{\underline{2}}7^{0}_{\underline{2}}10^{0}_{0}3^{1}_{0}6^{1}_{2}1^{2}_{2}3^{2}_{4}1^{3}_{6}$ 11^a_{128}: $1^{\underline{7}}_{\underline{16}}1^{\underline{6}}_{\underline{14}}1^{\underline{6}}_{\underline{12}}4^{\underline{5}}_{\underline{12}}1^{\underline{5}}_{\underline{10}}6^{\underline{4}}_{\underline{10}}5^{\underline{4}}_{\underline{8}}8^{\underline{3}}_{\underline{8}}5^{\underline{3}}_{\underline{6}}10^{\underline{2}}_{\underline{6}}8^{\underline{2}}_{\underline{4}}7^{\underline{1}}_{\underline{4}}10^{\underline{1}}_{\underline{2}}9^{0}_{\underline{2}}8^{0}_{0}5^{1}_{0}8^{1}_{2}3^{2}_{2}5^{2}_{4}1^{3}_{4}3^{3}_{6}1^{4}_{8}$	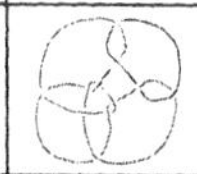	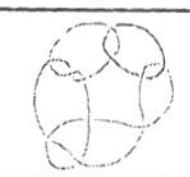
11^a_{129}: $1^{\underline{7}}_{\underline{18}}1^{\underline{6}}_{\underline{16}}1^{\underline{6}}_{\underline{14}}3^{\underline{5}}_{\underline{14}}1^{\underline{5}}_{\underline{12}}4^{\underline{4}}_{\underline{12}}4^{\underline{4}}_{\underline{10}}5^{\underline{3}}_{\underline{10}}3^{\underline{3}}_{\underline{8}}6^{\underline{2}}_{\underline{8}}5^{\underline{2}}_{\underline{6}}4^{\underline{1}}_{\underline{6}}6^{\underline{1}}_{\underline{4}}5^{0}_{\underline{4}}5^{0}_{\underline{2}}3^{1}_{\underline{2}}4^{1}_{0}2^{2}_{0}3^{2}_{2}1^{3}_{2}2^{3}_{4}1^{4}_{6}$ 11^a_{130}: $1^{\underline{2}}_{0}1^{\underline{1}}_{4}5^{0}_{4}2^{0}_{6}4^{1}_{6}3^{1}_{8}8^{2}_{8}4^{2}_{10}7^{3}_{10}8^{3}_{12}8^{4}_{12}7^{4}_{14}7^{5}_{14}8^{5}_{16}5^{6}_{16}7^{6}_{18}3^{7}_{18}5^{7}_{20}1^{8}_{20}3^{8}_{22}1^{9}_{24}$	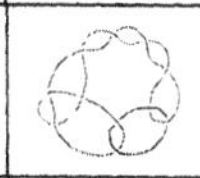	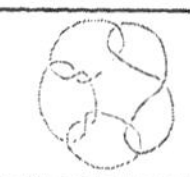
11^a_{131}: $1^{\underline{2}}_{\underline{2}}1^{\underline{1}}_{2}4^{0}_{2}2^{0}_{4}3^{1}_{4}2^{1}_{6}5^{2}_{6}3^{2}_{8}5^{3}_{8}5^{3}_{10}5^{4}_{10}5^{4}_{12}4^{5}_{12}5^{5}_{14}3^{6}_{14}4^{6}_{16}2^{7}_{16}3^{7}_{18}1^{8}_{18}2^{8}_{20}1^{9}_{22}$ 11^a_{132}: $1^{\underline{2}}_{\underline{4}}1^{\underline{1}}_{0}3^{0}_{0}2^{0}_{2}2^{1}_{2}1^{1}_{4}2^{2}_{4}2^{2}_{6}2^{3}_{6}2^{3}_{8}2^{4}_{8}2^{4}_{10}2^{5}_{10}2^{5}_{12}1^{6}_{12}2^{6}_{14}1^{7}_{14}1^{7}_{16}1^{8}_{16}1^{8}_{18}1^{9}_{20}$	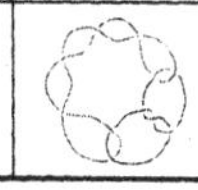	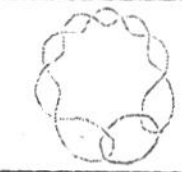

L: KH	1st line	2nd line
11^a_{133}: $1^{\underline{9}}_{\underline{22}}4^{\underline{8}}_{\underline{20}}1^{\underline{8}}_{\underline{18}}6^{\underline{7}}_{\underline{18}}4^{\underline{7}}_{\underline{16}}10^{\underline{6}}_{\underline{16}}6^{\underline{6}}_{\underline{14}}12^{\underline{5}}_{\underline{14}}10^{\underline{5}}_{\underline{12}}13^{\underline{4}}_{\underline{12}}13^{\underline{4}}_{\underline{10}}13^{\underline{3}}_{\underline{10}}12^{\underline{3}}_{\underline{8}}9^{\underline{2}}_{\underline{8}}13^{\underline{2}}_{\underline{6}}6^{\underline{1}}_{\underline{6}}9^{\underline{1}}_{\underline{4}}4^{0}_{\underline{4}}7^{0}_{\underline{2}}1^{1}_{\underline{2}}3^{1}_{0}1^{2}_{2}$	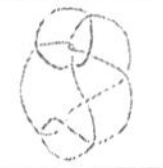	
11^a_{134}: $1^{\underline{8}}_{\underline{18}}3^{\underline{7}}_{\underline{16}}1^{\underline{7}}_{\underline{14}}5^{\underline{6}}_{\underline{14}}3^{\underline{6}}_{\underline{12}}8^{\underline{5}}_{\underline{12}}5^{\underline{5}}_{\underline{10}}10^{\underline{4}}_{\underline{10}}9^{\underline{4}}_{\underline{8}}11^{\underline{3}}_{\underline{8}}9^{\underline{3}}_{\underline{6}}11^{\underline{2}}_{\underline{6}}11^{\underline{2}}_{\underline{4}}7^{\underline{1}}_{\underline{4}}11^{\underline{1}}_{\underline{2}}7^{0}_{\underline{2}}8^{0}_{0}3^{1}_{0}6^{1}_{2}1^{2}_{2}3^{2}_{4}1^{3}_{6}$		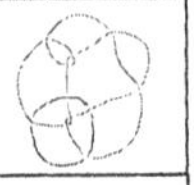
11^a_{135}: $1^{\underline{8}}_{\underline{20}}3^{\underline{7}}_{\underline{18}}1^{\underline{7}}_{\underline{16}}6^{\underline{6}}_{\underline{16}}3^{\underline{6}}_{\underline{14}}9^{\underline{5}}_{\underline{14}}6^{\underline{5}}_{\underline{12}}11^{\underline{4}}_{\underline{12}}10^{\underline{4}}_{\underline{10}}12^{\underline{3}}_{\underline{10}}10^{\underline{3}}_{\underline{8}}11^{\underline{2}}_{\underline{8}}12^{\underline{2}}_{\underline{6}}8^{\underline{1}}_{\underline{6}}11^{\underline{1}}_{\underline{4}}6^{0}_{\underline{4}}9^{0}_{\underline{2}}3^{1}_{\underline{2}}5^{1}_{0}1^{2}_{0}3^{2}_{2}1^{3}_{4}$	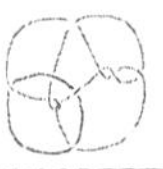	
11^a_{136}: $1^{\underline{8}}_{\underline{18}}2^{\underline{7}}_{\underline{16}}1^{\underline{7}}_{\underline{14}}4^{\underline{6}}_{\underline{14}}2^{\underline{6}}_{\underline{12}}5^{\underline{5}}_{\underline{12}}4^{\underline{5}}_{\underline{10}}8^{\underline{4}}_{\underline{10}}6^{\underline{4}}_{\underline{8}}8^{\underline{3}}_{\underline{8}}7^{\underline{3}}_{\underline{6}}8^{\underline{2}}_{\underline{6}}8^{\underline{2}}_{\underline{4}}6^{\underline{1}}_{\underline{4}}8^{\underline{1}}_{\underline{2}}5^{0}_{\underline{2}}7^{0}_{0}3^{1}_{0}4^{1}_{2}1^{2}_{2}3^{2}_{4}1^{3}_{6}$		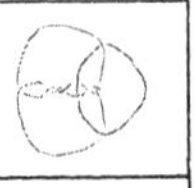
11^a_{137}: $1^{\underline{6}}_{\underline{12}}2^{\underline{5}}_{\underline{10}}1^{\underline{5}}_{\underline{8}}4^{\underline{4}}_{\underline{8}}3^{\underline{4}}_{\underline{6}}6^{\underline{3}}_{\underline{6}}3^{\underline{3}}_{\underline{4}}9^{\underline{2}}_{\underline{4}}6^{\underline{2}}_{\underline{2}}7^{\underline{1}}_{\underline{2}}9^{\underline{1}}_{0}10^{0}_{0}8^{0}_{2}7^{1}_{2}9^{1}_{4}5^{2}_{4}7^{2}_{6}3^{3}_{6}5^{3}_{8}1^{4}_{8}3^{4}_{10}1^{5}_{12}$		
11^a_{138}: $1^{\underline{6}}_{\underline{14}}3^{\underline{5}}_{\underline{12}}1^{\underline{5}}_{\underline{10}}6^{\underline{4}}_{\underline{10}}4^{\underline{4}}_{\underline{8}}8^{\underline{3}}_{\underline{8}}5^{\underline{3}}_{\underline{6}}11^{\underline{2}}_{\underline{6}}8^{\underline{2}}_{\underline{4}}10^{\underline{1}}_{\underline{4}}11^{\underline{1}}_{\underline{2}}11^{0}_{\underline{2}}11^{0}_{0}8^{1}_{0}10^{1}_{2}5^{2}_{2}8^{2}_{4}3^{3}_{4}5^{3}_{6}1^{4}_{6}3^{4}_{8}1^{5}_{10}$		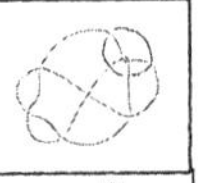
11^a_{139}: $1^{\underline{5}}_{\underline{10}}4^{\underline{4}}_{\underline{8}}1^{\underline{4}}_{\underline{6}}6^{\underline{3}}_{\underline{6}}4^{\underline{3}}_{\underline{4}}11^{\underline{2}}_{\underline{4}}7^{\underline{2}}_{\underline{2}}12^{\underline{1}}_{\underline{2}}10^{\underline{1}}_{0}14^{0}_{0}13^{0}_{2}12^{1}_{2}13^{1}_{4}10^{2}_{4}12^{2}_{6}6^{3}_{6}10^{3}_{8}3^{4}_{8}6^{4}_{10}1^{5}_{10}3^{5}_{12}1^{6}_{14}$	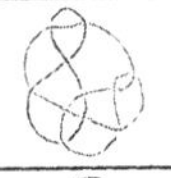	
11^a_{140}: $1^{\underline{7}}_{\underline{16}}3^{\underline{6}}_{\underline{14}}1^{\underline{6}}_{\underline{12}}6^{\underline{5}}_{\underline{12}}3^{\underline{5}}_{\underline{10}}10^{\underline{4}}_{\underline{10}}6^{\underline{4}}_{\underline{8}}12^{\underline{3}}_{\underline{8}}10^{\underline{3}}_{\underline{6}}14^{\underline{2}}_{\underline{6}}13^{\underline{2}}_{\underline{4}}13^{\underline{1}}_{\underline{4}}13^{\underline{1}}_{\underline{2}}11^{0}_{\underline{2}}14^{0}_{0}7^{1}_{0}10^{1}_{2}4^{2}_{2}7^{2}_{4}1^{3}_{4}4^{3}_{6}1^{4}_{8}$		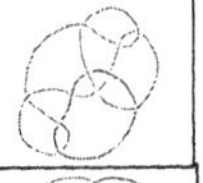
11^a_{141}: $1^{\underline{5}}_{\underline{10}}3^{\underline{4}}_{\underline{8}}1^{\underline{4}}_{\underline{6}}5^{\underline{3}}_{\underline{6}}3^{\underline{3}}_{\underline{4}}9^{\underline{2}}_{\underline{4}}6^{\underline{2}}_{\underline{2}}10^{\underline{1}}_{\underline{2}}8^{\underline{1}}_{0}12^{0}_{0}11^{0}_{2}10^{1}_{2}11^{1}_{4}9^{2}_{4}10^{2}_{6}5^{3}_{6}9^{3}_{8}3^{4}_{8}5^{4}_{10}1^{5}_{10}3^{5}_{12}1^{6}_{14}$	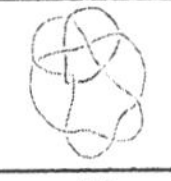	
11^a_{142}: $1^{\underline{5}}_{\underline{12}}3^{\underline{4}}_{\underline{10}}1^{\underline{4}}_{\underline{8}}5^{\underline{3}}_{\underline{8}}3^{\underline{3}}_{\underline{6}}9^{\underline{2}}_{\underline{6}}6^{\underline{2}}_{\underline{4}}10^{\underline{1}}_{\underline{4}}8^{\underline{1}}_{\underline{2}}11^{0}_{\underline{2}}11^{0}_{0}10^{1}_{0}10^{1}_{2}8^{2}_{2}10^{2}_{4}5^{3}_{4}8^{3}_{6}3^{4}_{6}5^{4}_{8}1^{5}_{8}3^{5}_{10}1^{6}_{12}$		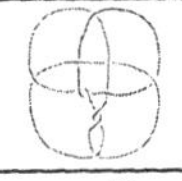
11^a_{143}: $1^{\underline{8}}_{\underline{20}}4^{\underline{7}}_{\underline{18}}1^{\underline{7}}_{\underline{16}}7^{\underline{6}}_{\underline{16}}4^{\underline{6}}_{\underline{14}}11^{\underline{5}}_{\underline{14}}7^{\underline{5}}_{\underline{12}}13^{\underline{4}}_{\underline{12}}11^{\underline{4}}_{\underline{10}}14^{\underline{3}}_{\underline{10}}13^{\underline{3}}_{\underline{8}}13^{\underline{2}}_{\underline{8}}15^{\underline{2}}_{\underline{6}}10^{\underline{1}}_{\underline{6}}12^{\underline{1}}_{\underline{4}}7^{0}_{\underline{4}}11^{0}_{\underline{2}}3^{1}_{\underline{2}}6^{1}_{0}1^{2}_{0}3^{2}_{2}1^{3}_{4}$	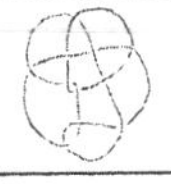	
11^a_{144}: $1^{\underline{5}}_{\underline{12}}3^{\underline{4}}_{\underline{10}}1^{\underline{4}}_{\underline{8}}5^{\underline{3}}_{\underline{8}}3^{\underline{3}}_{\underline{6}}8^{\underline{2}}_{\underline{6}}6^{\underline{2}}_{\underline{4}}9^{\underline{1}}_{\underline{4}}7^{\underline{1}}_{\underline{2}}10^{0}_{\underline{2}}10^{0}_{0}8^{1}_{0}9^{1}_{2}7^{2}_{2}8^{2}_{4}4^{3}_{4}7^{3}_{6}2^{4}_{6}4^{4}_{8}1^{5}_{8}2^{5}_{10}1^{6}_{12}$		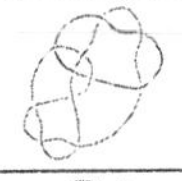
11^a_{145}: $1^{\underline{8}}_{\underline{20}}2^{\underline{7}}_{\underline{18}}1^{\underline{7}}_{\underline{16}}4^{\underline{6}}_{\underline{16}}2^{\underline{6}}_{\underline{14}}7^{\underline{5}}_{\underline{14}}4^{\underline{5}}_{\underline{12}}8^{\underline{4}}_{\underline{12}}7^{\underline{4}}_{\underline{10}}9^{\underline{3}}_{\underline{10}}8^{\underline{3}}_{\underline{8}}9^{\underline{2}}_{\underline{8}}10^{\underline{2}}_{\underline{6}}7^{\underline{1}}_{\underline{6}}8^{\underline{1}}_{\underline{4}}5^{0}_{\underline{4}}8^{0}_{\underline{2}}3^{1}_{\underline{2}}4^{1}_{0}1^{2}_{0}3^{2}_{2}1^{3}_{4}$	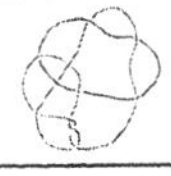	
11^a_{146}: $1^{\underline{5}}_{\underline{10}}2^{\underline{4}}_{\underline{8}}1^{\underline{4}}_{\underline{6}}3^{\underline{3}}_{\underline{6}}2^{\underline{3}}_{\underline{4}}6^{\underline{2}}_{\underline{4}}4^{\underline{2}}_{\underline{2}}6^{\underline{1}}_{\underline{2}}5^{\underline{1}}_{0}8^{0}_{0}7^{0}_{2}6^{1}_{2}7^{1}_{4}6^{2}_{4}6^{2}_{6}3^{3}_{6}6^{3}_{8}2^{4}_{8}3^{4}_{10}1^{5}_{10}2^{5}_{12}1^{6}_{14}$		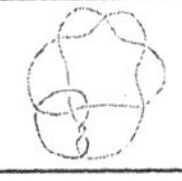
11^a_{147}: $1^{\underline{8}}_{\underline{22}}2^{\underline{7}}_{\underline{20}}1^{\underline{7}}_{\underline{18}}3^{\underline{6}}_{\underline{18}}2^{\underline{6}}_{\underline{16}}5^{\underline{5}}_{\underline{16}}3^{\underline{5}}_{\underline{14}}5^{\underline{4}}_{\underline{14}}5^{\underline{4}}_{\underline{12}}6^{\underline{3}}_{\underline{12}}5^{\underline{3}}_{\underline{10}}5^{\underline{2}}_{\underline{10}}7^{\underline{2}}_{\underline{8}}5^{\underline{1}}_{\underline{8}}4^{\underline{1}}_{\underline{6}}3^{0}_{\underline{6}}6^{0}_{\underline{4}}2^{1}_{\underline{4}}2^{1}_{\underline{2}}1^{2}_{\underline{2}}2^{2}_{0}1^{3}_{2}$	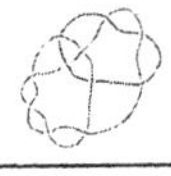	
11^a_{148}: $1^{\underline{5}}_{\underline{12}}2^{\underline{4}}_{\underline{10}}1^{\underline{4}}_{\underline{8}}4^{\underline{3}}_{\underline{8}}2^{\underline{3}}_{\underline{6}}7^{\underline{2}}_{\underline{6}}5^{\underline{2}}_{\underline{4}}8^{\underline{1}}_{\underline{4}}6^{\underline{1}}_{\underline{2}}9^{0}_{\underline{2}}9^{0}_{0}8^{1}_{0}8^{1}_{2}7^{2}_{2}8^{2}_{4}4^{3}_{4}7^{3}_{6}3^{4}_{6}4^{4}_{8}1^{5}_{8}3^{5}_{10}1^{6}_{12}$		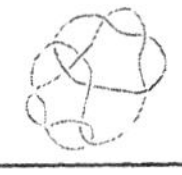
11^a_{149}: $1^{\underline{8}}_{\underline{20}}3^{\underline{7}}_{\underline{18}}1^{\underline{7}}_{\underline{16}}5^{\underline{6}}_{\underline{16}}3^{\underline{6}}_{\underline{14}}8^{\underline{5}}_{\underline{14}}5^{\underline{5}}_{\underline{12}}9^{\underline{4}}_{\underline{12}}8^{\underline{4}}_{\underline{10}}10^{\underline{3}}_{\underline{10}}9^{\underline{3}}_{\underline{8}}9^{\underline{2}}_{\underline{8}}11^{\underline{2}}_{\underline{6}}7^{\underline{1}}_{\underline{6}}8^{\underline{1}}_{\underline{4}}5^{0}_{\underline{4}}8^{0}_{\underline{2}}2^{1}_{\underline{2}}4^{1}_{0}1^{2}_{0}2^{2}_{2}1^{3}_{4}$		
11^a_{150}: $1^{\underline{7}}_{\underline{16}}4^{\underline{6}}_{\underline{14}}1^{\underline{6}}_{\underline{12}}8^{\underline{5}}_{\underline{12}}4^{\underline{5}}_{\underline{10}}12^{\underline{4}}_{\underline{10}}8^{\underline{4}}_{\underline{8}}15^{\underline{3}}_{\underline{8}}12^{\underline{3}}_{\underline{6}}17^{\underline{2}}_{\underline{6}}16^{\underline{2}}_{\underline{4}}15^{\underline{1}}_{\underline{4}}16^{\underline{1}}_{\underline{2}}13^{0}_{\underline{2}}16^{0}_{0}8^{1}_{0}12^{1}_{2}4^{2}_{2}8^{2}_{4}1^{3}_{4}4^{3}_{6}1^{4}_{8}$		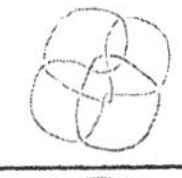
11^a_{151}: $1^{\underline{9}}_{\underline{22}}2^{\underline{8}}_{\underline{20}}1^{\underline{8}}_{\underline{18}}5^{\underline{7}}_{\underline{18}}2^{\underline{7}}_{\underline{16}}6^{\underline{6}}_{\underline{16}}5^{\underline{6}}_{\underline{14}}9^{\underline{5}}_{\underline{14}}6^{\underline{5}}_{\underline{12}}9^{\underline{4}}_{\underline{12}}9^{\underline{4}}_{\underline{10}}8^{\underline{3}}_{\underline{10}}9^{\underline{3}}_{\underline{8}}8^{\underline{2}}_{\underline{8}}9^{\underline{2}}_{\underline{6}}4^{\underline{1}}_{\underline{6}}7^{\underline{1}}_{\underline{4}}3^{0}_{\underline{4}}5^{0}_{\underline{2}}1^{1}_{\underline{2}}2^{1}_{0}1^{2}_{2}$		
11^a_{152}: $1^{\underline{11}}_{\underline{28}}2^{\underline{10}}_{\underline{26}}1^{\underline{10}}_{\underline{24}}5^{\underline{9}}_{\underline{24}}2^{\underline{9}}_{\underline{22}}7^{\underline{8}}_{\underline{22}}5^{\underline{8}}_{\underline{20}}9^{\underline{7}}_{\underline{20}}7^{\underline{7}}_{\underline{18}}10^{\underline{6}}_{\underline{18}}10^{\underline{6}}_{\underline{16}}10^{\underline{5}}_{\underline{16}}9^{\underline{5}}_{\underline{14}}7^{\underline{4}}_{\underline{14}}10^{\underline{4}}_{\underline{12}}5^{\underline{3}}_{\underline{12}}7^{\underline{3}}_{\underline{10}}3^{\underline{2}}_{\underline{10}}5^{\underline{2}}_{\underline{8}}3^{\underline{1}}_{\underline{6}}1^{0}_{\underline{6}}1^{0}_{\underline{4}}$		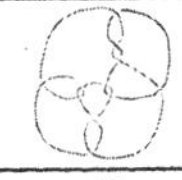
11^a_{153}: $1^{\underline{6}}_{\underline{12}}1^{\underline{6}}_{\underline{10}}2^{\underline{5}}_{\underline{10}}4^{\underline{4}}_{\underline{8}}2^{\underline{4}}_{\underline{6}}5^{\underline{3}}_{\underline{6}}4^{\underline{3}}_{\underline{4}}8^{\underline{2}}_{\underline{4}}5^{\underline{2}}_{\underline{2}}7^{\underline{1}}_{\underline{2}}8^{\underline{1}}_{0}9^{0}_{0}8^{0}_{2}6^{1}_{2}8^{1}_{4}5^{2}_{4}6^{2}_{6}2^{3}_{6}5^{3}_{8}1^{4}_{8}2^{4}_{10}1^{5}_{12}$	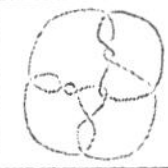	
11^a_{154}: $1^{\underline{6}}_{\underline{14}}1^{\underline{6}}_{\underline{12}}2^{\underline{5}}_{\underline{12}}3^{\underline{4}}_{\underline{10}}2^{\underline{4}}_{\underline{8}}5^{\underline{3}}_{\underline{8}}3^{\underline{3}}_{\underline{6}}7^{\underline{2}}_{\underline{6}}5^{\underline{2}}_{\underline{4}}6^{\underline{1}}_{\underline{4}}7^{\underline{1}}_{\underline{2}}8^{0}_{\underline{2}}7^{0}_{0}5^{1}_{0}7^{1}_{2}4^{2}_{2}5^{2}_{4}2^{3}_{4}4^{3}_{6}1^{4}_{6}2^{4}_{8}1^{5}_{10}$		
11^a_{155}: $1^{\underline{7}}_{\underline{16}}2^{\underline{6}}_{\underline{14}}1^{\underline{6}}_{\underline{12}}4^{\underline{5}}_{\underline{12}}2^{\underline{5}}_{\underline{10}}7^{\underline{4}}_{\underline{10}}4^{\underline{4}}_{\underline{8}}7^{\underline{3}}_{\underline{8}}7^{\underline{3}}_{\underline{6}}10^{\underline{2}}_{\underline{6}}8^{\underline{2}}_{\underline{4}}8^{\underline{1}}_{\underline{4}}9^{\underline{1}}_{\underline{2}}7^{0}_{\underline{2}}9^{0}_{0}5^{1}_{0}6^{1}_{2}2^{2}_{2}5^{2}_{4}1^{3}_{4}2^{3}_{6}1^{4}_{8}$	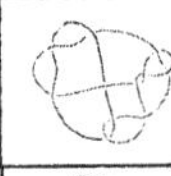	
11^a_{156}: $1^{\underline{4}}_{\underline{6}}2^{\underline{3}}_{\underline{4}}1^{\underline{3}}_{\underline{2}}5^{\underline{2}}_{\underline{2}}2^{\underline{2}}_{0}5^{\underline{1}}_{0}5^{\underline{1}}_{2}9^{0}_{2}6^{0}_{4}8^{1}_{4}8^{1}_{6}8^{2}_{6}9^{2}_{8}7^{3}_{8}7^{3}_{10}4^{4}_{10}7^{4}_{12}2^{5}_{12}4^{5}_{14}1^{6}_{14}2^{6}_{16}1^{7}_{18}$		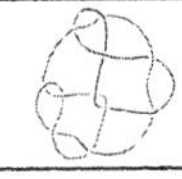
11^a_{157}: $1^{\underline{4}}_{\underline{8}}2^{\underline{3}}_{\underline{6}}1^{\underline{3}}_{\underline{4}}5^{\underline{2}}_{\underline{4}}2^{\underline{2}}_{\underline{2}}7^{\underline{1}}_{\underline{2}}5^{\underline{1}}_{0}10^{0}_{0}8^{0}_{2}10^{1}_{2}9^{1}_{4}10^{2}_{4}11^{2}_{6}8^{3}_{6}9^{3}_{8}5^{4}_{8}8^{4}_{10}3^{5}_{10}5^{5}_{12}1^{6}_{12}3^{6}_{14}1^{7}_{16}$	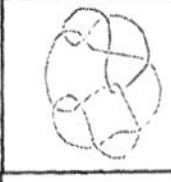	
11^a_{158}: $1^{\underline{6}}_{\underline{12}}2^{\underline{5}}_{\underline{10}}1^{\underline{5}}_{\underline{8}}5^{\underline{4}}_{\underline{8}}2^{\underline{4}}_{\underline{6}}6^{\underline{3}}_{\underline{6}}5^{\underline{3}}_{\underline{4}}10^{\underline{2}}_{\underline{4}}7^{\underline{2}}_{\underline{2}}9^{\underline{1}}_{\underline{2}}9^{\underline{1}}_{0}10^{0}_{0}10^{0}_{2}8^{1}_{2}9^{1}_{4}5^{2}_{4}8^{2}_{6}3^{3}_{6}5^{3}_{8}1^{4}_{8}3^{4}_{10}1^{5}_{12}$		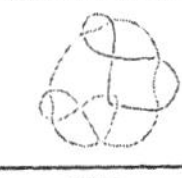
11^a_{159}: $1^{\underline{6}}_{\underline{14}}3^{\underline{5}}_{\underline{12}}1^{\underline{5}}_{\underline{10}}6^{\underline{4}}_{\underline{10}}3^{\underline{4}}_{\underline{8}}8^{\underline{3}}_{\underline{8}}6^{\underline{3}}_{\underline{6}}12^{\underline{2}}_{\underline{6}}9^{\underline{2}}_{\underline{4}}11^{\underline{1}}_{\underline{4}}11^{\underline{1}}_{\underline{2}}11^{0}_{\underline{2}}12^{0}_{0}9^{1}_{0}10^{1}_{2}5^{2}_{2}9^{2}_{4}3^{3}_{4}5^{3}_{6}1^{4}_{6}3^{4}_{8}1^{5}_{10}$		
11^a_{160}: $1^{\underline{7}}_{\underline{16}}2^{\underline{6}}_{\underline{14}}1^{\underline{6}}_{\underline{12}}5^{\underline{5}}_{\underline{12}}2^{\underline{5}}_{\underline{10}}7^{\underline{4}}_{\underline{10}}5^{\underline{4}}_{\underline{8}}9^{\underline{3}}_{\underline{8}}7^{\underline{3}}_{\underline{6}}11^{\underline{2}}_{\underline{6}}10^{\underline{2}}_{\underline{4}}9^{\underline{1}}_{\underline{4}}10^{\underline{1}}_{\underline{2}}9^{0}_{\underline{2}}10^{0}_{0}5^{1}_{0}8^{1}_{2}3^{2}_{2}5^{2}_{4}1^{3}_{4}3^{3}_{6}1^{4}_{8}$		
11^a_{161}: $1^{\underline{5}}_{\underline{10}}3^{\underline{4}}_{\underline{8}}1^{\underline{4}}_{\underline{6}}5^{\underline{3}}_{\underline{6}}3^{\underline{3}}_{\underline{4}}9^{\underline{2}}_{\underline{4}}6^{\underline{2}}_{\underline{2}}10^{\underline{1}}_{\underline{2}}8^{\underline{1}}_{0}12^{0}_{0}11^{0}_{2}10^{1}_{2}11^{1}_{4}9^{2}_{4}10^{2}_{6}5^{3}_{6}9^{3}_{8}3^{4}_{8}5^{4}_{10}1^{5}_{10}3^{5}_{12}1^{6}_{14}$		
11^a_{162}: $1^{\underline{9}}_{\underline{24}}2^{\underline{8}}_{\underline{22}}1^{\underline{8}}_{\underline{20}}4^{\underline{7}}_{\underline{20}}2^{\underline{7}}_{\underline{18}}5^{\underline{6}}_{\underline{18}}4^{\underline{6}}_{\underline{16}}7^{\underline{5}}_{\underline{16}}5^{\underline{5}}_{\underline{14}}6^{\underline{4}}_{\underline{14}}7^{\underline{4}}_{\underline{12}}6^{\underline{3}}_{\underline{12}}6^{\underline{3}}_{\underline{10}}5^{\underline{2}}_{\underline{10}}7^{\underline{2}}_{\underline{8}}3^{\underline{1}}_{\underline{8}}4^{\underline{1}}_{\underline{6}}2^{0}_{\underline{6}}4^{0}_{\underline{4}}1^{1}_{\underline{4}}1^{1}_{\underline{2}}1^{2}_{0}$		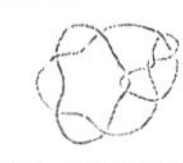

L: KH	1st line	2nd line
11^a_{163}: $1^{11}_{\underline{30}}2^{10}_{\underline{28}}1^{10}_{\underline{26}}4^{9}_{\underline{26}}2^{9}_{\underline{24}}6^{8}_{\underline{24}}4^{8}_{\underline{22}}7^{7}_{\underline{22}}6^{7}_{\underline{20}}7^{6}_{\underline{20}}8^{6}_{\underline{18}}8^{5}_{\underline{18}}6^{5}_{\underline{16}}4^{4}_{\underline{16}}8^{4}_{\underline{14}}4^{3}_{\underline{14}}4^{3}_{\underline{12}}2^{2}_{\underline{12}}4^{2}_{\underline{10}}2^{1}_{\underline{8}}1^{0}_{\underline{8}}1^{0}_{\underline{6}}$ 11^a_{164}: $1^{6}_{\underline{14}}1^{6}_{\underline{12}}2^{5}_{\underline{12}}3^{4}_{\underline{10}}2^{4}_{\underline{8}}5^{3}_{\underline{8}}3^{3}_{\underline{6}}6^{2}_{\underline{6}}5^{2}_{\underline{4}}6^{1}_{\underline{4}}6^{1}_{\underline{2}}7^{0}_{\underline{2}}7^{0}_{0}4^{1}_{0}6^{1}_{2}4^{2}_{2}4^{2}_{4}1^{3}_{4}4^{3}_{6}1^{4}_{6}1^{4}_{8}1^{5}_{10}$		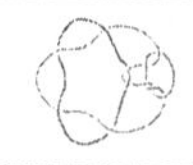
11^a_{165}: $1^{6}_{\underline{16}}1^{6}_{\underline{14}}2^{5}_{\underline{14}}2^{4}_{\underline{12}}2^{4}_{\underline{10}}4^{3}_{\underline{10}}2^{3}_{\underline{8}}4^{2}_{\underline{8}}4^{2}_{\underline{6}}4^{1}_{\underline{6}}4^{1}_{\underline{4}}5^{0}_{\underline{4}}5^{0}_{\underline{2}}3^{1}_{\underline{2}}4^{1}_{0}3^{2}_{0}3^{2}_{2}1^{3}_{2}3^{3}_{4}1^{4}_{4}1^{4}_{6}1^{5}_{8}$ 11^a_{166}: $1^{7}_{\underline{18}}2^{6}_{\underline{16}}1^{6}_{\underline{14}}4^{5}_{\underline{14}}2^{5}_{\underline{12}}6^{4}_{\underline{12}}4^{4}_{\underline{10}}6^{3}_{\underline{10}}6^{3}_{\underline{8}}8^{2}_{\underline{8}}7^{2}_{\underline{6}}6^{1}_{\underline{6}}7^{1}_{\underline{4}}5^{0}_{\underline{4}}7^{0}_{\underline{2}}4^{1}_{\underline{2}}4^{1}_{0}1^{2}_{0}4^{2}_{2}1^{3}_{2}1^{3}_{4}1^{4}_{6}$	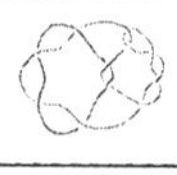	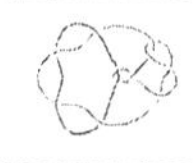
11^a_{167}: $1^{7}_{\underline{16}}3^{6}_{\underline{14}}1^{6}_{\underline{12}}6^{5}_{\underline{12}}3^{5}_{\underline{10}}7^{4}_{\underline{10}}6^{4}_{\underline{8}}11^{3}_{\underline{8}}7^{3}_{\underline{6}}10^{2}_{\underline{6}}11^{2}_{\underline{4}}9^{1}_{\underline{4}}10^{1}_{\underline{2}}9^{0}_{\underline{2}}10^{0}_{0}4^{1}_{0}8^{1}_{2}3^{2}_{2}5^{2}_{4}1^{3}_{4}2^{3}_{6}1^{4}_{8}$ 11^a_{168}: $1^{7}_{\underline{18}}3^{6}_{\underline{16}}1^{6}_{\underline{14}}5^{5}_{\underline{14}}3^{5}_{\underline{12}}8^{4}_{\underline{12}}5^{4}_{\underline{10}}10^{3}_{\underline{10}}8^{3}_{\underline{8}}10^{2}_{\underline{8}}11^{2}_{\underline{6}}10^{1}_{\underline{6}}9^{1}_{\underline{4}}8^{0}_{\underline{4}}11^{0}_{\underline{2}}5^{1}_{\underline{2}}7^{1}_{0}3^{2}_{0}5^{2}_{2}1^{3}_{2}3^{3}_{4}1^{4}_{6}$	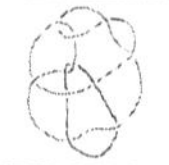	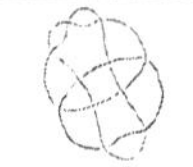
11^a_{169}: $1^{7}_{\underline{16}}3^{6}_{\underline{14}}1^{6}_{\underline{12}}5^{5}_{\underline{12}}3^{5}_{\underline{10}}9^{4}_{\underline{10}}5^{4}_{\underline{8}}10^{3}_{\underline{8}}9^{3}_{\underline{6}}12^{2}_{\underline{6}}11^{2}_{\underline{4}}11^{1}_{\underline{4}}11^{1}_{\underline{2}}9^{0}_{\underline{2}}12^{0}_{0}6^{1}_{0}8^{1}_{2}3^{2}_{2}6^{2}_{4}1^{3}_{4}3^{3}_{6}1^{4}_{8}$ 11^a_{170}: $1^{6}_{\underline{14}}3^{5}_{\underline{12}}1^{5}_{\underline{10}}7^{4}_{\underline{10}}3^{4}_{\underline{8}}11^{3}_{\underline{8}}7^{3}_{\underline{6}}13^{2}_{\underline{6}}11^{2}_{\underline{4}}15^{1}_{\underline{4}}13^{1}_{\underline{2}}14^{0}_{\underline{2}}16^{0}_{0}11^{1}_{0}13^{1}_{2}8^{2}_{2}12^{2}_{4}4^{3}_{4}7^{3}_{6}1^{4}_{6}4^{4}_{8}1^{5}_{10}$		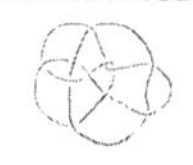
11^a_{171}: $1^{8}_{\underline{20}}4^{7}_{\underline{18}}1^{7}_{\underline{16}}8^{6}_{\underline{16}}4^{6}_{\underline{14}}12^{5}_{\underline{14}}8^{5}_{\underline{12}}15^{4}_{\underline{12}}12^{4}_{\underline{10}}16^{3}_{\underline{10}}15^{3}_{\underline{8}}15^{2}_{\underline{8}}17^{2}_{\underline{6}}12^{1}_{\underline{6}}14^{1}_{\underline{4}}8^{0}_{\underline{4}}13^{0}_{\underline{2}}4^{1}_{\underline{2}}7^{1}_{0}1^{2}_{0}4^{2}_{2}1^{3}_{4}$ 11^a_{172}: $1^{7}_{\underline{18}}2^{6}_{\underline{16}}1^{6}_{\underline{14}}4^{5}_{\underline{14}}2^{5}_{\underline{12}}6^{4}_{\underline{12}}4^{4}_{\underline{10}}7^{3}_{\underline{10}}6^{3}_{\underline{8}}8^{2}_{\underline{8}}8^{2}_{\underline{6}}7^{1}_{\underline{6}}7^{1}_{\underline{4}}6^{0}_{\underline{4}}8^{0}_{\underline{2}}4^{1}_{\underline{2}}5^{1}_{0}2^{2}_{0}4^{2}_{2}1^{3}_{2}2^{3}_{4}1^{4}_{6}$		
11^a_{173}: $1^{6}_{\underline{14}}2^{5}_{\underline{12}}1^{5}_{\underline{10}}4^{4}_{\underline{10}}2^{4}_{\underline{8}}6^{3}_{\underline{8}}4^{3}_{\underline{6}}8^{2}_{\underline{6}}7^{2}_{\underline{4}}8^{1}_{\underline{4}}7^{1}_{\underline{2}}8^{0}_{\underline{2}}9^{0}_{0}6^{1}_{0}7^{1}_{2}4^{2}_{2}6^{2}_{4}2^{3}_{4}4^{3}_{6}1^{4}_{6}2^{4}_{8}1^{5}_{10}$ 11^a_{174}: $1^{7}_{\underline{16}}2^{6}_{\underline{14}}1^{6}_{\underline{12}}4^{5}_{\underline{12}}2^{5}_{\underline{10}}7^{4}_{\underline{10}}4^{4}_{\underline{8}}7^{3}_{\underline{8}}7^{3}_{\underline{6}}10^{2}_{\underline{6}}8^{2}_{\underline{4}}8^{1}_{\underline{4}}9^{1}_{\underline{2}}7^{0}_{\underline{2}}9^{0}_{0}5^{1}_{0}6^{1}_{2}2^{2}_{2}5^{2}_{4}1^{3}_{4}2^{3}_{6}1^{4}_{8}$	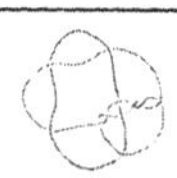	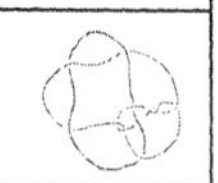
11^a_{175}: $1^{6}_{\underline{16}}2^{5}_{\underline{14}}1^{5}_{\underline{12}}4^{4}_{\underline{12}}2^{4}_{\underline{10}}5^{3}_{\underline{10}}4^{3}_{\underline{8}}7^{2}_{\underline{8}}6^{2}_{\underline{6}}7^{1}_{\underline{6}}6^{1}_{\underline{4}}6^{0}_{\underline{4}}8^{0}_{\underline{2}}6^{1}_{\underline{2}}5^{1}_{0}3^{2}_{0}6^{2}_{2}2^{3}_{2}3^{3}_{4}1^{4}_{4}2^{4}_{6}1^{5}_{8}$ 11^a_{176}: $1^{5}_{\underline{10}}3^{4}_{\underline{8}}1^{4}_{\underline{6}}6^{3}_{\underline{6}}3^{3}_{\underline{4}}9^{2}_{\underline{4}}6^{2}_{\underline{2}}11^{1}_{\underline{2}}9^{1}_{0}13^{0}_{0}12^{0}_{2}11^{1}_{2}12^{1}_{4}10^{2}_{4}12^{2}_{6}6^{3}_{6}9^{3}_{8}3^{4}_{8}6^{4}_{10}1^{5}_{10}3^{5}_{12}1^{6}_{14}$	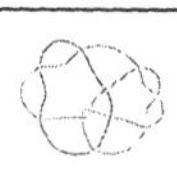	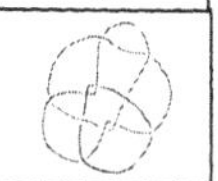
11^a_{177}: $1^{6}_{\underline{14}}3^{5}_{\underline{12}}1^{5}_{\underline{10}}6^{4}_{\underline{10}}3^{4}_{\underline{8}}10^{3}_{\underline{8}}6^{3}_{\underline{6}}11^{2}_{\underline{6}}10^{2}_{\underline{4}}13^{1}_{\underline{4}}11^{1}_{\underline{2}}12^{0}_{\underline{2}}14^{0}_{0}9^{1}_{0}11^{1}_{2}7^{2}_{2}10^{2}_{4}3^{3}_{4}6^{3}_{6}1^{4}_{6}3^{4}_{8}1^{5}_{10}$ 11^a_{178}: $1^{4}_{\underline{8}}3^{3}_{\underline{6}}1^{3}_{\underline{4}}7^{2}_{\underline{4}}3^{2}_{\underline{2}}9^{1}_{\underline{2}}7^{1}_{0}13^{0}_{0}10^{0}_{2}13^{1}_{2}12^{1}_{4}12^{2}_{4}14^{2}_{6}10^{3}_{6}11^{3}_{8}6^{4}_{8}10^{4}_{10}3^{5}_{10}6^{5}_{12}1^{6}_{12}3^{6}_{14}1^{7}_{16}$	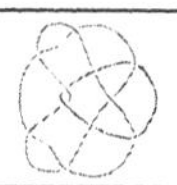	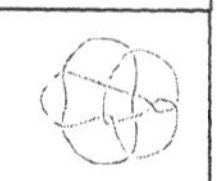
11^a_{179}: $1^{3}_{\underline{4}}3^{2}_{\underline{2}}1^{2}_{0}6^{1}_{0}3^{1}_{2}10^{0}_{2}7^{0}_{4}11^{1}_{4}9^{1}_{6}14^{2}_{6}12^{2}_{8}11^{3}_{8}13^{3}_{10}10^{4}_{10}11^{4}_{12}6^{5}_{12}10^{5}_{14}3^{6}_{14}6^{6}_{16}1^{7}_{16}3^{7}_{18}1^{8}_{20}$ 11^a_{180}: $1^{4}_{\underline{8}}4^{3}_{\underline{6}}1^{3}_{\underline{4}}8^{2}_{\underline{4}}4^{2}_{\underline{2}}11^{1}_{\underline{2}}8^{1}_{0}15^{0}_{0}12^{0}_{2}15^{1}_{2}14^{1}_{4}14^{2}_{4}16^{2}_{6}11^{3}_{6}13^{3}_{8}7^{4}_{8}11^{4}_{10}3^{5}_{10}7^{5}_{12}1^{6}_{12}3^{6}_{14}1^{7}_{16}$	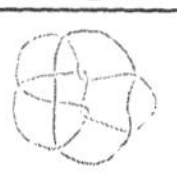	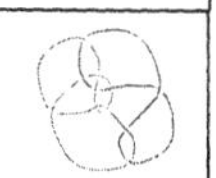
11^a_{181}: $1^{5}_{\underline{12}}3^{4}_{\underline{10}}1^{4}_{\underline{8}}6^{3}_{\underline{8}}3^{3}_{\underline{6}}10^{2}_{\underline{6}}7^{2}_{\underline{4}}12^{1}_{\underline{4}}9^{1}_{\underline{2}}13^{0}_{\underline{2}}13^{0}_{0}12^{1}_{0}12^{1}_{2}10^{2}_{2}12^{2}_{4}6^{3}_{4}10^{3}_{6}4^{4}_{6}6^{4}_{8}1^{5}_{8}4^{5}_{10}1^{6}_{12}$ 11^a_{182}: $1^{3}_{\underline{4}}3^{2}_{\underline{2}}1^{2}_{0}5^{1}_{0}3^{1}_{2}10^{0}_{2}6^{0}_{4}10^{1}_{4}9^{1}_{6}13^{2}_{6}11^{2}_{8}11^{3}_{8}12^{3}_{10}9^{4}_{10}11^{4}_{12}6^{5}_{12}9^{5}_{14}3^{6}_{14}6^{6}_{16}1^{7}_{16}3^{7}_{18}1^{8}_{20}$	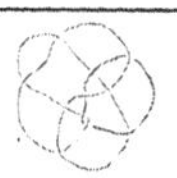	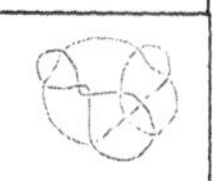
11^a_{183}: $1^{4}_{\underline{8}}3^{3}_{\underline{6}}1^{3}_{\underline{4}}6^{2}_{\underline{4}}3^{2}_{\underline{2}}9^{1}_{\underline{2}}6^{1}_{0}12^{0}_{0}10^{0}_{2}12^{1}_{2}11^{1}_{4}12^{2}_{4}13^{2}_{6}9^{3}_{6}11^{3}_{8}6^{4}_{8}9^{4}_{10}3^{5}_{10}6^{5}_{12}1^{6}_{12}3^{6}_{14}1^{7}_{16}$ 11^a_{184}: $1^{7}_{\underline{16}}4^{6}_{\underline{14}}1^{6}_{\underline{12}}7^{5}_{\underline{12}}4^{5}_{\underline{10}}10^{4}_{\underline{10}}7^{4}_{\underline{8}}13^{3}_{\underline{8}}10^{3}_{\underline{6}}14^{2}_{\underline{6}}14^{2}_{\underline{4}}12^{1}_{\underline{4}}13^{1}_{\underline{2}}11^{0}_{\underline{2}}13^{0}_{0}6^{1}_{0}10^{1}_{2}3^{2}_{2}6^{2}_{4}1^{3}_{4}3^{3}_{6}1^{4}_{8}$	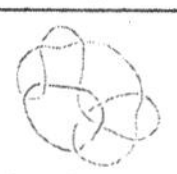	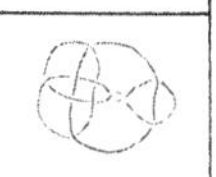
11^a_{185}: $1^{9}_{\underline{22}}3^{8}_{\underline{20}}1^{8}_{\underline{18}}4^{7}_{\underline{18}}3^{7}_{\underline{16}}8^{6}_{\underline{16}}5^{6}_{\underline{14}}9^{5}_{\underline{14}}7^{5}_{\underline{12}}9^{4}_{\underline{12}}9^{4}_{\underline{10}}9^{3}_{\underline{10}}9^{3}_{\underline{8}}7^{2}_{\underline{8}}9^{2}_{\underline{6}}4^{1}_{\underline{6}}7^{1}_{\underline{4}}3^{0}_{\underline{4}}5^{0}_{\underline{2}}1^{1}_{\underline{2}}2^{1}_{0}1^{2}_{2}$ 11^a_{186}: $1^{6}_{\underline{12}}1^{6}_{\underline{10}}2^{5}_{\underline{10}}3^{4}_{\underline{8}}2^{4}_{\underline{6}}4^{3}_{\underline{6}}3^{3}_{\underline{4}}7^{2}_{\underline{4}}4^{2}_{\underline{2}}5^{1}_{\underline{2}}7^{1}_{0}8^{0}_{0}6^{0}_{2}5^{1}_{2}7^{1}_{4}4^{2}_{4}5^{2}_{6}2^{3}_{6}4^{3}_{8}1^{4}_{8}2^{4}_{10}1^{5}_{12}$	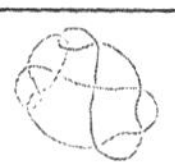	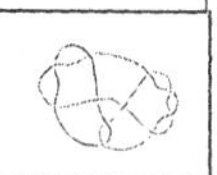
11^a_{187}: $1^{6}_{\underline{14}}1^{6}_{\underline{12}}3^{5}_{\underline{12}}4^{4}_{\underline{10}}3^{4}_{\underline{8}}6^{3}_{\underline{8}}4^{3}_{\underline{6}}9^{2}_{\underline{6}}6^{2}_{\underline{4}}7^{1}_{\underline{4}}9^{1}_{\underline{2}}9^{0}_{\underline{2}}8^{0}_{0}6^{1}_{0}8^{1}_{2}4^{2}_{2}6^{2}_{4}2^{3}_{4}4^{3}_{6}1^{4}_{6}2^{4}_{8}1^{5}_{10}$ 11^a_{188}: $1^{9}_{\underline{24}}1^{8}_{\underline{22}}1^{8}_{\underline{20}}3^{7}_{\underline{20}}1^{7}_{\underline{18}}3^{6}_{\underline{18}}3^{6}_{\underline{16}}5^{5}_{\underline{16}}3^{5}_{\underline{14}}4^{4}_{\underline{14}}5^{4}_{\underline{12}}4^{3}_{\underline{12}}4^{3}_{\underline{10}}4^{2}_{\underline{10}}5^{2}_{\underline{8}}2^{1}_{\underline{8}}3^{1}_{\underline{6}}2^{0}_{\underline{6}}3^{0}_{\underline{4}}1^{1}_{\underline{4}}1^{1}_{\underline{2}}1^{2}_{0}$	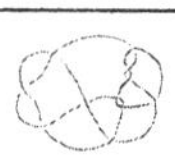	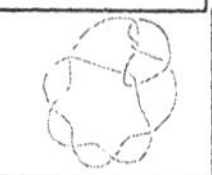
11^a_{189}: $1^{9}_{\underline{22}}1^{8}_{\underline{20}}1^{8}_{\underline{18}}4^{7}_{\underline{18}}1^{7}_{\underline{16}}5^{6}_{\underline{16}}4^{6}_{\underline{14}}7^{5}_{\underline{14}}5^{5}_{\underline{12}}8^{4}_{\underline{12}}7^{4}_{\underline{10}}7^{3}_{\underline{10}}8^{3}_{\underline{8}}7^{2}_{\underline{8}}8^{2}_{\underline{6}}4^{1}_{\underline{6}}6^{1}_{\underline{4}}3^{0}_{\underline{4}}5^{0}_{\underline{2}}1^{1}_{\underline{2}}2^{1}_{0}1^{2}_{2}$ 11^a_{190}: $1^{8}_{\underline{18}}2^{7}_{\underline{16}}1^{7}_{\underline{14}}4^{6}_{\underline{14}}3^{6}_{\underline{12}}6^{5}_{\underline{12}}3^{5}_{\underline{10}}7^{4}_{\underline{10}}6^{4}_{\underline{8}}7^{3}_{\underline{8}}7^{3}_{\underline{6}}8^{2}_{\underline{6}}7^{2}_{\underline{4}}5^{1}_{\underline{4}}8^{1}_{\underline{2}}5^{0}_{\underline{2}}6^{0}_{0}2^{1}_{0}4^{1}_{2}1^{2}_{2}2^{2}_{4}1^{3}_{6}$	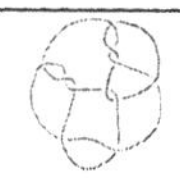	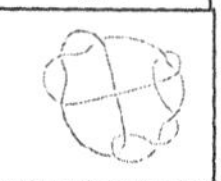
11^a_{191}: $1^{9}_{\underline{22}}2^{8}_{\underline{20}}1^{8}_{\underline{18}}4^{7}_{\underline{18}}2^{7}_{\underline{16}}5^{6}_{\underline{16}}4^{6}_{\underline{14}}7^{5}_{\underline{14}}5^{5}_{\underline{12}}7^{4}_{\underline{12}}7^{4}_{\underline{10}}6^{3}_{\underline{10}}7^{3}_{\underline{8}}6^{2}_{\underline{8}}7^{2}_{\underline{6}}3^{1}_{\underline{6}}5^{1}_{\underline{4}}2^{0}_{\underline{4}}4^{0}_{\underline{2}}1^{1}_{\underline{2}}1^{1}_{0}1^{2}_{2}$ 11^a_{192}: $1^{9}_{\underline{20}}1^{8}_{\underline{18}}1^{8}_{\underline{16}}2^{7}_{\underline{16}}1^{7}_{\underline{14}}3^{6}_{\underline{14}}2^{6}_{\underline{12}}3^{5}_{\underline{12}}3^{5}_{\underline{10}}4^{4}_{\underline{10}}3^{4}_{\underline{8}}3^{3}_{\underline{8}}4^{3}_{\underline{6}}4^{2}_{\underline{6}}4^{2}_{\underline{4}}2^{1}_{\underline{4}}3^{1}_{\underline{2}}2^{0}_{\underline{2}}3^{0}_{0}1^{1}_{0}1^{1}_{2}1^{2}_{4}$	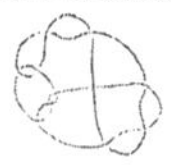	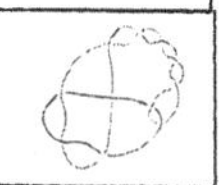

L: KH	1st line	2nd line
11^a_{193}: $1^{\underline{2}}_{0}1^{\underline{1}}_{2}1^{\underline{1}}_{4}5^{0}_{4}2^{0}_{6}5^{1}_{6}4^{1}_{8}9^{2}_{8}6^{2}_{10}8^{3}_{10}8^{3}_{12}9^{4}_{12}8^{4}_{14}7^{5}_{14}9^{5}_{16}5^{6}_{16}7^{6}_{18}3^{7}_{18}5^{7}_{20}1^{8}_{20}3^{8}_{22}1^{9}_{24}$ 11^a_{194}: $1^{\underline{2}}_{\underline{2}}1^{\underline{1}}_{0}1^{\underline{1}}_{2}4^{0}_{2}2^{0}_{4}4^{1}_{4}3^{1}_{6}7^{2}_{6}5^{2}_{8}6^{3}_{8}6^{3}_{10}6^{4}_{10}6^{4}_{12}5^{5}_{12}6^{5}_{14}3^{6}_{14}5^{6}_{16}2^{7}_{16}3^{7}_{18}1^{8}_{18}2^{8}_{20}1^{9}_{22}$		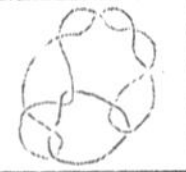
11^a_{195}: $1^{\underline{6}}_{\underline{14}}3^{\underline{5}}_{\underline{12}}1^{\underline{5}}_{\underline{10}}5^{\underline{4}}_{\underline{10}}3^{\underline{4}}_{\underline{8}}8^{\underline{3}}_{\underline{8}}5^{\underline{3}}_{\underline{6}}9^{\underline{2}}_{\underline{6}}8^{\underline{2}}_{\underline{4}}10^{\underline{1}}_{\underline{4}}9^{\underline{1}}_{\underline{2}}9^{0}_{\underline{2}}11^{0}_{0}7^{1}_{0}8^{1}_{2}5^{2}_{2}8^{2}_{4}2^{3}_{4}4^{3}_{6}1^{4}_{6}2^{4}_{8}1^{5}_{10}$ 11^a_{196}: $1^{\underline{6}}_{\underline{12}}2^{\underline{5}}_{\underline{10}}1^{\underline{5}}_{\underline{8}}3^{\underline{4}}_{\underline{8}}2^{\underline{4}}_{\underline{6}}5^{\underline{3}}_{\underline{6}}3^{\underline{3}}_{\underline{4}}6^{\underline{2}}_{\underline{4}}5^{\underline{2}}_{\underline{2}}6^{\underline{1}}_{\underline{2}}6^{\underline{1}}_{0}7^{0}_{0}7^{0}_{2}5^{1}_{2}6^{1}_{4}4^{2}_{4}6^{2}_{6}2^{3}_{6}3^{3}_{8}1^{4}_{8}2^{4}_{10}1^{5}_{12}$		
11^a_{197}: $1^{\underline{2}}_{\underline{2}}3^{\underline{1}}_{0}1^{\underline{1}}_{2}7^{0}_{2}4^{0}_{4}9^{1}_{4}6^{1}_{6}13^{2}_{6}10^{2}_{8}12^{3}_{8}12^{3}_{10}12^{4}_{10}12^{4}_{12}9^{5}_{12}12^{5}_{14}6^{6}_{14}9^{6}_{16}3^{7}_{16}6^{7}_{18}1^{8}_{18}3^{8}_{20}1^{9}_{22}$ 11^a_{198}: $1^{\underline{6}}_{\underline{12}}3^{\underline{5}}_{\underline{10}}1^{\underline{5}}_{\underline{8}}5^{\underline{4}}_{\underline{8}}3^{\underline{4}}_{\underline{6}}9^{\underline{3}}_{\underline{6}}5^{\underline{3}}_{\underline{4}}10^{\underline{2}}_{\underline{4}}9^{\underline{2}}_{\underline{2}}11^{\underline{1}}_{\underline{2}}10^{\underline{1}}_{0}12^{0}_{0}12^{0}_{2}8^{1}_{2}11^{1}_{4}7^{2}_{4}9^{2}_{6}3^{3}_{6}6^{3}_{8}1^{4}_{8}3^{4}_{10}1^{5}_{12}$	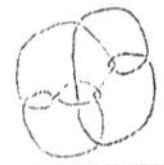	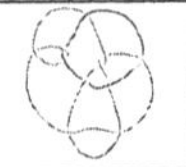
11^a_{199}: $1^{\underline{9}}_{\underline{24}}3^{\underline{8}}_{\underline{22}}1^{\underline{8}}_{\underline{20}}4^{\underline{7}}_{\underline{20}}3^{\underline{7}}_{\underline{18}}7^{\underline{6}}_{\underline{18}}5^{\underline{6}}_{\underline{16}}8^{\underline{5}}_{\underline{16}}6^{\underline{5}}_{\underline{14}}7^{\underline{4}}_{\underline{14}}8^{\underline{4}}_{\underline{12}}7^{\underline{3}}_{\underline{12}}7^{\underline{3}}_{\underline{10}}5^{\underline{2}}_{\underline{10}}7^{\underline{2}}_{\underline{8}}3^{\underline{1}}_{\underline{8}}5^{\underline{1}}_{\underline{6}}2^{0}_{\underline{6}}4^{0}_{\underline{4}}1^{1}_{\underline{4}}1^{1}_{\underline{2}}1^{2}_{0}$ 11^a_{200}: $1^{\underline{6}}_{\underline{14}}1^{\underline{6}}_{\underline{12}}2^{\underline{5}}_{\underline{12}}2^{\underline{4}}_{\underline{10}}2^{\underline{4}}_{\underline{8}}4^{\underline{3}}_{\underline{8}}2^{\underline{3}}_{\underline{6}}5^{\underline{2}}_{\underline{6}}4^{\underline{2}}_{\underline{4}}4^{\underline{1}}_{\underline{4}}5^{\underline{1}}_{\underline{2}}6^{0}_{\underline{2}}5^{0}_{0}3^{1}_{0}5^{1}_{2}3^{2}_{2}3^{2}_{4}1^{3}_{4}3^{3}_{6}1^{4}_{6}1^{4}_{8}1^{5}_{10}$		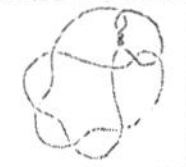
11^a_{201}: $1^{\underline{6}}_{\underline{16}}1^{\underline{6}}_{\underline{14}}3^{\underline{5}}_{\underline{14}}3^{\underline{4}}_{\underline{12}}3^{\underline{4}}_{\underline{10}}5^{\underline{3}}_{\underline{10}}3^{\underline{3}}_{\underline{8}}6^{\underline{2}}_{\underline{8}}5^{\underline{2}}_{\underline{6}}5^{\underline{1}}_{\underline{6}}6^{\underline{1}}_{\underline{4}}6^{0}_{\underline{4}}6^{0}_{\underline{2}}4^{1}_{\underline{2}}5^{1}_{0}3^{2}_{0}4^{2}_{2}1^{3}_{2}3^{3}_{4}1^{4}_{4}1^{4}_{6}1^{5}_{8}$ 11^a_{202}: $1^{\underline{9}}_{\underline{26}}1^{\underline{8}}_{\underline{24}}1^{\underline{8}}_{\underline{22}}2^{\underline{7}}_{\underline{22}}1^{\underline{7}}_{\underline{20}}2^{\underline{6}}_{\underline{20}}2^{\underline{6}}_{\underline{18}}3^{\underline{5}}_{\underline{18}}2^{\underline{5}}_{\underline{16}}2^{\underline{4}}_{\underline{16}}3^{\underline{4}}_{\underline{14}}3^{\underline{3}}_{\underline{14}}2^{\underline{3}}_{\underline{12}}2^{\underline{2}}_{\underline{12}}4^{\underline{2}}_{\underline{10}}2^{\underline{1}}_{\underline{10}}1^{\underline{1}}_{\underline{8}}1^{0}_{\underline{8}}3^{0}_{\underline{6}}1^{1}_{\underline{6}}1^{2}_{\underline{2}}$		
11^a_{203}: $1^{\underline{9}}_{\underline{24}}1^{\underline{8}}_{\underline{22}}1^{\underline{8}}_{\underline{20}}3^{\underline{7}}_{\underline{20}}1^{\underline{7}}_{\underline{18}}4^{\underline{6}}_{\underline{18}}3^{\underline{6}}_{\underline{16}}5^{\underline{5}}_{\underline{16}}4^{\underline{5}}_{\underline{14}}5^{\underline{4}}_{\underline{14}}5^{\underline{4}}_{\underline{12}}5^{\underline{3}}_{\underline{12}}5^{\underline{3}}_{\underline{10}}4^{\underline{2}}_{\underline{10}}6^{\underline{2}}_{\underline{8}}3^{\underline{1}}_{\underline{8}}3^{\underline{1}}_{\underline{6}}2^{0}_{\underline{6}}4^{0}_{\underline{4}}1^{1}_{\underline{4}}1^{1}_{\underline{2}}1^{2}_{0}$ 11^a_{204}: $1^{\underline{8}}_{\underline{20}}2^{\underline{7}}_{\underline{18}}1^{\underline{7}}_{\underline{16}}4^{\underline{6}}_{\underline{16}}3^{\underline{6}}_{\underline{14}}6^{\underline{5}}_{\underline{14}}3^{\underline{5}}_{\underline{12}}6^{\underline{4}}_{\underline{12}}6^{\underline{4}}_{\underline{10}}7^{\underline{3}}_{\underline{10}}6^{\underline{3}}_{\underline{8}}6^{\underline{2}}_{\underline{8}}7^{\underline{2}}_{\underline{6}}4^{\underline{1}}_{\underline{6}}6^{\underline{1}}_{\underline{4}}4^{0}_{\underline{4}}5^{0}_{\underline{2}}1^{1}_{\underline{2}}3^{1}_{0}1^{2}_{0}1^{2}_{2}1^{3}_{4}$	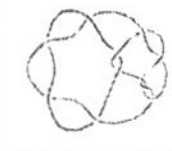	
11^a_{205}: $1^{\underline{9}}_{\underline{24}}2^{\underline{8}}_{\underline{22}}1^{\underline{8}}_{\underline{20}}4^{\underline{7}}_{\underline{20}}2^{\underline{7}}_{\underline{18}}5^{\underline{6}}_{\underline{18}}4^{\underline{6}}_{\underline{16}}6^{\underline{5}}_{\underline{16}}5^{\underline{5}}_{\underline{14}}6^{\underline{4}}_{\underline{14}}6^{\underline{4}}_{\underline{12}}5^{\underline{3}}_{\underline{12}}6^{\underline{3}}_{\underline{10}}4^{\underline{2}}_{\underline{10}}6^{\underline{2}}_{\underline{8}}3^{\underline{1}}_{\underline{8}}3^{\underline{1}}_{\underline{6}}1^{0}_{\underline{6}}4^{0}_{\underline{4}}1^{1}_{\underline{4}}1^{2}_{0}$ 11^a_{206}: $1^{\underline{9}}_{\underline{22}}1^{\underline{8}}_{\underline{20}}1^{\underline{8}}_{\underline{18}}2^{\underline{7}}_{\underline{18}}1^{\underline{7}}_{\underline{16}}3^{\underline{6}}_{\underline{16}}2^{\underline{6}}_{\underline{14}}3^{\underline{5}}_{\underline{14}}3^{\underline{5}}_{\underline{12}}4^{\underline{4}}_{\underline{12}}3^{\underline{4}}_{\underline{10}}3^{\underline{3}}_{\underline{10}}4^{\underline{3}}_{\underline{8}}3^{\underline{2}}_{\underline{8}}4^{\underline{2}}_{\underline{6}}2^{\underline{1}}_{\underline{6}}2^{\underline{1}}_{\underline{4}}1^{0}_{\underline{4}}3^{0}_{\underline{2}}1^{1}_{\underline{2}}1^{2}_{2}$	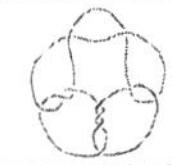	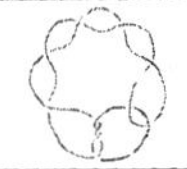
11^a_{207}: $1^{\underline{4}}_{\underline{8}}3^{\underline{3}}_{\underline{6}}1^{\underline{3}}_{\underline{4}}6^{\underline{2}}_{\underline{4}}3^{\underline{2}}_{\underline{2}}10^{\underline{1}}_{\underline{2}}6^{\underline{1}}_{0}13^{0}_{0}11^{0}_{2}13^{1}_{2}12^{1}_{4}14^{2}_{4}14^{2}_{6}10^{3}_{6}13^{3}_{8}7^{4}_{8}10^{4}_{10}4^{5}_{10}7^{5}_{12}1^{6}_{12}4^{6}_{14}1^{7}_{16}$ 11^a_{208}: $1^{\underline{7}}_{\underline{18}}2^{\underline{6}}_{\underline{16}}1^{\underline{6}}_{\underline{14}}4^{\underline{5}}_{\underline{14}}2^{\underline{5}}_{\underline{12}}5^{\underline{4}}_{\underline{12}}4^{\underline{4}}_{\underline{10}}8^{\underline{3}}_{\underline{10}}5^{\underline{3}}_{\underline{8}}6^{\underline{2}}_{\underline{8}}8^{\underline{2}}_{\underline{6}}7^{\underline{1}}_{\underline{6}}6^{\underline{1}}_{\underline{4}}6^{0}_{\underline{4}}8^{0}_{\underline{2}}3^{1}_{\underline{2}}5^{1}_{0}3^{2}_{0}4^{2}_{2}1^{3}_{2}2^{3}_{4}1^{4}_{6}$		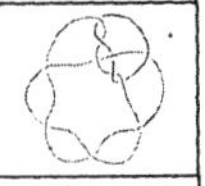
11^a_{209}: $1^{\underline{7}}_{\underline{16}}2^{\underline{6}}_{\underline{14}}1^{\underline{6}}_{\underline{12}}5^{\underline{5}}_{\underline{12}}2^{\underline{5}}_{\underline{10}}7^{\underline{4}}_{\underline{10}}5^{\underline{4}}_{\underline{8}}10^{\underline{3}}_{\underline{8}}7^{\underline{3}}_{\underline{6}}10^{\underline{2}}_{\underline{6}}10^{\underline{2}}_{\underline{4}}10^{\underline{1}}_{\underline{4}}10^{\underline{1}}_{\underline{2}}9^{0}_{\underline{2}}11^{0}_{0}5^{1}_{0}8^{1}_{2}4^{2}_{2}6^{2}_{4}1^{3}_{4}3^{3}_{6}1^{4}_{8}$ 11^a_{210}: $1^{\underline{5}}_{\underline{12}}3^{\underline{4}}_{\underline{10}}1^{\underline{4}}_{\underline{8}}6^{\underline{3}}_{\underline{8}}3^{\underline{3}}_{\underline{6}}9^{\underline{2}}_{\underline{6}}6^{\underline{2}}_{\underline{4}}11^{\underline{1}}_{\underline{4}}9^{\underline{1}}_{\underline{2}}12^{0}_{\underline{2}}12^{0}_{0}11^{1}_{0}11^{1}_{2}9^{2}_{2}12^{2}_{4}6^{3}_{4}8^{3}_{6}3^{4}_{6}6^{4}_{8}1^{5}_{8}3^{5}_{10}1^{6}_{12}$	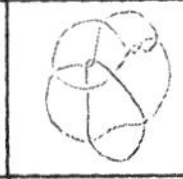	
11^a_{211}: $1^{\underline{8}}_{\underline{20}}2^{\underline{7}}_{\underline{18}}1^{\underline{7}}_{\underline{16}}5^{\underline{6}}_{\underline{16}}3^{\underline{6}}_{\underline{14}}8^{\underline{5}}_{\underline{14}}4^{\underline{5}}_{\underline{12}}9^{\underline{4}}_{\underline{12}}8^{\underline{4}}_{\underline{10}}10^{\underline{3}}_{\underline{10}}9^{\underline{3}}_{\underline{8}}10^{\underline{2}}_{\underline{8}}10^{\underline{2}}_{\underline{6}}7^{\underline{1}}_{\underline{6}}10^{\underline{1}}_{\underline{4}}6^{0}_{\underline{4}}8^{0}_{\underline{2}}3^{1}_{\underline{2}}5^{1}_{0}1^{2}_{0}3^{2}_{2}1^{3}_{4}$ 11^a_{212}: $1^{\underline{9}}_{\underline{24}}3^{\underline{8}}_{\underline{22}}1^{\underline{8}}_{\underline{20}}5^{\underline{7}}_{\underline{20}}3^{\underline{7}}_{\underline{18}}8^{\underline{6}}_{\underline{18}}6^{\underline{6}}_{\underline{16}}10^{\underline{5}}_{\underline{16}}7^{\underline{5}}_{\underline{14}}9^{\underline{4}}_{\underline{14}}10^{\underline{4}}_{\underline{12}}9^{\underline{3}}_{\underline{12}}9^{\underline{3}}_{\underline{10}}7^{\underline{2}}_{\underline{10}}9^{\underline{2}}_{\underline{8}}4^{\underline{1}}_{\underline{8}}7^{\underline{1}}_{\underline{6}}3^{0}_{\underline{6}}5^{0}_{\underline{4}}1^{1}_{\underline{4}}2^{1}_{\underline{2}}1^{2}_{0}$	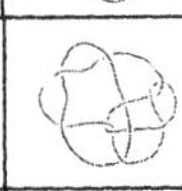	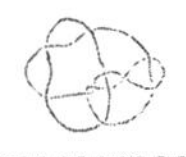
11^a_{213}: $1^{\underline{4}}_{\underline{6}}2^{\underline{3}}_{\underline{4}}1^{\underline{3}}_{\underline{2}}6^{\underline{2}}_{\underline{2}}3^{\underline{2}}_{0}7^{\underline{1}}_{0}5^{\underline{1}}_{2}11^{0}_{2}8^{0}_{4}10^{1}_{4}10^{1}_{6}11^{2}_{6}10^{2}_{8}8^{3}_{8}11^{3}_{10}6^{4}_{10}8^{4}_{12}3^{5}_{12}6^{5}_{14}1^{6}_{14}3^{6}_{16}1^{7}_{18}$ 11^a_{214}: $1^{\underline{6}}_{\underline{14}}1^{\underline{6}}_{\underline{12}}3^{\underline{5}}_{\underline{12}}5^{\underline{4}}_{\underline{10}}3^{\underline{4}}_{\underline{8}}8^{\underline{3}}_{\underline{8}}5^{\underline{3}}_{\underline{6}}11^{\underline{2}}_{\underline{6}}8^{\underline{2}}_{\underline{4}}10^{\underline{1}}_{\underline{4}}11^{\underline{1}}_{\underline{2}}12^{0}_{\underline{2}}11^{0}_{0}8^{1}_{0}11^{1}_{2}6^{2}_{2}8^{2}_{4}3^{3}_{4}6^{3}_{6}1^{4}_{6}3^{4}_{8}1^{5}_{10}$	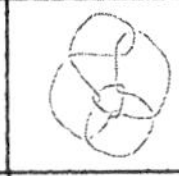	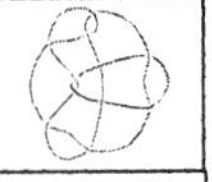
11^a_{215}: $1^{\underline{2}}_{0}2^{\underline{1}}_{2}1^{\underline{1}}_{4}6^{0}_{4}3^{0}_{6}6^{1}_{6}5^{1}_{8}10^{2}_{8}7^{2}_{10}9^{3}_{10}9^{3}_{12}9^{4}_{12}9^{4}_{14}7^{5}_{14}9^{5}_{16}5^{6}_{16}7^{6}_{18}2^{7}_{18}5^{7}_{20}1^{8}_{20}2^{8}_{22}1^{9}_{24}$ 11^a_{216}: $1^{\underline{11}}_{\underline{28}}3^{\underline{10}}_{\underline{26}}1^{\underline{10}}_{\underline{24}}5^{\underline{9}}_{\underline{24}}3^{\underline{9}}_{\underline{22}}8^{\underline{8}}_{\underline{22}}5^{\underline{8}}_{\underline{20}}10^{\underline{7}}_{\underline{20}}8^{\underline{7}}_{\underline{18}}10^{\underline{6}}_{\underline{18}}11^{\underline{6}}_{\underline{16}}11^{\underline{5}}_{\underline{16}}9^{\underline{5}}_{\underline{14}}7^{\underline{4}}_{\underline{14}}11^{\underline{4}}_{\underline{12}}5^{\underline{3}}_{\underline{12}}7^{\underline{3}}_{\underline{10}}3^{\underline{2}}_{\underline{10}}5^{\underline{2}}_{\underline{8}}3^{\underline{1}}_{\underline{6}}1^{0}_{\underline{6}}1^{0}_{\underline{4}}$	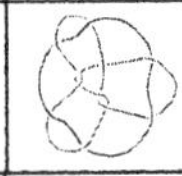	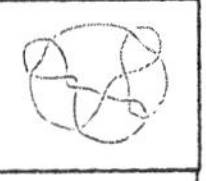
11^a_{217}: $1^{\underline{7}}_{\underline{16}}3^{\underline{6}}_{\underline{14}}1^{\underline{6}}_{\underline{12}}6^{\underline{5}}_{\underline{12}}3^{\underline{5}}_{\underline{10}}9^{\underline{4}}_{\underline{10}}6^{\underline{4}}_{\underline{8}}10^{\underline{3}}_{\underline{8}}9^{\underline{3}}_{\underline{6}}13^{\underline{2}}_{\underline{6}}11^{\underline{2}}_{\underline{4}}10^{\underline{1}}_{\underline{4}}12^{\underline{1}}_{\underline{2}}9^{0}_{\underline{2}}11^{0}_{0}6^{1}_{0}8^{1}_{2}2^{2}_{2}6^{2}_{4}1^{3}_{4}2^{3}_{6}1^{4}_{8}$ 11^a_{218}: $1^{\underline{2}}_{0}1^{\underline{1}}_{2}1^{\underline{1}}_{4}5^{0}_{4}2^{0}_{6}4^{1}_{6}4^{1}_{8}8^{2}_{8}5^{2}_{10}7^{3}_{10}7^{3}_{12}7^{4}_{12}7^{4}_{14}6^{5}_{14}7^{5}_{16}4^{6}_{16}6^{6}_{18}2^{7}_{18}4^{7}_{20}1^{8}_{20}2^{8}_{22}1^{9}_{24}$	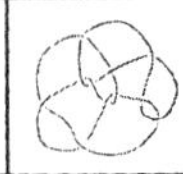	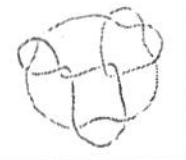
11^a_{219}: $1^{\underline{4}}_{\underline{6}}2^{\underline{3}}_{\underline{4}}1^{\underline{3}}_{\underline{2}}4^{\underline{2}}_{\underline{2}}2^{\underline{2}}_{0}6^{\underline{1}}_{0}4^{\underline{1}}_{2}8^{0}_{2}7^{0}_{4}8^{1}_{4}7^{1}_{6}9^{2}_{6}9^{2}_{8}6^{3}_{8}8^{3}_{10}5^{4}_{10}6^{4}_{12}2^{5}_{12}5^{5}_{14}1^{6}_{14}2^{6}_{16}1^{7}_{18}$ 11^a_{220}: $1^{\underline{7}}_{\underline{16}}2^{\underline{6}}_{\underline{14}}1^{\underline{6}}_{\underline{12}}6^{\underline{5}}_{\underline{12}}2^{\underline{5}}_{\underline{10}}8^{\underline{4}}_{\underline{10}}6^{\underline{4}}_{\underline{8}}10^{\underline{3}}_{\underline{8}}8^{\underline{3}}_{\underline{6}}13^{\underline{2}}_{\underline{6}}11^{\underline{2}}_{\underline{4}}10^{\underline{1}}_{\underline{4}}12^{\underline{1}}_{\underline{2}}10^{0}_{\underline{2}}11^{0}_{0}6^{1}_{0}9^{1}_{2}3^{2}_{2}6^{2}_{4}1^{3}_{4}3^{3}_{6}1^{4}_{8}$	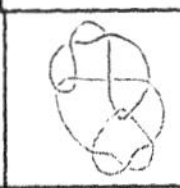	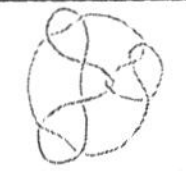
11^a_{221}: $1^{\underline{7}}_{\underline{16}}2^{\underline{6}}_{\underline{14}}1^{\underline{6}}_{\underline{12}}4^{\underline{5}}_{\underline{12}}2^{\underline{5}}_{\underline{10}}6^{\underline{4}}_{\underline{10}}4^{\underline{4}}_{\underline{8}}7^{\underline{3}}_{\underline{8}}6^{\underline{3}}_{\underline{6}}9^{\underline{2}}_{\underline{6}}8^{\underline{2}}_{\underline{4}}7^{\underline{1}}_{\underline{4}}8^{\underline{1}}_{\underline{2}}7^{0}_{\underline{2}}8^{0}_{0}4^{1}_{0}6^{1}_{2}2^{2}_{2}4^{2}_{4}1^{3}_{4}2^{3}_{6}1^{4}_{8}$ 11^a_{222}: $1^{\underline{2}}_{\underline{2}}1^{\underline{1}}_{0}1^{\underline{1}}_{2}5^{0}_{2}2^{0}_{4}5^{1}_{4}4^{1}_{6}8^{2}_{6}6^{2}_{8}8^{3}_{8}7^{3}_{10}7^{4}_{10}8^{4}_{12}6^{5}_{12}7^{5}_{14}4^{6}_{14}6^{6}_{16}2^{7}_{16}4^{7}_{18}1^{8}_{18}2^{8}_{20}1^{9}_{22}$	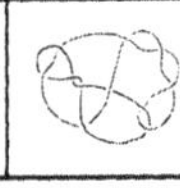	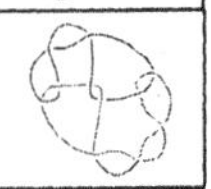

L: KH	1st line	2nd line
11^a_{223}: $1^{\underline{4}}_{\underline{8}}3^{\underline{3}}_{\underline{6}}1^{\underline{3}}_{\underline{4}}7^{\underline{2}}_{\underline{4}}3^{\underline{2}}_{\underline{2}}9^{\underline{1}}_{\underline{2}}7^{\underline{1}}_{0}13^{0}_{0}10^{0}_{2}13^{1}_{2}12^{1}_{4}12^{2}_{4}14^{2}_{6}10^{3}_{6}11^{3}_{8}6^{4}_{8}10^{4}_{10}3^{5}_{10}6^{5}_{12}1^{6}_{12}3^{6}_{14}1^{7}_{16}$ 11^a_{224}: $1^{\underline{4}}_{\underline{6}}2^{\underline{3}}_{\underline{4}}1^{\underline{3}}_{\underline{2}}5^{\underline{2}}_{\underline{2}}3^{\underline{2}}_{0}6^{\underline{1}}_{0}4^{\underline{1}}_{2}9^{0}_{2}7^{0}_{4}8^{1}_{4}8^{1}_{6}9^{2}_{6}8^{2}_{8}6^{3}_{8}9^{3}_{10}5^{4}_{10}6^{4}_{12}2^{5}_{12}5^{5}_{14}1^{6}_{14}2^{6}_{16}1^{7}_{18}$	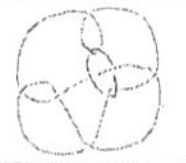	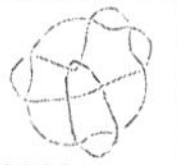
11^a_{225}: $1^{\underline{4}}_{\underline{8}}2^{\underline{3}}_{\underline{6}}1^{\underline{3}}_{\underline{4}}5^{\underline{2}}_{\underline{4}}3^{\underline{2}}_{\underline{2}}7^{\underline{1}}_{\underline{2}}4^{\underline{1}}_{0}9^{0}_{0}8^{0}_{2}9^{1}_{2}8^{1}_{4}9^{2}_{4}9^{2}_{6}6^{3}_{6}9^{3}_{8}5^{4}_{8}6^{4}_{10}2^{5}_{10}5^{5}_{12}1^{6}_{12}2^{6}_{14}1^{7}_{16}$ 11^a_{226}: $1^{\underline{11}}_{\underline{28}}3^{\underline{10}}_{\underline{26}}1^{\underline{10}}_{\underline{24}}7^{\underline{9}}_{\underline{24}}3^{\underline{9}}_{\underline{22}}10^{\underline{8}}_{\underline{22}}7^{\underline{8}}_{\underline{20}}13^{\underline{7}}_{\underline{20}}10^{\underline{7}}_{\underline{18}}14^{\underline{6}}_{\underline{18}}14^{\underline{6}}_{\underline{16}}14^{\underline{5}}_{\underline{16}}13^{\underline{5}}_{\underline{14}}10^{\underline{4}}_{\underline{14}}14^{\underline{4}}_{\underline{12}}7^{\underline{3}}_{\underline{12}}10^{\underline{3}}_{\underline{10}}4^{\underline{2}}_{\underline{10}}7^{\underline{2}}_{\underline{8}}4^{\underline{1}}_{\underline{6}}1^{0}_{\underline{6}}1^{0}_{\underline{4}}$	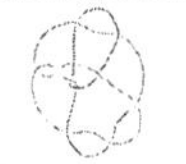	
11^a_{227}: $1^{\underline{8}}_{\underline{20}}2^{\underline{7}}_{\underline{18}}1^{\underline{7}}_{\underline{16}}6^{\underline{6}}_{\underline{16}}3^{\underline{6}}_{\underline{14}}9^{\underline{5}}_{\underline{14}}5^{\underline{5}}_{\underline{12}}11^{\underline{4}}_{\underline{12}}9^{\underline{4}}_{\underline{10}}12^{\underline{3}}_{\underline{10}}11^{\underline{3}}_{\underline{8}}12^{\underline{2}}_{\underline{8}}12^{\underline{2}}_{\underline{6}}9^{\underline{1}}_{\underline{6}}12^{\underline{1}}_{\underline{4}}7^{0}_{\underline{4}}10^{0}_{\underline{2}}4^{1}_{\underline{2}}6^{1}_{0}1^{2}_{0}4^{2}_{2}1^{3}_{4}$ 11^a_{228}: $1^{\underline{4}}_{\underline{8}}2^{\underline{3}}_{\underline{6}}1^{\underline{3}}_{\underline{4}}6^{\underline{2}}_{\underline{4}}3^{\underline{2}}_{\underline{2}}8^{\underline{1}}_{\underline{2}}5^{\underline{1}}_{0}11^{0}_{0}9^{0}_{2}11^{1}_{2}10^{1}_{4}11^{2}_{4}11^{2}_{6}8^{3}_{6}11^{3}_{8}6^{4}_{8}8^{4}_{10}3^{5}_{10}6^{5}_{12}1^{6}_{12}3^{6}_{14}1^{7}_{16}$		
11^a_{229}: $1^{\underline{6}}_{\underline{14}}1^{\underline{6}}_{\underline{12}}3^{\underline{5}}_{\underline{12}}5^{\underline{4}}_{\underline{10}}3^{\underline{4}}_{\underline{8}}8^{\underline{3}}_{\underline{8}}5^{\underline{3}}_{\underline{6}}10^{\underline{2}}_{\underline{6}}8^{\underline{2}}_{\underline{4}}10^{\underline{1}}_{\underline{4}}10^{\underline{1}}_{\underline{2}}11^{0}_{\underline{2}}11^{0}_{0}7^{1}_{0}10^{1}_{2}6^{2}_{2}7^{2}_{4}2^{3}_{4}6^{3}_{6}1^{4}_{6}2^{4}_{8}1^{5}_{10}$ 11^a_{230}: $1^{\underline{8}}_{\underline{20}}2^{\underline{7}}_{\underline{18}}1^{\underline{7}}_{\underline{16}}5^{\underline{6}}_{\underline{16}}3^{\underline{6}}_{\underline{14}}8^{\underline{5}}_{\underline{14}}4^{\underline{5}}_{\underline{12}}8^{\underline{4}}_{\underline{12}}8^{\underline{4}}_{\underline{10}}10^{\underline{3}}_{\underline{10}}8^{\underline{3}}_{\underline{8}}9^{\underline{2}}_{\underline{8}}10^{\underline{2}}_{\underline{6}}6^{\underline{1}}_{\underline{6}}9^{\underline{1}}_{\underline{4}}6^{0}_{\underline{4}}7^{0}_{\underline{2}}2^{1}_{\underline{2}}5^{1}_{0}1^{2}_{0}2^{2}_{2}1^{3}_{4}$	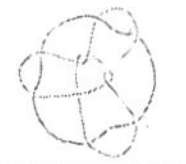	
11^a_{231}: $1^{\underline{9}}_{\underline{22}}2^{\underline{8}}_{\underline{20}}1^{\underline{8}}_{\underline{18}}5^{\underline{7}}_{\underline{18}}2^{\underline{7}}_{\underline{16}}8^{\underline{6}}_{\underline{16}}6^{\underline{6}}_{\underline{14}}10^{\underline{5}}_{\underline{14}}7^{\underline{5}}_{\underline{12}}11^{\underline{4}}_{\underline{12}}10^{\underline{4}}_{\underline{10}}10^{\underline{3}}_{\underline{10}}11^{\underline{3}}_{\underline{8}}9^{\underline{2}}_{\underline{8}}10^{\underline{2}}_{\underline{6}}5^{\underline{1}}_{\underline{6}}9^{\underline{1}}_{\underline{4}}4^{0}_{\underline{4}}6^{0}_{\underline{2}}1^{1}_{\underline{2}}3^{1}_{0}1^{2}_{2}$ 11^a_{232}: $1^{\underline{5}}_{\underline{10}}4^{\underline{4}}_{\underline{8}}1^{\underline{4}}_{\underline{6}}7^{\underline{3}}_{\underline{6}}4^{\underline{3}}_{\underline{4}}11^{\underline{2}}_{\underline{4}}7^{\underline{2}}_{\underline{2}}13^{\underline{1}}_{\underline{2}}11^{\underline{1}}_{0}15^{0}_{0}14^{0}_{2}13^{1}_{2}14^{1}_{4}11^{2}_{4}14^{2}_{6}7^{3}_{6}10^{3}_{8}3^{4}_{8}7^{4}_{10}1^{5}_{10}3^{5}_{12}1^{6}_{14}$		
11^a_{233}: $1^{\underline{2}}_{\underline{2}}3^{\underline{1}}_{0}1^{\underline{1}}_{2}7^{0}_{2}4^{0}_{4}10^{1}_{4}6^{1}_{6}13^{2}_{6}11^{2}_{8}13^{3}_{8}12^{3}_{10}13^{4}_{10}13^{4}_{12}9^{5}_{12}13^{5}_{14}7^{6}_{14}9^{6}_{16}3^{7}_{16}7^{7}_{18}1^{8}_{18}3^{8}_{20}1^{9}_{22}$ 11^a_{234}: $1^{\underline{4}}_{\underline{6}}3^{\underline{3}}_{\underline{4}}1^{\underline{3}}_{\underline{2}}5^{\underline{2}}_{\underline{2}}3^{\underline{2}}_{0}7^{\underline{1}}_{0}5^{\underline{1}}_{2}10^{0}_{2}8^{0}_{4}9^{1}_{4}9^{1}_{6}10^{2}_{6}10^{2}_{8}7^{3}_{8}9^{3}_{10}5^{4}_{10}7^{4}_{12}2^{5}_{12}5^{5}_{14}1^{6}_{14}2^{6}_{16}1^{7}_{18}$	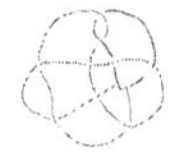	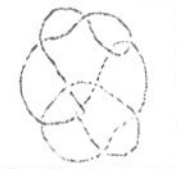
11^a_{235}: $1^{\underline{11}}_{\underline{28}}2^{\underline{10}}_{\underline{26}}1^{\underline{10}}_{\underline{24}}5^{\underline{9}}_{\underline{24}}2^{\underline{9}}_{\underline{22}}6^{\underline{8}}_{\underline{22}}5^{\underline{8}}_{\underline{20}}9^{\underline{7}}_{\underline{20}}6^{\underline{7}}_{\underline{18}}9^{\underline{6}}_{\underline{18}}10^{\underline{6}}_{\underline{16}}9^{\underline{5}}_{\underline{16}}8^{\underline{5}}_{\underline{14}}7^{\underline{4}}_{\underline{14}}9^{\underline{4}}_{\underline{12}}4^{\underline{3}}_{\underline{12}}7^{\underline{3}}_{\underline{10}}3^{\underline{2}}_{\underline{10}}4^{\underline{2}}_{\underline{8}}3^{\underline{1}}_{\underline{6}}1^{0}_{\underline{6}}1^{0}_{\underline{4}}$ 11^a_{236}: $1^{\underline{11}}_{\underline{26}}2^{\underline{10}}_{\underline{24}}1^{\underline{10}}_{\underline{22}}4^{\underline{9}}_{\underline{22}}2^{\underline{9}}_{\underline{20}}5^{\underline{8}}_{\underline{20}}4^{\underline{8}}_{\underline{18}}7^{\underline{7}}_{\underline{18}}5^{\underline{7}}_{\underline{16}}7^{\underline{6}}_{\underline{16}}8^{\underline{6}}_{\underline{14}}7^{\underline{5}}_{\underline{14}}6^{\underline{5}}_{\underline{12}}6^{\underline{4}}_{\underline{12}}7^{\underline{4}}_{\underline{10}}3^{\underline{3}}_{\underline{10}}6^{\underline{3}}_{\underline{8}}3^{\underline{2}}_{\underline{8}}3^{\underline{2}}_{\underline{6}}3^{\underline{1}}_{\underline{4}}1^{0}_{\underline{4}}1^{0}_{\underline{2}}$	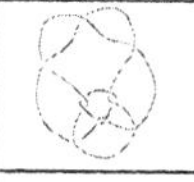	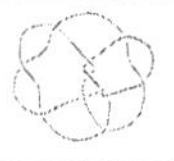
11^a_{237}: $1^{\underline{5}}_{\underline{10}}2^{\underline{4}}_{\underline{8}}1^{\underline{4}}_{\underline{6}}6^{\underline{3}}_{\underline{6}}2^{\underline{3}}_{\underline{4}}9^{\underline{2}}_{\underline{4}}6^{\underline{2}}_{\underline{2}}11^{\underline{1}}_{\underline{2}}9^{\underline{1}}_{0}14^{0}_{0}12^{0}_{2}12^{1}_{2}13^{1}_{4}11^{2}_{4}13^{2}_{6}7^{3}_{6}10^{3}_{8}4^{4}_{8}7^{4}_{10}1^{5}_{10}4^{5}_{12}1^{6}_{14}$ 11^a_{238}: $1^{\underline{4}}_{\underline{6}}2^{\underline{3}}_{\underline{4}}1^{\underline{3}}_{\underline{2}}6^{\underline{2}}_{\underline{2}}2^{\underline{2}}_{0}7^{\underline{1}}_{0}6^{\underline{1}}_{2}11^{0}_{2}8^{0}_{4}11^{1}_{4}10^{1}_{6}11^{2}_{6}12^{2}_{8}9^{3}_{8}10^{3}_{10}6^{4}_{10}9^{4}_{12}3^{5}_{12}6^{5}_{14}1^{6}_{14}3^{6}_{16}1^{7}_{18}$		
11^a_{239}: $1^{\underline{8}}_{\underline{18}}2^{\underline{7}}_{\underline{16}}1^{\underline{7}}_{\underline{14}}6^{\underline{6}}_{\underline{14}}3^{\underline{6}}_{\underline{12}}7^{\underline{5}}_{\underline{12}}5^{\underline{5}}_{\underline{10}}10^{\underline{4}}_{\underline{10}}7^{\underline{4}}_{\underline{8}}10^{\underline{3}}_{\underline{8}}10^{\underline{3}}_{\underline{6}}10^{\underline{2}}_{\underline{6}}10^{\underline{2}}_{\underline{4}}8^{\underline{1}}_{\underline{4}}10^{\underline{1}}_{\underline{2}}6^{0}_{\underline{2}}9^{0}_{0}3^{1}_{0}5^{1}_{2}1^{2}_{2}3^{2}_{4}1^{3}_{6}$ 11^a_{240}: $1^{\underline{6}}_{\underline{12}}1^{\underline{6}}_{\underline{10}}2^{\underline{5}}_{\underline{10}}5^{\underline{4}}_{\underline{8}}2^{\underline{4}}_{\underline{6}}5^{\underline{3}}_{\underline{6}}5^{\underline{3}}_{\underline{4}}10^{\underline{2}}_{\underline{4}}5^{\underline{2}}_{\underline{2}}8^{\underline{1}}_{\underline{2}}10^{\underline{1}}_{0}10^{0}_{0}9^{0}_{2}8^{1}_{2}9^{1}_{4}5^{2}_{4}8^{2}_{6}3^{3}_{6}5^{3}_{8}1^{4}_{8}3^{4}_{10}1^{5}_{12}$		
11^a_{241}: $1^{\underline{6}}_{\underline{14}}1^{\underline{6}}_{\underline{12}}2^{\underline{5}}_{\underline{12}}4^{\underline{4}}_{\underline{10}}2^{\underline{4}}_{\underline{8}}5^{\underline{3}}_{\underline{8}}4^{\underline{3}}_{\underline{6}}9^{\underline{2}}_{\underline{6}}5^{\underline{2}}_{\underline{4}}7^{\underline{1}}_{\underline{4}}9^{\underline{1}}_{\underline{2}}9^{0}_{\underline{2}}8^{0}_{0}7^{1}_{0}8^{1}_{2}4^{2}_{2}7^{2}_{4}3^{3}_{4}4^{3}_{6}1^{4}_{6}3^{4}_{8}1^{5}_{10}$ 11^a_{242}: $1^{\underline{2}}_{\underline{2}}1^{\underline{1}}_{0}1^{\underline{1}}_{2}5^{0}_{2}2^{0}_{4}6^{1}_{4}4^{1}_{6}9^{2}_{6}7^{2}_{8}9^{3}_{8}8^{3}_{10}9^{4}_{10}9^{4}_{12}7^{5}_{12}9^{5}_{14}5^{6}_{14}7^{6}_{16}3^{7}_{16}5^{7}_{18}1^{8}_{18}3^{8}_{20}1^{9}_{22}$		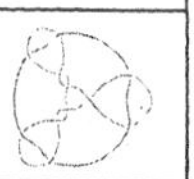
11^a_{243}: $1^{\underline{6}}_{\underline{14}}3^{\underline{5}}_{\underline{12}}1^{\underline{5}}_{\underline{10}}6^{\underline{4}}_{\underline{10}}3^{\underline{4}}_{\underline{8}}10^{\underline{3}}_{\underline{8}}6^{\underline{3}}_{\underline{6}}11^{\underline{2}}_{\underline{6}}10^{\underline{2}}_{\underline{4}}13^{\underline{1}}_{\underline{4}}11^{\underline{1}}_{\underline{2}}12^{0}_{\underline{2}}14^{0}_{0}9^{1}_{0}11^{1}_{2}7^{2}_{2}10^{2}_{4}3^{3}_{4}6^{3}_{6}1^{4}_{6}3^{4}_{8}1^{5}_{10}$ 11^a_{244}: $1^{\underline{6}}_{\underline{14}}1^{\underline{6}}_{\underline{12}}4^{\underline{5}}_{\underline{12}}7^{\underline{4}}_{\underline{10}}4^{\underline{4}}_{\underline{8}}10^{\underline{3}}_{\underline{8}}7^{\underline{3}}_{\underline{6}}15^{\underline{2}}_{\underline{6}}10^{\underline{2}}_{\underline{4}}13^{\underline{1}}_{\underline{4}}15^{\underline{1}}_{\underline{2}}15^{0}_{\underline{2}}14^{0}_{0}11^{1}_{0}14^{1}_{2}7^{2}_{2}11^{2}_{4}4^{3}_{4}7^{3}_{6}1^{4}_{6}4^{4}_{8}1^{5}_{10}$	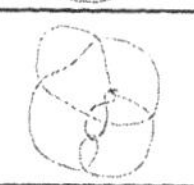	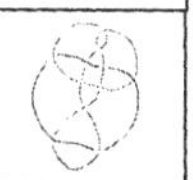
11^a_{245}: $1^{\underline{9}}_{\underline{22}}4^{\underline{8}}_{\underline{20}}1^{\underline{8}}_{\underline{18}}7^{\underline{7}}_{\underline{18}}4^{\underline{7}}_{\underline{16}}11^{\underline{6}}_{\underline{16}}8^{\underline{6}}_{\underline{14}}14^{\underline{5}}_{\underline{14}}10^{\underline{5}}_{\underline{12}}14^{\underline{4}}_{\underline{12}}14^{\underline{4}}_{\underline{10}}13^{\underline{3}}_{\underline{10}}14^{\underline{3}}_{\underline{8}}11^{\underline{2}}_{\underline{8}}13^{\underline{2}}_{\underline{6}}6^{\underline{1}}_{\underline{6}}11^{\underline{1}}_{\underline{4}}4^{0}_{\underline{4}}7^{0}_{\underline{2}}1^{1}_{\underline{2}}3^{1}_{0}1^{2}_{2}$ 11^a_{246}: $1^{\underline{7}}_{\underline{16}}3^{\underline{6}}_{\underline{14}}1^{\underline{6}}_{\underline{12}}5^{\underline{5}}_{\underline{12}}3^{\underline{5}}_{\underline{10}}10^{\underline{4}}_{\underline{10}}6^{\underline{4}}_{\underline{8}}12^{\underline{3}}_{\underline{8}}9^{\underline{3}}_{\underline{6}}13^{\underline{2}}_{\underline{6}}12^{\underline{2}}_{\underline{4}}12^{\underline{1}}_{\underline{4}}13^{\underline{1}}_{\underline{2}}11^{0}_{\underline{2}}13^{0}_{0}7^{1}_{0}10^{1}_{2}4^{2}_{2}7^{2}_{4}1^{3}_{4}4^{3}_{6}1^{4}_{8}$		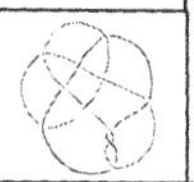
11^a_{247}: $1^{\underline{5}}_{\underline{10}}3^{\underline{4}}_{\underline{8}}1^{\underline{4}}_{\underline{6}}5^{\underline{3}}_{\underline{6}}3^{\underline{3}}_{\underline{4}}8^{\underline{2}}_{\underline{4}}5^{\underline{2}}_{\underline{2}}9^{\underline{1}}_{\underline{2}}8^{\underline{1}}_{0}12^{0}_{0}11^{0}_{2}10^{1}_{2}10^{1}_{4}8^{2}_{4}10^{2}_{6}5^{3}_{6}8^{3}_{8}3^{4}_{8}5^{4}_{10}1^{5}_{10}3^{5}_{12}1^{6}_{14}$ 11^a_{248}: $1^{\underline{5}}_{\underline{10}}3^{\underline{4}}_{\underline{8}}1^{\underline{4}}_{\underline{6}}5^{\underline{3}}_{\underline{6}}3^{\underline{3}}_{\underline{4}}9^{\underline{2}}_{\underline{4}}5^{\underline{2}}_{\underline{2}}10^{\underline{1}}_{\underline{2}}9^{\underline{1}}_{0}13^{0}_{0}12^{0}_{2}12^{1}_{2}11^{1}_{4}9^{2}_{4}12^{2}_{6}6^{3}_{6}9^{3}_{8}4^{4}_{8}6^{4}_{10}1^{5}_{10}4^{5}_{12}1^{6}_{14}$	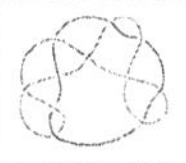	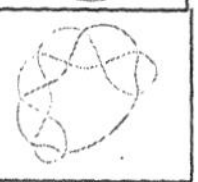
11^a_{249}: $1^{\underline{5}}_{\underline{10}}4^{\underline{4}}_{\underline{8}}1^{\underline{4}}_{\underline{6}}7^{\underline{3}}_{\underline{6}}4^{\underline{3}}_{\underline{4}}11^{\underline{2}}_{\underline{4}}7^{\underline{2}}_{\underline{2}}13^{\underline{1}}_{\underline{2}}11^{\underline{1}}_{0}16^{0}_{0}15^{0}_{2}14^{1}_{2}14^{1}_{4}11^{2}_{4}14^{2}_{6}7^{3}_{6}11^{3}_{8}4^{4}_{8}7^{4}_{10}1^{5}_{10}4^{5}_{12}1^{6}_{14}$ 11^a_{250}: $1^{\underline{7}}_{\underline{18}}3^{\underline{6}}_{\underline{16}}1^{\underline{6}}_{\underline{14}}4^{\underline{5}}_{\underline{14}}3^{\underline{5}}_{\underline{12}}7^{\underline{4}}_{\underline{12}}5^{\underline{4}}_{\underline{10}}9^{\underline{3}}_{\underline{10}}6^{\underline{3}}_{\underline{8}}8^{\underline{2}}_{\underline{8}}9^{\underline{2}}_{\underline{6}}8^{\underline{1}}_{\underline{6}}8^{\underline{1}}_{\underline{4}}7^{0}_{\underline{4}}9^{0}_{\underline{2}}4^{1}_{\underline{2}}6^{1}_{0}3^{2}_{0}4^{2}_{2}1^{3}_{2}3^{3}_{4}1^{4}_{6}$	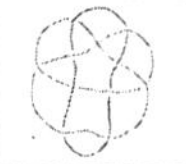	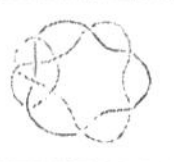
11^a_{251}: $1^{\underline{5}}_{\underline{10}}3^{\underline{4}}_{\underline{8}}1^{\underline{4}}_{\underline{6}}5^{\underline{3}}_{\underline{6}}3^{\underline{3}}_{\underline{4}}8^{\underline{2}}_{\underline{4}}5^{\underline{2}}_{\underline{2}}9^{\underline{1}}_{\underline{2}}8^{\underline{1}}_{0}12^{0}_{0}11^{0}_{2}10^{1}_{2}10^{1}_{4}8^{2}_{4}10^{2}_{6}5^{3}_{6}8^{3}_{8}3^{4}_{8}5^{4}_{10}1^{5}_{10}3^{5}_{12}1^{6}_{14}$ 11^a_{252}: $1^{\underline{7}}_{\underline{18}}3^{\underline{6}}_{\underline{16}}1^{\underline{6}}_{\underline{14}}5^{\underline{5}}_{\underline{14}}3^{\underline{5}}_{\underline{12}}8^{\underline{4}}_{\underline{12}}6^{\underline{4}}_{\underline{10}}10^{\underline{3}}_{\underline{10}}7^{\underline{3}}_{\underline{8}}10^{\underline{2}}_{\underline{8}}10^{\underline{2}}_{\underline{6}}9^{\underline{1}}_{\underline{6}}10^{\underline{1}}_{\underline{4}}8^{0}_{\underline{4}}10^{0}_{\underline{2}}5^{1}_{\underline{2}}7^{1}_{0}3^{2}_{0}5^{2}_{2}1^{3}_{2}3^{3}_{4}1^{4}_{6}$		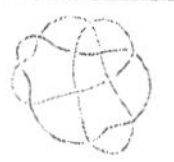

L: *KH*	1st line	2nd line
11^a_{253}: $1^{\underline{7}}_{\underline{18}}2^{\underline{6}}_{\underline{16}}1^{\underline{6}}_{\underline{14}}3^{\underline{5}}_{\underline{14}}2^{\underline{5}}_{\underline{12}}5^{\underline{4}}_{\underline{12}}4^{\underline{4}}_{\underline{10}}6^{\underline{3}}_{\underline{10}}4^{\underline{3}}_{\underline{8}}6^{\underline{2}}_{\underline{8}}6^{\underline{2}}_{\underline{6}}5^{\underline{1}}_{\underline{6}}6^{\underline{1}}_{\underline{4}}5^{0}_{\underline{4}}6^{0}_{\underline{2}}3^{1}_{\underline{2}}4^{1}_{0}2^{2}_{0}3^{2}_{2}1^{3}_{2}2^{3}_{4}1^{4}_{6}$ 11^a_{254}: $1^{\underline{8}}_{\underline{22}}2^{\underline{7}}_{\underline{20}}1^{\underline{7}}_{\underline{18}}3^{\underline{6}}_{\underline{18}}2^{\underline{6}}_{\underline{16}}4^{\underline{5}}_{\underline{16}}3^{\underline{5}}_{\underline{14}}5^{\underline{4}}_{\underline{14}}5^{\underline{4}}_{\underline{12}}6^{\underline{3}}_{\underline{12}}4^{\underline{3}}_{\underline{10}}4^{\underline{2}}_{\underline{10}}6^{\underline{2}}_{\underline{8}}4^{\underline{1}}_{\underline{8}}4^{\underline{1}}_{\underline{6}}3^{0}_{\underline{6}}5^{0}_{\underline{4}}2^{1}_{\underline{4}}2^{1}_{\underline{2}}1^{2}_{\underline{2}}2^{2}_{0}1^{3}_{2}$	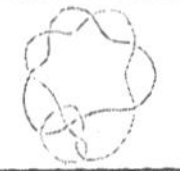	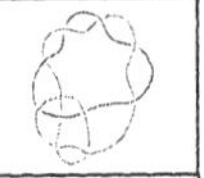
11^a_{255}: $1^{\underline{5}}_{\underline{10}}2^{\underline{4}}_{\underline{8}}1^{\underline{4}}_{\underline{6}}4^{\underline{3}}_{\underline{6}}2^{\underline{3}}_{\underline{4}}6^{\underline{2}}_{\underline{4}}4^{\underline{2}}_{\underline{2}}6^{\underline{1}}_{\underline{2}}6^{\underline{1}}_{0}10^{0}_{0}8^{0}_{2}7^{1}_{2}8^{1}_{4}6^{2}_{4}7^{2}_{6}4^{3}_{6}6^{3}_{8}2^{4}_{8}4^{4}_{10}1^{5}_{10}2^{5}_{12}1^{6}_{14}$ 11^a_{256}: $1^{\underline{7}}_{\underline{16}}2^{\underline{6}}_{\underline{14}}1^{\underline{6}}_{\underline{12}}4^{\underline{5}}_{\underline{12}}2^{\underline{5}}_{\underline{10}}8^{\underline{4}}_{\underline{10}}5^{\underline{4}}_{\underline{8}}9^{\underline{3}}_{\underline{8}}7^{\underline{3}}_{\underline{6}}11^{\underline{2}}_{\underline{6}}9^{\underline{2}}_{\underline{4}}9^{\underline{1}}_{\underline{4}}11^{\underline{1}}_{\underline{2}}9^{0}_{\underline{2}}10^{0}_{0}6^{1}_{0}8^{1}_{2}3^{2}_{2}6^{2}_{4}1^{3}_{4}3^{3}_{6}1^{4}_{8}$	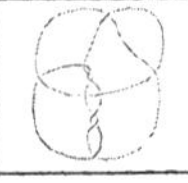	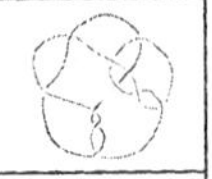
11^a_{257}: $1^{\underline{8}}_{\underline{20}}2^{\underline{7}}_{\underline{18}}1^{\underline{7}}_{\underline{16}}4^{\underline{6}}_{\underline{16}}2^{\underline{6}}_{\underline{14}}7^{\underline{5}}_{\underline{14}}4^{\underline{5}}_{\underline{12}}9^{\underline{4}}_{\underline{12}}8^{\underline{4}}_{\underline{10}}10^{\underline{3}}_{\underline{10}}8^{\underline{3}}_{\underline{8}}9^{\underline{2}}_{\underline{8}}10^{\underline{2}}_{\underline{6}}7^{\underline{1}}_{\underline{6}}9^{\underline{1}}_{\underline{4}}6^{0}_{\underline{4}}8^{0}_{\underline{2}}3^{1}_{\underline{2}}5^{1}_{0}1^{2}_{0}3^{2}_{2}1^{3}_{4}$ 11^a_{258}: $1^{\underline{8}}_{\underline{20}}1^{\underline{8}}_{\underline{18}}2^{\underline{7}}_{\underline{18}}3^{\underline{6}}_{\underline{16}}2^{\underline{6}}_{\underline{14}}5^{\underline{5}}_{\underline{14}}3^{\underline{5}}_{\underline{12}}7^{\underline{4}}_{\underline{12}}5^{\underline{4}}_{\underline{10}}7^{\underline{3}}_{\underline{10}}7^{\underline{3}}_{\underline{8}}7^{\underline{2}}_{\underline{8}}7^{\underline{2}}_{\underline{6}}5^{\underline{1}}_{\underline{6}}7^{\underline{1}}_{\underline{4}}5^{0}_{\underline{4}}6^{0}_{\underline{2}}2^{1}_{\underline{2}}4^{1}_{0}1^{2}_{0}2^{2}_{2}1^{3}_{4}$	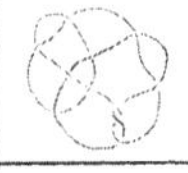	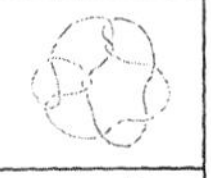
11^a_{259}: $1^{\underline{9}}_{\underline{24}}1^{\underline{8}}_{\underline{22}}1^{\underline{8}}_{\underline{20}}3^{\underline{7}}_{\underline{20}}1^{\underline{7}}_{\underline{18}}4^{\underline{6}}_{\underline{18}}3^{\underline{6}}_{\underline{16}}5^{\underline{5}}_{\underline{16}}4^{\underline{5}}_{\underline{14}}6^{\underline{4}}_{\underline{14}}6^{\underline{4}}_{\underline{12}}5^{\underline{3}}_{\underline{12}}5^{\underline{3}}_{\underline{10}}4^{\underline{2}}_{\underline{10}}5^{\underline{2}}_{\underline{8}}3^{\underline{1}}_{\underline{8}}4^{\underline{1}}_{\underline{6}}2^{0}_{\underline{6}}4^{0}_{\underline{4}}1^{1}_{\underline{4}}1^{1}_{\underline{2}}1^{2}_{0}$ 11^a_{260}: $1^{\underline{9}}_{\underline{22}}1^{\underline{8}}_{\underline{20}}1^{\underline{8}}_{\underline{18}}3^{\underline{7}}_{\underline{18}}1^{\underline{7}}_{\underline{16}}5^{\underline{6}}_{\underline{16}}3^{\underline{6}}_{\underline{14}}5^{\underline{5}}_{\underline{14}}5^{\underline{5}}_{\underline{12}}7^{\underline{4}}_{\underline{12}}6^{\underline{4}}_{\underline{10}}6^{\underline{3}}_{\underline{10}}6^{\underline{3}}_{\underline{8}}5^{\underline{2}}_{\underline{8}}6^{\underline{2}}_{\underline{6}}3^{\underline{1}}_{\underline{6}}5^{\underline{1}}_{\underline{4}}2^{0}_{\underline{4}}4^{0}_{\underline{2}}1^{1}_{\underline{2}}1^{1}_{0}1^{2}_{2}$	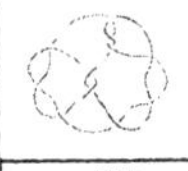	
11^a_{261}: $1^{\underline{7}}_{\underline{16}}2^{\underline{6}}_{\underline{14}}1^{\underline{6}}_{\underline{12}}5^{\underline{5}}_{\underline{12}}2^{\underline{5}}_{\underline{10}}7^{\underline{4}}_{\underline{10}}5^{\underline{4}}_{\underline{8}}10^{\underline{3}}_{\underline{8}}7^{\underline{3}}_{\underline{6}}11^{\underline{2}}_{\underline{6}}10^{\underline{2}}_{\underline{4}}9^{\underline{1}}_{\underline{4}}11^{\underline{1}}_{\underline{2}}10^{0}_{\underline{2}}11^{0}_{0}6^{1}_{0}8^{1}_{2}3^{2}_{2}6^{2}_{4}1^{3}_{4}3^{3}_{6}1^{4}_{8}$ 11^a_{262}: $1^{\underline{5}}_{\underline{10}}2^{\underline{4}}_{\underline{8}}1^{\underline{4}}_{\underline{6}}4^{\underline{3}}_{\underline{6}}2^{\underline{3}}_{\underline{4}}7^{\underline{2}}_{\underline{4}}4^{\underline{2}}_{\underline{2}}7^{\underline{1}}_{\underline{2}}7^{\underline{1}}_{0}11^{0}_{0}9^{0}_{2}9^{1}_{2}9^{1}_{4}7^{2}_{4}9^{2}_{6}5^{3}_{6}7^{3}_{8}3^{4}_{8}5^{4}_{10}1^{5}_{10}3^{5}_{12}1^{6}_{14}$		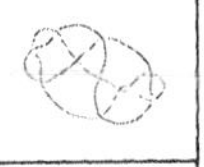
11^a_{263}: $1^{\underline{2}}_{0}1^{\underline{1}}_{2}1^{\underline{1}}_{4}4^{0}_{4}2^{0}_{6}4^{1}_{6}3^{1}_{8}6^{2}_{8}4^{2}_{10}5^{3}_{10}6^{3}_{12}7^{4}_{12}6^{4}_{14}5^{5}_{14}6^{5}_{16}3^{6}_{16}5^{6}_{18}2^{7}_{18}3^{7}_{20}1^{8}_{20}2^{8}_{22}1^{9}_{24}$ 11^a_{264}: $1^{\underline{4}}_{\underline{6}}2^{\underline{3}}_{\underline{4}}1^{\underline{3}}_{\underline{2}}4^{\underline{2}}_{\underline{2}}2^{\underline{2}}_{0}5^{\underline{1}}_{0}4^{\underline{1}}_{2}7^{0}_{2}6^{0}_{4}8^{1}_{4}6^{1}_{6}7^{2}_{6}8^{2}_{8}5^{3}_{8}7^{3}_{10}5^{4}_{10}6^{4}_{12}2^{5}_{12}4^{5}_{14}1^{6}_{14}2^{6}_{16}1^{7}_{18}$	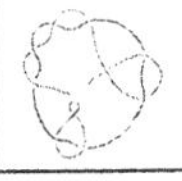	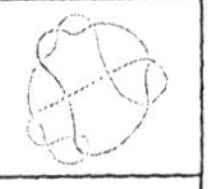
11^a_{265}: $1^{\underline{2}}_{\underline{2}}2^{\underline{1}}_{0}1^{\underline{1}}_{2}5^{0}_{2}3^{0}_{4}7^{1}_{4}4^{1}_{6}9^{2}_{6}7^{2}_{8}8^{3}_{8}9^{3}_{10}9^{4}_{10}9^{4}_{12}7^{5}_{12}8^{5}_{14}4^{6}_{14}7^{6}_{16}2^{7}_{16}4^{7}_{18}1^{8}_{18}2^{8}_{20}1^{9}_{22}$ 11^a_{266}: $1^{\underline{2}}_{\underline{2}}2^{\underline{1}}_{0}1^{\underline{1}}_{2}5^{0}_{2}3^{0}_{4}7^{1}_{4}4^{1}_{6}9^{2}_{6}7^{2}_{8}9^{3}_{8}9^{3}_{10}10^{4}_{10}10^{4}_{12}7^{5}_{12}9^{5}_{14}5^{6}_{14}7^{6}_{16}3^{7}_{16}5^{7}_{18}1^{8}_{18}3^{8}_{20}1^{9}_{22}$		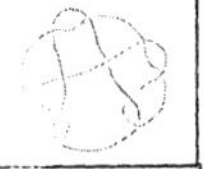
11^a_{267}: $1^{\underline{6}}_{\underline{12}}2^{\underline{5}}_{\underline{10}}1^{\underline{5}}_{\underline{8}}5^{\underline{4}}_{\underline{8}}3^{\underline{4}}_{\underline{6}}6^{\underline{3}}_{\underline{6}}4^{\underline{3}}_{\underline{4}}9^{\underline{2}}_{\underline{4}}6^{\underline{2}}_{\underline{2}}7^{\underline{1}}_{\underline{2}}9^{\underline{1}}_{0}10^{0}_{0}8^{0}_{2}7^{1}_{2}9^{1}_{4}4^{2}_{4}7^{2}_{6}3^{3}_{6}4^{3}_{8}1^{4}_{8}3^{4}_{10}1^{5}_{12}$ 11^a_{268}: $1^{\underline{7}}_{\underline{16}}3^{\underline{6}}_{\underline{14}}1^{\underline{6}}_{\underline{12}}5^{\underline{5}}_{\underline{12}}3^{\underline{5}}_{\underline{10}}9^{\underline{4}}_{\underline{10}}6^{\underline{4}}_{\underline{8}}10^{\underline{3}}_{\underline{8}}8^{\underline{3}}_{\underline{6}}11^{\underline{2}}_{\underline{6}}10^{\underline{2}}_{\underline{4}}10^{\underline{1}}_{\underline{4}}11^{\underline{1}}_{\underline{2}}9^{0}_{\underline{2}}11^{0}_{0}5^{1}_{0}8^{1}_{2}3^{2}_{2}5^{2}_{4}1^{3}_{4}3^{3}_{6}1^{4}_{8}$		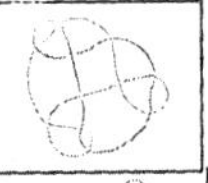
11^a_{269}: $1^{\underline{8}}_{\underline{20}}1^{\underline{8}}_{\underline{18}}3^{\underline{7}}_{\underline{18}}5^{\underline{6}}_{\underline{16}}3^{\underline{6}}_{\underline{14}}7^{\underline{5}}_{\underline{14}}5^{\underline{5}}_{\underline{12}}9^{\underline{4}}_{\underline{12}}7^{\underline{4}}_{\underline{10}}9^{\underline{3}}_{\underline{10}}9^{\underline{3}}_{\underline{8}}9^{\underline{2}}_{\underline{8}}9^{\underline{2}}_{\underline{6}}6^{\underline{1}}_{\underline{6}}9^{\underline{1}}_{\underline{4}}5^{0}_{\underline{4}}7^{0}_{\underline{2}}2^{1}_{\underline{2}}4^{1}_{0}1^{2}_{0}2^{2}_{2}1^{3}_{4}$ 11^a_{270}: $1^{\underline{9}}_{\underline{24}}2^{\underline{8}}_{\underline{22}}1^{\underline{8}}_{\underline{20}}5^{\underline{7}}_{\underline{20}}2^{\underline{7}}_{\underline{18}}6^{\underline{6}}_{\underline{18}}5^{\underline{6}}_{\underline{16}}7^{\underline{5}}_{\underline{16}}6^{\underline{5}}_{\underline{14}}8^{\underline{4}}_{\underline{14}}8^{\underline{4}}_{\underline{12}}7^{\underline{3}}_{\underline{12}}7^{\underline{3}}_{\underline{10}}5^{\underline{2}}_{\underline{10}}7^{\underline{2}}_{\underline{8}}3^{\underline{1}}_{\underline{8}}5^{\underline{1}}_{\underline{6}}2^{0}_{\underline{6}}4^{0}_{\underline{4}}1^{1}_{\underline{4}}1^{1}_{\underline{2}}1^{2}_{0}$		
11^a_{271}: $1^{\underline{9}}_{\underline{22}}2^{\underline{8}}_{\underline{20}}1^{\underline{8}}_{\underline{18}}4^{\underline{7}}_{\underline{18}}2^{\underline{7}}_{\underline{16}}5^{\underline{6}}_{\underline{16}}4^{\underline{6}}_{\underline{14}}6^{\underline{5}}_{\underline{14}}5^{\underline{5}}_{\underline{12}}8^{\underline{4}}_{\underline{12}}7^{\underline{4}}_{\underline{10}}6^{\underline{3}}_{\underline{10}}7^{\underline{3}}_{\underline{8}}5^{\underline{2}}_{\underline{8}}6^{\underline{2}}_{\underline{6}}3^{\underline{1}}_{\underline{6}}5^{\underline{1}}_{\underline{4}}2^{0}_{\underline{4}}4^{0}_{\underline{2}}1^{1}_{\underline{2}}1^{1}_{0}1^{2}_{2}$ 11^a_{272}: $1^{\underline{4}}_{\underline{8}}2^{\underline{3}}_{\underline{6}}1^{\underline{3}}_{\underline{4}}5^{\underline{2}}_{\underline{4}}2^{\underline{2}}_{\underline{2}}6^{\underline{1}}_{\underline{2}}5^{\underline{1}}_{0}9^{0}_{0}8^{0}_{2}9^{1}_{2}7^{1}_{4}8^{2}_{4}9^{2}_{6}6^{3}_{6}8^{3}_{8}4^{4}_{8}6^{4}_{10}2^{5}_{10}4^{5}_{12}1^{6}_{12}2^{6}_{14}1^{7}_{16}$	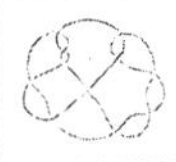	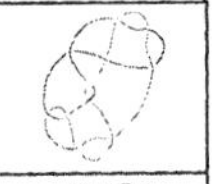
11^a_{273}: $1^{\underline{11}}_{\underline{30}}2^{\underline{10}}_{\underline{28}}1^{\underline{10}}_{\underline{26}}3^{\underline{9}}_{\underline{26}}2^{\underline{9}}_{\underline{24}}5^{\underline{8}}_{\underline{24}}4^{\underline{8}}_{\underline{22}}6^{\underline{7}}_{\underline{22}}4^{\underline{7}}_{\underline{20}}5^{\underline{6}}_{\underline{20}}6^{\underline{6}}_{\underline{18}}6^{\underline{5}}_{\underline{18}}5^{\underline{5}}_{\underline{16}}3^{\underline{4}}_{\underline{16}}6^{\underline{4}}_{\underline{14}}3^{\underline{3}}_{\underline{14}}3^{\underline{3}}_{\underline{12}}2^{\underline{2}}_{\underline{12}}3^{\underline{2}}_{\underline{10}}2^{\underline{1}}_{\underline{8}}1^{0}_{\underline{8}}1^{0}_{\underline{6}}$ 11^a_{274}: $1^{\underline{6}}_{\underline{14}}1^{\underline{5}}_{\underline{12}}1^{\underline{5}}_{\underline{10}}3^{\underline{4}}_{\underline{10}}2^{\underline{4}}_{\underline{8}}4^{\underline{3}}_{\underline{8}}2^{\underline{3}}_{\underline{6}}4^{\underline{2}}_{\underline{6}}4^{\underline{2}}_{\underline{4}}5^{\underline{1}}_{\underline{4}}4^{\underline{1}}_{\underline{2}}5^{0}_{\underline{2}}6^{0}_{0}3^{1}_{0}4^{1}_{2}3^{2}_{2}3^{2}_{4}1^{3}_{4}3^{3}_{6}1^{4}_{6}1^{4}_{8}1^{5}_{10}$	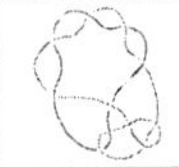	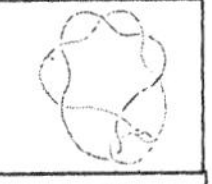
11^a_{275}: $1^{\underline{6}}_{\underline{16}}1^{\underline{5}}_{\underline{14}}1^{\underline{5}}_{\underline{12}}3^{\underline{4}}_{\underline{12}}2^{\underline{4}}_{\underline{10}}4^{\underline{3}}_{\underline{10}}2^{\underline{3}}_{\underline{8}}3^{\underline{2}}_{\underline{8}}4^{\underline{2}}_{\underline{6}}4^{\underline{1}}_{\underline{6}}3^{\underline{1}}_{\underline{4}}4^{0}_{\underline{4}}5^{0}_{\underline{2}}3^{1}_{\underline{2}}3^{1}_{0}2^{2}_{0}3^{2}_{2}1^{3}_{2}2^{3}_{4}1^{4}_{4}1^{4}_{6}1^{5}_{8}$ 11^a_{276}: $1^{\underline{8}}_{\underline{22}}1^{\underline{8}}_{\underline{20}}2^{\underline{7}}_{\underline{20}}2^{\underline{6}}_{\underline{18}}2^{\underline{6}}_{\underline{16}}3^{\underline{5}}_{\underline{16}}2^{\underline{5}}_{\underline{14}}3^{\underline{4}}_{\underline{14}}3^{\underline{4}}_{\underline{12}}4^{\underline{3}}_{\underline{12}}3^{\underline{3}}_{\underline{10}}3^{\underline{2}}_{\underline{10}}4^{\underline{2}}_{\underline{8}}2^{\underline{1}}_{\underline{8}}3^{\underline{1}}_{\underline{6}}3^{0}_{\underline{6}}3^{0}_{\underline{4}}1^{1}_{\underline{4}}2^{1}_{\underline{2}}1^{2}_{\underline{2}}1^{2}_{0}1^{3}_{2}$	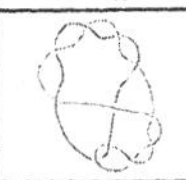	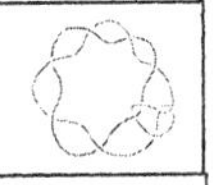
11^a_{277}: $1^{\underline{9}}_{\underline{26}}1^{\underline{8}}_{\underline{24}}1^{\underline{8}}_{\underline{22}}2^{\underline{7}}_{\underline{22}}1^{\underline{7}}_{\underline{20}}2^{\underline{6}}_{\underline{20}}2^{\underline{6}}_{\underline{18}}2^{\underline{5}}_{\underline{18}}2^{\underline{5}}_{\underline{16}}3^{\underline{4}}_{\underline{16}}3^{\underline{4}}_{\underline{14}}3^{\underline{3}}_{\underline{14}}2^{\underline{3}}_{\underline{12}}1^{\underline{2}}_{\underline{12}}3^{\underline{2}}_{\underline{10}}2^{\underline{1}}_{\underline{10}}1^{\underline{1}}_{\underline{8}}1^{0}_{\underline{8}}3^{0}_{\underline{6}}1^{1}_{\underline{6}}1^{2}_{\underline{2}}$ 11^a_{278}: $1^{\underline{9}}_{\underline{24}}1^{\underline{8}}_{\underline{22}}1^{\underline{8}}_{\underline{20}}2^{\underline{7}}_{\underline{20}}1^{\underline{7}}_{\underline{18}}3^{\underline{6}}_{\underline{18}}2^{\underline{6}}_{\underline{16}}3^{\underline{5}}_{\underline{16}}3^{\underline{5}}_{\underline{14}}4^{\underline{4}}_{\underline{14}}4^{\underline{4}}_{\underline{12}}3^{\underline{3}}_{\underline{12}}3^{\underline{3}}_{\underline{10}}2^{\underline{2}}_{\underline{10}}3^{\underline{2}}_{\underline{8}}2^{\underline{1}}_{\underline{8}}2^{\underline{1}}_{\underline{6}}1^{0}_{\underline{6}}3^{0}_{\underline{4}}1^{1}_{\underline{4}}1^{2}_{0}$	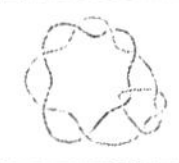	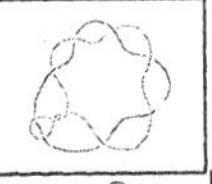
11^a_{279}: $1^{\underline{7}}_{\underline{18}}2^{\underline{6}}_{\underline{16}}1^{\underline{6}}_{\underline{14}}4^{\underline{5}}_{\underline{14}}2^{\underline{5}}_{\underline{12}}5^{\underline{4}}_{\underline{12}}4^{\underline{4}}_{\underline{10}}7^{\underline{3}}_{\underline{10}}5^{\underline{3}}_{\underline{8}}7^{\underline{2}}_{\underline{8}}7^{\underline{2}}_{\underline{6}}6^{\underline{1}}_{\underline{6}}7^{\underline{1}}_{\underline{4}}6^{0}_{\underline{4}}8^{0}_{\underline{2}}4^{1}_{\underline{2}}4^{1}_{0}2^{2}_{0}4^{2}_{2}1^{3}_{2}2^{3}_{4}1^{4}_{6}$ 11^a_{280}: $1^{\underline{5}}_{\underline{12}}2^{\underline{4}}_{\underline{10}}1^{\underline{4}}_{\underline{8}}4^{\underline{3}}_{\underline{8}}2^{\underline{3}}_{\underline{6}}6^{\underline{2}}_{\underline{6}}4^{\underline{2}}_{\underline{4}}6^{\underline{1}}_{\underline{4}}6^{\underline{1}}_{\underline{2}}8^{0}_{\underline{2}}8^{0}_{0}7^{1}_{0}6^{1}_{2}5^{2}_{2}7^{2}_{4}3^{3}_{4}5^{3}_{6}2^{4}_{6}3^{4}_{8}1^{5}_{8}2^{5}_{10}1^{6}_{12}$	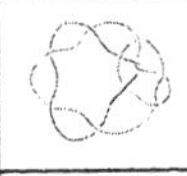	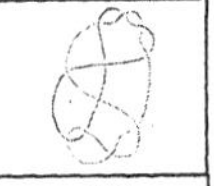
11^a_{281}: $1^{\underline{7}}_{\underline{16}}4^{\underline{6}}_{\underline{14}}1^{\underline{6}}_{\underline{12}}7^{\underline{5}}_{\underline{12}}4^{\underline{5}}_{\underline{10}}11^{\underline{4}}_{\underline{10}}8^{\underline{4}}_{\underline{8}}13^{\underline{3}}_{\underline{8}}10^{\underline{3}}_{\underline{6}}14^{\underline{2}}_{\underline{6}}13^{\underline{2}}_{\underline{4}}12^{\underline{1}}_{\underline{4}}14^{\underline{1}}_{\underline{2}}11^{0}_{\underline{2}}13^{0}_{0}6^{1}_{0}10^{1}_{2}3^{2}_{2}6^{2}_{4}1^{3}_{4}3^{3}_{6}1^{4}_{8}$ 11^a_{282}: $1^{\underline{5}}_{\underline{10}}2^{\underline{4}}_{\underline{8}}1^{\underline{4}}_{\underline{6}}4^{\underline{3}}_{\underline{6}}2^{\underline{3}}_{\underline{4}}5^{\underline{2}}_{\underline{4}}4^{\underline{2}}_{\underline{2}}7^{\underline{1}}_{\underline{2}}5^{\underline{1}}_{0}8^{0}_{0}8^{0}_{2}7^{1}_{2}7^{1}_{4}6^{2}_{4}7^{2}_{6}3^{3}_{6}6^{3}_{8}3^{4}_{8}4^{4}_{10}1^{5}_{10}2^{5}_{12}1^{6}_{14}$	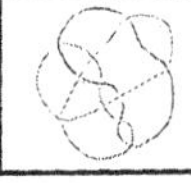	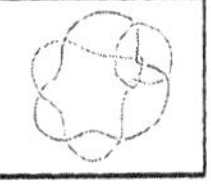

L: KH	1st line	2nd line
11^a_{283}: $1^{\underline{5}}_{\underline{12}}3^{\underline{4}}_{\underline{10}}1^{\underline{4}}_{\underline{8}}6^{\underline{3}}_{\underline{8}}3^{\underline{3}}_{\underline{6}}8^{\underline{2}}_{\underline{6}}6^{\underline{2}}_{\underline{4}}10^{\underline{1}}_{\underline{4}}8^{\underline{1}}_{\underline{2}}10^{0}_{\underline{2}}11^{0}_{0}10^{1}_{0}9^{1}_{2}7^{2}_{2}10^{2}_{4}4^{3}_{4}7^{3}_{6}3^{4}_{6}5^{4}_{8}1^{5}_{8}2^{5}_{10}1^{6}_{12}$ 11^a_{284}: $1^{\underline{3}}_{\underline{4}}2^{\underline{2}}_{\underline{2}}1^{\underline{2}}_{0}4^{\underline{1}}_{0}2^{\underline{1}}_{2}7^{0}_{2}5^{0}_{4}7^{1}_{4}6^{1}_{6}9^{2}_{6}7^{2}_{8}7^{3}_{8}9^{3}_{10}7^{4}_{10}8^{4}_{12}4^{5}_{12}6^{5}_{14}2^{6}_{14}4^{6}_{16}1^{7}_{16}2^{7}_{18}1^{8}_{20}$	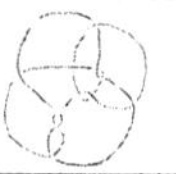	
11^a_{285}: $1^{\underline{6}}_{\underline{14}}2^{\underline{5}}_{\underline{12}}1^{\underline{5}}_{\underline{10}}4^{\underline{4}}_{\underline{10}}2^{\underline{4}}_{\underline{8}}7^{\underline{3}}_{\underline{8}}4^{\underline{3}}_{\underline{6}}8^{\underline{2}}_{\underline{6}}7^{\underline{2}}_{\underline{4}}8^{\underline{1}}_{\underline{4}}8^{\underline{1}}_{\underline{2}}9^{0}_{\underline{2}}10^{0}_{0}7^{1}_{0}7^{1}_{2}4^{2}_{2}7^{2}_{4}2^{3}_{4}4^{3}_{6}1^{4}_{6}2^{4}_{8}1^{5}_{10}$ 11^a_{286}: $1^{\underline{6}}_{\underline{16}}2^{\underline{5}}_{\underline{14}}1^{\underline{5}}_{\underline{12}}3^{\underline{4}}_{\underline{12}}2^{\underline{4}}_{\underline{10}}5^{\underline{3}}_{\underline{10}}3^{\underline{3}}_{\underline{8}}6^{\underline{2}}_{\underline{8}}5^{\underline{2}}_{\underline{6}}6^{\underline{1}}_{\underline{6}}6^{\underline{1}}_{\underline{4}}6^{0}_{\underline{4}}8^{0}_{\underline{2}}6^{1}_{\underline{2}}4^{1}_{0}3^{2}_{0}6^{2}_{2}2^{3}_{2}3^{3}_{4}1^{4}_{4}2^{4}_{6}1^{5}_{8}$	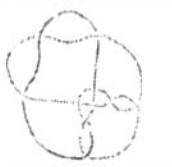	
11^a_{287}: $1^{\underline{4}}_{\underline{6}}2^{\underline{3}}_{\underline{4}}1^{\underline{3}}_{\underline{2}}4^{\underline{2}}_{\underline{2}}2^{\underline{2}}_{0}6^{\underline{1}}_{0}4^{\underline{1}}_{2}8^{0}_{2}7^{0}_{4}8^{1}_{4}7^{1}_{6}8^{2}_{6}8^{2}_{8}6^{3}_{8}8^{3}_{10}5^{4}_{10}7^{4}_{12}2^{5}_{12}4^{5}_{14}1^{6}_{14}2^{6}_{16}1^{7}_{18}$ 11^a_{288}: $1^{\underline{4}}_{\underline{8}}2^{\underline{3}}_{\underline{6}}1^{\underline{3}}_{\underline{4}}4^{\underline{2}}_{\underline{4}}2^{\underline{2}}_{\underline{2}}6^{\underline{1}}_{\underline{2}}4^{\underline{1}}_{0}8^{0}_{0}7^{0}_{2}9^{1}_{2}7^{1}_{4}7^{2}_{4}9^{2}_{6}6^{3}_{6}7^{3}_{8}5^{4}_{8}7^{4}_{10}2^{5}_{10}4^{5}_{12}1^{6}_{12}2^{6}_{14}1^{7}_{16}$		
11^a_{289}: $1^{\underline{4}}_{\underline{6}}2^{\underline{3}}_{\underline{4}}1^{\underline{3}}_{\underline{2}}5^{\underline{2}}_{\underline{2}}2^{\underline{2}}_{0}5^{\underline{1}}_{0}5^{\underline{1}}_{2}10^{0}_{2}7^{0}_{4}9^{1}_{4}8^{1}_{6}9^{2}_{6}9^{2}_{8}7^{3}_{8}9^{3}_{10}5^{4}_{10}7^{4}_{12}3^{5}_{12}5^{5}_{14}1^{6}_{14}3^{6}_{16}1^{7}_{18}$ 11^a_{290}: $1^{\underline{7}}_{\underline{16}}3^{\underline{6}}_{\underline{14}}1^{\underline{6}}_{\underline{12}}5^{\underline{5}}_{\underline{12}}3^{\underline{5}}_{\underline{10}}9^{\underline{4}}_{\underline{10}}6^{\underline{4}}_{\underline{8}}11^{\underline{3}}_{\underline{8}}8^{\underline{3}}_{\underline{6}}12^{\underline{2}}_{\underline{6}}11^{\underline{2}}_{\underline{4}}10^{\underline{1}}_{\underline{4}}12^{\underline{1}}_{\underline{2}}10^{0}_{\underline{2}}11^{0}_{0}6^{1}_{0}9^{1}_{2}3^{2}_{2}6^{2}_{4}1^{3}_{4}3^{3}_{6}1^{4}_{8}$		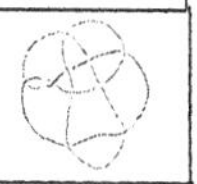
11^a_{291}: $1^{\underline{5}}_{\underline{12}}3^{\underline{4}}_{\underline{10}}1^{\underline{4}}_{\underline{8}}5^{\underline{3}}_{\underline{8}}3^{\underline{3}}_{\underline{6}}9^{\underline{2}}_{\underline{6}}5^{\underline{2}}_{\underline{4}}11^{\underline{1}}_{\underline{4}}9^{\underline{1}}_{\underline{2}}12^{0}_{\underline{2}}13^{0}_{0}12^{1}_{0}10^{1}_{2}9^{2}_{2}12^{2}_{4}6^{3}_{4}9^{3}_{6}4^{4}_{6}6^{4}_{8}1^{5}_{8}4^{5}_{10}1^{6}_{12}$ 11^a_{292}: $1^{\underline{5}}_{\underline{12}}2^{\underline{4}}_{\underline{10}}1^{\underline{4}}_{\underline{8}}5^{\underline{3}}_{\underline{8}}2^{\underline{3}}_{\underline{6}}7^{\underline{2}}_{\underline{6}}5^{\underline{2}}_{\underline{4}}7^{\underline{1}}_{\underline{4}}7^{\underline{1}}_{\underline{2}}10^{0}_{\underline{2}}9^{0}_{0}8^{1}_{0}8^{1}_{2}6^{2}_{2}8^{2}_{4}4^{3}_{4}6^{3}_{6}2^{4}_{6}4^{4}_{8}1^{5}_{8}2^{5}_{10}1^{6}_{12}$	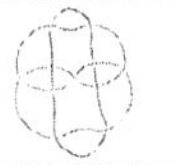	
11^a_{293}: $1^{\underline{6}}_{\underline{16}}2^{\underline{5}}_{\underline{14}}1^{\underline{5}}_{\underline{12}}4^{\underline{4}}_{\underline{12}}2^{\underline{4}}_{\underline{10}}6^{\underline{3}}_{\underline{10}}4^{\underline{3}}_{\underline{8}}6^{\underline{2}}_{\underline{8}}6^{\underline{2}}_{\underline{6}}7^{\underline{1}}_{\underline{6}}6^{\underline{1}}_{\underline{4}}7^{0}_{\underline{4}}9^{0}_{\underline{2}}6^{1}_{\underline{2}}5^{1}_{0}3^{2}_{0}6^{2}_{2}2^{3}_{2}3^{3}_{4}1^{4}_{4}2^{4}_{6}1^{5}_{8}$ 11^a_{294}: $1^{\underline{8}}_{\underline{20}}1^{\underline{8}}_{\underline{18}}2^{\underline{7}}_{\underline{18}}3^{\underline{6}}_{\underline{16}}2^{\underline{6}}_{\underline{14}}5^{\underline{5}}_{\underline{14}}3^{\underline{5}}_{\underline{12}}6^{\underline{4}}_{\underline{12}}5^{\underline{4}}_{\underline{10}}6^{\underline{3}}_{\underline{10}}6^{\underline{3}}_{\underline{8}}7^{\underline{2}}_{\underline{8}}6^{\underline{2}}_{\underline{6}}4^{\underline{1}}_{\underline{6}}7^{\underline{1}}_{\underline{4}}4^{0}_{\underline{4}}5^{0}_{\underline{2}}2^{1}_{\underline{2}}3^{1}_{0}1^{2}_{0}2^{2}_{2}1^{3}_{4}$	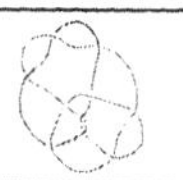	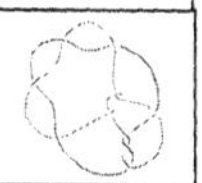
11^a_{295}: $1^{\underline{8}}_{\underline{18}}1^{\underline{8}}_{\underline{16}}2^{\underline{7}}_{\underline{16}}4^{\underline{6}}_{\underline{14}}2^{\underline{6}}_{\underline{12}}5^{\underline{5}}_{\underline{12}}4^{\underline{5}}_{\underline{10}}8^{\underline{4}}_{\underline{10}}5^{\underline{4}}_{\underline{8}}8^{\underline{3}}_{\underline{8}}8^{\underline{3}}_{\underline{6}}9^{\underline{2}}_{\underline{6}}8^{\underline{2}}_{\underline{4}}6^{\underline{1}}_{\underline{4}}9^{\underline{1}}_{\underline{2}}6^{0}_{\underline{2}}7^{0}_{0}3^{1}_{0}5^{1}_{2}1^{2}_{2}3^{2}_{4}1^{3}_{6}$ 11^a_{296}: $1^{\underline{7}}_{\underline{16}}1^{\underline{6}}_{\underline{14}}1^{\underline{6}}_{\underline{12}}4^{\underline{5}}_{\underline{12}}1^{\underline{5}}_{\underline{10}}5^{\underline{4}}_{\underline{10}}4^{\underline{4}}_{\underline{8}}6^{\underline{3}}_{\underline{8}}5^{\underline{3}}_{\underline{6}}8^{\underline{2}}_{\underline{6}}6^{\underline{2}}_{\underline{4}}6^{\underline{1}}_{\underline{4}}8^{\underline{1}}_{\underline{2}}7^{0}_{\underline{2}}8^{0}_{0}4^{1}_{0}5^{1}_{2}2^{2}_{2}4^{2}_{4}1^{3}_{4}2^{3}_{6}1^{4}_{8}$		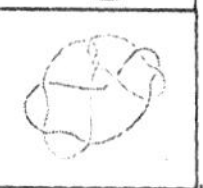
11^a_{297}: $1^{\underline{9}}_{\underline{22}}2^{\underline{8}}_{\underline{20}}1^{\underline{8}}_{\underline{18}}4^{\underline{7}}_{\underline{18}}2^{\underline{7}}_{\underline{16}}6^{\underline{6}}_{\underline{16}}4^{\underline{6}}_{\underline{14}}7^{\underline{5}}_{\underline{14}}6^{\underline{5}}_{\underline{12}}9^{\underline{4}}_{\underline{12}}8^{\underline{4}}_{\underline{10}}8^{\underline{3}}_{\underline{10}}8^{\underline{3}}_{\underline{8}}6^{\underline{2}}_{\underline{8}}8^{\underline{2}}_{\underline{6}}4^{\underline{1}}_{\underline{6}}6^{\underline{1}}_{\underline{4}}3^{0}_{\underline{4}}5^{0}_{\underline{2}}1^{1}_{\underline{2}}2^{1}_{0}1^{2}_{2}$ 11^a_{298}: $1^{\underline{7}}_{\underline{14}}1^{\underline{6}}_{\underline{12}}1^{\underline{6}}_{\underline{10}}3^{\underline{5}}_{\underline{10}}1^{\underline{5}}_{\underline{8}}4^{\underline{4}}_{\underline{8}}3^{\underline{4}}_{\underline{6}}4^{\underline{3}}_{\underline{6}}4^{\underline{3}}_{\underline{4}}7^{\underline{2}}_{\underline{4}}4^{\underline{2}}_{\underline{2}}4^{\underline{1}}_{\underline{2}}7^{\underline{1}}_{0}6^{0}_{0}6^{0}_{2}4^{1}_{2}4^{1}_{4}2^{2}_{4}4^{2}_{6}1^{3}_{6}2^{3}_{8}1^{4}_{10}$	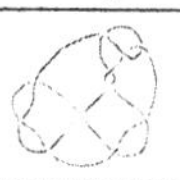	
11^a_{299}: $1^{\underline{2}}_{0}1^{\underline{1}}_{2}1^{\underline{1}}_{4}4^{0}_{4}2^{0}_{6}4^{1}_{6}3^{1}_{8}7^{2}_{8}4^{2}_{10}6^{3}_{10}7^{3}_{12}7^{4}_{12}7^{4}_{14}6^{5}_{14}6^{5}_{16}4^{6}_{16}6^{6}_{18}2^{7}_{18}4^{7}_{20}1^{8}_{20}2^{8}_{22}1^{9}_{24}$ 11^a_{300}: $1^{\underline{4}}_{\underline{6}}2^{\underline{3}}_{\underline{4}}1^{\underline{3}}_{\underline{2}}4^{\underline{2}}_{\underline{2}}2^{\underline{2}}_{0}6^{\underline{1}}_{0}4^{\underline{1}}_{2}8^{0}_{2}7^{0}_{4}9^{1}_{4}7^{1}_{6}9^{2}_{6}9^{2}_{8}6^{3}_{8}9^{3}_{10}6^{4}_{10}7^{4}_{12}3^{5}_{12}5^{5}_{14}1^{6}_{14}3^{6}_{16}1^{7}_{18}$		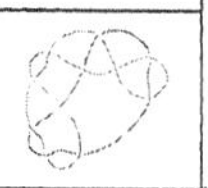
11^a_{301}: $1^{\underline{4}}_{\underline{6}}1^{\underline{3}}_{\underline{4}}1^{\underline{3}}_{\underline{2}}3^{\underline{2}}_{\underline{2}}1^{\underline{2}}_{0}3^{\underline{1}}_{0}3^{\underline{1}}_{2}6^{0}_{2}5^{0}_{4}5^{1}_{4}4^{1}_{6}5^{2}_{6}5^{2}_{8}4^{3}_{8}5^{3}_{10}3^{4}_{10}4^{4}_{12}1^{5}_{12}3^{5}_{14}1^{6}_{14}1^{6}_{16}1^{7}_{18}$ 11^a_{302}: $1^{\underline{4}}_{\underline{8}}1^{\underline{3}}_{\underline{6}}1^{\underline{3}}_{\underline{4}}3^{\underline{2}}_{\underline{4}}1^{\underline{2}}_{\underline{2}}3^{\underline{1}}_{\underline{2}}3^{\underline{1}}_{0}6^{0}_{0}5^{0}_{2}6^{1}_{2}4^{1}_{4}4^{2}_{4}6^{2}_{6}4^{3}_{6}4^{3}_{8}3^{4}_{8}4^{4}_{10}1^{5}_{10}3^{5}_{12}1^{6}_{12}1^{6}_{14}1^{7}_{16}$	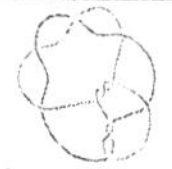	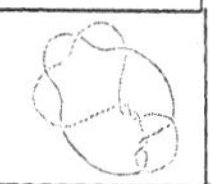
11^a_{303}: $1^{\underline{6}}_{\underline{14}}2^{\underline{5}}_{\underline{12}}1^{\underline{5}}_{\underline{10}}5^{\underline{4}}_{\underline{10}}3^{\underline{4}}_{\underline{8}}8^{\underline{3}}_{\underline{8}}4^{\underline{3}}_{\underline{6}}9^{\underline{2}}_{\underline{6}}8^{\underline{2}}_{\underline{4}}9^{\underline{1}}_{\underline{4}}9^{\underline{1}}_{\underline{2}}10^{0}_{\underline{2}}10^{0}_{0}7^{1}_{0}9^{1}_{2}5^{2}_{2}7^{2}_{4}2^{3}_{4}5^{3}_{6}1^{4}_{6}2^{4}_{8}1^{5}_{10}$ 11^a_{304}: $1^{\underline{8}}_{\underline{22}}1^{\underline{8}}_{\underline{20}}2^{\underline{7}}_{\underline{20}}2^{\underline{6}}_{\underline{18}}2^{\underline{6}}_{\underline{16}}4^{\underline{5}}_{\underline{16}}2^{\underline{5}}_{\underline{14}}4^{\underline{4}}_{\underline{14}}4^{\underline{4}}_{\underline{12}}4^{\underline{3}}_{\underline{12}}4^{\underline{3}}_{\underline{10}}4^{\underline{2}}_{\underline{10}}4^{\underline{2}}_{\underline{8}}3^{\underline{1}}_{\underline{8}}4^{\underline{1}}_{\underline{6}}3^{0}_{\underline{6}}4^{0}_{\underline{4}}1^{1}_{\underline{4}}2^{1}_{\underline{2}}1^{2}_{\underline{2}}1^{2}_{0}1^{3}_{2}$	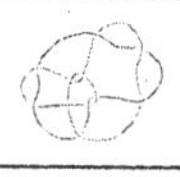	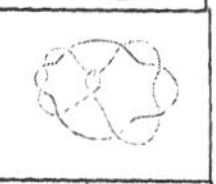
11^a_{305}: $1^{\underline{3}}_{\underline{4}}2^{\underline{2}}_{\underline{2}}1^{\underline{2}}_{0}3^{\underline{1}}_{0}2^{\underline{1}}_{2}6^{0}_{2}4^{0}_{4}7^{1}_{4}5^{1}_{6}8^{2}_{6}7^{2}_{8}6^{3}_{8}8^{3}_{10}7^{4}_{10}7^{4}_{12}4^{5}_{12}6^{5}_{14}2^{6}_{14}4^{6}_{16}1^{7}_{16}2^{7}_{18}1^{8}_{20}$ 11^a_{306}: $1^{\underline{11}}_{\underline{28}}3^{\underline{10}}_{\underline{26}}1^{\underline{10}}_{\underline{24}}5^{\underline{9}}_{\underline{24}}3^{\underline{9}}_{\underline{22}}8^{\underline{8}}_{\underline{22}}6^{\underline{8}}_{\underline{20}}10^{\underline{7}}_{\underline{20}}7^{\underline{7}}_{\underline{18}}9^{\underline{6}}_{\underline{18}}10^{\underline{6}}_{\underline{16}}10^{\underline{5}}_{\underline{16}}9^{\underline{5}}_{\underline{14}}7^{\underline{4}}_{\underline{14}}10^{\underline{4}}_{\underline{12}}4^{\underline{3}}_{\underline{12}}7^{\underline{3}}_{\underline{10}}3^{\underline{2}}_{\underline{10}}4^{\underline{2}}_{\underline{8}}3^{\underline{1}}_{\underline{6}}1^{0}_{\underline{6}}1^{0}_{\underline{4}}$	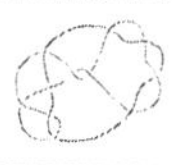	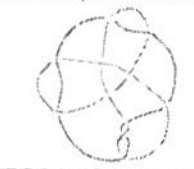
11^a_{307}: $1^{\underline{8}}_{\underline{20}}1^{\underline{8}}_{\underline{18}}3^{\underline{7}}_{\underline{18}}5^{\underline{6}}_{\underline{16}}3^{\underline{6}}_{\underline{14}}7^{\underline{5}}_{\underline{14}}5^{\underline{5}}_{\underline{12}}10^{\underline{4}}_{\underline{12}}7^{\underline{4}}_{\underline{10}}9^{\underline{3}}_{\underline{10}}10^{\underline{3}}_{\underline{8}}10^{\underline{2}}_{\underline{8}}9^{\underline{2}}_{\underline{6}}7^{\underline{1}}_{\underline{6}}10^{\underline{1}}_{\underline{4}}5^{0}_{\underline{4}}8^{0}_{\underline{2}}3^{1}_{\underline{2}}4^{1}_{0}1^{2}_{0}3^{2}_{2}1^{3}_{4}$ 11^a_{308}: $1^{\underline{9}}_{\underline{24}}2^{\underline{8}}_{\underline{22}}1^{\underline{8}}_{\underline{20}}5^{\underline{7}}_{\underline{20}}2^{\underline{7}}_{\underline{18}}7^{\underline{6}}_{\underline{18}}5^{\underline{6}}_{\underline{16}}8^{\underline{5}}_{\underline{16}}7^{\underline{5}}_{\underline{14}}9^{\underline{4}}_{\underline{14}}9^{\underline{4}}_{\underline{12}}9^{\underline{3}}_{\underline{12}}8^{\underline{3}}_{\underline{10}}6^{\underline{2}}_{\underline{10}}9^{\underline{2}}_{\underline{8}}4^{\underline{1}}_{\underline{8}}6^{\underline{1}}_{\underline{6}}3^{0}_{\underline{6}}5^{0}_{\underline{4}}1^{1}_{\underline{4}}2^{1}_{\underline{2}}1^{2}_{0}$		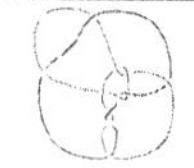
11^a_{309}: $1^{\underline{7}}_{\underline{16}}3^{\underline{6}}_{\underline{14}}1^{\underline{6}}_{\underline{12}}7^{\underline{5}}_{\underline{12}}3^{\underline{5}}_{\underline{10}}9^{\underline{4}}_{\underline{10}}7^{\underline{4}}_{\underline{8}}12^{\underline{3}}_{\underline{8}}9^{\underline{3}}_{\underline{6}}13^{\underline{2}}_{\underline{6}}12^{\underline{2}}_{\underline{4}}11^{\underline{1}}_{\underline{4}}13^{\underline{1}}_{\underline{2}}11^{0}_{\underline{2}}13^{0}_{0}6^{1}_{0}9^{1}_{2}3^{2}_{2}6^{2}_{4}1^{3}_{4}3^{3}_{6}1^{4}_{8}$ 11^a_{310}: $1^{\underline{4}}_{\underline{6}}2^{\underline{3}}_{\underline{4}}1^{\underline{3}}_{\underline{2}}4^{\underline{2}}_{\underline{2}}2^{\underline{2}}_{0}6^{\underline{1}}_{0}4^{\underline{1}}_{2}8^{0}_{2}7^{0}_{4}8^{1}_{4}7^{1}_{6}8^{2}_{6}8^{2}_{8}6^{3}_{8}8^{3}_{10}5^{4}_{10}7^{4}_{12}2^{5}_{12}4^{5}_{14}1^{6}_{14}2^{6}_{16}1^{7}_{18}$	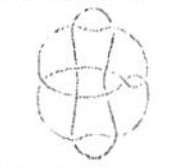	
11^a_{311}: $1^{\underline{7}}_{\underline{16}}3^{\underline{6}}_{\underline{14}}1^{\underline{6}}_{\underline{12}}5^{\underline{5}}_{\underline{12}}3^{\underline{5}}_{\underline{10}}9^{\underline{4}}_{\underline{10}}6^{\underline{4}}_{\underline{8}}11^{\underline{3}}_{\underline{8}}8^{\underline{3}}_{\underline{6}}12^{\underline{2}}_{\underline{6}}11^{\underline{2}}_{\underline{4}}10^{\underline{1}}_{\underline{4}}12^{\underline{1}}_{\underline{2}}10^{0}_{\underline{2}}11^{0}_{0}6^{1}_{0}9^{1}_{2}3^{2}_{2}6^{2}_{4}1^{3}_{4}3^{3}_{6}1^{4}_{8}$ 11^a_{312}: $1^{\underline{7}}_{\underline{18}}2^{\underline{6}}_{\underline{16}}1^{\underline{6}}_{\underline{14}}4^{\underline{5}}_{\underline{14}}2^{\underline{5}}_{\underline{12}}5^{\underline{4}}_{\underline{12}}4^{\underline{4}}_{\underline{10}}6^{\underline{3}}_{\underline{10}}5^{\underline{3}}_{\underline{8}}7^{\underline{2}}_{\underline{8}}6^{\underline{2}}_{\underline{6}}5^{\underline{1}}_{\underline{6}}7^{\underline{1}}_{\underline{4}}5^{0}_{\underline{4}}7^{0}_{\underline{2}}4^{1}_{\underline{2}}3^{1}_{0}1^{2}_{0}4^{2}_{2}1^{3}_{2}1^{3}_{4}1^{4}_{6}$	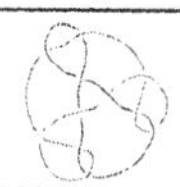	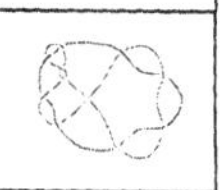

L: KH	1st line	2nd line
11^a_{313}: $1^{\underline{5}}_{\underline{12}}2^{\underline{4}}_{\underline{10}}1^{\underline{4}}_{\underline{8}}4^{\underline{3}}_{\underline{8}}2^{\underline{3}}_{\underline{6}}6^{\underline{2}}_{\underline{6}}4^{\underline{2}}_{\underline{4}}7^{\underline{1}}_{\underline{4}}6^{\underline{1}}_{\underline{2}}9^{0}_{\underline{2}}9^{0}_{0}7^{1}_{0}7^{1}_{2}6^{2}_{2}7^{2}_{4}4^{3}_{4}6^{3}_{6}2^{4}_{6}4^{4}_{8}1^{5}_{8}2^{5}_{10}1^{6}_{12}$ 11^a_{314}: $1^{\underline{3}}_{\underline{4}}3^{\underline{2}}_{\underline{2}}1^{\underline{2}}_{0}6^{\underline{1}}_{0}3^{\underline{1}}_{2}10^{0}_{2}7^{0}_{4}12^{1}_{4}9^{1}_{6}14^{2}_{6}12^{2}_{8}11^{3}_{8}14^{3}_{10}11^{4}_{10}12^{4}_{12}7^{5}_{12}10^{5}_{14}3^{6}_{14}7^{6}_{16}1^{7}_{16}3^{7}_{18}1^{8}_{20}$		
11^a_{315}: $1^{\underline{11}}_{\underline{30}}2^{\underline{10}}_{\underline{28}}1^{\underline{10}}_{\underline{26}}4^{\underline{9}}_{\underline{26}}2^{\underline{9}}_{\underline{24}}7^{\underline{8}}_{\underline{24}}5^{\underline{8}}_{\underline{22}}8^{\underline{7}}_{\underline{22}}6^{\underline{7}}_{\underline{20}}7^{\underline{6}}_{\underline{20}}8^{\underline{6}}_{\underline{18}}8^{\underline{5}}_{\underline{18}}7^{\underline{5}}_{\underline{16}}5^{\underline{4}}_{\underline{16}}8^{\underline{4}}_{\underline{14}}4^{\underline{3}}_{\underline{14}}5^{\underline{3}}_{\underline{12}}2^{\underline{2}}_{\underline{12}}4^{\underline{2}}_{\underline{10}}2^{\underline{1}}_{\underline{8}}1^{0}_{\underline{8}}1^{0}_{\underline{6}}$ 11^a_{316}: $1^{\underline{9}}_{\underline{24}}2^{\underline{8}}_{\underline{22}}1^{\underline{8}}_{\underline{20}}4^{\underline{7}}_{\underline{20}}2^{\underline{7}}_{\underline{18}}6^{\underline{6}}_{\underline{18}}4^{\underline{6}}_{\underline{16}}8^{\underline{5}}_{\underline{16}}6^{\underline{5}}_{\underline{14}}8^{\underline{4}}_{\underline{14}}9^{\underline{4}}_{\underline{12}}8^{\underline{3}}_{\underline{12}}7^{\underline{3}}_{\underline{10}}6^{\underline{2}}_{\underline{10}}8^{\underline{2}}_{\underline{8}}4^{\underline{1}}_{\underline{8}}6^{\underline{1}}_{\underline{6}}3^{0}_{\underline{6}}5^{0}_{\underline{4}}1^{1}_{\underline{4}}2^{1}_{\underline{2}}1^{2}_{0}$	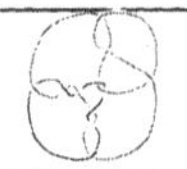	
11^a_{317}: $1^{\underline{7}}_{\underline{16}}3^{\underline{6}}_{\underline{14}}1^{\underline{6}}_{\underline{12}}7^{\underline{5}}_{\underline{12}}3^{\underline{5}}_{\underline{10}}10^{\underline{4}}_{\underline{10}}7^{\underline{4}}_{\underline{8}}14^{\underline{3}}_{\underline{8}}10^{\underline{3}}_{\underline{6}}15^{\underline{2}}_{\underline{6}}14^{\underline{2}}_{\underline{4}}13^{\underline{1}}_{\underline{4}}15^{\underline{1}}_{\underline{2}}13^{0}_{\underline{2}}15^{0}_{0}8^{1}_{0}11^{1}_{2}4^{2}_{2}8^{2}_{4}1^{3}_{4}4^{3}_{6}1^{4}_{8}$ 11^a_{318}: $1^{\underline{7}}_{\underline{20}}1^{\underline{6}}_{\underline{18}}1^{\underline{6}}_{\underline{16}}2^{\underline{5}}_{\underline{16}}1^{\underline{5}}_{\underline{14}}2^{\underline{4}}_{\underline{14}}2^{\underline{4}}_{\underline{12}}3^{\underline{3}}_{\underline{12}}2^{\underline{3}}_{\underline{10}}3^{\underline{2}}_{\underline{10}}3^{\underline{2}}_{\underline{8}}2^{\underline{1}}_{\underline{8}}3^{\underline{1}}_{\underline{6}}3^{0}_{\underline{6}}4^{0}_{\underline{4}}3^{1}_{\underline{4}}1^{1}_{\underline{2}}3^{2}_{0}1^{3}_{0}1^{4}_{4}$		
11^a_{319}: $1^{\underline{7}}_{\underline{18}}1^{\underline{6}}_{\underline{16}}1^{\underline{6}}_{\underline{14}}3^{\underline{5}}_{\underline{14}}1^{\underline{5}}_{\underline{12}}4^{\underline{4}}_{\underline{12}}3^{\underline{4}}_{\underline{10}}4^{\underline{3}}_{\underline{10}}4^{\underline{3}}_{\underline{8}}5^{\underline{2}}_{\underline{8}}4^{\underline{2}}_{\underline{6}}3^{\underline{1}}_{\underline{6}}5^{\underline{1}}_{\underline{4}}4^{0}_{\underline{4}}5^{0}_{\underline{2}}3^{1}_{\underline{2}}2^{1}_{0}3^{2}_{2}1^{3}_{2}1^{4}_{6}$ 11^a_{320}: $1^{\underline{8}}_{\underline{20}}2^{\underline{7}}_{\underline{18}}1^{\underline{7}}_{\underline{16}}5^{\underline{6}}_{\underline{16}}2^{\underline{6}}_{\underline{14}}7^{\underline{5}}_{\underline{14}}5^{\underline{5}}_{\underline{12}}8^{\underline{4}}_{\underline{12}}8^{\underline{4}}_{\underline{10}}10^{\underline{3}}_{\underline{10}}7^{\underline{3}}_{\underline{8}}8^{\underline{2}}_{\underline{8}}10^{\underline{2}}_{\underline{6}}6^{\underline{1}}_{\underline{6}}8^{\underline{1}}_{\underline{4}}5^{0}_{\underline{4}}7^{0}_{\underline{2}}2^{1}_{\underline{2}}4^{1}_{0}1^{2}_{0}2^{2}_{2}1^{3}_{4}$		
11^a_{321}: $1^{\underline{7}}_{\underline{16}}3^{\underline{6}}_{\underline{14}}1^{\underline{6}}_{\underline{12}}6^{\underline{5}}_{\underline{12}}3^{\underline{5}}_{\underline{10}}8^{\underline{4}}_{\underline{10}}6^{\underline{4}}_{\underline{8}}10^{\underline{3}}_{\underline{8}}8^{\underline{3}}_{\underline{6}}11^{\underline{2}}_{\underline{6}}10^{\underline{2}}_{\underline{4}}9^{\underline{1}}_{\underline{4}}11^{\underline{1}}_{\underline{2}}9^{0}_{\underline{2}}11^{0}_{0}5^{1}_{0}7^{1}_{2}2^{2}_{2}5^{2}_{4}1^{3}_{4}2^{3}_{6}1^{4}_{8}$ 11^a_{322}: $1^{\underline{5}}_{\underline{12}}3^{\underline{4}}_{\underline{10}}1^{\underline{4}}_{\underline{8}}6^{\underline{3}}_{\underline{8}}3^{\underline{3}}_{\underline{6}}9^{\underline{2}}_{\underline{6}}6^{\underline{2}}_{\underline{4}}10^{\underline{1}}_{\underline{4}}9^{\underline{1}}_{\underline{2}}12^{0}_{\underline{2}}12^{0}_{0}11^{1}_{0}10^{1}_{2}8^{2}_{2}11^{2}_{4}5^{3}_{4}8^{3}_{6}3^{4}_{6}5^{4}_{8}1^{5}_{8}3^{5}_{10}1^{6}_{12}$	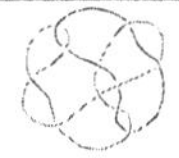	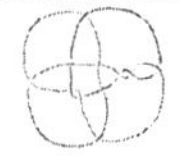
11^a_{323}: $1^{\underline{6}}_{\underline{14}}3^{\underline{5}}_{\underline{12}}1^{\underline{5}}_{\underline{10}}6^{\underline{4}}_{\underline{10}}3^{\underline{4}}_{\underline{8}}10^{\underline{3}}_{\underline{8}}6^{\underline{3}}_{\underline{6}}12^{\underline{2}}_{\underline{6}}10^{\underline{2}}_{\underline{4}}12^{\underline{1}}_{\underline{4}}12^{\underline{1}}_{\underline{2}}13^{0}_{\underline{2}}14^{0}_{0}10^{1}_{0}11^{1}_{2}6^{2}_{2}10^{2}_{4}3^{3}_{4}6^{3}_{6}1^{4}_{6}3^{4}_{8}1^{5}_{10}$ 11^a_{324}: $1^{\underline{6}}_{\underline{16}}1^{\underline{5}}_{\underline{14}}1^{\underline{5}}_{\underline{12}}3^{\underline{4}}_{\underline{12}}2^{\underline{4}}_{\underline{10}}4^{\underline{3}}_{\underline{10}}2^{\underline{3}}_{\underline{8}}4^{\underline{2}}_{\underline{8}}4^{\underline{2}}_{\underline{6}}5^{\underline{1}}_{\underline{6}}4^{\underline{1}}_{\underline{4}}4^{0}_{\underline{4}}6^{0}_{\underline{2}}4^{1}_{\underline{2}}3^{1}_{0}3^{2}_{0}4^{2}_{2}1^{3}_{2}3^{3}_{4}1^{4}_{4}1^{4}_{6}1^{5}_{8}$		
11^a_{325}: $1^{\underline{8}}_{\underline{22}}2^{\underline{7}}_{\underline{20}}1^{\underline{7}}_{\underline{18}}4^{\underline{6}}_{\underline{18}}2^{\underline{6}}_{\underline{16}}5^{\underline{5}}_{\underline{16}}4^{\underline{5}}_{\underline{14}}6^{\underline{4}}_{\underline{14}}6^{\underline{4}}_{\underline{12}}8^{\underline{3}}_{\underline{12}}5^{\underline{3}}_{\underline{10}}5^{\underline{2}}_{\underline{10}}8^{\underline{2}}_{\underline{8}}5^{\underline{1}}_{\underline{8}}5^{\underline{1}}_{\underline{6}}4^{0}_{\underline{6}}6^{0}_{\underline{4}}2^{1}_{\underline{4}}3^{1}_{\underline{2}}1^{2}_{\underline{2}}2^{2}_{0}1^{3}_{2}$ 11^a_{326}: $1^{\underline{6}}_{\underline{14}}3^{\underline{5}}_{\underline{12}}1^{\underline{5}}_{\underline{10}}6^{\underline{4}}_{\underline{10}}3^{\underline{4}}_{\underline{8}}9^{\underline{3}}_{\underline{8}}6^{\underline{3}}_{\underline{6}}11^{\underline{2}}_{\underline{6}}9^{\underline{2}}_{\underline{4}}12^{\underline{1}}_{\underline{4}}11^{\underline{1}}_{\underline{2}}12^{0}_{\underline{2}}14^{0}_{0}9^{1}_{0}10^{1}_{2}6^{2}_{2}9^{2}_{4}3^{3}_{4}6^{3}_{6}1^{4}_{6}3^{4}_{8}1^{5}_{10}$		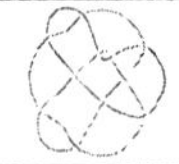
11^a_{327}: $1^{\underline{6}}_{\underline{14}}1^{\underline{5}}_{\underline{12}}1^{\underline{5}}_{\underline{10}}4^{\underline{4}}_{\underline{10}}2^{\underline{4}}_{\underline{8}}6^{\underline{3}}_{\underline{8}}3^{\underline{3}}_{\underline{6}}6^{\underline{2}}_{\underline{6}}6^{\underline{2}}_{\underline{4}}7^{\underline{1}}_{\underline{4}}6^{\underline{1}}_{\underline{2}}7^{0}_{\underline{2}}8^{0}_{0}5^{1}_{0}6^{1}_{2}4^{2}_{2}5^{2}_{4}1^{3}_{4}4^{3}_{6}1^{4}_{6}1^{4}_{8}1^{5}_{10}$ 11^a_{328}: $1^{\underline{6}}_{\underline{12}}3^{\underline{5}}_{\underline{10}}1^{\underline{5}}_{\underline{8}}5^{\underline{4}}_{\underline{8}}3^{\underline{4}}_{\underline{6}}7^{\underline{3}}_{\underline{6}}5^{\underline{3}}_{\underline{4}}9^{\underline{2}}_{\underline{4}}7^{\underline{2}}_{\underline{2}}8^{\underline{1}}_{\underline{2}}9^{\underline{1}}_{0}10^{0}_{0}10^{0}_{2}7^{1}_{2}8^{1}_{4}4^{2}_{4}7^{2}_{6}2^{3}_{6}4^{3}_{8}1^{4}_{8}2^{4}_{10}1^{5}_{12}$	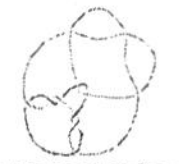	
11^a_{329}: $1^{\underline{6}}_{\underline{14}}2^{\underline{5}}_{\underline{12}}1^{\underline{5}}_{\underline{10}}5^{\underline{4}}_{\underline{10}}3^{\underline{4}}_{\underline{8}}7^{\underline{3}}_{\underline{8}}4^{\underline{3}}_{\underline{6}}8^{\underline{2}}_{\underline{6}}7^{\underline{2}}_{\underline{4}}9^{\underline{1}}_{\underline{4}}8^{\underline{1}}_{\underline{2}}9^{0}_{\underline{2}}10^{0}_{0}6^{1}_{0}8^{1}_{2}5^{2}_{2}6^{2}_{4}2^{3}_{4}5^{3}_{6}1^{4}_{6}2^{4}_{8}1^{5}_{10}$ 11^a_{330}: $1^{\underline{4}}_{\underline{6}}3^{\underline{3}}_{\underline{4}}1^{\underline{3}}_{\underline{2}}6^{\underline{2}}_{\underline{2}}3^{\underline{2}}_{0}7^{\underline{1}}_{0}6^{\underline{1}}_{2}12^{0}_{2}9^{0}_{4}11^{1}_{4}10^{1}_{6}11^{2}_{6}11^{2}_{8}8^{3}_{8}11^{3}_{10}6^{4}_{10}8^{4}_{12}3^{5}_{12}6^{5}_{14}1^{6}_{14}3^{6}_{16}1^{7}_{18}$		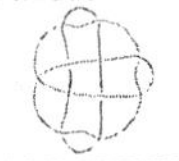
11^a_{331}: $1^{\underline{8}}_{\underline{22}}1^{\underline{8}}_{\underline{20}}3^{\underline{7}}_{\underline{20}}3^{\underline{6}}_{\underline{18}}3^{\underline{6}}_{\underline{16}}5^{\underline{5}}_{\underline{16}}3^{\underline{5}}_{\underline{14}}6^{\underline{4}}_{\underline{14}}5^{\underline{4}}_{\underline{12}}6^{\underline{3}}_{\underline{12}}6^{\underline{3}}_{\underline{10}}6^{\underline{2}}_{\underline{10}}6^{\underline{2}}_{\underline{8}}4^{\underline{1}}_{\underline{8}}6^{\underline{1}}_{\underline{6}}4^{0}_{\underline{6}}5^{0}_{\underline{4}}2^{1}_{\underline{4}}3^{1}_{\underline{2}}1^{2}_{\underline{2}}2^{2}_{0}1^{3}_{2}$ 11^a_{332}: $1^{\underline{11}}_{\underline{30}}3^{\underline{10}}_{\underline{28}}1^{\underline{10}}_{\underline{26}}4^{\underline{9}}_{\underline{26}}3^{\underline{9}}_{\underline{24}}7^{\underline{8}}_{\underline{24}}5^{\underline{8}}_{\underline{22}}8^{\underline{7}}_{\underline{22}}6^{\underline{7}}_{\underline{20}}7^{\underline{6}}_{\underline{20}}8^{\underline{6}}_{\underline{18}}8^{\underline{5}}_{\underline{18}}7^{\underline{5}}_{\underline{16}}4^{\underline{4}}_{\underline{16}}8^{\underline{4}}_{\underline{14}}4^{\underline{3}}_{\underline{14}}4^{\underline{3}}_{\underline{12}}2^{\underline{2}}_{\underline{12}}4^{\underline{2}}_{\underline{10}}2^{\underline{1}}_{\underline{8}}1^{0}_{\underline{8}}1^{0}_{\underline{6}}$	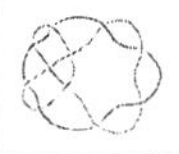	
11^a_{333}: $1^{\underline{6}}_{\underline{14}}3^{\underline{5}}_{\underline{12}}1^{\underline{5}}_{\underline{10}}6^{\underline{4}}_{\underline{10}}4^{\underline{4}}_{\underline{8}}9^{\underline{3}}_{\underline{8}}5^{\underline{3}}_{\underline{6}}11^{\underline{2}}_{\underline{6}}9^{\underline{2}}_{\underline{4}}11^{\underline{1}}_{\underline{4}}11^{\underline{1}}_{\underline{2}}12^{0}_{\underline{2}}12^{0}_{0}8^{1}_{0}11^{1}_{2}6^{2}_{2}8^{2}_{4}3^{3}_{4}6^{3}_{6}1^{4}_{6}3^{4}_{8}1^{5}_{10}$ 11^a_{334}: $1^{\underline{6}}_{\underline{14}}2^{\underline{5}}_{\underline{12}}1^{\underline{5}}_{\underline{10}}6^{\underline{4}}_{\underline{10}}3^{\underline{4}}_{\underline{8}}9^{\underline{3}}_{\underline{8}}5^{\underline{3}}_{\underline{6}}11^{\underline{2}}_{\underline{6}}9^{\underline{2}}_{\underline{4}}11^{\underline{1}}_{\underline{4}}11^{\underline{1}}_{\underline{2}}12^{0}_{\underline{2}}12^{0}_{0}9^{1}_{0}11^{1}_{2}6^{2}_{2}9^{2}_{4}3^{3}_{4}6^{3}_{6}1^{4}_{6}3^{4}_{8}1^{5}_{10}$	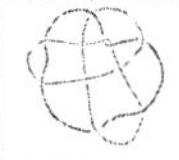	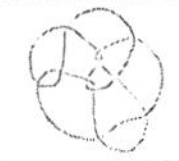
11^a_{335}: $1^{\underline{8}}_{\underline{20}}1^{\underline{8}}_{\underline{18}}4^{\underline{7}}_{\underline{18}}7^{\underline{6}}_{\underline{16}}4^{\underline{6}}_{\underline{14}}10^{\underline{5}}_{\underline{14}}7^{\underline{5}}_{\underline{12}}13^{\underline{4}}_{\underline{12}}10^{\underline{4}}_{\underline{10}}13^{\underline{3}}_{\underline{10}}13^{\underline{3}}_{\underline{8}}13^{\underline{2}}_{\underline{8}}13^{\underline{2}}_{\underline{6}}9^{\underline{1}}_{\underline{6}}13^{\underline{1}}_{\underline{4}}7^{0}_{\underline{4}}10^{0}_{\underline{2}}3^{1}_{\underline{2}}6^{1}_{0}1^{2}_{0}3^{2}_{2}1^{3}_{4}$ 11^a_{336}: $1^{\underline{11}}_{\underline{28}}4^{\underline{10}}_{\underline{26}}1^{\underline{10}}_{\underline{24}}6^{\underline{9}}_{\underline{24}}4^{\underline{9}}_{\underline{22}}10^{\underline{8}}_{\underline{22}}7^{\underline{8}}_{\underline{20}}12^{\underline{7}}_{\underline{20}}9^{\underline{7}}_{\underline{18}}11^{\underline{6}}_{\underline{18}}12^{\underline{6}}_{\underline{16}}12^{\underline{5}}_{\underline{16}}11^{\underline{5}}_{\underline{14}}8^{\underline{4}}_{\underline{14}}12^{\underline{4}}_{\underline{12}}5^{\underline{3}}_{\underline{12}}8^{\underline{3}}_{\underline{10}}3^{\underline{2}}_{\underline{10}}5^{\underline{2}}_{\underline{8}}3^{\underline{1}}_{\underline{6}}1^{0}_{\underline{6}}1^{0}_{\underline{4}}$		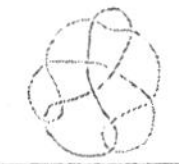
11^a_{337}: $1^{\underline{11}}_{\underline{28}}3^{\underline{10}}_{\underline{26}}1^{\underline{10}}_{\underline{24}}6^{\underline{9}}_{\underline{24}}3^{\underline{9}}_{\underline{22}}9^{\underline{8}}_{\underline{22}}7^{\underline{8}}_{\underline{20}}11^{\underline{7}}_{\underline{20}}8^{\underline{7}}_{\underline{18}}11^{\underline{6}}_{\underline{18}}11^{\underline{6}}_{\underline{16}}11^{\underline{5}}_{\underline{16}}11^{\underline{5}}_{\underline{14}}8^{\underline{4}}_{\underline{14}}11^{\underline{4}}_{\underline{12}}5^{\underline{3}}_{\underline{12}}8^{\underline{3}}_{\underline{10}}3^{\underline{2}}_{\underline{10}}5^{\underline{2}}_{\underline{8}}3^{\underline{1}}_{\underline{6}}1^{0}_{\underline{6}}1^{0}_{\underline{4}}$ 11^a_{338}: $1^{\underline{8}}_{\underline{20}}2^{\underline{7}}_{\underline{18}}1^{\underline{7}}_{\underline{16}}5^{\underline{6}}_{\underline{16}}2^{\underline{6}}_{\underline{14}}8^{\underline{5}}_{\underline{14}}5^{\underline{5}}_{\underline{12}}10^{\underline{4}}_{\underline{12}}9^{\underline{4}}_{\underline{10}}12^{\underline{3}}_{\underline{10}}9^{\underline{3}}_{\underline{8}}10^{\underline{2}}_{\underline{8}}12^{\underline{2}}_{\underline{6}}8^{\underline{1}}_{\underline{6}}10^{\underline{1}}_{\underline{4}}7^{0}_{\underline{4}}9^{0}_{\underline{2}}3^{1}_{\underline{2}}6^{1}_{0}1^{2}_{0}3^{2}_{2}1^{3}_{4}$	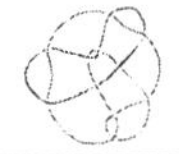	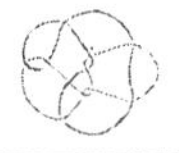
11^a_{339}: $1^{\underline{4}}_{\underline{6}}2^{\underline{3}}_{\underline{4}}1^{\underline{3}}_{\underline{2}}5^{\underline{2}}_{\underline{2}}2^{\underline{2}}_{0}5^{\underline{1}}_{0}5^{\underline{1}}_{2}9^{0}_{2}7^{0}_{4}9^{1}_{4}7^{1}_{6}8^{2}_{6}9^{2}_{8}6^{3}_{8}8^{3}_{10}5^{4}_{10}6^{4}_{12}2^{5}_{12}5^{5}_{14}1^{6}_{14}2^{6}_{16}1^{7}_{18}$ 11^a_{340}: $1^{\underline{4}}_{\underline{8}}2^{\underline{3}}_{\underline{6}}1^{\underline{3}}_{\underline{4}}6^{\underline{2}}_{\underline{4}}2^{\underline{2}}_{\underline{2}}7^{\underline{1}}_{\underline{2}}6^{\underline{1}}_{0}10^{0}_{0}9^{0}_{2}11^{1}_{2}8^{1}_{4}9^{2}_{4}11^{2}_{6}7^{3}_{6}9^{3}_{8}5^{4}_{8}7^{4}_{10}2^{5}_{10}5^{5}_{12}1^{6}_{12}2^{6}_{14}1^{7}_{16}$		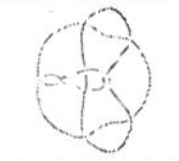
11^a_{341}: $1^{\underline{7}}_{\underline{16}}2^{\underline{6}}_{\underline{14}}1^{\underline{6}}_{\underline{12}}5^{\underline{5}}_{\underline{12}}2^{\underline{5}}_{\underline{10}}8^{\underline{4}}_{\underline{10}}5^{\underline{4}}_{\underline{8}}10^{\underline{3}}_{\underline{8}}8^{\underline{3}}_{\underline{6}}12^{\underline{2}}_{\underline{6}}10^{\underline{2}}_{\underline{4}}10^{\underline{1}}_{\underline{4}}12^{\underline{1}}_{\underline{2}}10^{0}_{\underline{2}}12^{0}_{0}7^{1}_{0}8^{1}_{2}3^{2}_{2}7^{2}_{4}1^{3}_{4}3^{3}_{6}1^{4}_{8}$ 11^a_{342}: $1^{\underline{9}}_{\underline{22}}2^{\underline{8}}_{\underline{20}}1^{\underline{8}}_{\underline{18}}5^{\underline{7}}_{\underline{18}}2^{\underline{7}}_{\underline{16}}8^{\underline{6}}_{\underline{16}}5^{\underline{6}}_{\underline{14}}9^{\underline{5}}_{\underline{14}}8^{\underline{5}}_{\underline{12}}11^{\underline{4}}_{\underline{12}}10^{\underline{4}}_{\underline{10}}10^{\underline{3}}_{\underline{10}}10^{\underline{3}}_{\underline{8}}8^{\underline{2}}_{\underline{8}}10^{\underline{2}}_{\underline{6}}5^{\underline{1}}_{\underline{6}}8^{\underline{1}}_{\underline{4}}3^{0}_{\underline{4}}6^{0}_{\underline{2}}1^{1}_{\underline{2}}2^{1}_{0}1^{2}_{2}$		

L: KH	1st line	2nd line
11^a_{343}: $1^{\underline{8}}_{\underline{20}}2^{\underline{7}}_{\underline{18}}1^{\underline{7}}_{\underline{16}}4^{\underline{6}}_{\underline{16}}2^{\underline{6}}_{\underline{14}}8^{\underline{5}}_{\underline{14}}4^{\underline{5}}_{\underline{12}}8^{\underline{4}}_{\underline{12}}8^{\underline{4}}_{\underline{10}}11^{\underline{3}}_{\underline{10}}8^{\underline{3}}_{\underline{8}}9^{\underline{2}}_{\underline{8}}11^{\underline{2}}_{\underline{6}}7^{\underline{1}}_{\underline{6}}9^{\underline{1}}_{\underline{4}}7^{0}_{\underline{4}}9^{0}_{\underline{2}}3^{1}_{\underline{2}}5^{1}_{0}1^{2}_{0}3^{2}_{2}1^{3}_{4}$ 11^a_{344}: $1^{\underline{7}}_{\underline{16}}1^{\underline{6}}_{\underline{14}}1^{\underline{6}}_{\underline{12}}4^{\underline{5}}_{\underline{12}}1^{\underline{5}}_{\underline{10}}6^{\underline{4}}_{\underline{10}}5^{\underline{4}}_{\underline{8}}8^{\underline{3}}_{\underline{8}}5^{\underline{3}}_{\underline{6}}10^{\underline{2}}_{\underline{6}}8^{\underline{2}}_{\underline{4}}7^{\underline{1}}_{\underline{4}}10^{\underline{1}}_{\underline{2}}9^{0}_{\underline{2}}8^{0}_{0}5^{1}_{0}8^{1}_{2}3^{2}_{2}5^{2}_{4}1^{3}_{4}3^{3}_{6}1^{4}_{8}$		
11^a_{345}: $1^{\underline{6}}_{\underline{12}}1^{\underline{5}}_{\underline{10}}1^{\underline{5}}_{\underline{8}}5^{\underline{4}}_{\underline{8}}2^{\underline{4}}_{\underline{6}}6^{\underline{3}}_{\underline{6}}4^{\underline{3}}_{\underline{4}}9^{\underline{2}}_{\underline{4}}6^{\underline{2}}_{\underline{2}}9^{\underline{1}}_{\underline{2}}9^{\underline{1}}_{0}10^{0}_{0}10^{0}_{2}8^{1}_{2}9^{1}_{4}6^{2}_{4}8^{2}_{6}3^{3}_{6}6^{3}_{8}1^{4}_{8}3^{4}_{10}1^{5}_{12}$ 11^a_{346}: $1^{\underline{4}}_{\underline{6}}1^{\underline{3}}_{\underline{4}}1^{\underline{3}}_{\underline{2}}5^{\underline{2}}_{\underline{2}}1^{\underline{2}}_{0}5^{\underline{1}}_{0}5^{\underline{1}}_{2}9^{0}_{2}7^{0}_{4}9^{1}_{4}7^{1}_{6}8^{2}_{6}9^{2}_{8}7^{3}_{8}8^{3}_{10}5^{4}_{10}7^{4}_{12}2^{5}_{12}5^{5}_{14}1^{6}_{14}2^{6}_{16}1^{7}_{18}$		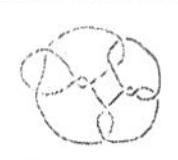
11^a_{347}: $1^{\underline{2}}_{0}1^{\underline{1}}_{4}5^{0}_{4}2^{0}_{6}4^{1}_{6}3^{1}_{8}8^{2}_{8}4^{2}_{10}7^{3}_{10}8^{3}_{12}8^{4}_{12}7^{4}_{14}7^{5}_{14}8^{5}_{16}5^{6}_{16}7^{6}_{18}3^{7}_{18}5^{7}_{20}1^{8}_{20}3^{8}_{22}1^{9}_{24}$ 11^a_{348}: $1^{\underline{7}}_{\underline{16}}5^{\underline{6}}_{\underline{14}}1^{\underline{6}}_{\underline{12}}8^{\underline{5}}_{\underline{12}}5^{\underline{5}}_{\underline{10}}12^{\underline{4}}_{\underline{10}}9^{\underline{4}}_{\underline{8}}15^{\underline{3}}_{\underline{8}}11^{\underline{3}}_{\underline{6}}15^{\underline{2}}_{\underline{6}}15^{\underline{2}}_{\underline{4}}13^{\underline{1}}_{\underline{4}}15^{\underline{1}}_{\underline{2}}12^{0}_{\underline{2}}14^{0}_{0}6^{1}_{0}11^{1}_{2}3^{2}_{2}6^{2}_{4}1^{3}_{4}3^{3}_{6}1^{4}_{8}$	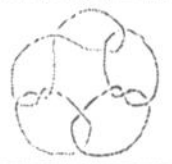	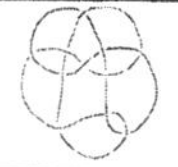
11^a_{349}: $1^{\underline{6}}_{\underline{14}}3^{\underline{5}}_{\underline{12}}1^{\underline{5}}_{\underline{10}}7^{\underline{4}}_{\underline{10}}3^{\underline{4}}_{\underline{8}}11^{\underline{3}}_{\underline{8}}7^{\underline{3}}_{\underline{6}}13^{\underline{2}}_{\underline{6}}11^{\underline{2}}_{\underline{4}}14^{\underline{1}}_{\underline{4}}13^{\underline{1}}_{\underline{2}}14^{0}_{\underline{2}}16^{0}_{0}11^{1}_{0}12^{1}_{2}7^{2}_{2}11^{2}_{4}3^{3}_{4}7^{3}_{6}1^{4}_{6}3^{4}_{8}1^{5}_{10}$ 11^a_{350}: $1^{\underline{6}}_{\underline{12}}2^{\underline{5}}_{\underline{10}}1^{\underline{5}}_{\underline{8}}6^{\underline{4}}_{\underline{8}}3^{\underline{4}}_{\underline{6}}7^{\underline{3}}_{\underline{6}}5^{\underline{3}}_{\underline{4}}11^{\underline{2}}_{\underline{4}}7^{\underline{2}}_{\underline{2}}10^{\underline{1}}_{\underline{2}}11^{\underline{1}}_{0}11^{0}_{0}11^{0}_{2}9^{1}_{2}10^{1}_{4}6^{2}_{4}9^{2}_{6}3^{3}_{6}6^{3}_{8}1^{4}_{8}3^{4}_{10}1^{5}_{12}$	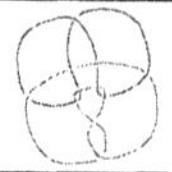	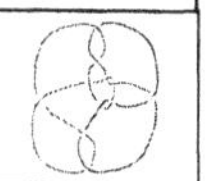
11^a_{351}: $1^{\underline{7}}_{\underline{16}}3^{\underline{6}}_{\underline{14}}1^{\underline{6}}_{\underline{12}}7^{\underline{5}}_{\underline{12}}3^{\underline{5}}_{\underline{10}}11^{\underline{4}}_{\underline{10}}7^{\underline{4}}_{\underline{8}}14^{\underline{3}}_{\underline{8}}11^{\underline{3}}_{\underline{6}}16^{\underline{2}}_{\underline{6}}14^{\underline{2}}_{\underline{4}}14^{\underline{1}}_{\underline{4}}16^{\underline{1}}_{\underline{2}}13^{0}_{\underline{2}}16^{0}_{0}9^{1}_{0}11^{1}_{2}4^{2}_{2}9^{2}_{4}1^{3}_{4}4^{3}_{6}1^{4}_{8}$ 11^a_{352}: $1^{\underline{8}}_{\underline{20}}4^{\underline{7}}_{\underline{18}}1^{\underline{7}}_{\underline{16}}8^{\underline{6}}_{\underline{16}}4^{\underline{6}}_{\underline{14}}11^{\underline{5}}_{\underline{14}}8^{\underline{5}}_{\underline{12}}14^{\underline{4}}_{\underline{12}}12^{\underline{4}}_{\underline{10}}15^{\underline{3}}_{\underline{10}}13^{\underline{3}}_{\underline{8}}13^{\underline{2}}_{\underline{8}}15^{\underline{2}}_{\underline{6}}10^{\underline{1}}_{\underline{6}}13^{\underline{1}}_{\underline{4}}7^{0}_{\underline{4}}11^{0}_{\underline{2}}3^{1}_{\underline{2}}6^{1}_{0}1^{2}_{0}3^{2}_{2}1^{3}_{4}$	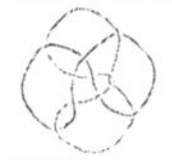	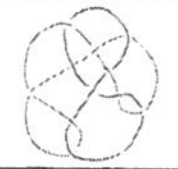
11^a_{353}: $1^{\underline{6}}_{\underline{12}}3^{\underline{5}}_{\underline{10}}1^{\underline{5}}_{\underline{8}}6^{\underline{4}}_{\underline{8}}3^{\underline{4}}_{\underline{6}}9^{\underline{3}}_{\underline{6}}6^{\underline{3}}_{\underline{4}}12^{\underline{2}}_{\underline{4}}9^{\underline{2}}_{\underline{2}}12^{\underline{1}}_{\underline{2}}12^{\underline{1}}_{0}13^{0}_{0}14^{0}_{2}10^{1}_{2}11^{1}_{4}7^{2}_{4}10^{2}_{6}3^{3}_{6}7^{3}_{8}1^{4}_{8}3^{4}_{10}1^{5}_{12}$ 11^a_{354}: $1^{\underline{7}}_{\underline{18}}2^{\underline{6}}_{\underline{16}}1^{\underline{6}}_{\underline{14}}5^{\underline{5}}_{\underline{14}}2^{\underline{5}}_{\underline{12}}8^{\underline{4}}_{\underline{12}}5^{\underline{4}}_{\underline{10}}9^{\underline{3}}_{\underline{10}}8^{\underline{3}}_{\underline{8}}10^{\underline{2}}_{\underline{8}}9^{\underline{2}}_{\underline{6}}9^{\underline{1}}_{\underline{6}}10^{\underline{1}}_{\underline{4}}8^{0}_{\underline{4}}11^{0}_{\underline{2}}6^{1}_{\underline{2}}6^{1}_{0}2^{2}_{0}6^{2}_{2}1^{3}_{2}2^{3}_{4}1^{4}_{6}$		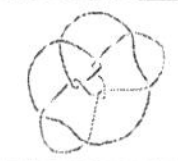
11^a_{355}: $1^{\underline{9}}_{\underline{24}}1^{\underline{8}}_{\underline{22}}1^{\underline{8}}_{\underline{20}}2^{\underline{7}}_{\underline{20}}1^{\underline{7}}_{\underline{18}}4^{\underline{6}}_{\underline{18}}3^{\underline{6}}_{\underline{16}}4^{\underline{5}}_{\underline{16}}3^{\underline{5}}_{\underline{14}}4^{\underline{4}}_{\underline{14}}4^{\underline{4}}_{\underline{12}}4^{\underline{3}}_{\underline{12}}4^{\underline{3}}_{\underline{10}}3^{\underline{2}}_{\underline{10}}4^{\underline{2}}_{\underline{8}}2^{\underline{1}}_{\underline{8}}3^{\underline{1}}_{\underline{6}}2^{0}_{\underline{6}}3^{0}_{\underline{4}}1^{1}_{\underline{4}}1^{1}_{\underline{2}}1^{2}_{0}$ 11^a_{356}: $1^{\underline{7}}_{\underline{16}}1^{\underline{6}}_{\underline{14}}1^{\underline{6}}_{\underline{12}}3^{\underline{5}}_{\underline{12}}1^{\underline{5}}_{\underline{10}}5^{\underline{4}}_{\underline{10}}3^{\underline{4}}_{\underline{8}}5^{\underline{3}}_{\underline{8}}5^{\underline{3}}_{\underline{6}}8^{\underline{2}}_{\underline{6}}6^{\underline{2}}_{\underline{4}}6^{\underline{1}}_{\underline{4}}7^{\underline{1}}_{\underline{2}}6^{0}_{\underline{2}}7^{0}_{0}4^{1}_{0}5^{1}_{2}2^{2}_{2}4^{2}_{4}1^{3}_{4}2^{3}_{6}1^{4}_{8}$	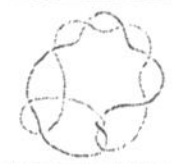	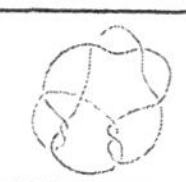
11^a_{357}: $1^{\underline{9}}_{\underline{22}}1^{\underline{8}}_{\underline{20}}1^{\underline{8}}_{\underline{18}}3^{\underline{7}}_{\underline{18}}1^{\underline{7}}_{\underline{16}}6^{\underline{6}}_{\underline{16}}4^{\underline{6}}_{\underline{14}}6^{\underline{5}}_{\underline{14}}5^{\underline{5}}_{\underline{12}}8^{\underline{4}}_{\underline{12}}6^{\underline{4}}_{\underline{10}}7^{\underline{3}}_{\underline{10}}8^{\underline{3}}_{\underline{8}}6^{\underline{2}}_{\underline{8}}7^{\underline{2}}_{\underline{6}}4^{\underline{1}}_{\underline{6}}6^{\underline{1}}_{\underline{4}}3^{0}_{\underline{4}}5^{0}_{\underline{2}}1^{1}_{\underline{2}}2^{1}_{0}1^{2}_{2}$ 11^a_{358}: $1^{\underline{5}}_{\underline{8}}1^{\underline{4}}_{\underline{6}}1^{\underline{4}}_{\underline{4}}2^{\underline{3}}_{\underline{4}}1^{\underline{3}}_{\underline{2}}4^{\underline{2}}_{\underline{2}}2^{\underline{2}}_{0}3^{\underline{1}}_{0}4^{\underline{1}}_{2}6^{0}_{2}4^{0}_{4}4^{1}_{4}5^{1}_{6}5^{2}_{6}5^{2}_{8}3^{3}_{8}4^{3}_{10}2^{4}_{10}3^{4}_{12}1^{5}_{12}2^{5}_{14}1^{6}_{16}$	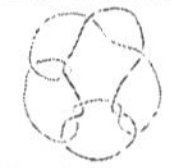	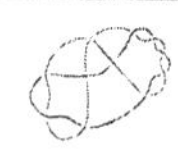
11^a_{359}: $1^{\underline{5}}_{\underline{10}}2^{\underline{4}}_{\underline{8}}1^{\underline{4}}_{\underline{6}}4^{\underline{3}}_{\underline{6}}2^{\underline{3}}_{\underline{4}}6^{\underline{2}}_{\underline{4}}4^{\underline{2}}_{\underline{2}}7^{\underline{1}}_{\underline{2}}6^{\underline{1}}_{0}9^{0}_{0}8^{0}_{2}7^{1}_{2}8^{1}_{4}7^{2}_{4}8^{2}_{6}4^{3}_{6}6^{3}_{8}2^{4}_{8}4^{4}_{10}1^{5}_{10}2^{5}_{12}1^{6}_{14}$ 11^a_{360}: $1^{\underline{2}}_{2}1^{\underline{1}}_{6}2^{0}_{6}1^{0}_{8}1^{1}_{8}1^{1}_{10}2^{2}_{10}1^{2}_{12}1^{3}_{12}2^{3}_{14}2^{4}_{14}1^{4}_{16}1^{5}_{16}2^{5}_{18}2^{6}_{18}2^{6}_{20}1^{7}_{20}1^{7}_{22}1^{8}_{22}1^{8}_{24}1^{9}_{26}$	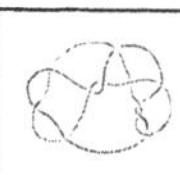	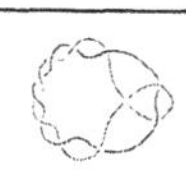
11^a_{361}: $1^{\underline{2}}_{0}1^{\underline{1}}_{2}1^{\underline{1}}_{4}4^{0}_{4}2^{0}_{6}5^{1}_{6}3^{1}_{8}6^{2}_{8}5^{2}_{10}7^{3}_{10}6^{3}_{12}7^{4}_{12}7^{4}_{14}5^{5}_{14}7^{5}_{16}5^{6}_{16}6^{6}_{18}2^{7}_{18}4^{7}_{20}1^{8}_{20}2^{8}_{22}1^{9}_{24}$ 11^a_{362}: $1^{\underline{2}}_{0}1^{\underline{1}}_{2}1^{\underline{1}}_{4}3^{0}_{4}2^{0}_{6}4^{1}_{6}2^{1}_{8}5^{2}_{8}4^{2}_{10}5^{3}_{10}5^{3}_{12}6^{4}_{12}5^{4}_{14}4^{5}_{14}6^{5}_{16}4^{6}_{16}5^{6}_{18}2^{7}_{18}3^{7}_{20}1^{8}_{20}2^{8}_{22}1^{9}_{24}$	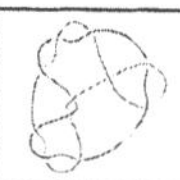	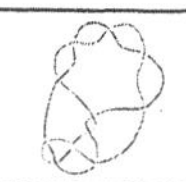
11^a_{363}: $1^{\underline{2}}_{\underline{2}}2^{\underline{1}}_{0}1^{\underline{1}}_{2}5^{0}_{2}3^{0}_{4}7^{1}_{4}4^{1}_{6}8^{2}_{6}7^{2}_{8}9^{3}_{8}8^{3}_{10}8^{4}_{10}9^{4}_{12}6^{5}_{12}8^{5}_{14}5^{6}_{14}7^{6}_{16}2^{7}_{16}4^{7}_{18}1^{8}_{18}2^{8}_{20}1^{9}_{22}$ 11^a_{364}: $1^{\underline{4}}_{\underline{6}}1^{\underline{3}}_{\underline{4}}1^{\underline{3}}_{\underline{2}}2^{\underline{2}}_{\underline{2}}1^{\underline{2}}_{0}2^{\underline{1}}_{0}2^{\underline{1}}_{2}4^{0}_{2}3^{0}_{4}3^{1}_{4}3^{1}_{6}4^{2}_{6}4^{2}_{8}3^{3}_{8}3^{3}_{10}2^{4}_{10}3^{4}_{12}1^{5}_{12}2^{5}_{14}1^{6}_{14}1^{6}_{16}1^{7}_{18}$	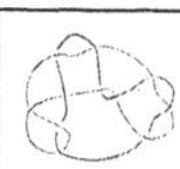	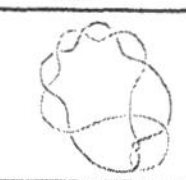
11^a_{365}: $1^{\underline{6}}_{\underline{14}}1^{\underline{5}}_{\underline{12}}1^{\underline{5}}_{\underline{10}}2^{\underline{4}}_{\underline{10}}1^{\underline{4}}_{\underline{8}}3^{\underline{3}}_{\underline{8}}2^{\underline{3}}_{\underline{6}}4^{\underline{2}}_{\underline{6}}4^{\underline{2}}_{\underline{4}}4^{\underline{1}}_{\underline{4}}3^{\underline{1}}_{\underline{2}}4^{0}_{\underline{2}}5^{0}_{0}3^{1}_{0}3^{1}_{2}2^{2}_{2}3^{2}_{4}1^{3}_{4}2^{3}_{6}1^{4}_{6}1^{4}_{8}1^{5}_{10}$ 11^a_{366}: $1^{\underline{6}}_{\underline{12}}1^{\underline{5}}_{\underline{10}}1^{\underline{5}}_{\underline{8}}3^{\underline{4}}_{\underline{8}}1^{\underline{4}}_{\underline{6}}4^{\underline{3}}_{\underline{6}}3^{\underline{3}}_{\underline{4}}6^{\underline{2}}_{\underline{4}}5^{\underline{2}}_{\underline{2}}6^{\underline{1}}_{\underline{2}}5^{\underline{1}}_{0}7^{0}_{0}7^{0}_{2}5^{1}_{2}6^{1}_{4}4^{2}_{4}5^{2}_{6}2^{3}_{6}4^{3}_{8}1^{4}_{8}2^{4}_{10}1^{5}_{12}$	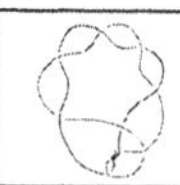	
11^a_{367}: $1^{\underline{8}}_{\underline{22}}1^{\underline{7}}_{\underline{20}}1^{\underline{7}}_{\underline{18}}2^{\underline{6}}_{\underline{18}}2^{\underline{6}}_{\underline{16}}3^{\underline{5}}_{\underline{16}}1^{\underline{5}}_{\underline{14}}2^{\underline{4}}_{\underline{14}}3^{\underline{4}}_{\underline{12}}3^{\underline{3}}_{\underline{12}}2^{\underline{3}}_{\underline{10}}2^{\underline{2}}_{\underline{10}}3^{\underline{2}}_{\underline{8}}2^{\underline{1}}_{\underline{8}}2^{\underline{1}}_{\underline{6}}2^{0}_{\underline{6}}3^{0}_{\underline{4}}1^{1}_{\underline{4}}1^{1}_{\underline{2}}1^{2}_{\underline{2}}1^{2}_{0}1^{3}_{2}$ 11^a_{368}: $1^{\underline{8}}_{\underline{20}}1^{\underline{7}}_{\underline{18}}1^{\underline{7}}_{\underline{16}}3^{\underline{6}}_{\underline{16}}2^{\underline{6}}_{\underline{14}}5^{\underline{5}}_{\underline{14}}2^{\underline{5}}_{\underline{12}}5^{\underline{4}}_{\underline{12}}5^{\underline{4}}_{\underline{10}}6^{\underline{3}}_{\underline{10}}5^{\underline{3}}_{\underline{8}}6^{\underline{2}}_{\underline{8}}6^{\underline{2}}_{\underline{6}}4^{\underline{1}}_{\underline{6}}6^{\underline{1}}_{\underline{4}}4^{0}_{\underline{4}}5^{0}_{\underline{2}}2^{1}_{\underline{2}}3^{1}_{0}1^{2}_{0}2^{2}_{2}1^{3}_{4}$	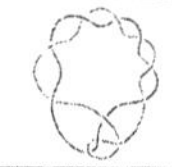	
11^a_{369}: $1^{\underline{7}}_{\underline{18}}1^{\underline{6}}_{\underline{16}}1^{\underline{6}}_{\underline{14}}3^{\underline{5}}_{\underline{14}}1^{\underline{5}}_{\underline{12}}4^{\underline{4}}_{\underline{12}}3^{\underline{4}}_{\underline{10}}4^{\underline{3}}_{\underline{10}}4^{\underline{3}}_{\underline{8}}6^{\underline{2}}_{\underline{8}}5^{\underline{2}}_{\underline{6}}4^{\underline{1}}_{\underline{6}}5^{\underline{1}}_{\underline{4}}4^{0}_{\underline{4}}5^{0}_{\underline{2}}3^{1}_{\underline{2}}3^{1}_{0}1^{2}_{0}3^{2}_{2}1^{3}_{2}1^{3}_{4}1^{4}_{6}$ 11^a_{370}: $1^{\underline{8}}_{\underline{18}}1^{\underline{7}}_{\underline{16}}1^{\underline{7}}_{\underline{14}}4^{\underline{6}}_{\underline{14}}2^{\underline{6}}_{\underline{12}}6^{\underline{5}}_{\underline{12}}3^{\underline{5}}_{\underline{10}}7^{\underline{4}}_{\underline{10}}6^{\underline{4}}_{\underline{8}}8^{\underline{3}}_{\underline{8}}7^{\underline{3}}_{\underline{6}}9^{\underline{2}}_{\underline{6}}8^{\underline{2}}_{\underline{4}}6^{\underline{1}}_{\underline{4}}9^{\underline{1}}_{\underline{2}}6^{0}_{\underline{2}}7^{0}_{0}3^{1}_{0}5^{1}_{2}1^{2}_{2}3^{2}_{4}1^{3}_{6}$	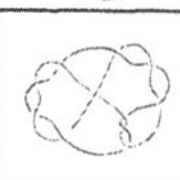	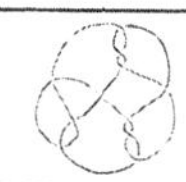
11^a_{371}: $1^{\underline{2}}_{\underline{2}}1^{\underline{1}}_{0}1^{\underline{1}}_{2}3^{0}_{2}2^{0}_{4}4^{1}_{4}2^{1}_{6}4^{2}_{6}4^{2}_{8}5^{3}_{8}4^{3}_{10}4^{4}_{10}5^{4}_{12}3^{5}_{12}4^{5}_{14}3^{6}_{14}4^{6}_{16}1^{7}_{16}2^{7}_{18}1^{8}_{18}1^{8}_{20}1^{9}_{22}$ 11^a_{372}: $1^{\underline{7}}_{\underline{20}}1^{\underline{6}}_{\underline{18}}1^{\underline{6}}_{\underline{16}}2^{\underline{5}}_{\underline{16}}1^{\underline{5}}_{\underline{14}}2^{\underline{4}}_{\underline{14}}2^{\underline{4}}_{\underline{12}}2^{\underline{3}}_{\underline{12}}2^{\underline{3}}_{\underline{10}}3^{\underline{2}}_{\underline{10}}3^{\underline{2}}_{\underline{8}}2^{\underline{1}}_{\underline{8}}2^{\underline{1}}_{\underline{6}}2^{0}_{\underline{6}}3^{0}_{\underline{4}}2^{1}_{\underline{4}}1^{1}_{\underline{2}}2^{2}_{0}1^{3}_{0}1^{4}_{4}$		

L: KH	1st line	2nd line
11^a_{373}: $1^{\underline{7}}_{\underline{16}}2^{\underline{6}}_{\underline{14}}1^{\underline{6}}_{\underline{12}}5^{\underline{5}}_{\underline{12}}2^{\underline{5}}_{\underline{10}}8^{\underline{4}}_{\underline{10}}5^{\underline{4}}_{\underline{8}}9^{\underline{3}}_{\underline{8}}8^{\underline{3}}_{\underline{6}}12^{\underline{2}}_{\underline{6}}10^{\underline{2}}_{\underline{4}}10^{\underline{1}}_{\underline{4}}11^{\underline{1}}_{\underline{2}}9^{0}_{\underline{2}}11^{0}_{0}6^{1}_{0}8^{1}_{2}3^{2}_{2}6^{2}_{4}1^{3}_{4}3^{3}_{6}1^{4}_{8}$ 11^a_{374}: $1^{\underline{8}}_{\underline{22}}1^{\underline{7}}_{\underline{20}}1^{\underline{7}}_{\underline{18}}3^{\underline{6}}_{\underline{18}}2^{\underline{6}}_{\underline{16}}4^{\underline{5}}_{\underline{16}}2^{\underline{5}}_{\underline{14}}3^{\underline{4}}_{\underline{14}}4^{\underline{4}}_{\underline{12}}5^{\underline{3}}_{\underline{12}}3^{\underline{3}}_{\underline{10}}3^{\underline{2}}_{\underline{10}}5^{\underline{2}}_{\underline{8}}3^{\underline{1}}_{\underline{8}}3^{\underline{1}}_{\underline{6}}3^{0}_{\underline{6}}4^{0}_{\underline{4}}1^{1}_{\underline{4}}2^{1}_{\underline{2}}1^{2}_{\underline{2}}1^{2}_{0}1^{3}_{2}$		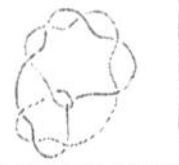
11^a_{375}: $1^{\underline{6}}_{\underline{14}}2^{\underline{5}}_{\underline{12}}1^{\underline{5}}_{\underline{10}}5^{\underline{4}}_{\underline{10}}2^{\underline{4}}_{\underline{8}}7^{\underline{3}}_{\underline{8}}5^{\underline{3}}_{\underline{6}}9^{\underline{2}}_{\underline{6}}8^{\underline{2}}_{\underline{4}}10^{\underline{1}}_{\underline{4}}8^{\underline{1}}_{\underline{2}}9^{0}_{\underline{2}}11^{0}_{0}7^{1}_{0}8^{1}_{2}5^{2}_{2}7^{2}_{4}2^{3}_{4}5^{3}_{6}1^{4}_{6}2^{4}_{8}1^{5}_{10}$ 11^a_{376}: $1^{\underline{8}}_{\underline{20}}1^{\underline{7}}_{\underline{18}}1^{\underline{7}}_{\underline{16}}4^{\underline{6}}_{\underline{16}}2^{\underline{6}}_{\underline{14}}6^{\underline{5}}_{\underline{14}}3^{\underline{5}}_{\underline{12}}6^{\underline{4}}_{\underline{12}}6^{\underline{4}}_{\underline{10}}8^{\underline{3}}_{\underline{10}}6^{\underline{3}}_{\underline{8}}7^{\underline{2}}_{\underline{8}}8^{\underline{2}}_{\underline{6}}5^{\underline{1}}_{\underline{6}}7^{\underline{1}}_{\underline{4}}5^{0}_{\underline{4}}6^{0}_{\underline{2}}2^{1}_{\underline{2}}4^{1}_{0}1^{2}_{0}2^{2}_{2}1^{3}_{4}$	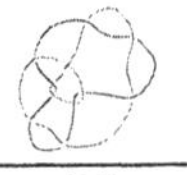	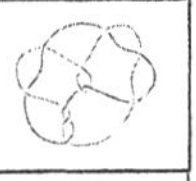
11^a_{377}: $1^{\underline{4}}_{\underline{6}}2^{\underline{3}}_{\underline{4}}1^{\underline{3}}_{\underline{2}}4^{\underline{2}}_{\underline{2}}2^{\underline{2}}_{0}5^{\underline{1}}_{0}4^{\underline{1}}_{2}8^{0}_{2}6^{0}_{4}7^{1}_{4}7^{1}_{6}8^{2}_{6}8^{2}_{8}6^{3}_{8}7^{3}_{10}4^{4}_{10}6^{4}_{12}2^{5}_{12}4^{5}_{14}1^{6}_{14}2^{6}_{16}1^{7}_{18}$ 11^a_{378}: $1^{\underline{4}}_{\underline{10}}2^{\underline{3}}_{\underline{8}}1^{\underline{3}}_{\underline{6}}4^{\underline{2}}_{\underline{6}}2^{\underline{2}}_{\underline{4}}5^{\underline{1}}_{\underline{4}}4^{\underline{1}}_{\underline{2}}6^{0}_{\underline{2}}6^{0}_{0}7^{1}_{0}5^{1}_{2}5^{2}_{2}8^{2}_{4}5^{3}_{4}4^{3}_{6}3^{4}_{6}5^{4}_{8}1^{5}_{8}3^{5}_{10}1^{6}_{10}1^{6}_{12}1^{7}_{14}$	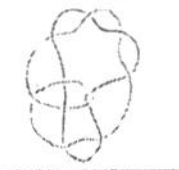	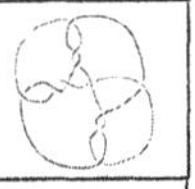
11^a_{379}: $1^{\underline{7}}_{\underline{18}}3^{\underline{6}}_{\underline{16}}1^{\underline{6}}_{\underline{14}}6^{\underline{5}}_{\underline{14}}3^{\underline{5}}_{\underline{12}}8^{\underline{4}}_{\underline{12}}6^{\underline{4}}_{\underline{10}}10^{\underline{3}}_{\underline{10}}8^{\underline{3}}_{\underline{8}}11^{\underline{2}}_{\underline{8}}11^{\underline{2}}_{\underline{6}}9^{\underline{1}}_{\underline{6}}10^{\underline{1}}_{\underline{4}}8^{0}_{\underline{4}}10^{0}_{\underline{2}}5^{1}_{\underline{2}}7^{1}_{0}2^{2}_{0}5^{2}_{2}1^{3}_{2}2^{3}_{4}1^{4}_{6}$ 11^a_{380}: $1^{\underline{7}}_{\underline{18}}2^{\underline{6}}_{\underline{16}}1^{\underline{6}}_{\underline{14}}5^{\underline{5}}_{\underline{14}}2^{\underline{5}}_{\underline{12}}7^{\underline{4}}_{\underline{12}}5^{\underline{4}}_{\underline{10}}8^{\underline{3}}_{\underline{10}}7^{\underline{3}}_{\underline{8}}10^{\underline{2}}_{\underline{8}}9^{\underline{2}}_{\underline{6}}8^{\underline{1}}_{\underline{6}}9^{\underline{1}}_{\underline{4}}7^{0}_{\underline{4}}9^{0}_{\underline{2}}5^{1}_{\underline{2}}6^{1}_{0}2^{2}_{0}5^{2}_{2}1^{3}_{2}2^{3}_{4}1^{4}_{6}$	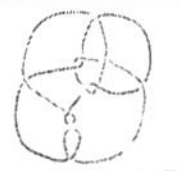	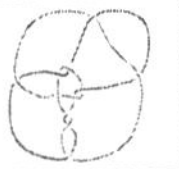
11^a_{381}: $1^{\underline{5}}_{\underline{10}}3^{\underline{4}}_{\underline{8}}1^{\underline{4}}_{\underline{6}}7^{\underline{3}}_{\underline{6}}3^{\underline{3}}_{\underline{4}}10^{\underline{2}}_{\underline{4}}7^{\underline{2}}_{\underline{2}}12^{\underline{1}}_{\underline{2}}10^{\underline{1}}_{0}15^{0}_{0}13^{0}_{2}12^{1}_{2}14^{1}_{4}11^{2}_{4}13^{2}_{6}7^{3}_{6}10^{3}_{8}3^{4}_{8}7^{4}_{10}1^{5}_{10}3^{5}_{12}1^{6}_{14}$ 11^a_{382}: $1^{\underline{6}}_{\underline{14}}2^{\underline{5}}_{\underline{12}}1^{\underline{5}}_{\underline{10}}4^{\underline{4}}_{\underline{10}}2^{\underline{4}}_{\underline{8}}6^{\underline{3}}_{\underline{8}}4^{\underline{3}}_{\underline{6}}8^{\underline{2}}_{\underline{6}}7^{\underline{2}}_{\underline{4}}8^{\underline{1}}_{\underline{4}}7^{\underline{1}}_{\underline{2}}8^{0}_{\underline{2}}9^{0}_{0}6^{1}_{0}7^{1}_{2}4^{2}_{2}6^{2}_{4}2^{3}_{4}4^{3}_{6}1^{4}_{6}2^{4}_{8}1^{5}_{10}$	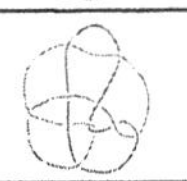	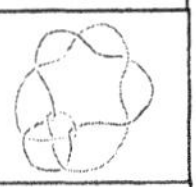
11^a_{383}: $1^{\underline{5}}_{\underline{10}}1^{\underline{4}}_{\underline{8}}1^{\underline{4}}_{\underline{6}}3^{\underline{3}}_{\underline{6}}1^{\underline{3}}_{\underline{4}}4^{\underline{2}}_{\underline{4}}3^{\underline{2}}_{\underline{2}}4^{\underline{1}}_{\underline{2}}4^{\underline{1}}_{0}7^{0}_{0}5^{0}_{2}4^{1}_{2}6^{1}_{4}5^{2}_{4}5^{2}_{6}3^{3}_{6}4^{3}_{8}1^{4}_{8}3^{4}_{10}1^{5}_{10}1^{5}_{12}1^{6}_{14}$ 11^a_{384}: $1^{\underline{2}}_{\underline{2}}2^{\underline{1}}_{0}1^{\underline{1}}_{2}6^{0}_{2}3^{0}_{4}9^{1}_{4}5^{1}_{6}10^{2}_{6}9^{2}_{8}12^{3}_{8}10^{3}_{10}11^{4}_{10}12^{4}_{12}8^{5}_{12}11^{5}_{14}7^{6}_{14}9^{6}_{16}3^{7}_{16}6^{7}_{18}1^{8}_{18}3^{8}_{20}1^{9}_{22}$	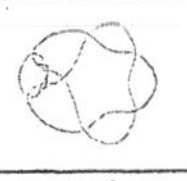	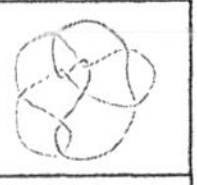
11^a_{385}: $1^{\underline{4}}_{\underline{7}}1^{\underline{4}}_{\underline{5}}2^{\underline{3}}_{\underline{5}}5^{\underline{2}}_{\underline{3}}2^{\underline{2}}_{\underline{1}}5^{\underline{1}}_{\underline{1}}5^{\underline{1}}_{1}12^{0}_{1}8^{0}_{3}10^{1}_{3}9^{1}_{5}11^{2}_{5}10^{2}_{7}7^{3}_{7}11^{3}_{9}7^{4}_{9}7^{4}_{11}3^{5}_{11}7^{5}_{13}1^{6}_{13}3^{6}_{15}1^{7}_{17}$ 11^a_{386}: $1^{\underline{6}}_{\underline{13}}1^{\underline{5}}_{\underline{11}}1^{\underline{5}}_{\underline{9}}6^{\underline{4}}_{\underline{9}}4^{\underline{4}}_{\underline{7}}8^{\underline{3}}_{\underline{7}}3^{\underline{3}}_{\underline{5}}11^{\underline{2}}_{\underline{5}}8^{\underline{2}}_{\underline{3}}9^{\underline{1}}_{\underline{3}}11^{\underline{1}}_{\underline{1}}13^{0}_{\underline{1}}10^{0}_{1}8^{1}_{1}12^{1}_{3}7^{2}_{3}8^{2}_{5}3^{3}_{5}7^{3}_{7}1^{4}_{7}3^{4}_{9}1^{5}_{11}$	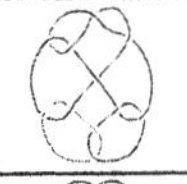	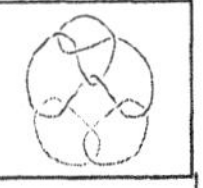
11^a_{387}: $1^{\underline{11}}_{\underline{29}}1^{\underline{10}}_{\underline{27}}1^{\underline{10}}_{\underline{25}}5^{\underline{9}}_{\underline{25}}1^{\underline{9}}_{\underline{23}}5^{\underline{8}}_{\underline{23}}5^{\underline{8}}_{\underline{21}}10^{\underline{7}}_{\underline{21}}5^{\underline{7}}_{\underline{19}}8^{\underline{6}}_{\underline{19}}10^{\underline{6}}_{\underline{17}}8^{\underline{5}}_{\underline{17}}8^{\underline{5}}_{\underline{15}}9^{\underline{4}}_{\underline{15}}11^{\underline{4}}_{\underline{13}}5^{\underline{3}}_{\underline{13}}6^{\underline{3}}_{\underline{11}}3^{\underline{2}}_{\underline{11}}5^{\underline{2}}_{\underline{9}}3^{\underline{1}}_{\underline{7}}1^{0}_{\underline{7}}1^{0}_{\underline{5}}$ 11^a_{388}: $1^{\underline{11}}_{\underline{27}}1^{\underline{10}}_{\underline{25}}1^{\underline{10}}_{\underline{23}}5^{\underline{9}}_{\underline{23}}1^{\underline{9}}_{\underline{21}}5^{\underline{8}}_{\underline{21}}5^{\underline{8}}_{\underline{19}}8^{\underline{7}}_{\underline{19}}5^{\underline{7}}_{\underline{17}}8^{\underline{6}}_{\underline{17}}8^{\underline{6}}_{\underline{15}}7^{\underline{5}}_{\underline{15}}8^{\underline{5}}_{\underline{13}}9^{\underline{4}}_{\underline{13}}10^{\underline{4}}_{\underline{11}}4^{\underline{3}}_{\underline{11}}6^{\underline{3}}_{\underline{9}}3^{\underline{2}}_{\underline{9}}4^{\underline{2}}_{\underline{7}}3^{\underline{1}}_{\underline{5}}1^{0}_{\underline{5}}1^{0}_{\underline{3}}$	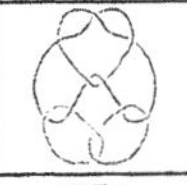	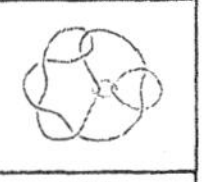
11^a_{389}: $1^{\underline{6}}_{\underline{13}}1^{\underline{5}}_{\underline{11}}1^{\underline{5}}_{\underline{9}}5^{\underline{4}}_{\underline{9}}2^{\underline{4}}_{\underline{7}}6^{\underline{3}}_{\underline{7}}4^{\underline{3}}_{\underline{5}}10^{\underline{2}}_{\underline{5}}6^{\underline{2}}_{\underline{3}}8^{\underline{1}}_{\underline{3}}10^{\underline{1}}_{\underline{1}}11^{0}_{\underline{1}}11^{0}_{1}9^{1}_{1}8^{1}_{3}5^{2}_{3}9^{2}_{5}3^{3}_{5}5^{3}_{7}1^{4}_{7}3^{4}_{9}1^{5}_{11}$ 11^a_{390}: $1^{\underline{8}}_{\underline{19}}1^{\underline{7}}_{\underline{17}}1^{\underline{7}}_{\underline{15}}5^{\underline{6}}_{\underline{15}}1^{\underline{6}}_{\underline{13}}5^{\underline{5}}_{\underline{13}}5^{\underline{5}}_{\underline{11}}11^{\underline{4}}_{\underline{11}}8^{\underline{4}}_{\underline{9}}11^{\underline{3}}_{\underline{9}}8^{\underline{3}}_{\underline{7}}10^{\underline{2}}_{\underline{7}}11^{\underline{2}}_{\underline{5}}8^{\underline{1}}_{\underline{5}}10^{\underline{1}}_{\underline{3}}7^{0}_{\underline{3}}9^{0}_{\underline{1}}4^{1}_{\underline{1}}6^{1}_{1}1^{2}_{1}4^{2}_{3}1^{3}_{5}$	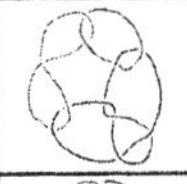	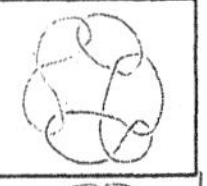
11^a_{391}: $1^{\underline{11}}_{\underline{29}}1^{\underline{10}}_{\underline{27}}1^{\underline{10}}_{\underline{25}}5^{\underline{9}}_{\underline{25}}1^{\underline{9}}_{\underline{23}}4^{\underline{8}}_{\underline{23}}5^{\underline{8}}_{\underline{21}}8^{\underline{7}}_{\underline{21}}4^{\underline{7}}_{\underline{19}}7^{\underline{6}}_{\underline{19}}8^{\underline{6}}_{\underline{17}}7^{\underline{5}}_{\underline{17}}7^{\underline{5}}_{\underline{15}}7^{\underline{4}}_{\underline{15}}10^{\underline{4}}_{\underline{13}}4^{\underline{3}}_{\underline{13}}4^{\underline{3}}_{\underline{11}}3^{\underline{2}}_{\underline{11}}4^{\underline{2}}_{\underline{9}}3^{\underline{1}}_{\underline{7}}1^{0}_{\underline{7}}1^{0}_{\underline{5}}$ 11^a_{392}: $1^{\underline{11}}_{\underline{27}}1^{\underline{10}}_{\underline{25}}1^{\underline{10}}_{\underline{23}}5^{\underline{9}}_{\underline{23}}1^{\underline{9}}_{\underline{21}}5^{\underline{8}}_{\underline{21}}5^{\underline{8}}_{\underline{19}}9^{\underline{7}}_{\underline{19}}5^{\underline{7}}_{\underline{17}}10^{\underline{6}}_{\underline{17}}9^{\underline{6}}_{\underline{15}}8^{\underline{5}}_{\underline{15}}10^{\underline{5}}_{\underline{13}}10^{\underline{4}}_{\underline{13}}11^{\underline{4}}_{\underline{11}}6^{\underline{3}}_{\underline{11}}7^{\underline{3}}_{\underline{9}}4^{\underline{2}}_{\underline{9}}6^{\underline{2}}_{\underline{7}}4^{\underline{1}}_{\underline{5}}1^{0}_{\underline{5}}1^{0}_{\underline{3}}$	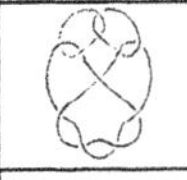	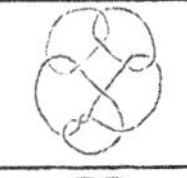
11^a_{393}: $1^{\underline{4}}_{\underline{5}}1^{\underline{4}}_{\underline{3}}1^{\underline{3}}_{\underline{3}}4^{\underline{2}}_{\underline{1}}1^{\underline{2}}_{1}1^{\underline{1}}_{1}4^{\underline{1}}_{3}8^{0}_{3}4^{0}_{5}4^{1}_{5}5^{1}_{7}7^{2}_{7}4^{2}_{9}3^{3}_{9}7^{3}_{11}4^{4}_{11}3^{4}_{13}2^{5}_{13}4^{5}_{15}1^{6}_{15}2^{6}_{17}1^{7}_{19}$ 11^a_{394}: $1^{\underline{4}}_{\underline{7}}1^{\underline{4}}_{\underline{5}}2^{\underline{3}}_{\underline{5}}5^{\underline{2}}_{\underline{3}}2^{\underline{2}}_{\underline{1}}5^{\underline{1}}_{\underline{1}}5^{\underline{1}}_{1}11^{0}_{1}8^{0}_{3}8^{1}_{3}8^{1}_{5}10^{2}_{5}8^{2}_{7}6^{3}_{7}10^{3}_{9}5^{4}_{9}6^{4}_{11}2^{5}_{11}5^{5}_{13}1^{6}_{13}2^{6}_{15}1^{7}_{17}$	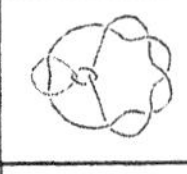	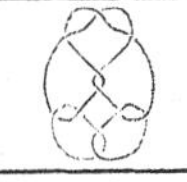
11^a_{395}: $1^{\underline{9}}_{\underline{23}}1^{\underline{8}}_{\underline{21}}1^{\underline{8}}_{\underline{19}}4^{\underline{7}}_{\underline{19}}1^{\underline{7}}_{\underline{17}}5^{\underline{6}}_{\underline{17}}4^{\underline{6}}_{\underline{15}}7^{\underline{5}}_{\underline{15}}5^{\underline{5}}_{\underline{13}}10^{\underline{4}}_{\underline{13}}10^{\underline{4}}_{\underline{11}}8^{\underline{3}}_{\underline{11}}7^{\underline{3}}_{\underline{9}}7^{\underline{2}}_{\underline{9}}8^{\underline{2}}_{\underline{7}}4^{\underline{1}}_{\underline{7}}7^{\underline{1}}_{\underline{5}}4^{0}_{\underline{5}}5^{0}_{\underline{3}}1^{1}_{\underline{3}}3^{1}_{\underline{1}}1^{2}_{1}$ 11^a_{396}: $1^{\underline{4}}_{\underline{7}}1^{\underline{4}}_{\underline{5}}1^{\underline{3}}_{\underline{5}}4^{\underline{2}}_{\underline{3}}1^{\underline{2}}_{\underline{1}}2^{\underline{1}}_{\underline{1}}4^{\underline{1}}_{1}10^{0}_{1}5^{0}_{3}7^{1}_{3}7^{1}_{5}8^{2}_{5}7^{2}_{7}6^{3}_{7}8^{3}_{9}5^{4}_{9}6^{4}_{11}3^{5}_{11}5^{5}_{13}1^{6}_{13}3^{6}_{15}1^{7}_{17}$	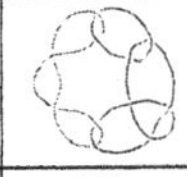	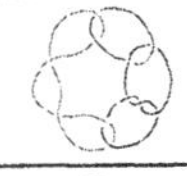
11^a_{397}: $1^{\underline{4}}_{\underline{9}}1^{\underline{4}}_{\underline{7}}1^{\underline{3}}_{\underline{7}}3^{\underline{2}}_{\underline{5}}1^{\underline{2}}_{\underline{3}}1^{\underline{1}}_{\underline{3}}3^{\underline{1}}_{\underline{1}}6^{0}_{\underline{1}}4^{0}_{1}4^{1}_{1}3^{1}_{3}4^{2}_{3}4^{2}_{5}3^{3}_{5}4^{3}_{7}2^{4}_{7}3^{4}_{9}2^{5}_{9}2^{5}_{11}1^{6}_{11}2^{6}_{13}1^{7}_{15}$ 11^a_{398}: $1^{\underline{6}}_{\underline{13}}4^{\underline{5}}_{\underline{11}}1^{\underline{5}}_{\underline{9}}6^{\underline{4}}_{\underline{9}}4^{\underline{4}}_{\underline{7}}11^{\underline{3}}_{\underline{7}}6^{\underline{3}}_{\underline{5}}12^{\underline{2}}_{\underline{5}}11^{\underline{2}}_{\underline{3}}13^{\underline{1}}_{\underline{3}}12^{\underline{1}}_{\underline{1}}14^{0}_{\underline{1}}15^{0}_{1}9^{1}_{1}12^{1}_{3}8^{2}_{3}11^{2}_{5}3^{3}_{5}6^{3}_{7}1^{4}_{7}3^{4}_{9}1^{5}_{11}$	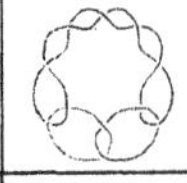	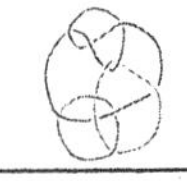
11^a_{399}: $1^{\underline{5}}_{\underline{11}}3^{\underline{4}}_{\underline{9}}1^{\underline{4}}_{\underline{7}}4^{\underline{3}}_{\underline{7}}3^{\underline{3}}_{\underline{5}}10^{\underline{2}}_{\underline{5}}6^{\underline{2}}_{\underline{3}}9^{\underline{1}}_{\underline{3}}8^{\underline{1}}_{\underline{1}}12^{0}_{\underline{1}}11^{0}_{1}10^{1}_{1}10^{1}_{3}8^{2}_{3}10^{2}_{5}5^{3}_{5}8^{3}_{7}3^{4}_{7}5^{4}_{9}1^{5}_{9}3^{5}_{11}1^{6}_{13}$ 11^a_{400}: $1^{\underline{5}}_{\underline{9}}3^{\underline{4}}_{\underline{7}}1^{\underline{4}}_{\underline{5}}4^{\underline{3}}_{\underline{5}}3^{\underline{3}}_{\underline{3}}10^{\underline{2}}_{\underline{3}}6^{\underline{2}}_{\underline{1}}8^{\underline{1}}_{\underline{1}}8^{\underline{1}}_{1}13^{0}_{1}10^{0}_{3}9^{1}_{3}11^{1}_{5}9^{2}_{5}9^{2}_{7}5^{3}_{7}9^{3}_{9}3^{4}_{9}5^{4}_{11}1^{5}_{11}3^{5}_{13}1^{6}_{15}$	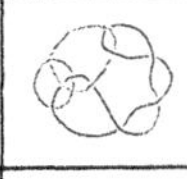	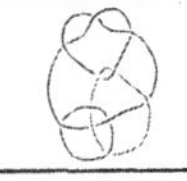
11^a_{401}: $1^{\underline{7}}_{\underline{17}}4^{\underline{6}}_{\underline{15}}1^{\underline{6}}_{\underline{13}}8^{\underline{5}}_{\underline{13}}4^{\underline{5}}_{\underline{11}}12^{\underline{4}}_{\underline{11}}8^{\underline{4}}_{\underline{9}}15^{\underline{3}}_{\underline{9}}12^{\underline{3}}_{\underline{7}}17^{\underline{2}}_{\underline{7}}17^{\underline{2}}_{\underline{5}}15^{\underline{1}}_{\underline{5}}15^{\underline{1}}_{\underline{3}}13^{0}_{\underline{3}}17^{0}_{\underline{1}}8^{1}_{\underline{1}}11^{1}_{1}4^{2}_{1}8^{2}_{3}1^{3}_{3}4^{3}_{5}1^{4}_{7}$ 11^a_{402}: $1^{\underline{6}}_{\underline{13}}3^{\underline{5}}_{\underline{11}}1^{\underline{5}}_{\underline{9}}6^{\underline{4}}_{\underline{9}}3^{\underline{4}}_{\underline{7}}9^{\underline{3}}_{\underline{7}}6^{\underline{3}}_{\underline{5}}12^{\underline{2}}_{\underline{5}}10^{\underline{2}}_{\underline{3}}12^{\underline{1}}_{\underline{3}}11^{\underline{1}}_{\underline{1}}13^{0}_{\underline{1}}14^{0}_{1}9^{1}_{1}11^{1}_{3}7^{2}_{3}10^{2}_{5}3^{3}_{5}6^{3}_{7}1^{4}_{7}3^{4}_{9}1^{5}_{11}$	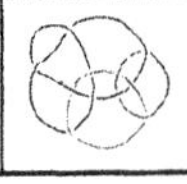	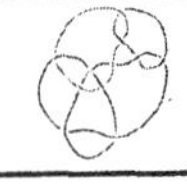

L: KH	1st line	2nd line
11^a_{403}: $1^{\underline{5}}_{\underline{9}}3^{\underline{4}}_{\underline{7}}1^{\underline{4}}_{\underline{5}}5^{\underline{3}}_{\underline{5}}3^{\underline{3}}_{\underline{3}}10^{\underline{2}}_{\underline{3}}6^{\underline{2}}_{\underline{1}}9^{\underline{1}}_{\underline{1}}9^{\underline{1}}_{1}14^{0}_{1}11^{0}_{3}10^{1}_{3}12^{1}_{5}10^{2}_{5}11^{2}_{7}6^{3}_{7}9^{3}_{9}3^{4}_{9}6^{4}_{11}1^{5}_{11}3^{5}_{13}1^{6}_{15}$	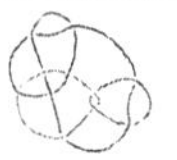	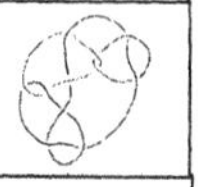
11^a_{404}: $1^{\underline{8}}_{\underline{19}}2^{\underline{7}}_{\underline{17}}1^{\underline{7}}_{\underline{15}}6^{\underline{6}}_{\underline{15}}4^{\underline{6}}_{\underline{13}}8^{\underline{5}}_{\underline{13}}4^{\underline{5}}_{\underline{11}}11^{\underline{4}}_{\underline{11}}9^{\underline{4}}_{\underline{9}}10^{\underline{3}}_{\underline{9}}10^{\underline{3}}_{\underline{7}}12^{\underline{2}}_{\underline{7}}10^{\underline{2}}_{\underline{5}}7^{\underline{1}}_{\underline{5}}12^{\underline{1}}_{\underline{3}}7^{0}_{\underline{3}}8^{0}_{\underline{1}}3^{1}_{\underline{1}}6^{1}_{1}1^{2}_{1}3^{2}_{3}1^{3}_{5}$		
11^a_{405}: $1^{\underline{9}}_{\underline{23}}2^{\underline{8}}_{\underline{21}}1^{\underline{8}}_{\underline{19}}5^{\underline{7}}_{\underline{19}}2^{\underline{7}}_{\underline{17}}6^{\underline{6}}_{\underline{17}}5^{\underline{6}}_{\underline{15}}9^{\underline{5}}_{\underline{15}}6^{\underline{5}}_{\underline{13}}9^{\underline{4}}_{\underline{13}}10^{\underline{4}}_{\underline{11}}8^{\underline{3}}_{\underline{11}}8^{\underline{3}}_{\underline{9}}8^{\underline{2}}_{\underline{9}}10^{\underline{2}}_{\underline{7}}4^{\underline{1}}_{\underline{7}}6^{\underline{1}}_{\underline{5}}3^{0}_{\underline{5}}5^{0}_{\underline{3}}1^{1}_{\underline{3}}2^{1}_{\underline{1}}1^{2}_{1}$	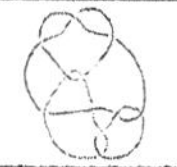	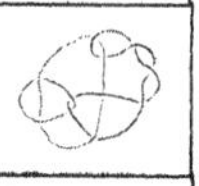
11^a_{406}: $1^{\underline{9}}_{\underline{21}}2^{\underline{8}}_{\underline{19}}1^{\underline{8}}_{\underline{17}}4^{\underline{7}}_{\underline{17}}2^{\underline{7}}_{\underline{15}}6^{\underline{6}}_{\underline{15}}4^{\underline{6}}_{\underline{13}}7^{\underline{5}}_{\underline{13}}6^{\underline{5}}_{\underline{11}}9^{\underline{4}}_{\underline{11}}8^{\underline{4}}_{\underline{9}}7^{\underline{3}}_{\underline{9}}8^{\underline{3}}_{\underline{7}}8^{\underline{2}}_{\underline{7}}9^{\underline{2}}_{\underline{5}}4^{\underline{1}}_{\underline{5}}6^{\underline{1}}_{\underline{3}}3^{0}_{\underline{3}}5^{0}_{\underline{1}}1^{1}_{\underline{1}}2^{1}_{1}1^{2}_{3}$		
11^a_{407}: $1^{\underline{4}}_{\underline{7}}1^{\underline{4}}_{\underline{5}}2^{\underline{3}}_{\underline{5}}5^{\underline{2}}_{\underline{3}}2^{\underline{2}}_{\underline{1}}5^{\underline{1}}_{\underline{1}}5^{\underline{1}}_{1}11^{0}_{1}6^{0}_{3}8^{1}_{3}10^{1}_{5}11^{2}_{5}10^{2}_{7}8^{3}_{7}9^{3}_{9}5^{4}_{9}8^{4}_{11}3^{5}_{11}5^{5}_{13}1^{6}_{13}3^{6}_{15}1^{7}_{17}$		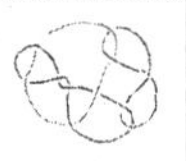
11^a_{408}: $1^{\underline{6}}_{\underline{13}}2^{\underline{5}}_{\underline{11}}1^{\underline{5}}_{\underline{9}}5^{\underline{4}}_{\underline{9}}3^{\underline{4}}_{\underline{7}}8^{\underline{3}}_{\underline{7}}4^{\underline{3}}_{\underline{5}}9^{\underline{2}}_{\underline{5}}8^{\underline{2}}_{\underline{3}}10^{\underline{1}}_{\underline{3}}9^{\underline{1}}_{\underline{1}}11^{0}_{\underline{1}}11^{0}_{1}7^{1}_{1}10^{1}_{3}7^{2}_{3}9^{2}_{5}3^{3}_{5}5^{3}_{7}1^{4}_{7}3^{4}_{9}1^{5}_{11}$		
11^a_{409}: $1^{\underline{5}}_{\underline{9}}2^{\underline{4}}_{\underline{7}}1^{\underline{4}}_{\underline{5}}5^{\underline{3}}_{\underline{5}}2^{\underline{3}}_{\underline{3}}6^{\underline{2}}_{\underline{3}}5^{\underline{2}}_{\underline{1}}8^{\underline{1}}_{\underline{1}}6^{\underline{1}}_{1}10^{0}_{1}10^{0}_{3}8^{1}_{3}8^{1}_{5}8^{2}_{5}9^{2}_{7}4^{3}_{7}7^{3}_{9}3^{4}_{9}4^{4}_{11}3^{5}_{13}1^{6}_{13}1^{6}_{15}$		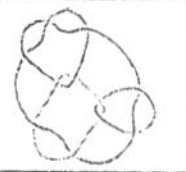
11^a_{410}: $1^{\underline{4}}_{\underline{5}}1^{\underline{3}}_{\underline{3}}1^{\underline{3}}_{\underline{1}}5^{\underline{2}}_{\underline{1}}2^{\underline{2}}_{1}3^{\underline{1}}_{1}4^{\underline{1}}_{3}9^{0}_{3}5^{0}_{5}5^{1}_{5}7^{1}_{7}8^{2}_{7}6^{2}_{9}5^{3}_{9}7^{3}_{11}4^{4}_{11}5^{4}_{13}2^{5}_{13}4^{5}_{15}1^{6}_{15}2^{6}_{17}1^{7}_{19}$		
11^a_{411}: $1^{\underline{4}}_{\underline{7}}1^{\underline{3}}_{\underline{5}}1^{\underline{3}}_{\underline{3}}5^{\underline{2}}_{\underline{3}}2^{\underline{2}}_{\underline{1}}4^{\underline{1}}_{\underline{1}}4^{\underline{1}}_{1}9^{0}_{1}6^{0}_{3}7^{1}_{3}7^{1}_{5}8^{2}_{5}8^{2}_{7}6^{3}_{7}7^{3}_{9}4^{4}_{9}6^{4}_{11}2^{5}_{11}4^{5}_{13}1^{6}_{13}2^{6}_{15}1^{7}_{17}$		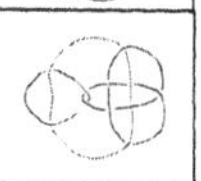
11^a_{412}: $1^{\underline{8}}_{\underline{19}}3^{\underline{7}}_{\underline{17}}1^{\underline{7}}_{\underline{15}}7^{\underline{6}}_{\underline{15}}3^{\underline{6}}_{\underline{13}}11^{\underline{5}}_{\underline{13}}7^{\underline{5}}_{\underline{11}}14^{\underline{4}}_{\underline{11}}12^{\underline{4}}_{\underline{9}}15^{\underline{3}}_{\underline{9}}13^{\underline{3}}_{\underline{7}}16^{\underline{2}}_{\underline{7}}17^{\underline{2}}_{\underline{5}}11^{\underline{1}}_{\underline{5}}14^{\underline{1}}_{\underline{3}}9^{0}_{\underline{3}}12^{0}_{\underline{1}}4^{1}_{\underline{1}}8^{1}_{1}1^{2}_{1}4^{2}_{3}1^{3}_{5}$		
11^a_{413}: $1^{\underline{11}}_{\underline{29}}2^{\underline{10}}_{\underline{27}}1^{\underline{10}}_{\underline{25}}5^{\underline{9}}_{\underline{25}}2^{\underline{9}}_{\underline{23}}7^{\underline{8}}_{\underline{23}}5^{\underline{8}}_{\underline{21}}9^{\underline{7}}_{\underline{21}}7^{\underline{7}}_{\underline{19}}10^{\underline{6}}_{\underline{19}}11^{\underline{6}}_{\underline{17}}10^{\underline{5}}_{\underline{17}}8^{\underline{5}}_{\underline{15}}7^{\underline{4}}_{\underline{15}}11^{\underline{4}}_{\underline{13}}5^{\underline{3}}_{\underline{13}}6^{\underline{3}}_{\underline{11}}3^{\underline{2}}_{\underline{11}}5^{\underline{2}}_{\underline{9}}3^{\underline{1}}_{\underline{7}}1^{0}_{\underline{7}}1^{0}_{\underline{5}}$	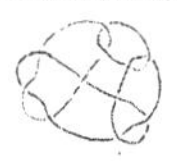	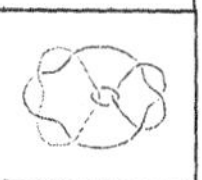
11^a_{414}: $1^{\underline{4}}_{\underline{5}}1^{\underline{4}}_{\underline{3}}1^{\underline{3}}_{\underline{3}}4^{\underline{2}}_{\underline{1}}1^{\underline{2}}_{1}2^{\underline{1}}_{1}4^{\underline{1}}_{3}8^{0}_{3}3^{0}_{5}4^{1}_{5}7^{1}_{7}8^{2}_{7}6^{2}_{9}5^{3}_{9}6^{3}_{11}4^{4}_{11}5^{4}_{13}2^{5}_{13}4^{5}_{15}1^{6}_{15}2^{6}_{17}1^{7}_{19}$		
11^a_{415}: $1^{\underline{4}}_{\underline{7}}1^{\underline{4}}_{\underline{5}}2^{\underline{3}}_{\underline{5}}4^{\underline{2}}_{\underline{3}}2^{\underline{2}}_{\underline{1}}5^{\underline{1}}_{\underline{1}}4^{\underline{1}}_{1}9^{0}_{1}6^{0}_{3}7^{1}_{3}8^{1}_{5}10^{2}_{5}9^{2}_{7}6^{3}_{7}8^{3}_{9}5^{4}_{9}6^{4}_{11}2^{5}_{11}5^{5}_{13}1^{6}_{13}2^{6}_{15}1^{7}_{17}$	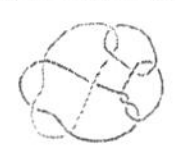	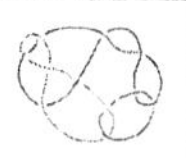
11^a_{416}: $1^{\underline{7}}_{\underline{17}}2^{\underline{6}}_{\underline{15}}1^{\underline{6}}_{\underline{13}}4^{\underline{5}}_{\underline{13}}2^{\underline{5}}_{\underline{11}}8^{\underline{4}}_{\underline{11}}5^{\underline{4}}_{\underline{9}}8^{\underline{3}}_{\underline{9}}7^{\underline{3}}_{\underline{7}}11^{\underline{2}}_{\underline{7}}10^{\underline{2}}_{\underline{5}}9^{\underline{1}}_{\underline{5}}9^{\underline{1}}_{\underline{3}}8^{0}_{\underline{3}}10^{0}_{\underline{1}}5^{1}_{\underline{1}}7^{1}_{1}3^{2}_{1}5^{2}_{3}1^{3}_{3}3^{3}_{5}1^{4}_{7}$		
11^a_{417}: $1^{\underline{4}}_{\underline{7}}1^{\underline{4}}_{\underline{5}}1^{\underline{3}}_{\underline{5}}3^{\underline{2}}_{\underline{3}}1^{\underline{2}}_{\underline{1}}2^{\underline{1}}_{\underline{1}}3^{\underline{1}}_{1}7^{0}_{1}3^{0}_{3}4^{1}_{3}6^{1}_{5}7^{2}_{5}6^{2}_{7}5^{3}_{7}5^{3}_{9}3^{4}_{9}5^{4}_{11}2^{5}_{11}3^{5}_{13}1^{6}_{13}2^{6}_{15}1^{7}_{17}$	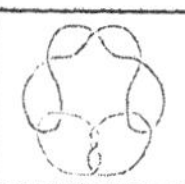	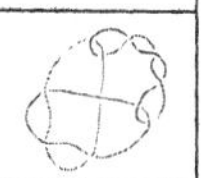
11^a_{418}: $1^{\underline{4}}_{\underline{9}}1^{\underline{4}}_{\underline{7}}2^{\underline{3}}_{\underline{7}}3^{\underline{2}}_{\underline{5}}2^{\underline{2}}_{\underline{3}}4^{\underline{1}}_{\underline{3}}3^{\underline{1}}_{\underline{1}}7^{0}_{\underline{1}}5^{0}_{1}5^{1}_{1}6^{1}_{3}7^{2}_{3}7^{2}_{5}5^{3}_{5}5^{3}_{7}3^{4}_{7}5^{4}_{9}2^{5}_{9}3^{5}_{11}1^{6}_{11}2^{6}_{13}1^{7}_{15}$		
11^a_{419}: $1^{\underline{4}}_{\underline{7}}1^{\underline{3}}_{\underline{5}}1^{\underline{3}}_{\underline{3}}5^{\underline{2}}_{\underline{3}}2^{\underline{2}}_{\underline{1}}5^{\underline{1}}_{\underline{1}}4^{\underline{1}}_{1}10^{0}_{1}7^{0}_{3}8^{1}_{3}8^{1}_{5}10^{2}_{5}9^{2}_{7}7^{3}_{7}9^{3}_{9}5^{4}_{9}7^{4}_{11}3^{5}_{11}5^{5}_{13}1^{6}_{13}3^{6}_{15}1^{7}_{17}$	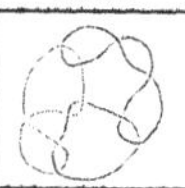	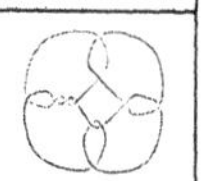
11^a_{420}: $1^{\underline{5}}_{\underline{9}}1^{\underline{4}}_{\underline{7}}1^{\underline{4}}_{\underline{5}}4^{\underline{3}}_{\underline{5}}1^{\underline{3}}_{\underline{3}}5^{\underline{2}}_{\underline{3}}4^{\underline{2}}_{\underline{1}}6^{\underline{1}}_{\underline{1}}5^{\underline{1}}_{1}9^{0}_{1}8^{0}_{3}7^{1}_{3}7^{1}_{5}7^{2}_{5}8^{2}_{7}4^{3}_{7}6^{3}_{9}3^{4}_{9}4^{4}_{11}3^{5}_{13}1^{6}_{13}1^{6}_{15}$		
11^a_{421}: $1^{\underline{11}}_{\underline{27}}2^{\underline{10}}_{\underline{25}}1^{\underline{10}}_{\underline{23}}6^{\underline{9}}_{\underline{23}}2^{\underline{9}}_{\underline{21}}8^{\underline{8}}_{\underline{21}}6^{\underline{8}}_{\underline{19}}10^{\underline{7}}_{\underline{19}}8^{\underline{7}}_{\underline{17}}13^{\underline{6}}_{\underline{17}}12^{\underline{6}}_{\underline{15}}11^{\underline{5}}_{\underline{15}}11^{\underline{5}}_{\underline{13}}10^{\underline{4}}_{\underline{13}}12^{\underline{4}}_{\underline{11}}6^{\underline{3}}_{\underline{11}}9^{\underline{3}}_{\underline{9}}4^{\underline{2}}_{\underline{9}}6^{\underline{2}}_{\underline{7}}4^{\underline{1}}_{\underline{5}}1^{0}_{\underline{5}}1^{0}_{\underline{3}}$	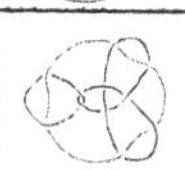	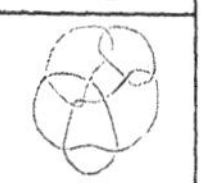
11^a_{422}: $1^{\underline{9}}_{\underline{23}}2^{\underline{8}}_{\underline{21}}1^{\underline{8}}_{\underline{19}}6^{\underline{7}}_{\underline{19}}2^{\underline{7}}_{\underline{17}}8^{\underline{6}}_{\underline{17}}6^{\underline{6}}_{\underline{15}}11^{\underline{5}}_{\underline{15}}8^{\underline{5}}_{\underline{13}}12^{\underline{4}}_{\underline{13}}12^{\underline{4}}_{\underline{11}}11^{\underline{3}}_{\underline{11}}11^{\underline{3}}_{\underline{9}}10^{\underline{2}}_{\underline{9}}13^{\underline{2}}_{\underline{7}}6^{\underline{1}}_{\underline{7}}8^{\underline{1}}_{\underline{5}}4^{0}_{\underline{5}}7^{0}_{\underline{3}}1^{1}_{\underline{3}}3^{1}_{\underline{1}}1^{2}_{1}$		
11^a_{423}: $1^{\underline{4}}_{\underline{7}}1^{\underline{4}}_{\underline{5}}3^{\underline{3}}_{\underline{5}}6^{\underline{2}}_{\underline{3}}3^{\underline{2}}_{\underline{1}}8^{\underline{1}}_{\underline{1}}6^{\underline{1}}_{1}13^{0}_{1}9^{0}_{3}11^{1}_{3}12^{1}_{5}14^{2}_{5}13^{2}_{7}9^{3}_{7}12^{3}_{9}7^{4}_{9}9^{4}_{11}3^{5}_{11}7^{5}_{13}1^{6}_{13}3^{6}_{15}1^{7}_{17}$	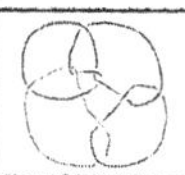	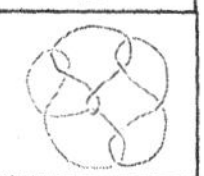
11^a_{424}: $1^{\underline{8}}_{\underline{19}}1^{\underline{7}}_{\underline{17}}1^{\underline{7}}_{\underline{15}}6^{\underline{6}}_{\underline{15}}3^{\underline{6}}_{\underline{13}}7^{\underline{5}}_{\underline{13}}4^{\underline{5}}_{\underline{11}}11^{\underline{4}}_{\underline{11}}8^{\underline{4}}_{\underline{9}}10^{\underline{3}}_{\underline{9}}10^{\underline{3}}_{\underline{7}}12^{\underline{2}}_{\underline{7}}10^{\underline{2}}_{\underline{5}}8^{\underline{1}}_{\underline{5}}12^{\underline{1}}_{\underline{3}}7^{0}_{\underline{3}}9^{0}_{\underline{1}}4^{1}_{\underline{1}}6^{1}_{1}1^{2}_{1}4^{2}_{3}1^{3}_{5}$		
11^a_{425}: $1^{\underline{4}}_{\underline{7}}1^{\underline{3}}_{\underline{5}}1^{\underline{3}}_{\underline{3}}6^{\underline{2}}_{\underline{3}}2^{\underline{2}}_{\underline{1}}6^{\underline{1}}_{\underline{1}}5^{\underline{1}}_{1}11^{0}_{1}8^{0}_{3}10^{1}_{3}9^{1}_{5}11^{2}_{5}11^{2}_{7}8^{3}_{7}10^{3}_{9}6^{4}_{9}8^{4}_{11}3^{5}_{11}6^{5}_{13}1^{6}_{13}3^{6}_{15}1^{7}_{17}$	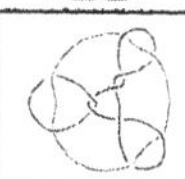	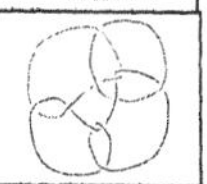
11^a_{426}: $1^{\underline{7}}_{\underline{17}}3^{\underline{6}}_{\underline{15}}1^{\underline{6}}_{\underline{13}}7^{\underline{5}}_{\underline{13}}3^{\underline{5}}_{\underline{11}}9^{\underline{4}}_{\underline{11}}7^{\underline{4}}_{\underline{9}}13^{\underline{3}}_{\underline{9}}9^{\underline{3}}_{\underline{7}}13^{\underline{2}}_{\underline{7}}14^{\underline{2}}_{\underline{5}}12^{\underline{1}}_{\underline{5}}12^{\underline{1}}_{\underline{3}}11^{0}_{\underline{3}}14^{0}_{\underline{1}}6^{1}_{\underline{1}}9^{1}_{1}4^{2}_{1}7^{2}_{3}1^{3}_{3}3^{3}_{5}1^{4}_{7}$		
11^a_{427}: $1^{\underline{6}}_{\underline{15}}1^{\underline{6}}_{\underline{13}}3^{\underline{5}}_{\underline{13}}4^{\underline{4}}_{\underline{11}}3^{\underline{4}}_{\underline{9}}6^{\underline{3}}_{\underline{9}}4^{\underline{3}}_{\underline{7}}10^{\underline{2}}_{\underline{7}}7^{\underline{2}}_{\underline{5}}7^{\underline{1}}_{\underline{5}}9^{\underline{1}}_{\underline{3}}10^{0}_{\underline{3}}9^{0}_{\underline{1}}7^{1}_{\underline{1}}8^{1}_{1}4^{2}_{1}7^{2}_{3}3^{3}_{3}4^{3}_{5}1^{4}_{5}3^{4}_{7}1^{5}_{9}$	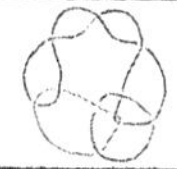	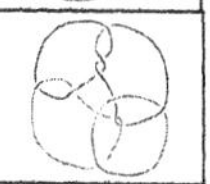
11^a_{428}: $1^{\underline{6}}_{\underline{13}}1^{\underline{6}}_{\underline{11}}3^{\underline{5}}_{\underline{11}}5^{\underline{4}}_{\underline{9}}3^{\underline{4}}_{\underline{7}}7^{\underline{3}}_{\underline{7}}5^{\underline{3}}_{\underline{5}}12^{\underline{2}}_{\underline{5}}8^{\underline{2}}_{\underline{3}}9^{\underline{1}}_{\underline{3}}11^{\underline{1}}_{\underline{1}}13^{0}_{\underline{1}}11^{0}_{1}8^{1}_{1}11^{1}_{3}6^{2}_{3}8^{2}_{5}3^{3}_{5}6^{3}_{7}1^{4}_{7}3^{4}_{9}1^{5}_{11}$		
11^a_{429}: $1^{\underline{3}}_{\underline{5}}3^{\underline{2}}_{\underline{3}}1^{\underline{2}}_{\underline{1}}5^{\underline{1}}_{\underline{1}}3^{\underline{1}}_{1}10^{0}_{1}7^{0}_{3}11^{1}_{3}8^{1}_{5}12^{2}_{5}12^{2}_{7}11^{3}_{7}11^{3}_{9}9^{4}_{9}11^{4}_{11}5^{5}_{11}9^{5}_{13}4^{6}_{13}6^{6}_{15}1^{7}_{15}3^{7}_{17}1^{8}_{19}$		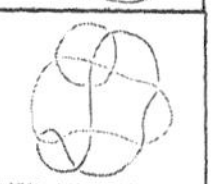
11^a_{430}: $1^{\underline{6}}_{\underline{13}}4^{\underline{5}}_{\underline{11}}1^{\underline{5}}_{\underline{9}}6^{\underline{4}}_{\underline{9}}4^{\underline{4}}_{\underline{7}}9^{\underline{3}}_{\underline{7}}6^{\underline{3}}_{\underline{5}}12^{\underline{2}}_{\underline{5}}9^{\underline{2}}_{\underline{3}}10^{\underline{1}}_{\underline{3}}12^{\underline{1}}_{\underline{1}}14^{0}_{\underline{1}}14^{0}_{1}9^{1}_{1}10^{1}_{3}5^{2}_{3}9^{2}_{5}3^{3}_{5}5^{3}_{7}1^{4}_{7}3^{4}_{9}1^{5}_{11}$		
11^a_{431}: $1^{\underline{5}}_{\underline{9}}3^{\underline{4}}_{\underline{7}}1^{\underline{4}}_{\underline{5}}4^{\underline{3}}_{\underline{5}}3^{\underline{3}}_{\underline{3}}8^{\underline{2}}_{\underline{3}}4^{\underline{2}}_{\underline{1}}6^{\underline{1}}_{\underline{1}}8^{\underline{1}}_{1}12^{0}_{1}10^{0}_{3}9^{1}_{3}8^{1}_{5}7^{2}_{5}9^{2}_{7}4^{3}_{7}7^{3}_{9}3^{4}_{9}4^{4}_{11}1^{5}_{11}3^{5}_{13}1^{6}_{15}$	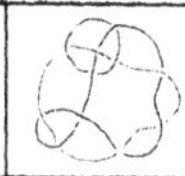	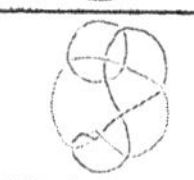
11^a_{432}: $1^{\underline{7}}_{\underline{17}}4^{\underline{6}}_{\underline{15}}1^{\underline{6}}_{\underline{13}}6^{\underline{5}}_{\underline{13}}4^{\underline{5}}_{\underline{11}}10^{\underline{4}}_{\underline{11}}8^{\underline{4}}_{\underline{9}}12^{\underline{3}}_{\underline{9}}8^{\underline{3}}_{\underline{7}}12^{\underline{2}}_{\underline{7}}12^{\underline{2}}_{\underline{5}}10^{\underline{1}}_{\underline{5}}12^{\underline{1}}_{\underline{3}}10^{0}_{\underline{3}}12^{0}_{\underline{1}}5^{1}_{\underline{1}}8^{1}_{1}3^{2}_{1}5^{2}_{3}1^{3}_{3}3^{3}_{5}1^{4}_{7}$		

L: KH	1st line	2nd line
11^a_{433}: $1^{\underline{7}}_{\underline{15}}3^{\underline{6}}_{\underline{13}}1^{\underline{6}}_{\underline{11}}3^{\underline{5}}_{\underline{11}}3^{\underline{5}}_{\underline{9}}8^{\underline{4}}_{\underline{9}}5^{\underline{4}}_{\underline{7}}8^{\underline{3}}_{\underline{7}}6^{\underline{3}}_{\underline{5}}10^{\underline{2}}_{\underline{5}}8^{\underline{2}}_{\underline{3}}7^{\underline{1}}_{\underline{3}}10^{\underline{1}}_{\underline{1}}9^{0}_{\underline{1}}9^{0}_{1}5^{1}_{1}7^{1}_{3}3^{2}_{3}5^{2}_{5}1^{3}_{5}3^{3}_{7}1^{4}_{9}$ 11^a_{434}: $1^{\underline{5}}_{\underline{11}}3^{\underline{4}}_{\underline{9}}1^{\underline{4}}_{\underline{7}}6^{\underline{3}}_{\underline{7}}3^{\underline{3}}_{\underline{5}}10^{\underline{2}}_{\underline{5}}6^{\underline{2}}_{\underline{3}}10^{\underline{1}}_{\underline{3}}10^{\underline{1}}_{\underline{1}}14^{0}_{\underline{1}}14^{0}_{1}12^{1}_{1}10^{1}_{3}9^{2}_{3}12^{2}_{5}5^{3}_{5}9^{3}_{7}3^{4}_{7}5^{4}_{9}1^{5}_{9}3^{5}_{11}1^{6}_{13}$	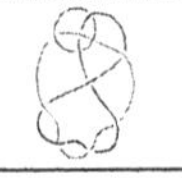	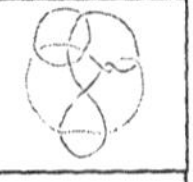
11^a_{435}: $1^{\underline{4}}_{\underline{7}}1^{\underline{4}}_{\underline{5}}3^{\underline{3}}_{\underline{5}}6^{\underline{2}}_{\underline{3}}3^{\underline{2}}_{\underline{1}}7^{\underline{1}}_{\underline{1}}6^{\underline{1}}_{1}12^{0}_{1}9^{0}_{3}10^{1}_{3}10^{1}_{5}11^{2}_{5}10^{2}_{7}7^{3}_{7}11^{3}_{9}6^{4}_{9}8^{4}_{11}2^{5}_{11}5^{5}_{13}1^{6}_{13}2^{6}_{15}1^{7}_{17}$ 11^a_{436}: $1^{\underline{9}}_{\underline{23}}2^{\underline{8}}_{\underline{21}}1^{\underline{8}}_{\underline{19}}6^{\underline{7}}_{\underline{19}}2^{\underline{7}}_{\underline{17}}7^{\underline{6}}_{\underline{17}}6^{\underline{6}}_{\underline{15}}9^{\underline{5}}_{\underline{15}}7^{\underline{5}}_{\underline{13}}10^{\underline{4}}_{\underline{13}}11^{\underline{4}}_{\underline{11}}9^{\underline{3}}_{\underline{11}}8^{\underline{3}}_{\underline{9}}7^{\underline{2}}_{\underline{9}}9^{\underline{2}}_{\underline{7}}3^{\underline{1}}_{\underline{7}}7^{\underline{1}}_{\underline{5}}3^{0}_{\underline{5}}5^{0}_{\underline{3}}1^{1}_{\underline{3}}1^{1}_{\underline{1}}1^{2}_{1}$		
11^a_{437}: $1^{\underline{9}}_{\underline{21}}2^{\underline{8}}_{\underline{19}}1^{\underline{8}}_{\underline{17}}4^{\underline{7}}_{\underline{17}}2^{\underline{7}}_{\underline{15}}5^{\underline{6}}_{\underline{15}}4^{\underline{6}}_{\underline{13}}5^{\underline{5}}_{\underline{13}}5^{\underline{5}}_{\underline{11}}8^{\underline{4}}_{\underline{11}}7^{\underline{4}}_{\underline{9}}6^{\underline{3}}_{\underline{9}}6^{\underline{3}}_{\underline{7}}5^{\underline{2}}_{\underline{7}}6^{\underline{2}}_{\underline{5}}2^{\underline{1}}_{\underline{5}}5^{\underline{1}}_{\underline{3}}3^{0}_{\underline{3}}4^{0}_{\underline{1}}1^{1}_{\underline{1}}1^{1}_{1}1^{2}_{3}$ 11^a_{438}: $1^{\underline{4}}_{\underline{7}}3^{\underline{3}}_{\underline{5}}1^{\underline{3}}_{\underline{3}}5^{\underline{2}}_{\underline{3}}3^{\underline{2}}_{\underline{1}}7^{\underline{1}}_{\underline{1}}5^{\underline{1}}_{1}11^{0}_{1}9^{0}_{3}11^{1}_{3}9^{1}_{5}10^{2}_{5}11^{2}_{7}7^{3}_{7}10^{3}_{9}7^{4}_{9}9^{4}_{11}3^{5}_{11}5^{5}_{13}1^{6}_{13}3^{6}_{15}1^{7}_{17}$	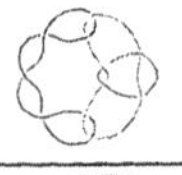	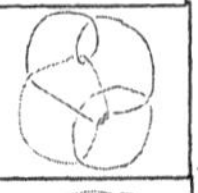
11^a_{439}: $1^{\underline{3}}_{\underline{3}}3^{\underline{2}}_{\underline{1}}1^{\underline{2}}_{1}4^{\underline{1}}_{1}3^{\underline{1}}_{3}9^{0}_{3}6^{0}_{5}9^{1}_{5}7^{1}_{7}12^{2}_{7}9^{2}_{9}8^{3}_{9}12^{3}_{11}10^{4}_{11}10^{4}_{13}6^{5}_{13}8^{5}_{15}3^{6}_{15}6^{6}_{17}1^{7}_{17}3^{7}_{19}1^{8}_{21}$ 11^a_{440}: $1^{\underline{9}}_{\underline{23}}2^{\underline{8}}_{\underline{21}}1^{\underline{8}}_{\underline{19}}4^{\underline{7}}_{\underline{19}}2^{\underline{7}}_{\underline{17}}6^{\underline{6}}_{\underline{17}}4^{\underline{6}}_{\underline{15}}7^{\underline{5}}_{\underline{15}}6^{\underline{5}}_{\underline{13}}10^{\underline{4}}_{\underline{13}}10^{\underline{4}}_{\underline{11}}8^{\underline{3}}_{\underline{11}}7^{\underline{3}}_{\underline{9}}6^{\underline{2}}_{\underline{9}}8^{\underline{2}}_{\underline{7}}4^{\underline{1}}_{\underline{7}}6^{\underline{1}}_{\underline{5}}3^{0}_{\underline{5}}5^{0}_{\underline{3}}1^{1}_{\underline{3}}2^{1}_{\underline{1}}1^{2}_{1}$	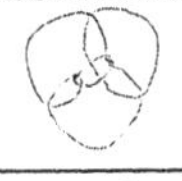	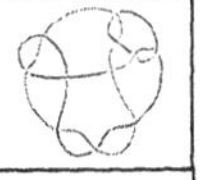
11^a_{441}: $1^{\underline{7}}_{\underline{15}}2^{\underline{6}}_{\underline{13}}1^{\underline{6}}_{\underline{11}}4^{\underline{5}}_{\underline{11}}2^{\underline{5}}_{\underline{9}}7^{\underline{4}}_{\underline{9}}5^{\underline{4}}_{\underline{7}}8^{\underline{3}}_{\underline{7}}6^{\underline{3}}_{\underline{5}}11^{\underline{2}}_{\underline{5}}8^{\underline{2}}_{\underline{3}}7^{\underline{1}}_{\underline{3}}11^{\underline{1}}_{\underline{1}}10^{0}_{\underline{1}}10^{0}_{1}6^{1}_{1}7^{1}_{3}3^{2}_{3}6^{2}_{5}1^{3}_{5}3^{3}_{7}1^{4}_{9}$ 11^a_{442}: $1^{\underline{4}}_{\underline{5}}1^{\underline{4}}_{\underline{3}}1^{\underline{3}}_{\underline{3}}3^{\underline{2}}_{\underline{1}}1^{\underline{2}}_{1}1^{\underline{1}}_{1}3^{\underline{1}}_{3}5^{0}_{3}2^{0}_{5}3^{1}_{5}4^{1}_{7}5^{2}_{7}3^{2}_{9}2^{3}_{9}5^{3}_{11}4^{4}_{11}4^{4}_{13}2^{5}_{13}2^{5}_{15}1^{6}_{15}2^{6}_{17}1^{7}_{19}$	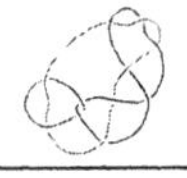	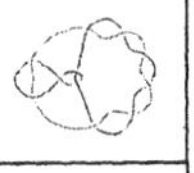
11^a_{443}: $1^{\underline{4}}_{\underline{7}}1^{\underline{4}}_{\underline{5}}2^{\underline{3}}_{\underline{5}}4^{\underline{2}}_{\underline{3}}2^{\underline{2}}_{\underline{1}}5^{\underline{1}}_{\underline{1}}4^{\underline{1}}_{1}8^{0}_{1}6^{0}_{3}7^{1}_{3}7^{1}_{5}8^{2}_{5}7^{2}_{7}5^{3}_{7}8^{3}_{9}5^{4}_{9}7^{4}_{11}2^{5}_{11}3^{5}_{13}1^{6}_{13}2^{6}_{15}1^{7}_{17}$ 11^a_{444}: $1^{\underline{4}}_{\underline{7}}1^{\underline{4}}_{\underline{5}}2^{\underline{3}}_{\underline{5}}4^{\underline{2}}_{\underline{3}}2^{\underline{2}}_{\underline{1}}5^{\underline{1}}_{\underline{1}}4^{\underline{1}}_{1}9^{0}_{1}6^{0}_{3}9^{1}_{3}8^{1}_{5}9^{2}_{5}9^{2}_{7}6^{3}_{7}9^{3}_{9}7^{4}_{9}8^{4}_{11}3^{5}_{11}5^{5}_{13}1^{6}_{13}3^{6}_{15}1^{7}_{17}$	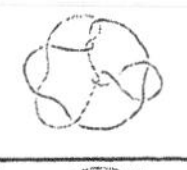	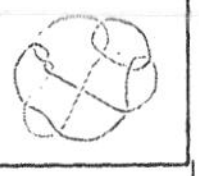
11^a_{445}: $1^{\underline{4}}_{\underline{7}}2^{\underline{3}}_{\underline{5}}1^{\underline{3}}_{\underline{3}}5^{\underline{2}}_{\underline{3}}2^{\underline{2}}_{\underline{1}}6^{\underline{1}}_{\underline{1}}5^{\underline{1}}_{1}11^{0}_{1}9^{0}_{3}11^{1}_{3}8^{1}_{5}10^{2}_{5}11^{2}_{7}7^{3}_{7}10^{3}_{9}7^{4}_{9}8^{4}_{11}3^{5}_{11}6^{5}_{13}1^{6}_{13}3^{6}_{15}1^{7}_{17}$ 11^a_{446}: $1^{\underline{7}}_{\underline{17}}2^{\underline{6}}_{\underline{15}}1^{\underline{6}}_{\underline{13}}4^{\underline{5}}_{\underline{13}}2^{\underline{5}}_{\underline{11}}6^{\underline{4}}_{\underline{11}}5^{\underline{4}}_{\underline{9}}7^{\underline{3}}_{\underline{9}}5^{\underline{3}}_{\underline{7}}8^{\underline{2}}_{\underline{7}}7^{\underline{2}}_{\underline{5}}6^{\underline{1}}_{\underline{5}}8^{\underline{1}}_{\underline{3}}7^{0}_{\underline{3}}9^{0}_{\underline{1}}4^{1}_{\underline{1}}4^{1}_{1}2^{2}_{1}4^{2}_{3}1^{3}_{3}2^{3}_{5}1^{4}_{7}$	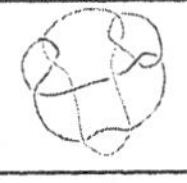	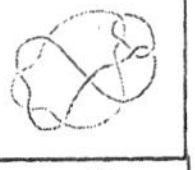
11^a_{447}: $1^{\underline{5}}_{\underline{9}}2^{\underline{4}}_{\underline{7}}1^{\underline{4}}_{\underline{5}}4^{\underline{3}}_{\underline{5}}2^{\underline{3}}_{\underline{3}}6^{\underline{2}}_{\underline{3}}4^{\underline{2}}_{\underline{1}}5^{\underline{1}}_{\underline{1}}6^{\underline{1}}_{1}10^{0}_{1}8^{0}_{3}7^{1}_{3}7^{1}_{5}6^{2}_{5}7^{2}_{7}3^{3}_{7}6^{3}_{9}3^{4}_{9}4^{4}_{11}1^{5}_{11}2^{5}_{13}1^{6}_{15}$ 11^a_{448}: $1^{\underline{4}}_{\underline{5}}1^{\underline{3}}_{\underline{3}}1^{\underline{3}}_{\underline{1}}4^{\underline{2}}_{\underline{1}}1^{\underline{2}}_{1}2^{\underline{1}}_{1}4^{\underline{1}}_{3}7^{0}_{3}5^{0}_{5}5^{1}_{5}4^{1}_{7}6^{2}_{7}5^{2}_{9}3^{3}_{9}6^{3}_{11}4^{4}_{11}4^{4}_{13}2^{5}_{13}3^{5}_{15}1^{6}_{15}2^{6}_{17}1^{7}_{19}$	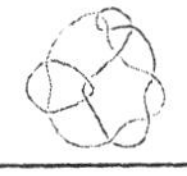	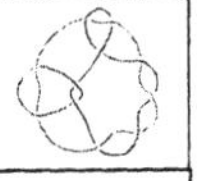
11^a_{449}: $1^{\underline{6}}_{\underline{13}}2^{\underline{5}}_{\underline{11}}1^{\underline{5}}_{\underline{9}}5^{\underline{4}}_{\underline{9}}3^{\underline{4}}_{\underline{7}}6^{\underline{3}}_{\underline{7}}4^{\underline{3}}_{\underline{5}}8^{\underline{2}}_{\underline{5}}6^{\underline{2}}_{\underline{3}}7^{\underline{1}}_{\underline{3}}8^{\underline{1}}_{\underline{1}}10^{0}_{\underline{1}}10^{0}_{1}6^{1}_{1}7^{1}_{3}4^{2}_{3}6^{2}_{5}2^{3}_{5}4^{3}_{7}1^{4}_{7}2^{4}_{9}1^{5}_{11}$ 11^a_{450}: $1^{\underline{4}}_{\underline{7}}2^{\underline{3}}_{\underline{5}}1^{\underline{3}}_{\underline{3}}5^{\underline{2}}_{\underline{3}}2^{\underline{2}}_{\underline{1}}6^{\underline{1}}_{\underline{1}}5^{\underline{1}}_{1}10^{0}_{1}9^{0}_{3}9^{1}_{3}7^{1}_{5}9^{2}_{5}9^{2}_{7}6^{3}_{7}9^{3}_{9}5^{4}_{9}7^{4}_{11}2^{5}_{11}4^{5}_{13}1^{6}_{13}2^{6}_{15}1^{7}_{17}$		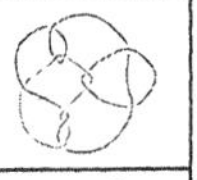
11^a_{451}: $1^{\underline{4}}_{\underline{9}}1^{\underline{4}}_{\underline{7}}2^{\underline{3}}_{\underline{7}}3^{\underline{2}}_{\underline{5}}2^{\underline{2}}_{\underline{3}}4^{\underline{1}}_{\underline{3}}3^{\underline{1}}_{\underline{1}}5^{0}_{\underline{1}}5^{0}_{1}6^{1}_{1}4^{1}_{3}5^{2}_{3}6^{2}_{5}3^{3}_{5}5^{3}_{7}4^{4}_{7}5^{4}_{9}2^{5}_{9}2^{5}_{11}1^{6}_{11}2^{6}_{13}1^{7}_{15}$ 11^a_{452}: $1^{\underline{6}}_{\underline{11}}1^{\underline{5}}_{\underline{9}}1^{\underline{5}}_{\underline{7}}4^{\underline{4}}_{\underline{7}}2^{\underline{4}}_{\underline{5}}3^{\underline{3}}_{\underline{5}}3^{\underline{3}}_{\underline{3}}7^{\underline{2}}_{\underline{3}}3^{\underline{2}}_{\underline{1}}4^{\underline{1}}_{\underline{1}}7^{\underline{1}}_{1}8^{0}_{1}7^{0}_{3}5^{1}_{3}5^{1}_{5}4^{2}_{5}5^{2}_{7}2^{3}_{7}4^{3}_{9}1^{4}_{9}2^{4}_{11}1^{5}_{13}$	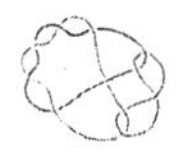	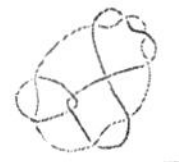
11^a_{453}: $1^{\underline{5}}_{\underline{9}}2^{\underline{4}}_{\underline{7}}1^{\underline{4}}_{\underline{5}}4^{\underline{3}}_{\underline{5}}2^{\underline{3}}_{\underline{3}}7^{\underline{2}}_{\underline{3}}4^{\underline{2}}_{\underline{1}}7^{\underline{1}}_{\underline{1}}7^{\underline{1}}_{1}11^{0}_{1}10^{0}_{3}8^{1}_{3}8^{1}_{5}8^{2}_{5}8^{2}_{7}4^{3}_{7}8^{3}_{9}3^{4}_{9}5^{4}_{11}1^{5}_{11}2^{5}_{13}1^{6}_{15}$ 11^a_{454}: $1^{\underline{6}}_{\underline{13}}2^{\underline{5}}_{\underline{11}}1^{\underline{5}}_{\underline{9}}6^{\underline{4}}_{\underline{9}}3^{\underline{4}}_{\underline{7}}9^{\underline{3}}_{\underline{7}}5^{\underline{3}}_{\underline{5}}12^{\underline{2}}_{\underline{5}}9^{\underline{2}}_{\underline{3}}11^{\underline{1}}_{\underline{3}}12^{\underline{1}}_{\underline{1}}14^{0}_{\underline{1}}14^{0}_{1}10^{1}_{1}11^{1}_{3}7^{2}_{3}10^{2}_{5}3^{3}_{5}7^{3}_{7}1^{4}_{7}3^{4}_{9}1^{5}_{11}$	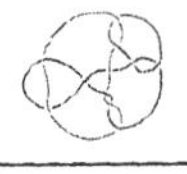	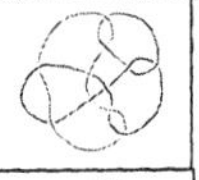
11^a_{455}: $1^{\underline{4}}_{\underline{9}}1^{\underline{4}}_{\underline{7}}2^{\underline{3}}_{\underline{7}}3^{\underline{2}}_{\underline{5}}2^{\underline{2}}_{\underline{3}}5^{\underline{1}}_{\underline{3}}3^{\underline{1}}_{\underline{1}}7^{0}_{\underline{1}}6^{0}_{1}7^{1}_{1}6^{1}_{3}6^{2}_{3}7^{2}_{5}5^{3}_{5}6^{3}_{7}5^{4}_{7}7^{4}_{9}2^{5}_{9}3^{5}_{11}1^{6}_{11}2^{6}_{13}1^{7}_{15}$ 11^a_{456}: $1^{\underline{5}}_{\underline{11}}2^{\underline{4}}_{\underline{9}}1^{\underline{4}}_{\underline{7}}5^{\underline{3}}_{\underline{7}}2^{\underline{3}}_{\underline{5}}8^{\underline{2}}_{\underline{5}}5^{\underline{2}}_{\underline{3}}8^{\underline{1}}_{\underline{3}}8^{\underline{1}}_{\underline{1}}11^{0}_{\underline{1}}11^{0}_{1}10^{1}_{1}8^{1}_{3}7^{2}_{3}10^{2}_{5}4^{3}_{5}7^{3}_{7}3^{4}_{7}5^{4}_{9}1^{5}_{9}2^{5}_{11}1^{6}_{13}$	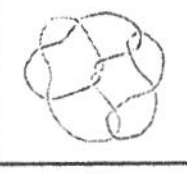	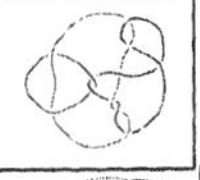
11^a_{457}: $1^{\underline{7}}_{\underline{17}}2^{\underline{6}}_{\underline{15}}1^{\underline{6}}_{\underline{13}}5^{\underline{5}}_{\underline{13}}2^{\underline{5}}_{\underline{11}}9^{\underline{4}}_{\underline{11}}6^{\underline{4}}_{\underline{9}}11^{\underline{3}}_{\underline{9}}8^{\underline{3}}_{\underline{7}}12^{\underline{2}}_{\underline{7}}11^{\underline{2}}_{\underline{5}}10^{\underline{1}}_{\underline{5}}12^{\underline{1}}_{\underline{3}}11^{0}_{\underline{3}}13^{0}_{\underline{1}}7^{1}_{\underline{1}}8^{1}_{1}3^{2}_{1}7^{2}_{3}1^{3}_{3}3^{3}_{5}1^{4}_{7}$ 11^a_{458}: $1^{\underline{7}}_{\underline{17}}4^{\underline{6}}_{\underline{15}}1^{\underline{6}}_{\underline{13}}7^{\underline{5}}_{\underline{13}}4^{\underline{5}}_{\underline{11}}10^{\underline{4}}_{\underline{11}}8^{\underline{4}}_{\underline{9}}13^{\underline{3}}_{\underline{9}}9^{\underline{3}}_{\underline{7}}13^{\underline{2}}_{\underline{7}}13^{\underline{2}}_{\underline{5}}11^{\underline{1}}_{\underline{5}}13^{\underline{1}}_{\underline{3}}11^{0}_{\underline{3}}14^{0}_{\underline{1}}6^{1}_{\underline{1}}8^{1}_{1}3^{2}_{1}6^{2}_{3}1^{3}_{3}3^{3}_{5}1^{4}_{7}$	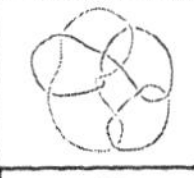	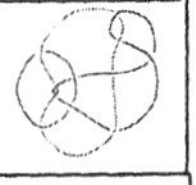
11^a_{459}: $1^{\underline{6}}_{\underline{13}}3^{\underline{5}}_{\underline{11}}1^{\underline{5}}_{\underline{9}}6^{\underline{4}}_{\underline{9}}4^{\underline{4}}_{\underline{7}}8^{\underline{3}}_{\underline{7}}5^{\underline{3}}_{\underline{5}}11^{\underline{2}}_{\underline{5}}8^{\underline{2}}_{\underline{3}}9^{\underline{1}}_{\underline{3}}11^{\underline{1}}_{\underline{1}}13^{0}_{\underline{1}}12^{0}_{1}8^{1}_{1}10^{1}_{3}5^{2}_{3}8^{2}_{5}3^{3}_{5}5^{3}_{7}1^{4}_{7}3^{4}_{9}1^{5}_{11}$ 11^a_{460}: $1^{\underline{3}}_{\underline{5}}4^{\underline{2}}_{\underline{3}}1^{\underline{2}}_{\underline{1}}6^{\underline{1}}_{\underline{1}}4^{\underline{1}}_{1}10^{0}_{1}8^{0}_{3}12^{1}_{3}8^{1}_{5}11^{2}_{5}12^{2}_{7}11^{3}_{7}11^{3}_{9}8^{4}_{9}12^{4}_{11}5^{5}_{11}7^{5}_{13}3^{6}_{13}5^{6}_{15}3^{7}_{17}1^{8}_{17}1^{8}_{19}$	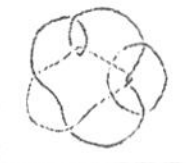	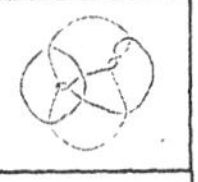
11^a_{461}: $1^{\underline{3}}_{\underline{3}}3^{\underline{2}}_{\underline{1}}1^{\underline{2}}_{1}3^{\underline{1}}_{1}3^{\underline{1}}_{3}7^{0}_{3}5^{0}_{5}8^{1}_{5}5^{1}_{7}8^{2}_{7}8^{2}_{9}7^{3}_{9}8^{3}_{11}7^{4}_{11}8^{4}_{13}4^{5}_{13}6^{5}_{15}3^{6}_{15}4^{6}_{17}3^{7}_{19}1^{8}_{19}1^{8}_{21}$ 11^a_{462}: $1^{\underline{8}}_{\underline{21}}1^{\underline{8}}_{\underline{19}}4^{\underline{7}}_{\underline{19}}6^{\underline{6}}_{\underline{17}}4^{\underline{6}}_{\underline{15}}8^{\underline{5}}_{\underline{15}}6^{\underline{5}}_{\underline{13}}11^{\underline{4}}_{\underline{13}}9^{\underline{4}}_{\underline{11}}11^{\underline{3}}_{\underline{11}}10^{\underline{3}}_{\underline{9}}10^{\underline{2}}_{\underline{9}}11^{\underline{2}}_{\underline{7}}7^{\underline{1}}_{\underline{7}}10^{\underline{1}}_{\underline{5}}6^{0}_{\underline{5}}9^{0}_{\underline{3}}3^{1}_{\underline{3}}4^{1}_{\underline{1}}1^{2}_{\underline{1}}3^{2}_{1}1^{3}_{3}$	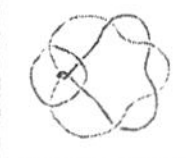	

L: KH	1st line	2nd line
11^a_{463}: $1^{\underline{3}}_{\underline{5}}3^{\underline{2}}_{\underline{3}}1^{\underline{2}}_{\underline{1}}4^{\underline{1}}_{\underline{1}}3^{\underline{1}}_{1}8^{0}_{1}6^{0}_{3}9^{1}_{3}6^{1}_{5}8^{2}_{5}9^{2}_{7}9^{3}_{7}8^{3}_{9}6^{4}_{9}10^{4}_{11}4^{5}_{11}5^{5}_{13}3^{6}_{13}4^{6}_{15}3^{7}_{17}1^{8}_{17}1^{8}_{19}$ 11^a_{464}: $1^{\underline{5}}_{\underline{11}}2^{\underline{4}}_{\underline{9}}1^{\underline{4}}_{\underline{7}}4^{\underline{3}}_{\underline{7}}2^{\underline{3}}_{\underline{5}}8^{\underline{2}}_{\underline{5}}4^{\underline{2}}_{\underline{3}}8^{\underline{1}}_{\underline{3}}8^{\underline{1}}_{\underline{1}}13^{0}_{\underline{1}}12^{0}_{1}11^{1}_{1}9^{1}_{3}8^{2}_{3}11^{2}_{5}6^{3}_{5}8^{3}_{7}4^{4}_{7}6^{4}_{9}1^{5}_{9}4^{5}_{11}1^{6}_{13}$		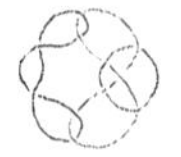
11^a_{465}: $1^{\underline{4}}_{\underline{7}}1^{\underline{3}}_{\underline{5}}1^{\underline{3}}_{\underline{3}}5^{\underline{2}}_{\underline{3}}1^{\underline{2}}_{\underline{1}}4^{\underline{1}}_{\underline{1}}5^{\underline{1}}_{1}11^{0}_{1}8^{0}_{3}10^{1}_{3}7^{1}_{5}9^{2}_{5}10^{2}_{7}7^{3}_{7}9^{3}_{9}6^{4}_{9}7^{4}_{11}3^{5}_{11}6^{5}_{13}1^{6}_{13}3^{6}_{15}1^{7}_{17}$ 11^a_{466}: $1^{\underline{8}}_{\underline{19}}2^{\underline{7}}_{\underline{17}}1^{\underline{7}}_{\underline{15}}5^{\underline{6}}_{\underline{15}}2^{\underline{6}}_{\underline{13}}7^{\underline{5}}_{\underline{13}}5^{\underline{5}}_{\underline{11}}9^{\underline{4}}_{\underline{11}}9^{\underline{4}}_{\underline{9}}10^{\underline{3}}_{\underline{9}}7^{\underline{3}}_{\underline{7}}9^{\underline{2}}_{\underline{7}}10^{\underline{2}}_{\underline{5}}6^{\underline{1}}_{\underline{5}}9^{\underline{1}}_{\underline{3}}6^{0}_{\underline{3}}8^{0}_{\underline{1}}2^{1}_{\underline{1}}4^{1}_{1}1^{2}_{1}2^{2}_{3}1^{3}_{5}$		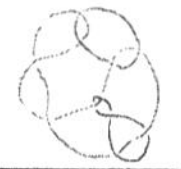
11^a_{467}: $1^{\underline{6}}_{\underline{13}}2^{\underline{5}}_{\underline{11}}1^{\underline{5}}_{\underline{9}}6^{\underline{4}}_{\underline{9}}3^{\underline{4}}_{\underline{7}}8^{\underline{3}}_{\underline{7}}5^{\underline{3}}_{\underline{5}}11^{\underline{2}}_{\underline{5}}9^{\underline{2}}_{\underline{3}}11^{\underline{1}}_{\underline{3}}10^{\underline{1}}_{\underline{1}}12^{0}_{\underline{1}}12^{0}_{1}8^{1}_{1}11^{1}_{3}7^{2}_{3}9^{2}_{5}3^{3}_{5}6^{3}_{7}1^{4}_{7}3^{4}_{9}1^{5}_{11}$ 11^a_{468}: $1^{\underline{9}}_{\underline{23}}2^{\underline{8}}_{\underline{21}}1^{\underline{8}}_{\underline{19}}4^{\underline{7}}_{\underline{19}}2^{\underline{7}}_{\underline{17}}6^{\underline{6}}_{\underline{17}}5^{\underline{6}}_{\underline{15}}7^{\underline{5}}_{\underline{15}}5^{\underline{5}}_{\underline{13}}8^{\underline{4}}_{\underline{13}}8^{\underline{4}}_{\underline{11}}6^{\underline{3}}_{\underline{11}}7^{\underline{3}}_{\underline{9}}6^{\underline{2}}_{\underline{9}}7^{\underline{2}}_{\underline{7}}3^{\underline{1}}_{\underline{7}}5^{\underline{1}}_{\underline{5}}2^{0}_{\underline{5}}4^{0}_{\underline{3}}1^{1}_{\underline{3}}1^{1}_{\underline{1}}1^{2}_{1}$	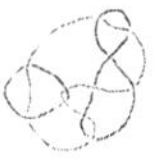	
11^a_{469}: $1^{\underline{9}}_{\underline{21}}2^{\underline{8}}_{\underline{19}}1^{\underline{8}}_{\underline{17}}4^{\underline{7}}_{\underline{17}}2^{\underline{7}}_{\underline{15}}7^{\underline{6}}_{\underline{15}}5^{\underline{6}}_{\underline{13}}7^{\underline{5}}_{\underline{13}}6^{\underline{5}}_{\underline{11}}10^{\underline{4}}_{\underline{11}}8^{\underline{4}}_{\underline{9}}7^{\underline{3}}_{\underline{9}}9^{\underline{3}}_{\underline{7}}8^{\underline{2}}_{\underline{7}}8^{\underline{2}}_{\underline{5}}4^{\underline{1}}_{\underline{5}}7^{\underline{1}}_{\underline{3}}3^{0}_{\underline{3}}5^{0}_{\underline{1}}1^{1}_{\underline{1}}2^{1}_{1}1^{2}_{3}$ 11^a_{470}: $1^{\underline{9}}_{\underline{23}}2^{\underline{8}}_{\underline{21}}1^{\underline{8}}_{\underline{19}}5^{\underline{7}}_{\underline{19}}2^{\underline{7}}_{\underline{17}}8^{\underline{6}}_{\underline{17}}6^{\underline{6}}_{\underline{15}}9^{\underline{5}}_{\underline{15}}7^{\underline{5}}_{\underline{13}}11^{\underline{4}}_{\underline{13}}10^{\underline{4}}_{\underline{11}}9^{\underline{3}}_{\underline{11}}10^{\underline{3}}_{\underline{9}}8^{\underline{2}}_{\underline{9}}10^{\underline{2}}_{\underline{7}}5^{\underline{1}}_{\underline{7}}7^{\underline{1}}_{\underline{5}}3^{0}_{\underline{5}}6^{0}_{\underline{3}}1^{1}_{\underline{3}}2^{1}_{\underline{1}}1^{2}_{1}$	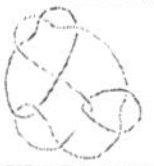	
11^a_{471}: $1^{\underline{5}}_{\underline{11}}3^{\underline{4}}_{\underline{9}}1^{\underline{4}}_{\underline{7}}6^{\underline{3}}_{\underline{7}}3^{\underline{3}}_{\underline{5}}9^{\underline{2}}_{\underline{5}}6^{\underline{2}}_{\underline{3}}11^{\underline{1}}_{\underline{3}}9^{\underline{1}}_{\underline{1}}13^{0}_{\underline{1}}13^{0}_{1}11^{1}_{1}11^{1}_{3}10^{2}_{3}13^{2}_{5}6^{3}_{5}8^{3}_{7}3^{4}_{7}6^{4}_{9}1^{5}_{9}3^{5}_{11}1^{6}_{13}$ 11^a_{472}: $1^{\underline{5}}_{\underline{9}}2^{\underline{4}}_{\underline{7}}1^{\underline{4}}_{\underline{5}}4^{\underline{3}}_{\underline{5}}2^{\underline{3}}_{\underline{3}}7^{\underline{2}}_{\underline{3}}4^{\underline{2}}_{\underline{1}}7^{\underline{1}}_{\underline{1}}7^{\underline{1}}_{1}11^{0}_{1}9^{0}_{3}8^{1}_{3}9^{1}_{5}9^{2}_{5}10^{2}_{7}5^{3}_{7}7^{3}_{9}3^{4}_{9}5^{4}_{11}1^{5}_{11}3^{5}_{13}1^{6}_{15}$	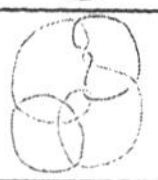	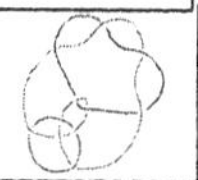
11^a_{473}: $1^{\underline{6}}_{\underline{13}}2^{\underline{5}}_{\underline{11}}1^{\underline{5}}_{\underline{9}}5^{\underline{4}}_{\underline{9}}2^{\underline{4}}_{\underline{7}}6^{\underline{3}}_{\underline{7}}5^{\underline{3}}_{\underline{5}}10^{\underline{2}}_{\underline{5}}8^{\underline{2}}_{\underline{3}}9^{\underline{1}}_{\underline{3}}8^{\underline{1}}_{\underline{1}}10^{0}_{\underline{1}}11^{0}_{1}7^{1}_{1}8^{1}_{3}5^{2}_{3}7^{2}_{5}2^{3}_{5}5^{3}_{7}1^{4}_{7}2^{4}_{9}1^{5}_{11}$ 11^a_{474}: $1^{\underline{4}}_{\underline{5}}1^{\underline{3}}_{\underline{3}}1^{\underline{3}}_{\underline{1}}4^{\underline{2}}_{\underline{1}}1^{\underline{2}}_{1}3^{\underline{1}}_{1}4^{\underline{1}}_{3}8^{0}_{3}5^{0}_{5}5^{1}_{5}6^{1}_{7}8^{2}_{7}7^{2}_{9}5^{3}_{9}6^{3}_{11}4^{4}_{11}5^{4}_{13}2^{5}_{13}4^{5}_{15}1^{6}_{15}2^{6}_{17}1^{7}_{19}$		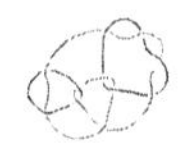
11^a_{475}: $1^{\underline{6}}_{\underline{13}}1^{\underline{5}}_{\underline{11}}1^{\underline{5}}_{\underline{9}}6^{\underline{4}}_{\underline{9}}3^{\underline{4}}_{\underline{7}}7^{\underline{3}}_{\underline{7}}4^{\underline{3}}_{\underline{5}}11^{\underline{2}}_{\underline{5}}7^{\underline{2}}_{\underline{3}}9^{\underline{1}}_{\underline{3}}11^{\underline{1}}_{\underline{1}}12^{0}_{\underline{1}}11^{0}_{1}9^{1}_{1}10^{1}_{3}6^{2}_{3}9^{2}_{5}3^{3}_{5}6^{3}_{7}1^{4}_{7}3^{4}_{9}1^{5}_{11}$ 11^a_{476}: $1^{\underline{6}}_{\underline{13}}2^{\underline{5}}_{\underline{11}}1^{\underline{5}}_{\underline{9}}6^{\underline{4}}_{\underline{9}}2^{\underline{4}}_{\underline{7}}8^{\underline{3}}_{\underline{7}}6^{\underline{3}}_{\underline{5}}12^{\underline{2}}_{\underline{5}}10^{\underline{2}}_{\underline{3}}12^{\underline{1}}_{\underline{3}}10^{\underline{1}}_{\underline{1}}13^{0}_{\underline{1}}14^{0}_{1}9^{1}_{1}11^{1}_{3}7^{2}_{3}9^{2}_{5}3^{3}_{5}7^{3}_{7}1^{4}_{7}3^{4}_{9}1^{5}_{11}$	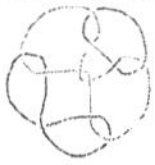	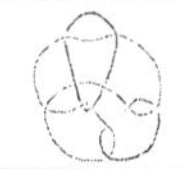
11^a_{477}: $1^{\underline{8}}_{\underline{21}}1^{\underline{7}}_{\underline{19}}1^{\underline{7}}_{\underline{17}}4^{\underline{6}}_{\underline{17}}3^{\underline{6}}_{\underline{15}}6^{\underline{5}}_{\underline{15}}2^{\underline{5}}_{\underline{13}}5^{\underline{4}}_{\underline{13}}6^{\underline{4}}_{\underline{11}}7^{\underline{3}}_{\underline{11}}5^{\underline{3}}_{\underline{9}}6^{\underline{2}}_{\underline{9}}7^{\underline{2}}_{\underline{7}}4^{\underline{1}}_{\underline{7}}6^{\underline{1}}_{\underline{5}}5^{0}_{\underline{5}}6^{0}_{\underline{3}}2^{1}_{\underline{3}}3^{1}_{\underline{1}}1^{2}_{\underline{1}}2^{2}_{1}1^{3}_{3}$ 11^a_{478}: $1^{\underline{8}}_{\underline{19}}1^{\underline{7}}_{\underline{17}}1^{\underline{7}}_{\underline{15}}5^{\underline{6}}_{\underline{15}}3^{\underline{6}}_{\underline{13}}7^{\underline{5}}_{\underline{13}}3^{\underline{5}}_{\underline{11}}8^{\underline{4}}_{\underline{11}}7^{\underline{4}}_{\underline{9}}9^{\underline{3}}_{\underline{9}}8^{\underline{3}}_{\underline{7}}10^{\underline{2}}_{\underline{7}}9^{\underline{2}}_{\underline{5}}6^{\underline{1}}_{\underline{5}}10^{\underline{1}}_{\underline{3}}7^{0}_{\underline{3}}8^{0}_{\underline{1}}3^{1}_{\underline{1}}5^{1}_{1}1^{2}_{1}3^{2}_{3}1^{3}_{5}$	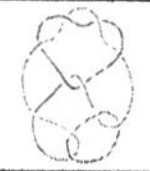	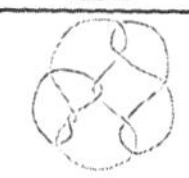
11^a_{479}: $1^{\underline{7}}_{\underline{15}}2^{\underline{6}}_{\underline{13}}1^{\underline{6}}_{\underline{11}}5^{\underline{5}}_{\underline{11}}2^{\underline{5}}_{\underline{9}}8^{\underline{4}}_{\underline{9}}5^{\underline{4}}_{\underline{7}}8^{\underline{3}}_{\underline{7}}8^{\underline{3}}_{\underline{5}}13^{\underline{2}}_{\underline{5}}10^{\underline{2}}_{\underline{3}}9^{\underline{1}}_{\underline{3}}11^{\underline{1}}_{\underline{1}}10^{0}_{\underline{1}}11^{0}_{1}6^{1}_{1}8^{1}_{3}3^{2}_{3}6^{2}_{5}1^{3}_{5}3^{3}_{7}1^{4}_{9}$ 11^a_{480}: $1^{\underline{7}}_{\underline{17}}2^{\underline{6}}_{\underline{15}}1^{\underline{6}}_{\underline{13}}5^{\underline{5}}_{\underline{13}}2^{\underline{5}}_{\underline{11}}8^{\underline{4}}_{\underline{11}}5^{\underline{4}}_{\underline{9}}9^{\underline{3}}_{\underline{9}}8^{\underline{3}}_{\underline{7}}12^{\underline{2}}_{\underline{7}}11^{\underline{2}}_{\underline{5}}10^{\underline{1}}_{\underline{5}}10^{\underline{1}}_{\underline{3}}9^{0}_{\underline{3}}12^{0}_{\underline{1}}6^{1}_{\underline{1}}7^{1}_{1}3^{2}_{1}6^{2}_{3}1^{3}_{3}3^{3}_{5}1^{4}_{7}$		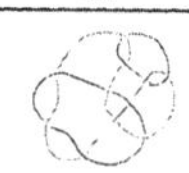
11^a_{481}: $1^{\underline{4}}_{\underline{7}}1^{\underline{4}}_{\underline{5}}2^{\underline{3}}_{\underline{5}}5^{\underline{2}}_{\underline{3}}2^{\underline{2}}_{\underline{1}}5^{\underline{1}}_{\underline{1}}5^{\underline{1}}_{1}12^{0}_{1}7^{0}_{3}10^{1}_{3}10^{1}_{5}11^{2}_{5}10^{2}_{7}8^{3}_{7}11^{3}_{9}7^{4}_{9}9^{4}_{11}4^{5}_{11}6^{5}_{13}1^{6}_{13}4^{6}_{15}1^{7}_{17}$ 11^a_{482}: $1^{\underline{8}}_{\underline{19}}2^{\underline{7}}_{\underline{17}}1^{\underline{7}}_{\underline{15}}5^{\underline{6}}_{\underline{15}}2^{\underline{6}}_{\underline{13}}8^{\underline{5}}_{\underline{13}}5^{\underline{5}}_{\underline{11}}11^{\underline{4}}_{\underline{11}}10^{\underline{4}}_{\underline{9}}13^{\underline{3}}_{\underline{9}}9^{\underline{3}}_{\underline{7}}12^{\underline{2}}_{\underline{7}}13^{\underline{2}}_{\underline{5}}8^{\underline{1}}_{\underline{5}}12^{\underline{1}}_{\underline{3}}9^{0}_{\underline{3}}10^{0}_{\underline{1}}4^{1}_{\underline{1}}7^{1}_{1}1^{2}_{1}4^{2}_{3}1^{3}_{5}$		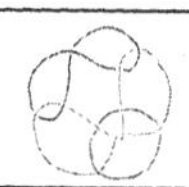
11^a_{483}: $1^{\underline{3}}_{\underline{5}}4^{\underline{2}}_{\underline{3}}1^{\underline{2}}_{\underline{1}}7^{\underline{1}}_{\underline{1}}4^{\underline{1}}_{1}11^{0}_{1}9^{0}_{3}13^{1}_{3}9^{1}_{5}14^{2}_{5}13^{2}_{7}11^{3}_{7}14^{3}_{9}11^{4}_{9}13^{4}_{11}6^{5}_{11}9^{5}_{13}3^{6}_{13}6^{6}_{15}1^{7}_{15}3^{7}_{17}1^{8}_{19}$ 11^a_{484}: $1^{\underline{5}}_{\underline{9}}3^{\underline{4}}_{\underline{7}}1^{\underline{4}}_{\underline{5}}5^{\underline{3}}_{\underline{5}}3^{\underline{3}}_{\underline{3}}9^{\underline{2}}_{\underline{3}}5^{\underline{2}}_{\underline{1}}7^{\underline{1}}_{\underline{1}}9^{\underline{1}}_{1}14^{0}_{1}11^{0}_{3}10^{1}_{3}10^{1}_{5}8^{2}_{5}10^{2}_{7}5^{3}_{7}8^{3}_{9}3^{4}_{9}5^{4}_{11}1^{5}_{11}3^{5}_{13}1^{6}_{15}$	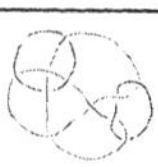	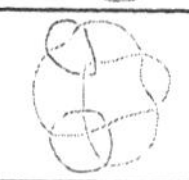
11^a_{485}: $1^{\underline{6}}_{\underline{13}}5^{\underline{5}}_{\underline{11}}1^{\underline{5}}_{\underline{9}}8^{\underline{4}}_{\underline{9}}5^{\underline{4}}_{\underline{7}}12^{\underline{3}}_{\underline{7}}8^{\underline{3}}_{\underline{5}}16^{\underline{2}}_{\underline{5}}12^{\underline{2}}_{\underline{3}}14^{\underline{1}}_{\underline{3}}16^{\underline{1}}_{\underline{1}}18^{0}_{\underline{1}}18^{0}_{1}12^{1}_{1}14^{1}_{3}7^{2}_{3}12^{2}_{5}4^{3}_{5}7^{3}_{7}1^{4}_{7}4^{4}_{9}1^{5}_{11}$ 11^a_{486}: $1^{\underline{3}}_{\underline{5}}3^{\underline{2}}_{\underline{3}}1^{\underline{2}}_{\underline{1}}6^{\underline{1}}_{\underline{1}}3^{\underline{1}}_{1}11^{0}_{1}8^{0}_{3}13^{1}_{3}9^{1}_{5}14^{2}_{5}13^{2}_{7}11^{3}_{7}14^{3}_{9}12^{4}_{9}13^{4}_{11}7^{5}_{11}10^{5}_{13}3^{6}_{13}7^{6}_{15}1^{7}_{15}3^{7}_{17}1^{8}_{19}$	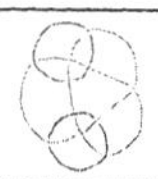	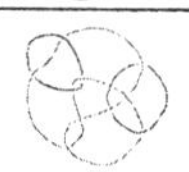
11^a_{487}: $1^{\underline{7}}_{\underline{17}}4^{\underline{6}}_{\underline{15}}1^{\underline{6}}_{\underline{13}}9^{\underline{5}}_{\underline{13}}4^{\underline{5}}_{\underline{11}}11^{\underline{4}}_{\underline{11}}9^{\underline{4}}_{\underline{9}}16^{\underline{3}}_{\underline{9}}11^{\underline{3}}_{\underline{7}}16^{\underline{2}}_{\underline{7}}16^{\underline{2}}_{\underline{5}}14^{\underline{1}}_{\underline{5}}16^{\underline{1}}_{\underline{3}}14^{0}_{\underline{3}}18^{0}_{\underline{1}}8^{1}_{\underline{1}}10^{1}_{1}4^{2}_{1}8^{2}_{3}1^{3}_{3}4^{3}_{5}1^{4}_{7}$ 11^a_{488}: $1^{\underline{2}}_{\underline{1}}2^{\underline{1}}_{1}1^{\underline{1}}_{3}6^{0}_{3}4^{0}_{5}8^{1}_{5}4^{1}_{7}11^{2}_{7}8^{2}_{9}9^{3}_{9}11^{3}_{11}12^{4}_{11}11^{4}_{13}8^{5}_{13}10^{5}_{15}6^{6}_{15}8^{6}_{17}2^{7}_{17}6^{7}_{19}1^{8}_{19}2^{8}_{21}1^{9}_{23}$		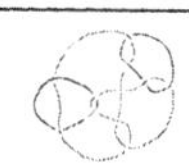
11^a_{489}: $1^{\underline{7}}_{\underline{15}}2^{\underline{6}}_{\underline{13}}1^{\underline{6}}_{\underline{11}}5^{\underline{5}}_{\underline{11}}2^{\underline{5}}_{\underline{9}}7^{\underline{4}}_{\underline{9}}6^{\underline{4}}_{\underline{7}}9^{\underline{3}}_{\underline{7}}6^{\underline{3}}_{\underline{5}}9^{\underline{2}}_{\underline{5}}9^{\underline{2}}_{\underline{3}}8^{\underline{1}}_{\underline{3}}9^{\underline{1}}_{\underline{1}}9^{0}_{\underline{1}}10^{0}_{1}4^{1}_{1}7^{1}_{3}3^{2}_{3}4^{2}_{5}3^{3}_{7}1^{4}_{7}1^{4}_{9}$ 11^a_{490}: $1^{\underline{4}}_{\underline{7}}1^{\underline{4}}_{\underline{5}}3^{\underline{3}}_{\underline{5}}6^{\underline{2}}_{\underline{3}}3^{\underline{2}}_{\underline{1}}7^{\underline{1}}_{\underline{1}}6^{\underline{1}}_{1}12^{0}_{1}9^{0}_{3}10^{1}_{3}10^{1}_{5}11^{2}_{5}10^{2}_{7}7^{3}_{7}11^{3}_{9}6^{4}_{9}8^{4}_{11}2^{5}_{11}5^{5}_{13}1^{6}_{13}2^{6}_{15}1^{7}_{17}$	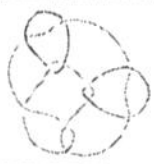	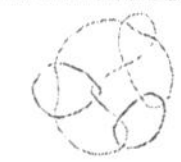
11^a_{491}: $1^{\underline{9}}_{\underline{23}}2^{\underline{8}}_{\underline{21}}1^{\underline{8}}_{\underline{19}}6^{\underline{7}}_{\underline{19}}2^{\underline{7}}_{\underline{17}}7^{\underline{6}}_{\underline{17}}6^{\underline{6}}_{\underline{15}}9^{\underline{5}}_{\underline{15}}7^{\underline{5}}_{\underline{13}}10^{\underline{4}}_{\underline{13}}11^{\underline{4}}_{\underline{11}}9^{\underline{3}}_{\underline{11}}8^{\underline{3}}_{\underline{9}}7^{\underline{2}}_{\underline{9}}9^{\underline{2}}_{\underline{7}}3^{\underline{1}}_{\underline{7}}7^{\underline{1}}_{\underline{5}}3^{0}_{\underline{5}}5^{0}_{\underline{3}}1^{1}_{\underline{3}}1^{1}_{\underline{1}}1^{2}_{1}$ 11^a_{492}: $1^{\underline{5}}_{\underline{11}}3^{\underline{4}}_{\underline{9}}1^{\underline{4}}_{\underline{7}}6^{\underline{3}}_{\underline{7}}3^{\underline{3}}_{\underline{5}}9^{\underline{2}}_{\underline{5}}6^{\underline{2}}_{\underline{3}}10^{\underline{1}}_{\underline{3}}9^{\underline{1}}_{\underline{1}}15^{0}_{\underline{1}}14^{0}_{1}11^{1}_{1}11^{1}_{3}9^{2}_{3}11^{2}_{5}6^{3}_{5}9^{3}_{7}3^{4}_{7}6^{4}_{9}1^{5}_{9}3^{5}_{11}1^{6}_{13}$		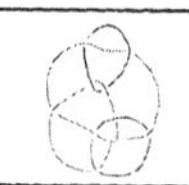

L: KH	1st line	2nd line
11^a_{493}: $1^{\underline{3}}_{\underline{3}}3^{\underline{2}}_{\underline{1}}1^{\underline{2}}_{1}5^{\underline{1}}_{1}3^{\underline{1}}_{3}10^{0}_{3}7^{0}_{5}10^{1}_{5}8^{1}_{7}14^{2}_{7}10^{2}_{9}9^{3}_{9}14^{3}_{11}11^{4}_{11}11^{4}_{13}7^{5}_{13}9^{5}_{15}3^{6}_{15}7^{6}_{17}1^{7}_{17}3^{7}_{19}1^{8}_{21}$ 11^a_{494}: $1^{\underline{7}}_{\underline{17}}2^{\underline{6}}_{\underline{15}}1^{\underline{6}}_{\underline{13}}6^{\underline{5}}_{\underline{13}}2^{\underline{5}}_{\underline{11}}7^{\underline{4}}_{\underline{11}}6^{\underline{4}}_{\underline{9}}10^{\underline{3}}_{\underline{9}}7^{\underline{3}}_{\underline{7}}11^{\underline{2}}_{\underline{7}}10^{\underline{2}}_{\underline{5}}9^{\underline{1}}_{\underline{5}}11^{\underline{1}}_{\underline{3}}10^{0}_{\underline{3}}13^{0}_{\underline{1}}6^{1}_{\underline{1}}6^{1}_{1}3^{2}_{1}6^{2}_{3}1^{3}_{3}3^{3}_{5}1^{4}_{7}$	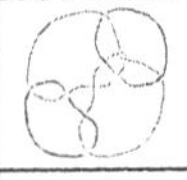	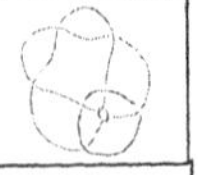
11^a_{495}: $1^{\underline{7}}_{\underline{15}}2^{\underline{6}}_{\underline{13}}1^{\underline{6}}_{\underline{11}}5^{\underline{5}}_{\underline{11}}2^{\underline{5}}_{\underline{9}}7^{\underline{4}}_{\underline{9}}5^{\underline{4}}_{\underline{7}}9^{\underline{3}}_{\underline{7}}7^{\underline{3}}_{\underline{5}}12^{\underline{2}}_{\underline{5}}9^{\underline{2}}_{\underline{3}}8^{\underline{1}}_{\underline{3}}12^{\underline{1}}_{\underline{1}}11^{0}_{\underline{1}}12^{0}_{1}7^{1}_{1}7^{1}_{3}3^{2}_{3}7^{2}_{5}1^{3}_{5}3^{3}_{7}1^{4}_{9}$ 11^a_{496}: $1^{\underline{4}}_{\underline{5}}1^{\underline{3}}_{\underline{3}}1^{\underline{3}}_{\underline{1}}5^{\underline{2}}_{\underline{1}}1^{\underline{2}}_{1}2^{\underline{1}}_{1}5^{\underline{1}}_{3}9^{0}_{3}6^{0}_{5}6^{1}_{5}5^{1}_{7}7^{2}_{7}6^{2}_{9}4^{3}_{9}7^{3}_{11}4^{4}_{11}4^{4}_{13}2^{5}_{13}4^{5}_{15}1^{6}_{15}2^{6}_{17}1^{7}_{19}$	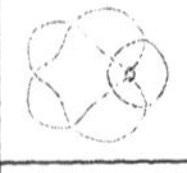	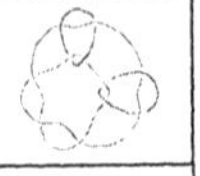
11^a_{497}: $1^{\underline{6}}_{\underline{13}}2^{\underline{5}}_{\underline{11}}1^{\underline{5}}_{\underline{9}}6^{\underline{4}}_{\underline{9}}4^{\underline{4}}_{\underline{7}}7^{\underline{3}}_{\underline{7}}4^{\underline{3}}_{\underline{5}}9^{\underline{2}}_{\underline{5}}7^{\underline{2}}_{\underline{3}}8^{\underline{1}}_{\underline{3}}9^{\underline{1}}_{\underline{1}}11^{0}_{\underline{1}}10^{0}_{1}6^{1}_{1}9^{1}_{3}5^{2}_{3}6^{2}_{5}2^{3}_{5}5^{3}_{7}1^{4}_{7}2^{4}_{9}1^{5}_{11}$ 11^a_{498}: $1^{\underline{7}}_{\underline{15}}2^{\underline{6}}_{\underline{13}}1^{\underline{6}}_{\underline{11}}5^{\underline{5}}_{\underline{11}}2^{\underline{5}}_{\underline{9}}7^{\underline{4}}_{\underline{9}}6^{\underline{4}}_{\underline{7}}10^{\underline{3}}_{\underline{7}}6^{\underline{3}}_{\underline{5}}11^{\underline{2}}_{\underline{5}}10^{\underline{2}}_{\underline{3}}9^{\underline{1}}_{\underline{3}}11^{\underline{1}}_{\underline{1}}10^{0}_{\underline{1}}11^{0}_{1}6^{1}_{1}8^{1}_{3}4^{2}_{3}6^{2}_{5}4^{3}_{7}1^{4}_{7}1^{4}_{9}$	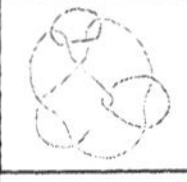	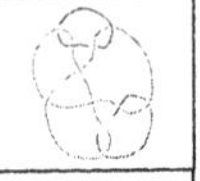
11^a_{499}: $1^{\underline{4}}_{\underline{7}}1^{\underline{3}}_{\underline{5}}1^{\underline{3}}_{\underline{3}}6^{\underline{2}}_{\underline{3}}1^{\underline{2}}_{\underline{1}}5^{\underline{1}}_{\underline{1}}6^{\underline{1}}_{1}11^{0}_{1}9^{0}_{3}10^{1}_{3}7^{1}_{5}9^{2}_{5}10^{2}_{7}7^{3}_{7}9^{3}_{9}5^{4}_{9}7^{4}_{11}2^{5}_{11}5^{5}_{13}1^{6}_{13}2^{6}_{15}1^{7}_{17}$ 11^a_{500}: $1^{\underline{4}}_{\underline{7}}1^{\underline{4}}_{\underline{5}}3^{\underline{3}}_{\underline{5}}6^{\underline{2}}_{\underline{3}}3^{\underline{2}}_{\underline{1}}8^{\underline{1}}_{\underline{1}}6^{\underline{1}}_{1}14^{0}_{1}10^{0}_{3}13^{1}_{3}12^{1}_{5}14^{2}_{5}13^{2}_{7}9^{3}_{7}14^{3}_{9}9^{4}_{9}10^{4}_{11}4^{5}_{11}8^{5}_{13}1^{6}_{13}4^{6}_{15}1^{7}_{17}$	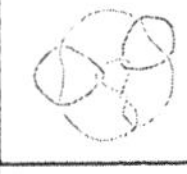	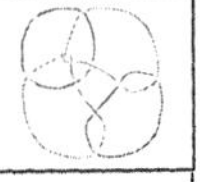
11^a_{501}: $1^{\underline{6}}_{\underline{13}}1^{\underline{6}}_{\underline{11}}3^{\underline{5}}_{\underline{11}}5^{\underline{4}}_{\underline{9}}3^{\underline{4}}_{\underline{7}}6^{\underline{3}}_{\underline{7}}5^{\underline{3}}_{\underline{5}}11^{\underline{2}}_{\underline{5}}7^{\underline{2}}_{\underline{3}}8^{\underline{1}}_{\underline{3}}10^{\underline{1}}_{\underline{1}}11^{0}_{\underline{1}}10^{0}_{1}7^{1}_{1}9^{1}_{3}5^{2}_{3}7^{2}_{5}2^{3}_{5}5^{3}_{7}1^{4}_{7}2^{4}_{9}1^{5}_{11}$ 11^a_{502}: $1^{\underline{11}}_{\underline{29}}2^{\underline{10}}_{\underline{27}}1^{\underline{10}}_{\underline{25}}5^{\underline{9}}_{\underline{25}}2^{\underline{9}}_{\underline{23}}6^{\underline{8}}_{\underline{23}}5^{\underline{8}}_{\underline{21}}8^{\underline{7}}_{\underline{21}}6^{\underline{7}}_{\underline{19}}9^{\underline{6}}_{\underline{19}}10^{\underline{6}}_{\underline{17}}8^{\underline{5}}_{\underline{17}}7^{\underline{5}}_{\underline{15}}6^{\underline{4}}_{\underline{15}}9^{\underline{4}}_{\underline{13}}4^{\underline{3}}_{\underline{13}}5^{\underline{3}}_{\underline{11}}2^{\underline{2}}_{\underline{11}}4^{\underline{2}}_{\underline{9}}2^{\underline{1}}_{\underline{7}}1^{0}_{\underline{7}}1^{0}_{\underline{5}}$	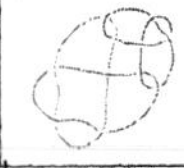	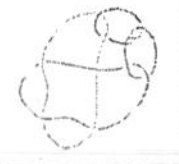
11^a_{503}: $1^{\underline{6}}_{\underline{15}}1^{\underline{6}}_{\underline{13}}3^{\underline{5}}_{\underline{13}}4^{\underline{4}}_{\underline{11}}3^{\underline{4}}_{\underline{9}}6^{\underline{3}}_{\underline{9}}4^{\underline{3}}_{\underline{7}}8^{\underline{2}}_{\underline{7}}7^{\underline{2}}_{\underline{5}}7^{\underline{1}}_{\underline{5}}7^{\underline{1}}_{\underline{3}}8^{0}_{\underline{3}}9^{0}_{\underline{1}}5^{1}_{\underline{1}}6^{1}_{1}4^{2}_{1}5^{2}_{3}1^{3}_{3}4^{3}_{5}1^{4}_{5}1^{4}_{7}1^{5}_{9}$ 11^a_{504}: $1^{\underline{11}}_{\underline{31}}2^{\underline{10}}_{\underline{29}}1^{\underline{10}}_{\underline{27}}4^{\underline{9}}_{\underline{27}}2^{\underline{9}}_{\underline{25}}5^{\underline{8}}_{\underline{25}}4^{\underline{8}}_{\underline{23}}6^{\underline{7}}_{\underline{23}}5^{\underline{7}}_{\underline{21}}6^{\underline{6}}_{\underline{21}}8^{\underline{6}}_{\underline{19}}7^{\underline{5}}_{\underline{19}}4^{\underline{5}}_{\underline{17}}3^{\underline{4}}_{\underline{17}}8^{\underline{4}}_{\underline{15}}4^{\underline{3}}_{\underline{15}}2^{\underline{3}}_{\underline{13}}1^{\underline{2}}_{\underline{13}}4^{\underline{2}}_{\underline{11}}1^{\underline{1}}_{\underline{9}}1^{0}_{\underline{9}}1^{0}_{\underline{7}}$	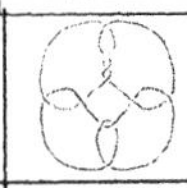	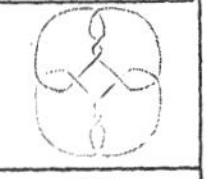
11^a_{505}: $1^{\underline{4}}_{\underline{7}}4^{\underline{3}}_{\underline{5}}1^{\underline{3}}_{\underline{3}}7^{\underline{2}}_{\underline{3}}4^{\underline{2}}_{\underline{1}}10^{\underline{1}}_{\underline{1}}7^{\underline{1}}_{1}14^{0}_{1}11^{0}_{3}13^{1}_{3}13^{1}_{5}14^{2}_{5}15^{2}_{7}10^{3}_{7}12^{3}_{9}7^{4}_{9}11^{4}_{11}3^{5}_{11}6^{5}_{13}1^{6}_{13}3^{6}_{15}1^{7}_{17}$ 11^a_{506}: $1^{\underline{5}}_{\underline{9}}4^{\underline{4}}_{\underline{7}}1^{\underline{4}}_{\underline{5}}6^{\underline{3}}_{\underline{5}}4^{\underline{3}}_{\underline{3}}11^{\underline{2}}_{\underline{3}}7^{\underline{2}}_{\underline{1}}11^{\underline{1}}_{\underline{1}}10^{\underline{1}}_{1}15^{0}_{1}13^{0}_{3}11^{1}_{3}13^{1}_{5}11^{2}_{5}12^{2}_{7}6^{3}_{7}10^{3}_{9}3^{4}_{9}6^{4}_{11}1^{5}_{11}3^{5}_{13}1^{6}_{15}$	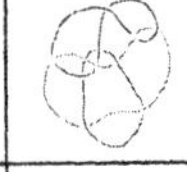	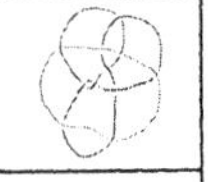
11^a_{507}: $1^{\underline{3}}_{\underline{5}}3^{\underline{2}}_{\underline{3}}1^{\underline{2}}_{\underline{1}}5^{\underline{1}}_{\underline{1}}3^{\underline{1}}_{1}9^{0}_{1}6^{0}_{3}10^{1}_{3}8^{1}_{5}12^{2}_{5}12^{2}_{7}10^{3}_{7}10^{3}_{9}9^{4}_{9}11^{4}_{11}5^{5}_{11}8^{5}_{13}3^{6}_{13}5^{6}_{15}1^{7}_{15}3^{7}_{17}1^{8}_{19}$ 11^a_{508}: $1^{\underline{4}}_{\underline{9}}3^{\underline{3}}_{\underline{7}}1^{\underline{3}}_{\underline{5}}5^{\underline{2}}_{\underline{5}}3^{\underline{2}}_{\underline{3}}8^{\underline{1}}_{\underline{3}}5^{\underline{1}}_{\underline{1}}10^{0}_{\underline{1}}9^{0}_{1}10^{1}_{1}9^{1}_{3}10^{2}_{3}12^{2}_{5}8^{3}_{5}8^{3}_{7}5^{4}_{7}9^{4}_{9}3^{5}_{9}4^{5}_{11}1^{6}_{11}3^{6}_{13}1^{7}_{15}$	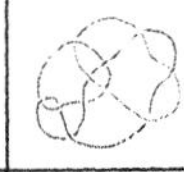	
11^a_{509}: $1^{\underline{5}}_{\underline{11}}4^{\underline{4}}_{\underline{9}}1^{\underline{4}}_{\underline{7}}8^{\underline{3}}_{\underline{7}}4^{\underline{3}}_{\underline{5}}13^{\underline{2}}_{\underline{5}}9^{\underline{2}}_{\underline{3}}15^{\underline{1}}_{\underline{3}}12^{\underline{1}}_{\underline{1}}18^{0}_{\underline{1}}17^{0}_{1}15^{1}_{1}16^{1}_{3}13^{2}_{3}16^{2}_{5}8^{3}_{5}12^{3}_{7}4^{4}_{7}8^{4}_{9}1^{5}_{9}4^{5}_{11}1^{6}_{13}$ 11^a_{510}: $1^{\underline{4}}_{\underline{5}}1^{\underline{3}}_{\underline{3}}1^{\underline{3}}_{\underline{1}}4^{\underline{2}}_{\underline{1}}2^{\underline{2}}_{1}3^{\underline{1}}_{1}3^{\underline{1}}_{3}7^{0}_{3}4^{0}_{5}4^{1}_{5}6^{1}_{7}7^{2}_{7}5^{2}_{9}4^{3}_{9}6^{3}_{11}4^{4}_{11}5^{4}_{13}2^{5}_{13}3^{5}_{15}1^{6}_{15}2^{6}_{17}1^{7}_{19}$		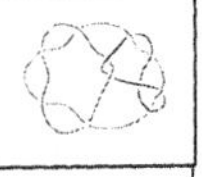
11^a_{511}: $1^{\underline{4}}_{\underline{7}}2^{\underline{3}}_{\underline{5}}1^{\underline{3}}_{\underline{3}}5^{\underline{2}}_{\underline{3}}3^{\underline{2}}_{\underline{1}}6^{\underline{1}}_{\underline{1}}4^{\underline{1}}_{1}9^{0}_{1}7^{0}_{3}8^{1}_{3}8^{1}_{5}9^{2}_{5}9^{2}_{7}6^{3}_{7}8^{3}_{9}5^{4}_{9}7^{4}_{11}2^{5}_{11}4^{5}_{13}1^{6}_{13}2^{6}_{15}1^{7}_{17}$ 11^a_{512}: $1^{\underline{9}}_{\underline{23}}3^{\underline{8}}_{\underline{21}}1^{\underline{8}}_{\underline{19}}5^{\underline{7}}_{\underline{19}}3^{\underline{7}}_{\underline{17}}8^{\underline{6}}_{\underline{17}}6^{\underline{6}}_{\underline{15}}10^{\underline{5}}_{\underline{15}}7^{\underline{5}}_{\underline{13}}10^{\underline{4}}_{\underline{13}}11^{\underline{4}}_{\underline{11}}9^{\underline{3}}_{\underline{11}}9^{\underline{3}}_{\underline{9}}8^{\underline{2}}_{\underline{9}}10^{\underline{2}}_{\underline{7}}4^{\underline{1}}_{\underline{7}}7^{\underline{1}}_{\underline{5}}3^{0}_{\underline{5}}5^{0}_{\underline{3}}1^{1}_{\underline{3}}2^{1}_{\underline{1}}1^{2}_{1}$	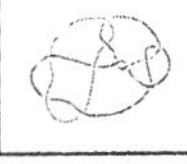	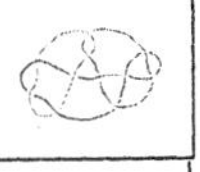
11^a_{513}: $1^{\underline{6}}_{\underline{13}}1^{\underline{6}}_{\underline{11}}2^{\underline{5}}_{\underline{11}}3^{\underline{4}}_{\underline{9}}2^{\underline{4}}_{\underline{7}}4^{\underline{3}}_{\underline{7}}3^{\underline{3}}_{\underline{5}}7^{\underline{2}}_{\underline{5}}4^{\underline{2}}_{\underline{3}}5^{\underline{1}}_{\underline{3}}7^{\underline{1}}_{\underline{1}}8^{0}_{\underline{1}}7^{0}_{1}5^{1}_{1}6^{1}_{3}4^{2}_{3}6^{2}_{5}2^{3}_{5}3^{3}_{7}1^{4}_{7}2^{4}_{9}1^{5}_{11}$ 11^a_{514}: $1^{\underline{6}}_{\underline{15}}1^{\underline{6}}_{\underline{13}}3^{\underline{5}}_{\underline{13}}4^{\underline{4}}_{\underline{11}}3^{\underline{4}}_{\underline{9}}6^{\underline{3}}_{\underline{9}}4^{\underline{3}}_{\underline{7}}8^{\underline{2}}_{\underline{7}}6^{\underline{2}}_{\underline{5}}7^{\underline{1}}_{\underline{5}}8^{\underline{1}}_{\underline{3}}8^{0}_{\underline{3}}9^{0}_{\underline{1}}6^{1}_{\underline{1}}6^{1}_{1}4^{2}_{1}7^{2}_{3}2^{3}_{3}3^{3}_{5}1^{4}_{5}2^{4}_{7}1^{5}_{9}$	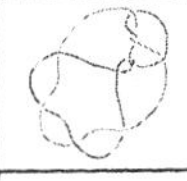	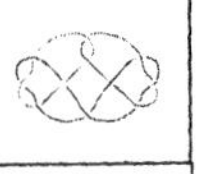
11^a_{515}: $1^{\underline{5}}_{\underline{9}}2^{\underline{4}}_{\underline{7}}1^{\underline{4}}_{\underline{5}}3^{\underline{3}}_{\underline{5}}2^{\underline{3}}_{\underline{3}}6^{\underline{2}}_{\underline{3}}4^{\underline{2}}_{\underline{1}}6^{\underline{1}}_{\underline{1}}5^{\underline{1}}_{1}8^{0}_{1}7^{0}_{3}6^{1}_{3}7^{1}_{5}7^{2}_{5}7^{2}_{7}3^{3}_{7}6^{3}_{9}3^{4}_{9}4^{4}_{11}1^{5}_{11}2^{5}_{13}1^{6}_{15}$ 11^a_{516}: $1^{\underline{7}}_{\underline{15}}3^{\underline{6}}_{\underline{13}}1^{\underline{6}}_{\underline{11}}6^{\underline{5}}_{\underline{11}}3^{\underline{5}}_{\underline{9}}9^{\underline{4}}_{\underline{9}}6^{\underline{4}}_{\underline{7}}10^{\underline{3}}_{\underline{7}}9^{\underline{3}}_{\underline{5}}14^{\underline{2}}_{\underline{5}}12^{\underline{2}}_{\underline{3}}10^{\underline{1}}_{\underline{3}}12^{\underline{1}}_{\underline{1}}11^{0}_{\underline{1}}12^{0}_{1}6^{1}_{1}9^{1}_{3}3^{2}_{3}6^{2}_{5}1^{3}_{5}3^{3}_{7}1^{4}_{9}$	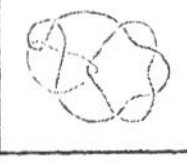	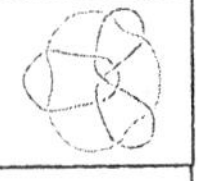
11^a_{517}: $1^{\underline{6}}_{\underline{13}}4^{\underline{5}}_{\underline{11}}1^{\underline{5}}_{\underline{9}}8^{\underline{4}}_{\underline{9}}4^{\underline{4}}_{\underline{7}}10^{\underline{3}}_{\underline{7}}8^{\underline{3}}_{\underline{5}}16^{\underline{2}}_{\underline{5}}12^{\underline{2}}_{\underline{3}}14^{\underline{1}}_{\underline{3}}14^{\underline{1}}_{\underline{1}}15^{0}_{\underline{1}}16^{0}_{1}11^{1}_{1}13^{1}_{3}7^{2}_{3}11^{2}_{5}3^{3}_{5}7^{3}_{7}1^{4}_{7}3^{4}_{9}1^{5}_{11}$ 11^a_{518}: $1^{\underline{6}}_{\underline{11}}1^{\underline{6}}_{\underline{9}}1^{\underline{5}}_{\underline{9}}3^{\underline{4}}_{\underline{7}}1^{\underline{4}}_{\underline{5}}2^{\underline{3}}_{\underline{5}}3^{\underline{3}}_{\underline{3}}6^{\underline{2}}_{\underline{3}}2^{\underline{2}}_{\underline{1}}3^{\underline{1}}_{\underline{1}}6^{\underline{1}}_{1}7^{0}_{1}5^{0}_{3}4^{1}_{3}5^{1}_{5}4^{2}_{5}5^{2}_{7}2^{3}_{7}3^{3}_{9}1^{4}_{9}2^{4}_{11}1^{5}_{13}$	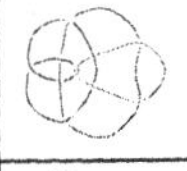	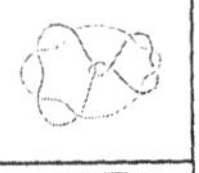
11^a_{519}: $1^{\underline{6}}_{\underline{13}}3^{\underline{5}}_{\underline{11}}1^{\underline{5}}_{\underline{9}}6^{\underline{4}}_{\underline{9}}3^{\underline{4}}_{\underline{7}}8^{\underline{3}}_{\underline{7}}6^{\underline{3}}_{\underline{5}}13^{\underline{2}}_{\underline{5}}10^{\underline{2}}_{\underline{3}}11^{\underline{1}}_{\underline{3}}11^{\underline{1}}_{\underline{1}}13^{0}_{\underline{1}}13^{0}_{1}9^{1}_{1}11^{1}_{3}6^{2}_{3}9^{2}_{5}3^{3}_{5}6^{3}_{7}1^{4}_{7}3^{4}_{9}1^{5}_{11}$ 11^a_{520}: $1^{\underline{6}}_{\underline{13}}5^{\underline{5}}_{\underline{11}}1^{\underline{5}}_{\underline{9}}10^{\underline{4}}_{\underline{9}}5^{\underline{4}}_{\underline{7}}13^{\underline{3}}_{\underline{7}}10^{\underline{3}}_{\underline{5}}20^{\underline{2}}_{\underline{5}}15^{\underline{2}}_{\underline{3}}18^{\underline{1}}_{\underline{3}}18^{\underline{1}}_{\underline{1}}19^{0}_{\underline{1}}20^{0}_{1}14^{1}_{1}17^{1}_{3}9^{2}_{3}14^{2}_{5}4^{3}_{5}9^{3}_{7}1^{4}_{7}4^{4}_{9}1^{5}_{11}$	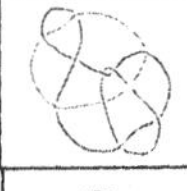	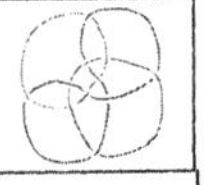
11^a_{521}: $1^{\underline{3}}_{\underline{3}}3^{\underline{2}}_{\underline{1}}1^{\underline{2}}_{1}4^{\underline{1}}_{1}3^{\underline{1}}_{3}8^{0}_{3}5^{0}_{5}8^{1}_{5}7^{1}_{7}10^{2}_{7}10^{2}_{9}9^{3}_{9}8^{3}_{11}7^{4}_{11}9^{4}_{13}4^{5}_{13}7^{5}_{15}3^{6}_{15}4^{6}_{17}3^{7}_{19}1^{8}_{19}1^{8}_{21}$ 11^a_{522}: $1^{\underline{5}}_{\underline{11}}3^{\underline{4}}_{\underline{9}}1^{\underline{4}}_{\underline{7}}6^{\underline{3}}_{\underline{7}}3^{\underline{3}}_{\underline{5}}10^{\underline{2}}_{\underline{5}}7^{\underline{2}}_{\underline{3}}12^{\underline{1}}_{\underline{3}}9^{\underline{1}}_{\underline{1}}13^{0}_{\underline{1}}13^{0}_{1}12^{1}_{1}12^{1}_{3}10^{2}_{3}13^{2}_{5}6^{3}_{5}9^{3}_{7}4^{4}_{7}7^{4}_{9}1^{5}_{9}3^{5}_{11}1^{6}_{13}$	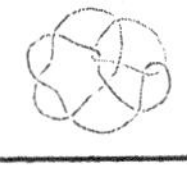	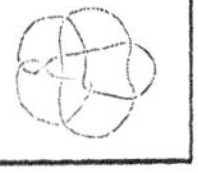

L: KH	1st line	2nd line
11^a_{523}: $1^{\underline{6}}_{\underline{13}}1^{\underline{6}}_{\underline{11}}3^{\underline{5}}_{\underline{11}}5^{\underline{4}}_{\underline{9}}3^{\underline{4}}_{\underline{7}}7^{\underline{3}}_{\underline{7}}5^{\underline{3}}_{\underline{5}}11^{\underline{2}}_{\underline{5}}7^{\underline{2}}_{\underline{3}}9^{\underline{1}}_{\underline{3}}11^{\underline{1}}_{\underline{1}}12^{0}_{\underline{1}}11^{0}_{1}8^{1}_{1}10^{1}_{3}6^{2}_{3}9^{2}_{5}3^{3}_{5}5^{3}_{7}1^{4}_{7}3^{4}_{9}1^{5}_{11}$ 11^a_{524}: $1^{\underline{5}}_{\underline{11}}4^{\underline{4}}_{\underline{9}}1^{\underline{4}}_{\underline{7}}7^{\underline{3}}_{\underline{7}}4^{\underline{3}}_{\underline{5}}11^{\underline{2}}_{\underline{5}}8^{\underline{2}}_{\underline{3}}14^{\underline{1}}_{\underline{3}}10^{\underline{1}}_{\underline{1}}14^{0}_{\underline{1}}15^{0}_{1}13^{1}_{1}13^{1}_{3}11^{2}_{3}14^{2}_{5}6^{3}_{5}10^{3}_{7}4^{4}_{7}7^{4}_{9}1^{5}_{9}3^{5}_{11}1^{6}_{13}$	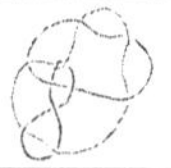	
11^a_{525}: $1^{\underline{9}}_{\underline{21}}3^{\underline{8}}_{\underline{19}}1^{\underline{8}}_{\underline{17}}5^{\underline{7}}_{\underline{17}}3^{\underline{7}}_{\underline{15}}8^{\underline{6}}_{\underline{15}}6^{\underline{6}}_{\underline{13}}9^{\underline{5}}_{\underline{13}}7^{\underline{5}}_{\underline{11}}11^{\underline{4}}_{\underline{11}}10^{\underline{4}}_{\underline{9}}8^{\underline{3}}_{\underline{9}}10^{\underline{3}}_{\underline{7}}9^{\underline{2}}_{\underline{7}}9^{\underline{2}}_{\underline{5}}4^{\underline{1}}_{\underline{5}}8^{\underline{1}}_{\underline{3}}3^{0}_{\underline{3}}5^{0}_{\underline{1}}1^{1}_{\underline{1}}2^{1}_{1}1^{2}_{3}$ 11^a_{526}: $1^{\underline{7}}_{\underline{17}}3^{\underline{6}}_{\underline{15}}1^{\underline{6}}_{\underline{13}}7^{\underline{5}}_{\underline{13}}3^{\underline{5}}_{\underline{11}}10^{\underline{4}}_{\underline{11}}7^{\underline{4}}_{\underline{9}}12^{\underline{3}}_{\underline{9}}10^{\underline{3}}_{\underline{7}}15^{\underline{2}}_{\underline{7}}14^{\underline{2}}_{\underline{5}}12^{\underline{1}}_{\underline{5}}13^{\underline{1}}_{\underline{3}}11^{0}_{\underline{3}}14^{0}_{\underline{1}}7^{1}_{\underline{1}}9^{1}_{1}3^{2}_{1}7^{2}_{3}1^{3}_{3}3^{3}_{5}1^{4}_{7}$	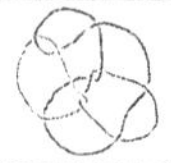	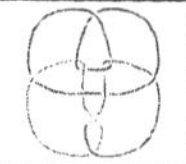
11^a_{527}: $1^{\underline{8}}_{\underline{19}}2^{\underline{7}}_{\underline{17}}1^{\underline{7}}_{\underline{15}}6^{\underline{6}}_{\underline{15}}3^{\underline{6}}_{\underline{13}}8^{\underline{5}}_{\underline{13}}5^{\underline{5}}_{\underline{11}}11^{\underline{4}}_{\underline{11}}9^{\underline{4}}_{\underline{9}}11^{\underline{3}}_{\underline{9}}10^{\underline{3}}_{\underline{7}}12^{\underline{2}}_{\underline{7}}12^{\underline{2}}_{\underline{5}}8^{\underline{1}}_{\underline{5}}11^{\underline{1}}_{\underline{3}}7^{0}_{\underline{3}}9^{0}_{\underline{1}}3^{1}_{\underline{1}}6^{1}_{1}1^{2}_{1}3^{2}_{3}1^{3}_{5}$ 11^a_{528}: $1^{\underline{6}}_{\underline{11}}2^{\underline{5}}_{\underline{9}}1^{\underline{5}}_{\underline{7}}5^{\underline{4}}_{\underline{7}}2^{\underline{4}}_{\underline{5}}4^{\underline{3}}_{\underline{5}}5^{\underline{3}}_{\underline{3}}10^{\underline{2}}_{\underline{3}}6^{\underline{2}}_{\underline{1}}6^{\underline{1}}_{\underline{1}}8^{\underline{1}}_{1}9^{0}_{1}8^{0}_{3}6^{1}_{3}7^{1}_{5}4^{2}_{5}6^{2}_{7}2^{3}_{7}4^{3}_{9}1^{4}_{9}2^{4}_{11}1^{5}_{13}$	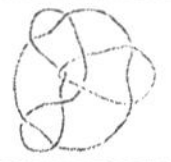	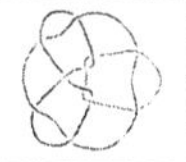
11^a_{529}: $1^{\underline{11}}_{\underline{27}}4^{\underline{10}}_{\underline{25}}1^{\underline{10}}_{\underline{23}}6^{\underline{9}}_{\underline{23}}4^{\underline{9}}_{\underline{21}}11^{\underline{8}}_{\underline{21}}7^{\underline{8}}_{\underline{19}}12^{\underline{7}}_{\underline{19}}10^{\underline{7}}_{\underline{17}}14^{\underline{6}}_{\underline{17}}14^{\underline{6}}_{\underline{15}}13^{\underline{5}}_{\underline{15}}12^{\underline{5}}_{\underline{13}}10^{\underline{4}}_{\underline{13}}13^{\underline{4}}_{\underline{11}}6^{\underline{3}}_{\underline{11}}10^{\underline{3}}_{\underline{9}}4^{\underline{2}}_{\underline{9}}6^{\underline{2}}_{\underline{7}}4^{\underline{1}}_{\underline{5}}1^{0}_{\underline{5}}1^{0}_{\underline{3}}$ 11^a_{530}: $1^{\underline{6}}_{\underline{13}}1^{\underline{6}}_{\underline{11}}3^{\underline{5}}_{\underline{11}}6^{\underline{4}}_{\underline{9}}3^{\underline{4}}_{\underline{7}}8^{\underline{3}}_{\underline{7}}6^{\underline{3}}_{\underline{5}}12^{\underline{2}}_{\underline{5}}8^{\underline{2}}_{\underline{3}}11^{\underline{1}}_{\underline{3}}12^{\underline{1}}_{\underline{1}}13^{0}_{\underline{1}}13^{0}_{1}9^{1}_{1}11^{1}_{3}7^{2}_{3}10^{2}_{5}3^{3}_{5}6^{3}_{7}1^{4}_{7}3^{4}_{9}1^{5}_{11}$	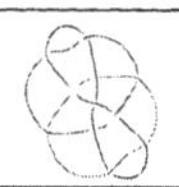	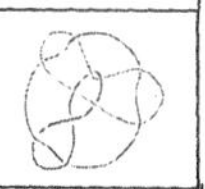
11^a_{531}: $1^{\underline{11}}_{\underline{28}}1^{\underline{10}}_{\underline{26}}1^{\underline{10}}_{\underline{24}}5^{\underline{9}}_{\underline{24}}1^{\underline{9}}_{\underline{22}}6^{\underline{8}}_{\underline{22}}6^{\underline{8}}_{\underline{20}}11^{\underline{7}}_{\underline{20}}5^{\underline{7}}_{\underline{18}}11^{\underline{6}}_{\underline{18}}11^{\underline{6}}_{\underline{16}}8^{\underline{5}}_{\underline{16}}11^{\underline{5}}_{\underline{14}}13^{\underline{4}}_{\underline{14}}14^{\underline{4}}_{\underline{12}}7^{\underline{3}}_{\underline{12}}7^{\underline{3}}_{\underline{10}}4^{\underline{2}}_{\underline{10}}7^{\underline{2}}_{\underline{8}}4^{\underline{1}}_{\underline{6}}1^{0}_{\underline{6}}1^{0}_{\underline{4}}$ 11^a_{532}: $1^{\underline{6}}_{\underline{12}}1^{\underline{5}}_{\underline{8}}6^{\underline{4}}_{\underline{8}}4^{\underline{4}}_{\underline{6}}5^{\underline{3}}_{\underline{6}}2^{\underline{3}}_{\underline{4}}11^{\underline{2}}_{\underline{4}}5^{\underline{2}}_{\underline{2}}5^{\underline{1}}_{\underline{2}}11^{\underline{1}}_{0}14^{0}_{0}9^{0}_{2}7^{1}_{2}10^{1}_{4}7^{2}_{4}7^{2}_{6}3^{3}_{6}7^{3}_{8}1^{4}_{8}3^{4}_{10}1^{5}_{12}$	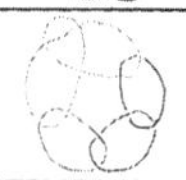	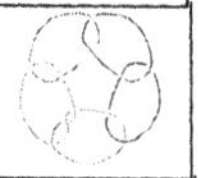
11^a_{533}: $1^{\underline{6}}_{\underline{14}}1^{\underline{5}}_{\underline{10}}5^{\underline{4}}_{\underline{10}}4^{\underline{4}}_{\underline{8}}4^{\underline{3}}_{\underline{8}}1^{\underline{3}}_{\underline{6}}7^{\underline{2}}_{\underline{6}}4^{\underline{2}}_{\underline{4}}4^{\underline{1}}_{\underline{4}}7^{\underline{1}}_{\underline{2}}10^{0}_{\underline{2}}8^{0}_{0}5^{1}_{0}6^{1}_{2}4^{2}_{2}5^{2}_{4}3^{3}_{4}4^{3}_{6}1^{4}_{6}3^{4}_{8}1^{5}_{10}$ 11^a_{534}: $1^{\underline{9}}_{\underline{22}}2^{\underline{8}}_{\underline{20}}1^{\underline{8}}_{\underline{18}}5^{\underline{7}}_{\underline{18}}2^{\underline{7}}_{\underline{16}}9^{\underline{6}}_{\underline{16}}6^{\underline{6}}_{\underline{14}}9^{\underline{5}}_{\underline{14}}8^{\underline{5}}_{\underline{12}}14^{\underline{4}}_{\underline{12}}12^{\underline{4}}_{\underline{10}}10^{\underline{3}}_{\underline{10}}11^{\underline{3}}_{\underline{8}}11^{\underline{2}}_{\underline{8}}13^{\underline{2}}_{\underline{6}}6^{\underline{1}}_{\underline{6}}8^{\underline{1}}_{\underline{4}}4^{0}_{\underline{4}}7^{0}_{\underline{2}}1^{1}_{\underline{2}}3^{1}_{0}1^{2}_{2}$	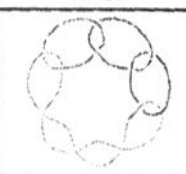	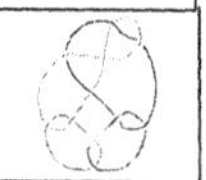
11^a_{535}: $1^{\underline{6}}_{\underline{12}}1^{\underline{5}}_{\underline{8}}5^{\underline{4}}_{\underline{8}}3^{\underline{4}}_{\underline{6}}4^{\underline{3}}_{\underline{6}}2^{\underline{3}}_{\underline{4}}8^{\underline{2}}_{\underline{4}}5^{\underline{2}}_{\underline{2}}5^{\underline{1}}_{\underline{2}}7^{\underline{1}}_{0}10^{0}_{0}6^{0}_{2}4^{1}_{2}9^{1}_{4}7^{2}_{4}7^{2}_{6}3^{3}_{6}4^{3}_{8}1^{4}_{8}3^{4}_{10}1^{5}_{12}$ 11^a_{536}: $1^{\underline{6}}_{\underline{14}}1^{\underline{5}}_{\underline{12}}1^{\underline{5}}_{\underline{10}}5^{\underline{4}}_{\underline{10}}4^{\underline{4}}_{\underline{8}}7^{\underline{3}}_{\underline{8}}2^{\underline{3}}_{\underline{6}}8^{\underline{2}}_{\underline{6}}8^{\underline{2}}_{\underline{4}}8^{\underline{1}}_{\underline{4}}7^{\underline{1}}_{\underline{2}}10^{0}_{\underline{2}}9^{0}_{0}5^{1}_{0}9^{1}_{2}7^{2}_{2}8^{2}_{4}3^{3}_{4}4^{3}_{6}1^{4}_{6}3^{4}_{8}1^{5}_{10}$	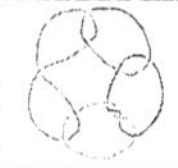	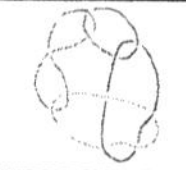
11^a_{537}: $1^{\underline{6}}_{\underline{12}}1^{\underline{5}}_{\underline{10}}1^{\underline{5}}_{\underline{8}}6^{\underline{4}}_{\underline{8}}2^{\underline{4}}_{\underline{6}}5^{\underline{3}}_{\underline{6}}5^{\underline{3}}_{\underline{4}}12^{\underline{2}}_{\underline{4}}8^{\underline{2}}_{\underline{2}}8^{\underline{1}}_{\underline{2}}9^{\underline{1}}_{0}13^{0}_{0}11^{0}_{2}7^{1}_{2}10^{1}_{4}7^{2}_{4}8^{2}_{6}3^{3}_{6}6^{3}_{8}1^{4}_{8}3^{4}_{10}1^{5}_{12}$ 11^a_{538}: $1^{\underline{9}}_{\underline{22}}2^{\underline{8}}_{\underline{20}}1^{\underline{8}}_{\underline{18}}5^{\underline{7}}_{\underline{18}}2^{\underline{7}}_{\underline{16}}9^{\underline{6}}_{\underline{16}}5^{\underline{6}}_{\underline{14}}8^{\underline{5}}_{\underline{14}}9^{\underline{5}}_{\underline{12}}14^{\underline{4}}_{\underline{12}}14^{\underline{4}}_{\underline{10}}11^{\underline{3}}_{\underline{10}}8^{\underline{3}}_{\underline{8}}9^{\underline{2}}_{\underline{8}}11^{\underline{2}}_{\underline{6}}4^{\underline{1}}_{\underline{6}}9^{\underline{1}}_{\underline{4}}4^{0}_{\underline{4}}6^{0}_{\underline{2}}1^{1}_{\underline{2}}2^{1}_{0}1^{2}_{2}$		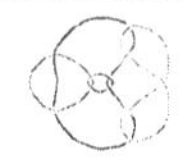
11^a_{539}: $1^{\underline{6}}_{\underline{12}}1^{\underline{5}}_{\underline{10}}1^{\underline{5}}_{\underline{8}}6^{\underline{4}}_{\underline{8}}3^{\underline{4}}_{\underline{6}}6^{\underline{3}}_{\underline{6}}4^{\underline{3}}_{\underline{4}}12^{\underline{2}}_{\underline{4}}6^{\underline{2}}_{\underline{2}}7^{\underline{1}}_{\underline{2}}12^{\underline{1}}_{0}15^{0}_{0}13^{0}_{2}9^{1}_{2}9^{1}_{4}7^{2}_{4}9^{2}_{6}3^{3}_{6}7^{3}_{8}1^{4}_{8}3^{4}_{10}1^{5}_{12}$ 11^a_{540}: $1^{\underline{8}}_{\underline{20}}1^{\underline{8}}_{\underline{18}}2^{\underline{7}}_{\underline{18}}4^{\underline{6}}_{\underline{16}}2^{\underline{6}}_{\underline{14}}5^{\underline{5}}_{\underline{14}}4^{\underline{5}}_{\underline{12}}11^{\underline{4}}_{\underline{12}}9^{\underline{4}}_{\underline{10}}10^{\underline{3}}_{\underline{10}}7^{\underline{3}}_{\underline{8}}9^{\underline{2}}_{\underline{8}}10^{\underline{2}}_{\underline{6}}6^{\underline{1}}_{\underline{6}}9^{\underline{1}}_{\underline{4}}8^{0}_{\underline{4}}9^{0}_{\underline{2}}3^{1}_{\underline{2}}5^{1}_{0}1^{2}_{0}3^{2}_{2}1^{3}_{4}$	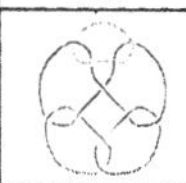	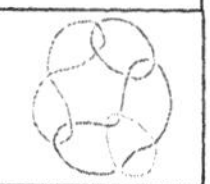
11^a_{541}: $1^{\underline{7}}_{\underline{16}}2^{\underline{6}}_{\underline{14}}1^{\underline{6}}_{\underline{12}}5^{\underline{5}}_{\underline{12}}2^{\underline{5}}_{\underline{10}}10^{\underline{4}}_{\underline{10}}7^{\underline{4}}_{\underline{8}}12^{\underline{3}}_{\underline{8}}8^{\underline{3}}_{\underline{6}}15^{\underline{2}}_{\underline{6}}12^{\underline{2}}_{\underline{4}}10^{\underline{1}}_{\underline{4}}15^{\underline{1}}_{\underline{2}}15^{0}_{\underline{2}}16^{0}_{0}9^{1}_{0}9^{1}_{2}4^{2}_{2}9^{2}_{4}1^{3}_{4}4^{3}_{6}1^{4}_{8}$ 11^a_{542}: $1^{\underline{3}}_{\underline{4}}3^{\underline{2}}_{\underline{2}}1^{\underline{2}}_{0}5^{\underline{1}}_{0}3^{\underline{1}}_{2}12^{0}_{2}7^{0}_{4}10^{1}_{4}10^{1}_{6}17^{2}_{6}14^{2}_{8}12^{3}_{8}13^{3}_{10}12^{4}_{10}14^{4}_{12}7^{5}_{12}10^{5}_{14}4^{6}_{14}7^{6}_{16}1^{7}_{16}4^{7}_{18}1^{8}_{20}$	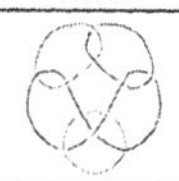	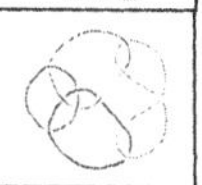
11^a_{543}: $1^{\underline{11}}_{\underline{28}}3^{\underline{10}}_{\underline{26}}1^{\underline{10}}_{\underline{24}}5^{\underline{9}}_{\underline{24}}3^{\underline{9}}_{\underline{22}}9^{\underline{8}}_{\underline{22}}6^{\underline{8}}_{\underline{20}}9^{\underline{7}}_{\underline{20}}8^{\underline{7}}_{\underline{18}}13^{\underline{6}}_{\underline{18}}13^{\underline{6}}_{\underline{16}}10^{\underline{5}}_{\underline{16}}9^{\underline{5}}_{\underline{14}}9^{\underline{4}}_{\underline{14}}12^{\underline{4}}_{\underline{12}}5^{\underline{3}}_{\underline{12}}7^{\underline{3}}_{\underline{10}}3^{\underline{2}}_{\underline{10}}5^{\underline{2}}_{\underline{8}}3^{\underline{1}}_{\underline{6}}1^{0}_{\underline{6}}1^{0}_{\underline{4}}$ 11^a_{544}: $1^{\underline{2}}_{0}1^{\underline{1}}_{2}1^{\underline{1}}_{4}6^{0}_{4}3^{0}_{6}5^{1}_{6}4^{1}_{8}11^{2}_{8}7^{2}_{10}7^{3}_{10}9^{3}_{12}12^{4}_{12}9^{4}_{14}6^{5}_{14}10^{5}_{16}7^{6}_{16}8^{6}_{18}3^{7}_{18}5^{7}_{20}1^{8}_{20}3^{8}_{22}1^{9}_{24}$	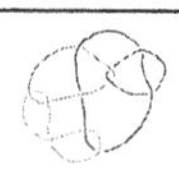	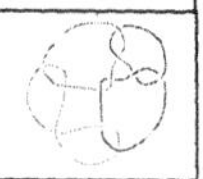
11^a_{545}: $1^{\underline{8}}_{\underline{20}}2^{\underline{7}}_{\underline{18}}1^{\underline{7}}_{\underline{16}}6^{\underline{6}}_{\underline{16}}4^{\underline{6}}_{\underline{14}}9^{\underline{5}}_{\underline{14}}4^{\underline{5}}_{\underline{12}}12^{\underline{4}}_{\underline{12}}11^{\underline{4}}_{\underline{10}}11^{\underline{3}}_{\underline{10}}10^{\underline{3}}_{\underline{8}}14^{\underline{2}}_{\underline{8}}13^{\underline{2}}_{\underline{6}}8^{\underline{1}}_{\underline{6}}12^{\underline{1}}_{\underline{4}}8^{0}_{\underline{4}}10^{0}_{\underline{2}}4^{1}_{\underline{2}}6^{1}_{0}1^{2}_{0}4^{2}_{2}1^{3}_{4}$ 11^a_{546}: $1^{\underline{8}}_{\underline{18}}1^{\underline{8}}_{\underline{16}}2^{\underline{7}}_{\underline{16}}6^{\underline{6}}_{\underline{14}}2^{\underline{6}}_{\underline{12}}5^{\underline{5}}_{\underline{12}}6^{\underline{5}}_{\underline{10}}13^{\underline{4}}_{\underline{10}}7^{\underline{4}}_{\underline{8}}8^{\underline{3}}_{\underline{8}}11^{\underline{3}}_{\underline{6}}14^{\underline{2}}_{\underline{6}}12^{\underline{2}}_{\underline{4}}8^{\underline{1}}_{\underline{4}}10^{\underline{1}}_{\underline{2}}7^{0}_{\underline{2}}9^{0}_{0}3^{1}_{0}6^{1}_{2}1^{2}_{2}3^{2}_{4}1^{3}_{6}$	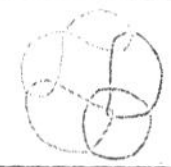	
11^a_{547}: $1^{\underline{11}}_{\underline{28}}3^{\underline{10}}_{\underline{26}}1^{\underline{10}}_{\underline{24}}6^{\underline{9}}_{\underline{24}}3^{\underline{9}}_{\underline{22}}11^{\underline{8}}_{\underline{22}}7^{\underline{8}}_{\underline{20}}11^{\underline{7}}_{\underline{20}}10^{\underline{7}}_{\underline{18}}16^{\underline{6}}_{\underline{18}}15^{\underline{6}}_{\underline{16}}13^{\underline{5}}_{\underline{16}}12^{\underline{5}}_{\underline{14}}11^{\underline{4}}_{\underline{14}}15^{\underline{4}}_{\underline{12}}7^{\underline{3}}_{\underline{12}}9^{\underline{3}}_{\underline{10}}4^{\underline{2}}_{\underline{10}}7^{\underline{2}}_{\underline{8}}4^{\underline{1}}_{\underline{6}}1^{0}_{\underline{6}}1^{0}_{\underline{4}}$ 11^a_{548}: $1^{\underline{8}}_{\underline{19}}1^{\underline{8}}_{\underline{17}}1^{\underline{7}}_{\underline{17}}5^{\underline{6}}_{\underline{15}}1^{\underline{6}}_{\underline{13}}1^{\underline{5}}_{\underline{13}}5^{\underline{5}}_{\underline{11}}15^{\underline{4}}_{\underline{11}}11^{\underline{4}}_{\underline{9}}10^{\underline{3}}_{\underline{9}}5^{\underline{3}}_{\underline{7}}11^{\underline{2}}_{\underline{7}}10^{\underline{2}}_{\underline{5}}4^{\underline{1}}_{\underline{5}}11^{\underline{1}}_{\underline{3}}11^{0}_{\underline{3}}9^{0}_{\underline{1}}4^{1}_{\underline{1}}6^{1}_{1}1^{2}_{1}4^{2}_{3}1^{3}_{5}$	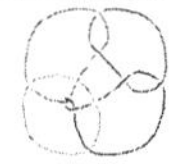	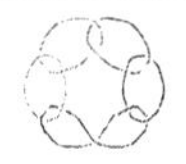

3.13. **11 Crossing Non-Alternating Links.**

L: KH	1st line	2nd line
11^n_{1}: $1^{\underline{7}}_{\underline{20}}1^{\underline{6}}_{\underline{18}}2^{\underline{6}}_{\underline{16}}3^{\underline{5}}_{\underline{16}}1^{\underline{5}}_{\underline{14}}1^{\underline{5}}_{\underline{12}}2^{\underline{4}}_{\underline{14}}4^{\underline{4}}_{\underline{12}}1^{\underline{4}}_{\underline{10}}3^{\underline{3}}_{\underline{12}}3^{\underline{3}}_{\underline{10}}3^{\underline{2}}_{\underline{10}}3^{\underline{2}}_{\underline{8}}1^{\underline{2}}_{\underline{6}}1^{\underline{1}}_{\underline{8}}3^{\underline{1}}_{\underline{6}}2^{0}_{\underline{6}}2^{0}_{\underline{4}}1^{1}_{\underline{4}}1^{1}_{\underline{2}}1^{2}_{0}$ 11^n_{2}: $1^{\underline{7}}_{\underline{16}}1^{\underline{6}}_{\underline{14}}1^{\underline{6}}_{\underline{12}}3^{\underline{5}}_{\underline{12}}1^{\underline{5}}_{\underline{10}}3^{\underline{4}}_{\underline{10}}3^{\underline{4}}_{\underline{8}}4^{\underline{3}}_{\underline{8}}3^{\underline{3}}_{\underline{6}}5^{\underline{2}}_{\underline{6}}4^{\underline{2}}_{\underline{4}}2^{\underline{1}}_{\underline{4}}5^{\underline{1}}_{\underline{2}}4^{0}_{\underline{2}}4^{0}_{0}2^{1}_{0}2^{1}_{2}2^{2}_{4}$	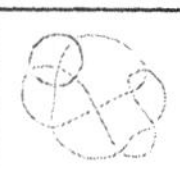	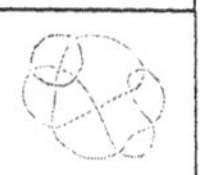

L: KH	1st line	2nd line
11^n_3: $1^{\underline{7}}_{\underline{16}}1^{\underline{6}}_{\underline{14}}1^{\underline{6}}_{\underline{12}}3^{\underline{5}}_{\underline{12}}1^{\underline{5}}_{\underline{10}}3^{\underline{4}}_{\underline{10}}3^{\underline{4}}_{\underline{8}}3^{\underline{3}}_{\underline{8}}3^{\underline{3}}_{\underline{6}}4^{\underline{2}}_{\underline{6}}3^{\underline{2}}_{\underline{4}}2^{\underline{1}}_{\underline{4}}4^{\underline{1}}_{\underline{2}}3^{0}_{\underline{2}}4^{0}_{0}1^{1}_{0}1^{1}_{2}1^{2}_{4}$		
11^n_4: $2^{\underline{6}}_{\underline{16}}2^{\underline{5}}_{\underline{14}}2^{\underline{5}}_{\underline{12}}4^{\underline{4}}_{\underline{12}}3^{\underline{4}}_{\underline{10}}5^{\underline{3}}_{\underline{10}}3^{\underline{3}}_{\underline{8}}4^{\underline{2}}_{\underline{8}}5^{\underline{2}}_{\underline{6}}4^{\underline{1}}_{\underline{6}}4^{\underline{1}}_{\underline{4}}3^{0}_{\underline{4}}5^{0}_{\underline{2}}2^{1}_{\underline{2}}2^{1}_{0}1^{2}_{0}2^{2}_{2}1^{3}_{4}$		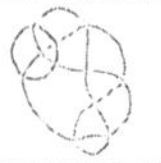
11^n_5: $1^{\underline{6}}_{\underline{12}}1^{\underline{5}}_{\underline{10}}1^{\underline{5}}_{\underline{8}}2^{\underline{4}}_{\underline{8}}1^{\underline{4}}_{\underline{6}}3^{\underline{3}}_{\underline{6}}2^{\underline{3}}_{\underline{4}}1^{\underline{2}}_{\underline{6}}3^{\underline{2}}_{\underline{4}}3^{\underline{2}}_{\underline{2}}4^{\underline{1}}_{\underline{2}}3^{\underline{1}}_{0}1^{0}_{\underline{2}}4^{0}_{0}4^{0}_{2}1^{1}_{0}2^{1}_{2}2^{1}_{4}2^{2}_{4}2^{2}_{6}1^{3}_{8}$	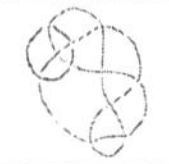	
11^n_6: $1^{\underline{4}}_{\underline{10}}1^{\underline{3}}_{\underline{8}}1^{\underline{3}}_{\underline{6}}3^{\underline{2}}_{\underline{6}}1^{\underline{2}}_{\underline{4}}3^{\underline{1}}_{\underline{4}}3^{\underline{1}}_{\underline{2}}4^{0}_{\underline{2}}5^{0}_{0}1^{0}_{2}4^{1}_{0}3^{1}_{2}2^{2}_{2}4^{2}_{4}2^{3}_{4}2^{3}_{6}1^{4}_{6}2^{4}_{8}1^{5}_{10}$		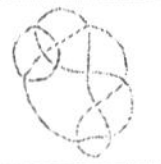
11^n_7: $2^{\underline{7}}_{\underline{20}}2^{\underline{6}}_{\underline{18}}2^{\underline{6}}_{\underline{16}}3^{\underline{5}}_{\underline{16}}2^{\underline{5}}_{\underline{14}}4^{\underline{4}}_{\underline{14}}4^{\underline{4}}_{\underline{12}}4^{\underline{3}}_{\underline{12}}3^{\underline{3}}_{\underline{10}}3^{\underline{2}}_{\underline{10}}4^{\underline{2}}_{\underline{8}}2^{\underline{1}}_{\underline{8}}3^{\underline{1}}_{\underline{6}}2^{0}_{\underline{6}}3^{0}_{\underline{4}}1^{1}_{\underline{4}}1^{1}_{\underline{2}}1^{2}_{0}$	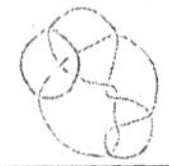	
11^n_8: $1^{\underline{7}}_{\underline{16}}1^{\underline{6}}_{\underline{14}}1^{\underline{6}}_{\underline{12}}2^{\underline{5}}_{\underline{12}}1^{\underline{5}}_{\underline{10}}3^{\underline{4}}_{\underline{10}}2^{\underline{4}}_{\underline{8}}1^{\underline{3}}_{\underline{10}}3^{\underline{3}}_{\underline{8}}3^{\underline{3}}_{\underline{6}}4^{\underline{2}}_{\underline{6}}3^{\underline{2}}_{\underline{4}}2^{\underline{1}}_{\underline{4}}3^{\underline{1}}_{\underline{2}}1^{0}_{\underline{4}}3^{0}_{\underline{2}}3^{0}_{0}1^{1}_{0}1^{1}_{2}1^{2}_{4}$		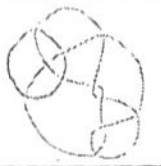
11^n_9: $1^{\underline{5}}_{\underline{14}}1^{\underline{5}}_{\underline{12}}1^{\underline{4}}_{\underline{12}}1^{\underline{4}}_{\underline{10}}1^{\underline{4}}_{\underline{8}}2^{\underline{3}}_{\underline{10}}2^{\underline{3}}_{\underline{8}}3^{\underline{2}}_{\underline{8}}3^{\underline{2}}_{\underline{6}}1^{\underline{2}}_{\underline{4}}2^{\underline{1}}_{\underline{6}}3^{\underline{1}}_{\underline{4}}1^{\underline{1}}_{\underline{2}}3^{0}_{\underline{4}}4^{0}_{\underline{2}}1^{0}_{0}2^{1}_{\underline{2}}2^{1}_{0}1^{2}_{0}2^{2}_{2}1^{3}_{2}1^{3}_{4}1^{4}_{6}$	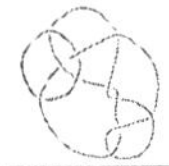	
11^n_{10}: $1^{\underline{9}}_{\underline{26}}1^{\underline{8}}_{\underline{24}}1^{\underline{8}}_{\underline{22}}3^{\underline{7}}_{\underline{22}}2^{\underline{7}}_{\underline{20}}2^{\underline{6}}_{\underline{20}}3^{\underline{6}}_{\underline{18}}1^{\underline{6}}_{\underline{16}}3^{\underline{5}}_{\underline{18}}2^{\underline{5}}_{\underline{16}}2^{\underline{4}}_{\underline{16}}4^{\underline{4}}_{\underline{14}}1^{\underline{4}}_{\underline{12}}2^{\underline{3}}_{\underline{14}}2^{\underline{3}}_{\underline{12}}1^{\underline{2}}_{\underline{12}}2^{\underline{2}}_{\underline{10}}1^{\underline{1}}_{\underline{8}}1^{0}_{\underline{8}}1^{0}_{\underline{6}}$		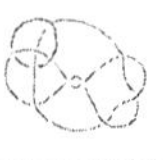
11^n_{11}: $1^{\underline{9}}_{\underline{22}}1^{\underline{8}}_{\underline{20}}1^{\underline{8}}_{\underline{18}}3^{\underline{7}}_{\underline{18}}1^{\underline{7}}_{\underline{16}}2^{\underline{6}}_{\underline{16}}3^{\underline{6}}_{\underline{14}}4^{\underline{5}}_{\underline{14}}2^{\underline{5}}_{\underline{12}}3^{\underline{4}}_{\underline{12}}4^{\underline{4}}_{\underline{10}}3^{\underline{3}}_{\underline{10}}3^{\underline{3}}_{\underline{8}}2^{\underline{2}}_{\underline{8}}3^{\underline{2}}_{\underline{6}}2^{\underline{1}}_{\underline{4}}2^{0}_{\underline{4}}2^{0}_{\underline{2}}$		
11^n_{12}: $1^{\underline{9}}_{\underline{22}}1^{\underline{8}}_{\underline{20}}1^{\underline{8}}_{\underline{18}}2^{\underline{7}}_{\underline{18}}1^{\underline{7}}_{\underline{16}}2^{\underline{6}}_{\underline{16}}2^{\underline{6}}_{\underline{14}}1^{\underline{5}}_{\underline{16}}3^{\underline{5}}_{\underline{14}}2^{\underline{5}}_{\underline{12}}3^{\underline{4}}_{\underline{12}}3^{\underline{4}}_{\underline{10}}1^{\underline{3}}_{\underline{12}}2^{\underline{3}}_{\underline{10}}2^{\underline{3}}_{\underline{8}}2^{\underline{2}}_{\underline{8}}2^{\underline{2}}_{\underline{6}}1^{\underline{1}}_{\underline{4}}1^{0}_{\underline{6}}2^{0}_{\underline{4}}1^{0}_{\underline{2}}$		
11^n_{13}: $1^{\underline{9}}_{\underline{24}}1^{\underline{8}}_{\underline{20}}2^{\underline{7}}_{\underline{20}}1^{\underline{6}}_{\underline{18}}2^{\underline{6}}_{\underline{16}}2^{\underline{5}}_{\underline{16}}1^{\underline{5}}_{\underline{14}}2^{\underline{4}}_{\underline{14}}3^{\underline{4}}_{\underline{12}}1^{\underline{4}}_{\underline{10}}1^{\underline{3}}_{\underline{12}}2^{\underline{3}}_{\underline{10}}1^{\underline{2}}_{\underline{10}}1^{\underline{2}}_{\underline{8}}1^{\underline{1}}_{\underline{6}}1^{0}_{\underline{6}}1^{0}_{\underline{4}}$	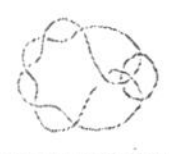	
11^n_{14}: $1^{\underline{9}}_{\underline{20}}1^{\underline{8}}_{\underline{16}}2^{\underline{7}}_{\underline{16}}1^{\underline{6}}_{\underline{14}}2^{\underline{6}}_{\underline{12}}2^{\underline{5}}_{\underline{12}}1^{\underline{5}}_{\underline{10}}2^{\underline{4}}_{\underline{10}}2^{\underline{4}}_{\underline{8}}1^{\underline{3}}_{\underline{8}}2^{\underline{3}}_{\underline{6}}2^{\underline{2}}_{\underline{6}}1^{\underline{2}}_{\underline{4}}2^{\underline{1}}_{\underline{2}}2^{0}_{\underline{2}}2^{0}_{0}$		
11^n_{15}: $1^{\underline{9}}_{\underline{20}}1^{\underline{8}}_{\underline{16}}1^{\underline{7}}_{\underline{16}}1^{\underline{6}}_{\underline{14}}1^{\underline{6}}_{\underline{12}}1^{\underline{5}}_{\underline{14}}1^{\underline{5}}_{\underline{12}}1^{\underline{5}}_{\underline{10}}2^{\underline{4}}_{\underline{10}}1^{\underline{4}}_{\underline{8}}1^{\underline{3}}_{\underline{10}}1^{\underline{3}}_{\underline{8}}1^{\underline{3}}_{\underline{6}}1^{\underline{2}}_{\underline{8}}2^{\underline{2}}_{\underline{6}}1^{\underline{2}}_{\underline{4}}1^{\underline{1}}_{\underline{4}}1^{\underline{1}}_{\underline{2}}1^{0}_{\underline{4}}2^{0}_{\underline{2}}1^{0}_{0}$		
11^n_{16}: $1^{\underline{9}}_{\underline{26}}1^{\underline{9}}_{\underline{24}}1^{\underline{8}}_{\underline{24}}1^{\underline{8}}_{\underline{22}}1^{\underline{8}}_{\underline{20}}2^{\underline{7}}_{\underline{22}}2^{\underline{7}}_{\underline{20}}2^{\underline{6}}_{\underline{20}}3^{\underline{6}}_{\underline{18}}1^{\underline{6}}_{\underline{16}}3^{\underline{5}}_{\underline{18}}2^{\underline{5}}_{\underline{16}}1^{\underline{5}}_{\underline{14}}1^{\underline{4}}_{\underline{16}}4^{\underline{4}}_{\underline{14}}1^{\underline{4}}_{\underline{12}}2^{\underline{3}}_{\underline{14}}1^{\underline{3}}_{\underline{12}}1^{\underline{2}}_{\underline{12}}2^{\underline{2}}_{\underline{10}}1^{\underline{1}}_{\underline{8}}1^{0}_{\underline{8}}1^{0}_{\underline{6}}$		
11^n_{17}: $1^{\underline{9}}_{\underline{22}}1^{\underline{8}}_{\underline{20}}1^{\underline{8}}_{\underline{18}}3^{\underline{7}}_{\underline{18}}1^{\underline{7}}_{\underline{16}}3^{\underline{6}}_{\underline{16}}3^{\underline{6}}_{\underline{14}}5^{\underline{5}}_{\underline{14}}3^{\underline{5}}_{\underline{12}}3^{\underline{4}}_{\underline{12}}5^{\underline{4}}_{\underline{10}}4^{\underline{3}}_{\underline{10}}3^{\underline{3}}_{\underline{8}}3^{\underline{2}}_{\underline{8}}4^{\underline{2}}_{\underline{6}}3^{\underline{1}}_{\underline{4}}2^{0}_{\underline{4}}2^{0}_{\underline{2}}$		
11^n_{18}: $1^{\underline{9}}_{\underline{22}}2^{\underline{8}}_{\underline{20}}1^{\underline{8}}_{\underline{18}}4^{\underline{7}}_{\underline{18}}2^{\underline{7}}_{\underline{16}}3^{\underline{6}}_{\underline{16}}4^{\underline{6}}_{\underline{14}}6^{\underline{5}}_{\underline{14}}3^{\underline{5}}_{\underline{12}}4^{\underline{4}}_{\underline{12}}6^{\underline{4}}_{\underline{10}}4^{\underline{3}}_{\underline{10}}4^{\underline{3}}_{\underline{8}}3^{\underline{2}}_{\underline{8}}4^{\underline{2}}_{\underline{6}}3^{\underline{1}}_{\underline{4}}2^{0}_{\underline{4}}2^{0}_{\underline{2}}$		
11^n_{19}: $1^{\underline{9}}_{\underline{26}}1^{\underline{9}}_{\underline{24}}1^{\underline{8}}_{\underline{22}}1^{\underline{8}}_{\underline{20}}2^{\underline{7}}_{\underline{22}}1^{\underline{7}}_{\underline{20}}1^{\underline{6}}_{\underline{20}}3^{\underline{6}}_{\underline{18}}1^{\underline{6}}_{\underline{16}}2^{\underline{5}}_{\underline{18}}1^{\underline{5}}_{\underline{16}}1^{\underline{5}}_{\underline{14}}1^{\underline{4}}_{\underline{16}}3^{\underline{4}}_{\underline{14}}1^{\underline{4}}_{\underline{12}}1^{\underline{3}}_{\underline{14}}1^{\underline{3}}_{\underline{12}}1^{\underline{2}}_{\underline{12}}1^{\underline{2}}_{\underline{10}}1^{\underline{1}}_{\underline{8}}1^{0}_{\underline{8}}1^{0}_{\underline{6}}$	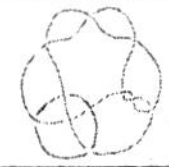	
11^n_{20}: $1^{\underline{9}}_{\underline{22}}1^{\underline{8}}_{\underline{18}}3^{\underline{7}}_{\underline{18}}2^{\underline{6}}_{\underline{16}}3^{\underline{6}}_{\underline{14}}4^{\underline{5}}_{\underline{14}}2^{\underline{5}}_{\underline{12}}3^{\underline{4}}_{\underline{12}}4^{\underline{4}}_{\underline{10}}3^{\underline{3}}_{\underline{10}}3^{\underline{3}}_{\underline{8}}3^{\underline{2}}_{\underline{8}}3^{\underline{2}}_{\underline{6}}3^{\underline{1}}_{\underline{4}}2^{0}_{\underline{4}}2^{0}_{\underline{2}}$		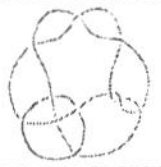
11^n_{21}: $1^{\underline{9}}_{\underline{22}}1^{\underline{8}}_{\underline{20}}1^{\underline{8}}_{\underline{18}}4^{\underline{7}}_{\underline{18}}1^{\underline{7}}_{\underline{16}}3^{\underline{6}}_{\underline{16}}4^{\underline{6}}_{\underline{14}}6^{\underline{5}}_{\underline{14}}3^{\underline{5}}_{\underline{12}}4^{\underline{4}}_{\underline{12}}6^{\underline{4}}_{\underline{10}}4^{\underline{3}}_{\underline{10}}4^{\underline{3}}_{\underline{8}}4^{\underline{2}}_{\underline{8}}4^{\underline{2}}_{\underline{6}}4^{\underline{1}}_{\underline{4}}2^{0}_{\underline{4}}2^{0}_{\underline{2}}$	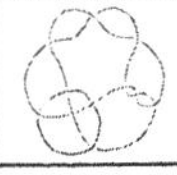	
11^n_{22}: $1^{\underline{5}}_{\underline{14}}2^{\underline{4}}_{\underline{12}}2^{\underline{4}}_{\underline{10}}1^{\underline{4}}_{\underline{8}}2^{\underline{3}}_{\underline{10}}2^{\underline{3}}_{\underline{8}}4^{\underline{2}}_{\underline{8}}3^{\underline{2}}_{\underline{6}}3^{\underline{1}}_{\underline{6}}4^{\underline{1}}_{\underline{4}}1^{\underline{1}}_{\underline{2}}3^{0}_{\underline{4}}4^{0}_{\underline{2}}3^{1}_{\underline{2}}2^{1}_{0}1^{2}_{0}3^{2}_{2}1^{3}_{2}1^{3}_{4}1^{4}_{6}$		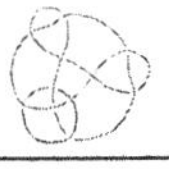
11^n_{23}: $1^{\underline{5}}_{\underline{10}}2^{\underline{4}}_{\underline{8}}1^{\underline{4}}_{\underline{6}}2^{\underline{3}}_{\underline{6}}2^{\underline{3}}_{\underline{4}}5^{\underline{2}}_{\underline{4}}2^{\underline{2}}_{\underline{2}}3^{\underline{1}}_{\underline{2}}5^{\underline{1}}_{0}5^{0}_{0}5^{0}_{2}4^{1}_{2}3^{1}_{4}2^{2}_{4}4^{2}_{6}1^{3}_{6}2^{3}_{8}1^{4}_{10}$	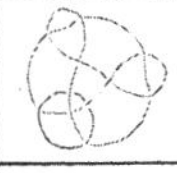	
11^n_{24}: $1^{\underline{5}}_{\underline{10}}2^{\underline{4}}_{\underline{8}}1^{\underline{4}}_{\underline{6}}2^{\underline{3}}_{\underline{6}}2^{\underline{3}}_{\underline{4}}3^{\underline{2}}_{\underline{4}}2^{\underline{2}}_{\underline{2}}3^{\underline{1}}_{\underline{2}}3^{\underline{1}}_{0}2^{0}_{\underline{2}}4^{0}_{0}4^{0}_{2}3^{1}_{2}2^{1}_{4}1^{2}_{2}1^{2}_{4}2^{2}_{6}1^{3}_{4}1^{3}_{6}1^{3}_{8}1^{4}_{8}1^{5}_{8}1^{6}_{12}$		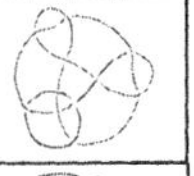
11^n_{25}: $2^{\underline{4}}_{\underline{10}}3^{\underline{3}}_{\underline{8}}2^{\underline{3}}_{\underline{6}}6^{\underline{2}}_{\underline{6}}3^{\underline{2}}_{\underline{4}}5^{\underline{1}}_{\underline{4}}6^{\underline{1}}_{\underline{2}}7^{0}_{\underline{2}}7^{0}_{0}7^{1}_{0}5^{1}_{2}3^{2}_{2}7^{2}_{4}3^{3}_{4}3^{3}_{6}1^{4}_{6}3^{4}_{8}1^{5}_{10}$	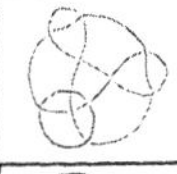	
11^n_{26}: $1^{\underline{9}}_{\underline{24}}1^{\underline{8}}_{\underline{20}}4^{\underline{7}}_{\underline{20}}2^{\underline{6}}_{\underline{18}}4^{\underline{6}}_{\underline{16}}3^{\underline{5}}_{\underline{16}}2^{\underline{5}}_{\underline{14}}4^{\underline{4}}_{\underline{14}}4^{\underline{4}}_{\underline{12}}2^{\underline{3}}_{\underline{12}}3^{\underline{3}}_{\underline{10}}2^{\underline{2}}_{\underline{10}}2^{\underline{2}}_{\underline{8}}2^{\underline{1}}_{\underline{6}}1^{0}_{\underline{6}}1^{0}_{\underline{4}}$		
11^n_{27}: $1^{\underline{9}}_{\underline{20}}1^{\underline{8}}_{\underline{16}}3^{\underline{7}}_{\underline{16}}1^{\underline{6}}_{\underline{14}}3^{\underline{6}}_{\underline{12}}2^{\underline{5}}_{\underline{12}}1^{\underline{5}}_{\underline{10}}3^{\underline{4}}_{\underline{10}}2^{\underline{4}}_{\underline{8}}1^{\underline{3}}_{\underline{10}}1^{\underline{3}}_{\underline{8}}3^{\underline{3}}_{\underline{6}}3^{\underline{2}}_{\underline{6}}1^{\underline{2}}_{\underline{4}}2^{\underline{1}}_{\underline{2}}1^{0}_{\underline{4}}2^{0}_{\underline{2}}1^{0}_{0}$		
11^n_{28}: $1^{\underline{9}}_{\underline{20}}1^{\underline{8}}_{\underline{16}}1^{\underline{7}}_{\underline{16}}1^{\underline{6}}_{\underline{14}}1^{\underline{6}}_{\underline{12}}3^{\underline{5}}_{\underline{14}}1^{\underline{5}}_{\underline{10}}1^{\underline{4}}_{\underline{12}}4^{\underline{4}}_{\underline{10}}4^{\underline{3}}_{\underline{10}}1^{\underline{3}}_{\underline{8}}1^{\underline{3}}_{\underline{6}}3^{\underline{2}}_{\underline{8}}4^{\underline{2}}_{\underline{6}}1^{\underline{1}}_{\underline{6}}3^{\underline{1}}_{\underline{4}}4^{0}_{\underline{4}}3^{0}_{\underline{2}}1^{1}_{\underline{2}}2^{1}_{0}1^{2}_{2}$		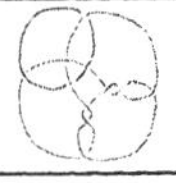
11^n_{29}: $2^{\underline{9}}_{\underline{24}}1^{\underline{8}}_{\underline{22}}2^{\underline{8}}_{\underline{20}}4^{\underline{7}}_{\underline{20}}1^{\underline{7}}_{\underline{18}}3^{\underline{6}}_{\underline{18}}4^{\underline{6}}_{\underline{16}}4^{\underline{5}}_{\underline{16}}3^{\underline{5}}_{\underline{14}}4^{\underline{4}}_{\underline{14}}5^{\underline{4}}_{\underline{12}}2^{\underline{3}}_{\underline{12}}3^{\underline{3}}_{\underline{10}}2^{\underline{2}}_{\underline{10}}2^{\underline{2}}_{\underline{8}}2^{\underline{1}}_{\underline{6}}1^{0}_{\underline{6}}1^{0}_{\underline{4}}$	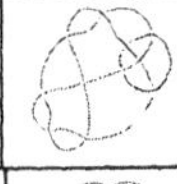	
11^n_{30}: $1^{\underline{9}}_{\underline{20}}1^{\underline{8}}_{\underline{16}}2^{\underline{7}}_{\underline{16}}1^{\underline{6}}_{\underline{14}}2^{\underline{6}}_{\underline{12}}1^{\underline{5}}_{\underline{14}}2^{\underline{5}}_{\underline{12}}1^{\underline{5}}_{\underline{10}}3^{\underline{4}}_{\underline{10}}2^{\underline{4}}_{\underline{8}}1^{\underline{3}}_{\underline{10}}1^{\underline{3}}_{\underline{8}}2^{\underline{3}}_{\underline{6}}1^{\underline{2}}_{\underline{8}}3^{\underline{2}}_{\underline{6}}1^{\underline{2}}_{\underline{4}}1^{\underline{1}}_{\underline{4}}2^{\underline{1}}_{\underline{2}}1^{0}_{\underline{4}}2^{0}_{\underline{2}}1^{0}_{0}$		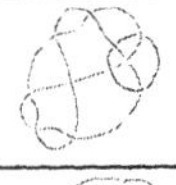
11^n_{31}: $2^{\underline{7}}_{\underline{18}}1^{\underline{6}}_{\underline{16}}2^{\underline{6}}_{\underline{14}}5^{\underline{5}}_{\underline{14}}1^{\underline{5}}_{\underline{12}}4^{\underline{4}}_{\underline{12}}5^{\underline{4}}_{\underline{10}}5^{\underline{3}}_{\underline{10}}4^{\underline{3}}_{\underline{8}}5^{\underline{2}}_{\underline{8}}5^{\underline{2}}_{\underline{6}}2^{\underline{1}}_{\underline{6}}5^{\underline{1}}_{\underline{4}}4^{0}_{\underline{4}}4^{0}_{\underline{2}}1^{1}_{\underline{2}}2^{1}_{0}1^{2}_{2}$	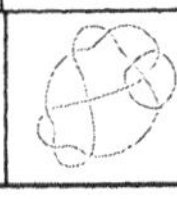	
11^n_{32}: $1^{\underline{5}}_{\underline{14}}3^{\underline{4}}_{\underline{12}}2^{\underline{4}}_{\underline{10}}3^{\underline{3}}_{\underline{10}}2^{\underline{3}}_{\underline{8}}4^{\underline{2}}_{\underline{8}}3^{\underline{2}}_{\underline{6}}3^{\underline{1}}_{\underline{6}}4^{\underline{1}}_{\underline{4}}4^{0}_{\underline{4}}4^{0}_{\underline{2}}3^{1}_{\underline{2}}3^{1}_{0}1^{2}_{0}3^{2}_{2}1^{3}_{2}1^{3}_{4}1^{4}_{6}$		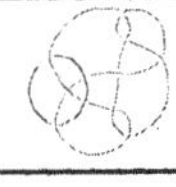

L: KH	1st line	2nd line
11^n_{33}: $1^{\underline{5}}_{\underline{10}}1^{\underline{4}}_{\underline{8}}1^{\underline{4}}_{\underline{6}}1^{\underline{3}}_{\underline{6}}1^{\underline{3}}_{\underline{4}}1^{\underline{2}}_{\underline{6}}2^{\underline{2}}_{\underline{4}}1^{\underline{2}}_{\underline{2}}1^{\underline{1}}_{\underline{4}}2^{\underline{1}}_{\underline{2}}2^{\underline{1}}_{0}3^{0}_{\underline{2}}4^{0}_{0}2^{0}_{2}2^{1}_{0}3^{1}_{2}1^{1}_{4}2^{2}_{2}2^{2}_{4}1^{2}_{6}2^{3}_{4}2^{3}_{6}1^{4}_{6}2^{4}_{8}1^{5}_{8}1^{5}_{10}1^{6}_{12}$ 11^n_{34}: $1^{\underline{5}}_{\underline{10}}3^{\underline{4}}_{\underline{8}}1^{\underline{4}}_{\underline{6}}4^{\underline{3}}_{\underline{6}}3^{\underline{3}}_{\underline{4}}8^{\underline{2}}_{\underline{4}}4^{\underline{2}}_{\underline{2}}6^{\underline{1}}_{\underline{2}}8^{\underline{1}}_{0}9^{0}_{0}8^{0}_{2}7^{1}_{2}7^{1}_{4}4^{2}_{4}7^{2}_{6}3^{3}_{6}4^{3}_{8}3^{4}_{10}$	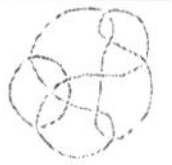	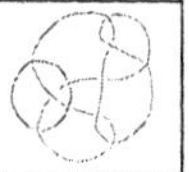
11^n_{35}: $1^{\underline{7}}_{\underline{16}}3^{\underline{6}}_{\underline{14}}1^{\underline{6}}_{\underline{12}}4^{\underline{5}}_{\underline{12}}3^{\underline{5}}_{\underline{10}}6^{\underline{4}}_{\underline{10}}4^{\underline{4}}_{\underline{8}}6^{\underline{3}}_{\underline{8}}6^{\underline{3}}_{\underline{6}}6^{\underline{2}}_{\underline{6}}6^{\underline{2}}_{\underline{4}}5^{\underline{1}}_{\underline{4}}6^{\underline{1}}_{\underline{2}}4^{0}_{\underline{2}}7^{0}_{0}2^{1}_{0}2^{1}_{2}2^{2}_{4}$ 11^n_{36}: $2^{\underline{4}}_{\underline{10}}1^{\underline{4}}_{\underline{8}}3^{\underline{3}}_{\underline{8}}1^{\underline{3}}_{\underline{6}}4^{\underline{2}}_{\underline{6}}3^{\underline{2}}_{\underline{4}}4^{\underline{1}}_{\underline{4}}4^{\underline{1}}_{\underline{2}}5^{0}_{\underline{2}}5^{0}_{0}4^{1}_{0}4^{1}_{2}3^{2}_{2}4^{2}_{4}1^{3}_{4}3^{3}_{6}1^{4}_{6}1^{4}_{8}1^{5}_{10}$	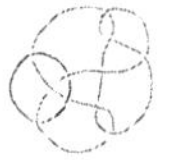	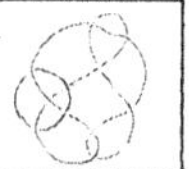
11^n_{37}: $1^{\underline{2}}_{\underline{4}}2^{\underline{1}}_{\underline{2}}1^{\underline{1}}_{0}5^{0}_{0}4^{0}_{2}5^{1}_{2}3^{1}_{4}5^{2}_{4}5^{2}_{6}4^{3}_{6}5^{3}_{8}4^{4}_{8}4^{4}_{10}2^{5}_{10}4^{5}_{12}1^{6}_{12}2^{6}_{14}1^{7}_{16}$ 11^n_{38}: $1^{\underline{7}}_{\underline{20}}1^{\underline{6}}_{\underline{18}}1^{\underline{6}}_{\underline{16}}3^{\underline{5}}_{\underline{16}}1^{\underline{5}}_{\underline{14}}3^{\underline{4}}_{\underline{14}}4^{\underline{4}}_{\underline{12}}1^{\underline{4}}_{\underline{10}}3^{\underline{3}}_{\underline{12}}3^{\underline{3}}_{\underline{10}}3^{\underline{2}}_{\underline{10}}3^{\underline{2}}_{\underline{8}}2^{\underline{1}}_{\underline{8}}3^{\underline{1}}_{\underline{6}}2^{0}_{\underline{6}}3^{0}_{\underline{4}}1^{1}_{\underline{4}}1^{1}_{\underline{2}}1^{2}_{0}$		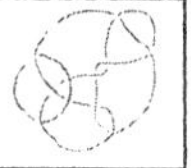
11^n_{39}: $1^{\underline{7}}_{\underline{16}}1^{\underline{6}}_{\underline{14}}1^{\underline{6}}_{\underline{12}}2^{\underline{5}}_{\underline{12}}1^{\underline{5}}_{\underline{10}}2^{\underline{4}}_{\underline{10}}2^{\underline{4}}_{\underline{8}}1^{\underline{3}}_{\underline{10}}2^{\underline{3}}_{\underline{8}}2^{\underline{3}}_{\underline{6}}1^{\underline{2}}_{\underline{8}}3^{\underline{2}}_{\underline{6}}2^{\underline{2}}_{\underline{4}}1^{\underline{1}}_{\underline{6}}2^{\underline{1}}_{\underline{4}}2^{\underline{1}}_{\underline{2}}2^{0}_{\underline{4}}3^{0}_{\underline{2}}3^{0}_{0}1^{1}_{\underline{2}}2^{1}_{0}1^{2}_{0}1^{2}_{2}1^{3}_{2}1^{3}_{4}1^{4}_{6}$ 11^n_{40}: $2^{\underline{9}}_{\underline{26}}2^{\underline{8}}_{\underline{24}}2^{\underline{8}}_{\underline{22}}4^{\underline{7}}_{\underline{22}}2^{\underline{7}}_{\underline{20}}3^{\underline{6}}_{\underline{20}}4^{\underline{6}}_{\underline{18}}4^{\underline{5}}_{\underline{18}}3^{\underline{5}}_{\underline{16}}3^{\underline{4}}_{\underline{16}}5^{\underline{4}}_{\underline{14}}3^{\underline{3}}_{\underline{14}}2^{\underline{3}}_{\underline{12}}1^{\underline{2}}_{\underline{12}}3^{\underline{2}}_{\underline{10}}1^{\underline{1}}_{\underline{8}}1^{0}_{\underline{8}}1^{0}_{\underline{6}}$		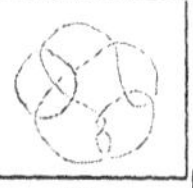
11^n_{41}: $1^{\underline{7}}_{\underline{20}}1^{\underline{6}}_{\underline{18}}1^{\underline{6}}_{\underline{16}}4^{\underline{5}}_{\underline{16}}2^{\underline{5}}_{\underline{14}}3^{\underline{4}}_{\underline{14}}4^{\underline{4}}_{\underline{12}}1^{\underline{4}}_{\underline{10}}4^{\underline{3}}_{\underline{12}}3^{\underline{3}}_{\underline{10}}3^{\underline{2}}_{\underline{10}}4^{\underline{2}}_{\underline{8}}2^{\underline{1}}_{\underline{8}}3^{\underline{1}}_{\underline{6}}3^{0}_{\underline{6}}4^{0}_{\underline{4}}1^{1}_{\underline{4}}1^{1}_{\underline{2}}1^{2}_{0}$ 11^n_{42}: $2^{\underline{9}}_{\underline{24}}1^{\underline{8}}_{\underline{22}}2^{\underline{8}}_{\underline{20}}3^{\underline{7}}_{\underline{20}}1^{\underline{7}}_{\underline{18}}3^{\underline{6}}_{\underline{18}}3^{\underline{6}}_{\underline{16}}3^{\underline{5}}_{\underline{16}}3^{\underline{5}}_{\underline{14}}3^{\underline{4}}_{\underline{14}}4^{\underline{4}}_{\underline{12}}2^{\underline{3}}_{\underline{12}}2^{\underline{3}}_{\underline{10}}1^{\underline{2}}_{\underline{10}}2^{\underline{2}}_{\underline{8}}1^{\underline{1}}_{\underline{6}}1^{0}_{\underline{6}}1^{0}_{\underline{4}}$	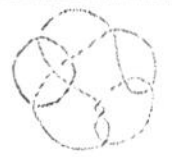	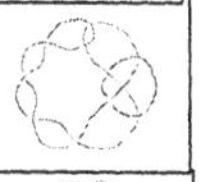
11^n_{43}: $2^{\underline{7}}_{\underline{18}}2^{\underline{6}}_{\underline{16}}2^{\underline{6}}_{\underline{14}}5^{\underline{5}}_{\underline{14}}2^{\underline{5}}_{\underline{12}}4^{\underline{4}}_{\underline{12}}5^{\underline{4}}_{\underline{10}}5^{\underline{3}}_{\underline{10}}4^{\underline{3}}_{\underline{8}}5^{\underline{2}}_{\underline{8}}5^{\underline{2}}_{\underline{6}}2^{\underline{1}}_{\underline{6}}5^{\underline{1}}_{\underline{4}}3^{0}_{\underline{4}}4^{0}_{\underline{2}}1^{1}_{\underline{2}}1^{1}_{0}1^{2}_{2}$ 11^n_{44}: $1^{\underline{9}}_{\underline{28}}1^{\underline{9}}_{\underline{26}}1^{\underline{8}}_{\underline{24}}1^{\underline{8}}_{\underline{22}}1^{\underline{7}}_{\underline{24}}1^{\underline{7}}_{\underline{22}}1^{\underline{6}}_{\underline{20}}1^{\underline{6}}_{\underline{18}}1^{\underline{5}}_{\underline{20}}2^{\underline{4}}_{\underline{16}}1^{\underline{4}}_{\underline{14}}1^{\underline{3}}_{\underline{16}}1^{\underline{2}}_{\underline{12}}1^{0}_{\underline{10}}1^{0}_{\underline{8}}$		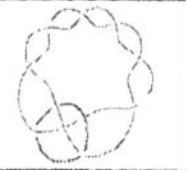
11^n_{45}: $1^{\underline{9}}_{\underline{24}}1^{\underline{8}}_{\underline{20}}2^{\underline{7}}_{\underline{20}}1^{\underline{6}}_{\underline{18}}2^{\underline{6}}_{\underline{16}}3^{\underline{5}}_{\underline{16}}1^{\underline{5}}_{\underline{14}}1^{\underline{4}}_{\underline{14}}3^{\underline{4}}_{\underline{12}}2^{\underline{3}}_{\underline{12}}1^{\underline{3}}_{\underline{10}}1^{\underline{2}}_{\underline{10}}2^{\underline{2}}_{\underline{8}}1^{\underline{1}}_{\underline{6}}2^{0}_{\underline{6}}2^{0}_{\underline{4}}$ 11^n_{46}: $1^{\underline{9}}_{\underline{24}}1^{\underline{8}}_{\underline{22}}1^{\underline{8}}_{\underline{20}}3^{\underline{7}}_{\underline{20}}1^{\underline{7}}_{\underline{18}}2^{\underline{6}}_{\underline{18}}3^{\underline{6}}_{\underline{16}}5^{\underline{5}}_{\underline{16}}2^{\underline{5}}_{\underline{14}}2^{\underline{4}}_{\underline{14}}5^{\underline{4}}_{\underline{12}}3^{\underline{3}}_{\underline{12}}2^{\underline{3}}_{\underline{10}}2^{\underline{2}}_{\underline{10}}3^{\underline{2}}_{\underline{8}}2^{\underline{1}}_{\underline{6}}2^{0}_{\underline{6}}2^{0}_{\underline{4}}$	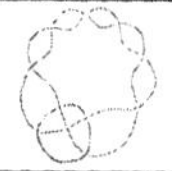	
11^n_{47}: $2^{\underline{9}}_{\underline{26}}1^{\underline{8}}_{\underline{24}}2^{\underline{8}}_{\underline{22}}4^{\underline{7}}_{\underline{22}}1^{\underline{7}}_{\underline{20}}2^{\underline{6}}_{\underline{20}}4^{\underline{6}}_{\underline{18}}3^{\underline{5}}_{\underline{18}}2^{\underline{5}}_{\underline{16}}3^{\underline{4}}_{\underline{16}}4^{\underline{4}}_{\underline{14}}2^{\underline{3}}_{\underline{14}}2^{\underline{3}}_{\underline{12}}1^{\underline{2}}_{\underline{12}}2^{\underline{2}}_{\underline{10}}1^{\underline{1}}_{\underline{8}}1^{0}_{\underline{8}}1^{0}_{\underline{6}}$ 11^n_{48}: $1^{\underline{9}}_{\underline{22}}1^{\underline{8}}_{\underline{18}}2^{\underline{7}}_{\underline{18}}1^{\underline{6}}_{\underline{16}}2^{\underline{6}}_{\underline{14}}1^{\underline{5}}_{\underline{16}}2^{\underline{5}}_{\underline{14}}1^{\underline{5}}_{\underline{12}}3^{\underline{4}}_{\underline{12}}2^{\underline{4}}_{\underline{10}}1^{\underline{3}}_{\underline{12}}1^{\underline{3}}_{\underline{10}}2^{\underline{3}}_{\underline{8}}2^{\underline{2}}_{\underline{8}}1^{\underline{2}}_{\underline{6}}1^{\underline{1}}_{\underline{4}}1^{0}_{\underline{6}}2^{0}_{\underline{4}}1^{0}_{\underline{2}}$	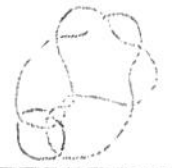	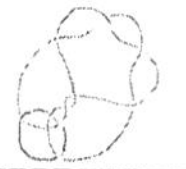
11^n_{49}: $2^{\underline{7}}_{\underline{20}}1^{\underline{6}}_{\underline{18}}2^{\underline{6}}_{\underline{16}}4^{\underline{5}}_{\underline{16}}1^{\underline{5}}_{\underline{14}}2^{\underline{4}}_{\underline{14}}4^{\underline{4}}_{\underline{12}}4^{\underline{3}}_{\underline{12}}2^{\underline{3}}_{\underline{10}}3^{\underline{2}}_{\underline{10}}4^{\underline{2}}_{\underline{8}}1^{\underline{1}}_{\underline{8}}3^{\underline{1}}_{\underline{6}}3^{0}_{\underline{6}}3^{0}_{\underline{4}}1^{1}_{\underline{4}}1^{1}_{\underline{2}}1^{2}_{0}$ 11^n_{50}: $1^{\underline{2}}_{\underline{4}}1^{\underline{1}}_{\underline{2}}1^{\underline{1}}_{0}5^{0}_{0}3^{0}_{2}4^{1}_{2}3^{1}_{4}4^{2}_{4}4^{2}_{6}4^{3}_{6}4^{3}_{8}3^{4}_{8}4^{4}_{10}2^{5}_{10}3^{5}_{12}1^{6}_{12}2^{6}_{14}1^{7}_{16}$	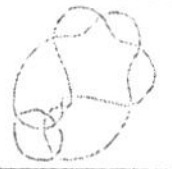	
11^n_{51}: $1^{\underline{4}}_{\underline{6}}1^{\underline{3}}_{\underline{4}}1^{\underline{3}}_{\underline{2}}4^{\underline{2}}_{\underline{2}}1^{\underline{2}}_{0}2^{\underline{1}}_{0}4^{\underline{1}}_{2}6^{0}_{2}4^{0}_{4}4^{1}_{4}4^{1}_{6}3^{2}_{6}4^{2}_{8}3^{3}_{8}3^{3}_{10}1^{4}_{10}3^{4}_{12}1^{5}_{14}$ 11^n_{52}: $1^{\underline{6}}_{\underline{12}}1^{\underline{5}}_{\underline{10}}1^{\underline{5}}_{\underline{8}}3^{\underline{4}}_{\underline{8}}1^{\underline{4}}_{\underline{6}}2^{\underline{3}}_{\underline{6}}3^{\underline{3}}_{\underline{4}}3^{\underline{2}}_{\underline{4}}2^{\underline{2}}_{\underline{2}}3^{\underline{1}}_{\underline{2}}3^{\underline{1}}_{0}1^{0}_{\underline{2}}3^{0}_{0}4^{0}_{2}2^{1}_{2}1^{1}_{4}2^{2}_{6}$		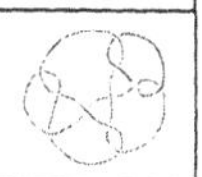
11^n_{53}: $1^{\underline{3}}_{\underline{8}}2^{\underline{2}}_{\underline{6}}1^{\underline{2}}_{\underline{4}}2^{\underline{1}}_{\underline{4}}2^{\underline{1}}_{\underline{2}}4^{0}_{\underline{2}}4^{0}_{0}1^{0}_{2}3^{1}_{0}3^{1}_{2}2^{2}_{2}3^{2}_{4}2^{3}_{4}2^{3}_{6}1^{4}_{6}2^{4}_{8}1^{5}_{10}$ 11^n_{54}: $1^{\underline{4}}_{\underline{6}}1^{\underline{4}}_{\underline{4}}1^{\underline{3}}_{\underline{4}}2^{\underline{2}}_{\underline{2}}1^{\underline{2}}_{0}1^{\underline{1}}_{0}2^{\underline{1}}_{2}1^{0}_{0}4^{0}_{2}1^{0}_{4}1^{1}_{2}1^{1}_{4}3^{1}_{6}1^{2}_{4}3^{2}_{6}1^{2}_{8}1^{3}_{6}2^{3}_{8}2^{3}_{10}1^{4}_{8}1^{4}_{10}1^{4}_{12}1^{5}_{12}1^{6}_{12}1^{7}_{16}$		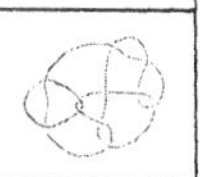
11^n_{55}: $2^{\underline{7}}_{\underline{18}}3^{\underline{6}}_{\underline{16}}2^{\underline{6}}_{\underline{14}}5^{\underline{5}}_{\underline{14}}3^{\underline{5}}_{\underline{12}}6^{\underline{4}}_{\underline{12}}6^{\underline{4}}_{\underline{10}}6^{\underline{3}}_{\underline{10}}5^{\underline{3}}_{\underline{8}}6^{\underline{2}}_{\underline{8}}6^{\underline{2}}_{\underline{6}}3^{\underline{1}}_{\underline{6}}6^{\underline{1}}_{\underline{4}}3^{0}_{\underline{4}}4^{0}_{\underline{2}}1^{1}_{\underline{2}}2^{1}_{0}1^{2}_{2}$ 11^n_{56}: $1^{\underline{4}}_{\underline{10}}1^{\underline{4}}_{\underline{8}}2^{\underline{3}}_{\underline{8}}3^{\underline{2}}_{\underline{6}}2^{\underline{2}}_{\underline{4}}3^{\underline{1}}_{\underline{4}}3^{\underline{1}}_{\underline{2}}5^{0}_{\underline{2}}4^{0}_{0}3^{1}_{0}4^{1}_{2}3^{2}_{2}3^{2}_{4}2^{3}_{4}3^{3}_{6}1^{4}_{6}2^{4}_{8}1^{5}_{10}$		
11^n_{57}: $1^{\underline{6}}_{\underline{12}}1^{\underline{5}}_{\underline{10}}1^{\underline{5}}_{\underline{8}}3^{\underline{4}}_{\underline{8}}2^{\underline{4}}_{\underline{6}}2^{\underline{3}}_{\underline{6}}2^{\underline{3}}_{\underline{4}}3^{\underline{2}}_{\underline{4}}2^{\underline{2}}_{\underline{2}}1^{\underline{1}}_{\underline{4}}2^{\underline{1}}_{\underline{2}}3^{\underline{1}}_{0}1^{0}_{\underline{2}}4^{0}_{0}2^{0}_{2}1^{1}_{0}1^{1}_{2}2^{1}_{4}1^{2}_{2}1^{2}_{4}1^{2}_{6}1^{3}_{6}1^{4}_{6}1^{5}_{10}$ 11^n_{58}: $2^{\underline{9}}_{\underline{24}}3^{\underline{8}}_{\underline{22}}2^{\underline{8}}_{\underline{20}}6^{\underline{7}}_{\underline{20}}3^{\underline{7}}_{\underline{18}}6^{\underline{6}}_{\underline{18}}6^{\underline{6}}_{\underline{16}}7^{\underline{5}}_{\underline{16}}6^{\underline{5}}_{\underline{14}}6^{\underline{4}}_{\underline{14}}8^{\underline{4}}_{\underline{12}}5^{\underline{3}}_{\underline{12}}5^{\underline{3}}_{\underline{10}}3^{\underline{2}}_{\underline{10}}5^{\underline{2}}_{\underline{8}}3^{\underline{1}}_{\underline{6}}1^{0}_{\underline{6}}1^{0}_{\underline{4}}$	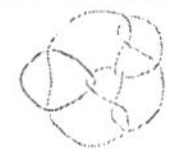	
11^n_{59}: $1^{\underline{4}}_{\underline{6}}1^{\underline{4}}_{\underline{4}}1^{\underline{2}}_{\underline{2}}1^{\underline{1}}_{\underline{2}}1^{\underline{1}}_{2}4^{0}_{0}3^{0}_{2}3^{1}_{2}3^{1}_{4}1^{1}_{6}4^{2}_{4}3^{2}_{6}3^{3}_{6}4^{3}_{8}3^{4}_{8}3^{4}_{10}2^{5}_{10}3^{5}_{12}1^{6}_{12}2^{6}_{14}1^{7}_{16}$ 11^n_{60}: $1^{\underline{7}}_{\underline{16}}2^{\underline{6}}_{\underline{14}}1^{\underline{6}}_{\underline{12}}4^{\underline{5}}_{\underline{12}}2^{\underline{5}}_{\underline{10}}5^{\underline{4}}_{\underline{10}}5^{\underline{4}}_{\underline{8}}5^{\underline{3}}_{\underline{8}}4^{\underline{3}}_{\underline{6}}6^{\underline{2}}_{\underline{6}}5^{\underline{2}}_{\underline{4}}3^{\underline{1}}_{\underline{4}}6^{\underline{1}}_{\underline{2}}4^{0}_{\underline{2}}4^{0}_{0}1^{1}_{0}3^{1}_{2}1^{2}_{4}$	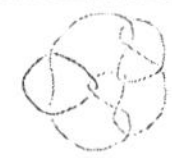	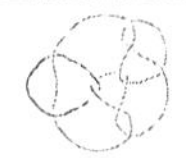
11^n_{61}: $1^{\underline{6}}_{\underline{12}}1^{\underline{5}}_{\underline{10}}1^{\underline{5}}_{\underline{8}}4^{\underline{4}}_{\underline{8}}2^{\underline{4}}_{\underline{6}}2^{\underline{3}}_{\underline{6}}3^{\underline{3}}_{\underline{4}}4^{\underline{2}}_{\underline{4}}2^{\underline{2}}_{\underline{2}}3^{\underline{1}}_{\underline{2}}4^{\underline{1}}_{0}1^{0}_{\underline{2}}3^{0}_{0}3^{0}_{2}2^{1}_{2}2^{1}_{4}2^{2}_{6}$ 11^n_{62}: $1^{\underline{11}}_{\underline{28}}1^{\underline{10}}_{\underline{26}}1^{\underline{10}}_{\underline{24}}3^{\underline{9}}_{\underline{24}}1^{\underline{9}}_{\underline{22}}2^{\underline{8}}_{\underline{22}}3^{\underline{8}}_{\underline{20}}3^{\underline{7}}_{\underline{20}}2^{\underline{7}}_{\underline{18}}3^{\underline{6}}_{\underline{18}}3^{\underline{6}}_{\underline{16}}1^{\underline{5}}_{\underline{18}}2^{\underline{5}}_{\underline{16}}3^{\underline{5}}_{\underline{14}}3^{\underline{4}}_{\underline{14}}3^{\underline{4}}_{\underline{12}}1^{\underline{3}}_{\underline{14}}1^{\underline{3}}_{\underline{10}}1^{\underline{2}}_{\underline{10}}1^{0}_{\underline{8}}1^{0}_{\underline{6}}$	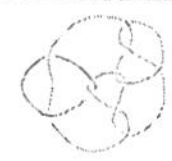	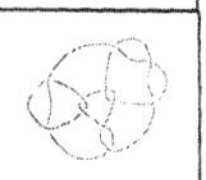

L: KH	1st line	2nd line
11^n_{63}: $2^{\underline{4}}_{\underline{8}}1^{\underline{4}}_{\underline{6}}3^{\underline{3}}_{\underline{6}}1^{\underline{3}}_{\underline{4}}5^{\underline{2}}_{\underline{4}}3^{\underline{2}}_{\underline{2}}5^{\underline{1}}_{\underline{2}}5^{\underline{1}}_{0}7^{0}_{0}6^{0}_{2}5^{1}_{2}6^{1}_{4}5^{2}_{4}5^{2}_{6}2^{3}_{6}5^{3}_{8}1^{4}_{8}2^{4}_{10}1^{5}_{12}$ 11^n_{64}: $1^{\underline{9}}_{\underline{22}}1^{\underline{8}}_{\underline{18}}1^{\underline{7}}_{\underline{18}}1^{\underline{6}}_{\underline{16}}1^{\underline{6}}_{\underline{14}}2^{\underline{5}}_{\underline{16}}1^{\underline{5}}_{\underline{12}}1^{\underline{4}}_{\underline{14}}4^{\underline{4}}_{\underline{12}}1^{\underline{4}}_{\underline{10}}2^{\underline{3}}_{\underline{12}}1^{\underline{3}}_{\underline{10}}1^{\underline{3}}_{\underline{8}}2^{\underline{2}}_{\underline{10}}2^{\underline{2}}_{\underline{8}}1^{\underline{1}}_{\underline{8}}2^{\underline{1}}_{\underline{6}}2^{0}_{\underline{6}}2^{0}_{\underline{4}}1^{1}_{\underline{4}}1^{1}_{\underline{2}}1^{2}_{0}$	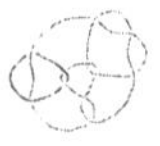	
11^n_{65}: $1^{\underline{11}}_{\underline{26}}1^{\underline{10}}_{\underline{22}}1^{\underline{9}}_{\underline{22}}1^{\underline{8}}_{\underline{20}}1^{\underline{8}}_{\underline{18}}2^{\underline{7}}_{\underline{20}}1^{\underline{7}}_{\underline{16}}1^{\underline{6}}_{\underline{18}}3^{\underline{6}}_{\underline{16}}2^{\underline{5}}_{\underline{16}}1^{\underline{5}}_{\underline{14}}1^{\underline{5}}_{\underline{12}}2^{\underline{4}}_{\underline{14}}3^{\underline{4}}_{\underline{12}}2^{\underline{3}}_{\underline{12}}1^{\underline{3}}_{\underline{10}}1^{\underline{2}}_{\underline{10}}2^{\underline{2}}_{\underline{8}}1^{\underline{1}}_{\underline{6}}1^{0}_{\underline{6}}1^{0}_{\underline{4}}$ 11^n_{66}: $2^{\underline{6}}_{\underline{16}}3^{\underline{5}}_{\underline{14}}2^{\underline{5}}_{\underline{12}}6^{\underline{4}}_{\underline{12}}4^{\underline{4}}_{\underline{10}}7^{\underline{3}}_{\underline{10}}5^{\underline{3}}_{\underline{8}}6^{\underline{2}}_{\underline{8}}7^{\underline{2}}_{\underline{6}}6^{\underline{1}}_{\underline{6}}6^{\underline{1}}_{\underline{4}}5^{0}_{\underline{4}}7^{0}_{\underline{2}}3^{1}_{\underline{2}}4^{1}_{0}1^{2}_{0}3^{2}_{2}1^{3}_{4}$		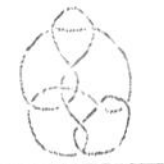
11^n_{67}: $3^{\underline{9}}_{\underline{24}}3^{\underline{8}}_{\underline{22}}3^{\underline{8}}_{\underline{20}}7^{\underline{7}}_{\underline{20}}3^{\underline{7}}_{\underline{18}}7^{\underline{6}}_{\underline{18}}7^{\underline{6}}_{\underline{16}}7^{\underline{5}}_{\underline{16}}7^{\underline{5}}_{\underline{14}}7^{\underline{4}}_{\underline{14}}8^{\underline{4}}_{\underline{12}}5^{\underline{3}}_{\underline{12}}6^{\underline{3}}_{\underline{10}}3^{\underline{2}}_{\underline{10}}5^{\underline{2}}_{\underline{8}}3^{\underline{1}}_{\underline{6}}1^{0}_{\underline{6}}1^{0}_{\underline{4}}$ 11^n_{68}: $1^{\underline{11}}_{\underline{28}}1^{\underline{10}}_{\underline{24}}2^{\underline{9}}_{\underline{24}}1^{\underline{8}}_{\underline{22}}2^{\underline{8}}_{\underline{20}}1^{\underline{7}}_{\underline{22}}2^{\underline{7}}_{\underline{20}}1^{\underline{7}}_{\underline{18}}3^{\underline{6}}_{\underline{18}}2^{\underline{6}}_{\underline{16}}1^{\underline{5}}_{\underline{18}}2^{\underline{5}}_{\underline{14}}3^{\underline{4}}_{\underline{14}}1^{\underline{4}}_{\underline{12}}1^{\underline{3}}_{\underline{14}}1^{\underline{3}}_{\underline{10}}1^{\underline{2}}_{\underline{10}}1^{0}_{\underline{8}}1^{0}_{\underline{6}}$	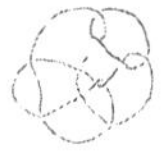	
11^n_{69}: $2^{\underline{4}}_{\underline{8}}1^{\underline{4}}_{\underline{6}}2^{\underline{3}}_{\underline{6}}1^{\underline{3}}_{\underline{4}}4^{\underline{2}}_{\underline{4}}2^{\underline{2}}_{\underline{2}}3^{\underline{1}}_{\underline{2}}4^{\underline{1}}_{0}6^{0}_{0}4^{0}_{2}4^{1}_{2}5^{1}_{4}3^{2}_{4}4^{2}_{6}2^{3}_{6}3^{3}_{8}1^{4}_{8}2^{4}_{10}1^{5}_{12}$ 11^n_{70}: $1^{\underline{7}}_{\underline{20}}1^{\underline{6}}_{\underline{18}}1^{\underline{6}}_{\underline{16}}2^{\underline{5}}_{\underline{16}}1^{\underline{5}}_{\underline{14}}2^{\underline{4}}_{\underline{14}}3^{\underline{4}}_{\underline{12}}3^{\underline{3}}_{\underline{12}}1^{\underline{3}}_{\underline{10}}2^{\underline{2}}_{\underline{10}}3^{\underline{2}}_{\underline{8}}1^{\underline{1}}_{\underline{8}}2^{\underline{1}}_{\underline{6}}2^{0}_{\underline{6}}2^{0}_{\underline{4}}1^{1}_{\underline{4}}1^{1}_{\underline{2}}1^{2}_{0}$		
11^n_{71}: $1^{\underline{11}}_{\underline{26}}1^{\underline{10}}_{\underline{22}}2^{\underline{9}}_{\underline{22}}1^{\underline{8}}_{\underline{20}}2^{\underline{8}}_{\underline{18}}1^{\underline{7}}_{\underline{20}}1^{\underline{7}}_{\underline{18}}1^{\underline{7}}_{\underline{16}}3^{\underline{6}}_{\underline{16}}1^{\underline{6}}_{\underline{14}}2^{\underline{5}}_{\underline{16}}2^{\underline{5}}_{\underline{12}}1^{\underline{4}}_{\underline{14}}4^{\underline{4}}_{\underline{12}}1^{\underline{4}}_{\underline{10}}1^{\underline{3}}_{\underline{12}}1^{\underline{3}}_{\underline{10}}1^{\underline{3}}_{\underline{8}}1^{\underline{2}}_{\underline{10}}1^{\underline{2}}_{\underline{8}}1^{\underline{1}}_{\underline{6}}1^{0}_{\underline{6}}1^{0}_{\underline{4}}$ 11^n_{72}: $2^{\underline{4}}_{\underline{10}}1^{\underline{4}}_{\underline{8}}3^{\underline{3}}_{\underline{8}}1^{\underline{3}}_{\underline{6}}5^{\underline{2}}_{\underline{6}}3^{\underline{2}}_{\underline{4}}4^{\underline{1}}_{\underline{4}}5^{\underline{1}}_{\underline{2}}6^{0}_{\underline{2}}5^{0}_{0}5^{1}_{0}5^{1}_{2}3^{2}_{2}5^{2}_{4}2^{3}_{4}3^{3}_{6}1^{4}_{6}2^{4}_{8}1^{5}_{10}$	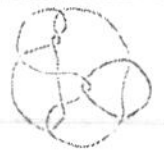	
11^n_{73}: $1^{\underline{7}}_{\underline{18}}2^{\underline{6}}_{\underline{16}}1^{\underline{6}}_{\underline{14}}3^{\underline{5}}_{\underline{14}}2^{\underline{5}}_{\underline{12}}5^{\underline{4}}_{\underline{12}}4^{\underline{4}}_{\underline{10}}5^{\underline{3}}_{\underline{10}}4^{\underline{3}}_{\underline{8}}4^{\underline{2}}_{\underline{8}}5^{\underline{2}}_{\underline{6}}3^{\underline{1}}_{\underline{6}}4^{\underline{1}}_{\underline{4}}3^{0}_{\underline{4}}4^{0}_{\underline{2}}1^{1}_{\underline{2}}2^{1}_{0}1^{2}_{2}$ 11^n_{74}: $1^{\underline{4}}_{\underline{4}}1^{\underline{4}}_{\underline{2}}1^{\underline{2}}_{0}1^{\underline{1}}_{4}2^{0}_{2}2^{0}_{4}1^{1}_{4}1^{1}_{6}1^{1}_{8}2^{2}_{6}1^{2}_{8}1^{3}_{8}2^{3}_{10}2^{4}_{10}1^{4}_{12}2^{5}_{14}1^{6}_{14}1^{7}_{18}$	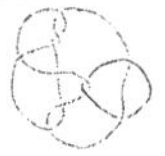	
11^n_{75}: $1^{\underline{4}}_{\underline{6}}1^{\underline{4}}_{\underline{4}}1^{\underline{3}}_{\underline{4}}1^{\underline{2}}_{\underline{2}}1^{\underline{2}}_{0}1^{\underline{1}}_{0}1^{\underline{1}}_{2}2^{0}_{0}3^{0}_{2}1^{0}_{4}2^{1}_{2}1^{1}_{4}2^{1}_{6}1^{2}_{4}3^{2}_{6}1^{3}_{6}1^{3}_{8}1^{3}_{10}2^{4}_{8}1^{4}_{10}2^{5}_{12}1^{6}_{12}1^{7}_{16}$ 11^n_{76}: $3^{\underline{4}}_{\underline{10}}1^{\underline{4}}_{\underline{8}}5^{\underline{3}}_{\underline{8}}2^{\underline{3}}_{\underline{6}}6^{\underline{2}}_{\underline{6}}5^{\underline{2}}_{\underline{4}}7^{\underline{1}}_{\underline{4}}6^{\underline{1}}_{\underline{2}}8^{0}_{\underline{2}}8^{0}_{0}6^{1}_{0}7^{1}_{2}5^{2}_{2}6^{2}_{4}2^{3}_{4}5^{3}_{6}1^{4}_{6}2^{4}_{8}1^{5}_{10}$	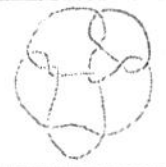	
11^n_{77}: $1^{\underline{5}}_{\underline{14}}4^{\underline{4}}_{\underline{12}}2^{\underline{4}}_{\underline{10}}4^{\underline{3}}_{\underline{10}}3^{\underline{3}}_{\underline{8}}5^{\underline{2}}_{\underline{8}}4^{\underline{2}}_{\underline{6}}5^{\underline{1}}_{\underline{6}}5^{\underline{1}}_{\underline{4}}5^{0}_{\underline{4}}6^{0}_{\underline{2}}4^{1}_{\underline{2}}4^{1}_{0}2^{2}_{0}4^{2}_{2}1^{3}_{2}2^{3}_{4}1^{4}_{6}$ 11^n_{78}: $1^{\underline{9}}_{\underline{22}}1^{\underline{8}}_{\underline{20}}1^{\underline{8}}_{\underline{18}}4^{\underline{7}}_{\underline{18}}1^{\underline{7}}_{\underline{16}}2^{\underline{6}}_{\underline{16}}4^{\underline{6}}_{\underline{14}}5^{\underline{5}}_{\underline{14}}2^{\underline{5}}_{\underline{12}}4^{\underline{4}}_{\underline{12}}5^{\underline{4}}_{\underline{10}}3^{\underline{3}}_{\underline{10}}4^{\underline{3}}_{\underline{8}}3^{\underline{2}}_{\underline{8}}3^{\underline{2}}_{\underline{6}}3^{\underline{1}}_{\underline{4}}2^{0}_{\underline{4}}2^{0}_{\underline{2}}$	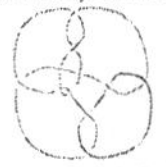	
11^n_{79}: $2^{\underline{4}}_{\underline{10}}1^{\underline{4}}_{\underline{8}}3^{\underline{3}}_{\underline{8}}1^{\underline{3}}_{\underline{6}}4^{\underline{2}}_{\underline{6}}3^{\underline{2}}_{\underline{4}}5^{\underline{1}}_{\underline{4}}4^{\underline{1}}_{\underline{2}}6^{0}_{\underline{2}}6^{0}_{0}4^{1}_{0}5^{1}_{2}4^{2}_{2}4^{2}_{4}2^{3}_{4}4^{3}_{6}1^{4}_{6}2^{4}_{8}1^{5}_{10}$ 11^n_{80}: $1^{\underline{9}}_{\underline{24}}1^{\underline{8}}_{\underline{22}}1^{\underline{8}}_{\underline{20}}4^{\underline{7}}_{\underline{20}}1^{\underline{7}}_{\underline{18}}3^{\underline{6}}_{\underline{18}}4^{\underline{6}}_{\underline{16}}5^{\underline{5}}_{\underline{16}}3^{\underline{5}}_{\underline{14}}3^{\underline{4}}_{\underline{14}}5^{\underline{4}}_{\underline{12}}4^{\underline{3}}_{\underline{12}}3^{\underline{3}}_{\underline{10}}2^{\underline{2}}_{\underline{10}}4^{\underline{2}}_{\underline{8}}2^{\underline{1}}_{\underline{6}}2^{0}_{\underline{6}}2^{0}_{\underline{4}}$	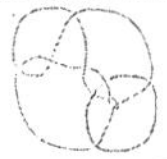	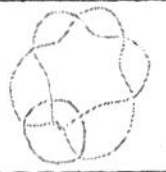
11^n_{81}: $2^{\underline{7}}_{\underline{20}}2^{\underline{6}}_{\underline{18}}2^{\underline{6}}_{\underline{16}}4^{\underline{5}}_{\underline{16}}2^{\underline{5}}_{\underline{14}}3^{\underline{4}}_{\underline{14}}4^{\underline{4}}_{\underline{12}}5^{\underline{3}}_{\underline{12}}3^{\underline{3}}_{\underline{10}}3^{\underline{2}}_{\underline{10}}5^{\underline{2}}_{\underline{8}}2^{\underline{1}}_{\underline{8}}3^{\underline{1}}_{\underline{6}}3^{0}_{\underline{6}}4^{0}_{\underline{4}}1^{1}_{\underline{4}}1^{1}_{\underline{2}}1^{2}_{0}$ 11^n_{82}: $2^{\underline{9}}_{\underline{24}}2^{\underline{8}}_{\underline{22}}2^{\underline{8}}_{\underline{20}}5^{\underline{7}}_{\underline{20}}2^{\underline{7}}_{\underline{18}}5^{\underline{6}}_{\underline{18}}5^{\underline{6}}_{\underline{16}}5^{\underline{5}}_{\underline{16}}5^{\underline{5}}_{\underline{14}}5^{\underline{4}}_{\underline{14}}6^{\underline{4}}_{\underline{12}}4^{\underline{3}}_{\underline{12}}4^{\underline{3}}_{\underline{10}}2^{\underline{2}}_{\underline{10}}4^{\underline{2}}_{\underline{8}}2^{\underline{1}}_{\underline{6}}1^{0}_{\underline{6}}1^{0}_{\underline{4}}$	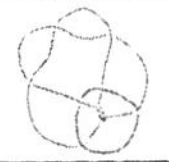	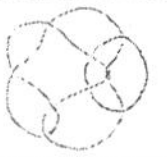
11^n_{83}: $2^{\underline{7}}_{\underline{18}}1^{\underline{6}}_{\underline{16}}2^{\underline{6}}_{\underline{14}}4^{\underline{5}}_{\underline{14}}1^{\underline{5}}_{\underline{12}}4^{\underline{4}}_{\underline{12}}4^{\underline{4}}_{\underline{10}}4^{\underline{3}}_{\underline{10}}4^{\underline{3}}_{\underline{8}}4^{\underline{2}}_{\underline{8}}4^{\underline{2}}_{\underline{6}}2^{\underline{1}}_{\underline{6}}4^{\underline{1}}_{\underline{4}}3^{0}_{\underline{4}}4^{0}_{\underline{2}}1^{1}_{\underline{2}}1^{1}_{0}1^{2}_{2}$ 11^n_{84}: $1^{\underline{4}}_{\underline{10}}1^{\underline{4}}_{\underline{8}}1^{\underline{3}}_{\underline{8}}3^{\underline{2}}_{\underline{6}}1^{\underline{2}}_{\underline{4}}2^{\underline{1}}_{\underline{4}}3^{\underline{1}}_{\underline{2}}4^{0}_{\underline{2}}3^{0}_{0}3^{1}_{0}3^{1}_{2}2^{2}_{2}3^{2}_{4}2^{3}_{4}2^{3}_{6}1^{4}_{6}2^{4}_{8}1^{5}_{10}$	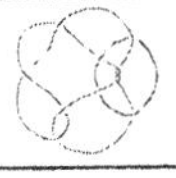	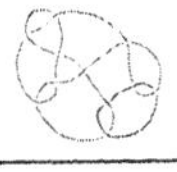
11^n_{85}: $1^{\underline{9}}_{\underline{22}}1^{\underline{8}}_{\underline{20}}1^{\underline{8}}_{\underline{18}}4^{\underline{7}}_{\underline{18}}1^{\underline{7}}_{\underline{16}}4^{\underline{6}}_{\underline{16}}4^{\underline{6}}_{\underline{14}}5^{\underline{5}}_{\underline{14}}4^{\underline{5}}_{\underline{12}}6^{\underline{4}}_{\underline{12}}6^{\underline{4}}_{\underline{10}}4^{\underline{3}}_{\underline{10}}5^{\underline{3}}_{\underline{8}}4^{\underline{2}}_{\underline{8}}4^{\underline{2}}_{\underline{6}}1^{\underline{1}}_{\underline{6}}4^{\underline{1}}_{\underline{4}}1^{0}_{\underline{4}}2^{0}_{\underline{2}}$ 11^n_{86}: $1^{\underline{4}}_{\underline{6}}1^{\underline{4}}_{\underline{4}}1^{\underline{3}}_{\underline{4}}3^{\underline{2}}_{\underline{2}}1^{\underline{2}}_{0}1^{\underline{1}}_{0}3^{\underline{1}}_{2}5^{0}_{2}2^{0}_{4}2^{1}_{4}4^{1}_{6}3^{2}_{6}2^{2}_{8}2^{3}_{8}3^{3}_{10}1^{4}_{10}2^{4}_{12}1^{5}_{14}$		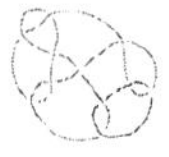
11^n_{87}: $1^{\underline{4}}_{\underline{10}}1^{\underline{4}}_{\underline{8}}1^{\underline{3}}_{\underline{10}}1^{\underline{2}}_{\underline{8}}2^{\underline{2}}_{\underline{6}}1^{\underline{1}}_{\underline{6}}1^{\underline{1}}_{\underline{4}}1^{\underline{1}}_{\underline{2}}2^{0}_{\underline{4}}2^{0}_{\underline{2}}1^{0}_{0}1^{1}_{\underline{2}}2^{1}_{0}1^{2}_{0}1^{2}_{2}1^{3}_{2}1^{3}_{4}1^{4}_{6}$ 11^n_{88}: $3^{\underline{4}}_{\underline{10}}5^{\underline{3}}_{\underline{8}}3^{\underline{3}}_{\underline{6}}7^{\underline{2}}_{\underline{6}}5^{\underline{2}}_{\underline{4}}8^{\underline{1}}_{\underline{4}}7^{\underline{1}}_{\underline{2}}9^{0}_{\underline{2}}10^{0}_{0}8^{1}_{0}7^{1}_{2}5^{2}_{2}8^{2}_{4}3^{3}_{4}5^{3}_{6}1^{4}_{6}3^{4}_{8}1^{5}_{10}$		
11^n_{89}: $1^{\underline{8}}_{\underline{18}}1^{\underline{7}}_{\underline{16}}1^{\underline{7}}_{\underline{14}}1^{\underline{6}}_{\underline{14}}1^{\underline{6}}_{\underline{12}}2^{\underline{5}}_{\underline{12}}1^{\underline{5}}_{\underline{10}}3^{\underline{4}}_{\underline{12}}2^{\underline{4}}_{\underline{10}}2^{\underline{4}}_{\underline{8}}2^{\underline{3}}_{\underline{10}}3^{\underline{3}}_{\underline{8}}1^{\underline{3}}_{\underline{6}}2^{\underline{2}}_{\underline{8}}3^{\underline{2}}_{\underline{6}}1^{\underline{2}}_{\underline{4}}3^{\underline{1}}_{\underline{6}}2^{\underline{1}}_{\underline{4}}1^{\underline{1}}_{\underline{2}}2^{0}_{\underline{4}}4^{0}_{\underline{2}}2^{1}_{\underline{2}}1^{1}_{0}1^{2}_{0}2^{2}_{2}1^{3}_{4}$ 11^n_{90}: $1^{\underline{8}}_{\underline{18}}1^{\underline{7}}_{\underline{16}}1^{\underline{7}}_{\underline{14}}3^{\underline{6}}_{\underline{14}}1^{\underline{6}}_{\underline{12}}2^{\underline{5}}_{\underline{12}}3^{\underline{5}}_{\underline{10}}4^{\underline{4}}_{\underline{10}}3^{\underline{4}}_{\underline{8}}3^{\underline{3}}_{\underline{8}}3^{\underline{3}}_{\underline{6}}2^{\underline{2}}_{\underline{6}}3^{\underline{2}}_{\underline{4}}2^{\underline{1}}_{\underline{4}}2^{\underline{1}}_{\underline{2}}1^{0}_{\underline{2}}3^{0}_{0}$		
11^n_{91}: $1^{\underline{9}}_{\underline{20}}1^{\underline{8}}_{\underline{16}}1^{\underline{7}}_{\underline{16}}1^{\underline{6}}_{\underline{14}}1^{\underline{6}}_{\underline{12}}2^{\underline{5}}_{\underline{14}}1^{\underline{5}}_{\underline{10}}1^{\underline{4}}_{\underline{12}}3^{\underline{4}}_{\underline{10}}3^{\underline{3}}_{\underline{10}}1^{\underline{3}}_{\underline{8}}1^{\underline{3}}_{\underline{6}}2^{\underline{2}}_{\underline{8}}3^{\underline{2}}_{\underline{6}}1^{\underline{1}}_{\underline{6}}2^{\underline{1}}_{\underline{4}}3^{0}_{\underline{4}}3^{0}_{\underline{2}}1^{1}_{\underline{2}}1^{1}_{0}1^{2}_{2}$ 11^n_{92}: $1^{\underline{3}}_{\underline{6}}3^{\underline{2}}_{\underline{4}}1^{\underline{2}}_{\underline{2}}3^{\underline{1}}_{\underline{2}}3^{\underline{1}}_{0}7^{0}_{0}5^{0}_{2}6^{1}_{2}5^{1}_{4}5^{2}_{4}6^{2}_{6}4^{3}_{6}5^{3}_{8}3^{4}_{8}4^{4}_{10}1^{5}_{10}3^{5}_{12}1^{6}_{14}$	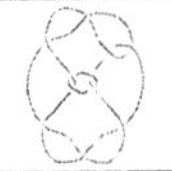	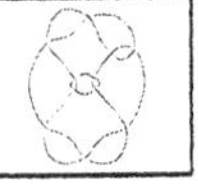

L: KH	1st line	2nd line
11^n_{93}: $1^{\underline{4}}_{\underline{10}}3^{\underline{3}}_{\underline{8}}1^{\underline{3}}_{\underline{6}}4^{\underline{2}}_{\underline{6}}3^{\underline{2}}_{\underline{4}}4^{\underline{1}}_{\underline{4}}4^{\underline{1}}_{\underline{2}}6^{0}_{\underline{2}}6^{0}_{0}5^{1}_{0}4^{1}_{2}3^{2}_{2}5^{2}_{4}2^{3}_{4}3^{3}_{6}1^{4}_{6}2^{4}_{8}1^{5}_{10}$ 11^n_{94}: $1^{\underline{4}}_{\underline{6}}1^{\underline{3}}_{\underline{4}}1^{\underline{3}}_{\underline{2}}2^{\underline{2}}_{\underline{2}}1^{\underline{2}}_{0}2^{\underline{1}}_{0}2^{\underline{1}}_{2}1^{0}_{0}4^{0}_{2}3^{0}_{4}1^{1}_{2}2^{1}_{4}2^{1}_{6}1^{2}_{4}3^{2}_{6}2^{2}_{8}1^{3}_{6}2^{3}_{8}2^{3}_{10}1^{4}_{8}1^{4}_{10}1^{4}_{12}1^{5}_{12}1^{6}_{12}1^{7}_{16}$	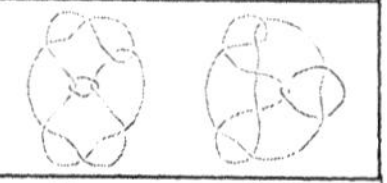	
11^n_{95}: $1^{\underline{11}}_{\underline{28}}1^{\underline{10}}_{\underline{26}}1^{\underline{10}}_{\underline{24}}2^{\underline{9}}_{\underline{24}}1^{\underline{9}}_{\underline{22}}2^{\underline{8}}_{\underline{22}}2^{\underline{8}}_{\underline{20}}1^{\underline{7}}_{\underline{22}}3^{\underline{7}}_{\underline{20}}2^{\underline{7}}_{\underline{18}}3^{\underline{6}}_{\underline{18}}3^{\underline{6}}_{\underline{16}}1^{\underline{5}}_{\underline{18}}1^{\underline{5}}_{\underline{16}}2^{\underline{5}}_{\underline{14}}3^{\underline{4}}_{\underline{14}}2^{\underline{4}}_{\underline{12}}1^{\underline{3}}_{\underline{14}}1^{\underline{3}}_{\underline{10}}1^{\underline{2}}_{\underline{10}}1^{0}_{\underline{8}}1^{0}_{\underline{6}}$ 11^n_{96}: $1^{\underline{6}}_{\underline{16}}2^{\underline{5}}_{\underline{14}}1^{\underline{5}}_{\underline{12}}3^{\underline{4}}_{\underline{12}}3^{\underline{4}}_{\underline{10}}4^{\underline{3}}_{\underline{10}}2^{\underline{3}}_{\underline{8}}4^{\underline{2}}_{\underline{8}}4^{\underline{2}}_{\underline{6}}3^{\underline{1}}_{\underline{6}}4^{\underline{1}}_{\underline{4}}3^{0}_{\underline{4}}4^{0}_{\underline{2}}2^{1}_{\underline{2}}2^{1}_{0}1^{2}_{0}2^{2}_{2}1^{3}_{4}$	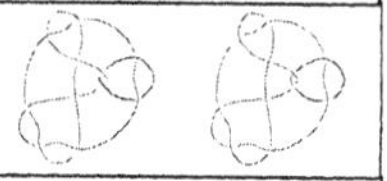	
11^n_{97}: $1^{\underline{5}}_{\underline{12}}3^{\underline{4}}_{\underline{10}}2^{\underline{4}}_{\underline{8}}3^{\underline{3}}_{\underline{8}}2^{\underline{3}}_{\underline{6}}5^{\underline{2}}_{\underline{6}}3^{\underline{2}}_{\underline{4}}4^{\underline{1}}_{\underline{4}}5^{\underline{1}}_{\underline{2}}5^{0}_{\underline{2}}5^{0}_{0}3^{1}_{0}4^{1}_{2}2^{2}_{2}3^{2}_{4}1^{3}_{4}2^{3}_{6}1^{4}_{8}$ 11^n_{98}: $1^{\underline{4}}_{\underline{4}}1^{\underline{4}}_{\underline{2}}1^{\underline{2}}_{0}1^{\underline{1}}_{4}2^{0}_{2}2^{0}_{4}2^{1}_{4}1^{1}_{6}1^{1}_{8}3^{2}_{6}2^{2}_{8}1^{3}_{8}3^{3}_{10}3^{4}_{10}1^{4}_{12}1^{5}_{12}3^{5}_{14}1^{6}_{14}1^{6}_{16}1^{7}_{18}$	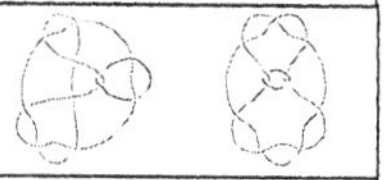	
11^n_{99}: $1^{\underline{4}}_{\underline{6}}1^{\underline{4}}_{\underline{4}}1^{\underline{3}}_{\underline{4}}1^{\underline{2}}_{\underline{2}}1^{\underline{2}}_{0}1^{\underline{1}}_{0}1^{\underline{1}}_{2}1^{0}_{0}3^{0}_{2}1^{0}_{4}1^{1}_{2}2^{1}_{6}1^{2}_{4}3^{2}_{6}1^{3}_{6}1^{3}_{8}2^{3}_{10}1^{4}_{8}1^{4}_{10}1^{5}_{12}1^{6}_{12}1^{7}_{16}$ 11^n_{100}: $2^{\underline{4}}_{\underline{12}}1^{\underline{4}}_{\underline{10}}2^{\underline{3}}_{\underline{10}}1^{\underline{3}}_{\underline{8}}2^{\underline{2}}_{\underline{8}}2^{\underline{2}}_{\underline{6}}2^{\underline{1}}_{\underline{6}}2^{\underline{1}}_{\underline{4}}3^{0}_{\underline{4}}3^{0}_{\underline{2}}2^{1}_{\underline{2}}2^{1}_{0}1^{2}_{0}2^{2}_{2}1^{3}_{2}1^{3}_{4}1^{4}_{6}$	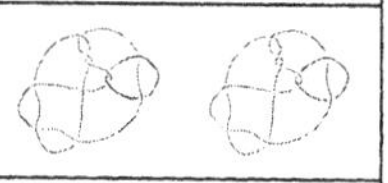	
11^n_{101}: $2^{\underline{6}}_{\underline{14}}2^{\underline{5}}_{\underline{12}}2^{\underline{5}}_{\underline{10}}5^{\underline{4}}_{\underline{10}}3^{\underline{4}}_{\underline{8}}6^{\underline{3}}_{\underline{8}}4^{\underline{3}}_{\underline{6}}6^{\underline{2}}_{\underline{6}}6^{\underline{2}}_{\underline{4}}5^{\underline{1}}_{\underline{4}}6^{\underline{1}}_{\underline{2}}5^{0}_{\underline{2}}6^{0}_{0}3^{1}_{0}4^{1}_{2}1^{2}_{2}3^{2}_{4}1^{3}_{6}$ 11^n_{102}: $3^{\underline{7}}_{\underline{18}}3^{\underline{6}}_{\underline{16}}3^{\underline{6}}_{\underline{14}}7^{\underline{5}}_{\underline{14}}3^{\underline{5}}_{\underline{12}}6^{\underline{4}}_{\underline{12}}7^{\underline{4}}_{\underline{10}}7^{\underline{3}}_{\underline{10}}6^{\underline{3}}_{\underline{8}}7^{\underline{2}}_{\underline{8}}7^{\underline{2}}_{\underline{6}}3^{\underline{1}}_{\underline{6}}7^{\underline{1}}_{\underline{4}}4^{0}_{\underline{4}}5^{0}_{\underline{2}}1^{1}_{\underline{2}}2^{1}_{0}1^{2}_{2}$	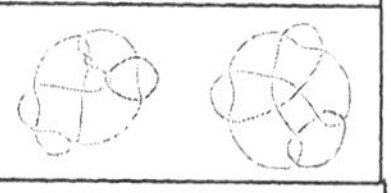	
11^n_{103}: $1^{\underline{11}}_{\underline{26}}1^{\underline{10}}_{\underline{22}}1^{\underline{9}}_{\underline{22}}1^{\underline{8}}_{\underline{20}}1^{\underline{8}}_{\underline{18}}3^{\underline{7}}_{\underline{20}}1^{\underline{7}}_{\underline{16}}1^{\underline{6}}_{\underline{18}}4^{\underline{6}}_{\underline{16}}3^{\underline{5}}_{\underline{16}}1^{\underline{5}}_{\underline{14}}1^{\underline{5}}_{\underline{12}}3^{\underline{4}}_{\underline{14}}4^{\underline{4}}_{\underline{12}}2^{\underline{3}}_{\underline{12}}2^{\underline{3}}_{\underline{10}}2^{\underline{2}}_{\underline{10}}2^{\underline{2}}_{\underline{8}}2^{\underline{1}}_{\underline{6}}1^{0}_{\underline{6}}1^{0}_{\underline{4}}$ 11^n_{104}: $1^{\underline{7}}_{\underline{18}}2^{\underline{6}}_{\underline{16}}1^{\underline{6}}_{\underline{14}}4^{\underline{5}}_{\underline{14}}2^{\underline{5}}_{\underline{12}}5^{\underline{4}}_{\underline{12}}5^{\underline{4}}_{\underline{10}}6^{\underline{3}}_{\underline{10}}4^{\underline{3}}_{\underline{8}}5^{\underline{2}}_{\underline{8}}6^{\underline{2}}_{\underline{6}}3^{\underline{1}}_{\underline{6}}5^{\underline{1}}_{\underline{4}}4^{0}_{\underline{4}}4^{0}_{\underline{2}}1^{1}_{\underline{2}}3^{1}_{0}1^{2}_{2}$	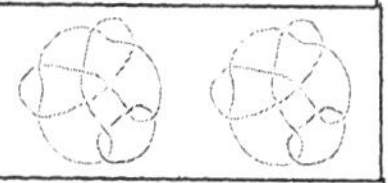	
11^n_{105}: $2^{\underline{5}}_{\underline{14}}3^{\underline{4}}_{\underline{12}}2^{\underline{4}}_{\underline{10}}5^{\underline{3}}_{\underline{10}}3^{\underline{3}}_{\underline{8}}5^{\underline{2}}_{\underline{8}}5^{\underline{2}}_{\underline{6}}5^{\underline{1}}_{\underline{6}}5^{\underline{1}}_{\underline{4}}6^{0}_{\underline{4}}7^{0}_{\underline{2}}4^{1}_{\underline{2}}4^{1}_{0}2^{2}_{0}4^{2}_{2}1^{3}_{2}2^{3}_{4}1^{4}_{6}$ 11^n_{106}: $1^{\underline{9}}_{\underline{22}}1^{\underline{8}}_{\underline{20}}1^{\underline{8}}_{\underline{18}}2^{\underline{7}}_{\underline{18}}1^{\underline{7}}_{\underline{16}}2^{\underline{6}}_{\underline{16}}2^{\underline{6}}_{\underline{14}}1^{\underline{5}}_{\underline{16}}2^{\underline{5}}_{\underline{14}}2^{\underline{5}}_{\underline{12}}1^{\underline{4}}_{\underline{14}}4^{\underline{4}}_{\underline{12}}3^{\underline{4}}_{\underline{10}}1^{\underline{3}}_{\underline{12}}2^{\underline{3}}_{\underline{10}}2^{\underline{3}}_{\underline{8}}1^{\underline{2}}_{\underline{10}}2^{\underline{2}}_{\underline{8}}1^{\underline{2}}_{\underline{6}}1^{\underline{1}}_{\underline{8}}1^{\underline{1}}_{\underline{6}}1^{\underline{1}}_{\underline{4}}1^{0}_{\underline{6}}2^{0}_{\underline{4}}1^{1}_{\underline{4}}1^{2}_{0}$	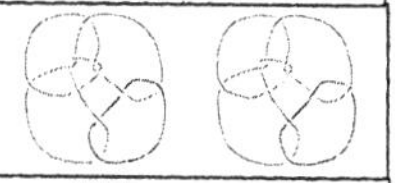	
11^n_{107}: $3^{0}_{2}2^{0}_{4}5^{1}_{4}1^{1}_{6}6^{2}_{6}5^{2}_{8}6^{3}_{8}6^{3}_{10}8^{4}_{10}6^{4}_{12}5^{5}_{12}8^{5}_{14}5^{6}_{14}5^{6}_{16}2^{7}_{16}5^{7}_{18}1^{8}_{18}2^{8}_{20}1^{9}_{22}$ 11^n_{108}: $1^{\underline{4}}_{\underline{6}}1^{\underline{4}}_{\underline{4}}1^{\underline{3}}_{\underline{4}}1^{\underline{2}}_{\underline{2}}1^{\underline{2}}_{0}1^{\underline{1}}_{0}1^{\underline{1}}_{2}3^{0}_{0}3^{0}_{2}1^{0}_{4}2^{1}_{2}2^{1}_{4}2^{1}_{6}2^{2}_{4}3^{2}_{6}2^{3}_{6}2^{3}_{8}1^{3}_{10}2^{4}_{8}2^{4}_{10}1^{5}_{10}2^{5}_{12}1^{6}_{12}1^{6}_{14}1^{7}_{16}$	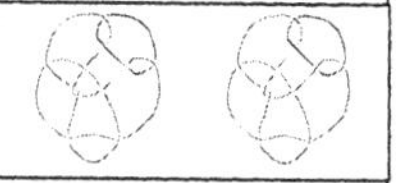	
11^n_{109}: $1^{\underline{11}}_{\underline{26}}1^{\underline{10}}_{\underline{22}}1^{\underline{9}}_{\underline{22}}1^{\underline{8}}_{\underline{20}}1^{\underline{8}}_{\underline{18}}1^{\underline{7}}_{\underline{20}}1^{\underline{7}}_{\underline{16}}2^{\underline{6}}_{\underline{16}}2^{\underline{5}}_{\underline{16}}1^{\underline{5}}_{\underline{12}}1^{\underline{4}}_{\underline{14}}3^{\underline{4}}_{\underline{12}}1^{\underline{4}}_{\underline{10}}1^{\underline{3}}_{\underline{12}}1^{\underline{3}}_{\underline{10}}1^{\underline{2}}_{\underline{10}}1^{\underline{2}}_{\underline{8}}1^{\underline{1}}_{\underline{6}}1^{0}_{\underline{6}}1^{0}_{\underline{4}}$ 11^n_{110}: $3^{\underline{4}}_{\underline{10}}1^{\underline{4}}_{\underline{8}}4^{\underline{3}}_{\underline{8}}2^{\underline{3}}_{\underline{6}}6^{\underline{2}}_{\underline{6}}4^{\underline{2}}_{\underline{4}}6^{\underline{1}}_{\underline{4}}6^{\underline{1}}_{\underline{2}}7^{0}_{\underline{2}}7^{0}_{0}6^{1}_{0}6^{1}_{2}4^{2}_{2}6^{2}_{4}2^{3}_{4}4^{3}_{6}1^{4}_{6}2^{4}_{8}1^{5}_{10}$	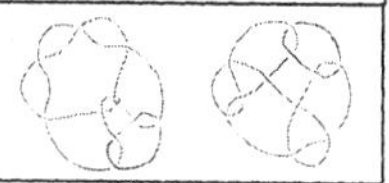	
11^n_{111}: $1^{\underline{9}}_{\underline{22}}1^{\underline{8}}_{\underline{20}}1^{\underline{8}}_{\underline{18}}1^{\underline{7}}_{\underline{18}}1^{\underline{7}}_{\underline{16}}2^{\underline{6}}_{\underline{16}}1^{\underline{6}}_{\underline{14}}1^{\underline{5}}_{\underline{16}}1^{\underline{5}}_{\underline{14}}2^{\underline{5}}_{\underline{12}}1^{\underline{4}}_{\underline{14}}3^{\underline{4}}_{\underline{12}}2^{\underline{4}}_{\underline{10}}1^{\underline{3}}_{\underline{12}}2^{\underline{3}}_{\underline{10}}1^{\underline{3}}_{\underline{8}}1^{\underline{2}}_{\underline{10}}1^{\underline{2}}_{\underline{8}}1^{\underline{2}}_{\underline{6}}1^{\underline{1}}_{\underline{8}}1^{\underline{1}}_{\underline{6}}1^{0}_{\underline{6}}2^{0}_{\underline{4}}1^{1}_{\underline{4}}1^{2}_{0}$ 11^n_{112}: $1^{\underline{4}}_{\underline{6}}1^{\underline{3}}_{\underline{2}}1^{\underline{2}}_{\underline{2}}1^{\underline{1}}_{0}1^{\underline{1}}_{2}3^{0}_{0}2^{0}_{2}1^{0}_{4}2^{1}_{2}2^{1}_{4}2^{2}_{4}2^{2}_{6}1^{2}_{8}2^{3}_{6}2^{3}_{8}2^{4}_{8}2^{4}_{10}1^{5}_{10}2^{5}_{12}1^{6}_{12}1^{6}_{14}1^{7}_{16}$	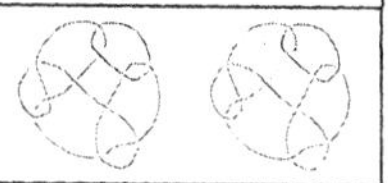	
11^n_{113}: $1^{\underline{7}}_{\underline{16}}2^{\underline{6}}_{\underline{14}}1^{\underline{6}}_{\underline{12}}4^{\underline{5}}_{\underline{12}}2^{\underline{5}}_{\underline{10}}5^{\underline{4}}_{\underline{10}}4^{\underline{4}}_{\underline{8}}5^{\underline{3}}_{\underline{8}}5^{\underline{3}}_{\underline{6}}6^{\underline{2}}_{\underline{6}}5^{\underline{2}}_{\underline{4}}4^{\underline{1}}_{\underline{4}}6^{\underline{1}}_{\underline{2}}4^{0}_{\underline{2}}6^{0}_{0}2^{1}_{0}2^{1}_{2}2^{2}_{4}$ 11^n_{114}: $1^{\underline{4}}_{\underline{6}}1^{\underline{3}}_{\underline{4}}1^{\underline{3}}_{\underline{2}}4^{\underline{2}}_{\underline{2}}1^{\underline{2}}_{0}3^{\underline{1}}_{0}4^{\underline{1}}_{2}6^{0}_{2}5^{0}_{4}5^{1}_{4}4^{1}_{6}4^{2}_{6}5^{2}_{8}3^{3}_{8}4^{3}_{10}2^{4}_{10}3^{4}_{12}2^{5}_{14}$	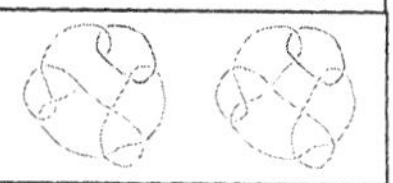	
11^n_{115}: $1^{\underline{6}}_{\underline{12}}1^{\underline{5}}_{\underline{10}}1^{\underline{5}}_{\underline{8}}4^{\underline{4}}_{\underline{8}}2^{\underline{4}}_{\underline{6}}3^{\underline{3}}_{\underline{6}}3^{\underline{3}}_{\underline{4}}4^{\underline{2}}_{\underline{4}}3^{\underline{2}}_{\underline{2}}4^{\underline{1}}_{\underline{2}}4^{\underline{1}}_{0}1^{0}_{\underline{2}}4^{0}_{0}4^{0}_{2}2^{1}_{2}3^{1}_{4}1^{2}_{4}2^{2}_{6}1^{3}_{8}$ 11^n_{116}: $1^{\underline{9}}_{\underline{22}}3^{\underline{8}}_{\underline{20}}1^{\underline{8}}_{\underline{18}}4^{\underline{7}}_{\underline{18}}3^{\underline{7}}_{\underline{16}}7^{\underline{6}}_{\underline{16}}4^{\underline{6}}_{\underline{14}}7^{\underline{5}}_{\underline{14}}7^{\underline{5}}_{\underline{12}}7^{\underline{4}}_{\underline{12}}8^{\underline{4}}_{\underline{10}}7^{\underline{3}}_{\underline{10}}6^{\underline{3}}_{\underline{8}}4^{\underline{2}}_{\underline{8}}7^{\underline{2}}_{\underline{6}}2^{\underline{1}}_{\underline{6}}4^{\underline{1}}_{\underline{4}}1^{0}_{\underline{4}}3^{0}_{\underline{2}}$	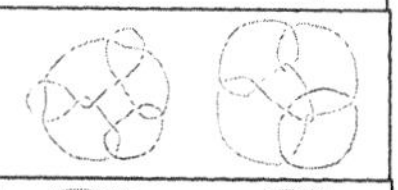	
11^n_{117}: $1^{\underline{8}}_{\underline{18}}2^{\underline{7}}_{\underline{16}}1^{\underline{7}}_{\underline{14}}3^{\underline{6}}_{\underline{14}}2^{\underline{6}}_{\underline{12}}5^{\underline{5}}_{\underline{12}}3^{\underline{5}}_{\underline{10}}5^{\underline{4}}_{\underline{10}}6^{\underline{4}}_{\underline{8}}5^{\underline{3}}_{\underline{8}}4^{\underline{3}}_{\underline{6}}5^{\underline{2}}_{\underline{6}}5^{\underline{2}}_{\underline{4}}2^{\underline{1}}_{\underline{4}}5^{\underline{1}}_{\underline{2}}3^{0}_{\underline{2}}3^{0}_{0}2^{1}_{2}$ 11^n_{118}: $2^{\underline{2}}_{\underline{4}}3^{\underline{1}}_{\underline{2}}2^{\underline{1}}_{0}6^{0}_{0}5^{0}_{2}7^{1}_{2}4^{1}_{4}6^{2}_{4}7^{2}_{6}5^{3}_{6}6^{3}_{8}5^{4}_{8}5^{4}_{10}2^{5}_{10}5^{5}_{12}1^{6}_{12}2^{6}_{14}1^{7}_{16}$	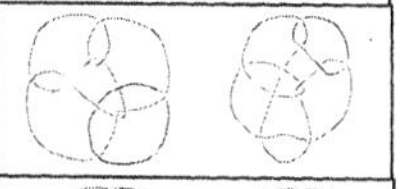	
11^n_{119}: $3^{0}_{2}2^{0}_{4}3^{1}_{4}1^{1}_{6}4^{2}_{6}3^{2}_{8}3^{3}_{8}4^{3}_{10}4^{4}_{10}3^{4}_{12}2^{5}_{12}4^{5}_{14}2^{6}_{14}2^{6}_{16}2^{7}_{18}$ 11^n_{120}: $1^{\underline{6}}_{\underline{12}}1^{\underline{5}}_{\underline{10}}1^{\underline{5}}_{\underline{8}}2^{\underline{4}}_{\underline{8}}2^{\underline{4}}_{\underline{6}}2^{\underline{3}}_{\underline{6}}1^{\underline{3}}_{\underline{4}}1^{\underline{2}}_{\underline{6}}2^{\underline{2}}_{\underline{4}}2^{\underline{2}}_{\underline{2}}1^{\underline{1}}_{\underline{4}}2^{\underline{1}}_{\underline{2}}2^{\underline{1}}_{0}2^{0}_{\underline{2}}4^{0}_{0}1^{0}_{2}2^{1}_{0}1^{1}_{2}2^{1}_{4}1^{2}_{2}2^{2}_{4}1^{3}_{4}1^{3}_{6}1^{4}_{6}1^{4}_{8}1^{5}_{10}$	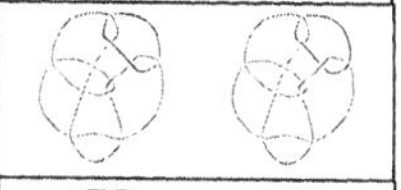	
11^n_{121}: $1^{\underline{4}}_{\underline{6}}1^{\underline{4}}_{\underline{4}}1^{\underline{2}}_{\underline{2}}1^{\underline{1}}_{\underline{2}}1^{\underline{1}}_{2}3^{0}_{0}3^{0}_{2}2^{1}_{2}2^{1}_{4}1^{1}_{6}3^{2}_{4}2^{2}_{6}1^{3}_{6}3^{3}_{8}2^{4}_{8}1^{4}_{10}1^{5}_{10}2^{5}_{12}1^{6}_{14}$ 11^n_{122}: $1^{\underline{4}}_{\underline{8}}1^{\underline{4}}_{\underline{6}}1^{\underline{2}}_{\underline{4}}1^{\underline{1}}_{\underline{4}}1^{\underline{1}}_{0}2^{0}_{\underline{2}}3^{0}_{0}1^{1}_{0}1^{1}_{2}1^{1}_{4}2^{2}_{2}1^{2}_{4}1^{3}_{4}2^{3}_{6}1^{4}_{6}1^{4}_{8}1^{5}_{8}1^{5}_{10}1^{6}_{12}$	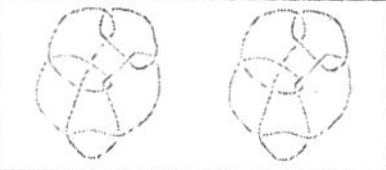	

L: KH	1st line	2nd line
11^n_{123}: $1^{\underline{9}}_{\underline{24}}2^{\underline{8}}_{\underline{22}}1^{\underline{8}}_{\underline{20}}5^{\underline{7}}_{\underline{20}}2^{\underline{7}}_{\underline{18}}4^{\underline{6}}_{\underline{18}}5^{\underline{6}}_{\underline{16}}6^{\underline{5}}_{\underline{16}}4^{\underline{5}}_{\underline{14}}5^{\underline{4}}_{\underline{14}}7^{\underline{4}}_{\underline{12}}4^{\underline{3}}_{\underline{12}}4^{\underline{3}}_{\underline{10}}3^{\underline{2}}_{\underline{10}}4^{\underline{2}}_{\underline{8}}3^{\underline{1}}_{\underline{6}}1^{0}_{\underline{6}}1^{0}_{\underline{4}}$	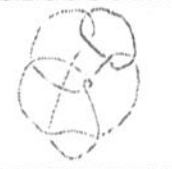	
11^n_{124}: $1^{\underline{3}}_{\underline{8}}3^{\underline{2}}_{\underline{6}}1^{\underline{2}}_{\underline{4}}3^{\underline{1}}_{\underline{4}}3^{\underline{1}}_{\underline{2}}6^{0}_{\underline{2}}5^{0}_{0}1^{0}_{2}5^{1}_{0}5^{1}_{2}4^{2}_{2}5^{2}_{4}4^{3}_{4}4^{3}_{6}2^{4}_{6}4^{4}_{8}1^{5}_{8}2^{5}_{10}1^{6}_{12}$		
11^n_{125}: $1^{\underline{7}}_{\underline{16}}2^{\underline{6}}_{\underline{14}}1^{\underline{6}}_{\underline{12}}2^{\underline{5}}_{\underline{12}}2^{\underline{5}}_{\underline{10}}4^{\underline{4}}_{\underline{10}}3^{\underline{4}}_{\underline{8}}3^{\underline{3}}_{\underline{8}}3^{\underline{3}}_{\underline{6}}1^{\underline{2}}_{\underline{8}}3^{\underline{2}}_{\underline{6}}3^{\underline{2}}_{\underline{4}}3^{\underline{1}}_{\underline{4}}3^{\underline{1}}_{\underline{2}}1^{0}_{\underline{4}}2^{0}_{\underline{2}}3^{0}_{0}1^{1}_{\underline{2}}1^{1}_{0}1^{1}_{2}1^{2}_{2}1^{3}_{2}1^{4}_{6}$		
11^n_{126}: $2^{\underline{2}}_{\underline{4}}2^{\underline{1}}_{\underline{2}}2^{\underline{1}}_{0}5^{0}_{0}3^{0}_{2}5^{1}_{2}4^{1}_{4}5^{2}_{4}6^{2}_{6}5^{3}_{6}4^{3}_{8}3^{4}_{8}5^{4}_{10}2^{5}_{10}3^{5}_{12}1^{6}_{12}2^{6}_{14}1^{7}_{16}$		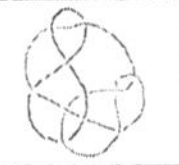
11^n_{127}: $1^{\underline{6}}_{\underline{12}}1^{\underline{5}}_{\underline{10}}1^{\underline{5}}_{\underline{8}}2^{\underline{4}}_{\underline{8}}1^{\underline{4}}_{\underline{6}}2^{\underline{3}}_{\underline{6}}2^{\underline{3}}_{\underline{4}}1^{\underline{2}}_{\underline{6}}3^{\underline{2}}_{\underline{4}}2^{\underline{2}}_{\underline{2}}2^{\underline{1}}_{\underline{2}}2^{\underline{1}}_{0}1^{0}_{\underline{2}}2^{0}_{0}2^{0}_{2}1^{1}_{2}1^{1}_{4}1^{2}_{6}$	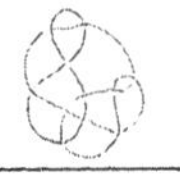	
11^n_{128}: $1^{\underline{4}}_{\underline{6}}1^{\underline{3}}_{\underline{4}}1^{\underline{3}}_{\underline{2}}3^{\underline{2}}_{\underline{2}}1^{\underline{2}}_{0}2^{\underline{1}}_{0}3^{\underline{1}}_{2}4^{0}_{2}3^{0}_{4}3^{1}_{4}3^{1}_{6}3^{2}_{6}4^{2}_{8}2^{3}_{8}2^{3}_{10}1^{4}_{10}2^{4}_{12}1^{5}_{14}$		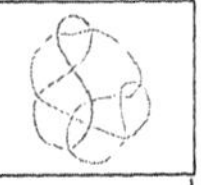
11^n_{129}: $1^{\underline{4}}_{\underline{10}}1^{\underline{3}}_{\underline{8}}1^{\underline{3}}_{\underline{6}}4^{\underline{2}}_{\underline{6}}2^{\underline{2}}_{\underline{4}}3^{\underline{1}}_{\underline{4}}3^{\underline{1}}_{\underline{2}}4^{0}_{\underline{2}}4^{0}_{0}4^{1}_{0}3^{1}_{2}2^{2}_{2}4^{2}_{4}2^{3}_{4}2^{3}_{6}1^{4}_{6}2^{4}_{8}1^{5}_{10}$	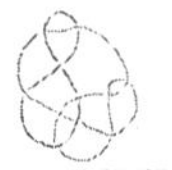	
11^n_{130}: $1^{\underline{8}}_{\underline{20}}2^{\underline{7}}_{\underline{18}}1^{\underline{7}}_{\underline{16}}3^{\underline{6}}_{\underline{16}}2^{\underline{6}}_{\underline{14}}5^{\underline{5}}_{\underline{14}}3^{\underline{5}}_{\underline{12}}4^{\underline{4}}_{\underline{12}}5^{\underline{4}}_{\underline{10}}5^{\underline{3}}_{\underline{10}}4^{\underline{3}}_{\underline{8}}4^{\underline{2}}_{\underline{8}}6^{\underline{2}}_{\underline{6}}2^{\underline{1}}_{\underline{6}}3^{\underline{1}}_{\underline{4}}2^{0}_{\underline{4}}3^{0}_{\underline{2}}1^{1}_{0}$		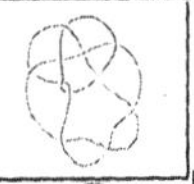
11^n_{131}: $1^{\underline{2}}_{\underline{4}}1^{\underline{2}}_{\underline{2}}2^{\underline{1}}_{\underline{2}}3^{0}_{0}3^{0}_{2}3^{1}_{2}2^{1}_{4}4^{2}_{4}3^{2}_{6}2^{3}_{6}4^{3}_{8}3^{4}_{8}2^{4}_{10}1^{5}_{10}3^{5}_{12}1^{6}_{12}1^{6}_{14}1^{7}_{16}$	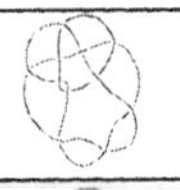	
11^n_{132}: $1^{\underline{9}}_{\underline{22}}1^{\underline{8}}_{\underline{18}}1^{\underline{7}}_{\underline{18}}1^{\underline{6}}_{\underline{16}}1^{\underline{6}}_{\underline{14}}1^{\underline{5}}_{\underline{16}}1^{\underline{5}}_{\underline{14}}1^{\underline{5}}_{\underline{12}}2^{\underline{4}}_{\underline{12}}1^{\underline{4}}_{\underline{10}}1^{\underline{3}}_{\underline{12}}1^{\underline{3}}_{\underline{8}}2^{\underline{2}}_{\underline{8}}1^{\underline{2}}_{\underline{6}}1^{0}_{\underline{6}}1^{0}_{\underline{4}}$		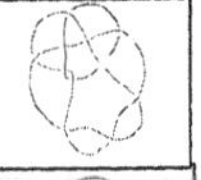
11^n_{133}: $1^{\underline{9}}_{\underline{26}}1^{\underline{8}}_{\underline{22}}1^{\underline{7}}_{\underline{22}}1^{\underline{6}}_{\underline{20}}2^{\underline{6}}_{\underline{18}}1^{\underline{5}}_{\underline{20}}1^{\underline{5}}_{\underline{18}}1^{\underline{4}}_{\underline{16}}1^{\underline{4}}_{\underline{14}}1^{\underline{3}}_{\underline{16}}1^{\underline{2}}_{\underline{12}}1^{0}_{\underline{10}}1^{0}_{\underline{8}}$	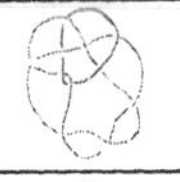	
11^n_{134}: $1^{\underline{6}}_{\underline{12}}1^{\underline{5}}_{\underline{10}}1^{\underline{5}}_{\underline{8}}2^{\underline{4}}_{\underline{8}}1^{\underline{4}}_{\underline{6}}2^{\underline{3}}_{\underline{6}}2^{\underline{3}}_{\underline{4}}3^{\underline{2}}_{\underline{4}}3^{\underline{2}}_{\underline{2}}2^{\underline{1}}_{\underline{2}}2^{\underline{1}}_{0}3^{0}_{0}3^{0}_{2}1^{1}_{2}2^{1}_{4}1^{2}_{4}1^{2}_{6}1^{3}_{8}$		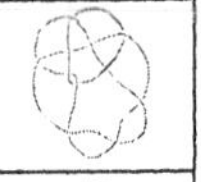
11^n_{135}: $1^{\underline{2}}_{\underline{4}}1^{\underline{2}}_{\underline{2}}1^{\underline{1}}_{\underline{2}}2^{0}_{0}2^{0}_{2}2^{1}_{2}1^{1}_{4}2^{2}_{4}2^{2}_{6}1^{3}_{6}2^{3}_{8}2^{4}_{8}1^{4}_{10}2^{5}_{12}1^{6}_{12}1^{7}_{16}$	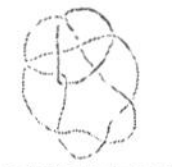	
11^n_{136}: $1^{\underline{6}}_{\underline{12}}1^{\underline{6}}_{\underline{10}}1^{\underline{5}}_{\underline{10}}1^{\underline{4}}_{\underline{8}}1^{\underline{4}}_{\underline{6}}1^{\underline{3}}_{\underline{6}}1^{\underline{3}}_{\underline{4}}2^{\underline{2}}_{\underline{4}}1^{\underline{2}}_{\underline{2}}2^{\underline{1}}_{0}3^{0}_{0}1^{0}_{2}2^{1}_{4}1^{2}_{4}1^{3}_{8}$		
11^n_{137}: $1^{\underline{8}}_{\underline{20}}3^{\underline{7}}_{\underline{18}}1^{\underline{7}}_{\underline{16}}4^{\underline{6}}_{\underline{16}}3^{\underline{6}}_{\underline{14}}6^{\underline{5}}_{\underline{14}}4^{\underline{5}}_{\underline{12}}6^{\underline{4}}_{\underline{12}}6^{\underline{4}}_{\underline{10}}6^{\underline{3}}_{\underline{10}}6^{\underline{3}}_{\underline{8}}5^{\underline{2}}_{\underline{8}}7^{\underline{2}}_{\underline{6}}3^{\underline{1}}_{\underline{6}}4^{\underline{1}}_{\underline{4}}2^{0}_{\underline{4}}4^{0}_{\underline{2}}1^{1}_{0}$	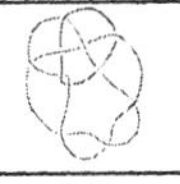	
11^n_{138}: $1^{\underline{8}}_{\underline{18}}1^{\underline{7}}_{\underline{16}}1^{\underline{7}}_{\underline{14}}1^{\underline{6}}_{\underline{14}}1^{\underline{6}}_{\underline{12}}2^{\underline{5}}_{\underline{12}}1^{\underline{5}}_{\underline{10}}1^{\underline{4}}_{\underline{10}}2^{\underline{4}}_{\underline{8}}1^{\underline{3}}_{\underline{8}}1^{\underline{3}}_{\underline{6}}2^{\underline{2}}_{\underline{6}}2^{\underline{2}}_{\underline{4}}1^{\underline{1}}_{\underline{2}}1^{0}_{\underline{2}}1^{0}_{0}$		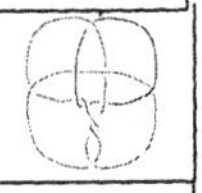
11^n_{139}: $1^{\underline{2}}_{\underline{4}}1^{\underline{2}}_{\underline{2}}2^{0}_{0}1^{0}_{2}1^{1}_{0}1^{1}_{4}2^{2}_{4}1^{3}_{4}1^{3}_{8}1^{4}_{6}1^{4}_{8}1^{5}_{10}1^{6}_{10}1^{7}_{14}$	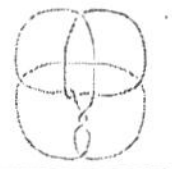	
11^n_{140}: $2^{\underline{7}}_{\underline{18}}1^{\underline{6}}_{\underline{16}}2^{\underline{6}}_{\underline{14}}3^{\underline{5}}_{\underline{14}}1^{\underline{5}}_{\underline{12}}3^{\underline{4}}_{\underline{12}}3^{\underline{4}}_{\underline{10}}2^{\underline{3}}_{\underline{10}}3^{\underline{3}}_{\underline{8}}3^{\underline{2}}_{\underline{8}}3^{\underline{2}}_{\underline{6}}1^{\underline{1}}_{\underline{6}}2^{\underline{1}}_{\underline{4}}1^{0}_{\underline{4}}2^{0}_{\underline{2}}$		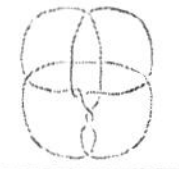
11^n_{141}: $1^{\underline{7}}_{\underline{22}}2^{\underline{6}}_{\underline{18}}1^{\underline{6}}_{\underline{16}}2^{\underline{5}}_{\underline{18}}1^{\underline{5}}_{\underline{16}}1^{\underline{4}}_{\underline{16}}2^{\underline{4}}_{\underline{14}}1^{\underline{4}}_{\underline{12}}1^{\underline{3}}_{\underline{14}}1^{\underline{3}}_{\underline{12}}1^{\underline{2}}_{\underline{12}}1^{\underline{2}}_{\underline{10}}1^{\underline{1}}_{\underline{8}}1^{0}_{\underline{8}}1^{0}_{\underline{6}}$	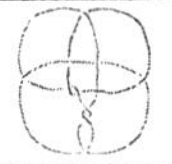	
11^n_{142}: $1^{\underline{6}}_{\underline{14}}1^{\underline{5}}_{\underline{12}}1^{\underline{5}}_{\underline{10}}3^{\underline{4}}_{\underline{10}}1^{\underline{4}}_{\underline{8}}3^{\underline{3}}_{\underline{8}}3^{\underline{3}}_{\underline{6}}4^{\underline{2}}_{\underline{6}}4^{\underline{2}}_{\underline{4}}4^{\underline{1}}_{\underline{4}}3^{\underline{1}}_{\underline{2}}3^{0}_{\underline{2}}5^{0}_{0}2^{1}_{0}2^{1}_{2}1^{2}_{2}2^{2}_{4}1^{3}_{6}$		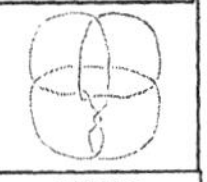
11^n_{143}: $1^{\underline{2}}_{\underline{6}}1^{\underline{2}}_{\underline{4}}1^{\underline{1}}_{\underline{4}}1^{0}_{\underline{2}}2^{0}_{0}2^{1}_{0}1^{2}_{2}2^{2}_{4}1^{3}_{4}1^{3}_{6}2^{4}_{6}1^{4}_{8}2^{5}_{10}1^{6}_{10}1^{7}_{14}$	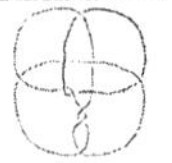	
11^n_{144}: $1^{\underline{6}}_{\underline{14}}1^{\underline{6}}_{\underline{12}}2^{\underline{5}}_{\underline{12}}2^{\underline{4}}_{\underline{10}}2^{\underline{4}}_{\underline{8}}3^{\underline{3}}_{\underline{8}}2^{\underline{3}}_{\underline{6}}4^{\underline{2}}_{\underline{6}}3^{\underline{2}}_{\underline{4}}2^{\underline{1}}_{\underline{4}}4^{\underline{1}}_{\underline{2}}4^{0}_{\underline{2}}3^{0}_{0}1^{1}_{0}3^{1}_{2}1^{2}_{2}1^{2}_{4}1^{3}_{6}$		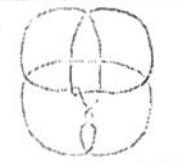
11^n_{145}: $1^{\underline{6}}_{\underline{16}}3^{\underline{5}}_{\underline{14}}1^{\underline{5}}_{\underline{12}}4^{\underline{4}}_{\underline{12}}3^{\underline{4}}_{\underline{10}}5^{\underline{3}}_{\underline{10}}4^{\underline{3}}_{\underline{8}}6^{\underline{2}}_{\underline{8}}6^{\underline{2}}_{\underline{6}}5^{\underline{1}}_{\underline{6}}5^{\underline{1}}_{\underline{4}}4^{0}_{\underline{4}}6^{0}_{\underline{2}}3^{1}_{\underline{2}}3^{1}_{0}1^{2}_{0}3^{2}_{2}1^{3}_{4}$	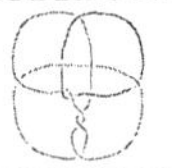	
11^n_{146}: $1^{\underline{5}}_{\underline{10}}2^{\underline{4}}_{\underline{8}}1^{\underline{4}}_{\underline{6}}3^{\underline{3}}_{\underline{6}}2^{\underline{3}}_{\underline{4}}5^{\underline{2}}_{\underline{4}}4^{\underline{2}}_{\underline{2}}4^{\underline{1}}_{\underline{2}}4^{\underline{1}}_{0}6^{0}_{0}5^{0}_{2}3^{1}_{2}5^{1}_{4}3^{2}_{4}3^{2}_{6}1^{3}_{6}3^{3}_{8}1^{4}_{10}$		
11^n_{147}: $1^{\underline{7}}_{\underline{20}}1^{\underline{6}}_{\underline{18}}1^{\underline{6}}_{\underline{16}}3^{\underline{5}}_{\underline{16}}1^{\underline{5}}_{\underline{14}}2^{\underline{4}}_{\underline{14}}3^{\underline{4}}_{\underline{12}}3^{\underline{3}}_{\underline{12}}2^{\underline{3}}_{\underline{10}}3^{\underline{2}}_{\underline{10}}4^{\underline{2}}_{\underline{8}}2^{\underline{1}}_{\underline{8}}2^{\underline{1}}_{\underline{6}}2^{0}_{\underline{6}}3^{0}_{\underline{4}}1^{1}_{\underline{4}}1^{1}_{\underline{2}}1^{2}_{0}$		
11^n_{148}: $1^{\underline{6}}_{\underline{12}}1^{\underline{6}}_{\underline{10}}1^{\underline{4}}_{\underline{8}}1^{\underline{3}}_{\underline{8}}1^{\underline{3}}_{\underline{4}}1^{\underline{2}}_{\underline{6}}2^{\underline{2}}_{\underline{4}}1^{\underline{1}}_{\underline{4}}1^{\underline{1}}_{\underline{2}}1^{\underline{1}}_{0}2^{0}_{\underline{2}}2^{0}_{0}1^{1}_{2}1^{2}_{2}1^{3}_{6}$		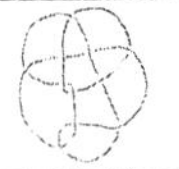
11^n_{149}: $1^{\underline{9}}_{\underline{22}}2^{\underline{8}}_{\underline{20}}1^{\underline{8}}_{\underline{18}}2^{\underline{7}}_{\underline{18}}2^{\underline{7}}_{\underline{16}}4^{\underline{6}}_{\underline{16}}3^{\underline{6}}_{\underline{14}}4^{\underline{5}}_{\underline{14}}3^{\underline{5}}_{\underline{12}}3^{\underline{4}}_{\underline{12}}4^{\underline{4}}_{\underline{10}}3^{\underline{3}}_{\underline{10}}3^{\underline{3}}_{\underline{8}}2^{\underline{2}}_{\underline{8}}3^{\underline{2}}_{\underline{6}}2^{\underline{1}}_{\underline{4}}1^{0}_{\underline{4}}1^{0}_{\underline{2}}$		
11^n_{150}: $1^{\underline{9}}_{\underline{22}}1^{\underline{8}}_{\underline{20}}1^{\underline{8}}_{\underline{18}}2^{\underline{7}}_{\underline{18}}1^{\underline{7}}_{\underline{16}}2^{\underline{6}}_{\underline{16}}2^{\underline{6}}_{\underline{14}}3^{\underline{5}}_{\underline{14}}2^{\underline{5}}_{\underline{12}}2^{\underline{4}}_{\underline{12}}3^{\underline{4}}_{\underline{10}}2^{\underline{3}}_{\underline{10}}2^{\underline{3}}_{\underline{8}}2^{\underline{2}}_{\underline{8}}3^{\underline{2}}_{\underline{6}}1^{\underline{1}}_{\underline{4}}1^{0}_{\underline{4}}1^{0}_{\underline{2}}$		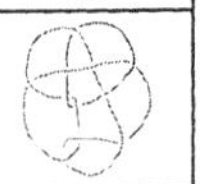
11^n_{151}: $1^{\underline{9}}_{\underline{26}}1^{\underline{8}}_{\underline{24}}1^{\underline{8}}_{\underline{22}}2^{\underline{7}}_{\underline{22}}1^{\underline{7}}_{\underline{20}}2^{\underline{6}}_{\underline{20}}3^{\underline{6}}_{\underline{18}}3^{\underline{5}}_{\underline{18}}1^{\underline{5}}_{\underline{16}}1^{\underline{4}}_{\underline{16}}3^{\underline{4}}_{\underline{14}}2^{\underline{3}}_{\underline{14}}1^{\underline{3}}_{\underline{12}}1^{\underline{2}}_{\underline{12}}2^{\underline{2}}_{\underline{10}}1^{\underline{1}}_{\underline{8}}1^{0}_{\underline{8}}1^{0}_{\underline{6}}$	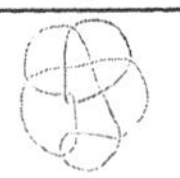	
11^n_{152}: $1^{\underline{5}}_{\underline{10}}2^{\underline{4}}_{\underline{8}}1^{\underline{4}}_{\underline{6}}2^{\underline{3}}_{\underline{6}}2^{\underline{3}}_{\underline{4}}4^{\underline{2}}_{\underline{4}}3^{\underline{2}}_{\underline{2}}3^{\underline{1}}_{\underline{2}}3^{\underline{1}}_{0}4^{0}_{0}4^{0}_{2}2^{1}_{2}3^{1}_{4}2^{2}_{4}2^{2}_{6}2^{3}_{8}$		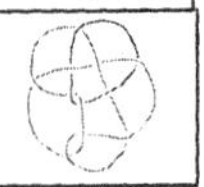

L: KH	1st line	2nd line
11^n_{153}: $1^{\underline{9}}_{\underline{22}}1^{\underline{8}}_{\underline{20}}1^{\underline{8}}_{\underline{18}}3^{\underline{7}}_{\underline{18}}1^{\underline{7}}_{\underline{16}}3^{\underline{6}}_{\underline{16}}3^{\underline{6}}_{\underline{14}}4^{\underline{5}}_{\underline{14}}3^{\underline{5}}_{\underline{12}}4^{\underline{4}}_{\underline{12}}4^{\underline{4}}_{\underline{10}}3^{\underline{3}}_{\underline{10}}4^{\underline{3}}_{\underline{8}}3^{\underline{2}}_{\underline{8}}4^{\underline{2}}_{\underline{6}}1^{\underline{1}}_{\underline{6}}2^{\underline{1}}_{\underline{4}}1^{0}_{\underline{4}}2^{0}_{\underline{2}}$ 11^n_{154}: $1^{\underline{9}}_{\underline{24}}1^{\underline{8}}_{\underline{22}}1^{\underline{8}}_{\underline{20}}3^{\underline{7}}_{\underline{20}}1^{\underline{7}}_{\underline{18}}3^{\underline{6}}_{\underline{18}}3^{\underline{6}}_{\underline{16}}4^{\underline{5}}_{\underline{16}}3^{\underline{5}}_{\underline{14}}3^{\underline{4}}_{\underline{14}}4^{\underline{4}}_{\underline{12}}3^{\underline{3}}_{\underline{12}}3^{\underline{3}}_{\underline{10}}2^{\underline{2}}_{\underline{10}}4^{\underline{2}}_{\underline{8}}1^{\underline{1}}_{\underline{8}}1^{\underline{1}}_{\underline{6}}1^{0}_{\underline{6}}2^{0}_{\underline{4}}$		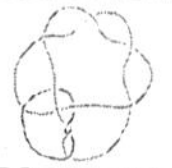
11^n_{155}: $1^{\underline{9}}_{\underline{24}}2^{\underline{8}}_{\underline{22}}1^{\underline{8}}_{\underline{20}}4^{\underline{7}}_{\underline{20}}2^{\underline{7}}_{\underline{18}}4^{\underline{6}}_{\underline{18}}4^{\underline{6}}_{\underline{16}}6^{\underline{5}}_{\underline{16}}4^{\underline{5}}_{\underline{14}}4^{\underline{4}}_{\underline{14}}6^{\underline{4}}_{\underline{12}}4^{\underline{3}}_{\underline{12}}4^{\underline{3}}_{\underline{10}}3^{\underline{2}}_{\underline{10}}5^{\underline{2}}_{\underline{8}}1^{\underline{1}}_{\underline{8}}2^{\underline{1}}_{\underline{6}}1^{0}_{\underline{6}}2^{0}_{\underline{4}}$ 11^n_{156}: $1^{\underline{6}}_{\underline{14}}1^{\underline{6}}_{\underline{12}}2^{\underline{5}}_{\underline{12}}3^{\underline{4}}_{\underline{10}}2^{\underline{4}}_{\underline{8}}4^{\underline{3}}_{\underline{8}}3^{\underline{3}}_{\underline{6}}5^{\underline{2}}_{\underline{6}}4^{\underline{2}}_{\underline{4}}4^{\underline{1}}_{\underline{4}}5^{\underline{1}}_{\underline{2}}5^{0}_{\underline{2}}5^{0}_{0}2^{1}_{0}4^{1}_{2}2^{2}_{2}2^{2}_{4}2^{3}_{6}$	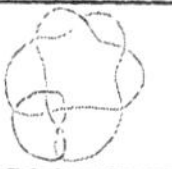	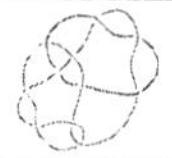
11^n_{157}: $1^{\underline{4}}_{\underline{8}}2^{\underline{3}}_{\underline{6}}1^{\underline{3}}_{\underline{4}}4^{\underline{2}}_{\underline{4}}3^{\underline{2}}_{\underline{2}}5^{\underline{1}}_{\underline{2}}3^{\underline{1}}_{0}6^{0}_{0}6^{0}_{2}5^{1}_{2}5^{1}_{4}5^{2}_{4}5^{2}_{6}2^{3}_{6}5^{3}_{8}2^{4}_{8}2^{4}_{10}2^{5}_{12}$ 11^n_{158}: $1^{\underline{8}}_{\underline{18}}1^{\underline{7}}_{\underline{16}}1^{\underline{7}}_{\underline{14}}2^{\underline{6}}_{\underline{14}}2^{\underline{6}}_{\underline{12}}2^{\underline{5}}_{\underline{12}}1^{\underline{5}}_{\underline{10}}2^{\underline{4}}_{\underline{10}}2^{\underline{4}}_{\underline{8}}1^{\underline{3}}_{\underline{8}}2^{\underline{3}}_{\underline{6}}2^{\underline{2}}_{\underline{6}}1^{\underline{2}}_{\underline{4}}2^{\underline{1}}_{\underline{2}}1^{0}_{\underline{2}}1^{0}_{0}$	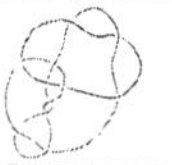	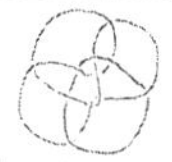
11^n_{159}: $1^{\underline{2}}_{0}1^{\underline{1}}_{2}1^{\underline{1}}_{4}2^{0}_{4}2^{0}_{6}2^{1}_{6}1^{1}_{8}1^{2}_{6}3^{2}_{8}2^{2}_{10}1^{3}_{10}2^{3}_{12}1^{4}_{10}2^{4}_{12}1^{4}_{14}1^{5}_{14}2^{5}_{16}$ 11^n_{160}: $1^{\underline{4}}_{\underline{10}}2^{\underline{3}}_{\underline{8}}1^{\underline{3}}_{\underline{6}}4^{\underline{2}}_{\underline{6}}3^{\underline{2}}_{\underline{4}}4^{\underline{1}}_{\underline{4}}3^{\underline{1}}_{\underline{2}}5^{0}_{\underline{2}}5^{0}_{0}4^{1}_{0}4^{1}_{2}3^{2}_{2}4^{2}_{4}2^{3}_{4}3^{3}_{6}1^{4}_{6}2^{4}_{8}1^{5}_{10}$	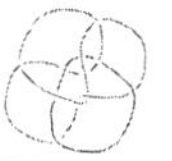	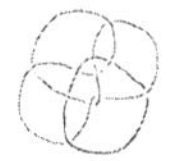
11^n_{161}: $1^{\underline{8}}_{\underline{18}}2^{\underline{7}}_{\underline{16}}1^{\underline{7}}_{\underline{14}}2^{\underline{6}}_{\underline{14}}2^{\underline{6}}_{\underline{12}}4^{\underline{5}}_{\underline{12}}2^{\underline{5}}_{\underline{10}}3^{\underline{4}}_{\underline{10}}4^{\underline{4}}_{\underline{8}}3^{\underline{3}}_{\underline{8}}3^{\underline{3}}_{\underline{6}}4^{\underline{2}}_{\underline{6}}4^{\underline{2}}_{\underline{4}}1^{\underline{1}}_{\underline{4}}3^{\underline{1}}_{\underline{2}}2^{0}_{\underline{2}}2^{0}_{0}1^{1}_{2}$ 11^n_{162}: $1^{\underline{6}}_{\underline{12}}1^{\underline{6}}_{\underline{10}}1^{\underline{4}}_{\underline{8}}1^{\underline{3}}_{\underline{8}}1^{\underline{3}}_{\underline{4}}2^{\underline{2}}_{\underline{6}}2^{\underline{2}}_{\underline{4}}2^{\underline{1}}_{\underline{4}}2^{\underline{1}}_{\underline{2}}1^{\underline{1}}_{0}3^{0}_{\underline{2}}3^{0}_{0}2^{1}_{0}2^{1}_{2}2^{2}_{2}2^{2}_{4}1^{3}_{4}2^{3}_{6}1^{4}_{6}1^{4}_{8}1^{5}_{10}$	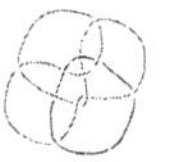	
11^n_{163}: $1^{\underline{9}}_{\underline{22}}2^{\underline{8}}_{\underline{20}}1^{\underline{8}}_{\underline{18}}3^{\underline{7}}_{\underline{18}}2^{\underline{7}}_{\underline{16}}5^{\underline{6}}_{\underline{16}}4^{\underline{6}}_{\underline{14}}5^{\underline{5}}_{\underline{14}}4^{\underline{5}}_{\underline{12}}5^{\underline{4}}_{\underline{12}}5^{\underline{4}}_{\underline{10}}4^{\underline{3}}_{\underline{10}}5^{\underline{3}}_{\underline{8}}3^{\underline{2}}_{\underline{8}}4^{\underline{2}}_{\underline{6}}1^{\underline{1}}_{\underline{6}}3^{\underline{1}}_{\underline{4}}1^{0}_{\underline{4}}2^{0}_{\underline{2}}$ 11^n_{164}: $1^{\underline{8}}_{\underline{18}}1^{\underline{7}}_{\underline{16}}1^{\underline{7}}_{\underline{14}}3^{\underline{6}}_{\underline{14}}2^{\underline{6}}_{\underline{12}}3^{\underline{5}}_{\underline{12}}2^{\underline{5}}_{\underline{10}}3^{\underline{4}}_{\underline{10}}3^{\underline{4}}_{\underline{8}}3^{\underline{3}}_{\underline{8}}3^{\underline{3}}_{\underline{6}}3^{\underline{2}}_{\underline{6}}3^{\underline{2}}_{\underline{4}}1^{\underline{1}}_{\underline{4}}3^{\underline{1}}_{\underline{2}}2^{0}_{\underline{2}}2^{0}_{0}1^{1}_{2}$	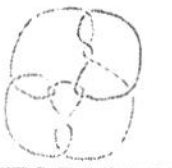	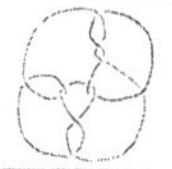
11^n_{165}: $1^{\underline{11}}_{\underline{28}}1^{\underline{10}}_{\underline{24}}1^{\underline{9}}_{\underline{24}}1^{\underline{8}}_{\underline{22}}1^{\underline{8}}_{\underline{20}}1^{\underline{7}}_{\underline{22}}1^{\underline{7}}_{\underline{18}}3^{\underline{6}}_{\underline{18}}1^{\underline{6}}_{\underline{16}}2^{\underline{5}}_{\underline{18}}1^{\underline{5}}_{\underline{16}}1^{\underline{5}}_{\underline{14}}1^{\underline{4}}_{\underline{16}}2^{\underline{4}}_{\underline{14}}1^{\underline{4}}_{\underline{12}}1^{\underline{3}}_{\underline{14}}1^{\underline{3}}_{\underline{12}}1^{\underline{2}}_{\underline{12}}1^{\underline{2}}_{\underline{10}}1^{\underline{1}}_{\underline{8}}1^{0}_{\underline{8}}1^{0}_{\underline{6}}$ 11^n_{166}: $1^{\underline{6}}_{\underline{14}}1^{\underline{6}}_{\underline{12}}1^{\underline{5}}_{\underline{12}}1^{\underline{4}}_{\underline{10}}1^{\underline{4}}_{\underline{8}}1^{\underline{3}}_{\underline{8}}1^{\underline{3}}_{\underline{6}}2^{\underline{2}}_{\underline{6}}1^{\underline{2}}_{\underline{4}}1^{\underline{1}}_{\underline{6}}2^{\underline{1}}_{\underline{2}}3^{0}_{\underline{2}}1^{0}_{0}1^{1}_{\underline{2}}1^{1}_{2}1^{2}_{0}1^{2}_{2}1^{3}_{4}1^{4}_{4}1^{5}_{8}$	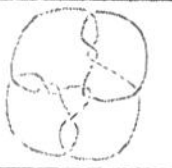	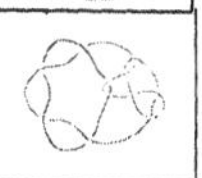
11^n_{167}: $1^{\underline{9}}_{\underline{24}}2^{\underline{8}}_{\underline{22}}1^{\underline{8}}_{\underline{20}}3^{\underline{7}}_{\underline{20}}2^{\underline{7}}_{\underline{18}}5^{\underline{6}}_{\underline{18}}4^{\underline{6}}_{\underline{16}}5^{\underline{5}}_{\underline{16}}4^{\underline{5}}_{\underline{14}}4^{\underline{4}}_{\underline{14}}5^{\underline{4}}_{\underline{12}}4^{\underline{3}}_{\underline{12}}4^{\underline{3}}_{\underline{10}}2^{\underline{2}}_{\underline{10}}4^{\underline{2}}_{\underline{8}}1^{\underline{1}}_{\underline{8}}2^{\underline{1}}_{\underline{6}}1^{0}_{\underline{6}}2^{0}_{\underline{4}}$ 11^n_{168}: $1^{\underline{8}}_{\underline{20}}1^{\underline{7}}_{\underline{18}}1^{\underline{7}}_{\underline{16}}3^{\underline{6}}_{\underline{16}}2^{\underline{6}}_{\underline{14}}4^{\underline{5}}_{\underline{14}}2^{\underline{5}}_{\underline{12}}3^{\underline{4}}_{\underline{12}}4^{\underline{4}}_{\underline{10}}4^{\underline{3}}_{\underline{10}}3^{\underline{3}}_{\underline{8}}3^{\underline{2}}_{\underline{8}}4^{\underline{2}}_{\underline{6}}1^{\underline{1}}_{\underline{6}}3^{\underline{1}}_{\underline{4}}2^{0}_{\underline{4}}2^{0}_{\underline{2}}1^{1}_{0}$	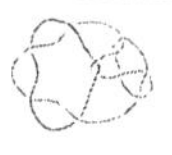	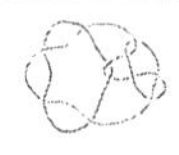
11^n_{169}: $1^{\underline{11}}_{\underline{30}}1^{\underline{10}}_{\underline{28}}1^{\underline{10}}_{\underline{26}}2^{\underline{9}}_{\underline{26}}1^{\underline{9}}_{\underline{24}}2^{\underline{8}}_{\underline{24}}2^{\underline{8}}_{\underline{22}}2^{\underline{7}}_{\underline{22}}2^{\underline{7}}_{\underline{20}}2^{\underline{6}}_{\underline{20}}3^{\underline{6}}_{\underline{18}}1^{\underline{5}}_{\underline{20}}2^{\underline{5}}_{\underline{18}}1^{\underline{5}}_{\underline{16}}1^{\underline{4}}_{\underline{16}}2^{\underline{4}}_{\underline{14}}1^{\underline{3}}_{\underline{16}}1^{\underline{2}}_{\underline{12}}1^{0}_{\underline{10}}1^{0}_{\underline{8}}$ 11^n_{170}: $1^{\underline{2}}_{\underline{4}}1^{\underline{2}}_{\underline{2}}3^{\underline{1}}_{\underline{2}}4^{0}_{0}4^{0}_{2}4^{1}_{2}3^{1}_{4}6^{2}_{4}4^{2}_{6}3^{3}_{6}6^{3}_{8}4^{4}_{8}3^{4}_{10}2^{5}_{10}4^{5}_{12}1^{6}_{12}2^{6}_{14}1^{7}_{16}$	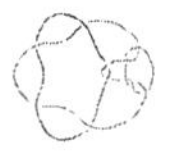	
11^n_{171}: $1^{\underline{4}}_{\underline{8}}3^{\underline{3}}_{\underline{6}}1^{\underline{3}}_{\underline{4}}5^{\underline{2}}_{\underline{4}}4^{\underline{2}}_{\underline{2}}6^{\underline{1}}_{\underline{2}}4^{\underline{1}}_{0}8^{0}_{0}7^{0}_{2}6^{1}_{2}7^{1}_{4}6^{2}_{4}6^{2}_{6}3^{3}_{6}6^{3}_{8}2^{4}_{8}3^{4}_{10}2^{5}_{12}$ 11^n_{172}: $1^{\underline{7}}_{\underline{16}}2^{\underline{6}}_{\underline{14}}1^{\underline{6}}_{\underline{12}}4^{\underline{5}}_{\underline{12}}2^{\underline{5}}_{\underline{10}}4^{\underline{4}}_{\underline{10}}4^{\underline{4}}_{\underline{8}}6^{\underline{3}}_{\underline{8}}4^{\underline{3}}_{\underline{6}}5^{\underline{2}}_{\underline{6}}6^{\underline{2}}_{\underline{4}}4^{\underline{1}}_{\underline{4}}5^{\underline{1}}_{\underline{2}}4^{0}_{\underline{2}}5^{0}_{0}1^{1}_{0}3^{1}_{2}1^{2}_{2}2^{2}_{4}$	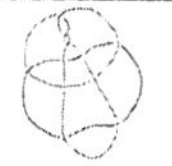	
11^n_{173}: $1^{\underline{4}}_{\underline{8}}3^{\underline{3}}_{\underline{6}}1^{\underline{3}}_{\underline{4}}5^{\underline{2}}_{\underline{4}}3^{\underline{2}}_{\underline{2}}6^{\underline{1}}_{\underline{2}}5^{\underline{1}}_{0}8^{0}_{0}7^{0}_{2}7^{1}_{2}7^{1}_{4}6^{2}_{4}8^{2}_{6}4^{3}_{6}5^{3}_{8}2^{4}_{8}4^{4}_{10}2^{5}_{12}$ 11^n_{174}: $2^{\underline{6}}_{\underline{18}}1^{\underline{6}}_{\underline{16}}2^{\underline{5}}_{\underline{16}}1^{\underline{5}}_{\underline{14}}2^{\underline{4}}_{\underline{14}}2^{\underline{4}}_{\underline{12}}3^{\underline{3}}_{\underline{12}}2^{\underline{3}}_{\underline{10}}2^{\underline{2}}_{\underline{10}}3^{\underline{2}}_{\underline{8}}2^{\underline{1}}_{\underline{8}}2^{\underline{1}}_{\underline{6}}2^{0}_{\underline{6}}3^{0}_{\underline{4}}1^{1}_{\underline{4}}1^{1}_{\underline{2}}1^{2}_{0}$		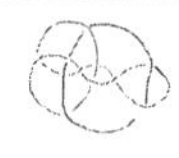
11^n_{175}: $1^{\underline{9}}_{\underline{22}}3^{\underline{8}}_{\underline{20}}1^{\underline{8}}_{\underline{18}}3^{\underline{7}}_{\underline{18}}3^{\underline{7}}_{\underline{16}}6^{\underline{6}}_{\underline{16}}4^{\underline{6}}_{\underline{14}}6^{\underline{5}}_{\underline{14}}5^{\underline{5}}_{\underline{12}}5^{\underline{4}}_{\underline{12}}6^{\underline{4}}_{\underline{10}}5^{\underline{3}}_{\underline{10}}5^{\underline{3}}_{\underline{8}}3^{\underline{2}}_{\underline{8}}5^{\underline{2}}_{\underline{6}}1^{\underline{1}}_{\underline{6}}3^{\underline{1}}_{\underline{4}}1^{0}_{\underline{4}}2^{0}_{\underline{2}}$ 11^n_{176}: $1^{\underline{8}}_{\underline{24}}1^{\underline{7}}_{\underline{22}}1^{\underline{7}}_{\underline{20}}1^{\underline{6}}_{\underline{20}}2^{\underline{6}}_{\underline{18}}1^{\underline{6}}_{\underline{16}}3^{\underline{5}}_{\underline{18}}2^{\underline{5}}_{\underline{16}}1^{\underline{4}}_{\underline{16}}3^{\underline{4}}_{\underline{14}}1^{\underline{4}}_{\underline{12}}2^{\underline{3}}_{\underline{14}}1^{\underline{3}}_{\underline{12}}1^{\underline{2}}_{\underline{12}}2^{\underline{2}}_{\underline{10}}1^{\underline{1}}_{\underline{8}}1^{0}_{\underline{8}}1^{0}_{\underline{6}}$		
11^n_{177}: $1^{\underline{3}}_{\underline{4}}3^{\underline{2}}_{\underline{2}}1^{\underline{2}}_{0}4^{\underline{1}}_{0}3^{\underline{1}}_{2}7^{0}_{2}5^{0}_{4}6^{1}_{4}6^{1}_{6}8^{2}_{6}7^{2}_{8}5^{3}_{8}7^{3}_{10}4^{4}_{10}5^{4}_{12}2^{5}_{12}4^{5}_{14}2^{6}_{16}$ 11^n_{178}: $1^{\underline{7}}_{\underline{16}}2^{\underline{6}}_{\underline{14}}1^{\underline{6}}_{\underline{12}}3^{\underline{5}}_{\underline{12}}2^{\underline{5}}_{\underline{10}}4^{\underline{4}}_{\underline{10}}3^{\underline{4}}_{\underline{8}}4^{\underline{3}}_{\underline{8}}4^{\underline{3}}_{\underline{6}}5^{\underline{2}}_{\underline{6}}5^{\underline{2}}_{\underline{4}}3^{\underline{1}}_{\underline{4}}4^{\underline{1}}_{\underline{2}}3^{0}_{\underline{2}}4^{0}_{0}1^{1}_{0}2^{1}_{2}1^{2}_{4}$		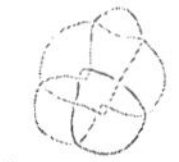
11^n_{179}: $1^{\underline{6}}_{\underline{14}}2^{\underline{5}}_{\underline{12}}1^{\underline{5}}_{\underline{10}}4^{\underline{4}}_{\underline{10}}2^{\underline{4}}_{\underline{8}}4^{\underline{3}}_{\underline{8}}4^{\underline{3}}_{\underline{6}}6^{\underline{2}}_{\underline{6}}5^{\underline{2}}_{\underline{4}}5^{\underline{1}}_{\underline{4}}5^{\underline{1}}_{\underline{2}}4^{0}_{\underline{2}}6^{0}_{0}3^{1}_{0}3^{1}_{2}1^{2}_{2}3^{2}_{4}1^{3}_{6}$ 11^n_{180}: $1^{\underline{6}}_{\underline{14}}1^{\underline{6}}_{\underline{12}}1^{\underline{4}}_{\underline{10}}1^{\underline{3}}_{\underline{10}}1^{\underline{3}}_{\underline{6}}1^{\underline{2}}_{\underline{8}}2^{\underline{2}}_{\underline{6}}1^{\underline{1}}_{\underline{6}}1^{\underline{1}}_{\underline{4}}1^{\underline{1}}_{\underline{2}}2^{0}_{\underline{4}}2^{0}_{\underline{2}}1^{0}_{0}1^{1}_{\underline{2}}2^{1}_{0}1^{2}_{0}1^{2}_{2}1^{3}_{2}1^{3}_{4}1^{4}_{6}$	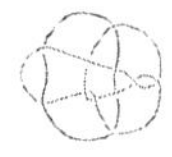	
11^n_{181}: $1^{\underline{6}}_{\underline{12}}1^{\underline{6}}_{\underline{10}}1^{\underline{4}}_{\underline{8}}1^{\underline{3}}_{\underline{8}}1^{\underline{3}}_{\underline{4}}2^{\underline{2}}_{\underline{6}}2^{\underline{2}}_{\underline{4}}1^{\underline{1}}_{\underline{4}}2^{\underline{1}}_{\underline{2}}1^{\underline{1}}_{0}3^{0}_{\underline{2}}2^{0}_{0}1^{1}_{0}2^{1}_{2}1^{2}_{2}1^{2}_{4}1^{3}_{4}1^{3}_{6}1^{4}_{8}$ 11^n_{182}: $1^{\underline{6}}_{\underline{14}}1^{\underline{6}}_{\underline{12}}3^{\underline{5}}_{\underline{12}}4^{\underline{4}}_{\underline{10}}3^{\underline{4}}_{\underline{8}}5^{\underline{3}}_{\underline{8}}4^{\underline{3}}_{\underline{6}}7^{\underline{2}}_{\underline{6}}5^{\underline{2}}_{\underline{4}}5^{\underline{1}}_{\underline{4}}7^{\underline{1}}_{\underline{2}}6^{0}_{\underline{2}}6^{0}_{0}3^{1}_{0}5^{1}_{2}2^{2}_{2}3^{2}_{4}2^{3}_{6}$		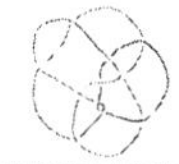

L: KH	1st line	2nd line
11^n_{183}: $1^{\underline{9}}_{\underline{22}}2^{\underline{8}}_{\underline{20}}1^{\underline{8}}_{\underline{18}}3^{\underline{7}}_{\underline{18}}2^{\underline{7}}_{\underline{16}}4^{\underline{6}}_{\underline{16}}3^{\underline{6}}_{\underline{14}}5^{\underline{5}}_{\underline{14}}4^{\underline{5}}_{\underline{12}}4^{\underline{4}}_{\underline{12}}5^{\underline{4}}_{\underline{10}}4^{\underline{3}}_{\underline{10}}4^{\underline{3}}_{\underline{8}}3^{\underline{2}}_{\underline{8}}5^{\underline{2}}_{\underline{6}}1^{\underline{1}}_{\underline{6}}2^{\underline{1}}_{\underline{4}}1^{0}_{\underline{4}}2^{0}_{\underline{2}}$ 11^n_{184}: $1^{\underline{8}}_{\underline{18}}1^{\underline{7}}_{\underline{16}}1^{\underline{7}}_{\underline{14}}2^{\underline{6}}_{\underline{14}}2^{\underline{6}}_{\underline{12}}3^{\underline{5}}_{\underline{12}}1^{\underline{5}}_{\underline{10}}2^{\underline{4}}_{\underline{10}}3^{\underline{4}}_{\underline{8}}2^{\underline{3}}_{\underline{8}}2^{\underline{3}}_{\underline{6}}3^{\underline{2}}_{\underline{6}}2^{\underline{2}}_{\underline{4}}3^{\underline{1}}_{\underline{2}}2^{0}_{\underline{2}}1^{0}_{0}1^{1}_{2}$		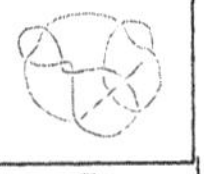
11^n_{185}: $1^{\underline{4}}_{\underline{10}}2^{\underline{3}}_{\underline{8}}1^{\underline{3}}_{\underline{6}}3^{\underline{2}}_{\underline{6}}3^{\underline{2}}_{\underline{4}}4^{\underline{1}}_{\underline{4}}2^{\underline{1}}_{\underline{2}}4^{0}_{\underline{2}}5^{0}_{0}3^{1}_{0}3^{1}_{2}3^{2}_{2}3^{2}_{4}1^{3}_{4}3^{3}_{6}1^{4}_{6}1^{4}_{8}1^{5}_{10}$ 11^n_{186}: $1^{\underline{5}}_{\underline{12}}2^{\underline{4}}_{\underline{10}}1^{\underline{4}}_{\underline{8}}3^{\underline{3}}_{\underline{8}}2^{\underline{3}}_{\underline{6}}5^{\underline{2}}_{\underline{6}}4^{\underline{2}}_{\underline{4}}4^{\underline{1}}_{\underline{4}}4^{\underline{1}}_{\underline{2}}5^{0}_{\underline{2}}5^{0}_{0}3^{1}_{0}4^{1}_{2}2^{2}_{2}3^{2}_{4}1^{3}_{4}2^{3}_{6}1^{4}_{8}$	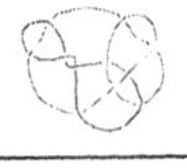	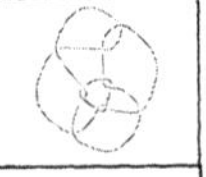
11^n_{187}: $2^{\underline{2}}_{\underline{2}}2^{\underline{1}}_{0}2^{\underline{1}}_{2}5^{0}_{2}3^{0}_{4}5^{1}_{4}4^{1}_{6}6^{2}_{6}6^{2}_{8}5^{3}_{8}5^{3}_{10}4^{4}_{10}5^{4}_{12}2^{5}_{12}4^{5}_{14}1^{6}_{14}2^{6}_{16}1^{7}_{18}$ 11^n_{188}: $2^{\underline{7}}_{\underline{18}}3^{\underline{6}}_{\underline{16}}2^{\underline{6}}_{\underline{14}}5^{\underline{5}}_{\underline{14}}3^{\underline{5}}_{\underline{12}}6^{\underline{4}}_{\underline{12}}5^{\underline{4}}_{\underline{10}}6^{\underline{3}}_{\underline{10}}6^{\underline{3}}_{\underline{8}}6^{\underline{2}}_{\underline{8}}7^{\underline{2}}_{\underline{6}}4^{\underline{1}}_{\underline{6}}5^{\underline{1}}_{\underline{4}}3^{0}_{\underline{4}}5^{0}_{\underline{2}}1^{1}_{\underline{2}}2^{1}_{0}1^{2}_{2}$	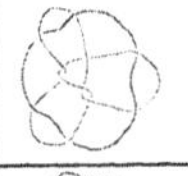	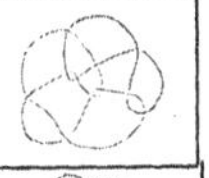
11^n_{189}: $1^{\underline{6}}_{\underline{12}}1^{\underline{6}}_{\underline{10}}1^{\underline{4}}_{\underline{8}}1^{\underline{3}}_{\underline{8}}1^{\underline{3}}_{\underline{4}}3^{\underline{2}}_{\underline{6}}2^{\underline{2}}_{\underline{4}}2^{\underline{1}}_{\underline{4}}3^{\underline{1}}_{\underline{2}}1^{\underline{1}}_{0}4^{0}_{\underline{2}}3^{0}_{0}3^{1}_{0}3^{1}_{2}2^{2}_{2}3^{2}_{4}2^{3}_{4}2^{3}_{6}1^{4}_{6}2^{4}_{8}1^{5}_{10}$ 11^n_{190}: $1^{\underline{9}}_{\underline{22}}2^{\underline{8}}_{\underline{20}}1^{\underline{8}}_{\underline{18}}4^{\underline{7}}_{\underline{18}}2^{\underline{7}}_{\underline{16}}5^{\underline{6}}_{\underline{16}}5^{\underline{6}}_{\underline{14}}6^{\underline{5}}_{\underline{14}}4^{\underline{5}}_{\underline{12}}6^{\underline{4}}_{\underline{12}}6^{\underline{4}}_{\underline{10}}4^{\underline{3}}_{\underline{10}}6^{\underline{3}}_{\underline{8}}4^{\underline{2}}_{\underline{8}}4^{\underline{2}}_{\underline{6}}1^{\underline{1}}_{\underline{6}}4^{\underline{1}}_{\underline{4}}1^{0}_{\underline{4}}2^{0}_{\underline{2}}$	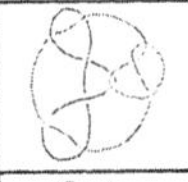	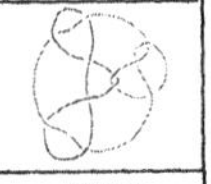
11^n_{191}: $1^{\underline{4}}_{\underline{6}}1^{\underline{3}}_{\underline{4}}1^{\underline{3}}_{\underline{2}}4^{\underline{2}}_{\underline{2}}2^{\underline{2}}_{0}2^{\underline{1}}_{0}3^{\underline{1}}_{2}5^{0}_{2}3^{0}_{4}3^{1}_{4}4^{1}_{6}3^{2}_{6}3^{2}_{8}2^{3}_{8}3^{3}_{10}1^{4}_{10}2^{4}_{12}1^{5}_{14}$ 11^n_{192}: $1^{\underline{5}}_{\underline{10}}2^{\underline{4}}_{\underline{8}}1^{\underline{4}}_{\underline{6}}3^{\underline{3}}_{\underline{6}}2^{\underline{3}}_{\underline{4}}4^{\underline{2}}_{\underline{4}}3^{\underline{2}}_{\underline{2}}4^{\underline{1}}_{\underline{2}}4^{\underline{1}}_{0}5^{0}_{0}5^{0}_{2}3^{1}_{2}4^{1}_{4}3^{2}_{4}4^{2}_{6}1^{3}_{6}2^{3}_{8}1^{4}_{10}$	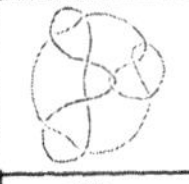	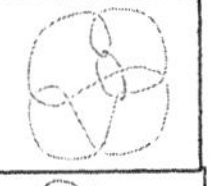
11^n_{193}: $1^{\underline{9}}_{\underline{22}}3^{\underline{8}}_{\underline{20}}1^{\underline{8}}_{\underline{18}}4^{\underline{7}}_{\underline{18}}3^{\underline{7}}_{\underline{16}}6^{\underline{6}}_{\underline{16}}5^{\underline{6}}_{\underline{14}}7^{\underline{5}}_{\underline{14}}5^{\underline{5}}_{\underline{12}}6^{\underline{4}}_{\underline{12}}7^{\underline{4}}_{\underline{10}}5^{\underline{3}}_{\underline{10}}6^{\underline{3}}_{\underline{8}}4^{\underline{2}}_{\underline{8}}5^{\underline{2}}_{\underline{6}}1^{\underline{1}}_{\underline{6}}4^{\underline{1}}_{\underline{4}}1^{0}_{\underline{4}}2^{0}_{\underline{2}}$ 11^n_{194}: $1^{\underline{6}}_{\underline{14}}2^{\underline{5}}_{\underline{12}}1^{\underline{5}}_{\underline{10}}4^{\underline{4}}_{\underline{10}}2^{\underline{4}}_{\underline{8}}5^{\underline{3}}_{\underline{8}}4^{\underline{3}}_{\underline{6}}6^{\underline{2}}_{\underline{6}}6^{\underline{2}}_{\underline{4}}6^{\underline{1}}_{\underline{4}}5^{\underline{1}}_{\underline{2}}5^{0}_{\underline{2}}7^{0}_{0}3^{1}_{0}4^{1}_{2}2^{2}_{2}3^{2}_{4}2^{3}_{6}$		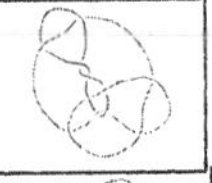
11^n_{195}: $1^{\underline{8}}_{\underline{20}}2^{\underline{7}}_{\underline{18}}1^{\underline{7}}_{\underline{16}}4^{\underline{6}}_{\underline{16}}3^{\underline{6}}_{\underline{14}}5^{\underline{5}}_{\underline{14}}3^{\underline{5}}_{\underline{12}}5^{\underline{4}}_{\underline{12}}5^{\underline{4}}_{\underline{10}}5^{\underline{3}}_{\underline{10}}5^{\underline{3}}_{\underline{8}}4^{\underline{2}}_{\underline{8}}5^{\underline{2}}_{\underline{6}}2^{\underline{1}}_{\underline{6}}4^{\underline{1}}_{\underline{4}}2^{0}_{\underline{4}}3^{0}_{\underline{2}}1^{1}_{0}$ 11^n_{196}: $2^{\underline{6}}_{\underline{14}}1^{\underline{6}}_{\underline{12}}3^{\underline{5}}_{\underline{12}}1^{\underline{5}}_{\underline{10}}5^{\underline{4}}_{\underline{10}}3^{\underline{4}}_{\underline{8}}5^{\underline{3}}_{\underline{8}}5^{\underline{3}}_{\underline{6}}7^{\underline{2}}_{\underline{6}}5^{\underline{2}}_{\underline{4}}5^{\underline{1}}_{\underline{4}}7^{\underline{1}}_{\underline{2}}5^{0}_{\underline{2}}6^{0}_{0}3^{1}_{0}4^{1}_{2}1^{2}_{2}3^{2}_{4}1^{3}_{6}$	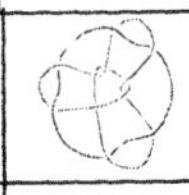	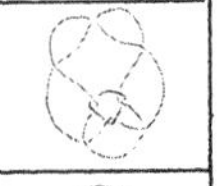
11^n_{197}: $1^{\underline{7}}_{\underline{16}}1^{\underline{6}}_{\underline{12}}1^{\underline{5}}_{\underline{12}}1^{\underline{4}}_{\underline{10}}1^{\underline{4}}_{\underline{8}}1^{\underline{3}}_{\underline{10}}1^{\underline{3}}_{\underline{6}}1^{\underline{2}}_{\underline{8}}3^{\underline{2}}_{\underline{6}}1^{\underline{2}}_{\underline{4}}1^{\underline{1}}_{\underline{6}}1^{\underline{1}}_{\underline{4}}1^{\underline{1}}_{\underline{2}}2^{0}_{\underline{4}}2^{0}_{\underline{2}}1^{0}_{0}1^{1}_{\underline{2}}2^{1}_{0}1^{2}_{0}1^{2}_{2}1^{3}_{2}1^{3}_{4}1^{4}_{6}$ 11^n_{198}: $1^{\underline{4}}_{\underline{10}}1^{\underline{4}}_{\underline{8}}1^{\underline{3}}_{\underline{8}}2^{\underline{2}}_{\underline{6}}1^{\underline{2}}_{\underline{4}}2^{\underline{1}}_{\underline{4}}2^{\underline{1}}_{\underline{2}}3^{0}_{\underline{2}}3^{0}_{0}2^{1}_{0}2^{1}_{2}2^{2}_{2}2^{2}_{4}1^{3}_{4}2^{3}_{6}1^{4}_{6}1^{4}_{8}1^{5}_{10}$	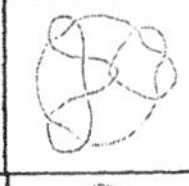	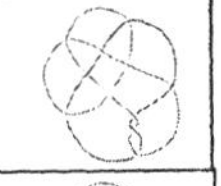
11^n_{199}: $1^{\underline{8}}_{\underline{22}}1^{\underline{8}}_{\underline{20}}1^{\underline{7}}_{\underline{22}}2^{\underline{6}}_{\underline{18}}2^{\underline{5}}_{\underline{18}}1^{\underline{5}}_{\underline{16}}1^{\underline{5}}_{\underline{14}}1^{\underline{4}}_{\underline{16}}2^{\underline{4}}_{\underline{14}}1^{\underline{4}}_{\underline{12}}1^{\underline{3}}_{\underline{14}}1^{\underline{3}}_{\underline{12}}1^{\underline{2}}_{\underline{12}}1^{\underline{2}}_{\underline{10}}1^{\underline{1}}_{\underline{8}}1^{0}_{\underline{8}}1^{0}_{\underline{6}}$ 11^n_{200}: $1^{\underline{8}}_{\underline{18}}1^{\underline{7}}_{\underline{16}}1^{\underline{7}}_{\underline{14}}2^{\underline{6}}_{\underline{14}}1^{\underline{6}}_{\underline{12}}3^{\underline{5}}_{\underline{12}}2^{\underline{5}}_{\underline{10}}3^{\underline{4}}_{\underline{10}}4^{\underline{4}}_{\underline{8}}3^{\underline{3}}_{\underline{8}}2^{\underline{3}}_{\underline{6}}3^{\underline{2}}_{\underline{6}}3^{\underline{2}}_{\underline{4}}1^{\underline{1}}_{\underline{4}}3^{\underline{1}}_{\underline{2}}2^{0}_{\underline{2}}2^{0}_{0}1^{1}_{2}$	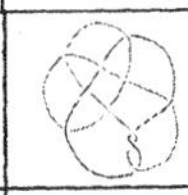	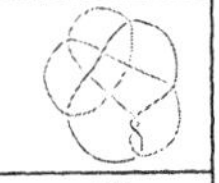
11^n_{201}: $1^{\underline{8}}_{\underline{18}}1^{\underline{8}}_{\underline{16}}1^{\underline{7}}_{\underline{16}}2^{\underline{6}}_{\underline{14}}1^{\underline{6}}_{\underline{12}}2^{\underline{5}}_{\underline{12}}2^{\underline{5}}_{\underline{10}}3^{\underline{4}}_{\underline{10}}2^{\underline{4}}_{\underline{8}}2^{\underline{3}}_{\underline{8}}3^{\underline{3}}_{\underline{6}}3^{\underline{2}}_{\underline{6}}2^{\underline{2}}_{\underline{4}}1^{\underline{1}}_{\underline{4}}3^{\underline{1}}_{\underline{2}}2^{0}_{\underline{2}}2^{0}_{0}1^{1}_{2}$ 11^n_{202}: $1^{\underline{6}}_{\underline{12}}1^{\underline{5}}_{\underline{10}}1^{\underline{5}}_{\underline{8}}1^{\underline{4}}_{\underline{8}}1^{\underline{4}}_{\underline{6}}2^{\underline{3}}_{\underline{6}}1^{\underline{3}}_{\underline{4}}2^{\underline{2}}_{\underline{4}}2^{\underline{2}}_{\underline{2}}1^{\underline{1}}_{\underline{2}}2^{\underline{1}}_{0}3^{0}_{0}3^{0}_{2}1^{1}_{2}1^{1}_{4}1^{2}_{4}1^{2}_{6}1^{3}_{8}$	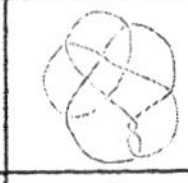	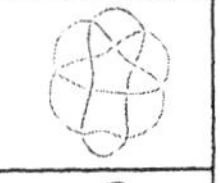
11^n_{203}: $1^{\underline{2}}_{\underline{4}}1^{\underline{1}}_{\underline{2}}1^{\underline{1}}_{0}4^{0}_{0}3^{0}_{2}4^{1}_{2}2^{1}_{4}3^{2}_{4}4^{2}_{6}3^{3}_{6}3^{3}_{8}3^{4}_{8}3^{4}_{10}1^{5}_{10}3^{5}_{12}1^{6}_{12}1^{6}_{14}1^{7}_{16}$ 11^n_{204}: $1^{\underline{8}}_{\underline{24}}1^{\underline{8}}_{\underline{22}}1^{\underline{7}}_{\underline{22}}1^{\underline{6}}_{\underline{18}}1^{\underline{5}}_{\underline{20}}1^{\underline{5}}_{\underline{18}}1^{\underline{4}}_{\underline{16}}1^{\underline{4}}_{\underline{14}}1^{\underline{3}}_{\underline{16}}1^{\underline{2}}_{\underline{12}}1^{0}_{\underline{10}}1^{0}_{\underline{8}}$	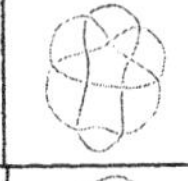	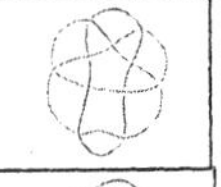
11^n_{205}: $1^{\underline{8}}_{\underline{18}}1^{\underline{8}}_{\underline{16}}1^{\underline{6}}_{\underline{14}}1^{\underline{5}}_{\underline{14}}1^{\underline{5}}_{\underline{10}}2^{\underline{4}}_{\underline{10}}1^{\underline{3}}_{\underline{10}}1^{\underline{3}}_{\underline{6}}1^{\underline{2}}_{\underline{8}}2^{\underline{2}}_{\underline{6}}1^{\underline{1}}_{\underline{4}}1^{\underline{1}}_{\underline{2}}1^{0}_{\underline{4}}1^{0}_{\underline{2}}$ 11^n_{206}: $1^{\underline{2}}_{\underline{4}}1^{\underline{1}}_{\underline{2}}1^{\underline{1}}_{0}3^{0}_{0}3^{0}_{2}3^{1}_{2}1^{1}_{4}3^{2}_{4}3^{2}_{6}2^{3}_{6}3^{3}_{8}2^{4}_{8}2^{4}_{10}1^{5}_{10}2^{5}_{12}1^{6}_{12}1^{6}_{14}1^{7}_{16}$	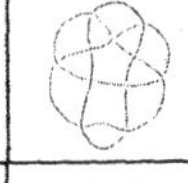	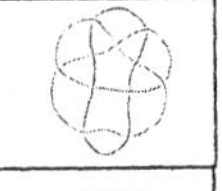
11^n_{207}: $1^{\underline{8}}_{\underline{20}}2^{\underline{7}}_{\underline{18}}1^{\underline{7}}_{\underline{16}}4^{\underline{6}}_{\underline{16}}2^{\underline{6}}_{\underline{14}}5^{\underline{5}}_{\underline{14}}4^{\underline{5}}_{\underline{12}}5^{\underline{4}}_{\underline{12}}6^{\underline{4}}_{\underline{10}}6^{\underline{3}}_{\underline{10}}4^{\underline{3}}_{\underline{8}}4^{\underline{2}}_{\underline{8}}6^{\underline{2}}_{\underline{6}}2^{\underline{1}}_{\underline{6}}4^{\underline{1}}_{\underline{4}}2^{0}_{\underline{4}}3^{0}_{\underline{2}}1^{1}_{0}$ 11^n_{208}: $1^{\underline{4}}_{\underline{12}}1^{\underline{4}}_{\underline{10}}1^{\underline{3}}_{\underline{10}}1^{\underline{2}}_{\underline{6}}1^{\underline{1}}_{\underline{6}}1^{0}_{\underline{4}}2^{0}_{\underline{2}}1^{1}_{\underline{2}}1^{2}_{0}1^{2}_{2}1^{3}_{4}1^{4}_{4}1^{5}_{8}$	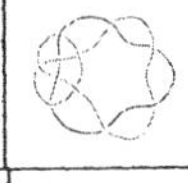	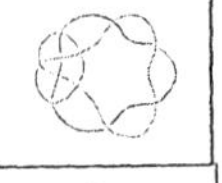
11^n_{209}: $1^{\underline{8}}_{\underline{20}}1^{\underline{7}}_{\underline{18}}1^{\underline{7}}_{\underline{16}}2^{\underline{6}}_{\underline{16}}1^{\underline{6}}_{\underline{14}}3^{\underline{5}}_{\underline{14}}2^{\underline{5}}_{\underline{12}}3^{\underline{4}}_{\underline{12}}4^{\underline{4}}_{\underline{10}}4^{\underline{3}}_{\underline{10}}2^{\underline{3}}_{\underline{8}}2^{\underline{2}}_{\underline{8}}4^{\underline{2}}_{\underline{6}}1^{\underline{1}}_{\underline{6}}2^{\underline{1}}_{\underline{4}}2^{0}_{\underline{4}}2^{0}_{\underline{2}}1^{1}_{0}$ 11^n_{210}: $1^{\underline{8}}_{\underline{20}}1^{\underline{8}}_{\underline{18}}1^{\underline{7}}_{\underline{18}}1^{\underline{6}}_{\underline{16}}1^{\underline{6}}_{\underline{14}}2^{\underline{5}}_{\underline{14}}1^{\underline{5}}_{\underline{12}}2^{\underline{4}}_{\underline{12}}2^{\underline{4}}_{\underline{10}}2^{\underline{3}}_{\underline{10}}2^{\underline{3}}_{\underline{8}}2^{\underline{2}}_{\underline{8}}2^{\underline{2}}_{\underline{6}}2^{\underline{1}}_{\underline{4}}2^{0}_{\underline{4}}1^{0}_{\underline{2}}1^{1}_{0}$	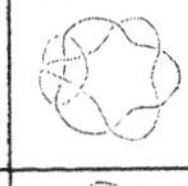	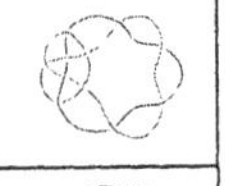
11^n_{211}: $1^{\underline{8}}_{\underline{20}}1^{\underline{8}}_{\underline{18}}2^{\underline{7}}_{\underline{18}}3^{\underline{6}}_{\underline{16}}2^{\underline{6}}_{\underline{14}}4^{\underline{5}}_{\underline{14}}3^{\underline{5}}_{\underline{12}}4^{\underline{4}}_{\underline{12}}4^{\underline{4}}_{\underline{10}}4^{\underline{3}}_{\underline{10}}4^{\underline{3}}_{\underline{8}}4^{\underline{2}}_{\underline{8}}4^{\underline{2}}_{\underline{6}}1^{\underline{1}}_{\underline{6}}4^{\underline{1}}_{\underline{4}}2^{0}_{\underline{4}}2^{0}_{\underline{2}}1^{1}_{0}$ 11^n_{212}: $1^{\underline{8}}_{\underline{20}}1^{\underline{8}}_{\underline{18}}2^{\underline{7}}_{\underline{18}}2^{\underline{6}}_{\underline{16}}2^{\underline{6}}_{\underline{14}}3^{\underline{5}}_{\underline{14}}2^{\underline{5}}_{\underline{12}}4^{\underline{4}}_{\underline{12}}3^{\underline{4}}_{\underline{10}}3^{\underline{3}}_{\underline{10}}4^{\underline{3}}_{\underline{8}}3^{\underline{2}}_{\underline{8}}3^{\underline{2}}_{\underline{6}}1^{\underline{1}}_{\underline{6}}3^{\underline{1}}_{\underline{4}}2^{0}_{\underline{4}}2^{0}_{\underline{2}}1^{1}_{0}$	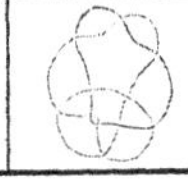	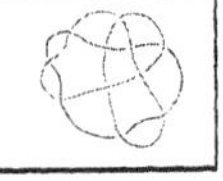

L: KH	1st line	2nd line
11^n_{213}: $1^{\underline{9}}_{\underline{24}}1^{\underline{8}}_{\underline{22}}1^{\underline{8}}_{\underline{20}}2^{\underline{7}}_{\underline{20}}1^{\underline{7}}_{\underline{18}}3^{\underline{6}}_{\underline{18}}2^{\underline{6}}_{\underline{16}}3^{\underline{5}}_{\underline{16}}3^{\underline{5}}_{\underline{14}}3^{\underline{4}}_{\underline{14}}4^{\underline{4}}_{\underline{12}}3^{\underline{3}}_{\underline{12}}2^{\underline{3}}_{\underline{10}}1^{\underline{2}}_{\underline{10}}3^{\underline{2}}_{\underline{8}}1^{\underline{1}}_{\underline{8}}1^{\underline{1}}_{\underline{6}}1^{0}_{\underline{6}}2^{0}_{\underline{4}}$ 11^n_{214}: $1^{\underline{9}}_{\underline{24}}2^{\underline{8}}_{\underline{22}}1^{\underline{8}}_{\underline{20}}4^{\underline{7}}_{\underline{20}}2^{\underline{7}}_{\underline{18}}5^{\underline{6}}_{\underline{18}}4^{\underline{6}}_{\underline{16}}5^{\underline{5}}_{\underline{16}}5^{\underline{5}}_{\underline{14}}5^{\underline{4}}_{\underline{14}}6^{\underline{4}}_{\underline{12}}5^{\underline{3}}_{\underline{12}}4^{\underline{3}}_{\underline{10}}2^{\underline{2}}_{\underline{10}}5^{\underline{2}}_{\underline{8}}1^{\underline{1}}_{\underline{8}}2^{\underline{1}}_{\underline{6}}1^{0}_{\underline{6}}2^{0}_{\underline{4}}$		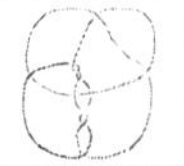
11^n_{215}: $1^{\underline{9}}_{\underline{22}}1^{\underline{8}}_{\underline{20}}1^{\underline{8}}_{\underline{18}}3^{\underline{7}}_{\underline{18}}1^{\underline{7}}_{\underline{16}}4^{\underline{6}}_{\underline{16}}3^{\underline{6}}_{\underline{14}}4^{\underline{5}}_{\underline{14}}4^{\underline{5}}_{\underline{12}}5^{\underline{4}}_{\underline{12}}5^{\underline{4}}_{\underline{10}}4^{\underline{3}}_{\underline{10}}4^{\underline{3}}_{\underline{8}}3^{\underline{2}}_{\underline{8}}4^{\underline{2}}_{\underline{6}}1^{\underline{1}}_{\underline{6}}3^{\underline{1}}_{\underline{4}}1^{0}_{\underline{4}}2^{0}_{\underline{2}}$ 11^n_{216}: $1^{\underline{3}}_{\underline{6}}2^{\underline{2}}_{\underline{4}}1^{\underline{2}}_{\underline{2}}2^{\underline{1}}_{\underline{2}}2^{\underline{1}}_{0}6^{0}_{0}4^{0}_{2}4^{1}_{2}4^{1}_{4}4^{2}_{4}4^{2}_{6}3^{3}_{6}4^{3}_{8}2^{4}_{8}3^{4}_{10}1^{5}_{10}2^{5}_{12}1^{6}_{14}$		
11^n_{217}: $1^{\underline{6}}_{\underline{14}}2^{\underline{5}}_{\underline{12}}1^{\underline{5}}_{\underline{10}}4^{\underline{4}}_{\underline{10}}2^{\underline{4}}_{\underline{8}}6^{\underline{3}}_{\underline{8}}4^{\underline{3}}_{\underline{6}}6^{\underline{2}}_{\underline{6}}6^{\underline{2}}_{\underline{4}}6^{\underline{1}}_{\underline{4}}6^{\underline{1}}_{\underline{2}}6^{0}_{\underline{2}}8^{0}_{0}4^{1}_{0}4^{1}_{2}2^{2}_{2}4^{2}_{4}2^{3}_{6}$ 11^n_{218}: $1^{\underline{2}}_{\underline{4}}1^{\underline{1}}_{0}2^{0}_{0}2^{0}_{2}1^{1}_{0}1^{1}_{2}2^{2}_{4}1^{2}_{6}1^{3}_{4}1^{3}_{8}1^{4}_{6}1^{4}_{8}1^{5}_{10}1^{6}_{10}1^{7}_{14}$	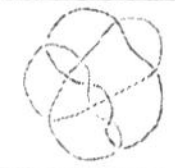	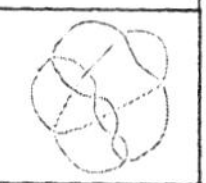
11^n_{219}: $1^{0}_{\underline{2}}2^{0}_{0}1^{0}_{2}1^{1}_{0}1^{2}_{4}1^{3}_{4}1^{4}_{6}1^{4}_{8}1^{5}_{10}1^{6}_{10}1^{7}_{14}$ 11^n_{220}: $1^{\underline{8}}_{\underline{24}}1^{\underline{8}}_{\underline{22}}2^{\underline{7}}_{\underline{22}}2^{\underline{6}}_{\underline{18}}2^{\underline{5}}_{\underline{18}}1^{\underline{4}}_{\underline{16}}2^{\underline{4}}_{\underline{14}}1^{\underline{3}}_{\underline{14}}1^{\underline{3}}_{\underline{12}}1^{\underline{2}}_{\underline{12}}1^{\underline{2}}_{\underline{10}}1^{\underline{1}}_{\underline{8}}1^{0}_{\underline{8}}1^{0}_{\underline{6}}$	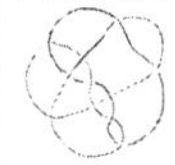	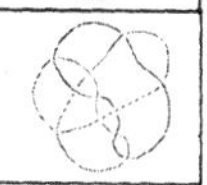
11^n_{221}: $2^{\underline{4}}_{\underline{10}}1^{\underline{4}}_{\underline{8}}4^{\underline{3}}_{\underline{8}}1^{\underline{3}}_{\underline{6}}5^{\underline{2}}_{\underline{6}}4^{\underline{2}}_{\underline{4}}5^{\underline{1}}_{\underline{4}}5^{\underline{1}}_{\underline{2}}7^{0}_{\underline{2}}6^{0}_{0}5^{1}_{0}6^{1}_{2}4^{2}_{2}5^{2}_{4}2^{3}_{4}4^{3}_{6}1^{4}_{6}2^{4}_{8}1^{5}_{10}$ 11^n_{222}: $1^{\underline{10}}_{\underline{28}}1^{\underline{9}}_{\underline{26}}1^{\underline{9}}_{\underline{24}}1^{\underline{8}}_{\underline{24}}2^{\underline{8}}_{\underline{22}}2^{\underline{7}}_{\underline{22}}1^{\underline{6}}_{\underline{20}}2^{\underline{6}}_{\underline{18}}1^{\underline{5}}_{\underline{20}}1^{\underline{5}}_{\underline{18}}1^{\underline{5}}_{\underline{16}}1^{\underline{4}}_{\underline{16}}1^{\underline{4}}_{\underline{14}}1^{\underline{3}}_{\underline{16}}1^{\underline{2}}_{\underline{12}}1^{0}_{\underline{10}}1^{0}_{\underline{8}}$		
11^n_{223}: $1^{\underline{7}}_{\underline{18}}1^{\underline{6}}_{\underline{16}}1^{\underline{6}}_{\underline{14}}3^{\underline{5}}_{\underline{14}}1^{\underline{5}}_{\underline{12}}4^{\underline{4}}_{\underline{12}}3^{\underline{4}}_{\underline{10}}4^{\underline{3}}_{\underline{10}}4^{\underline{3}}_{\underline{8}}4^{\underline{2}}_{\underline{8}}4^{\underline{2}}_{\underline{6}}3^{\underline{1}}_{\underline{6}}4^{\underline{1}}_{\underline{4}}3^{0}_{\underline{4}}5^{0}_{\underline{2}}2^{1}_{\underline{2}}1^{1}_{0}2^{2}_{2}$ 11^n_{224}: $1^{\underline{10}}_{\underline{26}}1^{\underline{9}}_{\underline{24}}1^{\underline{9}}_{\underline{22}}2^{\underline{8}}_{\underline{22}}2^{\underline{8}}_{\underline{20}}2^{\underline{7}}_{\underline{20}}1^{\underline{7}}_{\underline{18}}2^{\underline{6}}_{\underline{18}}2^{\underline{6}}_{\underline{16}}1^{\underline{5}}_{\underline{18}}2^{\underline{5}}_{\underline{16}}2^{\underline{5}}_{\underline{14}}2^{\underline{4}}_{\underline{14}}2^{\underline{4}}_{\underline{12}}1^{\underline{3}}_{\underline{14}}1^{\underline{3}}_{\underline{10}}1^{\underline{2}}_{\underline{10}}1^{0}_{\underline{8}}1^{0}_{\underline{6}}$		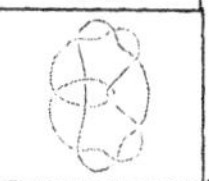
11^n_{225}: $1^{\underline{7}}_{\underline{18}}2^{\underline{6}}_{\underline{16}}1^{\underline{6}}_{\underline{14}}4^{\underline{5}}_{\underline{14}}2^{\underline{5}}_{\underline{12}}4^{\underline{4}}_{\underline{12}}4^{\underline{4}}_{\underline{10}}5^{\underline{3}}_{\underline{10}}4^{\underline{3}}_{\underline{8}}5^{\underline{2}}_{\underline{8}}5^{\underline{2}}_{\underline{6}}3^{\underline{1}}_{\underline{6}}5^{\underline{1}}_{\underline{4}}3^{0}_{\underline{4}}5^{0}_{\underline{2}}2^{1}_{\underline{2}}1^{1}_{0}2^{2}_{2}$ 11^n_{226}: $1^{\underline{7}}_{\underline{16}}1^{\underline{6}}_{\underline{12}}1^{\underline{5}}_{\underline{12}}1^{\underline{4}}_{\underline{12}}1^{\underline{4}}_{\underline{10}}1^{\underline{4}}_{\underline{8}}1^{\underline{3}}_{\underline{10}}2^{\underline{3}}_{\underline{8}}1^{\underline{3}}_{\underline{6}}1^{\underline{2}}_{\underline{8}}2^{\underline{2}}_{\underline{6}}1^{\underline{2}}_{\underline{4}}1^{\underline{1}}_{\underline{6}}1^{\underline{1}}_{\underline{4}}1^{\underline{1}}_{\underline{2}}1^{0}_{\underline{4}}3^{0}_{\underline{2}}1^{0}_{0}1^{1}_{\underline{2}}1^{2}_{2}$	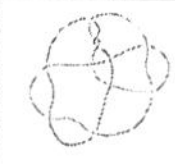	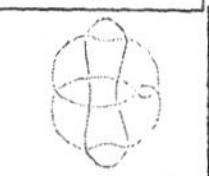
11^n_{227}: $1^{\underline{4}}_{\underline{6}}1^{\underline{3}}_{\underline{4}}1^{\underline{3}}_{\underline{2}}3^{\underline{2}}_{\underline{2}}1^{\underline{2}}_{0}2^{\underline{1}}_{0}3^{\underline{1}}_{2}5^{0}_{2}4^{0}_{4}3^{1}_{4}3^{1}_{6}3^{2}_{6}3^{2}_{8}2^{3}_{8}3^{3}_{10}1^{4}_{10}2^{4}_{12}1^{5}_{14}$ 11^n_{228}: $1^{\underline{6}}_{\underline{14}}1^{\underline{5}}_{\underline{12}}1^{\underline{5}}_{\underline{10}}2^{\underline{4}}_{\underline{10}}2^{\underline{4}}_{\underline{8}}2^{\underline{3}}_{\underline{8}}1^{\underline{3}}_{\underline{6}}2^{\underline{2}}_{\underline{6}}2^{\underline{2}}_{\underline{4}}1^{\underline{1}}_{\underline{6}}1^{\underline{1}}_{\underline{4}}2^{\underline{1}}_{\underline{2}}3^{0}_{\underline{2}}2^{0}_{0}1^{1}_{\underline{2}}1^{1}_{2}1^{2}_{0}1^{2}_{2}1^{3}_{4}1^{4}_{4}1^{5}_{8}$		
11^n_{229}: $2^{\underline{8}}_{\underline{24}}1^{\underline{8}}_{\underline{22}}2^{\underline{7}}_{\underline{22}}1^{\underline{7}}_{\underline{20}}1^{\underline{6}}_{\underline{20}}2^{\underline{6}}_{\underline{18}}3^{\underline{5}}_{\underline{18}}1^{\underline{5}}_{\underline{16}}1^{\underline{4}}_{\underline{16}}3^{\underline{4}}_{\underline{14}}2^{\underline{3}}_{\underline{14}}1^{\underline{3}}_{\underline{12}}1^{\underline{2}}_{\underline{12}}2^{\underline{2}}_{\underline{10}}1^{\underline{1}}_{\underline{8}}1^{0}_{\underline{8}}1^{0}_{\underline{6}}$ 11^n_{230}: $1^{\underline{6}}_{\underline{14}}1^{\underline{5}}_{\underline{12}}1^{\underline{5}}_{\underline{10}}4^{\underline{4}}_{\underline{10}}2^{\underline{4}}_{\underline{8}}5^{\underline{3}}_{\underline{8}}3^{\underline{3}}_{\underline{6}}5^{\underline{2}}_{\underline{6}}5^{\underline{2}}_{\underline{4}}5^{\underline{1}}_{\underline{4}}5^{\underline{1}}_{\underline{2}}5^{0}_{\underline{2}}6^{0}_{0}3^{1}_{0}4^{1}_{2}2^{2}_{2}3^{2}_{4}2^{3}_{6}$	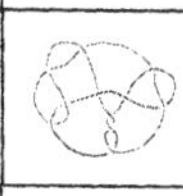	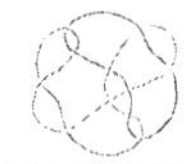
11^n_{231}: $1^{\underline{3}}_{\underline{4}}2^{\underline{2}}_{\underline{2}}1^{\underline{2}}_{0}3^{\underline{1}}_{0}2^{\underline{1}}_{2}5^{0}_{2}4^{0}_{4}4^{1}_{4}4^{1}_{6}5^{2}_{6}4^{2}_{8}3^{3}_{8}5^{3}_{10}3^{4}_{10}4^{4}_{12}1^{5}_{12}2^{5}_{14}1^{6}_{16}$ 11^n_{232}: $1^{\underline{2}}_{\underline{6}}2^{\underline{1}}_{\underline{4}}1^{\underline{1}}_{\underline{2}}2^{0}_{\underline{2}}3^{0}_{0}1^{0}_{2}3^{1}_{0}2^{1}_{2}2^{2}_{2}3^{2}_{4}2^{3}_{4}2^{3}_{6}2^{4}_{6}3^{4}_{8}1^{5}_{8}1^{5}_{10}1^{6}_{12}$	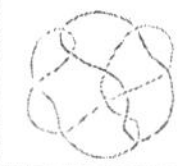	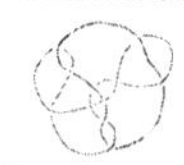
11^n_{233}: $1^{\underline{11}}_{\underline{30}}2^{\underline{10}}_{\underline{28}}1^{\underline{10}}_{\underline{26}}2^{\underline{9}}_{\underline{26}}2^{\underline{9}}_{\underline{24}}3^{\underline{8}}_{\underline{24}}3^{\underline{8}}_{\underline{22}}3^{\underline{7}}_{\underline{22}}2^{\underline{7}}_{\underline{20}}2^{\underline{6}}_{\underline{20}}3^{\underline{6}}_{\underline{18}}1^{\underline{5}}_{\underline{20}}2^{\underline{5}}_{\underline{18}}2^{\underline{5}}_{\underline{16}}1^{\underline{4}}_{\underline{16}}2^{\underline{4}}_{\underline{14}}1^{\underline{3}}_{\underline{16}}1^{\underline{2}}_{\underline{12}}1^{0}_{\underline{10}}1^{0}_{\underline{8}}$ 11^n_{234}: $2^{\underline{5}}_{\underline{12}}3^{\underline{4}}_{\underline{10}}2^{\underline{4}}_{\underline{8}}5^{\underline{3}}_{\underline{8}}3^{\underline{3}}_{\underline{6}}7^{\underline{2}}_{\underline{6}}5^{\underline{2}}_{\underline{4}}5^{\underline{1}}_{\underline{4}}7^{\underline{1}}_{\underline{2}}7^{0}_{\underline{2}}7^{0}_{0}5^{1}_{0}5^{1}_{2}2^{2}_{2}5^{2}_{4}1^{3}_{4}2^{3}_{6}1^{4}_{8}$	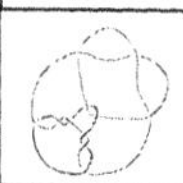	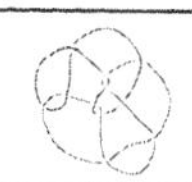
11^n_{235}: $1^{\underline{8}}_{\underline{20}}1^{\underline{8}}_{\underline{18}}1^{\underline{6}}_{\underline{16}}2^{\underline{5}}_{\underline{16}}1^{\underline{5}}_{\underline{12}}1^{\underline{4}}_{\underline{14}}3^{\underline{4}}_{\underline{12}}2^{\underline{3}}_{\underline{12}}1^{\underline{3}}_{\underline{10}}1^{\underline{3}}_{\underline{8}}2^{\underline{2}}_{\underline{10}}2^{\underline{2}}_{\underline{8}}1^{\underline{1}}_{\underline{8}}2^{\underline{1}}_{\underline{6}}2^{0}_{\underline{6}}2^{0}_{\underline{4}}1^{1}_{\underline{4}}1^{1}_{\underline{2}}1^{2}_{0}$ 11^n_{236}: $1^{\underline{10}}_{\underline{26}}1^{\underline{9}}_{\underline{22}}2^{\underline{8}}_{\underline{22}}1^{\underline{8}}_{\underline{20}}1^{\underline{7}}_{\underline{22}}1^{\underline{7}}_{\underline{20}}1^{\underline{7}}_{\underline{18}}2^{\underline{6}}_{\underline{18}}1^{\underline{6}}_{\underline{16}}2^{\underline{5}}_{\underline{18}}2^{\underline{5}}_{\underline{16}}1^{\underline{5}}_{\underline{14}}1^{\underline{4}}_{\underline{16}}2^{\underline{4}}_{\underline{14}}2^{\underline{4}}_{\underline{12}}1^{\underline{3}}_{\underline{14}}1^{\underline{3}}_{\underline{12}}1^{\underline{2}}_{\underline{12}}1^{\underline{2}}_{\underline{10}}1^{\underline{1}}_{\underline{8}}1^{0}_{\underline{8}}1^{0}_{\underline{6}}$	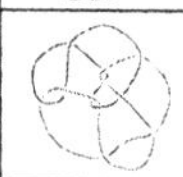	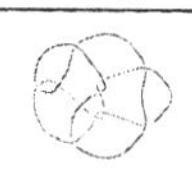
11^n_{237}: $1^{\underline{8}}_{\underline{20}}1^{\underline{8}}_{\underline{18}}1^{\underline{7}}_{\underline{18}}1^{\underline{6}}_{\underline{16}}1^{\underline{6}}_{\underline{14}}1^{\underline{5}}_{\underline{16}}1^{\underline{5}}_{\underline{14}}1^{\underline{5}}_{\underline{12}}1^{\underline{4}}_{\underline{14}}3^{\underline{4}}_{\underline{12}}1^{\underline{4}}_{\underline{10}}1^{\underline{3}}_{\underline{12}}1^{\underline{3}}_{\underline{10}}2^{\underline{3}}_{\underline{8}}1^{\underline{2}}_{\underline{10}}2^{\underline{2}}_{\underline{8}}1^{\underline{1}}_{\underline{8}}1^{\underline{1}}_{\underline{6}}1^{\underline{1}}_{\underline{4}}1^{0}_{\underline{6}}2^{0}_{\underline{4}}1^{1}_{\underline{4}}1^{2}_{0}$ 11^n_{238}: $3^{\underline{4}}_{\underline{10}}1^{\underline{4}}_{\underline{8}}4^{\underline{3}}_{\underline{8}}2^{\underline{3}}_{\underline{6}}6^{\underline{2}}_{\underline{6}}4^{\underline{2}}_{\underline{4}}6^{\underline{1}}_{\underline{4}}6^{\underline{1}}_{\underline{2}}7^{0}_{\underline{2}}7^{0}_{0}6^{1}_{0}6^{1}_{2}4^{2}_{2}6^{2}_{4}2^{3}_{4}4^{3}_{6}1^{4}_{6}2^{4}_{8}1^{5}_{10}$	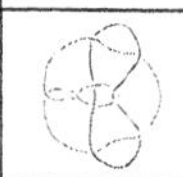	
11^n_{239}: $1^{\underline{6}}_{\underline{12}}1^{\underline{5}}_{\underline{8}}2^{\underline{4}}_{\underline{8}}1^{\underline{4}}_{\underline{6}}1^{\underline{3}}_{\underline{6}}1^{\underline{3}}_{\underline{4}}1^{\underline{2}}_{\underline{6}}1^{\underline{2}}_{\underline{4}}1^{\underline{2}}_{\underline{2}}1^{\underline{1}}_{\underline{4}}2^{\underline{1}}_{\underline{2}}1^{\underline{1}}_{0}2^{0}_{\underline{2}}3^{0}_{0}1^{0}_{2}2^{1}_{0}1^{1}_{2}1^{1}_{4}1^{2}_{2}2^{2}_{4}1^{3}_{4}1^{3}_{6}1^{4}_{6}1^{4}_{8}1^{5}_{10}$ 11^n_{240}: $1^{\underline{4}}_{\underline{6}}1^{\underline{3}}_{\underline{2}}1^{\underline{2}}_{\underline{2}}1^{\underline{1}}_{0}1^{\underline{1}}_{2}3^{0}_{0}2^{0}_{2}1^{0}_{4}2^{1}_{2}2^{1}_{4}2^{2}_{4}2^{2}_{6}1^{2}_{8}2^{3}_{6}2^{3}_{8}2^{4}_{8}2^{4}_{10}1^{5}_{10}2^{5}_{12}1^{6}_{12}1^{6}_{14}1^{7}_{16}$	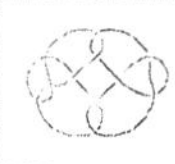	
11^n_{241}: $1^{\underline{4}}_{\underline{6}}1^{\underline{3}}_{\underline{4}}1^{\underline{3}}_{\underline{2}}4^{\underline{2}}_{\underline{2}}1^{\underline{2}}_{0}3^{\underline{1}}_{0}4^{\underline{1}}_{2}6^{0}_{2}5^{0}_{4}5^{1}_{4}4^{1}_{6}4^{2}_{6}5^{2}_{8}3^{3}_{8}4^{3}_{10}2^{4}_{10}3^{4}_{12}2^{5}_{14}$ 11^n_{242}: $1^{\underline{7}}_{\underline{16}}2^{\underline{6}}_{\underline{14}}1^{\underline{6}}_{\underline{12}}4^{\underline{5}}_{\underline{12}}2^{\underline{5}}_{\underline{10}}5^{\underline{4}}_{\underline{10}}4^{\underline{4}}_{\underline{8}}5^{\underline{3}}_{\underline{8}}5^{\underline{3}}_{\underline{6}}6^{\underline{2}}_{\underline{6}}5^{\underline{2}}_{\underline{4}}4^{\underline{1}}_{\underline{4}}6^{\underline{1}}_{\underline{2}}4^{0}_{\underline{2}}6^{0}_{0}2^{1}_{0}2^{1}_{2}2^{2}_{4}$	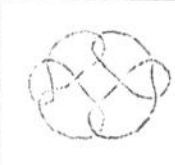	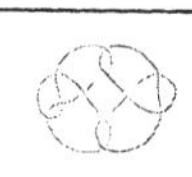

L: *KH*	1st line	2nd line
11^n_{243}: $1^{\underline{6}}_{\underline{12}}1^{\underline{5}}_{\underline{10}}1^{\underline{5}}_{\underline{8}}4^{\underline{4}}_{\underline{8}}2^{\underline{4}}_{\underline{6}}3^{\underline{3}}_{\underline{6}}3^{\underline{3}}_{\underline{4}}4^{\underline{2}}_{\underline{4}}3^{\underline{2}}_{\underline{2}}4^{\underline{1}}_{\underline{2}}4^{\underline{1}}_{0}1^{0}_{\underline{2}}4^{0}_{0}4^{0}_{2}2^{1}_{2}3^{1}_{4}1^{2}_{4}2^{2}_{6}1^{3}_{8}$ 11^n_{244}: $1^{\underline{6}}_{\underline{14}}1^{\underline{5}}_{\underline{12}}1^{\underline{5}}_{\underline{10}}1^{\underline{4}}_{\underline{10}}1^{\underline{4}}_{\underline{8}}1^{\underline{3}}_{\underline{10}}2^{\underline{3}}_{\underline{8}}1^{\underline{3}}_{\underline{6}}1^{\underline{2}}_{\underline{8}}2^{\underline{2}}_{\underline{6}}2^{\underline{2}}_{\underline{4}}1^{\underline{1}}_{\underline{6}}2^{\underline{1}}_{\underline{4}}1^{\underline{1}}_{\underline{2}}2^{0}_{\underline{4}}3^{0}_{\underline{2}}3^{0}_{0}1^{1}_{\underline{2}}2^{1}_{0}1^{2}_{0}1^{2}_{2}1^{3}_{2}1^{3}_{4}1^{4}_{6}$		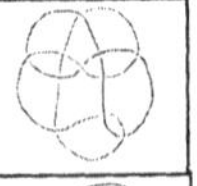
11^n_{245}: $2^{\underline{4}}_{\underline{10}}4^{\underline{3}}_{\underline{8}}2^{\underline{3}}_{\underline{6}}5^{\underline{2}}_{\underline{6}}4^{\underline{2}}_{\underline{4}}6^{\underline{1}}_{\underline{4}}5^{\underline{1}}_{\underline{2}}7^{0}_{\underline{2}}8^{0}_{0}6^{1}_{0}5^{1}_{2}4^{2}_{2}6^{2}_{4}2^{3}_{4}4^{3}_{6}1^{4}_{6}2^{4}_{8}1^{5}_{10}$ 11^n_{246}: $1^{\underline{7}}_{\underline{16}}3^{\underline{6}}_{\underline{14}}1^{\underline{6}}_{\underline{12}}4^{\underline{5}}_{\underline{12}}3^{\underline{5}}_{\underline{10}}6^{\underline{4}}_{\underline{10}}5^{\underline{4}}_{\underline{8}}6^{\underline{3}}_{\underline{8}}5^{\underline{3}}_{\underline{6}}6^{\underline{2}}_{\underline{6}}6^{\underline{2}}_{\underline{4}}4^{\underline{1}}_{\underline{4}}6^{\underline{1}}_{\underline{2}}4^{0}_{\underline{2}}5^{0}_{0}1^{1}_{0}3^{1}_{2}1^{2}_{4}$		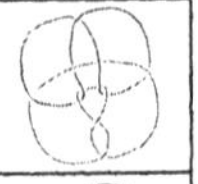
11^n_{247}: $1^{\underline{5}}_{\underline{10}}1^{\underline{4}}_{\underline{8}}1^{\underline{4}}_{\underline{6}}1^{\underline{3}}_{\underline{6}}1^{\underline{3}}_{\underline{4}}1^{\underline{2}}_{\underline{6}}2^{\underline{2}}_{\underline{4}}1^{\underline{2}}_{\underline{2}}1^{\underline{1}}_{\underline{4}}2^{\underline{1}}_{\underline{2}}2^{\underline{1}}_{0}2^{0}_{\underline{2}}4^{0}_{0}2^{0}_{2}2^{1}_{0}2^{1}_{2}1^{1}_{4}1^{2}_{2}2^{2}_{4}1^{2}_{6}1^{3}_{4}1^{3}_{6}1^{4}_{6}1^{4}_{8}1^{5}_{10}$ 11^n_{248}: $1^{\underline{4}}_{\underline{8}}3^{\underline{3}}_{\underline{6}}1^{\underline{3}}_{\underline{4}}6^{\underline{2}}_{\underline{4}}3^{\underline{2}}_{\underline{2}}6^{\underline{1}}_{\underline{2}}6^{\underline{1}}_{0}9^{0}_{0}8^{0}_{2}8^{1}_{2}7^{1}_{4}6^{2}_{4}8^{2}_{6}4^{3}_{6}6^{3}_{8}2^{4}_{8}4^{4}_{10}2^{5}_{12}$	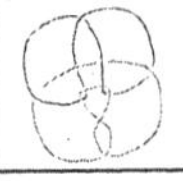	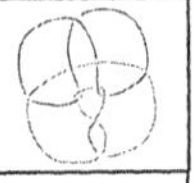
11^n_{249}: $3^{\underline{4}}_{\underline{10}}1^{\underline{4}}_{\underline{8}}5^{\underline{3}}_{\underline{8}}2^{\underline{3}}_{\underline{6}}7^{\underline{2}}_{\underline{6}}5^{\underline{2}}_{\underline{4}}7^{\underline{1}}_{\underline{4}}7^{\underline{1}}_{\underline{2}}9^{0}_{\underline{2}}8^{0}_{0}7^{1}_{0}8^{1}_{2}5^{2}_{2}7^{2}_{4}3^{3}_{4}5^{3}_{6}1^{4}_{6}3^{4}_{8}1^{5}_{10}$ 11^n_{250}: $1^{\underline{4}}_{\underline{6}}2^{\underline{3}}_{\underline{4}}1^{\underline{3}}_{\underline{2}}3^{\underline{2}}_{\underline{2}}2^{\underline{2}}_{0}4^{\underline{1}}_{0}3^{\underline{1}}_{2}5^{0}_{2}5^{0}_{4}4^{1}_{4}4^{1}_{6}4^{2}_{6}4^{2}_{8}2^{3}_{8}4^{3}_{10}2^{4}_{10}3^{4}_{12}1^{5}_{14}$	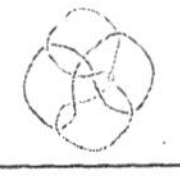	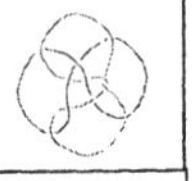
11^n_{251}: $1^{\underline{5}}_{\underline{12}}2^{\underline{4}}_{\underline{10}}1^{\underline{4}}_{\underline{8}}5^{\underline{3}}_{\underline{8}}2^{\underline{3}}_{\underline{6}}6^{\underline{2}}_{\underline{6}}5^{\underline{2}}_{\underline{4}}5^{\underline{1}}_{\underline{4}}6^{\underline{1}}_{\underline{2}}8^{0}_{\underline{2}}7^{0}_{0}5^{1}_{0}6^{1}_{2}3^{2}_{2}5^{2}_{4}2^{3}_{4}3^{3}_{6}2^{4}_{8}$ 11^n_{252}: $1^{\underline{10}}_{\underline{24}}1^{\underline{10}}_{\underline{22}}1^{\underline{8}}_{\underline{20}}1^{\underline{7}}_{\underline{20}}1^{\underline{7}}_{\underline{16}}1^{\underline{6}}_{\underline{18}}2^{\underline{6}}_{\underline{16}}2^{\underline{5}}_{\underline{16}}1^{\underline{5}}_{\underline{14}}1^{\underline{5}}_{\underline{12}}1^{\underline{4}}_{\underline{14}}2^{\underline{4}}_{\underline{12}}1^{\underline{3}}_{\underline{12}}1^{\underline{3}}_{\underline{10}}1^{\underline{2}}_{\underline{10}}1^{\underline{2}}_{\underline{8}}1^{\underline{1}}_{\underline{6}}1^{0}_{\underline{6}}1^{0}_{\underline{4}}$	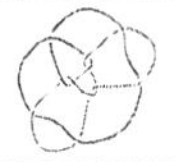	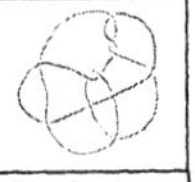
11^n_{253}: $1^{\underline{10}}_{\underline{26}}1^{\underline{10}}_{\underline{24}}1^{\underline{8}}_{\underline{22}}1^{\underline{7}}_{\underline{22}}1^{\underline{7}}_{\underline{18}}2^{\underline{6}}_{\underline{18}}2^{\underline{5}}_{\underline{18}}1^{\underline{5}}_{\underline{16}}1^{\underline{5}}_{\underline{14}}1^{\underline{4}}_{\underline{16}}2^{\underline{4}}_{\underline{14}}1^{\underline{4}}_{\underline{12}}1^{\underline{3}}_{\underline{14}}1^{\underline{3}}_{\underline{12}}1^{\underline{2}}_{\underline{12}}1^{\underline{2}}_{\underline{10}}1^{\underline{1}}_{\underline{8}}1^{0}_{\underline{8}}1^{0}_{\underline{6}}$ 11^n_{254}: $1^{\underline{10}}_{\underline{28}}1^{\underline{10}}_{\underline{26}}1^{\underline{9}}_{\underline{26}}1^{\underline{8}}_{\underline{24}}1^{\underline{8}}_{\underline{22}}1^{\underline{7}}_{\underline{22}}1^{\underline{7}}_{\underline{20}}1^{\underline{6}}_{\underline{20}}1^{\underline{6}}_{\underline{18}}1^{\underline{5}}_{\underline{20}}1^{\underline{5}}_{\underline{18}}1^{\underline{5}}_{\underline{16}}1^{\underline{4}}_{\underline{16}}1^{\underline{4}}_{\underline{14}}1^{\underline{3}}_{\underline{16}}1^{\underline{2}}_{\underline{12}}1^{0}_{\underline{10}}1^{0}_{\underline{8}}$		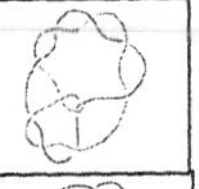
11^n_{255}: $3^{\underline{7}}_{\underline{19}}3^{\underline{6}}_{\underline{17}}3^{\underline{6}}_{\underline{15}}6^{\underline{5}}_{\underline{15}}3^{\underline{5}}_{\underline{13}}5^{\underline{4}}_{\underline{13}}7^{\underline{4}}_{\underline{11}}7^{\underline{3}}_{\underline{11}}4^{\underline{3}}_{\underline{9}}5^{\underline{2}}_{\underline{9}}7^{\underline{2}}_{\underline{7}}2^{\underline{1}}_{\underline{7}}5^{\underline{1}}_{\underline{5}}4^{0}_{\underline{5}}5^{0}_{\underline{3}}1^{1}_{\underline{3}}1^{1}_{\underline{1}}1^{2}_{1}$ 11^n_{256}: $1^{\underline{4}}_{\underline{7}}1^{\underline{4}}_{\underline{5}}1^{\underline{3}}_{\underline{5}}2^{\underline{2}}_{\underline{3}}1^{\underline{2}}_{\underline{1}}1^{\underline{1}}_{\underline{1}}2^{\underline{1}}_{1}1^{0}_{\underline{1}}5^{0}_{1}3^{0}_{3}1^{1}_{1}1^{1}_{3}2^{1}_{5}1^{2}_{3}4^{2}_{5}1^{2}_{7}1^{3}_{5}1^{3}_{7}3^{3}_{9}1^{4}_{7}1^{4}_{9}1^{5}_{11}1^{6}_{11}1^{7}_{15}$	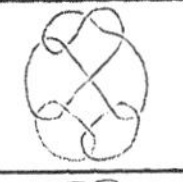	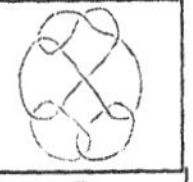
11^n_{257}: $1^{\underline{9}}_{\underline{21}}1^{\underline{8}}_{\underline{19}}1^{\underline{8}}_{\underline{17}}4^{\underline{7}}_{\underline{17}}1^{\underline{7}}_{\underline{15}}2^{\underline{6}}_{\underline{15}}4^{\underline{6}}_{\underline{13}}4^{\underline{5}}_{\underline{13}}2^{\underline{5}}_{\underline{11}}5^{\underline{4}}_{\underline{11}}5^{\underline{4}}_{\underline{9}}1^{\underline{3}}_{\underline{11}}3^{\underline{3}}_{\underline{9}}4^{\underline{3}}_{\underline{7}}4^{\underline{2}}_{\underline{7}}3^{\underline{2}}_{\underline{5}}3^{\underline{1}}_{\underline{3}}1^{0}_{\underline{5}}3^{0}_{\underline{3}}2^{0}_{\underline{1}}$ 11^n_{258}: $3^{\underline{5}}_{\underline{13}}3^{\underline{4}}_{\underline{11}}3^{\underline{4}}_{\underline{9}}7^{\underline{3}}_{\underline{9}}3^{\underline{3}}_{\underline{7}}6^{\underline{2}}_{\underline{7}}7^{\underline{2}}_{\underline{5}}6^{\underline{1}}_{\underline{5}}6^{\underline{1}}_{\underline{3}}7^{0}_{\underline{3}}9^{0}_{\underline{1}}5^{1}_{\underline{1}}4^{1}_{1}2^{2}_{1}5^{2}_{3}2^{3}_{5}1^{4}_{5}1^{4}_{7}$	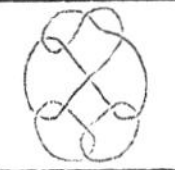	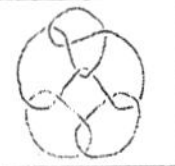
11^n_{259}: $1^{\underline{3}}_{\underline{5}}3^{\underline{2}}_{\underline{3}}1^{\underline{2}}_{\underline{1}}2^{\underline{1}}_{\underline{1}}3^{\underline{1}}_{1}6^{0}_{1}5^{0}_{3}4^{1}_{3}3^{1}_{5}4^{2}_{5}4^{2}_{7}2^{3}_{7}4^{3}_{9}2^{4}_{9}3^{4}_{11}1^{5}_{13}$ 11^n_{260}: $1^{\underline{4}}_{\underline{9}}3^{\underline{3}}_{\underline{7}}1^{\underline{3}}_{\underline{5}}4^{\underline{2}}_{\underline{5}}3^{\underline{2}}_{\underline{3}}4^{\underline{1}}_{\underline{3}}4^{\underline{1}}_{\underline{1}}6^{0}_{\underline{1}}7^{0}_{1}5^{1}_{1}3^{1}_{3}3^{2}_{3}5^{2}_{5}1^{3}_{5}3^{3}_{7}1^{4}_{7}2^{4}_{9}$	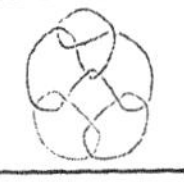	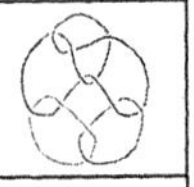
11^n_{261}: $1^{\underline{2}}_{\underline{1}}1^{\underline{1}}_{1}1^{\underline{1}}_{3}4^{0}_{3}4^{0}_{5}2^{1}_{5}1^{1}_{7}4^{2}_{7}2^{2}_{9}1^{3}_{7}1^{3}_{9}4^{3}_{11}1^{4}_{9}5^{4}_{11}1^{4}_{13}1^{5}_{11}3^{5}_{15}1^{6}_{13}1^{6}_{15}1^{7}_{17}1^{8}_{17}1^{9}_{21}$ 11^n_{262}: $1^{\underline{11}}_{\underline{27}}1^{\underline{10}}_{\underline{25}}1^{\underline{10}}_{\underline{23}}4^{\underline{9}}_{\underline{23}}1^{\underline{9}}_{\underline{21}}2^{\underline{8}}_{\underline{21}}4^{\underline{8}}_{\underline{19}}4^{\underline{7}}_{\underline{19}}2^{\underline{7}}_{\underline{17}}4^{\underline{6}}_{\underline{17}}4^{\underline{6}}_{\underline{15}}1^{\underline{5}}_{\underline{17}}1^{\underline{5}}_{\underline{15}}4^{\underline{5}}_{\underline{13}}6^{\underline{4}}_{\underline{13}}4^{\underline{4}}_{\underline{11}}1^{\underline{3}}_{\underline{13}}2^{\underline{3}}_{\underline{9}}1^{\underline{2}}_{\underline{9}}1^{0}_{\underline{7}}1^{0}_{\underline{5}}$	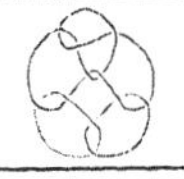	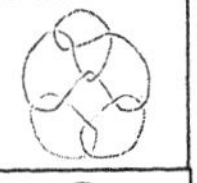
11^n_{263}: $1^{\underline{6}}_{\underline{15}}1^{\underline{5}}_{\underline{13}}1^{\underline{5}}_{\underline{11}}6^{\underline{4}}_{\underline{11}}4^{\underline{4}}_{\underline{9}}5^{\underline{3}}_{\underline{9}}3^{\underline{3}}_{\underline{7}}6^{\underline{2}}_{\underline{7}}5^{\underline{2}}_{\underline{5}}4^{\underline{1}}_{\underline{5}}6^{\underline{1}}_{\underline{3}}5^{0}_{\underline{3}}5^{0}_{\underline{1}}3^{1}_{\underline{1}}4^{1}_{1}1^{2}_{1}3^{2}_{3}1^{3}_{5}$ 11^n_{264}: $1^{\underline{6}}_{\underline{17}}2^{\underline{5}}_{\underline{15}}1^{\underline{5}}_{\underline{13}}4^{\underline{4}}_{\underline{13}}5^{\underline{4}}_{\underline{11}}1^{\underline{4}}_{\underline{9}}4^{\underline{3}}_{\underline{11}}2^{\underline{3}}_{\underline{9}}3^{\underline{2}}_{\underline{9}}4^{\underline{2}}_{\underline{7}}2^{\underline{1}}_{\underline{7}}3^{\underline{1}}_{\underline{5}}3^{0}_{\underline{5}}3^{0}_{\underline{3}}1^{1}_{\underline{3}}2^{1}_{\underline{1}}1^{2}_{1}$	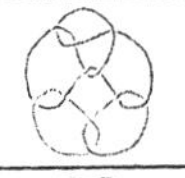	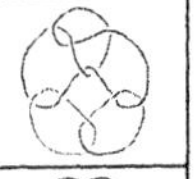
11^n_{265}: $3^{0}_{3}3^{0}_{5}4^{1}_{5}6^{2}_{7}4^{2}_{9}4^{3}_{9}6^{3}_{11}8^{4}_{11}5^{4}_{13}4^{5}_{13}7^{5}_{15}5^{6}_{15}4^{6}_{17}1^{7}_{17}5^{7}_{19}1^{8}_{19}1^{8}_{21}1^{9}_{23}$ 11^n_{266}: $1^{\underline{7}}_{\underline{17}}1^{\underline{6}}_{\underline{15}}1^{\underline{6}}_{\underline{13}}4^{\underline{5}}_{\underline{13}}1^{\underline{5}}_{\underline{11}}1^{\underline{4}}_{\underline{11}}4^{\underline{4}}_{\underline{9}}4^{\underline{3}}_{\underline{9}}1^{\underline{3}}_{\underline{7}}3^{\underline{2}}_{\underline{7}}4^{\underline{2}}_{\underline{5}}1^{\underline{1}}_{\underline{7}}1^{\underline{1}}_{\underline{5}}3^{\underline{1}}_{\underline{3}}4^{0}_{\underline{3}}4^{0}_{\underline{1}}1^{1}_{\underline{3}}1^{2}_{1}1^{4}_{3}1^{4}_{5}$	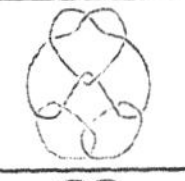	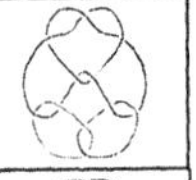
11^n_{267}: $1^{\underline{6}}_{\underline{11}}1^{\underline{5}}_{\underline{7}}4^{\underline{4}}_{\underline{7}}3^{\underline{4}}_{\underline{5}}1^{\underline{3}}_{\underline{5}}1^{\underline{3}}_{\underline{3}}3^{\underline{2}}_{\underline{3}}1^{\underline{2}}_{\underline{1}}1^{\underline{1}}_{\underline{1}}3^{\underline{1}}_{1}3^{0}_{\underline{1}}4^{0}_{1}1^{0}_{3}1^{1}_{1}2^{1}_{3}3^{1}_{5}1^{2}_{3}1^{2}_{5}1^{3}_{7}1^{4}_{7}1^{5}_{11}$ 11^n_{268}: $3^{0}_{3}3^{0}_{5}3^{1}_{5}4^{2}_{7}3^{2}_{9}3^{3}_{9}4^{3}_{11}7^{4}_{11}4^{4}_{13}2^{5}_{13}6^{5}_{15}4^{6}_{15}2^{6}_{17}1^{7}_{17}4^{7}_{19}1^{8}_{19}1^{8}_{21}1^{9}_{23}$	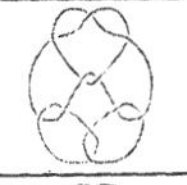	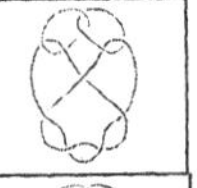
11^n_{269}: $1^{\underline{7}}_{\underline{17}}1^{\underline{6}}_{\underline{13}}2^{\underline{5}}_{\underline{13}}1^{\underline{4}}_{\underline{11}}2^{\underline{4}}_{\underline{9}}1^{\underline{3}}_{\underline{11}}3^{\underline{3}}_{\underline{9}}1^{\underline{3}}_{\underline{7}}2^{\underline{2}}_{\underline{7}}3^{\underline{2}}_{\underline{5}}1^{\underline{1}}_{\underline{7}}1^{\underline{1}}_{\underline{3}}4^{0}_{\underline{3}}3^{0}_{\underline{1}}1^{1}_{\underline{3}}1^{2}_{1}1^{4}_{3}1^{4}_{5}$ 11^n_{270}: $1^{\underline{6}}_{\underline{11}}1^{\underline{5}}_{\underline{7}}4^{\underline{4}}_{\underline{7}}3^{\underline{4}}_{\underline{5}}1^{\underline{3}}_{\underline{5}}1^{\underline{3}}_{\underline{3}}4^{\underline{2}}_{\underline{3}}1^{\underline{2}}_{\underline{1}}1^{\underline{1}}_{\underline{1}}4^{\underline{1}}_{1}1^{0}_{\underline{1}}4^{0}_{1}1^{0}_{3}1^{1}_{3}3^{1}_{5}1^{2}_{3}2^{2}_{5}1^{2}_{7}1^{3}_{7}2^{3}_{9}$	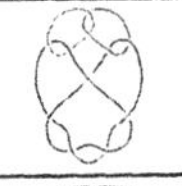	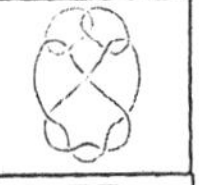
11^n_{271}: $3^{0}_{1}3^{0}_{3}5^{1}_{3}4^{2}_{5}5^{2}_{7}4^{3}_{7}4^{3}_{9}7^{4}_{9}5^{4}_{11}3^{5}_{11}6^{5}_{13}4^{6}_{13}3^{6}_{15}1^{7}_{15}4^{7}_{17}1^{8}_{17}1^{8}_{19}1^{9}_{21}$ 11^n_{272}: $1^{\underline{7}}_{\underline{15}}1^{\underline{6}}_{\underline{11}}2^{\underline{5}}_{\underline{11}}1^{\underline{4}}_{\underline{9}}2^{\underline{4}}_{\underline{7}}1^{\underline{3}}_{\underline{9}}1^{\underline{3}}_{\underline{7}}1^{\underline{3}}_{\underline{5}}3^{\underline{2}}_{\underline{5}}1^{\underline{2}}_{\underline{3}}2^{\underline{1}}_{\underline{5}}2^{\underline{1}}_{\underline{1}}1^{0}_{\underline{3}}5^{0}_{\underline{1}}3^{0}_{1}1^{1}_{\underline{1}}1^{1}_{1}1^{2}_{1}1^{2}_{3}1^{3}_{5}1^{4}_{5}1^{4}_{7}$	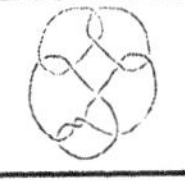	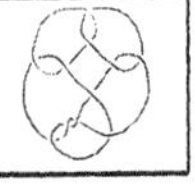

L: KH	1st line	2nd line
11^n_{273}: $1^{\underline{6}}_{\underline{13}}1^{\underline{5}}_{\underline{11}}1^{\underline{5}}_{\underline{9}}5^{\underline{4}}_{\underline{9}}4^{\underline{4}}_{\underline{7}}4^{\underline{3}}_{\underline{7}}2^{\underline{3}}_{\underline{5}}5^{\underline{2}}_{\underline{5}}4^{\underline{2}}_{\underline{3}}4^{\underline{1}}_{\underline{3}}5^{\underline{1}}_{\underline{1}}1^{0}_{\underline{3}}6^{0}_{\underline{1}}4^{0}_{1}1^{1}_{1}5^{1}_{3}2^{2}_{3}1^{2}_{5}2^{3}_{7}$ 11^n_{274}: $1^{\underline{6}}_{\underline{17}}2^{\underline{6}}_{\underline{15}}1^{\underline{6}}_{\underline{13}}2^{\underline{5}}_{\underline{15}}2^{\underline{4}}_{\underline{13}}4^{\underline{4}}_{\underline{11}}1^{\underline{4}}_{\underline{9}}3^{\underline{3}}_{\underline{11}}2^{\underline{3}}_{\underline{9}}1^{\underline{3}}_{\underline{7}}3^{\underline{2}}_{\underline{9}}4^{\underline{2}}_{\underline{7}}2^{\underline{1}}_{\underline{7}}3^{\underline{1}}_{\underline{5}}1^{\underline{1}}_{\underline{3}}3^{0}_{\underline{5}}3^{0}_{\underline{3}}1^{1}_{\underline{3}}2^{1}_{\underline{1}}1^{2}_{\underline{1}}1^{2}_{1}1^{3}_{3}$		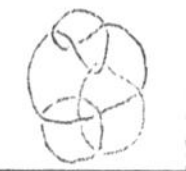
11^n_{275}: $1^{\underline{6}}_{\underline{13}}1^{\underline{6}}_{\underline{11}}2^{\underline{5}}_{\underline{11}}3^{\underline{4}}_{\underline{9}}2^{\underline{4}}_{\underline{7}}3^{\underline{3}}_{\underline{7}}3^{\underline{3}}_{\underline{5}}6^{\underline{2}}_{\underline{5}}4^{\underline{2}}_{\underline{3}}3^{\underline{1}}_{\underline{3}}5^{\underline{1}}_{\underline{1}}6^{0}_{\underline{1}}5^{0}_{1}2^{1}_{1}4^{1}_{3}2^{2}_{3}2^{2}_{5}2^{3}_{7}$ 11^n_{276}: $1^{\underline{9}}_{\underline{25}}1^{\underline{8}}_{\underline{21}}1^{\underline{7}}_{\underline{21}}2^{\underline{6}}_{\underline{19}}3^{\underline{6}}_{\underline{17}}1^{\underline{5}}_{\underline{19}}1^{\underline{5}}_{\underline{17}}2^{\underline{4}}_{\underline{15}}2^{\underline{4}}_{\underline{13}}1^{\underline{3}}_{\underline{15}}1^{\underline{2}}_{\underline{11}}1^{0}_{\underline{9}}1^{0}_{\underline{7}}$	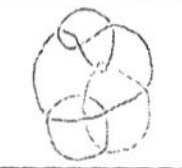	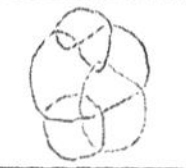
11^n_{277}: $1^{\underline{7}}_{\underline{19}}1^{\underline{6}}_{\underline{17}}2^{\underline{6}}_{\underline{15}}3^{\underline{5}}_{\underline{15}}1^{\underline{4}}_{\underline{13}}3^{\underline{4}}_{\underline{11}}2^{\underline{3}}_{\underline{11}}1^{\underline{3}}_{\underline{9}}2^{\underline{2}}_{\underline{9}}3^{\underline{2}}_{\underline{7}}1^{\underline{1}}_{\underline{5}}2^{0}_{\underline{5}}2^{0}_{\underline{3}}$ 11^n_{278}: $1^{\underline{4}}_{\underline{9}}1^{\underline{4}}_{\underline{7}}1^{\underline{2}}_{\underline{7}}3^{\underline{2}}_{\underline{5}}1^{\underline{2}}_{\underline{3}}1^{\underline{1}}_{\underline{5}}1^{\underline{1}}_{\underline{1}}1^{0}_{\underline{3}}3^{0}_{\underline{1}}1^{1}_{\underline{1}}1^{1}_{3}1^{2}_{1}1^{2}_{3}1^{3}_{5}1^{4}_{5}1^{5}_{9}$		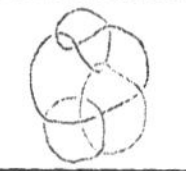
11^n_{279}: $2^{\underline{2}}_{\underline{3}}1^{\underline{2}}_{\underline{1}}1^{\underline{1}}_{\underline{1}}1^{\underline{1}}_{1}4^{0}_{1}3^{0}_{3}2^{1}_{3}2^{1}_{5}4^{2}_{5}3^{2}_{7}1^{3}_{7}3^{3}_{9}2^{4}_{9}1^{4}_{11}2^{5}_{13}$ 11^n_{280}: $1^{\underline{6}}_{\underline{13}}2^{\underline{5}}_{\underline{11}}1^{\underline{5}}_{\underline{9}}4^{\underline{4}}_{\underline{9}}4^{\underline{4}}_{\underline{7}}4^{\underline{3}}_{\underline{7}}2^{\underline{3}}_{\underline{5}}5^{\underline{2}}_{\underline{5}}4^{\underline{2}}_{\underline{3}}3^{\underline{1}}_{\underline{3}}5^{\underline{1}}_{\underline{1}}1^{0}_{\underline{3}}6^{0}_{\underline{1}}4^{0}_{1}1^{1}_{1}4^{1}_{3}1^{2}_{3}1^{2}_{5}1^{3}_{7}$	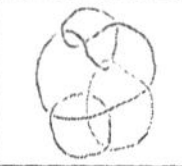	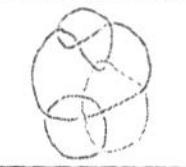
11^n_{281}: $1^{\underline{8}}_{\underline{21}}3^{\underline{7}}_{\underline{19}}1^{\underline{7}}_{\underline{17}}3^{\underline{6}}_{\underline{17}}3^{\underline{6}}_{\underline{15}}5^{\underline{5}}_{\underline{15}}3^{\underline{5}}_{\underline{13}}4^{\underline{4}}_{\underline{13}}7^{\underline{4}}_{\underline{11}}5^{\underline{3}}_{\underline{11}}2^{\underline{3}}_{\underline{9}}3^{\underline{2}}_{\underline{9}}5^{\underline{2}}_{\underline{7}}3^{\underline{1}}_{\underline{5}}2^{0}_{\underline{5}}2^{0}_{\underline{3}}$ 11^n_{282}: $1^{\underline{3}}_{\underline{5}}3^{\underline{2}}_{\underline{3}}1^{\underline{2}}_{\underline{1}}2^{\underline{1}}_{\underline{1}}3^{\underline{1}}_{1}7^{0}_{1}6^{0}_{3}4^{1}_{3}3^{1}_{5}5^{2}_{5}4^{2}_{7}2^{3}_{7}5^{3}_{9}2^{4}_{9}2^{4}_{11}2^{5}_{13}$		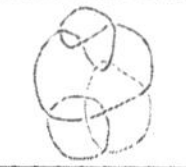
11^n_{283}: $1^{\underline{3}}_{\underline{7}}4^{\underline{2}}_{\underline{5}}3^{\underline{2}}_{\underline{3}}4^{\underline{1}}_{\underline{3}}2^{\underline{1}}_{\underline{1}}7^{0}_{\underline{1}}6^{0}_{1}5^{1}_{1}5^{1}_{3}6^{2}_{3}5^{2}_{5}3^{3}_{5}6^{3}_{7}3^{4}_{7}3^{4}_{9}1^{5}_{9}3^{5}_{11}1^{6}_{13}$ 11^n_{284}: $1^{\underline{3}}_{\underline{3}}2^{\underline{2}}_{\underline{1}}1^{\underline{2}}_{1}1^{\underline{1}}_{1}2^{\underline{1}}_{3}5^{0}_{3}3^{0}_{5}2^{1}_{5}3^{1}_{7}1^{2}_{5}5^{2}_{7}3^{2}_{9}2^{3}_{9}3^{3}_{11}1^{4}_{9}2^{4}_{11}2^{4}_{13}1^{5}_{13}2^{5}_{15}$	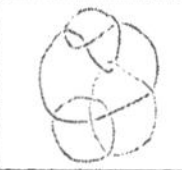	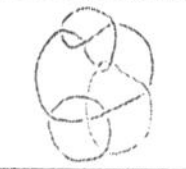
11^n_{285}: $1^{\underline{7}}_{\underline{19}}1^{\underline{6}}_{\underline{17}}1^{\underline{6}}_{\underline{15}}2^{\underline{5}}_{\underline{15}}1^{\underline{5}}_{\underline{13}}3^{\underline{4}}_{\underline{13}}3^{\underline{4}}_{\underline{11}}1^{\underline{4}}_{\underline{9}}2^{\underline{3}}_{\underline{11}}3^{\underline{3}}_{\underline{9}}3^{\underline{2}}_{\underline{9}}4^{\underline{2}}_{\underline{7}}1^{\underline{2}}_{\underline{5}}2^{\underline{1}}_{\underline{7}}2^{\underline{1}}_{\underline{5}}1^{0}_{\underline{5}}3^{0}_{\underline{3}}1^{1}_{\underline{3}}1^{2}_{1}$ 11^n_{286}: $1^{\underline{7}}_{\underline{15}}1^{\underline{6}}_{\underline{13}}1^{\underline{6}}_{\underline{11}}2^{\underline{5}}_{\underline{11}}1^{\underline{5}}_{\underline{9}}3^{\underline{4}}_{\underline{9}}2^{\underline{4}}_{\underline{7}}2^{\underline{3}}_{\underline{7}}3^{\underline{3}}_{\underline{5}}4^{\underline{2}}_{\underline{5}}3^{\underline{2}}_{\underline{3}}2^{\underline{1}}_{\underline{3}}3^{\underline{1}}_{\underline{1}}3^{0}_{\underline{1}}4^{0}_{1}1^{1}_{1}1^{1}_{3}1^{2}_{3}2^{2}_{5}$	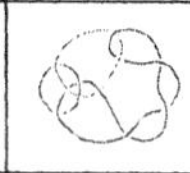	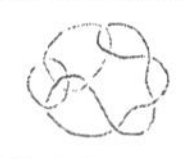
11^n_{287}: $1^{\underline{7}}_{\underline{15}}1^{\underline{6}}_{\underline{13}}1^{\underline{6}}_{\underline{11}}1^{\underline{5}}_{\underline{11}}1^{\underline{5}}_{\underline{9}}2^{\underline{4}}_{\underline{9}}1^{\underline{4}}_{\underline{7}}1^{\underline{3}}_{\underline{9}}2^{\underline{3}}_{\underline{7}}2^{\underline{3}}_{\underline{5}}1^{\underline{2}}_{\underline{7}}3^{\underline{2}}_{\underline{5}}2^{\underline{2}}_{\underline{3}}2^{\underline{1}}_{\underline{3}}2^{\underline{1}}_{\underline{1}}2^{0}_{\underline{3}}4^{0}_{\underline{1}}3^{0}_{1}1^{1}_{\underline{1}}1^{1}_{1}1^{2}_{3}1^{2}_{5}$ 11^n_{288}: $1^{\underline{7}}_{\underline{21}}1^{\underline{7}}_{\underline{19}}1^{\underline{6}}_{\underline{19}}1^{\underline{6}}_{\underline{17}}1^{\underline{6}}_{\underline{15}}2^{\underline{5}}_{\underline{17}}2^{\underline{5}}_{\underline{15}}2^{\underline{4}}_{\underline{15}}3^{\underline{4}}_{\underline{13}}2^{\underline{4}}_{\underline{11}}2^{\underline{3}}_{\underline{13}}2^{\underline{3}}_{\underline{11}}2^{\underline{2}}_{\underline{11}}4^{\underline{2}}_{\underline{9}}1^{\underline{2}}_{\underline{7}}2^{\underline{1}}_{\underline{9}}1^{\underline{1}}_{\underline{7}}1^{0}_{\underline{7}}3^{0}_{\underline{5}}1^{1}_{\underline{5}}1^{2}_{\underline{1}}$	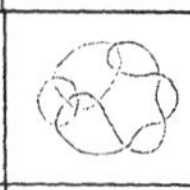	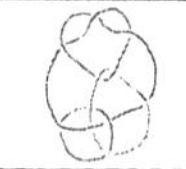
11^n_{289}: $1^{\underline{7}}_{\underline{17}}1^{\underline{6}}_{\underline{15}}1^{\underline{6}}_{\underline{13}}3^{\underline{5}}_{\underline{13}}1^{\underline{5}}_{\underline{11}}3^{\underline{4}}_{\underline{11}}3^{\underline{4}}_{\underline{9}}4^{\underline{3}}_{\underline{9}}3^{\underline{3}}_{\underline{7}}4^{\underline{2}}_{\underline{7}}5^{\underline{2}}_{\underline{5}}3^{\underline{1}}_{\underline{5}}3^{\underline{1}}_{\underline{3}}3^{0}_{\underline{3}}5^{0}_{\underline{1}}1^{1}_{\underline{1}}1^{1}_{1}1^{2}_{1}2^{2}_{3}$ 11^n_{290}: $1^{\underline{7}}_{\underline{17}}2^{\underline{6}}_{\underline{15}}1^{\underline{6}}_{\underline{13}}4^{\underline{5}}_{\underline{13}}2^{\underline{5}}_{\underline{11}}3^{\underline{4}}_{\underline{11}}4^{\underline{4}}_{\underline{9}}5^{\underline{3}}_{\underline{9}}3^{\underline{3}}_{\underline{7}}4^{\underline{2}}_{\underline{7}}5^{\underline{2}}_{\underline{5}}2^{\underline{1}}_{\underline{5}}4^{\underline{1}}_{\underline{3}}4^{0}_{\underline{3}}6^{0}_{\underline{1}}1^{1}_{\underline{1}}1^{2}_{3}$		
11^n_{291}: $1^{\underline{5}}_{\underline{13}}3^{\underline{4}}_{\underline{11}}2^{\underline{4}}_{\underline{9}}3^{\underline{3}}_{\underline{9}}2^{\underline{3}}_{\underline{7}}6^{\underline{2}}_{\underline{7}}5^{\underline{2}}_{\underline{5}}4^{\underline{1}}_{\underline{5}}4^{\underline{1}}_{\underline{3}}5^{0}_{\underline{3}}5^{0}_{\underline{1}}3^{1}_{\underline{1}}4^{1}_{1}2^{2}_{1}3^{2}_{3}1^{3}_{3}2^{3}_{5}1^{4}_{7}$ 11^n_{292}: $1^{\underline{5}}_{\underline{9}}2^{\underline{4}}_{\underline{7}}1^{\underline{4}}_{\underline{5}}2^{\underline{3}}_{\underline{5}}2^{\underline{3}}_{\underline{3}}5^{\underline{2}}_{\underline{3}}3^{\underline{2}}_{\underline{1}}3^{\underline{1}}_{\underline{1}}4^{\underline{1}}_{1}1^{0}_{\underline{1}}6^{0}_{1}4^{0}_{3}3^{1}_{3}4^{1}_{5}1^{2}_{3}3^{2}_{5}3^{2}_{7}1^{3}_{7}2^{3}_{9}1^{4}_{11}$	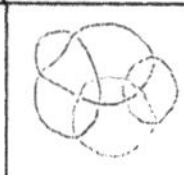	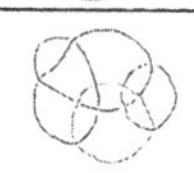
11^n_{293}: $1^{\underline{5}}_{\underline{9}}1^{\underline{4}}_{\underline{7}}1^{\underline{4}}_{\underline{5}}1^{\underline{3}}_{\underline{5}}1^{\underline{3}}_{\underline{3}}1^{\underline{2}}_{\underline{5}}2^{\underline{2}}_{\underline{3}}1^{\underline{2}}_{\underline{1}}1^{\underline{1}}_{\underline{3}}2^{\underline{1}}_{\underline{1}}2^{\underline{1}}_{1}5^{0}_{\underline{1}}6^{0}_{1}2^{0}_{3}3^{1}_{1}3^{1}_{3}1^{1}_{5}3^{2}_{3}3^{2}_{5}1^{2}_{7}2^{3}_{5}3^{3}_{7}2^{4}_{7}2^{4}_{9}1^{5}_{9}2^{5}_{11}1^{6}_{13}$ 11^n_{294}: $1^{\underline{2}}_{1}1^{\underline{1}}_{5}3^{0}_{5}2^{0}_{7}1^{1}_{7}1^{1}_{9}1^{2}_{7}3^{2}_{9}1^{2}_{11}1^{3}_{11}2^{3}_{13}2^{4}_{11}2^{4}_{13}1^{4}_{15}2^{5}_{15}2^{5}_{17}1^{6}_{15}1^{6}_{17}$	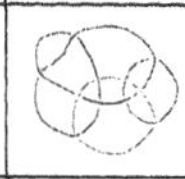	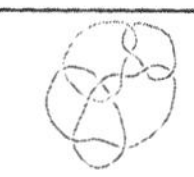
11^n_{295}: $2^{\underline{2}}_{\underline{3}}1^{\underline{2}}_{\underline{1}}2^{\underline{1}}_{\underline{1}}1^{\underline{1}}_{1}6^{0}_{1}4^{0}_{3}4^{1}_{3}4^{1}_{5}7^{2}_{5}5^{2}_{7}4^{3}_{7}6^{3}_{9}4^{4}_{9}4^{4}_{11}2^{5}_{11}4^{5}_{13}1^{6}_{13}2^{6}_{15}1^{7}_{17}$ 11^n_{296}: $1^{\underline{3}}_{\underline{7}}1^{\underline{2}}_{\underline{5}}1^{\underline{2}}_{\underline{3}}3^{\underline{1}}_{\underline{3}}1^{\underline{1}}_{\underline{1}}3^{0}_{\underline{1}}5^{0}_{1}4^{1}_{1}1^{1}_{3}3^{2}_{3}5^{2}_{5}2^{3}_{5}2^{3}_{7}2^{4}_{7}2^{4}_{9}2^{5}_{11}1^{6}_{11}1^{6}_{13}$	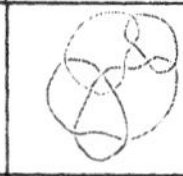	
11^n_{297}: $1^{\underline{7}}_{\underline{15}}1^{\underline{6}}_{\underline{13}}1^{\underline{6}}_{\underline{11}}3^{\underline{5}}_{\underline{11}}1^{\underline{5}}_{\underline{9}}2^{\underline{4}}_{\underline{9}}3^{\underline{4}}_{\underline{7}}3^{\underline{3}}_{\underline{7}}2^{\underline{3}}_{\underline{5}}4^{\underline{2}}_{\underline{5}}4^{\underline{2}}_{\underline{3}}1^{\underline{1}}_{\underline{3}}3^{\underline{1}}_{\underline{1}}4^{0}_{\underline{1}}3^{0}_{1}2^{1}_{3}1^{2}_{3}1^{2}_{5}$ 11^n_{298}: $2^{\underline{3}}_{\underline{7}}5^{\underline{2}}_{\underline{5}}3^{\underline{2}}_{\underline{3}}5^{\underline{1}}_{\underline{3}}4^{\underline{1}}_{\underline{1}}9^{0}_{\underline{1}}7^{0}_{1}7^{1}_{1}7^{1}_{3}7^{2}_{3}8^{2}_{5}5^{3}_{5}6^{3}_{7}3^{4}_{7}5^{4}_{9}1^{5}_{9}3^{5}_{11}1^{6}_{13}$		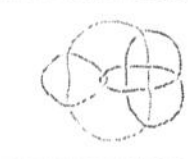
11^n_{299}: $1^{\underline{7}}_{\underline{15}}2^{\underline{6}}_{\underline{13}}1^{\underline{6}}_{\underline{11}}1^{\underline{5}}_{\underline{11}}2^{\underline{5}}_{\underline{9}}4^{\underline{4}}_{\underline{9}}2^{\underline{4}}_{\underline{7}}1^{\underline{3}}_{\underline{9}}2^{\underline{3}}_{\underline{7}}3^{\underline{3}}_{\underline{5}}2^{\underline{2}}_{\underline{7}}5^{\underline{2}}_{\underline{5}}3^{\underline{2}}_{\underline{3}}1^{\underline{1}}_{\underline{5}}3^{\underline{1}}_{\underline{3}}2^{\underline{1}}_{\underline{1}}3^{0}_{\underline{3}}3^{0}_{\underline{1}}2^{0}_{1}1^{1}_{\underline{1}}2^{1}_{1}1^{1}_{3}1^{2}_{1}1^{2}_{3}1^{3}_{3}1^{3}_{5}1^{4}_{7}$ 11^n_{300}: $2^{\underline{4}}_{\underline{11}}1^{\underline{4}}_{\underline{9}}2^{\underline{3}}_{\underline{9}}1^{\underline{3}}_{\underline{7}}4^{\underline{2}}_{\underline{7}}4^{\underline{2}}_{\underline{5}}3^{\underline{1}}_{\underline{5}}2^{\underline{1}}_{\underline{3}}4^{0}_{\underline{3}}4^{0}_{\underline{1}}2^{1}_{\underline{1}}3^{1}_{1}2^{2}_{1}2^{2}_{3}1^{3}_{3}2^{3}_{5}1^{4}_{7}$	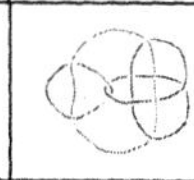	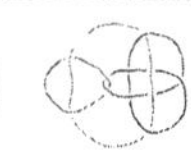
11^n_{301}: $3^{0}_{3}2^{0}_{5}3^{1}_{5}1^{1}_{7}6^{2}_{7}4^{2}_{9}5^{3}_{9}5^{3}_{11}7^{4}_{11}5^{4}_{13}4^{5}_{13}7^{5}_{15}5^{6}_{15}5^{6}_{17}2^{7}_{17}4^{7}_{19}1^{8}_{19}2^{8}_{21}1^{9}_{23}$ 11^n_{302}: $1^{\underline{5}}_{\underline{9}}1^{\underline{4}}_{\underline{5}}1^{\underline{3}}_{\underline{5}}1^{\underline{2}}_{\underline{3}}1^{\underline{2}}_{\underline{1}}1^{\underline{1}}_{\underline{3}}1^{\underline{1}}_{1}1^{0}_{\underline{1}}4^{0}_{1}1^{0}_{3}2^{1}_{1}1^{1}_{5}1^{2}_{3}3^{2}_{5}1^{2}_{7}1^{3}_{5}1^{3}_{7}1^{4}_{7}1^{4}_{9}1^{5}_{11}1^{6}_{11}1^{6}_{13}$	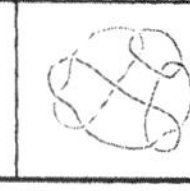	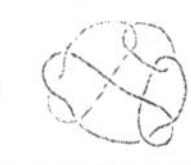

L: KH	1st line	2nd line
11^n_{303}: $1^{\underline{4}}_{\underline{5}}1^{\underline{4}}_{\underline{3}}1^{\underline{2}}_{\underline{1}}1^{\underline{1}}_{3}3^{0}_{1}2^{0}_{3}1^{1}_{3}2^{1}_{5}1^{1}_{7}4^{2}_{5}3^{2}_{7}2^{3}_{7}2^{3}_{9}2^{4}_{9}2^{4}_{11}1^{5}_{11}2^{5}_{13}1^{6}_{13}1^{6}_{15}1^{7}_{17}$ 11^n_{304}: $1^{\underline{7}}_{\underline{15}}1^{\underline{6}}_{\underline{13}}1^{\underline{6}}_{\underline{11}}3^{\underline{5}}_{\underline{11}}1^{\underline{5}}_{\underline{9}}3^{\underline{4}}_{\underline{9}}4^{\underline{4}}_{\underline{7}}4^{\underline{3}}_{\underline{7}}2^{\underline{3}}_{\underline{5}}4^{\underline{2}}_{\underline{5}}4^{\underline{2}}_{\underline{3}}2^{\underline{1}}_{\underline{3}}4^{\underline{1}}_{\underline{1}}4^{0}_{\underline{1}}3^{0}_{1}3^{1}_{3}2^{2}_{3}2^{2}_{5}$	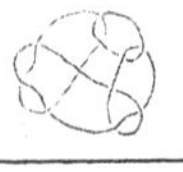	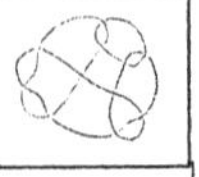
11^n_{305}: $1^{\underline{8}}_{\underline{19}}1^{\underline{7}}_{\underline{17}}1^{\underline{7}}_{\underline{15}}4^{\underline{6}}_{\underline{15}}3^{\underline{6}}_{\underline{13}}4^{\underline{5}}_{\underline{13}}2^{\underline{5}}_{\underline{11}}5^{\underline{4}}_{\underline{11}}5^{\underline{4}}_{\underline{9}}4^{\underline{3}}_{\underline{9}}4^{\underline{3}}_{\underline{7}}5^{\underline{2}}_{\underline{7}}4^{\underline{2}}_{\underline{5}}1^{\underline{1}}_{\underline{5}}5^{\underline{1}}_{\underline{3}}3^{0}_{\underline{3}}2^{0}_{\underline{1}}2^{1}_{1}$ 11^n_{306}: $1^{\underline{7}}_{\underline{21}}1^{\underline{6}}_{\underline{19}}2^{\underline{6}}_{\underline{17}}1^{\underline{6}}_{\underline{15}}3^{\underline{5}}_{\underline{17}}2^{\underline{5}}_{\underline{15}}2^{\underline{4}}_{\underline{15}}3^{\underline{4}}_{\underline{13}}1^{\underline{4}}_{\underline{11}}3^{\underline{3}}_{\underline{13}}2^{\underline{3}}_{\underline{11}}2^{\underline{2}}_{\underline{11}}4^{\underline{2}}_{\underline{9}}2^{\underline{1}}_{\underline{9}}1^{\underline{1}}_{\underline{7}}2^{0}_{\underline{7}}4^{0}_{\underline{5}}1^{1}_{\underline{5}}1^{2}_{\underline{1}}$	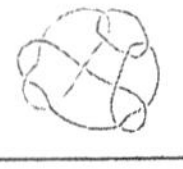	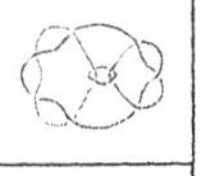
11^n_{307}: $2^{0}_{1}2^{0}_{3}4^{1}_{3}4^{2}_{5}5^{2}_{7}5^{3}_{7}3^{3}_{9}6^{4}_{9}5^{4}_{11}3^{5}_{11}6^{5}_{13}5^{6}_{13}4^{6}_{15}2^{7}_{15}4^{7}_{17}1^{8}_{17}2^{8}_{19}1^{9}_{21}$ 11^n_{308}: $1^{\underline{4}}_{\underline{7}}1^{\underline{4}}_{\underline{5}}2^{\underline{3}}_{\underline{5}}2^{\underline{2}}_{\underline{3}}2^{\underline{2}}_{\underline{1}}2^{\underline{1}}_{\underline{1}}2^{\underline{1}}_{1}1^{0}_{\underline{1}}5^{0}_{1}2^{0}_{3}1^{1}_{1}1^{1}_{3}4^{1}_{5}1^{2}_{3}5^{2}_{5}2^{2}_{7}1^{3}_{5}1^{3}_{7}2^{3}_{9}1^{4}_{7}1^{4}_{9}1^{4}_{11}1^{5}_{11}1^{6}_{11}1^{7}_{15}$	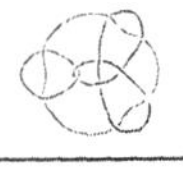	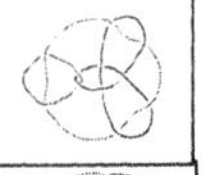
11^n_{309}: $1^{\underline{6}}_{\underline{13}}2^{\underline{5}}_{\underline{11}}1^{\underline{5}}_{\underline{9}}4^{\underline{4}}_{\underline{9}}3^{\underline{4}}_{\underline{7}}5^{\underline{3}}_{\underline{7}}3^{\underline{3}}_{\underline{5}}5^{\underline{2}}_{\underline{5}}5^{\underline{2}}_{\underline{3}}5^{\underline{1}}_{\underline{3}}5^{\underline{1}}_{\underline{1}}5^{0}_{\underline{1}}6^{0}_{1}2^{1}_{1}4^{1}_{3}3^{2}_{3}4^{2}_{5}1^{3}_{7}$ 11^n_{310}: $3^{\underline{2}}_{\underline{3}}1^{\underline{2}}_{\underline{1}}3^{\underline{1}}_{\underline{1}}2^{\underline{1}}_{1}7^{0}_{1}5^{0}_{3}6^{1}_{3}5^{1}_{5}8^{2}_{5}7^{2}_{7}5^{3}_{7}7^{3}_{9}5^{4}_{9}5^{4}_{11}2^{5}_{11}5^{5}_{13}1^{6}_{13}2^{6}_{15}1^{7}_{17}$	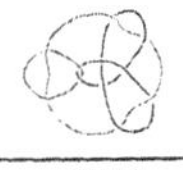	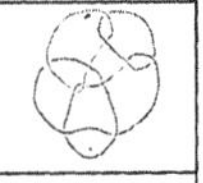
11^n_{311}: $1^{\underline{6}}_{\underline{11}}1^{\underline{5}}_{\underline{7}}2^{\underline{4}}_{\underline{7}}1^{\underline{4}}_{\underline{5}}1^{\underline{3}}_{\underline{5}}1^{\underline{3}}_{\underline{3}}3^{\underline{2}}_{\underline{5}}3^{\underline{2}}_{\underline{3}}1^{\underline{2}}_{\underline{1}}1^{\underline{1}}_{\underline{3}}2^{\underline{1}}_{\underline{1}}1^{\underline{1}}_{1}4^{0}_{\underline{1}}3^{0}_{1}1^{0}_{3}2^{1}_{1}3^{1}_{3}1^{1}_{5}2^{2}_{3}2^{2}_{5}1^{3}_{5}2^{3}_{7}1^{4}_{7}1^{4}_{9}1^{5}_{11}$ 11^n_{312}: $2^{\underline{7}}_{\underline{19}}3^{\underline{6}}_{\underline{17}}3^{\underline{6}}_{\underline{15}}6^{\underline{5}}_{\underline{15}}2^{\underline{5}}_{\underline{13}}5^{\underline{4}}_{\underline{13}}6^{\underline{4}}_{\underline{11}}6^{\underline{3}}_{\underline{11}}5^{\underline{3}}_{\underline{9}}6^{\underline{2}}_{\underline{9}}7^{\underline{2}}_{\underline{7}}3^{\underline{1}}_{\underline{7}}5^{\underline{1}}_{\underline{5}}4^{0}_{\underline{5}}5^{0}_{\underline{3}}1^{1}_{\underline{3}}2^{1}_{\underline{1}}1^{2}_{1}$	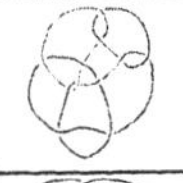	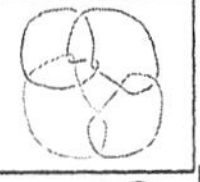
11^n_{313}: $1^{\underline{11}}_{\underline{27}}1^{\underline{10}}_{\underline{25}}1^{\underline{10}}_{\underline{23}}2^{\underline{9}}_{\underline{23}}1^{\underline{9}}_{\underline{21}}2^{\underline{8}}_{\underline{21}}2^{\underline{8}}_{\underline{19}}1^{\underline{7}}_{\underline{21}}2^{\underline{7}}_{\underline{19}}2^{\underline{7}}_{\underline{17}}1^{\underline{6}}_{\underline{19}}5^{\underline{6}}_{\underline{17}}3^{\underline{6}}_{\underline{15}}2^{\underline{5}}_{\underline{17}}1^{\underline{5}}_{\underline{15}}2^{\underline{5}}_{\underline{13}}1^{\underline{4}}_{\underline{15}}4^{\underline{4}}_{\underline{13}}2^{\underline{4}}_{\underline{11}}1^{\underline{3}}_{\underline{13}}1^{\underline{3}}_{\underline{11}}1^{\underline{3}}_{\underline{9}}1^{\underline{2}}_{\underline{11}}1^{\underline{2}}_{\underline{9}}1^{\underline{1}}_{\underline{7}}1^{0}_{\underline{7}}1^{0}_{\underline{5}}$ 11^n_{314}: $2^{\underline{7}}_{\underline{19}}3^{\underline{6}}_{\underline{17}}2^{\underline{6}}_{\underline{15}}4^{\underline{5}}_{\underline{15}}3^{\underline{5}}_{\underline{13}}6^{\underline{4}}_{\underline{13}}5^{\underline{4}}_{\underline{11}}5^{\underline{3}}_{\underline{11}}5^{\underline{3}}_{\underline{9}}5^{\underline{2}}_{\underline{9}}7^{\underline{2}}_{\underline{7}}4^{\underline{1}}_{\underline{7}}3^{\underline{1}}_{\underline{5}}2^{0}_{\underline{5}}5^{0}_{\underline{3}}1^{1}_{\underline{3}}1^{1}_{\underline{1}}1^{2}_{1}$	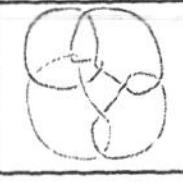	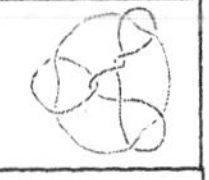
11^n_{315}: $1^{\underline{7}}_{\underline{15}}1^{\underline{6}}_{\underline{13}}1^{\underline{6}}_{\underline{11}}1^{\underline{5}}_{\underline{11}}1^{\underline{5}}_{\underline{9}}2^{\underline{4}}_{\underline{9}}1^{\underline{4}}_{\underline{7}}1^{\underline{3}}_{\underline{9}}1^{\underline{3}}_{\underline{7}}2^{\underline{3}}_{\underline{5}}1^{\underline{2}}_{\underline{7}}3^{\underline{2}}_{\underline{5}}2^{\underline{2}}_{\underline{3}}1^{\underline{1}}_{\underline{5}}2^{\underline{1}}_{\underline{3}}1^{\underline{1}}_{\underline{1}}2^{0}_{\underline{3}}3^{0}_{\underline{1}}2^{0}_{1}1^{1}_{\underline{1}}1^{1}_{1}1^{2}_{1}2^{2}_{3}1^{3}_{3}1^{4}_{7}$ 11^n_{316}: $3^{0}_{1}2^{0}_{3}3^{1}_{3}1^{1}_{5}4^{2}_{5}4^{2}_{7}4^{3}_{7}3^{3}_{9}4^{4}_{9}4^{4}_{11}2^{5}_{11}4^{5}_{13}3^{6}_{13}3^{6}_{15}1^{7}_{15}2^{7}_{17}1^{8}_{19}$		
11^n_{317}: $2^{\underline{6}}_{\underline{15}}1^{\underline{6}}_{\underline{13}}4^{\underline{5}}_{\underline{13}}1^{\underline{5}}_{\underline{11}}5^{\underline{4}}_{\underline{11}}4^{\underline{4}}_{\underline{9}}6^{\underline{3}}_{\underline{9}}5^{\underline{3}}_{\underline{7}}8^{\underline{2}}_{\underline{7}}7^{\underline{2}}_{\underline{5}}5^{\underline{1}}_{\underline{5}}7^{\underline{1}}_{\underline{3}}6^{0}_{\underline{3}}7^{0}_{\underline{1}}3^{1}_{\underline{1}}4^{1}_{1}1^{2}_{1}3^{2}_{3}1^{3}_{5}$ 11^n_{318}: $1^{\underline{9}}_{\underline{23}}3^{\underline{8}}_{\underline{21}}1^{\underline{8}}_{\underline{19}}5^{\underline{7}}_{\underline{19}}3^{\underline{7}}_{\underline{17}}6^{\underline{6}}_{\underline{17}}6^{\underline{6}}_{\underline{15}}9^{\underline{5}}_{\underline{15}}5^{\underline{5}}_{\underline{13}}6^{\underline{4}}_{\underline{13}}9^{\underline{4}}_{\underline{11}}6^{\underline{3}}_{\underline{11}}6^{\underline{3}}_{\underline{9}}5^{\underline{2}}_{\underline{9}}7^{\underline{2}}_{\underline{7}}1^{\underline{1}}_{\underline{7}}4^{\underline{1}}_{\underline{5}}2^{0}_{\underline{5}}3^{0}_{\underline{3}}$	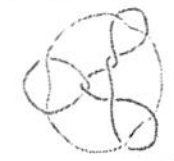	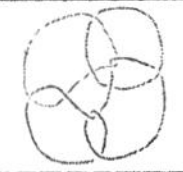
11^n_{319}: $1^{\underline{4}}_{\underline{7}}2^{\underline{3}}_{\underline{5}}1^{\underline{3}}_{\underline{3}}4^{\underline{2}}_{\underline{3}}3^{\underline{2}}_{\underline{1}}4^{\underline{1}}_{\underline{1}}3^{\underline{1}}_{1}7^{0}_{1}6^{0}_{3}4^{1}_{3}5^{1}_{5}6^{2}_{5}5^{2}_{7}2^{3}_{7}5^{3}_{9}2^{4}_{9}2^{4}_{11}2^{5}_{13}$ 11^n_{320}: $1^{\underline{6}}_{\underline{17}}1^{\underline{6}}_{\underline{15}}1^{\underline{5}}_{\underline{15}}1^{\underline{5}}_{\underline{13}}1^{\underline{5}}_{\underline{11}}2^{\underline{4}}_{\underline{13}}4^{\underline{4}}_{\underline{11}}2^{\underline{4}}_{\underline{9}}3^{\underline{3}}_{\underline{11}}2^{\underline{3}}_{\underline{9}}2^{\underline{2}}_{\underline{9}}4^{\underline{2}}_{\underline{7}}1^{\underline{2}}_{\underline{5}}2^{\underline{1}}_{\underline{7}}2^{\underline{1}}_{\underline{5}}1^{\underline{1}}_{\underline{3}}2^{0}_{\underline{5}}3^{0}_{\underline{3}}1^{1}_{\underline{3}}1^{1}_{\underline{1}}1^{2}_{\underline{1}}1^{2}_{1}1^{3}_{3}$	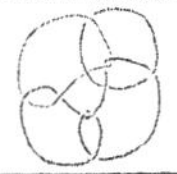	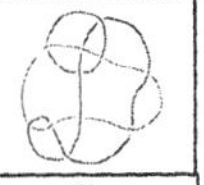
11^n_{321}: $1^{\underline{6}}_{\underline{13}}1^{\underline{5}}_{\underline{11}}1^{\underline{5}}_{\underline{9}}3^{\underline{4}}_{\underline{9}}2^{\underline{4}}_{\underline{7}}4^{\underline{3}}_{\underline{7}}2^{\underline{3}}_{\underline{5}}5^{\underline{2}}_{\underline{5}}4^{\underline{2}}_{\underline{3}}3^{\underline{1}}_{\underline{3}}5^{\underline{1}}_{\underline{1}}6^{0}_{\underline{1}}6^{0}_{1}3^{1}_{1}3^{1}_{3}2^{2}_{3}3^{2}_{5}2^{3}_{7}$ 11^n_{322}: $1^{\underline{6}}_{\underline{13}}2^{\underline{5}}_{\underline{11}}1^{\underline{5}}_{\underline{9}}4^{\underline{4}}_{\underline{9}}2^{\underline{4}}_{\underline{7}}4^{\underline{3}}_{\underline{7}}4^{\underline{3}}_{\underline{5}}7^{\underline{2}}_{\underline{5}}6^{\underline{2}}_{\underline{3}}5^{\underline{1}}_{\underline{3}}5^{\underline{1}}_{\underline{1}}6^{0}_{\underline{1}}7^{0}_{1}3^{1}_{1}4^{1}_{3}2^{2}_{3}3^{2}_{5}2^{3}_{7}$	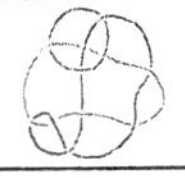	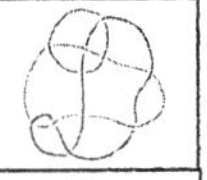
11^n_{323}: $1^{\underline{7}}_{\underline{21}}1^{\underline{7}}_{\underline{19}}1^{\underline{6}}_{\underline{19}}2^{\underline{6}}_{\underline{17}}1^{\underline{6}}_{\underline{15}}1^{\underline{5}}_{\underline{17}}1^{\underline{5}}_{\underline{15}}1^{\underline{5}}_{\underline{13}}2^{\underline{4}}_{\underline{15}}4^{\underline{4}}_{\underline{13}}2^{\underline{4}}_{\underline{11}}2^{\underline{3}}_{\underline{13}}2^{\underline{3}}_{\underline{11}}1^{\underline{2}}_{\underline{11}}2^{\underline{2}}_{\underline{9}}1^{\underline{2}}_{\underline{7}}1^{\underline{1}}_{\underline{9}}1^{\underline{1}}_{\underline{7}}1^{0}_{\underline{7}}2^{0}_{\underline{5}}1^{1}_{\underline{5}}1^{2}_{\underline{1}}$ 11^n_{324}: $1^{\underline{7}}_{\underline{17}}1^{\underline{6}}_{\underline{15}}1^{\underline{6}}_{\underline{13}}2^{\underline{5}}_{\underline{13}}1^{\underline{5}}_{\underline{11}}4^{\underline{4}}_{\underline{11}}3^{\underline{4}}_{\underline{9}}4^{\underline{3}}_{\underline{9}}3^{\underline{3}}_{\underline{7}}4^{\underline{2}}_{\underline{7}}4^{\underline{2}}_{\underline{5}}2^{\underline{1}}_{\underline{5}}4^{\underline{1}}_{\underline{3}}4^{0}_{\underline{3}}5^{0}_{\underline{1}}2^{1}_{\underline{1}}1^{1}_{1}2^{2}_{3}$	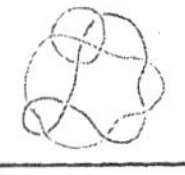	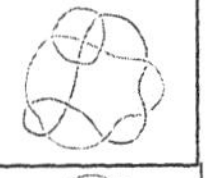
11^n_{325}: $1^{\underline{7}}_{\underline{17}}2^{\underline{6}}_{\underline{15}}1^{\underline{6}}_{\underline{13}}4^{\underline{5}}_{\underline{13}}2^{\underline{5}}_{\underline{11}}5^{\underline{4}}_{\underline{11}}4^{\underline{4}}_{\underline{9}}5^{\underline{3}}_{\underline{9}}5^{\underline{3}}_{\underline{7}}7^{\underline{2}}_{\underline{7}}7^{\underline{2}}_{\underline{5}}4^{\underline{1}}_{\underline{5}}5^{\underline{1}}_{\underline{3}}4^{0}_{\underline{3}}6^{0}_{\underline{1}}2^{1}_{\underline{1}}2^{1}_{1}2^{2}_{3}$ 11^n_{326}: $1^{\underline{5}}_{\underline{13}}1^{\underline{4}}_{\underline{11}}2^{\underline{4}}_{\underline{9}}1^{\underline{4}}_{\underline{7}}2^{\underline{3}}_{\underline{9}}1^{\underline{3}}_{\underline{7}}3^{\underline{2}}_{\underline{7}}3^{\underline{2}}_{\underline{5}}2^{\underline{1}}_{\underline{5}}3^{\underline{1}}_{\underline{3}}1^{\underline{1}}_{\underline{1}}4^{0}_{\underline{3}}5^{0}_{\underline{1}}1^{0}_{1}2^{1}_{\underline{1}}2^{1}_{1}1^{2}_{1}2^{2}_{3}1^{3}_{3}1^{3}_{5}1^{4}_{7}$	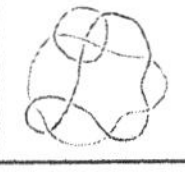	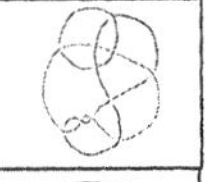
11^n_{327}: $1^{\underline{5}}_{\underline{9}}1^{\underline{4}}_{\underline{7}}1^{\underline{4}}_{\underline{5}}2^{\underline{3}}_{\underline{5}}1^{\underline{3}}_{\underline{3}}4^{\underline{2}}_{\underline{3}}2^{\underline{2}}_{\underline{1}}2^{\underline{1}}_{\underline{1}}4^{\underline{1}}_{1}6^{0}_{1}5^{0}_{3}3^{1}_{3}3^{1}_{5}3^{2}_{5}3^{2}_{7}1^{3}_{7}3^{3}_{9}1^{4}_{9}2^{4}_{11}$ 11^n_{328}: $1^{\underline{5}}_{\underline{9}}1^{\underline{4}}_{\underline{7}}1^{\underline{4}}_{\underline{5}}2^{\underline{3}}_{\underline{5}}1^{\underline{3}}_{\underline{3}}3^{\underline{2}}_{\underline{3}}2^{\underline{2}}_{\underline{1}}2^{\underline{1}}_{\underline{1}}3^{\underline{1}}_{1}1^{0}_{\underline{1}}5^{0}_{1}3^{0}_{3}2^{1}_{3}3^{1}_{5}1^{2}_{3}3^{2}_{5}3^{2}_{7}1^{3}_{7}1^{3}_{9}1^{4}_{11}$	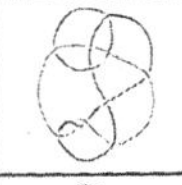	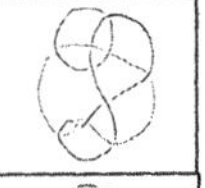
11^n_{329}: $1^{\underline{5}}_{\underline{15}}1^{\underline{4}}_{\underline{13}}2^{\underline{4}}_{\underline{11}}1^{\underline{4}}_{\underline{9}}2^{\underline{3}}_{\underline{11}}2^{\underline{3}}_{\underline{9}}2^{\underline{2}}_{\underline{9}}2^{\underline{2}}_{\underline{7}}1^{\underline{2}}_{\underline{5}}1^{\underline{1}}_{\underline{7}}2^{\underline{1}}_{\underline{5}}3^{0}_{\underline{5}}4^{0}_{\underline{3}}1^{0}_{\underline{1}}2^{1}_{\underline{3}}1^{1}_{\underline{1}}2^{2}_{1}1^{3}_{1}1^{4}_{5}$ 11^n_{330}: $1^{\underline{5}}_{\underline{11}}1^{\underline{4}}_{\underline{9}}1^{\underline{4}}_{\underline{7}}2^{\underline{3}}_{\underline{7}}1^{\underline{3}}_{\underline{5}}3^{\underline{2}}_{\underline{5}}2^{\underline{2}}_{\underline{3}}2^{\underline{1}}_{\underline{3}}3^{\underline{1}}_{\underline{1}}5^{0}_{\underline{1}}5^{0}_{1}3^{1}_{1}2^{1}_{3}1^{2}_{3}3^{2}_{5}1^{3}_{5}1^{3}_{7}1^{4}_{7}2^{4}_{9}$	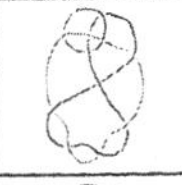	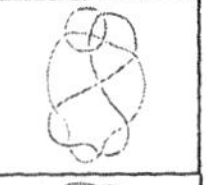
11^n_{331}: $1^{\underline{5}}_{\underline{11}}1^{\underline{4}}_{\underline{9}}1^{\underline{4}}_{\underline{7}}2^{\underline{3}}_{\underline{7}}1^{\underline{3}}_{\underline{5}}3^{\underline{2}}_{\underline{5}}2^{\underline{2}}_{\underline{3}}2^{\underline{1}}_{\underline{3}}3^{\underline{1}}_{\underline{1}}4^{0}_{\underline{1}}4^{0}_{1}2^{1}_{1}2^{1}_{3}2^{2}_{3}4^{2}_{5}1^{3}_{5}1^{4}_{9}$ 11^n_{332}: $1^{\underline{7}}_{\underline{19}}1^{\underline{6}}_{\underline{17}}1^{\underline{6}}_{\underline{15}}2^{\underline{5}}_{\underline{15}}1^{\underline{5}}_{\underline{13}}5^{\underline{4}}_{\underline{13}}5^{\underline{4}}_{\underline{11}}1^{\underline{4}}_{\underline{9}}3^{\underline{3}}_{\underline{11}}3^{\underline{3}}_{\underline{9}}3^{\underline{2}}_{\underline{9}}3^{\underline{2}}_{\underline{7}}2^{\underline{1}}_{\underline{7}}3^{\underline{1}}_{\underline{5}}2^{0}_{\underline{5}}3^{0}_{\underline{3}}1^{1}_{\underline{3}}1^{1}_{\underline{1}}1^{2}_{1}$	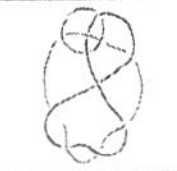	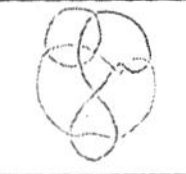

L: KH	1st line	2nd line
11^n_{333}: $1^{\underline{7}}_{\underline{15}}1^{\underline{6}}_{\underline{13}}1^{\underline{6}}_{\underline{11}}2^{\underline{5}}_{\underline{11}}1^{\underline{5}}_{\underline{9}}4^{\underline{4}}_{\underline{9}}3^{\underline{4}}_{\underline{7}}2^{\underline{3}}_{\underline{7}}3^{\underline{3}}_{\underline{5}}4^{\underline{2}}_{\underline{5}}2^{\underline{2}}_{\underline{3}}2^{\underline{1}}_{\underline{3}}4^{\underline{1}}_{\underline{1}}1^{0}_{\underline{3}}4^{0}_{\underline{1}}4^{0}_{1}1^{1}_{1}1^{1}_{3}1^{2}_{5}$ 11^n_{334}: $1^{\underline{7}}_{\underline{15}}1^{\underline{6}}_{\underline{13}}1^{\underline{6}}_{\underline{11}}1^{\underline{5}}_{\underline{11}}1^{\underline{5}}_{\underline{9}}2^{\underline{4}}_{\underline{9}}1^{\underline{4}}_{\underline{7}}1^{\underline{3}}_{\underline{9}}1^{\underline{3}}_{\underline{7}}2^{\underline{3}}_{\underline{5}}2^{\underline{2}}_{\underline{7}}4^{\underline{2}}_{\underline{5}}2^{\underline{2}}_{\underline{3}}1^{\underline{1}}_{\underline{5}}2^{\underline{1}}_{\underline{3}}1^{\underline{1}}_{\underline{1}}3^{0}_{\underline{3}}3^{0}_{\underline{1}}2^{0}_{1}1^{1}_{\underline{1}}2^{1}_{1}1^{2}_{1}1^{2}_{3}1^{3}_{3}1^{3}_{5}1^{4}_{7}$	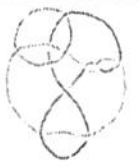	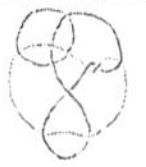
11^n_{335}: $3^{\underline{6}}_{\underline{15}}2^{\underline{6}}_{\underline{13}}5^{\underline{5}}_{\underline{13}}1^{\underline{5}}_{\underline{11}}7^{\underline{4}}_{\underline{11}}5^{\underline{4}}_{\underline{9}}7^{\underline{3}}_{\underline{9}}7^{\underline{3}}_{\underline{7}}10^{\underline{2}}_{\underline{7}}7^{\underline{2}}_{\underline{5}}6^{\underline{1}}_{\underline{5}}10^{\underline{1}}_{\underline{3}}7^{0}_{\underline{3}}8^{0}_{\underline{1}}4^{1}_{\underline{1}}5^{1}_{1}1^{2}_{1}4^{2}_{3}1^{3}_{5}$ 11^n_{336}: $1^{\underline{2}}_{\underline{5}}2^{\underline{1}}_{\underline{3}}1^{\underline{1}}_{\underline{1}}4^{0}_{\underline{1}}4^{0}_{1}1^{0}_{3}5^{1}_{1}3^{1}_{3}3^{2}_{3}5^{2}_{5}4^{3}_{5}3^{3}_{7}4^{4}_{7}6^{4}_{9}2^{5}_{9}2^{5}_{11}1^{6}_{11}2^{6}_{13}1^{7}_{15}$		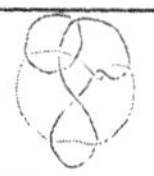
11^n_{337}: $1^{\underline{11}}_{\underline{27}}1^{\underline{10}}_{\underline{25}}1^{\underline{10}}_{\underline{23}}3^{\underline{9}}_{\underline{23}}1^{\underline{9}}_{\underline{21}}3^{\underline{8}}_{\underline{21}}4^{\underline{8}}_{\underline{19}}3^{\underline{7}}_{\underline{19}}2^{\underline{7}}_{\underline{17}}4^{\underline{6}}_{\underline{17}}3^{\underline{6}}_{\underline{15}}1^{\underline{5}}_{\underline{17}}1^{\underline{5}}_{\underline{15}}4^{\underline{5}}_{\underline{13}}5^{\underline{4}}_{\underline{13}}3^{\underline{4}}_{\underline{11}}1^{\underline{3}}_{\underline{13}}2^{\underline{3}}_{\underline{9}}1^{\underline{2}}_{\underline{9}}1^{0}_{\underline{7}}1^{0}_{\underline{5}}$ 11^n_{338}: $1^{\underline{8}}_{\underline{17}}1^{\underline{8}}_{\underline{15}}1^{\underline{6}}_{\underline{13}}1^{\underline{5}}_{\underline{9}}4^{\underline{4}}_{\underline{11}}3^{\underline{4}}_{\underline{9}}3^{\underline{3}}_{\underline{9}}2^{\underline{3}}_{\underline{7}}1^{\underline{3}}_{\underline{5}}4^{\underline{2}}_{\underline{7}}3^{\underline{2}}_{\underline{5}}2^{\underline{1}}_{\underline{5}}4^{\underline{1}}_{\underline{3}}4^{0}_{\underline{3}}3^{0}_{\underline{1}}2^{1}_{\underline{1}}3^{1}_{1}1^{2}_{1}2^{2}_{3}1^{3}_{5}$		
11^n_{339}: $1^{\underline{9}}_{\underline{21}}1^{\underline{8}}_{\underline{17}}1^{\underline{7}}_{\underline{17}}1^{\underline{6}}_{\underline{15}}1^{\underline{6}}_{\underline{13}}2^{\underline{5}}_{\underline{15}}1^{\underline{5}}_{\underline{11}}2^{\underline{4}}_{\underline{13}}5^{\underline{4}}_{\underline{11}}1^{\underline{4}}_{\underline{9}}3^{\underline{3}}_{\underline{11}}1^{\underline{3}}_{\underline{9}}1^{\underline{3}}_{\underline{7}}2^{\underline{2}}_{\underline{9}}3^{\underline{2}}_{\underline{7}}1^{\underline{1}}_{\underline{7}}2^{\underline{1}}_{\underline{5}}3^{0}_{\underline{5}}3^{0}_{\underline{3}}1^{1}_{\underline{3}}1^{1}_{\underline{1}}1^{2}_{1}$ 11^n_{340}: $3^{\underline{4}}_{\underline{9}}2^{\underline{4}}_{\underline{7}}4^{\underline{3}}_{\underline{7}}1^{\underline{3}}_{\underline{5}}7^{\underline{2}}_{\underline{5}}4^{\underline{2}}_{\underline{3}}5^{\underline{1}}_{\underline{3}}7^{\underline{1}}_{\underline{1}}9^{0}_{\underline{1}}7^{0}_{1}6^{1}_{1}7^{1}_{3}5^{2}_{3}6^{2}_{5}2^{3}_{5}5^{3}_{7}1^{4}_{7}2^{4}_{9}1^{5}_{11}$		
11^n_{341}: $2^{\underline{5}}_{\underline{13}}3^{\underline{4}}_{\underline{11}}3^{\underline{4}}_{\underline{9}}6^{\underline{3}}_{\underline{9}}2^{\underline{3}}_{\underline{7}}5^{\underline{2}}_{\underline{7}}6^{\underline{2}}_{\underline{5}}5^{\underline{1}}_{\underline{5}}5^{\underline{1}}_{\underline{3}}6^{0}_{\underline{3}}7^{0}_{\underline{1}}4^{1}_{\underline{1}}4^{1}_{1}2^{2}_{1}4^{2}_{3}2^{3}_{5}1^{4}_{5}1^{4}_{7}$ 11^n_{342}: $1^{\underline{6}}_{\underline{11}}1^{\underline{5}}_{\underline{7}}4^{\underline{4}}_{\underline{7}}2^{\underline{4}}_{\underline{5}}1^{\underline{3}}_{\underline{5}}2^{\underline{3}}_{\underline{3}}3^{\underline{2}}_{\underline{3}}1^{\underline{2}}_{\underline{1}}2^{\underline{1}}_{\underline{1}}3^{\underline{1}}_{1}3^{0}_{\underline{1}}4^{0}_{1}2^{0}_{3}1^{1}_{1}2^{1}_{3}2^{1}_{5}1^{2}_{3}1^{2}_{5}1^{2}_{7}1^{3}_{7}1^{4}_{7}1^{5}_{11}$	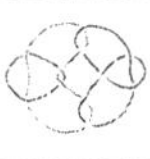	
11^n_{343}: $1^{\underline{6}}_{\underline{13}}1^{\underline{5}}_{\underline{9}}4^{\underline{4}}_{\underline{9}}2^{\underline{4}}_{\underline{7}}3^{\underline{3}}_{\underline{7}}2^{\underline{3}}_{\underline{5}}5^{\underline{2}}_{\underline{5}}3^{\underline{2}}_{\underline{3}}3^{\underline{1}}_{\underline{3}}5^{\underline{1}}_{\underline{1}}5^{0}_{\underline{1}}5^{0}_{1}3^{1}_{1}3^{1}_{3}2^{2}_{3}3^{2}_{5}2^{3}_{7}$ 11^n_{344}: $1^{\underline{7}}_{\underline{17}}2^{\underline{6}}_{\underline{15}}1^{\underline{6}}_{\underline{13}}4^{\underline{5}}_{\underline{13}}2^{\underline{5}}_{\underline{11}}5^{\underline{4}}_{\underline{11}}5^{\underline{4}}_{\underline{9}}5^{\underline{3}}_{\underline{9}}4^{\underline{3}}_{\underline{7}}6^{\underline{2}}_{\underline{7}}5^{\underline{2}}_{\underline{5}}3^{\underline{1}}_{\underline{5}}6^{\underline{1}}_{\underline{3}}4^{0}_{\underline{3}}6^{0}_{\underline{1}}2^{1}_{\underline{1}}1^{1}_{1}2^{2}_{3}$	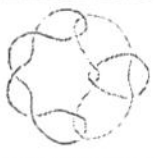	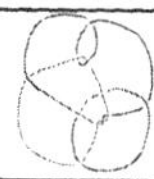
11^n_{345}: $1^{\underline{2}}_{1}1^{\underline{1}}_{5}3^{0}_{5}2^{0}_{7}1^{1}_{7}1^{1}_{9}2^{2}_{9}1^{2}_{11}1^{3}_{11}2^{3}_{13}2^{4}_{11}3^{4}_{13}1^{4}_{15}1^{5}_{13}1^{5}_{15}2^{5}_{17}1^{6}_{17}1^{8}_{19}1^{8}_{21}$ 11^n_{346}: $2^{\underline{5}}_{\underline{13}}4^{\underline{4}}_{\underline{11}}3^{\underline{4}}_{\underline{9}}6^{\underline{3}}_{\underline{9}}3^{\underline{3}}_{\underline{7}}7^{\underline{2}}_{\underline{7}}6^{\underline{2}}_{\underline{5}}5^{\underline{1}}_{\underline{5}}7^{\underline{1}}_{\underline{3}}8^{0}_{\underline{3}}8^{0}_{\underline{1}}5^{1}_{\underline{1}}5^{1}_{1}2^{2}_{1}5^{2}_{3}1^{3}_{3}2^{3}_{5}1^{4}_{7}$	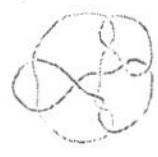	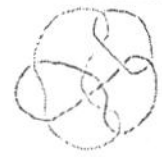
11^n_{347}: $1^{\underline{8}}_{\underline{19}}1^{\underline{8}}_{\underline{17}}1^{\underline{6}}_{\underline{15}}2^{\underline{5}}_{\underline{15}}1^{\underline{5}}_{\underline{11}}2^{\underline{4}}_{\underline{13}}4^{\underline{4}}_{\underline{11}}3^{\underline{3}}_{\underline{11}}1^{\underline{3}}_{\underline{9}}1^{\underline{3}}_{\underline{7}}2^{\underline{2}}_{\underline{9}}3^{\underline{2}}_{\underline{7}}1^{\underline{1}}_{\underline{7}}2^{\underline{1}}_{\underline{5}}3^{0}_{\underline{5}}3^{0}_{\underline{3}}1^{1}_{\underline{3}}1^{1}_{\underline{1}}1^{2}_{1}$ 11^n_{348}: $1^{\underline{9}}_{\underline{21}}1^{\underline{8}}_{\underline{19}}1^{\underline{8}}_{\underline{17}}1^{\underline{7}}_{\underline{17}}1^{\underline{7}}_{\underline{15}}2^{\underline{6}}_{\underline{15}}1^{\underline{6}}_{\underline{13}}1^{\underline{5}}_{\underline{15}}1^{\underline{5}}_{\underline{13}}2^{\underline{5}}_{\underline{11}}3^{\underline{4}}_{\underline{13}}5^{\underline{4}}_{\underline{11}}2^{\underline{4}}_{\underline{9}}2^{\underline{3}}_{\underline{11}}2^{\underline{3}}_{\underline{9}}1^{\underline{3}}_{\underline{7}}2^{\underline{2}}_{\underline{9}}2^{\underline{2}}_{\underline{7}}1^{\underline{2}}_{\underline{5}}1^{\underline{1}}_{\underline{7}}2^{\underline{1}}_{\underline{5}}2^{0}_{\underline{5}}2^{0}_{\underline{3}}1^{1}_{\underline{3}}1^{1}_{\underline{1}}1^{2}_{1}$	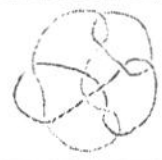	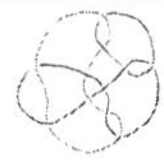
11^n_{349}: $1^{\underline{4}}_{\underline{7}}1^{\underline{4}}_{\underline{5}}2^{\underline{3}}_{\underline{5}}4^{\underline{2}}_{\underline{3}}2^{\underline{2}}_{\underline{1}}4^{\underline{1}}_{\underline{1}}4^{\underline{1}}_{1}6^{0}_{1}5^{0}_{3}5^{1}_{3}5^{1}_{5}5^{2}_{5}5^{2}_{7}2^{3}_{7}5^{3}_{9}3^{4}_{9}4^{4}_{11}1^{5}_{13}$ 11^n_{350}: $1^{\underline{6}}_{\underline{11}}1^{\underline{5}}_{\underline{7}}2^{\underline{4}}_{\underline{7}}1^{\underline{4}}_{\underline{5}}1^{\underline{3}}_{\underline{5}}1^{\underline{3}}_{\underline{3}}1^{\underline{2}}_{\underline{5}}1^{\underline{2}}_{\underline{3}}1^{\underline{2}}_{\underline{1}}2^{\underline{1}}_{\underline{1}}1^{\underline{1}}_{1}4^{0}_{\underline{1}}4^{0}_{1}1^{0}_{3}2^{1}_{1}1^{1}_{3}1^{1}_{5}1^{2}_{3}2^{2}_{5}1^{3}_{5}1^{3}_{7}1^{4}_{7}1^{4}_{9}1^{5}_{11}$	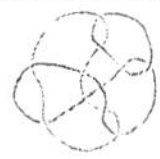	
11^n_{351}: $3^{\underline{4}}_{\underline{9}}1^{\underline{4}}_{\underline{7}}4^{\underline{3}}_{\underline{7}}2^{\underline{3}}_{\underline{5}}7^{\underline{2}}_{\underline{5}}4^{\underline{2}}_{\underline{3}}6^{\underline{1}}_{\underline{3}}7^{\underline{1}}_{\underline{1}}9^{0}_{\underline{1}}9^{0}_{1}7^{1}_{1}6^{1}_{3}5^{2}_{3}7^{2}_{5}2^{3}_{5}5^{3}_{7}1^{4}_{7}2^{4}_{9}1^{5}_{11}$ 11^n_{352}: $1^{\underline{8}}_{\underline{17}}1^{\underline{7}}_{\underline{13}}1^{\underline{6}}_{\underline{13}}1^{\underline{5}}_{\underline{11}}1^{\underline{5}}_{\underline{9}}4^{\underline{4}}_{\underline{11}}3^{\underline{4}}_{\underline{9}}1^{\underline{4}}_{\underline{7}}3^{\underline{3}}_{\underline{9}}2^{\underline{3}}_{\underline{7}}4^{\underline{2}}_{\underline{7}}3^{\underline{2}}_{\underline{5}}1^{\underline{2}}_{\underline{3}}2^{\underline{1}}_{\underline{5}}4^{\underline{1}}_{\underline{3}}4^{0}_{\underline{3}}3^{0}_{\underline{1}}2^{1}_{\underline{1}}3^{1}_{1}1^{2}_{1}2^{2}_{3}1^{3}_{5}$		
11^n_{353}: $2^{\underline{5}}_{\underline{13}}3^{\underline{4}}_{\underline{11}}2^{\underline{4}}_{\underline{9}}5^{\underline{3}}_{\underline{9}}3^{\underline{3}}_{\underline{7}}6^{\underline{2}}_{\underline{7}}5^{\underline{2}}_{\underline{5}}4^{\underline{1}}_{\underline{5}}6^{\underline{1}}_{\underline{3}}7^{0}_{\underline{3}}8^{0}_{\underline{1}}5^{1}_{\underline{1}}3^{1}_{1}1^{2}_{1}5^{2}_{3}1^{3}_{3}1^{3}_{5}1^{4}_{7}$ 11^n_{354}: $1^{\underline{9}}_{\underline{21}}1^{\underline{8}}_{\underline{19}}1^{\underline{8}}_{\underline{17}}1^{\underline{7}}_{\underline{17}}1^{\underline{7}}_{\underline{15}}3^{\underline{6}}_{\underline{15}}2^{\underline{6}}_{\underline{13}}1^{\underline{5}}_{\underline{15}}1^{\underline{5}}_{\underline{13}}2^{\underline{5}}_{\underline{11}}1^{\underline{4}}_{\underline{13}}4^{\underline{4}}_{\underline{11}}2^{\underline{4}}_{\underline{9}}1^{\underline{3}}_{\underline{11}}2^{\underline{3}}_{\underline{9}}2^{\underline{3}}_{\underline{7}}2^{\underline{2}}_{\underline{9}}3^{\underline{2}}_{\underline{7}}1^{\underline{2}}_{\underline{5}}1^{\underline{1}}_{\underline{7}}1^{\underline{1}}_{\underline{5}}1^{\underline{1}}_{\underline{3}}1^{0}_{\underline{5}}2^{0}_{\underline{3}}1^{1}_{\underline{3}}1^{2}_{1}$	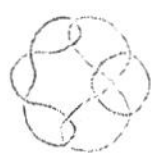	
11^n_{355}: $2^{\underline{6}}_{\underline{15}}1^{\underline{6}}_{\underline{13}}3^{\underline{5}}_{\underline{13}}1^{\underline{5}}_{\underline{11}}6^{\underline{4}}_{\underline{11}}4^{\underline{4}}_{\underline{9}}5^{\underline{3}}_{\underline{9}}5^{\underline{3}}_{\underline{7}}8^{\underline{2}}_{\underline{7}}6^{\underline{2}}_{\underline{5}}5^{\underline{1}}_{\underline{5}}7^{\underline{1}}_{\underline{3}}5^{0}_{\underline{3}}6^{0}_{\underline{1}}3^{1}_{\underline{1}}4^{1}_{1}1^{2}_{1}3^{2}_{3}1^{3}_{5}$ 11^n_{356}: $1^{\underline{5}}_{\underline{13}}3^{\underline{4}}_{\underline{11}}2^{\underline{4}}_{\underline{9}}4^{\underline{3}}_{\underline{9}}2^{\underline{3}}_{\underline{7}}5^{\underline{2}}_{\underline{7}}5^{\underline{2}}_{\underline{5}}5^{\underline{1}}_{\underline{5}}4^{\underline{1}}_{\underline{3}}5^{0}_{\underline{3}}6^{0}_{\underline{1}}3^{1}_{\underline{1}}4^{1}_{1}3^{2}_{1}4^{2}_{3}1^{3}_{3}2^{3}_{5}1^{4}_{7}$		
11^n_{357}: $1^{\underline{6}}_{\underline{11}}1^{\underline{5}}_{\underline{7}}3^{\underline{4}}_{\underline{7}}1^{\underline{4}}_{\underline{5}}1^{\underline{3}}_{\underline{5}}2^{\underline{3}}_{\underline{3}}3^{\underline{2}}_{\underline{3}}2^{\underline{2}}_{\underline{1}}2^{\underline{1}}_{\underline{1}}2^{\underline{1}}_{1}1^{0}_{\underline{1}}3^{0}_{1}2^{0}_{3}1^{1}_{3}2^{1}_{5}1^{2}_{3}2^{2}_{5}1^{2}_{7}1^{3}_{9}$ 11^n_{358}: $1^{\underline{6}}_{\underline{13}}1^{\underline{5}}_{\underline{11}}1^{\underline{5}}_{\underline{9}}3^{\underline{4}}_{\underline{9}}2^{\underline{4}}_{\underline{7}}3^{\underline{3}}_{\underline{7}}2^{\underline{3}}_{\underline{5}}4^{\underline{2}}_{\underline{5}}4^{\underline{2}}_{\underline{3}}3^{\underline{1}}_{\underline{3}}3^{\underline{1}}_{\underline{1}}4^{0}_{\underline{1}}4^{0}_{1}1^{1}_{1}3^{1}_{3}2^{2}_{3}2^{2}_{5}1^{3}_{7}$	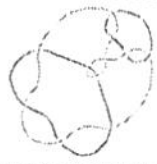	
11^n_{359}: $1^{\underline{6}}_{\underline{11}}1^{\underline{5}}_{\underline{7}}2^{\underline{4}}_{\underline{7}}1^{\underline{4}}_{\underline{5}}1^{\underline{3}}_{\underline{5}}1^{\underline{3}}_{\underline{3}}2^{\underline{2}}_{\underline{5}}2^{\underline{2}}_{\underline{3}}1^{\underline{2}}_{\underline{1}}1^{\underline{1}}_{\underline{3}}2^{\underline{1}}_{\underline{1}}1^{\underline{1}}_{1}3^{0}_{\underline{1}}3^{0}_{1}1^{0}_{3}2^{1}_{1}2^{1}_{3}1^{1}_{5}2^{2}_{3}3^{2}_{5}1^{3}_{5}1^{3}_{7}1^{4}_{7}1^{4}_{9}1^{5}_{11}$ 11^n_{360}: $1^{\underline{7}}_{\underline{15}}1^{\underline{6}}_{\underline{11}}1^{\underline{5}}_{\underline{11}}1^{\underline{4}}_{\underline{11}}1^{\underline{4}}_{\underline{9}}1^{\underline{4}}_{\underline{7}}1^{\underline{3}}_{\underline{9}}1^{\underline{3}}_{\underline{7}}1^{\underline{3}}_{\underline{5}}3^{\underline{2}}_{\underline{7}}4^{\underline{2}}_{\underline{5}}1^{\underline{2}}_{\underline{3}}2^{\underline{1}}_{\underline{5}}2^{\underline{1}}_{\underline{3}}1^{\underline{1}}_{\underline{1}}3^{0}_{\underline{3}}4^{0}_{\underline{1}}1^{0}_{1}2^{1}_{\underline{1}}2^{1}_{1}1^{2}_{1}2^{2}_{3}1^{3}_{3}1^{3}_{5}1^{4}_{7}$		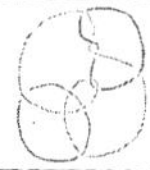
11^n_{361}: $1^{\underline{9}}_{\underline{21}}2^{\underline{8}}_{\underline{19}}1^{\underline{8}}_{\underline{17}}2^{\underline{7}}_{\underline{17}}2^{\underline{7}}_{\underline{15}}5^{\underline{6}}_{\underline{15}}4^{\underline{6}}_{\underline{13}}4^{\underline{5}}_{\underline{13}}3^{\underline{5}}_{\underline{11}}4^{\underline{4}}_{\underline{11}}4^{\underline{4}}_{\underline{9}}3^{\underline{3}}_{\underline{9}}4^{\underline{3}}_{\underline{7}}3^{\underline{2}}_{\underline{7}}3^{\underline{2}}_{\underline{5}}3^{\underline{1}}_{\underline{3}}2^{0}_{\underline{3}}2^{0}_{\underline{1}}$ 11^n_{362}: $1^{\underline{3}}_{\underline{5}}3^{\underline{2}}_{\underline{3}}1^{\underline{2}}_{\underline{1}}4^{\underline{1}}_{\underline{1}}3^{\underline{1}}_{1}7^{0}_{1}6^{0}_{3}8^{1}_{3}5^{1}_{5}6^{2}_{5}8^{2}_{7}6^{3}_{7}6^{3}_{9}4^{4}_{9}6^{4}_{11}1^{5}_{11}4^{5}_{13}2^{6}_{13}3^{6}_{15}$	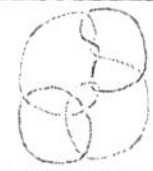	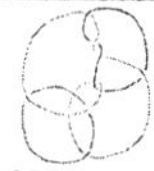

L: KH	1st line	2nd line
11^n_{363}: $1^{\underline{7}}_{\underline{17}} 1^{\underline{6}}_{\underline{15}} 1^{\underline{6}}_{\underline{13}} 2^{\underline{5}}_{\underline{13}} 1^{\underline{5}}_{\underline{11}} 2^{\underline{4}}_{\underline{11}} 2^{\underline{4}}_{\underline{9}} 2^{\underline{3}}_{\underline{9}} 2^{\underline{3}}_{\underline{7}} 1^{\underline{2}}_{\underline{9}} 3^{\underline{2}}_{\underline{7}} 4^{\underline{2}}_{\underline{5}} 2^{\underline{1}}_{\underline{5}} 1^{\underline{1}}_{\underline{3}} 1^{0}_{\underline{5}} 2^{0}_{\underline{3}} 3^{0}_{\underline{1}} 1^{1}_{\underline{3}} 1^{1}_{\underline{1}} 1^{2}_{1} 1^{3}_{1} 1^{4}_{5}$ 11^n_{364}: $1^{\underline{2}}_{\underline{3}} 1^{\underline{1}}_{\underline{1}} 1^{\underline{1}}_{1} 5^{0}_{1} 3^{0}_{3} 3^{1}_{3} 3^{1}_{5} 6^{2}_{5} 5^{2}_{7} 4^{3}_{7} 4^{3}_{9} 3^{4}_{9} 4^{4}_{11} 2^{5}_{11} 3^{5}_{13} 1^{6}_{13} 2^{6}_{15} 1^{7}_{17}$		
11^n_{365}: $1^{\underline{8}}_{\underline{19}} 1^{\underline{7}}_{\underline{17}} 1^{\underline{7}}_{\underline{15}} 4^{\underline{6}}_{\underline{15}} 3^{\underline{6}}_{\underline{13}} 4^{\underline{5}}_{\underline{13}} 2^{\underline{5}}_{\underline{11}} 4^{\underline{4}}_{\underline{11}} 4^{\underline{4}}_{\underline{9}} 4^{\underline{3}}_{\underline{9}} 4^{\underline{3}}_{\underline{7}} 4^{\underline{2}}_{\underline{7}} 4^{\underline{2}}_{\underline{5}} 1^{\underline{1}}_{\underline{5}} 4^{\underline{1}}_{\underline{3}} 3^{0}_{\underline{3}} 3^{0}_{\underline{1}} 1^{1}_{1}$ 11^n_{366}: $1^{\underline{2}}_{1} 1^{\underline{1}}_{5} 3^{0}_{5} 2^{0}_{7} 1^{1}_{7} 1^{1}_{9} 2^{2}_{9} 1^{2}_{11} 1^{3}_{11} 2^{3}_{13} 1^{4}_{11} 2^{4}_{13} 1^{4}_{15} 1^{5}_{15} 2^{5}_{17} 1^{6}_{15} 2^{6}_{17} 1^{6}_{19}$		
11^n_{367}: $2^{\underline{2}}_{\underline{7}} 2^{\underline{2}}_{\underline{5}} 1^{\underline{2}}_{\underline{3}} 1^{\underline{1}}_{\underline{5}} 1^{\underline{1}}_{\underline{3}} 2^{0}_{\underline{3}} 3^{0}_{\underline{1}} 1^{0}_{1} 2^{1}_{\underline{1}} 1^{1}_{1} 1^{2}_{1} 2^{2}_{3} 1^{3}_{3} 1^{3}_{5} 1^{4}_{5} 1^{4}_{7} 1^{5}_{9}$ 11^n_{368}: $1^{\underline{9}}_{\underline{21}} 1^{\underline{8}}_{\underline{19}} 1^{\underline{8}}_{\underline{17}} 1^{\underline{7}}_{\underline{17}} 1^{\underline{7}}_{\underline{15}} 2^{\underline{6}}_{\underline{15}} 1^{\underline{6}}_{\underline{13}} 1^{\underline{5}}_{\underline{15}} 1^{\underline{5}}_{\underline{13}} 2^{\underline{5}}_{\underline{11}} 2^{\underline{4}}_{\underline{13}} 4^{\underline{4}}_{\underline{11}} 2^{\underline{4}}_{\underline{9}} 2^{\underline{3}}_{\underline{11}} 2^{\underline{3}}_{\underline{9}} 1^{\underline{3}}_{\underline{7}} 1^{\underline{2}}_{\underline{9}} 2^{\underline{2}}_{\underline{7}} 1^{\underline{2}}_{\underline{5}} 1^{\underline{1}}_{\underline{7}} 1^{\underline{1}}_{\underline{5}} 2^{0}_{\underline{5}} 3^{0}_{\underline{3}} 1^{1}_{\underline{3}} 1^{2}_{1}$	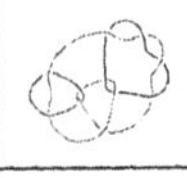	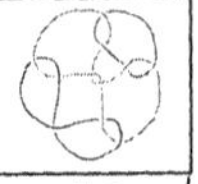
11^n_{369}: $1^{\underline{6}}_{\underline{15}} 2^{\underline{5}}_{\underline{13}} 1^{\underline{5}}_{\underline{11}} 5^{\underline{4}}_{\underline{11}} 4^{\underline{4}}_{\underline{9}} 6^{\underline{3}}_{\underline{9}} 3^{\underline{3}}_{\underline{7}} 6^{\underline{2}}_{\underline{7}} 6^{\underline{2}}_{\underline{5}} 4^{\underline{1}}_{\underline{5}} 6^{\underline{1}}_{\underline{3}} 6^{0}_{\underline{3}} 6^{0}_{\underline{1}} 3^{1}_{\underline{1}} 4^{1}_{1} 1^{2}_{1} 3^{2}_{3} 1^{3}_{5}$ 11^n_{370}: $1^{\underline{5}}_{\underline{13}} 4^{\underline{4}}_{\underline{11}} 3^{\underline{4}}_{\underline{9}} 5^{\underline{3}}_{\underline{9}} 2^{\underline{3}}_{\underline{7}} 6^{\underline{2}}_{\underline{7}} 5^{\underline{2}}_{\underline{5}} 4^{\underline{1}}_{\underline{5}} 6^{\underline{1}}_{\underline{3}} 7^{0}_{\underline{3}} 6^{0}_{\underline{1}} 4^{1}_{\underline{1}} 5^{1}_{1} 2^{2}_{1} 4^{2}_{3} 1^{3}_{3} 2^{3}_{5} 1^{4}_{7}$	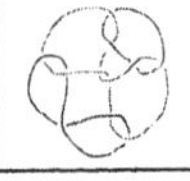	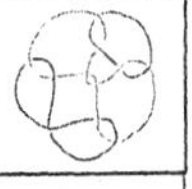
11^n_{371}: $2^{\underline{5}}_{\underline{13}} 4^{\underline{4}}_{\underline{11}} 3^{\underline{4}}_{\underline{9}} 5^{\underline{3}}_{\underline{9}} 3^{\underline{3}}_{\underline{7}} 7^{\underline{2}}_{\underline{7}} 6^{\underline{2}}_{\underline{5}} 6^{\underline{1}}_{\underline{5}} 6^{\underline{1}}_{\underline{3}} 6^{0}_{\underline{3}} 7^{0}_{\underline{1}} 4^{1}_{\underline{1}} 5^{1}_{1} 3^{2}_{1} 5^{2}_{3} 1^{3}_{3} 2^{3}_{5} 1^{4}_{7}$ 11^n_{372}: $2^{\underline{4}}_{\underline{11}} 1^{\underline{4}}_{\underline{9}} 3^{\underline{3}}_{\underline{9}} 1^{\underline{3}}_{\underline{7}} 4^{\underline{2}}_{\underline{7}} 4^{\underline{2}}_{\underline{5}} 5^{\underline{1}}_{\underline{5}} 3^{\underline{1}}_{\underline{3}} 5^{0}_{\underline{3}} 6^{0}_{\underline{1}} 4^{1}_{\underline{1}} 4^{1}_{1} 4^{2}_{1} 5^{2}_{3} 2^{3}_{3} 3^{3}_{5} 1^{4}_{5} 2^{4}_{7} 1^{5}_{9}$	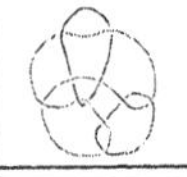	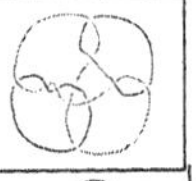
11^n_{373}: $3^{\underline{4}}_{\underline{9}} 1^{\underline{4}}_{\underline{7}} 3^{\underline{3}}_{\underline{7}} 2^{\underline{3}}_{\underline{5}} 6^{\underline{2}}_{\underline{5}} 4^{\underline{2}}_{\underline{3}} 6^{\underline{1}}_{\underline{3}} 5^{\underline{1}}_{\underline{1}} 7^{0}_{\underline{1}} 7^{0}_{1} 5^{1}_{1} 6^{1}_{3} 5^{2}_{3} 6^{2}_{5} 2^{3}_{5} 4^{3}_{7} 1^{4}_{7} 2^{4}_{9} 1^{5}_{11}$ 11^n_{374}: $1^{\underline{7}}_{\underline{19}} 2^{\underline{6}}_{\underline{17}} 1^{\underline{6}}_{\underline{15}} 3^{\underline{5}}_{\underline{15}} 2^{\underline{5}}_{\underline{13}} 5^{\underline{4}}_{\underline{13}} 5^{\underline{4}}_{\underline{11}} 1^{\underline{4}}_{\underline{9}} 5^{\underline{3}}_{\underline{11}} 4^{\underline{3}}_{\underline{9}} 3^{\underline{2}}_{\underline{9}} 5^{\underline{2}}_{\underline{7}} 3^{\underline{1}}_{\underline{7}} 3^{\underline{1}}_{\underline{5}} 3^{0}_{\underline{5}} 5^{0}_{\underline{3}} 1^{1}_{\underline{3}} 1^{1}_{\underline{1}} 1^{2}_{1}$	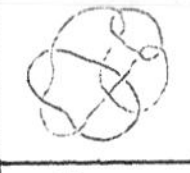	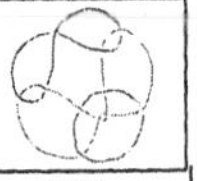
11^n_{375}: $1^{\underline{9}}_{\underline{25}} 1^{\underline{8}}_{\underline{23}} 2^{\underline{8}}_{\underline{21}} 1^{\underline{8}}_{\underline{19}} 3^{\underline{7}}_{\underline{21}} 1^{\underline{7}}_{\underline{19}} 2^{\underline{6}}_{\underline{19}} 4^{\underline{6}}_{\underline{17}} 4^{\underline{5}}_{\underline{17}} 2^{\underline{5}}_{\underline{15}} 1^{\underline{5}}_{\underline{13}} 3^{\underline{4}}_{\underline{15}} 6^{\underline{4}}_{\underline{13}} 1^{\underline{4}}_{\underline{11}} 2^{\underline{3}}_{\underline{13}} 2^{\underline{3}}_{\underline{11}} 2^{\underline{2}}_{\underline{11}} 2^{\underline{2}}_{\underline{9}} 2^{\underline{1}}_{\underline{7}} 1^{0}_{\underline{7}} 1^{0}_{\underline{5}}$ 11^n_{376}: $1^{\underline{7}}_{\underline{19}} 1^{\underline{6}}_{\underline{15}} 2^{\underline{5}}_{\underline{15}} 2^{\underline{4}}_{\underline{13}} 4^{\underline{4}}_{\underline{11}} 1^{\underline{4}}_{\underline{9}} 2^{\underline{3}}_{\underline{11}} 1^{\underline{3}}_{\underline{9}} 1^{\underline{2}}_{\underline{9}} 2^{\underline{2}}_{\underline{7}} 1^{\underline{1}}_{\underline{5}} 2^{0}_{\underline{5}} 2^{0}_{\underline{3}}$		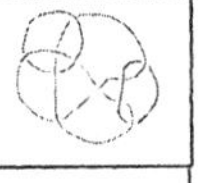
11^n_{377}: $1^{\underline{9}}_{\underline{21}} 1^{\underline{8}}_{\underline{19}} 1^{\underline{8}}_{\underline{17}} 3^{\underline{7}}_{\underline{17}} 1^{\underline{7}}_{\underline{15}} 3^{\underline{6}}_{\underline{15}} 3^{\underline{6}}_{\underline{13}} 4^{\underline{5}}_{\underline{13}} 3^{\underline{5}}_{\underline{11}} 5^{\underline{4}}_{\underline{11}} 6^{\underline{4}}_{\underline{9}} 3^{\underline{3}}_{\underline{9}} 3^{\underline{3}}_{\underline{7}} 4^{\underline{2}}_{\underline{7}} 3^{\underline{2}}_{\underline{5}} 4^{\underline{1}}_{\underline{3}} 2^{0}_{\underline{3}} 2^{0}_{\underline{1}}$ 11^n_{378}: $1^{\underline{7}}_{\underline{15}} 1^{\underline{6}}_{\underline{11}} 2^{\underline{5}}_{\underline{11}} 1^{\underline{4}}_{\underline{9}} 2^{\underline{4}}_{\underline{7}} 1^{\underline{3}}_{\underline{7}} 1^{\underline{3}}_{\underline{5}} 2^{\underline{2}}_{\underline{5}} 1^{\underline{2}}_{\underline{3}} 2^{\underline{1}}_{\underline{1}} 1^{0}_{\underline{3}} 4^{0}_{\underline{1}} 3^{0}_{1}$	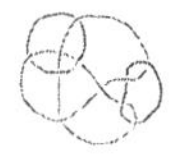	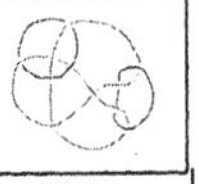
11^n_{379}: $1^{\underline{4}}_{\underline{11}} 1^{\underline{4}}_{\underline{9}} 1^{\underline{3}}_{\underline{11}} 2^{\underline{3}}_{\underline{9}} 1^{\underline{2}}_{\underline{7}} 2^{\underline{2}}_{\underline{5}} 1^{\underline{1}}_{\underline{7}} 3^{0}_{\underline{3}} 2^{0}_{\underline{1}} 1^{1}_{\underline{3}} 1^{2}_{1} 1^{4}_{3} 1^{4}_{5}$ 11^n_{380}: $1^{\underline{9}}_{\underline{21}} 1^{\underline{8}}_{\underline{19}} 1^{\underline{8}}_{\underline{17}} 3^{\underline{7}}_{\underline{17}} 1^{\underline{7}}_{\underline{15}} 2^{\underline{6}}_{\underline{15}} 3^{\underline{6}}_{\underline{13}} 3^{\underline{5}}_{\underline{13}} 2^{\underline{5}}_{\underline{11}} 1^{\underline{4}}_{\underline{13}} 5^{\underline{4}}_{\underline{11}} 4^{\underline{4}}_{\underline{9}} 1^{\underline{3}}_{\underline{11}} 2^{\underline{3}}_{\underline{9}} 3^{\underline{3}}_{\underline{7}} 3^{\underline{2}}_{\underline{7}} 2^{\underline{2}}_{\underline{5}} 2^{\underline{1}}_{\underline{3}} 1^{0}_{\underline{5}} 2^{0}_{\underline{3}} 1^{0}_{\underline{1}}$	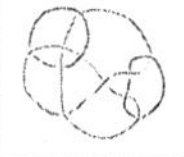	
11^n_{381}: $1^{\underline{7}}_{\underline{15}} 1^{\underline{6}}_{\underline{11}} 1^{\underline{5}}_{\underline{11}} 1^{\underline{4}}_{\underline{9}} 1^{\underline{4}}_{\underline{7}} 2^{\underline{3}}_{\underline{9}} 1^{\underline{3}}_{\underline{7}} 1^{\underline{3}}_{\underline{5}} 1^{\underline{2}}_{\underline{7}} 3^{\underline{2}}_{\underline{5}} 1^{\underline{2}}_{\underline{3}} 1^{\underline{1}}_{\underline{3}} 1^{\underline{1}}_{\underline{1}} 3^{0}_{\underline{3}} 4^{0}_{\underline{1}} 1^{0}_{1} 1^{1}_{\underline{1}} 1^{2}_{3}$ 11^n_{382}: $2^{\underline{3}}_{\underline{7}} 2^{\underline{2}}_{\underline{5}} 2^{\underline{2}}_{\underline{3}} 3^{\underline{1}}_{\underline{3}} 2^{\underline{1}}_{\underline{1}} 3^{0}_{\underline{1}} 5^{0}_{1} 4^{1}_{1} 1^{1}_{3} 2^{2}_{3} 4^{2}_{5} 2^{3}_{7} 2^{4}_{7} 2^{4}_{9}$	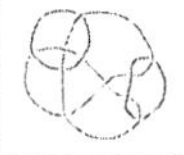	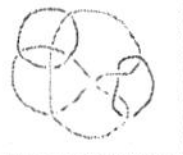
11^n_{383}: $1^{\underline{7}}_{\underline{15}} 2^{\underline{6}}_{\underline{13}} 1^{\underline{6}}_{\underline{11}} 1^{\underline{5}}_{\underline{11}} 2^{\underline{5}}_{\underline{9}} 5^{\underline{4}}_{\underline{9}} 3^{\underline{4}}_{\underline{7}} 1^{\underline{3}}_{\underline{9}} 3^{\underline{3}}_{\underline{7}} 3^{\underline{3}}_{\underline{5}} 1^{\underline{2}}_{\underline{7}} 5^{\underline{2}}_{\underline{5}} 3^{\underline{2}}_{\underline{3}} 3^{\underline{1}}_{\underline{3}} 4^{\underline{1}}_{\underline{1}} 2^{0}_{\underline{3}} 4^{0}_{\underline{1}} 2^{0}_{1} 1^{1}_{\underline{1}} 1^{1}_{1} 2^{1}_{3} 1^{2}_{3} 1^{2}_{5}$ 11^n_{384}: $2^{\underline{3}}_{\underline{7}} 4^{\underline{2}}_{\underline{5}} 2^{\underline{2}}_{\underline{3}} 3^{\underline{1}}_{\underline{3}} 4^{\underline{1}}_{\underline{1}} 8^{0}_{\underline{1}} 7^{0}_{1} 6^{1}_{1} 4^{1}_{3} 5^{2}_{3} 6^{2}_{5} 3^{3}_{5} 5^{3}_{7} 2^{4}_{7} 3^{4}_{9} 1^{5}_{9} 2^{5}_{11} 1^{6}_{13}$	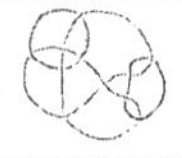	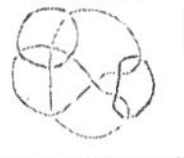
11^n_{385}: $1^{\underline{7}}_{\underline{21}} 1^{\underline{7}}_{\underline{19}} 1^{\underline{6}}_{\underline{19}} 2^{\underline{6}}_{\underline{17}} 1^{\underline{6}}_{\underline{15}} 2^{\underline{5}}_{\underline{17}} 1^{\underline{5}}_{\underline{15}} 1^{\underline{5}}_{\underline{13}} 1^{\underline{4}}_{\underline{15}} 4^{\underline{4}}_{\underline{13}} 2^{\underline{4}}_{\underline{11}} 3^{\underline{3}}_{\underline{13}} 2^{\underline{3}}_{\underline{11}} 1^{\underline{2}}_{\underline{11}} 3^{\underline{2}}_{\underline{9}} 1^{\underline{2}}_{\underline{7}} 1^{\underline{1}}_{\underline{9}} 1^{\underline{1}}_{\underline{7}} 2^{0}_{\underline{7}} 3^{0}_{\underline{5}} 1^{1}_{\underline{5}} 1^{2}_{\underline{1}}$ 11^n_{386}: $1^{\underline{7}}_{\underline{17}} 1^{\underline{6}}_{\underline{15}} 1^{\underline{6}}_{\underline{13}} 3^{\underline{5}}_{\underline{13}} 1^{\underline{5}}_{\underline{11}} 3^{\underline{4}}_{\underline{11}} 3^{\underline{4}}_{\underline{9}} 5^{\underline{3}}_{\underline{9}} 3^{\underline{3}}_{\underline{7}} 4^{\underline{2}}_{\underline{7}} 5^{\underline{2}}_{\underline{5}} 2^{\underline{1}}_{\underline{5}} 4^{\underline{1}}_{\underline{3}} 5^{0}_{\underline{3}} 6^{0}_{\underline{1}} 2^{1}_{\underline{1}} 1^{1}_{1} 2^{2}_{3}$	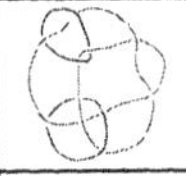	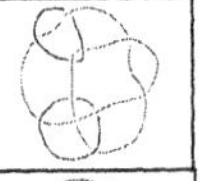
11^n_{387}: $1^{\underline{6}}_{\underline{13}} 3^{\underline{5}}_{\underline{11}} 1^{\underline{5}}_{\underline{9}} 4^{\underline{4}}_{\underline{9}} 3^{\underline{4}}_{\underline{7}} 5^{\underline{3}}_{\underline{7}} 4^{\underline{3}}_{\underline{5}} 6^{\underline{2}}_{\underline{5}} 5^{\underline{2}}_{\underline{3}} 5^{\underline{1}}_{\underline{3}} 6^{\underline{1}}_{\underline{1}} 1^{0}_{\underline{3}} 7^{0}_{\underline{1}} 8^{0}_{1} 3^{1}_{1} 3^{1}_{3} 1^{2}_{3} 3^{2}_{5} 1^{3}_{7}$ 11^n_{388}: $3^{\underline{5}}_{\underline{13}} 3^{\underline{4}}_{\underline{11}} 3^{\underline{4}}_{\underline{9}} 6^{\underline{3}}_{\underline{9}} 3^{\underline{3}}_{\underline{7}} 6^{\underline{2}}_{\underline{7}} 6^{\underline{2}}_{\underline{5}} 4^{\underline{1}}_{\underline{5}} 6^{\underline{1}}_{\underline{3}} 7^{0}_{\underline{3}} 8^{0}_{\underline{1}} 4^{1}_{\underline{1}} 3^{1}_{1} 1^{2}_{1} 4^{2}_{3} 1^{3}_{5}$	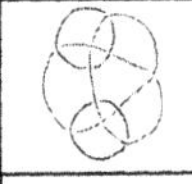	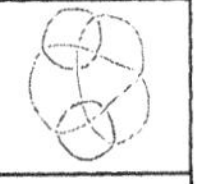
11^n_{389}: $1^{\underline{9}}_{\underline{25}} 2^{\underline{8}}_{\underline{23}} 2^{\underline{8}}_{\underline{21}} 5^{\underline{7}}_{\underline{21}} 1^{\underline{7}}_{\underline{19}} 3^{\underline{6}}_{\underline{19}} 5^{\underline{6}}_{\underline{17}} 4^{\underline{5}}_{\underline{17}} 3^{\underline{5}}_{\underline{15}} 5^{\underline{4}}_{\underline{15}} 6^{\underline{4}}_{\underline{13}} 3^{\underline{3}}_{\underline{13}} 3^{\underline{3}}_{\underline{11}} 2^{\underline{2}}_{\underline{11}} 3^{\underline{2}}_{\underline{9}} 2^{\underline{1}}_{\underline{7}} 1^{0}_{\underline{7}} 1^{0}_{\underline{5}}$ 11^n_{390}: $1^{\underline{6}}_{\underline{15}} 2^{\underline{5}}_{\underline{13}} 1^{\underline{5}}_{\underline{11}} 1^{\underline{4}}_{\underline{11}} 3^{\underline{4}}_{\underline{9}} 3^{\underline{3}}_{\underline{9}} 2^{\underline{2}}_{\underline{7}} 3^{\underline{2}}_{\underline{5}} 1^{\underline{1}}_{\underline{7}} 1^{\underline{1}}_{\underline{5}} 2^{\underline{1}}_{\underline{3}} 3^{0}_{\underline{3}} 3^{0}_{\underline{1}} 1^{1}_{\underline{3}} 1^{2}_{1} 1^{4}_{3} 1^{4}_{5}$	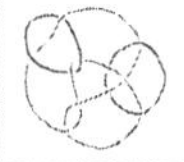	
11^n_{391}: $1^{\underline{9}}_{\underline{21}} 1^{\underline{8}}_{\underline{17}} 1^{\underline{7}}_{\underline{17}} 1^{\underline{6}}_{\underline{17}} 1^{\underline{6}}_{\underline{15}} 1^{\underline{6}}_{\underline{13}} 3^{\underline{5}}_{\underline{15}} 1^{\underline{5}}_{\underline{13}} 1^{\underline{5}}_{\underline{11}} 3^{\underline{4}}_{\underline{13}} 6^{\underline{4}}_{\underline{11}} 1^{\underline{4}}_{\underline{9}} 5^{\underline{3}}_{\underline{11}} 2^{\underline{3}}_{\underline{9}} 1^{\underline{3}}_{\underline{7}} 3^{\underline{2}}_{\underline{9}} 5^{\underline{2}}_{\underline{7}} 2^{\underline{1}}_{\underline{7}} 3^{\underline{1}}_{\underline{5}} 4^{0}_{\underline{5}} 4^{0}_{\underline{3}} 1^{1}_{\underline{3}} 2^{1}_{\underline{1}} 1^{2}_{1}$ 11^n_{392}: $1^{\underline{5}}_{\underline{11}} 3^{\underline{4}}_{\underline{9}} 1^{\underline{4}}_{\underline{7}} 6^{\underline{3}}_{\underline{7}} 3^{\underline{3}}_{\underline{5}} 7^{\underline{2}}_{\underline{5}} 6^{\underline{2}}_{\underline{3}} 9^{\underline{1}}_{\underline{3}} 7^{\underline{1}}_{\underline{1}} 8^{0}_{\underline{1}} 11^{0}_{1} 8^{1}_{1} 6^{1}_{3} 5^{2}_{3} 8^{2}_{5} 1^{3}_{5} 5^{3}_{7} 2^{4}_{7} 3^{4}_{9}$		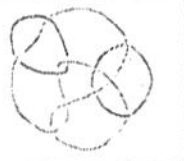

L: KH	1st line	2nd line
11^n_{393}: $1^{\underline{7}}_{\underline{15}}2^{\underline{6}}_{\underline{13}}1^{\underline{6}}_{\underline{11}}2^{\underline{5}}_{\underline{11}}2^{\underline{5}}_{\underline{9}}5^{\underline{4}}_{\underline{9}}4^{\underline{4}}_{\underline{7}}3^{\underline{3}}_{\underline{7}}3^{\underline{3}}_{\underline{5}}1^{\underline{2}}_{\underline{7}}4^{\underline{2}}_{\underline{5}}3^{\underline{2}}_{\underline{3}}3^{\underline{1}}_{\underline{3}}4^{\underline{1}}_{\underline{1}}2^{0}_{\underline{3}}4^{0}_{\underline{1}}3^{0}_{1}1^{1}_{\underline{1}}1^{1}_{1}2^{1}_{3}1^{2}_{3}1^{3}_{3}1^{4}_{7}$ 11^n_{394}: $1^{\underline{5}}_{\underline{13}}3^{\underline{4}}_{\underline{11}}3^{\underline{4}}_{\underline{9}}1^{\underline{4}}_{\underline{7}}3^{\underline{3}}_{\underline{9}}2^{\underline{3}}_{\underline{7}}5^{\underline{2}}_{\underline{7}}4^{\underline{2}}_{\underline{5}}4^{\underline{1}}_{\underline{5}}5^{\underline{1}}_{\underline{3}}1^{\underline{1}}_{\underline{1}}5^{0}_{\underline{3}}6^{0}_{\underline{1}}3^{1}_{\underline{1}}3^{1}_{1}2^{2}_{1}3^{2}_{3}1^{3}_{3}2^{3}_{5}1^{4}_{7}$		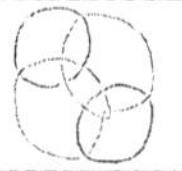
11^n_{395}: $1^{\underline{5}}_{\underline{9}}2^{\underline{4}}_{\underline{7}}1^{\underline{4}}_{\underline{5}}2^{\underline{3}}_{\underline{5}}2^{\underline{3}}_{\underline{3}}5^{\underline{2}}_{\underline{3}}2^{\underline{2}}_{\underline{1}}3^{\underline{1}}_{\underline{1}}5^{\underline{1}}_{1}1^{0}_{\underline{1}}7^{0}_{1}6^{0}_{3}4^{1}_{3}3^{1}_{5}1^{2}_{3}3^{2}_{5}4^{2}_{7}2^{3}_{7}3^{3}_{9}1^{4}_{11}$ 11^n_{396}: $1^{\underline{5}}_{\underline{9}}1^{\underline{4}}_{\underline{7}}1^{\underline{4}}_{\underline{5}}2^{\underline{3}}_{\underline{5}}1^{\underline{3}}_{\underline{3}}1^{\underline{2}}_{\underline{5}}2^{\underline{2}}_{\underline{3}}2^{\underline{2}}_{\underline{1}}3^{\underline{1}}_{\underline{1}}2^{\underline{1}}_{1}5^{0}_{\underline{1}}6^{0}_{1}3^{0}_{3}2^{1}_{1}3^{1}_{3}2^{1}_{5}2^{2}_{3}3^{2}_{5}1^{2}_{7}2^{3}_{5}2^{3}_{7}1^{3}_{9}1^{4}_{7}2^{4}_{9}1^{5}_{9}1^{5}_{11}1^{6}_{13}$	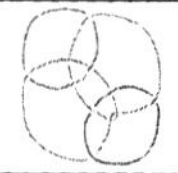	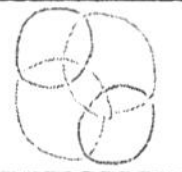
11^n_{397}: $1^{\underline{6}}_{\underline{11}}1^{\underline{5}}_{\underline{7}}3^{\underline{4}}_{\underline{7}}2^{\underline{4}}_{\underline{5}}1^{\underline{3}}_{\underline{5}}1^{\underline{3}}_{\underline{3}}1^{\underline{2}}_{\underline{5}}2^{\underline{2}}_{\underline{3}}1^{\underline{2}}_{\underline{1}}2^{\underline{1}}_{\underline{1}}2^{\underline{1}}_{1}4^{0}_{\underline{1}}4^{0}_{1}1^{0}_{3}2^{1}_{1}2^{1}_{3}2^{1}_{5}1^{2}_{3}2^{2}_{5}1^{3}_{5}1^{3}_{7}1^{4}_{7}1^{4}_{9}1^{5}_{11}$ 11^n_{398}: $1^{\underline{11}}_{\underline{27}}1^{\underline{10}}_{\underline{25}}1^{\underline{10}}_{\underline{23}}3^{\underline{9}}_{\underline{23}}1^{\underline{9}}_{\underline{21}}3^{\underline{8}}_{\underline{21}}4^{\underline{8}}_{\underline{19}}3^{\underline{7}}_{\underline{19}}2^{\underline{7}}_{\underline{17}}4^{\underline{6}}_{\underline{17}}3^{\underline{6}}_{\underline{15}}1^{\underline{5}}_{\underline{17}}1^{\underline{5}}_{\underline{15}}4^{\underline{5}}_{\underline{13}}5^{\underline{4}}_{\underline{13}}3^{\underline{4}}_{\underline{11}}1^{\underline{3}}_{\underline{13}}2^{\underline{3}}_{\underline{9}}1^{\underline{2}}_{\underline{9}}1^{0}_{\underline{7}}1^{0}_{\underline{5}}$	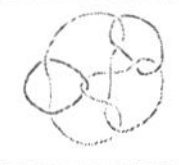	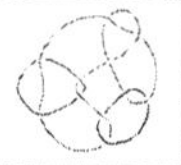
11^n_{399}: $1^{\underline{9}}_{\underline{21}}1^{\underline{8}}_{\underline{17}}1^{\underline{7}}_{\underline{17}}1^{\underline{6}}_{\underline{15}}1^{\underline{6}}_{\underline{13}}2^{\underline{5}}_{\underline{15}}1^{\underline{5}}_{\underline{11}}2^{\underline{4}}_{\underline{13}}5^{\underline{4}}_{\underline{11}}1^{\underline{4}}_{\underline{9}}3^{\underline{3}}_{\underline{11}}1^{\underline{3}}_{\underline{9}}1^{\underline{3}}_{\underline{7}}2^{\underline{2}}_{\underline{9}}3^{\underline{2}}_{\underline{7}}1^{\underline{1}}_{\underline{7}}2^{\underline{1}}_{\underline{5}}3^{0}_{\underline{5}}3^{0}_{\underline{3}}1^{1}_{\underline{3}}1^{1}_{\underline{1}}1^{2}_{1}$ 11^n_{400}: $1^{\underline{8}}_{\underline{17}}1^{\underline{8}}_{\underline{15}}1^{\underline{6}}_{\underline{13}}1^{\underline{5}}_{\underline{9}}4^{\underline{4}}_{\underline{11}}3^{\underline{4}}_{\underline{9}}3^{\underline{3}}_{\underline{9}}2^{\underline{3}}_{\underline{7}}1^{\underline{3}}_{\underline{5}}4^{\underline{2}}_{\underline{7}}3^{\underline{2}}_{\underline{5}}2^{\underline{1}}_{\underline{5}}4^{\underline{1}}_{\underline{3}}4^{0}_{\underline{3}}3^{0}_{\underline{1}}2^{1}_{\underline{1}}3^{1}_{1}1^{2}_{1}2^{2}_{3}1^{3}_{5}$		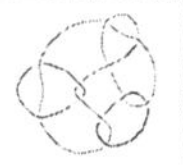
11^n_{401}: $3^{\underline{4}}_{\underline{9}}2^{\underline{4}}_{\underline{7}}4^{\underline{3}}_{\underline{7}}1^{\underline{3}}_{\underline{5}}7^{\underline{2}}_{\underline{5}}4^{\underline{2}}_{\underline{3}}5^{\underline{1}}_{\underline{3}}7^{\underline{1}}_{\underline{1}}9^{0}_{\underline{1}}7^{0}_{1}6^{1}_{1}7^{1}_{3}5^{2}_{3}6^{2}_{5}2^{3}_{5}5^{3}_{7}1^{4}_{7}2^{4}_{9}1^{5}_{11}$ 11^n_{402}: $2^{\underline{5}}_{\underline{13}}3^{\underline{4}}_{\underline{11}}3^{\underline{4}}_{\underline{9}}6^{\underline{3}}_{\underline{9}}2^{\underline{3}}_{\underline{7}}5^{\underline{2}}_{\underline{7}}6^{\underline{2}}_{\underline{5}}5^{\underline{1}}_{\underline{5}}5^{\underline{1}}_{\underline{3}}6^{0}_{\underline{3}}7^{0}_{\underline{1}}4^{1}_{\underline{1}}4^{1}_{1}2^{2}_{1}4^{2}_{3}2^{3}_{5}1^{4}_{5}1^{4}_{7}$	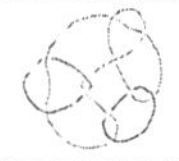	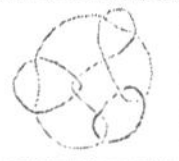
11^n_{403}: $1^{\underline{6}}_{\underline{11}}1^{\underline{5}}_{\underline{7}}4^{\underline{4}}_{\underline{7}}2^{\underline{4}}_{\underline{5}}1^{\underline{3}}_{\underline{5}}2^{\underline{3}}_{\underline{3}}3^{\underline{2}}_{\underline{3}}1^{\underline{2}}_{\underline{1}}2^{\underline{1}}_{\underline{1}}3^{\underline{1}}_{1}3^{0}_{\underline{1}}4^{0}_{1}2^{0}_{3}1^{1}_{1}2^{1}_{3}2^{1}_{5}1^{2}_{3}1^{2}_{5}1^{2}_{7}1^{3}_{7}1^{4}_{7}1^{5}_{11}$ 11^n_{404}: $1^{\underline{7}}_{\underline{17}}1^{\underline{6}}_{\underline{13}}2^{\underline{5}}_{\underline{13}}1^{\underline{4}}_{\underline{13}}1^{\underline{4}}_{\underline{11}}2^{\underline{4}}_{\underline{9}}1^{\underline{3}}_{\underline{11}}4^{\underline{3}}_{\underline{9}}1^{\underline{3}}_{\underline{7}}2^{\underline{2}}_{\underline{9}}2^{\underline{2}}_{\underline{7}}3^{\underline{2}}_{\underline{5}}1^{\underline{1}}_{\underline{7}}2^{\underline{1}}_{\underline{5}}1^{\underline{1}}_{\underline{3}}2^{0}_{\underline{5}}5^{0}_{\underline{3}}3^{0}_{\underline{1}}2^{1}_{\underline{3}}1^{1}_{\underline{1}}2^{2}_{1}1^{3}_{1}1^{4}_{5}$		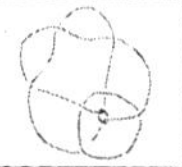
11^n_{405}: $1^{\underline{5}}_{\underline{11}}2^{\underline{4}}_{\underline{9}}1^{\underline{4}}_{\underline{7}}4^{\underline{3}}_{\underline{7}}2^{\underline{3}}_{\underline{5}}4^{\underline{2}}_{\underline{5}}4^{\underline{2}}_{\underline{3}}6^{\underline{1}}_{\underline{3}}4^{\underline{1}}_{\underline{1}}6^{0}_{\underline{1}}8^{0}_{1}5^{1}_{1}4^{1}_{3}3^{2}_{3}5^{2}_{5}1^{3}_{5}3^{3}_{7}2^{4}_{7}3^{4}_{9}$ 11^n_{406}: $1^{\underline{7}}_{\underline{15}}1^{\underline{6}}_{\underline{11}}2^{\underline{5}}_{\underline{11}}1^{\underline{4}}_{\underline{11}}1^{\underline{4}}_{\underline{9}}2^{\underline{4}}_{\underline{7}}2^{\underline{3}}_{\underline{7}}1^{\underline{3}}_{\underline{5}}2^{\underline{2}}_{\underline{7}}2^{\underline{2}}_{\underline{5}}1^{\underline{2}}_{\underline{3}}1^{\underline{1}}_{\underline{5}}2^{\underline{1}}_{\underline{3}}2^{\underline{1}}_{\underline{1}}2^{0}_{\underline{3}}5^{0}_{\underline{1}}3^{0}_{1}2^{1}_{\underline{1}}1^{1}_{1}2^{2}_{3}1^{3}_{3}1^{4}_{7}$	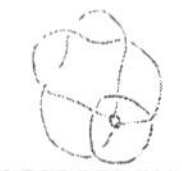	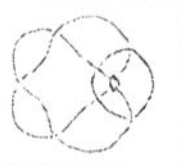
11^n_{407}: $1^{\underline{5}}_{\underline{9}}2^{\underline{4}}_{\underline{7}}1^{\underline{4}}_{\underline{5}}3^{\underline{3}}_{\underline{5}}2^{\underline{3}}_{\underline{3}}5^{\underline{2}}_{\underline{3}}3^{\underline{2}}_{\underline{1}}5^{\underline{1}}_{\underline{1}}5^{\underline{1}}_{1}6^{0}_{1}7^{0}_{3}5^{1}_{3}4^{1}_{5}4^{2}_{5}5^{2}_{7}1^{3}_{7}4^{3}_{9}2^{4}_{9}3^{4}_{11}$ 11^n_{408}: $1^{\underline{7}}_{\underline{15}}1^{\underline{6}}_{\underline{13}}1^{\underline{6}}_{\underline{11}}2^{\underline{5}}_{\underline{11}}1^{\underline{5}}_{\underline{9}}2^{\underline{4}}_{\underline{9}}2^{\underline{4}}_{\underline{7}}2^{\underline{3}}_{\underline{7}}2^{\underline{3}}_{\underline{5}}1^{\underline{2}}_{\underline{7}}2^{\underline{2}}_{\underline{5}}2^{\underline{2}}_{\underline{3}}2^{\underline{1}}_{\underline{3}}2^{\underline{1}}_{\underline{1}}2^{0}_{\underline{3}}4^{0}_{\underline{1}}4^{0}_{1}1^{1}_{\underline{1}}1^{1}_{1}1^{2}_{3}1^{3}_{3}1^{4}_{7}$	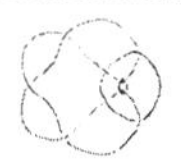	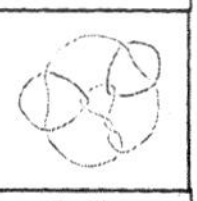
11^n_{409}: $2^{\underline{5}}_{\underline{11}}3^{\underline{4}}_{\underline{9}}2^{\underline{4}}_{\underline{7}}5^{\underline{3}}_{\underline{7}}3^{\underline{3}}_{\underline{5}}8^{\underline{2}}_{\underline{5}}5^{\underline{2}}_{\underline{3}}5^{\underline{1}}_{\underline{3}}8^{\underline{1}}_{\underline{1}}9^{0}_{\underline{1}}9^{0}_{1}6^{1}_{1}5^{1}_{3}3^{2}_{3}6^{2}_{5}1^{3}_{5}3^{3}_{7}1^{4}_{9}$ 11^n_{410}: $3^{\underline{7}}_{\underline{19}}4^{\underline{6}}_{\underline{17}}3^{\underline{6}}_{\underline{15}}6^{\underline{5}}_{\underline{15}}4^{\underline{5}}_{\underline{13}}7^{\underline{4}}_{\underline{13}}8^{\underline{4}}_{\underline{11}}8^{\underline{3}}_{\underline{11}}5^{\underline{3}}_{\underline{9}}6^{\underline{2}}_{\underline{9}}8^{\underline{2}}_{\underline{7}}3^{\underline{1}}_{\underline{7}}6^{\underline{1}}_{\underline{5}}4^{0}_{\underline{5}}5^{0}_{\underline{3}}1^{1}_{\underline{3}}2^{1}_{\underline{1}}1^{2}_{1}$	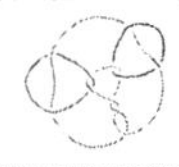	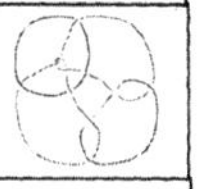
11^n_{411}: $1^{\underline{11}}_{\underline{27}}1^{\underline{10}}_{\underline{25}}1^{\underline{10}}_{\underline{23}}1^{\underline{9}}_{\underline{23}}1^{\underline{9}}_{\underline{21}}3^{\underline{8}}_{\underline{21}}2^{\underline{8}}_{\underline{19}}3^{\underline{7}}_{\underline{21}}2^{\underline{7}}_{\underline{19}}2^{\underline{7}}_{\underline{17}}1^{\underline{6}}_{\underline{19}}5^{\underline{6}}_{\underline{17}}2^{\underline{6}}_{\underline{15}}2^{\underline{5}}_{\underline{17}}2^{\underline{5}}_{\underline{15}}2^{\underline{5}}_{\underline{13}}3^{\underline{4}}_{\underline{15}}5^{\underline{4}}_{\underline{13}}1^{\underline{4}}_{\underline{11}}2^{\underline{3}}_{\underline{13}}1^{\underline{3}}_{\underline{11}}1^{\underline{3}}_{\underline{9}}1^{\underline{2}}_{\underline{11}}2^{\underline{2}}_{\underline{9}}1^{\underline{1}}_{\underline{7}}1^{0}_{\underline{7}}1^{0}_{\underline{5}}$ 11^n_{412}: $1^{\underline{9}}_{\underline{21}}1^{\underline{8}}_{\underline{19}}1^{\underline{8}}_{\underline{17}}3^{\underline{7}}_{\underline{17}}1^{\underline{7}}_{\underline{15}}3^{\underline{6}}_{\underline{15}}4^{\underline{6}}_{\underline{13}}4^{\underline{5}}_{\underline{13}}2^{\underline{5}}_{\underline{11}}4^{\underline{4}}_{\underline{11}}4^{\underline{4}}_{\underline{9}}2^{\underline{3}}_{\underline{9}}4^{\underline{3}}_{\underline{7}}4^{\underline{2}}_{\underline{7}}3^{\underline{2}}_{\underline{5}}3^{\underline{1}}_{\underline{3}}2^{0}_{\underline{3}}2^{0}_{\underline{1}}$	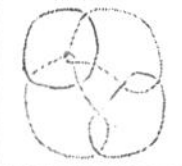	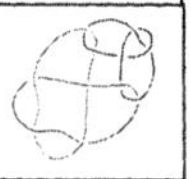
11^n_{413}: $1^{\underline{6}}_{\underline{11}}1^{\underline{6}}_{\underline{9}}1^{\underline{4}}_{\underline{7}}1^{\underline{3}}_{\underline{3}}3^{\underline{2}}_{\underline{5}}2^{\underline{2}}_{\underline{3}}1^{\underline{1}}_{\underline{3}}2^{\underline{1}}_{\underline{1}}1^{\underline{1}}_{1}4^{0}_{\underline{1}}3^{0}_{1}2^{1}_{1}2^{1}_{3}2^{2}_{3}2^{2}_{5}1^{3}_{5}2^{3}_{7}1^{4}_{7}1^{4}_{9}1^{5}_{11}$ 11^n_{414}: $1^{\underline{7}}_{\underline{15}}1^{\underline{6}}_{\underline{11}}1^{\underline{5}}_{\underline{11}}1^{\underline{4}}_{\underline{9}}1^{\underline{4}}_{\underline{7}}1^{\underline{3}}_{\underline{9}}1^{\underline{3}}_{\underline{5}}1^{\underline{2}}_{\underline{7}}4^{\underline{2}}_{\underline{5}}1^{\underline{2}}_{\underline{3}}2^{\underline{1}}_{\underline{5}}1^{\underline{1}}_{\underline{1}}1^{0}_{\underline{3}}3^{0}_{\underline{1}}1^{0}_{1}1^{1}_{\underline{1}}1^{1}_{1}1^{2}_{1}1^{2}_{3}1^{3}_{5}1^{4}_{5}1^{4}_{7}$	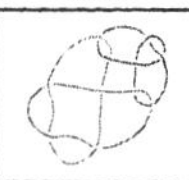	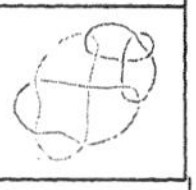
11^n_{415}: $3^{\underline{2}}_{\underline{3}}2^{\underline{2}}_{\underline{1}}3^{\underline{1}}_{\underline{1}}1^{\underline{1}}_{1}6^{0}_{1}4^{0}_{3}5^{1}_{3}5^{1}_{5}7^{2}_{5}5^{2}_{7}4^{3}_{7}7^{3}_{9}5^{4}_{9}5^{4}_{11}2^{5}_{11}4^{5}_{13}1^{6}_{13}2^{6}_{15}1^{7}_{17}$ 11^n_{416}: $2^{\underline{7}}_{\underline{19}}2^{\underline{6}}_{\underline{17}}3^{\underline{6}}_{\underline{15}}5^{\underline{5}}_{\underline{15}}1^{\underline{5}}_{\underline{13}}4^{\underline{4}}_{\underline{13}}5^{\underline{4}}_{\underline{11}}4^{\underline{3}}_{\underline{11}}4^{\underline{3}}_{\underline{9}}5^{\underline{2}}_{\underline{9}}5^{\underline{2}}_{\underline{7}}2^{\underline{1}}_{\underline{7}}4^{\underline{1}}_{\underline{5}}3^{0}_{\underline{5}}4^{0}_{\underline{3}}1^{1}_{\underline{3}}1^{1}_{\underline{1}}1^{2}_{1}$	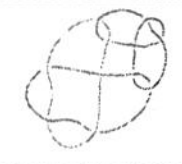	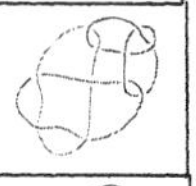
11^n_{417}: $1^{\underline{4}}_{\underline{5}}1^{\underline{3}}_{\underline{1}}4^{\underline{2}}_{\underline{1}}2^{\underline{2}}_{1}1^{\underline{1}}_{1}2^{\underline{1}}_{3}4^{0}_{3}2^{0}_{5}2^{1}_{5}3^{1}_{7}1^{2}_{5}3^{2}_{7}2^{2}_{9}2^{3}_{9}3^{3}_{11}1^{4}_{9}2^{4}_{11}1^{4}_{13}1^{5}_{15}$ 11^n_{418}: $1^{\underline{8}}_{\underline{19}}1^{\underline{7}}_{\underline{17}}1^{\underline{7}}_{\underline{15}}3^{\underline{6}}_{\underline{15}}2^{\underline{6}}_{\underline{13}}3^{\underline{5}}_{\underline{13}}2^{\underline{5}}_{\underline{11}}4^{\underline{4}}_{\underline{11}}4^{\underline{4}}_{\underline{9}}3^{\underline{3}}_{\underline{9}}3^{\underline{3}}_{\underline{7}}4^{\underline{2}}_{\underline{7}}4^{\underline{2}}_{\underline{5}}1^{\underline{1}}_{\underline{5}}3^{\underline{1}}_{\underline{3}}2^{0}_{\underline{3}}2^{0}_{\underline{1}}1^{1}_{1}$		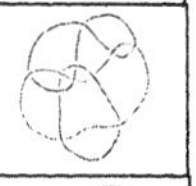
11^n_{419}: $2^{\underline{2}}_{\underline{3}}1^{\underline{2}}_{\underline{1}}1^{\underline{1}}_{\underline{1}}1^{\underline{1}}_{1}3^{0}_{1}2^{0}_{3}2^{1}_{3}2^{1}_{5}3^{2}_{5}3^{2}_{7}1^{3}_{7}2^{3}_{9}2^{4}_{9}2^{4}_{11}1^{5}_{13}$ 11^n_{420}: $1^{\underline{6}}_{\underline{17}}1^{\underline{6}}_{\underline{15}}1^{\underline{5}}_{\underline{17}}1^{\underline{5}}_{\underline{15}}1^{\underline{4}}_{\underline{15}}2^{\underline{4}}_{\underline{13}}2^{\underline{4}}_{\underline{11}}1^{\underline{3}}_{\underline{13}}1^{\underline{3}}_{\underline{11}}1^{\underline{2}}_{\underline{11}}2^{\underline{2}}_{\underline{9}}1^{\underline{2}}_{\underline{7}}1^{\underline{1}}_{\underline{9}}1^{\underline{1}}_{\underline{7}}1^{0}_{\underline{7}}2^{0}_{\underline{5}}1^{1}_{\underline{5}}1^{2}_{\underline{1}}$	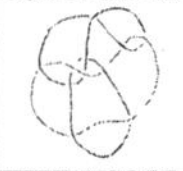	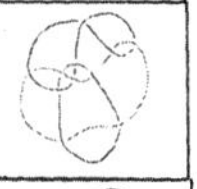
11^n_{421}: $1^{\underline{5}}_{\underline{9}}1^{\underline{4}}_{\underline{7}}1^{\underline{4}}_{\underline{5}}1^{\underline{3}}_{\underline{5}}1^{\underline{3}}_{\underline{3}}3^{\underline{2}}_{\underline{3}}2^{\underline{2}}_{\underline{1}}1^{\underline{1}}_{\underline{1}}2^{\underline{1}}_{1}3^{0}_{1}2^{0}_{3}1^{1}_{3}2^{1}_{5}2^{2}_{5}2^{2}_{7}1^{3}_{9}1^{4}_{9}1^{4}_{11}$ 11^n_{422}: $1^{\underline{1}}_{\underline{1}}3^{0}_{1}2^{0}_{3}3^{1}_{3}2^{1}_{5}5^{2}_{5}5^{2}_{7}5^{3}_{7}3^{3}_{9}3^{4}_{9}5^{4}_{11}3^{5}_{11}3^{5}_{13}2^{6}_{13}3^{6}_{15}2^{7}_{17}1^{8}_{17}1^{8}_{19}$	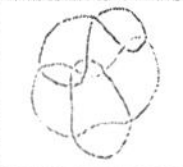	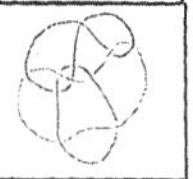

L: KH	1st line	2nd line
11^n_{423}: $1^{\underline{5}}_{\underline{9}}2^{\underline{4}}_{\underline{7}}1^{\underline{4}}_{\underline{5}}1^{\underline{3}}_{\underline{5}}2^{\underline{3}}_{\underline{3}}4^{\underline{2}}_{\underline{3}}2^{\underline{2}}_{\underline{1}}2^{\underline{1}}_{\underline{1}}3^{\underline{1}}_{1}1^{0}_{\underline{1}}4^{0}_{1}3^{0}_{3}2^{1}_{3}2^{1}_{5}1^{2}_{3}2^{2}_{5}2^{2}_{7}1^{3}_{9}$	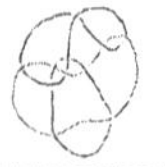	
11^n_{424}: $1^{\underline{7}}_{\underline{17}}2^{\underline{6}}_{\underline{15}}1^{\underline{6}}_{\underline{13}}4^{\underline{5}}_{\underline{13}}2^{\underline{5}}_{\underline{11}}5^{\underline{4}}_{\underline{11}}4^{\underline{4}}_{\underline{9}}5^{\underline{3}}_{\underline{9}}5^{\underline{3}}_{\underline{7}}7^{\underline{2}}_{\underline{7}}7^{\underline{2}}_{\underline{5}}4^{\underline{1}}_{\underline{5}}5^{\underline{1}}_{\underline{3}}4^{0}_{\underline{3}}6^{0}_{\underline{1}}2^{1}_{\underline{1}}2^{1}_{1}2^{2}_{3}$		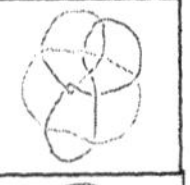
11^n_{425}: $1^{\underline{6}}_{\underline{13}}1^{\underline{6}}_{\underline{11}}2^{\underline{5}}_{\underline{11}}2^{\underline{4}}_{\underline{9}}2^{\underline{4}}_{\underline{7}}2^{\underline{3}}_{\underline{7}}2^{\underline{3}}_{\underline{5}}4^{\underline{2}}_{\underline{5}}2^{\underline{2}}_{\underline{3}}1^{\underline{1}}_{\underline{3}}4^{\underline{1}}_{\underline{1}}4^{0}_{\underline{1}}3^{0}_{1}1^{1}_{1}2^{1}_{3}1^{2}_{3}2^{2}_{5}$	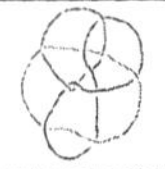	
11^n_{426}: $1^{\underline{6}}_{\underline{11}}1^{\underline{5}}_{\underline{7}}1^{\underline{4}}_{\underline{7}}1^{\underline{3}}_{\underline{5}}1^{\underline{3}}_{\underline{3}}2^{\underline{2}}_{\underline{5}}2^{\underline{2}}_{\underline{3}}1^{\underline{2}}_{\underline{1}}1^{\underline{1}}_{\underline{1}}3^{0}_{\underline{1}}2^{0}_{1}1^{0}_{3}1^{1}_{3}1^{2}_{3}1^{3}_{7}$		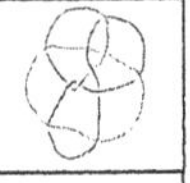
11^n_{427}: $2^{\underline{6}}_{\underline{15}}1^{\underline{6}}_{\underline{13}}2^{\underline{5}}_{\underline{13}}1^{\underline{5}}_{\underline{11}}5^{\underline{4}}_{\underline{11}}3^{\underline{4}}_{\underline{9}}4^{\underline{3}}_{\underline{9}}4^{\underline{3}}_{\underline{7}}6^{\underline{2}}_{\underline{7}}5^{\underline{2}}_{\underline{5}}4^{\underline{1}}_{\underline{5}}5^{\underline{1}}_{\underline{3}}4^{0}_{\underline{3}}5^{0}_{\underline{1}}2^{1}_{\underline{1}}3^{1}_{1}1^{2}_{1}2^{2}_{3}1^{3}_{5}$	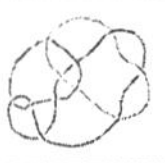	
11^n_{428}: $1^{\underline{4}}_{\underline{9}}1^{\underline{3}}_{\underline{5}}1^{\underline{2}}_{\underline{7}}3^{\underline{2}}_{\underline{5}}1^{\underline{2}}_{\underline{3}}1^{\underline{1}}_{\underline{5}}1^{\underline{1}}_{\underline{3}}1^{\underline{1}}_{\underline{1}}1^{0}_{\underline{3}}3^{0}_{\underline{1}}2^{0}_{1}1^{1}_{\underline{1}}1^{1}_{1}1^{2}_{1}1^{2}_{3}1^{2}_{5}1^{3}_{5}1^{4}_{5}1^{5}_{9}$		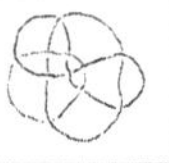
11^n_{429}: $1^{\underline{6}}_{\underline{17}}1^{\underline{6}}_{\underline{15}}2^{\underline{5}}_{\underline{15}}3^{\underline{4}}_{\underline{13}}3^{\underline{4}}_{\underline{11}}3^{\underline{3}}_{\underline{11}}2^{\underline{3}}_{\underline{9}}4^{\underline{2}}_{\underline{9}}4^{\underline{2}}_{\underline{7}}3^{\underline{1}}_{\underline{7}}3^{\underline{1}}_{\underline{5}}3^{0}_{\underline{5}}4^{0}_{\underline{3}}2^{1}_{\underline{3}}2^{1}_{\underline{1}}1^{2}_{\underline{1}}2^{2}_{1}1^{3}_{3}$	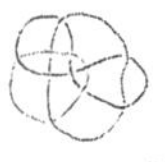	
11^n_{430}: $1^{\underline{6}}_{\underline{13}}2^{\underline{5}}_{\underline{11}}1^{\underline{5}}_{\underline{9}}3^{\underline{4}}_{\underline{9}}2^{\underline{4}}_{\underline{7}}3^{\underline{3}}_{\underline{7}}3^{\underline{3}}_{\underline{5}}6^{\underline{2}}_{\underline{5}}5^{\underline{2}}_{\underline{3}}3^{\underline{1}}_{\underline{3}}4^{\underline{1}}_{\underline{1}}5^{0}_{\underline{1}}5^{0}_{1}2^{1}_{1}3^{1}_{3}1^{2}_{3}2^{2}_{5}1^{3}_{7}$		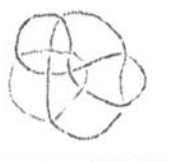
11^n_{431}: $1^{\underline{1}}_{\underline{1}}4^{0}_{1}2^{0}_{3}4^{1}_{3}3^{1}_{5}7^{2}_{5}6^{2}_{7}6^{3}_{7}5^{3}_{9}6^{4}_{9}7^{4}_{11}4^{5}_{11}5^{5}_{13}3^{6}_{13}4^{6}_{15}1^{7}_{15}3^{7}_{17}1^{8}_{19}$	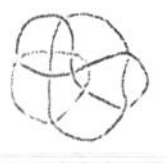	
11^n_{432}: $1^{\underline{2}}_{\underline{1}}1^{\underline{1}}_{3}2^{0}_{3}1^{0}_{5}1^{1}_{3}1^{1}_{5}1^{1}_{7}1^{2}_{5}4^{2}_{7}2^{2}_{9}2^{3}_{7}1^{3}_{9}1^{3}_{11}1^{4}_{9}3^{4}_{11}1^{4}_{13}1^{5}_{11}1^{5}_{13}1^{5}_{15}1^{6}_{13}1^{6}_{15}1^{7}_{17}1^{8}_{17}1^{8}_{19}$		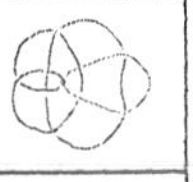
11^n_{433}: $1^{\underline{2}}_{1}1^{\underline{1}}_{5}2^{0}_{5}1^{0}_{7}1^{1}_{7}1^{1}_{9}2^{2}_{7}3^{2}_{9}1^{2}_{11}1^{3}_{9}1^{3}_{11}1^{3}_{13}2^{4}_{11}2^{4}_{13}1^{4}_{15}1^{5}_{13}2^{5}_{15}1^{5}_{17}1^{6}_{15}1^{6}_{17}1^{7}_{19}1^{8}_{19}1^{8}_{21}$	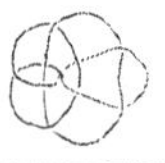	
11^n_{434}: $1^{\underline{3}}_{\underline{7}}3^{\underline{2}}_{\underline{5}}2^{\underline{2}}_{\underline{3}}4^{\underline{1}}_{\underline{3}}2^{\underline{1}}_{\underline{1}}5^{0}_{\underline{1}}5^{0}_{1}5^{1}_{1}4^{1}_{3}5^{2}_{3}6^{2}_{5}3^{3}_{5}4^{3}_{7}3^{4}_{7}4^{4}_{9}1^{5}_{9}2^{5}_{11}1^{6}_{13}$		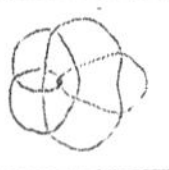
11^n_{435}: $3^{\underline{2}}_{\underline{3}}1^{\underline{2}}_{\underline{1}}4^{\underline{1}}_{\underline{1}}2^{\underline{1}}_{1}7^{0}_{1}5^{0}_{3}7^{1}_{3}6^{1}_{5}9^{2}_{5}8^{2}_{7}6^{3}_{7}8^{3}_{9}6^{4}_{9}7^{4}_{11}3^{5}_{11}5^{5}_{13}1^{6}_{13}3^{6}_{15}1^{7}_{17}$	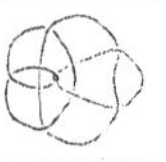	
11^n_{436}: $1^{\underline{5}}_{\underline{11}}1^{\underline{4}}_{\underline{9}}1^{\underline{4}}_{\underline{7}}1^{\underline{3}}_{\underline{7}}1^{\underline{3}}_{\underline{5}}1^{\underline{2}}_{\underline{7}}2^{\underline{2}}_{\underline{5}}1^{\underline{2}}_{\underline{3}}1^{\underline{1}}_{\underline{5}}2^{\underline{1}}_{\underline{3}}2^{\underline{1}}_{\underline{1}}2^{0}_{\underline{3}}5^{0}_{\underline{1}}4^{0}_{1}2^{1}_{\underline{1}}2^{1}_{1}1^{2}_{1}2^{2}_{3}1^{2}_{5}1^{3}_{3}1^{3}_{5}1^{4}_{5}1^{4}_{7}1^{5}_{9}$		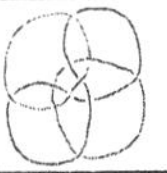
11^n_{437}: $1^{\underline{8}}_{\underline{19}}1^{\underline{8}}_{\underline{17}}1^{\underline{7}}_{\underline{17}}1^{\underline{6}}_{\underline{15}}1^{\underline{6}}_{\underline{13}}1^{\underline{5}}_{\underline{15}}1^{\underline{5}}_{\underline{13}}1^{\underline{5}}_{\underline{11}}1^{\underline{4}}_{\underline{13}}3^{\underline{4}}_{\underline{11}}1^{\underline{4}}_{\underline{9}}1^{\underline{3}}_{\underline{11}}1^{\underline{3}}_{\underline{9}}2^{\underline{3}}_{\underline{7}}2^{\underline{2}}_{\underline{9}}4^{\underline{2}}_{\underline{7}}1^{\underline{2}}_{\underline{5}}1^{\underline{1}}_{\underline{7}}1^{\underline{1}}_{\underline{5}}1^{\underline{1}}_{\underline{3}}1^{0}_{\underline{5}}2^{0}_{\underline{3}}1^{1}_{\underline{3}}1^{2}_{1}$	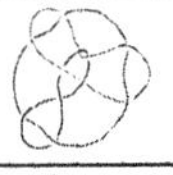	
11^n_{438}: $4^{0}_{2}4^{0}_{4}6^{1}_{4}7^{2}_{6}6^{2}_{8}4^{3}_{8}7^{3}_{10}11^{4}_{10}8^{4}_{12}5^{5}_{12}7^{5}_{14}5^{6}_{14}5^{6}_{16}1^{7}_{16}5^{7}_{18}1^{8}_{18}1^{8}_{20}1^{9}_{22}$		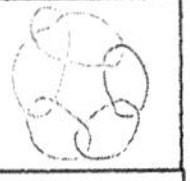
11^n_{439}: $1^{\underline{5}}_{\underline{10}}1^{\underline{4}}_{\underline{6}}1^{\underline{3}}_{\underline{6}}1^{\underline{2}}_{\underline{4}}1^{\underline{2}}_{\underline{2}}3^{\underline{1}}_{\underline{4}}1^{\underline{1}}_{0}1^{0}_{\underline{2}}7^{0}_{0}4^{0}_{2}4^{1}_{0}1^{1}_{2}1^{2}_{2}4^{2}_{4}1^{3}_{4}1^{3}_{6}4^{4}_{6}5^{4}_{8}1^{5}_{8}1^{6}_{12}$		
11^n_{440}: $1^{\underline{4}}_{\underline{6}}1^{\underline{4}}_{\underline{4}}1^{\underline{2}}_{\underline{2}}1^{\underline{1}}_{2}6^{0}_{0}7^{0}_{2}4^{1}_{2}1^{1}_{6}4^{2}_{4}4^{2}_{6}1^{3}_{6}4^{3}_{8}5^{4}_{8}2^{4}_{10}1^{5}_{10}4^{5}_{12}1^{6}_{12}1^{6}_{14}1^{7}_{16}$		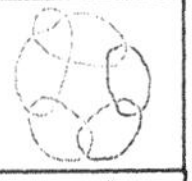
11^n_{441}: $1^{\underline{8}}_{\underline{18}}1^{\underline{8}}_{\underline{16}}1^{\underline{7}}_{\underline{16}}5^{\underline{6}}_{\underline{14}}1^{\underline{6}}_{\underline{12}}1^{\underline{5}}_{\underline{12}}5^{\underline{5}}_{\underline{10}}11^{\underline{4}}_{\underline{10}}7^{\underline{4}}_{\underline{8}}4^{\underline{3}}_{\underline{8}}5^{\underline{3}}_{\underline{6}}7^{\underline{2}}_{\underline{6}}4^{\underline{2}}_{\underline{4}}3^{\underline{1}}_{\underline{4}}7^{\underline{1}}_{\underline{2}}1^{0}_{\underline{4}}4^{0}_{\underline{2}}3^{0}_{0}3^{1}_{2}$	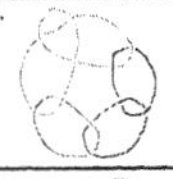	
11^n_{442}: $3^{\underline{2}}_{\underline{4}}1^{\underline{2}}_{\underline{2}}3^{\underline{1}}_{\underline{2}}2^{\underline{1}}_{0}7^{0}_{0}6^{0}_{2}6^{1}_{2}4^{1}_{4}8^{2}_{4}9^{2}_{6}5^{3}_{6}5^{3}_{8}5^{4}_{8}6^{4}_{10}2^{5}_{10}4^{5}_{12}1^{6}_{12}2^{6}_{14}1^{7}_{16}$		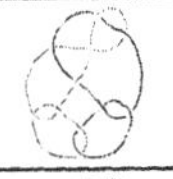
11^n_{443}: $1^{\underline{5}}_{\underline{10}}1^{\underline{4}}_{\underline{8}}1^{\underline{4}}_{\underline{6}}1^{\underline{3}}_{\underline{6}}1^{\underline{3}}_{\underline{4}}3^{\underline{2}}_{\underline{4}}2^{\underline{2}}_{\underline{2}}1^{\underline{1}}_{\underline{4}}1^{\underline{1}}_{\underline{2}}2^{\underline{1}}_{0}1^{0}_{\underline{2}}5^{0}_{0}4^{0}_{2}1^{1}_{0}2^{1}_{2}1^{1}_{4}2^{2}_{2}4^{2}_{4}3^{2}_{6}1^{3}_{4}1^{3}_{6}1^{4}_{6}2^{4}_{8}1^{5}_{8}1^{6}_{12}$	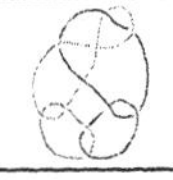	
11^n_{444}: $1^{\underline{8}}_{\underline{18}}1^{\underline{7}}_{\underline{16}}1^{\underline{7}}_{\underline{14}}5^{\underline{6}}_{\underline{14}}2^{\underline{6}}_{\underline{12}}2^{\underline{5}}_{\underline{12}}4^{\underline{5}}_{\underline{10}}8^{\underline{4}}_{\underline{10}}5^{\underline{4}}_{\underline{8}}4^{\underline{3}}_{\underline{8}}5^{\underline{3}}_{\underline{6}}1^{\underline{2}}_{\underline{8}}7^{\underline{2}}_{\underline{6}}6^{\underline{2}}_{\underline{4}}3^{\underline{1}}_{\underline{4}}4^{\underline{1}}_{\underline{2}}1^{0}_{\underline{4}}3^{0}_{\underline{2}}3^{0}_{0}2^{1}_{2}$		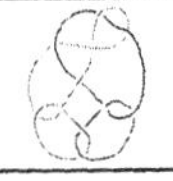
11^n_{445}: $2^{\underline{2}}_{\underline{4}}1^{\underline{1}}_{\underline{2}}2^{\underline{1}}_{0}8^{0}_{0}7^{0}_{2}1^{0}_{4}7^{1}_{2}3^{1}_{4}5^{2}_{4}7^{2}_{6}4^{3}_{6}5^{3}_{8}6^{4}_{8}6^{4}_{10}2^{5}_{10}4^{5}_{12}1^{6}_{12}2^{6}_{14}1^{7}_{16}$	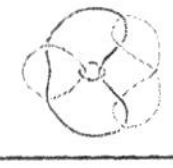	
11^n_{446}: $1^{\underline{5}}_{\underline{10}}1^{\underline{4}}_{\underline{8}}1^{\underline{4}}_{\underline{6}}1^{\underline{3}}_{\underline{6}}1^{\underline{3}}_{\underline{4}}2^{\underline{2}}_{\underline{4}}1^{\underline{2}}_{\underline{2}}1^{\underline{1}}_{\underline{4}}1^{\underline{1}}_{\underline{2}}2^{\underline{1}}_{0}3^{0}_{\underline{2}}7^{0}_{0}5^{0}_{2}2^{1}_{0}2^{1}_{2}1^{2}_{2}2^{2}_{4}1^{2}_{6}1^{3}_{4}1^{3}_{6}2^{4}_{6}3^{4}_{8}1^{5}_{8}1^{6}_{12}$		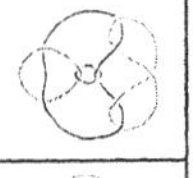
11^n_{447}: $1^{\underline{8}}_{\underline{18}}1^{\underline{7}}_{\underline{16}}1^{\underline{7}}_{\underline{14}}5^{\underline{6}}_{\underline{14}}1^{\underline{6}}_{\underline{12}}2^{\underline{5}}_{\underline{12}}5^{\underline{5}}_{\underline{10}}10^{\underline{4}}_{\underline{10}}8^{\underline{4}}_{\underline{8}}5^{\underline{3}}_{\underline{8}}4^{\underline{3}}_{\underline{6}}6^{\underline{2}}_{\underline{6}}5^{\underline{2}}_{\underline{4}}3^{\underline{1}}_{\underline{4}}6^{\underline{1}}_{\underline{2}}1^{0}_{\underline{4}}4^{0}_{\underline{2}}4^{0}_{0}2^{1}_{2}$	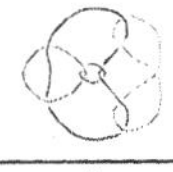	
11^n_{448}: $1^{\underline{4}}_{\underline{10}}2^{\underline{4}}_{\underline{8}}1^{\underline{4}}_{\underline{6}}1^{\underline{3}}_{\underline{8}}3^{\underline{2}}_{\underline{6}}2^{\underline{2}}_{\underline{4}}1^{\underline{1}}_{\underline{4}}3^{\underline{1}}_{\underline{2}}1^{\underline{1}}_{0}7^{0}_{\underline{2}}7^{0}_{0}2^{0}_{2}3^{1}_{0}3^{1}_{2}2^{2}_{2}3^{2}_{4}2^{3}_{4}2^{3}_{6}1^{4}_{6}2^{4}_{8}1^{5}_{10}$		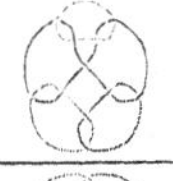
11^n_{449}: $1^{\underline{8}}_{\underline{18}}1^{\underline{8}}_{\underline{16}}1^{\underline{7}}_{\underline{16}}4^{\underline{6}}_{\underline{14}}1^{\underline{6}}_{\underline{12}}1^{\underline{5}}_{\underline{12}}4^{\underline{5}}_{\underline{10}}8^{\underline{4}}_{\underline{10}}5^{\underline{4}}_{\underline{8}}3^{\underline{3}}_{\underline{8}}4^{\underline{3}}_{\underline{6}}5^{\underline{2}}_{\underline{6}}3^{\underline{2}}_{\underline{4}}2^{\underline{1}}_{\underline{4}}5^{\underline{1}}_{\underline{2}}1^{0}_{\underline{4}}4^{0}_{\underline{2}}4^{0}_{0}1^{1}_{2}$	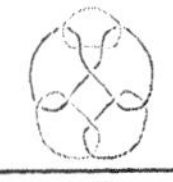	
11^n_{450}: $4^{\underline{4}}_{\underline{10}}2^{\underline{4}}_{\underline{8}}6^{\underline{3}}_{\underline{8}}2^{\underline{3}}_{\underline{6}}8^{\underline{2}}_{\underline{6}}6^{\underline{2}}_{\underline{4}}6^{\underline{1}}_{\underline{4}}8^{\underline{1}}_{\underline{2}}12^{0}_{\underline{2}}12^{0}_{0}8^{1}_{0}6^{1}_{2}5^{2}_{2}8^{2}_{4}2^{3}_{4}5^{3}_{6}1^{4}_{6}2^{4}_{8}1^{5}_{10}$		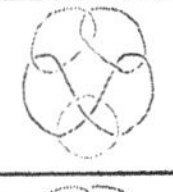
11^n_{451}: $1^{\underline{7}}_{\underline{16}}2^{\underline{6}}_{\underline{14}}1^{\underline{6}}_{\underline{12}}3^{\underline{5}}_{\underline{12}}2^{\underline{5}}_{\underline{10}}4^{\underline{4}}_{\underline{10}}5^{\underline{4}}_{\underline{8}}4^{\underline{3}}_{\underline{8}}2^{\underline{3}}_{\underline{6}}1^{\underline{2}}_{\underline{8}}4^{\underline{2}}_{\underline{6}}4^{\underline{2}}_{\underline{4}}2^{\underline{1}}_{\underline{4}}4^{\underline{1}}_{\underline{2}}2^{0}_{\underline{4}}6^{0}_{\underline{2}}6^{0}_{0}1^{1}_{\underline{2}}1^{1}_{0}1^{2}_{2}1^{3}_{2}1^{4}_{6}$	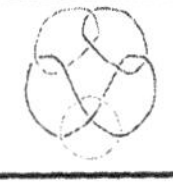	
11^n_{452}: $1^{\underline{6}}_{\underline{12}}1^{\underline{5}}_{\underline{10}}1^{\underline{5}}_{\underline{8}}5^{\underline{4}}_{\underline{8}}3^{\underline{4}}_{\underline{6}}2^{\underline{3}}_{\underline{6}}3^{\underline{3}}_{\underline{4}}5^{\underline{2}}_{\underline{4}}2^{\underline{2}}_{\underline{2}}3^{\underline{1}}_{\underline{2}}5^{\underline{1}}_{0}4^{0}_{\underline{2}}8^{0}_{0}5^{0}_{2}1^{1}_{0}2^{1}_{2}2^{1}_{4}1^{2}_{2}1^{2}_{4}2^{2}_{6}1^{3}_{6}1^{4}_{6}1^{5}_{10}$		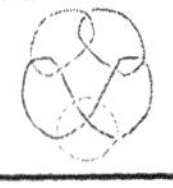

L: KH	1st line	2nd line
11^n_{453}: $1^{\underline{9}}_{\underline{26}}2^{\underline{8}}_{\underline{24}}2^{\underline{8}}_{\underline{22}}1^{\underline{8}}_{\underline{20}}2^{\underline{7}}_{\underline{22}}2^{\underline{7}}_{\underline{20}}4^{\underline{6}}_{\underline{20}}6^{\underline{6}}_{\underline{18}}2^{\underline{6}}_{\underline{16}}4^{\underline{5}}_{\underline{18}}2^{\underline{5}}_{\underline{16}}2^{\underline{4}}_{\underline{16}}6^{\underline{4}}_{\underline{14}}1^{\underline{4}}_{\underline{12}}3^{\underline{3}}_{\underline{14}}1^{\underline{3}}_{\underline{12}}1^{\underline{2}}_{\underline{12}}3^{\underline{2}}_{\underline{10}}1^{\underline{1}}_{\underline{8}}1^{0}_{\underline{8}}1^{0}_{\underline{6}}$ 11^n_{454}: $1^{\underline{9}}_{\underline{22}}2^{\underline{8}}_{\underline{20}}1^{\underline{8}}_{\underline{18}}2^{\underline{7}}_{\underline{18}}2^{\underline{7}}_{\underline{16}}5^{\underline{6}}_{\underline{16}}4^{\underline{6}}_{\underline{14}}4^{\underline{5}}_{\underline{14}}3^{\underline{5}}_{\underline{12}}5^{\underline{4}}_{\underline{12}}6^{\underline{4}}_{\underline{10}}3^{\underline{3}}_{\underline{10}}3^{\underline{3}}_{\underline{8}}4^{\underline{2}}_{\underline{8}}5^{\underline{2}}_{\underline{6}}2^{\underline{1}}_{\underline{4}}2^{0}_{\underline{4}}2^{0}_{\underline{2}}$		
11^n_{455}: $1^{\underline{3}}_{\underline{8}}3^{\underline{2}}_{\underline{6}}3^{\underline{2}}_{\underline{4}}1^{\underline{2}}_{\underline{2}}4^{\underline{1}}_{\underline{4}}2^{\underline{1}}_{\underline{2}}4^{0}_{\underline{2}}6^{0}_{0}1^{0}_{2}5^{1}_{0}3^{1}_{2}4^{2}_{2}7^{2}_{4}3^{3}_{4}2^{3}_{6}3^{4}_{6}5^{4}_{8}1^{5}_{8}1^{5}_{10}1^{6}_{12}$ 11^n_{456}: $1^{\underline{7}}_{\underline{16}}2^{\underline{6}}_{\underline{14}}1^{\underline{6}}_{\underline{12}}2^{\underline{5}}_{\underline{12}}2^{\underline{5}}_{\underline{10}}4^{\underline{4}}_{\underline{10}}3^{\underline{4}}_{\underline{8}}3^{\underline{3}}_{\underline{8}}3^{\underline{3}}_{\underline{6}}1^{\underline{2}}_{\underline{8}}6^{\underline{2}}_{\underline{6}}6^{\underline{2}}_{\underline{4}}1^{\underline{1}}_{\underline{6}}2^{\underline{1}}_{\underline{4}}2^{\underline{1}}_{\underline{2}}4^{0}_{\underline{2}}4^{0}_{0}1^{1}_{\underline{2}}1^{1}_{2}1^{2}_{2}1^{4}_{4}1^{4}_{6}$		
11^n_{457}: $1^{\underline{9}}_{\underline{24}}4^{\underline{8}}_{\underline{22}}2^{\underline{8}}_{\underline{20}}3^{\underline{7}}_{\underline{20}}3^{\underline{7}}_{\underline{18}}8^{\underline{6}}_{\underline{18}}7^{\underline{6}}_{\underline{16}}6^{\underline{5}}_{\underline{16}}4^{\underline{5}}_{\underline{14}}6^{\underline{4}}_{\underline{14}}8^{\underline{4}}_{\underline{12}}4^{\underline{3}}_{\underline{12}}4^{\underline{3}}_{\underline{10}}3^{\underline{2}}_{\underline{10}}4^{\underline{2}}_{\underline{8}}3^{\underline{1}}_{\underline{6}}1^{0}_{\underline{6}}1^{0}_{\underline{4}}$ 11^n_{458}: $3^{0}_{2}2^{0}_{4}4^{1}_{4}1^{1}_{6}7^{2}_{6}6^{2}_{8}6^{3}_{8}5^{3}_{10}9^{4}_{10}8^{4}_{12}5^{5}_{12}7^{5}_{14}6^{6}_{14}7^{6}_{16}3^{7}_{16}4^{7}_{18}1^{8}_{18}3^{8}_{20}1^{9}_{22}$		
11^n_{459}: $1^{\underline{4}}_{\underline{6}}1^{\underline{4}}_{\underline{4}}1^{\underline{3}}_{\underline{4}}2^{\underline{2}}_{\underline{2}}1^{\underline{2}}_{0}1^{\underline{1}}_{0}2^{\underline{1}}_{2}3^{0}_{0}5^{0}_{2}1^{0}_{4}1^{1}_{2}2^{1}_{4}3^{1}_{6}4^{2}_{4}6^{2}_{6}2^{2}_{8}2^{3}_{6}2^{3}_{8}1^{3}_{10}2^{4}_{8}3^{4}_{10}1^{4}_{12}1^{5}_{10}1^{5}_{12}1^{6}_{12}1^{6}_{14}1^{7}_{16}$	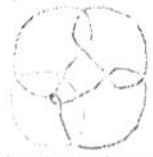	

REFERENCES

[BN1] D. Bar-Natan, *On Khovanov's Categorification of the Jones polynomial,* Algebraic and Geometric Topology **2-16** (2002) 337–370, `http://www.ma.huji.ac.il/~drorbn/papers/Categorification/`, arXiv:math.GT/0201043.

[BN2] D. Bar-Natan, *Khovanov's Homology for Tangles and Cobordisms*, in preparation.

[HT] J. Hoste and M. Thistlethwaite, *Knotscape*, `http://dowker.math.utk.edu/knotscape.html`

[Ja] M. Jacobsson, *An Invariant of Link Cobordisms from Khovanov's Homology Theory,* arXiv:math.GT/0206303.

[Kh1] M. Khovanov, *A Categorification of the Jones Polynomial,* arXiv:math.QA/9908171.

[Kh2] ——, *A Functor-Valued Invariant of Tangles,* arXiv:math.QA/0103190.

[Kh3] ——, *An Invariant of Tangle Cobordisms,* arXiv:math.QA/0207264.

[Kh4] ——, *sl*(3) *Link Homology,* arXiv:math.QA/0304375.

[OS] P. Ozsvath and Z. Szabo, *Holomorphic Disks and Knot Invariants,* arXiv:math.GT/0209056.

[Ro] D. Rolfsen, *Knots and Links*, Publish or Perish, Mathematics Lecture Series **7**, Wilmington 1976.

[Sc] R. Scharein, *KnotPlot,* `http://www.cs.ubc.ca/nest/imager/contributions/scharein/KnotPlot.html`

DEPARTMENT OF MATHEMATICS, UNIVERSITY OF TORONTO, TORONTO ONTARIO M5S 3G3, CANADA
E-mail address: `drorbn@math.toronto.edu`
URL: `http://www.math.toronto.edu/~drorbn`

Perturbative Quantum Field Theory and L_∞-algebras

Lucian M. Ionescu (lmiones@ilstu.edu)
Illinois State University

Abstract. L_∞−morphisms are investigated from the point of view of perturbative quantum field theory, as generalizations of Feynman expansions. Ideas from TQFT and Hopf algebra approach to renormalization are exploited.

It is proved that the algebra of graphs with Kontsevich graph homology differential and Kreimer's coproduct is a DG-coalgebra.

The weights of the corresponding expansions are proved to be cycles of the DG-coalgebra of Feynman graphs, leading to graph cohomology via the cobar construction. Moreover, the moduli space of L-infinity morphims (partition functions/QFTs) is isomorphic to the cohomology of Feynman graphs.

The weights constructed via integrals over configuration spaces represent a prototypical example of "Feynman integrals". The present cohomological point of view aims to construct the coefficients of formality morphisms using an algebraic machinery, as an alternative to the analytical approach using integrals over configuration spaces. It is also expected to yield a categorical formulation for the Feynman path integral quantization, which is presently sketched in the context of L_∞-algebras.

Keywords: L_∞-algebra, configuration spaces, renormalization, QFT

Primary: 18G55: Secondary: 81Q30, 81T18

1. Introduction

L_∞-morphisms may be represented as perturbation series over a class of Feynman graphs, as Kontsevich showed in [1]. In this article ideas from TQFT and the Hopf algebra approach to renormalization are used to study such "Feynman-Taylor" expansions, and their physical interpretation, supplementing [2].

Applying the cobar construction to graph homology, we prove that such L_∞-morphisms correspond to the cohomology classes of the corresponding DG-coalgebra of Feynman graphs (Theorem 2.3).

Examples of such cocycles (periods/weights) are provided by integrals over the compactification of the configuration spaces of a given manifold with boundary. These are "Feynman integrals" corresponding to a given propagator and a certain class of Feynman graphs.

Part of the motivation for this work is provided by the author's hope that combinatorial examples of such cycles may be constructed based on the Hopf algebra of trees. Applied to deformation quantization, this algebraic approach would provide the coefficients of the (local)

J.M. Bryden (ed.), Advances in Topological Quantum Field Theory, 243–252.

star-product of a Poisson manifold, with applications to the Hausdorff series of a Lie algebra [3]. It also reinforces the statement that the related process of renormalization is essentially an algebraic process, independent on the regularization and renormalization schemes [4, 5, 6].

This work is based on an analysis of [1], aiming to extract an axiomatic "interface" from Kontsevich implementation of the formality morphism, which was based on integrals over configuration spaces (see [7] for additional details). As declared in [1], p.4, and announced in [8], p.147, Kontsevich's formula for the star-product [1], is the mathematical implementation of an open string theory, namely a Poisson sigma model on the disk [2]. The coefficients of the terms of the 2-point functions are integrals over compactified configuration spaces. Reminiscent of renormalization of Feynman path integrals, some intrinsic properties are extracted, and the algebraic condition establishing the L_∞-morphism is reinterpreted as a certain "Forest Formula" ([4, 6]), by using the coalgebra structure of renormalization.

As already stated, the main property of the compactification of configuration spaces is the coalgebra structure of their boundaries. As a consequence, the periods of a closed form over the codimension one boundary define a cocycle of the cobar DG-algebra of Feynman graphs. Since these cocycles correspond to L_∞-morphisms, the success of using integrals over configuration spaces to implement formality morphisms, and in particular deformation quantization formulas, is explained.

It is known that the cobar construction allows to compute the homology of loop spaces ([9], p.81). Its appearance in the process of understanding the "loop structure" of Feynman graphs is hardly surprising. An intrinsic approach to their study, in the context of A_∞-algebras, must be from the perspective of codifferentials on the tensor algebra, suggesting the application of the cobar construction to graph homology.

We hope that our approach will contribute to the understanding of the cycles obtained from an A_∞-algebra [10], p.13, also providing one more instance in the plethora of "partition functions" obtained via a "state-sum model" on a "generalized cobordism category".

The paper is organized as follows. The results on L_∞-morphisms expanded as a perturbation series over a class of "Feynman" graphs are explained in §2. The coefficients of L_∞-morphisms satisfy a certain cocycle condition and L_∞-morphisms, modulo homotopy, are represented by cohomology classes of the DG-coalgebra of Feynman graphs.

A mathematical interface to perturbative QFT is sketched in section 3 With this interpretation in mind, the previous result is a classification of the corresponding QFTs (partition functions) in terms of the cohomology of Feynman graphs.

Acknowledgments I would like to express my gratitude for the excellent research conditions at I.H.E.S., where this project was conceived under the influence of, and benefitting from stimulating discussions with Maxim Kontsevich.

The referee's and editor's comments leading to a better structure of the paper are equally appreciated.

2. Cohomology of Feynman graphs and L_∞-morphisms

We investigate when a graded map between L_∞-algebras represented as a *Feynman expansion* over a given class of graphs ("partition function") is an L_∞-morphism. The goal is to understand the coefficients of formality morphisms and Kontsevich deformation quantization formula, from the perspective of perturbative QFT. We will prove that the obstruction for a pre-L_∞-morphism ([8], p.142) to be a morphism is of cohomological nature and point to its relation with renormalization.

2.1. Feynman graphs

Consider a graded class of *Feynman graphs* (e.g. 1-dimensional CW-complexes or finite graded category, i.e. both objects and Homs are finite in each degree), and g the vector space over some field k of characteristic zero, with homogeneous generators $\Gamma \in \mathcal{G}_n$ (e.g. "admissible graphs" [1], p.22).

A Feynman graph will be thought off both as an object in a category of Feynman graphs (categorical point of view), as well as a cobordism between their *boundary vertices* (TQFT point of view). The main assumption needed is the existence of subgraphs and quotients.

While the concept of *subgraph* γ of Γ is clear, being modeled after that of a subcategory, we will define the *quotient* of Γ by the subgraph γ (in the same vein), as the graph Γ' obtained by collapsing γ (vertices and internal edges) to a vertex of the quotient (e.g. see [4], p.11).

Remark 1. When γ contains "external legs", i.e. edges with 1-valent vertices belonging to the boundary of the Feynman graph when thought of as a cobordism, we will say that γ *meets the boundary* of Γ. In this case the vertex of the quotient obtained by collapsing γ will be part of the boundary (of Γ/γ) too. In other words the boundary of the quotient is the quotient of the boundary (compare [1], p.27).

Formal definitions will be introduced elsewhere.

DEFINITION 2.1. *A subgraph γ of $\Gamma \in \mathcal{G}$ is* normal *iff the corresponding* quotient *Γ/γ belongs to the same class of Feynman graphs $\mathcal{G}$.*

DEFINITION 2.2. *An* extension $\gamma \hookrightarrow \Gamma \twoheadrightarrow \gamma'$ *in* $\mathcal{G}$ *is a triple (as displayed) determined by a subgraph* γ *of* Γ, *such that the quotient* γ' *is in* $\mathcal{G}$. *The extension is a* full extension *if* γ *is a* full subgraph, *i.e. together with two vertices of* Γ *contains all the corresponding connecting arrows (the respective "Hom").*

Edges will play the role of simple objects.

DEFINITION 2.3. *A subgraph consisting of a single edge is called a* simple subgraph.

Example 2.1. As a first example consider the class $\mathcal{G}_a$ of *admissible graphs* provided in [1]. Denote by $\mathcal{G}$ the larger class of graphs, including those for which edges from boundary points may point towards internal vertices (essentially all finite graph-like one-sided "cobordisms": $\emptyset \to [m]$). Then the normal subgraphs relative to the class $\mathcal{G}_a$ are precisely the subgraphs corresponding to the "bad-edge" case ([1], p.27), i.e. those for which the quotient is still an admissible graph.

Another example is the class of Feynman graphs of ϕ^3-theory. In this context a subgraph of a 3-valent graph collapses to a 3-valent vertex precisely when it is a normal subgraph in our sense.

There is a natural pre-Lie operation based on the operation of insertion of a graph at an internal vertex of another graph [6], [11], p14, addressed next. It is essentially a sum over extensions corresponding to two given graphs.

DEFINITION 2.4. *The* extension product $\star : g \otimes g \to g$, $g = k[\mathcal{G}]$ *($k = \mathbb{R}$ or $\mathbb{C}$), is the bilinear operation which on generators equals the sum over all possible extensions of one graph by the other one:*

$$\gamma' \star \gamma = \sum_{\gamma \to \Gamma \to \gamma'} \pm \Gamma. \tag{1}$$

It is essentially the "superposition of $Ext^1(\gamma', \gamma)$". As noted in [11], p.14, it is a pre-Lie operation, endowing g with a canonical Lie bracket (loc. cit. $\mathcal{L}_{FG}$).

Let $H = T(g)$ be the tensor algebra with (reduced) coproduct:

$$\Delta\Gamma = \sum_{\gamma \to \Gamma \to \gamma'} \gamma \otimes \gamma', \tag{2}$$

where the sum is over all non-trivial normal subgraphs of Γ, with non-trivial quotient (compare with condition (7) [4], p.11).

Remark 2. The two operations introduced are in a sense "opposite" to one another, since the coproduct unfolds a given graph into its constituents, while the product assembles two constituents in all possible ways. For the moment we will not dwell on the resulting algebraic structure.

With the appearance of a Lie bracket and a comultiplication, we should be looking for a differential (towards a DG-structure).

Consider the graph homology differential [12], p.109:

$$d\Gamma = \sum_{e \in E_\Gamma} \pm \Gamma/\gamma_e, \tag{3}$$

where the sum is over the edges of Γ, γ_e is the one-edge graph, and Γ/γ_e is the quotient (forget about the signs for now).

THEOREM 2.1. (H, d, Δ) *is a differential graded coalgebra.*

Proof. That it is a coalgebra results from [4], p.12. All we need to prove is that d is a coderivation:

$$\Delta d = (d \otimes id + id \otimes d)\Delta.$$

Comparing the two sides (with signs omitted):

$$\begin{aligned} LHS &= \textstyle\sum_{e\in\Gamma} \sum_{\bar\gamma\subset\Gamma/e\to\bar\gamma'} \bar\gamma \otimes \bar\gamma' \\ &= \textstyle\sum_{e\in\Gamma} (\sum_{e/e\in\bar\gamma\subset\Gamma/e\to\bar\gamma'} \bar\gamma \otimes \bar\gamma' + \sum_{e/e\notin\bar\gamma\subset\Gamma/e\to\bar\gamma'} \bar\gamma \otimes \bar\gamma'), \end{aligned}$$

and

$$\begin{aligned} RHS &= \textstyle\sum_{\gamma\subset\Gamma\to\gamma'} (\sum_{e\in\gamma} \gamma/e \otimes \gamma' + \sum_{e\in\gamma'} \gamma \otimes \gamma'/e) \\ &= \textstyle\sum_{e\in\Gamma} (\sum_{e\in\gamma\subset\Gamma\to\gamma'} \gamma/e \otimes \gamma' + \sum_{e\notin\gamma\subset\Gamma\to\gamma'} \gamma \otimes \gamma'/e), \end{aligned}$$

with a correspondence uniquely defined by $e \in \gamma \to \bar\gamma$, i.e. $\bar\gamma = \gamma/e$ and $e \in \gamma' \to \bar\gamma'$, i.e. $\bar\gamma' = \gamma'/e$ respectively, concludes the proof.

The boundary of the codimension one strata of the configuration spaces (see [1], p.22) suggests to consider its cobar construction $C(H) = T(s^{-1}\bar{H})$ ([13], p.366, [14], p.171). Moreover, this is the natural set up for DG(L)A-infinity structures (e.g. [15]).

The total differential is $D = d + \bar\Delta$, where the "coalgebra part" $\bar\Delta$ is the graded derivation:

$$\bar\Delta\Gamma = \sum_{\gamma\to\Gamma\to\gamma'} \gamma \otimes \gamma', \tag{4}$$

corresponding to the reduced coproduct Δ given by equation 2.

DEFINITION 2.5. *The cobar construction* $(C(H), D)$ *of the DG-coalgebra* (H, d, Δ) *of Feynman graphs is called the* Feynman cobar construction *on* $\mathcal{G}$.

Taking the homology of its dual relative some field k, with dual differential δ, yields the following

DEFINITION 2.6. *The cohomology of the DG-coalgebra of Feynman diagrams* $\mathcal{G}$ *is:*

$$H^\bullet(\mathcal{G}; k) = H_\bullet(Hom_{Calg}(C(H), k), \delta),$$

where (H, d, Δ) *is the DG-coalgebra of Feynman graphs.*

We will see in Section 2.3 that it characterizes L_∞-morphisms represented as Feynman expansions.

2.2. Feynman-Taylor coefficients

Let (g_1, Q_1) and (g_2, Q_2) be L_∞-algebras, and $f : g_1 \to g_2$ a pre-L_∞ morphism ([1], p.11) with associated morphism of graded cocommutative coalgebras $\mathcal{F}_* : C(g_1[1]) \to C(g_2[1])$, thought of as the Feynman expansion of a partition function:

$$\mathcal{F}_* = \sum \mathcal{F}_n, \quad \mathcal{F}_n(a) = \sum_{\Gamma \in \mathcal{G}_n} < \Gamma, a >, \quad a \in g_1^n,$$

where the "pairing" $< \ , \ >$ corresponds to a morphism $B : H \to Hom(g_1, g_2)$.

DEFINITION 2.7. *A morphism* $B : H \to Hom(g_1, g_2)$ *is called* a generalized Feynman integral. *Its value* $< \Gamma, a >$ *will be called a* Feynman-Taylor coefficient.

Characters $W : H \to \mathbf{R}$ *act on Feynman integrals:*

$$U = W \cdot B, \quad U(\Gamma) = W(\Gamma)B(\Gamma), \ \Gamma \in \mathcal{G}.$$

An example of a generalized Feynman integral is U_Γ defined in [1], p.23, using the pairing between polyvector fields and functions on $\mathbb{R}^n$. An example of (pre)L_∞-morphisms associated with graphs is provided by $U_n = \sum_{\Gamma \in \mathcal{G}_n} W_\Gamma B_\Gamma$, the formality morphism of [1], p.24.

2.3. L_∞-morphisms

Before addressing the general case of L_∞-algebras, we will characterize formality morphisms of DGLAs (e.g. polyvector fields and polydifferential operators).

THEOREM 2.2. *Let* $(g_1, 0, [\ ,\ ]_{SN})$ *and* $(g_2, d_2, [\ ,\ ])$ *be two DGLAs, and* $U = W \cdot B : g_1 \to g_2$ *a pre-*L_∞*-morphism as above. Then*

(i) $[U, Q] = \delta(W)U$*, where* Q *denotes the appropriate* L_∞*-structure.*

(ii) U *is an* L_∞*-morphism iff the character* W *is a cocycle of the DG-coalgebra of Feynman graphs:* $\delta W = 0$.

Proof. The proof in the general case is essentially the proof from [1].

Alternatively, a direct argument at the level of the associated cobar constructions for g_i may be considered: W is a cocycle iff $W : C(H) \to \mathbb{R}$ is a DG-coalgebra map, where $\mathbb{R}$ has trivial differential [16], p.111.

DEFINITION 2.8. *A character* W *is called a* weight *if it is such a cocycle.*

We claim that the above result holds for arbitrary $L_\infty-algebras$. Moreover L_∞-morphisms can be expanded over a suitable class of Feynman graphs, and their moduli space corresponds to the cohomology group of the corresponding DG-coalgebra of Feynman graphs.

THEOREM 2.3. *("Feynman-Taylor")*

Let $\mathcal{G}$ *be the class of Kontsevich graphs and* g_1, g_2 *two* L_∞*-algebras.*

In the homotopy category of L_∞*-algebras,* L_∞*-morphisms correspond to the cohomology of the corresponding Feynman DG-coalgebra:*

$$\mathcal{H}o(g_1, g_2) = H^\bullet(\mathcal{G}; k).$$

The basic examples (formality morphisms) are provided by cocycles constructed using integrals over compactification of configuration spaces (periods ([17], p.26).

Remark 3. The initial motivation for the present approach was to find an algebraic construction for such cocycles. The idea consists in defining an algebraic version of the "configuration functor" $S : H \to C_\bullet(M)$, a top closed form $\omega : H \to \Omega^\bullet(M)$ with the integration pairing $< S, \omega >$. Their properties suggests that the general setup consist from a chain map $S : (H, d) \to (C_\bullet, \partial)$, and a cocycle ω in some dual cohomological complex $C^\bullet$, with $< \partial S, \omega > = < S, d\omega >$ $(= 0)$, so that the "Stokes theorem" holds. Then $W = < S, \omega >$ would be such a cocycle.

A physical interpretation will be suggested next, and investigated elsewhere.

3. Relations to perturbative QFT

The result of the previous section is important from the physical point of view. It is known that the Kontsevich cocycle W representing the coefficients of the formality morphism is based on a non-linear sigma model on the disk [2], and the (formality) L-infinity morphism is a partition function of a QFT. The integrals over configuration spaces W_Γ are Feynman integrals for a specific propagator.

From this point of view, Theorem 2.3 classifies QFTs determined by their partition functions in terms of Feynman integrals.

In this sense the prototypical example of Kontsevich cocycle is an algebraic framework for ("post-renormalization") Feynman integrals, as sketched in what follows (see also [7]). The relation with the Connes-Kreimer algebraic approach to renormalization [4] will be addressed elsewhere (see [18] for some additional details).

Let H be the Hopf algebra of a class of Feynman graphs $\mathcal{G}$. If Γ is such a graph, then positions are attached to its vertices (configurations), while momenta are attached to edges in the two dual representations (Feynman rules in position and momentum spaces).

This duality is *represented* by a pairing between a "configuration functor" and a "Lagrangian" (e.g. ω determined by its value on an edge, i.e. by a propagator). Together with the pairing (typically integration) representing the action, they are thought of as part of the Feynman model of the *state space of a quantum system.*

As already argued in [19], this "Feynman picture" is more general than the "Riemannian picture" based on manifolds, since it models in a more direct way the observable aspects of quantum phenomena ("interactions" modeled by a class of graphs), without the assumption of a continuity (or even the existence) of the interaction or propagation process in an ambient "space-time", the later being clearly only an artificial model useful to relate with the classical physics, i.e. convenient for "quantization purposes".

Now an *action* on $\mathcal{G}$ ("S_{int}"), is a character $W : H \to \mathbf{R}$ which is a cocycle in the associated DG-coalgebra $(T(H^*), D)$.

A source of such actions is provided by a morphism of complexes $S : H \to C_\bullet(M)$ ("configuration functor"), where M is some "space", $C_\bullet(M)$ is a complex ("configurations on M"), endowed with a pairing $\int : C_\bullet(M) \times C^\bullet(M) \to \mathbf{R}$, where $C^\bullet(M)$ is some dual complex ("forms on configuration spaces"), i.e. such that "Stokes theorem" holds $< \partial S, \omega >=< S, d\omega >$.

A *Lagrangian* on the class $\mathcal{G}$ of Feynman graphs is a k-linear map $\omega : H \to C^\bullet(M)$ associating to any Feynman graph Γ a closed "volume form" on $S(\Gamma)$ vanishing on the boundaries, i.e. for any subgraph $\gamma \to \Gamma$

(viewed as a subobject) meeting the boundary of $\Gamma : [s] \to [t]$ (viewed as a cobordism), $\omega(\gamma) = 0$. Then the associated action is $W =< S, \omega >$.

As mentioned above, a prototypical "configuration functor" is given by the compactification of configuration spaces $C_{n,m}$ described in [1]. The second condition for a Lagrangian emulates the vanishing on the boundary of the angle-form α (see [1], p.22). The coefficient $W(\Gamma)$ is then a *period* of the quadruple $(C_\Gamma, \partial C_\Gamma, \wedge_{k=1}^{|E_\Gamma|}\alpha(z_{i_k}, z_{j_k})$ ([17], p.24; see also "effective periods" p.27).

In conclusion, the intent of the present article is to isolate some algebraic properties of, and to establish a perhaps simpler "interface" to a mathematical model for the Feynman path integral quantization based on homotopical algebra:

$$\text{"} \int \mathcal{D}\gamma \; e^{S[\gamma]} \quad \text{"} => \sum_n \sum_\gamma U_n(\gamma).$$

The left hand side is a conceptual framework which need not be implemented using analytical tools (integrals, measures, etc.), but most likely with algebraic tools, e.g. the state sum models yielding TQFTs, and more general still, as representations of generalized cobordism categories.

References

1. M. Kontsevich: Deformation quantization of Poisson manifolds I, q-alg/9709040.
2. A. S. Cattaneo and G. Felder: A path integral approach to the Kontsevich quantization formula, math.QA/9902090.
3. L. M. Ionescu, A combinatorial approach to Kontsevich coefficients in deformation quantization, in preparation.
4. A. Connes and D. Kreimer: Renormalization in quantum field theory and the Riemann-Hilbert problem I, hep-th/9912092.
5. D. Kreimer: Combinatorics of (perturbative) Quantum Field Theory, hep-th/0010059.
6. D. Kreimer: Structures in Feynman Graphs - Hopf algebras and symmetries, hep-th/0202110.
7. L. M. Ionescu, Perturbative quantum field theory and configuration space integrals, hep-th0307062.
8. M. Kontsevich: Formality conjecture, Deformation theory and symplectic geometry (Ascona, 1996), 139–156, Math. Phys. Stud., 20, Kluwer Acad. Publ., Dordrecht, 1997.
9. M. J. F. Adams: On the cobar construction, Colloque de topologie algbrique, Louvain, 1956, pp. 81–87. Georges Thone, Lige; Masson & Cie, Paris, 1957.
10. M. Penkava: Infinity algebras and the homology of graph complexes, q-alg/9601018, v1.
11. A. Connes and D. Kreimer: Insertion and Elimination: the doubly infinite Lie algebra of Feynman graphs, hep-th/0201157.

12. M. Kontsevich: Feynman diagrams and low-dimensional topology, Joseph, A. (ed.) et al., First European congress of mathematics (ECM), Paris, France, July 6-10, 1992. Volume II: Invited lectures (Part 2). Basel: Birkhuser. Prog. Math. 120, 97-121 (1994).
13. V.K.A.M. Gugenheim, L.A. Lambe and J.D. Stasheff: Perturbation theory in differential homological algebra II, Ill. J. Math. Vol.35, No.3, Fall 1991, p.357-373.
14. M. Markl: A cohomology theory for A(m)-algebras and applications, JPAA 83 (1992) 141-175.
15. B. Keller: An introduction to $A-\infty$-algebras, math.RA/9910179.
16. J. D. S. Jones, Lectures on operads, Contemporary Mathematics, Vol. **315**, 2002, p.89-130.
17. M. Kontsevich: Operads and Motives in Deformation Quantization, math.QA/9904055.
18. L. M. Ionescu, A Hopf algebra deformation approach to renormalization, hep-th/0306112.
19. L. M. Ionescu: Remarks on quantum physics and noncommutative geometry, math.HO/0006024.

Address for Offprints: Department of Mathematics, Illinois State University, Normal IL 61790-4520.

A Linking Form Conjecture for 3-Manifolds

J. Bryden [1] and F. Deloup [2]

Abstract. Kawauchi and Kojima have shown that for any linking pairing (G, ϕ) on a finite abelian group G there is a closed, connected, oriented 3-manifold, M, with $H_1(M) = G$ and linking form $\lambda_M \cong \phi$. Our object is to refine this theorem by proving that any linking pairing on a finite abelian group can be realized as the linking form of an oriented Seifert manifold which is a rational homology sphere. In particular, since such Seifert manifolds are irreducible, any linking pairing on a finite abelian group would then be isomorphic to the linking form of an irreducible 3-manifold. We refer to this as the linking form conjecture.

Keywords: Seifert manifolds, linking form

Mathematics Subject Classification 2000:- Primary: 57N65: Secondary: 57N27, 20J06, 20K10, 81Q30

§1 Introduction

In this paper M will denote an oriented Seifert fibred manifold with oriented orbit surface. Using the standard notation introduced by Seifert (S),

$$M \cong (O, o; 0 \mid e : (a_1, b_1), \ldots, (a_m, b_m)) \ .$$

Here the orbit surface of M has genus $g = 0$, e is the Euler number, m is the number of singular fibres and, for each i, (a_i, b_i) is a pair of relatively prime integers that characterize the twisting of the i-th singular fibre. In addition to Seifert's original paper, good expositions of the basic facts about Seifert fibred manifolds can be found in (H), (Mon), (O), (ST).

For any prime p, let $\nu_p(B)$ denote the p-valuation of the positive integer B, that is, $\nu_p(B)$ is the largest power of p that divides B. Adopt the convention that $\nu_p(0) = \infty$. Suppose that s is the maximal p-valuation of the Seifert invariants $a_1, \ldots, a_m$ and t is a non negative integer with $0 \leq t \leq s$. Then for each t, let $a_{t,1}, \ldots, a_{t,r_t}$ denote the Seifert invariants which satisfy the condition $\nu_p(a_{t,i}) = t$, $1 \leq i \leq r_t$. This imposes an ordering on the Seifert invariants since $\nu_p(a_{t,i}) < \nu_p(a_{l,j})$ when $t < l$. Thus the invariants $a_1, \ldots, a_m$ and their p-valuations can be listed as follows:

$$\begin{array}{lll} a_{s,1}, \ \ldots, \ a_{s,r_s}, & \nu_p(a_{s,i}) = s, & 1 \leq i \leq r_s, \\ \quad\vdots & \quad\vdots & \\ a_{t,1}, \ \ldots, \ a_{t,r_t}, & \nu_p(a_{t,i}) = t, & 1 \leq i \leq r_t, \\ \quad\vdots & \quad\vdots & \\ a_{0,1}, \ \ldots, \ a_{0,r_0}, & \nu_p(a_{0,i}) = 0, & 1 \leq i \leq r_0 . \end{array}$$

1 Partially supported by NSERC operating grant RGP203233.
2 Supported by E.U. Marie Curie Fellowship HMPF-CT-2001-01174.

J.M. Bryden (ed.), Advances in Topological Quantum Field Theory, 253–265.

Both notations a_j and $a_{t,i}$ will be used interchangeably as needed.

Let $n = \sum_{i=1}^{s} r_i$. Then the Seifert invariants $a_1, a_2, \ldots a_m$ can be reordered so they satisfy $0 \neq \nu_p(a_1) \leq \nu_p(a_2) \leq \ldots \leq \nu_p(a_n)$ and $\nu_p(a_{n+1}) = \nu_p(a_{n+2}) = \cdots = \nu_p(a_m) = 0$, as explained above. This fact is used when applying the results of (BLPZ). As a final notational convention, let $A = \prod_{i=1}^{n} a_i$, $A_j = a_j^{-1}A \in \mathbf{Z}$ and $C = \sum b_i A_i$.

It was proved by Kawauchi and Kojima in (KK) that if (G, ϕ) is a linking pairing on a finite abelian group, then there is a closed, connected, oriented 3-manifold M with $H_1(M) \cong G$ and whose linking form λ_M is isomorphic to ϕ. The 3-manifold M, corresponding to (G, ϕ), is a connected sum of the following three types of irreducible 3-manifolds: (i) lens spaces, (ii) 3-manifolds for which there is a PL embedding into S^4, and (iii) fibres of fibred 2-knots that are embedded in S^4.

The object of this paper is to give a preliminary report and an overview of the techniques that will enable us to address the following conjecture for 3-manifolds that would be a refinement of Kawauchi and Kojima's result.

Linking form conjecture: All isomorphism classes of linking pairings of finite abelian groups can be realized as the linking form of a Seifert manifold which is a rational homology sphere.

Since such Seifert manifolds are irreducible, the linking form conjecture would then imply that any linking pairing is isomorphic to the linking form of an irreducible 3-manifold rather than the linking form of a connected sum of 3-manifolds, as was proved in (KK).

Our intention is to show that given any linking pairing (G, ϕ) on a finite abelian group G, there is a Seifert manifold, which is also a rational homology sphere, whose linking form has the same block sum decomposition as the linking pairing. This will be discussed in §3.

The weaker form of this conjecture, where the manifold M is only required to be irreducible, can be proved using results of Gordon-Luecke and Myers.

Proposition 1. *Any linking pairing on a finite abelian group $\lambda : G \times G \to \mathbf{Q}/\mathbf{Z}$ can be realized as the linking pairing of an irreducible 3-manifold.*

Proof. By applying Kawauchi-Kojima's theorem, there is a (possibly not irreducible) 3-manifold realizing the given linking pairing. It follows from the work of Gordon-Luecke (GL) and Myers (My) that we can perform surgery on some null-homotopic knot in M so that the resulting 3-manifold is irreducible, with the same linking pairing. □

One important application of this work is to the abelian WRT-type invariants constructed in (De1), (De2). These invariants were first described by Turaev in (T1), although a special case of this class of invariants was examined in (MOO) and used to give a description of the Dijkgraaf-Witten invariants discussed in (DW).

§2 Cohomology of the Seifert Manifolds

For an oriented Seifert manifold $M \cong (O, o; 0 \mid e : (a_1, b_1), \ldots, (a_n, b_n))$, abelianization of the fundamental group gives the presentation:

$$H_1(M) \approx \langle s_j, h \mid a_j s_j + b_j h = 0, \text{for } j = 1, \ldots, n;\ \Sigma s_j - eh = 0 \rangle .$$

It then follows that $H_1(M)$ is a finite abelian group unless $Ae + C = 0$ (cf. (BLPZ)), in which case

$$H_1(M) \approx \mathbf{Z} \oplus \text{Tors } H_1(M) .$$

Furthermore when $Ae + C \neq 0$ it turns out that for some integer q

$$\text{Tors}_p H_1(M) \cong H_1(M) \otimes \mathbf{Z}/p^q \cong H^1(M; \mathbf{Z}/p^q) , \qquad (1.1).$$

The first goal is to calculate the integer q.

Let $F_p(G)$ denote the p-component of an abelian group G, that is, $F_p(G)$ is the quotient group obtained by factoring out the subgroup of all torsion elements having order prime to p. It follows from (BLPZ) Theorem 1, that for a Seifert manifold $M = (O, o, 0|e, (a_1, b_1), \ldots (a_n, b_n))$

$$F_p(H_1(M)) = \mathbf{Z}/p^c \oplus \mathbf{Z}/p^{\nu(a_1)} \oplus \cdots \oplus \mathbf{Z}/p^{\nu(a_{n-2})} , \qquad (1.2)$$

where $c = \nu_p(Ae+C) - \nu_p(A) + \nu(a_{n-1}) + \nu(a_n)$. Here we make the following conventions, $\mathbf{Z}/p^0 = \{0\}$, $\mathbf{Z}/p^\infty = \mathbf{Z}$

Equations (1.1) and (1.2) will allow us to analyze the linking forms of oriented Seifert manifolds using the product structure in cohomology. From this point on we assume that $Ae + C \neq 0$. In this case (1.2) yields:

$$\text{Tors}_p H_1(M) = \mathbf{Z}/p^c \oplus \mathbf{Z}/p^{\nu(a_1)} \oplus \cdots \oplus \mathbf{Z}/p^{\nu(a_{n-2})} .$$

The following Theorem gives a relation between the number c defined above and the maximal p-valuation of the Seifert invariants $a_1, \ldots, a_m$ of the given Seifert manifold. This result will allow us to distinguish two cases in the computation of the cohomology of these manifolds.

Theorem 1. *Suppose that the Seifert manifold M is a rational homology sphere. Then $c = \nu_p(a_{n-1}) = \nu_p(a_n)$ if and only if*

$$b_{j_0} A'_{j_0} + \ldots + b_n A'_n \not\equiv 0 \bmod p .$$

where $j_0 = \min\{1 \leq j \leq n | \nu_p(a_n) = \nu_p(a_j)\}$, $A'_j = \prod a'_k$ *and* $a'_k p^{\nu_p(a_k)} = a_k$.

Proof. Since M is a rational homology sphere, rational abelianization of the fundamental group gives the relation

$$e = \sum_{1 \leq j \leq n} b_j / a_j \in \mathbf{Q} .$$

Then,

$$\begin{aligned} Ae + C &= 2Ae \\ &= 2 \sum_{1 \le j \le n} b_j A_j \\ &= 2 \sum_{1 \le j \le n} b_j p^{\sum_{k \neq j} \nu_p(a_k)} A'_j \\ &= 2\, p^{\nu_p(a_1) + \ldots + \nu_p(a_{n-1})} \sum_{1 \le j \le n} b_j p^{\sum_{k \neq j} \nu_p(a_k) - \sum_{l=1}^{n-1} \nu_p(a_l)} A'_j \\ &= 2\, p^{\nu_p(a_1) + \ldots + \nu_p(a_{n-1})} \sum_{1 \le j \le n} b_j p^{\nu_p(a_n) - \nu_p(a_j)} A'_j \,. \end{aligned}$$

Now set $J = \{1 \le j \le n | \nu_p(a_n) = \nu_p(a_j)\}$, and let $j_0 = \min J$. Then the equation for $Ae + C$ from above can be expressed in the form

$$Ae + C = 2\, p^{\nu_p(a_1) + \ldots + \nu_p(a_{n-1})} \left(\sum_{1 \le j \le j_0} b_j p^{\nu_p(a_n) - \nu_p(a_j)} A'_j + \sum_{j_0 \le j \le n} b_j A'_j \right) .$$

This implies that

$$\nu_p(Ae + C) = \begin{cases} \nu_p(a_1) + \ldots + \nu_p(a_{n-1}) + \nu_p(X) \,, & \text{if } p > 2 \,, \\ 1 + \nu_p(a_1) + \ldots + \nu_p(a_{n-1}) + \nu_p(X) \,, & \text{if } p = 2 \,, \end{cases}$$

where $X = \sum_{1 \le j \le j_0} b_j p^{\nu_p(a_n) - \nu_p(a_j)} A'_j + \sum_{j_0 \le j \le n} b_j A'_j$. In particular,

$$\nu_p(Ae + C) \ge \begin{cases} \sum_{j=1}^{n-1} \nu_p(a_j) \,, & \text{if } p > 2 \,, \\ 1 + \sum_{j=1}^{n-1} \nu_p(a_j) \,, & \text{if } p = 2 \,. \end{cases} \tag{1.3}$$

Observe that (1.3) is an equality if and only if $c = \nu_p(a_{n-1})$. That is, $\sum_{j_0 \le j \le n} b_j A'_j$ is coprime to p. □

Recall from above that the maximal p-valuation of the Seifert invariants $a_{t,j}$, $1 \le t \le s$, $1 \le j \le r_t$, of M is s. Theorem 1 gives an important condition that ensures that the integer q in (1.1) is equal to s and so $\mathrm{Tor}_p H^2(M) \cong H^2(M; \mathbf{Z}/p^s)$.

In order to describe the mod p^c cohomology ring of $M = (O, o; 0|e, (a_1, b_1), \ldots (a_n, b_n))$, first recall the description of the CW structure of the oriented Seifert manifolds with orbit surface S^2 from (BHZZ1), (BHZZ2). The equivariant chain complex for the universal cover $\tilde{M}$ is constructed by lifting the cell structure of M and consists of the free $\mathbf{Z}[\pi_1(M)]$-modules C_i, $i = 0, 1, 2, 3$, with free generators:

$$\begin{array}{lll} 0: & \sigma_0^0, \ldots, \sigma_m^0; & (C_0) \\ 1: & \sigma_1^1, \ldots, \sigma_m^1; \rho_0^1, \ldots, \rho_m^1; \eta_0^1, \ldots, \eta_m^1; & (C_1) \\ 2: & \sigma_1^2, \ldots, \sigma_m^2; \rho_0^2, \ldots, \rho_m^2; \mu_0^2, \ldots, \mu_m^2; \delta^2; & (C_2) \\ 3: & \sigma_0^3, \ldots, \sigma_m^3; \delta^3. & (C_3) \end{array}$$

We thus obtain the free $\mathbf{Z}[\pi_1(M)]$-resolution,

$$\mathcal{C}: \qquad 0 \to C_3 \xrightarrow{\partial_3} C_2 \xrightarrow{\partial_2} C_1 \xrightarrow{\partial_1} C_0 \xrightarrow{\varepsilon} \mathbf{Z} \to 0\,,$$

of $\mathbf{Z}$. The differentials of this resolution were described in (BHZZ2), (BZ) and (BZ2). There is also a description of the differentials in (B).

The generators of this free resolution correspond to cells in the universal cover, $\tilde{M}$, lifted from M in the following way: M is decomposed into $m+1$ solid tori $V_0, \ldots, V_m$ and its central part $B(m+1) \times S^1$, where $B(m+1) = \overline{S^2 - D_0^2 \cup \cdots \cup D_m^2}$ is the closure of the sphere minus $m+1$ disks. The solid tori $V_i = D_i^2 \times S^1$, $i = 1 \ldots m$, are regular neighborhoods of the singular fibers $\rho_1^1, \ldots, \rho_m^1$ and $\eta_1^1, \ldots, \eta_m^1$ are the crossing curves on $\partial V_i = S^1 \times S^1$, while ρ_0^1 is an ordinary fibre of the ordinary solid torus V_0 with crossing curve η_0^1. Now let σ_i^0 be a point on the fibre ρ_i^1 and define the 1-cells σ_i^1, $i = 1, \ldots, m$, to be paths from σ_0^0 to σ_i^0. Next there are three families of 2-cells in addition to the open 2-cell $\delta^2 = S^2 - D_0^2 \cup \cdots \cup D_m^2$ which are, (i) $\rho_i^2 = \partial V_i \setminus \rho_i^1 \cup \eta_i^1$, $i = 1, \ldots, m$, (ii) σ_i^2 are the cells with boundaries η_0^1, σ_i^1, η_i^1, $i = 1, \ldots, m$, and (iii) for each $i = 1, \ldots, m$ there are cells μ_i^2 interior to to the fibred solid tori V_i which are defined as the image of a map $\varphi: D^2 \to V_i$ such that the restriction of φ to the interior of D^2 is an embedding and $\varphi(\partial D^2) \subset \rho_i^1 \cup \eta_i^1$. Finally, there is a 3-cell $\delta^3 = \delta^2 \times \eta_0^1$ and a family of 3-cells $\sigma_i^3 = \operatorname{Int} V_i \setminus \mu_i^2$ for $i = 1 \ldots m$. Because the closures of these cells are the images of closed disks, they can be lifted to the universal covering space, beginning with the 0-cells, 1-cells, etc.. These cells give the generators listed in C_0, C_1, C_2, C_3.

Each Seifert invariant (a_i, b_i), $1 \le i \le m$, of a singular fibre, has a corresponding fibred solid torus V_i, $1 \le i \le m$, which has a cellular decomposition in terms of the cells σ_i^0, ρ_i^1, η_i^1, ρ_i^2, μ_i^2, σ_i^3, described above. Furthermore, the cells σ_i^1 and σ_i^2 are attached to ∂V_i. We will use the following notational device in order to describe the cohomology classes of M: a fibred solid torus $V_{t,j}$ of a singular fibre of M with corresponding Seifert invariant $(a_{t,j}, b_{t,j})$ has a cellular decomposition into the cells $\sigma_{t,i}^1$, $\rho_{t,i}^1$, $\eta_{t,i}^1$, $\sigma_{t,i}^2$, $\rho_{t,i}^2$, $\mu_{t,i}^2$ and further the cells $\sigma_{t,j}^1$ and $\sigma_{t,j}^2$ are attached to the boundary of $V_{t,j}$.

Since an irreducible 3-manifold M with infinite fundamental group is an Eilenberg-MacLane space, it follows that $H^*(M; \mathbf{Z}/p^s) \cong H^*(\pi_1(M); \mathbf{Z}/p^s)$. That is, the cohomology of M is isomorphic to the group cohomology of $\pi_1(M)$ (cf. (M)). In particular, any Seifert manifold with infinite fundamental group is irreducible and so must be an Eilenberg-MacLane space. This is true for any Seifert manifold which is a rational homology sphere. Thus for such a Seifert manifold M, $H^*(M; \mathbf{Z}/p^s) \cong H^*(\pi_1(M); \mathbf{Z}/p^s)$ can be found from the homology of the cochain complex:

$$Hom(C_0; \mathbf{Z}/p^s) \xrightarrow{\partial^0} Hom(C_1; \mathbf{Z}/p^s) \xrightarrow{\partial^1} Hom(C_2; \mathbf{Z}/p^s) \xrightarrow{\partial^2} Hom(C_3; \mathbf{Z}/p^s) \to 0$$

(cf. (BHZZ2), (BZ2), (B)). For any generator α of C_i, let $\hat{\alpha}$ denote the dual generator of $Hom(C_i; \mathbf{Z}/p^s)$; that is, $\hat{\alpha}(\alpha) = 1$, and $\hat{\alpha}(\beta) = 0$ for any other generator β of C_i, for $i = 0, 1, 2, 3$.

The next Theorem describes the $\mathbf{Z}/p^s$ cohomology ring of Seifert manifolds that are rational homology spheres. In Section 3 we will require that the maximal p-valuation s of the Seifert invariants a_i of M equals the number c described in (1.1). However, the proof of Theorem 1 does not depend on this requirement. As a final remark, note that since any Seifert invariant (a_i, b_i) of M satisfies the condition g.c.d. $(a_i, b_i) = 1$, b_i has a multiplicative inverse in $\mathbf{Z}/p^s$.

Theorem 2. *Let $M := (O, o; 0 \mid e : (a_1, b_1), \ldots, (a_m, b_m))$ be a Seifert manifold for which $Ae + C \neq 0$. As above let s denote the maximal p-valuation of the Seifert invariants a_i. If $n > 1$, then as a graded group,*

$$H^*(M; \mathbf{Z}/p^s) = \langle 1, \alpha_{t,i}, \beta_{t,i}, \gamma \mid 1 \leq i \leq r_t, 1 \leq t < s;\ \ 2 \leq i \leq r_s,\ for\ t = s \rangle ,$$

with $\alpha_{t,i}$ in degree 1, $\beta_{t,i}$ in degree 2, and γ in degree 3. There is exactly one relation given by

$$\beta_{s,1} = -\sum_{\substack{1 \leq t \leq s \\ 1 \leq i \leq r_t \\ t=s, s \neq 1}} \beta_{t,i} .$$

Let δ_{jk} denote the Kronecker delta. The non-zero cup products in $H^*(M; \mathbf{Z}/p^s)$ are given by the following.

1. *Let $p = 2$ and suppose that either $1 \leq t, l < s$ and $1 \leq i \leq r_t$, $1 \leq j \leq r_l$, or at least one of t, l is equal to s, in which case either $2 \leq i \leq r_s$ or $2 \leq j \leq r_s$. Then*

$$\alpha_{t,i} \cdot \alpha_{l,j} = 2^{2s-t-l} \left[\binom{a_{s,1}}{2} b_{s,1}^{-1} \beta_{s,1} + \delta_{t,l} \delta_{ij} \binom{a_{t,i}}{2} b_{t,j}^{-1} \beta_{t,i} \right] .$$

2. *If $1 \leq t, l < s$ and $1 \leq i \leq r_t$, $1 \leq j \leq r_l$, or either t, l is equal to s in which case either $2 \leq i \leq r_s$ or $2 \leq j \leq r_s$, then for any prime p,*

$$\alpha_{t,i} \cdot \beta_{l,j} = -\delta_{tl} \delta_{ij} p^{s-t} \gamma .$$

Additionally, the mod p^s Bockstein, B_{p^s}, on $H^1(M; \mathbf{Z}/p^s)$ is given by

$$B_{p^s}(\alpha_{t,i}) = \frac{a_{t,i} b_{t,i}^{-1} \beta_{t,i} - a_{s,1} b_{s,1}^{-1} \beta_{s,1}}{p^t} \in H^2(M; \mathbf{Z}/p^s) .$$

Sketch of Proof: The coboundaries of the cochain complex are given in (BHZZ2) and (BZ2). It follows that the generators in cohomology are $\alpha_{t,i} = p^{s-t}[\hat{\rho}^1_{t,i} - \hat{\rho}^1_{s,1}]$ in dimension 1, $\beta_{t,i} = [b_{t,i} \hat{\mu}^2_{t,i}]$ in dimension 2 and $\gamma = \hat{\delta}^3$, which is the class dual to the mod p^s reduction of the fundamental class of M. The cup products and Bockstein maps were determined for a special case of this theorem in Appendix A (BZ2) and the general results are given in (B). □

Remark 1. *Given any finite abelian group G, it is clear that by simply varying the parameters $a_{t,j}$, r_t we can always find an appropriate Seifert manifold M that satisfies the condition of Theorem 1 (which means that $s = c$) and so that $H^2(M; \mathbf{Z}/p^s) \cong Tors_p H^2(M) \cong Tors_p G$ for all p. Thus it follows that $H^2(M) \cong G$.*

The next theorem describes the cohomology ring of a Seifert manifold that is also a rational homology sphere, when all the Seifert invariants satisfy the condition $\nu_p(a_1) = \cdots = \nu_p(a_m) = 0$. This means that $a_i \not\equiv 0 \,(\mathrm{mod}\, p^k)$, for all k and $1 \leq i \leq m$. Since g.c.d. $(a_i, b_i) = 1$, we can order the Seifert invariants and define r so that $b_1, \ldots, b_r \not\equiv 0 \,(\mathrm{mod}\, p^c)$, $b_{r+1}, \ldots, b_m \equiv 0 \,(\mathrm{mod}\, p^c)$.

Theorem 3. *Let $M := (O, o; 0 \mid e : (a_1, b_1), \ldots, (a_m, b_m))$. Suppose that $Ae + C \neq 0$ and $n = 0$, that is, $\nu_p(a_1) = \cdots = \nu_p(a_m) = 0$. Suppose that $b_1, \ldots, b_r \not\equiv 0 \ (\mathrm{mod}\ p^q)$ while $b_{r+1}, \ldots, b_n \equiv 0 \ (\mathrm{mod}\ p^q)$. Then as a graded group*

$$H^*(M; \mathbf{Z}/p^q) = \begin{cases} \langle 1, \alpha, \beta, \gamma \rangle, & \text{if } Ae + C \equiv 0 \, (\mathrm{mod}\ p^q), \\ \langle 1, \gamma \rangle, & \text{if } Ae + C \not\equiv 0 \, (\mathrm{mod}\ p^q), \end{cases}$$

where $\deg(\alpha) = 1$, $\deg(\beta) = 2$, $\deg(\gamma) = 3$.

When $p > 2$ there is only one non-trivial cup product, which is:

$$\alpha \cdot \beta = -\gamma, \qquad \text{if } \ Ae + C \equiv 0 \, (\mathrm{mod}\ p^q).$$

Furthermore, the mod p^q Bockstein on $H^1(M; \mathbf{Z}/p^q)$ is given by,

$$B_{p^q}(\alpha) = -\frac{A^{-1}}{p^q} \left[\sum_{i=r+1}^{m} b_i A_i + Ae + C \right] \beta \in H^2(M; \mathbf{Z}/p^q), \qquad \text{if } \ Ae + C \equiv 0 \, (\mathrm{mod}\ p^q).$$

Sketch of Proof: Observe that $\mathrm{Im}\,(\partial^0) = \left\langle \hat{\sigma}_j^1 \mid 1 \leq j \leq m, \right\rangle$, and furthermore that if $x = \sum_{j=0}^{m} \hat{\eta}_j^1 - \sum_{j=1}^{r} b_j a_j^{-1} \hat{\rho}_j^1 - e\hat{\rho}_0^1, \hat{\sigma}_j^1$, then

$$\mathrm{Ker}(\partial^1) = \begin{cases} \left\langle x, \ \hat{\sigma}_j^1, : 1 \leq j \leq m \right\rangle, & \text{if } Ae + C \equiv 0 \ (\mathrm{mod}\ p^q), \\ \left\langle \hat{\sigma}_j^1 : 1 \leq j \leq m \right\rangle, & \text{if } Ae + C \not\equiv 0 \ (\mathrm{mod}\ p^q). \end{cases}$$

It is clear that in the case where $Ae + C \not\equiv 0 \ (\mathrm{mod}\ p^q)$, $H^1(M; \mathbf{Z}/p^q) \cong H^2(M; \mathbf{Z}/p^q) \cong 0$. When $Ae + C \equiv 0 \ (\mathrm{mod}\ p^q)$ set $\alpha = \left[\sum_{j=0}^{m} \hat{\eta}_j^1 - \sum_{j=1}^{r} b_j a_j^{-1} \hat{\rho}_j^1 - e\hat{\rho}_0^1 \right]$ and $\beta = \left[\hat{\delta}^2 \right]$. As in Theorem 2, γ is the dual of the mod p^q reduction of the fundamental class. The remaining details of this proof are a generalization of Theorem 1.3 and Case (2), Theorem 4.1 (BZ2). □

Remark 2. *There is one final case that occurs when $Ae + C \neq 0$ and there is precisely one Seifert invariant a_i, $i = 1, \ldots, m$, that has non-zero p-valuation, for at least one prime p. The above results do not deal with this case. However this case is not of importance when considering primes greater than 2, unless the Seifert manifold is a lens space.*

§3 Linking Forms

In (W) and (KK) the structure of the monoid $\mathcal{N}$ of isomorphism classes of linking pairings of finite abelian groups was completely determined. It is clear that $\mathcal{N} \cong \oplus \mathcal{N}_p$, where $\mathcal{N}_p$ is the monoid of isomorphism classes of linking pairings on p-groups. When $p > 2$ the generators of $\mathcal{N}_p$ are (p^{-k}) and $(n(p)p^{-k})$ where $n(p)$ is a fixed quadratic non-residue mod p. These are linking pairings on $\mathbf{Z}/p^k$. When $p = 2$ the generators of $\mathcal{N}_2$ are: $(n2^{-k})$, for $k \geq 1$, which is a linking pairing on $\mathbf{Z}/2^k$, and

$$E_0^k = \begin{pmatrix} 0 & 2^{-k} \\ 2^{-k} & 0 \end{pmatrix}, \qquad E_1^k = \begin{pmatrix} 2^{1-k} & 2^{-k} \\ 2^{-k} & 2^{1-k} \end{pmatrix},$$

which are linking pairings on $\mathbf{Z}/2^k \oplus \mathbf{Z}/2^k$.

Furthermore, if $\bar{G}_p^k$ denotes the subgroup of a finite abelian group G generated by elements of order p^s for $s \leq k$ when p is an odd prime, then the linking pairing $\phi: G \times G \to \mathbf{Q}/\mathbf{Z}$ determines a linking pairing

$$\tilde{\phi}_p^k: \tilde{G}_p^k \times \tilde{G}_p^k \to \mathbf{Q}/\mathbf{Z}$$

for $\tilde{G}_p^k := \frac{\bar{G}_p^k}{\bar{G}_p^{k-1} + p\bar{G}_p^{k+1}}$, defined by:

$$\tilde{\phi}_p^k([x],[y]) = p^{k-1}\phi(x,y),$$

$x, y \in \bar{G}_p^k$, k a positive integer. Wall (W) proved that the series $\{(\rho_p^k, \sigma_p^k)\}$, where $\rho_p^k = \dim_{\mathbf{Z}/p}(\tilde{G}_p^k)$ and $\sigma_p^k = \left(\frac{\det \tilde{\phi}_p^k}{p}\right)$, is a complete minimal system of invariants of linking forms on groups of odd order. There is also a complete minimal system of invariants $\{(\rho_2^k, \sigma_2^k)\}$ of linking pairings for 2-groups (cf. (KK), (De3)).

For an oriented 3-manifold M, the (usual) linking form

$$\lambda: \text{Tors } H^2(M) \otimes \text{Tors } H^2(M) \to \mathbf{Q}/\mathbf{Z}$$

of the manifold is defined as follows: Given $x, y \in \text{Tors } H^2(M)$, define

$$\lambda(x,y) = \left\langle x \cup B^{-1}y, [M] \right\rangle,$$

where $B: H^1(M; \mathbf{Q}/\mathbf{Z}) \to H^2(M)$ here denotes the $\mathbf{Q}/\mathbf{Z}$-Bockstein. Equivalently, if $Nx = Ny = 0$ in $\text{Tors } H^2(M)$ for some integer $N > 1$, then

$$\lambda(x,y) = \frac{1}{N}\left\langle x \cup B_N^{-1}y, [M] \right\rangle,$$

where $B_N: H^1(M; \mathbf{Z}/N) \to H^2(M)$ denotes the mod N Bockstein (cf. (T2)).

Instead of studying the linking form directly, we will define a new linking pairing on $H^1(M; \mathbf{Z}/p^c)$ in order to prove the linking form conjecture for abelian groups of odd order.

Definition 1. *Define a linking pairing*

$$\hat{\lambda}_M^p: H^1(M; \mathbf{Z}/p^c) \otimes H^1(M; \mathbf{Z}/p^c) \to \mathbf{Q}/\mathbf{Z}$$

on an oriented 3*-manifold* M *by* $\hat{\lambda}_M^p(x,y) = \frac{1}{p^c}\langle x \cup B_{p^c}(y), [M]\rangle$, *for* $x, y \in H^1(M; \mathbf{Z}/p^c)$.

As stipulated previously the Seifert manifold $M = (O, o; 0 \mid e : (a_1, b_1), \ldots, (a_m, b_m))$ satisfies the condition $Ae + C \neq 0$. We also want the condition of Theorem 1 to be satisfied so that $s = c$. In this case $H_1(M) \cong H^2(M)$ is a torsion group and since $\text{Tors}_p H_1(M) \cong H_1(M) \otimes \mathbf{Z}/p^s \cong H^1(M; \mathbf{Z}/p^s)$ we can define a linking pairing $\hat{\lambda}_M: H^2(M) \otimes H^2(M) \to \mathbf{Q}/\mathbf{Z}$ in terms of the pairings $\hat{\lambda}_M^p$ by setting $\hat{\lambda}_M := \oplus_p \hat{\lambda}_M^p$.

Let Λ^p be the matrix of the linking pairing $\hat{\lambda}_M^p$ with respect to the basis given in Theorem 2. Then Λ^p has the following form:

$$\Lambda^p = \begin{pmatrix} \Lambda_{1,1} & \Lambda_{1,2} & \dots & \Lambda_{1,s} \\ \Lambda_{2,1} & \Lambda_{2,2} & \dots & \Lambda_{2,s} \\ \vdots & \vdots & \vdots & \vdots \\ \Lambda_{s,1} & \Lambda_{s,2} & \dots & \Lambda_{s,s} \end{pmatrix}$$

where each $\Lambda_{l,t}$ is an $r_l \times r_t$ matrix, except in the cases when $l = s$ or $t = s$. In these cases it is an $r_s - 1 \times r_t$ or $r_l \times r_s - 1$-matrix respectively. (This follows because there are only $r_s - 1$ generators that arise from level s.)

Now suppose that under basis change Λ^p is given in its block diagonal form, $\Lambda^p = \operatorname{diag}(\Lambda_{1,1}, \dots \Lambda_{t,t}, \dots \Lambda_{s,s})$. In order to verify our conjecture it is necessary and sufficient to show that each diagonal block $\Lambda_{t,t}$ gives an arbitrary pairing on either $(\mathbf{Z}/p^t)^k$, for some k or $(\mathbf{Z}/2^t \oplus \mathbf{Z}/2^t)^l$, for some l. For then Λ^p would be an arbitrary linking pairing on $\text{Tors}_p H_1(M)$ and hence all isomorphism classes of linking pairings on finite abelian groups could be realized by the pairing $\hat{\lambda}_M := \oplus_p \hat{\lambda}_M^p$, for some M. Since the (usual) linking form of any closed, connected, oriented 3-manifold must belong to one of these isomorphism classes, the linking form conjecture would follow as well.

Recall that since g.c.d. $(a_{t,j}, b_{t,j}) = 1$, the elements $b_{t,j}$ are invertible modulo p^c. Throughout the remainder of the paper let $c_{l,j}$ denote $-b_{t,j}^{-1}$ modulo p^c for convenience. Theorem 2 will now be used to calculate the linking matrix Λ^p.

Theorem 4. *The matrix blocks $\Lambda_{l,t}$ of the matrix Λ^p of the linking pairing $\hat{\lambda}_M^p$, for $M = (O, o; 0 \mid e : (a_1, b_1), \dots, (a_m, b_m))$ when $Ae + C \neq 0$, have the following form for any prime p:*

1. *When $l \neq t$,*

$$\Lambda_{l,t} = \frac{1}{p^{t+l}} \begin{pmatrix} a_{s,1}c_{s,1} & a_{s,1}c_{s,1} & \dots & a_{s,1}c_{s,1} \\ a_{s,1}c_{s,1} & a_{s,1}c_{s,1} & \dots & a_{s,1}c_{s,1} \\ \vdots & \vdots & \vdots & \vdots \\ a_{s,1}c_{s,1} & a_{s,1}c_{s,1} & \dots & a_{s,1}c_{s,1} \end{pmatrix}.$$

2. *When $l = t \neq s$,*

$$\Lambda_{t,t} = \frac{1}{p^{2t}} \begin{pmatrix} a_{s,1}c_{s,1} + a_{t,1}c_{t,1} & a_{s,1}c_{s,1} & \dots & a_{s,1}c_{s,1} \\ a_{s,1}c_{s,1} & a_{s,1}c_{s,1} + a_{t,2}c_{t,2} & \dots & a_{s,1}c_{s,1} \\ \vdots & \vdots & \vdots & \vdots \\ a_{s,1}c_{s,1} & a_{s,1}c_{s,1} & \dots & a_{s,1}c_{s,1} + a_{t,r_t}c_{t,r_t} \end{pmatrix}.$$

3. *When $l = t = s$,*

$$\Lambda_{s,s} = \frac{1}{p^{2s}} \begin{pmatrix} a_{s,1}c_{s,1} + a_{s,2}c_{s,2} & a_{s,1}c_{s,1} & \dots & a_{s,1}c_{s,1} \\ a_{s,1}c_{s,1} & a_{s,1}c_{s,1} + a_{s,3}c_{s,3} & \dots & a_{s,1}c_{s,1} \\ \vdots & \vdots & \vdots & \vdots \\ a_{s,1}c_{s,1} & a_{s,1}c_{s,1} & \dots & a_{s,1}c_{s,1} + a_{s,r_s}c_{s,r_s} \end{pmatrix}.$$

The block sum diagonalization of Λ^p depends only on the diagonal blocks of Λ^p. This follows from the fact that the invariants (ρ_p^m, σ_p^m) of the diagonal blocks of Λ^p determine the linking form completely. By the theory of linking pairings for odd primes p,

$$\Lambda_{t,t} = \begin{cases} r_t\left(\frac{1}{p^t}\right), & \text{if } \left(\frac{\det \Lambda_{t,t}}{p^t}\right) = 1, \\ (r_t - 1)\left(\frac{1}{p^t}\right) \oplus \left(\frac{\det \Lambda_{t,t}}{p^t}\right), & \text{if } \left(\frac{\det \Lambda_{t,t}}{p^t}\right) = -1. \end{cases} \tag{2.1}$$

Lemma 1. *Let p be an odd prime. Then det $\Lambda_{t,t}$ is a square mod p^t if and only if $\prod_{j=1}^{r_t} a'_{t,j}c_{t,j}$ is as well, for all $1 \le t \le s$.*

Proof. First observe that for a square matrix $A = (a_{i,j})$, defined so that $a_{i,i} = x_i$, for $1 \le i \le n$ and $a_{i,j} = a$, for $1 \le i \ne j \le n$,

$$\det\ (A) = -af'(a) + f(a),$$

where $f(t) = \prod_{i=1}^n (x_i - t)$. It follows by applying this fact to $\Lambda_{t,t}$ that

$$\det \Lambda_{t,t} = -a'_{s,1}c_{s,1}\left(-\prod_{j=1}^{r_t} a'_{t,j}c_{t,j} \cdot \sum_{i=1}^{r_t} \frac{1}{a'_{t,i}c_{t,i}}\right) + \prod_{j=1}^{r_t} a'_{t,j}c_{t,j},$$

where $a'_{t,j} = \frac{a_{t,j}}{p^t} \in \mathbf{Z}$ and $a'_{s,1} = \frac{a_{s,1}}{p^t} \in p^{s-t}\mathbf{Z}$.

The result now follows from the fact that x is a square mod p^k if and only if x is a square mod p^{k+1} for all $k \in \mathbf{N}$. (Necessity of this statement is proved by induction on k, while sufficiency is clear.) □

Theorem 5. *For any prime $p > 2$*

$$\Lambda_{t,t} = \begin{cases} r_t\left(\frac{1}{p^t}\right), & \textit{if } \prod_{j=1}^{r_t} a'_{t,j}c_{t,j} \textit{ is a square mod } p, \\ (r_t - 1)\left(\frac{1}{p^t}\right) \oplus \left(\frac{\prod_{j=1}^{r_t} a'_{t,j}c_{t,j}}{p^t}\right), & \textit{if } \prod_{j=1}^{r_t} a'_{t,j}c_{t,j} \textit{ is a non-square mod } p. \end{cases}$$

Proof. The proof follows immediately by applying Lemma 1 to equation 2.1. □

The following remark shows how to obtain the block sum decomposition of any linking pairing on abelian groups of odd order.

Remark 3. *Let Λ^p be a linking matrix with block sum decomposition*

$$\text{diag}\,(\Lambda_{1,1}, \ldots, \Lambda_{t,t}, \ldots, \Lambda_{s,s}).$$

Each diagonal block, $\Lambda_{t,t}$, represents a linking pairing on $\bar{G}_t = (\mathbf{Z}/p^t)^{r_t}$ as described in Theorem 5, for any prime $p > 2$. The isomorphism type of $\Lambda_{t,t}$ is symmetric in the Seifert invariants $(a_{t,i}, b_{t,i})$, where $1 \le i \le r_t$ and $1 \le t \le s$. (When $t = s$ this fact is not completely obvious.) Furthermore, under basis change

$$\Lambda_{t,t} = \begin{pmatrix} \frac{1}{p^t} & 0 & \ldots & 0 & 0 \\ 0 & \frac{1}{p^t} & \ldots & 0 & 0 \\ \vdots & \vdots & \vdots & \vdots & \vdots \\ 0 & 0 & \ldots & \frac{1}{p^t} & 0 \\ 0 & 0 & \ldots & 0 & \frac{S}{p^t} \end{pmatrix},$$

where S is either 1 or $\prod_{j=1}^{r_t} a'_{t,j}c_{t,j}$.

Let G be a finite abelian group with $Tors_pG = \oplus_t\bar{G}_t$. Remark 1 shows that there is a Seifert manifold M, satisfying the condition of Theorem 1, with $H^2(M;\mathbf{Z}/p^s) \cong Tors_pG \cong \oplus_t\bar{G}_t$, for all p. Next observe that all non-squares $\prod_{j=1}^{r_t} a'_{t,j}c_{t,j}$ mod p in Theorem 5 give isomorphic linking pairings. It is now clear that each diagonal block $\Lambda_{t,t}$ gives an arbitrary linking pairing on $\bar{G}_t$, by simply varying the parameters r_t, $a_{t,i}$, of $\Lambda_{t,t}$.

Since this can be done for all t, the matrix Λ^p gives an arbitrary linking pairing on the p-torsion of $H^2(M)$. (Recall that Λ^p is the matrix of the linking pairing $\hat{\lambda}^p_M$.) It now follows that an arbitrary linking pairing $\hat{\lambda}_M$ can be constructed on $H_1(M) \cong H^2(M)$ as the orthogonal sum $\hat{\lambda}_M = \oplus_p\hat{\lambda}^p_M$. This proves the conjecture for abelian groups of odd order.

Remark 4. *Observe that throughout this section we have only considered the case when $s = c$. In this case $Tors_pH^2(M) \cong H^2(M;\mathbf{Z}/p^s)$. This means that the p-torsion of $H^2(M)$ depends only on the p-valuations of the Seifert invariants $a_{t,j}$. The case when $s \neq c$ depends additionally on the Euler number e and on the $b_{t,i}$. However this case, which is more complicated, is not required when determining the isomorphism classes of linking pairings for groups of odd order.*

For completeness it should also be observed that certain linking pairings also arise from the case considered in Theorem 3, when $p > 2$.

Theorem 6. *Given a Seifert manifold $M := (O, o; 0 \mid e : (a_1, b_1), \dots, (a_m, b_m))$, suppose that $n = 0$, that is, each $a_i \not\equiv 0 \pmod{p^t}$ for all $1 \le i \le n$ and $1 \le t \le s$. The linking form of M is determined by Theorem 3 when $q = c$ and gives only one case, which occurs when $Ae + C \equiv 0 \pmod{p^c}$. The linking form in this case is given by the 1×1 matrix $\left(\frac{A^{-1}}{p^c}\left[\sum_{i=r+1}^m b_iA_i + Ae + C\right]\right)$ and so represents a linking pairing on $\mathbf{Z}/p^c$.*

When $p = 2$, the complete minimal system of invariants $\{(\rho_2^k, \sigma_2^k)\}$ of linking pairings for 2-groups has a somewhat different description than in the case for the odd primes. We will use the characterization of these invariants given by Deloup in (De3) and summarize this construction as follows: Given a quadratic form $q: G \to \mathbf{Q}/\mathbf{Z}$ on a finite abelian group G, define a Gauss sum associated to q by,

$$\Gamma(G, q) = |G|^{-\frac{1}{2}} \sum_{x \in G} e^{2\pi i q(x)}.$$

For a linking pairing $\phi: G \times G \to \mathbf{Q}/\mathbf{Z}$, define

$$\tau_2^k(\phi) = \Gamma(G, 2^{k-1}q_\phi),$$

where $q_\phi(x) = \phi(x, x)$.

Definition 2. *For a linking form ϕ on a 2-group G, let*

$$\sigma_2^m(\phi) = \begin{cases} \infty, & \text{if } \tau_2^m(\phi) = 0, \\ \frac{1}{2\pi}\text{Arg } \tau_2^m(\phi), & \text{otherwise}. \end{cases}$$

The results given above clearly show that all linking pairings of the form $(n2^{-k})$ on $\mathbf{Z}/2^k$, for $k \geq 1$, do exist. However two problems remain. The first is to distinguish when the linking form on the 2-group is hyperbolic or not. It follows from Corollary 5.2 (KK) that a linking pairing on a finite 2-group is hyperbolic if and only if all the ranks ρ_2^k of the homogeneous groups of exponent 2^k, for all $k = 1, 2, 3, \ldots$ are even and all the signatures σ_2^k are 0. Although this gives a criterion for distinguishing hyperbolic linking forms from the linkings over $\mathbf{Z}/2^k$, it is still difficult to implement. Furthermore, the main problem, which is to distinguish between the hyperbolic linking pairings must still be resolved.

Summary: Our object is to show that any linking pairing on a finite abelian group is isomorphic to the linking form on an oriented irreducible 3-manifold and in particular that it is isomorphic to the linking form of a Seifert manifold that is a rational homology sphere. In this paper we have shown that any linking matrix on an abelian group of odd order has a block sum decomposition that corresponds to the linking form of a Seifert manifold, M, that is a rational homology sphere for which $\mathrm{Tor}_2 H_1(M) = 0$. Our approach is to first calculate the matrix Λ^p of a linking pairing that we define on the p-torsion of first homology of M. We calculate this by using the product structure in cohomology. Next we show that by varying the parameters of the manifold, M, we obtain an arbitrary linking pairing on the p-torsion of $H_1(M)$. Since we can do this for all $p > 2$ we obtain an arbitrary linking pairing on all abelian groups of odd order and hence obtain all isomorphism classes of linking pairings on abelian groups of odd order. Since the linking form of the manifold M belongs to one of these isomorphism classes, this proves the conjecture for abelian groups of odd order.

We hope to show in a later paper that this result can be extended to all finite abelian groups by showing that Seifert manifolds satisfying the hypotheses of Theorem 2 also account for arbitrary linking pairings on $(\mathbf{Z}/2^t)^k$ or $(\mathbf{Z}/2^t \oplus \mathbf{Z}/2^t)^l$.

Acknowledgements: Both authors would like to thank Dror Bar Natan and Ruth Lawrence for their invitation to the Einstein Institute for Mathematics at Hebrew University Jerusalem, Israel where the research for this paper was conducted. We would also like to thank the institute itself for its support and for providing a stimulating atmosphere.

References

Bryden, J. *Cohomology rings of oriented Seifert manifolds with mod p^s coefficients*, Advances in Topological Quantum Field Theory, NATO Science Series, Kluwer, (2004).

Bryden, J.; Hayat-Legrand, C.; Zieschang, H.; Zvengrowski, P. *L'anneau de cohomologie d'une variété de Seifert* , C. R. Acad. Sci. Paris, 324, (1) (1997) 323-326.

Bryden, J.; Hayat-Legrand, C.; Zieschang, H.; Zvengrowski, P. *The cohomology ring of a class of Seifert manifolds* , Top. and its Appl. 105 (2) (2000) 123-156.

Bryden, J.; Pigott, B.; Lawson, T.; Zvengrowski, P. *The integral homology of the oriented Seifert manifolds* Top. and Its Appl., 127 (1-2) (2003) 259-276 .

Bryden, J.; Zvengrowski, P. *The cohomology algebras of oriented Seifert manifolds and applications to Lusternik-Schnirelmann category* , Homotopy and Geometry, Banach Center Publications, Vol. 45 (1998) 25-39.

Bryden, J.; Zvengrowski, P. *The cohomology ring of the oriented Seifert manifolds II*, Top. and Its Appl., 127 (1-2) (2003) 213-257 .

Deloup, F. *Linking forms, reciprocity for Gauss sums and invariants of* 3-*manifolds*, Trans. of the AMS, 35 (5) (1999) 1895-1918.

Deloup, F. *An explicit construction of an abelian topological quantum field theory in dimension 3*, Top. and its Appl. 127 (1-2) (2003) 199-211.

Deloup, F. *Une description combinatoire du monode des enlacements* C. R. Math. Acad. Sci. Paris 337 (4) (2003), 227–232 .

Dijkgraaf, R.; Witten, E. *Topological gauge theories and group cohomology*, Commun. Math. Phys. 129 (1990) 393-429.

Hempel, J. *3-Manifolds* , Vol. 86, Annals of Math Studies, Princeton Univ. Press, Princeton, New Jersey (1976).

Gordon, C. McA.; Luecke, J. *Reducible manifolds and Dehn surgery*, Topology 35 (2) (1996), 385–409.

Kawauchi, A.; Kojima, S. *Algebraic classification of linking pairings on* 3-*manifolds*, Math. Ann. 253 (1980) 29-42.

MacLane, S. *Homology*. Springer Verlag, Berlin(1963).

Montesinos, J.M. *Classical Tessellations and Three-Manifolds* , Springer-Verlag, Berlin-Heidelberg-New York (1987).

Murakami, H.; Ohtsuki, T.; Okada, M. *Invariants of three-manifolds derived from linking matrices of framed links*, Osaka J. Math. 29 (1992) 545-572.

Myers, Robert *Simple knots in compact, orientable* 3-*manifolds*, Trans. Amer. Math. Soc. 273 (1) (1982) 75–91.

Orlik, P. *Seifert Manifolds*, Lecture Notes in Math. 291, Springer-Verlag, Berlin-Heidelberg-New York (1972).

Seifert, H. *Topologie dreidimensionaler gefaserter Räume* , Acta. Math. 60 (1932) 147-238.

Seifert, H.; Threlfall, W. *A Textbook of Topology*, Academic Press, London (1980).

Turaev, V. *Cohomology rings, linking forms and invariants of spin structures of three-dimensional manifolds*, Math USSR Sbornik, 48 No.1 65-79 (1984).

Turaev , V. *Quantum invariants of knots and* 3-*manifolds*, de Gruyter Studies in Mathematics (1994).

Wall, C.T.C. *Quadratic forms on finite groups , and related topics*, Topology, 2 (1964) 281-298.

Address for Offprints:

John Bryden
Department of Mathematics and Statistics
University of Calgary
and
Department of Mathematics and Statistics
Southern Illinois University
Edwardsville, IL 62025
email: jbryden@siue.edu

F. Deloup
Laboratoire Emile Picard
Université Paul Sabatier
Toulouse, France
email: deloup@math.huji.ac.il

Mappings of nonzero degree between 3-manifolds: a new obstruction[1]

Dale Rolfsen (*rolfsen@math.ubc.edu*)
Department of Mathematics
University of British Columbia
Vancouver. B.C., Canada

Abstract. My purpose here is to discuss a new technique which can be applied to 3-manifold theory. Orderability properties of fundamental groups may be used to provide a method of analyzing mappings between closed orientable 3-manifolds. In particular, an obstruction to the existence of nonzero degree maps is developed – a part of joint work with S. Boyer and B. Wiest [1]. We also discuss numerous examples to which the obstruction applies.

Keywords: 3-manifolds, degree of mappings, ordered groups

Mathematics Subject Classification 2000:- Primary: 57M27 : Secondary: 20F60

1. Definitions and statement of results.

An important question in the theory of manifolds is the following: given two closed orientable manifolds, M^n and N^n, does a degree one map $f : M^n \to N^n$ exist? The answer is always "yes" if N is the sphere S^n or a homotopy sphere, but overall it is a subtle problem to analyze this partial ordering among manifolds. More generally, one may ask if such a map f exists with nonzero degree.

My purpose in this exposition is to explain a new approach to this question for the case $n = 3$. It involves orderability of the fundamental groups and results that appear in [1]. After preliminaries and the main result, I will describe a number of manifolds to which this criterion, our main theorem, applies.

A group G is said to be *left-orderable* (LO) if there exists a strict total ordering of its elements such that $g < h \Rightarrow fg < fh$ for all $f, g, h \in G$. LO groups are easily seen to be torsion-free, but the converse is not true (examples will be discussed below). It should be noted that a group is LO if and only if it has a (possibly different) right-invariant ordering. Groups which are LO have many pleasant properties; for example their group rings have no zero divisors. Further information on orderable groups may be found in [6], [8] and [9]. We also note that Farrell [5] has shown that if X is a reasonable space (Hausdorff, paracompact and possessing a universal cover), then $\pi_1(X)$ is LO if and only if the universal cover $\tilde{X}$ of X embeds in $X \times [0,1]$ so that the projection $X \times [0,1] \to X$ restricts to the covering map.

[1] This research was partially supported by NSERC research grant 8-82-02

J.M. Bryden (ed.), Advances in Topological Quantum Field Theory, 267–273.

Many interesting nonabelian groups are LO; in particular the fundamental groups of many 3-manifolds. It is shown in [1] that for each of the eight 3-manifold geometries, there exist manifolds modelled on that geometry whose fundamental groups are LO and other examples which are not LO.

We now state the main result, which may be interpreted as an obstruction to the existence of nonzero degree maps.

THEOREM 1.1. *Suppose M^3 and N^3 are closed, connected, oriented 3-manifolds. Assume M is irreducible, that $\pi_1(N^3)$ is LO, but $\pi_1(M^3)$ is not LO. Then any map $M^3 \to N^3$ has degree zero.*

I would like to thank Steve Boyer and Bert Wiest for their collaboration with me on orderable 3-manifold groups, which made this presentation possible. Parts of the present paper appear in our work [1].

2. Proof of Theorem 1.1

One of the principal results of [1] is the following criterion for left-orderability of a 3-manifold group, which we state only for the case of orientable manifolds.

THEOREM 2.1. *Suppose that M is a compact, connected, orientable, irreducible 3-manifold. Then $\pi_1(M)$ is LO if and only if either $\pi_1(M)$ is trivial or there exists a non-trivial homomorphism from $\pi_1(M)$ to some LO group. In particular, if $H_1(M)$ is infinite, $\pi_1(M)$ is LO.* □

LEMMA 2.2. *Suppose M^n and N^n are closed connected oriented manifolds and $f : M \to N$ is a map of nonzero degree. Then the index of $f_*\pi_1(M)$ in $\pi_1(N)$ is finite.*

Proof: Consider the covering space $p : \tilde{N} \to N$ corresponding to the $f_*\pi_1(M)$. Then f lifts to $\tilde{f} : M \to \tilde{N}$, so we have $f = p \circ \tilde{f}$. If the index were infinite, $\tilde{N}$ would be noncompact and hence have zero homology in dimension three. But this would imply that f, which factors through $\tilde{N}$, has degree zero. □

We can now prove the Theorem 1.1. Assume the hypotheses and suppose there is a map $f : M \to N$ of nonzero degree. By the lemma, since its image has finite index, $f_* : \pi_1(M) \to \pi_1(N)$ is a nontrivial homomorphism to the LO group $\pi_1(N)$. Theorem 3.2 then implies that $\pi_1(M)$ is LO, a contradiction. □

3. Examples and applications

The goal of this section is to present a number of examples to which Theorem 1.1 can be applied. We begin with several examples of 3-manifold groups which are torsion-free, yet not LO.

Example 1: Consider the Klein bottle K^2 and its fundamental group which we also denote by K, without the superscript,

$$K = \pi_1(K^2) = \langle x, y; x^{-1}yx = y^{-1} \rangle.$$

It is a well-known example of an LO group which cannot possess any ordering which is invariant under multiplication on both sides. It is LO, since it fits into an exact sequence

$$1 \to \mathbb{Z} \to K \to \mathbb{Z} \to 1$$

and left-orderability is preserved under extensions (by a straightforward argument). In any LO group we clearly have $1 < y \Leftrightarrow y^{-1} < 1$. If K had a 2-sided invariant ordering, the order would also be invariant under conjugation, and we would be led by the defining relation to the contradiction $y > 1 \Leftrightarrow y < 1$.

Now we will use this example to construct a group G which is torsion-free, but not LO. In an LO group, define $|x| = x$ if $x \geq 1$ and $|x| = x^{-1}$ if $x < 1$. The notation $x << y$ means that all powers of x are less than y.

LEMMA 3.1. *In any left-ordering of the group K presented above, $|y| << |x|$.*

Proof: There are four cases, according to whether the generators are positive or negative. Consider the case $x > 1$ and $y > 1$. Then, noting that $y^{-1}x = xy$ in K, for any positive integer n we have $y^{-n}x = xy^n > 1$, and left-invariance implies $x > y^n$. The other cases are similar. □

Now consider the subgroup H of K generated by x^2 and y; clearly $H \cong \mathbb{Z} \oplus \mathbb{Z}$ and H has index two in K. Let $\bar{K}$ be another copy of K, with corresponding generators $\bar{x}, \bar{y}$ and $\bar{H} = \langle \bar{x}^2, \bar{y} \rangle$ be the corresponding copy of H. The trick is to identify H and $\bar{H}$ with a twist, via

$$x^2 \leftrightarrow \bar{y}, \quad y \leftrightarrow \bar{x}^2$$

and form the corresponding free product with amalgamation:

$$G := K *_H \bar{K}.$$

PROPOSITION 3.2. *G is torsion-free, but not LO.*

Proof: It is well-known that amalgams of torsion-free groups are also torsion free. If there were a left-invariant ordering of G, it would restrict to left-orderings of both K and $\bar{K}$. By the lemma, $|y| << |x| < |x^2|$, but our identifications would then imply that $|\bar{y}| > |\bar{x}^2| > |\bar{x}|$, contradicting the same lemma applied to $\bar{K}$. □

3.1. 3-MANIFOLDS WITH π_1 NOT LO

Example 1, continued: Next we point out that the group G we have just constructed is the fundamental group of a closed, orientable 3-manifold. Indeed, consider the orientation double cover $p : T^2 \to K^2$ and let $P = Cyl(p)$ be its mapping cylinder. P is the twisted I bundle over the Klein bottle, an orientable 3-manifold with boundary T^2. The fundamental group of P is our group K, with $\pi_1(T^2)$ corresponding to the subgroup H as described above.

Let $\bar{P}$ be another copy of P and identify their torus boundaries, but with the "twist" which interchanges longitude and meridian of the two tori, to form the closed, orientable 3-manifold

$$M_G := P \cup_{T^2} \bar{P}.$$

By the Seifert-Van Kampen theorem, we see that $\pi_1(M_G) \cong G$. It can be readily verified that M_G is irreducible, and is a Haken manifold, as the T^2 in the middle is incompressible. In fact, M_G is a graph manifold in the sense of Waldhausen [12], the graph having two vertices and a single edge.

J. Przytycki has pointed out during this conference that the manifold M_G can also be described simply as the 2-fold branched cyclic covering of S^3, branched over the Borromean rings.

Example 2: A very similar example to the above was considered in [1]. It is constructed by sewing the complements of two trefoil knots in S^3 together along their torus boundaries, but with the twist interchanging the meridian with a curve corresponding to the Seifert fibre of the trefoil's complement. That the group of this 3-manifold is not LO is proven in much the same way as above; (see [1] for details.) This example is also an irreducible graph-manifold.

Example 3: Examples of (irreducible) hyperbolic closed 3-manifolds whose groups are not LO have recently been developed by Roberts, Shareshian and Stein [11]. There is one such manifold $M^3_{p,q,m}$ for each choice of a negative integer m and relatively prime positive integers p and q. Its fundamental group has the following presentation:

$$\pi_1(M^3_{p,q,m}) \cong \langle t, a, b : t^{-1}at = aba^{m-1}, t^{-1}bt = a^{-1}, t^{-p} = (aba^{-1}b^{-1})^q \rangle.$$

They argue that this group is not LO by showing that it does not act effectively on the real numbers by order-preserving homeomorphisms. It is well known that countable groups are LO if and only if they embed in $Homeo_+(\mathbb{R})$. They use this result (in a strengthened form) to conclude that these manifolds cannot possess a 2-dimensional foliation without Reeb components. Note that these examples have finite first homology, as can be easily verified by abelianizing $\pi_1(M^3_{p,q,m})$.

Example 4: Calegari and Dunfield [2] have also investigated orderability of a number of hyperbolic manifolds, including the Weeks manifold W^3, whose first homology group is $\mathbb{Z}_5 \times \mathbb{Z}_5$. This is the one of smallest volume among all known hyperbolic 3-manifolds. It can be obtained by surgery on a Whitehead link in S^3 with coefficients 5 and 5/2 and has a fundamental group with presentation

$$\pi_1(W) \cong \langle a, b : ababab^{-1}a^2b^{-1} = bababa^{-1}b^2a^{-1} = 1 \rangle.$$

Their proof that this group is non-LO is worth repeating here, as it is short and illustrates an algorithmic approach to showing a group is not LO. First, note that if a group G is LO, then the positive cone $P = \{g \in G : 1 < g\}$ is closed under multiplication and $G \setminus 1$ is the disjoint union of P and P^{-1}. (In fact, the existence of such a P is easily seen to be equivalent to left-orderability.) We will show that such a P cannot exist for $\pi_1(W)$. Suppose such P does exist. Without loss of generality, we can assume $a \in P$ (reverse the ordering if necessary).

Case $b \in P$. Subcase $ab^{-1} \in P$: Then $abab(ab^{-1})a(ab^{-1})$ belongs to P, a contradiction, since it equals the identity. Subcase $ba^{-1} \in P$ leads to a similar contradiction, symmetrically.

Case $b^{-1} \in P$. Note that

$$b^{-1}ab^{-2}ab^{-1}a^2b^{-1} = b^{-1}(bababa^{-1}b^2a^{-1})^{-1}b(ababab^{-1}a^2b^{-1}) = 1$$

but in this case we would have $b^{-1}ab^{-2}ab^{-1}a^2b^{-1} \in P$, again leading to a contradiction. So $\pi_1(W)$ cannot be LO.

In [2] a census of 128 hyperbolic 3-manifolds of small hyperbolic volume are tabulated, and 44 were shown to have non-LO fundamental groups.

Example 5: Further examples of irreducible 3-manifolds with torsion-free, but non-LO, groups have very recently been described by Dabkowski, Przytycki and Togha [3]. These manifolds are constructed as certain branched coverings of S^3, with branch set a particular knot or link, including torus links, pretzel links and 2-bridge links. A particular family of examples are the n-fold branched cyclic covers of the torus link of type $(2, 2k)$, with antiparallel orientation, where n, k are arbitrary positive integers. Other infinite families of examples are constructed in [3], including our Example 1 as a special case.

3.2. 3-MANIFOLDS WITH LEFT-ORDERABLE π_1

There are many 3-manifold groups which *are* LO. Theorem 3.2 provides many examples, for example irreducible manifolds (with or without boundary) which have positive first Betti number.

As explained in [1], the fact that the universal cover of $PSL_2(\mathbb{R})$ is left-orderable also implies (as corollary to Theorem 3.2) that many manifolds with first Betti number zero also have LO fundamental group.

COROLLARY 3.3. *With the exception of Poincaré 's dodecahedral space, with π_1 of order 120, every homology sphere which is Seifert-fibred has LO fundamental group.*

Among orientable Seifert fibred spaces, the ones with LO fundamental group have been characterized in [1], as follows.

THEOREM 3.4. *A closed orientable Seifert fibred space M^3 has LO fundamental group if and only if one of the following holds:*

(i) $M^3 \cong S^3$,

(ii) $H_1(M)$ *is infinite, or*

(iii) $\pi_1(M)$ *is infinite, the base orbifold is a 2-sphere (with possibly some cone points), and* M *admits a foliation which is everywhere transverse to the fibres.* □

The Seifert manifolds satisfying condition (iii) have been explicitly characterized in [4, 7, 10]. In the Calegari-Dunfield tabulation, they show that several of the hyperbolic manifolds with finite homology and small volume have left-orderable fundamental groups.

3.3. Conclusion

All the examples in 3.1 describe 3-manifolds which are orientable, irreducible and have infinite, torsion-free fundamental groups which fail to be left orderable.

As an application of Theorem 1.1, take M^3 any one of these manifolds and N^3 any of the manifolds described in 3.2, which have left-orderable fundamental group. Then there is no map $M \to N$ of nonzero degree.

In closing, I would like to discuss the existence of orderable subgroups of finite index. Our work [1] led Boyer, Wiest and myself to ask if the fundamental group of any (irreducible) 3-manifold may be *virtually* left-orderable:

VLO Question: If M^3 is an irreducible compact 3-manifold, does $\pi_1(M)$ contain a finite-index subgroup which is LO?

This is related to the virtual Betti number conjecture, that every such M^3 has a finite-sheeted covering which has positive first Betti number. Such a finite-sheeted covering would have LO fundamental group, by Corollary 3.3, and so $\pi_1(M)$ would be virtually left-orderable.

We do not know the answer to the VLO question, even for hyperbolic 3-manifolds. It was shown in [1] that the fundamental groups of 3-manifolds with any of the other seven geometries, the answer is "yes." That is, if M^3 is Seifert-fibred or a Sol manifold, then $\pi_1(M)$ is virtually LO. Indeed $\pi_1(M)$ is virtually bi-orderable, in that it has a finite index subgroup which supports an ordering invariant under multiplication from both sides.

VO Question: If M^3 is an irreducible compact 3-manifold, does $\pi_1(M)$ contain a finite-index subgroup which is bi-orderable?

An affirmative answer to this last question would imply the venerable virtual Haken conjecture, because it happens that any nontrivial group which is bi-orderable must have infinite abelianization, see [1]. Summarizing:

PROPOSITION 3.5. *"Yes" to the VO question implies the virtual first Betti number conjecture, which in turn implies an affirmative answer to the VLO question.*

For this reason, an affirmative answer to the VO question or negative answer to the VLO question would be very big news.

References

1. S. Boyer, D. Rolfsen and B. Wiest, *Orderable 3-manifold groups*, preprint 2001.
2. D. Calegari and N. Dunfield, *Laminations and groups of homeomorphisms of the circle*, to appear in Invent. Math.
3. M. Dabkowski, J. Przytycki and A. Togha, *Non-left-orderable 3-manifold groups*, preprint.
4. D. Eisenbud, U. Hirsch, W. Neumann, *Transverse foliations on Seifert bundles and self-homeomorphisms of the circle*, Comm. Math. Helv. **56** (1981), 638–660.
5. F. T. Farrell, *Right-orderable deck transformation groups*, Rocky Mountain J. Math. **6**(1976), no. 3, 441–447.
6. A. M. W. Glass, *Partially Ordered groups*, Series in Algebra, vol. 7, World Scientific,London, 1999.
7. M. Jankins, W. Neumann, *Rotation numbers and products of circle homomorphisms*, Math. Ann. **271** (1985), 381–400.
8. Valeriĭ M. Kopitov and Nikolaĭ Ya. Medvedev, *Right-Ordered Groups*, Plenum Publishing Corporation, New York, 1996.
9. Roberta Botto Mura and Akbar Rhemtulla, *Orderable groups*, Lecture Notes in Pure and Applied Mathematics, vol. 27, Marcel Dekker, New York, 1977.
10. R. Naimi, *Foliations transverse to fibers of Seifert manifolds*, Comm. Math. Helv. **69**(1994), 155–162.
11. R. Roberts, J. Shareshian, M. Stein, *Infinitely many hyperbolic 3-manifolds which contain no Reebless foliation*, to appear in J.A.M.S.
12. F. Waldhausen, *Eine Klasse von 3-dimensionalen Mannigfaltigkeiten, I and II*, Invent. Math. **3**(1967), 308 – 333 and **4**(1967), 87 – 117.

On braid groups, homotopy groups, and modular forms

F. R. Cohen *

Abstract. The purpose of this article is to list some connections between braid groups, homotopy groups, representations of braid groups, associated fibre bundles, and their cohomological properties. Several related problems are posed.

Keywords: braid groups, homotopy groups, and representations

Mathematics Subject Classification 2000:- Primary: 20F36, 55Q40: Secondary: 32S22, 55N15, 55P35, 57M99

§1 Introduction

This article gives a description of certain features of braid groups, and how these features fit together. The sections of this article are as follows.

1: Introduction

2: Braid groups, and classical homotopy groups

3: Representations and associated bundles for braid groups

4: Cohomology classes, representations, and their connections to modular forms

5: A sample computation

6: On bundles obtained from representations in section 2, their connections to modular forms as well as wild speculation

The author would like to thank John Bryden for organizing this interesting, and provocative conference as well as fostering the mathematical connections at this conference.

§2 Braid groups, and classical homotopy groups

There is a close connection between braid groups, the homotopy groups of the 2-sphere, Vassiliev invariants, and the unstable Adams spectral sequence or Bousfield-Kan spectral sequence. This section is an exposition of how these different structures fit together in a natural way based on joint work of J. Berrick, Y. L. Wong, J. Wu, and the author (CW; BCWW; W).

One way in which braid groups arise in homotopy theory is through the structure of a simplicial group Γ_* a collection of groups

$$\Gamma_0, \Gamma_1, \cdots, \Gamma_n, \cdots$$

* Partially supported by the NSF.

J.M. Bryden (ed.), Advances in Topological Quantum Field Theory, 275–288.

together with face operations

$$d_i : \Gamma_n \to \Gamma_{n-1},$$

and degeneracy operations

$$s_i : \Gamma_n \to \Gamma_{n+1},$$

for $0 \leq i \leq n$. These homomorphisms are required to satisfy the standard simplicial identities.

An example of a simplicial group is given by $\Gamma_n = P_{n+1}$, Artin's $(n+1)$-st pure braid group in degree n, and is elucidated in (CW). The face operations are given by deletion of a strand, while the degeneracies are gotten by "doubling" of a strand. This simplicial group is denoted AP_*.

Recall that P_{n+1} is generated by symbols $A_{i,j}$ for $1 \leq i < j \leq n+1$. Artin's relations are listed in (MKS). A reformulation of those relations is given by the following equalities of commutators for which $[a,b] = a^{-1} \cdot b^{-1} \cdot a \cdot b$.

1. $[A_{r,s}, A_{i,k}] = 1$ for either $r < s < i < k$ or $i < k < r < s$,
2. $[A_{k,s}, A_{i,k}] = [A_{i,s}^{-1}, A_{i,k}]$ for $i < k < s$,
3. $[A_{r,k}, A_{i,k}] = [A_{i,k}^{-1}, A_{i,r}^{-1}]$ for $i < r < k$, and
4. $[A_{r,s}, A_{i,k}] = [[A_{i,s}^{-1}, A_{i,r}^{-1}], A_{i,k}]$ for $i < r < k < s$.

The origin of the face, and degeneracy maps below is gotten by omitting a strand for a pure braid in the case of face maps, and doubling a strand for a pure braid in the case of degeneracy maps. The results are stated, but the computations are omitted. The face operations in the simplicial group AP_* are given by the following formulas for $d_t(A_{i,j})$.

1. $A_{i-1,j-1}$ if $t+1 < i$,
2. 1 if $t+1 = i$,
3. $A_{i,j-1}$ if $i < t+1 < j$,
4. 1 if $t+1 = j$, and
5. $A_{i,j}$ if $t+1 > j$.

The degeneracy operations $s_t(A_{i,j})$ are defined as follows:

1. $A_{i+1,j+1}$ if $t+1 < i$,
2. $A_{i,j+1} \cdot A_{i+1,j+1}$ if $t+1 = i$,
3. $A_{i,j+1}$ if $i < t+1 < j$,
4. $A_{i,j} \cdot A_{i,j+1}$ if $t+1 = j$, and
5. $A_{i,j}$ if $t+1 > j$.

Simplicial groups admit structures analogous to those of topological spaces, but in a more rigid setting. For example, Moore (Mo) defined the homotopy groups of a simplicial group Γ_* by

$$\pi_n\Gamma_* = Z_n/d_0(C_{n+1})$$

where

$$Z_n = \cap_{0\leq i\leq n} ker[d_i : \Gamma_n \to \Gamma_{n-1}],$$

and

$$C_{n+1} = \cap_{1\leq i\leq n+1} ker[d_i : \Gamma_{n+1} \to \Gamma_n].$$

These homotopy groups are isomorphic to the classical homotopy groups of the geometric realization of Γ_*.

In addition, the loop-space of a connected simplicial group Γ_* denoted $\Omega\Gamma_*$ is defined next as in (Mo) where $\Omega\Gamma_n$, the simplicial loop space in degree n, is the kernel of

$$d_0 : \Gamma_{n+1} \to \Gamma_n.$$

Define face and degeneracy operations $\bar{d}_i$, and $\bar{s}_i$ given by d_{i+1}, and s_{i+1} respectively for $i \geq 0$ by restriction to the subgroup $\Omega\Gamma_n$ in Γ_{n+1}. This formulation endows $\Omega\Gamma_*$ with the structure of a simplicial group. The groups $\pi_{n+1}\Omega\Gamma_*$, and $\pi_n\Gamma_*$ are naturally isomorphic as long as Γ_* is connected (Mo).

Recall Milnor's free group construction (M) given by $F[K_*]$ for a simplicial set K_* with a base-point $*$ in degree 0. Define $F[K_*]$ in degree n to be the free group generated by the n-simplices K_n modulo the single relation $s_0{}^n(*) = 1$. One feature of $F[K_*]$ is that the geometric realization $|F[K_*]|$ is homotopy equivalent to $\Omega\Sigma|K_*|$ in case K_* is reduced, that is K_0 is a single point (M). Let $\Delta[1]$ denote the simplicial 1-simplex with S^1 the simplicial circle.

Theorem 1. *The (simplicial) loop space of AP_* is isomorphic to $F[\Delta[1]]$ and is thus contractible. Hence $\pi_n AP_*$ is the trivial group for all n.*

A proof of this theorem arises by a direct comparison of $F[\Delta[1]]$, and the (simplicial) loop space of AP_*. An explicit map from $F[\Delta[1]]$ to $\Omega(AP_*)$ which realizes this isomorphism is given in (CW).

In addition, a second result gives that Milnor's free group construction $F[S^1]$ the simplicial loop-space of S^2 is embedded naturally in AP_*. Here recall that the simplicial circle S^1 in degree n is given by the simplicies $< 0^i, 1^{n+1-i} >$ for $0 \leq i \leq n$ with $< 0^{n+1} >=< 1^{n+1} >$ as the base-point.

Theorem 2. *There exists a unique morphism of simplicial groups*

$$\Theta : F[S^1] \to AP_*$$

with $\Theta(< 0, 1 >) = A_{2,1}$. The map Θ is an embedding. Hence the homotopy groups of $F[S^1]$ are natural sub-quotients of AP_*, and the geometric realization of quotient simplicial set $AP_*/F[S^1]$ is homotopy equivalent to the 2-sphere. Furthermore, the smallest sub-simplicial group of AP_* which contains the element $\Theta(< 0, 1 >) = A_{1,2}$ is isomorphic to $F[S^1]$.

The method of proof of this theorem is by a comparison of Lie algebras. That is consider the restriction of the homomorphism Θ to $\Theta_n : F_n \to P_{n+1}$. Next consider the morphism of associated graded Lie algebras obtained by filtering by the descending central series

$$E_0^*(\Theta_n) : E_0^*(F_n) \to E_0^*(P_{n+1}).$$

Here F_n denotes a free group on n generators, $F[S^1]$ in degree n. Using the structure of the Lie algebra $E_0^*(P_{n+1})$ analyzed by T. Kohno (K; K1), as well as Falk, and Randell (FR), a computation gives that $E_0^*(\Theta_n)$ is an embedding. That Θ_n is an embedding follows from the next proposition.

Recall that a discrete group Γ is said to be residually nilpotent group if

$$\bigcap_{i\geq 1} \Gamma^i(\Pi) = \{identity\}$$

where $\Gamma^i(\Pi)$ denotes the i-th stage of the descending central series for Π. Examples of residually nilpotent groups are free groups, and P_n.

Proposition 2.1. *1. Assume that Π is a residually nilpotent group. Let*

$$\rho : \Pi \to G$$

be a homomorphism of discrete groups such that the morphism of associated graded Lie algebras

$$E_0^*(\rho) : E_0^*(\Pi) \to E_0^*(G)$$

is a monomorphism. Then ρ is a monomorphism.

2. If Π is a free group, and $E_0^*(\rho)$ is a monomorphism, then ρ is a monomorphism.

One other feature of this proof is that the filtration arising from descending central series provides the method of Bousfield-Kan to construct the unstable Adams spectral sequence with the modifications obtained by using the mod-p descending central series. On the level of homotopy groups, this is precisely the E^0-term of the unstable Adams spectral sequence. On the other-hand, the Lie algebra above gives the Vassiliev invariants of pure braids via (K; K1). This point of view is that the Vassiliev invariants of pure braids specialize to give the E^0-term of the unstable Adams spectral sequence for the 2-sphere.

Replacing pure braid group P_n by the pure braid group of the 2-sphere results in a different, but similar construction $AP_*(S^2)$. The projection maps induce analogous maps to those of the face operations for a simplicial group, but degeneracies do not exist in this case.

Nevertheless, this analogue $AP_*(S^2)$, a Δ-group rather than a simplicial group, has homotopy sets analogous to homotopy groups defined as the set of left cosets above

$$\pi_n AP_*(S^2) = Z_n / d_0(C_{n+1}).$$

In the case of the pure braid group for S^2, these sets are groups as long as $n \geq 4$ with the following result from (BCWW).

Theorem 3. *If $n \geq 4$, then there is an isomorphism of groups*

$$\pi_n AP_*(S^2) \to \pi_n(S^2).$$

An analogue for all spheres arises at once by taking coproducts of simplicial groups $AP_* \vee AP_*$ which in degree n is given by the free product $P_{n+1} \amalg P_{n+1}$. Then the smallest simplicial subgroup which contains $P_2 \amalg P_2$ in degree 1 has geometric realization given by $\Omega(S^2 \vee S^2)$. Recall that ΩS^n is a retract of $\Omega(S^2 \vee S^2)$ for any $n \geq 2$ by the Hilton-Milnor theorem.

For the remainder of this section restrict to the simplicial group $\Gamma_* = F[S^1]$. It would be interesting to understand the group extension

$$1 \to d_0(C_{n+1}) \to C_n \to C_n/d_0(C_{n+1}) \to 1$$

for $C_n = \cap_{1 \leq i \leq n} ker[d_i : \Gamma_n \to \Gamma_{n-1}]$ on the level of pure braid groups. In particular, the Serre exact sequence for homology gives an exact sequence

$$\cdots \to H_1(d_0(C_{n+1}))_{C_n} \to H_1(C_n) \to H_1(C_n/d_0(C_{n+1})) \to 0$$

for which $H_1(d_0(C_{n+1}))_{C_n}$ denotes the module of coinvariants. Since the group C_n is free, the group $H_1(C_n)$ is free abelian.

The map $H_1(d_0(C_{n+1}))_{C_n} \to H_1(C_n)$ appears to be complicated combinatorially. Computations of this map are similar to those encountered with the classical partition function. As a rough analogy, it is natural to ask whether there is a generating function keeping track of these combinatorics as in the case of the classical partition function. It is also natural to ask whether there are associated functions which reflect the combinatorics of the map $H_1(d_0(C_{n+1}))_{C_n} \to H_1(C_n)$ in a way similar to that of the Dedekind η function and the generating function for the classical partition function. A specific computation is given in section 5 below.

§3 Representations of braid groups, and associated bundles

Given a discrete group Γ together with a topological group G, consider the space of all homomorphisms

$$Hom(\Gamma, G).$$

The quotient space modulo the conjugation action by inner automorphisms

$$Rep(\Gamma, G)$$

has similar features, but will not be addressed here. In case Γ is the pure braid group, the topology of these spaces have natural properties discussed in this section as given in joint work of A. Adem, D. Cohen, and the author (ACC; ACC2).

For example, consider the universal n-plane bundle

$$EO(n) \times_{O(n)} \mathbb{R}^n \to BO(n).$$

If G is any subgroup of the classical orthogonal group $O(n)$, then the pull-back of the universal bundle is that bundle given by

$$EO(n) \times_G \mathbb{R}^n \to EO(n)/G$$

with fibre $\mathbb{R}^n$ where $BG = EO(n)/G$. In addition, there is a bundle obtained over $B\Gamma$ obtained pulling back the universal bundle via a representation $\rho : \Gamma \to G$. That bundle is $E\Gamma \times_\Gamma \mathbb{R}^n \to B\Gamma$ in which Γ acts on $\mathbb{R}^n$ via ρ. In addition, there are natural evaluation maps

$$e : Hom(\Gamma, G) \to [B\Gamma, BG]$$

for which $[B\Gamma, BG]$ denotes the set of pointed homotopy classes of maps. The evaluation map e is defined by sending an element f to $e(f)$ the homotopy class of the induced map on the level of classifying spaces.

In case Γ is a discrete group, the classifying space $B\Gamma$ is $K(\Gamma, 1)$. There are two natural questions which arise here inspired by work of Toshitake Kohno on Vassiliev invariants of pure braids in case Γ is the pure braid group (K; K1).

1. For which representations $\rho : \Gamma \to G$ are the resulting bundles isomorphic to the trivial bundle ?

2. What is the topology of the space $Hom(\Gamma, G)$, and $Rep(\Gamma, G)$ when G is one of the classical Lie groups such as $O(n)$, $GL(n, \mathbb{R})$, or $PGL(n, \mathbb{R})$, or $Sp(2n, \mathbb{R})$?

The first crude question above, whether the bundle itself is a product as a bundle, admits further questions associated to properties satisfied by flat connections (K; K1). It may be interesting to see what additional information is encoded in representations which give rise to trivial bundles.

Recall that the pure braid group on k strands is the fundamental group of the complement of a complex hyperplane arrangement $\mathcal{A} \subset \mathbb{C}^k$, $M(\mathcal{A})$, given by the configuration space of ordered k-tuples of distinct points in the complex line, the braid arrangement. In the case of the pure braid groups, the initial question as to whether the associated bundles are trivial admits the following answer (ACC).

Theorem 4. *Let $\Gamma = \pi_1(M(\mathcal{A}))$ be the fundamental group of the complement of a complex hyperplane arrangement $\mathcal{A} \subset \mathbb{C}^k$. Then the vector bundle over $M(\mathcal{A})$ associated to a representation $\rho : \Gamma \to O(n)$ is trivial if and only if this representation lifts to $Spin(n)$. If $\gamma : \Gamma \to U(n)$ is any unitary representation, then the vector bundle over $M(\mathcal{A})$ associated to a representation γ is trivial.*

The basic methods of proof are similar in spirit to those for the cohomology of finite groups. That is, in case H is a finite group with p-Sylow subgroup H_p, the natural inclusion $H_p \to H$ induces a monomorphism in mod-p cohomology. A standard technique is to analyze the image of the restriction map. The method used to obtain Theorem 4 above for analyzing the bundles is analogous.

In particular, given $\Gamma = \pi_1(M(\mathcal{A}))$, a group Π together, a homomorphism $\Pi \to \Gamma$ is constructed such that the induced map on K-theory is a monomorphism, and the bundles obtained from representations of Π are computable from characterisitc classes. Thus the groups Π detect non-triviality of bundles arising from representations. This procedure is carried out in the special cases of $\Gamma = \pi_1(M(\mathcal{A}))$ the fundamental group of the complement of a complex hyperplane arrangement with $\mathcal{A} \subset \mathbb{C}^k$ by

1. constructing explicit maps of tori to $M(\mathcal{A})$,

2. extending these maps to a map of the bouquet of these tori, and

3. proving that the K-theory of this bouquet detects non-trivial bundles which arise from representations.

In addition, the fundamental group of the bouquet in part 2 above is a free product of free abelian groups (not necessarily a free group). The determination of which bundles arise from representations of free abelian groups follows directly from a computation of the characteristic classes for the representation. The property that free products of free abelian groups detect non-triviality of bundles arising from representations of $\Gamma = \pi_1(M(\mathcal{A}))$ is given in the article (ACC) via a certain cofibre sequence.

In subsequent extensions of this result, the following feature arises (ACC2). If Γ is the pure braid group, or more generally the fundamental group of the complement of a complex hyperplane arrangement $\mathcal{A} \subset \mathbb{C}^k$, then the question of whether a bundle is trivial is closely tied to the structure of the maximal abelian subgroups A, those abelian subgroups which are not proper subgroups of any other abelian subgroup. One subsequent result is as follows (ACC2) in which properties of free products of free abelian groups are central to the proof.

Theorem 5. *Assume that*

1. *$\rho : \Gamma \to G$ is any representation of the fundamental group of the complement of a complex hyperplane arrangement $\mathcal{A} \subset \mathbb{C}^k$, and*

2. *G is any topological group with the property that every maximal abelian subgroup is path-connected.*

Then the element $B\rho : B\Gamma \to BG$ regarded as an element of $[B\Gamma, BG]$ is the trivial element.

If G is a compact Lie group with a connected maximal torus, then the hypotheses of the theorem are satisfied. Examples are $SU(n)$, $U(n)$, but not $SO(n)$, or $O(n)$.

A related question is to ask about the subgroup of the real K-theory of $B\Gamma$ which these bundles generate. This subgroup denoted $KO^0_{rep}(B\Gamma)$ is addressed in (ACC).

Theorem 6. *Let Γ be the fundamental group of the complement of a $K(\Gamma, 1)$ arrangement and let ζ_1 and ζ_2 be arbitrary classes in $H^1(\Gamma; \mathbb{Z}/2\mathbb{Z})$ and $H^2(\Gamma; \mathbb{Z}/2\mathbb{Z})$. Then there is a finite dimensional orthogonal representation of Γ with first and second Stiefel-Whitney classes given by ζ_1 and ζ_2 respectively. Moreover for these groups Γ, the Stiefel-Whitney classes induce an isomorphism of groups*

$$KO^0_{rep}(B\Gamma) \cong H^1(\Gamma, \mathbb{Z}/2) \oplus H^2(\Gamma, \mathbb{Z}/2).$$

The previous theorem provides a lower bound for the set of path-components of $Hom(\Gamma, G)$. It is natural to ask for conditions which imply that the map

$$e : Hom(\Gamma, G) \to [B\Gamma, BG]$$

induces an isomorphism on the level of sets

$$E : \pi_0 Hom(\Gamma, G) \to [B\Gamma, BG].$$

There are two notable examples where this last map is an isomorphism:

1. Work in (Li; HL) implies that if Γ denotes the fundamental group of a closed, compact, orientable Riemann surface of genus at least 2, and G denotes a connected, compact, semi-simple Lie group, then E is an isomorphism of sets.

2. Deep work of Miller, and Lannes gives that E is an isomorphism of sets in case Γ is an elementary abelian p-group, and G is a compact Lie group.

In addition, $\pi_0 Hom(\Gamma, G)$ is a single point when Γ is a finitely generated free abelian group of rank $k > 0$, and G is $U(n)$. However in case n is much larger than k with $k > 1$, $[B\Gamma, BG]$ is of infinite cardinality. Thus E fails to be an isomorphism of sets in these cases.

The last part of this section addresses one related example of Γ with $G = SO(3)$. Let $Conf(S^2, k)$ denote the configuration space of ordered k-tuples of distinct points in the 2-sphere S^2. The symmetric group on k letters Σ_k acts naturally by permutation of coordinates. Write $Conf(S^2, k)/\Sigma_k$ for the quotient.

Then $SO(3)$, and $PGL(2, \mathbb{C})$ act naturally on the space of complex lines through the origin in complex 2-space $S^2 = \mathbb{CP}^1$. Hence $SO(3)$ as well as $PGL(2, \mathbb{C})$ act diagonally on $Conf(S^2, k)/\Sigma_k$. Form the Borel construction

$$ESO(3) \times_{SO(3)} Conf(S^2, k)/\Sigma_k$$

together with the natural projection map

$$p : ESO(3) \times_{SO(3)} Conf(S^2, k)/\Sigma_k \to ESO(3)/SO(3) = BSO(3).$$

It was proven in (C2) that $ESO(3) \times_{SO(3)} Conf(S^2, k)/\Sigma_k$ is $K(\Gamma_0^k, 1)$ if $k \geq 3$ where Γ_0^k is the mapping class group for genus zero surfaces with k punctures as described below in section 6. Thus the projection map above is a map

$$p : B\Gamma_0^k \to BSO(3)$$

for $k \geq 3$.

If $k \geq 2$, then this map $B\Gamma_0^{2k} \to BSO(3)$ satisfies the condition that the mod-2 cohomology of $BSO(3)$ injects in that for $B\Gamma_0^{2k}$, and so w_1, and w_2 both non-zero (C2). However, the map p is not homotopic to $B\rho$ for any representation $\rho : \Gamma_0^k \to SO(3)$ for $k \geq 3$ (BC).

It is natural to ask whether the composite

$$B\Gamma_0^6 \to BSO(3) \to BPGL(2, \mathbb{C})$$

induced by the inclusion of the maximal compact subgroup $SO(3) \to PGL(2, \mathbb{C})$ is homotopic to a map $B\rho'$ induced by a representation $\rho' : \Gamma_0^6 \to PGL(2, \mathbb{C})$.

The next question is whether these representations admit useful applications. Part of this question is addressed in the next 3 sections in which representations take values in $GL(n, \mathbb{Z})$.

§4 Cohomology classes, representations, and their connections to modular forms

Let

$$\rho : \Gamma \to GL(V)$$

be a fixed representation of a discrete group Γ for a vector space V over a field $\mathbb{F}$. There are associated representations given by

$$\Theta : \Gamma \to Aut(\Phi(V))$$

where

$$\Phi(V)$$

is a functor with source, and target given by the category of modules over $\mathbb{F}[\Gamma]$. Examples of such functors occur ubiquitously in nature such as the symmetric algebra $S[V]$, the tensor algebra $T[V]$, the exterior algebra $E[V]$, or their tensor products are naturally modules over $\mathbb{F}[\Gamma]$.

Recall that $S[V]$ is a direct sum of vector spaces

$$\oplus_{k \geq 0} Sym^k[V]$$

for which $Sym^k[V]$ denotes the homogeneous polynomials of classical degree k. Thus $Sym^k[V]$ admits the structure of $\mathbb{F}[\Gamma]$-module obtained from the multiplicative extension of the action on V.

Either the homology or the cohomology groups of Γ with coefficients in $Sym^k[V]$,

$$H_*(\Gamma, Sym^k[V]),$$

or

$$H^*(\Gamma, Sym^k[V])$$

have useful properties. One classical application arises with work of Eichler, and Shimura (E; S) concerning the ring of modular forms. A later application is to the cohomology of the classifying space of the group of orientation preserving diffeomorphisms of a torus $BDiff^+(S \times S^1)$ in work of Furusawa, Tezuka, and Yagita (FTY). A third application is a determination of the cohomology for certain mapping class groups Γ_1^k as given in (C).

A classical application due to Eichler, and Shimura (E; S) identifies the ring of modular forms based on the $SL(2,\mathbb{Z})$-action on the upper 1/2-plane. Recall that the group $SL(2,\mathbb{Z})$ acts on the upper 1/2-plane $\mathbb{H}^2$ by fractional linear transformations. For each positive integer k, a complex holomorphic function

$$f : \mathbb{H}^2 \to \mathbb{C}$$

is called an (integral) modular form of weight k with respect to $SL(2,\mathbb{Z})$ provided the following properties are satisfied:

1. The function satisfies

$$f(\{az+b\}/\{cz+d\}) = (cz+d)^k f(z)$$

for every matrix

$$A = \begin{pmatrix} a & b \\ c & d \end{pmatrix}$$

in $SL(2,\mathbb{Z})$.

2. The function f is holomorphic everywhere including the natural extension at ∞.
3. Furthermore, f is called a cusp form if $f(\infty) = 0$.
4. The vector space of modular forms of weight k is denoted $M_k(SL(2,\mathbb{Z}))$.

The Eichler-Shimura isomorphism (E; S; FTY; Se) in the special case below is an $\mathbb{R}$-linear isomorphism of real vector spaces given by

$$H^1(SL(2,\mathbb{Z}); Sym^{2k}(V_2^{\mathbb{R}})) \rightarrow M_{2k+2}(SL(2,\mathbb{Z}))$$

for $k \geq 0$, and

$$H^1(SL(2,\mathbb{Z}); Sym^{2k+1}(V_2^{\mathbb{R}})) = M_{2k+1}(SL(2,\mathbb{Z})) = \{0\}.$$

There are modifications for subgroups Γ of $SL(2,\mathbb{Z})$ giving modular forms $M_k(\Gamma)$ in case $\mathbb{H}^2/\Gamma$ has finite volume (E; S; FTY).

One connection between the ring of modular forms, representations of braid groups, and the homomorphisms $\Theta_n : F_n \rightarrow P_{n+1}$ is addressed in the next two sections.

§5 A sample computation

The purpose of this section, joint work with J. Wu, is to give a connection between the maps

$$\Theta_n : F_n \rightarrow P_{n+1}$$

of section 2, and bundles arising from representations of the braid groups as described in sections 3. The main computation here concerns the composite of $\Theta_2 : F_2 \rightarrow P_3$, with a classical map $\phi : B_3 \rightarrow SL(2,\mathbb{Z})$ arising from representations of the braid groups in mapping class groups. The composite map $\phi \circ \Theta_2$ is a special case of maps

$$\Lambda_n : F_n \rightarrow Sp(2g,\mathbb{Z})$$

for $n = 2g,\ 2g+1$ which uses the natural topology arising from the mapping class group as described next.

Consider the mapping class group Γ_g, the group of isotopy classes of orientation preserving diffeomorphisms for a closed orientable surface S_g of genus g. Next consider the symplectic representation obtained by evaluating a diffeomorphism on the first homology group of the surface

$$\Phi_g : \Gamma_g \rightarrow Sp(2g,\mathbb{Z}).$$

There are maps $\phi_g : B_{2g+2} \rightarrow \Gamma_g$ obtained from the centralizer of the hyperelliptic involution which are constructed by Dehn twists along a "necklace" of circles on S_g analogous to the constructions given on pages 183-188 in (B). Restrict to $g = 1$ to obtain

$$B_3 \rightarrow B_4 \rightarrow SL(2,\mathbb{Z})$$

to obtain the composite Λ_2.

Next, consider maps Λ_{2g+1} defined to be the composite

$$F_{2g+1} \xrightarrow{\Theta_{2g+1}} P_{2g+2} \xrightarrow{inclusion} B_{2g+2} \xrightarrow{\phi_g} Sp(2g,\mathbb{Z}).$$

There are analogous maps Λ_{2g} defined by the composite

$$F_{2g} \xrightarrow{\Theta_{2g}} P_{2g+1} \xrightarrow{inclusion} P_{2g+2} \xrightarrow{\phi_g} Sp(2g,\mathbb{Z}).$$

Similar maps

$$F_n \xrightarrow{\Lambda_n} Sp(2g,\mathbb{Z}) \xrightarrow{reduction} PSp(2g,\mathbb{Z})$$

obtained by composition with the natural reduction map to $Sp(2g,\mathbb{Z}) \to PSp(2g,\mathbb{Z})$ will also be denoted by Λ_n.

In the special case of $n = 2$, these maps restrict to $\Lambda_2 : F_2 \to PSL(2,\mathbb{Z})$. Thus the maps Λ_n may be regarded as extensions of Λ_2. To state one property of this map, recall that the principal congruence subgroup of level n in $PSL(2,\mathbb{Z})$, denoted $\bar{\Gamma}(2)$ here, is the kernel of the reduction map $PSL(2,\mathbb{Z}) \to PSL(2,\mathbb{Z}/n\mathbb{Z})$.

Theorem 7. *The homomorphism $\Lambda_2 : F_2 \to PSL(2,\mathbb{Z})$ maps F_2 isomorphically onto the principal congruence subgroup of level 2, $\bar{\Gamma}(2)$ in $PSL(2,\mathbb{Z})$.*

Proof. This proof, a direct calculation, is listed next. The braid group B_3 is generated by two elements σ_i for $i = 1,2$ while the symplectic representation above satisfies the following for the matrices $\Phi(\sigma_i)$:

$$\Phi(\sigma_1) = \begin{pmatrix} 1 & 1 \\ 0 & 1 \end{pmatrix}, \quad \Phi(\sigma_2) = \begin{pmatrix} 1 & 0 \\ -1 & 1 \end{pmatrix}.$$

Furthermore, the formulas below hold in B_3.

1. $A_{1,2} = \sigma_1^2$,
2. $A_{2,3} = \sigma_2^2$, and
3. $A_{1,3} = \sigma_2 \cdot \sigma_1^2 \cdot \sigma_2^{-1}$.

The image of these elements in $SL(2,\mathbb{Z})$ is given by the formulas

$$\Phi(A_{1,2}) = \begin{pmatrix} 1 & 2 \\ 0 & 1 \end{pmatrix},$$

$$\Phi(A_{2,3}) = \begin{pmatrix} 1 & 0 \\ -2 & 1 \end{pmatrix},$$

and

$$\Phi(A_{1,3}) = \begin{pmatrix} 3 & 2 \\ -2 & -1 \end{pmatrix}.$$

Recall the map $\Theta_2 : F_2 \to P_3$ of section 2 for which

1. $\Theta_2(x_1) = A_{1,3} \cdot A_{2,3}$, and
2. $\Theta_2(x_2) = A_{1,2} \cdot A_{1,3}$.

Thus the image of F_2 in $SL(2,\mathbb{Z})$ is specified by the following values for $\Lambda_2(x_i)$:

$$\Lambda_2(x_1) = \begin{pmatrix} -1 & 2 \\ 0 & -1 \end{pmatrix},$$

and

$$\Lambda_2(x_2) = \begin{pmatrix} -1 & 0 \\ -2 & -1 \end{pmatrix}.$$

Notice that the values of $\Lambda_2(x_i)$ for $i = 1, 2$ give a basis for the free group on 2 generators given by $\bar{\Gamma}(2)$ (S; N), and the theorem follows. □

One property of the maps $F_{2g+1} \to PSp(2g,\mathbb{Z})$ is listed next.

Theorem 8. *The image of the composite*

$$F_{2g+1} \xrightarrow{\Theta_{2g+1}} P_{2g+2} \xrightarrow{inclusion} B_{2g+2} \xrightarrow{\Phi} Sp(2g,\mathbb{Z}) \xrightarrow{reduction} PSp(2g,\mathbb{Z})$$

lies in the principal congruence subgroup given by the kernel of the mod-2 reduction map

$$PSp(2g,\mathbb{Z}) \longrightarrow PSp(2g,\mathbb{Z}/2\mathbb{Z}).$$

Proof. The natural map factoring through Γ_g,

$$B_{2g+2} \to Sp(2g,\mathbb{Z})$$

composed with the mod-2 reduction map $Sp(2g,\mathbb{Z}) \to Sp(2g,\mathbb{Z}/2\mathbb{Z})$ factors through the symmetric group on $2g+2$ letters Σ_{2g+2}. The theorem follows. □

The representations $\Lambda_n : F_n \to Sp(2g,\mathbb{Z})$ for $n = 2g,\ 2g+1$ provide bundles

$$E\Gamma \times_\Gamma (\mathbb{CP}^\infty)^{2g}$$

over $B\Gamma$ with $\Gamma = F_n$, and with fibre $(\mathbb{CP}^\infty)^{2g}$. The next section addresses the restriction of Λ_n to the case $n = 2$ which touches on these associated bundles, as well as modular forms.

§6 On bundles obtained from representations in section 2, their connections to modular forms as well as wild speculation

If $\Gamma = SL(2,\mathbb{Z})$, the Eichler-Shimura isomorphism gives that the real cohomology of the bundle $E\Gamma \times_\Gamma (\mathbb{CP}^\infty)^2$ is isomorphic, additively, and up to a degree shift, to the ring of modular forms based on the $SL(2,\mathbb{Z})$ action on the upper 1/2-plane (S; FTY; R). In addition, the real cohomology of the bundle $E\Gamma \times_{\Gamma_0(2)} (\mathbb{CP}^\infty)^2$ was considered by G. Nishida, and T. Ratliff (N; R) in work related to that of P. Landweber (L) for which $\Gamma_0(2)$ denotes the level 2 congruence subgroup of $SL(2,\mathbb{Z})$.

Let $\Gamma(N)$ denote the principal congruence subgroup of level N in $SL(2,\mathbb{Z})$. As pointed out in (N; R), there is an extension

$$1 \longrightarrow \Gamma(2) \longrightarrow \Gamma_0(2) \longrightarrow \mathbb{Z}/2\mathbb{Z} \longrightarrow 1.$$

Let $P\Gamma(2)$ denote $\Gamma(2)$ modulo its' center. The computation given by Theorem 7 of section 5 gives that the induced map $\Lambda_2 : F_2 \to P\Gamma(2)$ is an isomorphism. The next result is a corollary of the work in (N; R), as well as Theorem 7 here.

Theorem 9. *There is an isomorphism*

$$M_{2k+2}(\Gamma_0(2)) \to H^{4k+1}(EF_2 \times_{F_2} (\mathbb{CP}^\infty)^2; \mathbb{R})$$

for all $k > 0$ for which $EF_2 \times_{F_2} (\mathbb{CP}^\infty)^2$ is the bundle obtained from the representation

$$\Lambda_2 : F_2 \to SL(2,\mathbb{Z})$$

with the natural action of $SL(2,\mathbb{Z})$ on $(\mathbb{CP}^\infty)^2$.

This theorem in conjunction with Theorem 2 of section 2 gives a complicated way to detect different multiples of the classical Hopf map $\eta : S^3 \to S^2$ as follows. Restrict to the action of the infinite cyclic group Γ generated by the commutator $[x_1, x_2]$ as given in the proof of Theorem 7. A direct computation of the commutator gives that

$$\Lambda_2([x_1,x_2]) = \begin{pmatrix} 13 & 8 \\ 8 & 5 \end{pmatrix}.$$

In addition, the natural map $H^*(F_2; H^*(\mathbb{CP}^\infty)^2) \to H^*(\Gamma; H^*(\mathbb{CP}^\infty)^2)$ is non-trivial while $[x_1, x_2]$ is a cycle which represents η. Crossed homomorphisms corresponding to $H^1(\Gamma; H^*(\mathbb{CP}^\infty)^2)$ detect different multiples of η by a direct computation.

Furthermore, the bundle $EF_2 \times_{F_2} (\mathbb{CP}^\infty)^2$ is precisely that in (L; N; R) addressing the structure of Landweber's connective elliptic cohomology theory Ell^*. Ratliff's work in (R) gives a stable splitting of this bundle with one stable summand having real cohomology given exactly by the ring of modular forms corresponding to $\Gamma_0(2)$.

Given that the homotopy groups of the 2-sphere are natural subquotients of braid groups via $F[S^1]$ regarded as a simplicial subgroup of AP_* as described in section 2, it is reasonable to try to measure features of these braids. One possible feature is given by the cohomology of the spaces

$$H^1(\Gamma; H^*(\mathbb{CP}^\infty)^{2g})$$

for $\Gamma = F_n$ with $n = 2g,\ 2g+1$ for representations $\Lambda_n : F_n \to Sp(2g,\mathbb{Z})$. The example given by the commutator $[x_1, x_2]$ suggests considering crossed homomorphisms of F_n into some choice of representation which encodes natural combinatorial properties of $F[S^1]$. This last paragraph is the "wild speculation" as given in the title of this section.

References

Adem, A. Cohen, F. R., Cohen, D., *On representations and K-theory of the braid groups*, Math. Annalen, 326 (2003), no. 3, 515–542.

Adem, A. Cohen, F. R., Cohen, D., in preparation.

Benson, D. J., Cohen, F. R., *Mapping class groups of low genus and their cohomology*, Memoirs of the American Mathematical Society, **443**(1991).

Berrick, J., Cohen, F. R., Wong, Y. L., and Wu, J., *Configurations, braids, and homotopy groups*, submitted.

Birman, J., *Braids, Links and Mapping Class Groups*, Ann. of Math. Studies, **82**(1975), Princeton Univ. Press, Princeton, N.J..

Cohen, F. R., *On genus one mapping class groups, function spaces, and modular forms*, Cont. Math. **279**(2001), 103-128.

Cohen, F. R., *On the hyperelliptic mapping class groups, $SO(3)$, and $Spin^c(3)$* , American J. Math., **115**(1993), 389–434.

Cohen, F. R., and J. Wu, *On braid groups, free groups, and the loop space of the 2-sphere*, Progress in Mathematics, **215**(2003), 93-105, Birkhaüser, and *Braid groups, free groups, and the loop space of the 2-sphere*, preprint.

Eichler, M., *Eine Verallgemeinerung der Abelschen Integrale*, Math. Zeit., **67**(1957), 267-298.

Fadell, E., and Neuwirth, L., *Configuration spaces*, Math. Scand. **10**(1962), 119-126.

Falk, M., and Randell, R., *The lower central series of a fiber-type arrangement*, Invent. Math. **82**(1985), 77-88.

Furusawa, M., Tezuka, M., and Yagita, N., *On the cohomology of classifying spaces of torus bundles, and automorphic forms*, J. London Math. Soc., (2)**37**(1988), 528-543.

N. Ho, C. M. Liu, *Connected components of the space of surface group representations*, arXiv:math.SG/0303255 v1 20 Mar 2003.

Kohno, T., *Linear representations of braid groups and classical Yang-Baxter equations*, Cont. Math., **78**(1988), 339-363.

Kohno, T., *Vassiliev invariants and de Rham complex on the space of knots*, in: Symplectic Geometry and Quantization, Contemp. Math., **179**(1994), Amer. Math. Soc., Providence, RI, 123-138.

Landweber, P., *Elliptic cohomology and modular forms*, Elliptic Curves, in: Modular Forms in Algebraic Topology, Princeton, 1986, Springer-Verlag, New York, 1988, 55-68.

Li, J., *The space of surface group representations*, Manuscript. Math. **78**(1993), no. 3, 223-243.

Magnus, W., Karrass, A., and Solitar, D., *Combinatorial Group Theory*, Dover Publications, Inc., 1966.

Milnor, J., *On the construction F[K], In: A student's Guide to Algebraic Topology*, J.F. Adams, editor, Lecture Notes of the London Mathematical Society, **4**(1972), 119-136.

Moore, J. C., *Homotopie des complexes monoïdaux*, Séminaire H. Cartan, (1954/55).

Nishida, G., *Modular forms and the double transfer for BT^2*, Japan Journal of Mathematics, **17**(1991), 187-201.

Ratliff, T. C., *Congruence subgroups,elliptic cohomology, and the Eichler-Shimura map*, Journal of Pure and Applied Algebra, **109**(1996), 295-322.

Serre, J. P., *A Course in Arithmetic*, Springer-Verlag Graduate Texts in Mathematics, **7**(1970).

Shimura, G., *Introduction to the arithmetic theory of automorphic forms*, Publications of the Mathematical Society of Japan 11, Iwanami Shoten, Tokyo; University Press, Princeton, 1971.

Wu, J. *On combinatorial descriptions of the homotopy groups of certain spaces*, Math. Proc. Camb. Philos. Soc., **130**(2001), no.3, 489-513.

Address for Offprints:

F. R. Cohen
Department of Mathematics
University of Rochester
Rochester, NY 14627
email: cohf@math.rochester.edu

A note on symplectic circle actions and Massey products

Z. Stepień [1] and A. Tralle [2]

Abstract. In this note we show that the property of having only vanishing Massey products in the equivariant cohomology is inherited by the set of fixed points of symplectic circle actions on closed symplectic manifolds. This result can be considered in a more general context of characterizing homotopic properties of symplectic Lie group actions.

Keywords: Symplectic circle action, Massey product

Mathematics Subject Classification 2000: Primary: 53D35: Secondary: 53C15, 55P62

§1 Introduction

In this article we prove the following theorem.

Theorem 1. *Let (M,ω) be a closed symplectic manifold endowed with a symplectic circle action $S^1 \times M \to M$. Let F be any connected component of the fixed point set of this action. If there exists a non-vanishing Massey product in $H^*(F)$, then the same is valid for the equivariant cohomology $H^*_{S^1}(M)$.*

Of course the above theorem can be rephrased as follows. If (M,ω) is a closed symplectic manifold endowed with a circle action $S^1 \times M \to M$ compatible with ω, such that all the triple Massey products in the equivariant cohomology $H^*_{S^1}(M)$ vanish, then any connected component F of the fixed point set M^{S^1} also has only vanishing Massey products in $H^*(F)$.

Recently, there has been an increasing interest in the problems of compatibility of circle actions with symplectic forms, motivated by problems posed by Taubes (Ta) and Baldridge (Ba) (see (HW), (W)). For example, Baldridge asked the following question: *do there exist closed symplectic (M,ω) with free circle actions not compatible with any symplectic form on them?* It appears, that this question is difficult to answer, and in fact it is even difficult to construct an example of (M,ω) endowed with circle action (not necessarily free) which is not compatible with any symplectic structure. In (Au) examples of closed 4-manifolds with circle actions not compatible with any symplectic structures were given. However, it is not known, if these manifolds admit symplectic structures. Hence, it is natural to look for at least necessary conditions on actions ensuring their symplecticness. Theorem 1 gives a condition of this kind in terms of secondary cohomology operations. It is worth mentioning that necessary conditions for symplecticness were discussed in other works, for example, in (A), (G).

1, 2 Partially supported by the Polish Committee for the Scientific Research (KBN), grant 2P03A 036 24.

J.M. Bryden (ed.), Advances in Topological Quantum Field Theory, 289–295.

§2 Preliminaries and notation

Massey products. Here we recall briefly the notion of Massey products in the form suitable for our considerations. Note that in this paper all algebras and cohomologies are considered over the reals. To get a more detailed exposition of this topic in this form, one can consult (RT). Note however, that we use the term "non-vanishing" or "nonzero" Massey product instead of "essential" Massey product in (RT). Let there be given a commutative differential graded algebra (A, d). The cohomology algebra of (A, d) is denoted by $H^*(A)$. If $a \in A$ is a cocycle, we write $[a]$ for the corresponding cohomology class. For a homogeneous element $a \in A$ of degree p we use the notation $\bar{a} = (-1)^p a$. Assume that we are given a triple of the cohomology classes $[a], [b], [c]$ such that $[a][b] = [b][c] = 0$. Consider $x \in A$ and $y \in A$ such that $dx = \bar{a}b$ and $dy = \bar{b}c$. One can check that the element $\bar{a}y + \bar{x}c$ is a cocycle and therefore determines the cohomology class $[\bar{a}y + \bar{x}c] \in H^*(A)$. Note that this class depends on the choice of x and y. By definition the *set* of all cohomology classes $[\bar{a}y + \bar{x}c]$ is denoted by $\langle [a], [b], [c] \rangle$ and is called the *(triple) Massey product.*

Definition 1. *We say that the set $\langle [a], [b], [c] \rangle$ is defined if $[a][b] = [b][c] = 0$ and does not vanish if the set of all cohomology classes $[\bar{a}y + \bar{x}c]$ does not contain zero. In the opposite case we say that $\langle [a], [b], [c] \rangle$ vanishes.*

One can easily describe the indeterminacy in the definition of the triple Massey product and formulate it in the following way (see (RT), Prop. 1.5). Let $\langle [a], [b], [c] \rangle$ be a defined triple Massey product in $H^*(A)$. Denote by $([a], [c])$ the ideal in $H^*(A)$ generated by elements $[a]$ and $[c]$. The product $\langle [a], [b], [c] \rangle$ does not vanish if and only of there exists a cohomology class $x \in \langle [a], [b], [c] \rangle$ such that $x \notin ([a], [c])$.

We also need the following formulas (see (RT), Prop. 1.4):

$$\xi\langle a_1, a_2, a_3 \rangle \subset \langle \xi a_1, a_2, a_3 \rangle$$

$$\xi\langle a_1, a_2, a_3 \rangle \subset \langle a_1, \xi a_2, a_3 \rangle$$

$$\xi\langle a_1, a_2, a_3 \rangle \subset \langle a_1, a_2, \xi a_3 \rangle \tag{2.1}$$

which are valid for any $a_1, a_2, a_3 \in H^*(A)$ and for any ξ represented by a central element in A.

Let $f : (A, d) \to (A', d')$ be a morphism of differential graded algebras. Then

$$f^*\langle [a], [b], [c] \rangle \subset \langle f^*[a], f^*[b], f^*[c] \rangle \tag{2.2}$$

(see Prop. 1.3 in (RT)).

Cartan model. Consider now the case of a G-manifold, i.e. a smooth manifold endowed with a smooth action of a Lie group. We use the *equivariant* cohomology of the G-manifold, i.e. the cohomology of the total space of the Borel fibration:

$$M \to EG \times_G M \to BG$$

associated with the universal principal G-bundle $G \to EG \to BG$ over the classifying space BG of the Lie group G. Thus, $H_G^*(M) = H^*(EG \times_G M)$ (see (Au), (GS)).

In the proof of Theorem 1 we will calculate Massey products with respect to the *Cartan model.* Recall that for any G-manifold M one can associate the following differential graded algebra. Consider $\Omega^*_G(M) = (\Omega^*(M) \otimes S(\mathfrak{g}^*))^G$, where $\Omega^*(M)$ is the de Rham algebra of M, and $S(\mathfrak{g}^*)$ is a symmetric algebra over the dual to the Lie algebra $\mathfrak{g}$ of G. Then G acts on $\mathfrak{g}^*$ by the coadjoint representation and hence there is a natural G-action on the tensor product $\Omega^*(M) \otimes S(\mathfrak{g}^*)$. We consider the subalgebra $\Omega^*_G(M)$ of fixed points of the given action. The details of this construction can be found in (BV), (GS), (JK), (McDS). We use the fact that there is a natural differential $D : \Omega^*_G(M) \to \Omega^*_G(M)$ and that

$$H^*(\Omega^*_G(M), D) \cong H^*_G(M).$$

§3 Proof of Theorem 1

Consider the triple (M, ω, G). Recall that if the G-action is symplectic, the fixed point set M^G is a symplectic submanifold and it is a finite disjoint union of connected closed symplectic submanifolds:

$$M^G = \cup_{i=1}^p F_i.$$

Choose one of the connected components, say F_1, and denote it for brevity as F. We have a symplectic embedding

$$i_F : F \to M.$$

Consider F as a symplectic G-manifold with a trivial action of G. In particular, $H^*_G(F) = H^*(F) \otimes S(\mathfrak{g}^*)^G$. In case $G = S^1$ we have

$$H^*_{S^1}(F) = H^*(F) \otimes \mathbb{R}[h] \tag{3.1}$$

where $\mathbb{R}[h]$ denotes the free polynomial algebra with one generator h of degree 2. Passing to the Borel fibrations we can write the following commutative diagram

$$\begin{array}{ccc} F & \xrightarrow{i_F} & M \\ \downarrow & & \downarrow \\ EG \times_G F & \xrightarrow{(i_F)_G} & EG \times_G M \\ \downarrow & & \downarrow \\ BG & \xrightarrow{=} & BG \end{array}$$

On the cohomology level we will get the maps

$$(i_F)^*_G : H^*_G(M) \to H^*_G(F) \text{and } i^*_F : H^*(M) \to H^*(F).$$

Let ν denote the normal bundle of the symplectic embedding i_F, and let $E(\nu)$ be the total space of it. Since G acts on this bundle fiberwise, one can define the *equivariant* normal bundle with totoal space $E(\nu_G) = EG \times_G E(\nu)$. In this way we obtain the vector bundle

$$EG \times_G E(\nu) \to E_G \times_G F. \tag{3.2}$$

In particular, the Euler class of the vector bundle (3.2) is called the *equivariant Euler class of* ν. Throughout the paper, it is denoted by $\chi \in H^*_G(F)$.

The following facts can be found in (McDS), (pp. 192-193) or in (JK) and can be summarized in the following propositions.

Proposition 1. *Let $G = S^1$ act on a closed symplectic manifold (M, ω) and let F denote the chosen connected component of M^G. Then:*

(i) the normal bundle ν splits into the sum of complex line bundles

$$\nu = \oplus_{j=1}^m L_j, \; m = codim_M F$$

invariant with respect to S^1-action. The circle group acts on each L_j with weight $k_j \neq 0$,

(ii) the equivariant Euler class has the form

$$\chi = \prod_{j=1}^m (c_1(L_j) + k_j h), \tag{3.3}$$

where $c_1(L_j)$ denotes the first Chern class of L_j.

Proposition 2. *(Au),(K) Assume that G is a torus and acts symplectically on a closed symplectic manifold (M, ω). Then*

(i) there exists a linear map (the Gysin homomorphism) $((i_F)_G)_* : H^*_G(F) \to H^*_G(M)$ with the property:

$$(i_F)^*_G((i_F)_G)_*(x) = \chi x$$

for any $x \in H^*_G(F)$,

(ii) χ is not a zero divisor in $H^*_G(F)$.

The Gysin map was introduced in the cited papers (Au),(JK) in topological terms. We need an alternative way of describing it in the language of the Cartan complex. This was done in (GS), Chapter 10. Let τ denote the equivariant Thom form of the equivariant normal bundle (GS). Recall that $\tau \in \Omega^*_G(\nu)_c$, where $\Omega^*_G(\nu)_c$ denotes the Cartan complex of equivariant differential forms on the normal bundle ν *with compact supports.* If one identifies ν with the tubular neighbourhood of F in M, one can extend τ onto M by zero. Consider the natural projection $\pi : \nu \to F$. For any $\theta \in \Omega^*_G(F)$, define $i_*\theta$ as

$$i_*\theta = \pi^*\theta \wedge \tau.$$

Extending $i_*\theta$ onto the whole M by zero, and passing to equivariant cohomology, we get the *Gysin map*

$$((i_F)_G)_* : H^*_G(F) \to H^*_G(M)$$

written as

$$((i_F)_G)_*\theta = j^*(\pi^*\theta \wedge \tau), \tag{3.4}$$

where j^* denotes the extension onto M by zero (identified with $j^* : H^*_G(M, M \setminus N) \cong H^*_G(\nu)_c \to H^*_G(M)$ induced by the corresponding map of pairs (cf. (AB)). Here one considers equivariant forms and cohomologies with compact supports: $H^*_G(\nu)_c = H^*(\Omega^*_G(\nu)_c)$.

The proof of Theorem 1 will now follow from the two lemmas below. Note that in both lemmas we keep the same notation and we assume that G is a torus acting on (M, ω) symplectically.

Lemma 1. *Let there be given a defined Massey product $\langle u, v, w\rangle \subset H^*_G(F)$. Then there is a defined Massey product in $H^*_G(F)$:*

$$\langle \chi u, \chi v, \chi w\rangle. \tag{3.5}$$

If the Massey product (3.5) does not contain zero, then the Massey product

$$\langle ((i_F)_G)_*u, ((i_F)_G)_*v, ((i_F)_G)_*w\rangle \subset H^*_G(M) \tag{3.6}$$

is defined and does not contain zero.

Proof. To avoid clumsy notation let us temporarily denote $((i_F)_G)_*$ as $(i_F)_*$ and $(i_F)^*_G$ as $(i_F)^*$. Using the equality $i^*_F\theta = j^*(\pi^*\theta \wedge \tau)$ (see (3.4)) we write

$$(i_F)_*u(i_F)_*v = j^*(\pi^*u \wedge \tau)j^*(\pi^*v \wedge \tau) = j^*(\pi^*(uv) \wedge \tau \wedge \tau) = 0,$$

since $uv = 0$. Hence, (3.6) is defined. The following formulae show that (3.6) does not contain zero:

$$i^*_F\langle (i_F)_*u, (i_F)_*v, (i_F)_*w\rangle \subset \langle i^*_F(i_F)_*u, i^*_F(i_F)_*v, i^*_F(i_F)_*w\rangle = \langle \chi u, \chi v, \chi w\rangle.$$

Here we used (2.2) and Proposition 2.

Lemma 2. *Assume $G = S^1$ and that $\langle u, v, w\rangle$ is a non-vanishing triple Massey product in $H^*(F) \subset H^*_{S^1}(F) = H^*(F) \otimes \mathbb{R}[h]$. Then*

$$\langle \chi u, \chi v, \chi w\rangle \neq 0.$$

Proof. According to Section 2, we need only to prove that there exists an element z in the set $\langle \chi u, \chi v, \chi w\rangle$ such that $z \notin (\chi u, \chi w)$. Use the equivariant cohomology $H^*_{S^1}(F)$ calculated with respect to the Cartan model. Take a non-trivial Massey product $\langle u, v, w\rangle$ considered as a non-trivial Massey product in the equivariant cohomology (one can easily check by straightforward calculation that $\langle u, v, w\rangle$ cannot become zero in the tensor product $H^*(F) \otimes \mathbb{R}[h]$ by writing the corresponding cocycles in the Cartan model). From Lemma 1, $\langle \chi u, \chi v, \chi w\rangle$ is defined. Since $\langle u, v, w\rangle \neq 0$, there exists $x \in \langle u, v, w\rangle$ such that $x \notin (u, w)$. Note that $\chi^3 x \in \langle \chi u, \chi v, \chi w\rangle$ (by (2.1)). Assume that

$$\langle \chi u, \chi v, \chi w\rangle = 0.$$

It means that any $z \in \langle \chi u, \chi v, \chi w\rangle$ belongs to the ideal $(\chi u, \chi w) \subset H^*_{S^1}(F)$. In particular, $\chi^3 x \in (\chi u, \chi w)$. Hence

$$\chi^3 x = \chi u a + \chi w b,\ a, b \in H^*_{S^1}(F).$$

Therefore

$$\chi(\chi^2 x - ua - wb) = 0.$$

Recalling that χ is not a zero divisor (Proposition 2 (ii)) one can write

$$\chi^2 x = ua + wb.$$

Taking into consideration that $u, w \in H^*(F) \subset H^*_{S^1}(F)$ and the expression for the Euler class (3.3) one obtains

$$\prod_{j=1}^{m}(c_1(L_j) + k_j h)^2 x = u(a_0 + a_1 h + ... + a_{2m}h^{2m}) + w(b_0 + b_1 h + ... + b_{2m}h^{2m}).$$

Note that h is a free generator, and $k_1 \cdots k_m \neq 0$. Using this and comparing the coefficients of h^{2m} on both sides of the latter equation yields $x \in (u, w)$, a contradiction. Finally, $z = \chi^3 x$ is the required element.

Now we can complete the proof of Theorem 1. If $\langle u, v, w\rangle$ is a non-trivial Massey triple product in $H^*(F)$, Lemma 2 implies that $\langle \chi u, \chi v, \chi w\rangle$ is a non-vanishing triple Massey product in $H^*_{S^1}(F)$. From Lemma 1 we get a nontrivial triple Massey product in $H^*_G(M)$ expressed by formula (3.6).

References

C. Allday *Examples of circle actions on symplectic spaces*, in: Homotopy and Geometry (J. Oprea and A. Tralle, eds.), Banach Center Publ. **45** (1998), 87-90.

M. Atiyach, R. Bott *The moment map and equivariant cohomology*, Topology, **23** (1984), 1-28

M. Audin, *The topology of torus actions on symplectic manifolds*, Birkhäuser, Basel, 1991

S. Baldridge *Seiberg-Witten invariants and 4-manifolds with free circle actions*, Comm. Cont. Math. **3** (2001), 341-353.

M. Berline and M. Vergne *Zeros d'un champes de vecteurs et classes characteristique equivariantes*, Duke Math. J. **50** (1983), 539-549.

V.L. Ginzburg *Some remarks on symplectic actions of compact groups* Math. Z. **210**, (1992), 625-640.

V. Guillemin and S. Sternberg *Supersymmetry and equivariant de Rham theory*, Springer, Berlin, 1999.

B. Hajduk, R. Walczak *Symplectic forms invariant under free circle actions* preprint, ArXiv: math.SG/0312465

L. Jeffrey and F. Kirwan *Applications of equivariant cohomology to symplectic geometry and moduli spaces*, in: Symplectic Geometry, Y. Eliashberg and L. Traynor, eds. Providence, RI, 1999, 1-18

F. Kirwan *Cohomology of quotients in symplectic and algebraic geometry*, Princeton, 1984

G. Lupton and J. Oprea, *Cohomologically symplectic spaces. Toral actions and the Gottlieb group*, Trans. Amer. Math. Soc. **347** (1995), 261-288.

D. McDuff *The moment map for circle actions on symplectic manifolds* J. Geom. Phys. **5** (1988), 149-160

D. McDuff and D. Salamon *Introduction to symplectic topology* Oxford, 1998

Y. Rudyak and A. Tralle *On Thom spaces, Massey products and non-formal symplectic manifolds*, Internat. Math. Res. Notices **10** (2000), 495-513.

C. Taubes *The geometry of the Seiberg-Witten invariants* Proc. ICM Berlin, Doc. Math. J. Extra volume **2** (1998), 493-504.

R. Walczak *Existence of symplectic structures on torus bundles over surfaces* Preprint, ArXiv: math.SG/0310261

Address for Offprints:

Aleksy Tralle
Department of Mathematics and Information Technology
University of Warmia and Mazury
Olsztyn, Poland
email: tralle@matman.uwm.edu.pl

Zofia Stepień
Institute of Mathematics
Szczecin Technical University, Poland
email: stepien@arcadia.tuniv.szczecin.pl

Realization of Primitive Branched Coverings over Closed Surfaces *

S. A. Bogatyi †, D. L. Gonçalves ‡, E. A. Kudryavtseva § and H. Zieschang ¶

Abstract. Let V be a closed surface, $H \subseteq \pi_1(V)$ a subgroup of finite index ℓ and $\mathcal{D} = [A_1, \ldots, A_m]$ a collection of partitions of a given number $d \geq 2$ with positive defect $v(\mathcal{D})$. *When does there exist a connected branched covering $f : W \to V$ of order d with branch data $\mathcal{D}$ and $f_\#(\pi_1(W)) = H$?*

We show that, for a surface V different from the sphere and the projective plane and $\ell = 1$, the corresponding branched covering exists (the data $\mathcal{D}$ is realizable) if and only if the data $\mathcal{D}$ fulfills the Hurwitz congruence $v(\mathcal{D}) \equiv 0 \mod 2$. In the case $\ell > 1$, the corresponding branched covering exists if and only if $v(\mathcal{D}) \equiv 0 \mod 2$, the number d/ℓ is an integer, and each partition $A_i \in \mathcal{D}$ splits into the union of ℓ partitions of the number d/ℓ.

The realization problem for the projective plane and $\ell = 1$ has been solved in (Edmonds-Kulkarni-Stong, 1984). The case of the sphere is treated in (Berstein-Edmonds, 1979; Berstein-Edmonds, 1984; Husemoller, 1962; Edmonds-Kulkarni-Stong, 1984).

AMS: Primary: 55M20, Secondary: 57M12, 20F99

Introduction

In his study of surface homeomorphisms, J. Nielsen introduced in 1927 a new homotopical invariant (nowadays called Nielsen number) for a mapping which gives a lower bound for the number of fixed points of mappings homotopic to the given one (Nielsen, 1927). Later the Nielsen coincidence number $NC(f_1, f_2)$ was introduced for pairs of mappings $f_1, f_2 \colon W^n \to V^n$ between n-dimensional orientable closed manifolds and it was proved that the Nielsen number is a lower bound for the

* This work was done at the Ruhr-Universität Bochum in June/July 2001 (supported by the DFG-project "Niedrigdimensionale Topologie und geometrische Methoden in der Gruppentheorie") and at the Moscow State Lomonosov-University in May/June 2002.

† The author was supported by the RFBI grant 00-01-00289.

‡ The visits of the Ruhr-Universität Bochum and the Moscow State Lomonosov-University were supported by the "Projeto Temático Topologia Algébrica e Geométrica – FAPESP (Fundaçâo de amparo a pesquisa do Estado de São Paulo)".

§ The author was partially supported by the "Support of Leading Scientifical Schools" 00-15-96059, by the RFBI grant 01-01-00583, and the INRIA project 01-07.

¶ The work of this author in Moscow was supported by the Stiftungsinitiative Johann Gottfried Herder.

J.M. Bryden (ed.), Advances in Topological Quantum Field Theory, 297–316.

number of geometrically different coincidence points for all $\tilde{f}_1$ and $\tilde{f}_2$ homotopic to f_1 and f_2, respectively. H. Schirmer (Schirmer, 1955) proved that for $n \geq 3$ this bound is strong in the following sense: given two arbitrary continuous mappings $f_1, f_2\colon W^n \to V^n$, $n \geq 3$ then f_1 can be deformed into a continuous mapping $g\colon W^n \to V^n$ such that

$$|\operatorname{coin}(g, f_2)| = NC(f_1, f_2).$$

However the original problem, namely the coincidence or fixed point problem for surfaces, is not solved and there is not known a calculation of the minimal number of coincidence or fixed points of mappings within given homotopy classes.

In the papers (Gonçalves-Zieschang, 2001; Gonçalves-Kudryavtseva-Zieschang, 2002) (for orientable closed and for all closed surfaces) there are given conditions under which the Nielsen coincidence number of a given mapping with a constant map turns out to be a strong lower bound for the cardinality of the preimage of a given point for maps within the considered homotopy class. In (Bogatyi-Gonçalves-Zieschang, 2001; Bogatyi-Gonçalves-Kudryavtseva-Zieschang, 2004) an explicit simple formula is given for the minimal number of points mapped to the given one for mappings of a given homotopy class.

All four mentioned above articles of the authors are based on the construction of suitable branched coverings and the deep theorem of Gabai-Kazez on the classification of mappings between surfaces (Gabai-Kasez, 1987). If we want to find an "economical" representative in a homotopy class $[f]$, it suffices, according to the Gabai-Kazez theorem, to construct an "economical" mapping with a given image of the fundamental group. In the above articles of the authors, branched coverings show up to be "economical" mappings. In (Husemoller, 1962; Ezell, 1978; Berstein-Edmonds, 1979; Berstein-Edmonds, 1984; Edmonds-Kulkarni-Stong, 1984), the problem of the existence of a branched covering over a given closed surface with an arbitrary given branch data is discussed, and the complete solution is obtained for all surfaces except the sphere.

In (Bogatyi-Gonçalves-Zieschang, 2001) there has been formulated the *refined realization problem* to construct a branched covering with given branch data and a given image $H \subset \pi_1(V^2, *)$ of the fundamental group of W^2 (if the image is the full fundamental group, the mapping is called *primitive*). For the calculation of the minimal number of preimages in a given homotopy class, it was sufficient to solve the problem in very special cases and this was done in the above papers of the authors. In particular, there have been obtained realizations over closed orientable surfaces S_g of genus $g \geq 2$ and over the non-orientable surfaces N_g of genus $g \geq 4$. Any mapping into the sphere is primitive, therefore the problems of primitive and usual realizations over the sphere coincide. In (Edmonds-Kulkarni-Stong, 1984), it was decided which sets of partitions of a number d can be realized as branch data of a connected branched covering over the projective plane, and it was shown that all such data can be realized by primitive coverings. The realization problem over the sphere and the projective plane (with trivial image of the fundamental group) remains open and is not considered in this article. The paper (Berstein-Edmonds, 1979, Proposition 5.2) contains some concrete examples of primitive branched coverings over the torus.

However, the refined question remained open to describe the branch data which can be realized by a branched covering with a given image of the fundamental group (or, in particular, with an induced epimorphism between the fundamental groups). More precisely, the realization problem remained open only over the torus $T = S_1$ in the orientable case and over the Klein bottle $K = N_2$ and N_3 in the nonorientable case; remember that the sphere and the projective plane are excluded from our consideration. *In this paper we give a full solution of the refined realization problem over closed surfaces different from the sphere and the projective plane,* see Theorem 4.2. In particular (Proposition 4.1), we prove that the branch data of any branched covering which is not a usual covering can also be realized by a primitive connected branched covering.

The paper is divided into five sections. In Section 1 we recall the notions of branch data and their defect; this leads us to the concept of *virtual branch data.* Furthermore we remind the known results on the refined realization problem. In Section 2 we define special branched coverings over a surface and introduce a gluing operation among them. In Section 3 we construct special primitive branched coverings realizing all possible branch data over the torus and the Klein bottle with one branch point. This is done by explicit constructions for the cases when the preimage of the branch point consists of one or two points. Then, using the gluing operation defined in Section 2, we construct special primitive branched coverings with one branch point over the torus and the Klein bottle. In Section 4 we consider the general case with arbitrary many branch points and obtain the main result Theorem 4.2. The arguments in the sections 3-4 are of geometrical nature and several times "one has to see" some properties. For a full proof one has to determine the stars at the different vertices of the covering surface as was done in (Gonçalves-Zieschang, 2001); this is performed in Section 5 for "generic" cases. In a forthcoming paper (Bogatyi-Gonçalves-Kudryavtseva-Zieschang, 2003) we use the pure algebraic approach of Hurwitz (Hurwitz, 1891) to study and construct branched coverings based on calculations in symmetric groups.

1. Preliminaries

1.1 Definition Given a branched covering $f\colon W \to V$ between surfaces, we have the *branch data* as defined for example in (Bogatyi-Gonçalves-Zieschang, 2001). The branch data $\mathcal{D} = [A_1, \ldots, A_m]$ contain the branching orders $A_i = [d_{i1}, \ldots, d_{ir_i}]$ of the r_i different points over the i-th branch point, $1 \leq i \leq m$. The number

$$v(\mathcal{D}) = \sum_{i=1}^{m} \sum_{j=1}^{r_i} (d_{ij} - 1) = \sum_{i=1}^{m} \left(\sum_{j=1}^{r_i} d_{ij} - r_i \right)$$

is called the *defect* of the branched covering or of the branch data. The branch data satisfy certain necessary conditions, namely, if f is a d-fold branched covering which is not a usual covering then

(i) $d_{i1} + \ldots + d_{ir_i} = d$, $d_{ij} \geq 1$ for $i = 1, \ldots, m$, that is, A_i is a *partition* of d,

(ii) $v(\mathcal{D}) \equiv 0 \mod 2$ and

(iii) $v(\mathcal{D}) \neq 0$.

For a system $\mathcal{D}$ of partitions of a number d with the above properties we use the notion of *virtual branch data.*

As mentioned in the introduction, for any surface different from the sphere and the projective plane, arbitrary virtual branch data can be realized by connected branched coverings (Husemoller, 1962; Ezell, 1978; Berstein-Edmonds, 1979; Berstein-Edmonds, 1984; Edmonds-Kulkarni-Stong, 1984). We are looking for a *primitive* connected branched covering over such a surface. More general, we are looking for a branched covering such that the image of the fundamental group of the covering space is a given subgroup of finite index in the fundamental group of the base. By (Bogatyi-Gonçalves-Zieschang, 2001, Proposition 5.8), for $V = S_g$, the orientable closed surface of genus g, every virtual branch data can be realized by a primitive connected branched covering if the surface S_g has genus $g \geq 2$ (or Euler characteristic ≤ -2). In a similar fashion, in (Bogatyi-Gonçalves-Kudryavtseva-Zieschang, 2004, 4.1), it is proved that for $V = N_g$, the nonorientable closed surface of genus g, every virtual branch data can be realized as required if the surface N_g has genus $g \geq 4$ (or Euler characteristic ≤ -2). These results were obtained using the Hurwitz realization theorem (Bogatyi-Gonçalves-Zieschang, 2001, 5.4). So, the problem of realizing virtual branch data by a primitive branched covering was not solved only for the torus, the Klein bottle and the nonorientable closed surface N_3 of genus 3. In this paper we consider these cases and construct, in particular, primitive branched coverings corresponding to given virtual branch data. To do so, we define a gluing operation on the family of so-called special branched coverings.

2. Gluing Branched Coverings

In the following, the surfaces V and W are assumed to be closed and connected.

For a twosided simple closed curve $\gamma \subset W$, denote by $W(\gamma)$ the surface with boundary obtained by cutting W along γ. Then $\partial W(\gamma) = \gamma(1) \cup \gamma(2)$ where $\gamma(i)$, $i \in \{1,2\}$ are copies of γ and the natural projection $W(\gamma) \to W$ homeomorphically maps each $\gamma(i)$ to γ.

2.1 Definition A continuous mapping is called *primitive* if the induced homomorphism between the fundamental groups is surjective. A quadrupel (W, γ, f, V) is called a *special branched covering* if $f\colon W \to V$ is a branched covering, W connected, and $\gamma \subset W$ is a simple twosided closed curve which is homeomorphically mapped onto a non-separating simple loop $f(\gamma) \subset V \setminus B_f$; here B_f is the set of branch points of f. If the composition $\bar{f}\colon W(\gamma) \to W \to V$ is primitive the quadrupel (W, γ, f, V) is called a *special primitive branched covering.*

2.2 Lemma *Let V be an orientable closed surface different from the sphere or let V be the Klein bottle. Given special branched coverings (W_1, γ_1, f_1, V), (W_2, γ_2, f_2, V), there is a branched covering $f_2'\colon W_2 \to V$ admitting the same branch set and branch data as f_2 such that $f_1(\gamma_1) = f_2'(\gamma_2)$ and (W_2, γ_2, f_2', V) is a special branched covering.*

Proof. Since the two curves $f_1(\gamma_1)$, $f_2(\gamma_2)$ are twosided simple closed curves and do not separate, there is a surface homeomorphism $\psi\colon V \to V$ which maps $f_2(\gamma_2)$ to $f_1(\gamma_1)$ (for this well known fact see, e.g., (Zieschang-Vogt-Coldewey, 1980, Theorem 3.5.4)) and leaves the branch set fixed. Then the composite $f_2' = \psi \circ f_2$ has the required properties.

For nonorientable surfaces, there are several but few types of twosided non-seperating simple closed curves; in the case of the Klein bottle the curve is even uniquely determined up to isotopy. □

Now we define an operation among pairs of special branched coverings one of which is special primitive.

2.3 Definition of the Gluing Operation Let (W_1, γ_1, f_1, V) be a special primitive branched covering and (W_2, γ_2, f_2, V) a special branched covering where $f_1(\gamma_1) = f_2(\gamma_2)$ is a non-separating simple twosided loop γ. Cutting V along γ we get a surface with two boundary components γ^+, γ^-. Doing the same with W_i, γ_i we obtain a surface $W_i(\gamma_i)$ with boundary components γ_i^+, γ_i^-, the "preimages" of γ^+, γ^-. Now we identify γ_i^+ with γ_j^-, $i \neq j$ and denote the obtained surface by W_3. Define $f_1 \# f_2 \colon W_3 \to V$ as the map which restriction to $W_i(\gamma_i)$ is the projection $W_i(\gamma_i) \to W_i$ followed by f_i, $i = 1, 2$. Hence, $f_1 \# f_2$ is a branched covering and $(W_3, \gamma, f_1 \# f_2, V)$ is a special primitive branched covering where the curve γ is one of the curves along which the surfaces are glued together.

2.4 Lemma *Let V be a closed orientable surface different from the sphere or let V be the Klein bottle. Let (W_1, γ_1, f_1, V) be a special primitive branched covering and (W_2, γ_2, f_2, V) a special branched covering with one branch point where $f_1 \colon W_1 \to V$, $f_2 \colon W_2 \to V$ have branch data $\mathcal{D}_1 = [B]$, $B = [d_1, \ldots, d_r]$ and $\mathcal{D}_2 = [C]$, $C = [d'_1, \ldots, d'_s]$, respectively.*

Then there is a special primitive branched covering (W, γ, f, V) with one branch point and branch data

$$\mathcal{D} = [A], \quad A = B \sqcup C = [d_1, \ldots, d_r, d'_1, \ldots, d'_s].$$

Proof. In according to the homogeneity of the surface and Lemma 2.2, we may assume that the coverings f_1 and f_2 have the same branch point and $f_1(\gamma_1) = f_2(\gamma_2)$. As $f_1(\gamma_1) = f_2(\gamma_2)$, we can apply the gluing operation 2.3. Then the map $f_1 \# f_2 \colon W_3 \to T$ provides a special branched cover which has branch data $A = [d_1, \ldots, d_r, d'_1, \ldots, d'_s]$. □

By a similar gluing operation, the following more general lemma can be obtained.

2.5 Lemma *Let V be a closed orientable surface different from the sphere or let V be the Klein bottle. Let (W_1, γ_1, f_1, V) be a special primitive branched covering of order d_1 and (W_2, γ_2, f_2, V) a special branched covering of order d_2 with m branch points where $f_1 \colon W_1 \to V$, $f_2 \colon W_2 \to V$ have branch data*

$$\mathcal{D}_1 = [B_1, \ldots, B_m], \ B_i = [d_{i1}, \ldots, d_{ir_i}] \ (1 \leq i \leq m) \quad \text{and}$$
$$\mathcal{D}_2 = [C_1, \ldots, C_m], \ C_i = [d'_{i1}, \ldots, d'_{is_i}] \ (1 \leq i \leq m),$$

respectively. Then there is a special primitive branched covering (W, γ, f, V) of order $d_1 + d_2$ with branch data

$$\mathcal{D} = [A_1, \ldots, A_m], \quad A_i = B_i \sqcup C_i = [d_{i1}, \ldots, d_{ir_i}, d'_{i1}, \ldots, d'_{is_i}]$$

for $1 \leq i \leq m$. □

This result is also true for the case that some of the B_j or C_j are "trivial", namely consisting only of 1's; in other words, no branching happens at this place. Therefore the branched coverings may also have different numbers and positions of the branch points in the target. In particular, the second branched covering may be unbranched.

3. Realizing Special Primitive Branched Coverings Over the Torus and the Klein Bottle

In the sections 3-4 we show that, for any closed surface V, $\chi(V) \leq 0$ and any virtual branch data $\mathcal{D} = [A_1, \ldots, A_m]$, there is a primitive branched covering $f\colon W \to V$ with m branch points in V realizing the given virtual branch data. Actually, there is a simple closed curve γ such that (W, γ, f, V) is a special primitive branched covering. In this section we will assume that the target is either the torus T or the Klein bottle K. We will see in the next section that our general realization question (not only for the primitive case) for $\chi(V) < 0$ directly follows from the torus case.

3.1 Proposition *For a surface* $V \in \{T, K\}$ *and any odd number* $d = 2k+1$, *there is a special primitive branched covering* (W, γ, f, V) *with branch data* $\mathcal{D} = [A]$, $A = [d]$.

Proof. For the torus the result can be obtained from (Gonçalves-Zieschang, 2001), but we give a full proof here. The corresponding branched covering f is given in Fig. 5.1, a corrected copy of (Gonçalves-Zieschang, 2001, Figure 3). Take as γ the curve a_{k+1}; it is closed and $f|_\gamma\colon \gamma \to a$ is a homeomorphism. The loops a_{k+1} and b_{2k+1} are disjoint and mapped onto a and b; hence, f is special primitive.

The case of the Klein bottle can be solved in a similar form using Figure 5.3 with the disjoint curves $\gamma = b_2$ and $b_{2k+1}a_{k+1}b_1$. □

3.2 Proposition *For a surface* $V \in \{T, K\}$ *and any integers* k, ℓ *such that* $0 \leq \ell \leq k-2$, *there is a special primitive branched covering* (W, γ, f, V) *with branch data* $\mathcal{D} = [A]$, $A = [k+\ell, k-\ell]$.

Proof. First we consider the torus. From Figure 5.2, we obtain a branched covering f with branch data $\mathcal{D} = [k+\ell, k-\ell]$. We can see that the curve a_k (or b_{k+1}) is a simple loop in W which projects homeomorphically to a simple closed curve a (or b, respectively). The curves b_{k+1} and a_k are disjoint. So (W, a_k, f, T) is a special primitive branched covering.

For the Klein bottle K, let $f\colon W \to K$ be the branched covering described in Fig. 5.4. The simple closed curve b_{k+1} is mapped homeomorphically to the non-separating twosided loop b. The curve $b_{2k}a_k b_1$ is disjoint from b_{k+1} and its image belongs to the class $[bab] = [a] \in \pi_1(K)$. Hence, (W, b_{k+1}, f, K) is special primitive. □

Now we show the main result of this section.

3.3 Theorem *For a surface $V \in \{T, K\}$ and arbitrary virtual branch data $\mathcal{D} = [A]$, $A = [d_1, \ldots, d_r]$, there is a special primitive branched covering (W, γ, f, V) with the branch data $\mathcal{D}$.*

Proof by induction on r. We assume that $d_1 \geq d_2 \geq \ldots \geq d_r$. If $r = 1$, the result coincides with Proposition 3.1.

If $r \geq 2$, we will distinguish two cases.

Case 1: there is an odd number d_j (for instance, $d_2 = 1$ for $r = 2$). Consider two numbers

$$d' = d_1 + \ldots + d_{j-1} + d_{j+1} + \ldots + d_r \quad \text{and} \quad d'' = d_j$$

and their partitions $[d_1, \ldots, d_{j-1}, d_{j+1}, \ldots, d_r]$, $[d_j]$. Here we put $j = r$ if $d_r = 1$, in order to obtain a nontrivial partition of the number d'. Using the induction hypothesis, take a special primitive branched covering realizing the first partition and a special branched covering realizing the second partition. The induction hypothesis is fulfilled, since the defect $v([d_j]) = d_j - 1$ of the partition $[d_j]$ is even and

$$v(A) = v([d_1, \ldots, d_{j-1}, d_{j+1}, \ldots, d_r]) + v([d_j]),$$

so both considered partitions have an even defect. The first one is nontrivial, and we can apply Lemma 2.4 to the special branched coverings under consideration.

Case 2: all d_j are even. Then their sum is also even and, thus, r too.

If $r = 2$, the result follows from Proposition 3.2 for $k = (d_1 + d_2)/2$ and $\ell = (d_1 - d_2)/2$. So suppose that $r \geq 3$. Then $r \geq 4$ and $d_r \geq 2$. Now we apply Lemma 2.4 to the special primitive branched coverings realizing the partitions $[d_1, \ldots, d_{r-2}]$ and $[d_{r-1}, d_r]$ of the numbers $d - d_{r-1} - d_r = d_1 + \ldots + d_{r-2}$ and $d_{r-1} + d_r$ which exist by induction hypothesis. □

For the relation of branched coverings with representations on the symmetric group Σ_d corresponding to the Hurwitz realization theorem see, for example, (Bogatyi-Gonçalves-Zieschang, 2001). Now we get the following algebraic corollary from Theorem 3.3:

3.4 Corollary *For $V \in \{T, K\}$ and a nontrivial even permutation $\hat{c} \in \Sigma_d$ there are permutations $\hat{a}, \hat{b} \in \Sigma_d$ with the following properties:*

- *$\hat{c} = R(\hat{a}, \hat{b})$ where R is the defining relation in $\pi_1(V) = \langle a, b \mid R(a, b)\rangle$,*
- *the subgroup of Σ_d generated by $\hat{a}, \hat{b}$ acts transitively on $\{1, 2, \ldots, d\}$,*
- *the corresponding cover over V admitting 1 branch point is primitive.*

Proof. Consider the decomposition of $\hat{c}$ as a product of cycles and let $d_1, d_2, \ldots, d_r$ be the lengths of the cycles. Put $A = [d_1, \ldots, d_r]$. From the fact that the permutation is even and nontrivial, it follows that $v(A) = d_1 + \ldots + d_r - r \equiv 0 \mod 2$ and $v(A) \neq 0$. By Theorem 3.3, there is a connected and primitive branched covering $f'\colon W \to V$ realizing the virtual branch data $A = [d_1, \ldots, d_r]$. By the Hurwitz realization theorem, this branched covering corresponds to a representation $\varphi\colon \pi_1(V \setminus B_{f'}) = \langle a, b \mid \rangle \to \Sigma_d$ such that the subgroup $H' = \langle \hat{a}', \hat{b}' \rangle$ of Σ_d generated by $\hat{a}' = \varphi(a)$, $\hat{b}' = \varphi(b)$ acts transitively on $\{1, 2, \ldots, d\}$. But $\hat{c}' = R(\hat{a}', \hat{b}')$ is a permutation which has the same cyclic decomposition as $\hat{c}$. This implies that $\hat{c}$ and $\hat{c}'$ are conjugate. Take $\hat{a}$ and $\hat{b}$ as the corresponding conjugates of $\hat{a}'$ and $\hat{b}'$, so $\hat{c} = R(\hat{a}, \hat{b})$. Then the subgroup $H = \langle \hat{a}, \hat{b} \rangle$ acts also transitively on $\{1, 2, \ldots, d\}$ and the corresponding branched covering is primitive. □

From Corollary 3.4 it follows that, *for any nontrivial even permutation $\hat{c} \in \Sigma_d$, there are permutations $\hat{a}$, $\hat{b}$ such that $\hat{c} = [\hat{a}, \hat{b}]$ and the subgroup of Σ_d generated by $\hat{a}, \hat{b}$ acts transitively on $\{1, 2, \ldots, d\}$*. This is a weaker form of a well-known fact (Husemoller, 1962, Proposition 4), (Ezell, 1978, Lemma 3.2), and (Edmonds-Kulkarni-Stong, 1984, Lemma 3.2) where the permutation $\hat{a}$ is even a d-cycle.

4. The General Case

Let us recall that, for virtual branch data $\mathcal{D}$, the defect $v(\mathcal{D})$ is even and does not vanish.

4.1 Proposition *For any closed surface V different from the sphere and the projective plane, any virtual branch data $\mathcal{D}$ can be realized by a connected primitive branched covering over V.*

Proof. The case $d = 2$ can easily be treated. Now we assume that $d \geq 3$.

Case $\chi(V) = 0$: Now V is the torus or the Klein bottle. Consider virtual branch data $\mathcal{D} = [A_1, \ldots, A_m]$ with $A_i = [d_{i1}, \ldots, d_{ir_i}]$, $\sum_{j=1}^{r_i} d_{ij} = d$. For each i consider a permutation σ_i which is the product of r_i disjoint cyclic permutations of lengths $d_{i1}, \ldots, d_{ir_i}$. We may assume – after perhaps changing one of the σ_i, for details see (Bogatyi-Gonçalves-Kudryavtseva-Zieschang, 2003) – that $\sigma = \prod_{i=1}^{m} \sigma_i$ is not the identity. Take a disk D^2 with m-point subset $B \subset \mathring{D}^2$ and consider the representation of the fundamental group of $D^2 \setminus B$ on Σ_d sending m obvious generators to $\sigma_1, \ldots, \sigma_m$ and, hence, the homotopy class of ∂D^2 to σ. Let $p_1 \colon W_1 \to D^2$ be the branched covering related to this representation.

Let $A = [d_1, \ldots, d_r]$ be the collection of the lengths of cycles appearing in the decomposition of σ into the product of disjoint cycles. Observe that the partition A of d satisfies the Hurwitz congruence $v(A) \equiv 0 \mod 2$. In fact, one easily checks that $v(A)$ has the same parity as the permutation σ; therefore

$$v(A) \equiv \sum_{i=1}^{m} v(A_i) \equiv v(\mathcal{D}) \equiv 0 \mod 2.$$

By Theorem 3.3, there is a connected primitive branched covering $p_2 \colon W_2 \to V$ with one branch point which realizes the partition A as branch data. We identify D^2 with a small disk in V around the branch point.

It follows from the construction that the restrictions of these two branched coverings to ∂D are equivalent in the following sense: there exists a homeomorphism $h \colon p_1^{-1}(\partial D^2) \to p_2^{-1}(\partial D^2)$ respecting the projections p_i. Therefore we can construct a connected branched covering W_3 over V by gluing W_1 and $W_2 \setminus p_2^{-1}(\mathring{D}^2)$ with respect to h, and define the projection $p_3 \colon W_3 \to V$ as

$$p_3|_{W_1} = p_1 \quad \text{and} \quad p_3|_{W_2 \setminus p_2^{-1}(\mathring{D}^2)} = p_2|_{W_2 \setminus p_2^{-1}(\mathring{D}^2)}.$$

The obtained branched covering has the required branch data. Since the inclusion $W_2 \setminus p_2^{-1}(\mathring{D}^2) \to W_2$ is primitive and the covering $p_2 \colon W_2 \to V$ is primitive by construction, it follows that $W_2 \setminus p_2^{-1}(\mathring{D}^2) \to V$ is primitive and consequently $p_3 \colon W_3 \to V$ is also primitive and the result follows.

Case $\chi(V) < 0$: Observe that V can be obtained by gluing two surfaces V_1 and V_2 where $V_1 = T \setminus \mathring{D}^2$, the torus minus an open disk, and V_2 is a compact surface with one boundary component. Take the branched covering $f_0\colon W_0 \to T$ constructed in the above case. We may assume that D^2 lies in the complement of the set of branch points. Further we remove $f_0^{-1}(\mathring{D}^2)$ from W_0 and obtain a branched covering $f_1\colon W_1 \to V_1$ with $W_1 = W_0 \setminus f_0^{-1}(\mathring{D}^2)$. Let W_2 be the disjoint union of d copies of V_2 and let $f_2\colon W_2 \to V_2$ be the covering which is the identity on each copy of V_2. Let W be obtained by gluing W_1 and W_2 along the boundary respecting the projections $f_1|_{\partial W_1}$ and $f_2|_{\partial W_2}$. Then we obtain a branch covering $f\colon W \to V$ with the branch data $[A_1, \ldots, A_m]$ for $f|_{W_1} = f_1|_{W_1}$ and $f|_{W_2} = f_2|_{W_2}$. Let us show that f is primitive.

Take base points $*_V \in \partial V_1 = \partial V_2$ and $*_W \in \partial W_1$. Let x be an element from $\pi_1(V, *_V)$ which can be represented by a loop ξ lying in V_2. The lift of ξ starting at $*_W$ is also closed; hence, $x \in f_\#(\pi_1(W, *_W))$. It remains to show that any element $g \in \pi_1(V_1, v)$ belongs to $f_\#(\pi_1(W, *_W))$. Consider the curve $\beta = \partial V_1$ and its homotopy class $B \in \pi_1(V, *_V)$. It follows from the primitivity of $f_0\colon W_0 \to T$ that there exist $h \in \pi_1(W_1)$, $h_1, \ldots, h_\ell \in \pi_1(V_1, *_V)$ and $\varepsilon_1, \ldots, \varepsilon_\ell \in \mathbb{Z}$ such that $g = f_\#(h) \prod_{i=1}^{\ell} h_i B^{\varepsilon_i} h_i^{-1}$. Let γ_i be a representative of h_i. Since any lifting of β is closed, the lifting of the curve $\gamma_i \beta^{\varepsilon_i} \gamma_i^{-1}$ is also closed and, thus, $h_i B^{\varepsilon_i} h_i^{-1} \in f_\#(\pi_1(W, *_W))$ what implies $g \in f_\#(\pi_1(W, *_W))$. □

Now we formulate the main result. We say that the *subgroup H of $\pi_1(V)$ corresponds to the branched covering $f\colon W \to V$* if $H = f_\#(\pi_1(W))$.

4.2 Theorem *Let V be a closed surface different from the sphere and the projective plane, $H \subset \pi_1(V)$ a subgroup, and let $\mathcal{D} = [A_1, \ldots, A_m]$ be some virtual branch data with order d. Then the following two assertions are equivalent.*

(a) *The subgroup H corresponds to some connected branched covering realizing the branch data $\mathcal{D}$.*

(b) *H is a subgroup of finite index ℓ, $\ell | d$. For each $i \in \{1, \ldots, m\}$ there exist ℓ partitions*

$$B_{i1} = [d_{i11}, \ldots, d_{i1r_{i1}}], \ldots, B_{i\ell} = [d_{i\ell 1}, \ldots, d_{i\ell r_{i\ell}}]$$

of the number d/ℓ such that

$$A_i = B_{i1} \sqcup \ldots \sqcup B_{i\ell} = [d_{i11}, \ldots, d_{i1r_{i1}}, \ldots, d_{i\ell 1}, \ldots, d_{i\ell r_{i\ell}}].$$

Proof. (a) $\Longrightarrow$ (b): Let $f\colon W \to V$ be a branched covering with $f_\#(\pi_1(W)) = H$. Consider the unbranched covering $p\colon \bar{V} \to V$ corresponding to the subgroup H. Then f lifts to $\bar{f}\colon W \to \bar{V}$. Now, for any branch point $x \in B_f$, the union of the branch data (with respect to $\bar{f}$) of ℓ points $\{y_1, \ldots, y_\ell\} = p^{-1}(x)$ gives the branch data for f at x.

(b) $\Longrightarrow$ (a): Let $p\colon \bar{V} \to V$ be the unbranched covering which corresponds to the subgroup H. Consider the virtual branch data $\bar{\mathcal{D}} = [B_{11}, \ldots, B_{1\ell}, \ldots, B_{m1}, \ldots, B_{m\ell}]$. Since $\chi(\bar{V}) = \ell \cdot \chi(V) \leq 0$, the surface $\bar{V}$ is different from the sphere and the projective plane. It follows from Proposition 4.1 that there is a connected primitive branched covering $h\colon W \to \bar{V}$ which realizes $\bar{\mathcal{D}}$. Therefore $p \circ h$ is the required covering. □

Let us call the procedure described in (b) *a splitting of a partition.* Putting $m = 1$ we immediately obtain the following consequence.

4.3 Corollary *Let V be a closed surface different from the sphere and projective plane and $H \subset \pi_1(V)$ be a subgroup of finite index ℓ. Let $A = [d_1, \ldots, d_r]$ be a partition of d, $d - r$ even, which can be splitted into ℓ partitions of the number $\bar{d} = \frac{d}{\ell}$. Then the virtual branch data $[A]$ can be realized by a branched covering with exactly one branch point such that H corresponds to this covering.* □

5. Some Special Branched Coverings in Figures

The figures below illustrate the proofs of the propositions 3.1, 3.2 showing primitive branched coverings over the torus and the Klein bottle with minimal numbers of roots. Actually, each figure presents a covering over the surface minus a small open disk around the branch point. The edges denoted by the same symbols are identified. In order to demonstrate that the pictures determine unbranched coverings, we calculate the stars at the different vertices of the cover. The branching appears after attaching disks to the boundary components.

Each figure contains several series of polygons where the subscript of any edge or vertex monotonely varies by 1; the first and the last polygon of each series is explicitely shown, while the others are omitted and replaced by dots. The series are allowed to be empty.

Figure 5.1 illustrates a special primitive branched covering which realizes the branch data $[A]$, $A = [2k + 1]$ over the torus ($k \geq 1$).

The polygons of the cover have edges a_i, b_i, c_i, δ_i where each symbol except of δ_i appears twice and i runs from 1 to $2k + 1$. This figure contains two series each consisting of k polygons. Next we calculate the stars at the different vertices:

$$\text{at } A_j: \ c_j^+,\ a_j^+,\ b_{k+j+1}^-,\ a_{k+j+1}^-,\ b_{k+j+1}^+,\ c_j^+, \quad 1 \le j \le k;$$

$$\text{at } A_{k+1}: \ c_{k+1}^+,\ a_{k+1}^+,\ b_1^-,\ a_{k+1}^-,\ b_{k+1}^+,\ c_{k+1}^+;$$

$$\text{at } A_{k+j}: \ c_{k+j}^+,\ a_{k+j}^+,\ b_{k-j+3}^-,\ a_{k-j+2}^-,\ b_{k-j+2}^+,\ c_{k+j}^+, \quad 1 < j \le k+1.$$

Here a_i^+ denotes the end of a_i, while a_i^- denotes the beginning of it; the similar notations are used for $b_i^+, b_i^-, c_i^+, c_i^-$. Over the boundary component we have the only boundary component $\delta_1 \dots \delta_{2k+1}$.

Figure 5.2 illustrates a special primitive branched covering which realizes the branch data $[A]$, $A = [k+\ell, k-\ell]$ over the torus $(0 \le \ell \le k-2)$.

The polygons of the cover have edges a_i, b_i, c_i, δ_i where each symbol except of δ_i appears twice and i runs from 1 to $2k$. This figure contains three series consisting of $k-1$, ℓ, and $k-\ell-2$ polygons, respectively; the second and the third series are allowed to be empty (if $\ell = 0$ or $\ell = k-2$). The stars at the vertices are:

$$\text{at } A_j: \ c_j^+,\ a_j^+,\ b_{k+j}^-,\ a_{k+j}^-,\ b_{k+j}^+,\ c_j^+, \quad 1 \le j < k;$$

$$\text{at } A_k: \ c_k^+,\ a_k^+,\ b_{2k}^-,\ a_k^-,\ b_k^+,\ c_k^+;$$

$$\text{at } A_{k+j}: \ c_{k+j}^+,\ a_{k+j}^+,\ b_{k-j+1}^-,\ a_{k-j}^-,\ b_{k-j}^+,\ c_{k+j}^+, \quad 1 \le j < \ell;$$

$$\text{at } A_{k+\ell}: \ c_{k+\ell}^+,\ a_{k+\ell}^+,\ b_{k-\ell+1}^-,\ a_{k-\ell}^-,\ b_{k-\ell}^+,\ c_{k+\ell}^+;$$

$$\text{at } A_{k+\ell+1}: \ c_{2k}^+,\ a_{2k}^+,\ b_{k-\ell}^-,\ a_{k-\ell-1}^-,\ b_{k-\ell-1}^+,\ c_{2k}^+;$$

$$\text{at } A_{k+j+1}: \ c_{k+j}^+,\ a_{k+j}^+,\ b_{k-j}^-,\ a_{k-j-1}^-,\ b_{k-j-1}^+,\ c_{k+j}^+, \quad \ell < j < k-1;$$

$$\text{at } A_{2k}: \ c_{2k-1}^+,\ a_{2k-1}^+,\ b_1^-,\ a_{2k}^-,\ b_{2k}^+,\ c_{2k-1}^+$$

for $\ell \ge 1$. For $\ell = 0$ one has to drop the third and the fourth row. Over the boundary component we have two boundary components: $\delta_1 \dots \delta_{k+\ell}$ and $\delta_{k+\ell+1} \dots \delta_{2k}$.

Figure 5.3 illustrates a special primitive branched covering which realizes the branch data $[A]$, $A = [2k+1]$ over the Klein bottle $(k \ge 1)$.

The polygons of the cover have edges a_i, b_i, c_i, δ_i where each symbol except of δ_i appears twice and i runs from 1 to $2k+1$. This figure contains two series consisting of $k-1$ and k polygons, respectively; the first series is allowed to be empty (if $k = 1$). The stars at the vertices are:

$$\text{at } A_j: \; c_j^+, \; a_j^+, \; b_{j+1}^-, \; a_{k+j+1}^-, \; b_{j+1}^+, \; c_j^+, \quad 1 \le j \le k;$$

$$\text{at } A_{k+1}: \; c_{k+1}^+, \; a_{k+1}^+, \; b_1^-, \; a_2^-, \; b_{k+2}^+, \; c_{k+1}^+;$$

$$\text{at } A_{k+j}: \; c_{k+j}^+, \; a_{k+j}^+, \; b_{k+j}^-, \; a_{j+1}^-, \; b_{k+j+1}^+, \; c_{k+j}^+, \quad 1 < j \le k;$$

$$\text{at } A_{2k+1}: \; c_{2k+1}^+, \; a_{2k+1}^+, \; b_{2k+1}^-, \; a_1^-, \; b_1^+, \; c_{2k+1}^+.$$

Over the boundary component we have the only boundary component $\delta_1 \dots \delta_{2k+1}$.

Figure 5.4 illustrates a special primitive branched covering which realizes the branch data $[A]$, $A = [k+\ell, k-\ell]$ over the Klein bottle $(0 \le \ell \le k-2)$.

The polygons of the cover have edges a_i, b_i, c_i, δ_i where each symbol except of δ_i appears twice and i runs from 1 to $2k$. This figure contains three series consisting of $k-1$, ℓ, and $k-\ell-3$ polygons, respectively; the second and the third series are allowed to be empty (if $\ell = 0$ or $\ell \ge k-3$). In the case $\ell = k-2$, the third series together with its preceding and following polygon must be replaced by a polygon with boundary $c_{2k-1}a_{2k-1}^{-1}b_{2k-1}a_{2k-1}b_{2k}c_{2k}^{-1}\delta_{2k-1}$ and vertices $A_{2k}, A_{k-1}, A_{k-1}, A_{2k}, A_{2k-1}$. The stars are:

$$\text{at } A_j: \; c_j^+, \; a_j^+, \; b_{k+j}^-, \; a_{k+j}^-, \; b_{k+j}^+, \; c_j^+, \quad 1 \le j < k;$$

$$\text{at } A_k: \; c_k^+, \; a_k^+, \; b_1^-, \; a_1^-, \; b_2^+, \; c_k^+;$$

$$\text{at } A_{k+j}: \; c_{k+j}^+, \; a_{k+j}^+, \; b_{j+1}^-, \; a_{j+1}^-, \; b_{j+2}^+, \; c_{k+j}^+, \quad 1 \le j \le \ell;$$

$$\text{at } A_{k+\ell+1}: \; c_{2k}^+, \; a_{2k}^+, \; b_{\ell+2}^-, \; a_{\ell+2}^-, \; b_{\ell+3}^+, \; c_{2k}^+;$$

$$\text{at } A_{k+j+1}: \; c_{k+j}^+, \; a_{k+j}^+, \; b_{j+2}^-, \; a_{j+2}^-, \; b_{j+3}^+, \; c_{k+j}^+, \quad \ell < j \le k-3;$$

$$\text{at } A_{2k-1}: \; c_{2k-2}^+, \; a_{2k-2}^+, \; b_k^-, \; a_k^-, \; b_{2k}^+, \; c_{2k-2}^+;$$

$$\text{at } A_{2k}: \; c_{2k-1}^+, \; a_{2k-1}^+, \; b_{2k}^-, \; a_{2k}^-, \; b_1^+, \; c_{2k-1}^+$$

for $\ell \le k-3$. For $\ell = k-2$ we have to replace the fourth, the fifth, and the sixth row by the following row

$$\text{at } A_{2k-1}: \; c_{2k}^+, \; a_{2k}^+, \; b_k^-, \; a_k^-, \; b_{2k}^+, \; c_{2k}^+.$$

Over the boundary component we have two boundary components: $\delta_1 \dots \delta_{k+\ell}$ and $\delta_{k+\ell+1} \dots \delta_{2k}$.

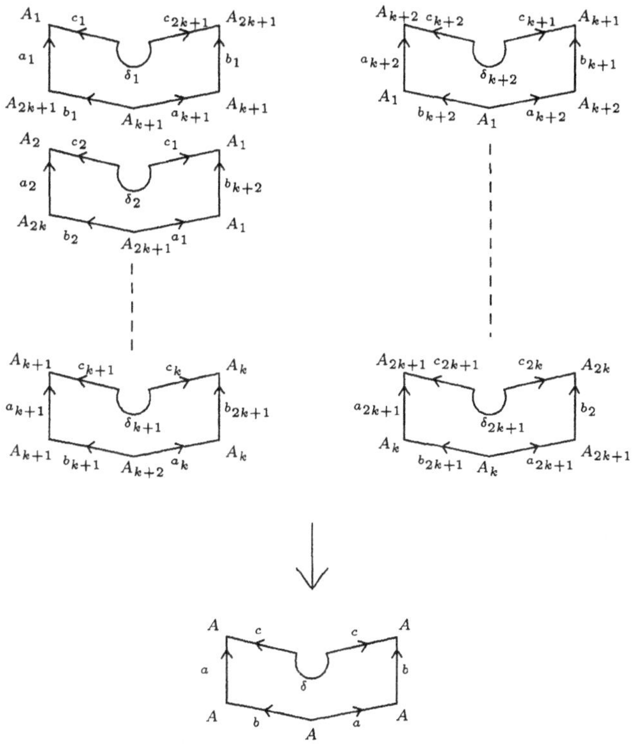

Fig. 5.1: Branched covering over the torus with 1 preimage of the branch point.

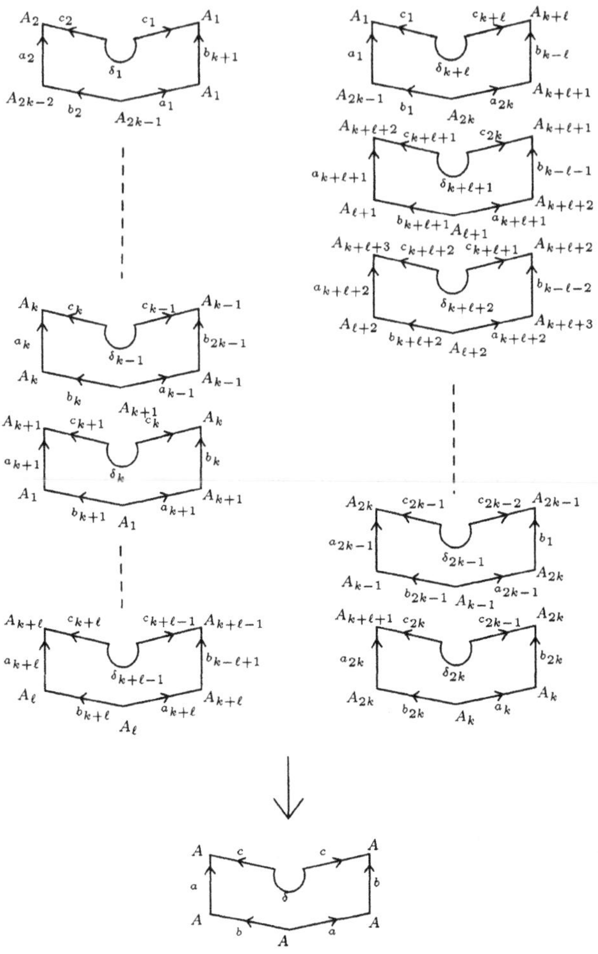

Fig. 5.2: Branched covering over the torus with 2 preimages of the branch point.

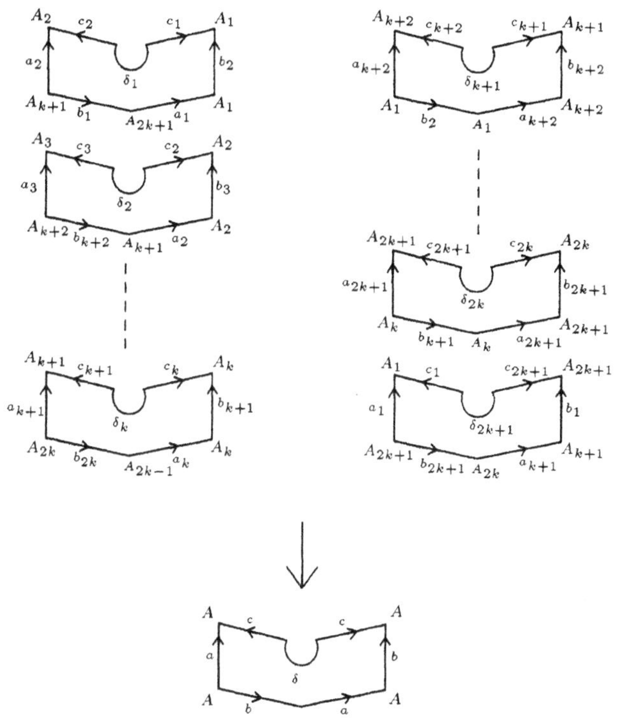

Fig. 5.3: Branched covering over the Klein bottle with 1 preimage of the branch point.

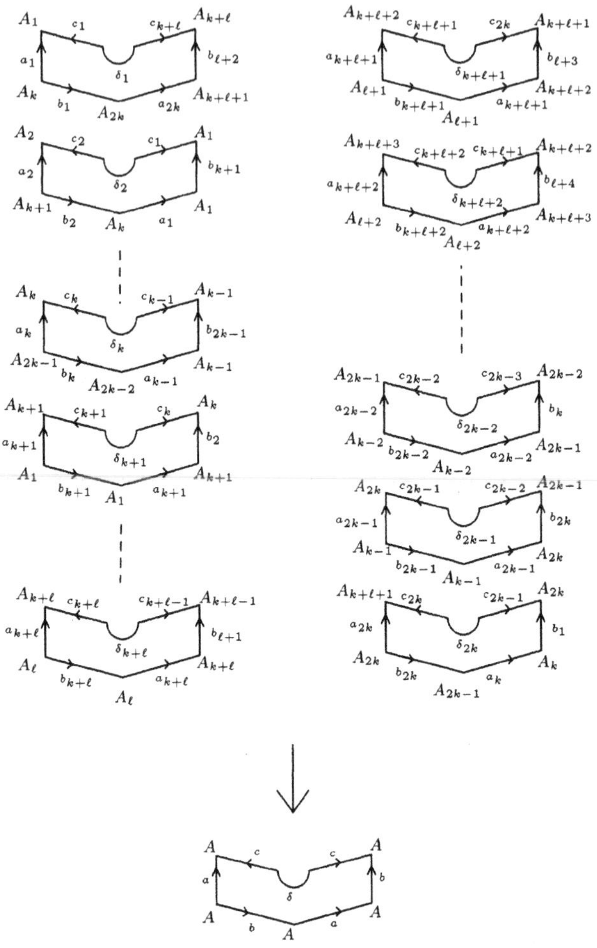

Fig. 5.4: Branched covering over the Klein bottle with 2 preimages of the branch point.

References

Berstein, I. and A. L. Edmonds. *On the construction of branched coverings of low-dimensional manifolds.* Trans. Amer. Math. Soc., **247**:87–124, 1979.

Berstein, I. and A. L. Edmonds. *On the classification of generic branched coverings of surfaces.* Illinois. J. Math., **28**:64–82, 1984.

Bogatyi, S., D. L. Gonçalves, E. A. Kudryavtseva, and H. Zieschang. *Construction of branched coverings over closed surfaces following the Hurwitz approach* J. of Math., **2**:184–197, 2003.

Bogatyi, S., D. L. Gonçalves, E. A. Kudryavtseva, and H. Zieschang. *Minimal number of preimages under mappings between surfaces.* Math. Z., **5**:13–19, 2004.

Bogatyi, S., D. L. Gonçalves, and H. Zieschang. *The minimal number of roots of surface mappings and quadratic equations in free products.* Math. Z., **236**:419–452, 2001.

Ezell, C.L. *Branch point structure of covering maps onto nonorientable surfaces.* Trans. Amer. Math. Soc., **243**:123–133, 1978.

Edmonds, A.L., R. S. Kulkarni, and R. E. Stong. *Realizability of branched coverings of surfaces.* Trans. Amer. Math. Soc., **282**:773–790, 1984.

Gabai, D. and W. H. Kazez. *The classification of maps of surfaces.* Invent. math., **90**:219–242, 1987.

Gonçalves, D. L., E. A. Kudryavtseva, and H. Zieschang. *Roots of mappings on nonorientable surfaces and equations in free groups.* manuscr. math., **107**:311–341, 2002.

Gonçalves, D. L. and H. Zieschang. *Equations in free groups and coincidence of mappings on surfaces.* Math. Z., **237**:1–29, 2001.

Hurwitz, A. *Über Riemannsche Flächen mit gegebenen Verzweigungspunkten.* Math. Ann., **39**:1–60, 1891.

Husemoller, D.H. *Ramified coverings of Riemann surfaces.* Duke Math. J., **29**:167–174, 1962.

Nielsen, J. *Untersuchungen zur Topologie der geschlossenen zweiseitigen Flächen.* Acta Math., **50**:189–358, 1927. Engl. transl. in: Jakob Nielsen, *Collected Mathematical Papers,* pp. 223-341. Birkhäuser, Basel, 1986.

Schirmer, H. *Mindestzahlen von Koinzidenzpunkten.* J. reine angew. Math., **194**:21–39, 1955.

Zieschang, H., E. Vogt, and H. D. Coldewey. *Surfaces and planar discontinuous groups.* Springer-Verlag, Lecture Notes Math. **835**, Berlin-Heidelberg-New York, 1980.

Address for Offprints:
Semeon Bogatyi
Mechanics-Mathematics Faculty
Moscow State Lomonosov-University
119992 Moscow - Russia
e-mail: bogatyi@mech.math.msu.su

Daciberg L. Gonçalves
Departamento de Matemática - IME-USP
Caixa Postal 66281 - Agência Cidade de São Paulo
05311-970 - São Paulo - SP - Brasil
e-mail: dlgoncal@ime.usp.br

Elena Kudryavtseva
Mechanics-Mathematics Faculty
Moscow State Lomonosov-University
119992 Moscow - Russia
e-mail: eakudr@mech.math.msu.su

Heiner Zieschang
Institut für Mathematik
Ruhr-Universität Bochum
44780 Bochum - Germany
e-mail: heiner.zieschang@ruhr-uni-bochum.de
and
Mechanics-Mathematics Faculty
Moscow State Lomonosov-University
119992 Moscow - Russia
e-mail: zieschan@mech.math.msu.su

Cohomology Rings of Oriented Seifert Manifolds with mod p^s Coefficients

J. Bryden

Abstract. This paper discusses the cohomology rings of the orientable Seifert manifolds with $\mathbf{Z}/p^c$ coefficients, where s is the maximal p-valuation of the Seifert invariants and $c \leq s$. The primary motivation for this work lies in its application to the classification of isomorphism classes of linking forms of closed, connected, oriented 3-manifolds and to abelian WRT-type invariants.

Keywords: Seifert manifolds, cup products, cohomology ring, Bockstein map

Mathematics Subject Classification 2000:- Primary: 57M25 : Secondary: 20F38, 20J05, 57N65, 57N27, 20J06, 20K10, 81Q30

Dedicated to the memory of my mother, E. Bryden, who passed away on Nov 17, 2002.

§1 Introduction

This is one in a series of papers (cf. (BHZZ1), (BHZZ2), (BZ2), (BLPZ)) whose purpose is to study the properties of the Seifert manifolds having infinite fundamental group. One of the applications of this work is to the classification of linking forms of 3-manifolds (cf. (BD)) and the description of abelian WRT-type invariants (cf. (De1), (De2), (De3)) that depend on the linking form of the 3-manifold. The object of this paper is to determine the cup products and Bockstein maps for the cohomology of orientable Seifert manifolds of the form

$$M \cong (O, o, g \mid e : (a_1, b_1), \ldots, (a_m, b_m)) ,$$

with $\mathbf{Z}/p^s$ coefficients. Seifert's notation will be used to describe this class of manifolds and can be found in (S).

Aside from Seifert's original paper (S), other references which describe the Seifert fibred manifolds can be found in (H), (Mon), (O), (ST). In (BHZZ1), (BHZZ2), (BZ), and (BZ2) the cohomology ring and Bockstein maps of $H^*(M; \mathbb{Z}/p)$, for any prime p, have been found for any orientable Seifert manifold M. Furthermore, these four papers also discuss applications to degree one maps and the Lusternik-Schnirelmann category of these manifolds amongst other topics. In (BLPZ) a presentation for the p-components $F_p(H_1(M))$ of $H_1(M; \mathbf{Z})$ is given for all Seifert manifolds. From this presentation the unique cyclic decomposition of $F_p(H_1(M))$ is determined.

The main result of this paper, which is stated below, describes the cohomology ring structure and Bockstein maps of $H^*(M; \mathbb{Z}/p^s)$ where s is the maximal p-valuation of the Seifert invariants $a_1, \ldots, a_m$, for Seifert manifolds of the form given above.

For any prime p, let $\nu_p(B)$ denote the p-valuation of the positive integer B. Suppose that s is the maximal p-valuation of the Seifert invariants $a_1, \ldots, a_m$ and t is an integer

This research was partially supported by NSERC operating grant RGP203233.

J.M. Bryden (ed.), Advances in Topological Quantum Field Theory, 317–326.

with $0 \leq t \leq s$. Then for each t, let $a_{t,1}, \ldots, a_{t,r_t}$ denote the Seifert invariants which satisfy the condition $\nu_p(a_{t,i}) = t$, $1 \leq i \leq r_t$ and set $n = \sum_{i=1}^{s} r_s$.

Main Theorem 1. *Let $M := (O, o; g \mid e : (a_1, b_1), \ldots, (a_m, b_m))$ and let s denote the maximal p-valuation of the Seifert invariants a_i. If $n > 1$, then as a graded group*

$$H^*(M; \mathbf{Z}/p^s) = < 1, \alpha_{t,i}, \theta_k, \theta'_k, \beta_{t,i}, \phi_k, \phi'_k, \gamma \mid 1 \leq i \leq r_t, 1 \leq t < s; \\ 2 \leq i \leq r_s, \text{ for } t = s; \; 1 \leq k \leq g >,$$

with generators $\alpha_{t,i}$, θ_k, θ'_k, in degree 1, $\beta_{t,i}$, ϕ_k, ϕ'_k, in degree 2, and γ in degree 3. Moreover, there is exactly one relation given by

$$\beta_{s,1} = -\sum \beta_{t,i} \, .$$

Let δ_{jk} denote the Kronecker delta. Then the non-zero cup products in $H^*(M; \mathbf{Z}/p^s)$ are given by the following.

1. *Let $p = 2$. If $1 \leq t, l < s$ and $1 \leq i \leq r_t$, $1 \leq j \leq r_l$, or at least one of t, l is equal to s in which case either $2 \leq i \leq r_s$ or $2 \leq j \leq r_s$, then*

$$\alpha_{t,i} \cdot \alpha_{l,j} = p^{2s-t-l} \left[\binom{a_{s,1}}{2} b_{s,1}^{-1} \beta_{s,1} + \delta_{t,l} \delta_{ij} \binom{a_{t,i}}{2} b_{t,j}^{-1} \beta_{t,i} \right] .$$

2. *If $1 \leq t, l < s$ and $1 \leq i \leq r_t$, $1 \leq j \leq r_l$, or either t, l is equal to s in which case either $2 \leq i \leq r_s$ or $2 \leq j \leq r_s$, then for any prime p,*

$$\alpha_{t,i} \cdot \beta_{l,j} = -p^{s-t} \delta_{tl} \delta_{ij} \gamma \, , \qquad \theta_l \cdot \phi'_l = \gamma = \theta'_l \cdot \phi_l \, .$$

Additionally, the mod p^s Bockstein B_{p^s} on $H^1(M; \mathbf{Z}/p^s)$ is given by

$$B_{p^s}(\alpha_{t,i}) = \frac{a_{t,i} b_{t,i}^{-1} \beta_{t,i} - a_{s,1} b_{s,1}^{-1} \beta_{s,1}}{p^t} \in H^2(M; \mathbf{Z}/p^s) \, ,$$

$$B_{p^s}(\theta_l) = B_{p^s}(\theta'_l) = 0 \, .$$

Furthermore, Remarks 3 and 4 show how to obtain the cohomology ring structure and Bockstein maps in other cases.

§2 Description of the Equivariant Chain Complex

In order to describe the mod p^s cohomology ring of $M = (O, o; g|e, (a_1, b_1), \ldots (a_n, b_n))$ we first recall the description of the CW structure of the oriented Seifert manifolds given in (BHZZ2) and (BZ2) and then use this information to construct the equivariant chain complex from which the cohomology ring structure is determined. There is also a brief account explaining how to construct the CW structure for manifolds which have the form $M = (O, o; 0|e, (a_1, b_1), \ldots (a_n, b_n))$ given in (BD).

First decompose M into $m+1$ solid tori $V_0, \ldots, V_m$ and its central part $B(m+1) \times S^1$, where $B(m+1) = \overline{S^2 \setminus D_0^2 \cup \cdots \cup D_m^2}$. Each fibred solid torus V_i, $1 \leq i \leq m$, corresponds to a singular fibre that has Seifert invariant (a_i, b_i), $1 \leq i \leq m$. The fibred solid torus

V_0 is an ordinary torus and has Seifert invariant $(a_0, b_0) = (1, e)$. Each fibred solid torus V_i, $1 \leq i \leq m$, can itself be decomposed as a cell complex as follows: let σ_i^0 denote a point on the fibre ρ_i^1, and let η_i^1 be a crossing curve of ρ_i^1, that is, η_i^1 is chosen so that it along with ρ_i^1 are linearly independent and generate $H_1(\partial V_i)$. Next there are 2-cells $\rho_i^2 = \partial V_0 \setminus \rho_i^1 \cup \eta_i^1$, and 2-cells μ_i^2 interior to V_i, plus 3-cells $\sigma_i^3 = \text{int}\,(V_i) \setminus \mu_i^2$. Furthermore there are two cells attached to ∂V_i. The first is a 1-cell σ_i^1 which is simply a path from σ_0^0 to σ_i^0 and a 2-cell σ_i^2 that sits over σ_i^1. Finally there are a number of cells that are independent of the cellular decomposition of the solid tori V_i. The 2-cell δ^2 is the sphere minus $m+1$ discs and $\delta^3 = \delta^2 \times \eta_0^1$. There is also a family of 1 and 2-cells ν_l^1, ω_l^1, ν_l^2, ω_l^2, for $l = 1, \ldots, g$, that are associated to the orbit surface of M.

We will use the following notational device in order to describe the cohomology classes of M: a fibred solid torus $V_{t,j}$ with singular fibre $\rho_{t,j}^1$, and corresponding Seifert invariant $(a_{t,j}, b_{t,j})$, has a cellular decomposition into the cells $\sigma_{t,i}^0$, $\sigma_{t,i}^1$, $\rho_{t,i}^1$, $\eta_{t,i}^1$, $\sigma_{t,i}^2$, $\rho_{t,i}^2$, $\mu_{t,i}^2$ and further the cells $\sigma_{t,j}^1$ and $\sigma_{t,j}^2$ are attached to the boundary of $V_{t,j}$. The ordinary solid torus with Seifert invariant $(1, e)$ is denoted by $V_{0,0}$ and has an analogous cell decomposition. The cells δ^2, δ^3, ν_l^1, ω_l^1, ν_l^2, ω_l^2, for $l = 1, \ldots, g$, are not dependent on the cellular decomposition of the tori $V_{t,j}$ and so do not require the double index notation.

The equivariant chain complex $\mathcal{C}$ for the universal cover $\tilde{M}$ consists of free $\mathbf{Z}[\pi_1(M)]$-modules C_i in dimensions $i = 0, 1, 2, 3$, whose generators correspond to cells in dimensions $0, 1, 2, 3$, that are lifted from the cells of M. This gives a free $\mathbf{Z}[\pi_1(M)]$-resolution,

$$\mathcal{C}: \qquad 0 \to C_3 \xrightarrow{\partial_3} C_2 \xrightarrow{\partial_2} C_1 \xrightarrow{\partial_1} C_0 \xrightarrow{\varepsilon} \mathbf{Z} \to 0\,,$$

of $\mathbf{Z}$, with free generators:

$$\begin{array}{lll}
0: & \sigma_{0,0}^0, \ldots, \sigma_{0,r_0}^0, \sigma_{1,1}^0, \ldots, \sigma_{1,r_1}^0, \ldots, \sigma_{s,1}^0, \ldots, \sigma_{s,r_s}^0; & (C_0)\\
1: & \sigma_{0,1}^1, \ldots, \sigma_{s,r_s}^1; \rho_{0,0}^1, \ldots, \rho_{s,r_s}^1; \eta_{0,0}^1, \ldots, \eta_{s,r_s}^1; \nu_1^1, \omega_1^1, \ldots \nu_g^1, \omega_g^1; & (C_1)\\
2: & \sigma_{0,1}^2, \ldots, \sigma_{s,r_s}^2; \rho_{0,0}^2, \ldots, \rho_{s,r_s}^2; \mu_{0,0}^2, \ldots, \mu_{s,r_s}^2; \delta^2; \nu_1^2, \omega_1^2, \ldots \nu_g^2, \omega_g^2; & (C_2)\\
3: & \sigma_{0,0}^3, \ldots, \sigma_{s,r_s}^3; \delta^3. & (C_3)
\end{array}$$

In each case (using the notation in the previous paragraph) when $1 \leq t \leq s$, then $1 \leq j \leq r_t$. However, if $t = 0$, then $0 \leq j \leq r_t$, except in the case of $\sigma_{t,j}^1$ and $\sigma_{t,j}^2$.

The boundary map ∂ of the chain complex $\mathcal{C}$ is determined by an incidence function. The general construction is described in (CF). In order to define ∂ in this case, we use the following conventions and definitions.

In addition to the list of generators given in (C_1), (C_2), set $\sigma_{0,0}^1 = 0, \sigma_{0,0}^2 = 0$. With this convention it follows that if $t = 0$, then $0 \leq j \leq r_t$ for $\sigma_{t,j}^1$, $\sigma_{t,j}^2$ as well.

Next recall that the standard presentation of the fundamental group of M is defined as follows (cf. (H)):

$$\pi_1(M) = \left\langle x_{t,j}, v_1, w_1, \ldots, v_g, w_g, h \mid [x_{t,j}, h], x_{t,j}^{a_{t,j}} h^{b_{t,j}}, [v_j, h], [w_j, h], \right\rangle .$$

Geometrically the generators $x_{t,j}$, $0 \leq t \leq s$, $1 \leq j \leq r_t$, correspond to the singular fibres of M with Seifert invariants $(a_{t,j}, b_{t,j})$. Normally these elements are denoted $s_1, \ldots, s_m$ when there are m singular fibres (cf. (S), (H), (BZ2)). However, to be consistent with the notation defined above we use the double index notation here too. The generators v_j, w_j,

$1 \le j \le g$ are associated to the orbit surface of M which has genus g. There are precisely two generators for each torus in the connected sum. Now define the elements

$$y_{t,j} = \prod_{\substack{0\le l\le t \\ 0\le k\le j,\, t=0 \\ 1\le k\le j,\, t\neq 0}} x_{l,k}\,, \qquad y_{s,r_s+l} = y_{s,r_s}\prod_{k=1}^{l}[v_k, w_k]\,.$$

In order for this notation to be consistent, adopt the conventions $y_{0,-1} = 1$, $x_{0,0} = y_{0,0} = h^{-e}$ ($x_{0,0}$ corresponds to the ordinary fibre of $V_{0,0}$), $y_{t,0} = y_{t-1,r_{t-1}}$ for $1 \le t \le s$ and note that $y_{s,r_s+g} = 1$. (This notation is adapted from (BZ2).)

Given relatively prime integers $a_{t,j} > 0$, $b_{t,j} > 0$, choose integers $c_{t,j} > 0$, $d_{t,j} > 0$ so that

$$\begin{pmatrix} a_{t,j} & b_{t,j} \\ c_{t,j} & d_{t,j} \end{pmatrix} = 1$$

and let $z_{t,j} = x_{t,j}^{c_{t,j}} h^{d_{t,j}}$. Furthermore define two Laurent polynomials

$$F_{t,j} = \frac{z_{t,j}^{a_{t,j}} - 1}{z_{t,j} - 1}, \qquad G_{t,j} = \frac{1 - z_{t,j}^{-b_{t,j}}}{z_{t,j} - 1}\,.$$

Lastly, define the chains:

$$\begin{aligned}
\pi^1_{t,j} &:= y_{t,j-1}\left(\sigma^1_{t,j} + \rho^1_{t,j}\right) - y_{t,j}\sigma^1_{t,j} \in C_1,\\
\pi^1_{s,r_s+j} &:= y_{s,r_s+j-1}\left(1 - v_j w_j v_j^{-1}\right)\nu^1_j + \left(y_{s,r_s+j-1}v_j - y_{s,r_s+j}\right)\omega^1_j \in C_1\,,\\
\pi^2_{t,j} &:= -y_{t,j-1}\left(\sigma^2_{t,j} + \rho^2_{t,j}\right) + y_{t,j}\sigma^2_{t,j}, \in C_2\\
\pi^2_{s,r_s+j} &:= y_{s,r_s+j-1}\left(v_j w_j v_j^{-1} - 1\right)\nu^2_j + \left(y_{s,r_s+j} - y_{s,r_s+j-1}v_j\right)\omega^2_j \in C_2\,.
\end{aligned}$$

As explained above the boundary map ∂ is constructed from an incidence function. This construction is carried out explicitly in (BHZZ2) and (BZ2). Although, the boundary in degree 2 was determined using the Fox calculus (cf. (F)) which uses the standard presentation of the fundamental group of M given above. The boundary map is defined on the generators of the chain complex as follows:

$$\begin{aligned}
\partial\sigma^1_{t,j} &= \sigma^0_{t,j} - \sigma^0_{0,0}, \quad 0\le t\le s;\, 1\le j\le r_t\,, && (R_{1,1})\\
\partial\rho^1_{t,j} &= (x_{t,j}-1)\sigma^0_{t,j}, \quad 0\le t\le s;\, 1\le j\le r_t\,, t\neq 0;\, 0\le j\le r_0\,, && (R_{1,2})\\
\partial\eta^1_{t,j} &= (h-1)\sigma^0_{t,j}, \quad 0\le t\le s;\, 1\le j\le r_t\,, t\neq 0;\, 0\le j\le r_0\,, && (R_{1,3})\\
\partial\nu^1_j &= (v_j-1)\sigma^0_{0,0}, \quad \partial\omega^1_j = (w_j-1)\sigma^0_{0,0}, \quad 1\le j\le g && (R_{1,4})
\end{aligned}$$

$$\begin{aligned}
\partial\sigma^2_{t,j} &= \eta^1_{0,0} - \eta^1_{t,j} + (h-1)\sigma^1_{t,j}, \quad 0\le t\le s; 1\le j\le r_t\,, && (R_{2,1})\\
\partial\rho^2_{t,j} &= (1-x_{t,j})\eta^1_{t,j} + (h-1)\rho^1_{t,j}, \quad 0\le t\le s;\, 1\le j\le r_t\,, t\neq 0;\, 0\le j\le r_0\,, && (R_{2,2})\\
\partial\nu^2_j &= (1-v_j)\eta^1_{0,0} + (h-1)\nu^1_j, \quad \partial\omega^2_j = (1-w_j)\eta^1_{0,0} + (h-1)\omega^1_j, \quad 1\le j\le g && (R_{2,3})\\
\partial\delta^2 &= \textstyle\sum_{t,j}\pi^1_{t,j} + \sum_j \pi^1_{s,r_s+j} && (R_{2,4})\\
\partial\mu^2_{t,j} &= F_{t,j}\cdot\rho^1_{t,j} + G_{t,j}\cdot\eta^1_{t,j}, \quad 0\le t\le s;\, 1\le j\le r_t\,, t\neq 0;\, 0\le j\le r_0\,, && (R_{2,5})
\end{aligned}$$

$$\begin{aligned}
\partial\sigma^3_{t,j} &= \rho^2_{t,j} + (1-z_{t,j})\mu^2_{t,j}, \quad 0\le t\le s;\, 1\le j\le r_t\,, t\neq 0;\, 0\le j\le r_0\,, && (R_{3,1})\\
\partial\delta^3 &= (1-h)\delta^2 - \textstyle\sum_{t,j}\pi^2_{t,j} - \sum_j \pi^2_{s,r_s+j} && (R_{3,2}).
\end{aligned}$$

An irreducible 3-manifold M with infinite fundamental group is an Eilenberg-MacLane space, and therefore $H^*(M;\mathbf{Z}/p^s) \cong H^*(\pi_1(M);\mathbf{Z}/p^s)$. That is, the cohomology of M is isomorphic to the group cohomology of $\pi_1(M)$ (cf. (M)). Since Seifert manifolds with infinite fundamental group are irreducible, they are Eilenberg MacLane spaces. Thus the cohomology of these Seifert manifolds can be viewed from the point of view of group cohomology.

For such Seifert manifolds the group cohomology and the cup products can now be obtained from the equivariant chain complex via the cochain complex:

$$Hom(C_0;\mathbf{Z}/p^s)\xrightarrow{\partial^0}Hom(C_1;\mathbf{Z}/p^s)\xrightarrow{\partial^1}Hom(C_2;\mathbf{Z}/p^s)\xrightarrow{\partial^2}Hom(C_3;\mathbf{Z}/p^s)\to 0$$

(cf. (BHZZ2), (BZ2)). For any generator α of C_i, let $\hat{\alpha}$ denote the dual generator of $Hom(C_i;\mathbf{Z}/p^s)$; that is, $\hat{\alpha}(\alpha) = 1$, and $\hat{\alpha}(\beta) = 0$ for any other generator β of C_i, for $i = 0, 1, 2, 3$.

The fundamental reason for reformulating the problem in terms of group cohomology is to calculate the cup product structure, which is a complex task. The main difficulty is the construction of a chain approximation to the diagonal. Such a diagonal approximation, Δ was constructed in (BHZZ2) and (BZ2) for all orientable Seifert manifolds. Recall that once the diagonal approximation has been found the cup products can be found in the following manner: let $A = [\hat{u}] \in H^i(\pi_1(M);\mathbb{Z}/p^s)$, $B = [\hat{v}] \in H^j(\pi_1(M);\mathbb{Z}/p^s)$, where $\hat{u}$, $\hat{v}$ are cocycles. Define an $i+j$-cocycle $\hat{\mu} \smile \hat{\nu}$ by:

$$(\hat{u} \smile \hat{v})(z) = \times(\hat{u}\otimes\hat{v})(\Delta z)$$

where $\times: \mathbb{Z}/p\otimes\mathbb{Z}/p \to \mathbb{Z}/p$ is multiplication, $z \in C_{i+j}$ is an $(i+j)$-chain. Then set $A\cdot B = [\hat{\mu} \smile \hat{\nu}]$ (cf. (CE)).

§3 Cohomology of Oriented Seifert manifolds

The method described in §1 will now be applied to Seifert manifolds of the form $M := (O, o; g \mid e : (a_1, b_1), \ldots, (a_m, b_m))$ when $n > 1$. The result for the case when $n = 0$ is given in Remark 3. A result similar to Theorem 1 will be stated for Seifert manifolds $M := (O, n; k \mid e : (a_1, b_1), \ldots, (a_m, b_m))$ when $n > 1$ in Remark 4.

Theorem 1. *Let $M := (O, o; g \mid e : (a_1, b_1), \ldots, (a_m, b_m))$ and let s denote the maximal p-valuation of the Seifert invariants a_i. If $n > 1$, then as a graded group*

$$H^*(M;\mathbf{Z}/p^s) = <1, \alpha_{t,i}, \theta_k, \theta'_k, \beta_{t,i}, \phi_k, \phi'_k, \gamma \mid 1 \le i \le r_t, 1 \le t < s; \\ 2 \le i \le r_s,\ for\ t = s;\ 1 \le k \le g>,$$

with generators $\alpha_{t,i}$, θ_k, θ'_k, in degree 1, $\beta_{t,i}$, ϕ_k, ϕ'_k, in degree 2, and γ in degree 3. Moreover, there is exactly one relation given by

$$\beta_{s,1} = -\sum_{\substack{1\le t\le s\\ 1\le i\le r_t\\ t=s, s\neq 1}} \beta_{t,i}\,.$$

These generators are defined as follows:

$$1 = \left[\textstyle\sum_{t,j}\hat{\sigma}^0_{t,j}\right], \quad \alpha_{t,i} = p^{s-t}[\hat{\rho}^1_{t,i} - \hat{\rho}^1_{s,1}], \quad \beta_{t,i} = [\hat{\sigma}^2_{t,i}] = [b_{t,i}\,\hat{\mu}^2_{t,i}],$$
$$\theta_k = [\hat{\nu}^1_k], \quad \phi_k = [\hat{\nu}^2_k],$$
$$\theta'_k = [\hat{\omega}^1_k], \quad \phi'_k = [\hat{\omega}^2_k], \quad \gamma = [\delta^3]\,.$$

Let δ_{jk} denote the Kronecker delta. Then the non-zero cup products in $H^*(M;\mathbf{Z}/p^s)$ are given by the following.

1. *Let $p=2$. If $1\le t,l<s$ and $1\le i\le r_t$, $1\le j\le r_l$, or at least one of t,l is equal to s in which case either $2\le i\le r_s$ or $2\le j\le r_s$, then*

$$\alpha_{t,i}\cdot\alpha_{l,j}=p^{2s-t-l}\left[\binom{a_{s,1}}{2}b_{s,1}^{-1}\beta_{s,1}+\delta_{t,l}\delta_{ij}\binom{a_{t,i}}{2}b_{t,j}^{-1}\beta_{t,i}\right].$$

2. *If $1\le t,l<s$ and $1\le i\le r_t$, $1\le j\le r_l$, or either t,l is equal to s in which case either $2\le i\le r_s$ or $2\le j\le r_s$, then for any prime p,*

$$\alpha_{t,i}\cdot\beta_{l,j}=-\delta_{tl}\delta_{ij}p^{s-t}\gamma\,,\qquad \theta_l\cdot\phi'_l=\gamma=\theta'_l\cdot\phi_l\,.$$

Additionally, the mod p^s Bockstein B_{p^s} on $H^1(M;\mathbf{Z}/p^s)$ is given by

$$\begin{aligned}B_{p^s}(\alpha_{t,i})&=\frac{a_{t,i}b_{t,i}^{-1}\beta_{t,i}-a_{s,1}b_{s,1}^{-1}\beta_{s,1}}{p^t}\in H^2(M;\mathbf{Z}/p^s)\,,\\ B_{p^s}(\theta_l)&=B_{p^s}(\theta'_l)=0\,.\end{aligned}$$

Proof. The calculations for $H^0(M;\mathbf{Z}/p^s)$ and $H^3(M;\mathbf{Z}/p^s)$ are clear. The calculation of $H^1(M;\mathbf{Z}/p^s)$ and $H^2(M;\mathbf{Z}/p^s)$ are similar to those given in (BZ2) for $H^1(M;\mathbf{Z}/p)$ and $H^2(M;\mathbf{Z}/p)$. However, a further refinement is necessary in order to identify the generators in this case and then to determine the cup products and Bockstein maps.

As in (BZ2) the coboundaries of the generators in dimension 1 are:

$$\begin{aligned}&\partial^1\hat\sigma^1_{t,j}=0,\\ &\partial^1\hat\eta^1_{0,0}=\textstyle\sum_{t,j}\hat\sigma^2_{t,j}+e\cdot\hat\mu^2_{0,0}, &&\partial^1\hat\rho^1_{0,0}=\hat\delta^2+\hat\mu^2_{0,0},\\ &\partial^1\hat\eta^1_{t,j}=-\hat\sigma^2_{t,j}+b_{t,j}\hat\mu^2_{t,j}, &&\partial^1\hat\rho^1_{t,j}=\begin{cases}\hat\delta^2, & t=s\,,\\ \hat\delta^2+a_{t,j}\hat\mu^2_{t,j}, & 1\le t<s\,,\end{cases}\\ &\partial^1\hat\nu^1_k=0,\quad 1\le k\le g\,, &&\partial^1\hat\omega^1_k=0,\quad 1\le k\le g.\end{aligned}\tag{2.1}$$

When $t=s$, observe that $\partial^1(\alpha_{s,i})=\partial^1(\hat\rho^1_{s,i}-\hat\rho^1_{s,1})=\hat\delta^2-\hat\delta^2=0$. Further, for $t\ne s$, $\partial^1(\alpha_{t,i})=\partial^1\left[p^{s-t}\left(\hat\rho^1_{t,i}-\hat\rho^1_{s,1}\right)\right]=p^{s-t}\left(\hat\delta^2+a_{t,i}\hat\mu^2_{t,i}-\hat\delta^2\right)=p^{s-t}a_{t,i}\hat\mu^2_{t,i}=0$. Thus,

$$\begin{aligned}\mathrm{Ker}\,(\partial^1)=<\hat\sigma^1_{t,i},p^{s-t}\left(\hat\rho^1_{t,i}-\hat\rho^1_{s,1}\right),\hat\nu^1_k,\ \hat\omega^1_k\mid 1\le i\le r_t,1\le t<s;\\ 2\le i\le r_s,\ \text{for } t=s;\ 1\le k\le g>,\end{aligned}$$

and since $\mathrm{Im}(\partial^0)=\left\langle\hat\sigma^1_{t,j}\mid 1\le i\le r_t,1\le t\le s\right\rangle$, it follows that

$$H^1(M;\mathbb{Z}/p^s)=\langle\alpha_{t,j},\theta_k,\theta'_k\mid 1\le i\le r_t,1\le t<s;\ 2\le i\le r_s,\ \text{for } t=s;\ 1\le k\le g\rangle\,.$$

The coboundary in dimension 2 is again given in (BZ2) and is defined by,

$$\begin{aligned}&\partial^2\hat\sigma^2_{t,j}=0, &&\partial^2\hat\mu^2_{t,j}=0,\\ &\partial^2\hat\rho^2_{t,j}=\hat\sigma^3_{t,j}+\hat\delta^3, &&\partial^2\hat\nu^2_k=0,\\ &\partial^2\hat\delta^2=0, &&\partial^2\hat\omega^2_k=0\,.\end{aligned}$$

Hence,

$$\operatorname{Ker}\left(\partial^2\right) = \left\langle \hat{\delta}^2, \hat{\sigma}^2_{t,j}, \hat{\mu}^2_{t,j}, \hat{\nu}^2_k, \hat{\omega}^2_k \mid 1 \le i \le r_t, 1 \le t \le s;\ 1 \le k \le g \right\rangle .$$

The equivalence relation induced on Ker ∂^2 by (2.1) gives $H^2(M;\mathbb{Z}/p^s)$. These relations imply that $b_{t,j}\hat{\mu}^2_{t,j} \sim \hat{\sigma}^2_{t,j}$ with $b_{t,j} \neq 0$ for $1 \le j \le r_t$, $1 \le t \le s$. Furthermore $-\hat{\delta}^2 \sim \hat{\mu}^2_{0,0} \sim a_{0,1}\hat{\mu}^2_{0,1} \sim \ldots \sim a_{0,r_0}\hat{\mu}^2_{0,r_0}$. Since $\partial^1\hat{\rho}^1_{s,j} = \hat{\delta}^2$, for $1 \le j \le r_s$, and $\nu_p(a_{0,1}) = \cdots = \nu_p(a_{0,r_0}) = 0$, $\hat{\delta}^2 \sim \hat{\mu}^2_{0,0} \sim \hat{\mu}^2_{0,1} \sim \ldots \sim \hat{\mu}^2_{0,r_0} \sim 0$. As a consequence $\hat{\sigma}^2_{0,j} \sim 0$ for $1 \le j \le r_0$. Finally observe that $\sum_{j,t}\hat{\sigma}^2_{t,j} + e\cdot 0 \sim \sum_{j,t}\hat{\sigma}^2_{t,j} \sim 0$ is the only relation amongst these elements. This shows that $-\left[\hat{\sigma}^1_{s,1}\right] = \sum_{t,j}\left[\hat{\sigma}^2_{t,j}\right]$. That is, $\beta_{s,1} = -\sum_{t,j}\beta_{t,j}$. Thus

$$H^2(M;\mathbb{Z}/p^s) = \langle \beta_{t,i}, \varphi_k, \varphi'_k \mid 1 \le i \le r_t, 1 \le t < s;\ 2 \le i \le r_s,\ \text{for } t = s;\ 1 \le k \le g \rangle .$$

The cup products are determined using the methods of (BHZZ2) and (BZ2). Apart from the calculation of $\alpha_{t,i} \cdot \alpha_{l,j}$, there is no essential difference from the calculations given in (BZ2).

First consider the case when $t = l = s$. By definition we have

$$\alpha_{s,i} \cdot \alpha_{s,j} := \left[\left(\hat{\rho}^1_{s,i} - \hat{\rho}^1_{s,1}\right) \smile \left(\hat{\rho}^1_{s,j} - \hat{\rho}^1_{s,1}\right)\right] .$$

A special case of this theorem is proved in Theorem A.1 (BZ2). In particular the calculation of $\alpha_{s,i} \cdot \alpha_{s,j}$ is given there. Thus,

$$\begin{aligned}\alpha_{s,i} \cdot \alpha_{s,j} &= \binom{a_{s,1}}{2} b^{-1}_{s,1}[b_{s,1}\hat{\mu}^2_{s,1}] + \delta_{i,j}\binom{a_{s,j}}{2} b^{-1}_{s,j}[b_{s,j}\hat{\mu}^2_{s,j}] \\ &= \binom{a_{s,1}}{2} b^{-1}_{s,1}\beta_{s,1} + \delta_{i,j}\binom{a_{s,j}}{2} b^{-1}_{s,j}\beta_{s,j} .\end{aligned}$$

When $p > 2$ it is clear that this expression is zero.

Next consider the product $\alpha_{t,i} \cdot \alpha_{l,j}$ when one or both of t, l are not equal to s. Since $\alpha_{t,i} = p^{s-t}\left[\hat{\rho}^1_{t,i} - \hat{\rho}^1_{s,1}\right]$, the class $\alpha_{t,i}$ is simply a multiple of the expression $\left[\hat{\rho}^1_{t,i} - \hat{\rho}^1_{s,1}\right]$. Hence

$$\begin{aligned}\alpha_{t,i} \cdot \alpha_{l,j} &= \left[p^{s-t}\left(\hat{\rho}^1_{t,i} - \hat{\rho}^1_{s,1}\right) \smile p^{s-l}\left(\hat{\rho}^1_{l,j} - \hat{\rho}^1_{s,1}\right)\right] \\ &= p^{2s-t-l}\left[\left(\hat{\rho}^1_{t,i} - \hat{\rho}^1_{s,1}\right) \smile \left(\hat{\rho}^1_{l,j} - \hat{\rho}^1_{s,1}\right)\right] .\end{aligned}$$

But now a similar calculation to that given in Theorem A.1 (BZ2) shows that

$$\alpha_{t,i} \cdot \alpha_{l,j} = p^{2s-t-l}\left[\binom{a_{s,1}}{2} b^{-1}_{s,1}\beta_{s,1} + \delta_{t,l}\delta_{ij}\binom{a_{t,i}}{2} b^{-1}_{t,j}\beta_{t,i}\right] .$$

Once again observe that when $p > 2$ this expression is zero.

Finally to describe the Bockstein map $B_{p^s}\colon H^1(M;\mathbb{Z}/p^s) \to H^2(M;\mathbb{Z}/p^s)$, recall the method described in (BZ2). First lift the cocycle $p^{s-t}\left[\hat{\rho}^1_{t,i} - \hat{\rho}^1_{s,1}\right] \in C^1(M;\mathbf{Z}/p^s)$ to a $\mathbf{Z}$-cochain $p^{s-t}\left[\hat{\rho}^1_{t,i} - \hat{\rho}^1_{s,1}\right] \in C^1(M;\mathbf{Z})$. Now apply the $\mathbf{Z}$-coboundary map to $\alpha_{t,i}$ to obtain

$$\begin{aligned}\partial\left(p^{s-t}\left(\hat{\rho}^1_{t,i} - \hat{\rho}^1_{s,1}\right)\right) &= p^{s-t}\partial\left(\left(\hat{\rho}^1_{t,i} - \hat{\rho}^1_{s,1}\right)\right) \\ &= p^{s-t}\left(\hat{\delta}^2 + a_{t,i}\hat{\mu}^2_{t,i} - \hat{\delta}^2 - a_{s,1}\hat{\mu}^2_{s,1}\right) \\ &= p^{s-t}\left(a_{t,i}\hat{\mu}^2_{t,i} - a_{s,1}\hat{\mu}^2_{s,1}\right)\end{aligned}$$

Now pull back along the map $(\times p^{s-t})^*$ to find a 2-cochain D that satisfies $(\times p^{s-t})^*(D) = \partial(\alpha_{t,i})$. Then define $B_{p^s}(\alpha_{t,i}) = \mu^*_{p^s}(D)$, where $\mu_{p^s}\colon \mathbf{Z} \to \mathbf{Z}/p^s$ is the canonical projection. Thus,

$$\begin{aligned}
B_{p^s}(\alpha_{t,i}) &= \mu^*_{p^s}(D)\\
&= p^{-s}p^{s-t}\left(a_{t,i}\hat{\mu}^2_{t,i} - a_{s,1}\hat{\mu}^2_{s,1}\right)\\
&= p^{-t}\left(a_{t,i}b^{-1}_{t,i}\left[b_{t,i}\hat{\mu}^2_{t,i}\right] - a_{s,1}b_{s,1}\left[b_{s,1}\hat{\mu}^2_{s,1}\right]\right)\\
&= \frac{a_{t,i}b^{-1}_{t,i}\beta_{t,i} - a_{s,1}b^{-1}_{s,1}\beta_{s,1}}{p^t} \in H^2(M;\mathbf{Z}/p^s),
\end{aligned}$$

The calculations of $B_{p^s}(\theta_l)$ and $B_{p^s}(\theta'_l)$ are similar. □

Remark 1. *The cyclic decomposition of $H^1(M,\mathbf{Z}/p^s)$ and $H^2(M,\mathbf{Z}/p^s)$ is easy to compute and using the notation defined above is given by,*

$$H^1(M,\mathbf{Z}/p^s) \cong H^2(M,\mathbf{Z}/p^s) \cong (\mathbf{Z}/p)^{r_1} \oplus \cdots \oplus \left(\mathbf{Z}/p^{s-1}\right)^{r_{s-1}} \oplus (\mathbf{Z}/p^s)^{2g+r_s-1}.$$

Remark 2. *As above let s denote the maximal p-valuation of the Seifert invariants a_j and let c be a positive integer. If $c < s$ then the generators of $H^*(M;\mathbf{Z}/p^c)$ which are different from those defined in Theorem 1 are:*

$$\alpha_{t,i} = \begin{cases} p^{c-t}\left[\hat{\rho}^1_{t,i} - \hat{\rho}^1_{s,1}\right], & \textit{for } 1 \le t < c,\\ \left[\hat{\rho}^1_{t,i} - \hat{\rho}^1_{s,1}\right], & \textit{for } c \le t \le s.\end{cases}$$

The remaining generators of $H^*(M;\mathbf{Z}/p^c)$ are defined in exactly the same way as those in Theorem 1.

Remark 3. *Let $M := (O,o;0 \mid e : (a_1,b_1),\ldots,(a_m,b_m))$. Suppose that $Ae + C \ne 0$ and $n = 0$, that is, $\nu_p(a_1) = \cdots = \nu_p(a_m) = 0$. Suppose that $b_1,\ldots,b_r \not\equiv 0\,(\mathrm{mod}\ p^q)$ while $b_{r+1},\ldots,b_n \equiv 0\,(\mathrm{mod}\ p^q)$. Then as a graded group*

$$H^*(M;\mathbf{Z}/p^q) = \begin{cases} \langle 1;\, \alpha,\theta_l,\theta'_l;\, \beta,\phi_l,\phi'_l;\gamma \mid 1 \le l \le g\rangle, & \textit{if } Ae + C \equiv 0\,(\mathrm{mod}\ p^q)\\ \langle 1;\, \theta_l,\theta'_l;\, \phi_l,\phi_l;\, \gamma \mid 1 \le l \le g\rangle, & \textit{if } Ae + C \not\equiv 0\,(\mathrm{mod}\ p^q)\end{cases},$$

where $deg(\alpha) = deg(\theta'_l) = deg(\theta'_l) = 1$, $deg(\beta) = deg(\phi_l) = deg(\phi'_l) = 2$, $deg(\gamma) = 3$. The generators θ_l, θ'_l, ϕ_l, ϕ'_l are defined in exactly the same way as they are defined in Theorem 1 in both cases. Furthermore, when $Ae + C \equiv 0\,(\mathrm{mod}\ p^q)$ set

$$\alpha = \left[\sum_{j=0}^{m}\hat{\eta}^1_j - \sum_{j=1}^{r} b_j a_j^{-1}\hat{\rho}^1_j - e\hat{\rho}^1_0\right], \qquad \beta = \left[\hat{\delta}^2\right].$$

When $p > 2$, the non-zero cup products are (in either case, when the classes are defined):

$$\begin{aligned}
\alpha\cdot\theta_l &= \phi_l, & \alpha\cdot\theta'_l &= \phi'_l, & \theta_l\cdot\theta'_l &= \beta\\
\alpha\cdot\beta &= -\gamma, & \theta_l\cdot\phi'_l, &= \theta'_l\cdot\phi_l = \gamma.
\end{aligned}$$

Additionally, when $Ae + C \equiv 0 \,(mod\ p^q)$ the mod p^q Bockstein on $H^1(M; \mathbf{Z}/p^q)$ is given by,

$$B_{p^q}(\alpha) = -\frac{A^{-1}}{p^q}\left[\sum_{i=r+1}^{m} b_i A_i + Ae + C\right]\beta \in H^2(M; \mathbf{Z}/p^q),$$

$$B_{p^q}(\theta_l) = B_{p^q}(\theta_l) = 0\,.$$

Remark 4. *There is an analogue to Theorem 1 for Seifert manifolds of the form $M := (O, n; k \mid e : (a_1, b_1), \ldots, (a_m, b_m))$. In this case when $n > 1$,*

$$H^*(M; \mathbf{Z}/p^s) = < 1, \alpha_{t,i}, \theta_k, \beta_{t,i}, \phi_k,\ \gamma \mid 1 \le i \le r_t, 1 \le t < s; \quad 2 \le i \le r_s,\ for\ t = s;\ 1 \le k \le g >,$$

with $\alpha_{t,i}$, θ_k, in degree 1, $\beta_{t,i}$, ϕ_k, in degree 2, and γ in degree 3. Moreover, there is exactly one relation given by

$$\beta_{s,1} = -\sum_{\substack{1 \le t \le s \\ 1 \le i \le r_t \\ t=s, s \ne 1}} \beta_{t,i} - 2\sum_{k=1}^{g} \phi_k\,.$$

These generators are defined as follows:

$$1 = \left[\textstyle\sum_{t,j} \hat{\sigma}^0_{t,j}\right], \qquad \alpha_{t,i} = p^{s-t}[\hat{\rho}^1_{t,i} - \hat{\rho}^1_{s,1}], \qquad \beta_{t,i} = [\hat{\sigma}^2_{t,i}] = [b_{t,i}\,\hat{\mu}^2_{t,i}],$$
$$\theta_k = [\hat{\nu}^1_k], \qquad \phi_k = [\hat{\nu}^2_k], \qquad \gamma = [\delta^3]\,.$$

The non-zero cup products in $H^*(M; \mathbf{Z}/p^s)$ are given by the following.

1. *Let $p = 2$. If $1 \le t, l < s$ and $1 \le i \le r_t$, $1 \le j \le r_l$, or at least one of t, l, is equal to s in which case either $2 \le i \le r_s$ or $2 \le j \le r_s$, then*

$$\alpha_{t,i} \cdot \alpha_{l,j} = 2^{2s-t-l}\left[\binom{a_{s,1}}{2} b_{s,1}^{-1}\beta_{s,1} + \delta_{t,l}\delta_{ij}\binom{a_{t,i}}{2} b_{t,j}^{-1}\beta_{t,i}\right].$$

2. *If $1 \le t, l < s$ and $1 \le i \le r_t$, $1 \le j \le r_l$, or either t, l is equal to s in which case either $2 \le i \le r_s$ or $2 \le j \le r_s$, then for any prime p,*

$$\alpha_{t,i} \cdot \beta_{l,j} = -\delta_{tl}\delta_{ij}p^{s-t}\gamma\,, \qquad \theta_l \cdot \phi_l = -\gamma\,.$$

Additionally, the mod p^s Bockstein B_{p^s} on $H^1(M; \mathbf{Z}/p^s)$, is given by

$$B_{p^s}(\alpha_{t,i}) = \frac{a_{t,i}b_{t,i}^{-1}\beta_{t,i} - a_{s,1}b_{s,1}^{-1}\beta_{s,1}}{p^t} \in H^2(M; \mathbf{Z}/p^s)\,,$$

$$B_{p^s}(\theta_l) = 0\,.$$

Acknowledgements: I would like to thank Dror Bar Natan and Ruth Lawrence for their invitation to the Einstein Institute for Mathematics at Hebrew University Jerusalem, Israel where the research for this paper was completed. I would also like to thank the institute itself for its support and for providing a stimulating atmosphere.

References

Bryden, J.; Deloup, F. *A linking form conjecture for* 3*-manifolds*, Advances in Topological Quantum Field Theory, NATO Science Series, Kluwer, (2004).

Bryden, J.; Hayat-Legrand, C.; Zieschang, H.; Zvengrowski, P. *L'anneau de cohomologie d'une variété de Seifert* , C. R. Acad. Sci. Paris, 324, (1) (1997) 323-326.

Bryden, J.; Hayat-Legrand, C.; Zieschang, H.; Zvengrowski, P. *The cohomology ring of a class of Seifert manifolds* , Top. and its Appl. 105 (2) (2000) 123-156.

Bryden, J.; Pigott, B.; Lawson, T.; Zvengrowski, P. *The integral homology of the oriented Seifert manifolds* Top. and Its Appl., 127 (1-2) (2003) 259-276.

Bryden, J.; Zvengrowski, P. *The cohomology algebras of oriented Seifert manifolds and applications to Lusternik-Schnirelmann category* , Homotopy and Geometry, Banach Center Publications, Vol. 45 (1998) 25-39.

Bryden, J.; Zvengrowski, P. *The cohomology ring of the oriented Seifert manifolds II*, Top. and Its Appl., 127 (1-2) (2003) 213-257.

Cartan, E.; Eilenberg, S. *Homological Algebra*, Princeton University Press, Princeton New Jersey (1957).

Cooke, G.; Finney, R. *Homology of Cell Complexes*, Princeton University Press, Princeton New Jersey (1976).

Deloup, F. *Explicit formulas for abelian quantum invariants of links in* 3*-manifolds*, Ph.D. thesis, Columbia University (1998).

Deloup, F. *Linking forms, reciprocity for Gauss sums and invariants of* 3*-manifolds*, Trans. AMS, 35 (5) (1999) 1895-1918.

Deloup, F. *An explicit construction of an abelian topological quantum field theory in dimension 3*, Top. and its Appl. 127 (1-2) (2003) 199-211.

Fox, R. *Free differential calculus 1. Derivations in the free group ring*, Ann. of Math. 57 (1953) 457-560.

Hempel, J. *3-Manifolds* , Vol. 86, Annals of Math Studies, Princeton Univ. Press, Princeton, New Jersey (1976).

MacLane, S. *Homology*. Springer Verlag, Berlin (1963).

Montesinos, J.M. *Classical Tesselations and Three-Manifolds* , Springer-Verlag, Berlin-Heidelberg-New York (1987).

Orlik, P. *Seifert Manifolds*, Lecture Notes in Math. 291, Springer-Verlag, Berlin-Heidelberg-New York (1972).

Seifert, H. *Topologie dreidimensionaler gefaserter Räume* , Acta. Math. 60 (1932) 147-238.

Seifert, H.; Threlfall, W. *A Textbook of Topology*, Academic Press, London (1980).

Address for Offprints:

John Bryden
Department of Mathematics and Statistics
University of Calgary
and
Department of Mathematics and Statistics
Southern Illinois University
Edwardsville, IL 62025
email: jbryden@siue.edu

ON CYCLIC COVERS OF THE RIEMANN SPHERE AND A RELATED CLASS OF CURVES

S. KALLEL , D. SJERVE , Y. SONG

This note consists of two short parts. The first part is semi-expository and summarizes some known and less well-known classification results about moduli and automorphisms of prime cyclic covers of the Riemann sphere. In the second part, we restrict attention to those curves affording fixed-point free induced actions on their vector space of holomorphic differentials. These curves correspond to those with all cyclic actions ramifying over the sphere. We describe them completely in terms of their affine equations.

1. An Illustrative Example

A Riemann surface in the sense of Riemann is the collection of all branches of a multi-valued algebraic function $w = f(z)$ obtained by solving an irreducible polynomial equation

$$P(z,w) = a_0(z)w^n + a_1(z)w^{n-1} + \cdots + a_n(z) = 0,$$

where the $a_i(z)$ are polynomials in z. In the sense of Poincaré however, a Riemann surface (of genus $g > 1$) is the quotient of the upper half plane $\mathcal{H}$ by a discrete torsion free subgroup Γ of $\mathrm{Aut}(\mathcal{H})$ (i.e. a Fuchsian group). A standard difficulty in the theory of Riemann surfaces is going back and forth between Riemann's approach and Poincaré's (for recent work in that direction see [2]). Below is a leisurely example of how this correspondence sometimes work.

Let C be a closed Riemann surface (or curve for short) of genus $g = 2$ uniformized by a Fuchsian group Γ with fundamental domain in the

J.M. Bryden (ed.), Advances in Topological Quantum Field Theory, 327–353.

hyperbolic plane given by a regular polygon as shown in figure 1, with 8 sides labeled $a_1, \ldots, a_8$ and ordered clockwise. All angles are equal to $2\pi/8$ and all sides have equal length. The group Γ is generated by elements $\gamma_1, \ldots \gamma_4$, where γ_i is defined by the conditions $\gamma_i(P) \cap int(P) = \emptyset$, $\gamma_i(a_i) = a_{i+4}$ if $i = 1, 3$, and $\gamma_i(a_{i+4}) = a_i$ if $i = 2, 4$. The polygon $P(2)$ is *canonical* in the sense of Schaller [24]. That Γ has fundamental domain $P(2)$ is a consequence of a classical theorem of Poincaré (cf. [24], section 3). Note that all the vertices Q_i are identified under the action of Γ and they map to a unique Q on the surface $C = \mathcal{H}/\Gamma$.

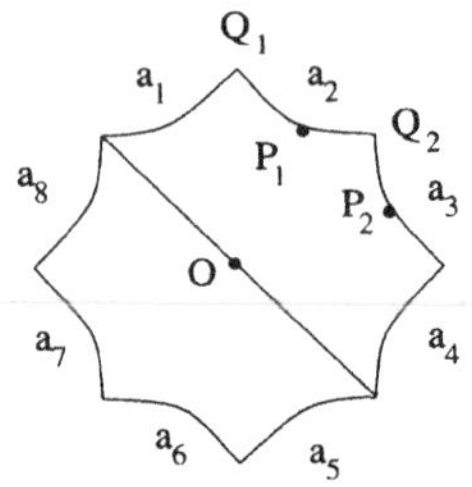

Figure 1. Fundamental polygon $P(2)$ for a genus 2 (compact) curve

Rotation about the center of the polygon O gives an action of the cyclic group $G = \mathbb{Z}_8$ on $P(2)$ respecting identifications, and hence an action on C. By closing down sides, the quotient surface of C under the action of G is obviously $\mathbb{P}^1$. Consider then the sequence of quotient maps

$$P(2) \xrightarrow{q} C \xrightarrow{\pi} \mathbb{P}^1$$

Naturally the origin O (or its image $q(O)$ in C) is a fixed point of the action. It is hence a *ramification point* of π and its image in $\mathbb{P}^1$ a *branching* point. Since Q_1 and all of its translates $Q_i = T^i(Q_1)$ get identified under q, their image in C is also a fixed point of G. Similarly P_i and $P_{4+i} = T^4 P_i$ get identified under q so that $q(P_i)$ is a fixed point for the subgroup of order 2 in G. These are the only fixed points of the action and we get in total six ramification points on the surface.

On the other hand, there are only three branched points in $\mathbb{P}^1$ given by $\pi(q(O))$, $\pi(q(Q_1))$ and $\pi(q(P_1))$. The ramification about the points $q(O), q(Q_1)$ and $q(P_1)$ is such that there are 8 sheets coming together at O, the same number at Q_1 and only 2 sheets at P_1. That is the "ramification numbers" of the action are $2, 8, 8$. This is naturally all consistent with the Riemann Hurwitz formula relating the genus g of C to the genus h of C/G

and the ramification numbers[1] n_y of each point y in the branch locus B; i.e.

$$2 - 2g = |G| \left(2 - 2h - \sum_{y \in B} \left(1 - \frac{1}{n_y} \right) \right)$$

Next we make use of Galois theory for coverings to describe a form for the affine equation $f(x,y) = 0, (x,y) \in \mathbb{C}^2$ that the curve satisfies (this form is not unique). Given a curve C, we consider its field of meromorphic functions $\mathcal{M}(C) = \{f : C \longrightarrow \mathbb{C} \text{ meromorphic}\}$. Note that $\mathcal{M}(\mathbb{P}^1)$ is $\mathbb{C}(x)$ the field of rational functions in one variable. Since $\pi : C \longrightarrow \mathbb{P}^1$ is a covering, we get by precomposition a map $\mathbb{C}(x) \longrightarrow \mathcal{M}(C)$ which is an inclusion of fields. This exhibits $\mathcal{M}(C)$ as a field extension of $\mathbb{C}(x)$ which in fact is Galois if the original cover is Galois. In the case at hand, $\mathbb{P}^1$ is the quotient of C by the action of $G = \mathbb{Z}_8$ and the cover is by definition Galois.

The fact that our curve is dimension one (complex), $\mathcal{M}(C)$ is necessarily an algebraic function field in one variable (Siegel), or more precisely it is a finite field extension of $\mathbb{C}(x)$. We can write $\mathcal{M}(C) = \mathbb{C}(x,y)$ with x, y satisfying an equation $F(x,y) = a_0(x)y^n + a_1(x)y^{n-1} + \cdots + a_n(x)$ with $a_i(x) \in \mathbb{C}(x)$, and n is the degree of the covering. Now theorem 6.2 in [18] (chap. VIII,§6) states that if K is a degree n cyclic field extension of k, n prime to char(k) and k containing a primitive n-th root, then there is $\alpha \in K$ such that $K = k(\alpha)$ and α satisfies an equation $Y^n - a = 0$ for $a \in k$. In our case, $k = \mathbb{C}(x)$, $n = 8$ and $a = f(x)$ so that $\mathcal{M}(C) = \mathbb{C}(x,y)$ with $y^8 = f(x)$. The curve C is the locus of this polynomial and the cyclic quotient $C \to \mathbb{P}^1$ is the restriction to C of the projection $\mathbb{C}^2 \to \mathbb{C}$, $(x,y) \mapsto x$. Note that $F(x,y) = y^8 - f(x)$ is necessarily irreducible since we started with a connected Riemann surface.

We can set our three branched points[2] to be $0, 1$ and -1. These points correspond to the zeros of $f(x)$, so that our equation becomes $y^8 = x^a(x+1)^b(x-1)^c$. The numbers a, b, c relate to the ramification indexes in an interesting way. Notice that in a small neighborhood of either O, Q_1 or P_1, the action of G is rotation by a multiple of $2\pi/8$. These multiples can be chosen to be a, b, c (respectively) and are such that

$$a + b + c \equiv 0 \bmod 8 \;, 0 < a, b, c < 8 \;, \quad [a, 8] = [b, 8] = 1 \;, [c, 8] = 4$$

[1]These numbers correspond to the order of the stabilizer subgroups at the fixed points, and these subgroups are necessarily cyclic.

[2]Recall that the action of $Aut(\mathbb{P}^1)$ on the Riemann sphere is 3-transitive and hence any choice for the 3 branched points yield isomorphic coverings.

The first congruence is necessary to avoid having ramification over the point at infinity in $\mathbb{P}^1 = \mathbb{C} \cup \{\infty\}$ (there are only 3 branched points as we pointed out). The triple (a, b, c) is well defined up to a multiple k prime to 8 (since of course the numbering of the sheets of $C \longrightarrow \mathbb{P}^1$ is only well-defined up to permutation). If we choose $a = 1$, then necessarily $b = 3$ and $c = 4$. Our surface has affine equation

$$y^8 = x(x-1)^3(x+1)^4$$

This curve was studied by Kulkarni for instance (cf. [14], [15]) who determined its full automorphism group $Aut(C) = GL_2(\mathbb{F}_3)$ (the general linear group over the finite field $\mathbb{F}_3$). Exceptionally, this curve is completely determined by the fact that it is of genus two and that it admits a $\mathbb{Z}_8$-action.

Notice that since our curve is hyperelliptic (with involution rotation by π about the axis through Q_4 and Q_8 in figure 1), we could have searched for an equation of the form $y^2 = g(x)$. But the reduced group of automorphisms of this curve; i.e. the quotient by the central involution, is $GL_2(\mathbb{F}_3)/\mathbb{Z}_2 = S_4$ (see Coxeter-Moser, p:96), and by the classication result of Bolza (§3.1), another equation for C is $y^2 = x(x^4 - 1)$.

2. Synopsis

We say C is p-elliptic if it admits an action of the cyclic group $\mathbb{Z}_p$ with quotient $\mathbb{P}^1$. When p is a prime, these curves admit affine equations of the form

$$w^p = (z - e_1)^{a_1} \times (z - e_2)^{a_2} \times \cdots \times (z - e_r)^{a_r} \tag{1}$$

where $e_1, \ldots, e_r$ are distinct complex numbers, and r is related to p by the formula $2g = (r-2)(p-1)$. We can assume (without any loss of generality) that $a_1, \ldots, a_r$ are integers satisfying $1 \leq a_i \leq p-1$ and $\sum_{i=1}^r a_i \equiv 0 \pmod{p}$. The projection $\pi : (w, z) \mapsto z$ is of course *branched* over the e_i's. The last condition ensures that there is no branching over ∞ in $\mathbb{P}^1 = \mathbb{C} \cup \{\infty\}$.

In the first part of this note we collect and slightly expand on several disseminated facts about the automorphism groups and moduli of p-elliptic curves. Interestingly for example, and combining results in [11] and [14], we obtain

Proposition 1 *Suppose $g > (p-1)^2$, and p prime. Then a p-elliptic curve C_g admits a unique normal subgroup $\mathbb{Z}_p$ which acts with quotient the Riemann sphere.*

This is the analog for $p > 2$ of the known fact that hyperelliptic curves ($p = 2$) admit a *unique* hyperelliptic involution.

For a cyclic n-fold cover of the line (n not necessarily prime), it is believed that the branch data fully determines the curve (up to isomorphism). Nakajo [21] and Gabino-Diez [12] verified this directly for p prime. A calculation of Lloyd [19] on the other hand gives a count for the distinct isomorphism classes of p-elliptic curves of a given genus. The following, of which we give a slightly novel proof, summarizes the situation

Proposition 2 : *Assume* $2g = (r-2)(p-1)$. *Then the moduli space* $\mathbf{M}(g,p)$ *of genus* g *prime Galois covers of the sphere splits into* N_r *disjoint copies of* $C_{r-2}(\mathbb{C})$, *the configuration space of unordered* $r-2$ *tuples of distinct complex numbers, where* N_r *is obtained from the generating function*

$$\sum_r N_r x^r = \frac{1}{p-1}\left\{\frac{1}{p}\left[\frac{1}{(1-x)^{p-1}} + (p-1)\frac{(1-x)}{(1-x^p)}\right] + \sum_{\substack{ll'=p-1\\ l\neq 1}} \phi(l)(1-x^l)^{-l'}\right\}$$

REMARKS.
(a) In the case $p = 2$ ($r = 2g+2$), the series reduces to $1+x^2+x^4+\cdots$ and hence $N_r = 1$ (for any given $g > 1$). The moduli space is connected and coincides with the set of conformal classes of hyperelliptic curves (classical).
(b) When $r = 3$, $g = (p-1)/2$ and the curve is isomorphic to one with equation $y^p = x^a(x-1)$ for some a, $1 \leq a \leq \frac{p-1}{2}$ (see [14]). Such curves are usually called *Lefschetz*. The following count of conformal classes of distinct Lefschetz curves can be deduced from Proposition 2 and appears for instance in [22]:

$$N_3 = \begin{cases} (p+1)/6, \text{if } p \equiv 2 \ (mod\ 3) \\ (p+5)/6, \text{if } p \equiv 1 \ (mod\ 3) \end{cases}$$

Recall that any genus g curve C admits g-independent holomorphic one forms (Riemann). Let V be this vector space, and suppose $G \times C \to C$ is an action of the finite group G on the curve C. We say that the induced action on V is fixed point free if $+1$ is not an eigenvalue of any non-identity element $g \in G$. Such an element does act fixed point freely on $V^* = V\backslash 0$.

It has been observed in [13] that the associated linear representation $\rho : G \to GL(V)$ is fixed point free if and only if the genus of each orbit surface $C/\mathbb{Z}_p$ is zero for every subgroup $\mathbb{Z}_p \subseteq G$, where p is a prime dividing the order of G. A group with this property is said to have a "genus-zero action" on C. Such a property imposes strong Sylow conditions on G, which in fact have allowed the authors in [13] to completely classify these groups.

Theorem 1 *[13] The groups admitting genus-zero actions on surfaces of genus $g > 1$ are: the cyclic groups $\mathbb{Z}_{p^e}$ for primes $p \geq 2$ and exponents $e \geq 1$; the cyclic groups $\mathbb{Z}_{pq}$ for distinct primes p, q; the generalized quaternion groups $Q(2^n)$ for $n \geq 3$; and the ZM groups $G_{p,4}(-1)$ for odd primes p.*

For the definition of ZM (Zassenhaus metacyclic) groups see §5 or [27]. Our goal here is to give equations and complete listing for the surfaces admitting fixed point free actions on their vector space of differentials. We can summarize our calculations in one main theorem which we split into the following three propositions (cf. sections 4.1 and 4.2):

Proposition 3 *Let C be a Riemann surface of genus $g > 1$ admitting a genus zero action by $G = \mathbb{Z}_{pq}$. Then*

1. *Either $g = \frac{1}{2}(p-1)(q-1)$, and C is isomorphic to the Fermat curve $w^p = z^q - 1$,*
2. *or $g = (p-1)(q-1)$, and C is isomorphic to the surface with the equation $w^p = \dfrac{z^q - 1}{z^q - \lambda}$ (with $\lambda \neq 0$ and $\lambda^q \neq 1$).*

Proposition 4 *Let C be a Riemann surface of genus g admitting a genus zero action by $G = G_{p,4}(-1) = \langle x, y | x^p = 1,\ y^4 = 1,\ yxy^{-1} = x^{-1}\rangle$. Then $g = p - 1$ and C is isomorphic to the surface $w^2 = z^{2p} - 1$.*

Similar results are obtained for the cyclic groups $\mathbb{Z}_{p^e}$, $e > 1$, and for the quaternionic groups as summarized towards the end of the paper. Here's a sample:

Proposition 5 *$Q_8 = \langle A, B \mid A^4 = 1, A^2 = B^2, BAB^{-1} = A^{-1}\rangle$ admits a genus zero action on the genus 4 surface*

$$w^2 = z(z^4 - 1)(z^4 + 1) = z(z^8 - 1)$$

as follows: $A(x, y) = (-x, iy)$ and $B(x, y) = (-1/x, y/x^5)$. The induced action on the space of holomorphic differentials is fixed-point free.

3. Part I : On Cyclic Covers of the Sphere, their Automorphisms and Moduli

3.1. AUTOMORPHISMS

One is tempted to classify all finite groups that can arise as automorphisms of p-elliptic curves $C_p : y^p = f(x)$ in terms of p and the polynomial f. We assume the genus g of C to be bigger than 2 so that the group $Aut(C)$ of all automorphisms of C is finite. It is clear that for generic curves, $Aut(C)$ is reduced to just $\mathbb{Z}_p$ (cf. [8] for instance). The answer is also known for both small and "sufficiently" large genera.

For curves of genus $g = 2$ and $g = 3$ the classification of the automorphism groups is completely known (whether the curve is p-elliptic or not). For $g = 3$ see [4], [17], [20]. Genus two curves are necessarily hyperelliptic (i.e. 2-elliptic) and have affine equations of the form $y^2 = f(x)$, $(x, y) \in \mathbb{C}^2$. The involution $(x, y) \mapsto (x, -y)$ is always central; see [10]. O. Bolza (1888) seems to have been first to determine the groups acting on genus two curves and write equations for them (see [1], chapter 1 for example). One of Bolza's curves has full automorphism group $GL_2(\mathbb{F}_3)$ (reduced group $GL_2(\mathbb{F}_3)/\mathbb{Z}_2 = S_4$) and has affine equation $w^2 = z(z^4-1)$. There is on the other hand a single isomorphism class of curves with automorphisms $\mathbb{Z}_{10}$; it has equation $w^2 = z^5 - 1$. The other possible full automorphism groups that can occur are $\mathbb{Z}_2 \oplus \mathbb{Z}_2$, D_8 and D_{12} (see [23] for example).

Proposition 1 addresses the problem of determining $\mathrm{Aut}(C)$ when the number of branchpoints of f in the defining equation $C : y^p = f(x)$ is large enough (so is the genus).

PROOF (of Proposition 1) Let $C : y^p = f(x)$ be a p-elliptic curve, and denote by r the number of distinct roots of f (i.e. the branchpoints). This is related to g and p by the formula $2g = (r-2)(p-1)$. According to theorem 4 of [14], the p-cyclic subgroup G acting on a p-elliptic curve $C : y^p = f(x)$ becomes normal in $\mathrm{Aut}(C)$ as soon as r exceeds $2p$. On the other hand, a theorem of Gabino-Diez [11] asserts that any other cyclic subgroup G' of order p acting on C with quotient the Riemann sphere must be conjugate to G. Since G is normal, G' must coincide with G and hence the uniqueness statement. ∎

REMARKS AND COROLLARY:
(a) The theorem above implies that for p-elliptic curves with large enough genus $g > (p-1)^2$, $\mathrm{Aut}(C)$ is an extension of $\mathbb{Z}_p$ by a polyhedral group (i.e.

a finite subgroup of SO(3)). The polyhedral groups are: the finite cyclic groups, the dihedral groups D_{2n} of order $2n$, the tetrahedral group A_4, the octahedral group S_4 and the icosahedral group A_5.
(b) The theorem is true for "general curves" of genus $g > p-1$, or equivalently for the number of branched points r exceeding 4. (A curve $y^p = f(x)$ is general for an open dense choice of branched points in S^2). This is essentially the statement of the main lemma in [8].
(c) Finally the conclusion of the theorem is not anymore true for smaller genus since the Fermat curve $x^p + y^p = 1$ has genus $(p-1)(p-2)/2$ but affords two distinct cyclic p-actions ramifying over the sphere (the obvious ones) and which are conjugate under the involution $(x,y) \mapsto (y,x)$.

Proposition 6 *The following cyclic p-covers admit actions by the corresponding polyhedral groups:*

1. *Dihedral* D_{2n}: $y^p = x^n - 1$, $p|n$.
2. *Octohedral (and Tetrahedral):* $y^{2p} = x^{2p} - 1$, *p odd.*

PROOF: The action of D_{2n} is easy enough (cf. [14]). The involution in this case acts on $y^p = x^n - 1$ by $(x,y) \mapsto (x/y^m, 1/y)$, with $m = n/p$. For the tetrahedral group A_4 we use the $(2,3,3)$ triangle group presentation

$$\langle R, S, T \mid R^2 = S^3 = T^3 = RST = 1 \rangle .$$

Then an action of A_4 on $y^{2p} = x^{2p} - 1$, whenever p is odd, is given by

$$R(x,y) = (x, -y), \; S(x,y) = (\imath/y, \imath x/y), \; T(x,y) = (-y/x, \imath/x).$$

This action extends to an action of

$$S_4 = \langle R, S, T \mid R^2 = S^3 = T^4 = RST = 1 \rangle$$

as follows: $R(x,y) = (x/y, 1/y)$, $S(x,y) = (y/x, \imath/x)$, $T(x,y) = (\imath y, \imath x)$. ∎

It turns out that it is possible to completely classify Aut(C) when the number of branched points is less than 4; in which case we are dealing with surfaces of the form

$$C_n(a,b,c) : y^n = x^a(x-1)^b(x+1)^c, \quad a+b+c \equiv 0 \; (mod\; n) \qquad (2)$$

In [14], a classification of all groups $Aut(C_n(a,b,c))$ in terms of n, a, b and c is given. Here one assumes $n \geq 4$ (so that $g \geq 2$) and $1 \leq a, b, c < n$. When n is not prime, one also assumes $GCD(n,a,b,c) = 1$ to ensure connectedness. The following definition is needed for the rest of the paper.

DEFINITION. The mod-n Nielsen class of a tuple $(a_1, \ldots, a_k)$ of integers consists of all $(a'_1, \ldots, a'_k)$ such that $(a'_1, \ldots, a'_k) = (ka_{\tau(1)}, \ldots, ka_{\tau(k)})$ for some permutation $\tau \in S_k$ and k prime to n. We write

$$(a'_1, \ldots, a'_k) \sim_\tau (a_1, \ldots, a_k) \operatorname{mod}(n)$$

(and $\sim$ instead of $\sim_\tau$ when there is no need for explicit mention of τ). Note that triples in the same class yield isomorphic curves in (2) (a classic observation of Nielsen). We will see shortly and similarly that Nielsen classes together with the cross ratios of the branched points is all that determines the isomorphism type of Galois p-covers.

Theorem 2 *[14] Suppose n is odd. Then $Aut(C_n(a,b,c))$ is determined by the Nielsen class of (a,b,c) as follows:*

1. *$Aut(C)$ is $\mathbb{Z}_2 \oplus \mathbb{Z}_n$ if $(a,b,c) \sim (1,1,n-2)$.*
2. *$Aut(C)$ is the metacyclic group $\mathbb{Z}_n : \mathbb{Z}_2$ if $(a,b,c) \sim (1,b,n-1-b)$, where $GCD(b,n) = 1, b \neq 1, 8 \not| n, b^2 \equiv 1 \pmod n$.*
3. *$Aut(C)$ is the metacyclic group $\mathbb{Z}_n : \mathbb{Z}_3$ if $(a,b,c) \sim (1,b,b^2)$, where $b \neq 1$ and $GCD(b,n) = 1$.*
4. *$Aut(C)$ is $PSL_2(7)$ if $n = 7$ and $(a,b,c) \sim (1,2,4)$. C is the Klein curve.*
5. *$Aut(C) = \mathbb{Z}_n$ in all other cases.*

The notation $A : B$ denotes a non-split extension of A by B. The situation for n even is equally well understood [14].

3.2. THE MODULI SPACE

As opposed to the full moduli space of genus g closed Riemann surfaces, the moduli space of cyclic coverings of the line has a very simple description.

Theorem 3 (EQUIVALENCE OF CYCLIC p-COVERINGS)

1. *([8]) Fix some branch set $B = \{q_1, \ldots, q_r\}$ in a curve X, and consider two $\mathbb{Z}_p$-coverings C_1 and C_2 over X, branched over the q_i's with multiplicities $(k_1, \ldots, k_r)$ and $(l_1, \ldots, l_r)$ respectively. Then C_1 is isomorphic to C_2 (as branched covers) if and only if $(k_1, \ldots, k_r) \sim (l_1, \ldots, l_r)$ mod-p.*

2. *([21]) Now Suppose $X = \mathbb{P}^1$; C_1 is branched over $B_1 \subset X$, with branch points q_i of multiplicities k_i, $1 \leq i \leq r$; and C_2 is branched over $B_2 \subset X$ with branch points p_i of multiplicities l_i, $1 \leq i \leq r$. Then C_1 is isomorphic to C_2 if and only if there is $\sigma \in PGL_2(\mathbb{C})$ and $\tau \in \Sigma_r$ such that $p_{\tau(i)} = \sigma q_i$ and $(k_1, \dots, k_r) \sim_\tau (l_1, \dots, l_r)$.*

The sufficiency part of these assertions is a direct consequence of Riemann's extension theorem. Riemann's theorem asserts that a ramified cover $\pi : X \longrightarrow Y$ branched over $B \subset Y$ is determined by the étale cover $\pi' : X - \pi^{-1}(B) \longrightarrow Y - B$ and by the way the sheets of π "come together" at the ramification points. This is specified by the monodromy around the branch points. More precisely

(Riemann) Suppose X is a curve, $B \subset X$ is a finite subset and $x_0 \notin B$. Then there exists a correspondence

Degree d (algebraic) branched covers *$C \longrightarrow X$, branched over B,* *modulo covering transformations*	$\longleftrightarrow$	Transitive representations $\pi_1(X-B,x_0) \longrightarrow \Sigma_d$, *modulo* *equivalence of representations*

By elementary covering space theory, the étale cover is uniquely determined (up to equivalence of covers) by the kernel of the monodromy $\rho : \pi_1(X - B, x_0) \longrightarrow \Sigma_d$. If the cover is connected, then the image of ρ (the monodromy sugbroup) acts transitively on the cover. The "existence" part stipulates then that an étale cover of curves $\pi' : C - B_0 \longrightarrow X - B$, B a finite set covered by B_0, always extends to an analytic (and hence algebraic) ramified cover $C \longrightarrow X$. Part(1) and Part(2) (only if part) of theorem 3 is an immediate consequence of the RET after observing that for $\mathbb{Z}_p$-Galois coverings over $\mathbb{P}^1$, we can replace Σ_p by $\mathbb{Z}_p$, and that automorphisms of p-cyclic groups are given by raising elements to a power k prime to p. The permutation τ entering in the Nielsen class comes from the fact again that the sheets of the covering can only be numbered up to a permutation.

Part (2) of the above theorem as observed by Nakajo (see also [11]) asserts that any abstract isomorphism of prime cyclic covers of the line is in fact an equivalence of branched coverings. We can see this statement through the eyes of uniformization theory as follows.

Let Π denote the fundamental group of a p-elliptic curve C, and P a fundamental polygon for C in the upper half-plane $\mathcal{H}$. The polygon P then affords a $\mathbb{Z}_p$ symmetry (see §1), and in fact there is another Fuchsian group

$\Gamma \subset \mathrm{PSL}_2(\mathbb{R})$ and a short exact sequence (called a skep)

$$1 \longrightarrow \Pi \longrightarrow \Gamma \xrightarrow{\ \theta\ } \mathbb{Z}_p \longrightarrow 1$$

uniformizing the action. The group Γ has *signature* $(0|\overbrace{p,\ldots,p}^{r})$. This means that as an abstract group it has the presentation

4.2 $$\langle x_1, x_2, \ldots, x_r \mid x_1^p = x_2^p = \cdots = x_r^p = x_1 x_2 \cdots x_r = 1 \rangle .$$

The x_i are called the elliptic generators. Describing the $\mathbb{Z}_p$ action on C is done in a standard way and comes down to choosing a realization of the abstract group Γ as a Fuchsian group (i.e embedding it in $\mathrm{PSL}_2(\mathbb{R})$ up to conjugation) and then specifying the epimorphism $\theta : \Gamma \to \mathbb{Z}_p$ with torsion free kernel Π. Pick a generator $T \in \mathbb{Z}_p$. Any epimorphism $\theta : \Gamma \to \mathbb{Z}_p$ is described on the elliptic elements by:

$$\theta(x_i) = T^{a_i}, \text{ where } 1 \le a_i \le p-1,\ 1 \le i \le r, \text{ and } \sum_{i=1}^{r} a_i \equiv 0\ (mod\ p).$$

The conditions guarantee that θ is a well defined epimorphism with torsion free kernel. The number of elliptic generators r corresponds to the number of fixed points of $T : C \to C$.

An isomorphism of Fuchsian groups is an abstract isomorphism of groups that is induced from an element of $\mathrm{PSL}_2(\mathbb{R})$ (that is both groups can be embedded in $\mathrm{PSL}_2(\mathbb{R})$ as conjugates). We say that two skeps θ_1 and θ_2 are equivalent if the corresponding extensions are equivalent; that is if there are vertical isomorphisms of Fuchsian groups making the diagram commute

$$\begin{array}{ccccc} \Pi_1 & \longrightarrow & \Gamma_1 & \xrightarrow{\theta_1} & \mathbb{Z}_p \\ \downarrow \lambda & & \downarrow \alpha & & \downarrow \beta \\ \Pi_2 & \longrightarrow & \Gamma_2 & \xrightarrow{\theta_2} & \mathbb{Z}_p \end{array}$$

Lemma 1 : *There is a 1-1 correspondence between equivalence classes of skeps $\Gamma \to \mathbb{Z}_p$ and isomorphism classes of p-elliptic curves.*

PROOF: Consider an isomorphism class of skeps where α is induced from $\tau : \mathcal{H} \to \mathcal{H}$. Here $\ker \theta_i = \Pi$ as abstract groups and λ is an automorphism of Π. Both copies of Π (embedded in $\mathrm{PSL}_2(\mathbb{R})$) are conjugate by $\tau \in \mathrm{PSL}_2(\mathbb{R})$

and hence $\mathcal{H}/\Pi$ is a well defined Riemann surface (up to isomorphism). The correspondence so indicated is well-defined.

To see that it is bijective, start with two isomorphic p-elliptic curves $f : C_1 \xrightarrow{\cong} C_2$ and $\mathbb{Z}_p$ actions $S_i : \mathbb{Z}_p \times C_i \longrightarrow C_i$. Consider the diagram of branched coverings

$$\begin{array}{ccc} \mathcal{H} & \dashrightarrow & \mathcal{H} \\ \downarrow & & \downarrow \\ C_1 & \xrightarrow{f} & C_2 \\ \downarrow{\scriptstyle \pi_1} & & \downarrow{\scriptstyle \pi_2} \\ \Sigma & \overset{A}{\dashrightarrow} & \Sigma \end{array}$$

where π_i are the projections $C_i \rightarrow C_i/\mathbb{Z}_p = \Sigma$. If the diagram extends at the bottom, i.e. if there is a linear fractional transformation $A : \Sigma \rightarrow \Sigma$ making the bottom diagram commute, then the diagram extends at the top as well, and the commutative diagram so obtained readily proves the lemma. It is not possible in general to compress f to A for this means that necessarily $\pi_2(f(g_1(x)) = \pi_2(f(x))$ (that is that $f(g_1(x)) = g_2^k f(x)$ for some k prime to p; here we write $g_i^k(x) = S_i(g^k, x)$ where g is the generator of $\mathbb{Z}_p$). However one can replace f by an isomorphism f' with such a property (and hence a commuting diagram as above exists with f replaced by f'). To this end, we pull back the action of $\mathbb{Z}_p$ on C_2 via f to an action on C_1; $S : \mathbb{Z}_p \times C_1 \rightarrow C_1$ determined by $S(g, x) := g_s(x) = f^{-1}g_2(f(x))$. But the two actions S and S_1 are necessarily conjugate in $\mathrm{Aut}(C_1)$ according to a beautiful result [11] (see remark below), and hence there is $h \in \mathrm{Aut}C_1$ such that $h^{-1}Sh = S_1$; i.e. such that $h^{-1}g_s h(x) = g_1(x)$. Define $f' = f \circ h$ from C_1 to C_2. Then we check

$$\begin{aligned} \pi_2(f'(g_1(x)) &= \pi_2[fhg_1(x)] = \pi_2[fg_s h(x)] = \pi_2[g_2 f(h(x))] \\ &= \pi_2[fh(x)] = \pi_2[f'(x)] \end{aligned}$$

which precisely states that f' descends to a holomorphic 1-1 map (necessarily a fractional linear transformation) A defined by $A(x) = \pi_2(f'\pi^{-1}(x))$. The claim follows. ∎

PROOF (of Proposition 2) According to Lemma 1, we see that an isomorphism class of p-elliptic curves corresponds to an isomorphism class of skeps $\Gamma(0, \underbrace{p, \ldots, p}_{r}) \rightarrow \mathbb{Z}_p$ for some r. Such a class is determined by the orbit of r-points in Σ under the action of $\mathrm{PGL}_2(\mathbb{C})$ and by the automorphism $\beta : \mathbb{Z}_p \rightarrow \mathbb{Z}_p$. Since automorphisms of $\mathbb{Z}_p$ are always of the form $\tau \longrightarrow \tau^k$

for some $1 \leq k < p$, β is uniquely determined by some integer k prime to p, as pointed out earlier.

On the other hand, an automorphism of Γ as in 3.2 must be given on generators by $\theta(x_i) = \lambda_i x_j \lambda_i^{-1}$ (see [19]). A diagram of skeps as in 2.5 then determines a permutation on the generators x_i of Γ and an integer k. It follows that for a given projective class of r points in S^2, there are N_r distinct isomorphism classes of skeps $\Gamma \to \mathbb{Z}_p$ where N_r is the number of equivalence classes of tuples of distint integers $(k_1, \ldots, k_r)$, $1 \leq k_i \leq r$ as in Theorem 3 above. Using generating series, Lloyd [19] was now able to compute the numbers N_r for all r and his result is summarized in Proposition 2. ∎

Remark A pivotal result in the proof of Lemma 1 is the result of Gabino-Diez that if two p-cyclic groups (p prime) act on C with quotient $\mathbb{P}^1$, then necessarily the two groups are conjugate in $\mathrm{Aut}(C)$. It is interesting to compare this to a result of Nielsen which states that two orientation preserving periodic maps on a topological surface C are conjugate (in $Homeo^+(C)$) if and only if they have the same period and the same fixed point data.

4. Part II: Fixed point free representations

This part occupies the rest of the paper. If a finite group G acts on a Riemann surface C, it is often very useful to consider the induced action on the complex vector space $V = H^0(C, w_c)$ of holomorphic differentials. The complex dimension of V is the genus g. Thus we have an associated linear representation $\rho : G \to GL_g(\mathbb{C})$ which is faithful whenever the quotient $C/G \cong \mathbb{P}^1$ and $g \geq 2$. Thus the action of G on C induces an embedding

$$\mathrm{Aut}(C) \hookrightarrow GL_g(\mathbb{C})$$

These differentials are usually computed as follows. Let C be the (smooth) curve with affine equation $p(x, y) = 0$, $x, y \in \mathbb{C}$. Then a basis of differentials is given by

$$w_{r,s} = \frac{x^r y^s}{p'_y} dx \mid r, s \geq 0, r + s \leq n - 3$$

where $n = \deg p_y$ and p'_y is the partial with respect to y. A count of differentials gives $g = (1/2)(n-1)(n-2)$ as is well-known for smooth curves in $\mathbb{P}^2$. For the Fermat curve for instance ($x^n + y^n = 1$), we get the forms $w_{r,s} = \frac{x^{r-1} y^{s-1}}{y^{n-1}}$, $1 \leq r, s \leq n$.

Example The special case of curves $C : y^n = f(x)$ is treated in [7] for example. When $f(x) = \prod(x - e_i)$ is a polynomial with distinct roots and $\deg f = kn$, then the $w_{ij} = x^i y^{j-n+1} dx$, $i + kj \leq k(n-1) - 2$ form a basis of $V = H^0(C, w_c)$. If $n = p$ is a prime then the action of $\mathbb{Z}_p$ at a fixed point is given by rotation by $2\pi/p$. This immediately gives the action on $H^0(C, w_c)$: for $w = f(z)dz$, $\sigma^*(w) = f(\zeta z)\zeta dz$, where σ is a generator for $\mathbb{Z}_p$ and ζ a primitive p-th root of unity.

A finite group G acts on C with the genus zero property (GZP) if for any non-trivial subgroup $H \subset G$, $C/H = \mathbb{P}^1$. This is equivalent to the induced action $G \times H^0(C, w_c) \longrightarrow H^0(C, w_c)$ being fixed-point free (see §2). It turns out that groups of this type are rare. When $g > 1$ for example, the only abelian groups having this property are $\mathbb{Z}_{p^e}$ and $\mathbb{Z}_{pq}$ where p and q are distinct primes.

The case $g = 1$ is easy. There are four finite groups which can act on a torus C with the GZP, namely $\mathbb{Z}_2, \mathbb{Z}_3, \mathbb{Z}_4$, and $\mathbb{Z}_6$. If C admits a genus zero action by $\mathbb{Z}_3$ then it also admits one by $\mathbb{Z}_6$. In fact all genus zero actions on C contain the hyperelliptic involution (which in terms of the equation $w^2 = 4z^3 - g_2 z - g_3$ corresponds to $(z, w) \to (z, -w)$). Let $S : C \to C$ denote an automorphism of order 4 or 6.

Lemma 2 *For the groups $G \cong \mathbb{Z}_4$ or $G \cong \mathbb{Z}_6$ there is a unique torus with a genus zero action by G. The tori and actions are given by*

1. *For $G \cong \mathbb{Z}_4 :\ w^2 = 4z^3 - z,\ S(z, w) = (-z, \imath w)$.*
2. *For $G \cong \mathbb{Z}_6 :\ w^2 = 4z^3 - 1, S(z, w) = (\zeta z, -w)$ where $\zeta = e^{2\pi\imath/3}$.*

PROOF: If C has an automorphism of order 4 then the lattice Λ must admit a rotation of order 4, that is $\imath\Lambda = \Lambda$. From this it follows that $g_3 = 0$, so the torus must have the equation $w^2 = 4z^3 - g_2 z$ for some $g_2 \in \mathbb{C}$, $g_2 \neq 0$. Moreover, multiplication by $\imath : \Lambda \to \Lambda$ then corresponds to $(z, w) \to (-z, \imath w)$. The torus is unique because the elliptic modular invariant $J(\Lambda) = 1$ for any lattice Λ satisfying $\imath\Lambda = \Lambda$, and therefore there is no loss of generality in assuming $g_2 = 1$.

Similarly if C has an automorphism of order 3, $\zeta\Lambda = \Lambda$. Here $g_2 = 0$, $J(\Lambda) = 0$, and we can take $g_3 = 1$. ∎

More generally now, and as was observed in [13], if G has the GZP than necessarily its Sylow p-subgroups (for $p > 2$) are cyclic. Such groups include the cyclic, dihedral and generalized quaternionic groups. As it turns out, only the cyclic groups $\mathbb{Z}_{p^e}$ and $\mathbb{Z}_{pq}$, for p and q distinct primes, admit genus zero actions on Riemann surfaces of genus $g > 1$. The only genus zero action the dihedral group has is on the Riemann sphere.

Finite groups having all Sylow subgroups cyclic are called the Zassenhauss metacyclic groups $G_{m,n}(r)$ (or ZM for short) and are described as follows:

$G_{m,n}(r)$ is the group presented as follows:

$$\begin{aligned} \text{generators}: &\quad A, B; \\ \text{relations}: &\quad A^m = 1,\ B^n = 1,\ BAB^{-1} = A^r; \\ \text{conditions}: &\quad GCD((r-1)n, m) = 1 \text{ and } r^n \equiv 1\ (mod\ m). \end{aligned}$$

These are the groups described by extensions

$$1 \to \mathbb{Z}_m\{A\} \to G_{m,n}(r) \to \mathbb{Z}_n\{B\} \to 1,$$

where the cyclic group generated by A, $\mathbb{Z}\{A\}$, is the commutator subgroup of G. Only $G_{p,4}(-1)$ (p is an odd prime) turns out to act with genus zero (Theorem 1).

In the next few paragraphs we give representative equations for Riemann surfaces of genus $g \geq 1$ admitting genus zero actions by the cyclic groups, the metacyclic group $G_{p,4}(-1)$ and the quaternion groups.

4.1. THE CYCLIC CASE

In this section we analyze actions $G \times C \to C$, where G is cyclic and $g > 1$. According to Theorem 1 either $G \cong \mathbb{Z}_{p^e}$ or $G \cong \mathbb{Z}_{pq}$, where p and q are distinct primes. Let $S \in G$ denote a generator. The following was proved in [13]:

Proposition 7 *Suppose p, q are distinct primes. Then the genus-zero actions of $\mathbb{Z}_{pq}$ have signature and corresponding genus given by*

1. *$sig(\Gamma) = (0|pq, pq)$, in which case $g = 0$.*
2. *$sig(\Gamma) = (0|p, q, pq)$, in which case $g = \frac{1}{2}(p-1)(q-1)$.*
3. *$sig(\Gamma) = (0|p, p, q, q)$, in which case $g = (p-1)(q-1)$.*

According to the theorem above there are 2 possibilities we need consider in this section; either $g = \frac{1}{2}(p-1)(q-1)$ or $g = (p-1)(q-1)$.

Proposition 8 *Let C denote a surface of genus $g = \frac{1}{2}(p-1)(q-1)$ admitting a genus-zero action by $G = \mathbb{Z}_{pq}$. Then C is isomorphic to the Fermat curve $w^p = z^q - 1$, and the action is $S(z,w) = (\eta z, \zeta w)$, where η, ζ are respectively a primitive q^{th} root of unity and a primitive p^{th} root of unity.*

PROOF: The automorphism $T = S^q : C \to C$ has order p and the quotient of the action is $\mathbb{P}^1$. As pointed out earlier, this implies that C is the curve associated to an equation $w^p = \prod_{i=1}^r (z-e_i)^{a_i}$, where $e_i \neq e_j$ if $i \neq j$, $1 \leq a_i \leq p-1$ and $\sum_{i=1}^r a_i = kp$. We write $T(z,w) = (z, \kappa w)$ for some primitive p^{th} root of unity κ. Applying Lemma 3 in the appendix we have $S(z,w) = (Z,W)$ where

$$Z = A(z) = \frac{az+b}{cz+d}, \quad W = \frac{\mu}{(cz+d)^k} w,$$

$$\text{and } \mu \text{ satisfies } \mu^p = \prod_{i=1}^r (a - ce_i)^{a_i}.$$

A has order q since $S^q(z,w) = (z, \kappa w) = (A^q(z), \kappa w)$, and $A \neq I$. Moreover, A must permute the e_i and, if $A(e_i) = e_j$ then $a_i = a_j$. The Riemann-Hurwitz formula gives $r = q+1$, and therefore the only possibility is that one of the e_i is fixed by A and the other q form a complete cycle under the action of A. For argument's sake let's assume

$$A(e_1) = e_2, A(e_2) = e_3, \cdots, A(e_{q-1}) = e_q, A(e_q) = e_1 \text{ and } A(e_{q+1}) = e_{q+1}.$$

It follows that $a_1 = a_2 = \cdots = a_q = a$ and $qa + a_{q+1} = kp$.

Up to conjugation we can assume A has the form $A(z) = \lambda z$, where λ is a primitive q^{th} root of unity. That is $A = \begin{bmatrix} \xi & 0 \\ 0 & \xi^{-1} \end{bmatrix}$, where $\xi^2 = \lambda$. Therefore $e_{q+1} = 0$ and $\{e_i\}_{1 \leq i \leq q} = \{\lambda^i e\}_{1 \leq i \leq q}$ for some $e \neq 0$. It follows that the equation for C is

$$w^p = z^{a_{q+1}}(z-e)^a(z-\lambda e)^a \cdots (z - \lambda^{q-1}e)^a = z^{a_{q+1}}(z^q - e^q)^a,$$

and $S : C \to C$ is given by

$$S(z,w) = (Z,W), \text{ where } Z = \lambda z \text{ and } W = \mu\xi^k w$$
$$\text{for some } \mu \text{ satisfying } \mu^p = \xi^{kp}.$$

Therefore, $\mu = \rho\xi^k$ for some p^{th} root of unity ρ. In fact ρ must be a primitive p^{th} root of unity, for otherwise S would not have order pq. Therefore $S(z,w) = (\lambda z, \rho\xi^{2k}w) = (\lambda z, \rho\lambda^k w)$.

The projection $\psi : C \to \Sigma$, $\psi : (z,w) \to z$, has no branching over ∞ since $qa + a_{q+1} = kp$. If we make the change of variables $x = z^{-1}$, $y = z^{-k}w$ then the equation becomes $y^p = (1 - e^q x^q)^a$ and there is now branching over ∞. The formula for the action by G is $S(x,y) = (\lambda^{-1}x, \rho y)$.

We can make another change of variables so that C is $v^p = (u^q - 1)^a$ and $S(u,v) = (\lambda^{-1}u, \rho v)$. Finally we make the change of variables

$$w = \frac{v^l}{(u^q-1)^m}, \; z = u \text{ where } l, m \text{ are chosen so that } la - mp = 1.$$

Then C has the equation $w^p = z^q - 1$ and $S(z,w) = (\lambda^{-1}z, \rho^k w) = (\eta z, \zeta w)$. ∎

In much the same manner we can prove the following result:

Proposition 9 *Let C denote a surface of genus $g = (p-1)(q-1)$ admitting a genus zero action by $G = \mathbb{Z}_{pq}$, where p and q are distinct primes. Then C is isomorphic to the curve with the equation $w^p = \dfrac{z^q - 1}{z^q - \lambda}$, where $\lambda \neq 0$ and $\lambda^q \neq 1$. Moreover, under this isomorphism, $S(z,w) = (\eta z, \zeta w)$, where η, ζ are respectively a primitive q^{th} root of unity and a primitive p^{th} root of unity.*

Now we consider genus-zero actions of $G = \mathbb{Z}_{p^e}$, $e \geq 2$. Let S be a generator of G, let $T = S^{p^{e-1}}$ and set H equal to the subgroup of order p generated by T. Suppose $G \times C \to C$ is a genus-zero action on a surface of genus g. There is a short exact sequence $1 \to \Pi \to \Gamma \xrightarrow{\theta} G \to 1$, where the signatures of Π and Γ are $(g|-)$ and $(0|\overbrace{p, \cdots, p}^{r}, p^e, p^e)$ respectively, and the genus is $g = \frac{1}{2}r(p^e - p^{e-1})$.

Proposition 10 *C and the action are given by*

$$w^p = z^a \prod_{i=1}^{r} \left(z^{p^{e-1}} - f_i\right)^{a_i}, \quad S(z,w) = (\eta z, \lambda w),$$

where

1. $1 \leq a \leq p-1$ *and* $1 \leq a_i \leq p-1$ *for* $1 \leq i \leq r$.
2. $a + a_1 + \cdots + a_r \not\equiv 0$ *(mod p).*
3. η *is a primitive root of unity of order* p^{e-1} *and* λ *is any complex number so that* $\lambda^p = \eta^a$.
4. *the* f_i *are distinct non-zero complex numbers.*

PROOF: As before C is given by

$$w^p = \prod_{i=1}^{n} (z - e_i)^{a_i}, \text{ where } 1 \leq a_i \leq p-1 \text{ and } \sum_{i=1}^{n} a_i = kp.$$

The projection map $\psi : C \to \Sigma$, $\psi : (z,w) \to z$, is branched over the n points $\{e_i\}_{1 \leq i \leq n}$, with all branching of order $p-1$. It is not branched over ∞. By Riemann-Hurwitz we have $g = 1 - p + \frac{1}{2} n(p-1)$. On the other hand $g = \frac{1}{2} r(p^e - p^{e-1})$ and therefore $n = 2 + rp^{e-1}$. Thus C and the action of T can be described by the equations:

$$w^p = \prod_{i=1}^{2+rp^{e-1}} (z - e_i)^{a_i}, \; T(z,w) = (z, \zeta w),$$

where ζ is a primitive p^{th} root of unity. From Lemma 3 we have $S(z,w) = (A(z), W)$, where $A(z) = \dfrac{az+b}{cz+d}$, $W = \dfrac{\mu}{(cz+d)^k} w$ and μ is a complex number so that

$$\mu^p = \prod_{i=1}^{2+rp^{e-1}} (a - ce_i)^{a_i}.$$

Moreover A will have order p^{e-1}.

Now A permutes the $2 + rp^{e-1}$ points e_i and therefore must fix two of them, say $e_{1+rp^{e-1}}$ and $e_{2+rp^{e-1}}$. This is because the order of A is p^{e-1}. The remaining e_i must fall into r orbits, each of length p^{e-1}. We may assume that $e_1, \cdots, e_r$ are representatives of the orbits. Moreover, if $A(e_i) = e_j$, then $a_i = a_j$, and therefore

$$w^p = (z - e_{1+rp^{e-1}})^{a_{1+rp^{e-1}}} (z - e_{2+rp^{e-1}})^{a_{2+rp^{e-1}}} \prod_{i=1}^{r} \prod_{j=1}^{p^{e-1}} (z - A^j(e_i))^{a_i},$$

$$\mu^p = (a - ce_{1+rp^{e-1}})^{a_{1+rp^{e-1}}} (a - ce_{2+rp^{e-1}})^{a_{2+rp^{e-1}}} \prod_{i=1}^{r} \prod_{j=1}^{p^{e-1}} (a - cA^j(e_i))^{a_i}$$

Now make the change of variables

$$
\begin{aligned}
x &= L(z) = \frac{z - e_{1+rp^{e-1}}}{z - e_{2+rp^{e-1}}}, \; y = \frac{(x-1)^k w}{\nu}, \text{ where} \\
\nu^p &= (e_{2+rp^{e-1}} - e_{1+rp^{e-1}})^{a_{1+rp^{e-1}}+a_{2+rp^{e-1}}} \prod_{i=1}^{r} \prod_{j=1}^{p^{e-1}} \left(e_{2+rp^{e-1}} - A^j(e_i)\right)^{a_i}
\end{aligned}
$$

Then LAL^{-1} has fixed points $x = 0, \infty$ and therefore $LAL^{-1}(x) = \eta x$, where η is a primitive p^{e-1} root of unity. Now it is routine to check that in these variables the surface C has the equation

$$
\begin{aligned}
y^p &= x^{a_{1+rp^{e-1}}} \prod_{i=1}^{r} \prod_{j=1}^{p^{e-1}} \left(x - LA^j(e_i)\right)^{a_i} = x^a \prod_{i=1}^{r} \prod_{j=1}^{p^{e-1}} \left(x - LA^j L^{-1}(e_i')\right)^{a_i} \\
&= x^a \prod_{i=1}^{r} \prod_{j=1}^{p^{e-1}} \left(x - \eta^j e_i'\right)^{a_i} = x^a \prod_{i=1}^{r} \left(x^{p^{e-1}} - f_i\right)^{a_i}, \text{ where} \\
a &= a_{1+rp^{e-1}}, \; e_i' = L(e_i) \text{ and } f_i = L(e_i)^{p^{e-1}}, \; 1 \le i \le r.
\end{aligned}
$$

One can also check that $S(x,y) = (\eta x, \lambda y)$. ∎

4.2. THE METACYCLIC CASE

In this section we give equations for genus-zero actions by the metacyclic group $G = G_{p,4}(-1)$ presented by

$$G = \langle X, Y | X^p = 1, \; Y^4 = 1, \; YXY^{-1} = X^{-1} \rangle .$$

It was shown in [13] that any surface C admitting a genus-zero action by G is associated to a short exact sequence

$$1 \to \Pi \to \Gamma \xrightarrow{\theta} G \to 1$$

where the genus of C is $g = p - 1$ and the signature of Γ is $(0|4, 4, p)$.

Proposition 11 *The surface C is equivalent to the surface $w^2 = z^{2p} - 1$ and the action is given by $X(z,w) = (\zeta z, \epsilon w)$, $Y(z,w) = \left(\frac{1}{z}, \frac{\epsilon w}{(\imath z)^p}\right)$, where ζ is a primitive p^{th} root of unity and $\epsilon = \pm 1$.*

PROOF: C is hyperelliptic, with hyperelliptic involution Y^2, and therefore we can present the surface and action of Y^2 by

$$w^2 = \prod_{i=1}^{2g+2} (z - e_i) = \prod_{i=1}^{2p} (z - e_i), \; Y^2(z, w) = (z, -w),$$

where the e_i are distinct complex numbers.

The automorphism X then has the form

$$\begin{aligned} X(z,w) &= (A(z), W), \text{ where } A(z) = \frac{az+b}{cz+d} \text{ has order } p, \text{ and} \\ W &= \frac{\mu}{(cz+d)^p} w \text{ for some } \mu \text{ satisfying } \mu^2 = \prod_{i=1}^{2p} (a - ce_i). \end{aligned}$$

Since A has order p and permutes the $2p$ numbers e_i, $1 \le i \le 2p$, we see that the e_i must fall into 2 orbits with respect to the action of A. Suppose e_1 and e_2 are in different orbits. Then we have

$$\begin{aligned} \mu^2 &= \prod_{i=1}^{2p} (a - ce_i) = \prod_{j=1}^{p} \left(a - cA^{-j}(e_1)\right) \prod_{j=1}^{p} \left(a - cA^{-j}(e_2)\right) \\ &= \epsilon^2 = 1 \text{ (by Lemma 5 applied to } A^{-1}). \end{aligned}$$

Moreover, $X^p(z,w) = \left(z, \frac{\mu^p}{\epsilon^p} w\right) = (z,w)$, and therefore $\mu = \epsilon$. The same considerations apply to the automorphism Y, that is

$$\begin{aligned} Y(z,w) &= (B(z), V), \text{ where } B(z) = \frac{\alpha z + \beta}{\gamma z + \delta} \text{ has order 2, and} \\ V &= \frac{\lambda}{(\gamma z + \delta)^p} w \text{ for some } \lambda \text{ satisfying } \lambda^2 = \prod_{i=1}^{2p} (\alpha - \gamma e_i). \end{aligned}$$

We must have $\delta = -\alpha$ and so $Y(z,w) = \left(B(z), \frac{\lambda}{(\gamma z - \alpha)^p} w\right)$.

The automorphism B permutes the e_i, $1 \le i \le 2p$, and in fact fixes 2 of them and pairs off the remaining $2p - 2$. To see this note that

$$(\gamma B(z) - \alpha)(\gamma z - \alpha) = -1, \text{ and therefore } Y^2(z,w) = (z, -\lambda^2 w)$$

Thus $\lambda^2 = 1$ since Y^2 is the hyperelliptic involution. If $B(e_i) \ne e_i$ then the contribution of $\{e_i, B(e_i)\}$ to λ^2 would be $(\alpha - \gamma e_i)(\alpha - \gamma B(e_i)) = -1$.

Thus if B did not fix any of the e_i we would have $\lambda^2 = (-1)^p = -1$, a contradiction. Therefore, B must fix some of the e_i, and in fact 2 of them. Finally, the fixed points of B are $\dfrac{\alpha \pm \imath}{\gamma}$ and their contribution to λ^2 is $\left(\alpha - \gamma\dfrac{\alpha+\imath}{\gamma}\right)\left(\alpha - \gamma\dfrac{\alpha-\imath}{\gamma}\right) = 1$.

We may alter A within its conjugacy class and therefore there is no loss of generality in assuming $A = \begin{bmatrix} \xi & 0 \\ 0 & \xi^{-1} \end{bmatrix}$, where $\xi^2 = \zeta$ is a primitive p^{th} root of unity. The relation $XYX = Y$ holds in G and therefore, by calculation we see that $\alpha = 0$. This relation implies that $ABA = B$, and from this we see that the fixed points of B are in different orbits with respect to A.

For argument's sake suppose the fixed points of B are $e_1 = \dfrac{\imath}{\gamma}$, $e_2 = -\dfrac{\imath}{\gamma}$. By a change of variables we may assume $e_1 = 1$ and $e_2 = -1$, that is we may assume $\gamma = \imath$. Therefore

$$\{e_i\}_{1 \le i \le 2p} = \{\zeta^j | 1 \le j \le p\} \cup \{-\zeta^j | 1 \le j \le p\}$$

We then see that the equation for C is

$$\begin{aligned} w^2 &= \prod_{i=1}^{2p}(z-e_i) = \prod_{j=1}^{p}(z-\zeta^j)\prod_{j=1}^{p}(z+\zeta^j) \\ &= \prod_{j=1}^{p}(z^2-\zeta^{2j}) = z^{2p}-1. \end{aligned}$$

With these choices the generators of G are acting as stated in the theorem. ∎

4.3. THE QUATERNIONIC CASE

In this section we give equations for genus-zero actions by the generalized quaternion groups $Q = Q(2^n)$, $n \ge 3$. A presentation of Q is

$$Q = \left\langle A, B \mid A^{2^{n-1}} = 1, A^{2^{n-2}} = B^2, BAB^{-1} = A^{-1} \right\rangle.$$

Genus-zero actions on Q have been fully described in [13]. In particular actions by $Q(2^n)$ have signature $(0|\overbrace{2,\ldots,2}^{r},4,4,2^{n-1})$, where r is odd. The genus is $g = 2^{n-2}(r+1)$.

Proposition 12 *Let C be a Riemann surface with a genus zero action by Q. Then C is conformally equivalent to the surface with the equation*

$$w^2 = z(z^{2^{n-1}} - \beta^{2^{n-1}})\prod_{i=1}^{r}\left(z^{2^{n-1}} - \left(e_i + \frac{1}{e_i}\beta^{2^{n-1}}\right)z^{2^{n-2}} + \beta^{2^{n-1}}\right),$$

where $\beta \neq 0$ and the e_i are distinct non-zero complex numbers $\neq \pm\beta^{2^{n-2}}$. The action of Q on C is given by:

$$\begin{aligned} A(z,w) &= (\eta z, \lambda w), \text{ where } \eta^{2^{n-2}} = 1 \text{ and } \lambda^2 = \eta, \\ B(z,w) &= \left(-\frac{\beta^2}{z}, \frac{\mu}{z^m}w\right), \text{ where } m = 2^{n-2}(r+1)+1 \text{ and } \mu = \pm\beta^m. \end{aligned}$$

PROOF: The subgroup $< A > \cong \mathbb{Z}_{2^{n-1}}$ has a genus-zero action on C and so according to Proposition 10, we have

$$w^2 = z\prod_{i=1}^{2(r+1)}\left(z^{2^{n-2}} - f_i\right), \quad A(z,w) = (\eta z, \lambda w) \tag{5.1}$$

where η is a primitive 2^{n-2} root of 1, $\lambda^2 = \eta$, and the f_i are distinct non-zero complex numbers. The curve C is hyperelliptic with corresponding involution $A^{2^{n-2}}(z,w) = (z,-w)$. The orbit surface $C/ < A^{2^{n-2}} > \cong \mathbb{P}^1$ and the quotient map $\psi : C \to \mathbb{P}^1$ can be identified with $\psi(z,w) = z$. Now the automorphism $B : C \to C$ commutes with the hyperelliptic involution and therefore there is a linear fractional transformation $N(z) = \dfrac{\alpha z + \beta}{\gamma z + \delta}$ of order 2 so that $\psi \circ B = N \circ \psi$.

Thus $B(z,w) = (N(z), W(z,w))$ for some $W(z,w)$. The automorphism B normalizes the subgroup $< A >$, the orbit surface $C/ < A > \cong \mathbb{P}^1$ and the quotient map $\phi : C \to \mathbb{P}^1$ can be identified with the map $\phi : C \to \mathbb{P}^1$, $\phi(z,w) = z^{2^{n-2}}$. Therefore there is a linear fractional transformation $N'(z) = \dfrac{\alpha' z + \beta'}{\gamma' z + \delta'}$ so that $\rho \circ N = N' \circ \rho$, where $\rho(z) = z^{2^{n-2}}$. Thus we must have

$$\left(\frac{\alpha z + \beta}{\gamma z + \delta}\right)^{2^{n-2}} = \frac{\alpha' z^{2^{n-2}} + \beta'}{\gamma' z^{2^{n-2}} + \delta'}.$$

This last equation implies either $\alpha' = \delta' = 0$ and $\alpha = \delta = 0$, or $\beta' = \gamma' = 0$ and $\beta = \gamma = 0$. If $\beta' = \gamma' = 0$ and $\beta = \gamma = 0$ then $N(z) = \dfrac{\alpha z}{\delta} = -z$ (since $\alpha + \delta = 0$), and so $B(z,w) = (-z, W)$. Substituting into the equation (5.1) for C we see that $W = \pm \imath w$. Checking this against the relation $ABA = B$ gives a contradiction and therefore $\alpha' = \delta' = 0$ and $\alpha = \delta = 0$. Thus $N(z) = \dfrac{\beta}{\gamma z} = -\dfrac{\beta^2}{z}$ since $\beta\gamma = -1$.

Now substitute $B(z, w) = \left(-\dfrac{\beta^2}{z}, W\right)$ into the equation for C:

$$W^2 = -\frac{\beta^2}{z} \prod_{i=1}^{2(r+1)} \left(\left(-\frac{\beta^2}{z}\right)^{2^{n-2}} - f_i \right) \tag{3}$$

$$= -\frac{\beta^2}{z^{2m-1}} \prod_{i=1}^{2(r+1)} \left(\beta^{2^{n-1}} - f_i z^{2^{n-2}} \right) \tag{4}$$

$$= -\left(\beta^2 \prod_{i=1}^{2(r+1)} f_i \right) z^{-2m} z \prod_{i=1}^{2(r+1)} \left(z^{2^{n-2}} - \frac{1}{f_i} \beta^{2^{n-1}} \right) \tag{5}$$

$$= \mu^2 z^{-2m} w^2, \text{ where } \mu^2 = -\beta^2 \prod_{i=1}^{2(r+1)} f_i. \tag{6}$$

To see the last line note that as sets we have

$$\{f_i \mid 1 \le i \le 2(r+1)\} = \left\{ \frac{1}{f_i} \beta^{2^{n-1}} \mid 1 \le i \le 2(r+1) \right\}$$

because both represent the Weirstrass points of C. Therefore $B(z, w) = \left(-\dfrac{\beta^2}{z}, \dfrac{\mu}{z^m} w\right)$.

Next we check to see what conditions are imposed by the relations in Q. For example $B^2(z, w) = (z, -w)$ implies $\mu^2 = \beta^{2m}$, so $\mu = \pm\beta^m$. The relation $A^{2^{n-1}} = 1$ is clearly satisfied, and one can check that the relation $ABA = B$ is also satisfied.

Finally we consider the conditions that follow from equations (5), (7) above and from and $\mu = \pm\beta^m$. First note that (5) and $\mu = \pm\beta^m$ imply $\prod_{i=1}^{2(r+1)} f_i = -\beta^{2m-2} = -\beta^{2^{n-1}(r+1)}$. Now suppose $f_i \neq \dfrac{1}{f_i} \beta^{2^{n-1}}$ for $1 \le i \le 2(r+1)$. Then equation (7) gives $\prod_{i=1}^{2(r+1)} f_i = \beta^{2^{n-1}(r+1)}$, a contradiction.

Thus there is at least one i, and in fact 2, so that $f_i = \frac{1}{f_i}\beta^{2^{n-1}}$. For these values of i we have $f_i = \pm\beta^{2^{n-2}}$. Let the other f_i be denoted $e_i, 1 \leq i \leq r$. Then substituting into (5.1) we get the equation in the statement of the theorem. ∎

5. Appendix

In this appendix we collect some technicalities and proofs. Consider genus-zero actions $G \times C \to C$, where G has a normal subgroup H of prime order p. Let $T \in H$ denote a generator and suppose $S \in G$ is an element of order n, not in H. Then C and the action of T are given by:

$$w^p = \prod_{1 \leq i \leq r} (z - e_1)^{a_1},\ T(z,w) = (z, \zeta w),\ 1 \leq a_i \leq p-1 \text{ and } \sum_{i=1}^{r} a_i = kp.$$

Lemma 3 *There exists an element* $A(z) = \frac{az+b}{cz+d}$ *in* $PSL_2(\mathbb{C})$ *so that*

$$\begin{aligned} S(z,w) &= (Z,W),\ \textit{where } Z = A(z) \textit{ and} \\ W &= \frac{\mu}{(cz+d)^k} w \textit{ for some } \mu \textit{ satisfying } \mu^p = \prod_{i=1}^{r} (a - ce_i)^{a_i}. \end{aligned}$$

Moreover $A^n = 1$, *A must permute the* e_i, $1 \leq i \leq r$, *and if* $A(e_i) = e_j$ *then* $a_i = a_j$.

Proof: Let $\psi : C \to \mathbb{P}^1$ be the quotient map $C \to C/H \cong \mathbb{P}^1$, that is $\psi(z,w) = z$. Since H is normal in G there is an element $A \in PSL_2(\mathbb{C})$ so that the following diagram is commutative:

$$\begin{array}{ccc} C & \xrightarrow{S} & C \\ \downarrow \psi & & \downarrow \psi \\ \mathbb{P}^1 & \xrightarrow{A} & \mathbb{P}^1 \end{array}$$

S has order n and therefore $A^n = 1$. The branching of $\psi : C \to \mathbb{P}^1$ is preserved by A, and therefore A must permute the e_i. This implies that $a - ce_i \neq 0$ for $1 \leq i \leq r$ (to see this note that $A(\infty) = a/c$). The orders of the branch points are also preserved and so $a_i = a_j$ if $A(e_i) = e_j$.

Since $(Z, W) \in C$ we have

$$\begin{aligned} W^p &= \prod_{i=1}^{r} (Z - e_i)^{a_i} = \prod_{i=1}^{r} \left(\frac{az+b}{cz+d} - e_i \right)^{a_i} \\ &= \frac{1}{(cz+d)^{kp}} \prod_{i=1}^{r} ((a - ce_i)z - (de_i - b))^{a_i} \\ &= \frac{\lambda}{(cz+d)^{kp}} \prod_{i=1}^{r} \left(z - A^{-1}(e_i)\right)^{a_i}, \text{ where } \lambda = \prod_{i=1}^{n} (a - ce_i)^{a_i}. \end{aligned}$$

Therefore $W^p = \dfrac{\lambda}{(cz+d)^{kp}} \displaystyle\prod_{i=1}^{r} (z - e_i)^{a_i} = \dfrac{\lambda}{(cz+d)^{kp}} w^p$. It follows that $W = \dfrac{\mu}{(cz+d)^k} w$, where μ is a complex number satisfying $\mu^p = \lambda = \prod_{i=1}^{r} (a - ce_i)^{a_i}$. ∎

The next lemma follows easily by induction.

Lemma 4 *In the notation above $S^l(z, w)$ equals*

$$\left(A^l(z), \frac{\mu}{\left(cA^{l-1}(z) + d\right)^k} \times \frac{\mu}{\left(cA^{l-2}(z) + d\right)^k} \times \cdots \times \frac{\mu}{(cz+d)^k} \times w \right).$$

For the next lemma let $\tilde{A} = \begin{bmatrix} a & b \\ c & d \end{bmatrix} \in SL_2(\mathbb{C})$ be a matrix representative of $A \in PSL_2(\mathbb{C})$ and assume the order of A is n. Then $\tilde{A}^n = \epsilon I$, where ϵ is ± 1.

Lemma 5

$$\left(cA^{n-1}(z) + d\right) \times \left(cA^{n-2}(z) + d\right) \times \cdots \times (cz + d) = \epsilon$$

for all $z \in \mathbb{C}$ for which the left hand side has no zero terms.

Proof: We have a telescoping product

$$(cz+d) \times (cA(z)+d) \times \cdots \times \left(cA^{n-1}(z) + d\right) =$$

$$(cz+d) \times \left(c\frac{az+b}{cz+d} + d \right) \times \cdots \times \left(cA^{n-1}(z) + d\right) =$$

$$(cz+d) \times \left(\frac{c(az+b) + d(cz+d)}{cz+d} \right) \times \cdots \times \left(cA^{n-1}(z) + d\right)$$

in which every numerator cancels the next denominator, leaving only the last numerator. Since $\tilde{A}^{n-1} = \epsilon \tilde{A}^{-1} = \begin{bmatrix} \epsilon d & -\epsilon b \\ -\epsilon c & \epsilon a \end{bmatrix}$ as matrices in $SL_2(\mathbb{C})$ we get

$$\begin{aligned} cA^{n-1}(z) + d &= c \times \frac{\epsilon dz - \epsilon b}{-\epsilon cz + \epsilon a} + d \\ &= \frac{c(\epsilon dz - \epsilon b) + d(-\epsilon cz + \epsilon a)}{-\epsilon cz + \epsilon a} \\ &= \frac{\epsilon}{-\epsilon cz + \epsilon a}. \end{aligned}$$

Thus the last numerator is ϵ. ∎

References

1. A. Aigon, *Transformations hyperboliques et courbes algebriques en genre 2 et 3*, thesis university of Montpellier II (2001). http://tel.ccsd.cnrs.fr/documents/archives0/00/00/11/54/
2. A. Aigon, R. Silhol, *Hyperbolic lego and algebraic curves in genus 3*, J. London Math. Soc. (2) **66** (2002), no. 3, 671–690.
3. T. Breuer, *Characters and automorphisms of Riemann surfaces*, London Math. Soc. Lect. Notes **280**, Cambridge U. Press. (2000).
4. A. Broughton, *Classifying finite group actions on surfaces of low genus*, J. Pure and applied algebra **69**, (1990), 233–270.
5. E. Bujalance, J.M. Gamboa, G. Gromadski, *The full automorphism groups of hyperelliptic Riemann surfaces*, manuscripta math. **79** (1993), 267–282.
6. Burnside, *On finite groups in which all the Sylow subgroups are cyclical*, Messenger of Mathematics, **35**, (1905), pp. 46–50.
7. C. Ciliberto, R. Miranda, *Gaussian maps for certain families of canonical curves*, Complex projective geometry (Trieste, 1989/Bergen, 1989), 106–127, London Math. Soc. Lecture Note Ser., 179, Cambridge Univ. Press, Cambridge, 1992.
8. M. Cornalba, *On the locus of curves with automorphisms* Ann. Mat. Pura Appl. (4) 149 (1987), 135–151.
9. P. Debes, *Methodes topologiques et analytiques en theorie de Galois*, workshop notes, St. Etienne, April 2000.
10. H. Farkas, I. Kra, *Riemann Surfaces*, Graduate Texts in Mathematics, 71(1980), Springer.
11. G. Gonzalez-Diez, *On prime galois coverings of the Riemann sphere*, Annali Mat. Pura. Appl.(IV), Vol. CLXVIII (1995), 1–15.
12. G. Gonzalez-Diez, *Loci of curves which are prime Galois coverings of the* $\mathbb{P}^1$, Proc. London Math. Soc. (3), **62** (1991), 469–489.
13. S. Kallel, D. Sjerve *Genus Zero Actions on Riemann Surfaces*, Kyushu J. Math., **54**, (2000), pp. 1-24.
14. S. Kallel, D. Sjerve *On automorphism groups of cyclic covers of the line*, preprint 2002. To appear in Math. Proc. Cambridge Math. Soc.
15. R. S. Kulkarni, *Isolated points in the branch locus of moduli spaces*, Annales Acade. Scient. Fennicae, series A. **16**, (1991), 71–81.
16. A. Kuribayashi *On Analytic Families of Compact Riemann Surfaces with Non-trivial Automorphisms* Nagoya Math J, **28**, (1966), pp. 119–165.

17. A. Kuribayashi, K. Komiya, *On the structure of the automorphism group of a compact surface of genus 3*, Bull. Facul. Sci. Eng. Chuo Univ. **23** (1980), 1–34.
18. S. Lang, "Abstract Algebra", second edition, Wiley and sons.
19. K. Lloyd, *Riemann surface transformation groups*, J. of Comb. Theory (A) **13**, (1972), 17–27.
20. K. Magaard, T. Shaska, S. Shpectorov, H. Volklein, *The locus of curves with prescribed automorphism group*, preprint April 2002.
21. N. Nakajo, *On the equivalence problem of cyclic branched coverings of the Riemann sphere*, Kyushu J. Math **53** (1999), 127–131.
22. G.Riera, R. Rodriguez, *Riemann surfaces and abelian varieties with an automorphism of prime order*, Duke. Math. J. **69**, 1 (1993), 199–217.
23. T. Shaska, *Computing the automorphism group of hyperelliptic curves*, Proceedings of the 2003 International Symposium on Symbolic and Algebraic Computation (Philadelphia), pg. 248-254, ACM, New York, 2003.
24. P.S. Schaller, *Teichmuller space and fundamental domains of fuchsian groups*, Enseigement Math. **45** (1999), 169–187.
25. D. Sjerve, Q. Yang, *The Eichler Trace of $\mathbb{Z}_p$ actions on Riemann Surfaces*, Can. J. Math., **50(3)**, 1998, pp. 620-637.
26. H. Völklein, *Groups as galois groups*, Cambridge Studies in Advanced Mathematics **53** (1996). ISBN 0-521-56280-5.
27. Wolf, *Spaces of Constant Curvature*, Publish or Perish, 1974.